D0982588

ROLLING BEARING ANALYSIS

ROLLING BEARING ANALYSIS

Third Edition

TEDRIC A. HARRIS

A Wiley-Interscience Publication

JOHN WILEY & SONS, INC.

New York · Chichester · Brisbane · Toronto · Singapore

Library of Congress Cataloging in Publication Data:

Harris, Tedric A.

Rolling bearing analysis / Tedric A. Harris. — 3rd ed.

p. cm.

"A Wiley-Interscience publication."

Includes bibliographical references.

1. Roller bearings. 2. Ball-bearings. I. Title.

TJ1071.H35 1990

621.8'22—dc20 90-33828

ISBN 0-471-51349-0 CIP

Printed in the United States of America

10 9 8 7 6 5 4 3 2 1

PREFACE

Ball and roller bearings, generically called rolling bearings, are commonly used machine elements. They are employed to permit rotary motion of or about shafts in simple commercial devices such as bicycles, roller skates, and electric motors. They are also utilized in complex engineering mechanisms including, among others, aircraft gas turbines, rolling mills, dental drill assemblies, gyroscopes, and power transmissions. Until the beginning of World War II, the design and application of these bearings could be considered more of an art than a science. Little was understood about the physical phenomena that occur during the operation of rolling bearings. Since 1945, the date that approximately marks the birth of the atomic age, scientific progress has occurred virtually at an exponential pace. Today, in the space age, continually increasing demands are being made of engineering equipment. To ascertain the effectiveness of rolling bearings in modern engineering applications, it is necessary to obtain a firm understanding of how these bearings perform under varied and often extremely demanding conditions of operation.

Most information and data pertaining to the performance of rolling bearings are presented within manufacturers' catalogs. These data are almost entirely empirical in nature, stemming either from the larger manufacturers' product testing or, more likely for smaller manufacturers, based on information contained in the American National Standards Institute (ANSI) or International Standards Organization (ISO) publications or other similar publications. Furthermore, these data rarely pertain to other than slow speed, simple loading, and nominal temperature applications. If the engineer wishes to examine bearing applications beyond these bounds, it is necessary to return to the basics of rolling and sliding motions over the concentrated contacts that occur in rolling bearings.

One of the first books written on this subject was *Ball and Roller Bearing Engineering* by Arvid Palmgren, Technical Director of ABSKF for many years. It explained, more completely than had been done previously, the concept of rolling bearing fatigue life. Palmgren, together with Gustav Lundberg of Chalmers Institute of Technology in Göteborg, Sweden, was the originator of the formulas on which current ANSI and ISO standards for the calculation of rolling bearing fatigue life are based. Also, A. Burton Jones' book *Analysis of Stresses and Deflections* gave a good explanation of the static loading of ball bearings. Jones, who worked in various technical capacities for New Departure Ball Bearings Division of General Motors, Marlin-Rockwell Corporation, and Fafnir Ball Bearing Company, and also as a consultant, pioneered the use of digital computers to analyze ball and roller bearing shaft-bearing-housing systems. The remainder of other early texts on rolling bearings were largely empirical in their approach to applications analysis.

Particularly since 1960, much research has been conducted into rolling bearings and rolling contact phenomena. The use of modern laboratory equipment such as electron microscopes, X-ray diffraction devices, and high-speed digital computers has shed new light on the mechanical, hydrodynamic, metallurgical, and chemical phenomena involved in rolling bearing operation. Many significant papers have been published by the various technical societies, for example, the American Society of Mechanical Engineers, the Institution of Mechanical Engineers, and the Society of Tribologists and Lubrication Engineers, analyzing the performance of rolling bearings in exceptional applications involving high-speed, heavy load, and extraordinary internal design and materials. Since 1960, substantial attention has been given to the mechanisms of rolling bearing lubrication and the rheology of lubricants. Notwithstanding the existence of the foregoing literature, there has been the need for a reference that presents a unified, up-to-date approach to the analysis of rolling bearing performance. That is my intention in presenting this book.

To accomplish this goal, I have attempted to review the most significant technical papers and texts covering the performance of rolling bearings, their constituent materials, and lubrication. The concepts and mathematical presentations contained in the reviewed technical literature have been condensed and simplified in this book for rapidity and ease of understanding. It is not to be construed, however, that this book supplies a complete bibliography on rolling bearings. Only those data that I have found most useful in practical analysis have been referenced. Several of the references cited are my own works since in some cases these are the original or only ones available on the particular subject. Considerable use has been made of data published by SKF Group companies. These provide valuable and often unique insight into rolling bearing metallurgy, fatigue endurance, and lubrication mechanics.

The format of *Rolling Bearing Analysis* is aimed at developing for the reader a basic understanding of rolling bearing operation. Thus, the initial chapters discuss the simplest elements of rolling bearings, such as basic bearing types, geometry, loading of single balls and rollers, and contact stresses and deformations. Then, the complex analysis of load distribution among the rolling elements, speeds and velocities, elastohydrodynamic lubrication, statistics of bearing endurance, fatigue life, friction, and temperature are considered. Several topics depend almost entirely upon the preceding discussions. As nearly as possible an attempt has been made to maintain a continuity of presentation. To amplify the discussion, numerical examples are presented in most chapters. In these examples, a single bearing of each type is examined. For instance, numerical examples deal with a 209 radial ball bearing, a 209 cylindrical roller bearing, a 218 angular-contact ball bearing, and a 22317 spherical roller bearing in each chapter. It is evident that analytical data for each bearing are being accumulated as the reader progresses through the book. In the first edition of *Rolling Bearing Analysis* the numerical examples were executed only in English system units (inches, pounds, seconds, °F, and so on). In this third edition, the examples are carried out in metric system units (millimeters, Newtons, seconds, °C, and so on); however, for reference purposes the results are given parenthetically in English system units.

The material covered herein spans many scientific disciplines, for example, geometry, mechanical elasticity, dynamics, statics, hydrodynamics, statistics, and heat transfer. Thus many mathematical symbols have been employed. In some cases, the same symbol has been chosen to represent various quantities in different chapters. To avoid confusion a list of symbols is presented at the beginning of each chapter. In the interest of clarity, however, certain symbols have been retained for singular usage. For example, D is always ball or roller diameter, d_m is pitch diameter, and α is contact angle.

Because of the several scientific disciplines that this book spans, the treatment of each topic may vary somewhat in scope and manner. Where it has been feasible, analytical solutions to various problems have been represented. On the other hand, empirical approaches to certain problems have been used where they have appeared to be more practical. The wedding of analytical and empirical techniques is particularly evident in the chapters covering lubrication, friction, and fatigue life.

Rolling Bearing Analysis, first and second editions, dealt principally with the methods used to predict the performance of ball and roller bearings in various applications, particularly applications beyond the scope of those covered in manufacturers' catalogs and engineering handbooks.

Only brief attention was given to topics such as materials; their properties, behavior, and survival under operating conditions; methods of testing and evaluation of test data; methods of lubrication and their usages; bearing-induced vibration; rotor dynamics; flange-roller contacts and roller skewing control. *Rolling Bearing Analysis*, third edition, endeavors to cover these subjects in greater detail and also covers the initial development of a substantially improved and more accurate fatigue life theory, information on which was not available for publication in the second edition.

The breadth of the material contained in this text, for credibility, can hardly be covered by the expertise of a single person. Therefore, to assist in compiling this text, I have utilized information from various experts in the field of ball and roller bearing technology. I cite the following contributors who, subject to editing and augmentation of the material by me, prepared most of the material in the indicated chapters:

Contributor(s)	Chapter	Topic
Robert A. Pallini	11	Rotor Dynamics and Critical Speeds
Donald R. Wensing and John R. Rumierz	16	Bearing Structural Materials
Colin G. Hingley and John R. Rumierz	17	Lubricants and Lubrication Techniques
Frank R. Morrison	19	Endurance Testing
Robert E. Maurer	22	Material Response to Rolling Contact
Neal DesRuisseaux	25	Vibration and Noise

Moreover, the material in the following chapters, subject only to editing, was prepared by the contributors indicated below:

Contributor	Chapter	Topic
John I. McCool	14	Microcontact Phenomena
	20	Statistical Methods to Analyze Endurance
Lavern D. Wedeven	24	Wear
Anthony DiGiorgio	26	Failure Investigation and Analysis

Additionally, Fred W. Brown and Mark A. Ragen made significant contributions toward new material in Chapters 6 and 7 while Luc Houpert, as cited in the second edition, contributed significantly to the material on elastohydrodynamic lubrication in Chapter 12.

As stated previously, the material presented herein exists substantially in other publications. The purpose of this text is to concentrate that knowledge in one place for the benefit of both the student and the rolling bearing user who need or want a broader understanding of the technical field and/or product. Moreover, the references provided at the end of each chapter enable the curious reader to go into further detail.

Several of the illustrations in this text have appeared in various SKF publications; for such illustrations, appropriate references are identified. Unless otherwise specified, photographs of bearings were taken from various SKF advertisement material spanning a three decade period. Bearing configurations in the photographs are used to illustrate basic bearing designs only; they are not to be construed as being the most recent designs. To obtain constantly improving performance, SKF rolling bearing internal designs have undergone significant, and sometimes not directly obvious, alterations during the time span of *Rolling Bearing Analysis* publication.

That a substantial number of illustrations are of SKF origin is the result of my association with the company in several different capacities for 30 years. In this regard, I am grateful to SKF for the engineering, managerial, and intellectual opportunities afforded me during this long association, not the least of which has been the opportunity for the publication of *Rolling Bearing Analysis.*

TEDRIC A. HARRIS

Lakewood, New York
January 1991

CONTENTS

1

ROLLING BEARING TYPES

INTRODUCTION TO ROLLING BEARINGS

After humans invented the wheel, it became obvious that less effort was required to move an object on rollers than to slide the same object over the same surface. Even after man discovered lubrication, which reduced the effort required in sliding, rolling motion was still less difficult when it could be used. It was therefore to be expected that bearings based on rolling motion would eventually be developed for use in complex machine mechanisms. Figure 1.1 depicts, in a rather simplistic manner, the evolution of rolling bearings. Dowson [1.1] provides a substantially complete presentation on the history of bearings and lubrication in general; however, the coverage of ball and roller bearings is also quite extensive. The universal acceptance of rolling bearings by design engineers was, at first, impeded by the inability of manufacturers to supply rolling bearings that could compete in endurance characteristics with hydrodynamic sliding bearings. This picture, however, has been favorably altered since the turn of the twentieth century and particularly within the last four decades by development of superior rolling bearings steels and by constant improvement in manufacturing techniques providing long-lived extremely accurate rolling bearing assemblies. Initially, this develop-

With the rollers used by the Assyrians to move massive stones in 1100 BC . . .

and later, with crude cart wheels, man strived to overcome friction's drag.

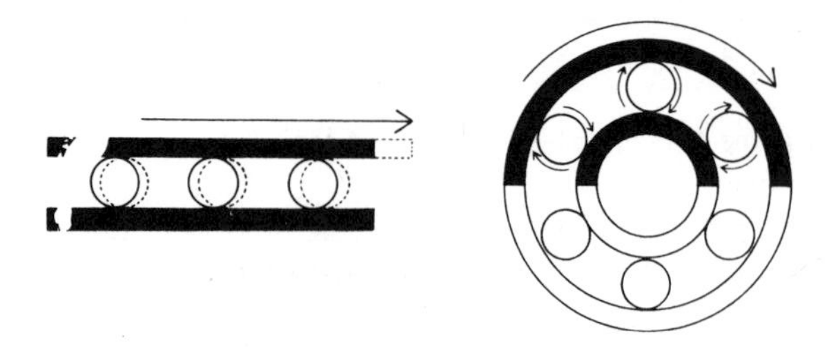

The simple ball bearing for 19th century bicycles marked man's first important victory.

FIGURE 1.1. The evolution of rolling bearings (courtesy of SKF).

ment was triggered by the bearing requirements for high speed aircraft gas turbines; however, the competition between ball and roller bearing manufacturers for worldwide markets increased substantially during the 1970s, and this has served to provide standard bearings of outstanding performance at relatively low cost to the consumer.

The term *rolling bearings* includes all forms of bearings that utilize the rolling action of balls and/or rollers to permit minimum friction, constrained motion of one body relative to another. Most rolling bearings are employed to permit rotation of a shaft relative to some fixed structure. Some rolling bearings permit translation; that is, relative linear motion, of a fixture in the direction provided by a stationary shaft, and a few rolling bearing designs permit a combination of relative linear and rotary motion between two bodies.

This book will concern itself predominantly with the standardized forms of ball and roller bearings that permit relative rotary motion between two machine elements. These bearings will always include a complement of balls and/or rollers that physically maintain the shaft and a usually stationary supporting structure, frequently called a *housing*, in a radially spaced apart relationship. Usually, a bearing may be obtained as a unit, which includes two steel rings each of which has a hardened raceway on which hardened steel balls or rollers roll. The balls or rollers, frequently called *rolling elements*, are usually held in an angularly spaced relationship by a *cage* that may also be called a *separator* or *retainer*.

As indicated above, balls, rollers, and rings of good quality, rolling bearings are normally manufactured from steel that has a capability of being hardened to a high degree, at least on the surface. In wide use throughout the ball bearing industry is AISI 52100, a steel moderately rich in chromium and easily hardened throughout (through-hardened) the mass of most bearing components to 61–65 Rockwell C scale hardness. This steel is also used in roller bearings by some manufacturers. Miniature ball bearing manufacturers whose bearings are used in sensitive instruments such as gyroscopes prefer to fabricate components from stainless steels such as AISI 440C. Roller bearing manufacturers frequently prefer to fabricate rolling components from case-hardening steels such as AISI 3310, 4620, and 8620. In all cases, at least the surfaces of the rolling components are extremely hard. During the late 1970s, it was established that certain light weight, high compressive strength ceramic materials such as silicon nitride might be used instead of steel for balls and rollers in bearing applications involving high speed, high temperature, or both. Bearings utilizing such ceramics can operate using dry lubricants or with no lubrication for short periods of time with minimal apparent danger of seizure.

Cage materials, as compared to materials for balls, rollers, and rings, are generally required to be relatively soft. They must also possess a good strength-to-weight ratio, and therefore materials as widely divergent in physical properties as mild steel, brass, bronze, aluminum, polyamide (nylon), polytetrafluoroethylene (Teflon or PTFE), and fiberglass are used. Most recently, plastic materials filled with carbon fibers are under consideration as well as polyamidimide, polyethersulfone, and polyether/etherketone for high temperature, high strength cage requirements.

In this age of continuing space exploration, many different kinds of bearings have come into use such as gas film bearings, foil bearings, spiral groove bearings, magnetic bearings, and externally pressurized (hydrostatic) bearings. Each of these bearings excels in some specialized field of application. For example, hydrostatic bearings are excellent for applications in which size is no problem, an ample supply of pressurized fluid is available, and extreme rigidity under heavy loading is required. Self-acting gas bearings may be used for applications in which loads are light, speeds are high, and a gaseous atmosphere exists. Rolling bearings, however, are not quite so limited in the scope of useful applications as are other bearings. Consequently, miniature ball bearings can be found in precision applications such as inertial guidance gyroscopes and high speed dental drills, large roller bearings such as that illustrated in Fig. 1.2 are utilized in metal rolling mill applications and even larger slewing bearings such as that shown in Fig. 1.3 are used in tunneling machines for the English Channel project. Figure 1.4 is a schematic diagram of the slewing bearing internal design.

Specifically, rolling bearings have the following advantages as compared to other bearing types:

FIGURE 1.2. A large spherical roller bearing assembly for a steel rolling mill application.

FIGURE 1.3. A slewing bearing for an English Channel tunneling machine.

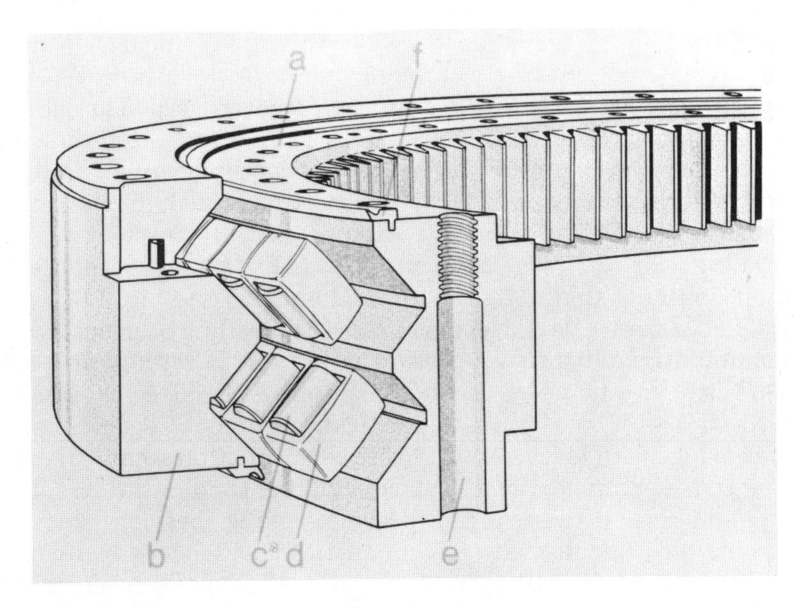

FIGURE 1.4. Schematic diagram of the internal design of the slewing bearing of Fig. 1.3. (*a*) inner ring, (*b*) outer ring, (*c*) rollers, (*d*) spacers, (*e*) threaded holes, (*f*) integral seals (SKF photograph).

1. Rolling bearings have much less friction torque than conventional hydrodynamic bearing types and therefore considerably less friction power loss, that is, less heat power loss.
2. Starting (static) friction torque of rolling bearings is only slightly greater than moving (kinetic) friction torque.
3. Rolling bearing deflection is less sensitive to load fluctuation than is deflection in a conventional hydrodynamic bearing.
4. Only a small quantity of lubricant is usually required for satisfactory operation of a rolling bearing, and an expensive lubricating system is thereby obviated. Some rolling bearings can be obtained with a self-contained lifelong lubricant supply.
5. A rolling bearing occupies a shorter axial length than does a conventional hydrodynamic bearing.
6. Within reasonable limits, changes in load, speed, and operating temperature have but slight effect on the satisfactory performance of a rolling bearing.
7. Most rolling bearings are designed to support combinations of radial and thrust load simultaneously.
8. The load and speed range to which a given rolling bearing may be subjected and still yield excellent performance is quite wide.

It is hardly proper to cite these advantages without mention of what has been considered the principal shortcoming of rolling bearings. Tallian [1.2] has defined three eras of modern rolling bearing development: an "empirical" era continuing through the 1920s, a "classical" period lasting through the early 1950s, and the "modern" era. Through the empirical, classical, and well into the modern era, it was said that even if rolling bearings are properly lubricated, properly mounted, protected from dirt and moisture, and otherwise properly operated, they will eventually fail because of fatigue of the surfaces in rolling contact. Historically, rolling bearings have been considered to have a statistical life distribution similar to that of light bulbs and humans (see Fig. 1.5).

Research in the 1960s [1.3] demonstrated that rolling bearings exhibit a minimum fatigue life; that is, "crib deaths" due to rolling contact fatigue are nonexistent when the aforementioned operating criteria are achieved. Moreover, modern manufacturing techniques enable production of bearings with extremely accurate component internal and external geometries and exceptionally smooth rolling contact surfaces, modern steel-making processes can provide rolling bearing steels of outstanding homogeneity with few impurities, and modern sealing and lubricant filtration methods act to minimize the incursion of harmful contaminants into the rolling contact zones. These phenomena can now be used in combination to virtually eliminate the occurrence of the rolling contact fatigue in almost any rolling bearing application, even some

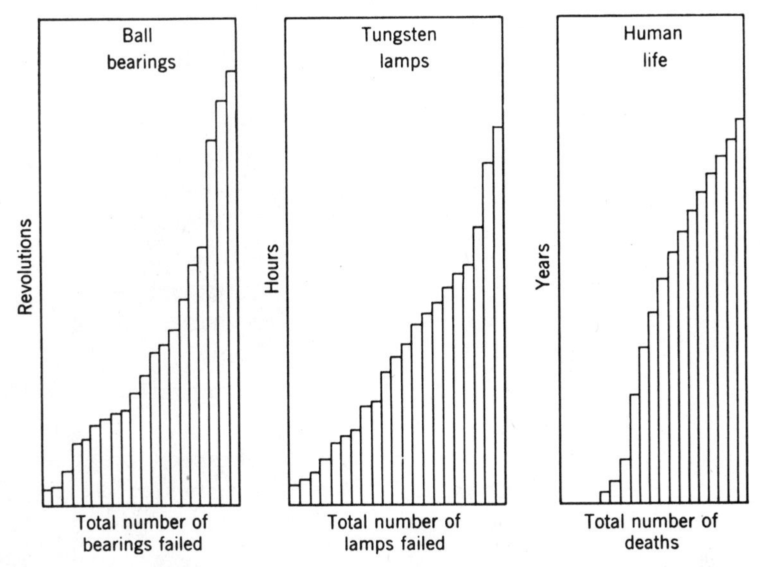

FIGURE 1.5. Comparison of rolling bearing fatigue life distribution with those of humans and light bulbs.

involving very heavy applied loading. In many lightly loaded applications, for example, most electric motors, fatigue life need not be a major design consideration.

There are many different kinds of rolling bearings, and before embarking on a discussion of the theory and analysis of their operation, it is necessary to become somewhat familiar with each type. In the succeeding pages a description is given for each of the most popular ball and roller bearings in current use.

BALL BEARINGS

Radial Ball Bearings

Single-Row Deep-Groove Conrad Assembly Ball Bearing. This ball bearing is shown in Fig. 1.6, and it is the most popular rolling bearing. The inner and outer raceway grooves have curvature radii between 51.5 and 53% of the ball diameter for most commercial bearings.

To assemble these bearings, the balls are inserted between the inner and outer rings as shown by Figures 1.7 and 1.8. The *assembly angle* ϕ is given as follows:

$$\phi = 2(Z - 1)\, D/d_m \tag{1.1}$$

FIGURE 1.6. A single-row, deep-groove, Conrad-assembly, radial ball bearing.

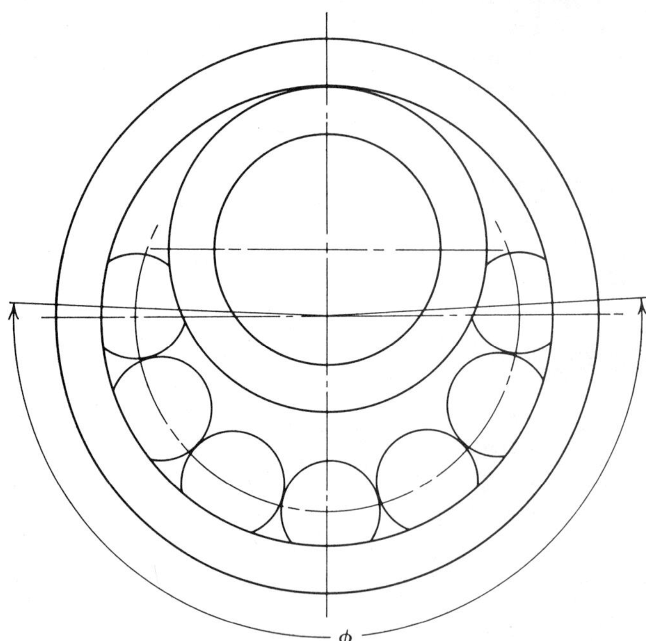

FIGURE 1.7. Diagram illustrating the method of assembly of a Conrad-type, deep-groove ball bearing.

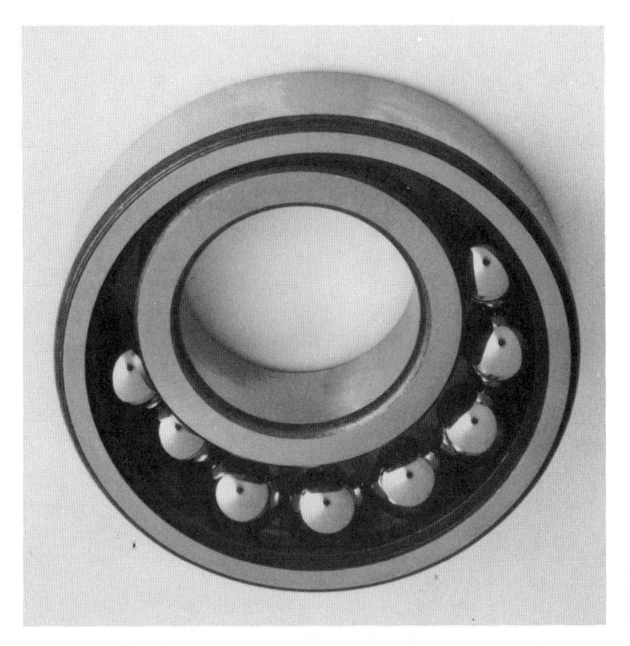

FIGURE 1.8. Photograph showing Conrad-type, ball bearing components just prior to snapping the inner ring to the position concentric with the outer ring.

in which Z is the number of balls, D is ball diameter, and d_m is pitch diameter. The inner ring is then snapped to a position concentric with the outer ring, the balls are separated uniformly, and a rivetted cage as shown in Fig. 1.6 or a plastic cage as illustrated by Fig. 16.25*a* is inserted to maintain the separation. Because of the high osculation and an appropriate ball diameter and ball complement to substantially *fill* the bearing pitch circle, the deep-groove ball bearing has comparatively high load-carrying capacity when accurately manufactured from good quality steel and operated in accordance with good lubrication and contaminant-exclusion practices. Although it is designed to carry radial load, it performs well under combined radial and thrust load and under thrust alone. With proper cage design, deep-groove ball bearings can withstand misaligning loads (moment loads) of small magnitude. By making the bearing outside surface a portion of a sphere as illustrated in Fig. 1.9, however, the bearing can be made externally self-aligning and, thus, incapable of supporting a moment load.

The deep-groove ball bearing can be readily adapted with seals as shown in Fig. 1.10 or shields as shown by Fig. 1.11 or both as illustrated by Fig. 1.12. These components function to keep lubricant in the bearing and exclude contaminants. Seals and shields come in many different configurations to serve general or selective applications; those shown in Figs.

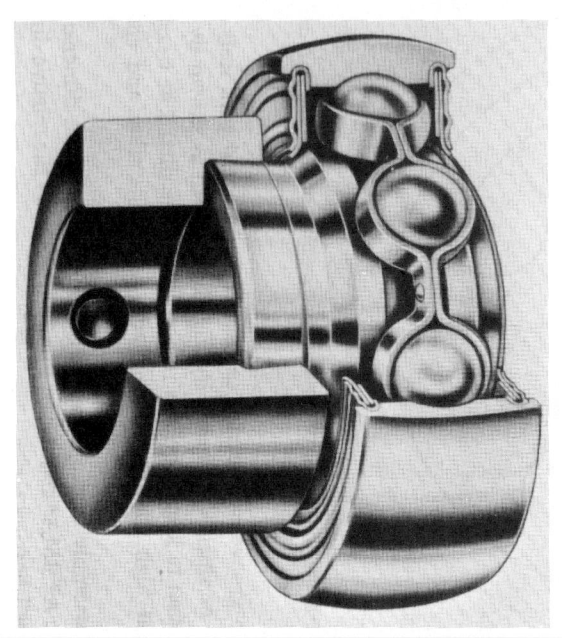

FIGURE 1.9. A single-row deep-groove ball bearing assembly having a sphered outer surface to make it externally aligning.

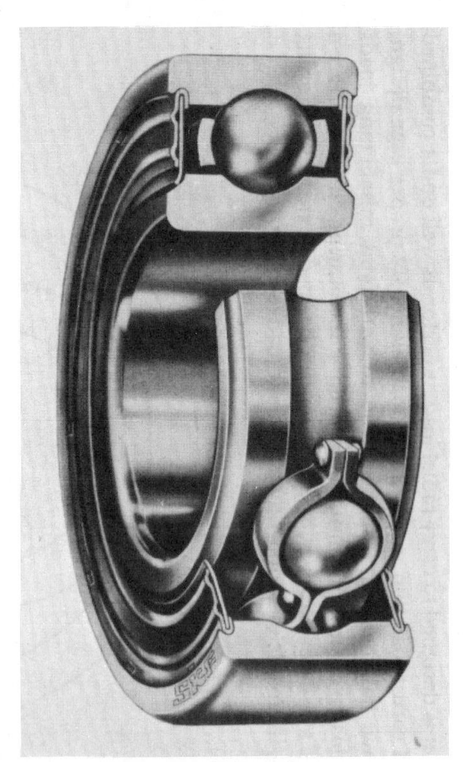

FIGURE 1.10. A single-row deep-groove ball bearing having two seals to retain lubricant (grease) and prevent ingress of dirt into the bearing.

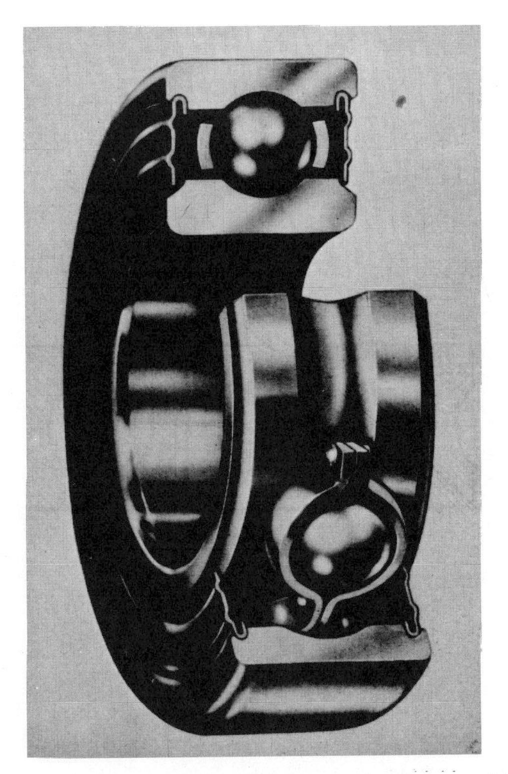

FIGURE 1.11. A single-row deep-groove ball bearing having two shields to exclude dirt from the bearing.

1.10–1.12 should be taken only as examples. In Chapter 17, seals are discussed in greater detail.

Deep-groove ball bearings perform well at high speeds provided adequate lubrication and cooling are available. Speed limits shown in manufacturers' catalogs generally pertain to bearing operation without the benefit of external cooling capability or special cooling techniques.

Conrad assembly bearings can be obtained in different dimension series according to ANSI and ISO* standards. Figure 1.13 shows the relative dimensions of various ball bearing series.

Single-Row Deep-Groove Filling-Slot Assembly Ball Bearings. This bearing as illustrated in Fig. 1.14 has a slot machined in the side wall of each of the inner and outer ring grooves to permit the assembly of more balls than the Conrad type does, and thus it has more radial load-carrying capacity. Because the slot disrupts the groove continuity, the

*American National Standards Institute and International Standards Organization.

FIGURE 1.12. A single-row deep-groove ball bearing assembly having shields and seals. The shields are used to exclude large particles of foreign matter. This bearing might be used in an agricultural application.

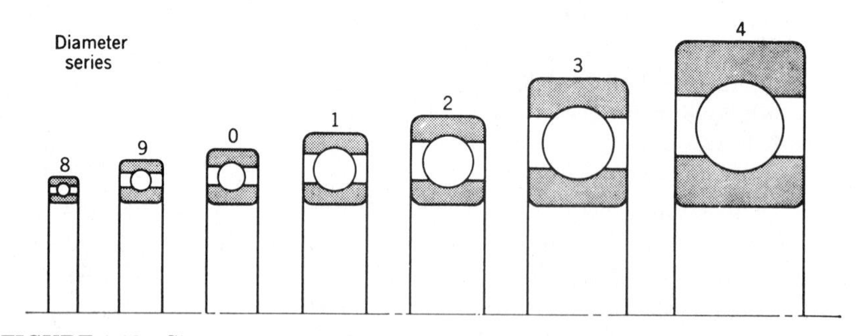

FIGURE 1.13. Size comparison of popular deep-groove ball bearing dimension series.

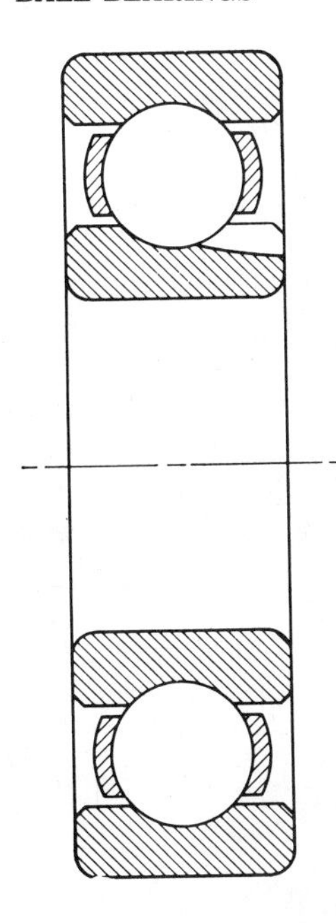

FIGURE 1.14. Diagram showing the filling slot in a single-row deep-groove filling-slot-type ball bearing assembly.

bearing is not recommended for thrust load applications. Otherwise, the bearing has characteristics similar to those of the Conrad type.

Double-Row Deep-Groove Ball Bearings. This ball bearing as shown in Fig. 1.15 has greater radial load-carrying capacity than the single-row types. Proper load sharing between the rows is a function of the geometrical accuracy of the grooves. Otherwise, these bearings behave similarly to single-row ball bearings.

Instrument Ball Bearings. In metric design, the standardized form of these bearings ranges in size from 1.5-mm (0.05906-in.) bore and 4-mm (0.15748-in.) o.d. to 9-mm (0.35433-in.) bore and 26-mm (1.02362-in.) o.d. See reference [1.4]. As detailed in reference [1.5], standardized form, inch design instrument ball bearings range from 0.635-mm (0.0250-in.) bore and 2.54-mm (0.100-in.) o.d. to 19.050-mm (0.7500-in.) bore and 41.275-mm (1.6250-in.) o.d. Additionally, instrument ball bearings have *extra thin series* that range up to 47.625-mm (1.8750-in.) o.d. and *thin series* that range up to 100-mm (3.93701-in.) o.d. Those bearings having less than 9-mm (0.3543-in.) o.d. are classified as miniature ball bearings according to [1.6]; such bearings can use balls as small as 0.6350-mm (0.0250-in.) diameter. Figure 1.16 illustrates this type of bearing. They are fabricated according to more stringent manufacturing standards, such as for cleanliness, than are any of the bearings previously de-

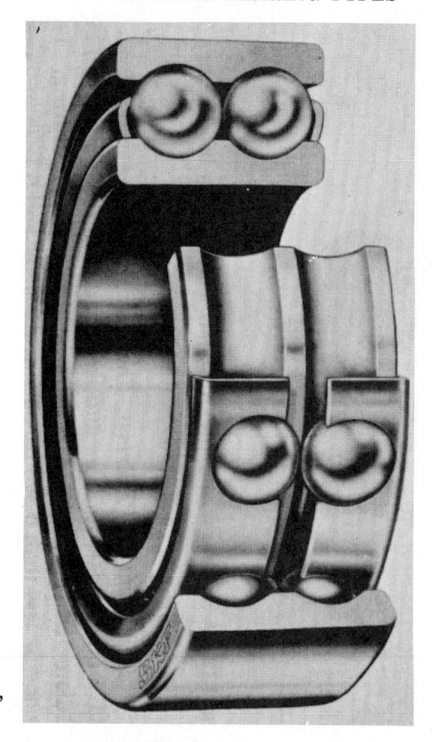

FIGURE 1.15. A double-row, deep-groove, radial ball bearing.

FIGURE 1.16. An instrument ball bearing assembly.

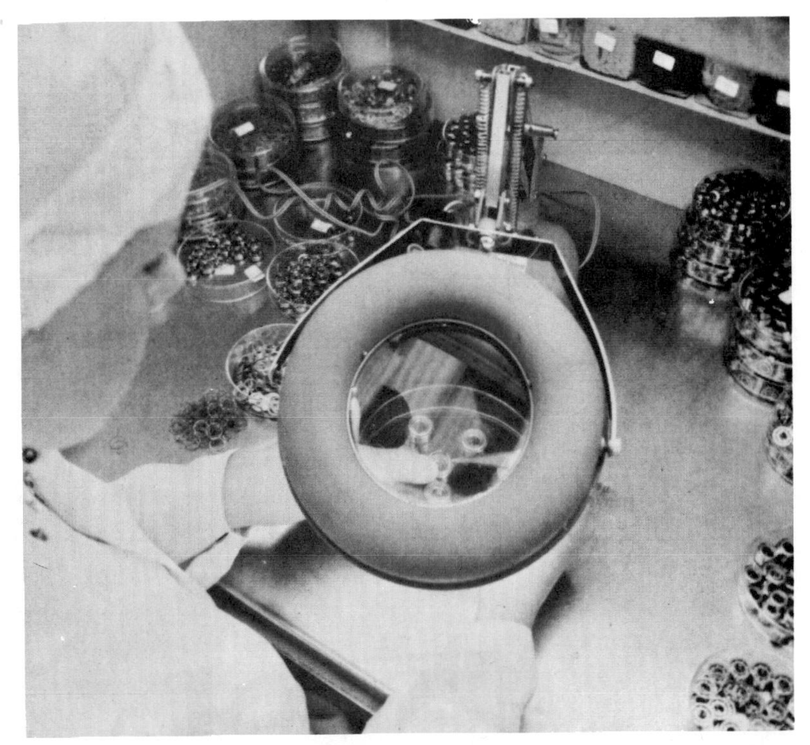

FIGURE 1.17. A delicate, final assembly operation on an instrument ball bearing assembly is performed under magnification in a "white room."

scribed. This is because minute particles of foreign matter can significantly increase the friction torque and negatively affect the smooth operation of the bearings. For this reason, they are assembled in a *white* room as illustrated in Fig. 1.17.

Groove radii of instrument ball bearings are usually not smaller than 57% of the ball diameter. The bearings are usually fabricated from stainless steels since corrosion particles will seriously deteriorate bearing performance.

Angular-Contact Ball Bearings

Single-Row Angular-Contact Ball Bearings. Angular-contact ball bearings as shown in Fig. 1.18 are designed to support combined radial and thrust loads or heavy thrust loads depending on the contact angle magnitude. The bearings having large contact angles can support heavier thrust loads. Figure 1.19 shows bearings having small and large contact angles. The bearings generally have groove curvature radii in the range of 52–53% of the ball diameter. The contact angle does not

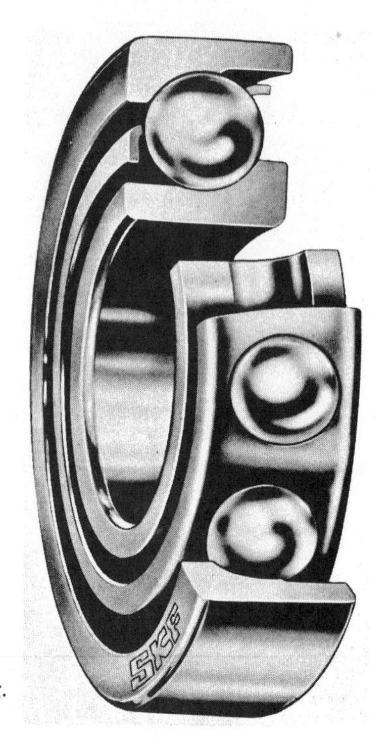

FIGURE 1.18. An angular-contact ball bearing.

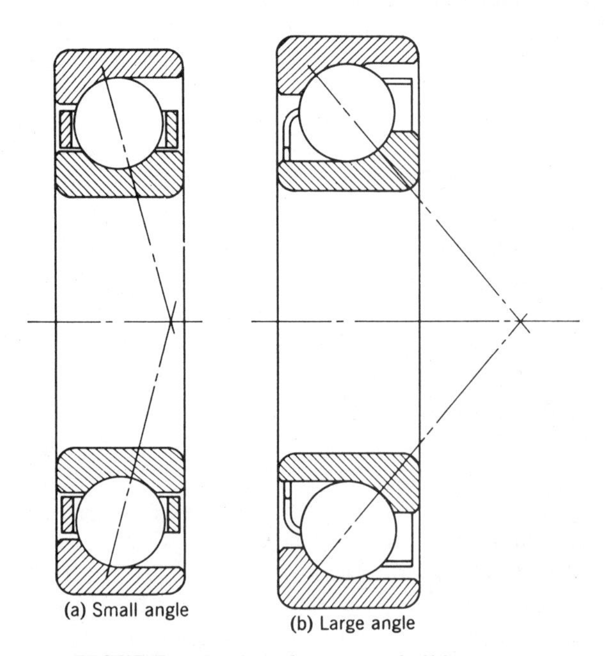

FIGURE 1.19. Angular-contact ball bearings.

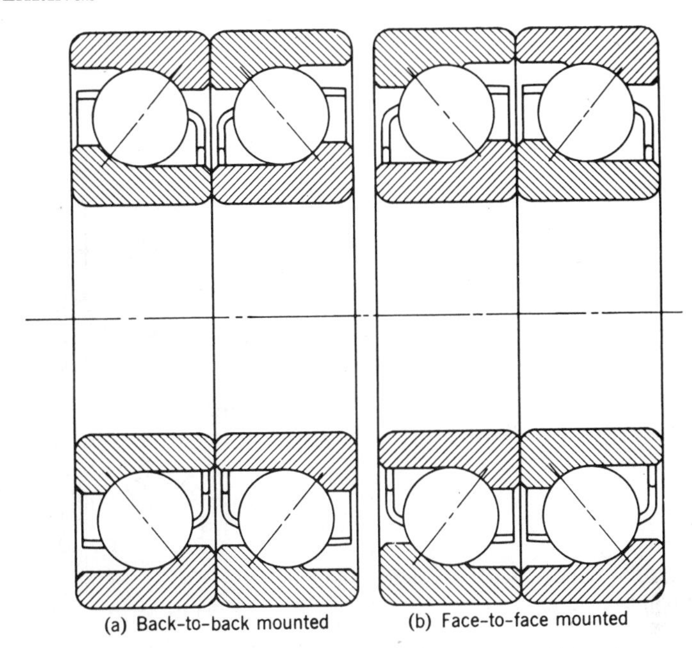

FIGURE 1.20. Duplex pairs of angular-contact ball bearings.

usually exceed 40°. The bearings are usually mounted in pairs with the free endplay removed as shown in Fig. 1.20. These sets may be preloaded against each other to stiffen the assembly in the axial direction. The bearings may also be mounted in tandem as illustrated in Fig. 1.21 to achieve greater thrust-carrying capacity.

Double-Row Angular-Contact Ball Bearings. These bearings as depicted in Fig. 1.22 can carry thrust load in either direction or a combination of radial and thrust load. Bearings of the rigid type are able to withstand moment loading effectively. Essentially, the bearings perform similarly to duplex pairs of single-row angular-contact ball bearings.

Self-Aligning Double-Row Ball Bearings. As illustrated in Fig. 1.23, the outer raceway of this bearing is a portion of a sphere. Thus, the bearings are internally self-aligning and cannot support a moment load. Because the balls do not conform well to the outer raceway (it is not grooved), the outer raceway has reduced load-carrying capacity. This is compensated somewhat by use of a very large ball complement that minimizes the load carried by each ball. The bearings are particularly useful in applications in which it is difficult to obtain exact parallelism between the shaft and housing bores. Figure 1.24 shows this bearing with a tapered sleeve and locknut adapter. With this arrangement the bearing does not require a locating shoulder on the shaft.

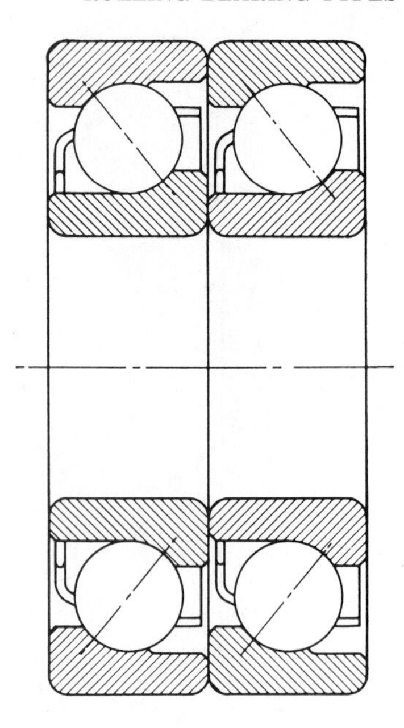

FIGURE 1.21. A tandem-mounted pair of angular-contact ball bearings.

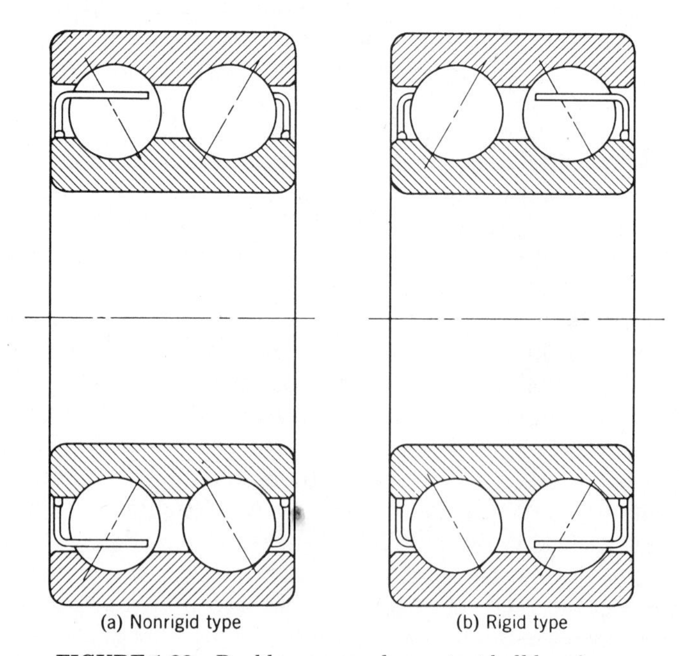

FIGURE 1.22. Double-row angular-contact ball bearings.

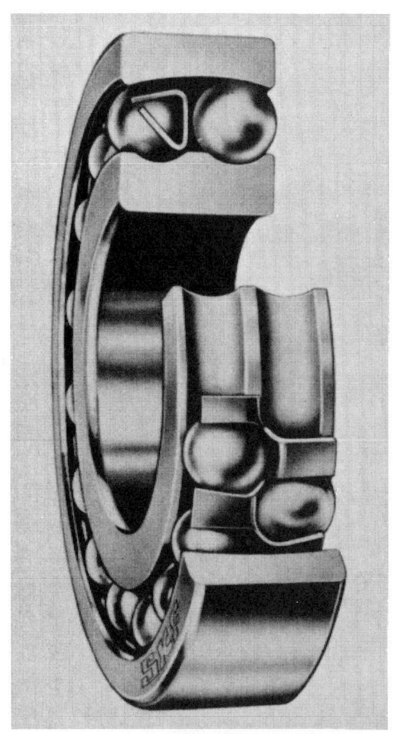

FIGURE 1.23. A double-row internally self-aligning ball bearing assembly.

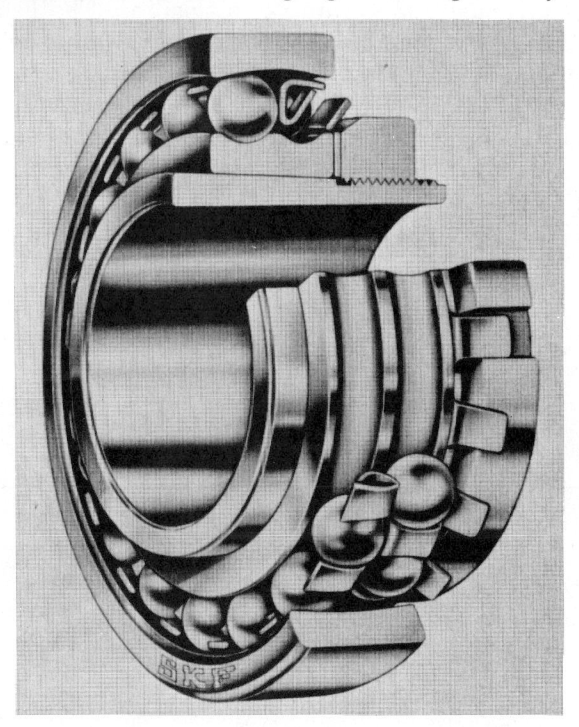

FIGURE 1.24. A double-row internally self-aligning ball bearing assembly with a tapered sleeve and locknut adapter for simplified mounting on a shaft of uniform diameter.

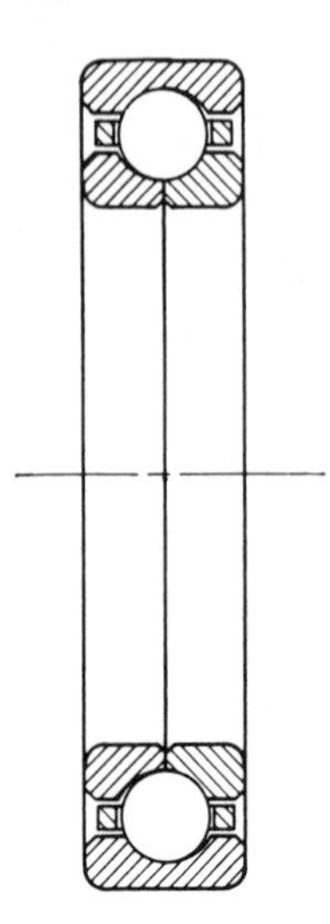

FIGURE 1.25. A split inner ring ball bearing assembly.

Split Inner Ring Ball Bearings. These bearings are illustrated in Fig. 1.25. As can be seen, the inner ring consists of two axial halves such that a heavy thrust load can be supported in either direction. They may also support, simultaneously, moderate radial loading. The bearings have found extensive use in supporting the thrust loads acting on high speed, gas turbine engine mainshafts. Figure 1.26 shows the compressor and turbine shaft ball bearing locations in a high-performance aircraft gas turbine engine. Obviously, both the inner and outer rings must be *locked up* on both axial sides to support a reversing thrust load. It is possible

FIGURE 1.26. High-performance gas turbine engine showing mainshaft ball bearing locations (courtesy of Pratt & Whitney Aircraft, United Technologies Corp.).

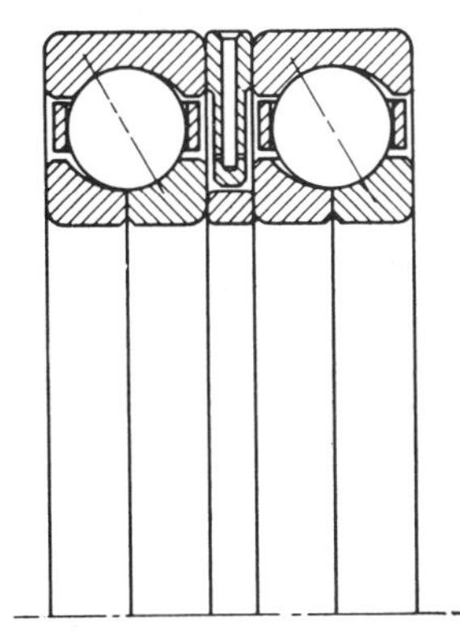

FIGURE 1.27. A tandem-mounted pair of split inner ring ball bearings.

with accurate *flush grinding* at the factory to utilize these bearings in tandem as shown in Fig. 1.27 to share a thrust load in a given direction.

Thrust Ball Bearings

The thrust ball bearing illustrated in Fig. 1.28 has a 90° contact angle; however, ball bearings whose contact angles exceed 45° are also classi-

FIGURE 1.28. A 90° contact angle thrust ball bearing assembly.

FIGURE 1.29. A 90° contact angle thrust ball bearing having a spherical seat to make it externally aligning.

fied as thrust bearings. As for radial ball bearings, thrust ball bearings are suitable for operation at high speeds. To achieve a degree of externally aligning ability, thrust ball bearings are sometimes mounted on spherical seats. This arrangement is demonstrated by Fig. 1.29. A thrust ball bearing whose contact angle is 90° cannot support any radial load.

ROLLER BEARINGS

General

Roller bearings are usually used for applications requiring exceptionally large load-supporting capability, which cannot be feasibly obtained using a ball bearing assembly. Roller bearings are usually much stiffer structures (less deflection per unit loading) and provide greater fatigue endurance than do ball bearings of a comparable size. In general, they also cost more to manufacture, and hence purchase, than comparable ball bearing assemblies. They usually require greater care in mounting than do ball bearing assemblies. Accuracy of alignment of shafts and housings can be a problem in all but self-aligning roller bearings.

Radial Roller Bearings

Cylindrical Roller Bearings. Cylindrical roller bearings as illustrated in Fig. 1.30 have exceptionally low friction torque characteristics that

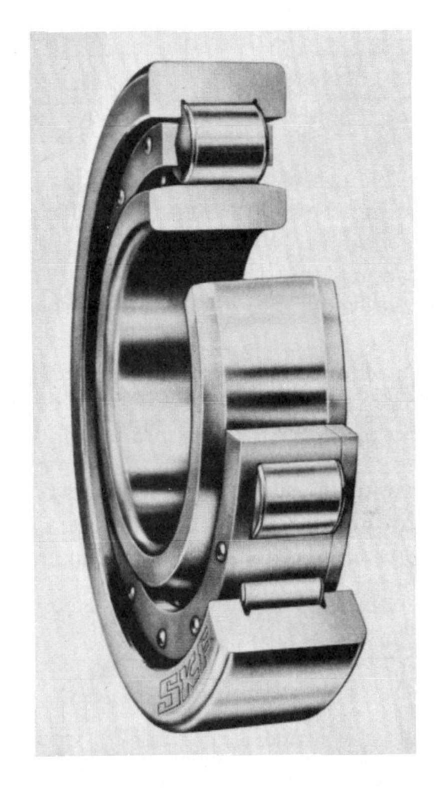

FIGURE 1.30. A radial cylindrical roller bearing.

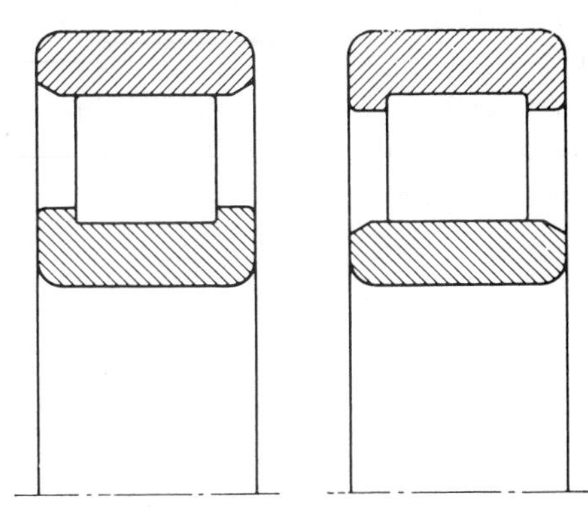

FIGURE 1.31. Cylindrical roller bearings without thrust flanges.

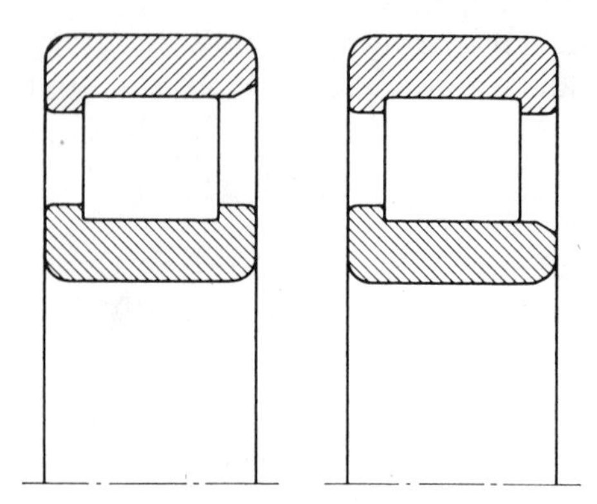

FIGURE 1.32. Cylindrical roller bearings having thrust flanges.

make them suitable for high speed operation. They also have high radial-load-carrying capacity. The usual cylindrical roller bearing is free to float axially. It has two roller-guiding flanges on one ring and none on the other, as shown in Fig. 1.31. By equipping the bearing with a guide flange on the opposing ring (illustrated by Fig. 1.32) the bearing can be made to support thrust load.

To prevent high stresses at the edges of the rollers the rollers are usually crowned as shown in Fig. 1.33. This crowning of rollers also gives the bearing protection against the effects of slight misalignment. The crown is ideally designed for only one condition of loading. Crowned raceways may be used in lieu of crowned rollers.

To achieve greater radial-load-carrying capacity, cylindrical roller bearings are frequently constructed of two or more rows of rollers rather than of longer rollers. This is done to reduce the tendency of the rollers

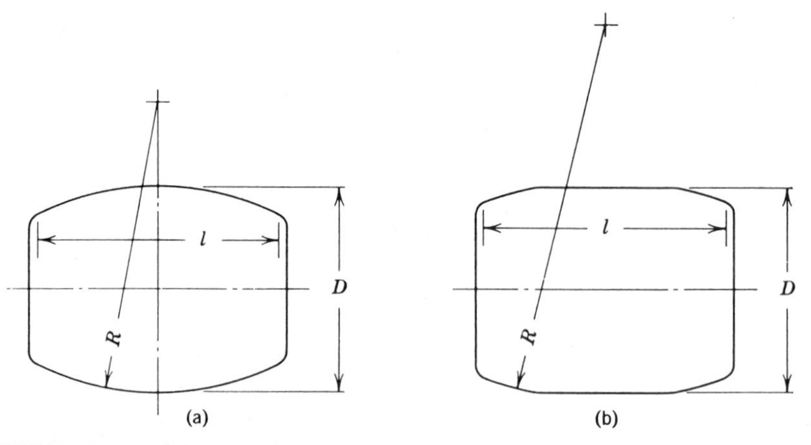

FIGURE 1.33. (*a*) Spherical roller (fully crowned); (*b*) partially crowned cylindrical roller (crown radius is greatly exaggerated for clarity).

FIGURE 1.34. A double-row, cylindrical roller bearing for precision machine tool spindle applications.

to skew. Figure 1.34 shows a small double-row cylindrical roller bearing designed for use in precision applications. Figure 1.35 illustrates a large multirow cylindrical roller bearing for a steel rolling mill application.

Needle Roller Bearings. A needle roller bearing is a cylindrical roller bearing having rollers of considerably greater length than diameter. This bearing is illustrated in Fig. 1.36. Because of the geometry of the rollers, they cannot be manufactured as accurately as other cylindrical rollers nor can they be guided as well. Consequently, needle roller bearings have relatively greater friction than other cylindrical roller bearings.

Needle roller bearings are designed to fit in applications in which radial space is at a premium. Sometimes to conserve space the needles bear directly on a hardened shaft. They are useful for applications in which oscillatory motion occurs or in which continuous rotation occurs but loading is light and intermittent. The bearings may be assembled without a cage, as shown in Fig. 1.37. In this full-complement-type bearing,

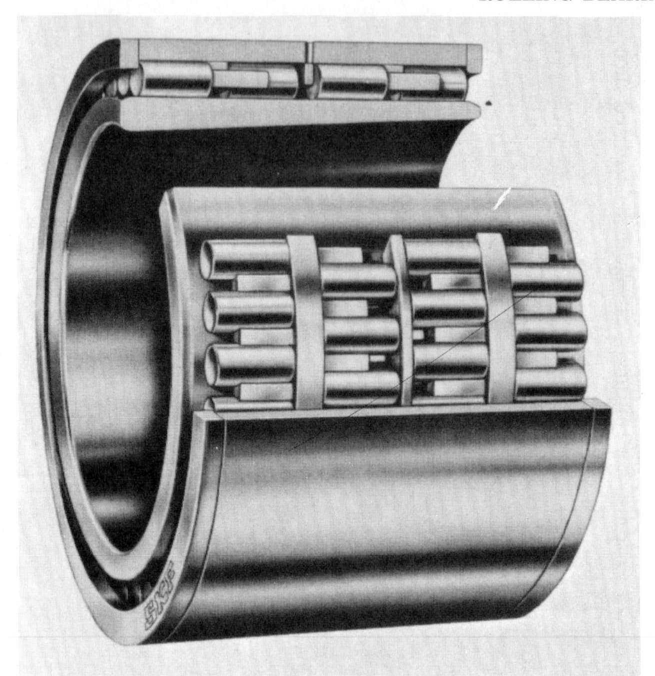

FIGURE 1.35. A multirow cylindrical roller bearing for a steel rolling mill application.

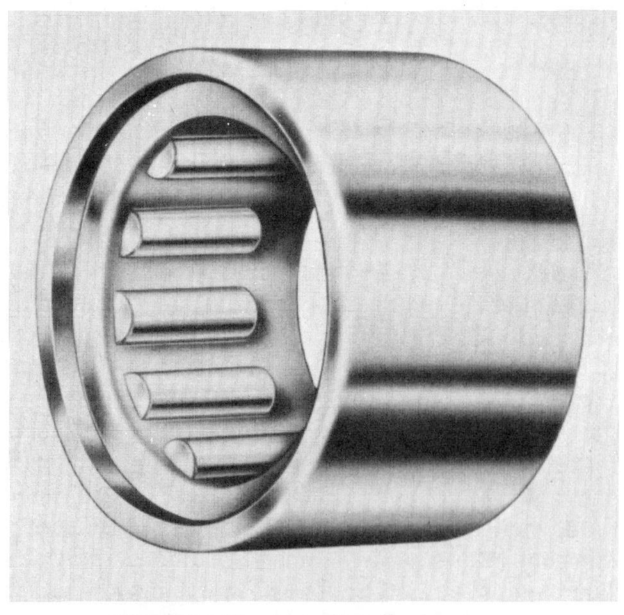

FIGURE 1.36. A needle roller bearing.

FIGURE 1.37. A full complement (no cage) needle roller bearing.

the rollers are frequently retained by turned-under flanges that are integral with the outer shell. The raceways are frequently hardened but not ground.

Tapered Roller Bearings

The single-row tapered roller bearing shown in Fig. 1.38 has the ability to carry combinations of large radial and thrust loads or to carry thrust load only. Because of the difference between the inner and outer raceway contact angles, there is a force component that drives the tapered rollers against the guide flange. Because of the relatively large sliding friction generated at this flange, the bearing is not suitable for high speed operation without special attention to cooling and/or lubrication.

Tapered roller bearing terminology differs somewhat from that pertaining to other roller bearings; the inner ring being called the *cone* and the outer ring the *cup*. Depending on the magnitude of the thrust load to be supported, the bearing may have a small or steep contact angle, as shown in Fig. 1.39. Since tapered roller bearing rings are separable, the bearings are mounted in pairs as indicated in Fig. 1.40, and one bearing is adjusted against the other. To achieve greater radial load-carrying capacity and eliminate problems of axial adjustment due to distance be-

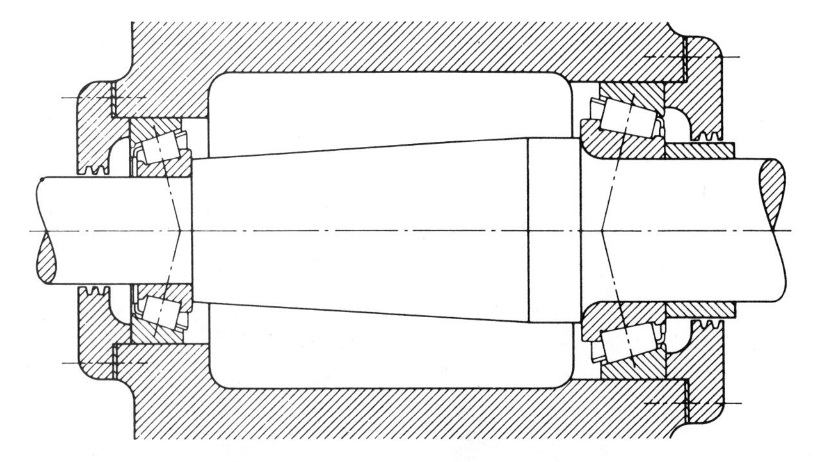

FIGURE 1.40. Typical mounting of tapered roller bearings.

tween bearings, tapered roller bearings may be combined as shown in Figs. 1.41 and 1.42 into two-row bearings. Double-row tapered roller bearings may also be combined into four-row or *quad* bearings for exceptionally heavy radial load applications such as rolling mills. Figure 1.43 shows a *quad* bearing having integral seals.

As with cylindrical roller bearings, tapered rollers or raceways are frequently crowned to relieve heavy stresses on the axial extremities of the rolling members.

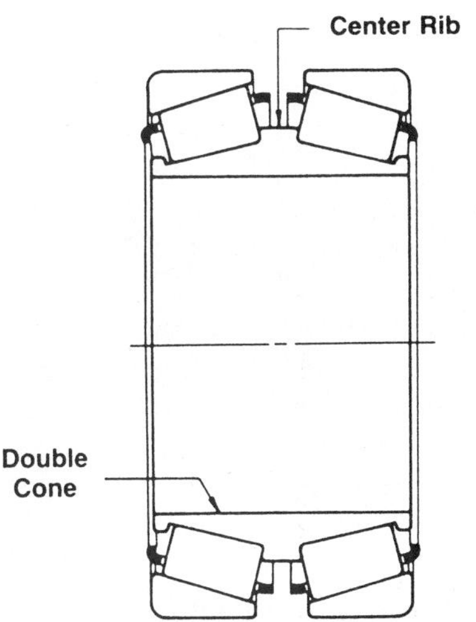

FIGURE 1.41. A double-row, double cone tapered roller bearing assembly.

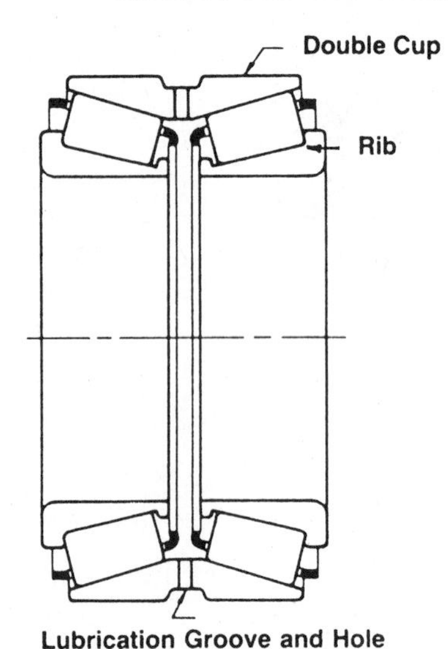

FIGURE 1.42. A double-row, double cup tapered roller bearing assembly.

FIGURE 1.43. A four-row tapered roller bearing with integral seals for a hot strip mill application.

Spherical Roller Bearings

Most spherical roller bearings have an outer raceway that is a portion of a sphere; hence, the bearings are internally self-aligning (see Fig. 1.44). Each roller has a curved generatrix in the direction transverse to rotation that conforms relatively closely to the inner and outer raceways. This gives the bearing high load-carrying capacity. Various executions of double-row, spherical roller bearings are shown in Fig. 1.45.

Figure 1.45*a* shows a bearing with asymmetrical rollers. This bearing, similar to tapered roller bearings, has force components that drive

FIGURE 1.44. A double-row spherical roller bearing assembly.

the rollers against the fixed central guide flange. Bearings such as illustrated in Fig. 1.45*b* and 1.45*c* have symmetrical (*barrel* or *hourglass* shape) rollers, and these force components are absent. Double-row bearings having *barrel-shape* symmetrical rollers frequently use an axially floating central flange as illustrated in Fig. 1.45*b*. This eliminates undercuts in the inner raceways and permits use of longer rollers, thus increasing the load-carrying capacity of the bearing. Roller guiding in such bearings tends to be accomplished by the raceways in conjunction with the cage. In a well-designed bearing, the roller-cage loads due to roller skewing motions may be minimized (see Chapter 13).

Because of the close osculation between rollers and raceways and curved generatrices, spherical roller bearings have inherently greater friction than cylindrical roller bearings. This is due to the degree of sliding that occurs in the roller-raceway contacts. Spherical roller bearings are therefore not readily suited for use in high-speed operations. They perform well in heavy duty applications such as rolling mills, paper mills, and power transmissions and in marine applications. Double-row bearings can carry combined radial and thrust loads; they cannot support moment loading. Radial, single-row spherical roller bearings have a basic

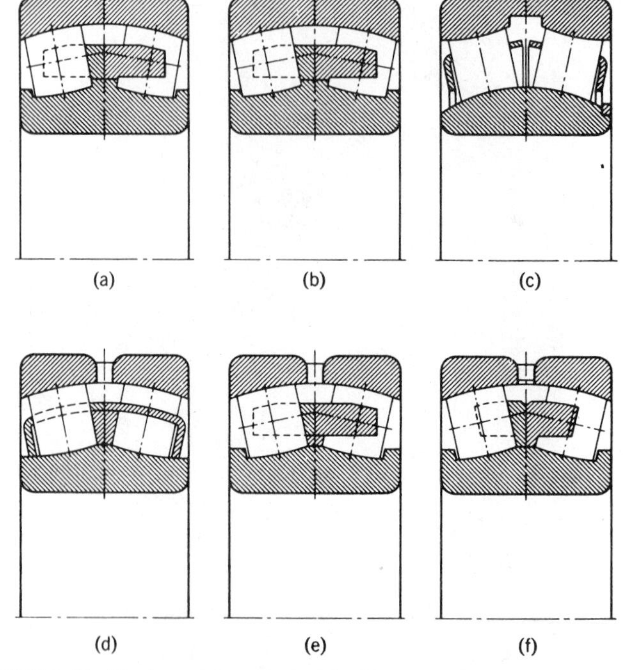

FIGURE 1.45. Various executions of double-row spherical roller bearings.

contact angle of 0°. Under thrust loading, this contact angle does not increase significantly; consequently, any amount of thrust loading magnifies roller-raceway loading substantially. Therefore, these bearings should not be used to carry combined radial and thrust loading.

Thrust Roller Bearings

Spherical Roller Thrust Bearings. The spherical roller thrust bearing shown in Fig. 1.46 has very high load-carrying capacity due to high osculation between the rollers and raceways. It can carry a combination thrust and radial load and is internally self-aligning. Because the rollers are asymmetrical, force components occur that drive the sphere ends of the rollers against a concave spherical guide flange. Thus, the bearings experience sliding friction at this flange and do not lend themselves readily to high-speed applications.

Cylindrical Roller Thrust Bearings. The cylindrical roller thrust bearings shown in Figs. 1.47 and 1.48 experience a large amount of sliding between the rollers and raceways because of the bearing geometry. Thus, the bearings are limited to slow speed operation. Sliding is reduced

FIGURE 1.38. A single-row tapered roller bearing.

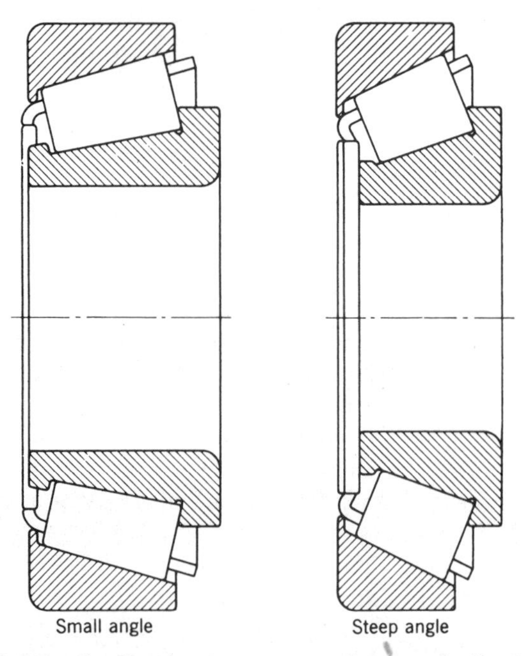

FIGURE 1.39. Small and steep contact angle tapered roller bearings.

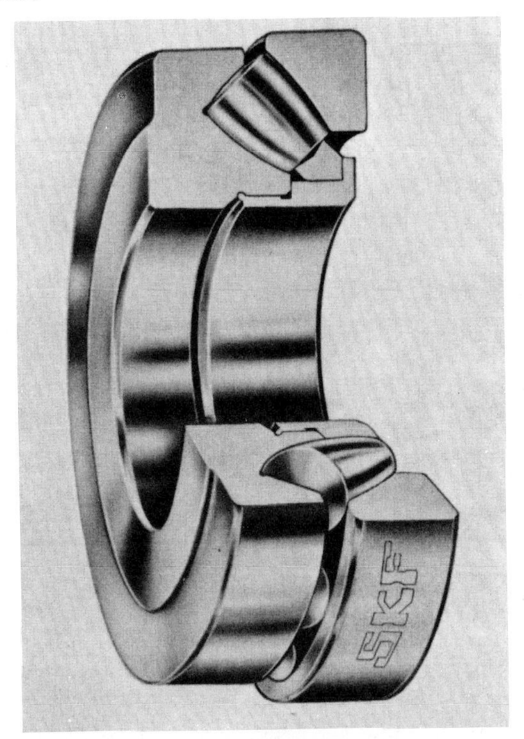

FIGURE 1.46. A spherical roller thrust bearing assembly.

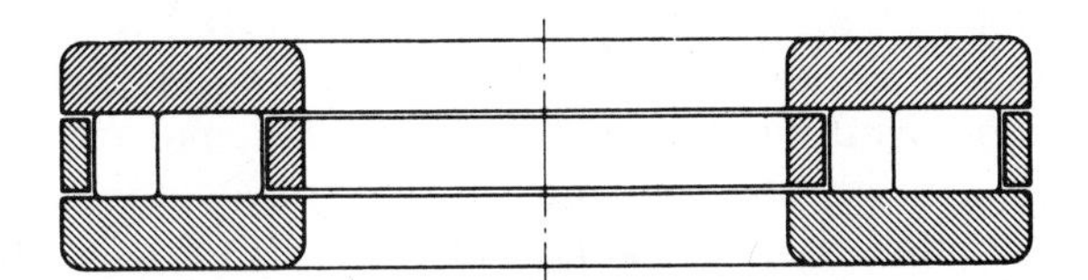

FIGURE 1.47. Cylindrical roller thrust bearing.

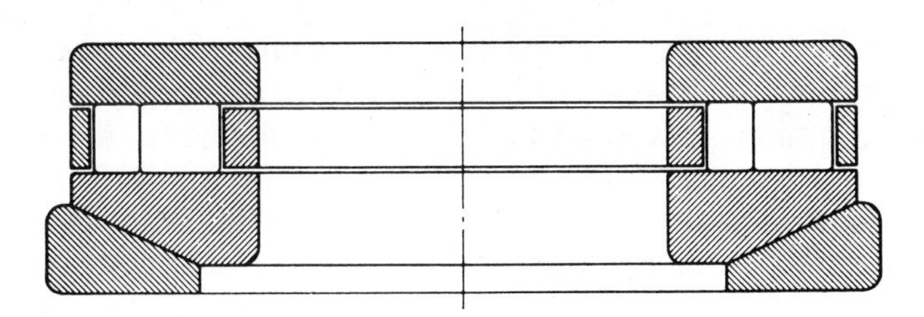

FIGURE 1.48. Cylindrical roller thrust bearing, externally aligning.

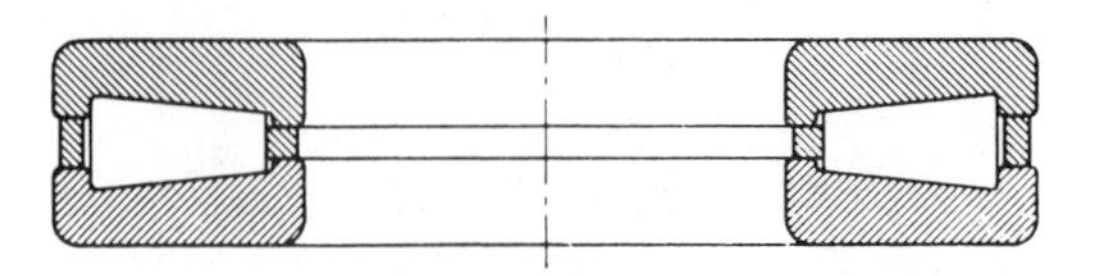

FIGURE 1.49. Tapered roller thrust bearing, both washers tapered.

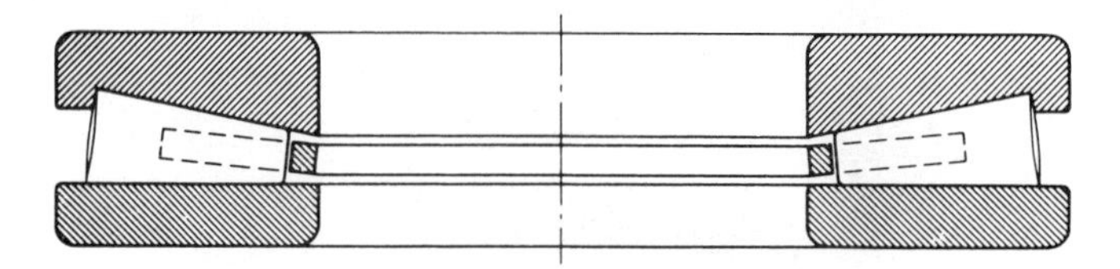

FIGURE 1.50. Tapered roller thrust bearing, one washer tapered.

somewhat by using several short rollers in each cage pocket rather than a single integral roller.

Tapered Roller Thrust Bearings. This bearing, which is illustrated in Figs. 1.49 and 1.50, has an inherent force component that forces each roller against the outboard flange. The sliding frictional forces generated at the contact between the rollers and flange limit the bearing to relatively slow speed applications.

Needle Roller Thrust Bearings. This bearing is similar to the cylindrical roller thrust bearing except that needle rollers are used in lieu of normal size rollers. Consequently, sliding is prevalent in this bearing to a large degree and loading must be light. The principal advantage of this bearing is that it requires only a narrow axial space. Figure 1.51 illustrates a needle roller-cage assembly that may be purchased in lieu of an entire bearing assembly.

LINEAR MOTION BEARINGS

Until relatively recent times, bearings for linear motions such as those used in machine tool "ways;" for example, V-ways, employed only lubricated sliding action. These sliding actions are subject to relatively high friction, wear, and subsequent loss of locational accuracy. Ball bushings were therefore introduced to overcome these problems. Operating on a hardened steel shaft, such units, as illustrated schematically in Fig. 1.52 provide many of the low friction minimal wear characteristics of radial rolling bearings.

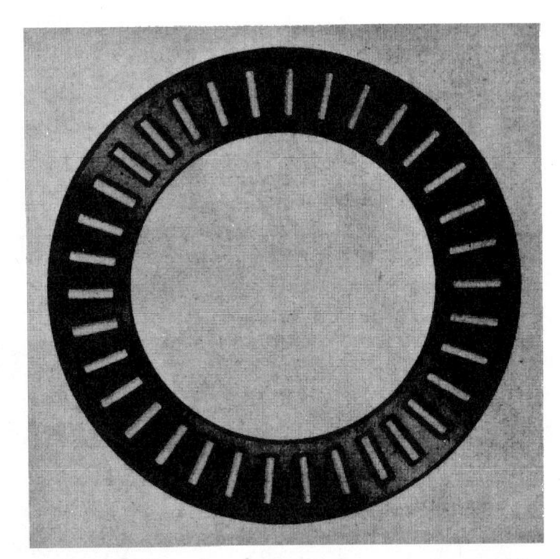

FIGURE 1.51. A needle roller-cage assembly.

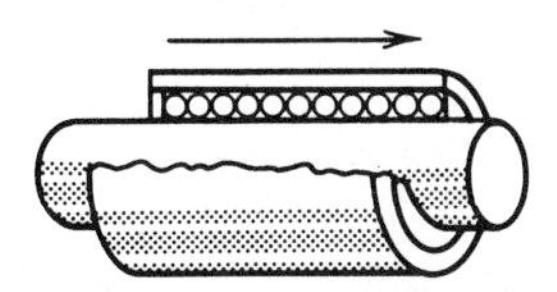

FIGURE 1.52. Schematic illustration of a ball bushing.

The ball bushing, which provides linear travel along the shaft limited only by built-in motion stoppers, contains three or more oblong circuits of recirculating balls. As illustrated in Fig. 1.53 one portion of the oblong ball complement supports load on the rolling balls while the remaining balls operate with clearance in the return track.

The ball retainer units can be fabricated relatively inexpensively of pressed steel or nylon (polyamid) material. Figure 1.54 is a photograph showing an actual unit with its components. Ball bushings of instrument quality are made to operate on shaft diameters as small as 3.18 mm (0.125 in.).

Ball bushings can be lubricated with medium-heavy weight oil or with a light grease to prevent wear and corrosion. For high linear speeds, light oils are recommended. Seals can be provided; however, friction is increased significantly.

As with radial ball bearings, life can be limited by subsurface-initiated fatigue of the rolling contact surfaces. A unit is usually designed to perform satisfactorily for several million units of linear travel. Since the hardened shaft is subject to surface fatigue and/or wear, provision can be made for rotating the bushing or shaft to bring new bearing surface into play.

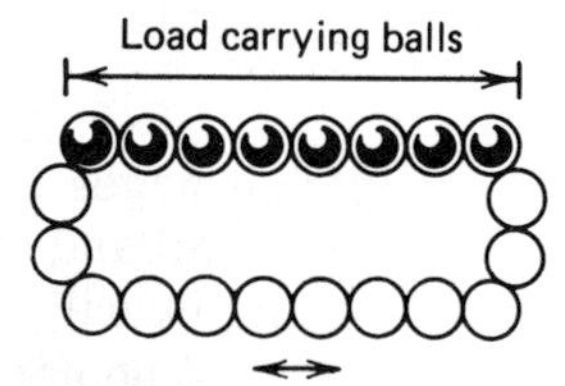

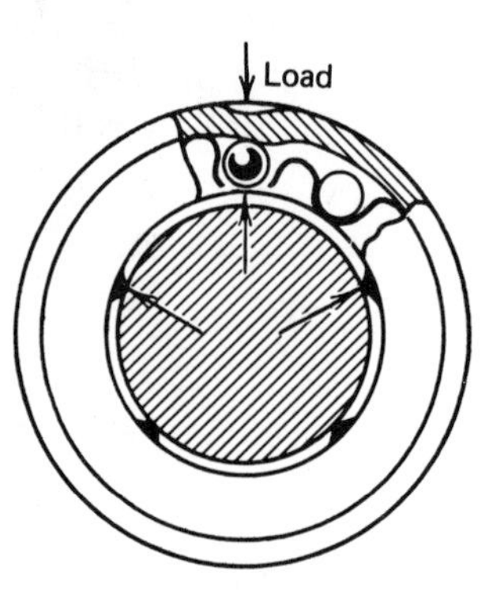

FIGURE 1.53. Schematic diagram of a ball bushing showing a recirculating ball set.

FIGURE 1.54. Linear ball bushing showing various components.

CLOSURE

The foregoing pages have illustrated and described various types and executions of ball and roller bearings. It is not to be construed that every type of rolling bearing has been described; discussion has been limited to only the most popular and basic forms. There are, for example, cylindrical roller bearing designs that use snap rings, instead of machined and ground flanges. Also being currently manufactured and sold are such designs as single-row spherical roller bearings, fractured outer ring ball bearings, hollow roller cylindrical roller bearings, and so on. These de-

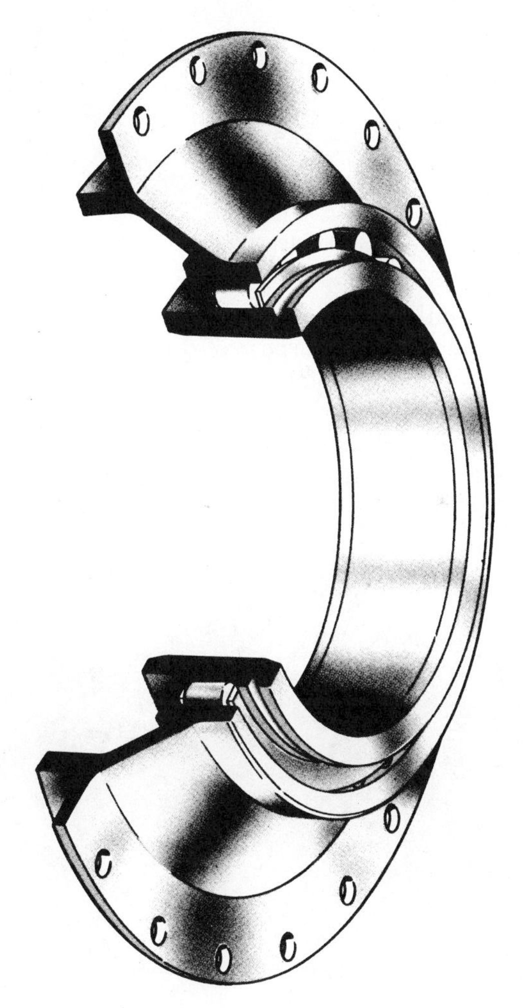

FIGURE 1.55. Aircraft gas turbine engine mainshaft roller bearing assembly showing (*a*) outer ring integral with bolting flange (from [1.7]).

FIGURE 1.55. (*Continued*) (*b*) Cutouts in outer ring flange to save weight.

signs are, however, only minor variations of the bearings that have already been described. It is the purpose of this book to be concerned with bearing performance, and therefore only the basic bearing designs need be considered.

Many variations of the illustrated bearing types are manufactured for specialized applications, such as automotive front and rear wheel bear-

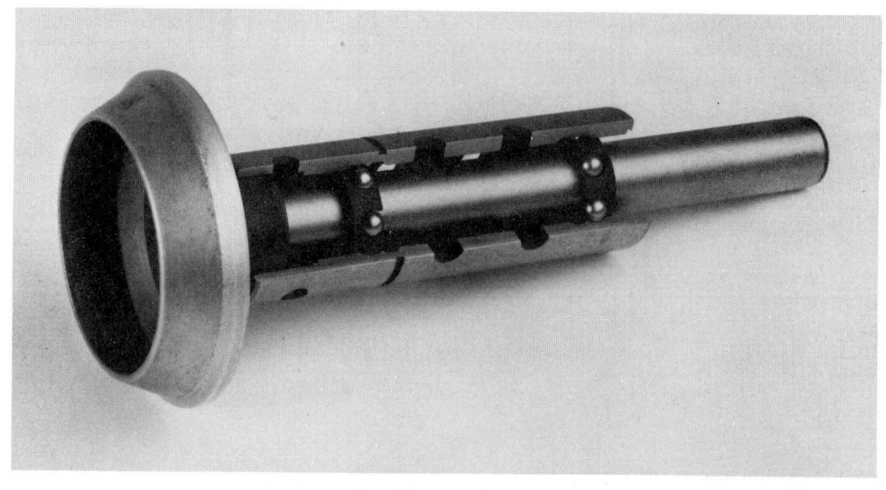

FIGURE 1.56. Textile spinning spindle unit.

ings, fan and water pump bearings, air-frame bearings, clutch-throwout bearings, and planet gear bearings, to name a few. Generally, in these specialized bearings, raceway and rolling element constructions are identical to those shown herein; but, the inner and outer ring structures have been uniquely designed to be accommodated in the specific application. This usually results in a saving of overall cost, weight, or space, or combinations thereof.

Ball and roller bearings used in aircraft gas turbine engine mainshaft

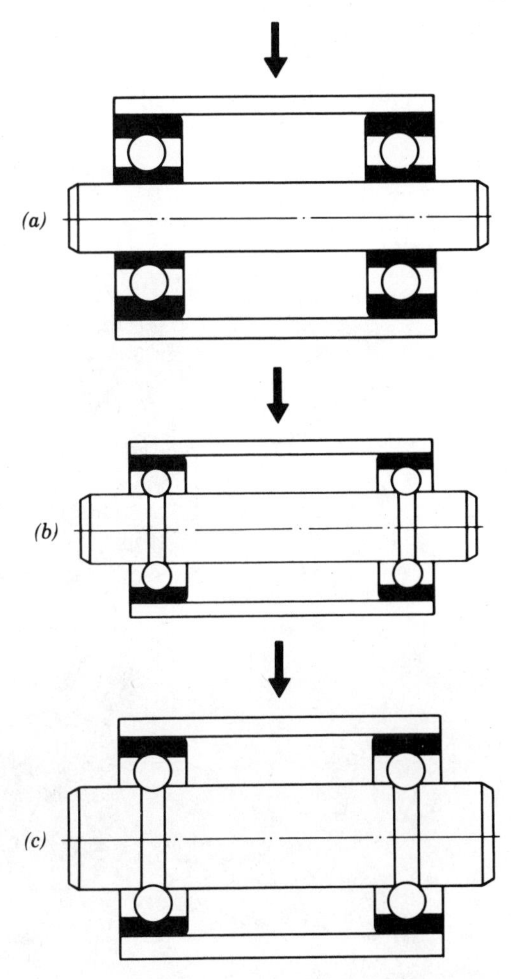

FIGURE 1.57. Compacting effects associated with integral bearing-spindle unit. (*a*) conventional discrete bearing unit; (*b*) integrated bearing unit with same shaft diameter, o.d. reduced 35%; (*c*) integrated bearing unit with same o.d., high stiffness. (SKF photograph).

applications such as those shown in Fig. 1.26 are bolted to the engine frame to minimize space and weight. Figure 1.55*a* illustrates the bearing and mounting arrangement. The bolting flanges integral with the outer rings are frequently scalloped or have cutouts to further minimize weight as demonstrated by the bearing photograph in Fig. 1.55*b*. Figure 1.56 shows a textile spinning spindle unit with integral ball bearings. Figure 1.57 illustrates the effects of compacting the design and the engineering options associated with the compact design.

Another bearing unit combines a front wheel-driven automobile wheel bearing with the constant velocity joint. Figure 1.58 shows a photograph of such a unit with a cutaway section.

It is apparent that bearing failure in such units means the replacement of the entire unit at considerable cost; however, the units are generally precisely assembled, aligned, and frequently sealed for life, eliminating causes of premature failure and providing general assurance of exceptionally long life.

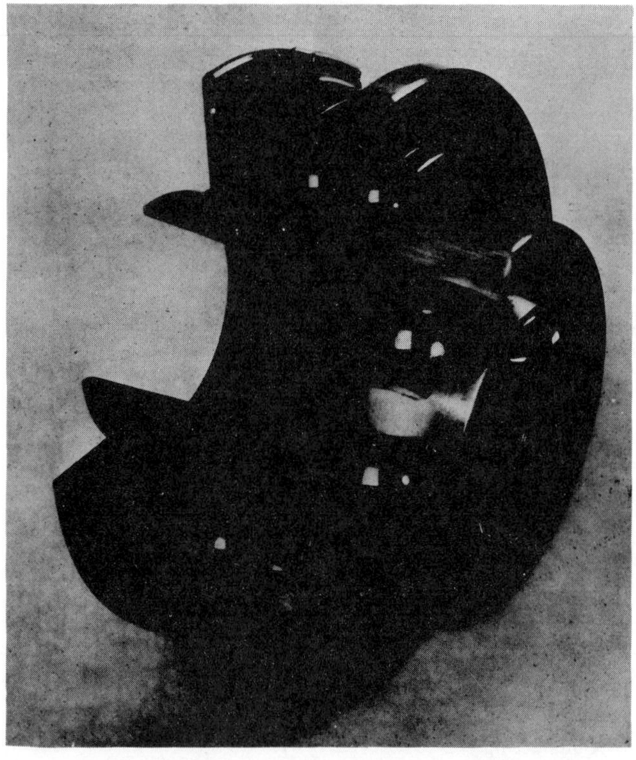

FIGURE 1.58. Hub bearing unit for a front wheel-driven automobile showing a double-row ball bearing integral with mounting flanges and a constant velocity joint.

REFERENCES

1.1. D. Dowson, *History of Tribology*, Longman, New York (1979).

1.2. T. Tallian, "Progress in Rolling Contact Technology," SKF Industries, Inc. Report AL690007, 1969.

1.3. T. Tallian, "Weibull Distribution of Rolling Contact Fatigue Life and Deviations Therefrom," *ASLE Trans.* **5**(1), 183–196 (1962).

1.4. American National Standards Institute, "Instrument Ball Bearings Metric Design," *ANSI/AFBMA Std 12.1-1985* (May 13, 1985).

1.5. American National Standards Institute, "Instrument Ball Bearings Inch Design," *ANSI/AFBMA Std 12.2-1985* (May 13, 1985).

1.6. American National Standards Institute, "Terminology for Anti-Friction Ball and Roller Bearings and Parts," *ANSI/AFBMA Std 1-1984* (April 13, 1984).

1.7. R. Spitzer, "New Case Hardening Steel Provides Greater Fracture Toughness," SKF Ball Bearing J., 234, 6–11, (Sept. 1989).

2

ROLLING BEARING MACROGEOMETRY

LIST OF SYMBOLS

Symbol	Description	Units
A	Distance between raceway groove curvature centers	mm (in.)
B	A/D	mm (in.)
d	Raceway diameter	mm (in.)
D	Ball or roller diameter	mm (in.)
d_m	Bearing pitch diameter	mm (in.)
f	r/D	
l	Roller effective length	mm (in.)
P_d	Bearing diametral clearance	mm (in.)
P_e	Free endplay	mm(in.)
r	Raceway groove curvature radius	mm (in.)
R	Roller contour radius	mm (in.)
S_d	Diametral play	mm (in.)
Z	Number of balls or rollers per row	
w_s	Shim width	mm (in.)

Symbol	Description	Units
α°	Free contact angle	°
α	Contact angle	°
α_s	Shim angle	°
γ	$D \cos \alpha/d_m$	°
θ	Misalignment angle	°
ρ	Curvature	mm^{-1} ($in.^{-1}$)
$F(\rho)$	Curvature difference	
$\Sigma\rho$	Curvature sum	mm^{-1} ($in.^{-1}$)
ϕ	Osculation	
ω	Angular velocity	rad/sec

SUBSCRIPTS

c	Refers to cage
o	Refers to outer raceway
i	Refers to inner raceway
r	Refers to roller

GENERAL

Although ball and roller bearings appear to be simple mechanisms, their internal geometries are quite complex. For example, a radial ball bearing subjected to thrust loading assumes angles of contact between the balls and raceways in accordance with the relative conformities of the balls to the raceways and the diametral clearance. On the other hand, the ability of the same bearing to support the thrust loading depends on the contact angles formed. The same diametral clearance or play produces an axial endplay that may or may not be tolerable to the bearing user. In later chapters it will be demonstrated how diametral clearance affects not only contact angles and endplay but also stresses, deflections, load distributions, and fatigue life. Stresses, deflections, load distribution, and life in roller bearings are also affected by clearance.

In the determination of stresses and deflections the relative conformities of balls and rollers to their contacting raceways are of vital interest. In this chapter the principal macrogeometric relationships governing the operation of ball and roller bearings shall be developed and examined.

BALL BEARINGS

Clearance and Pitch Diameter

The ball bearing can be illustrated in its most simple form as in Fig. 2.1. From Fig. 2.1 one can easily see that the bearing pitch diameter is ap-

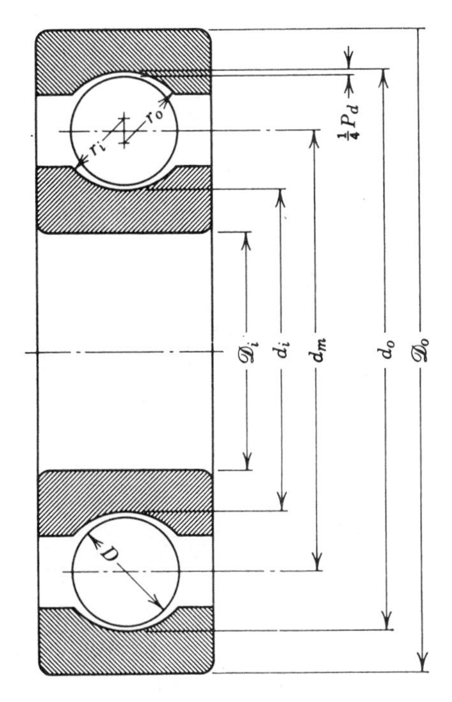

FIGURE 2.1. Radial ball bearing showing diametral clearance.

proximately equal to the mean of the bore and O.D. or

$$d_m \approx \tfrac{1}{2}(\text{bore} + \text{O.D.}) \tag{2.1}$$

More precisely, however, the bearing pitch diameter is the mean of the inner and outer ring raceway contact diameters. Therefore,

$$d_m = \tfrac{1}{2}(d_i + d_o) \tag{2.2}$$

Generally, ball bearings and other radial rolling bearings such as cylindrical roller bearings are designed with clearance. From Fig. 2.1, the diametral* clearance is as follows:

$$P_d = d_o - d_i - 2D \tag{2.3}$$

Table 2.1 taken from Reference 2.1 gives values of radial internal clearance for radial contact ball bearings under no load.

*Clearance is always measured on a diameter; however, because measurement takes place in a radial plane, it is commonly called radial clearance. This text will use diametral and radial clearance interchangeably.

TABLE 2.1. PART 1. Radial Internal Clearance Values for Radial Contact Ball Bearings

d mm		SYMBOL 2[a]		SYMBOL 0[a] (Normal)		SYMBOL 3[a]		SYMBOL 4[a]		SYMBOL 5[a]	
		Clearance Values in micrometers									
over	incl.	min.	max.	min.	max.	min.	max.	min.	max.	min.	max.
2.5	6	0	7	2	13	8	23	—	—	—	—
6	10	0	7	2	13	8	23	14	29	20	37
10	18	0	9	3	18	11	25	18	33	25	45
18	24	0	10	5	20	13	28	20	36	28	48
24	30	1	11	5	20	13	28	23	41	30	53
30	40	1	11	6	20	15	33	28	46	40	64
40	50	1	11	6	23	18	36	30	51	45	73
50	65	1	15	8	28	23	43	38	61	55	90
65	80	1	15	10	30	25	51	46	71	65	105
80	100	1	18	12	36	30	58	53	84	75	120
100	120	2	20	15	41	36	66	61	97	90	140
120	140	2	23	18	48	41	81	71	114	105	160
140	160	2	23	18	53	46	91	81	130	120	180
160	180	2	25	20	61	53	102	91	147	135	200
180	200	2	30	25	71	63	117	107	163	150	230

[a] These symbols relate to the ANSI/AFBMA Identification Code.

TABLE 2.1. PART 2

Clearance values in 0.0001 in.

d mm		SYMBOL 2[a]		SYMBOL 0[a] (Normal)		SYMBOL 3[a]		SYMBOL 4[a]		SYMBOL 5[a]	
over	incl.	min.	max.	min.	max.	min.	max.	min.	max.	min.	max.
2.5	6	0	3	1	5	3	9	—	—	—	—
6	10	0	3	1	5	3	9	6	11	8	15
10	18	0	3.5	1	7	4	10	7	13	10	18
18	24	0	4	2	8	5	11	8	14	11	19
24	30	0.5	4.5	2	8	5	11	9	16	12	21
30	40	0.5	4.5	2	8	6	13	11	18	16	25
40	50	0.5	4.5	2.5	9	7	14	12	20	18	29
50	65	0.5	6	3.5	11	9	17	15	24	22	35
65	80	0.5	6	4	12	10	20	18	28	26	41
80	100	0.5	7	4.5	14	12	23	21	33	30	47
100	120	1	8	6	16	14	26	24	38	35	55
120	140	1	9	7	19	16	32	28	45	41	63
140	160	1	9	7	21	18	36	32	51	47	71
160	180	1	10	8	24	21	40	36	58	53	79
180	200	1	12	10	28	25	46	42	64	59	91

[a]These symbols relate to the ANSI/AFBMA Identification Code.

Example 2.1. A 209 single-row radial deep-groove ball bearing has the following dimensions:

Inner raceway diameter, d_i	52.291 mm (2.0587 in.)
Outer raceway diameter, d_o	77.706 mm (3.0593 in.)
Ball diameter, D	12.7 mm (0.5 in.)
Number of balls, Z	9
Inner groove radius, r_i	6.6 mm (0.26 in.)
Outer groove radius, r_o	6.6 mm (0.26 in.)

Determine the bearing pitch diameter d_m and diametral clearance P_d.

$$d_m = \tfrac{1}{2}(d_i + d_o) \tag{2.2}$$

$$= \tfrac{1}{2}(52.3 + 77.7)$$

$$= 65 \text{ mm } (2.559 \text{ in.})$$

$$P_d = d_o - d_i - 2D \tag{2.3}$$

$$= 77.706 - 52.291 - 2 \times 12.7$$

$$= 0.015 \text{ mm } (0.0006 \text{ in.})$$

Osculation

The ability of a ball bearing to carry load depends in large measure on the osculation of the rolling elements and raceways. Osculation is the ratio of the radius of curvature of the rolling element to that of the raceway in a direction transverse to the direction of rolling. From Fig. 2.1 it can be seen that for a ball mating with a raceway, osculation is given by

$$\phi = \frac{D}{2r} \tag{2.4}$$

Letting $r = fD$, osculation is

$$\phi = \frac{1}{2f} \tag{2.5}$$

It is to be noted that the osculation is not necessarily identical for inner and outer contacts.

Example 2.2. Determine the osculation in the 209 radial ball bearing of Example 2.1.

$$f_i = \frac{r_i}{D} = \frac{6.6}{12.7} = 0.52$$

$$\phi_i = \frac{1}{2f_i} = \frac{1}{2 \times 0.52} = 0.962 \tag{2.5}$$

$$f_o = \frac{r_o}{D} = \frac{6.6}{12.7} = 0.52$$

$$\phi_o = \frac{1}{2f_o} = \frac{1}{2 \times 0.52} = 0.962 \tag{2.5}$$

Contact Angle and Endplay

Because a radial ball bearing is generally designed to have a diametral clearance in the no-load state, the bearing also can experience an axial play. Removal of this axial freedom causes the ball-raceway contact to assume an oblique angle with the radial plane; hence, a contact angle different from zero degrees will occur. Angular-contact ball bearings are specifically designed to operate under thrust load and the clearance built into the unloaded bearing along with the raceway groove curvatures determines the bearing free contact angle. Figure 2.2 shows the geometry of a radial ball bearing with axial play removed. From Fig. 2.2 it can be seen that the distance between the centers of curvature 0′ and 0″ of the inner and outer ring grooves is

$$A = r_o + r_i - D \tag{2.6}$$

Substituting $r = fD$ yields

$$A = (f_o + f_i - 1)D = BD \tag{2.7}$$

in which $B = f_o + f_i - 1$ is defined as the total curvature of the bearing.

Also from Fig. 2.2, it can be seen that the free contact angle is the angle made by the line passing through the points of contact of the ball and both raceways and a plane perpendicular to the bearing axis of rotation. The magnitude of the free contact angle can be described as follows:

$$\cos \alpha^\circ = \frac{\frac{1}{2}A - \frac{1}{4}P_d}{\frac{1}{2}A} \tag{2.8}$$

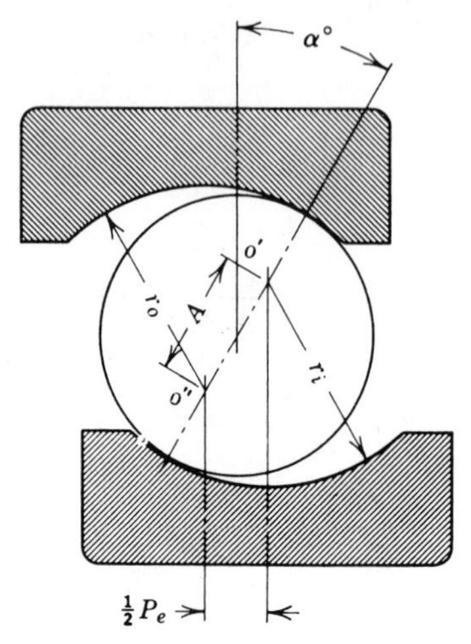

FIGURE 2.2. Radial ball bearing showing ball-raceway contact due to axial shift of inner and outer rings.

or

$$\alpha^\circ = \cos^{-1}\left(1 - \frac{P_d}{2A}\right) \tag{2.9}$$

Example 2.3. A 218 angular-contact ball bearing has dimensions as follows:

Inner raceway diameter, d_i	102.79 mm (4.047 in.)
Outer raceway diameter, d_o	147.73 mm (5.816 in.)
Ball diameter, D	22.23 mm (0.875 in.)
Inner groove radius, r_i	11.63 mm (0.4578 in.)
Outer groove radius, r_o	11.63 mm (0.4578 in.)

Determine the free contact angle of this bearing.

$$f_i = \frac{r_i}{D}$$

$$= \frac{11.63}{22.23} = 0.5232$$

$$f_o = \frac{r_o}{D}$$

$$= \frac{11.63}{22.23} = 0.5232$$

$$B = f_i + f_o - 1$$

$$= 0.5232 + 0.5232 - 1 = 0.0464$$

$$A = BD \tag{2.7}$$

$$= 0.0464 \times 22.23 = 1.031 \text{ mm } (0.0406 \text{ in.})$$

$$P_d = d_o - d_i - 2D \tag{2.3}$$

$$= 147.73 - 102.79 - 2 \times 22.23 = 0.48 \text{ mm } (0.019 \text{ in.})$$

$$\alpha^\circ = \cos^{-1}\left(1 - \frac{P_d}{2A}\right) \tag{2.9}$$

$$= \cos^{-1}\left(1 - \frac{0.48}{2 \times 1.031}\right) = 40^\circ$$

If in mounting the bearing an interference fit is used, then the diametral clearance must be diminished by the change in ring diameter to obtain the free contact angle. Hence

$$\alpha^\circ = \cos^{-1}\left(1 - \frac{P_d + \Delta P_d)}{2A}\right) \tag{2.10}$$

Because of diametral clearance a radial bearing is free to float axially under the condition of no load. This free endplay may be defined as the maximum relative axial movement of the inner ring with respect to the outer ring under zero load. From Fig. 2.2,

$$\tfrac{1}{2}P_e = A \sin \alpha^\circ \tag{2.11}$$

$$P_e = 2A \sin \alpha^\circ \tag{2.12}$$

Figure 2.3 shows the free contact angle and endplay versus P_d/D for single-row ball bearings.

Double-row angular-contact ball bearings are generally assembled with a certain amount of diametral play (smaller than diametral clearance). It can be determined that the free endplay for a double-row bearing is

$$P_e = 2A \sin \alpha^\circ - 2\left[A^2 - \left(A \cos \alpha^\circ + \frac{S_d}{2}\right)^2\right]^{1/2} \tag{2.13}$$

Split inner ring ball bearings as illustrated in Fig. 2.4 have inner rings that are ground with a shim between the ring halves. The width

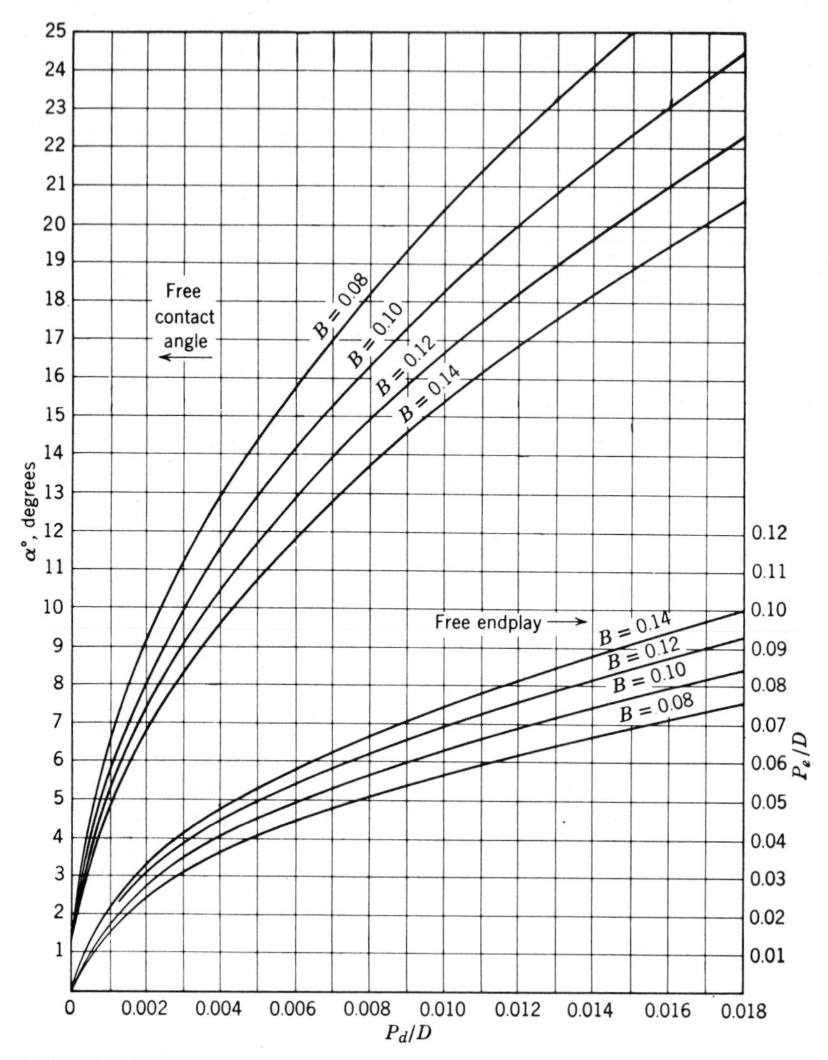

FIGURE 2.3. Free contact angle and endplay versus $B = f_o + f_i - 1$ for single-row ball bearings.

of this shim is associated with a shim angle that is obtained by removing the shim and abutting the ring halves.

From Fig. 2.5 it can be determined that the shim width is given by

$$w_s = (2r_i - D)\sin\alpha_s \tag{2.14}$$

Since $f_i = r_i/D$, equation (2.14) becomes

$$w_s = (2f_i - 1)D\sin\alpha_s \tag{2.15}$$

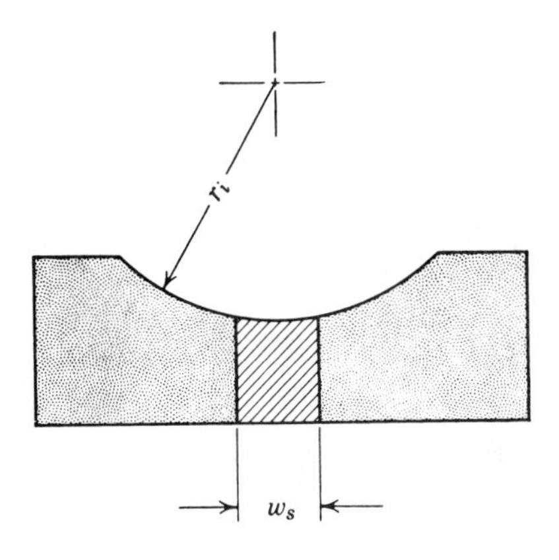

FIGURE 2.4. Inner rings of split inner ring ball bearing showing shim for grinding.

The shim angle α_s, and the assembled diametral play S_d of the bearing accordingly dictate the free contact angle. The effective clearance P_d of the bearing may be determined from Fig. 2.5 to be

$$P_d = S_d + (2f_i - 1)(1 - \cos \alpha_s)D \tag{2.16}$$

Thus, the bearing contact angle which is shown in Fig. 2.2 is given by

$$\alpha^\circ = \cos^{-1}\left(1 - \frac{S_d}{2BD} - \frac{(2f_i - 1)(1 - \cos \alpha_s)}{2B}\right) \tag{2.17}$$

Free Angle of Misalignment

Furthermore, diametral clearance can allow a ball bearing to misalign slightly under no load. The free angle of misalignment is defined as the maximum angle through which the axis of the inner ring can be rotated

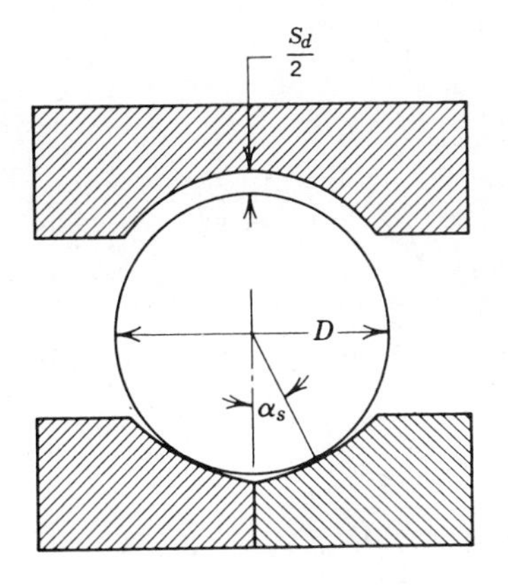

FIGURE 2.5. Split inner ring ball bearing assembly showing shim angle.

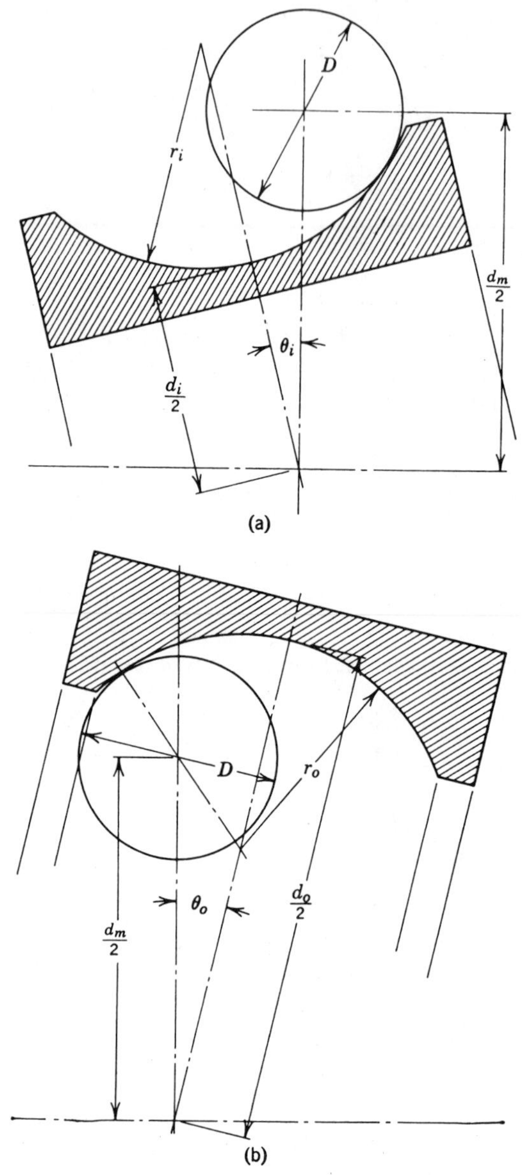

FIGURE 2.6. (*a*) Free misalignment of inner ring of single-row ball bearing, (*b*) free misalignment of outer ring of single-row ball bearing.

with respect to the axis of the outer ring before stressing bearing components. From Fig. 2.6 using the law of cosines it can be determined that

$$\cos \theta_{\mathrm{i}} = 1 - \frac{P_{\mathrm{d}}[(2f_{\mathrm{i}} - 1)D - (P_{\mathrm{d}}/4)]}{2d_{\mathrm{m}}[d_{\mathrm{m}} + (2f_{\mathrm{i}} - 1)D - (P_{\mathrm{d}}/2)]} \tag{2.18}$$

$$\cos\theta_o = 1 - \frac{P_d[(2f_o - 1)D - (P_d/4)]}{2d_m[d_m - (2f_o - 1)D + (P_d/2)]} \tag{2.19}$$

Therefore θ, the free angle of misalignment, is

$$\theta = \theta_i + \theta_o \tag{2.20}$$

Since the following trigonometric identity is true,

$$\cos\theta_i + \cos\theta_o = 2\cos\tfrac{1}{2}(\theta_i + \theta_o)\cos\tfrac{1}{2}(\theta_i - \theta_o) \tag{2.21}$$

and since $\theta_i - \theta_o$ approaches zero, therefore,

$$\theta = 2\cos^{-1}\left(\frac{\cos\theta_i + \cos\theta_o}{2}\right) \tag{2.22}$$

or

$$\theta = 2\cos^{-1}\left[1 - \frac{P_d}{4d_m}\left(\frac{(2f_i - 1)D - (P_d/4)}{d_m + (2f_i - 1)D - (P_d/2)} + \frac{(2f_o - 1)D - (P_d/4)}{d_m - (2f_o - 1)D + (P_d/2)}\right)\right] \tag{2.23}$$

Example 2.4. Determine the free contact angle $\alpha°$, free endplay P_e, and free angle of misalignment of the 209 radial ball bearing in Example 2.1.

$f_i = f_o = 0.52$ Ex. 2.2

$d_m = 65$ mm (2.559 in.) Ex. 2.1

$P_d = 0.015$ mm (0.0006 in.) Ex. 2.1

$$B = f_i + f_o - 1 = 0.52 + 0.52 - 1 = 0.04$$

$$A = BD = 0.04 \times 12.7 = 0.508 \tag{2.7}$$

$$\alpha° = \cos^{-1}\left(1 - \frac{P_d}{2A}\right) \tag{2.9}$$

$$= \cos^{-1}\left(1 - \frac{0.015}{2 \times 0.508}\right) = 9°52'$$

$$P_e = 2A \sin \alpha^\circ = 2 \times 0.508 \times \sin(9^\circ 52') = 0.174 \text{ mm } (0.0069 \text{ in.}) \tag{2.12}$$

$$\theta = 2\cos^{-1}\left[1 - \frac{P_d}{4d_m}\left(\frac{(2f_i - 1)D - (P_d/4)}{d_m + (2f_i - 1)D - (P_d/2)} + \frac{(2f_o - 1)D - (P_d/4)}{d_m - (2f_o - 1)D + (P_d/2)}\right)\right]$$

$$= 2\cos^{-1}\left[1 - \frac{0.015}{4 \times 65}\left(\frac{(2 \times 0.52 - 1) \times 12.7 - (0.015/4)}{65 + (2 \times 0.52 - 1) \times 12.7 - (0.015/2)} + \frac{(2 \times 0.52 - 1) \times 12.7 - (0.015/4)}{65 - (2 \times 0.52 - 1) \times 12.7 + (0.015/2)}\right)\right]$$

$$= 9'20'' \tag{2.23}$$

Note how small the free angle of misalignment is.

Curvature and Relative Curvature

Two bodies of revolution having different radii of curvature in a pair of principal planes passing through the contact between the bodies may contact each other at a single point under the condition of no applied load. Such a condition is called point contact. Figure 2.7 demonstrates this condition.

In Fig. 2.7 the upper body is denoted by I and the lower body by II. The principal planes are denoted by 1 and 2. Therefore, the radius of curvature of body I in plane 2 is denoted by $r_{\mathrm{I}2}$. Since r denotes radius of curvature, curvature is defined as

$$\rho = \frac{1}{r} \tag{2.24}$$

Although radius of curvature is always of positive sign, curvature may be positive or negative, convex surfaces being positive and concave surfaces negative.

To describe the contact between mating surfaces of revolution, the following definitions are used.

1. Curvature sum:

$$\sum \rho = \frac{1}{r_{\mathrm{I}1}} + \frac{1}{r_{\mathrm{I}2}} + \frac{1}{r_{\mathrm{II}1}} + \frac{1}{r_{\mathrm{II}2}} \tag{2.25}$$

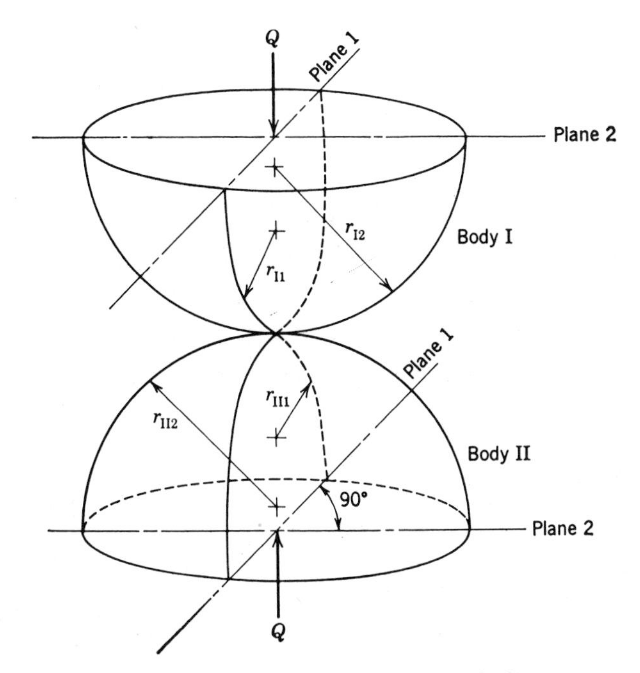

FIGURE 2.7. Geometry of contacting bodies.

2. Curvature difference:

$$F(\rho) = \frac{(\rho_{\mathrm{I}1} - \rho_{\mathrm{I}2}) + (\rho_{\mathrm{II}1} - \rho_{\mathrm{II}2})}{\sum \rho} \tag{2.26}$$

In equations (2.25) and (2.26) the sign convention for convex and concave surfaces is used. Furthermore, care must be exercised to see that $F(\rho)$ is positive.

By way of example, $F(\rho)$ is determined for a ball-inner raceway contact as follows (see Fig. 2.8):

$$r_{\mathrm{I}1} = \tfrac{1}{2}D$$

$$r_{\mathrm{I}2} = \tfrac{1}{2}D$$

$$r_{\mathrm{II}1} = \tfrac{1}{2}d_{\mathrm{i}} = \frac{1}{2}\left(\frac{d_{\mathrm{m}}}{\cos\alpha} - D\right)$$

$$r_{\mathrm{II}2} = f_{\mathrm{i}}D$$

Let

$$\gamma = \frac{D\cos\alpha}{d_{\mathrm{m}}} \tag{2.27}$$

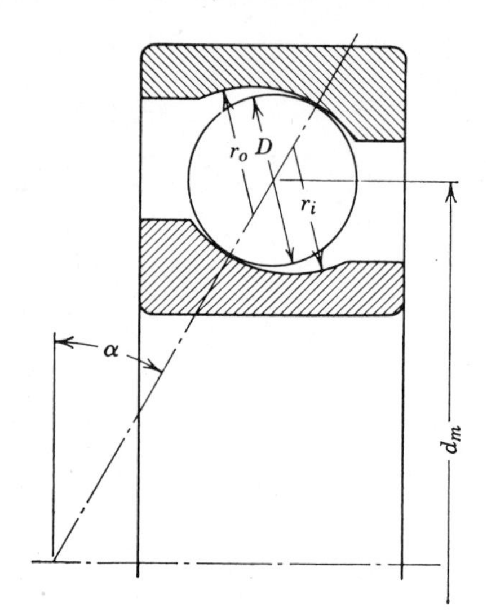

FIGURE 2.8. Ball bearing geometry.

Then

$$\rho_{\mathrm{I1}} = \rho_{\mathrm{I2}} = \frac{2}{D}$$

$$\rho_{\mathrm{II1}} = \frac{2}{D}\left(\frac{\gamma}{1-\gamma}\right)$$

$$\rho_{\mathrm{II2}} = -\frac{1}{f_{\mathrm{i}}D}$$

$$\sum \rho_{\mathrm{i}} = \frac{4}{D} - \frac{1}{f_{\mathrm{i}}D} + \frac{2}{D}\left(\frac{\gamma}{1-\gamma}\right) = \frac{1}{D}\left(4 - \frac{1}{f_{\mathrm{i}}} + \frac{2\gamma}{1-\gamma}\right) \tag{2.28}$$

$$F(\rho)_{\mathrm{i}} = \frac{\frac{2}{D}\left(\frac{\gamma}{1-\gamma}\right) - \left(-\frac{1}{f_{\mathrm{i}}D}\right)}{\sum \rho} = \frac{\frac{1}{f_{\mathrm{i}}} + \frac{2\gamma}{1-\gamma}}{4 - \frac{1}{f_{\mathrm{i}}} + \frac{2\gamma}{1-\gamma}} \tag{2.29}$$

For the ball-outer raceway contact $\rho_{\mathrm{I1}} = \rho_{\mathrm{I2}} = 2/D$ as above; however,

$$r_{\mathrm{II1}} = \frac{1}{2}\left(\frac{d_{\mathrm{m}}}{\cos\alpha} + D\right)$$

$$r_{\mathrm{II2}} = f_{\mathrm{o}}D$$

Therefore,

$$\rho_{\text{II}1} = -\frac{2}{D}\left(\frac{\gamma}{1+\gamma}\right)$$

$$\rho_{\text{II}2} = -\frac{1}{f_o D}$$

$$\sum \rho_o = \frac{1}{D}\left(4 - \frac{1}{f_o} - \frac{2\gamma}{1+\gamma}\right)$$

$$F(\rho)_o = \frac{\dfrac{1}{f_o} - \dfrac{2\gamma}{1+\gamma}}{4 - \dfrac{1}{f_o} - \dfrac{2\gamma}{1+\gamma}} \tag{2.30}$$

$F(\rho)$ is always a number between 0 and 1.

Example 2.5. Determine the values of curvature sum and curvature difference for the 209 radial ball bearing of Example 2.1, subjected to radial load only.

$$f_i = f_o = 0.52 \qquad \text{Ex. 2.2}$$

$$d_m = 65 \text{ mm } (2.559 \text{ in.}) \qquad \text{Ex. 2.1}$$

$$\gamma = \frac{D \cos \alpha}{d_m} \tag{2.27}$$

$$= \frac{12.7 \times \cos(0°)}{65} = 0.1954$$

$$\sum \rho_i = \frac{1}{D}\left(4 - \frac{1}{f_i} + \frac{2\gamma}{1-\gamma}\right) \tag{2.28}$$

$$= \frac{1}{12.7}\left(4 - \frac{1}{0.52} + \frac{2 \times 0.1954}{1 - 0.1954}\right) = 0.202 \text{ mm}^{-1} \ (5.126 \text{ in.}^{-1})$$

$$F(\rho)_i = \frac{\dfrac{1}{f_i} + \dfrac{2\gamma}{1-\gamma}}{4 - \dfrac{1}{f_i} + \dfrac{2\gamma}{1-\gamma}} \tag{2.29}$$

$$= \frac{\dfrac{1}{0.52} + \dfrac{2 \times 0.1954}{1 - 0.1954}}{4 - \dfrac{1}{0.52} + \dfrac{2 \times 0.1954}{1 - 0.1954}} = 0.9399$$

$$\sum \rho_o = \frac{1}{D}\left(4 - \frac{1}{f_o} - \frac{2\gamma}{1+\gamma}\right) \tag{2.30}$$

$$= \frac{1}{12.7}\left(4 - \frac{1}{0.52} - \frac{2 \times 0.1954}{1 + 0.1954}\right) = 0.1378 \text{ mm}^{-1} \text{ (3.500 in.}^{-1}\text{)}$$

$$F(\rho)_o = \frac{\dfrac{1}{f_o} - \dfrac{2\gamma}{1+\gamma}}{4 - \dfrac{1}{f_o} - \dfrac{2\gamma}{1+\gamma}} \tag{2.31}$$

$$= \frac{\dfrac{1}{0.52} - \dfrac{2 \times 0.1954}{1 + 0.1954}}{4 - \dfrac{1}{0.52} - \dfrac{2 \times 0.1954}{1 + 0.1954}} = 0.9120$$

Note that when $f_i = f_o$, $F(\rho)_i > F(\rho)_o$. This condition will be used to demonstrate in a later example that an elliptical area of contact of greater ellipticity generally exists at the inner raceway contact as opposed to the outer raceway contact.

Example 2.6. Determine the magnitude of curvature sum and curvature difference for the 218 angular-contact ball bearing of Example 2.3 subjected to light axial loading.

$$f_i = f_o = 0.5232 \qquad \text{Ex. 2.3}$$

$$d_m = \tfrac{1}{2}(d_i + d_o) \tag{2.2}$$

$$= (102.79 + 147.73) = 125.26 \text{ mm (4.932 in.)}$$

$$\alpha^\circ = 40^\circ * \qquad \text{Ex. 2.3}$$

$$\gamma = \frac{D \cos \alpha}{d_m} \tag{2.27}$$

$$= \frac{22.23 \times \cos(40^\circ)}{125.26} = 0.1359$$

$$\sum \rho_i = \frac{1}{D}\left(4 - \frac{1}{f_i} + \frac{2\gamma}{1-\gamma}\right) \tag{2.28}$$

$$= \frac{1}{22.23}\left(4 - \frac{1}{0.5232} + \frac{2 \times 0.1359}{1 - 0.1359}\right) = 0.108 \text{ mm}^{-1} \text{ (2.747 in.}^{-1}\text{)}$$

*It will be demonstrated in a later chapter that contact angle increases under thust (axial) loading. This will not be considered in this example.

$$F(\rho)_i = \frac{\dfrac{1}{f_i} + \dfrac{2\gamma}{1-\gamma}}{4 - \dfrac{1}{f_i} + \dfrac{2\gamma}{1-\gamma}} \tag{2.29}$$

$$= \frac{\dfrac{1}{0.5232} + \dfrac{2 \times 0.11359}{1 - 0.1359}}{4 - \dfrac{1}{0.5232} + \dfrac{2 \times 0.1359}{1 - 0.1359}} = 0.9260$$

$$\sum \rho_o = \frac{1}{D}\left(4 - \frac{1}{f_o} - \frac{2\gamma}{1+\gamma}\right) \tag{2.30}$$

$$= \frac{1}{22.23}\left(4 - \frac{1}{0.5232} - \frac{2 \times 0.1359}{1 + 0.1359}\right)$$

$$= 0.0832 \text{ mm}^{-1} \ (2.114 \text{ in.}^{-1})$$

$$F(\rho)_o = \frac{\dfrac{1}{f_o} - \dfrac{2\gamma}{1+\gamma}}{4 - \dfrac{1}{f_o} - \dfrac{2\gamma}{1+\gamma}} \tag{2.31}$$

$$= \frac{\dfrac{1}{0.5232} - \dfrac{2 \times 0.1359}{1 + 0.1359}}{4 - \dfrac{1}{0.5232} - \dfrac{2 \times 0.1359}{1 + 0.1359}} = 0.9038$$

Comparison of the $F(\rho)_i$ and $F(\rho)_o$ values in Examples 2.6 and 2.7 indicates that magnitudes in the neighborhood of 0.9 are to be expected for ball bearings. Larger magnitudes of f_i and f_o cause subsequently smaller values of $F(\rho)$.

ROLLER BEARINGS

Pitch Diameter and Clearance

Equations (2.1)–(2.3), are equally valid for roller bearings as well as ball bearings. Tables 2.2 and 2.3 give standard values of internal clearance for cylindrical and spherical roller bearings under no load. These data are excerpted from reference [2.1].

TABLE 2.2. PART 1. Radial Internal Clearance Values for Cylindrical Roller Bearings

Clearance values in micrometers

d mm		SYMBOL 2[a]				SYMBOL 0[a] (Normal)				SYMBOL 3[a]				SYMBOL 4[a]			
		Interchangeable[b]	Matched[c]	Matched[c]	Interchangeable[b]	Interchangeable[b]	Matched[c]	Matched[c]	Interchangeable[b]	Interchangeable[b]	Matched[c]	Matched[c]	Interchangeable[b]	Interchangeable[b]	Matched[c]	Matched[c]	Interchangeable[b]
Over	Incl.	Low	Low	High	High	Low	Low	High	High	Low	Low	High	High	Low	Low	High	High
	10	0	10	20	30	10	20	30	41	25	36	46	56	36	46	56	66
10	18	0	10	20	30	10	20	30	41	25	36	46	56	36	46	56	66
18	24	0	10	20	30	10	20	30	41	25	36	46	56	36	46	56	66
24	30	0	10	25	30	10	25	36	46	30	41	51	66	41	51	61	71
30	40	0	13	25	36	15	25	41	51	36	46	56	71	46	56	71	81
40	50	5	15	30	41	20	30	46	56	41	51	66	76	56	66	81	89
50	65	5	15	36	46	20	36	51	66	46	56	76	89	66	76	89	104
65	80	5	20	41	56	25	41	61	76	56	71	89	104	76	89	109	124
80	100	10	25	46	61	30	46	71	81	66	81	104	114	89	104	124	140
100	120	10	25	51	66	36	51	81	89	81	94	119	135	104	119	145	160
120	140	10	30	61	76	41	61	89	104	89	104	135	155	114	135	160	180
140	160	15	36	66	81	51	66	99	114	99	114	150	165	130	150	180	196
160	180	20	36	76	86	61	76	109	124	109	124	165	175	150	165	201	216
180	200	25	41	81	94	66	81	119	135	124	140	180	196	165	180	221	234
200	225	30	46	89	104	76	—	—	150	140	—	—	216	180	—	—	254
225	250	41	51	99	114	90	—	—	165	155	—	—	229	206	—	—	279
250	280	46	56	109	124	99	—	—	180	175	—	—	254	229	—	—	310
280	315	51	61	119	132	109	—	—	196	196	—	—	279	254	—	—	340
315	355	56	66	135	145	124	—	—	216	216	—	—	305	279	—	—	371
355	400	66	76	150	160	140	—	—	236	244	—	—	340	320	—	—	414
400	450	71	—	—	191	155	—	—	274	269	—	—	389	356	—	—	455
450	500	84	—	—	206	180	—	—	300	300	—	—	419	394	—	—	513

[a]These symbols relate to the ANSI/AFBMA Identification Code.
[b]The Term "Interchangeable" refers to such assembly of rings and rollers that the separable ring can be replaced by any other ring of the ANSI/AFBMA same design and manufacture.
[c]The term "Matched" refers to such assembly of rings and rollers that the separable ring can not be replaced by any other ring.

TABLE 2.2. PART 2.

Clearance values in 0.0001 inches

d mm		SYMBOL 2[a]				SYMBOL 0[a] (Normal)				SYMBOL 3[a]				SYMBOL 4[a]			
			Interchangeable[b] Matched[c]				Interchangeable[b] Matched[c]				Interchangeable[b] Matched[c]				Interchangeable[b] Matched[c]		
Over	Incl.	Low	Low	High	High	Low	Low	High	High	Low	Low	High	High	Low	Low	High	High
	10	0	4	8	12	4	8	12	16	10	14	18	22	14	18	22	26
10	18	0	4	8	12	4	8	12	16	10	14	18	22	14	18	22	26
18	24	0	4	8	12	4	8	12	16	10	14	18	22	14	18	22	26
24	30	0	4	10	12	4	10	14	18	12	16	20	26	16	20	24	28
30	40	0	5	10	14	6	10	16	20	14	18	22	28	18	22	28	32
40	50	2	6	12	16	8	12	18	22	16	20	26	30	22	26	32	35
50	65	2	6	14	18	8	14	20	26	18	22	30	35	26	30	35	41
65	80	2	8	16	22	10	16	24	30	22	28	35	41	30	35	43	49
80	100	4	10	18	24	12	18	28	32	26	32	41	45	35	41	49	55
100	120	4	10	20	26	14	20	32	35	32	37	47	53	41	47	57	63
120	140	4	12	24	30	16	24	35	41	35	41	53	61	45	53	63	71
140	160	6	14	26	32	20	26	39	45	39	45	59	65	51	59	71	77
160	180	8	14	30	34	24	30	43	49	43	49	65	69	59	65	79	85
180	200	10	16	32	37	26	32	47	53	49	55	71	77	65	71	87	92
200	225	12	18	35	41	30	—	—	59	55	—	—	85	71	—	—	100
225	250	16	20	39	45	35	—	—	65	61	—	—	90	81	—	—	110
250	280	18	22	43	49	39	—	—	71	69	—	—	100	90	—	—	122
280	315	20	24	47	52	43	—	—	77	77	—	—	110	100	—	—	134
315	355	22	26	53	57	49	—	—	85	85	—	—	120	110	—	—	146
355	400	26	30	59	63	55	—	—	93	96	—	—	134	126	—	—	163
400	450	28	—	—	75	61	—	—	108	106	—	—	153	140	—	—	179
450	500	33	—	—	81	71	—	—	118	118	—	—	165	155	—	—	202

[a]These symbols relate to the ANSI/AFBMA Identification Code.
[b]The Term "Interchangeable" refers to such assembly of rings and rollers that the separable ring can be replaced by any other ring of the ANSI/AFBMA same design and manufacture.
[c]The term "Matched" refers to such assembly of rings and rollers that the separable ring can not be replaced by any other ring.

TABLE 2.3. PART 1. Radial Internal Clearance Values for Self-Aligning Roller Bearings with Cylindrical Bore

d mm		*Clearance values in micrometers*									
		SYMBOL 2[a]		SYMBOL 0[a] (Normal)		SYMBOL 3[a]		SYMBOL 4[a]		SYMBOL 5[a]	
over	incl.	min.	max.	min.	max.	min.	max.	min.	max.	min.	max.
14	24	10	20	20	35	35	45	45	60	60	75
24	30	15	25	25	40	40	55	55	75	75	95
30	40	15	30	30	45	45	60	60	80	80	100
40	50	20	35	35	55	55	75	75	100	100	125
50	65	20	40	40	65	65	90	90	120	120	150
65	80	30	50	50	80	80	110	110	145	145	180
80	100	35	60	60	100	100	135	135	180	180	225
100	120	40	75	75	120	120	160	160	210	210	260
120	140	50	95	95	145	145	190	190	240	240	300
140	160	60	110	110	170	170	220	220	280	280	350
160	180	65	120	120	180	180	240	240	310	310	390
180	200	70	130	130	200	200	260	260	340	340	430
200	225	80	140	140	220	220	290	290	380	380	470
225	250	90	150	150	240	240	320	320	420	420	520
250	280	100	170	170	260	260	350	350	460	460	570
280	315	110	190	190	280	280	370	370	500	500	630
315	355	120	200	200	310	310	410	410	550	550	690
355	400	130	220	220	340	340	450	450	600	600	750
400	450	140	240	240	370	370	500	500	660	660	820
450	500	140	260	260	410	410	550	550	720	720	900
500	560	150	280	280	440	440	600	600	780	780	1,000
560	630	170	310	310	480	480	650	650	850	850	1,100
630	710	190	350	350	530	530	700	700	920	920	1,190
710	800	210	390	390	580	580	770	770	1,010	1,010	1,300
800	900	230	430	430	650	650	860	860	1,120	1,120	1,440
900	1,000	260	480	480	710	710	930	930	1,220	1,220	1,570

[a]These symbols relate to the ANSI/AFBMA Identification Code.

TABLE 2.3. PART 2.

Clearance values in 0.0001 inches

d mm		SYMBOL 2[a]		SYMBOL 0[a] (Normal)		SYMBOL 3[a]		SYMBOL 4[a]		SYMBOL 5[a]	
Over	Incl.	Low	High	Low	High	Low	High	Low	High	Low	High
14	24	4	8	8	14	14	18	18	24	24	30
24	30	6	10	10	16	16	22	22	30	30	37
30	40	6	12	12	18	18	24	24	31	31	39
40	50	8	14	14	22	22	30	30	39	39	49
50	65	8	16	16	26	26	35	35	47	47	59
65	80	12	20	20	31	31	43	43	57	57	71
80	100	14	24	24	39	39	53	53	71	71	89
100	120	16	30	30	47	47	63	63	83	83	102
120	140	20	37	37	57	57	75	75	94	94	118
140	160	24	43	43	67	67	87	87	110	110	138
160	180	26	47	47	71	71	94	94	122	122	154
180	200	28	51	51	79	79	102	102	134	134	169
200	225	31	55	55	87	87	114	114	150	150	185
225	250	35	59	59	94	94	126	126	165	165	205
250	280	39	67	67	102	102	138	138	181	181	224
280	315	43	75	75	110	110	146	146	197	197	248
315	355	47	79	79	122	122	161	161	217	217	272
355	400	51	87	87	134	134	177	177	236	236	295
400	450	55	94	94	146	146	197	197	260	260	323
450	500	55	102	102	161	161	217	217	283	283	354
500	560	59	110	110	173	173	236	236	307	307	394
560	630	67	122	122	189	189	256	256	335	335	433
630	710	75	138	138	209	209	276	276	362	362	469
710	800	83	154	154	228	228	303	303	398	398	512
800	900	91	169	169	256	256	339	339	441	441	567
900	1,000	102	189	189	280	280	366	366	480	480	618

[a]These symbols relate to the ANSI/AFBMA Identification Code.

Example 2.7. A 209 cylindrical roller bearing has the following dimensions:

Inner raceway diameter, d_i	54.991 mm (2.165 in.)
Outer raceway diameter, d_o	75.032 mm (2.954 in.)
Roller diameter, D	10 mm (0.3937 in.)
Roller effective length, l	9.601 mm* (0.3780 in.)
Roller total length, l_t	10 mm (0.3937 in.)
Number of rollers, Z	14

Determine the bearing pitch diameter d_m and diametral clearance P_d.

$$d_m = \tfrac{1}{2}(d_i + d_o) \tag{2.2}$$

$$= \tfrac{1}{2}(54.991 + 75.032) = 65.011 \text{ mm (2.559 in.)}$$

$$P_d = d_o - d_i - 2D \tag{2.3}$$

$$= 75.032 - 54.991 - 2 \times 10 = 0.041 \text{ mm (0.0016 in.)}$$

Osculation and Crowning

The term osculation also applies to roller bearings in that the roller and/or raceways may be crowned in the direction transverse to rolling. Such crowning consists of forming all or a portion of the roller or raceway contour parallel to the roller's axis of rotation of a relatively large radius arc. Figure 1.33*b* shows a partially crowned roller.

For a roller bearing, therefore, osculation is stated as follows:

$$\phi = \frac{R}{r} \tag{2.32}$$

in which R is the roller contour radius and r is the contour radius of the raceway.

*Roller effective length is the length presumed to be in contact with the raceway under loading. Generally.

$$l = l_t - 2r_c$$

in which r_c is the roller corner radius or the grinding undercut, whichever is larger. A combination of the corner radius and grinding undercut may also be required to estimate effective length.

Example 2.8. A 22317 two-row spherical roller bearing has the following dimensions:

Roller diameter, D	25 mm (0.9843 in.)
Number of rollers per row, Z	14
Roller effective length, l	20.762 mm (0.8154 in.)
Roller contour radius, R	79.959 mm (3.148 in.)
Inner raceway contour radius, r_i	81.585 mm (3.212 in.)
Outer raceway sphere radius, r_o	81.585 mm (3.212 in.)
Bearing pitch diameter, d_m	135.077 mm (5.318 in.)
Diametral play, S_d	0.102 mm (0.004 in.)

Determine the osculation at each raceway-roller contact.

$$\phi_i = \frac{R}{r_i} \tag{2.32}$$

$$= \frac{79.959}{81.585} = 0.98$$

$$\phi_o = \frac{R}{r_o} \tag{2.32}$$

$$= \frac{79.959}{81.585} = 0.98$$

Free Endplay and Contact Angle

The free contact angle of a straight contour roller bearing is not affected by the diametral play. Radial cylindrical roller bearings have a 0° contact angle and may take thrust load only by virtue of axial flanges. Tapered roller bearings must be subjected to a thrust load or the inner and outer rings will not remain assembled; therefore, tapered roller bearings do not exhibit free diametral play. Radial spherical roller bearings are, however, normally assembled with free diametral play and hence exhibit free endplay.

Diametral play S_d in a spherical roller bearing can be measured with a feeler gage such that from Fig. 2.9:

$$S_d = 2[r_o - (r_i + D)] \tag{2.33}$$

in which r_i is the radius from the bearing center to the contact point on the inner raceway. It may be seen from Fig. 2.9 that

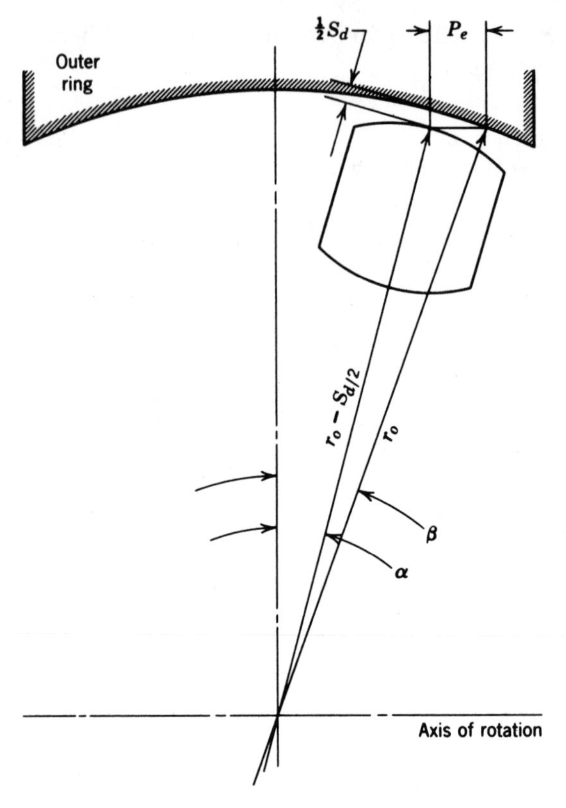

FIGURE 2.9. Spherical roller bearing schematic diagram showing diametral play and endplay.

$$r_o \cos \beta = \left(r_o - \frac{S_d}{2} \right) \cos \alpha \tag{2.34}$$

or

$$\beta = \cos^{-1} \left[\left(1 - \frac{S_d}{2r_o} \right) \cos \alpha \right] \tag{2.35}$$

Figure 2.9 also shows that

$$P_e = 4r_o(\sin \beta - \sin \alpha) + 2S_d \sin \alpha \tag{2.36}$$

Example 2.9. Estimate the magnitude of the free endplay P_e for the 22317 spherical roller bearing of Example 2.8 if the contact angle is nominally 12°.

$$\beta = \cos^{-1}\left[\left(1 - \frac{S_d}{2r_o}\right)\cos\alpha\right] \tag{2.35}$$

$$= \cos^{-1}\left[\left(1 - \frac{0.102}{2 \times 81.585}\right)\cos(12°)\right] = 12°10'$$

$$P_e = 4r_o(\sin\beta - \sin\alpha) + 2S_d\sin\alpha \tag{2.36}$$

$$= 4 \times 85.585\,(0.2108 - 0.2079) + 2 \times 0.102 \times 0.2079$$

$$= 0.956 \text{ mm } (0.0376 \text{ in.})$$

Curvature

For roller bearings with point contact between roller and raceway the equations for curvature sum and difference are as follows (see Fig. 2.10):

$$\sum\rho_i = \frac{2}{D} + \frac{1}{R} + \frac{2\gamma}{D(1-\gamma)} - \frac{1}{r_i} = \frac{1}{D}\left[\frac{2}{1-\gamma} + D\left(\frac{1}{R} - \frac{1}{r_i}\right)\right] \tag{2.37}$$

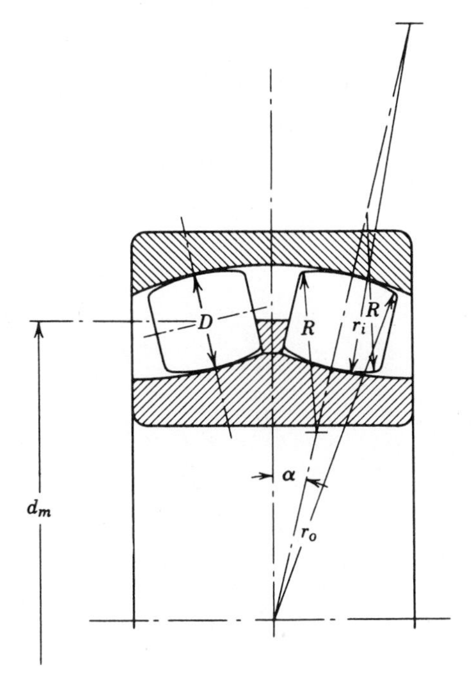

FIGURE 2.10. Spherical roller bearing geometry.

$$F(\rho)_{\mathrm{i}} = \frac{\frac{2}{D} - \frac{1}{R} + \frac{2\gamma}{D(1-\gamma)} - \left(-\frac{1}{r_{\mathrm{i}}}\right)}{\sum \rho_{\mathrm{i}}} = \frac{\frac{2}{1-\gamma} - D\left(\frac{1}{R} - \frac{1}{r_{\mathrm{i}}}\right)}{\frac{2}{1-\gamma} + D\left(\frac{1}{R} - \frac{1}{r_{\mathrm{i}}}\right)} \tag{2.38}$$

$$\sum \rho_{\mathrm{o}} = \frac{2}{D} + \frac{1}{R} - \frac{2\gamma}{D(1+\gamma)} - \frac{1}{r_{\mathrm{o}}} = \frac{1}{D}\left[\frac{2}{1+\gamma} + D\left(\frac{1}{R} - \frac{1}{r_{\mathrm{o}}}\right)\right] \tag{2.39}$$

$$F(\rho)_{\mathrm{o}} = \frac{\frac{2}{D} - \frac{1}{R} - \frac{2\gamma}{D(1+\gamma)} - \left(-\frac{1}{r_{\mathrm{o}}}\right)}{\sum \rho_{\mathrm{o}}}$$

$$= \frac{\frac{2}{1+\gamma} - D\left(\frac{1}{R} - \frac{1}{r_{o}}\right)}{\frac{2}{1+\gamma} + D\left(\frac{1}{R} - \frac{1}{r_{\mathrm{o}}}\right)} \tag{2.40}$$

It can be seen from equations (2.38) and (2.40), that $F(\rho) = 1$ if the roller contour radius R is identical to the raceway groove radius r. This condition also occurs for roller bearings having straight raceway and roller contours, in which case $R = r_{\mathrm{i}} = r_{\mathrm{o}} = \infty$.

Example 2.10. Calculate the magnitudes of curvature sum and difference for the 22317 spherical roller bearing of Example 2.8.

$$\gamma = \frac{D \cos \alpha}{d_{\mathrm{m}}} \tag{2.27}$$

$$= 25 \cos (12°)/135.077 = 0.1810$$

$$\frac{1}{R} - \frac{1}{r_{\mathrm{i}}} = \frac{1}{79.959} - \frac{1}{81.585} = 0.00025 \text{ mm}^{-1} \text{ } (0.0064 \text{ in.}^{-1})$$

$$\sum \rho_{\mathrm{i}} = \frac{1}{D}\left[\frac{2}{1-\gamma} + D\left(\frac{1}{R} - \frac{1}{r_{\mathrm{i}}}\right)\right] \tag{2.37}$$

$$= \frac{1}{25}\left[\frac{2}{1-0.1810} + 25 \times 0.00025\right]$$

$$= 0.09793 \text{ mm}^{-1} \text{ } (2.487 \text{ in.}^{-1})$$

$$F(\rho)_i = \frac{\dfrac{2}{1-\gamma} - D\left(\dfrac{1}{R} - \dfrac{1}{r_i}\right)}{\dfrac{2}{1-\gamma} + D\left(\dfrac{1}{R} - \dfrac{1}{r_i}\right)} \tag{2.38}$$

$$= \frac{2/(1 - 0.1810) - 25 \times 0.00025}{2/(1 - 0.1810) + 25 \times 0.00025} = 0.9951$$

$$\frac{1}{R} - \frac{1}{r_o} = \frac{1}{79.959} - \frac{1}{81.585}$$

$$= 0.00025 \text{ mm}^{-1} \ (0.0064 \text{ in.}^{-1})$$

$$\sum \rho_o = \frac{1}{D}\left[\frac{2}{1+\gamma} + D\left(\frac{1}{R} - \frac{1}{r_o}\right)\right] \tag{2.39}$$

$$= \frac{1}{25}\left(\frac{2}{1 + 0.1810} + 25 \times 0.00025\right)$$

$$= 0.068 \text{ mm}^{-1} \ (1.726 \text{ in.}^{-1})$$

$$F(\rho)_o = \frac{\dfrac{2}{1+\gamma} - D\left(\dfrac{1}{R} - \dfrac{1}{r_o}\right)}{\dfrac{2}{1+\gamma} + D\left(\dfrac{1}{R} - \dfrac{1}{r_o}\right)} \tag{2.40}$$

$$= \frac{2/(1 + 0.1810) - 25 \times 0.00025}{2/(1 + 0.1810) + 25 \times 0.00025} = 0.9929$$

Note that the magnitude of $F(\rho)$ approaches unity for spherical roller bearings. For cylindrical roller bearings and tapered roller bearings without crowned rollers or crowned raceways, $F(\rho)_i$ and $F(\rho)_o$ equal 1.

CLOSURE

The relationships developed in this chapter are based only on the macroshape of the rolling components of the bearing. When load is applied to the bearing, these contours may be altered somewhat; however, the undeformed geometry must be considered in the determination of the distorted shape.

Numerical examples developed in this chapter were of necessity very simple in format. The quantity of these simple examples is justified in that the results from the calculations will subsequently be used as starting points in more complex numerical examples involving stress, deflection, torque, and fatigue life.

REFERENCES

2.1. *American National Standard (ANSI/AFBMA) Std 20-1987*, "Radial Bearings of Ball, Cylindrical Roller and Spherical Roller Types, Metric Design," (October 28, 1987).

2.2. A. Jones, "Analysis of Stresses and Deflections," Vol. 1, New Departure Division, General Motors Corporation, Bristol, CT, p. 12 (1946).

3

INTERFERENCE FITTING AND CLEARANCE

LIST OF SYMBOLS

Symbol	Description	Units
B	Basic inner ring width	mm (in.)
B_s	Single width of an inner ring	mm (in.)
C	Basic outer ring width	mm (in.)
C_s	Single width of an outer ring	mm (in.)
d	Basic bore diameter	mm (in.)
d_i	Bearing inner raceway diameter	mm (in.)
d_o	Bearing outer raceway diameter	mm (in.)
d_s	Single diameter of a bore	mm (in.)
d_{mp}	Single plane mean bore diameter	mm (in.)
D	Basic outside diameter	mm (in.)
D_s	Single diameter of an outside surface	mm (in.)
D_{mp}	Single plane mean outside diameter	mm (in.)
$\mathfrak{D}$	Common diameter	mm (in.)
$\mathfrak{D}_h$	Basic housing bore	mm (in.)
$\mathfrak{D}_1$	Outside ring o.d.	mm (in.)

Symbol	Description	Units
$\mathcal{D}_2$	Inside ring i.d.	mm (in.)
$\mathcal{D}_s$	Basic shaft diameter	mm (in.)
E	Modulus of elasticity	N/mm^2 (psi)
I	Interference	mm (in.)
K_{ia}	Radial runout of assembled bearing inner ring	μm (in.)
K_{ea}	Radial runout of assembled bearing outer ring	μm (in.)
$\mathcal{L}$	Length	mm (in.)
P_d	Bearing clearance	mm (in.)
p	Pressure	N/mm^2 (psi)
$\mathcal{R}$	Ring radius	mm (in.)
$\mathcal{R}_i$	Inside radius of ring	mm (in.)
$\mathcal{R}_o$	Outside radius of ring	mm (in.)
S_d	Inner ring reference face runout with bore	μm (in.)
S_D	Outside cylindrical surface runout with outer ring reference face	μm (in.)
S_{ia}	Axial runout of assembled bearing inner ring	μm (in.)
S_{ea}	Axial runout of assembled bearing outer ring	μm (in.)
u	Radial deflection	mm (in.)
V_{dp}	Bore diameter variation in a single radial plane	μm (in.)
V_{dmp}	Mean bore diameter variation	μm (in.)
V_{Dmp}	Mean outside diameter variation	μm (in.)
V_{Dp}	Outside diameter variation in single radial plane	μm (in.)
Δ_{Bs}	Single inner ring width deviation from basic	μm (in.)
Δ_{Cs}	Single outer ring width deviation from basic	μm (in.)
Δ_{ds}	Single bore diameter deviation from basic	μm (in.)
Δ_{dmp}	Single plane mean bore diameter deviation from basic for a tapered bore small end	μm (in.)
Δ_{d1mp}	Single plane mean bore diameter deviation at large end of tapered bore	μm (in.)
Δ_{Ds}	Single outside diameter deviation from basic	μm (in.)

Symbol	Description	Units
Δ_{Dmp}	Single plane mean outside diameter deviation from basic	μm (in.)
Δ_h	Clearance reduction due to press fitting of bearing in housing	mm (in.)
Δ_s	Clearance reduction due to press fitting of bearing on shaft	mm (in.)
Δ_t	Clearance increase due to thermal expansion	mm (in.)
T	Temperature	°C (°F)
ϵ_r	Strain in radial direction	mm/mm (in./in.)
ϵ_t	Strain in tangential direction	mm/mm (in./in.)
Γ	Coefficient of linear expansion	mm/mm/°C (in./in./°F)
ξ	Poisson's ratio	
σ_r	Normal stress in radial direction	N/mm^2 (psi)
σ_t	Normal stress in tangential direction	N/mm^2 (psi)

GENERAL

Ball and roller bearings are usually mounted on shafts or in housings with interference fits. This is usually done to prevent fretting corrosion that could be produced by relative movement between the bearing inner ring bore and the shaft o.d. and/or the bearing outer ring o.d. and the housing bore. The interference fit of the bearing inner ring with the shaft is usually accomplished by pressing the former member over the latter. In some cases, however, the inner ring is heated to a controlled temperature in an oven or in an oil bath. Then the inner ring is slipped over the shaft and allowed to cool, thus accomplishing a shrink fit.

Press or shrink fitting of the inner ring on the shaft causes the inner ring to expand slightly. Similarly, press fitting of the outer ring in the housing causes the former member to shrink slightly, Thus, the bearing's diametral clearance will tend to decrease. Large amounts of interference in fitting practice can cause bearing clearance to vanish and even produce negative clearance or interference in the bearing.

Thermal conditions of bearing operation can also affect the diametral clearance. Heat generated by friction causes internal temperatures to rise. This in turn causes expansion of the shaft, housing, and bearing components. Depending on the shaft and housing materials and on the magnitude of thermal gradients across the bearing and these supporting structures, clearance can tend to increase or decrease. It is also apparent that the thermal environment in which a bearing operates may have a significant effect on clearance.

In Chapter 2 it was demonstrated that clearance significantly affects ball bearing contact angle. Subsequently, the effect of clearance on load distribution and life will be investigated. It is therefore clear that the mechanics of bearing fitting practice is an important part of this book.

ANSI AND ISO STANDARDS

Standards defining recommended practices for ball and roller bearing usage have been developed in the United States by the Anti-Friction Bearing Manufacturers Association (AFBMA). These standards have subsequently been adopted by the American National Standards Institute (ANSI) and are published under the title "American National Standard (ANSI/AFBMA)" with the corresponding standard number and date of approval. In general, corresponding standards are published by the International Standards Organization as "International Standard (ISO)" and the corresponding standard number. In this chapter, various bearing, shaft, and housing tolerance data are excerpted from the "American National Standards."

Reference [3.1] defines recommended practice in fitting bearing inner rings to shafts and outer rings in housings. These fits are recommended in terms of *light*, *normal*, and *heavy* loading, as defined by Table 3.1. Shaft tolerance ranges in the standard are designated by a lower case letter followed by a number, for example, g6, h5, and so on. Similarly, tolerance range symbols for housings consist of an upper case letter followed by a number, for example G7 or H6. Figure 3.1 shows the relative magnitudes of the shaft and bearing tolerance ranges. Correspondingly, Table 3.2 gives the ANSI recommended practice for fitting inner rings on shafts and Table 3.3 shows the shaft diameter tolerance limits corresponding to the recommended fit. Tables 3.4 and 3.5 yield similar data for fitting of bearing outer rings in housing bores.

At the time of preparation of this text, the United States is in the

TABLE 3.1. Ball and Roller Bearing Loads

Load Magnitude	Ball Bearings	Roller Bearings
Light	$<0.07C$[a]	$<0.08C$
Normal	$>0.07C$ $\leq 0.15C$	$>0.08C$ $\leq 0.18C$
Heavy	$>0.15C$	$>0.18C$

[a]C is the basic load rating of the bearing as given in bearing catalogs.

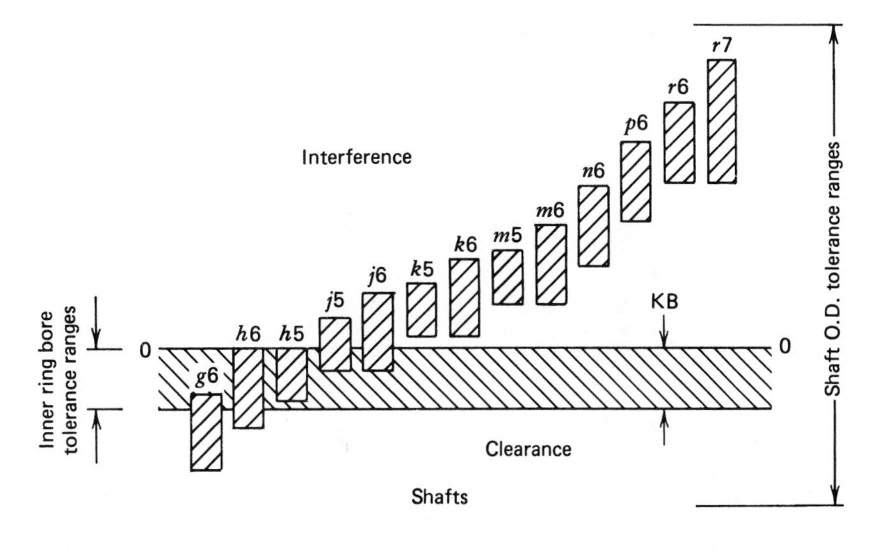

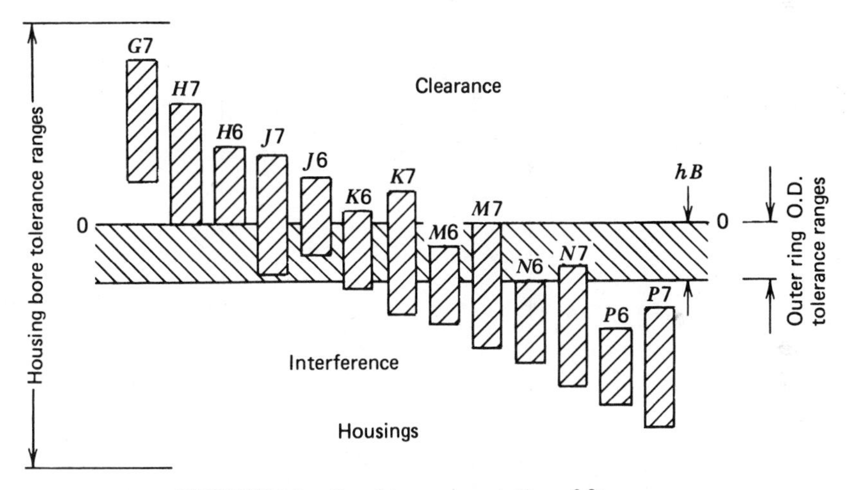

FIGURE 3.1. Graphic representation of fits.

At the time of preparation of this text, the United States is in the process of converting from English System units to Metric System units. Many of the ANSI/AFBMA Standards contain dimensions and tolerances primarily in millimeters and micrometers (microns) and secondarily in inches. Reference [3.1] is currently an exception in that shaft and housing tolerances are given in inches only. Subsequently, tabular data excerpted from the ANSI/AFBMA Standards are shown herein primarily in Metric System units and secondarily in English System units. Numerical examples are presented in similar fashion.

TABLE 3.2. Selection of Shaft Tolerance Classifications For Metric Radial Ball and Roller Bearings[a]

Design and Operating Conditions		Ball Bearings				Cylindrical Roller Bearings				Spherical Roller Bearings				Tolerance Classification[b]
		Basic Bearing Bore												
Rotational Conditions	Radial Loading	mm		in.		mm		in.		mm		in.		
		Over	Incl.	Over	Incl.	Over	Incl.	Over	Incl.	Over	Incl.	Over	Incl.	
Inner ring rotating in relation to load direction or	Light radial load	0	18	0	0.71									h5
		18	All	0.71	All	0	40	0	1.57	0	40	0	1.57	j6[c]
						40	140	1.57	5.52	40	100	1.57	3.94	k6[c]
						140	320	5.52	12.6	100	200	3.94	7.88	m6[c]
	Normal radial load	0	18	0	0.71									j5
		18	All	0.71	All	0	40	0	1.57	0	40	0	1.97	k5
						40	100	1.57	3.94	40	65	1.57	2.56	m5
						100	140	3.94	5.52	65	100	2.56	3.94	m6
						140	320	5.52	12.6	100	140	3.94	5.52	n6
										140	280	5.52	11.1	p6
										280	500	11.1	19.7	r6
										500	All	19.7	All	r7

Load direction is indeterminate	Heavy radial load		18 100	100 All	0.71 3.94	3.94 All	0 40 65 140	40 65 140 320	0 1.57 2.56 5.52	1.57 2.56 5.52 12.6	0 40 65 100 140 200	40 65 100 140 200 All	0 1.57 2.56 3.94 5.52 7.88	1.57 2.56 3.94 5.52 7.88 All	k5 m5 m6 n6 p6 r6 r7
Inner ring stationary in relation to load direction	All loads	Inner ring must be easily axially displaceable	All sizes												g6
		Inner ring need not be easily axially displaceable	All sizes												h6
Pure thrust (axial) load			All sizes												j6

[a]Tolerance class ABEC-1, RBEC-1.
[b]Tolerance classifications shown are for solid steel shafts. For hollow or nonferrous shafts tighter fits may be needed.
[c]If greater accuracy is needed, substitute j5, k5, and m5 for j6, k6, and m6, respectively.

TABLE 3.3. Shaft Diameter Tolerance Limits[a]

Basic bore diameter d				Tolerance classifications												
				g6	h6	h5	j5	j6	k5	k6	m5	m6	n6	p6	r6	r7
in.		mm		Allowable Deviations from Basic Bore Diameter												
Over	Incl.	Over	Incl.													
0.1181	0.2362	3	6	−0.0002 −0.0005	0 −0.0003	0 −0.0002	+0.0001 −0.0001					+0.0005 +0.0002				
0.2362	0.3937	6	10	−0.0002 −0.0006	0 −0.0004	0 −0.0003	+0.0002 −0.0001	+0.0003 −0.0001				+0.0006 +0.0002				
0.3937	0.7087	10	18	−0.0003 −0.0007	0 −0.0004	0 −0.0003	+0.0002 −0.0001	+0.0003 −0.0001	+0.0004 +0.0001	+0.0005 +0.0001	+0.0006 +0.0003	+0.0007 +0.0003				
0.7087	1.1811	18	30	−0.0003 −0.0008	0 −0.0005	0 −0.0004	+0.0002 −0.0002	+0.0003 −0.0002	+0.0005 +0.0001	+0.0006 +0.0001	+0.0007 +0.0003	+0.0008 +0.0003	+0.0011 +0.0006			
1.1811	1.9685	30	50	−0.0004 −0.0010	0 −0.0006	0 −0.0004	+0.0002 −0.0002	+0.0004 −0.0002	+0.0005 +0.0001	+0.0007 +0.0001	+0.0008 +0.0004	+0.0010 +0.0004	+0.0013 +0.0007	+0.0016 +0.0010		
1.9685	3.1496	50	80	−0.0004 −0.0011	0 −0.0007	0 −0.0005	+0.0002 −0.0003	+0.0004 −0.0003	+0.0006 +0.0001	+0.0008 +0.0001	+0.0010 +0.0005	+0.0012 +0.0005	+0.0015 +0.0008	+0.0021 +0.0014	+0.0023 +0.0016	
3.1496	4.7244	80	120	−0.0005 −0.0014	0 −0.0009	0 −0.0006	+0.0002 −0.0004	+0.0005 −0.0004	+0.0007 +0.0001	+0.0010 +0.0001	+0.0011 +0.0005	+0.0014 +0.0005	+0.0019 +0.0010	+0.0025 +0.0016	+0.0029 +0.0020	
4.7244	7.0866	120	180	−0.0006 −0.0016	0 −0.0010	0 −0.0007	+0.0003 −0.0004	+0.0006 −0.0004	+0.0008 +0.0001	+0.0011 +0.0001	+0.0013 +0.0006	+0.0016 +0.0006	+0.0022 +0.0012	+0.0028 +0.0018	+0.0035 +0.0025	

7.0866	9.8425	180	250	−0.0006 −0.0018	0 −0.0012	0 −0.0008	+0.0003 −0.0005	+0.0007 −0.0005	+0.0010 +0.0002	+0.0014 +0.0002	+0.0014 +0.0006	+0.0018 +0.0006	+0.0026 +0.0014	+0.0032 +0.0020	+0.0042 +0.0030	+0.0048 +0.0030
9.8425	12.4016	250	315	−0.0007 −0.0019	0 −0.0012	0 −0.0009	+0.0003 −0.0006	+0.0007 −0.0006	+0.0011 +0.0002	+0.0014 +0.0002	+0.0017 +0.0008	+0.0020 +0.0008	+0.0026 +0.0014	+0.0034 +0.0022	+0.0047 +0.0035	+0.0055 +0.0035
12.4016	15.7480	315	400	−0.0007 −0.0021	0 −0.0014	0 −0.0010	+0.0003 −0.0007	+0.0007 −0.0007	+0.0012 +0.0002	+0.0016 +0.0002	+0.0018 +0.0008	+0.0022 +0.0008	+0.0030 +0.0016	+0.0039 +0.0025	+0.0059 +0.0045	+0.0067 +0.0045
15.7480	19.6850	400	500	−0.0008 −0.0024	0 −0.0016	0 −0.0010	+0.0003 −0.0007	+0.0009 −0.0007	+0.0012 +0.0002	+0.0018 +0.0002	+0.0019 +0.0009	+0.0025 +0.0009	+0.0034 +0.0018	+0.0044 +0.0028	+0.0066 +0.0050	+0.0075 +0.0050
19.6850	22.0472	500	560		0 −0.0017	0 −0.0012	+0.0003 −0.0009	+0.0009 −0.0008	+0.0014 +0.0002	+0.0019 +0.0002	+0.0022 +0.0010	+0.0027 +0.0010	+0.0034 +0.0017	+0.0046 +0.0030	+0.0074 +0.0057	+0.0083 +0.0057
22.0472	24.8031	560	630												+0.0077 +0.0060	+0.0087 +0.0060
24.8031	27.9527	630	710		0 −0.0019	0 −0.0013	+0.0004 −0.0009	+0.0009 −0.0009	+0.0015 +0.0002	+0.0020 +0.0002	+0.0024 +0.0011	+0.0030 +0.0011	+0.0038 +0.0019	+0.0051 +0.0033	+0.0085 +0.0067	+0.0097 +0.0067
27.9527	31.4960	710	800												+0.0088 +0.0069	+0.0099 +0.0069
31.4960	35.4331	800	900		0 −0.0020	0 −0.0015	+0.0004 −0.0011	+0.0010 −0.0010	+0.0017 +0.0002	+0.0023 +0.0002	+0.0027 +0.0013	+0.0033 +0.0013	+0.0042 +0.0021	+0.0056 +0.0036	+0.0098 +0.0077	+0.0110 +0.0077
35.4331	39.3700	900	1000												+0.0102 +0.0081	+0.0114 +0.0081
39.3700	44.0945	1000	1120		0 −0.0023	0 −0.0016	+0.0004 −0.0012	+0.0011 −0.0011	+0.0018 +0.0002	+0.0025 +0.0002	+0.0030 +0.0014	+0.0037 +0.0014	+0.0046 +0.0024	+0.0062 +0.0039	+0.0112 +0.0089	+0.0126 +0.0089

[a]Tolerance limits in inches.

TABLE 3.4. Selection of Housing Tolerance Classifications[a]

Design and Operating Conditions				Tolerance Classification[b]
Rotational Conditions	Loading	Outer Ring Axial Displacement Limitations	Other Conditions	
Outer ring stationary in relation to load direction	Light, normal, and heavy	Outer ring must be easily axially displaceable	Heat input through shaft	G7
			Housing split axially	H7
			Housing not split axially	H6[c]
	Shock with temporary complete unloading	Outer ring need not be axially displaceable		J6[c]
Load direction is indeterminate	Light and normal			
	Normal and heavy		Split housing not recommended	K6[c]
	Heavy shock			M6[c]
Outer ring rotating in relation to load direction	Light			
	Normal and heavy			N6[c]
	Heavy		Thin wall housing not split	P6[c]

[a]For metric radial ball and roller bearings of tolerance class ABEC-1, RBEC-1.
[b]For cast iron or steel housings. For housing nonferrous alloys tighter fits may be needed.
[c]Where wider tolerances are permissible, use tolerance classifications P7, N7, M7, K7, J7, and H7 in place of P6, N6, M6, K6, J6, and H6, respectively.

TABLE 3.5. Housing Bore Tolerance Limits[a]

Basic o.d. *D*				Tolerance classifications												
				G7	H7	H6	J7	J6	K6	K7	M6	M7	N6	N7	P6	P7
in.		mm		Allowable Deviation from Basic o.d.												
Over	Incl.	Over	Incl.													
0.3937	0.7086	10	18	+0.0003 +0.0010	0 +0.0007	0 +0.0004	−0.0003 +0.0004	−0.0002 +0.0002	−0.0004 +0.0000	−0.0005 −0.0002	−0.0006 −0.0002	−0.0007 0	−0.0008 −0.0004	−0.0009 −0.0002	−0.0010 −0.0006	−0.0011 −0.0004
0.7086	1.1811	18	30	+0.0003 +0.0011	0 +0.0008	0 +0.0005	−0.0003 +0.0005	−0.0002 +0.0003	−0.0005 +0.0000	−0.0006 +0.0002	−0.0007 −0.0002	−0.0008 0	−0.0010 −0.0005	−0.0011 −0.0003	−0.0012 −0.0007	−0.0013 −0.0005
1.1811	1.9685	30	50	+0.0004 +0.0014	0 +0.0010	0 +0.0006	−0.0004 +0.0006	−0.0002 +0.0004	−0.0005 +0.0001	−0.0007 +0.0003	−0.0008 −0.0002	−0.0010 0	−0.0011 −0.0005	−0.0013 −0.0003	−0.0014 −0.0008	−0.0016 −0.0006
1.9685	3.1496	50	80	+0.0004 +0.0016	0 +0.0012	0 +0.0007	−0.0004 +0.0008	−0.0002 +0.0005	−0.0006 +0.0001	−0.0008 +0.0004	−0.0010 −0.0003	−0.0012 0	−0.0013 −0.0006	−0.0015 −0.0003	−0.0019 −0.0012	−0.0021 −0.0009
3.1496	4.7244	80	120	+0.0005 +0.0019	0 +0.0014	0 +0.0009	−0.0005 +0.0009	−0.0002 +0.0007	−0.0007 +0.0002	−0.0010 +0.0004	−0.0012 −0.0003	−0.0014 0	−0.0016 −0.0007	−0.0018 −0.0004	−0.0022 −0.0013	−0.0025 −0.0011
4.7244	7.0866	120	180	+0.0006 +0.0022	0 +0.0016	0 +0.0010	−0.0006 +0.0010	−0.0003 +0.0007	−0.0008 +0.0002	−0.0011 +0.0005	−0.0013 −0.0003	−0.0016 0	−0.0019 −0.0009	−0.0022 −0.0006	−0.0025 −0.0015	−0.0028 −0.0012
7.0866	9.8425	180	250	+0.0006 +0.0024	0 +0.0018	0 +0.0012	−0.0007 +0.0011	−0.0003 +0.0009	−0.0010 +0.0002	−0.0013 +0.0005	−0.0015 −0.0003	−0.0018 0	−0.0022 −0.0010	−0.0026 −0.0008	−0.0028 −0.0016	−0.0032 −0.0014

(continued on next page)

TABLE 3.5. Housing Bore Tolerance Limits[a] (continued)

9.8425	12.4016	250	315	+0.0007 +0.0027	0 +0.0020	0 +0.0012	−0.0007 +0.0013	−0.0003 +0.0009	−0.0010 +0.0002	−0.0014 +0.0006	−0.0016 −0.0004	−0.0020 0	−0.0023 −0.0011	−0.0028 −0.0008	−0.0031 −0.0019	−0.0034 −0.0014
12.4016	15.7480	315	400	+0.0007 +0.0029	0 +0.0022	0 +0.0014	−0.0007 +0.0015	−0.0003 +0.0011	−0.0012 +0.0002	−0.0016 +0.0006	−0.0018 −0.0004	−0.0022 0	−0.0026 −0.0012	−0.0030 −0.0008	−0.0035 −0.0021	−0.0039 −0.0017
15.7480	19.6850	400	500	+0.0008 +0.0033	0 +0.0025	0 +0.0016	−0.0009 +0.0016	−0.0003 +0.0013	−0.0012 +0.0004	−0.0018 +0.0007	−0.0020 −0.0004	−0.0025 0	−0.0028 −0.0012	−0.0034 −0.0009	−0.0038 −0.0022	−0.0044 −0.0019
19.6850	24.8031	500	630	+0.0009 +0.0035	0 +0.0027	0 +0.0017	−0.0009 +0.0018	−0.0003 +0.0014	−0.0014 +0.0003	−0.0019 +0.0008	−0.0022 −0.0005	−0.0027 0	−0.0029 −0.0012	−0.0034 −0.0007	−0.0041 −0.0024	−0.0046 −0.0020
24.8031	31.4960	630	800	+0.0009 +0.0039	0 +0.0030	0 +0.0019	−0.0009 +0.0020	−0.0004 +0.0015	−0.0015 +0.0004	−0.0020 +0.0009	−0.0024 −0.0006	−0.0030 0	−0.0032 −0.0014	−0.0038 −0.0008	−0.0046 −0.0027	−0.0051 −0.0021
31.4960	39.3700	800	1000	+0.0110 +0.0043	0 +0.0033	0 +0.0021	−0.0010 +0.0023	−0.0004 +0.0017	−0.0017 +0.0004	−0.0023 +0.0010	−0.0027 −0.0007	−0.0033 0	−0.0036 −0.0015	−0.0042 −0.0009	−0.0050 −0.0030	−0.0056 −0.0023
39.3700	49.2126	1000	1250	+0.0011 +0.0047	0 +0.0037	0 +0.0023	−0.0011 +0.0025	−0.0004 +0.0019	−0.0018 +0.0005	−0.0025 +0.0011	−0.0030 −0.0007	−0.0037 0	−0.0039 −0.0017	−0.0046 −0.0010	−0.0055 −0.0032	−0.0062 −0.0026
49.2126	62.9921	1250	1600	+0.0011 +0.0053	0 +0.0041	0 +0.0025	−0.0013 +0.0028	−0.0004 +0.0021	−0.0020 +0.0005	−0.0028 +0.0013	−0.0033 −0.0008	−0.0041 0	−0.0044 −0.0019	−0.0052 −0.0011	−0.0061 −0.0035	−0.0069 −0.0028

[a]Tolerance limits in inches.

TABLE 3.6. ANSI/AFBMA versus ISO Tolerance Classifications

ANSI/AFBMA	ISO
ABEC 1 or RBEC 1	Normal class
ABEC 3 or RBEC 3	Class 6
ABEC 5 or RBEC 5	Class 5
ABEC 7	Class 4
ABEC 9	Class 2

ANSI/AFBMA in references [3.2–3.9] also provide standards for tolerance ranges on bearing bore and o.d. for various types of radial bearings. Several of these bearing types, for example, tapered roller bearings, needle roller bearings, and instrument ball bearings, exist in too many variations to include all of the appropriate tolerance tables herein. On the other hand, reference [3.9] covers a wide range of standard radial ball and roller bearings; Tables 3.7–3.11 are taken from reference [3.9]. For radial ball bearings these tolerances are grouped in ABEC* classes 1, 3, 5, 7, and 9 according to accuracy of manufacturing. Accuracy improves and tolerance ranges narrow as the class number increases. Tables 3.7–3.11 give tolerance ranges for all ABEC classifications. Additionally, Tables 3.7–3.9 provide the tolerances or bore and o.d. for radial roller bearings as well as for ball bearings. The ABEC and RBEC† tolerance classes correspond in every respect to the precision classes endorsed by the ISO. Table 3.6 shows the correspondence between the ANSI/AFBMA and ISO classifications. It is further noted that inch tolerances given in Part II of Tables 3.7–3.11 are calculated from primary metric tolerances given in Part I of those tables.

To define the range of interference or looseness in the mounting of an inner ring on a shaft or an outer ring in a housing, it is necessary to consider combination of the shaft, housing, and bearing tolerances.

EFFECT OF INTERFERENCE FITTING ON CLEARANCE

The solution to this problem may be obtained by using elastic thick ring theory. Consider the ring of Fig. 3.2 subjected to an internal pressure p per unit length. The ring has a bore radius $\mathcal{R}_i$ and an outside radius $\mathcal{R}_o$. For the elemental area $\mathcal{R}\ d\mathcal{R}\ d\phi$ the summation of forces in the radial

*Annular Bearing Engineers' Committee of AFBMA.
†Roller Bearing Engineers' Committee of AFBMA.

TABLE 3.7. PART 1. Tolerance Class ABEC-1, RBEC-1. Metric Ball and Roller Bearings [except tapered roller bearings[e]] of Dimensions Conforming to the Basic Plan for Boundary Dimensions of Metric Radial Bearings Given in Table 1 of [3.9].

Inner Ring (Tolerance values in micrometers)

				V_{dp}						Δ_{Bs}		
				diameter series								
d mm		Δ_{dmp}		7, 8, 9	0, 1	2, 3, 4	V_{dmp}	K_{ia}	all	normal	modified[d]	V_{Bs}
over	incl.	high	low	max.			max.	max.	high	low		max.
a 0.6	2.5	0	−8	10	8	6	6	10	0	−40	—	12
2.5	10	0	−8	10	8	6	6	10	0	−120	−250	15
10	18	0	−8	10	8	6	6	10	0	−120	−250	20
18	30	0	−10	13	10	8	8	13	0	−120	−250	20
30	50	0	−12	15	12	9	9	15	0	−120	−250	20
50	80	0	−15	19	19	11	11	20	0	−150	−380	25
80	120	0	−20	25	25	15	15	25	0	−200	−380	25
120	180	0	−25	31	31	19	19	30	0	−250	−500	30
180	250	0	−30	38	38	23	23	40	0	−300	−500	30
250	315	0	−35	44	44	26	26	50	0	−350	−500	35
315	400	0	−40	50	50	30	30	60	0	−400	−630	40
400	500	0	−45	56	56	34	34	65	0	−450	—	50
500	630	0	−50	63	63	38	38	70	0	−500	—	60
630	800	0	−75	—	—	—	—	80	0	−750	—	70
800	1,000	0	−100	—	—	—	—	90	0	−1,000	—	80
1,000	1,250	0	−125	—	—	—	—	100	0	−1,250	—	100
1,250	1,600	0	−160	—	—	—	—	120	0	−1,600	—	120
1,600	2,000	0	−200	—	—	—	—	140	0	−2,000	—	140

Outer Ring (Tolerance values in micrometers)

D mm		Δ_{Dmp}		V_{Dp} [c] Open Bearings, diameter series 7, 8, 9	V_{Dp} [c] Open Bearings, diameter series 0, 1	V_{Dp} [c] Open Bearings, diameter series 2, 3, 4	V_{Dp} [c] Capped Bearings [b], diameter series 2, 3, 4	V_{Dmp} [c]	K_{ea}	Δ_{Cs}		V_{Cs}
over	incl.	high	low	max.	max.	max.	max.	max.	max.	high	low	max.
[a] 2.5	6	0	−8	10	8	6	10	6	15			
6	18	0	−8	10	8	6	10	6	15			
18	30	0	−9	12	9	7	12	7	15			
30	50	0	−11	14	11	8	16	8	20			
50	80	0	−13	16	13	10	20	10	25			
80	120	0	−15	19	19	11	26	11	35			
120	150	0	−18	23	23	14	30	14	40			
150	180	0	−25	31	31	19	38	19	45			
180	250	0	−30	38	38	23	—	23	50	Identical to Δ_{Bs} and V_{Bs} of inner ring of same bearing		
250	315	0	−35	44	44	26	—	26	60			
315	400	0	−40	50	50	30	—	30	70			
400	500	0	−45	56	56	34	—	34	80			
500	630	0	−50	63	63	38	—	38	100			
630	800	0	−75	94	94	55	—	55	120			
800	1,000	0	−100	125	125	75	—	75	140			
1,000	1,250	0	−125	—	—	—	—	—	160			
1,250	1,600	0	−160	—	—	—	—	—	190			
1,600	2,000	0	−200	—	—	—	—	—	220			
2,000	2,500	0	−250	—	—	—	—	—	250			

[a] This diameter is included in the group.
[b] No values have been established for diameter series 7, 8, 9, 0, and 1.
[c] Applies before mounting and after removal of internal or external snap ring.
[d] This refers to the rings of single bearings made for paired or stack mounting.
[e] For tapered roller bearing tolerances see [3.7, 3.8].

TABLE 3.7. PART 2. Tolerance Class ABEC-1, RBEC-1. Metric Ball and Roller Bearings [except tapered roller bearings[e]] of Dimensions Conforming to the Basic Plan for Boundary Dimensions of Metric Radial Bearings Given in Table 1 of [3.9].

Inner Ring (Tolerance values in .0001 in.)

d mm		Δ_{dmp}		V_{dp} diameter series 7, 8, 9	V_{dp} diameter series 0, 1	V_{dp} diameter series 2, 3, 4	V_{dmp}	K_{ia}	Δ_{Bs} all	Δ_{Bs} normal	Δ_{Bs} modified[d]	V_{Bs}
over	incl.	high	low	max.	max.	max.	max.	max.	high	low	low	max.
[a] 0.6	2.5	0	−3	4	3	2.5	2.5	4	0	−16	—	4.5
2.5	10	0	−3	4	3	2.5	2.5	4	0	−47	−98	6
10	18	0	−3	4	3	2.5	2.5	4	0	−47	−98	8
18	30	0	−4	5	4	3	3	5	0	−47	−98	8
30	50	0	−4.5	6	4.5	3.5	3.5	6	0	−47	−98	8
50	80	0	−6	7.5	7.5	4.5	4.5	8	0	−59	−150	10
80	120	0	−8	10	10	6	6	10	0	−79	−150	10
120	180	0	−10	12	12	7.5	7.5	12	0	−98	−197	12
180	250	0	−12	15	15	9	9	16	0	−118	−197	12
250	315	0	−14	17	17	10	10	20	0	−138	−197	14
315	400	0	−16	20	20	12	12	24	0	−157	−248	16
400	500	0	−18	22	22	13	13	26	0	−177	—	20
500	630	0	−20	25	25	15	15	28	0	−197	—	24
630	800	0	−30	—	—	—	—	31	0	−295	—	28
800	1,000	0	−39	—	—	—	—	35	0	−394	—	31
1,000	1,250	0	−49	—	—	—	—	39	0	−492	—	39
1,250	1,600	0	−63	—	—	—	—	47	0	−630	—	47
1,600	2,000	0	−79	—	—	—	—	55	0	−787	—	55

Outer Ring (Tolerance values in .0001 in.)

D mm		Δ_{Dmp}		V_{Dp} [c] Open Bearings, diameter series 7, 8, 9	V_{Dp} [c] Open Bearings, diameter series 0, 1	V_{Dp} [c] Open Bearings, diameter series 2, 3, 4	V_{Dp} [c] Capped Bearings [b], diameter series 2, 3, 4	V_{Dmp} [c]	K_{ea}	Δ_{Cs}		V_{Cs}
over	incl.	high	low	max.	max.	max.	max.	max.	max.	high	low	max.
[a] 2.5	6	0	−3	4	3	2.5	4	2.5	6			
6	18	0	−3	4	3	2.5	4	2.5	6			
18	30	0	−3.5	4.5	3.5	3	4.5	3	6			
30	50	0	−4.5	5.5	4.5	3	6.5	3	8			
50	80	0	−5	6.5	5	4	8	4	10			
80	120	0	−6	7.5	7.5	4.5	10	4.5	14			
120	150	0	−7	9	9	5.5	12	5.5	16			
150	180	0	−10	12	12	7.5	15	7.5	18			
180	250	0	−12	15	15	9	—	9	20	Identical to Δ_{Bs} and V_{Bs} of inner ring of same bearing		
250	315	0	−14	17	17	10	—	10	24			
315	400	0	−16	20	20	12	—	12	28			
400	500	0	−18	22	22	13	—	13	31			
500	630	0	−20	25	25	15	—	15	39			
630	800	0	−30	37	37	22	—	22	47			
800	1,000	0	−39	49	49	30	—	30	55			
1,000	1,250	0	−49	—	—	—	—	—	63			
1,250	1,600	0	−63	—	—	—	—	—	75			
1,600	2,000	0	−79	—	—	—	—	—	87			
2,000	2,500	0	−98	—	—	—	—	—	98			

[a] This diameter is included in the group.
[b] No values have been established for diameter series 7, 8, 9, 0, and 1.
[c] Applies before mounting and after removal of internal or external snap ring.
[d] This refers to the rings of single bearings made for paired or stack mounting.
[e] For tapered roller bearing tolerances see [3.7, 3.8].

TABLE 3.8. PART 1. Tolerance Class ABEC-3, RBEC-3. Metric Ball and Roller Bearings [except tapered roller bearings[e]] of Dimensions of Conforming to the Basic Plan for Boundary Dimensions of Metric Radial Bearings Given in Table 1 of [3.9].

Inner Ring (Tolerance values in micrometers)

	d mm		Δ_{dmp}	V_{dp} diameter series 7, 8, 9	V_{dp} diameter series 0, 1	V_{dp} diameter series 2, 3, 4	V_{dmp}	K_{ia}	Δ_{Bs} all	Δ_{Bs} normal	Δ_{Bs} modified[d]	V_{Bs}
	over	incl.	high	low	max.	max.	max.	max.	max.	high	low	low	max.
a	0.6	2.5	0	−7	9	7	5	5	5	0	−40	—	12
	2.5	10	0	−7	9	7	5	5	6	0	−120	−250	15
	10	18	0	−7	9	7	5	5	7	0	−120	−250	20
	18	30	0	−8	10	8	6	6	8	0	−120	−250	20
	30	50	0	−10	13	10	8	8	10	0	−120	−250	20
	50	80	0	−12	15	15	9	9	10	0	−150	−380	25
	80	120	0	−15	19	19	11	11	13	0	−200	−380	25
	120	180	0	−18	23	23	14	14	18	0	−250	−500	30
	180	250	0	−22	28	28	17	17	20	0	−300	−500	30
	250	315	0	−25	31	31	19	19	25	0	−350	−500	35
	315	400	0	−30	38	38	23	23	30	0	−400	−630	40
	400	500	0	−35	44	44	26	26	35	0	−450	—	45
	500	630	0	−40	50	50	30	30	40	0	−500	—	50

Outer Ring (Tolerance values in micrometers)

D mm		Δ_{Dmp}		V_{Dp} [c] Open Bearings, diameter series 7, 8, 9	0, 1	2, 3, 4	Capped Bearings [b] 0, 1, 2, 3, 4	V_{Dmp} [c]	K_{ea}	Δ_{Cs}		V_{Cs}
over	incl.	high	low	max.	max.	max.	max.	max.	max.	high	low	max.
[a] 2.5	6	0	−7	9	7	5	9	5	8			
6	18	0	−7	9	7	5	9	5	8			
18	30	0	−8	10	8	6	10	6	9			
30	50	0	−9	11	9	7	13	7	10			
50	80	0	−11	14	11	8	16	8	13			
80	120	0	−13	16	16	10	20	10	18			
120	150	0	−15	19	19	11	25	11	20	Identical to Δ_{Bs} and V_{Bs} of inner ring of same bearing		
150	180	0	−18	23	23	14	30	14	23			
180	250	0	−20	25	25	15	—	15	25			
250	315	0	−25	31	31	19	—	19	30			
315	400	0	−28	35	35	21	—	21	35			
400	500	0	−33	41	41	25	—	25	40			
500	630	0	−38	48	48	29	—	29	50			
630	800	0	−45	56	56	34	—	34	60			
800	1,000	0	−60	75	75	45	—	45	75			

[a] This diameter is included in the group.
[b] No values have been established for diameter series 7, 8, 9.
[c] Applies before mounting and after removal of internal or external snap ring.
[d] This refers to the rings of single bearings made for paired or stack mounting.
[e] For tapered roller bearing tolerances see [3.7, 3.8].

TABLE 3.8. PART 2. Tolerance Class ABEC-3, RBEC-3. Metric Ball and Roller Bearings [except tapered roller bearings[e]] of Dimensions Conforming to the Basic Plan for Boundary Dimensions of Metric Radial Bearings Given in Table 1 of [3.9].

Inner Ring (Tolerance values in .0001 in.)

				V_{dp}						Δ_{Bs}		
				diameter series								
d mm		Δ_{dmp}		7, 8, 9	0, 1	2, 3, 4	V_{dmp}	K_{ia}	all	normal	modified[d]	V_{Bs}
over	incl.	high	low		max.		max.	max.	high		low	max.
a 0.6	2.5	0	−3	3.5	3	2	2	2	0	−16	—	4.5
2.5	10	0	−3	3.5	3	2	2	2.5	0	−47	−98	6
10	18	0	−3	3.5	3	2	2	3	0	−47	−98	8
18	30	0	−3	4	3	2.5	2.5	3	0	−47	−98	8
30	50	0	−4	5	4	3	3	4	0	−47	−98	8
50	80	0	−4.5	6	6	3.5	3.5	4	0	−59	−150	10
80	120	0	−6	7.5	7.5	4.5	4.5	5	0	−79	−150	10
120	180	0	−7	9	9	5.5	5.5	7	0	−98	−197	12
180	250	0	−8.5	11	11	6.5	6.5	8	0	−118	−197	12
250	315	0	−10	12	12	7.5	7.5	10	0	−138	−197	14
315	400	0	−12	15	15	9	9	12	0	−157	−248	16
400	500	0	−14	17	17	10	10	14	0	−177	—	18
500	630	0	−16	20	20	12	12	16	0	−197	—	20

Outer Ring (Tolerance values in .0001 in.)

D mm		Δ_{Dmp}		V_{Dp} [c]				V_{Dmp} [c]	K_{ea}	Δ_{Cs}		V_{Cs}
				Open Bearings			Capped Bearings [b]					
				diameter series								
				7, 8, 9	0, 1	2, 3, 4	0, 1, 2, 3, 4					
over	incl.	high	low	max.				max.	max.	high	low	max.
[a] 2.5	6	0	−3	3.5	3	2	3.5	2	3			
6	18	0	−3	3.5	3	2	3.5	2	3			
18	30	0	−3	4	3	2.5	4	2.5	3.5			
30	50	0	−3.5	4.5	3.5	3	5	3	4			
50	80	0	−4.5	5.5	4.5	3	6.5	3	5			
80	120	0	−5	6.5	6.5	4	8	4	7			
120	150	0	−6	7.5	7.5	4.5	10	4.5	8	Identical to Δ_{Bs} and		
150	180	0	−7	9	9	5.5	12	5.5	9	V_{Bs} of inner ring of		
180	250	0	−8	10	10	6	—	6	10	same bearing		
250	315	0	−10	12	12	7.5	—	7.5	12			
315	400	0	−11	14	14	8.5	—	8.5	14			
400	500	0	−13	16	16	10	—	10	16			
500	630	0	−15	19	19	11	—	11	20			
630	800	0	−18	22	22	13	—	13	24			
800	1,000	0	−24	30	30	18	—	18	30			

[a] This diameter is included in the group.
[b] No values have been established for diameter series 7, 8, and 9.
[c] Applies before mounting and after removal of internal or external snap ring.
[d] This refers to the rings of single bearings made for paired or stack mounting.
[e] For tapered roller bearing tolerances see [3.7, 3.8].

TABLE 3.9. PART 1. Tolerance Class ABEC-5, RBEC-5. Metric Ball and Roller Bearings [except instrument bearings,[e] and tapered roller bearings[f]] of Dimensions Conforming to the Basic Plan for Boundary Dimensions of Metric Radial Bearings Given in Table 1 of [3.9].

Inner Ring (Tolerance values in micrometers)

d mm		Δ_{dmp}		V_{dp} diameter series 7, 8, 9	V_{dp} diameter series 0, 1, 2, 3, 4	V_{dmp}	K_{ia}	S_d	S_{ia}[c]	Δ_{Bs} all	Δ_{Bs} normal	Δ_{Bs} mod.[d]	V_{Bs}
over	incl.	high	low	max.	max.	max.	max.	max.	max.	high	low	low	max.
a 0.6	2.5	0	−5	5	4	3	4	7	7	0	−40	−250	5
2.5	10	0	−5	5	4	3	4	7	7	0	−40	−250	5
10	18	0	−5	5	4	3	4	7	7	0	−80	−250	5
18	30	0	−6	6	5	3	4	8	8	0	−120	−250	5
30	50	0	−8	8	6	4	5	8	8	0	−120	−250	5
50	80	0	−9	9	7	5	5	8	8	0	−150	−250	6
80	120	0	−10	10	8	5	6	9	9	0	−200	−380	7
120	180	0	−13	13	10	7	8	10	10	0	−250	−380	8
180	250	0	−15	15	12	8	10	11	13	0	−300	−500	10
250	315	0	−18	18	14	9	13	13	15	0	−350	−500	13
315	400	0	−23	23	18	12	15	15	20	0	−400	−630	15

Outer Ring (Tolerance values in micrometers)

D mm		Δ_{Dmp}		V_{Dp} [b] diameter series 7, 8, 9	V_{Dp} [b] diameter series 0, 1, 2, 3, 4	V_{Dmp}	K_{ea}	S_D	S_{ea} [c]	Δ_{Cs}		V_{Cs}
over	incl.	high	low	max.	max.	max.	max.	max.	max.	high	low	max.
[a] 2.5	6	0	−5	5	4	3	5	8	8	Identical to ΔB_s of inner ring of same bearing		5
6	18	0	−5	5	4	3	5	8	8			5
18	30	0	−6	6	5	3	6	8	8			5
30	50	0	−7	7	5	4	7	8	8			5
50	80	0	−9	9	7	5	8	8	10			6
80	120	0	−10	10	8	5	10	9	11			8
120	150	0	−11	11	8	6	11	10	13			8
150	180	0	−13	13	10	7	13	10	14			8
180	250	0	−15	15	11	8	15	11	15			10
250	315	0	−18	18	14	9	18	13	18			11
315	400	0	−20	20	15	10	20	13	20			13
400	500	0	−23	23	17	12	23	15	23			15
500	630	0	−28	28	21	14	25	18	25			18
630	800	0	−35	35	26	18	30	20	30			20

[a] This diameter is included in the group.
[b] No values have been established for sealed or shielded bearings.
[c] Applies to groove type ball bearings only.
[d] This refers to the rings of single bearings made for paired or stack mounting.
[e] For instrument ball bearing tolerances see [3.2, 3.3].
[f] For tapered roller bearing tolerances see [3.7, 3.8].

TABLE 3.9. PART 2. Tolerance Class ABEC-5, RBEC-5. Metric Ball and Roller Bearings [except instrument bearings,[e] and tapered roller bearings[f]] of Dimensions Conforming to the Basic Plan for Boundary Dimensions of Metric Radial Bearings Given in Table 1 of [3.9].

Inner Ring (Tolerance values in .0001 in.)

d mm		Δ_{dmp}		V_{dp} diameter series 7, 8, 9	V_{dp} diameter series 0, 1, 2, 3, 4	V_{dmp}	K_{ia}	S_d	S_{ia}[c]	Δ_{Bs} all	Δ_{Bs} normal	Δ_{Bs} mod.[d]	V_{Bs}
over	incl.	high	low	max.	max.	max.	max.	max.	max.	high	low	low	max.
[a] 0.6	2.5	0	−2	2	1.5	1	1.5	3	3	0	−16	−98	2
2.5	10	0	−2	2	1.5	1	1.5	3	3	0	−16	−98	2
10	18	0	−2	2	1.5	1	1.5	3	3	0	−31	−98	2
18	30	0	−2.5	2.5	2	1	1.5	3	3	0	−47	−98	2
30	50	0	−3	3	2.5	1.5	2	3	3	0	−47	−98	2
50	80	0	−3.5	3.5	3	2	2	3	3	0	−59	−98	2.5
80	120	0	−4	4	3	2	2.5	3.5	3.5	0	−79	−150	3
120	180	0	−5	5	4	3	3	4	4	0	−98	−150	3
180	250	0	−6	6	4.5	3	4	4.5	5	0	−118	−197	4
250	315	0	−7	7	5.5	3.5	5	5	6	0	−138	−197	5
315	400	0	−9	9	7	4.5	6	6	8	0	−157	−248	6

Outer Ring (Tolerance values in .0001 in.)

D mm		Δ_{Dmp}		V_{Dp} [b] diameter series 7, 8, 9	V_{Dp} [b] diameter series 0, 1, 2, 3, 4	V_{Dmp}	K_{ea}	S_D	S_{ea} [c]	Δ_{Cs}		V_{Cs}
over	incl.	high	low	max.	max.	max.	max.	max.	max.	high	low	max.
[a] 2.5	6	0	−2	2	1.5	1	2	3	3	Identical to ΔB_s of inner ring of same bearing		2
6	18	0	−2	2	1.5	1	2	3	3			2
18	30	0	−2.5	2.5	2	1	2.5	3	3			2
30	50	0	−3	3	2	1.5	3	3	3			2
50	80	0	−3.5	3.5	3	2	3	3	4			2.5
80	120	0	−4	4	3	2	4	3.5	4.5			3
120	150	0	−4.5	4.5	3	2.5	4.5	4	5			3
150	180	0	−5	5	4	3	5	4	5.5			3
180	250	0	−6	6	4.5	3	6	4.5	6			4
250	315	0	−7	7	5.5	3.5	7	5	7			4.5
315	400	0	−8	8	6	4	8	5	8			5
400	500	0	−9	9	6.5	4.5	9	6	9			6
500	630	0	−11	11	8.5	5.5	10	7	10			7
630	800	0	−14	14	10	7	12	8	12			8

[a] This diameter is included in the group.
[b] No values have been established for sealed or shielded bearings.
[c] Applies to groove type ball bearings only.
[d] This refers to the rings of single bearings made for paired or stack mounting.
[e] For instrument ball bearing tolerances see [3.2, 3.3].
[f] For tapered roller bearing tolerances see [3.7, 3.8].

TABLE 3.10. PART 1. Tolerance Class ABEC-7. Metric Ball [except instrument bearings [f]] of Dimensions Conforming to the Basic Plan for Boundary Dimensions of Metric Radial Bearings Given in Table 1 of [3.9].

Inner Ring (Tolerance values in micrometers)

d mm		Δ_{dmp}		Δ_{ds} [b]		V_{dp} diameter series 7, 8, 9	V_{dp} diameter series 0, 1, 2, 3, 4	V_{dmp}	K_{ia}	S_d	S_{ia} [d]	Δ_{Bs} all	Δ_{Bs} normal	Δ_{Bs} mod. [e]	V_{Bs}
over	incl.	high	low	high	low	max.		max.	max.	max.	max.	high	low		max.
[a] 0.6	2.5	0	−4	0	−4	4	3	2	2.5	3	3	0	−40	−250	2.5
2.5	10	0	−4	0	−4	4	3	2	2.5	3	3	0	−40	−250	2.5
10	18	0	−4	0	−4	4	3	2	2.5	3	3	0	−80	−250	2.5
18	30	0	−5	0	−5	5	4	2.5	3	4	4	0	−120	−250	2.5
30	50	0	−6	0	−6	6	5	3	4	4	4	0	−120	−250	3
50	80	0	−7	0	−7	7	5	3.5	4	5	5	0	−150	−250	4
80	120	0	−8	0	−8	8	6	4	5	5	5	0	−200	−380	4
120	180	0	−10	0	−10	10	8	5	6	6	7	0	−250	−380	5
180	250	0	−12	0	−12	12	9	6	8	7	8	0	−300	−500	6

Outer Ring (Tolerance values in micrometers)

D mm		Δ_{Dmp}		Δ_{Ds} [b]		V_{Dp} [c] diameter series 7, 8, 9	V_{Dp} [c] diameter series 0, 1, 2, 3, 4	V_{Dmp}	K_{ea}	S_D	S_{ea} [d]	Δ_{Cs}		V_{Cs}
over	incl.	high	low	high	low	max.	max.	max.	max.	max.	max.	high	low	max.
[a] 2.5	6	0	−4	0	−4	4	3	2	3	4	5			2.5
6	18	0	−4	0	−4	4	3	2	3	4	5			2.5
18	30	0	−5	0	−5	5	4	2.5	4	4	5			2.5
30	50	0	−6	0	−6	6	5	3	5	4	5	Identical to ΔB_s of inner ring of same bearing		2.5
50	80	0	−7	0	−7	7	5	3.5	5	4	5			3
80	120	0	−8	0	−8	8	6	4	6	5	6			4
120	150	0	−9	0	−9	9	7	5	7	5	7			5
150	180	0	−10	0	−10	10	8	5	8	5	8			5
180	250	0	−11	0	−11	11	8	6	10	7	10			7
250	315	0	−13	0	−13	13	10	7	11	8	10			7
315	400	0	−15	0	−15	15	11	8	13	10	13			8

[a] This diameter is included in the group.
[b] These deviations apply to diameter series 0, 1, 2, 3, and 4 only.
[c] No values have been established for sealed or shielded bearings.
[d] Applies to groove type ball bearings only.
[e] This refers to the rings of single bearings made for paired or stack mounting.
[f] For instrument ball bearing tolerances see [3.2, 3.3].

TABLE 3.10. PART 2. Tolerance Class ABEC-7. Metric Ball [except instrument bearings [f]] of Dimensions Conforming to the Basic Plan for Boundary Dimensions of Metric Radial Bearings Given in Table 1 of [3.9].

Inner Ring (Tolerance values in .0001 in.)

d mm		Δ_{dmp}		Δ_{ds} [b]		V_{dp} diameter series 7, 8, 9	V_{dp} diameter series 0, 1, 2, 3, 4	V_{dmp}	K_{ia}	S_d	S_{ia} [d]	Δ_{Bs} all	Δ_{Bs} normal	Δ_{Bs} mod. [e]	V_{Bs}
over	incl.	high	low	high	low	max.	max.	max.	max.	max.	max.	high	low	low	max.
[a] 0.6	2.5	0	−1.5	0	−1.5	1.5	1	1	1	1	1	0	−16	−98	1
2.5	10	0	−1.5	0	−1.5	1.5	1	1	1	1	1	0	−16	−98	1
10	18	0	−1.5	0	−1.5	1.5	1	1	1	1	1	0	−31	−98	1
18	30	0	−2	0	−2	2	1.5	1	1	1.5	1.5	0	−47	−98	1
30	50	0	−2.5	0	−2.5	2.5	2	1	1.5	1.5	1.5	0	−47	−98	1
50	80	0	−3	0	−3	3	2	1.5	1.5	2	2	0	−59	−98	1.5
80	120	0	−3	0	−3	3	2.5	1.5	2	2	2	0	−79	−150	1.5
120	180	0	−4	0	−4	4	3	2	2.5	2.5	3	0	−98	−150	2
180	250	0	−4.5	0	−4.5	4.5	3.5	2.5	3	3	3	0	−118	−197	2.5

Outer Ring (Tolerance values in .0001 in.)

D mm		Δ_{Dmp}		Δ_{Ds} [b]		V_{dp} [c] diameter series		V_{Dmp}	K_{ea}	S_D	S_{ea} [d]	Δ_{Cs}		V_{Cs}
						7, 8, 9	0, 1, 2, 3, 4							
over	incl.	high	low	high	low	max.		max.	max.	max.	max.	high	low	max.
a 2.5	6	0	−1.5	0	−1.5	1.5	1	1	1	1.5	2			1
6	18	0	−1.5	0	−1.5	1.5	1	1	1	1.5	2			1
18	30	0	−2	0	−2	2	1.5	1	1.5	1.5	2			1
30	50	0	−2.5	0	−2.5	2.5	2	1	2	1.5	2	Identical		1
50	80	0	−3	0	−3	3	2	1.5	2	1.5	2	to Δ_{Bs} of		1
80	120	0	−3	0	−3	3	2.5	1.5	2.5	2	2.5	inner		1.5
120	150	0	−3.5	0	−3.5	3.5	3	2	3	2	3	ring of		2
150	180	0	−4	0	−4	4	3	2	3	2	3	same		2
180	250	0	−4.5	0	−4.5	4.5	3	2.5	4	3	4	bearing		3
250	315	0	−5	0	−5	5	4	3	4.5	3	4			3
315	400	0	−6	0	−6	6	4.5	3	5	4	5			3

[a] This diameter is included in the group.
[b] These deviations apply to diameter series 0, 1, 2, 3, and 4 only.
[c] No values have been established for sealed or shielded bearings.
[d] Applies to groove type ball bearings only.
[e] This refers to the rings of single bearings made for paired or stack mounting.
[f] For instrument ball bearing tolerances see [3.2, 3.3].

TABLE 3.11. PART 1. Tolerance Class ABEC-9. Metric Ball [except instrument bearings[d]] of Dimensions Conforming to the Basic Plan for Boundary Dimensions of Metric Radial Bearings Given in Table 1 of [3.9].

Inner Ring (Tolerance values in micrometers)

d mm		Δ_{dmp}		Δ_{ds}		V_{dp}	V_{dmp}	K_{ia}	S_d	S_{ia} [c]	Δ_{Bs}		V_{Bs}
over	incl.	high	low	high	low	max.	max.	max.	max.	max.	high	low	max.
[a] 0.6	2.5	0	−2.5	0	−2.5	2.5	1.5	1.5	1.5	1.5	0	−40	1.5
2.5	10	0	−2.5	0	−2.5	2.5	1.5	1.5	1.5	1.5	0	−40	1.5
10	18	0	−2.5	0	−2.5	2.5	1.5	1.5	1.5	1.5	0	−80	1.5
18	30	0	−2.5	0	−2.5	2.5	1.5	2.5	1.5	2.5	0	−120	1.5
30	50	0	−2.5	0	−2.5	2.5	1.5	2.5	1.5	2.5	0	−120	1.5
50	80	0	−4	0	−4	4	2	2.5	1.5	2.5	0	−150	1.5
80	120	0	−5	0	−5	5	2.5	2.5	2.5	2.5	0	−200	2.5
120	150	0	−7	0	−7	7	3.5	2.5	2.5	2.5	0	−250	2.5
150	180	0	−7	0	−7	7	3.5	5	4	5	0	−300	4
180	250	0	−8	0	−8	8	4	5	5	5	0	−350	5

Outer Ring (Tolerance values in micrometers)

D mm		Δ_{Dmp}		Δ_{Ds}		V_{Dp} [b]	V_{Dmp}	K_{ea}	S_D	S_{ea} [c]	Δ_{Cs}		V_{Cs}
over	incl.	high	low	high	low	max.	max.	max.	max.	max.	high	low	max.
[a] 2.5	6	0	−2.5	0	−2.5	2.5	1.5	1.5	1.5	1.5			1.5
6	18	0	−2.5	0	−2.5	2.5	1.5	1.5	1.5	1.5			1.5
18	30	0	−4	0	−4	4	2	2.5	1.5	2.5	Identical to Δ_{Bs} of inner ring of same bearing		1.5
30	50	0	−4	0	−4	4	2	2.5	1.5	2.5			1.5
50	80	0	−4	0	−4	4	2	4	1.5	4			1.5
80	120	0	−5	0	−5	5	2.5	5	2.5	5			2.5
120	150	0	−5	0	−5	5	2.5	5	2.5	5			2.5
150	180	0	−7	0	−7	7	3.5	5	2.5	5			2.5
180	250	0	−8	0	−8	8	4	7	4	7			4
250	315	0	−8	0	−8	8	4	7	5	7			5
315	400	0	−10	0	−10	10	5	8	7	8			7

[a] This diameter is included in the group.
[b] No values have been established for sealed or shielded bearings.
[c] Applies to groove type ball bearings only.
[d] For instrument ball bearing tolerances see [3.2, 3.3].

TABLE 3.11. PART 2. Tolerance Class ABEC-9. Metric Ball [except instrument bearings[d]] of Dimensions Conforming to the Basic Plan for Boundary Dimensions of Metric Radial Bearings Given in Table 1 of [3.9].

Inner Ring (Tolerance values in .0001 in.)

d mm		Δ_{dmp}		Δ_{ds}		V_{dp}	V_{dmp}	K_{ia}	S_d	S_{ia}[c]	Δ_{Bs}		V_{Bs}
over	incl.	high	low	high	low	max.	max.	max.	max.	max.	high	low	max.
a 0.6	2.5	0	−1	0	−1	1	.5	.5	.5	.5	0	−16	.5
2.5	10	0	−1	0	−1	1	.5	.5	.5	.5	0	−16	.5
10	18	0	−1	0	−1	1	.5	.5	.5	.5	0	−31	.5
18	30	0	−1	0	−1	1	.5	1	.5	1	0	−47	.5
30	50	0	−1	0	−1	1	.5	1	.5	1	0	−47	.5
50	80	0	−1.5	0	−1.5	1.5	1	1	.5	1	0	−59	.5
80	120	0	−2	0	−2	2	1	1	1	1	0	−79	1
120	150	0	−3	0	−3	3	1.5	1	1	1	0	−98	1
150	180	0	−3	0	−3	3	1.5	2	1.5	2	0	−118	1.5
180	250	0	−3	0	−3	3	1.5	2	2	2	0	−138	2

Outer Ring (Tolerance values in .0001 in.)

D mm		Δ_{Dmp}		Δ_{Ds}		V_{Dp} [b]	V_{Dmp}	K_{ea}	S_D	S_{ea} [c]	Δ_{Cs}		V_{Cs}
over	incl.	high	low	high	low	max.	max.	max.	max.	max.	high	low	max.
[a] 2.5	6	0	−1	0	−1	1	.5	.5	.5	.5			.5
6	18	0	−1	0	−1	1	.5	.5	.5	.5			.5
18	30	0	−1.5	0	−1.5	1.5	1	1	.5	1	Identical to		.5
30	50	0	−1.5	0	−1.5	1.5	1	1	.5	1	Δ_{Bs} of inner		.5
50	80	0	−1.5	0	−1.5	1.5	1	1.5	.5	1.5	ring of		.5
80	120	0	−2	0	−2	2	1	2	1	2	same bear-		1
120	150	0	−2	0	−2	2	1	2	1	2	ing		1
150	180	0	−3	0	−3	3	1.5	2	1	2			1
180	250	0	−3	0	−3	3	1.5	3	1.5	3			1.5
250	315	0	−3	0	−3	3	1.5	3	2	3			2
315	400	0	−4	0	−4	4	2	3	3	3			3

[a] This diameter is included in the group.
[b] No values have been established for sealed or shielded bearings.
[c] Applies to groove type ball bearings only.
[d] For instrument ball bearing tolerances see [3.2, 3.3].

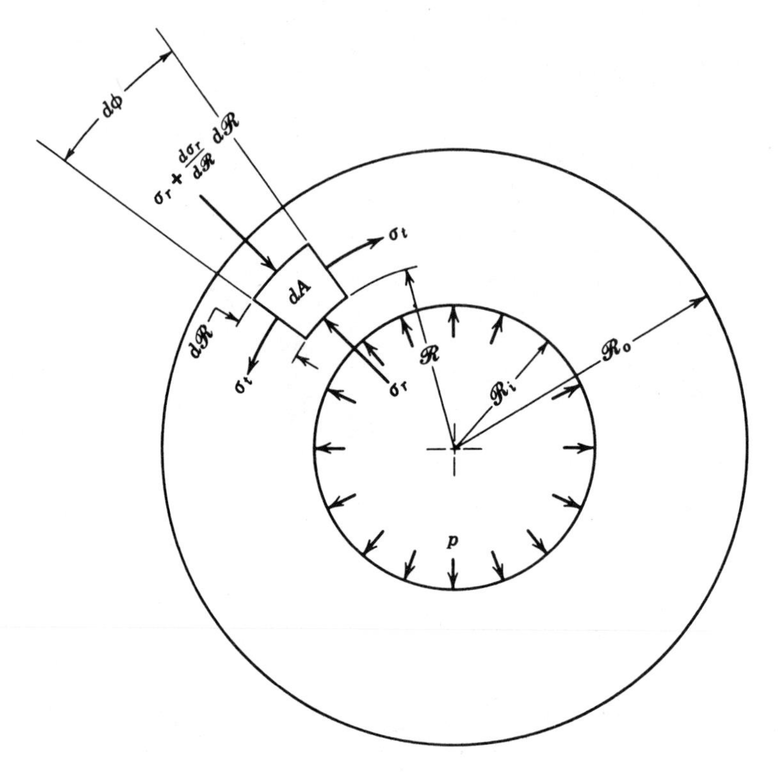

FIGURE 3.2. Thick ring loaded by internal pressure p.

direction is zero for static equilibrium:

$$\sigma_r \mathcal{R}\, d\phi + 2\sigma_t d\mathcal{R} \sin\frac{d\phi}{2} - \left(\sigma_r + \frac{d\sigma_r}{d\mathcal{R}}\, d\mathcal{R}\right)(\mathcal{R} + d\mathcal{R})\, d\phi = 0 \quad (3.1)$$

Since $d\phi$ is small, $\sin \frac{1}{2}\, d\phi = \frac{1}{2}\, d\phi$ and, neglecting small quantities of higher order,

$$\sigma_t - \sigma_r - \mathcal{R}\frac{d\sigma_r}{d\mathcal{R}} = 0 \quad (3.2)$$

Corresponding to the stress in the radial direction, there is an elongation u and the unit strain in the radial direction is

$$\epsilon_r = \frac{du}{d\mathcal{R}} \quad (3.3)$$

In the circumferential direction the unit strain is

$$\epsilon_r = \frac{u}{\mathcal{R}} \tag{3.4}$$

According to plane strain theory,

$$\epsilon_r = \frac{1}{E}(\sigma_r - \xi\sigma_t) \tag{3.5}$$

$$\epsilon_t = \frac{1}{E}(\sigma_t - \xi\sigma_r) \tag{3.6}$$

Combining equations (3.3–3.6) yields

$$\sigma_r = \frac{E}{1 - \xi^2}\left(\frac{du}{d\mathcal{R}} + \xi\frac{u}{\mathcal{R}}\right) \tag{3.7}$$

$$\sigma_t = \frac{E}{1 - \xi^2}\left(\frac{u}{\mathcal{R}} + \xi\frac{du}{d\mathcal{R}}\right) \tag{3.8}$$

Substituting equations (3.7) and (3.8) into (3.2) yields

$$\frac{d^2u}{d\mathcal{R}^2} + \frac{1}{\mathcal{R}}\frac{du}{d\mathcal{R}} - \frac{u}{\mathcal{R}^2} = 0 \tag{3.9}$$

The general solution to equation (3.9) is

$$u = c_1\mathcal{R} + c_2\mathcal{R}^{-1} \tag{3.10}$$

Substituting in equations (3.7) and (3.8) from (3.10) gives

$$\sigma_r = \frac{E}{1 - \xi^2}\left[c_1(1 + \xi) - c_2\frac{(1 - \xi)}{\mathcal{R}^2}\right] \tag{3.11}$$

$$\sigma_t = \frac{E}{1 - \xi^2}\left[c_1(1 + \xi) + c_2\frac{(1 - \xi)}{\mathcal{R}^2}\right] \tag{3.12}$$

At the boundary defined by $\mathcal{R} = \mathcal{R}_o$, $\sigma_r = p = 0$; therefore,

$$c_1(1 + \xi) = \frac{c_2}{\mathcal{R}_o^2}(1 - \xi) \tag{3.13}$$

At $\mathcal{R} = \mathcal{R}_i$, $\sigma_r = p$ and, therefore,

$$c_2 = \frac{p(1 + \xi)\mathcal{R}_i^2\mathcal{R}_o^2}{E(\mathcal{R}_o^2 - \mathcal{R}_i^2)} \tag{3.14}$$

Substituting equations (3.13) and (3.14) into (3.11) and (3.12) yields

$$\sigma_r = p\left[\frac{(\mathcal{R}_o/\mathcal{R})^2 - 1}{(\mathcal{R}_o/\mathcal{R}_i)^2 - 1}\right] \tag{3.15}$$

$$\sigma_t = p\left[\frac{(\mathcal{R}_o/\mathcal{R})^2 + 1}{(\mathcal{R}_o/\mathcal{R}_i)^2 - 1}\right] \tag{3.16}$$

Similarly, for a ring loaded by external pressure only,

$$\sigma_r = p\left[\frac{1 - (\mathcal{R}/\mathcal{R}_o)^2}{1 - (\mathcal{R}_i/\mathcal{R}_o)^2}\right] \tag{3.17}$$

$$\sigma_t = p\left[\frac{1 + (\mathcal{R}/\mathcal{R}_o)^2}{1 - (\mathcal{R}_i/\mathcal{R}_o)^2}\right] \tag{3.18}$$

From equations (3.15), (3.16), and (3.5), the increase in the internal radius of a ring loaded by internal pressure p is given by

$$u_i = \frac{p\mathcal{R}_i}{E}\left[\frac{(\mathcal{R}_o/\mathcal{R}_i)^2 + 1}{(\mathcal{R}_o/\mathcal{R}_i)^2 - 1} + \xi\right] \tag{3.19}$$

Similarly, the decrease in the external radius $\mathcal{R}_o$ of a ring loaded by external pressure p is given by

$$u_o = \frac{p\mathcal{R}_o}{E}\left[\frac{(\mathcal{R}_o/\mathcal{R}_i)^2 + 1}{(\mathcal{R}_o/\mathcal{R}_i)^2 - 1} - \xi\right] \tag{3.20}$$

If a ring having elastic modulus E_1, outside diameter $\mathfrak{D}_1$, and bore $\mathfrak{D}$ is mounted with a diametral interference I on a second ring having modulus E_2, outside diameter $\mathfrak{D}$, and bore $\mathfrak{D}_2$, then a common pressure p develops between the rings. The radial interference is the sum of the radial deflection of each ring due to pressure p. Hence the diametral interference is given by

$$I = 2(u_1 + u_2) \tag{3.21}$$

In terms of the common diameter $\mathfrak{D}$, therefore,

$$I = p\mathfrak{D}\left\{\frac{1}{E_1}\left[\frac{(\mathfrak{D}_1/\mathfrak{D})^2 + 1}{(\mathfrak{D}_1/\mathfrak{D})^2 - 1} + \xi_1\right] + \frac{1}{E_2}\left[\frac{(\mathfrak{D}/\mathfrak{D}_2)^2 + 1}{(\mathfrak{D}/\mathfrak{D}_2)^2 - 1} - \xi_2\right]\right\} \tag{3.22}$$

It can be seen that equation (3.22) can be used to determine p if I is known; thus,

$$p = \frac{\dfrac{I}{\mathfrak{D}}}{\dfrac{1}{E_1}\left[\dfrac{(\mathfrak{D}_1/\mathfrak{D})^2 + 1}{(\mathfrak{D}_1/\mathfrak{D})^2 - 1} + \xi_1\right] + \dfrac{1}{E_2}\left[\dfrac{(\mathfrak{D}/\mathfrak{D}_2)^2 + 1}{(\mathfrak{D}/\mathfrak{D}_2)^2 - 1} - \xi_2\right]} \tag{3.23}$$

If the external ring is a bearing inner ring of diameter $\mathfrak{D}_1$ and bore $\mathfrak{D}_s$ as shown in Fig. 3.3, then the increase in $\mathfrak{D}_1$ due to press fitting is

$$\Delta_s = \frac{2I\left(\dfrac{\mathfrak{D}_1}{\mathfrak{D}_s}\right)}{[(\mathfrak{D}_1/\mathfrak{D}_s)^2 - 1]\left\{\left[\dfrac{(\mathfrak{D}_1/\mathfrak{D}_s)^2 + 1}{(\mathfrak{D}_1/\mathfrak{D}_s)^2 - 1} + \xi_b\right] + \dfrac{E_b}{E_s}\left[\dfrac{(\mathfrak{D}_s/\mathfrak{D}_2)^2 + 1}{(\mathfrak{D}_s/\mathfrak{D}_2)^2 - 1} - \xi_s\right]\right\}} \tag{3.24}$$

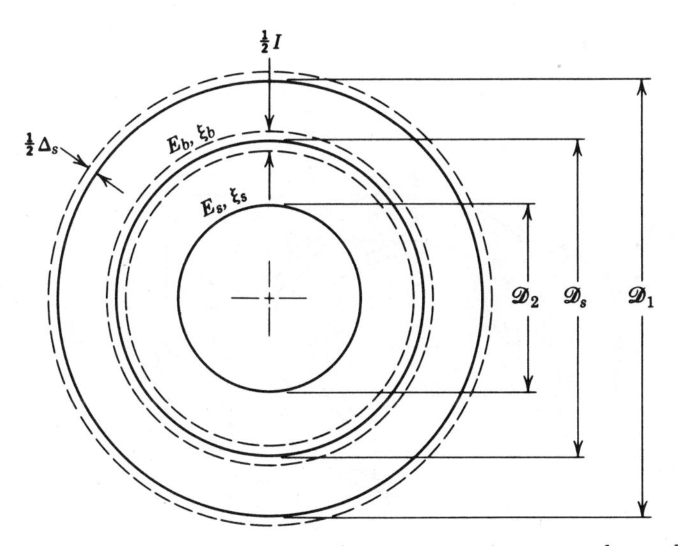

FIGURE 3.3. Schematic diagram of a bearing inner ring mounted on a shaft.

If the bearing inner ring and shaft are both fabricated from the same material, then

$$\Delta_s = I\left(\frac{\mathfrak{D}_1}{\mathfrak{D}_s}\right)\left[\frac{(\mathfrak{D}_s/\mathfrak{D}_2)^2 - 1}{(\mathfrak{D}_1/\mathfrak{D}_2)^2 - 1}\right] \tag{3.25}$$

For a bearing inner ring mounted on a solid shaft of the same material, diameter $\mathfrak{D}_2$ is zero and

$$\Delta_s = I\left(\frac{\mathfrak{D}_s}{\mathfrak{D}_1}\right) \tag{3.26}$$

By a similar process it is possible to determine the contraction of the bore of the internal ring of the assembly shown in Fig. 3.4. Thus,

$$\Delta_h = \frac{2I\left(\dfrac{\mathfrak{D}_h}{\mathfrak{D}_2}\right)}{[(\mathfrak{D}_h/\mathfrak{D}_2)^2 - 1]\left\{\dfrac{(\mathfrak{D}_h/\mathfrak{D}_2)^2 + 1}{(\mathfrak{D}_h/\mathfrak{D}_2)^2 - 1} - \xi_b + \dfrac{E_b}{E_h}\left[\dfrac{(\mathfrak{D}_1/\mathfrak{D}_h)^2 + 1}{(\mathfrak{D}_1/\mathfrak{D}_h)^2 - 1} + \xi_h\right]\right\}} \tag{3.27}$$

For a bearing outer ring pressed into a housing of the same material,

$$\Delta_h = I\left(\frac{\mathfrak{D}_h}{\mathfrak{D}_2}\right)\left[\frac{(\mathfrak{D}_1/\mathfrak{D}_h)^2 - 1}{(\mathfrak{D}_1/\mathfrak{D}_2)^2 - 1}\right] \tag{3.28}$$

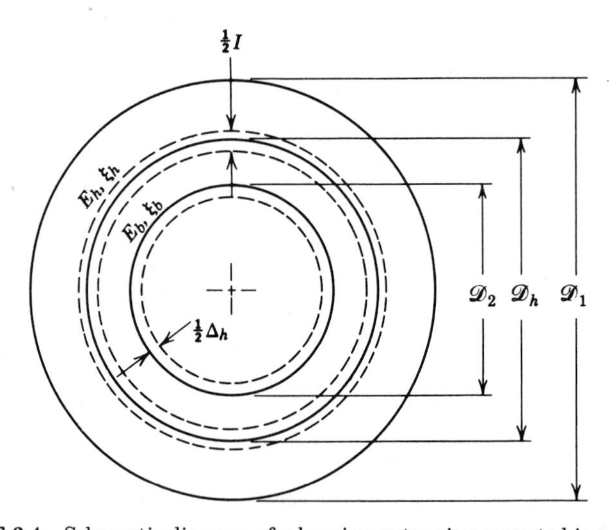

FIGURE 3.4. Schematic diagram of a bearing outer ring mounted in a housing.

If the housing is large compared to the ring dimensions, diameter $\mathfrak{D}_1$ approaches infinity and

$$\Delta_h = I\left(\frac{\mathfrak{D}_2}{\mathfrak{D}_h}\right) \tag{3.29}$$

Considering a bearing having a clearance P_d prior to mounting, the change in clearance after mounting is given by

$$\Delta P_d = -\Delta_s - \Delta_h \tag{3.30}$$

The preceding equation takes no account of differential thermal expansion.

PRESS FORCE

Since the pressure p between interfering surfaces is known, it is possible to estimate the amount of axial force necessary to accomplish or remove an interference fit. Because the area of shear is $\pi\mathfrak{D}B$, the axial force is given by

$$F_a = \mu\pi\mathfrak{D}Bp \tag{3.31}$$

in which μ is the coefficient of friction. According to Jones [3.8] the force required to press a steel ring on a solid steel shaft may be estimated by

$$F_a = 47100\ BI\left[1 - \left(\frac{\mathfrak{D}_s}{\mathfrak{D}_1}\right)^2\right] \tag{3.32}$$

This is based on a kinetic coefficient of friction $\mu = 0.15$. Similarly, the axial force required to press a steel bearing into a steel housing is given by

$$F_a = 47100\ CI\left[1 - \left(\frac{\mathfrak{D}_2}{\mathfrak{D}_h}\right)^2\right] \tag{3.33}$$

DIFFERENTIAL EXPANSION

Rolling bearings are usually fabricated from hardened steel and are generally mounted with press fits on steel shafts. In many applications such as in aircraft, however, the bearing may be mounted in a housing of a dissimilar material. Bearings are usually mounted at room temperature;

but they may operate at temperatures elevated ΔT above room temperature. The amount of temperature elevation may be determined by using the heat generation and heat transfer techniques indicated in Chapter 15. Under the influence of increased temperature, materials will expand linearly to the following equation:

$$u = \Gamma \mathcal{L}(T - T_a) \tag{3.34}$$

in which Γ is the coefficient of linear expansion in mm per mm per °C and $\mathcal{L}$ is a characteristic length.

Considering a bearing outer ring of outside diameter d_o at temperatures $T_o - T_a$ above ambient, the increase in ring outside circumference is given approximately by

$$u_{toc} = \Gamma_b \pi d_o(T_o - T_a) \tag{3.35}$$

Therefore the approximate increase in diameter is

$$u_{to} = \Gamma_b d_o(T_o - T_a) \tag{3.36}$$

The inner ring will undergo a similar expansion:

$$u_{ti} = \Gamma_b d_i(T_i - T_a) \tag{3.37}$$

Thus the net diametral expansion of the fit is given by

$$\Delta_T = \Gamma_b[d_o(T_o - T_a) - d_i(T_i - T_a)] \tag{3.38}$$

When the housing is fabricated from a material other than steel, the interference I between the housing and outer ring may either increase or decrease at elevated temperatures. Equation (3.39) gives the change in I with temperature:

$$\Delta I = (\Gamma_b - \Gamma_h)\mathcal{D}_h(T_o - T_a) \tag{3.39}$$

in which Γ_b and Γ_h are the coefficients of expansion of the bearing and housing, respectively. For dissimilar materials the housing is likely to expand more than the bearing, which tends to reduce any interference fit. Equation (3.30) therefore becomes

$$\Delta P_d = \Delta_T - \Delta_s - \Delta_h \tag{3.40}$$

If the shaft is not fabricated from the same material (usually steel) as the bearing, then a similar analysis applies.

EFFECT OF SURFACE FINISH

The interference I between a bore and o.d. is somewhat less than the apparent dimensional value due to the smoothing of the minute peaks and valleys of the surface. The schedule of Table 3.12 for reduction of I may be used.

It can be seen from Table 3.12 that for an accurately ground shaft mating with a similar bore, it may be expected that the reduction on the bore diameter would be 0.0020 mm (0.00008 in.) and on the shaft possibly 0.0041 mm (0.00016 in.) or a total reduction in I of 0.0061 mm (0.00024 in.).

Example 3.1. The 209 radial ball bearing of Example 2.4 is manufactured to ABEC 5 specifications. The bearing is mounted on a solid steel shaft with a k5 fit and in a rigid steel housing with a K6 fit. The nominal bearing bore is 45 mm (1.7717 in.), and the nominal o.d. is 85 mm (3.3465 in.). Determine the bearing contact angle and free endplay under light thrust loading.

Shaft tolerance range from Table 3.3 is 0.0025 mm (0.0001 in.) to 0.0127 mm (0.0005 in.) or 0.0076 mm (0.0003 in.) mean. Bearing bore mean tolerance from Table 3.9 is 0.004 mm (0.00016 in.), a negative value; that is, −0.004 mm (−0.00016 in.)

The mean interference on the shaft is

$$I = 0.0076 + 0.004 = 0.0116 \text{ mm } (0.0005 \text{ in.})$$

Assuming the bearing is mounted on a ground surface, the reduction in I due to surface finish is approximately 0.0020 mm (0.00008 in.) (see Table 3.12) for the bearing bore and shaft. Therefore

$$I = 0.0116 - 2 \times 0.0020 = 0.0076 \text{ mm } (0.00030 \text{ in.})$$

$$d_i = 52.3 \text{ mm } (2.0587 \text{ in.}) \qquad \text{Ex. 2.4}$$

$$\mathfrak{D}_1 = d_i$$

TABLE 3.12. Reduction in Interference due to Surface Condition

Finish	Reduction 0.0001 mm	Reduction 0.00001 in.
Accurately ground surface	20–51	(8–20)
Very smooth turned surface	61–142	(24–56)
Machine-reamed bores	102–239	(40–94)
Ordinary accurately turned surface	239–483	(94–190)

$$\Delta_s = I\left(\frac{\mathfrak{D}_s}{\mathfrak{D}_1}\right) \tag{3.26}$$

$$= 0.0076\left(\frac{45}{52.3}\right) = 0.0065 \text{ mm } (0.00026 \text{ in.})$$

Housing tolerance range from Table 3.4 is −0.0178 mm (−0.0007 in.) to 0.0051 mm (0.0002 in.) or −0.0064 mm (−0.00025 in.) mean. Bearing o.d. mean tolerance range from Table 3.9 is 0.005 mm (0.00020 in.), a negative value, that is, −0.005 mm (−0.00020 in.).

The mean interference in the housing is

$$I = 0.0064 - 0.005 = 0.0014 \text{ mm } (0.00006 \text{ in.})$$

Assuming the bearing housing bore is accurately ground, the reduction I due to surface finish is approximately 0.0020 mm (0.00008 in.) (see Table 3.12) for the housing bore and the bearing o.d. Thus, the net interference in the housing is virtually zero.

$$P_d = 0.015 \text{ mm } (0.0006 \text{ in}) \qquad \text{Ex. 2.4}$$

$$\Delta P_d = -\Delta_s - \Delta_h \tag{3.30}$$

$$= -0.0065 + 0 = -0.0065 \text{ mm } (-0.00026 \text{ in.})$$

$$A = 0.508 \text{ mm } (0.02 \text{ in.}) \qquad \text{Ex. 2.4}$$

$$\alpha^\circ = \cos^{-1}\left(1 - \frac{P_d + \Delta P_d}{2A}\right) \tag{2.10}$$

$$= \cos^{-1}\left(1 - \frac{0.015 - 0.0065}{2 \times 0.508}\right) = 7°25'$$

$$P_e = 2A \sin \alpha^\circ \tag{2.12}$$

$$= 2 \times 0.508 \times \sin(7°25') = 0.1312 \text{ mm } (0.0052 \text{ in.})$$

Comparison of these values of α° and P_e with those of Example 2.4 indicates the necessity of including the effect of interference fitting in the determination of clearance.

Example 3.2. The 218 angular-contact ball bearing of Example 2.3 has a 90 mm (3.5433 in.) bore, a 160 mm (6.2992 in.) o.d. and is manufactured to ABEC 7 tolerance limits. The bearing is mounted on a hollow steel shaft of 63.5 mm (2.5 in.) bore with a k6 fit and in a titanium housing having an effective o.d. of 203.2 mm (8 in.) with an M6 fit. Determine the free contact angle of the bearing.

Shaft tolerance range from Table 3.3 is +0.0025 mm (0.0001 in.) to +0.0254 mm (0.0010 in.), or a +0.0140 mm (0.00055 in.) mean.

Bore mean tolerance range from Table 3.10 is 0.004 mm (0.00016 in.), a negative value, that is, −0.004 mm (−0.00016 in.)

The mean interference on the shaft is

$$I = 0.0140 + 0.004 = 0.0180 \text{ mm } (0.00071 \text{ in.})$$

Assuming the bearing is mounted on a ground surface, the reduction in I due to surface finish is approximately 0.0020 mm (0.00008 in.) (see Table 3.12) for the bearing bore and shaft, therefore,

$$I = 0.0180 - 2 \times 0.0020 = 0.0140 \text{ mm } (0.00055 \text{ in.})$$

$$d_{\mathrm{i}} = 102.8 \text{ mm } (4.047 \text{ in.}) \qquad \text{Ex. 2.3}$$

$$\mathfrak{D}_1 = d_{\mathrm{i}}$$

$$\Delta_{\mathrm{s}} = I\left(\frac{\mathfrak{D}_1}{\mathfrak{D}_{\mathrm{s}}}\right)\left[\frac{(\mathfrak{D}_{\mathrm{s}}/\mathfrak{D}_2)^2 - 1}{(\mathfrak{D}_1/\mathfrak{D}_2)^2 - 1}\right] \qquad (3.25)$$

$$= 0.0140 \times \frac{102.8}{90}\left[\frac{(90/63.5)^2 - 1}{(102.8/63.5)^2 - 1}\right]$$

$$= 0.00995 \text{ mm } (0.00039 \text{ in.})$$

Housing tolerance range from Table 3.4 is −0.033 mm (−0.0013 in.) to −0.0076 mm (−0.0003 in.), or a −0.0203 mm (−0.0008 in.) mean.

Bearing mean o.d. tolerance range from Table 3.10 is 0.005 mm (0.0002 in.), a negative value, that is, −0.005 mm (−0.0002 in.).

The mean interference is the housing is

$$I = 0.0203 - 0.005 = 0.0153 \text{ mm } (0.0006 \text{ in.})$$

Assuming the housing is mounted on a ground surface, the reduction in I due to surface finish is approximately 0.0020 mm (0.00008 in.) (see Table 3.12) for the bearing o.d. and housing bore, therefore

$$I = 0.0153 - 2 \times 0.0020 = 0.0113 \text{ mm } (0.00044 \text{ in.})$$

$$d_{\mathrm{o}} = 147.7 \text{ mm } (5.816 \text{ in.}) \qquad \text{Ex. 2.3}$$

$$\mathfrak{D}_2 = d_{\mathrm{o}}$$

For steel

$$E = 206900 \text{ N/mm}^2 \ (30 \times 10^6 \text{ psi})$$

$$\xi = 0.3$$

For titanium

$$E = 103500 \text{ N/mm}^2 \ (15 \times 10^6 \text{ psi})$$

$$\xi = 0.33$$

$$\Delta_h = \frac{2I(\mathfrak{D}_h/\mathfrak{D}_2)}{[(\mathfrak{D}_h/\mathfrak{D}_2)^2 - 1]\left\{\dfrac{(\mathfrak{D}_h/\mathfrak{D}_2)^2 + 1}{(\mathfrak{D}_h/\mathfrak{D}_2)^2 - 1} - \xi_b + \dfrac{E_b}{E_h}\left[\dfrac{(\mathfrak{D}_1/\mathfrak{D}_h)^2 + 1}{(\mathfrak{D}_1/\mathfrak{D}_h)^2 - 1} + \xi_h\right]\right\}} \qquad (3.27)$$

$$= \frac{2 \times 0.0113 \times (160/147.7)}{\left[\left(\dfrac{160}{147.7}\right)^2 - 1\right]\left\{\dfrac{\left(\dfrac{160}{147.7}\right)^2 + 1}{\left(\dfrac{160}{147.7}\right)^2 - 1} - 0.3 + \dfrac{206900}{103500}\left[\dfrac{\left(\dfrac{203.2}{160}\right)^2 + 1}{\left(\dfrac{203.2}{160}\right)^2 - 1} + 0.33\right]\right\}}$$

$$= 0.0064 \text{ mm } (0.00025 \text{ in.})$$

$$\Delta P_d = \Delta_s - \Delta_h \qquad (3.30)$$

$$= -0.00995 - 0.0064 = -0.01635 \text{ mm } (-0.00064 \text{ in.})$$

$$P_d = 0.483 \text{ mm } (0.019 \text{ in.}) \qquad \text{Ex. 2.3}$$

$$\alpha^\circ = \cos^{-1}\left(1 - \frac{P_d + \Delta P_d}{2A}\right) \qquad (2.10)$$

$$= \cos^{-1}\left(1 - \frac{0.483 - 0.01635}{2 \times 1.031}\right) = 39°19'$$

Example 3.3. The inner ring of the 218 angular-contact ball bearing operates at a mean temperature of 148.9°C (300°F) and the outer ring is at 121.1°C (250°F). Considering that the bearing was assembled at 21.1°C (70°F) and considering the press fits of Example 3.2, what free contact angle will occur?

For steel

$$\Gamma = 11.7 \times 10^{-6} \text{ mm/mm/°C } (6.5 \times 10^{-6} \text{ in./in./°F})$$

For titanium

$$\Gamma = 8.5 \times 10^{-6} \text{ mm/mm/°C } (4.7 \times 10^{-6} \text{ in./in./°F})$$

$$d_i = 102.8 \text{ mm } (4.047 \text{ in.}) \qquad \text{Ex. 2.3}$$

$$d_o = 147.7 \text{ mm } (5.816 \text{ in.}) \qquad \text{Ex. 2.3}$$

Because of differential expansion,

$$\Delta_T = \Gamma_b[d_o(T_o - T_a) - d_i(T_i - T_a)] \qquad (3.38)$$

$$= 11.7 \times 10^{-6}(147.7 \times 100 - 102.8 \times 127.8)$$

$$= 0.0191 \text{ mm } (0.00075 \text{ in.})$$

The outer ring and housing have different rates of expansion

$$\mathfrak{D}_h = 160 \text{ mm } (6.2992 \text{ in.}) \qquad \text{Ex. 3.2}$$

$$\Delta I = (\Gamma_b - \Gamma_h)\mathfrak{D}_h(T_o - T_a) \qquad (3.39)$$

$$= (11.7 - 8.5) \times 10^{-6} \times 160 \times (121.1 - 21.1)$$

$$= 0.0508 \text{ mm } (0.0020 \text{ in.})$$

$$\Delta_h = 0.0064 \text{ mm } (0.00025 \text{ in.}) \qquad \text{Ex. 3.2}$$

$$I = 0.0113 \text{ mm } (0.00044 \text{ in.}) \qquad \text{Ex. 3.2}$$

$$I = I + \Delta I$$

$$= 0.0113 + 0.0508 = 0.062 \text{ mm } (0.00244 \text{ in.})$$

$$\Delta_h = \frac{0.062}{0.0113} \times 0.00635 = 0.0348 \text{ mm } (0.00137 \text{ in.})$$

$$\Delta_s = 0.00995 \text{ mm } (0.00039 \text{ in.}) \qquad \text{Ex. 3.2}$$

$$\Delta P_d = \Delta_T - \Delta_s - \Delta_h \qquad (3.40)$$

$$= 0.0191 - 0.00995 - 0.0348 = -0.0257 \text{ mm } (-0.00101 \text{ in.})$$

$$P_d = 0.483 \text{ mm } (0.019 \text{ in.}) \qquad \text{Ex. 2.3}$$

$$A = 1.031 \text{ mm } (0.0406 \text{ in.}) \qquad \text{Ex. 2.3}$$

$$\alpha^\circ = \cos^{-1}\left(1 - \frac{P_d + \Delta P_d}{2A}\right) \qquad (2.10)$$

$$= \cos^{-1}\left(1 - \frac{0.483 - 0.0257}{2 \times 1.031}\right) = 38°54'$$

Example 3.4. The 209 ball bearing of Example 3.1 has a nominal width of 19 mm (0.7480 in.). What force is required to accomplish the shaft press fit in Example 3.1?

$$I = 0.0076 \text{ mm } (0.00030 \text{ in.}) \qquad \text{Ex. 3.1}$$

$$\mathfrak{D}_s = 45 \text{ mm } (1.7717 \text{ in.}) \qquad \text{Ex. 3.1}$$

$$\mathfrak{D}_1 = 52.3 \text{ mm } (2.0587 \text{ in.}) \qquad \text{Ex. 3.1}$$

$$F_a = 4.71 \times 10^4 BI \left[1 - \left(\frac{\mathfrak{D}_s}{\mathfrak{D}_1}\right)^2\right] \tag{3.32}$$

$$= 4.71 \times 10^4 \times 19 \times 0.0076 \left[1 - \left(\frac{45}{52.3}\right)^2\right]$$

$$= 1766 \text{ N } (397 \text{ lb})$$

CLOSURE

The important effect of bearing fitting practice on diametral clearance has been demonstrated for ball bearings in the numerical examples. Because the ball bearing contact angle determines its ability to carry thrust load and the contact angle is dependent on clearance, the analysis of the fit-up is important in many applications. The numerical examples herein were based on mean tolerance conditions. In many cases, however, it is necessary to examine the extremes of fit.

Although only the effect of fit-up on contact angle has been examined, it is not to be construed that this is the only effect of significance. Later, the sensitivity of other phases of rolling bearing operation to clearance will be investigated.

The thermal conditions of operation have been shown to be of no less significance than the fit-up. In precision applications, the clearance must be evaluated under operating conditions.

Tables 3.7–3.11 contain tolerance limits on radial and axial runout as well as the tolerance limits on mean diameters. Runout affects bearing performance in subtle ways such as through vibration as discussed in Chapter 25.

REFERENCES

3.1. *American National Standard (ANSI/AFBMA) Std 7-1972*, "Shaft and Housing Fits for Metric Radial Ball and Roller Bearings (Except Tapered Roller Bearings) Conforming to Basic Boundary Plans," R1978 (May 2, 1978).

3.2. *American National Standard (ANSI/AFBMA) Std 12.1-1985*, "Instrument Ball Bearings Metric Design" (May 13, 1985).

3.3. *American National Standard ANSI/AFBMA Std 12.2-1985*, "Instrument Ball Bearings Inch Design" (May 13, 1985).

3.4. *American National Standard (ANSI/AFBMA) Std 16.2-1978*, "Airframe Ball, Roller and Needle Bearings-Inch Design" (November 21, 1982).

3.5. *American National Standard (ANSI/AFBMA) Std 18.8-1982*, "Needle Roller Bearings Radial, Metric Design" (December 2, 1982).

3.6. *American National Standard (ANSI/AFBMA) Std 18.2-1982*, "Needle Roller Bearings Radial, Inch Design" (May 14, 1982).

3.7. *American National Standard (ANSI) Std B3.19-1975*, "Tapered Roller Bearings—Radial, Inch Design" (August 4, 1975).

3.8. *American National Standard (ANSI/AFBMA) Std 19.1-1987*, "Tapered Roller Bearing—Radial, Metric Design" (October 19, 1987).

3.9. *American National Standard (ANSI/AFBMA) Std 20-1987*, "Radial Bearings of Ball, Cylindrical Roller and Spherical Roller Types, Metric Design" (October 28, 1987).

3.10. A. Jones, *Analysis of Stresses and Deflection*, Vol. 1, New Departure Division, General Motors Corporation, Bristol, CT, pp. 161–170 (1946).

4

BALL AND ROLLER LOADS

LIST OF SYMBOLS

Symbol	Description	Units
D	Ball or roller diameter	mm (in.)
d_m	Pitch diameter	mm (in.)
F	Force	N (lb)
F_c	Centrifugal force	N (lb)
g	Gravitational constant	mm/sec^2 (in./sec^2)
J	Mass moment of inertia	kg · mm^2 (in. · lb · sec^2)
l_t	Roller length	mm (in.)
M	Moment	N · mm (in. · lb)
M_g	Gyratory moment	N · mm (in. · lb)
m	Mass of ball or roller	kg (lb · sec^2/in.)
n	Rotational speed	rpm
n_m	Orbital ball or roller speed, cage speed	rpm
n_R	Ball or roller speed about its own axis	rpm

Symbol	Description	Units
Q	Ball or roller normal load	N (lb)
Q_r	Radial direction load on ball or roller	N (lb)
Q_a	Axial direction load on ball or roller	N (lb)
r	Radius	mm (in.)
U	Coordinate direction distance	mm (in.)
V	Coordinate direction distance	mm (in.)
W	Coordinate direction distance	mm (in.)
x	Coordinate direction distance	mm (in.)
$\ddot{x}$	Acceleration in x-direction	mm/sec^2 ($in./sec^2$)
x'	Coordinate direction distance	mm (in.)
y	Coordinate direction distance	mm (in.)
$\ddot{y}$	Acceleration in y-direction	mm/sec^2 ($in./sec^2$)
y'	Coordinate direction distance	mm (in.)
z	Coordinate direction distance	mm (in.)
$\ddot{z}$	Acceleration in z-direction	mm/sec^2 ($in./sec^2$)
z'	Coordinate direction distance	mm (in.)
α	Contact angle	°, rad
β	Angle between W axis and z' axis	rad
β'	Angle between projection of the U axis on the $x'y'$ plane and the x' axis	rad
γ	$D \cos \alpha / d_m$	
γ_s	Roller skewing angle	°, rad
ζ	Roller tilting angle	°, rad
θ	Angle	rad
ρ	Mass density	kg/mm^3 ($lb \cdot sec^2 \cdot in.^{-4}$)
ϕ	Angle in WV plane	rad
ψ	Angle in yz plane	rad
ω_m	Orbital angular velocity of ball or roller	rad/sec
ω_R	Angular velocity of ball or roller about its own axis	rad/sec

SUBSCRIPTS

a	Refers to axial direction
e	Refers to rotation about an eccentric axis
f	Refers to guide flange
i	Refers to inner raceway
j	refers to rolling element at location j
m	Refers to orbital rotation

Symbol	Description	Units
	SUBSCRIPTS	
o	Refers to outer raceway	
R	Refers to rolling element	
r	Refers to radial direction	

GENERAL

The loads carried by ball and roller bearings are transmitted through the rolling elements from one ring to the other. The magnitude of the loading carried by the individual ball or roller depends on the internal geometry of the bearing and on the type of load impressed on it. In addition to applied loading, rolling elements are subjected to dynamic loading due to speed effects. Bearing geometry also affects the dynamic loading. The object of this chapter is to define the rolling element loading in ball and roller bearings under varied conditions of bearing operation.

STATIC LOADING

A rolling element can support a normal load along the line of contact between the rolling element and the raceway (see Fig. 4.1). If a radial load Q_r is applied to the ball of Fig. 4.1, then the normal load supported

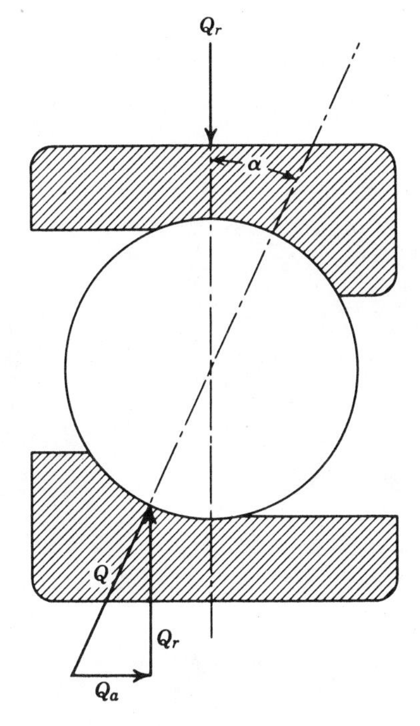

FIGURE 4.1. Radially loaded ball.

by the ball is

$$Q = \frac{Q_r}{\cos \alpha} \tag{4.1}$$

Hence a thrust load of magnitude

$$Q_a = Q \sin \alpha \tag{4.2}$$

or

$$Q_a = Q_r \tan \alpha \tag{4.3}$$

is induced in the assembly.

Example 4.1. The 209 radial ball bearing of Example 3.1 is subjected to a thrust (axial) load of 445 N (100 lb) per ball. What is the magnitude of the induced ball radial load assuming the contact angle* is not changed by the thrust load?

$$\alpha = 7° \, 25' \qquad \text{Ex. 3.1}$$

$$Q_a = Q_r \tan \alpha \qquad (4.3)$$

$$445 = Q_r \tan (7° \, 25')$$

$$Q_r = 3419 \text{ N (768 lb)}$$

This result indicates the degree of thrust load amplification at small contact angles.

Example 4.2. The 218 angular-contact ball bearing of Example 3.3 is subjected to a thrust load of 2225 N (500 lb) per ball. What is the magnitude of the normal ball load that is induced assuming the contact angle* is not changed by the thrust load?

$$\alpha = 38° \, 54' \qquad \text{Ex. 3.3}$$

$$Q_a = Q \sin \alpha \qquad (4.2)$$

$$2225 = Q \sin (38° \, 54')$$

$$Q = 3543 \text{ N (796.2 lb)}$$

It can be seen that a bearing with a 40° contact angle can support a thrust load better than a bearing with an 7° 25′ contact angle.

*This assumption is not accurate in this case and is made only to illustrate a point.

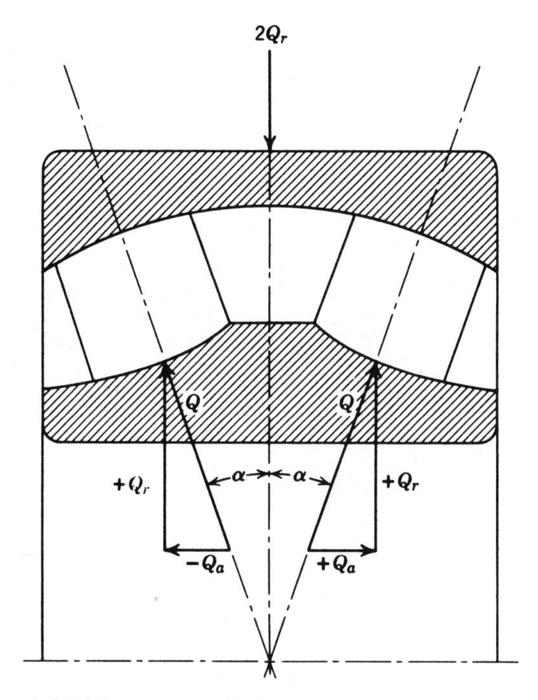

FIGURE 4.2. Radially loaded symmetrical rollers.

Equations (4.2) and (4.3) are also valid for spherical roller bearings using symmetrical rollers. For a double-row spherical roller bearing under an applied radial load the induced roller thrust loads are self-equilibrating (see Fig. 4.2).

Spherical roller bearings having asymmetrical rollers and tapered roller bearings usually have a fixed guide flange on the bearing inner ring. This flange as shown in Fig. 4.3 is subjected to loading through the roller ends. If a radial load Q_{ir} is applied to the assembly, the following loading occurs:

$$Q_{\text{i}} = \frac{Q_{\text{ir}}}{\cos \alpha_{\text{i}}} \tag{4.4}$$

$$Q_{\text{ia}} = Q_{\text{ir}} \tan \alpha_{\text{i}} \tag{4.5}$$

For static equilibrium the sum of forces in any direction is equal to zero; therefore,

$$Q_{\text{ir}} - Q_{\text{fr}} - Q_{\text{or}} = 0 \tag{4.6}$$

$$Q_{\text{ia}} + Q_{\text{fa}} - Q_{\text{oa}} = 0 \tag{4.7}$$

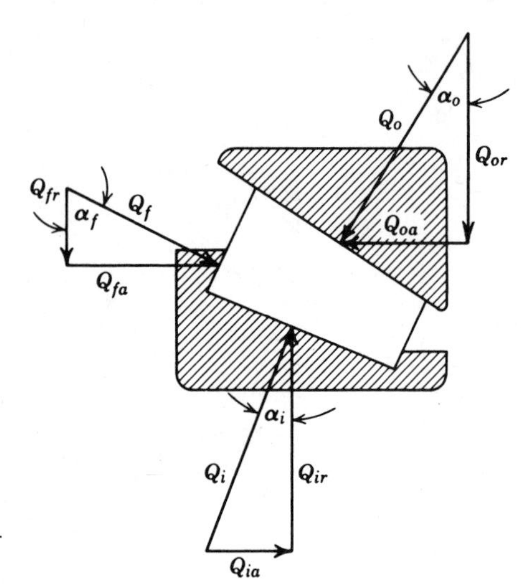

FIGURE 4.3. Radially loaded asymmetrical roller.

or

$$Q_{\mathrm{ir}} - Q_{\mathrm{f}} \cos \alpha_{\mathrm{f}} - Q_{\mathrm{o}} \cos \alpha_{\mathrm{o}} = 0 \tag{4.8}$$

$$Q_{\mathrm{ir}} \tan \alpha_{\mathrm{i}} + Q_{\mathrm{f}} \sin \alpha_{\mathrm{f}} - Q_{\mathrm{o}} \sin \alpha_{\mathrm{o}} = 0 \tag{4.9}$$

Solving equations (4.8) and (4.9) for Q_{o} and Q_{f} yields

$$Q_{\mathrm{o}} = Q_{\mathrm{ir}} \frac{(\sin \alpha_{\mathrm{f}} + \tan \alpha_{\mathrm{i}} \cos \alpha_{\mathrm{f}})}{\sin (\alpha_{\mathrm{o}} + \alpha_{\mathrm{f}})} \tag{4.10}$$

$$Q_{\mathrm{f}} = Q_{\mathrm{ir}} \frac{(\sin \alpha_{\mathrm{o}} - \tan \alpha_{\mathrm{i}} \cos \alpha_{\mathrm{o}})}{\sin (\alpha_{\mathrm{o}} + \alpha_{\mathrm{f}})} \tag{4.11}$$

The thrust load induced by the applied radial load is

$$Q_{\mathrm{oa}} = Q_{\mathrm{ir}} \frac{\sin \alpha_{\mathrm{o}}}{\sin (\alpha_{\mathrm{o}} + \alpha_{\mathrm{f}})} (\sin \alpha_{\mathrm{f}} + \tan \alpha_{\mathrm{i}} \cos \alpha_{\mathrm{f}}) \tag{4.12}$$

Under an applied thrust load Q_{ia} the following equations of load obtain, considering static equilibrium:

$$Q_{\mathrm{o}} = Q_{\mathrm{ia}} \frac{(\cos \alpha_{\mathrm{f}} + \operatorname{ctn} \alpha_{\mathrm{i}} \sin \alpha_{\mathrm{f}})}{\sin (\alpha_{\mathrm{o}} + \alpha_{\mathrm{f}})} \tag{4.13}$$

$$Q_{\mathrm{f}} = Q_{\mathrm{ia}} \frac{(\operatorname{ctn} \alpha_{\mathrm{i}} \sin \alpha_{\mathrm{o}} - \cos \alpha_{\mathrm{o}})}{\sin (\alpha_{\mathrm{o}} + \alpha_{\mathrm{f}})} \tag{4.14}$$

Example 4.3. A 90000 series steep-angle tapered roller bearing has the following dimensions:

$$\alpha_i = 22°$$

$$\alpha_o = 29°$$

$$D = 22.86 \text{ mm } (0.9000 \text{ in.}) \text{ (mean)}$$

$$l = 30.48 \text{ mm } (1.2 \text{ in.})$$

$$d_m = 142.24 \text{ mm } (5.600 \text{ in.}) \text{ (mean)}$$

$$\alpha_f = 64° \text{ (with bearing axis)}$$

If the most heavily loaded roller supports a thrust of 22,250 N (5000 lb), what is the magnitude of the maximum load on the guide flange?

$$Q_f = Q_{ia} \frac{(\text{ctn } \alpha_i \sin \alpha_o - \cos \alpha_o)}{\sin(\alpha_o + \alpha_f)} \tag{4.14}$$

$$= 22250 \frac{(\text{ctn } 22° \sin 29° - \cos 29°)}{\sin(29° + 64°)} = 7245 \text{ N } (1628 \text{ lb})$$

Compare this load with the maximum normal load on the cone.

$$Q_i = \frac{Q_{ia}}{\sin \alpha_i} \qquad \text{Fig. 4.3}$$

$$= \frac{22250}{\sin 22°} = 59{,}410 \text{ N } (13{,}350 \text{ lb})$$

DYNAMIC LOADING

Body Forces Due to Rolling Element Rotations

The development of equations in this section is based on the motions occurring in an angular-contact ball bearing because it is the most general form of rolling bearing. Subsequently, the equations developed can be so restricted as to apply to other ball bearings and also to roller bearings.

Figure 4.4 illustrates the instantaneous position of a particle of mass in a ball of an angular-contact ball bearing operating at high rotational speed about an axis x. To simplify the analysis the following coordinate axes systems are introduced:

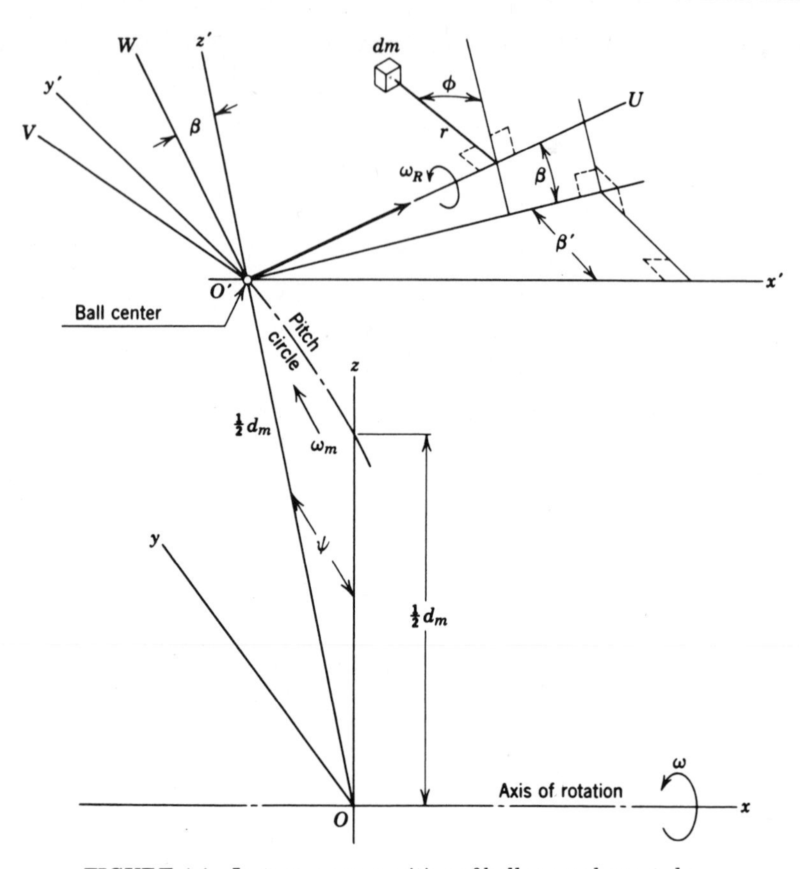

FIGURE 4.4. Instantaneous position of ball mass element dm.

x, y, z	A fixed set of Cartesian coordinates with the x axis coincident with the bearing rotational axis.
x', y', z'	A set of Cartesian coordinates with the x' axis parallel to the x axis of the fixed set. This set of coordinates has its origin O' at the ball center and rotates at orbital speed about the fixed x axis at radius $\frac{1}{2}\, d_{\mathrm{m}}$.
U, V, W	A set of Cartesian coordinates with origin at the ball center O' and rotating at orbital speed ω_{m}. The U axis is collinear with the axis of rotation of the ball about its own center. The W axis is in the plane of the U axis and z' axis; the angle between the W axis and z' axis is β.
U, r, ϕ	A set of polar coordinates rotating with the ball.

In addition to the foregoing coordinate systems, the following symbols are introduced:

β' The angle between the projection of the U axis on the $x'y'$ plane and the x' axis.

ψ The angle between the z axis and z' axis, that is, the angular position of the ball on the pitch circle.

Consider that an element of mass dm in the ball has the following instantaneous location in the system of rotating coordinates: U, r, ϕ. Since

$$\begin{aligned} U &= U \\ V &= r \sin \phi \\ W &= r \cos \phi \end{aligned} \tag{4.15}$$

and

$$\begin{aligned} x' &= U \cos \beta \cos \beta' - V \sin \beta' - W \sin \beta \cos \beta' \\ y' &= U \cos \beta \sin \beta' + V \cos \beta' - W \sin \beta \sin \beta' \\ z' &= U \sin \beta + W \cos \beta \end{aligned} \tag{4.16}$$

and

$$\begin{aligned} x &= x' \\ y &= \tfrac{1}{2} d_m \sin \psi + y' \cos \psi + z' \sin \psi \\ z &= \tfrac{1}{2} d_m \cos \psi - y' \sin \psi + z' \cos \psi \end{aligned} \tag{4.17}$$

therefore, by substitution of equations (4.15) into (4.16) and thence into (4.17), the following expressions relating the instantaneous position of the element of mass dm to the fixed system of Cartesian coordinates can be formulated.

$$x = U \cos \beta \cos \beta' - r(\sin \beta' \sin \phi + \sin \beta \cos \beta' \cos \phi) \tag{4.18}$$

$$\begin{aligned} y = {} & \frac{d_m}{2} \sin \psi + U(\cos \beta \sin \beta' \cos \psi + \sin \beta \sin \psi) \\ & + r(\cos \beta \sin \phi \cos \psi + \cos \beta \cos \phi \sin \psi \\ & \quad - \sin \beta \sin \beta' \cos \phi \cos \psi) \end{aligned} \tag{4.19}$$

$$\begin{aligned} z = {} & \frac{d_m}{2} \cos \psi + U(-\cos \beta \sin \beta' \sin \psi + \sin \beta \cos \psi) \\ & + r(-\cos \beta' \sin \phi \sin \psi + \cos \beta \cos \phi \cos \psi \\ & \quad + \sin \beta \sin \beta' \cos \phi \sin \psi) \end{aligned} \tag{4.20}$$

In accordance with Newton's second law of motion, the following relationships can be determined if the rolling element position angle ψ is arbitrarily set equal to 0°:

$$dF_{x'} = \ddot{x} dm \tag{4.21}$$

$$dF_{y'} = \ddot{y} dm \tag{4.22}$$

$$dF_{z'} = \ddot{z} dm \tag{4.23}$$

$$\begin{aligned} dM'_z = \{&-\ddot{x}[U \cos\beta \sin\beta' + r(\cos\beta' \sin\phi - \sin\beta \sin\beta' \cos\phi)] \\ &+ \ddot{y}[U \cos\beta \cos\beta' - r(\sin\beta' \sin\phi + \sin\beta \cos\beta' \cos\phi)]\}\, dm \end{aligned} \tag{4.24}$$

$$\begin{aligned} dM'_y = \{&\ddot{x}[U \sin\beta + r \cos\beta \cos\phi] \\ &- \ddot{z}[U \cos\beta \cos\beta' - r(\sin\beta' \sin\phi + \sin\beta \cos\beta' \cos\phi)]\}\, dm \end{aligned} \tag{4.25}$$

The net moment about the x axis must be zero for constant speed motion. At each ball location (ψ, β), ω_R (rotational speed $d\phi/dt$ of the ball about its own axis $U - 0'$) and ω_m (orbital speed $d\psi/dt$ of the ball about the bearing axis x) are constant; therefore, at $\psi = 0$:

$$\ddot{x} = \frac{d^2x}{dt^2} = r\omega_R^2(\sin\beta' \sin\phi + \sin\beta \cos\beta' \cos\phi) \tag{4.26}$$

$$\begin{aligned} \ddot{y} = \frac{d^2y}{dt^2} = &-2\omega_R\omega_m r \cos\beta \sin\phi \\ &+ \omega_m^2[-U \cos\beta \sin\beta' + r(-\cos\beta' \sin\phi + \sin\beta \cos\phi \sin\beta')] \\ &+ \omega_R^2 r(-\cos\beta' \cos\phi + \sin\beta \sin\beta' \sin\phi) \end{aligned} \tag{4.27}$$

$$\begin{aligned} \ddot{z} = \frac{d^2z}{dt^2} = &-2\omega_R\omega_m r(\cos\beta' \cos\phi + \sin\beta \sin\beta' \sin\phi) \\ &- \omega_m^2\left(\frac{d_m}{2} + U \sin\beta + r \cos\beta \cos\phi\right) \\ &- \omega_R^2 r \cos\beta \cos\phi \end{aligned} \tag{4.28}$$

Substitution of equations (4.26) to (4.28) into equations (4.21) to (4.25) and placing the latter into integral format yields

$$F_{x'} = -\rho \int_{-r_R}^{+r_R} \int_0^{(r_R^2 - U^2)^{1/2}} \int_0^{2\pi} \ddot{x} r \, dr \, dU \, d\phi \tag{4.29}$$

$$F_{y'} = -\rho \int_{-r_R}^{+r_R} \int^{(r_R^2 - U^2)^{1/2}} \int_0^{2\pi} \ddot{y} r \, dr \, dU \, d\phi \tag{4.30}$$

$$F_{z'} = -\rho \int_{-r_R}^{+r_R} \int_0^{(r_R^2 - U^2)^{1/2}} \int_0^{2\pi} \ddot{z} r \, dr \, dU \, d\phi \tag{4.31}$$

$$\begin{aligned} M_{z'} = -\rho \int_{-r_R}^{+r_R} \int_0^{(r_R^2 - U^2)^{1/2}} \int_0^{2\pi} & \{-\ddot{x}[U \cos\beta \sin\beta' \\ & + r(\cos\beta' \sin\phi - \sin\beta \sin\beta' \cos\phi)] \\ & + \ddot{y}[U \cos\beta \cos\beta' - r(\sin\beta' \sin\phi \\ & + \sin\beta \cos\beta' \cos\phi)]\} r \, dr \, dU \, d\phi \end{aligned} \tag{4.32}$$

$$\begin{aligned} M_{y'} = -\rho \int_{-r_R}^{+r_R} \int_0^{(r_R^2 - U^2)^{1/2}} \int_0^{2\pi} & \{\ddot{x}[U \sin\beta + r \cos\beta \cos\phi] \\ & - \ddot{z}[U \cos\beta \cos\beta' - r(\sin\beta' \sin\phi \\ & + \sin\beta \cos\beta' \cos\phi)]\} r \, dr \, dU \, d\phi \end{aligned} \tag{4.33}$$

In equations (4.29)–(4.33) ρ is the mass density of the ball material and r_R is the ball radius.

Performing the integrations indicated by equations (4.29)–(4.33) establishes that the net forces in the x' and y' directions are zero and that

$$F_{z'} = \tfrac{1}{2} m d_m \omega_m^2 \tag{4.34}$$

$$M_{y'} = J \omega_R \omega_m \sin\beta \tag{4.35}$$

$$M_{z'} = -J \omega_R \omega_m \cos\beta \sin\beta' \tag{4.36}$$

in which m is the mass of the ball and J is the mass moment of inertia. m and J are defined as follows:

$$m = \tfrac{1}{6} \rho \pi D^3 \tag{4.37}$$

$$J = \tfrac{1}{60} \rho \pi D^5 \tag{4.38}$$

Centrifugal Force

Rotation about the Bearing Axis. Substituting equation (4.37) into (4.34) and recognizing that

$$\omega_m = \frac{2\pi n_m}{60} \tag{4.39}$$

equation (4.40) yielding ball centrifugal force is obtained:

$$F_c = \frac{\pi^3 \rho}{10800g} D^3 n_m^2 d_m \tag{4.40}$$

For steel balls,

$$F_c = 2.26 \times 10^{-11} D^3 n_m^2 d_m \tag{4.41}$$

For an applied thrust load per ball Q_{ia} and a ball centrifugal load F_c directed radially outward, the ball loading is as shown in Fig. 4.5. For conditions of equilibrium, assuming the bearing rings are not flexible,

$$Q_{ia} - Q_{oa} = 0 \tag{4.42}$$

$$Q_{ir} + F_c - Q_{or} = 0 \tag{4.43}$$

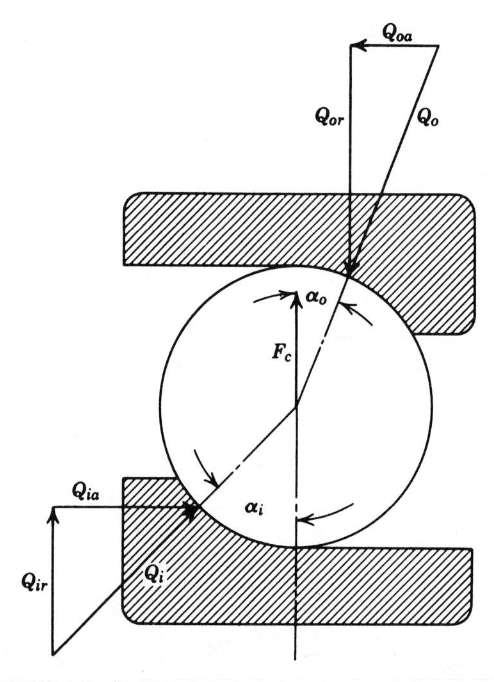

FIGURE 4.5. Ball under thrust load and centrifugal load.

or

$$Q_{ia} - Q_o \sin \alpha_o = 0 \tag{4.44}$$

$$Q_{ia} \operatorname{ctn} \alpha_i + F_c - Q_o \cos \alpha_o = 0 \tag{4.45}$$

In equations (4.44) and (4.45), Q_o and α_o are unknown; therefore, these equations must be solved simultaneously for Q_o and α_o. Thus,

$$\alpha_o = \operatorname{ctn}^{-1}\left(\operatorname{ctn} \alpha_i + \frac{F_c}{Q_{ia}}\right) \tag{4.46}$$

$$Q_o = \left[1 + \left(\operatorname{ctn} \alpha_i + \frac{F_c}{Q_{ia}}\right)^2\right]^{1/2} Q_{ia} \tag{4.47}$$

Further,

$$Q_i = \frac{Q_{ia}}{\sin \alpha_i} \tag{4.48}$$

It is apparent from equation (4.46) that because of centrifugal force F_c, $\alpha_o < \alpha_i$. The quantity α_i is the contact angle under thrust load and it is greater than the free contact angle $\alpha°$. This condition will be discussed in Chapter 6.

Example 4.4. If the 218 angular-contact ball bearing of Example 4.2 has a cage speed of 5000 rpm, what outer ring contact angle will obtain?

$Q_a = 2225$ N (500 lb) Ex. 4.2

$\alpha_i = 38° 54'$ Ex. 3.3

$D = 22.23$ mm (0.875 in.) Ex. 2.3

$d_m = 125.3$ mm (4.932 in.) Ex. 2.6

$$F_c = 2.26 \times 10^{-11} D^3 n_m^2 d_m \tag{4.41}$$

$$= 2.26 \times 10^{-11}(22.23)^3\,(5000)^2 \times 125.3 = 775.2 \text{ N } (174.2 \text{ lb})$$

$$\alpha_o = \operatorname{ctn}^{-1}\left(\operatorname{ctn} \alpha_i + \frac{F_c}{Q_{ia}}\right) \tag{4.46}$$

$$= \operatorname{ctn}^{-1}\left(\operatorname{ctn} 38°54' + \frac{775.2}{2225}\right) = 32°12'$$

Compare the normal ball loads at the inner and outer raceways.

$$Q_i = 3543 \text{ N } (796.2 \text{ lb}) \qquad \text{Ex. 4.2}$$

$$Q_o = \left[1 + \left(\text{ctn } \alpha_i + \frac{F_c}{Q_{ia}}\right)^2\right]^{1/2} Q_{ia} \tag{4.47}$$

$$= \left[1 + \left(\text{ctn } 38°54' + \frac{775.2}{2225}\right)^2\right]^{1/2} \times 2225 = 4176 \text{ N } (938.5 \text{ lb})$$

Ball thrust bearings with nominal contact angle $\alpha = 90°$ operating at high speed and light load tend to permit the balls to override the land on both rings. The contact angle thus deviates from 90° in the same direction on both raceways (see Fig. 4.6).

From Fig. 4.6,

$$Q = \frac{F_c}{2 \cos \alpha} \tag{4.49}$$

$$\alpha = \tan^{-1}\left(\frac{2Q_a}{F_c}\right) \tag{4.50}$$

Example 4.5. A 90° nominal contact angle ball thrust bearing has a cage speed of 5000 rpm. The bearing that has a 127 mm (5 in.) pitch

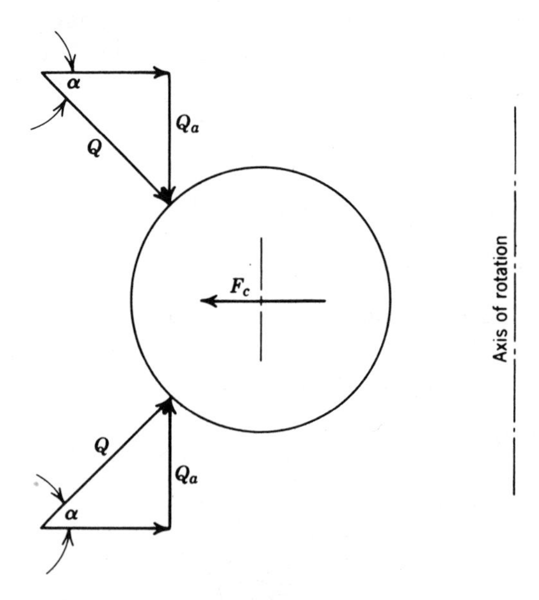

FIGURE 4.6. Ball loading in a ball thrust bearing.

diameter and 12.7 mm (0.5 in.) balls, supports a thrust load of 445 N (100 lb) per ball. What contact angle obtains?

$$F_c = 2.26 \times 10^{-11} D^3 n_m^2 d_m \quad (4.41)$$

$$= 2.26 \times 10^{-11} (12.7)^3 (5000)^2 \times 127$$

$$= 146.7 \text{ N } (32.97 \text{ lb})$$

$$\alpha = \tan^{-1}\left(\frac{2Q_a}{F_c}\right) \quad (4.50)$$

$$= \tan^{-1}\left(\frac{2 \times 445}{146.7}\right)$$

$$= 80°38'$$

Equation (4.34) is not restrictive as to geometry and since the mass of a cylindrical (or nearly cylindrical) roller is given by

$$m = \tfrac{1}{4} \rho \pi D^2 l_t \quad (4.51)$$

the centrifugal force for a steel roller rotating about a bearing axis is given by

$$F_c = 3.39 \times 10^{-11} D^2 l_t d_m n_m^2 \quad (4.52)$$

Consider a double-row spherical roller bearing having symmetrical rollers subjected to a radial load and rotating at high speed. Since the contact angle at the outer ring is fixed and since induced thrust loads are self-equilibrating, the roller centrifugal force induces a larger thrust component on the outer raceway than on the inner.

For a tapered roller bearing, however, roller centrifugal force alters the distribution of load between the outer raceway and central guide flange. Figure 4.7 demonstrates this condition for an applied thrust load Q_{ia}. For equilibrium to exist,

$$Q_{ia} - Q_{fa} - Q_{oa} = 0 \quad (4.53)$$

$$Q_{ir} - Q_{fr} + F_c - Q_{or} = 0 \quad (4.54)$$

or

$$Q_{ia} + Q_f \sin \alpha_f - Q_o \sin \alpha_o = 0 \quad (4.55)$$

$$Q_{ia} \operatorname{ctn} \alpha_i - Q_f \cos \alpha_f + F_c - Q_o \cos \alpha_o = 0 \quad (4.56)$$

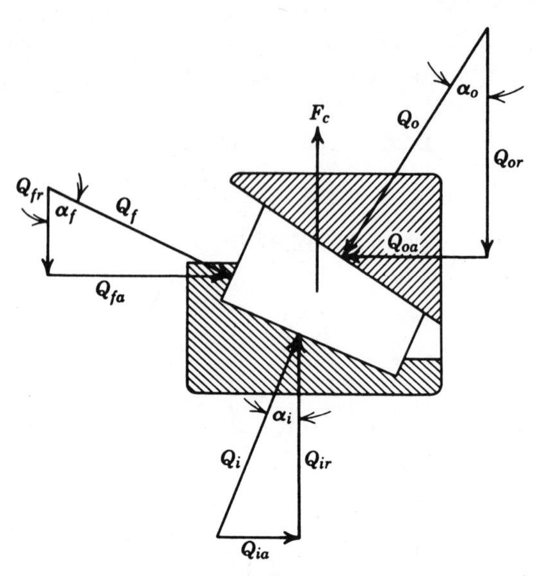

FIGURE 4.7. Asymmetrical roller under thrust load and centrifugal load.

Solving equations (4.55) and (4.56) simultaneously yields

$$Q_o = \frac{Q_{ia}(\text{ctn}\ \alpha_i \sin \alpha_f + \cos \alpha_f) + F_c \sin \alpha_f}{\sin(\alpha_o + \alpha_f)} \tag{4.57}$$

$$Q_f = \frac{Q_{ia}(\text{ctn}\ \alpha_i \sin \alpha_o + \cos \alpha_o) + F_c \sin \alpha_o}{\sin(\alpha_o + \alpha_f)} \tag{4.58}$$

Example 4.6. If the tapered roller bearing of Example 4.3 has a cage speed of 1700 rpm, what is the guide flange load due to the most heavily loaded roller?

$$Q_{ia} = 22{,}250 \text{ N (5000 lb)} \qquad \text{Ex. 4.3}$$

$$D = 22.86 \text{ mm (0.9000 in.)} \qquad \text{Ex. 4.3}$$

$$d_m = 142.2 \text{ mm (5.600 in.)} \qquad \text{Ex. 4.3}$$

$$l_t = 30.48 \text{ mm (1.2 in.)} \qquad \text{Ex. 4.3}$$

$$F_c = 3.39 \times 10^{-11} D^2 l_t d_m n_m^2 \tag{4.52}$$

$$= 3.39 \times 10^{-11}(22.86)^2 \times 30.48 \times 142.2(1700)^2$$

$$= 221.9 \text{ N (49.86 lb)}$$

$$Q_{\mathrm{f}} = \frac{Q_{\mathrm{ia}}(\mathrm{ctn}\ \alpha_{\mathrm{i}} \sin \alpha_{\mathrm{o}} - \cos \alpha_{\mathrm{o}}) + F_{\mathrm{c}} \sin \alpha_{\mathrm{o}}}{\sin(\alpha_{\mathrm{o}} + \alpha_{\mathrm{f}})} \tag{4.58}$$

$$= \frac{22250(\mathrm{ctn}\ 22^\circ \sin 29^\circ - \cos 29^\circ) + 221.9 \sin 29^\circ}{\sin(29^\circ + 64^\circ)}$$

$$= 7351 \text{ N } (1652 \text{ lb})$$

What is the maximum normal load at the outer raceway?

$$Q_{\mathrm{o}} = \frac{Q_{\mathrm{ia}}(\mathrm{ctn}\ \alpha_{\mathrm{i}} \sin \alpha_{\mathrm{f}} + \cos \alpha_{\mathrm{f}}) + F_{\mathrm{c}} \sin \alpha_{\mathrm{f}}}{\sin(\alpha_{\mathrm{o}} + \alpha_{\mathrm{f}})} \tag{4.57}$$

$$= \frac{22250(\mathrm{ctn}\ 22^\circ \sin 64^\circ + \cos 64^\circ) + 221.9 \sin 64^\circ}{\sin(29^\circ + 64^\circ)}$$

$$= 59{,}500 \text{ N } (13{,}370 \text{ lb})$$

Care must be exercised in operating a tapered roller bearing at very high speed. At some critical speed related to the magnitude of the applied load, the force at the inner ring raceway contact approaches zero and the entire axial load is carried at the flange-roller end contact. Since this contact has only sliding motion, very high friction results with attendant high heat generation.

Rotation about an Eccentric Axis. The foregoing section dealt with rolling element centrifugal loading when the bearing rotates about its own axis. This is the usual case. In planetary gear transmissions, however, the planet gear bearings rotate about the input and/or output shaft axes as well as about their own axes. Hence an additional inertial or centrifugal force is induced in the rolling element. Figure 4.8 shows a schematic diagram of such a system. From Fig. 4.8 it can be seen that the instantaneous radius of rotation is by the law of cosines

$$r = (r_{\mathrm{m}}^2 + r_{\mathrm{e}}^2 - 2r_{\mathrm{m}}r_{\mathrm{e}} \cos \psi)^{1/2} \tag{4.59}$$

Therefore, the corresponding centrifugal force is

$$F_{\mathrm{ce}} = m\omega_{\mathrm{e}}^2(r_{\mathrm{m}}^2 + r_{\mathrm{e}}^2 - 2r_{\mathrm{m}}r_{\mathrm{c}} \cos \psi)^{1/2} \tag{4.60}$$

This force F_{ce} is maximum at $\psi = 180^\circ$ and at that angle is algebraically additive to F_{c}. At $\psi = 0$ the total centrifugal force is $F_{\mathrm{c}} - F_{\mathrm{ce}}$. The angle between F_{c} and F_{ce} as derived from the law of cosines is

$$\theta = \cos^{-1}\left(\frac{r_{\mathrm{m}}}{r} - \frac{r_{\mathrm{e}}}{r} \cos \psi\right) \tag{4.61}$$

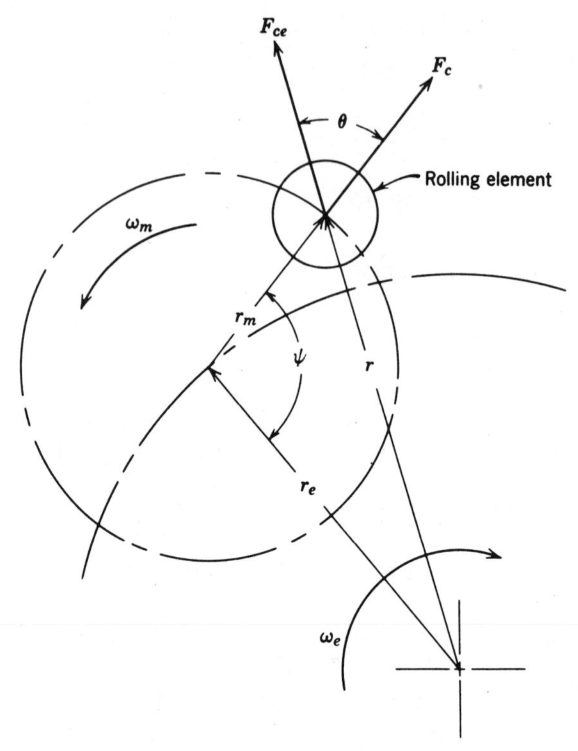

FIGURE 4.8. Rolling element centrifugal forces due to bearing rotation about an eccentric axis.

F_{ce} can be resolved into a radial force and a tangential force as follows:

$$F_{cer} = F_{ce} \cos \theta \tag{4.62}$$

$$F_{cet} = F_{ce} \sin \theta \tag{4.63}$$

Hence the total instantaneous radial centrifugal force acting on the rolling element is

$$F_{cr} = m\omega_m^2 r_m + m\omega_e^2(r_m - r_e \cos \psi) \tag{4.64}$$

or

$$F_{cr} = \frac{W}{g}[r_m(\omega_m^2 + \omega_e^2) - r_e\omega_e^2 \cos \psi] \tag{4.65}$$

in which the positive direction is that taken by the constant component F_c. For steel ball and roller elements the following equations are respectively valid:

$$F_{cr} = 2.26 \times 10^{-11} D^3 [d_m(n_m^2 + n_e^2) - d_e n_e^2 \cos \psi] \tag{4.66}$$

$$F_{cr} = 3.39 \times 10^{-11} D^2 l_t [d_m(n_m^2 + n_e^2) - d_e n_e^2 \cos \psi] \tag{4.67}$$

The instantaneous tangential component of eccentric centrifugal force is

$$F_{ct} = m\omega_e^2 r_e \sin \psi \tag{4.68}$$

For steel ball and roller elements, respectively, the following equations pertain:

$$F_{ct} = 2.26 \times 10^{-11} D^3 d_e n_e^2 \sin \psi \tag{4.69}$$

$$F_{ct} = 3.39 \times 10^{-11} D^2 l_t d_e n_e^2 \sin \psi \tag{4.70}$$

This tangential force alternates direction and tends to produce sliding between the rolling element and raceway. It is therefore resisted by a frictional force between the contacting surfaces.

The bearing cage also undergoes this eccentric motion and if it is supported on the rolling elements, it will impose additional load on the individual rolling elements. This cage load may be reduced by using a material of smaller mass density.

Gyroscopic Moment

It can usually be assumed with minimal loss of calculational accuracy that pivotal motion due to gyroscopic moment is negligible; then the angle β' is zero and equation (4.36) is of no consequence. The gyroscopic moment as defined by equation (4.35) is therefore resisted successfully by friction forces at the bearing raceways for ball bearings and by normal forces for roller bearings. Substituting equation (4.39) into (4.35) the following expression is obtained for ball bearings:

$$M_g = \tfrac{1}{60} \rho \pi D^5 \omega_R \omega_m \sin \beta \tag{4.71}$$

since

$$\omega_R = \frac{2\pi n_R}{60} \tag{4.72}$$

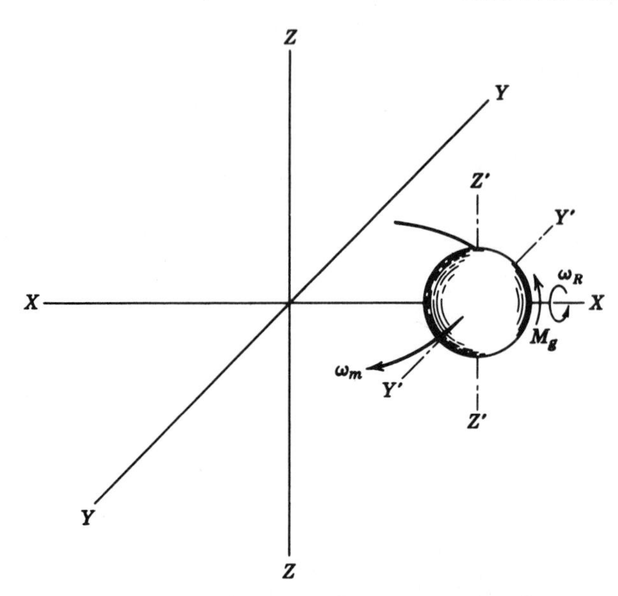

FIGURE 4.9. Gyroscopic moment due to simultaneous rotation about nonparallel axes.

and

$$\omega_{\mathrm{m}} = \frac{2\pi n_{\mathrm{m}}}{60} \tag{4.39}$$

The gyroscopic moment for steel ball bearings is given by

$$M_{\mathrm{g}} = 4.47 \times 10^{-12} D^5 n_{\mathrm{R}} n_{\mathrm{m}} \sin \beta \tag{4.73}$$

Figure 4.9 shows the direction of the gyroscopic moment in a ball bearing. Accordingly, Fig. 4.10 shows the ball loading due to the action of gyroscopic moment and centrifugal force on a thrust-loaded ball bearing.

Example 4.7. In the ball thrust bearing of Example 4.5, the balls rotate at a speed of 50,000 rpm about axes perpendicular to the bearing axis, that is, $\beta = 90°$. What is the gyroscopic moment that must be resisted by raceway friction forces and what coefficient of friction is required to resist gyroscopic pivotal rotation?

$$D = 12.7 \text{ mm } (0.5 \text{ in.}) \qquad \text{Ex. 4.5}$$

$$N_{\mathrm{m}} = 5000 \text{ rpm} \qquad \text{Ex. 4.5}$$

$$M_{\mathrm{g}} = 4.47 \times 10^{-12} D^5 n_{\mathrm{R}} n_{\mathrm{m}} \sin \beta \tag{4.73}$$

$$= 4.47 \times 10^{-12} (12.7)^5 \times 50{,}000 \times 5000 \times 1$$

$$= 0.369 \text{ N} \cdot \text{m } (3.265 \text{ in} \cdot \text{lb})$$

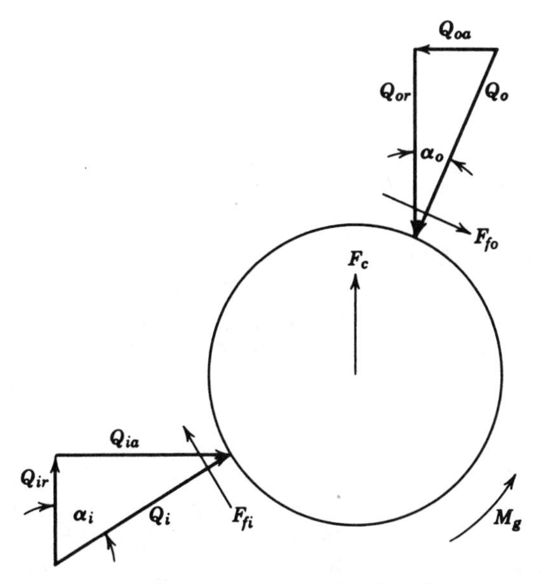

FIGURE 4.10. Forces acting on a high speed ball under applied thrust load.

$$F_c = 146.7 \text{ N } (32.97 \text{ lb}) \qquad \text{Ex. 4.5}$$

$$\alpha = 80°38' \qquad \text{Ex. 4.5}$$

$$Q = \frac{F_c}{2 \cos \alpha} \tag{4.49}$$

$$= \frac{146.7}{2 \cos (80°38')} = 450.8 \text{ N } (101.3 \text{ lb})$$

$$(F_{fi} + F_{fo}) \frac{D}{2} - M_g = 0 \qquad \text{Fig. 4.10}$$

$$\mu(Q_i + Q_o) \frac{D}{2} - M_g = 0$$

$$\mu(450.8 + 450.8) \times \frac{12.7}{2} - 0.369 \times 10^{-3} = 0$$

$$\mu = 0.0645$$

Gyroscopic moments also act on roller bearings. The rollers, however, are geometrically constrained from rotating; therefore, a gyroscopic moment of significant magnitude tends to alter the distribution of load across the roller contour. For steel rollers the gyroscopic moment is given by

$$M_g = 8.37 \times 10^{-12} D^4 l_t n_R n_m \sin \beta \tag{4.74}$$

ROLLER AXIAL LOADING IN RADIAL BEARINGS

Axial Loading Effects in Radial Roller Bearings

Axial loading of the rollers in a radial roller bearing can significantly affect bearing performance. In cylindrical and tapered roller bearings a roller axial load is reacted between the roller end and a guide flange. The combination of load and sliding in this contact can cause the bearing to generate excessive heat and, under certain conditions, lead to wear and smearing of the contacting surfaces. Conversely, with proper management of roller end-flange design and lubrication, bearing heat generation and axial load-carrying capacity can be optimized. Axial roller loads interact with roller radial forces and ring deflections to determine roller tilting and skewing motions that influence roller-raceway contact stresses and sliding velocity distributions. Roller tilting and skewing substantially affect roller bearing heat generation, friction torque, and rolling contact fatigue life. Many modern spherical roller bearing designs do not use flanges as the primary source of roller guidance. Such designs rely on proper management of roller tilting and skewing motions via roller-raceway traction forces to provide roller guidance while minimizing bearing heat generation and friction torque.

External Applied Roller Thrust Loading

Many radial roller bearing designs have the ability to carry applied axial or thrust load in addition to their predominant radial loading. Cylindrical roller bearings can carry thrust loading by virtue of flanges fixed to inner and outer rings. The angular-contact roller arrangement of tapered roller bearings reacts applied thrust load over the combination of roller-raceway contact surfaces and roller end-flange contacts. Spherical roller bearings react applied thrust loading through their roller-raceway contacts (and flanges if asymmetric rollers are employed). Axial loading of radial roller bearings causes alteration of internal load distribution and roller response that can significantly affect bearing performance.

Cylindrical roller bearings with fixed inner and outer ring flanges, as shown in Fig. 4.11*a*, can carry axial load through contacts between the

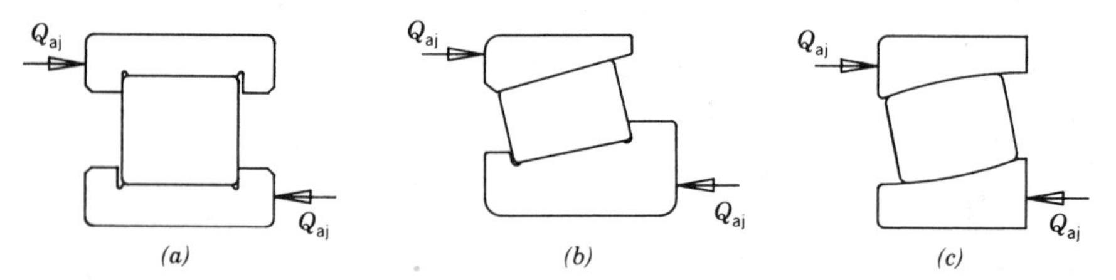

FIGURE 4.11. Applied thrust loading on radial roller bearings. (*a*) Cylindrical roller. (*b*) Tapered roller. (*c*) Spherical roller.

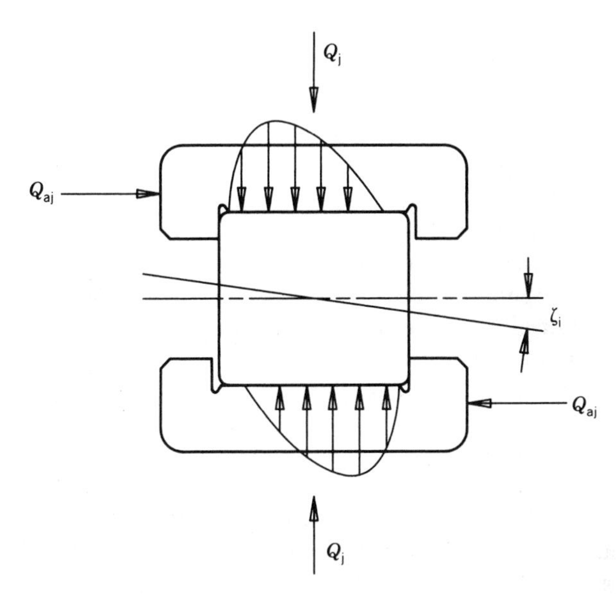

FIGURE 4.12. Roller tilting displacement and contact load distribution due to combined axial and radial load on a cylindrical roller bearing.

roller end faces and the flanges. The couple produced on the roller by the axial roller end-flange forces Q_{aj} results in a tilting ζ_j about the center of the roller. As the roller tilts, the roller-raceway load distribution shifts asymmetrically, as represented in Fig. 4.12. The roller-raceway load distribution of Fig. 4.12 can be compared with the ideal load distribution in Fig. 5.20*b*.

Roller Axial Loading Due to Roller Skewing

In roller bearings skewing is defined as an angular rotation of the roller axis (in a plane tangent to its orbital direction) with respect to the axis of the contacting ring. The magnitude of the skewing motion may be expressed as a skewing angle γ_s. The skewing angle of a roller at a given azimuth position might be different for inner and outer ring contacts if misalignment exists between inner and outer rings. In general, the skewing angle will vary with roller azimuth position ψ.

Skewing is caused by forces acting on the roller, which result in a moment loading about an axis oriented in the bearing radial direction and passing through the roller at its midpoint. These skewing forces are often due to asymmetrical distribution of tangential friction forces at the roller-raceway contacts arising from asymmetrical normal loading and/or sliding velocity distributions along the contacts. Normal and frictional forces acting on the roller due to contact with guide flanges (including roller applied axial loads) and cage may contribute to, or serve to limit, roller skewing motions. Operating conditions associated with asymmetrical load distributions include misalignment of inner and outer rings in cylindrical roller bearings and tapered roller bearings and applied roller axial loading in cylindrical roller bearings. In high-speed roller bearing operation the dynamic effects of roller mass unbalance and impact loading between roller and flange or cage can become significant causative factors in roller skewing. Forces that give rise to roller skewing motions may also, by virtue of their points of application or radial force components, be coupled to the roller tilting action. Quantification of roller skewing and coupled tilting behavior require computer programs designed for this purpose.

As an illustration of roller skewing, consider the cylindrical roller bearing roller subjected to radial and axial applied loads shown in Fig. 4.13. The applied axial load Q_{aj} causes the roller to tilt through the angle ζ_j. The tilting motion gives rise to the asymmetrical contact load distribution in Fig. 4.13*a*. Assuming no gross sliding or skidding in the application, the tilting also causes sliding motion to occur along the inner and outer raceway contacts. Roller deformations cause the sliding velocity on the uncrowned roller to be zero at only one point along each raceway contact. The tangential sliding velocity distributions for the unskewed roller are represented by Fig. 4.13*b*. Positive and negative values obtain corresponding to sliding in the z direction. Tangential friction forces in both contacts are related to the magnitude of the normal contact load and magnitude and direction of tangential sliding velocity. The asymmetrical tangential friction force distribution shown in Fig. 4.13*c* results. This distribution causes a skewing moment about the y axis. The skewing moment tends to cause the roller to skew and generate an axial friction force at both contacts, the resultant of which, Q_a, is shown in Fig. 4.13*d*. Also contributing to the roller skewing moment are the frictional forces generated at the flange contacts.

In the illustration the skewing moment and axial friction force must be reacted by the flange contacts. In principle, a skewing angle γ_s may be achieved whereby roller skewing moment and axial force are in equilibrium with the flange contact forces. Note that, in general, the locations of the roller-flange contacts, the roller-raceway normal load distribution, sliding velocities, and tangential and axial friction forces are all functions of the roller skewing angle. The skewing angle at which this force balance is obtained is known as an equilibrium skewing angle.

As bearing operational speed increases, dynamic effects become significant. Brown et al. [4.1] investigated the problem of roller axial load due to skewing in cylindrical roller bearings under high-speed conditions. Their analytical and experimental work highlighted the detrimental effects of unbalance forces due to roller corner radius runout. Analytical models were developed for roller impact loading on flange and cage. The effect of bearing design parameters was empirically correlated with observed roller end wear.

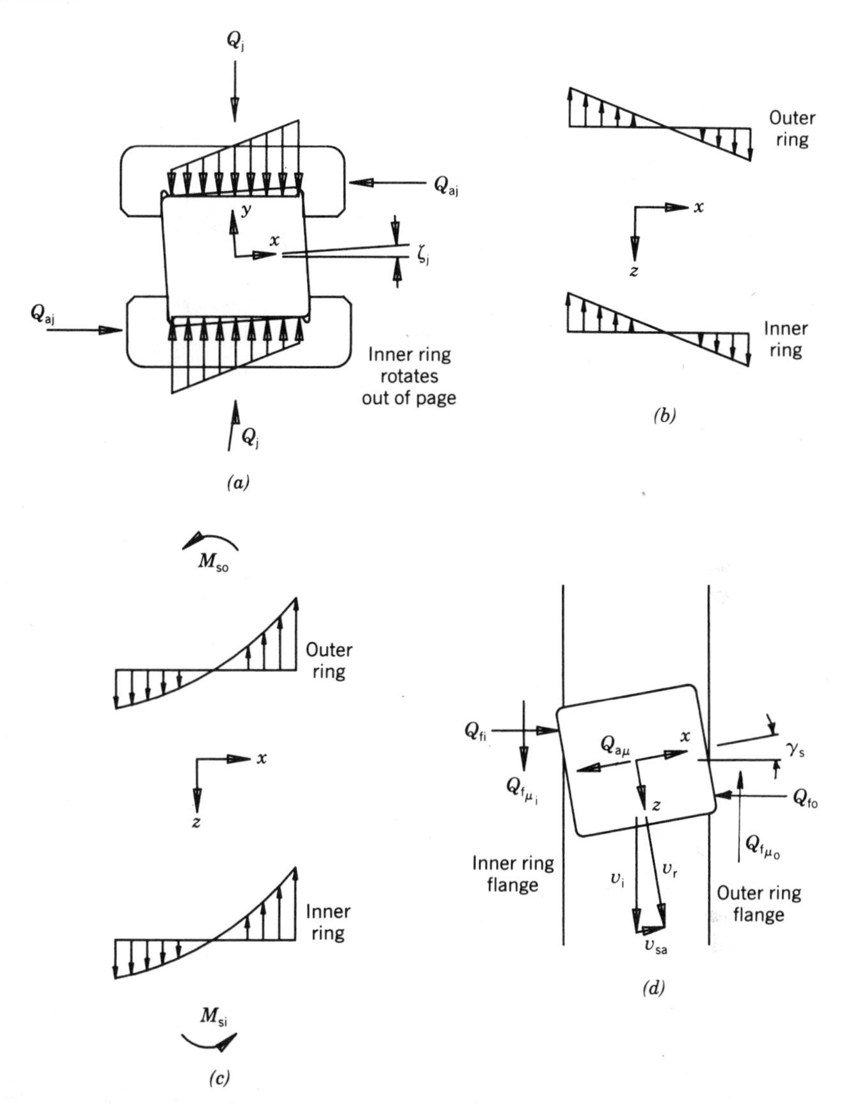

FIGURE 4.13. Roller loading and tangential sliding velocity in a cylindrical roller bearing with applied axial and radial load. (*a*) Applied roller loads and roller load distribution. (*b*) Tangential sliding velocity distribution. (*c*) Tangential traction force distribution and skewing moments. (*d*) Roller-flange contact forces and resultant axial friction force (view from outer ring).

CLOSURE

To analyze rolling bearing performance in a given application, it is usually necessary to determine the load on individual balls or rollers. How well the balls or rollers accept the applied and induced loads will determine bearing endurance. For example, light radial load applied to a 90°

contact angle thrust bearing can cause the bearing to fail rapidly. Similarly, thrust load applied to a 0° contact angle, radial ball bearing is largely magnified according to the final contact angle that obtains. In Chapters 6 and 8 this book will concern itself with the distribution of load among balls and rollers. It will be shown that the manner in which each rolling element accepts its load will determine in large measure the loading of all others. Moreover, in angular-contact ball bearings the ball loading can affect ball and cage speeds significantly. In high-speed roller bearings if roller loading is too light, rolling motion may be preempted by skidding. The material in this chapter is therefore fundamental to even the most rudimentary analysis of a rolling bearing application.

REFERENCE

4.1. P. Brown, D. Robinson, L. Dobek, and J. Miner, "Mainshaft High-Speed, Cylindrical Roller Bearings for Gas Turbine Engines," U.S. Navy Contract N000140-76-C-0383 Interim Report (1978).

5

CONTACT STRESS AND DEFORMATION

LIST OF SYMBOLS

Symbol	Description	Units
a	Semimajor axis of the projected contact ellipse	mm (in.)
a^*	Dimensionless semimajor axis of contact ellipse	
b	Semiminor axis of the projected contact ellipse	mm (in.)
b^*	Dimensionless semiminor axis of contact ellipse	
E	Modulus of elasticity	N/mm^2 (psi)
$\mathcal{E}$	Complete elliptic integral of the second kind	
$\mathcal{E}(\phi)$	Elliptic integral of the second kind	
$\mathfrak{F}$	Complete elliptic integral of the first kind	
$\mathfrak{F}(\phi)$	Elliptic integral of the first kind	
F	Force	N (lb)
G	Shear modulus of elasticity	N/mm^2 (psi)
l	Roller effective length	mm (in.)

Symbol	Description	Units
Q	Normal force between rolling element and raceway	N (lb)
r	Radius of curvature	mm (in.)
S	Principal stress	N/mm^2 (psi)
u	Deflection in x direction	mm (in.)
U	Arbitrary function	
v	Deflection in y direction	mm (in.)
V	Arbitrary function	
w	Deflection in z direction	mm (in.)
x	Principal direction distance	mm (in.)
X	Dimensionless parameter	
y	Principal direction distance	mm (in.)
Y	Dimensionless parameter	
z	Principal direction distance	mm (in.)
z_1	Depth to maximum shear stress at $x = 0$, $y = 0$	mm (in.)
z_0	Depth to maximum reversing shear stress $y \neq 0$, $x = 0$	mm (in.)
Z	Dimensionless parameter	
γ	Shear strain	
δ	Deformation	mm (in.)
δ^*	Dimensionless contact deformation	
ϵ	Linear strain	
ζ	z/b, roller tilting angle	°, rad
θ	Angle	rad
ϑ	Auxiliary angle	rad
κ	a/b	
λ	Parameter	
ξ	Poisson's ratio	
σ	Normal stress	N/mm^2 (psi)
τ	Shear stress	N/mm^2 (psi)
ν	Auxiliary angle	rad
ϕ	auxiliary angle	rad or °
$F(\rho)$	Curvature difference	
$\Sigma\rho$	Curvature sum	mm^{-1} ($in.^{-1}$)

SUBSCRIPTS

i	Refers to inner raceway
o	Refers to outer raceway
r	Refers to radial direction
x	Refers to x direction
y	Refers to y direction

Symbol	Description	Units
	SUBSCRIPTS	
z	Refers to z direction	
yz	Refers to yz plane	
xz	Refers to xz plane	
I	Refers to contact body I	
II	Refers to contact body II	

GENERAL

Loads acting between the rolling elements and raceways in rolling bearings develop only small areas of contact between the mating members. Consequently, although the elemental loading may only be moderate, stresses induced on the surfaces of the rolling elements and raceways are usually large. It is not uncommon for rolling bearings to operate continuously with normal stresses exceeding 1380 N/mm^2 (200,000 psi) compression on the rolling surfaces. In some applications and during endurance testing normal stresses on rolling surfaces may exceed 3449 N/mm^2 (500,000 psi) compression. Since the effective area over which load is supported rapidly increases with depth below a rolling surface, the high compressive stress occurring at the surface does not permeate the entire rolling member. Therefore, bulk failure of rolling members is generally not a significant factor in rolling bearing design; however, destruction of the rolling surfaces is. This chapter is therefore concerned only with the determination of surface stresses and stresses occurring near the surface. Contact deformations are caused by contact stresses. Because of the rigid nature of the rolling members, these deformations are generally of a low order of magnitude, for example 0.025 mm (0.001 in.) or less in steel bearings. It is the purpose of this chapter to develop relationships permitting the determination of contact stresses and deformations in rolling bearings.

THEORY OF ELASTICITY

The classical solution for the local stress and deformation of two elastic bodies apparently contacting at a single point was established by Hertz [5.1] in 1881. Today, contact stresses are frequently called Hertzian or simply Hertz stresses in recognition of his accomplishment.

To develop the mathematics of contact stresses, one must have a firm foundation in principles of mechanical elasticity. It is, however, not a purpose of this text to teach theory of elasticity and therefore only a rudimentary discussion of that discipline is presented herein to demon-

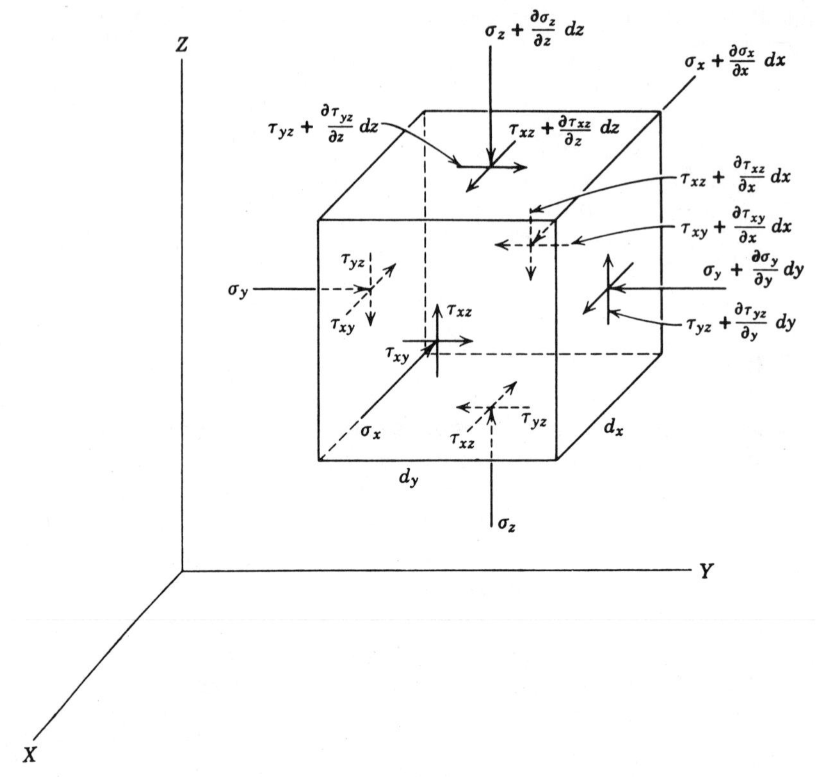

FIGURE 5.1. Stresses acting on an infinitesimal cube of material under load.

strate the complexity of contact stress problems. In that light consider an infinitesimal cube of an isotropic homogeneous elastic material subjected to the stresses shown in Fig. 5.1. Considering the stresses acting in the x direction and in the absence of body forces, static equilibrium requires that

$$\sigma_x \, dz \, dy + \tau_{xy} \, dx \, dz + \tau_{xz} \, dx \, dy - \left(\sigma_x + \frac{\partial \sigma_x}{\partial x} dx \right) dz \, dy$$
$$- \left(\tau_{xy} + \frac{\partial \tau_{xy}}{\partial y} dy \right) dx \, dz - \left(\tau_{xz} + \frac{\partial \tau_{xz}}{\partial z} dz \right) dx \, dy = 0 \quad (5.1)$$

Therefore,

$$\frac{\partial \sigma_x}{\partial x} + \frac{\partial \tau_{xy}}{\partial y} + \frac{\partial \tau_{xz}}{\partial z} = 0 \quad (5.2)$$

Similarly, for the y and z directions, respectively,

$$\frac{\partial \sigma_y}{\partial y} + \frac{\partial \tau_{xy}}{\partial x} + \frac{\partial \tau_{yz}}{\partial z} = 0 \tag{5.3}$$

$$\frac{\partial \sigma_z}{\partial z} + \frac{\partial \tau_{xz}}{\partial x} + \frac{\partial \tau_{yz}}{\partial y} = 0 \tag{5.4}$$

Equations (5.2)–(5.4) are the equations of equilibrium in Cartesian coordinates. Hooke's law for an elastic material states that within the proportional limit

$$\epsilon = \frac{\sigma}{E} \tag{5.5}$$

in which ϵ is strain and E is the modulus of elasticity of the strained material. If u, v, *and* w are the deflections in the x, y, and z directions, then

$$\begin{aligned} \epsilon_x &= \frac{\partial u}{\partial x} \\ \epsilon_y &= \frac{\partial v}{\partial y} \\ \epsilon_z &= \frac{\partial w}{\partial z} \end{aligned} \tag{5.6}$$

If instead of an elongation or compression the sides of the cube undergo relative rotation such that the sides in the deformed conditions are no longer mutually perpendicular, then the rotational strains are given as

$$\begin{aligned} \gamma_{xy} &= \frac{\partial u}{\partial y} + \frac{\partial v}{\partial x} \\ \gamma_{xz} &= \frac{\partial u}{\partial z} + \frac{\partial w}{\partial x} \\ \gamma_{yz} &= \frac{\partial v}{\partial z} + \frac{\partial w}{\partial y} \end{aligned} \tag{5.7}$$

When a tensile stress σ_x is applied to two faces of a cube, then in addition to an extension in the x direction, a contraction is produced in the

y and z directions as follows:

$$\epsilon_x = \frac{\sigma_x}{E}$$

$$\epsilon_y = -\frac{\xi\sigma_x}{E} \tag{5.8}$$

$$\epsilon_z = -\frac{\xi\sigma_x}{E}$$

In equation (5.8), ξ is Poisson's ratio; for steel $\xi \approx 0.3$.

Now the total strain in each principal direction due to the action of normal stresses σ_x, σ_y, and σ_z is the total of the individual strains. Hence

$$\epsilon_x = \frac{1}{E}[\sigma_x - \xi(\sigma_y + \sigma_z)]$$

$$\epsilon_y = \frac{1}{E}[\sigma_y - \xi(\sigma_x + \sigma_z)] \tag{5.9}$$

$$\epsilon_z = \frac{1}{E}[\sigma_z - \xi(\sigma_x + \sigma_y)]$$

Equations (5.9) were obtained by the method of superposition.

In accordance with Hooke's law, it can further be demonstrated that shear stress is related to shear strain as follows:

$$\gamma_{xy} = \frac{\tau_{xy}}{G}$$

$$\gamma_{xz} = \frac{\tau_{xz}}{G} \tag{5.10}$$

$$\gamma_{yz} = \frac{\tau_{yz}}{G}$$

in which G is the modulus of elasticity in shear and it is defined

$$G = \frac{E}{2(1 + \xi)} \tag{5.11}$$

One further defines the volume expansion of the cube as follows:

$$\epsilon = \epsilon_x + \epsilon_y + \epsilon_z \tag{5.12}$$

Combining equations (5.9), (5.11), and (5.12) one obtains for normal stresses

$$\sigma_x = 2G\left(\frac{\partial u}{\partial x} + \frac{\xi}{1 - 2\xi}\epsilon\right)$$

$$\sigma_y = 2G\left(\frac{\partial v}{\partial y} + \frac{\xi}{1 - 2\xi}\epsilon\right) \quad (5.13)$$

$$\sigma_z = 2G\left(\frac{\partial w}{\partial z} + \frac{\xi}{1 - 2\xi}\epsilon\right)$$

Finally, a set of "compatibility" conditions can be developed by differentiation of the strain relationships, both linear and rotational, and substituting in the equilibrium equations (5.2)–(5.4):

$$\nabla^2 u + \frac{1}{1 - 2\xi}\frac{\partial \epsilon}{\partial x} = 0$$

$$\nabla^2 v + \frac{1}{1 - 2\xi}\frac{\partial \epsilon}{\partial y} = 0 \quad (5.14)$$

$$\nabla^2 w + \frac{1}{1 - 2\xi}\frac{\partial \epsilon}{\partial z} = 0$$

in which

$$\nabla^2 = \frac{\partial^2}{\partial x^2} + \frac{\partial^2}{\partial y^2} + \frac{\partial^2}{\partial z^2} \quad (5.15)$$

Equations (5.14) represent a set of conditions that by using the known stresses acting on a body must be solved to determine the subsequent strains and internal stresses of that body. See Timoshenko and Goodier [5.2] for a detailed presentation of the foregoing.

SURFACE STRESSES AND DEFORMATIONS

Using polar coordinates rather than Cartesian, Boussinesq [5.3] in 1892 solved the simple radial distribution of stress within a semiinfinite solid as shown in Fig. 5.2. With the boundary condition of a surface free of shear stress, the following solution was obtained for radial stress:

$$\sigma_r = -\frac{2F\cos\theta}{\pi r} \quad (5.16)$$

It is apparent from equation (5.16) that as r approaches 0, σ_r becomes infinitely large. It is further apparent that this condition cannot exist without causing gross yielding or failure of the material at the surface.

Hertz reasoned that instead of a point or line contact, a small contact

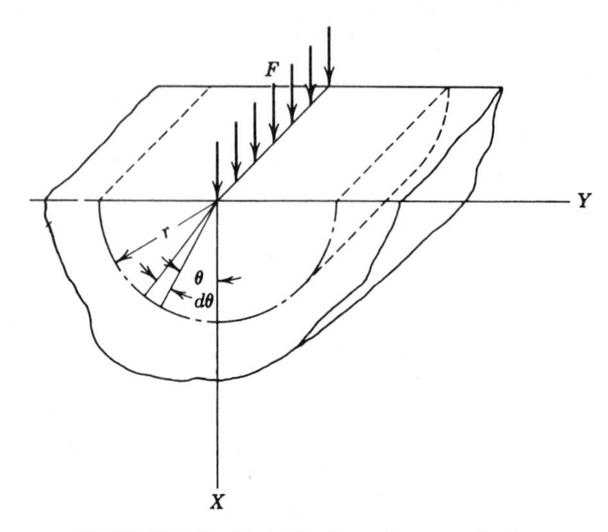

FIGURE 5.2. Model for Boussinesq analysis.

area must form, causing the load to be distributed over a surface, and thus alleviating the condition of infinite stress. In performing his analysis, he made the following assumptions:

1. The proportional limit of the material is not exceeded, that is, all deformation occurs in the elastic range.
2. Loading is perpendicular to the surface, that is, the effect of surface shear stresses is neglected.
3. The contact area dimensions are small compared to the radii of curvature of the bodies under load.
4. The radii of curvature of the contact areas are very large compared to the dimensions of these areas.

The solution of theoretical problems in elasticity is based on the assumption of a stress function or functions that singly or in combination fit the compatibility equations and the boundary conditions. For stress distribution in a semiinfinite elastic solid, Hertz introduced the assumptions:

$$X = \frac{x}{b}$$

$$Y = \frac{y}{b} \tag{5.17}$$

$$Z = \frac{z}{b}$$

in which b is an arbitrary fixed length and hence, X, Y, and Z are dimensionless parameters. Also,

$$\begin{aligned}\frac{u}{c} &= \frac{\partial U}{\partial X} - Z\frac{\partial V}{\partial X}\\ \frac{v}{c} &= \frac{\partial U}{\partial Y} - Z\frac{\partial V}{\partial Y}\\ \frac{w}{c} &= \frac{\partial U}{\partial Z} - Z\frac{\partial V}{\partial Z} + V\end{aligned} \tag{5.18}$$

in which c is an arbitrary length such that deformations u/c, v/c, and w/c are dimensionless. U and V are arbitrary functions of X and Y only such that

$$\begin{aligned}\nabla^2 U &= 0\\ \nabla^2 V &= 0\end{aligned} \tag{5.19}$$

Furthermore, b and c are related to U as follows:

$$\frac{b\epsilon}{c} = -2\frac{\partial^2 U}{\partial Z^2} \tag{5.20}$$

The foregoing assumptions, which are partly intuitive and partly based on experience, when combined with elasticity relationships (5.7), (5.10), (5.12), (5.13), and (5.14) yield the following expressions:

$$\begin{aligned}\frac{\sigma_x}{\sigma_0} &= Z\frac{\partial^2 V}{\partial X^2} - \frac{\partial^2 U}{\partial X^2} - 2\frac{\partial V}{\partial Z}\\ \frac{\sigma_y}{\sigma_0} &= Z\frac{\partial^2 V}{\partial Y^2} - \frac{\partial^2 U}{\partial Y^2} - 2\frac{\partial V}{\partial Z}\\ \frac{\sigma_z}{\sigma_0} &= Z\frac{\partial^2 V}{\partial Z^2} - \frac{\partial V}{\partial Z}\\ \frac{\tau_{xy}}{\sigma_0} &= Z\frac{\partial^2 V}{\partial X\partial Y} - \frac{\partial^2 U}{\partial X\partial Y}\\ \frac{\tau_{xz}}{\sigma_0} &= Z\frac{\partial^2 V}{\partial X\partial Z}\\ \frac{\tau_{yz}}{\sigma_0} &= Z\frac{\partial^2 V}{\partial Y\partial Z}\end{aligned} \tag{5.21}$$

in which

$$\sigma_0 = (-2Gc)/b \qquad \text{and} \qquad U = (1 - 2\xi) \int_z^\infty V(X, Y, \zeta)\, d\zeta$$

From the preceding formulas, the stresses and deformations may be determined for a semiinfinite body limited by the xy plane on which $\tau_{xz} = \tau_{yz} = 0$ and σ_z is finite on the surface, that is, at $z = 0$.

Hertz's last assumption was that the shape of the deformed surface was that of an ellipsoid of revolution. The function V was expressed as follows:

$$V = \frac{1}{2} \int_{\mathcal{S}_0}^\infty \frac{\left(1 - \dfrac{X^2}{\kappa^2 + \mathcal{S}^2} - \dfrac{Y^2}{1 + \mathcal{S}^2} - \dfrac{Z^2}{\mathcal{S}^2}\right)}{\sqrt{(\kappa^2 + \mathcal{S}^2)(1 + \mathcal{S}^2)}}\, \kappa\, d\mathcal{S} \tag{5.22}$$

in which $\mathcal{S}_0$ is the largest positive root of the equation

$$\frac{X^2}{\kappa^2 + \mathcal{S}_0^2} + \frac{Y^2}{1 + \mathcal{S}_0^2} + \frac{Z^2}{\mathcal{S}_0^2} = 1 \tag{5.23}$$

and

$$\kappa = a/b \tag{5.24}$$

Here, a and b are the semimajor and semiminor axes of the projected elliptical area of contact. For an elliptical contact area, the stress at the geometrical center is

$$\sigma_0 = -\frac{3Q}{2\pi ab} \tag{5.25}$$

The arbitrary length c is defined by

$$c = \frac{3Q}{4\pi Ga} \tag{5.26}$$

For the special case $\kappa = \infty$, then

$$\sigma_0 = -\frac{2Q}{\pi b} \tag{5.27}$$

$$c = \frac{Q}{\pi G} \tag{5.28}$$

Since the contact surface is assumed to be relatively small compared to the dimensions of the bodies, the distance between the bodies may be expressed as

$$z = \frac{x^2}{2r_x} + \frac{y^2}{2r_y} \tag{5.29}$$

in which r_x and r_y are the principal radii of curvature.

Introducing the auxiliary quantity $F(\rho)$ as determined by equation (2.26), this is found to be a function of the elliptical parameters a and b as follows:

$$F(\rho) = \frac{(\kappa^2 + 1)\mathcal{E} - 2\mathfrak{F}}{(\kappa^2 - 1)\mathcal{E}} \tag{5.30}$$

in which $\mathfrak{F}$ and $\mathcal{E}$ are the complete elliptic integrals of the first and second kind, respectively.

$$\mathfrak{F} = \int_0^{\pi/2} \left[1 - \left(1 - \frac{1}{\kappa^2}\right) \sin^2 \phi\right]^{-1/2} d\phi \tag{5.31}$$

$$\mathcal{E} = \int_0^{\pi/2} \left[1 - \left(1 - \frac{1}{\kappa^2}\right) \sin^2 \phi\right]^{1/2} d\phi \tag{5.32}$$

By assuming values of the elliptical eccentricity parameter κ, it is possible to calculate corresponding values of $F(\rho)$ and thus create a table of κ vs $F(\rho)$. Recall that $F(\rho)$ is a function of curvature of contacting bodies.

$$F(\rho) = \frac{(\rho_{\text{I}1} - \rho_{\text{I}2}) + (\rho_{\text{II}1} - \rho_{\text{II}2})}{\Sigma\rho} \tag{2.26}$$

It was further determined that

$$a = a^*\left[\frac{3Q}{2\Sigma\rho}\left(\frac{(1 - \xi_{\text{I}}^2)}{E_{\text{I}}} + \frac{(1 - \xi_{\text{II}}^2)}{E_{\text{II}}}\right)\right]^{1/3} \tag{5.33}$$

$$= 0.0236a^*\left(\frac{Q}{\Sigma\rho}\right)^{1/3} \quad \text{(for steel bodies)} \tag{5.34}$$

$$b = b^*\left[\frac{3Q}{2\Sigma\rho}\left(\frac{(1 - \xi_{\text{I}}^2)}{E_{\text{I}}} + \frac{(1 - \xi_{\text{II}}^2)}{E_{\text{II}}}\right)\right]^{1/3} \tag{5.35}$$

$$= 0.0236b^*\left(\frac{Q}{\Sigma\rho}\right)^{1/3} \quad \text{(for steel bodies)} \tag{5.36}$$

$$\delta = \delta^* \left[\frac{3Q}{2\Sigma\rho} \left(\frac{(1 - \xi_{\mathrm{I}}^2)}{E_{\mathrm{I}}} + \frac{(1 - \xi_{\mathrm{II}}^2)}{E_{\mathrm{II}}} \right) \right]^{2/3} \frac{\Sigma\rho}{2} \tag{5.37}$$

$$= 2.79 \times 10^{-4} \delta^* Q^{2/3} \Sigma\rho^{1/3} \quad \text{(for steel bodies)} \tag{5.38}$$

in which δ is the relative approach of remote points in the contacting bodies and

$$a^* = \left(\frac{2\kappa^2 \mathcal{E}}{\pi} \right)^{1/3} \tag{5.39}$$

$$b^* = \left(\frac{2\mathcal{E}}{\pi\kappa} \right)^{1/3} \tag{5.40}$$

$$\delta^* = \frac{2\mathfrak{F}}{\pi} \left(\frac{\pi}{2\kappa^2 \mathcal{E}} \right)^{1/3} \tag{5.41}$$

Values of the dimensionless quantities a^*, b^*, and δ^* as functions of $F(\rho)$ are given in Table 5.1. The values of Table 5.1 are also plotted in Figs. 5.3–5.5.

For an elliptical contact area, the maximum compressive stress occurs at the geometrical center. The magnitude of this stress is

$$\sigma_{\max} = \frac{3Q}{2\pi ab} \tag{5.42}$$

The normal stress at other points within the contact area is given by equation (5.43) in accordance with Fig. 5.6:

$$\sigma = \frac{3Q}{2\pi ab} \left[1 - \left(\frac{x}{a} \right)^2 - \left(\frac{y}{b} \right)^2 \right]^{1/2} \tag{5.43}$$

The foregoing equations (5.30)–(5.43) of surface stress and deformation apply to point contact.

Example 5.1. For the 218 angular-contact ball bearing of Example 4.2 determine the maximum normal contact stresses and contact deformations at the inner and outer raceways.

$Q = 3543$ N (796.2 lb)	Ex. 4.2
$\alpha = 38° \, 54'$	Ex. 3.3
$D = 22.23$ mm (0.875 in.)	Ex. 2.3
$d_m = 125.3$ mm (4.932 in.)	Ex. 2.6

TABLE 5.1. Dimensionless Contact Parameters

$F(\rho)$	a^*	b^*	δ^*
0	1	1	1
0.1075	1.0760	0.9318	0.9974
0.3204	1.2623	0.8114	0.9761
0.4795	1.4556	0.7278	0.9429
0.5916	1.6440	0.6687	0.9077
0.6716	1.8258	0.6245	0.8733
0.7332	2.011	0.5881	0.8394
0.7948	2.265	0.5480	0.7961
0.83495	2.494	0.5186	0.7602
0.87366	2.800	0.4863	0.7169
0.90999	3.233	0.4499	0.6636
0.93657	3.738	0.4166	0.6112
0.95738	4.395	0.3830	0.5551
0.97290	5.267	0.3490	0.4960
0.983797	6.448	0.3150	0.4352
0.990902	8.062	0.2814	0.3745
0.995112	10.222	0.2497	0.3176
0.997300	12.789	0.2232	0.2705
0.9981847	14.839	0.2072	0.2427
0.9989156	17.974	0.18822	0.2106
0.9994785	23.55	0.16442	0.17167
0.9998527	37.38	0.13050	0.11995
1	∞	0	0

$$\gamma = \frac{D \cos \alpha}{d_m} \tag{2.27}$$

$$= \frac{22.23 \cos (38^\circ\ 54')}{125.3} = 0.1381$$

Since this value is only slightly larger than the value of γ obtained in Example 2.6, it is sufficient for the purpose of this example to use values of $\Sigma\rho_i$, $F(\rho)_i$, $\Sigma\rho_o$, and $F(\rho)_o$ obtained in that example. Hence,

$$\Sigma\rho_i = 0.108 \text{ mm}^{-1} (2.747 \text{ in.}^{-1})$$

$$F(\rho)_i = 0.9260$$

$$\Sigma\rho_o = 0.0832 \text{ mm}^{-1} (2.114 \text{ in.}^{-1})$$

$$F(\rho)_o = 0.9038$$

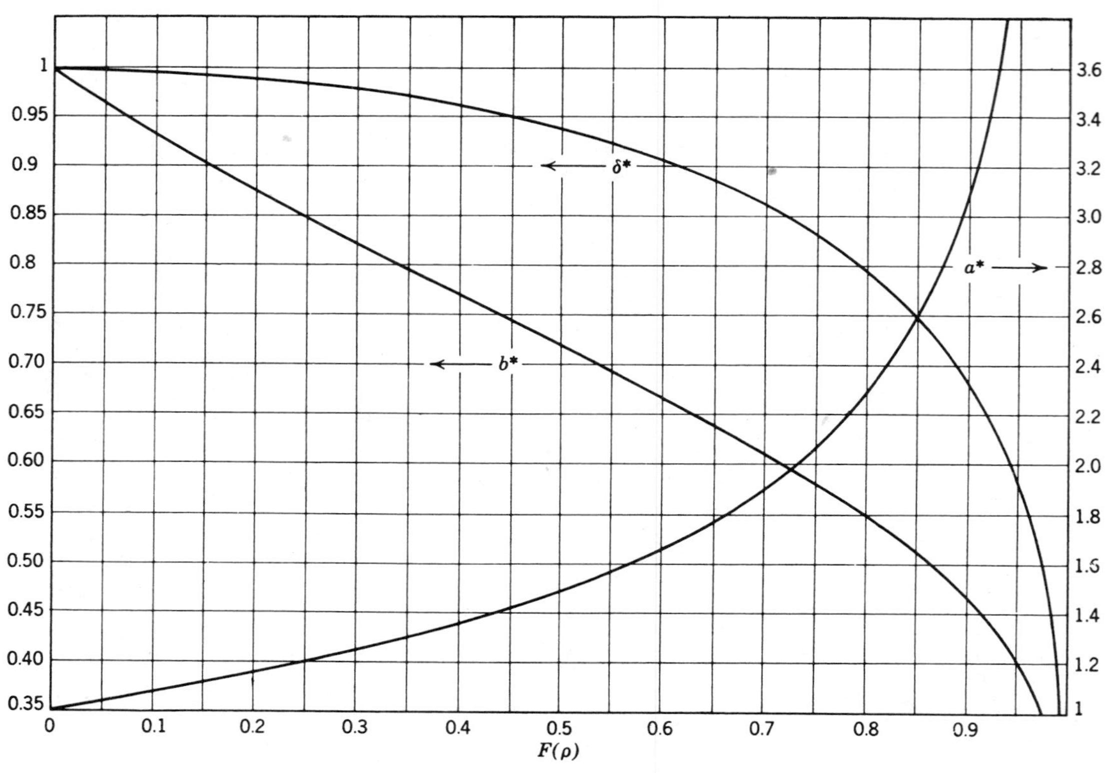

FIGURE 5.3. a^*, b^*, and δ^* vs $F(\rho)$.

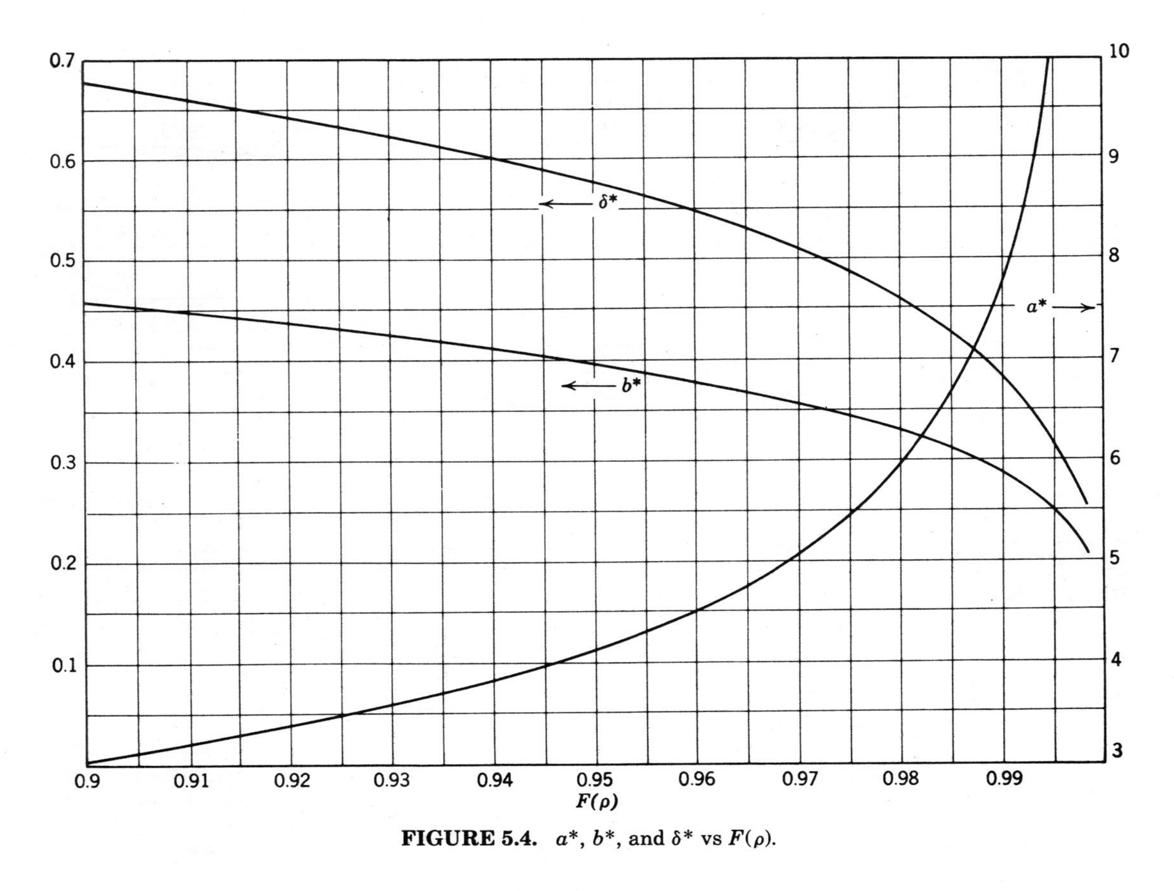

FIGURE 5.4. a^*, b^*, and δ^* vs $F(\rho)$.

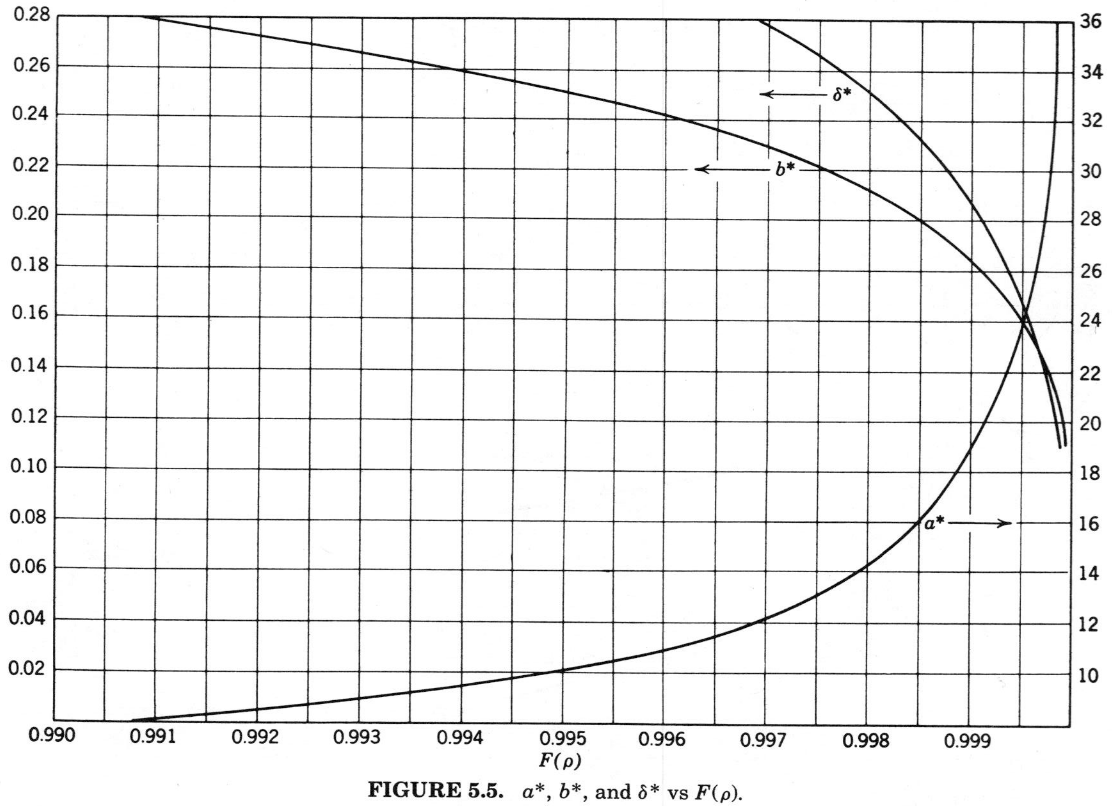

FIGURE 5.5. a^*, b^*, and δ^* vs $F(\rho)$.

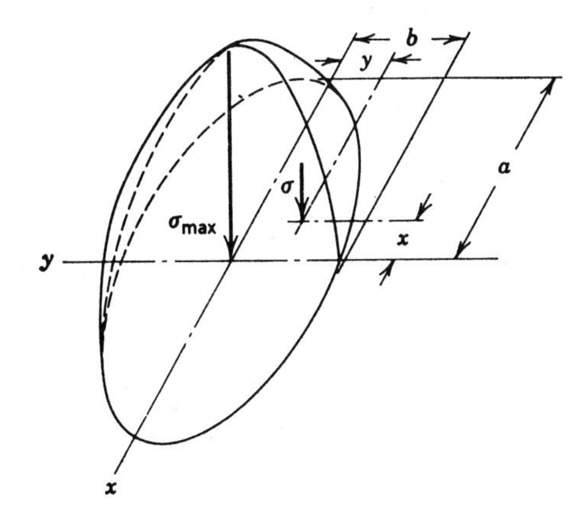

FIGURE 5.6. Ellipsoidal surface compressive stress distribution of point contact.

From Fig. 5.4,

$$a_\mathrm{i}^* = 3.50, \quad b_i^* = 0.430, \quad \delta_i^* = 0.630$$

$$a_\mathrm{i} = 0.0236 a_i^* \left(\frac{Q}{\Sigma \rho_i} \right)^{1/3} \tag{5.34}$$

$$= 0.0236 \times 3.50 \left(\frac{3543}{0.108} \right)^{1/3} = 2.64 \text{ mm (0.1040 in.)}$$

$$b_\mathrm{i} = 0.0236 b_i^* \left(\frac{Q}{\Sigma \rho_i} \right)^{1/3} \tag{5.36}$$

$$= 0.0236 \times 0.430 \times \left(\frac{3543}{0.108} \right)^{1/3} = 0.324 \text{ mm (0.01277 in.)}$$

$$\sigma_\mathrm{imax} = \frac{3Q}{2\pi a_i b_i} \tag{5.42}$$

$$= \frac{3 \times 3543}{2\pi \times 2.64 \times 0.324}$$

$$= 1976 \text{ N/mm}^2 \text{ (286,400 psi)} \quad \text{(compression)}$$

$$\delta_\mathrm{i} = 2.79 \times 10^{-4} \delta_i^* Q^{2/3} \Sigma \rho_i^{1/3} \tag{5.38}$$

$$= 2.79 \times 10^{-4} \times 0.630 \times (3543)^{2/3} (0.108)^{1/3}$$

$$= 0.0195 \text{ mm (0.000766 in.)}$$

At the outer raceway, from Fig. 5.4,

$$a_\mathrm{o}^* = 3.10, \quad b_o^* = 0.455, \quad \delta_\mathrm{o}^* = 0.672$$

$$a_o = 0.0236 a_o^* \left(\frac{Q}{\Sigma\rho_o}\right)^{1/3} \tag{5.34}$$

$$= 0.0236 \times 3.10 \times \left(\frac{3543}{0.0832}\right)^{1/3} = 2.56 \text{ mm } (0.1007 \text{ in.})$$

$$b_o = 0.0236 b_o^* \left(\frac{Q}{\Sigma\rho_o}\right)^{1/3} \tag{5.36}$$

$$= 0.0236 \times 0.455 \times \left(\frac{3543}{0.0832}\right)^{1/3} = 0.3754 \text{ mm } (0.01478 \text{ in.})$$

$$\sigma_{\text{omax}} = \frac{3Q}{2\pi a_o b_o} \tag{5.42}$$

$$= \frac{3 \times 3543}{2\pi \times 2.56 \times 0.3754}$$

$$= 1762 \text{ N/mm}^2 \text{ } (255{,}500 \text{ psi}) \quad \text{(compression)}$$

$$\delta_o = 2.79 \times 10^{-4} \delta_o^* Q^{2/3} \Sigma\rho_o^{1/3} \tag{5.38}$$

$$= 2.79 \times 10^{-4} \times 0.672 \times (3543)^{2/3}(0.0832)^{1/3}$$

$$= 0.01902 \text{ mm } (0.000749 \text{ in.})$$

Note that σ_{imax} is greater than σ_{omax}. This is true for most ball and roller bearings in static loading.

For ideal line contact to exist, the length of body I must equal that of body II. Then κ approaches infinity and the stress distribution in the contact area degenerates to a semicylindrical form as shown by Fig. 5.7. For this condition,

$$\sigma_{\max} = \frac{2Q}{\pi l b} \tag{5.44}$$

$$\sigma = \frac{2Q}{\pi l b}\left[1 - \left(\frac{y}{b}\right)^2\right]^{1/2} \tag{5.45}$$

$$b = \left[\frac{4Q}{\pi l \Sigma\rho}\left(\frac{(1 - \xi_{\text{I}}^2)}{E_{\text{I}}} + \frac{(1 - \xi_{\text{II}}^2)}{E_{\text{II}}}\right)\right]^{1/2} \tag{5.46}$$

For steel roller bearings the semiwidth of the contact surface may be approximated by

$$b = 3.35 \times 10^{-3}\left(\frac{Q}{l\Sigma\rho}\right)^{1/2} \tag{5.47}$$

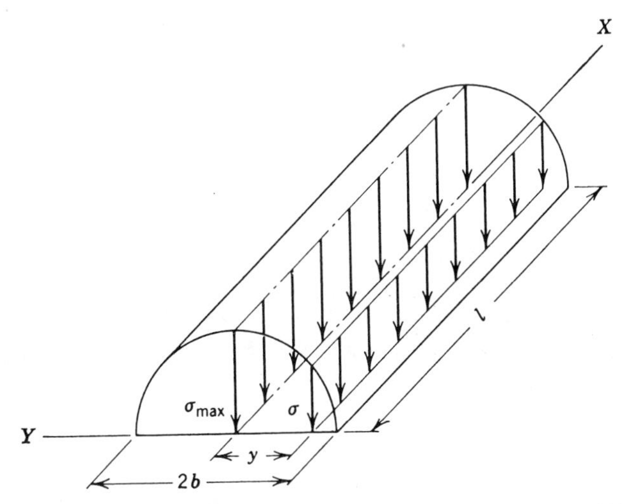

FIGURE 5.7. Semicylindrical surface compressive stress distribution of ideal line contact.

The contact deformation for a line-contact condition was determined subsequent to the Hertz analysis. Palmgren [5.3] gives for steel on steel,

$$\delta = 3.84 \times 10^{-5} \frac{Q^{0.9}}{l^{0.8}} \tag{5.48}$$

Several references exist concerning the determination and calculation of local surface stresses on contacting members. Some of these are by Hertz [5.1], Thomas and Hoersch [5.4], and Lundberg and Sjovall [5.5]. These references may be seen for more complete information on the solution of the elasticity problem.

Example 5.2. Estimate the maximum normal contact stress and deformation at the inner raceway of the 90000 series tapered roller bearing of Example 4.3:

$Q_i = 59{,}410$ N (13,350 lb) Ex. 4.3

$\alpha_i = 22°$ Ex. 4.3

$D = 22.86$ mm (0.9000 in.) Ex 4.3

$l = 30.48$ mm (1.2 in.) Ex 4.3

$d_m = 142.2$ mm (5.600 in.) Ex 4.3

$$\gamma_i = \frac{D \cos \alpha_i}{d_m} \tag{2.27}$$

$$= \frac{22.86 \times \cos(22°)}{142.2} = 0.1490$$

$$\Sigma\rho_i = \frac{1}{D}\left[\frac{2}{1 - \gamma_i} + D\left(\frac{1}{R} - \frac{1}{r_i}\right)\right] \tag{2.37}$$

Assuming that neither rollers nor raceways are crowned, that is, $R = r_i = \infty$ and ideal line contact occurs,

$$\Sigma\rho_i = \frac{1}{D(1 - \gamma_i)}$$

$$= \frac{2}{22.86(1 - 0.1490)} = 0.1028 \text{ mm}^{-1} \text{ (2.611 in.}^{-1}\text{)}$$

$$b_i = 3.35 \times 10^{-3}\left(\frac{Q_i}{l\Sigma\rho_i}\right)^{1/2} \tag{5.47}$$

$$= 3.35 \times 10^{-3}\left(\frac{59410}{30.40 \times 0.1028}\right)^{1/2} = 0.461 \text{ mm (0.01815 in.)}$$

$$\sigma_{imax} = \frac{2Q_i}{\pi l b_i} \tag{5.44}$$

$$= \frac{2 \times 59410}{\pi \times 30.48 \times 0.461} = 2692 \text{ N/mm}^2 \text{ (390,200 psi)}$$

This roller is very heavily loaded and probably requires crowning to avoid edge loading (see Figs. 5.21 and 5.22):

$$\delta_i = 3.84 \times 10^{-5} \frac{Q_i^{0.9}}{l^{0.8}}$$

$$= 3.84 \times 10^{-5} \times \frac{(59410)^{0.9}}{(30.48)^{0.8}} = 0.0491 \text{ mm (0.001932 in.)} \tag{5.48}$$

SUBSURFACE STRESSES

Hertz's analysis applied only to surface stresses caused by a concentrated force applied perpendicular to the surface. Experimental evidence indicates that failure of rolling bearings in surface fatigue as caused by the aforementioned load emanates from points below the stressed surface. Therefore, it is of interest to determine the magnitude of the subsurface stresses. Since the fatigue failure of the surfaces in contact is a statistical phenomenon dependent on the volume of material stressed (see Chapter 18), the depths at which significant stresses occur below the surface are also of interest.

Again, considering only stresses caused by a concentrated force normal to the surface, Jones [5.7] after Thomas and Hoersch [5.4] gives the following equations by which to calculate the principal stresses S_x, S_y, and S_z occurring along the Z axis at any depth below the contact surface. Since surface stress is maximum at the Z axis, therefore the principal stresses must attain maximum values there (see Fig. 5.8):

$$
\begin{aligned}
S_x &= \lambda(\Omega_x + \xi\Omega_x') \\
S_y &= \lambda(\Omega_y + \xi\Omega_y') \qquad (5.49) \\
S_z &= -\tfrac{1}{2}\lambda\left(\frac{1}{\nu} - \nu\right)
\end{aligned}
$$

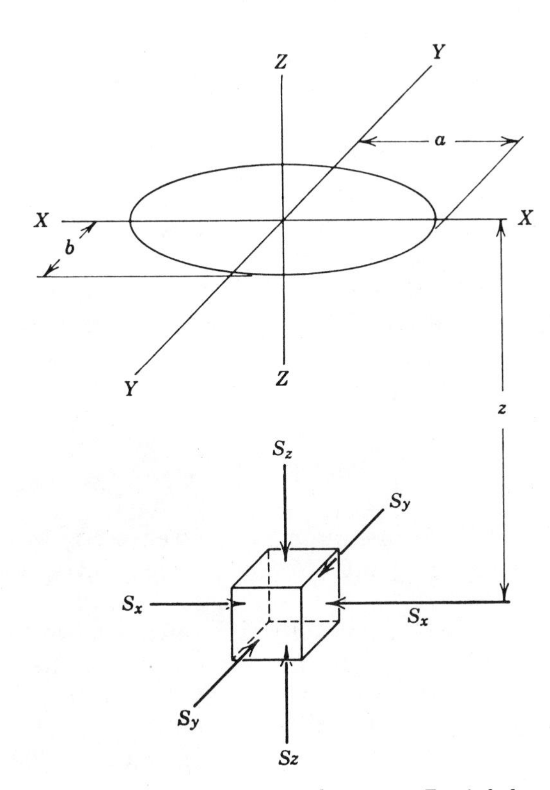

FIGURE 5.8. Principal stresses occurring on element on Z axis below contact surface.

in which

$$\lambda = \frac{b\Sigma\rho}{\left(\kappa - \frac{1}{\kappa}\right)\mathcal{E}\left(\frac{1-\xi_{\mathrm{I}}^2}{E_{\mathrm{I}}} + \frac{1-\xi_{\mathrm{II}}^2}{E_{\mathrm{II}}}\right)} \tag{5.50}$$

$$\nu = \left(\frac{1+\zeta^2}{\kappa^2+\zeta^2}\right)^{1/2} \tag{5.51}$$

$$\zeta = \frac{z}{b} \tag{5.52}$$

$$\Omega_x = -\tfrac{1}{2}(1-\nu) + \zeta[\mathfrak{F}(\phi) - \mathcal{E}(\phi)] \tag{5.53}$$

$$\Omega_x' = 1 - \kappa^2\nu + \zeta[\kappa^2\mathcal{E}(\phi) - \mathfrak{F}(\phi)] \tag{5.54}$$

$$\Omega_y = \frac{1}{2}\left(1 + \frac{1}{\nu}\right) - \kappa^2\nu + \zeta[\kappa^2\mathcal{E}(\phi) - \mathfrak{F}(\phi)] \tag{5.55}$$

$$\Omega_y' = -1 + \nu + \zeta[\mathfrak{F}(\phi) - \mathcal{E}(\phi)] \tag{5.56}$$

$$\mathfrak{F}(\phi) = \int_0^\phi \left[1 - \left(1 - \frac{1}{\kappa^2}\right)\sin^2\phi\right]^{-1/2} d\phi \tag{5.57}$$

$$\mathcal{E}(\phi) = \int_0^\phi \left[1 - \left(1 - \frac{1}{\kappa^2}\right)\sin^2\phi\right]^{1/2} d\phi \tag{5.58}$$

The principal stress indicated by the foregoing equations are graphically illustrated by Figs. 5.9–5.11.

Since each of the maximum principal stresses can be determined, it is further possible to evaluate the maximum shear stress on the z axis below the contact surface. By Mohr's circle (see reference [5.2]), the maximum shear stress is found to be

$$\tau_{yz} = \tfrac{1}{2}(S_z - S_y) \tag{5.59}$$

As shown by Fig. 5.12 the maximum shear stress occurs at various depths z, below the surface, being at $0.467b$ for simple point contact and $0.786b$ for line contact.

During the passage of a loaded rolling element over a point on the raceway surface, the maximum shear stress on the z axis varies between 0 and $\tau_{\max}$. If the element rolls in the direction of the y axis, then the shear stresses occurring in the yz plane below the contact surface assume values from negative to positive for values of y less than and greater than zero, respectively. Thus, the maximum variation of shear stress in the yz plane at any point for a given depth is $2\tau_{yz}$.

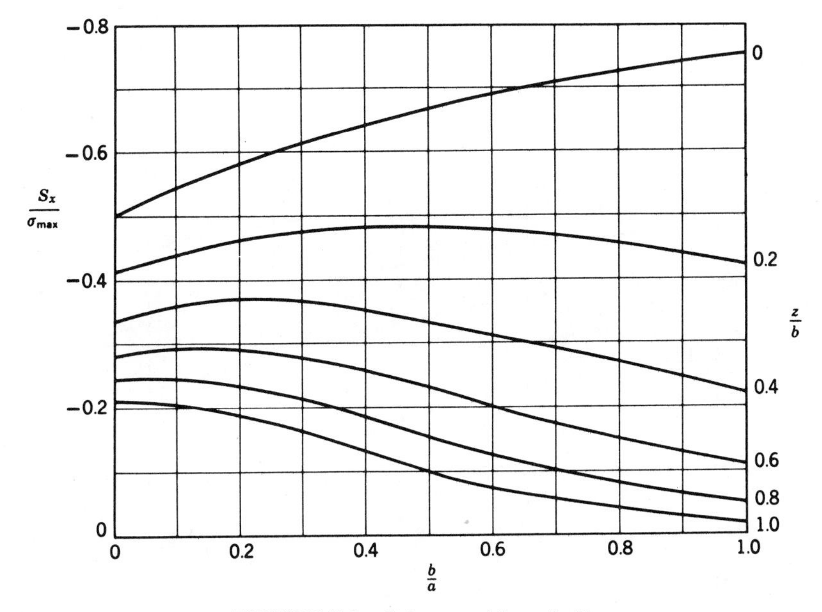

FIGURE 5.9. S_x/σ_{max} vs b/a and z/b.

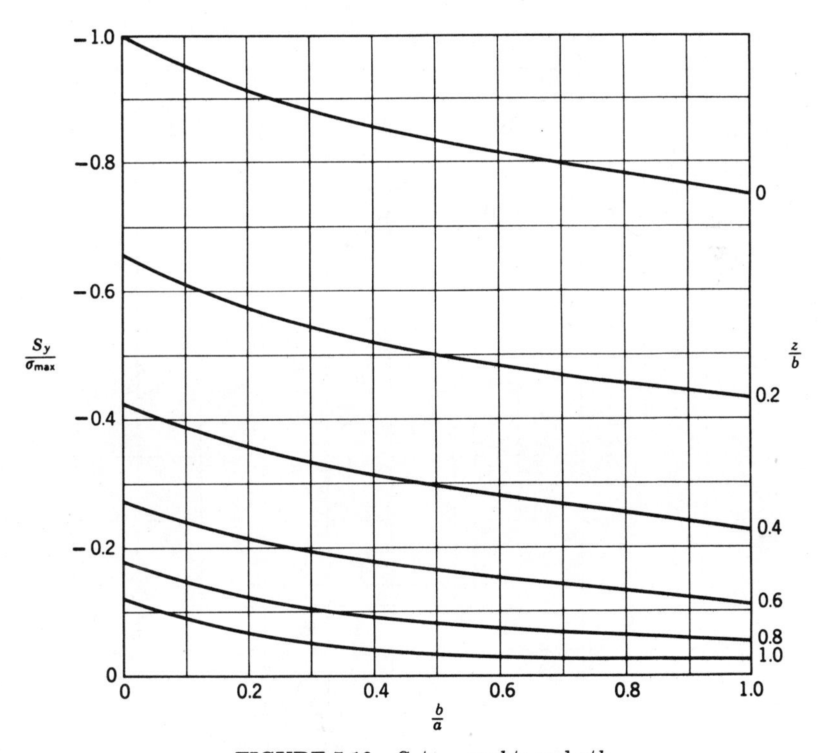

FIGURE 5.10. S_y/σ_{max} vs b/a and z/b.

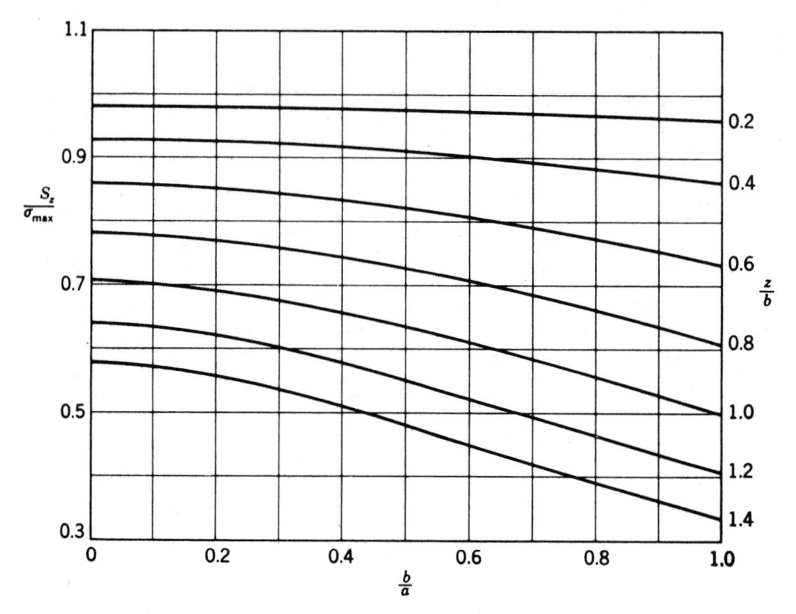

FIGURE 5.11. S_z/σ_{max} vs b/a and z/b.

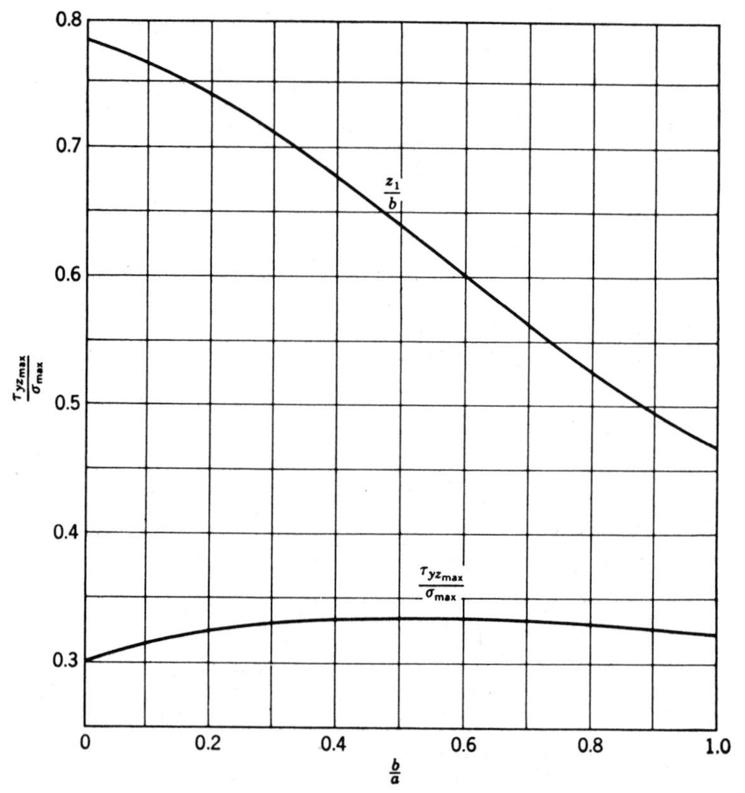

FIGURE 5.12. $\tau_{yz\max}/\sigma_{max}$ and z_1/b vs b/a.

Lundberg and Palmgren [5.8] show that

$$\tau_{yz} = \frac{3Q}{2\pi} \times \frac{\cos^2\phi \sin\phi \sin\vartheta}{a^2 \tan^2\vartheta + b^2 \cos^2\phi} \tag{5.60}$$

wherein

$$y = (b^2 + a^2 \tan^2\vartheta)^{1/2} \sin\phi \tag{5.61}$$

$$z = a \tan\vartheta \cos\phi \tag{5.62}$$

Here, ϑ and ϕ are auxiliary angles such that

$$\frac{\partial \tau_{yz}}{\partial \phi} = \frac{\partial \tau_{yz}}{\partial \vartheta} = 0$$

which defines the amplitude τ_0 of the shear stress. Further, ϑ and ϕ are related as follows:

$$\begin{aligned} \tan^2\phi &= t \\ \tan^2\vartheta &= t - 1 \end{aligned} \tag{5.63}$$

in which t is an auxiliary parameter such that

$$\frac{b}{a} = [(t^2 - 1)(2t - 1)]^{1/2} \tag{5.64}$$

Solving equations (5.60)–(5.64) simultaneously, it is shown in reference [5.8] that

$$\frac{2\tau_0}{\sigma_{\max}} = \frac{(2t - 1)^{1/2}}{t(t + 1)} \tag{5.65}$$

and

$$\zeta = \frac{1}{(t + 1)(2t - 1)^{1/2}} \tag{5.66}$$

Figure 5.13 shows the resulting distribution of shear stress at depth z_0 in the direction of rolling for $b/a = 0$, that is, a line contact.

Figure 5.14 shows the shear stress amplitude of equation (5.65) as a function of b/a. Also shown is the depth below the surface at which this shear stress occurs. Since the shear stress amplitude indicated by Fig. 5.14 is greater than that of Fig. 5.12, Lundberg and Palmgren [5.8] assumed this shear stress, called the maximum orthogonal shear stress, to

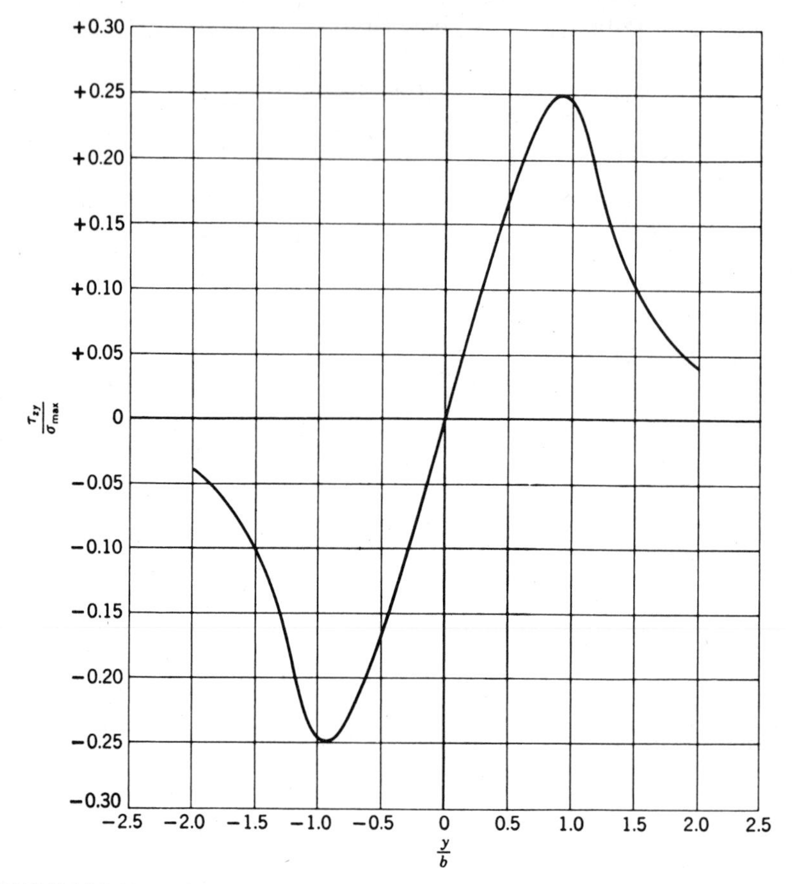

FIGURE 5.13. τ_{zy}/σ_{max} vs y/b for $b/a = 0$ and $z = z_0$ (concentrated normal load).

be significant in causing fatigue failure of the surfaces in rolling contact. As can be seen from Fig. 5.14, for a typical rolling bearing point contact of $b/a = 0.1$, the depth below the surface at which this stress occurs is approximately $0.49b$. Moreover, as seen by Fig. 5.13, this stress occurs at any instant under the extremities of the contact ellipse with regard to the direction of motion, that is, at $y = \pm 0.9b$.

Metallurgical research [5.16] based on plastic alterations detected in sub-surface material by transmission electron microscopic investigation gives indications that the subsurface depth at which significant amounts of material alteration occur is approximately $0.75b$. Assuming such plastic alteration is the forerunner of material failure, then it would appear that the maximum shear stress of Fig. 5.12 may be worthy of consideration as the significant stress causing failure. Figures 5.15 and 5.16 from reference [5.16] are photomicrographs showing the subsurface changes caused by constant rolling on the surface. As a further consideration

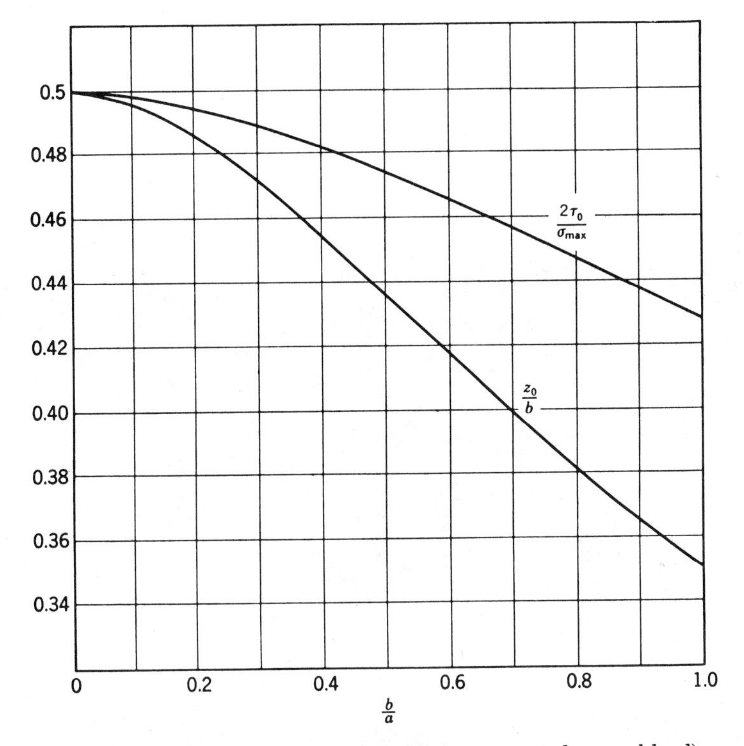

FIGURE 5.14. $2\tau_0/\sigma_{max}$ and z_0/b vs b/a (concentrated normal load).

some researchers consider the Von Mises distortion energy theory of failure [5.17] and the scalar stress level derived therefrom as the criterion for failure. The magnitude of the maximum Von Mises stress is roughly equivalent to that of the maximum orthogonal shear stress, that is, $0.5\,\sigma$, while the subsurface depth at which the maximum occurs is consistent with that of the maximum shear stress, that is, $0.7b$–$0.8b$.

Example 5.3. Determine the amplitude of the maximum orthogonal shear stress at the inner and outer raceways of the 218 angular-contact ball bearing of Example 5.1. Estimate the depths below the rolling surfaces at which these stresses occur.

$$a_i = 2.64 \text{ mm } (0.1040 \text{ in.}) \qquad \text{Ex. 5.1}$$

$$b_i = 0.324 \text{ mm } (0.01277 \text{ in.}) \qquad \text{Ex. 5.1}$$

$$\frac{b_i}{a_i} = \frac{0.324}{2.64} = 0.1227$$

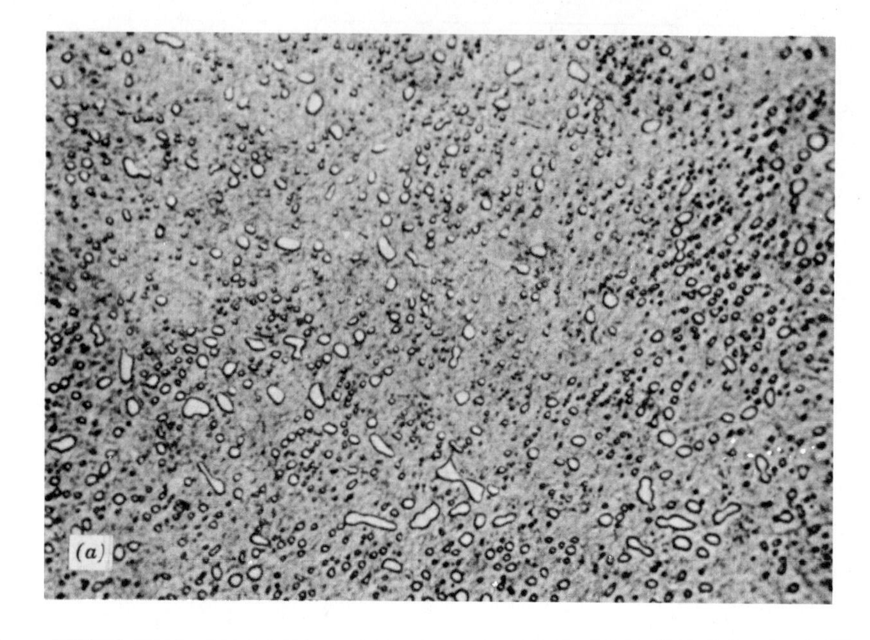

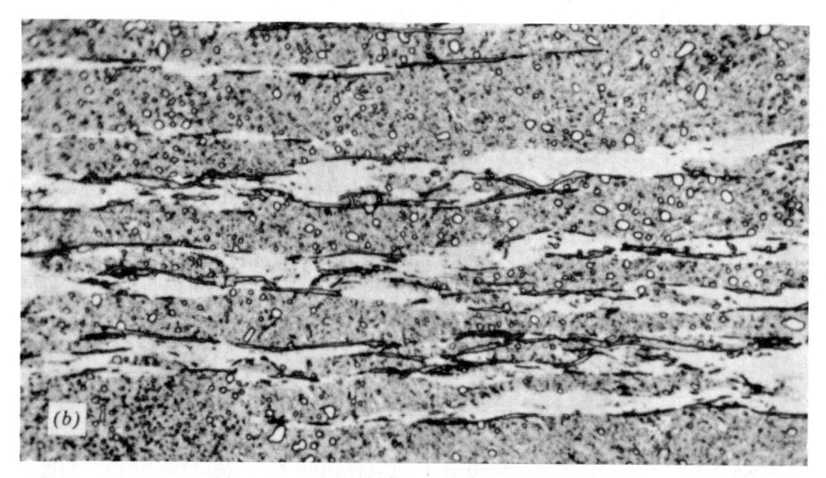

FIGURE 5.15. Subsurface metallurgical structure (1300 times magnification after picral etch) showing change due to repeated rolling under load. (*a*) Normal structure; (*b*) stress-cycled structure—white deformation bands and lenticular carbide formations are in evidence.

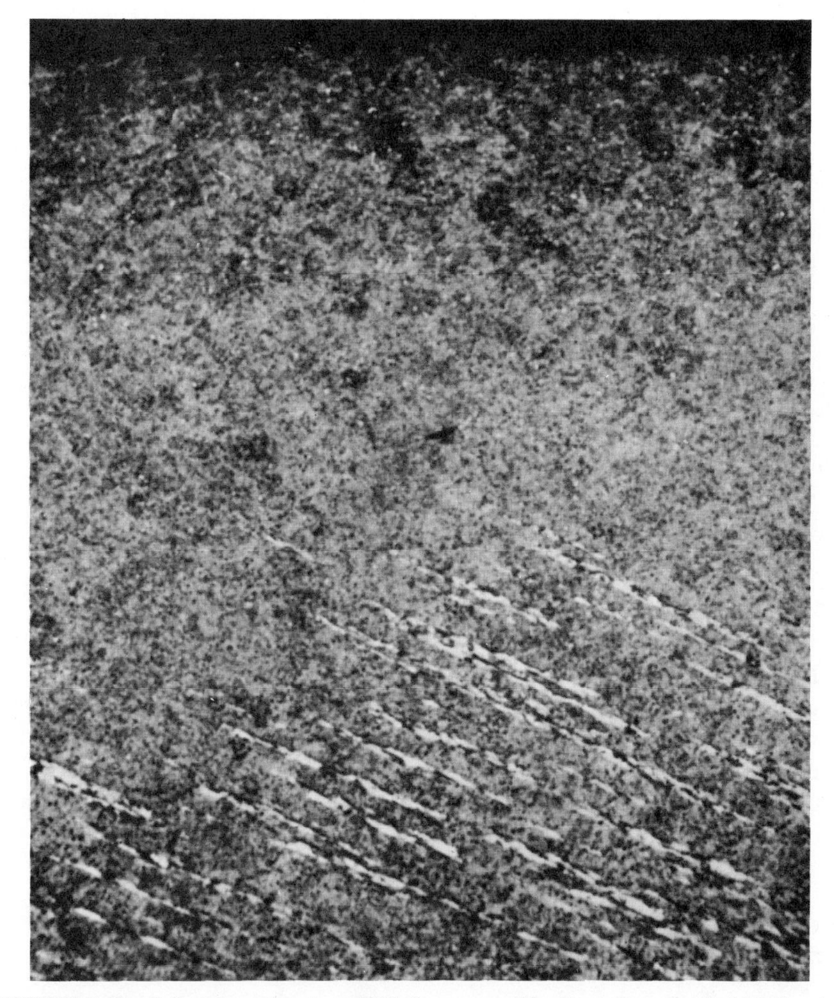

FIGURE 5.16. Subsurface structure (300 times magnification after picral etch) showing orientation of carbides to direction of rolling. Carbides are thought to be weakness locations at which fatigue failure is initiated.

From Fig. 5.14,

$$\frac{2\tau_{0i}}{\sigma_{imax}} = 0.498, \qquad \frac{z_{0i}}{b_i} = 0.493$$

$$\sigma_{imax} = 1976 \text{ N/mm}^2 \text{ (286,400 psi)}$$

$$\tau_{0i} = \frac{0.498 \times 1976}{2} = 491.9 \text{ N/mm}^2 \text{ (71,310 psi)}$$

$$a_o = 2.558 \text{ mm } (0.1007 \text{ in.}) \qquad \text{Ex. 5.1}$$

$$b_o = 0.375 \text{ mm } (0.01478 \text{ in.})$$

$$\frac{b_o}{a_o} = \frac{0.375}{2.558} = 0.1468 \qquad \text{Ex. 5.1}$$

Ex. 5.1

From Fig. 5.14,

$$\frac{2\tau_{0o}}{\sigma_{omax}} = 0.497, \qquad \frac{Z_{0o}}{b_o} = 0.491$$

$$\sigma_{omax} = 1762 \text{ N/mm}^2 \text{ } (255{,}500 \text{ psi}) \qquad \text{Ex. 5.1}$$

$$\tau_{0o} = \frac{0.497 \times 1762}{2} = 438 \text{ N/mm}^2 \text{ } (63{,}490 \text{ psi})$$

$$z_{0i} = 0.493 \times 0.324 = 0.160 \text{ mm } (0.00630 \text{ in.})$$

$$z_{0o} = 0.491 \times 0.375 = 0.184 \text{ mm } (0.00726 \text{ in.})$$

For case-hardened bearings the value of z_{0i} and z_{0o} can be used to estimate the required case depth. Note that the maximum shear stress at the center of contact occurs at $z_{1i} = 0.76b_i$ and $z_{1o} = 0.755b_o$ for the inner and outer raceways, respectively (see Fig. 5.12). Hence $z_{1i} = 0.246$ mm (0.00967 in.) and $z_{1o} = 0.281$ mm (0.01108 in.). It is more conservative to base case depth on these values. Case depth should exceed z_0 or z_1 by at least a factor of three.

EFFECT OF SURFACE SHEAR STRESS

In the determination of contact deformation vs load only the concentrated load applied normal to the surface need be considered for most applications. Moreover, in most rolling bearing applications, lubrication is at least adequate, and the sliding friction between rolling elements and raceways is negligible (see Chapter 13). This means that the shear stresses acting on the rolling elements and raceway surfaces in contact, that is, the elliptical areas of contact, are negligible compared to normal stresses.

For the determination of bearing endurance with regard to fatigue of the contacting rolling surfaces, the surface shear stress cannot be neglected and in many cases is the most significant factor in determining endurance of a rolling bearing in a given application. Methods of calculation of the surface shear stresses (traction stresses) will be discussed in Chapter 13. The means for determining the effect on the subsurface stresses of the combination of normal and tangential (traction) stresses applied at the surface are extremely complex requiring the use of digital computation. Among others, Zwirlein and Schlicht [5.18] have calculated subsurface stress fields based upon assumed ratios of surface shear stress to applied normal stress. Reference [5.18] assumes that the Von Mises stress is most significant with regard to fatigue failure and gives illustrations of this stress in Fig. 5.17.

Figure 5.18 also from reference [5.18] shows the depth at which the various stresses occur. Figure 5.18 shows that as the ratio of surface shear to normal stress increases, the maximum Von Mises stress moves closer to the surface. At a ratio value of $\tau/\sigma = 0.3$, the maximum Von Mises stress occurs at the surface. Various other investigators have found that if a shear stress is applied at the contact surface in addition to the normal stress, the maximum shear stress tends to increase and it is located closer to the surface (see references [5.9–5.12]). References [5.13–5.15] give indications of the effect of higher order surfaces on the contact stress solution The references cited above are intended not to be extensive, but to give only a representation of the field of knowledge.

The foregoing discussion pertained to the subsurface stress field caused by a concentrated normal load applied in combination with a uniform surface shear stress. The ratio of surface shear stress to normal stress is also called the coefficient of friction (see Chapter 13). Because of infinitesimally small irregularities in the basic surface geometries of the rolling contact bodies, neither uniform normal stress fields as shown by Figs. 5.6 and 5.7 nor a uniform shear stress field are likely to occur in practice. Sayles et al. [5.19] use the model shown by Fig. 5.19 in developing an "elastic conformity factor."

Kalker [5.20] has developed a mathematical model and computer program to calculate the subsurface stress distribution owing to an arbitrary distribution of shear and normal stresses over a surface in concentrated contact.

The foregoing does not infer that the analysis presented in this chapter is not adequate. To the contrary, experience has shown the method to be sufficiently accurate for the analysis of most rolling bearing applications.

TYPE OF CONTACT

Basically, two hypothetical types of contact can be defined under conditions of *zero* load. These are

1. *Point contact*, that is, two surfaces touch at a single point.
2. *Line contact*, that is, two surfaces touch along a straight or curved line of zero width.

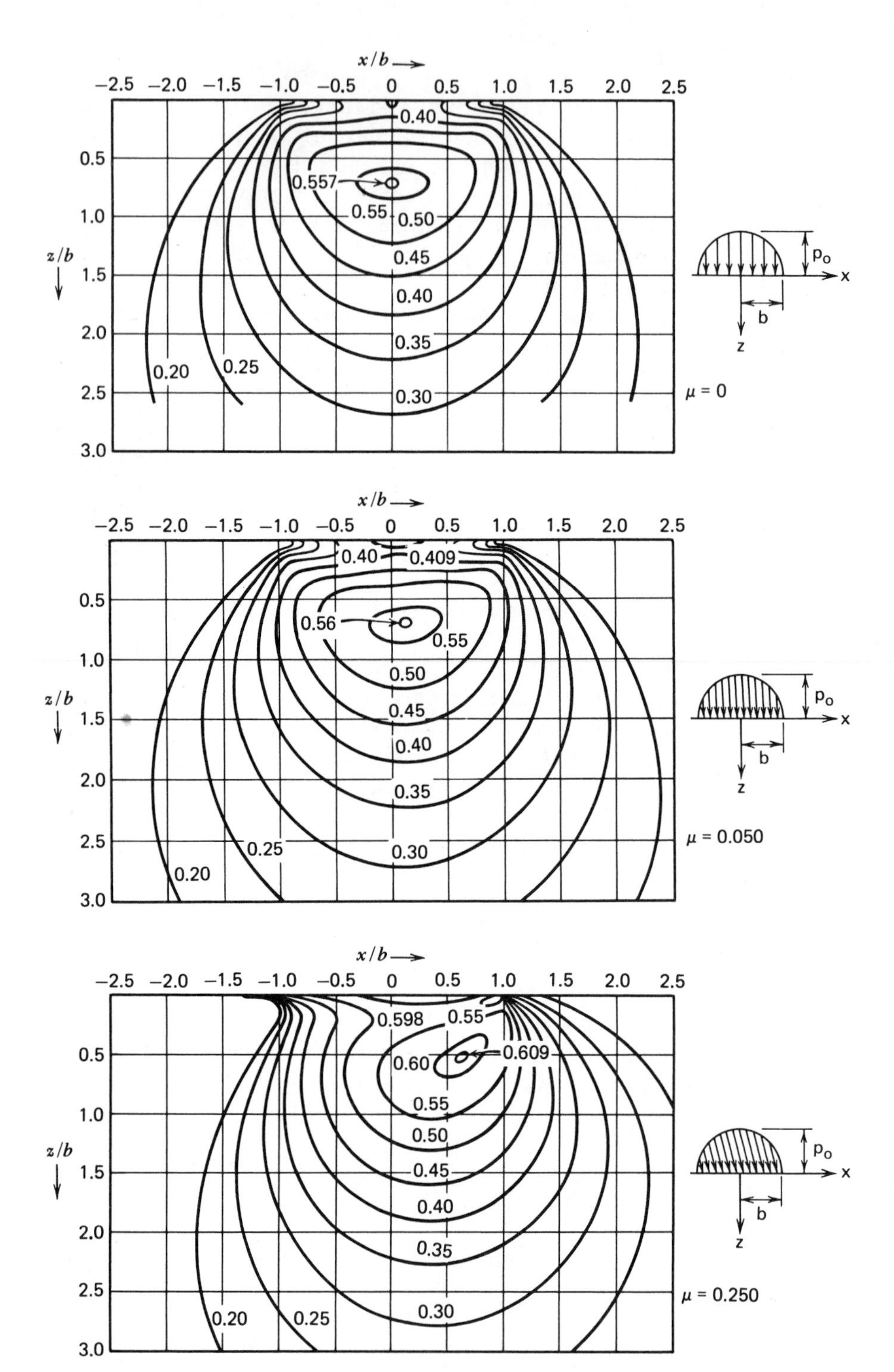

FIGURE 5.17. Lines of equal von Mises stress/normal applied stress for various surface shear stresses τ/normal applied stress σ.

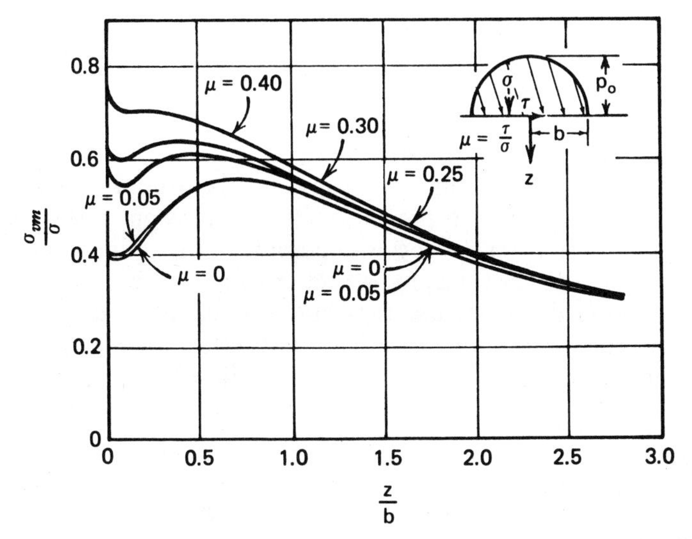

FIGURE 5.18. Material stressing (σ_{vm}/σ) vs depth for different amounts of surface shear stress (τ/σ).

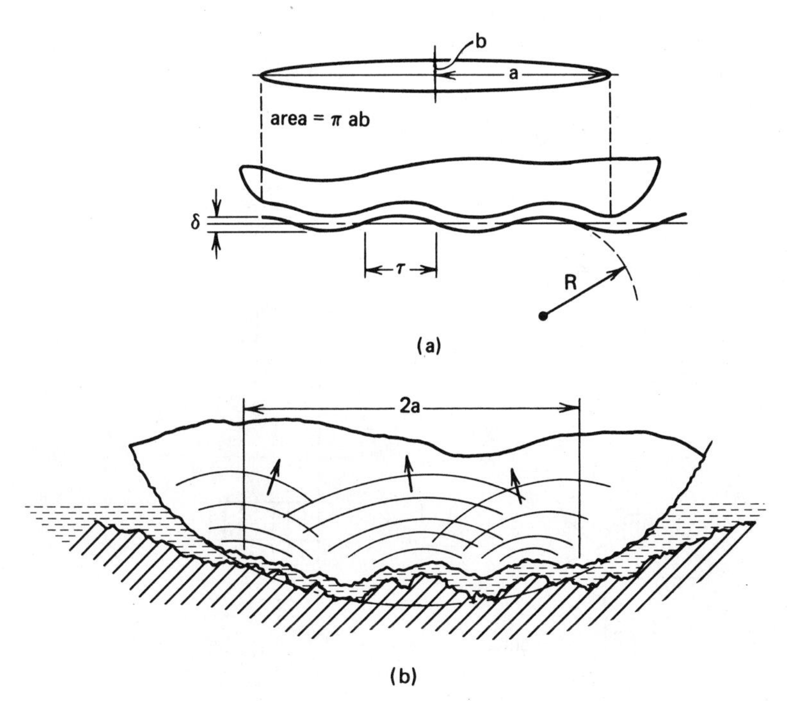

FIGURE 5.19. Models for less-than-ideal elastic conformity. (*a*) Hertzian contact model used in developing elastic conformity parameter. (*b*) Elastic conformity envisaged with real roughness would be preferential to certain asperity wavelengths. For convenience the figure shows only one compliant rolling element, whereas in practice if materials of similar modulus were employed the deformation would be shared.

Obviously, after load is applied to the contacting bodies the point expands to an ellipse and the line to a rectangle in *ideal line contact*, that is, the bodies have equal length. Figure 5.20 illustrates the surface compressive stress distribution which occurs in each case.

When a roller of finite length contacts a raceway of greater length, the axial stress distribution along the roller is altered from that of Fig. 5.20. Since the material in the raceway is in tension at the roller ends because of depression of the raceway outside of the roller ends, the roller end compressive stress tends to be higher than that in the center of contact. Figure 5.21 demonstrates this condition of *edge loading*.

To counteract this condition, cylindrical rollers (or the raceways) may be crowned as shown by Fig 1.33. The stress distribution is thereby made

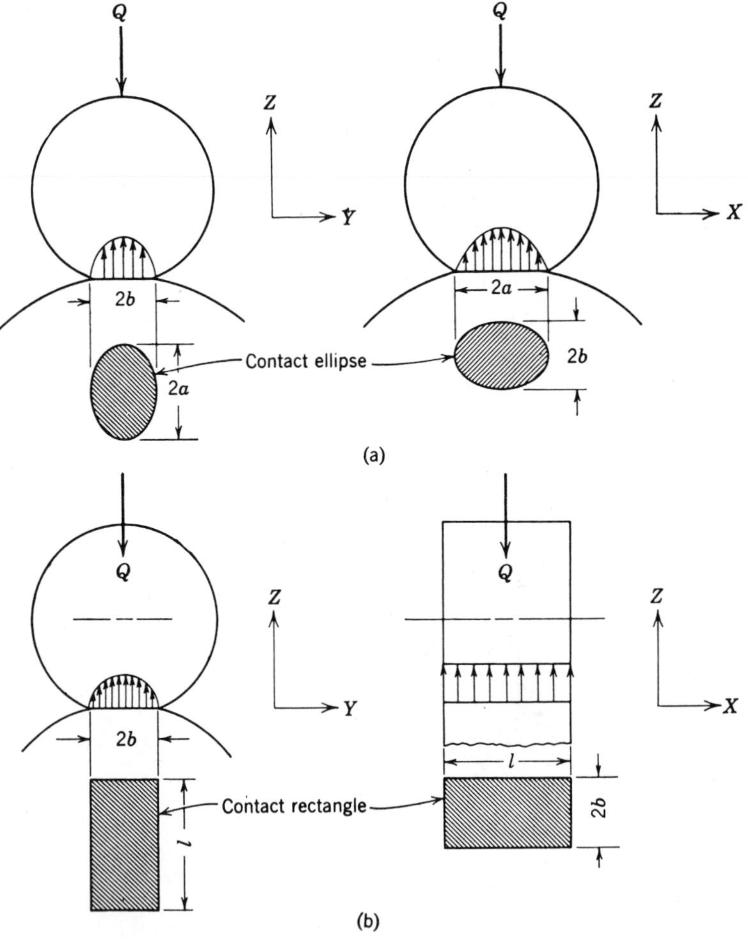

FIGURE 5.20. Surface compressive stress distribution. (*a*) Point contact; (*b*) ideal line contact.

more uniform depending upon applied load. If the applied load is increased significantly, *edge loading* will occur once again.

Lundberg et al. [5.8] have defined a condition of *modified line contact* for roller-raceway contact. Thus, when the major axis ($2a$) of the contact ellipse is greater than the effective roller length l but less than $1.5l$, *modified line contact* is said to exist. If $2a < l$, then point contact exists; if $2a > 1.5l$, then *line contact* exists with attendant *edge loading*. This condition may be ascertained approximately by the methods presented in the section "Surface Stresses and Deformations," using the roller crown radius for R in equations (2.37)–(2.40).

The analysis of contact stress and deformation presented in this sec-

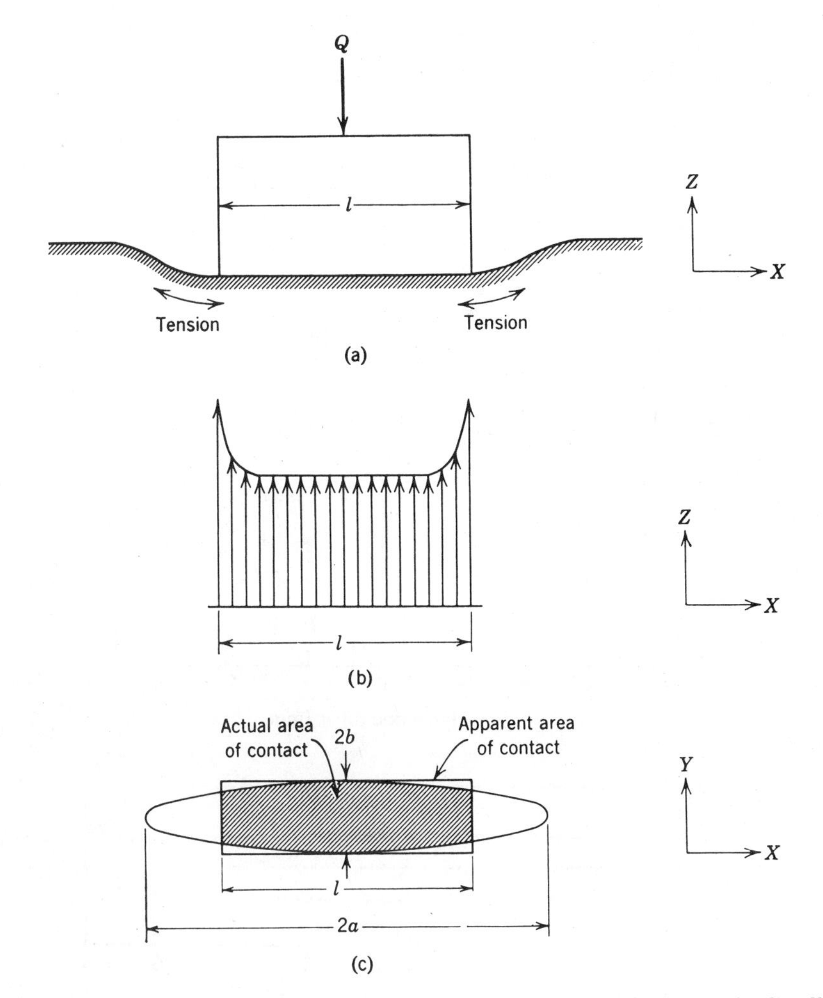

FIGURE 5.21. Line contact. (*a*) Roller contacting a surface of infinite length; (*b*) roller–raceway compressive stress distribution; (*c*) contact ellipse.

tion is based on the existence of an elliptical area of contact, except for the ideal roller under load, which has a rectangular contact. Since it is desirable to preclude *edge loading* and attendant high stress concentrations, roller bearing applications should be examined carefully according to the *modified line contact* criterion. Where that criterion is exceeded, redesign of roller and/or raceway curvatures may be necessitated.

Rigorous mathematical/numerical methods have ben developed to calculate the distribution and magnitude of surfaces stresses in any "line" contact situation, that is, including the effects of crowning of rollers, raceways, and combinations thereof (see references [5.21] and [5.22]. Additionally, finite element methods (FEM) have been employed [5.23] to perform the same analysis. In all cases, a rather great amount of time on a digital computer is required to solve a single contact situation. Figure 5.22 shows the result of an FEM analysis of a heavily loaded typical spherical roller on a raceway. Note the slight "dogbone" shape of the

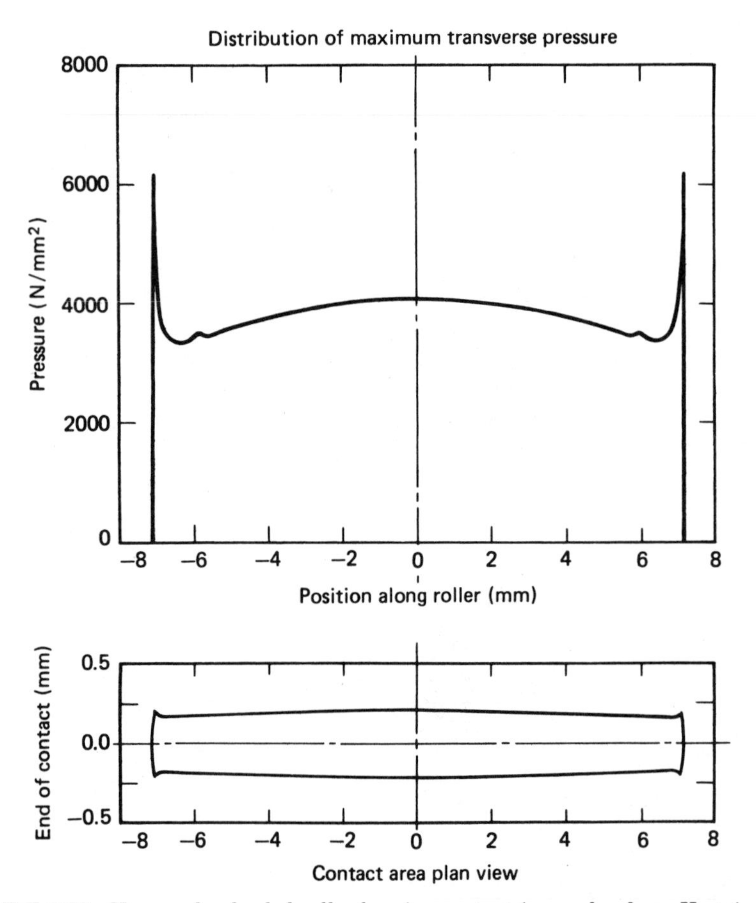

FIGURE 5.22. Heavy edge loaded roller bearing contact (example of non-Hertzian contact).

contact surface. Note also the slight pressure increase where the roller crown blends into the roller end geometry.

Example 5.4. The 22317 spherical roller bearing of Example 2.10 experiences a peak roller load of 2225 N (500 lb). Estimate the type of raceway–roller contact at each raceway

$$l = 20.71 \text{ mm } (0.8154 \text{ in.}) \qquad \text{Ex. 2.8}$$

$$\sum \rho_i = 0.0979 \text{ mm}^{-1} \ (2.487 \text{ in.}^{-1}) \qquad \text{Ex. 2.10}$$

$$F(\rho)_i = 0.9951 \qquad \text{Ex. 2.10}$$

$$\sum \rho_o = 0.068 \text{ mm}^{-1} \ (1.726 \text{ in.}^{-1}) \qquad \text{Ex. 2.10}$$

$$F(\rho)_o = 0.9929 \qquad \text{Ex. 2.10}$$

From Fig. 5.5,

$$a_i^* = 10.2$$

$$a_o^* = 8.8$$

$$a_i = 0.0236 a_i^* \left(\frac{Q}{\sum \rho_i} \right)^{1/3} \qquad (5.34)$$

$$= 0.0236 \times 10.2 \left(\frac{2225}{0.0979} \right)^{1/3} = 6.828 \text{ mm } (0.2688 \text{ in.})$$

$$2a_i = 13.64 \text{ mm } (0.537 \text{ in.})$$

Since $2a_i < l$ point contact occurs at the inner raceway

$$a_o = 0.0236 a_o^* \left(\frac{Q}{\sum \rho_o} \right)^{1/3} \qquad (5.34)$$

$$= 0.0236 \times 8.8 \left(\frac{2225}{0.068} \right)^{1/3} = 6.65 \text{ mm } (0.2618 \text{ in.})$$

$$2a_o = 13.3 \text{ mm } (0.5236 \text{ in.})$$

Since $2a_o < l$, point contact occurs at the outer raceway also.

Example 5.5. Estimate the type of contact that occurs at each raceway of a 22317 spherical roller bearing if the peak roller load is 22,250 N (5000 lb.). At 2225 N (500 lb.).

$$a_i = 6.828 \text{ mm } (0.2688 \text{ in.}) \qquad \text{Ex. 5.4}$$

$$a_o = 6.65 \text{ mm } (0.2618 \text{ in.}) \qquad \text{Ex. 5.4}$$

$$l = 20.71 \text{ mm } (0.8154 \text{ in.}) \qquad \text{Ex. 2.8}$$

$$a_i = 6.828\left(\frac{22250}{2225}\right)^{1/3} = 14.69 \text{ mm } (0.5785 \text{ in.})$$

$$2a_i = 29.39 \text{ mm } (1.157 \text{ in.})$$

$$1.5l = 1.5 \times 20.71 = 31.06 \text{ mm } (1.223 \text{ in.})$$

Since $l < 2a_i < 1.5l$, modified line contact occurs at the inner raceway.

$$a_o = 6.65\left(\frac{22250}{2225}\right)^{1/3} = 14.31 \text{ mm } (0.5632 \text{ in.})$$

$$2a_o = 28.6 \text{ mm } (1.126 \text{ in.})$$

Since $l < 2a_o < 1.5l$, modified line contact occurs at the outer raceway.

The circular crown shown in Fig. 1.33*a* resulted from the theory of Hertz [5.11] whereas the cylindrical/crowned profile of Fig. 1.33*b* resulted from the work of Lundberg et al [5.51]. As illustrated in Fig. 5.23, each of these surface profiles, while minimizing edge stresses, has its drawbacks. Under light loads, a circular crowned profile does not enjoy full use of the roller length, somewhat negating the use of rollers in lieu of balls to carry heavier loads with longer endurance (see Chapter 18). Under heavier loads, while edge stresses are avoided for most applications, contract stress in the center of the contact can greatly exceed that in a straight profile contact, again resulting in substantially reduced endurance characteristics.

Under light loads, the partially crowned roller of Fig. 1.33*b* as illustrated by Fig. 5.23*c* experiences less contact stress than does a fully crowned roller under the same loading. Under heavy loading the partially crowned roller also tends to outlast the fully crowned roller because of lower stress in the center of the contact; however, unless careful attention is paid to blending of the intersections of the "flat" (straight portion of the profiles) and the crown, stress concentrations can occur at the intersections with substantial reduction in endurance (see Chapter 18). When the roller axis is tilted relative to the bearing axis, both the fully crowned and partially crowned profiles tend to generate less edge stress under a given load as compared to the straight profile.

After many years of investigation and with the assistance of mathematical tools such as finite difference and finite element methods as prac-

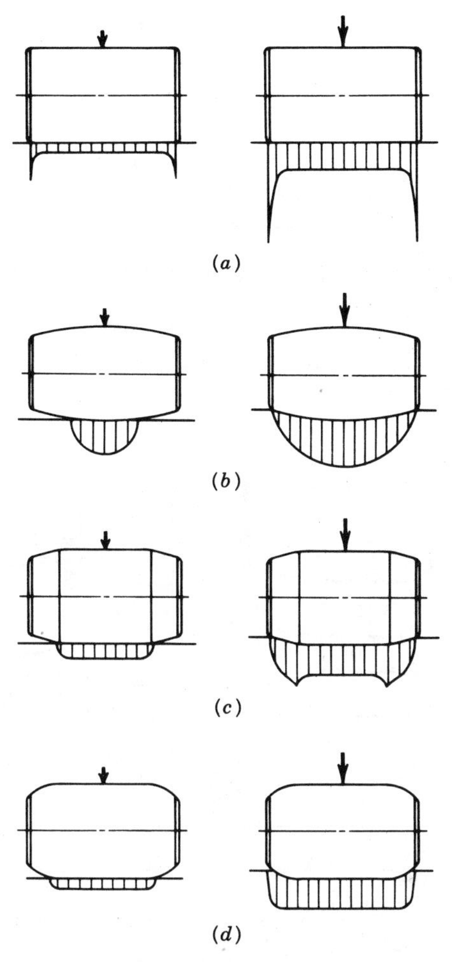

FIGURE 5.23. Roller–raceway contact load vs length and applied load; a comparison of straight, fully crowned, partially crowned, and logarithmic profiles (from [5.24]).

ticed using powerful computers, a "logarithmic" profile was developed [5.24] yielding a substantially optimized stress distribution under most conditions of loading (see Fig. 5.23*d*). The profile is so named because it can be expressed mathematically as a special logarithmic function. Under all loading conditions, the logarithmic profile uses more of the roller length than either the fully crowned or partially crowned roller profiles. Under misalignment, edge loading tends to be avoided under all but exceptionally heavy loads. Under specific loading (Q/lD) from 20 to 100 N/mm^2 (2900–14500 psi), Fig. 5.24 taken from [5.24] illustrates the contact stress distributions attendant to the various surface profiles discussed

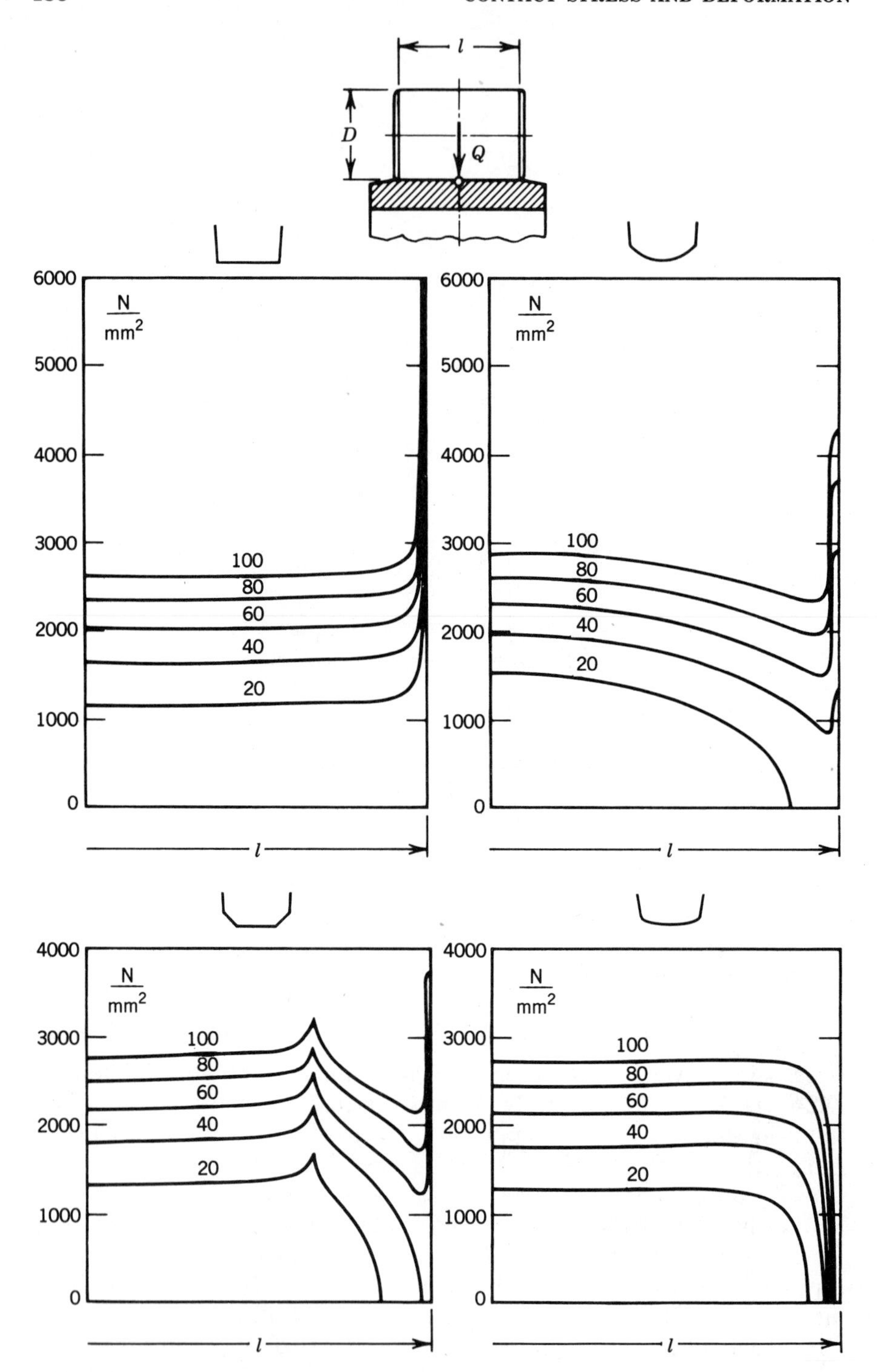

FIGURE 5.24. Compressive stress vs length and specific rollerload (Q/lD) for various roller (or raceway) profiles (from [5.24]).

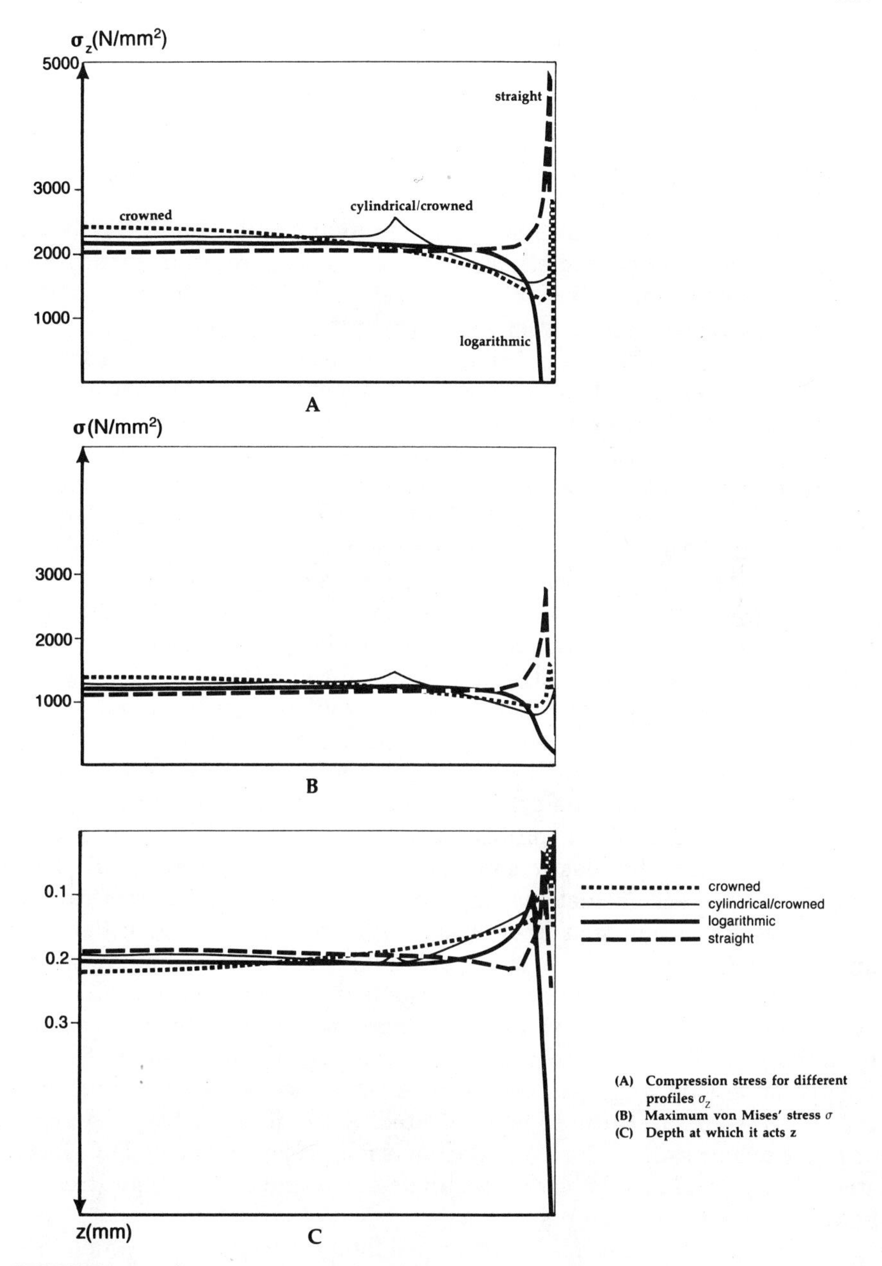

FIGURE 5.25. Comparison of surface compressive stress σ_z, maximum Von Mises stress σ_{vm}, and depth z to the maximum Von Mises stress for various roller (or raceway) profiles (from [5.24]).

herein. Figure 5.25, also from [5.24], compares the surface and subsurface stress characteristics for the various surface profiles.

ROLLER END–FLANGE CONTACT STRESS

The contact stresses arising between flange and roller ends may be estimated from the contact stress and deformation relationships previously presented. The roller ends are usually flat with corner radii blending into the crowned portion of the roller profile. The flange may also be a portion of a flat surface, in which case the flat roller ends mate directly with the flat flange surface. This is the usual design in cylindrical roller bearings. When it is desired to have the rollers carry thrust loads between the roller ends and the flange, sometimes the flange surface is designed as a portion of a cone. In this case, the roller corners contact the flange. The angle between the flange and a radial plane is called the layback angle. Alternatively, the roller end may be designed as a portion of a sphere that contacts the flange. The latter arrangement, that is a sphere-end roller contacting an angled flange, is conducive to improved lubrication while sacrificing some flange–roller guidance capability. In this case, some skewing control may have to be provided by the cage.

For the case of sphere-end rollers and angled flange geometry, the individual contact may be modeled as a sphere contacting a cylinder. For the purpose of calculation the sphere radius is set equal to the roller sphere end radius, and the cylinder radius can be approximated by the radius of curvature of the conical flange at the theoretical point of contact. By knowing the elastic contact load, roller–flange material properties, and contact geometries, the contact stress and deflection can be calculated. This approach is only approximate, because the roller end and flange do not meet the Hertzian half-space assumption. Also, the radius of curvature on the conical flange is not a constant but will vary across the contact width. This method applies only to contacts that are fully confined to the spherical roller end and conical portion of the flange. It is possible that improper geometry or excessive skewing could cause the elastic contact ellipse to be truncated by the flange edge, undercut, or roller corner radius. Such a situation is not properly modeled by Hertz stress theory and should be avoided in design because high edge stresses and poor lubrication can result.

The case of a flat end roller and angled flange contact is less amenable to simple contact stress evaluation. The nature of the contact surface on the roller, being at or near the intersection of the corner radius and end flat, is difficult to model adequately. The notion of an "effective" roller radius based on an assumed blend radius between roller corner and end flat is suitable for approximate calculations. A more precise contact stress distribution can be obtained by using finite element stress analysis technique if necessary.

CLOSURE

The information presented in this chapter is sufficient to make a determination of the contact stress level and elastic deformations occurring in a statically loaded rolling bearing. The model of a statically loaded bearing is somewhat distorted by surface tangential stresses induced by rolling and lubricant action. However, under the effects of moderate to heavy loading, the contact stresses calculated herein are sufficiently accurate for the rotating bearing as well as the static bearing. The same is true with regard to the effect of "edge stresses" on roller load distribution and hence deformation. These stresses subtend a rather small area and therefore do not influence the overall elastic load-deformation characteristic. In any event, from the simplified analytical methods presented in this chapter, a level of loading can be calculated against which to check other bearings at the same or different loads. The methods for calculation of elastic contact deformation are also sufficiently accurate, and these can be used to compare rolling bearing stiffness against stiffness of other bearing types.

REFERENCES

5.1. H. Hertz, "On the Contact of Rigid Elastic Solids and on Hardness," *Miscellaneous Papers*, MacMillan, London, pp. 163–183 (1896).

5.2. S. Timoshenko and J. Goodier, *Theory of Elasticity*, 2nd ed., McGraw-Hill, New York, pp. 1–27, 213–244 (1951).

5.3. J. Boussinesq, *Compt. Rend.* **114,** 1465 (1892).

5.4. H. R. Thomas, and V. A. Hoersch, "Stresses Due to the Pressure of One Elastic Solid upon Another," *Univ. Illinois Bull.* **212** (July 15, 1930).

5.5. G. Lundberg, and H. Sjovall, *Stress and Deformation in Elastic Contacts*, Publication No. 4, The Institute of Theory of Elasticity and Strength of Materials, Chalmers University of Technology, Gothenburg (1958).

5.6. N. Belyayev, "Work in the Theory of Elasticity and Plasticity," State Press for Technical Theoretical Literature, Moscow (1957).

5.7. A. B. Jones, *Analysis of Stresses and Deflections*, New Departure Engineering Data, Bristol, CT, pp. 12–22 (1946).

5.8. G. Lundberg and A. Palmgren, "Dynamic Capacity of Rolling Bearings," *Acta Polytech.* No. 7, R.S.A.E.E. (1947).

5.9. K. L. Johnson, "The Effects of an Oscillating Tangential Force at the Interface between Elastic Bodies in Contact," Ph.D. Thesis, University of Manchester (1954).

5.10. J. O. Smith and C. K. Liu, "Stresses Due to Tangential and Normal Loads on an Elastic Solid with Application to Some Contact Stress Problems," *ASME* paper 52 A-13 (December 1952).

5.11. E. I. Radzimovsky, "Stress Distribution and Strength Condition of Two Rolling Cylinders Pressed Together," *Univ. Illinois Engineer. Exp. Station Bull.* Series No. 408 (February 1953).

5.12. C. K. Liu, "Stress and Deformations Due to Tangential and Normal Loads on an Elastic Solid with Application to Contact Stress," Ph. D. thesis, University of Illinois (June 1950).

5.13. C. Cattaneo, "A Theory of Second Order Elastic Contact," *Univ. Roma Rend. Mat. Appl.* **6,** 505–512 (1947).

5.14. T. T. Loo, "A Second Approximation Solution on the Elastic Contact Problem," *Sci. Sinica* **7,** 1235–1246 (1958).

5.15. H. Deresiewicz, "A Note on Second Order Hertz Contact," *ASME J. Appl. Mech.* 141–142 (March 1961).

5.16. J. Martin, S. Borgese, and A. Eberhardt, "Microstructural Alterations of Rolling Bearing Steel undergoing Cyclic Stressing," *ASME Preprint* 65-WA CF-4 (November 1965).

5.17. M. F. Spotts, *Design of Machine Elements*, 3rd ed., Prentice-Hall, Englewood Cliffs, pp. 85, 115 (1961).

5.18. O. Zwirlein and H. Schlicht, "Werkstoffanstrengung bei Wälzbeanspruchung—Einfluss von Reibung und Eigenspannungen," *Z. Werkstofftech.* **11,** 1–14 (1980).

5.19. R. Sayles, G. deSilva, J. Leather, J. Anderson, and P. MacPherson, "Elastic Conformity in Hertzian Contacts," *Tribol. Int.* 315–322 (1981).

5.20. J. J. Kalker, "Numerical Calculation of the Elastic Field in a Half-Space due to an Arbitrary Load Distributed over a Bounded Region of the Surface," SKF Engineering and Research Center report NL82D002 (Appendix) (June, 1982).

5.21. K. Kunert, "Spannungsverteilung im Halbraum bei Elliptischer Flächenpressungsverteilung über einer Rechteckigen Druckfläche," *Forsch. Geb. Ingenieurwes.* **27,**(6), 165–174 (1961).

5.22. H. Reusner, Druckflächenbelastung und Overflächenverschiebung in Wälzkontakt von Rotätionskörpern, Dissertation, Schweinfurt, West Germany (1977).

5.23. B. Fredriksson, "Three-Dimensional Roller-Raceway Contact Stress Analysis," Advanced Engineering Corp. Report, Linköping, Sweden (1980).

5.24. H. Reusner, "The Logarithmic Roller Profile—the Key to Superior Performance of Cylindrical and Taper Roller Bearings," *Ball Bearing J.* No. 230, SKF (June 1987).

6

DISTRIBUTION OF LOAD IN STATICALLY LOADED BEARINGS

LIST OF SYMBOLS

Symbol	Description	Units
A	Distance between raceway groove curvature centers	mm (in.)
B	$f_i + f_o - 1$, total curvature	
c	Crown drop	mm (in.)
C	Influence coefficient	mm/N (in./lb)
D	Ball or roller diameter	mm (in.)
d_m	Bearing pitch diameter	mm (in.)
e	Eccentricity of loading	mm (in.)
E	Modulus of elasticity	N/mm^2 (psi)
f	r/D	
F	Applied load	N (lb)
h	Roller thrust couple moment arm	mm (in.)
i	Number of rows of rolling elements	
I	Ring section moment of inertia	mm^4 ($in.^4$)
J_a	Axial load integral	
J_r	Radial load integral	
J_m	Moment load integral	

Symbol	Description	Units
k	Number of laminae	
K	Load-deflection factor; axial load-deflection factor	
l	Roller length	mm (in.)
L	Distance between rows	mm (in.)
$\mathfrak{M}$	Moment applied to bearing	N · mm (in. · lb)
M	Moment	N · mm (in. · lb)
n	Load-deflection exponent; $n = 1.5$ for point contact; $n = 1.11$ for line contact	
P_d	Diametral clearance	mm (in.)
q	Load per unit length	N/mm (lb/in.)
Q	Roller or ball load	N (lb)
$\mathcal{R}$	Ring radius to neutral axis, radius of locus of raceway groove curvature centers	mm (in.)
r	Raceway groove curvature radius	mm (in.)
s	Distance between loci of inner and outer raceway groove curvature centers	mm (in.)
U	Strain energy	N · mm (in. · lb)
u	Ring radial deflection	mm (in.)
Z	Number of balls or rollers per row	
α°	Free contact angle	rad,°
α	Mounted contact angle	rad,°
β	$\tan^{-1} l/(d_m - D)$	rad,°
γ	$D \cos \alpha/d_m$	
δ	Deflection or contact deformation	mm (in.)
δ_1	Distance between inner and outer rings	mm (in.)
Δ	Contact deformation under ideal loading (no misalignment or roller tilt)	mm (in.)
ϵ	Load distribution factor	
ζ	Roller tilt angle	rad,°
η	$\tan^{-1} D/l$	rad,°
θ	Misalignment angle	rad,°
λ	Laminum position	
φ	Position angle	rad,°
ψ	Azimuth (position) angle	rad,°
$\Sigma\rho$	Curvature sum	mm^{-1} ($in.^{-1}$)
$\Delta\psi$	Angular distance between rolling elements	rad,°
$\mathcal{E}$	Error on variable	
$\mathcal{G}$	Error function	

SUBSCRIPTS

a	Refers to axial direction	
i	Refers to inner raceway	

Symbol	Description	Units
	SUBSCRIPTS	
i	Refers to ring angular position	
j	Refers to ball or roller position	
k	Refers to ball or roller position	
l	Refers to line contact	
m	Refers to ball or roller position	
M	Refers to moment loading	
n	Refers to direction collinear with normal load	
o	Refers to outer raceway	
p	Refers to point contact	
r	Refers to radial direction	
R	Refers to ball or roller	
s	Refers to squeezing or gear separating load	
t	Refers to gear tangential load	
1, 2	Refers to bearing row	
ψ	Refers to angular location	

GENERAL

Having determined in Chapter 4 how each ball or roller in a bearing carries load, it is possible to determine how the bearing load is distributed among the balls or rollers. To do this it is first necessary to develop load-deflection relationships for rolling elements contacting raceways. By using Chapters 2 and 5 these load-deflection relationships can be developed for any type of rolling element contacting any type of raceway. Hence, the material presented in this chapter is completely dependent on the previous chapters, and a quick review might be advantageous at this point.

Most rolling bearing applications involve steady-state rotation of either the inner or outer raceway or both; however, the speeds of rotation are usually not so great as to cause ball or roller centrifugal forces or gyroscopic moments of magnitude large enough to affect significantly the distribution of applied load among the rolling elements. Moreover, in most applications the frictional forces and moments acting on the rolling elements also should not significantly influence this load distribution. Consequently, in analyzing the distribution of rolling element loads, it is usually satisfactory to ignore these effects in most applications. In this chapter the load distribution of statically loaded ball and roller bearings will be investigated.

LOAD-DEFLECTION RELATIONSHIPS

From equation (5.38) it can be seen that for a given ball-raceway contact (point loading)

$$\delta \sim Q^{2/3} \tag{6.1}$$

Inverting equation (6.1) and expressing it in equation format yields

$$Q = K_{\mathrm{p}}\delta^{3/2} \tag{6.2}$$

Similarly, for a given roller-raceway contact (line contact)

$$Q = K_{\mathrm{l}}\delta^{10/9} \tag{6.3}$$

In general then

$$Q = K\delta^{n} \tag{6.4}$$

in which $n = 1.5$ for ball bearings and $n = 1.11$ for roller bearings.

The total normal approach between two raceways under load separated by a rolling element is the sum of the approaches between the rolling element and each raceway. Hence

$$\delta_{\mathrm{n}} = \delta_{\mathrm{i}} + \delta_{\mathrm{o}} \tag{6.5}$$

Therefore,

$$K_{\mathrm{n}} = \left[\frac{1}{(1/K_{\mathrm{i}})^{1/n} + (1/K_{\mathrm{o}})^{1/n}}\right]^{n} \tag{6.6}$$

and

$$Q = K_{\mathrm{n}}\delta^{n} \tag{6.7}$$

For steel ball-steel raceway contact,

$$K_{\mathrm{p}} = 2.15 \times 10^{5} \sum \rho^{-1/2}(\delta^{*})^{-3/2} \tag{6.8}$$

Similarly, for steel roller and raceway contact,

$$K_{\mathrm{l}} = 7.86 \times 10^{4}\, l^{8/9} \tag{6.9}$$

BEARINGS UNDER RADIAL LOAD

For a rigidly supported bearing subjected to radial load, the radial deflection at any rolling element angular position is given by

$$\delta_\psi = \delta_r \cos \psi - \tfrac{1}{2}P_d \tag{6.10}$$

in which δ_r is the ring radial shift, occurring at $\psi = 0$ and P_d is the diametral clearance. Figure 6.1 illustrates a radial bearing with clear-

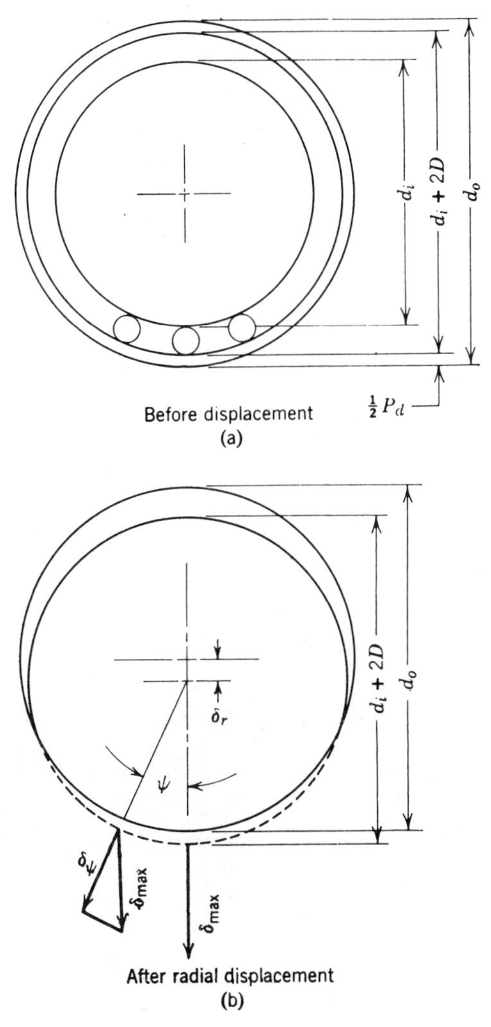

FIGURE 6.1. Bearing rings.

ance. Equation (6.10) may be rearranged in terms of maximum deformation as follows:

$$\delta_\psi = \delta_{max}\left[1 - \frac{1}{2\epsilon}(1 - \cos\psi)\right] \tag{6.11}$$

in which

$$\epsilon = \frac{1}{2}\left(1 - \frac{P_d}{2\delta_r}\right) \tag{6.12}$$

It is clear from equation (6.12) that the angular extent of the load zone is determined by the diametral clearance such that

$$\psi_1 = \cos^{-1}\left(\frac{P_d}{2\delta_r}\right) \tag{6.13}$$

For zero clearance, $\psi_1 = 90°$.

From equation (6.4),

$$\frac{Q_\psi}{Q_{max}} = \left(\frac{\delta_\psi}{\delta_{max}}\right)^n \tag{6.14}$$

Therefore, from (6.11) and (6.14),

$$Q_\psi = Q_{max}\left[1 - \frac{1}{2\epsilon}(1 - \cos\psi)\right]^n \tag{6.15}$$

For static equilibrium to exist, the applied radial load must equal the sum of the vertical components of the rolling element loads:

$$F_r = \sum_{\psi=0}^{\psi=\pm\psi_1} Q_\psi \cos\psi \tag{6.16}$$

or

$$F_r = Q_{max} \sum_{\psi=0}^{\psi=\pm\psi_1} \left[1 - \frac{1}{2\epsilon}(1 - \cos\psi)\right]^n \cos\psi \tag{6.17}$$

Equation (6.17) can also be written in integral form:

$$F_r = ZQ_{max} \times \frac{1}{2\pi}\int_{-\psi_1}^{+\psi_1}\left[1 - \frac{1}{2\epsilon}(1 - \cos\psi)\right]^n \cos\psi \, d\psi \tag{6.18}$$

TABLE 6.1. Load Distribution Integral $J_r(\epsilon)$

ϵ	Point Contact	Line Contact	ϵ	Point Contact	Line Contact
0	1/Z	1/Z	0.8	0.2559	0.2658
0.1	0.1156	0.1268	0.9	0.2576	0.2628
0.2	0.1590	0.1737	1.0	0.2546	0.2523
0.3	0.1892	0.2055	1.25	0.2289	0.2078
0.4	0.2117	0.2286	1.67	0.1871	0.1589
0.5	0.2288	0.2453	2.5	0.1339	0.1075
0.6	0.2416	0.2568	5.0	0.0711	0.0544
0.7	0.2505	0.2636	∞	0	0

or

$$F_r = ZQ_{max}J_r(\epsilon) \tag{6.19}$$

in which

$$J_r(\epsilon) = \frac{1}{2\pi}\int_{-\psi_l}^{+\psi_l}\left[1 - \frac{1}{2\epsilon}(1 - \cos\psi)\right]^n \cos\psi \, d\psi \tag{6.20}$$

The radial integral of equation (6.20) has been evaluated numerically for various values of ϵ. This is given in Table 6.1.

From equation (6.7),

$$Q_{max} = K_n\delta^n_{\psi=0} = K_n(\delta_r - \tfrac{1}{2}P_d)^n \tag{6.21}$$

Therefore,

$$F_r = ZK_n(\delta_r - \tfrac{1}{2}P_d)^n J_r(\epsilon) \tag{6.22}$$

For a given bearing with a given clearance under a given load, equation (6.22) may be solved by trial and error. A value of δ_r is first assumed and ϵ is calculated from equation (6.12). This yields $J_r(\epsilon)$ from Table 6.1. If equation (6.22) does not then balance, the process is repeated.

Figure 6.2 gives values of J_r vs ϵ.

Figure 6.3 shows radial load distribution for various values of ϵ. Here, ϵ is the ratio of the projected load zone on the bearing diameter.

For ball bearings under pure radial load and zero clearance Stribeck [6.1] concluded that

$$Q_{max} = \frac{4.37F_r}{Z\cos\alpha} \tag{6.23}$$

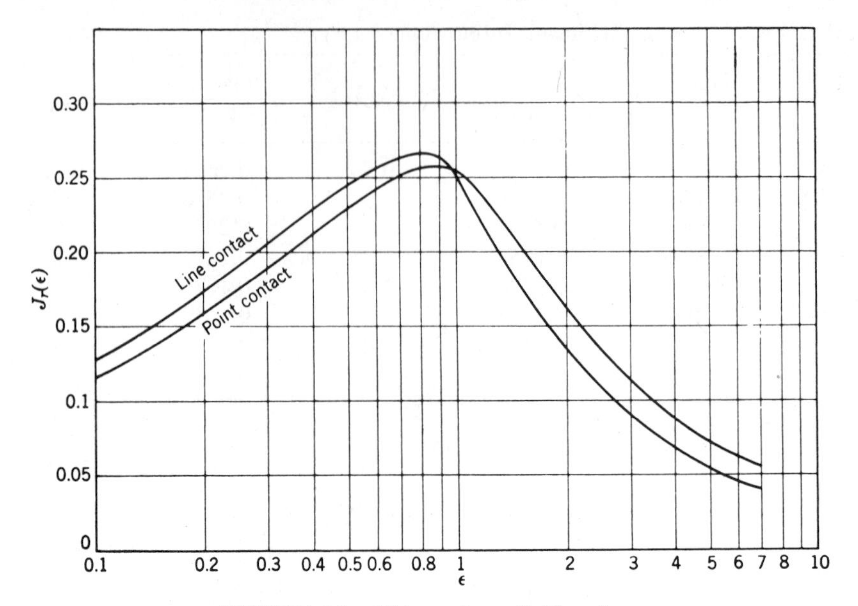

FIGURE 6.2. $J_r(\epsilon)$ vs ϵ for radial bearings.

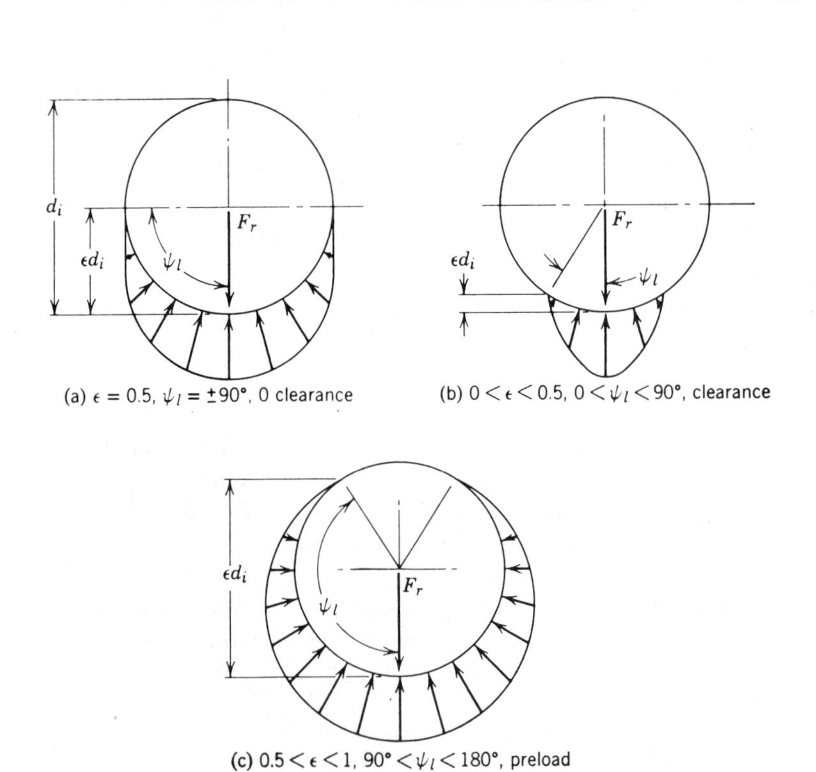

FIGURE 6.3. Rolling element load distribution for different amounts of clearance.

Accounting for nominal diametral clearance in the bearing, one may use the following approximation.

$$Q_{max} = \frac{5F_r}{Z \cos \alpha} \tag{6.24}$$

Example 6.1. The 209 radial ball bearing of Example 2.1 experiences a radial load of 8900 N (2000 lb). Determine the loading at each ball location.

$$Z = 9 \qquad \text{Ex. 2.1}$$

$$P_d = 0.0150 \text{ mm } (0.0006 \text{ in.}) \qquad \text{Ex. 2.1}$$

$$d_m = 65 \text{ mm } (2.559 \text{ in.}) \qquad \text{Ex. 2.1}$$

$$D = 12.7 \text{ mm } (0.5 \text{ in.}) \qquad \text{Ex. 2.1}$$

$$f_i = f_o = 0.52 \qquad \text{Ex. 2.2}$$

$$\gamma = 0.1954 \qquad \text{Ex. 2.5}$$

$$\sum \rho_i = 0.202 \text{ mm}^{-1} (5.126 \text{ in.}^{-1}) \qquad \text{Ex. 2.5}$$

$$F(\rho)_i = 0.9399 \qquad \text{Ex. 2.5}$$

$$\sum \rho_o = 0.138 \text{ mm}^{-1} (3.500 \text{ in.}^{-1}) \qquad \text{Ex. 2.5}$$

$$F(\rho)_o = 0.9120 \qquad \text{Ex. 2.5}$$

From Fig. 5.4,

$$\delta_i^* = 0.602, \qquad \delta_o^* = 0.658$$

$$K_{pi} = 2.15 \times 10^5 \sum \rho_i^{-1/2} (\delta_i^*)^{-3/2} \tag{6.8}$$

$$= 2.15 \times 10^5 \times (0.202)^{-1/2} (0.602)^{-3/2}$$

$$= 1.026 \times 10^6 \text{ N/mm}^{1.5} (2.951 \times 10^7 \text{ lb/in.}^{1.5})$$

$$K_{po} = 2.15 \times 10^5 \sum \rho_o^{-1/2} \delta_o^*)^{-3/2} \tag{6.8}$$

$$= 2.15 \times 10^5 (0.138)^{-1/2} (0.658)^{-3/2}$$

$$= 1.089 \times 10^6 \text{ N/mm}^{1.5} (3.132 \times 10^7 \text{ lb/in.}^{1.5})$$

$$K_n = \left[\frac{1}{(1/K_i)^{1/1.5} + (1/K_o)^{1/1.5}} \right]^{1.5} \tag{6.6}$$

$$= \left[\frac{1}{(1/1.026)^{0.667} + (1/1.089)^{0.667}} \right]^{1.5} \times 10^6$$

$$= 3.735 \times 10^5 \text{ N/mm}^{1.5} (1.075 \times 10^7 \text{ lb/in.}^{1.5})$$

$$F_r = ZK_n(\delta_r - \tfrac{1}{2}P_d)^{1.5} J_r(\epsilon) \tag{6.22}$$

$$8900 = 9 \times 3.735 \times 10^5 \left(\delta_r - \frac{0.0150}{2} \right)^{1.5} J_r(\epsilon)$$

$$(\delta_r - 0.0075)^{1.5} J_r(\epsilon) = 0.002654 \quad \text{(a)}$$

$$\epsilon = \frac{1}{2}\left(1 - \frac{P_d}{2\delta_r}\right) \quad (6.12)$$

$$= \frac{1}{2}\left(1 - \frac{0.015}{2\delta_r}\right)$$

$$= 0.5 - \frac{0.00375}{\delta_r} \quad \text{(b)}$$

Equations (a) and (b) may be solved by trial and error or iteration.

1. Assume $\epsilon = 0.4$
   From Fig. 6.2, $J_r(0.4) = 0.212$
2. $(\delta_r - 0.0075)^{1.5} \times 0.212 = 0.002654$ (a)
   $\delta_r = 0.0614$ mm (0.002418 in.)
3. $\epsilon = 0.5 - (0.00375/0.06141) = 0.439$ (b)
4. $0.439 \neq 0.400$ (assumed in step 1)
5. Assume $\epsilon = 0.435$
   From Fig. 6.2, $J_r(0.435) = 0.218$
6. $(\delta_r - 0.0075)^{1.5} \times 0.218 = 0.002654$ (a)
   $\delta_r = 0.06041$ mm (0.002379 in.)
7. $\epsilon = 0.5 - (0.00375/0.06041) = 0.437$ (b)
8. $0.437 \neq 0.435$, and so on

$$\delta_r = 0.06041 \text{ mm } (0.002379 \text{ in.})$$

$$\epsilon = 0.438$$

$$J_r(0.434) = 0.218$$

$$F_r = ZQ_{max} J_r(\epsilon) \quad (6.19)$$

$$8900 = 9 \times Q_{max} \times 0.218$$

$$Q_{max} = 4536 \text{ N } (1019 \text{ lb})$$

$$Q_\psi = Q_{max}\left[1 - \frac{1}{2\epsilon}(1 - \cos\psi)\right]^{1.5} \quad (6.15)$$

$$= 4536\left[1 - \frac{1}{2 \times 0.438}(1 - \cos\psi)\right]^{1.5}$$

$$= 4536\,(1.142\cos\psi - 0.142)^{1.5}$$

$$\Delta\psi = \frac{360°}{Z} = \frac{360°}{9} = 40°$$

ψ	$\cos\psi$	Q_ψ (N)
0	1	4536 (1019 lb)
±40	0.7660	2846 (638.6 lb)
±80	0.1737	61 (13.7 lb)
±120	−0.5000	0
±160	−0.9397	0

Example 6.2. By using Stribeck's equation (6.23), determine the load distribution for the 209 radial ball bearing of the preceding example.

$$Q_{\max} = \frac{4.37F_r}{Z\cos\alpha} \tag{6.23}$$

$$= \frac{4.37 \times 8900}{9 \times \cos(0°)}$$

$$= 4321 \text{ N } (971.2 \text{ lb})$$

Corresponding to equation (6.23), $\epsilon = 0.5$

$$Q_\psi = Q_{\max}\left[1 - \frac{1}{2\epsilon}(1 - \cos\psi)\right]^{1.5} \tag{6.15}$$

$$= 4321\left[1 - \frac{1}{2 \times 0.5}(1 - \cos\psi)\right]^{1.5} = 4321\cos^{1.5}\psi$$

ψ	Q_ψ (N)
0°	4321 (971.2 lb)
±40°	2897 (650.7 lb)
±80°	313 (70.1 lb)
±120°	0
±160°	0

For radial roller bearings with zero clearance under pure radial load, it can also be determined that

$$Q_{\max} = \frac{4.08F_r}{Z\cos\alpha} \tag{6.25}$$

Equation (6.24) is also a valid approximation for radial roller bearings with nominal clearance.

Example 6.3. The 209 cylindrical roller bearing of Example 2.7 supports a radial load of 4450 N (1000 lb). Determine the loading at each roller location and the extent of the load zone.

$$Z = 14 \qquad \text{Ex. 2.7}$$

$$d_m = 65 \text{ mm } (2.559 \text{ in.}) \qquad \text{Ex. 2.7}$$

$$P_d = 0.041 \text{ mm } (0.0016 \text{ in.}) \qquad \text{Ex. 2.7}$$

$$l = 9.6 \text{ mm } (0.3780 \text{ in.}) \qquad \text{Ex. 2.7}$$

$$K_l = 7.86 \times 10^4 l^{8/9} \qquad (6.9)$$

$$= 7.86 \times 10^4 (9.6)^{8/9}$$

$$= 5.869 \times 10^5 \text{ N/mm}^{1.11} \; (4.799 \times 10^6 \text{ lb/in.}^{1.11})$$

$$K_n = \left[\frac{1}{\left(\frac{1}{K_i}\right)^{1/1.11} + \left(\frac{1}{K_o}\right)^{1/1.11}} \right]^{1.11} \qquad (6.6)$$

$$= (0.5)^{1.11} \times 5.869 \times 10^5$$

$$= 2.720 \times 10^5 \text{ N/mm}^{1.11} \; (2.222 \times 10^6 \text{ lb/in.}^{1.11})$$

$$F_r = ZK_n(\delta_r - \tfrac{1}{2}P_d)^{1.11} J_r(\epsilon) \qquad (6.22)$$

$$4450 = 14 \times 2.720 \times 10^5 \left(\delta_r - \frac{0.041}{2}\right)^{1.11} J_r(\epsilon)$$

$$(\delta_r - 0.0205)^{1.11} J_r(\epsilon) = 0.001169 \qquad \text{(a)}$$

$$\epsilon = \frac{1}{2}\left(1 - \frac{P_d}{2\delta_r}\right) \qquad (6.12)$$

$$= \frac{1}{2}\left(1 - \frac{0.041}{2\delta_r}\right) = 0.5 - \frac{0.01025}{\delta_r} \qquad \text{(b)}$$

1. Assume $\epsilon = 0.3$
From Fig. 6.2, $J_r(0.3) = 0.206$

2. $(\delta_r - 0.0205)^{1.11} \times 0.206 = 0.001169$ (a)
$\delta_r = 0.03002$ mm (0.00118 in.)

3. $\epsilon = 0.5 - \dfrac{0.01025}{0.03002} = 0.159$ (b)

4. $0.159 \neq 0.3$ (assumed in step 1)

5. Assume $\epsilon = 0.158$
From Fig. 6.2, $J_r(0.158) = 0.154$

6. $(\delta_r - 0.0205)^{1.11} \times 0.154 = 0.001169$ (a)
$\delta_r = 0.03287$ mm (0.001294 in.)

7. $\epsilon = 0.5 - \dfrac{0.01025}{0.03287} = 0.1882$ (b)

8. $0.1882 \neq 0.156$, and so on

$$\delta_r = 0.0320 \text{ mm } (0.00126 \text{ in.})$$

$$\epsilon = 0.1824$$

$$J_r(0.1824) = 0.165$$

$$F_r = ZQ_{max}J_r(\epsilon) \tag{6.19}$$

$$4450 = 14 \times Q_{max} \times 0.165$$

$$Q_{max} = 1926 \text{ N } (432.8 \text{ lb})$$

$$Q_\psi = Q_{max}\left[1 - \frac{1}{2\epsilon}(1 - \cos\psi)\right]^{1.11} \tag{6.15}$$

$$= 1926\left[1 - \frac{1}{2 \times 0.1824}(1 - \cos\psi)\right]^{1.11}$$

$$= 1926\,(2.741\cos\psi - 1.741)^{1.11}$$

$$\Delta\psi = \frac{360°}{Z} = \frac{360°}{14} = 25.71°$$

ψ	$\cos\psi$	Q_ψ (N)
0°	1	1926 (432.8 lb)
25.71°	0.9010	1355 (304.7 lb)
51.42°	0.6237	0
77.13°	0.2227	0
102.84°	−0.2227	0
128.55°	−0.6237	0
154.26°	−0.9010	0
180°	−1	0

$$\psi_1 = \cos^{-1}\left(\frac{P_d}{2\delta_r}\right) \tag{6.13}$$

$$= \cos^{-1}\left(\frac{0.041}{2 \times 0.0320}\right) = \pm 50°\ 10'$$

Example 6.4. By using equation (6.24) to determine Q_{max} evaluate the load distribution which occurs for the 209 cylindrical roller bearing of the preceding example.

$$Q_{max} = \frac{5F_r}{Z \cos \alpha} \tag{6.24}$$

$$= \frac{5 \times 4450}{14 \times \cos (0°)} = 1589 \text{ N (357.1 lb)}$$

$$F_r = ZQ_{max}J_r(\epsilon) \tag{6.19}$$

$$4450 = 14 \times 1589 \times J_r(\epsilon)$$

$$J_r(\epsilon) = 0.2000$$

From Fig. 6.2, $\epsilon = 0.28$

$$Q_\psi = Q_{max}\left[1 - \frac{1}{2\epsilon}(1 - \cos \psi)\right]^{1.11} \tag{6.15}$$

$$= 1589\left[1 - \frac{1}{2 \times 0.28}(1 - \cos \psi)\right]^{1.11}$$

$$= 1589\,(1.786 \cos \psi - 0.786)^{1.11}$$

ψ	Q_ψ (N)
0	1589 (357.1 lb)
25.71°	1280 (287.6 lb)
51.42°	461 (103.6 lb)
77.13°	0
102.84°	0
128.55°	0
154.26°	0
180°	0

$$\psi_1 = \cos^{-1}\left(\frac{P_d}{2\delta_r}\right) \tag{6.13}$$

since

$$\epsilon = \frac{1}{2}\left(1 - \frac{P_d}{2\delta_r}\right) \tag{6.12}$$

$$\psi_1 = \cos^{-1}(1 - 2\epsilon)$$

$$= \cos^{-1}(1 - 2 \times 0.28) = \pm 63°\ 54'$$

For lightly loaded rolling bearings, the approximation of equation (6.24) is not adequate to determine maximum rolling element load and it should not be used in that instance.

BEARINGS UNDER THRUST LOAD

Centric Thrust Load

Thrust ball and roller bearings subjected to a centric thrust load have the load distributed equally among the rolling elements. Hence

$$Q = \frac{F_a}{Z \sin \alpha} \tag{6.26}$$

In equation (6.26), α is the contact angle that occurs in the loaded bearing. For thrust ball bearings whose contact angles are nominally less than 90°, the contact angle in the loaded bearing is greater than the initial contact angle α° that occurs in the nonloaded bearing. This phenomenon is discussed in detail in the following paragraphs.

Angular-Contact Ball Bearings

In the absence of centrifugal loading, the contact angles at inner and outer raceways are identical; however, they are greater than those in the unloaded condition. In the unloaded condition, contact angle is defined by

$$\cos \alpha^\circ = 1 - \frac{P_d}{2BD} \tag{6.27}$$

in which P_d is the mounted diametral clearance. A thrust load F_a applied to the inner ring as shown in Fig. 6.4 causes an axial deflection δ_a. This axial deflection is a component of a normal deflection along the line of contact such that from Fig. 6.4,

$$\delta_n = BD\left(\frac{\cos \alpha^\circ}{\cos \alpha} - 1\right) \tag{6.28}$$

Since $Q = K_n \delta_n^{1.5}$,

$$Q = K_n (BD)^{1.5}\left(\frac{\cos \alpha^\circ}{\cos \alpha} - 1\right)^{1.5} \tag{6.29}$$

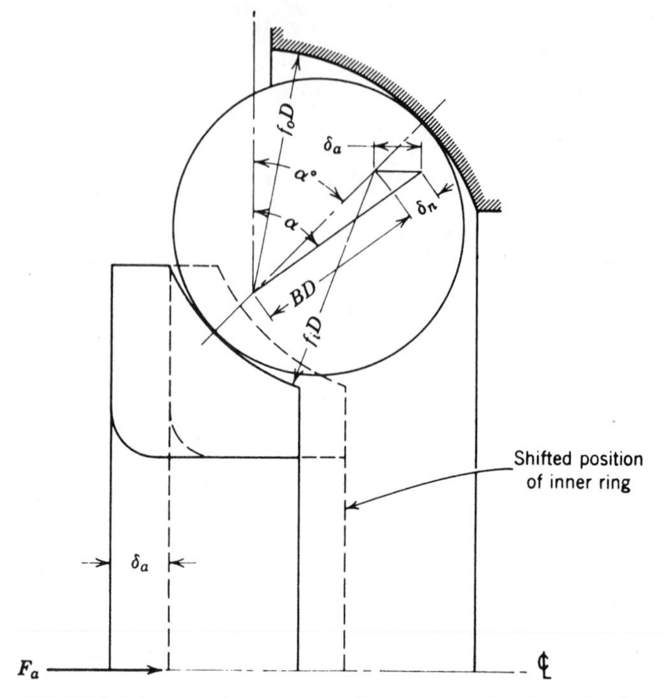

FIGURE 6.4. Angular-contact ball bearing under thrust load.

Substitution of equation (6.26) into (6.29) yields

$$\frac{F_a}{ZK_n(BD)^{1.5}} = \sin\alpha\left(\frac{\cos\alpha^\circ}{\cos\alpha} - 1\right)^{1.5} \tag{6.30}$$

Since K_n is a function of the final contact angle α, equation (6.30) must be solved by trial and error to yield an exact solution for α. Jones [6.2], however, defines an axial deflection constant K as follows:

$$K = \frac{B}{g(+\gamma) + g(-\gamma)} \tag{6.31}$$

in which $\gamma = (D\cos\alpha)/d_m$ and $g(+\gamma)$ refers to the inner raceway and $g(-\gamma)$ refers to the outer raceway. Jones [6.2] further indicates that the sum of $g(+\gamma)$ and $g(-\gamma)$ remains virtually constant for all contact angles being dependent only on total curvature B. The axial deflection constant K is related to K_n as follows:

$$K_n = \frac{KD^{0.5}}{B^{1.5}} \tag{6.32}$$

Hence,

$$\frac{F_a}{ZD^2K} = \sin\alpha \left(\frac{\cos\alpha^\circ}{\cos\alpha} - 1\right)^{1.5} \tag{6.33}$$

Taking K from Fig. 6.5, equation (6.33) may be solved numerically by the Newton–Raphson method. The iterative equation to be satisfied is

$$\alpha' = \alpha + \frac{\dfrac{F_a}{ZD^2K} - \sin\alpha\left(\dfrac{\cos\alpha^\circ}{\cos\alpha} - 1\right)^{1.5}}{\cos\alpha\left(\dfrac{\cos\alpha^\circ}{\cos\alpha} - 1\right)^{1.5} + 1.5\tan^2\alpha\left(\dfrac{\cos\alpha^\circ}{\cos\alpha} - 1\right)^{0.5}\cos\alpha^\circ} \tag{6.34}$$

Equation (6.34) is satisfied when $\alpha' - \alpha$ is essentially zero.

The axial deflection δ_a corresponding to δ_n may also be determined

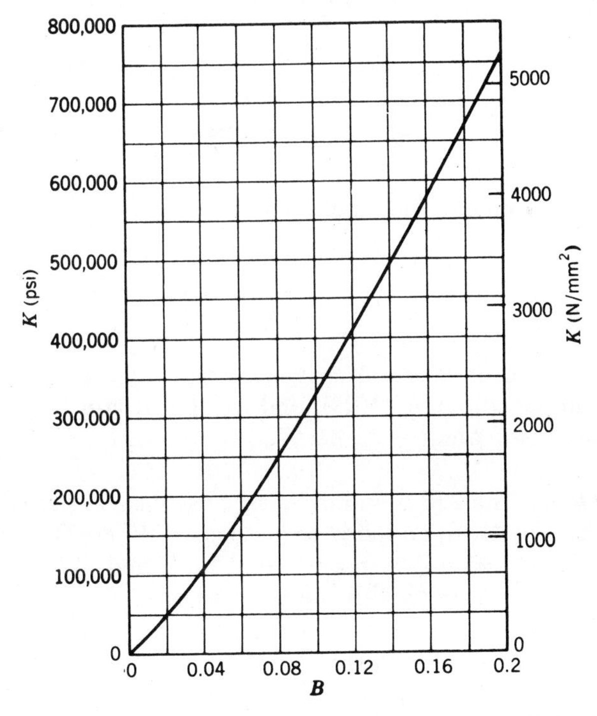

FIGURE 6.5. Axial deflection constant K vs total curvature B for ball bearings. $B = f_o + f_i - 1$, $f = r/D$ [6.2]).

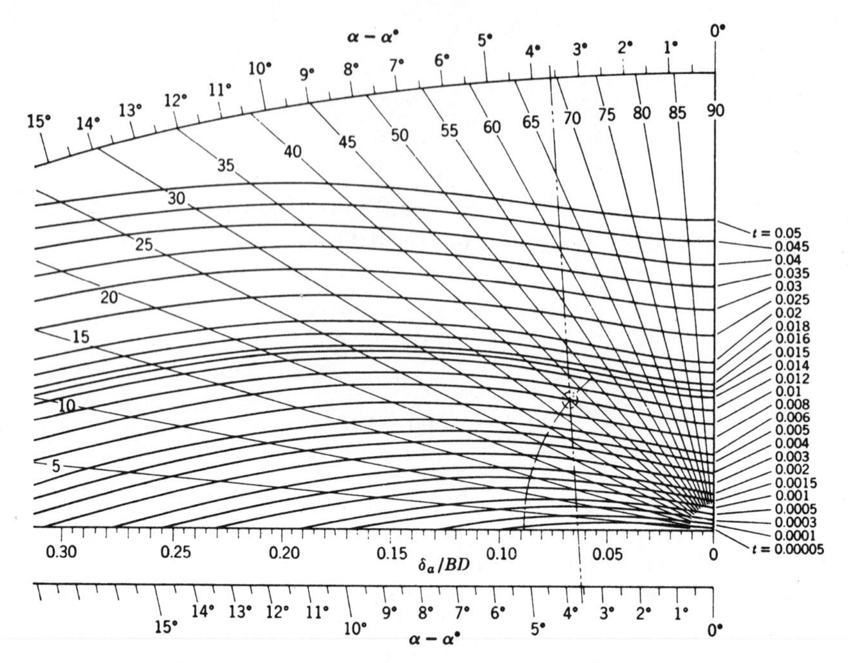

FIGURE 6.6. δ_a/BD and $\alpha = \alpha°$ vs $t = F_a/ZD^2K$ and $\alpha°$.

from Fig. 6.6 as follows:

$$\delta_a = (BD + \delta_n)\sin\alpha - BD\sin\alpha° \tag{6.35}$$

Substituting δ_n from equation (6.28) yields

$$\delta_a = \frac{BD\sin(\alpha - \alpha°)}{\cos\alpha} \tag{6.36}$$

Figure 6.6 presents a series of curves for the rapid calculation of the change in contact angle $(\alpha - \alpha°)$, and axial deflection as functions of initial contact angle and $t = F_a/ZD^2K$.

Example 6.5. The 218 angular-contact ball bearing of Example 2.3 experiences a statically applied thrust load of 17,800 N (4000 lb). Determine the contact angle, normal ball load, and axial deflection of the bearing considering a ball complement of 16.

$B = 0.0464$ — Ex. 2.3

$\alpha° = 40°$ — Ex. 2.3

$D = 22.23$ mm (0.875 in.) — Ex. 2.3

From Fig. 6.5, $K = 896.7$ N/mm^2 (130,000 psi)

$$\frac{F_a}{ZD^2K} = \sin\alpha\left(\frac{\cos\alpha^\circ}{\cos\alpha} - 1\right)^{1.5} \tag{6.33}$$

$$\frac{17800}{16(22.23)^2 \times 896.7} = \sin\alpha\left(\frac{\cos 40^\circ}{\cos\alpha} - 1\right)^{1.5}$$

$$0.002512 = \sin\alpha\left(\frac{0.7660}{\cos\alpha} - 1\right)^{1.5} \tag{a}$$

$$\alpha' = \alpha + \frac{\dfrac{F_a}{ZD^2K} - \sin\alpha\left(\dfrac{\cos\alpha^\circ}{\cos\alpha} - 1\right)^{1.5}}{\cos\alpha\left(\dfrac{\cos\alpha^\circ}{\cos\alpha} - 1\right)^{1.5} + 1.5\tan^2\alpha\left(\dfrac{\cos\alpha^\circ}{\cos\alpha} - 1\right)^{0.5}\cos\alpha^\circ} \tag{6.34}$$

$$= \alpha + \frac{0.002512 - \sin\alpha\left(\dfrac{0.7660}{\cos\alpha} - 1\right)^{1.5}}{\cos\alpha\left(\dfrac{0.7660}{\cos\alpha} - 1\right)^{1.5} + 1.149\tan^2\alpha\left(\dfrac{0.7660}{\cos\alpha} - 1\right)^{0.5}} \tag{b}$$

1. Assume $\alpha = 40.5^\circ$ or $\alpha = 0.7069$ rad,

$$\sin\alpha = 0.6495$$
$$\cos\alpha = 0.7604$$
$$\tan\alpha = 0.8541$$

2. Then we have

$$\alpha' = 0.7069 + \frac{0.002512 - 0.6495\left(\dfrac{0.7660}{0.7604} - 1\right)^{1.5}}{0.7604\left(\dfrac{0.7660}{0.7604} - 1\right)^{1.5} + 1.149(0.8541)^2\left(\dfrac{0.7660}{0.7604} - 1\right)^{0.5}} \tag{b}$$

$$= 0.7069 + \frac{-0.002103}{+0.0715} = 0.7363 \text{ rad}$$

3. Assume $\alpha = 0.7363$ rad $(\alpha = 42.19°)$

$$\cos\alpha = 0.7410$$
$$\sin\alpha = 0.6715$$
$$\tan\alpha = 0.9062$$

4. Then

$$\alpha' = 0.7363 + \frac{0.002512 - 0.6715\left(\dfrac{0.7660}{0.7410} - 1\right)^{1.5}}{0.7410\left(\dfrac{0.7660}{0.7410} - 1\right)^{1.5} + 1.149(0.9062)^2\left(\dfrac{0.7660}{0.7410} - 1\right)^{0.5}} \quad \text{(b)}$$

$$= 0.7363 + \frac{-0.00162}{+0.1777} = 0.7272 \text{ rad}$$

5. Assume $\alpha = 0.7272$ rad $(41.66°)$

$$\cos\alpha = 0.7470$$
$$\sin\alpha = 0.6648$$
$$\tan\alpha = 0.8899$$

$$\alpha' = 0.7272 + \frac{0.002512 - 0.6648\left(\dfrac{0.7660}{0.7470} - 1\right)^{1.5}}{0.7470\left(\dfrac{0.7660}{0.7470} - 1\right)^{1.5} + 1.149(0.8899)^2\left(\dfrac{0.7660}{0.7470} - 1\right)^{0.5}} \quad \text{(b)}$$

$$= 0.7272 + \frac{-0.00018}{+0.1476} = 0.7260 \text{ rad} \quad (41° \; 36')$$

This result is sufficiently accurate for the illustrative purpose intended here. A similar result could have been obtained by using Fig. 6.6 at $t = 0.0025$ and $\alpha° = 40°$.

$$Q = \frac{F_a}{Z\sin\alpha} \quad (6.26)$$

$$= \frac{17800}{16 \times \sin(41° \; 36')} = 1676 \text{ N } (376.6 \text{ lb})$$

$$\delta_a = \frac{BD \sin(\alpha - \alpha^\circ)}{\cos \alpha} \tag{6.36}$$

$$= \frac{0.0464 \times 22.23 \sin(1^\circ\ 36')}{\cos(41^\circ\ 36')} = 0.0386 \text{ mm } (0.00152 \text{ in.})$$

Eccentric Thrust Load

Single Direction Bearings. Figure 6.7 illustrates a single-row thrust bearing subjected to an eccentric thrust load. Taking $\psi = 0$ as the position of the maximum loaded rolling element, then

$$\delta_\psi = \delta_a + \tfrac{1}{2}\theta\, d_m \cos \psi \tag{6.37}$$

Also,

$$\delta_{max} = \delta_a + \tfrac{1}{2}\theta\, d_m \tag{6.38}$$

From equations (6.37) and (6.38), one may develop the familiar relationship

$$\delta_\psi = \delta_{max}\left[1 - \frac{1}{2\epsilon}(1 - \cos \psi)\right] \tag{6.39}$$

in which

$$\epsilon = \frac{1}{2}\left(1 + \frac{2\delta_a}{\theta d_m}\right) \tag{6.40}$$

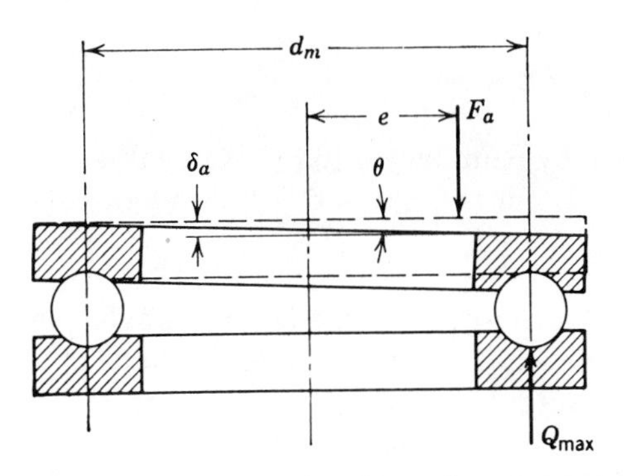

FIGURE 6.7. 90° ball thrust bearing under an eccentric load.

The extent of the zone of loading is defined by

$$\psi_1 = \cos^{-1}\left(\frac{-2\delta_a}{\theta d_m}\right) \tag{6.41}$$

As before,

$$Q_\psi = Q_{max}\left[1 - \frac{1}{2\epsilon}(1 - \cos\psi)\right]^n \tag{6.42}$$

The laws of static equilibrium require that

$$F_a = \sum_{\psi=0}^{\psi=\pm\pi} Q_\psi \sin\alpha \tag{6.43}$$

$$\mathfrak{M} = eF_a = \sum_{\psi=0}^{\psi=\pm\pi} \tfrac{1}{2}Q_\psi d_m \sin\alpha\cos\psi \tag{6.44}$$

Equations (6.43) and (6.44) may also be written in terms of thrust and moment integrals as follows:

$$F_a = ZQ_{max}J_a(\epsilon)\sin\alpha \tag{6.45}$$

in which

$$J_a(\epsilon) = \frac{1}{2\pi}\int_{-\psi_1}^{+\psi_1}\left[1 - \frac{1}{2\epsilon}(1 - \cos\psi)\right]^n d\psi \tag{6.46}$$

$$\mathfrak{M} = eF_a = \tfrac{1}{2}Q_{max}d_m J_m(\epsilon)\sin\alpha \tag{6.47}$$

in which

$$J_m(\epsilon) = \frac{1}{2\pi}\int_{-\psi_1}^{+\psi_1}\left[1 - \frac{1}{2\epsilon}(1 - \cos\psi)\right]^n \cos\psi\, d\psi \tag{6.48}$$

Table 6.2 as shown by Rumbarger [6.3] gives values of $J_a(\epsilon)$ and $J_m(\epsilon)$ as functions of $2e/d_m$. Figures 6.8 and 6.9 yield identical data in graphical format.

Example 6.6. Assuming that the contact angle remains constant at 41°36′, determine what the magnitude would be of the maximum ball load in the 218 angular-contact ball bearing of Example 6.5 if the 17,800 N (4000 lb) thrust load was applied at a point 50.8 mm (2 in.) distant from the bearing's axis of rotation.

TABLE 6.2. $J_a(\epsilon)$ and $J_m(\epsilon)$ for Single-Row Thrust Bearings

	Point Contact			Line Contact		
ϵ	$\frac{2e}{d_m}$	$J_m(\epsilon)$	$J_a(\epsilon)$	$\frac{2e}{d_m}$	$J_m(\epsilon)$	$J_a(\epsilon)$
0	1.0000	$1/z$	$1/z$	1.0000	$1/z$	$1/z$
0.1	0.9663	0.1156	0.1196	0.9613	0.1268	0.1319
0.2	0.9318	0.159	0.1707	0.9215	0.1737	0.1885
0.3	0.8964	0.1892	0.2110	0.8805	0.2055	0.2334
0.4	0.8601	0.2117	0.2462	0.8380	0.2286	0.2728
0.5	0.8225	0.2288	0.2782	0.7939	0.2453	0.3090
0.6	0.7835	0.2416	0.3084	0.7480	0.2568	0.3433
0.7	0.7427	0.2505	0.3374	0.6999	0.2636	0.3766
0.8	0.6995	0.2559	0.3658	0.6486	0.2658	0.4098
0.9	0.6529	0.2576	0.3945	0.5920	0.2628	0.4439
1.0	0.6000	0.2546	0.4244	0.5238	0.2523	0.4817
1.25	0.4338	0.2289	0.5044	0.3598	0.2078	0.5775
1.67	0.3088	0.1871	0.6060	0.2340	0.1589	0.6790
2.5	0.1850	0.1339	0.7240	0.1372	0.1075	0.7837
5.0	0.0831	0.0711	0.8558	0.0611	0.0544	0.8909
∞	0	0	1.0000	0	0	1.0000

$$d_m = 125.3 \text{ mm } (4.932 \text{ in.}) \qquad \text{Ex. 2.6}$$

$$\frac{2e}{d_m} = \frac{2 \times 50.8}{125.3} = 0.8110$$

From Fig. 6.8, $J_a = 0.285$, $J_m = 0.233$, $\epsilon = 0.525$

$$Z = 16 \qquad \text{Ex. 6.5}$$

$$F_a = ZQ_{max}J_a(\epsilon) \sin \alpha \qquad (6.45)$$

$$17800 = 16 \times Q_{max} \times 0.285 \times \sin (41°36')$$

$$Q_{max} = 5878 \text{ N } (1321 \text{ lb})$$

An identical result is obtained by using $J_m(\epsilon)$ in equation (6.47),

$$\psi_l = \cos^{-1}(1 - 2\epsilon) \qquad (6.40), (6.41)$$

$$= \cos^{-1}(1 - 2 \times 0.525) = \pm 92°52'$$

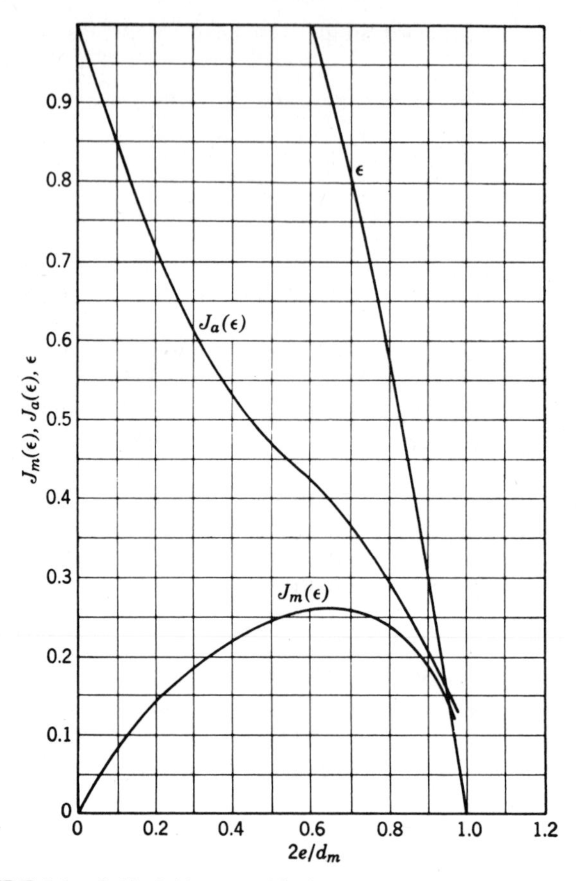

FIGURE 6.8. $J_m(\epsilon)$, $J_a(\epsilon)$, ϵ vs $2e/d_m$ for point-contact thrust bearings.

Figure 6.10 demonstrates a typical distribution of load in a 90° thrust bearing subjected to eccentric load.

Double Direction Bearings. The following relationships are valid for a two-row double-direction thrust bearing:

$$\delta_{a1} = -\delta_{a2} \tag{6.49}$$

$$\theta_1 = \theta_2 \tag{6.50}$$

It can also be shown that

$$\epsilon_1 + \epsilon_2 = 1 \tag{6.51}$$

and

$$\frac{\delta_{\max_2}}{\delta_{\max_1}} = \frac{\epsilon_2}{\epsilon_1} \tag{6.52}$$

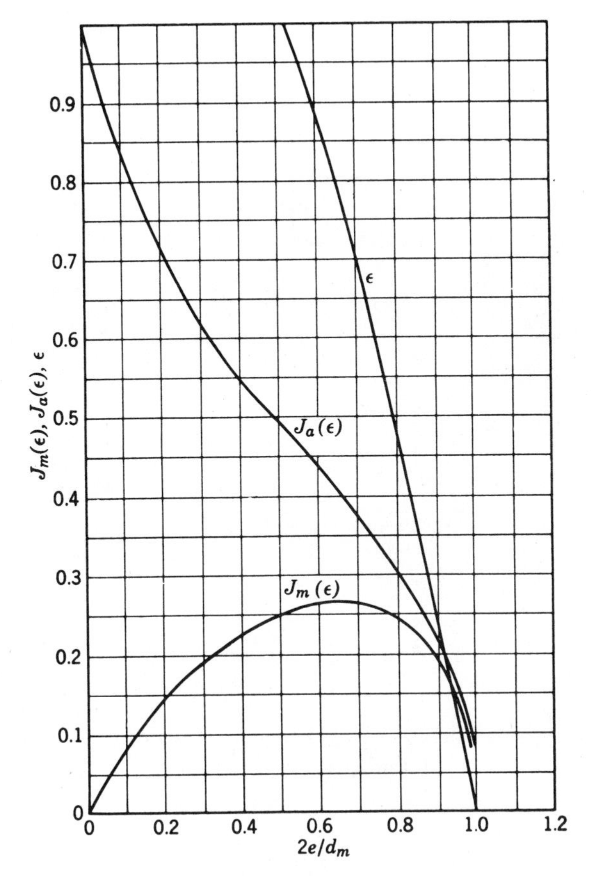

FIGURE 6.9. $J_m(\epsilon)$, $J_a(\epsilon)$, ϵ vs $2e/d_m$ for line-contact thrust bearings.

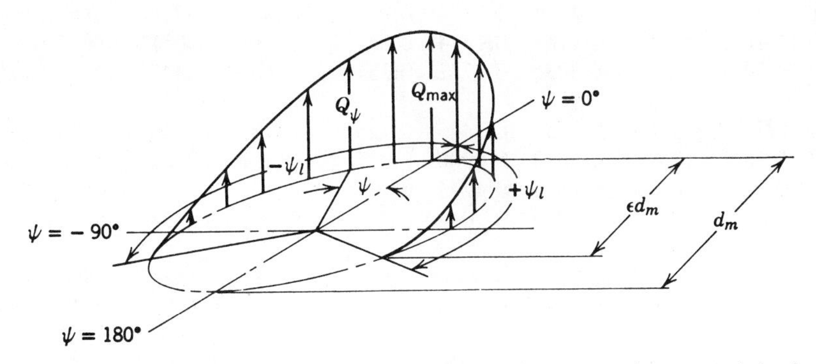

FIGURE 6.10. Load distribution in a 90° thrust bearing under eccentric load.

Considering equation (6.4), equation (6.52) becomes

$$\frac{Q_{\max 2}}{Q_{\max 1}} = \left(\frac{\epsilon_2}{\epsilon_1}\right)^n \tag{6.53}$$

In equation (6.53), $n = 1.5$ for ball bearings, and $n = 1.11$ for roller bearings. From conditions of equilibrium one may conclude that

$$F_a = F_{a1} - F_{a2} = ZQ_{\max 1} J_a \sin\alpha \tag{6.54}$$

in which

$$J_a = J_a(\epsilon_1) - \frac{Q_{\max 2}}{Q_{\max 1}} J_a(\epsilon_2) \tag{6.55}$$

$$\mathfrak{M} = \mathfrak{M}_1 + \mathfrak{M}_2 = \tfrac{1}{2} ZQ_{\max 1} d_m J_m \sin\alpha \tag{6.56}$$

in which

$$J_m = J_m(\epsilon_1) + \frac{Q_{\max 2}}{Q_{\max 1}} J_m(\epsilon_2) \tag{6.57}$$

Table 6.3 below gives values of J_a and J_m as functions of $2e/d_m$ for two-row bearings. Figures 6.11 and 6.12 give the same data in graphical format.

TABLE 6.3. J_a and J_m for Two-Row Thrust Bearings

		Point Contact				Line Contact			
ϵ_1	ϵ_2	$\frac{2e}{d_m}$	J_m	J_a	$\frac{Q_{\max 2}}{Q_{\max 1}}$	$\frac{2e}{d_m}$	J_m	J_a	$\frac{Q_{\max 2}}{Q_{\max 1}}$
0.50	0.50	∞	0.4577	0	1.000	∞	0.4906	0	1.000
0.51	0.49	25.72	0.4476	0.0174	0.941	28.50	0.4818	0.0169	0.955
0.60	0.40	2.046	0.3568	0.1744	0.544	2.389	0.4031	0.1687	0.640
0.70	0.30	1.092	0.3036	0.2782	0.281	1.210	0.3445	0.2847	0.394
0.80	0.20	0.800	0.2758	0.3445	0.125	0.823	0.3036	0.3688	0.218
0.90	0.10	0.671	0.2618	0.3900	0.037	0.634	0.2741	0.4321	0.089
1.0	0	0.600	0.2546	0.4244	0	0.524	0.2523	0.4817	0
1.25	0	0.434	0.2289	0.5044	0	0.360	0.2078	0.5775	0
1.67	0	0.309	0.1871	0.6060	0	0.234	0.1589	0.6790	0
2.5	0	0.185	0.1339	0.7240	0	0.137	0.1075	0.7837	0
5.0	0	0.083	0.0711	0.8558	0	0.061	0.0544	0.8909	0
	0	0	0	1.0000	0	0	0	1.0000	0

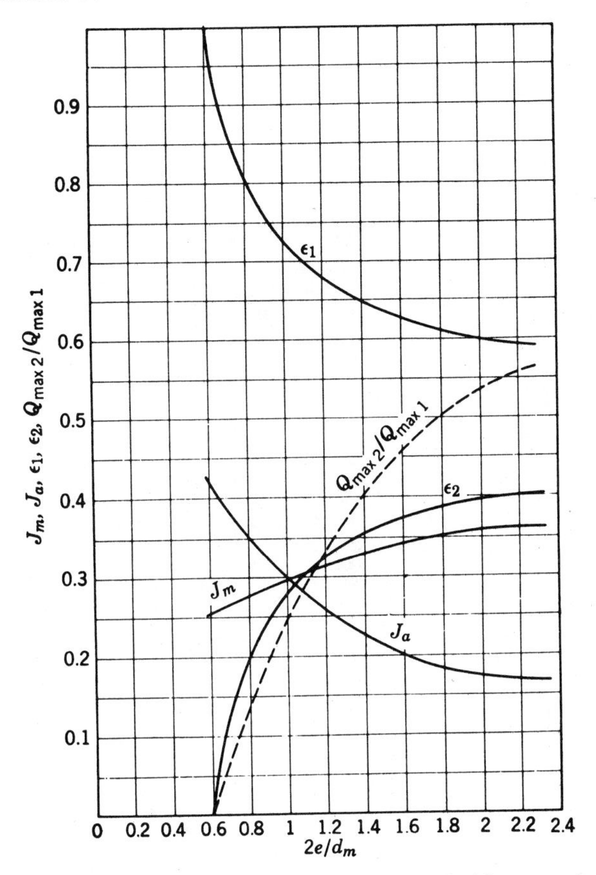

FIGURE 6.11. J_m, J_a, ϵ_1, ϵ_2, Q_{max2}/Q_{max1} vs $2e/d_m$ for double-row point-contact thrust bearings.

BEARINGS UNDER COMBINED RADIAL AND THRUST LOAD

Single-Row Bearings

If a rolling bearing without diametral clearance is subjected simultaneously to a radial load in the central plane of the rollers and a centric thrust load, then the inner and outer rings of the bearing will remain parallel and will be relatively displaced a distance δ_a in the axial direction and δ_r in the radial direction. At any angular position ψ measured from the most heavily loaded rolling element, the approach of the rings is

$$\delta_\psi = \delta_a \sin\alpha + \delta_r \cos\alpha \cos\psi \tag{6.58}$$

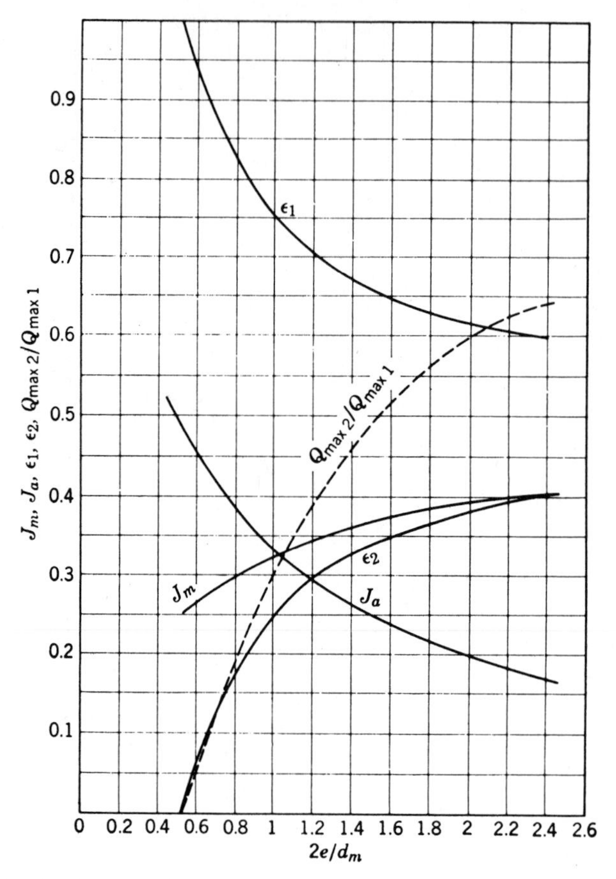

FIGURE 6.12. J_m, J_a, ϵ_1, ϵ_2, $Q_{max\,2}/Q_{max\,1}$ vs $2e/d_m$ for double-row line-contact thrust bearings.

Figure 6.13 illustrates this condition. At $\psi = 0$ maximum deflection occurs and is given by

$$\delta_{max} = \delta_a \sin \alpha + \delta_r \cos \alpha \tag{6.59}$$

Combining equations (6.58) and (6.59) yields

$$\delta_\psi = \delta_{max}\left[1 - \frac{1}{2\epsilon}(1 - \cos \psi)\right] \tag{6.60}$$

This expression is identical in form to equation (6.11), however,

$$\epsilon = \frac{1}{2}\left(1 + \frac{\delta_a \tan \alpha}{\delta_r}\right) \tag{6.61}$$

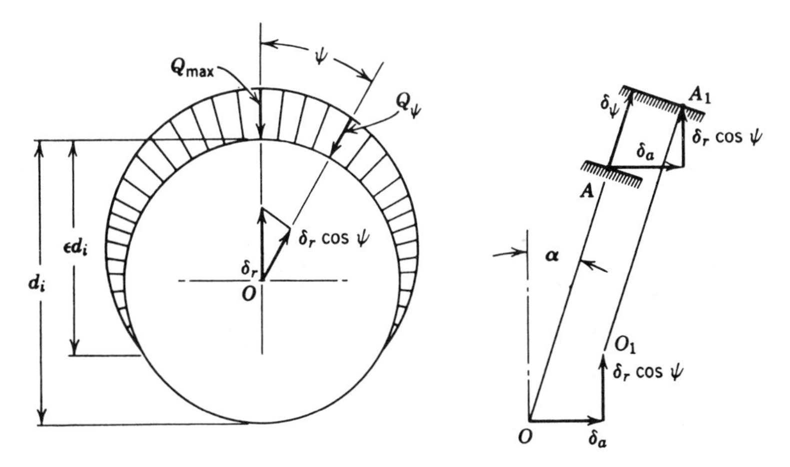

FIGURE 6.13. Rolling bearing displacements due to combined radial and axial loading.

It should also be apparent that

$$Q_\psi = Q_{\max}\left[1 - \frac{1}{2\epsilon}(1 - \cos\psi)\right]^n \tag{6.62}$$

As in equation (6.4), $n = 1.5$ for ball bearings and $n = 1.11$ for roller bearings.

For static equilibrium to exist, the summation of rolling element forces in each direction must equal the applied load in that direction.

$$F_r = \sum_{\psi=-\psi_1}^{\psi=+\psi_1} Q_\psi \cos\alpha \cos\psi \tag{6.63}$$

$$F_a = \sum_{\psi=-\psi_1}^{\psi=+\psi_1} Q_\psi \sin\alpha \tag{6.64}$$

in which the limiting angle is defined by

$$\psi_1 = \cos^{-1}\left(-\frac{\delta_a \tan\alpha}{\delta_r}\right) \tag{6.65}$$

Equations (6.63) and (6.64) may be rewritten in terms of a radial integral and thrust integral, respectively.

$$F_r = ZQ_{\max}J_r(\epsilon)\cos\alpha \tag{6.66}$$

where

$$J_r(\epsilon) = \frac{1}{2\pi}\int_{-\psi_1}^{+\psi_1}\left[1 - \frac{1}{2\epsilon}(1 - \cos\psi)\right]^n \cos\psi \, d\psi \tag{6.67}$$

and

$$F_a = ZQ_{\max} J_a(\epsilon) \sin\alpha \tag{6.68}$$

where

$$J_a(\epsilon) = \frac{1}{2\pi}\int_{-\psi_1}^{+\psi_1}\left[1 - \frac{1}{2\epsilon}(1 - \cos\psi)\right]^n d\psi \tag{6.69}$$

The integrals of equations (6.67) and (6.69) were introduced by Sjovall [6.4]. Table 6.4 gives values of these integrals for point and line contact as functions of $F_r \tan\alpha / F_a$.

TABLE 6.4. $J_r(\epsilon)$ and $J_a(\epsilon)$ for Single-Row Bearings

	Point Contact			Line Contact		
ϵ	$\dfrac{F_r \tan\alpha}{F_a}$	$J_r(\epsilon)$	$J_a(\epsilon)$	$\dfrac{F_r \tan\alpha}{F_a}$	$J_r(\epsilon)$	$J_a(\epsilon)$
0	1	$1/z$	$1/z$	1	$1/z$	$1/z$
0.2	0.9318	0.1590	0.1707	0.9215	0.1737	0.1885
0.3	0.8964	0.1892	0.2110	0.8805	0.2055	0.2334
0.4	0.8601	0.2117	0.2462	0.8380	0.2286	0.2728
0.5	0.8225	0.2288	0.2782	0.7939	0.2453	0.3090
0.6	0.7835	0.2416	0.3084	0.7480	0.2568	0.3433
0.7	0.7427	0.2505	0.3374	0.6999	0.2636	0.3766
0.8	0.6995	0.2559	0.3658	0.6486	0.2658	0.4098
0.9	0.6529	0.2576	0.3945	0.5920	0.2628	0.4439
1	0.6000	0.2546	0.4244	0.5238	0.2523	0.4817
1.25	0.4338	0.2289	0.5044	0.3598	0.2078	0.5775
1.67	0.3088	0.1871	0.6060	0.2340	0.1589	0.6790
2.5	0.1850	0.1339	0.7240	0.1372	0.1075	0.7837
5	0.0831	0.0711	0.8558	0.0611	0.0544	0.8909
∞	0	0	1	0	0	1

Note that the contact angle α is assumed identical for all loaded rollers. Thus the values of the integrals are approximate; however, they are sufficiently accurate for most calculations. Using these integrals,

$$Q_{\max} = \frac{F_r}{J_r(\epsilon) Z \cos \alpha} \tag{6.70}$$

or

$$Q_{\max} = \frac{F_a}{J_a(\epsilon) Z \sin \alpha} \tag{6.71}$$

Figures 6.14 and 6.15 also give values of J_r, J_a, and ϵ vs $F_r \tan \alpha / F_a$ for point and line contact, respectively.

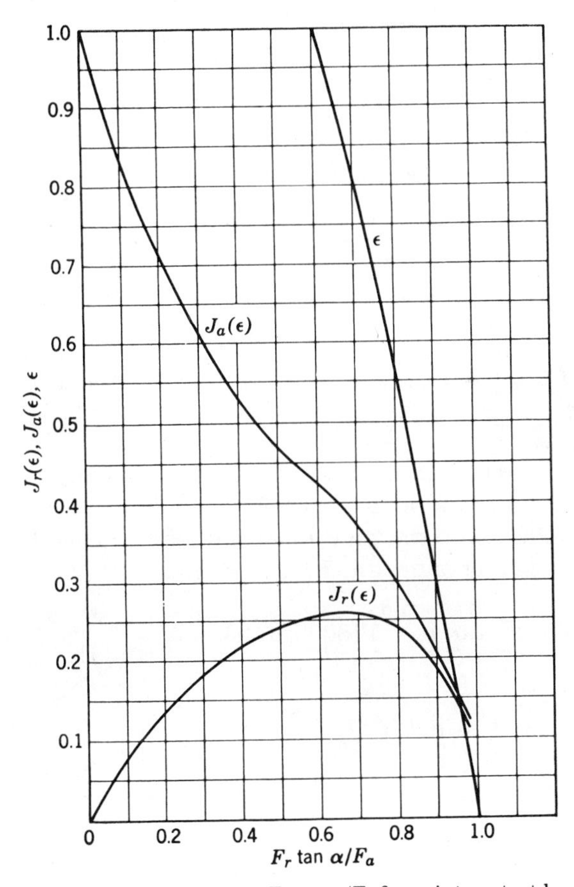

FIGURE 6.14. $J_r(\epsilon)$, $J_a(\epsilon)$, ϵ vs $F_r \tan \alpha / F_a$ for point-contact bearings.

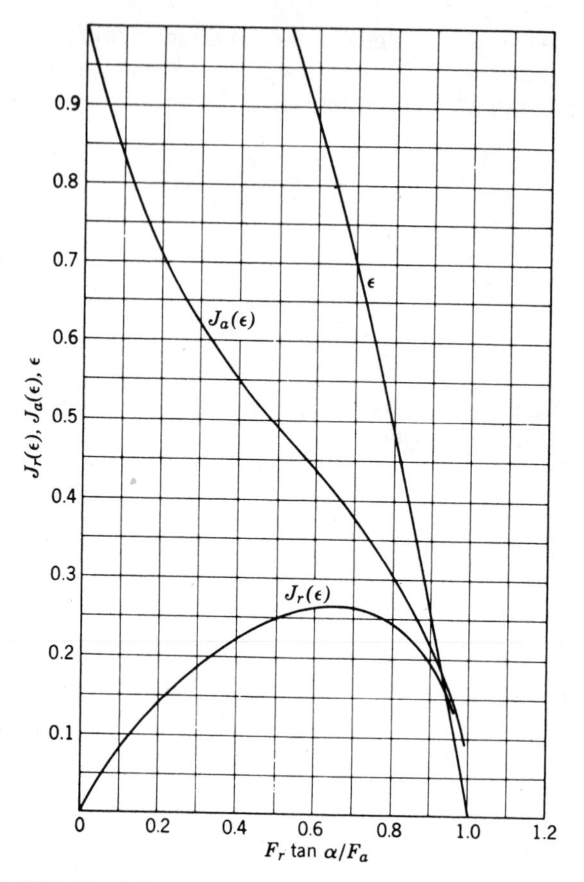

FIGURE 6.15. $J_r(\epsilon)$, $J_a(\epsilon)$, ϵ vs $F_r \tan \alpha / F_a$ for line-contact bearings.

Example 6.7. The 218 angular-contact ball bearing of Example 6.5 is subjected to a radial load of 17,800 N (4000 lb) and a thrust load of 17,800 N (4000 lb). Estimate the normal load on each ball if the contact angle can be presumed to remain constant at 40°.

$$\frac{F_r \tan \alpha}{F_a} = \frac{17800 \tan (40°)}{17800} = 0.8391$$

From Fig. 6.14, $J_r = 0.221$, $J_a = 0.263$, $\epsilon = 0.455$,

$$Z = 16 \qquad \text{Ex. 6.5}$$

$$Q_{max} = \frac{F_r}{J_r(\epsilon) Z \cos \alpha} \qquad (6.70)$$

$$= \frac{17800}{0.221 \times 16 \times \cos (40°)} = 6571 \text{ N (1477 lb)}$$

The same result could have been obtained by using equation (6.71).

$$\psi_1 = \cos^{-1}(1 - 2\epsilon) \qquad (6.61), (6.65)$$

$$= \cos^{-1}(1 - 2 \times 0.455) = \pm 84°47'$$

$$Q_\psi = Q_{max}\left[1 - \frac{1}{2\epsilon}(1 - \cos\psi)\right]^{1.5} \qquad (6.62)$$

$$= 6571\left[1 - \frac{1}{2 \times 0.455}(1 - \cos\psi)\right]^{1.5}$$

$$= 6571(1.099\cos\psi - 0.0989)^{1.5}$$

$$\Delta\psi = \frac{360}{Z} = \frac{360}{16} = 22.5°(22°30')$$

$\psi°$	$\cos\psi$	Q_ψ (N)
0	1	6571 (1477 lb)
22.5	0.9239	5765 (1296 lb)
45	0.7071	3670 (824.2 lb)
67.5	0.3827	1200 (269.6 lb)
90	0	0
112.5	−0.3827	0
135	−0.7071	0
157.5	−0.9239	0
180	−1	0

Double-Row Bearings

Let the indices 1 and 2 designate the rows of a two-row bearing having zero diametral clearance. Then

$$\delta_{r1} = \delta_{r2} = \delta_r \qquad (6.72)$$

$$\delta_{a1} = -\delta_{a2} \qquad (6.73)$$

Substituting these conditions into equations (6.59) and (6.60) yields

$$\frac{\delta_{max_2}}{\delta_{max_1}} = \frac{\epsilon_2}{\epsilon_1} \qquad (6.74)$$

$$\epsilon_1 + \epsilon_1 = 1 \qquad (6.75)$$

Equation (6.75) pertains only if both rows are loaded. If only one row is loaded, then

$$\epsilon_1 \geqq 1, \quad \epsilon_2 = 0 \qquad (6.76)$$

It is further clear from equation (6.4) that

$$\frac{Q_{\max 2}}{Q_{\max 1}} = \left(\frac{\epsilon_2}{\epsilon_1}\right)^n \tag{6.77}$$

The laws of static equilibrium dictate that

$$F_r = F_{r1} + F_{r2} \tag{6.78}$$

$$F_a = F_{a1} - F_{a2} \tag{6.79}$$

As before,

$$F_r = ZQ_{\max 1} J_r \cos \alpha \tag{6.80}$$

$$F_a = ZQ_{\max 1} J_a \sin \alpha \tag{6.81}$$

in which

$$J_r = J_r(\epsilon_1) + \frac{Q_{\max 2}}{Q_{\max 1}} J_r(\epsilon_2) \tag{6.82}$$

$$J_a = J_a(\epsilon_1) + \frac{Q_{\max 2}}{Q_{\max 1}} J_a(\epsilon_2) \tag{6.83}$$

Table 6.5 gives values of J_r and J_a as functions of $F_r \tan \alpha / F_a$. Figures 6.16 and 6.17 give the same data in graphical format for point and line contact, respectively.

TABLE 6.5. J_a and J_r for Double-Row Bearings

		Point Contact					Line Contact				
ϵ_1	ϵ_2	$\frac{F_r \tan \alpha}{F_a}$	J_r	J_a	$\frac{Q_{\max 2}}{Q_{\max 1}}$	$\frac{F_{r2}}{F_{r1}}$	$\frac{F_r \tan \alpha}{F_a}$	J_r	J_a	$\frac{Q_{\max 2}}{Q_{\max 1}}$	$\frac{F_{r2}}{F_{r1}}$
0.5	0.5	∞	0.4577	0	1	1	∞	0.4906	0	1	1
0.6	0.4	2.046	0.3568	0.1744	0.544	0.477	2.389	0.4031	0.1687	0.640	0.570
0.7	0.3	1.092	0.3036	0.2782	0.281	0.212	1.210	0.3445	0.2847	0.394	0.306
0.8	0.2	0.8005	0.2758	0.3445	0.125	0.078	0.8232	0.3036	0.3688	0.218	0.142
0.9	0.1	0.6713	0.2618	0.3900	0.037	0.017	0.6343	0.2741	0.4321	0.089	0.043
1.0	0	0.6000	0.2546	0.4244	0	0	0.5238	0.2523	0.4817	0	0

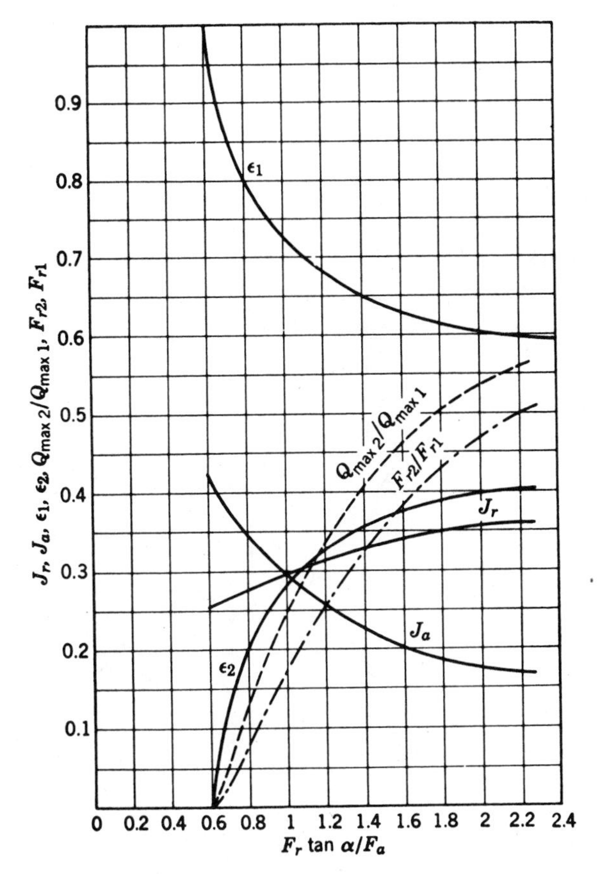

FIGURE 6.16. J_r, J_a, ϵ_1, ϵ_2, Q_{max2}/Q_{max1}, F_{r2}/F_{r1} vs $F_r \tan \alpha/F_a$ for double-row point-contact bearings.

Example 6.8. The 22317 two-row spherical roller bearing of Example 2.8 supports a 89,000 N (20,000 lb) radial load and a 22,250 N (5000 lb) thrust load simultaneously. Estimate the roller load distribution.

$$\alpha = 12° \qquad \text{Ex. 2.9}$$

$$Z = 14 \qquad \text{Ex. 2.8}$$

$$\frac{F_r \tan \alpha}{F_a} = \frac{89000 \times \tan(12°)}{22250} = 0.8502$$

From Fig. 6.17, $J_r = 0.303$, $J_a = 0.370$, $\epsilon_1 = 0.8$, $\epsilon_2 = 0.2$, $Q_{max_2}/Q_{max_1} = 0.220$, $F_{r2}/F_{r1} = 0.143$

$$F_r = ZQ_{max_1} J_r \cos \alpha \tag{6.80}$$

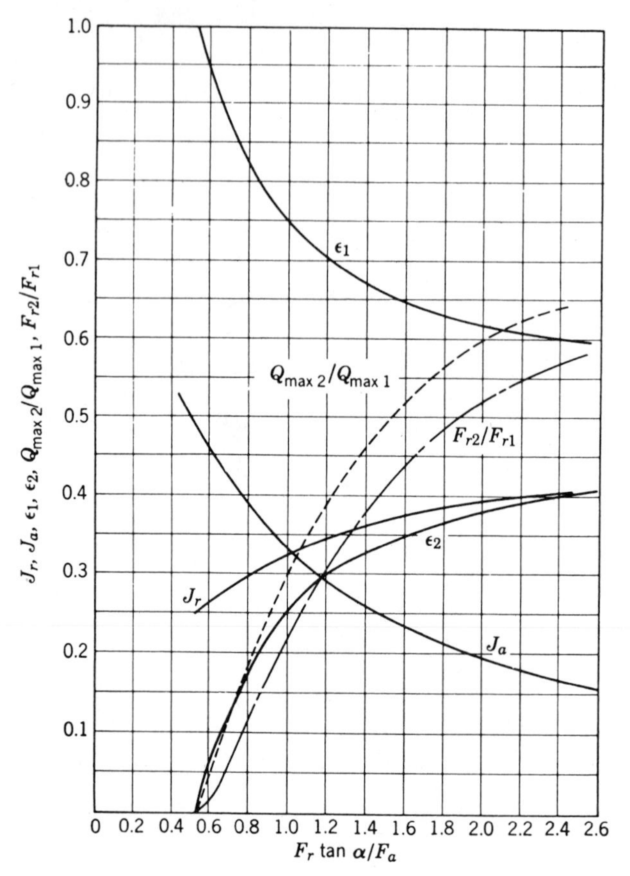

FIGURE 6.17. J_r, J_a, ϵ_1, ϵ_2, $Q_{\max 2}/Q_{\max 1}$, F_{r2}/F_{r1} vs $F_r \tan \alpha/F_a$ for double-row line-contact bearings.

$$89000 = 14 \times Q_{\max_1} \times 0.303 \times \cos(12^\circ)$$

$$Q_{\max_1} = 21450 \text{ N (4819 lb)}$$

$$Q_{\max_2} = \frac{Q_{\max_2}}{Q_{\max_1}} \times Q_{\max_1}$$

$$= 0.220 \times 21450 = 4719 \text{ N (1060 lb)}$$

$$Q_{\psi 1} = Q_{\max_1}\left[1 - \frac{1}{2\epsilon}(1 - \cos\psi)\right]^{1.11} \tag{6.62}$$

$$= 21450\left[1 - \frac{1}{2 \times 0.8}(1 - \cos\psi)\right]^{1.11}$$

$$= 21450(0.625 \cos\psi + 0.375)^{1.11}$$

$$\psi_{l_1} = \cos^{-1}(1 - 2\epsilon_1)$$

$$= \cos^{-1}(1 - 2 \times 0.8) = \pm 126°52'$$

$$Q_{\psi 2} = Q_{\max 2}\left[1 - \frac{1}{2\epsilon_2}(1 - \cos\psi)\right]^{1.11} \tag{6.62}$$

$$= 4719\left[1 - \frac{1}{2 \times 0.2}(1 - \cos\psi)\right]^{1.11} = 4719(2.5\cos\psi - 1.5)^{1.11}$$

$$\psi_{l_2} = \cos^{-1}(1 - 2\epsilon_2)$$

$$= \cos^{-1}(1 - 2 \times 0.2) = \pm 53°8'$$

$$\Delta\psi = \frac{360°}{Z} = \frac{360}{14} = 25.71°(25°43')$$

$\psi(°)$	$\cos\psi$	$Q_{\psi 1}$ (N)	$Q_{\psi 2}$ (N)
0	1	21450 (4819 lb)	4719 (1060 lb)
25.71	0.9010	19980 (4488 lb)	3442 (773 lb)
51.42	0.6237	15930 (3578 lb)	204 (46 lb)
77.13	0.2227	10250 (2299 lb)	0
102.84	−0.2227	4321 (964 lb)	0
128.55	−0.6237	0	0
154.26	−0.9010	0	0
180	−1	0	0

BALL BEARINGS UNDER COMBINED RADIAL, THRUST, AND MOMENT LOAD

If a ball is compressed by a load Q, then since the centers of curvature of the raceway grooves are fixed with respect to the corresponding raceway, the distance between the centers is increased by the amount of the normal approach between the raceways. From Fig. 6.18 it can be seen that

$$s = A + \delta_i + \delta_o \tag{6.84}$$

$$\delta_n = \delta_i + \delta_o = s - A \tag{6.85}$$

If a ball bearing that has a number of rolling elements situated symmetrically about a pitch circle is subjected to a combination of radial, thrust, and moment loads, the following relative displacements of inner and outer ring raceways may be defined:

δ_a relative axial displacement

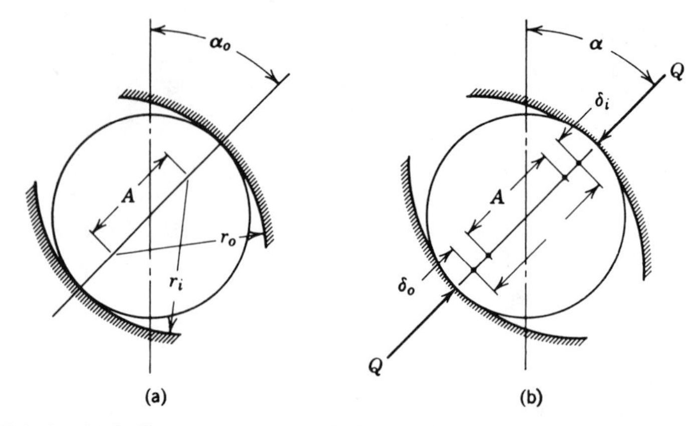

FIGURE 6.18. (*a*) Ball–raceway contact before loading; (*b*) ball–raceway contact under load.

δ_r relative radial displacement
θ relative angular misalignment

These relative displacements are shown in Fig. 6.19.

Consider a rolling bearing prior to the application of load. Figure 6.20 shows the positions of the loci of the centers of the inner and outer race-

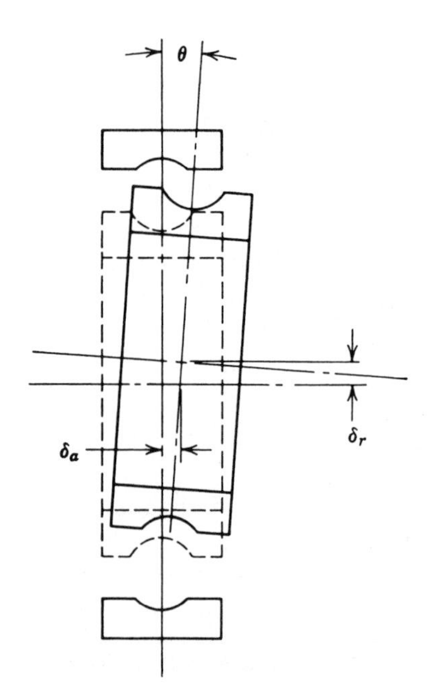

FIGURE 6.19. Displacements of an inner ring (outer ring fixed) due to combined radial, axial, and moment loading.

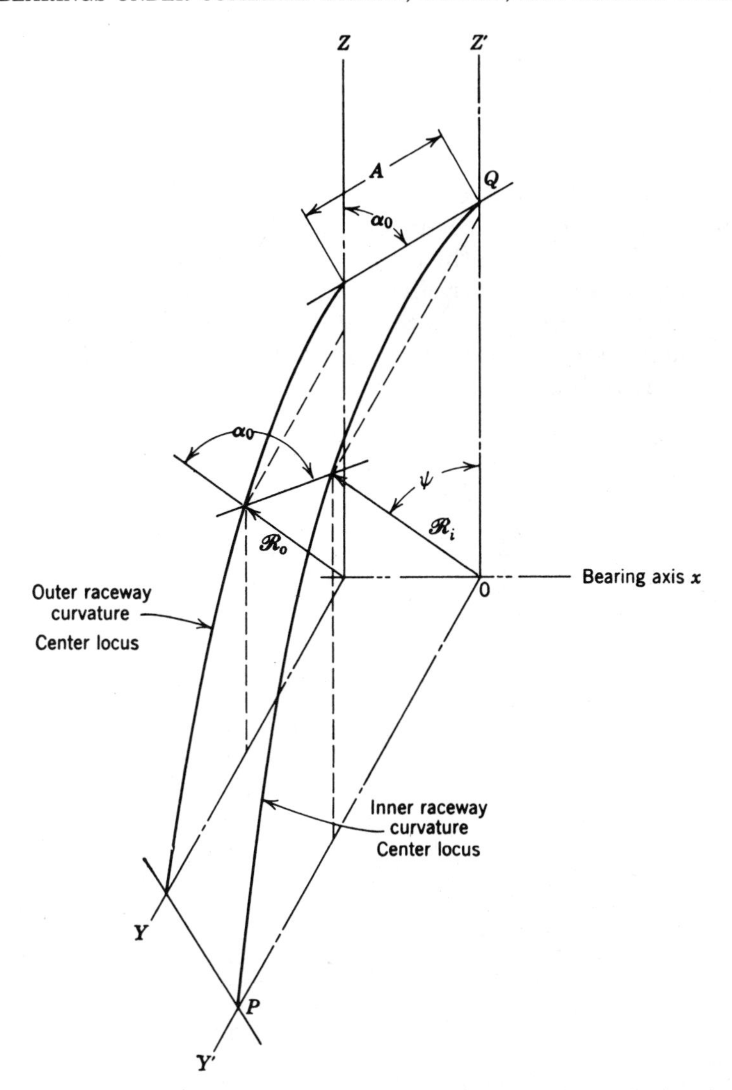

FIGURE 6.20. Loci of raceway groove curvature radii centers before loading (reprinted from [6.2]).

way groove curvature radii. It can be determined from Fig. 2.2 that the locus of the centers of the inner ring raceway groove curvature radii is expressed by

$$\mathcal{R}_i = \tfrac{1}{2}d_m + (r_i - 0.5D)\cos\alpha^\circ \tag{6.86}$$

in which α° is the free contact angle determined by bearing diametral clearance. From Fig. 6.20, then

$$\mathcal{R}_o = \mathcal{R}_i - A \cos \alpha^\circ \tag{6.87}$$

$$\mathcal{R}_i - \mathcal{R}_o = A \cos \alpha^\circ \tag{6.88}$$

In Fig. 6.20, ψ is the angle between the most heavily loaded rolling element and any other rolling element. Because of symmetry $0 \leq \psi \leq \pi$.

If the outer ring of the bearing is considered fixed in space as load is applied to the bearing, then the inner ring will be displaced and the locus of inner ring raceway groove radii centers will also be displaced as shown in Fig. 6.21. From Fig. 6.21 it can be determined that s, the distance between the centers of curvature of the inner and outer ring raceway grooves at any rolling element position ψ, is given by

$$s = [(A \sin \alpha^\circ + \delta_a + \mathcal{R}_i\, \theta \cos \psi)^2 + (A \cos \alpha^\circ + \delta_r \cos \psi)^2]^{1/2} \tag{6.89}$$

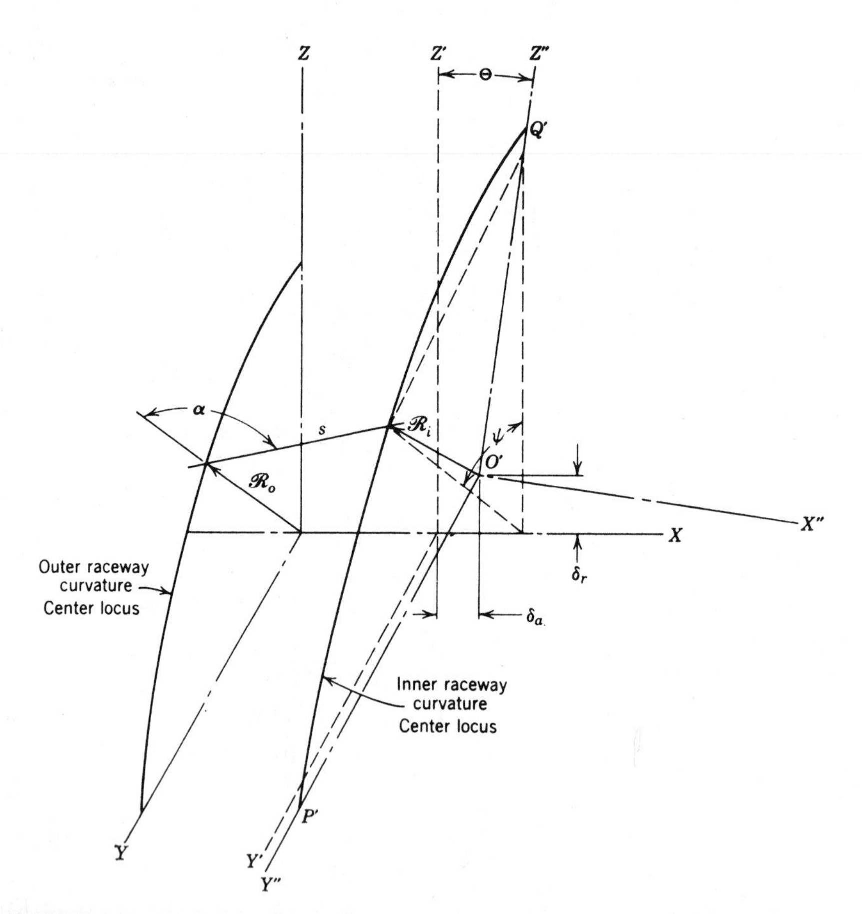

FIGURE 6.21. Loci of raceway groove curvature radii centers after displacement (reprinted from [6.2]).

or

$$s = A[(\sin\alpha^\circ + \bar{\delta}_a + \mathcal{R}_i\bar{\theta}\cos\psi)^2 + (\cos\alpha^\circ + \bar{\delta}_r\cos\psi)^2]^{1/2} \tag{6.90}$$

in which

$$\bar{\delta}_a = \frac{\delta_a}{A} \tag{6.91}$$

$$\bar{\delta}_r = \frac{\delta_r}{A} \tag{6.92}$$

$$\bar{\theta} = \frac{\theta}{A} \tag{6.93}$$

Substituting equation (6.90) into (6.85) yields

$$\delta_n = A\{[(\sin\alpha^\circ + \bar{\delta}_a + \mathcal{R}_i\bar{\theta}\cos\psi)^2 + (\cos\alpha^\circ + \bar{\delta}_r\cos\psi)^2]^{1/2} - 1\} \tag{6.94}$$

Substitution of equation (6.94) into (6.4) gives

$$Q = K_n A^n\{[(\sin\alpha^\circ + \bar{\delta}_a + \mathcal{R}_i\bar{\theta}\cos\psi)^2 + (\cos\alpha^\circ + \bar{\delta}_r\cos\psi)^2]^{1/2} - 1\}^n \tag{6.95}$$

At any rolling element position ψ, the operating contact angle is α. This contact angle can be determined from

$$\sin\alpha = \frac{\sin\alpha^\circ + \bar{\delta}_a + \mathcal{R}_i\bar{\theta}\cos\psi}{[(\sin\alpha^\circ + \bar{\delta}_a + \mathcal{R}_i\bar{\theta}\cos\psi)^2 + (\cos\alpha^\circ + \bar{\delta}_r\cos\psi)^2]^{1/2}} \tag{6.96}$$

or

$$\cos\alpha = \frac{\cos\alpha^\circ + \bar{\delta}_r\cos\psi}{[(\sin\alpha^\circ + \bar{\delta}_a + \mathcal{R}_i\bar{\theta}\cos\psi)^2 + (\cos\alpha^\circ + \bar{\delta}_r\cos\psi)^2]^{1/2}} \tag{6.97}$$

Equation (6.95) describes the normal load acting through contact angle α. This normal load may be resolved into axial and radial components as follows:

$$Q_a = Q\sin\alpha \tag{6.98}$$

$$Q_r = Q\cos\psi\cos\alpha \tag{6.99}$$

If the applied radial and thrust loads on the bearing are F_r and F_a, respectively, then for static equilibrium to exist:

$$F_a = \sum_{\psi=0}^{\psi=\pm\pi} Q_\psi \sin \alpha \tag{6.100}$$

$$F_r = \sum_{\psi=0}^{\psi=\pm\pi} Q_\psi \cos \psi \cos \alpha \tag{6.101}$$

Additionally, each of the thrust components produces a moment about the Y axis (moments about the Z axis are self-equilibrating), which is given by

$$\mathfrak{M} = \tfrac{1}{2} Q d_m \sin \alpha \cos \psi \tag{6.102}$$

For static equilibrium, the applied moment $\mathfrak{M}$ about the Y axis must equal the sum of the moments of each rolling element load about the Y axis, that is,

$$\mathfrak{M} = \tfrac{1}{2} d_m \sum_{\psi=0}^{\psi=\pm\pi} Q_\psi \cos \psi \sin \alpha \tag{6.103}$$

Combination of equations (6.95), (6.98), and (6.100) yields

$$F_a - K_n A^n \sum_{\psi=0}^{\psi=\pm\pi} \frac{\{[(\sin \alpha^\circ + \bar{\delta}_a + \mathcal{R}_i \bar{\theta} \cos \psi)^2 + (\cos \alpha^\circ + \bar{\delta}_r \cos \psi)^2]^{1/2} - 1\}^n \times (\sin \alpha^\circ + \bar{\delta}_a + \mathcal{R}_i \bar{\theta} \cos \psi)}{[(\sin \alpha^\circ + \bar{\delta}_a + \mathcal{R}_i \bar{\theta} \cos \psi)^2 + (\cos \alpha^\circ + \bar{\delta}_r \cos \psi)^2]^{1/2}} = 0 \tag{6.104}$$

Similarly,

$$F_r - K_n A^n \sum_{\psi=0}^{\psi=\pm\pi} \frac{\{[(\sin \alpha^\circ + \bar{\delta}_a + \mathcal{R}_i \bar{\theta} \cos \psi)^2 + (\cos \alpha^\circ + \bar{\delta}_r \cos \psi)^2]^{1/2} - 1\}^n \times (\cos \alpha^\circ + \bar{\delta}_r + \cos \psi) \cos \psi}{[(\sin \alpha^\circ + \bar{\delta}_a + \mathcal{R}_i \bar{\theta} \cos \psi)^2 + (\cos \alpha^\circ + \bar{\delta}_r \cos \psi)^2]^{1/2}} = 0 \tag{6.105}$$

and

$$\mathfrak{M} - \tfrac{1}{2} d_m K_n A^n \sum_{\psi=0}^{\psi=\pm\pi} \frac{\{[(\sin \alpha^\circ + \bar{\delta}_a + \mathcal{R}_i \bar{\theta} \cos \psi)^2 + (\cos \alpha^\circ + \bar{\delta}_r \cos \psi)^2]^{1/2} - 1\}^n \times (\sin \alpha^\circ + \bar{\delta}_a + \mathcal{R}_i \bar{\theta} \cos \psi) \cos \psi}{[(\sin \alpha^\circ + \bar{\delta}_a + \mathcal{R}_i \bar{\theta} \cos \psi)^2 + (\cos \alpha^\circ + \bar{\delta}_r \cos \psi)^2]^{1/2}} = 0 \tag{6.106}$$

The foregoing equations were developed by Jones [6.2].

Equations (6.104)–(6.106) are simultaneous nonlinear equations with unknowns $\bar{\delta}_a$, $\bar{\delta}_r$, and $\bar{\theta}$. They must be solved by the Newton–Raphson method as follows:

1. Equations (6.104)–(6.106) may be written:

$$\mathcal{G}_m(\delta_k) = 0 \qquad k = 1, 2, 3 \qquad m = 1, 2, 3 \tag{6.107}$$

in which $\delta_1 = \bar{\delta}_a$, $\delta_2 = \bar{\delta}_r$, $\delta_3 = \bar{\theta}$.

2. Then

$$\mathcal{G}_m + \sum_{k=1}^{k=3} \frac{\partial \mathcal{G}_m}{\partial \delta_k} \mathcal{E}_k = 0 \tag{6.108}$$

3. The initial assumption is made concerning values of δ_k. Then the equations (6.108) are solved for the errors $\mathcal{E}_k$.
4. New values of δ_k may now be calculated giving

$$\delta_k' = \delta_k^0 + \mathcal{E}_k \tag{6.109}$$

5. The procedure is repeated until $\mathcal{E}_k$ are sufficiently small that $\delta_k' \approx \delta_k^0 = \delta_k$.

Having solved for $\bar{\delta}_a$, $\bar{\delta}_r$, and $\bar{\theta}$, it is then possible to solve for the maximum rolling element load. This is obtained from equation (6.95) for $\psi = 0$.

$$Q_{\max} = K_n A^n \{[(\sin \alpha^\circ + \bar{\delta}_a + \mathcal{R}_i \bar{\theta})^2 + (\cos \alpha^\circ + \bar{\delta}_r)^2]^{1/2} - 1\}^n \tag{6.110}$$

As can be seen, the foregoing procedure is very tedious for hand calculation. Generally, a digital computer must be used. Where a combination of loads including a moment load on the bearing is involved, it would be necessary to apply the foregoing equations. In certain specific cases, however, simplified solutions have been developed to evaluate maximum rolling element load. These have been demonstrated in previous sections.

MISALIGNMENT OF RADIAL ROLLER BEARINGS

Although it is usually undesirable, radial cylindrical roller bearings and tapered roller bearings can support to a small extent the moment loading due to misalignment. The various types of misalignment are illustrated in Fig. 6.22. Clearly, spherical roller bearings are designed to exclude all moment loads on the bearings and therefore are not included in this discussion.

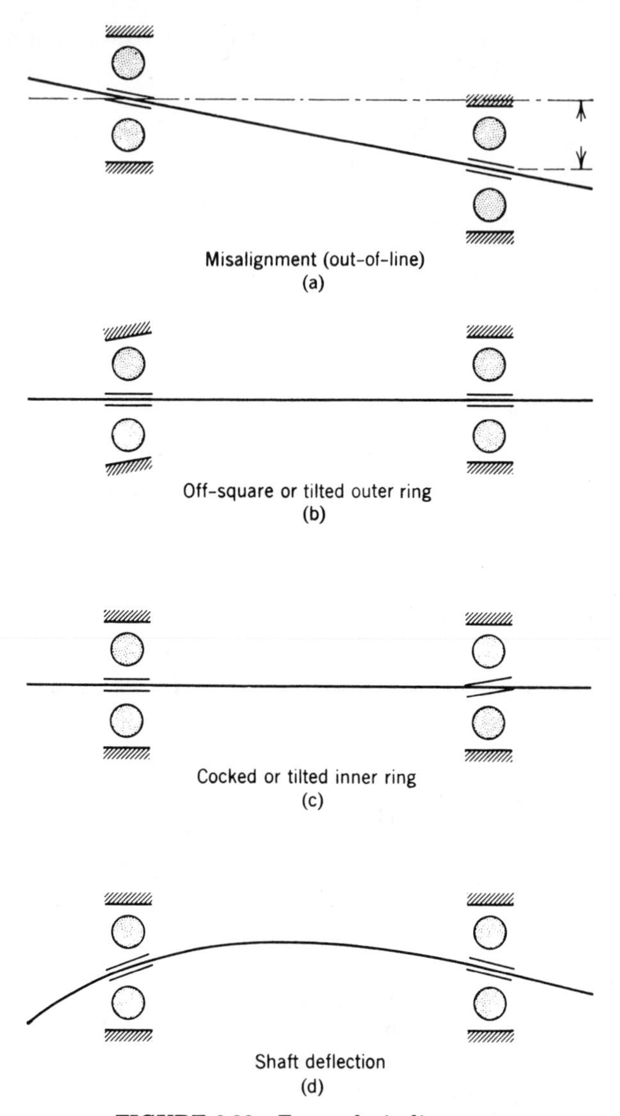

FIGURE 6.22. Types of misalignment.

Figure 6.23 illustrates the misalignment of a cylindrical roller bearing inner ring relative to the outer ring.

To commence the analysis, it is assumed that any roller-raceway contact can be subdivided into a number of laminae situated in planes parallel to the radial plane of the bearing. It is also assumed that shear effects between these laminae can be neglected owing to the small magnitudes of the contact deformations that develop. (Only contact deformations are considered.)

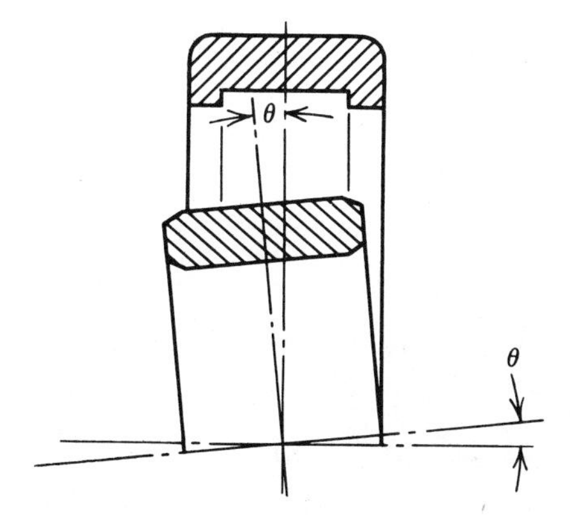

FIGURE 6.23. Misalignment of bearing rings.

As discussed in Chapter 5, equation (5.48) relates contact deformation to roller load as follows:

$$\delta = 3.84 \times 10^{-5} \frac{Q^{0.9}}{l^{0.8}} \tag{5.48}$$

Since the roller–raceway contact is subdivided into k laminae, each of width w, then $l = kw$. Substituting the roller load per unit length into equation (5.48), the following relationship can be obtained:

$$\delta = 3.84 \times 10^{-5} q^{0.9} (kw)^{0.1} \tag{6.111}$$

Equation (6.111) can be rearranged into the more useful format:

$$q = \frac{\delta^{1.11}}{1.24 \times 10^{-5} (kw)^{0.11}} \tag{6.112}$$

Equation (6.112) does not consider edge stresses; however, because these only obtain over a very small area, they can be neglected with little loss in accuracy when considering equilibrium of loading.

Roller Axial Load Distribution

For any roller–raceway contact the contact deformation δ can occur only after all clearance has been removed between the roller and raceway. Therefore, the crown drop must be factored into the analysis. In cylindrical roller bearings, usually the rollers are crowned and not the raceways although the latter condition is not altogether precluded. As shown in Chapter 1, roller crown is specified in terms of roller straight length, roller effective length, and crown radius. From Fig. 6.24, it can be deter-

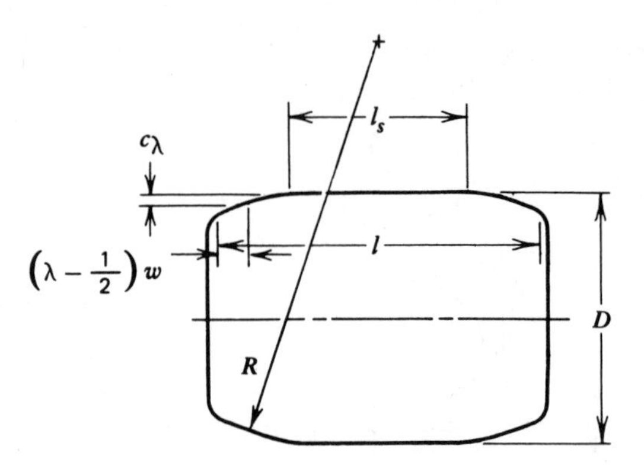

FIGURE 6.24. Crowned roller showing crown radius, straight length, and effective length.

mined that the crown drop (on roller diameter) is given by:

$$c_\lambda = \left[R^2 - \left(\frac{l_s}{2}\right)^2\right]^{1/2} - \left[R^2 - \left(\frac{l_s}{2} + \frac{l - l_s}{2} - (\lambda - \tfrac{1}{2})w\right)^2\right]^{1/2}$$

$$0 \leq (\lambda - \tfrac{1}{2})w \leq \frac{l - l_s}{2}$$

$$c_\lambda = 0$$

$$\frac{l - l_s}{2} \leq (\lambda - \tfrac{1}{2})w \leq \frac{l + l_s}{2} \tag{6.113}$$

$$c_\lambda = \left[R^2 - \left(\frac{l_s}{2}\right)^2\right]^{1/2} - \left[R^2 - \left((\lambda - \tfrac{1}{2})w - \frac{l}{2}\right)^2\right]^{1/2}$$

$$\frac{l + l_s}{2} \leq (\lambda - \tfrac{1}{2})w \leq l$$

where λ varies from 1 to k.

For the bearing misalignment shown in Fig. 6.23 the effective misalignment at a given roller location angle ψ_j is $\frac{1}{2}\theta \cos \psi_j$; for $0 \leq \psi_j \leq \pi/2$; $-\frac{1}{2}\theta \cos \psi_j$ for $\pi/2 \leq \psi_j \leq \pi$ (assuming symmetry about the $0 - \pi$ diameter. Hence the contact deformation at this roller location as a function of roller axial distance is given by (see Fig. 6.25):

$$\delta_{\lambda j} = \Delta_j \pm \tfrac{1}{2}\theta(\lambda - \tfrac{1}{2})w \cos \psi_j - c_\lambda \qquad \lambda = 1, k \tag{6.114}$$

Substitution of equation (6.114) into (6.112) yields

$$q_{\lambda j} = \frac{[\Delta_j \pm \frac{1}{2}\theta(\lambda - \frac{1}{2})w \cos \psi_j - c_\lambda]^{1.11}}{1.24 \times 10^{-5}(k_j w)^{0.11}} \tag{6.115}$$

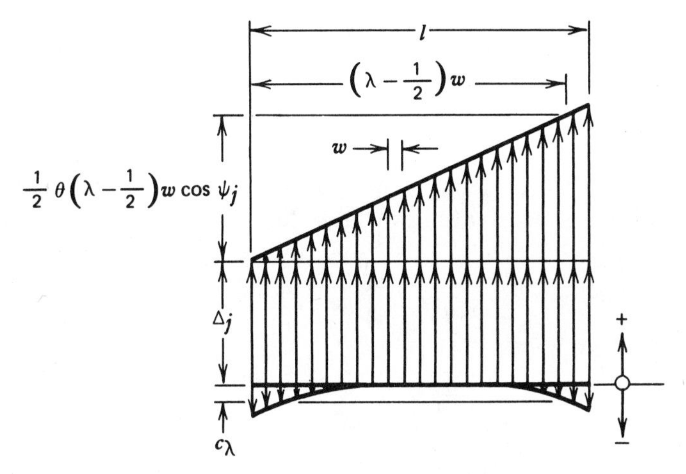

FIGURE 6.25. Components of roller deflection due to radial load, misalignment, and crowning.

where k_j is the number of laminae corresponding to the number of positive values of the numerator, that is, k_j refers to the number of laminae under load. At any roller location angle ψ_j, the total roller loading is

$$Q_j = \frac{w^{0.89}}{1.24 \times 10^{-5} k_j^{0.11}} \sum_{\lambda=1}^{\lambda=k_j} [\Delta_j \pm \tfrac{1}{2}\theta(\lambda - \tfrac{1}{2})w \cos \psi_j - c_\lambda]^{1.11} \quad (6.116)$$

Equations of Static Equilibrium

To determine the individual roller loading, it is necessary to satisfy the requirements of static equilibrium. Hence, for an applied radial load,

$$\frac{F_{\mathrm{r}}}{2} - \sum_{j=1}^{j=Z/2+1} \tau_j Q_j \cos \psi_j = 0 \qquad \tau_j = 0.5; \quad \psi_j = 0, \pi$$

$$\tau_j = 1; \quad \psi_j \neq 0, \pi \quad (6.117)$$

Substituting equation (6.116) into (6.117) yields

$$\frac{3.84 \times 10^{-5} F_{\mathrm{r}}}{w^{0.89}} - \sum_{j=1}^{j=Z/2+1} \frac{\tau_j \cos \psi_j}{k_j^{0.11}} \sum_{\lambda=1}^{\lambda=k_j} [\Delta_j \pm \tfrac{1}{2}\theta(\lambda - \tfrac{1}{2})w \cos \psi_j - c_\lambda]^{1.11}$$
$$= 0 \quad (6.118)$$

For an applied coplanar misaligning moment load, the equilibrium condition to be satisfied is

$$\frac{\mathfrak{M}}{2} - \sum_{j=1}^{j=Z/2+1} \tau_j Q_j e_j \cos\psi_j = 0 \qquad \tau_j = 0.5; \qquad \psi_j = 0, \pi$$

$$\tau_j = 1; \qquad \psi_j \neq 0, \pi \tag{6.119}$$

where e_j is the eccentricity of loading at each roller location. Accordingly, e_j, which is illustrated in Fig. 6.26, is given by

$$e_j = \frac{\sum_{\lambda=1}^{\lambda=k_j} q_{\lambda j}(\lambda - \frac{1}{2})w}{\sum_{\lambda=1}^{\lambda=k_j} q_{\lambda j}} - \frac{l}{2} \qquad j = 3, \frac{Z}{2} + 3 \tag{6.120}$$

Hence,

$$\frac{3.84 \times 10^{-5}\mathfrak{M}}{w^{0.89}} - \sum_{j=1}^{j=Z/2+1} \frac{\tau_j \cos\psi_j}{k_j^{0.11}} \times \left\{ \sum_{\lambda=1}^{\lambda=k_j} [\Delta_j \pm \tfrac{1}{2}\theta(\lambda - \tfrac{1}{2})w \cos\psi_j - c_\lambda]^{1.11}(\lambda - \tfrac{1}{2})w - \frac{l}{2}\sum_{\lambda=1}^{\lambda=k_j} [\Delta_j \pm \tfrac{1}{2}\theta(\lambda - \tfrac{1}{2})w \cos\psi_j - c_\lambda]^{1.11} \right\} = 0 \tag{6.121}$$

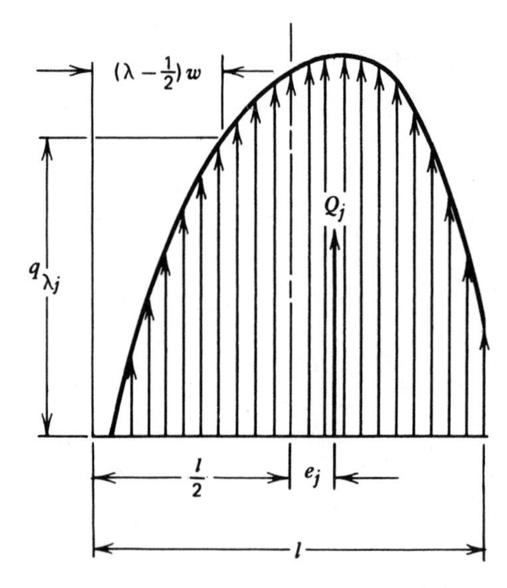

FIGURE 6.26. Load distribution for a misaligned crowned roller showing eccentricity of loading.

Deflection Equations

The remaining equations to be established are the radial deflection relationships. It is necessary here to determine the relative radial movement of the rings caused by the misalignment as well as that owing to radial loading. To assist in the first determination, Fig. 6.27 shows schematically an inner ring–roller assembly misaligned with respect to the outer ring. From this sketch, it is evident that one-half of the roller included angle is described by

$$\beta = \tan^{-1} \frac{l}{d_{\mathrm{m}} - D} \tag{6.122}$$

and

$$\sin \beta = \frac{l}{[(d_{\mathrm{m}} - D)^2 + l^2]^{1/2}} \tag{6.123}$$

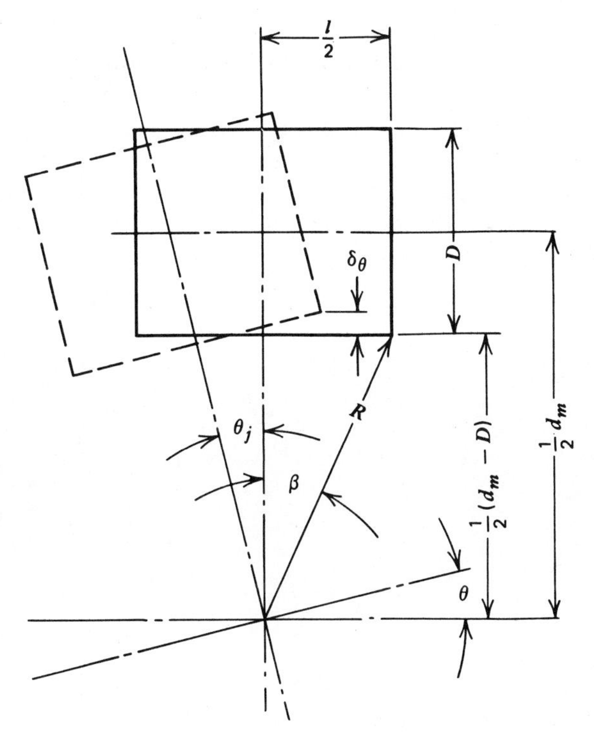

FIGURE 6.27. Schematic diagram of misaligned roller–inner ring assembly showing interference with outer ring.

The maximum radial interference between a roller and the outer ring owing to misalignment is given by

$$\delta_\theta = R\cos(\beta - \theta_j) - R\cos\beta \tag{6.124}$$

where

$$R = [(d_m - D)^2 + l^2]^{1/2} \tag{6.125}$$

In developing equations (6.124) and (6.125), the effect of crown drop was investigated and found to be negligible.

Expanding equation (6.124) in terms of the trigonometric identity further yields

$$\delta_\theta = R(\cos\beta\cos\theta_j + \sin\beta\sin\theta_j - \cos\beta) \tag{6.126}$$

Since θ_j is small, $\cos\theta_j \to 1$, and $\sin\theta_j \to \theta_j$. Moreover $\theta_j = \pm\theta\cos\psi_j$ and $\sin\beta = l/2R$, therefore,

$$\delta_\theta = \pm\tfrac{1}{2}l\theta\cos\psi_j \tag{6.127}$$

The shift of the inner ring center relative to the outer ring center owing to radial loading and clearance, and the subsequent relative radial movement at any roller location are shown in Fig. 6.28. The sum of the relative radial movement of the rings at each roller angular location minus the clearance is equal to the sum of the inner and outer raceway maximum contact deformations at the same angular location. Stating

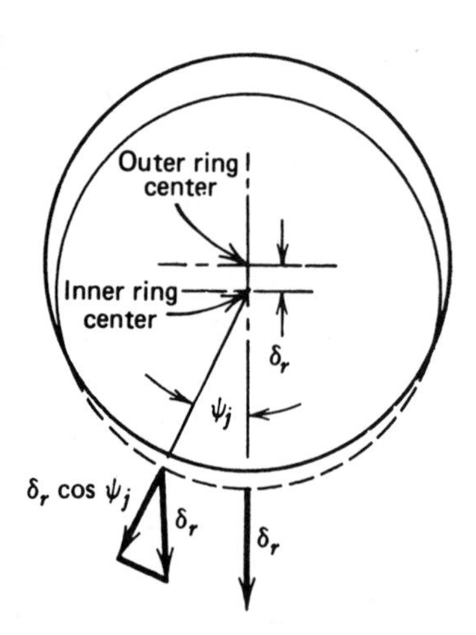

FIGURE 6.28. Displacement of ring centers caused by radial loading showing relative radial movement.

this relationship in equation format:

$$[\delta_r \pm \tfrac{1}{2}l\theta]\cos\psi_j - \frac{P_d}{2} - 2[\Delta_j \pm \tfrac{1}{2}\theta(\lambda - \tfrac{1}{2})w\cos\psi_j - c_\lambda]_{\max} = 0 \quad (6.128)$$

Equations (6.118), (6.121), and (6.128) constitute a set of $Z/2 + 3$ simultaneous nonlinear equations that can be solved for δ_r, θ, and Δ_j using numerical analysis techniques and digital computation. Thereafter, the variation of roller load per unit length, and subsequently the roller load, may be determined for each roller location using equations (6.115) and (6.116), respectively.

Using the foregoing method and digital computation, Harris [6.5] analyzed a 309 cylindrical roller bearing having the following dimensions and loading:

Number of rollers	12
Roller effective length	12.6 mm (0.496 in.)
Roller straight lengths	4.78, 7.70, 12.6 mm
Roller crown radius	1245 mm (49 in.)
Roller diameter	14 mm (0.551 in.)
Bearing pitch diameter	72.39 mm (2.85 in.)
Applied radial load	31,600 N (7100 lb)

For the above conditions, Fig. 6.29 shows the loading on various rollers for the bearing with ideally crowned rollers [$l = 12.6$ mm (0.496 in.) and with fully crowned rollers ($l = 0$).

Fig. 6.30 shows the effect or roller crowning on bearing radial deflection as a function of misalignment.

THRUST LOADING OF RADIAL CYLINDRICAL ROLLER BEARINGS

When radial cylindrical roller bearings have fixed flanges on both inner and outer rings, as shown in Fig. 1.30, they can carry some thrust load in addition to radial load. In fact, the greater the amount of radial load applied, the more thrust load that can be carried. As shown by Harris [6.12] and seen in Fig. 6.31, the thrust load causes each roller to tilt an amount ζ_j.

As in the previous section, it is assumed that a roller–raceway contact can be subdivided into laminae in planes parallel to the radial plane of the bearing. When a radial cylindrical roller bearing is subjected to applied thrust load, the inner ring shifts axially relative to the outer ring. Assuming deflections owing to roller–end-flange contacts are negligible,

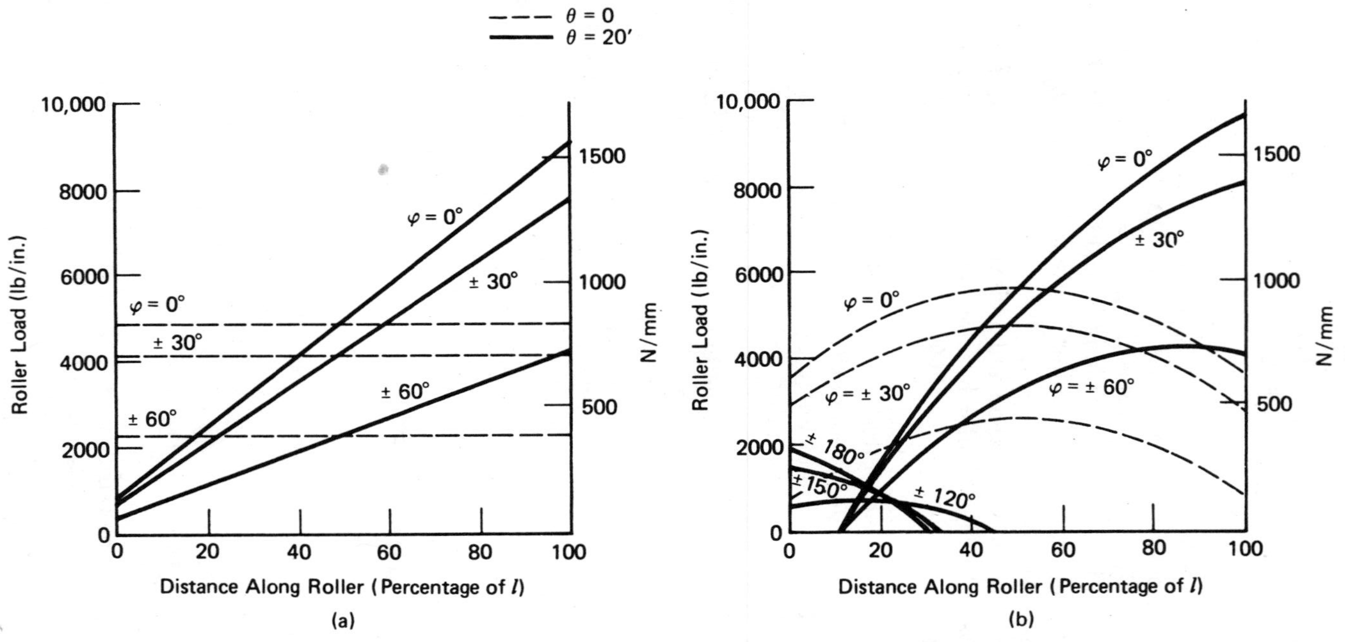

FIGURE 6.29. Roller loading vs axial and circumferential location—309 cylindrical roller bearing. (*a*) Ideally crowned rollers; (*b*) fully crowned rollers.

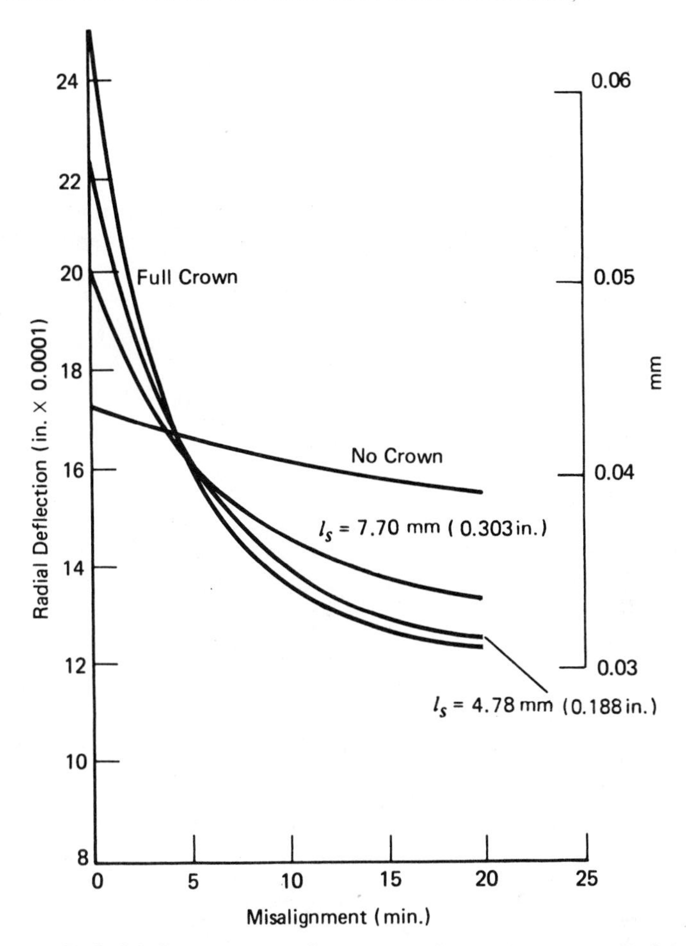

FIGURE 6.30. Radial deflection vs misalignment and crowning—309 cylindrical roller bearing at 31,600 N (7100 lb) radial load.

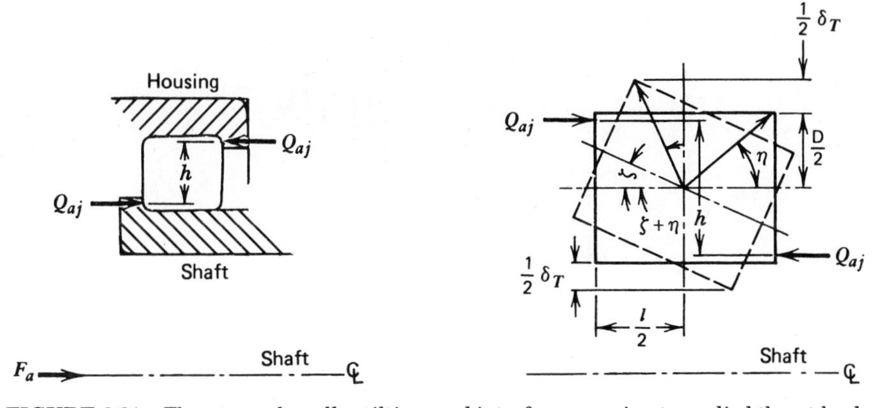

FIGURE 6.31. Thrust couple, roller tilting, and interference owing to applied thrust load.

then the interference at any axial location (laminum) is

$$\delta_{\lambda j} = \Delta_j + \zeta_j(\lambda - \tfrac{1}{2})w - c_\lambda \qquad \lambda = 1, k_j \tag{6.129}$$

where c_λ is given by equation (6.113). Figure 6.32 illustrates the component deflections in equation (6.129). Substituting equation (6.129) into (6.112) yields

$$q_{\lambda j} = \frac{[\Delta_j + \zeta_j(\lambda - \frac{1}{2})w - c_\lambda]^{1.11}}{1.24 \times 10^{-5}(k_j w)^{0.11}} \tag{6.130}$$

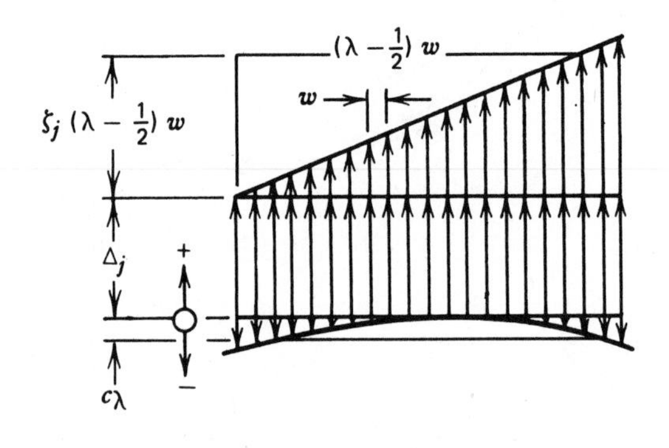

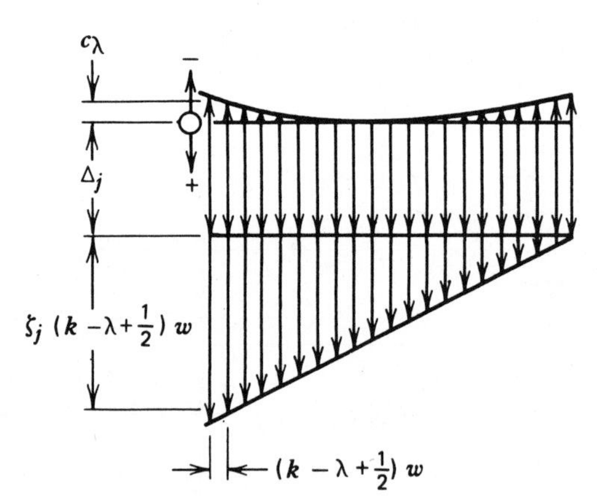

FIGURE 6.32. Components of roller–raceway deflection at opposing raceways due to radial load, thrust load, and crowning.

and at any azimuth ψ_j, the total roller loading is

$$Q_j = \frac{w^{0.89}}{1.24 \times 10^{-5} k_j^{0.11}} \sum_{\lambda=1}^{\lambda=k_j} [\Delta_j + \zeta_j(\lambda - \tfrac{1}{2})w - c_\lambda]^{1.11} \tag{6.131}$$

Equilibrium Equations

To determine roller loading, it is necessary to satisfy static equilibrium requirements. Hence, for applied radial load

$$\frac{F_r}{2} - \sum_{j=1}^{j=Z/2+1} \tau_j Q_j \cos\psi_j = 0 \qquad \begin{matrix} \tau_j = 0.5; & \psi_j = 0, \pi \\ \tau_j = 1; & \psi_j \neq 0, \pi \end{matrix} \tag{6.132}$$

Substituting equation (6.131) into (6.132) yields

$$\frac{3.84 \times 10^{-5} F_r}{w^{0.89}} - \sum_{j=1}^{j=Z/2+1} \frac{\tau_j \cos\psi_j}{k_j^{0.11}} \sum_{\lambda=1}^{\lambda=k_j} [\Delta_j + \zeta_j(\lambda - \tfrac{1}{2})w - c_\lambda]^{1.11} \tag{6.133}$$

For an applied centric thrust load, the equilibrium condition to be satisfied is

$$\frac{F_a}{2} - \sum_{j=1}^{j=Z/2+1} \tau_j Q_{aj} = 0 \tag{6.134}$$

At each roller location, the thrust couple is balanced by a radial load couple caused by the skewed axial load distribution. Thus, $hQ_{aj} = 2Q_j e_j$ and

$$\frac{F_a}{2} - \frac{2}{h} \sum_{j=1}^{j=Z/2+1} \tau_j Q_j e_j = 0 \qquad \begin{matrix} \tau_j = 0.5; & \psi_j = 0, \pi \\ \tau_j = 1; & \psi_j \neq 0, \pi \end{matrix} \tag{6.135}$$

where e_j is the eccentricity of loading indicated by Fig. 6.26 and defined by

$$e_j = \frac{\sum_{\lambda=1}^{\lambda=k_j} q_{\lambda j}(\lambda - \tfrac{1}{2})w}{\sum_{\lambda=1}^{\lambda=k_j} q_{\lambda j}} - \frac{l}{2} \tag{6.136}$$

Substitution of equations (6.131) and (6.136) into (6.135) yields

$$\frac{1.92 \times 10^{-5} F_{\mathrm{a}}}{w^{0.89}} - \sum_{j=1}^{j=Z/2+1} \frac{\tau_j}{k_j^{0.11}} \times \left\{ \sum_{\lambda=1}^{\lambda=k_j} [\Delta_j + \zeta_j(\lambda - \tfrac{1}{2})w - c_\lambda]^{1.11} (\lambda - \tfrac{1}{2})w - \frac{l}{2} \times \sum_{\lambda=1}^{\lambda=k_j} [\Delta_j + \zeta_j(\lambda - \tfrac{1}{2})w - c_\lambda]^{1.11} \right\} = 0 \qquad \begin{array}{ll} \tau_j = \tfrac{1}{2}; & \psi_j = 0, \pi \\ \tau_j = 1; & \psi_j \neq 0, \pi \end{array} \tag{6.137}$$

Deflection Equations

Radial deflection relationships remain to be established. It is here necessary to determine the relative radial movement of the bearing rings caused by the thrust loading as well as that due to radial loading. To assist in this derivation, Fig. 6.31 shows schematically a thrust-loaded roller–ring assembly. From this sketch, a roller angle is described by

$$\tan \eta = \frac{D}{l} \tag{6.138}$$

The maximum radial interference between a roller and both rings is given by

$$\delta_j = D \left[\frac{\sin (\zeta_j + \eta)}{\sin \eta} - 1 \right] \tag{6.139}$$

In developing equation (6.139) the effect of crown drop was found to be negligible. Expanding equation (6.139) in terms of the trigonometric identity and recognizing that ζ_j is small and $l = D \operatorname{ctn} \eta$, yields

$$\delta_{tj} = l \zeta_j \tag{6.140}$$

Whereas δ_{tj} is the radial deflection due to roller tilting, it can be similarly shown that axial deflection owing to roller tilting is

$$\delta_{aj} = D \zeta_j \tag{6.141}$$

Therefore, the radial interference caused by axial deflection is

$$\delta_{ra} = \delta_{a} \frac{l}{D} \tag{6.142}$$

The sum of the relative radial movements of the inner and outer rings at each roller azimuth minus the radial clearance is equal to the sum of the inner and outer raceway maximum contact deformations at the same azimuth, or

$$\delta_a \frac{l}{D} + \delta_r \cos \psi_j - \frac{P_d}{2} - 2[\Delta_j + \zeta_j(\lambda - \tfrac{1}{2})w - c_\lambda]_{max} = 0 \quad (6.143)$$

Equations (6.133), (6.137), and (6.143) are a set of simultaneous equations that can be solved for ζ_j, Δ_j, δ_r, and δ_a. Thereafter, the variation of roller load per unit length and roller load may be determined for each roller azimuth using equations (6.130) and (6.131), respectively. The axial load on each roller may be determined from

$$Q_{aj} = \frac{w^{0.89}}{3.84 \times 10^{-5} k_j^{0.11} h} \left\{ \sum_{\lambda=1}^{\lambda=k_j} [\Delta_j + \zeta_j(\lambda - \tfrac{1}{2})w - c_\lambda]^{1.11} (\lambda - \tfrac{1}{2})w - \frac{l}{2} \sum_{\lambda=1}^{\lambda=k_j} [\Delta_j + \zeta_j(\lambda - \tfrac{1}{2})w - c_\lambda]^{1.11} \right\} \quad (6.144)$$

RADIAL, THRUST, AND MOMENT LOADING OF RADIAL ROLLER BEARINGS

For radial cylindrical roller bearings, it is possible to apply combined loading. In general, the equations described for equilibrium of radial, thrust, and moment loading apply; however, the interference at any laminum in the roller–raceway contact is given by

$$\delta_{i\lambda j} = \Delta_j + w(\lambda - \tfrac{1}{2})[\pm \tfrac{1}{2}\theta \cos \psi_j + \nu_i \zeta_j] - c_\lambda \qquad \nu_1 = 1, \nu_2 = -1 \quad (6.145)$$

where $i = 1$ refers to the inner raceway, $i = 2$ the outer raceway, and the load per unit length is

$$q_{i\lambda j} = \frac{\{\Delta_j + (\lambda - \tfrac{1}{2})w[\pm \tfrac{1}{2}\theta \cos \psi_j + \nu_i \zeta_j] - c_\lambda\}^{1.11}}{1.24 \times 10^{-5} (k_j w)^{0.11}} \quad (6.146)$$

From an analytical point of view, cylindrical roller bearings may be considered a special form of tapered roller bearing; that is, a tapered roller bearing with a 0° contact angle. The equations used previously can be applied to tapered roller bearings provided the equilibrium equations are expanded to include the effect of the inner ring flange–roller end loading. This may also be accomplished under static loading conditions by considering the equilibrium of forces and moments on the outer ring (cone) as compared to the inner ring–flange arrangement (see Fig. 4.3).

As stated previously, spherical roller bearings are self-aligning and cannot therefore support moment loading. Moreover, in the absence of roller inertial loading and not considering friction, double-row spherical roller bearings having symmetrical (barrel-shaped) rollers will exhibit no tendency to tilt. Hence, the simpler analytical methods in previous sections may be applied and yield accurate results. For spherical roller bearings having asymmetrical rollers, for example, spherical roller thrust bearings (Fig. 1.45), roller tilting is not eliminated. In this case the roller may be considered a special type of tapered roller having fully crowned rollers. Then, the methods shown in the previous section can be employed for increased accuracy.

FLEXIBILY SUPPORTED ROLLING BEARINGS

Ring Deflections

The preceding discussion of distribution of load among the bearing rolling elements pertains to bearings having rigidly supported rings. Such bearings are assumed to be supported in infinitely stiff or rigid housings and on solid shafts of rigid material. The deflections considered in the determination of load distribution were contact deformations, that is, Hertzian deflections. This assumption is, in fact, an excellent approximation for most bearing applications.

In some radial bearing applications, however, the outer ring of the bearing may be supported at one or two angular positions only, and the shaft on which the inner ring is positioned may be hollow. The condition of two-point outer ring support, as shown in Figs. 6.33 and 6.34, occurs in the planet gear bearings of a planetary gear power transmission system, and was analyzed by Jones and Harris [6.6]. In certain rolling mill applications, the back-up roll bearings may be supported at only one point on the outer ring or possibly at two points as shown in Fig. 6.35. These conditions were analyzed by Harris [6.7]. In certain high speed radial bearings, to prevent skidding it is desirable to preload the rolling elements by using an elliptical raceway, thus achieving essentially two-point ring loading under conditions of light applied load. The case of a flexible outer ring and an elliptical inner ring was investigated by Harris and Broschard [6.8]. In each of the foregoing applications, the outer ring must be considered flexible to achieve a correct analysis of rolling element loading.

In many aircraft applications to conserve weight the power transmission shafting is made hollow. In these cases the inner ring deflections

FIGURE 6.33. Planet gear bearing.

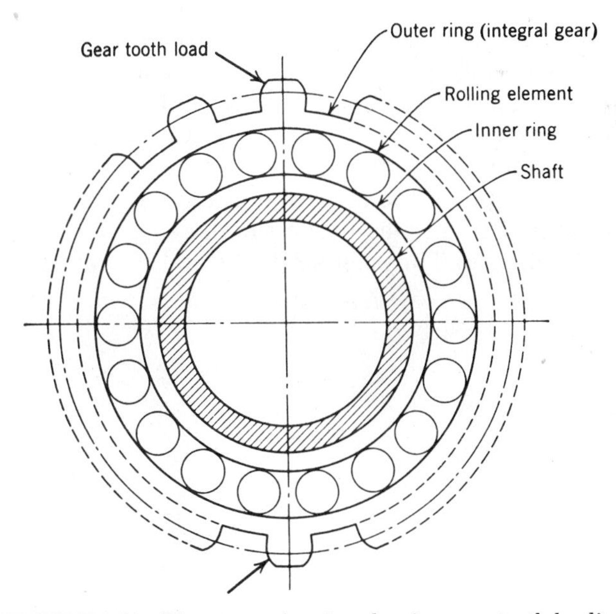

FIGURE 6.34. Planet gear bearing showing gear tooth loading.

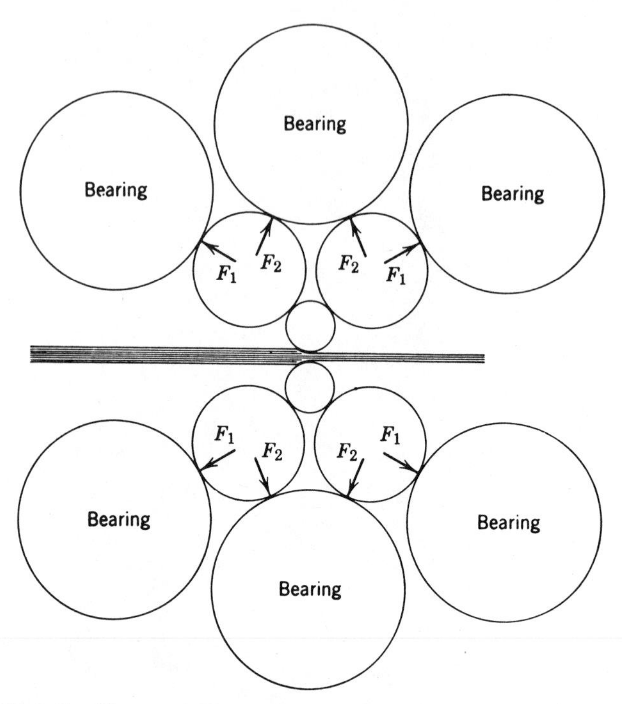

FIGURE 6.35. Cluster mill assembly showing back-up roll bearing loading.

will alter the load distribution from that considering only contact deformation.

To determine the load distribution among the rolling elements when one or both of the bearing rings is flexible, it is necessary to determine the deflections of a ring loaded at various points around its periphery. This analysis may be achieved by the application of classical energy methods for the bending of thin rings.

As a simple example of the method of analysis, consider a thin ring subjected to loads of equal magnitude equally spaced at angles $\Delta\psi$ (see Fig. 6.36). According to Timoshenko [6.9] the differential equation de-

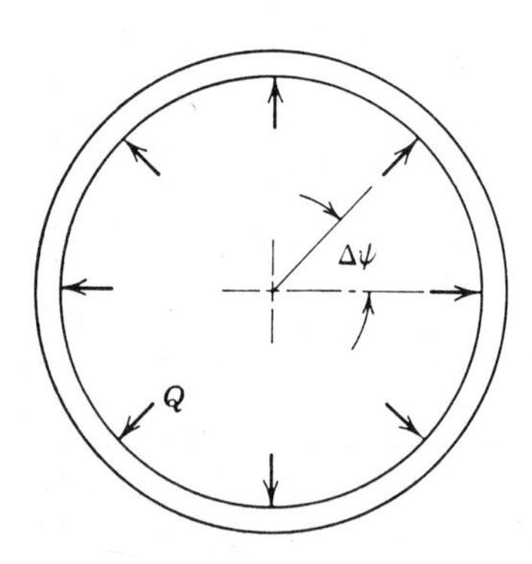

FIGURE 6.36. Thin ring loaded by equally spaced loads of equal magnitude.

scribing radial deflection u for bending of a thin bar with a circular center line is

$$\frac{d^2u}{d\phi^2} + u = -\frac{M\mathcal{R}^2}{EI} \tag{6.147}$$

in which I is the section moment of inertia in bending and E is the modulus of elasticity. It can be shown that the complete solution of equation (6.147) consists of a complementary solution and a particular solution. The complementary solution is

$$u_c = C_1 \sin\phi + C_2 \cos\phi \tag{6.148}$$

in which C_1 and C_2 are arbitrary constants.

Consider that the ring is cut at two positions: at the position of loading, $\phi = \frac{1}{2}\Delta\psi$, and at the position $\phi = 0$, midway between the loads. The loads required to maintain equilibrium over the section are shown in Fig. 6.37. From Fig. 6.37 it can be seen that since horizontal forces are balanced,

$$Q = 2F_0 \sin\phi \tag{6.149}$$

or

$$F_0 = \frac{Q}{2\sin\phi} \tag{6.150}$$

The moment at any angle ϕ between 0 and $\frac{1}{2}\Delta\psi$ is apparently

$$M = M_0 - F_0\mathcal{R}(1 - \cos\phi) \tag{6.151}$$

or

$$M = M_0 - \frac{Q\mathcal{R}}{2\sin\phi}(1 - \cos\phi) \tag{6.152}$$

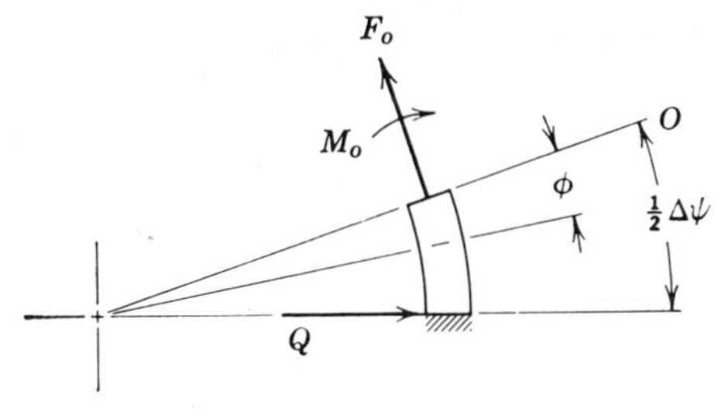

FIGURE 6.37. Loading of section of thin ring between $0 \le \phi \le \frac{1}{2}\Delta\psi$.

Since the section at $\phi = 0$ is midway between loads, it cannot rotate. According to Castigliano's theorem [6.9] the angular rotation at any section is

$$\theta = \frac{\partial U}{\partial M} \tag{6.153}$$

in which U is the strain energy in the beam at the position of loading. Timoshenko [6.9] shows that for a curved beam

$$U = \int_0^{\phi} \frac{M^2 \mathcal{R}}{2EI}\, d\phi \tag{6.154}$$

At $\phi = 0$, $M = M_0$ and since the section is constrained from rotation,

$$\frac{\partial U}{\partial M_0} = 0 = \frac{\mathcal{R}}{EI} \int_0^{1/2\Delta\psi} M \frac{\partial M}{\partial M_0}\, d\phi \tag{6.155}$$

Substituting equation (6.152) into (6.155) and integrating yields

$$M_0 = \frac{Q\mathcal{R}}{2}\left[\frac{1}{\sin\left(\frac{1}{2}\Delta\psi\right)} - \frac{2}{\Delta\psi}\right] \tag{6.156}$$

Hence,

$$M = \frac{Q\mathcal{R}}{2}\left[\frac{\cos\phi}{\sin\left(\frac{1}{2}\Delta\psi\right)} - \frac{2}{\Delta\psi}\right] \tag{6.157}$$

Equation (6.157) may be substituted for M in (6.147) such that the particular solution is

$$u_p = \frac{Q\mathcal{R}^3}{2EI}\left[\frac{\phi \sin\phi}{2\sin\left(\frac{1}{2}\Delta\psi\right)} - \frac{1}{\Delta\psi}\right] \tag{6.158}$$

The complete solution is

$$u = u_c + u_p = C_1 \sin\phi + C_2 \cos\phi + \frac{Q\mathcal{R}^3}{2EI}\left[\frac{\phi \sin\phi}{2\sin\left(\frac{1}{2}\Delta\psi\right)} - \frac{1}{\Delta\psi}\right] \tag{6.159}$$

Because the sections at $\phi = 0$ and $\phi = \frac{1}{2}\Delta\psi$ do not rotate, therefore

$$\left.\frac{du}{d\phi}\right|_{\phi=0} = 0; \qquad C_1 = 0$$

$$\left.\frac{du}{d\phi}\right|_{\phi=1/2\Delta\psi} = 0; \qquad C_2 = \frac{Q\mathcal{R}^3}{4EI\sin(\frac{1}{2}\Delta\psi)}[\tfrac{1}{2}\Delta\psi \operatorname{ctn}(\tfrac{1}{2}\Delta\psi) + 1]$$

Hence, the radial deflection at any angle ϕ between $\phi = 0$ and $\phi = \frac{1}{2}\Delta\psi$ is

$$u = \frac{Q\mathcal{R}^3}{2EI}\left\{\frac{2}{\Delta\psi} - \left[\frac{\Delta\psi\cos(\frac{1}{2}\Delta\psi)}{4\sin^2(\frac{1}{2}\Delta\psi)} + \frac{1}{2\sin(\frac{1}{2}\Delta\psi)}\right]\cos\phi - \frac{\phi\sin\phi}{2\sin(\frac{1}{2}\Delta\psi)}\right\} \tag{6.160}$$

Equation (6.160) may be expressed in another format as follows:

$$u = C_\phi Q \tag{6.161}$$

in which C_ϕ are influence coefficients dependent on angular position and ring dimensions.

$$C_\phi = \frac{\mathcal{R}^3}{2EI}\left\{\frac{2}{\Delta\psi} - \left[\frac{\Delta\psi\cos(\frac{1}{2}\Delta\psi)}{4\sin^2(\frac{1}{2}\Delta\psi)} + \frac{1}{2\sin(\frac{1}{2}\Delta\psi)}\right] \times \cos\phi - \frac{\phi\sin\phi}{2\sin(\frac{1}{2}\Delta\psi)}\right\} \tag{6.162}$$

Lutz [6.10] using procedures similar to those described above developed influence coefficients for various conditions of point loading of a thin ring. These coefficients have been expressed in infinite series format for the sake of simplicity of use.

For a thin ring loaded by forces of equal magnitude symmetrically located about a diameter as shown in Fig. 6.38, the following equation yields radial deflections:

$$_Qu_i = {}_QC_{ij}Q_j \tag{6.163}$$

in which

$$_QC_{ij} = \mp\frac{2\mathcal{R}^3}{\pi EI}\sum_{m=2}^{m=\infty}\frac{\cos m\psi_j\cos m\psi_i}{(m^2-1)^2} \tag{6.164}$$

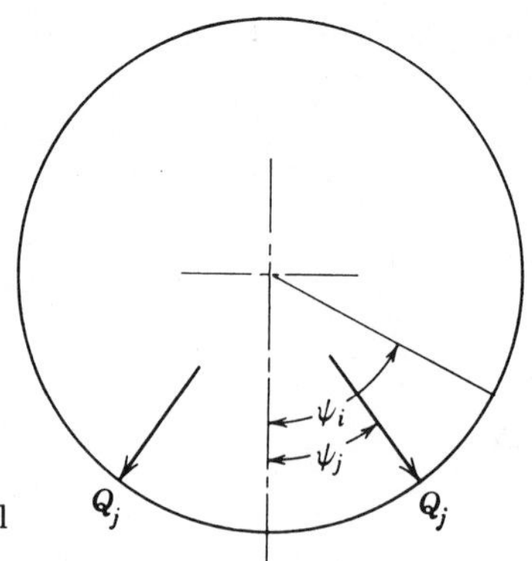

FIGURE 6.38. Thin ring loaded by forces of equal magnitude located symmetrically about a diameter.

The negative sign in (6.164) is used for internal loads and the positive sign is used for external loads. Equation (6.163) defines radial deflection at angle ψ_i caused by Q_j at position angle ψ_j. When rolling element loads Q_j are such that a rigid body translation δ_1 of the ring occurs, in the direction of an applied load, equations (6.163) are not self-sufficient in establishing a solution; however, a directional equilibrium equation may be used in conjunction with (6.163) to determine the translatory movement. Referring to Fig. 6.39 the appropriate equilibrium equation is as follows:

$$F_r \cos \psi_l - Q_j \cos \psi_j = 0 \tag{6.165}$$

In the planet gear bearing application demonstrated in Fig. 6.34 the gear tooth loads may be resolved into tangential forces, radial forces, and moment loads at $\psi = 90°$ (see Fig. 6.40). The ring radial deflections at

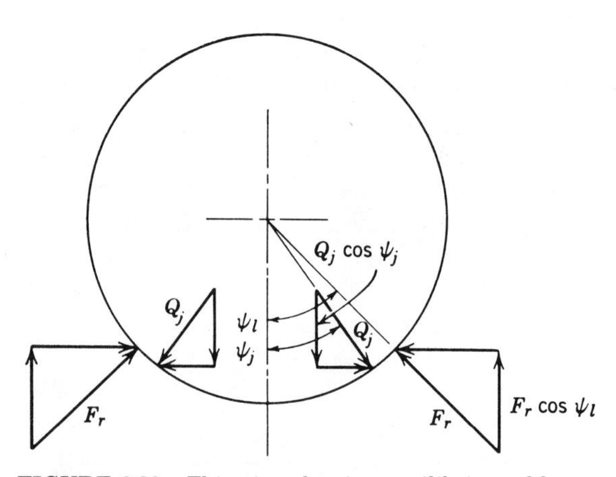

FIGURE 6.39. Thin ring showing equilibrium of forces.

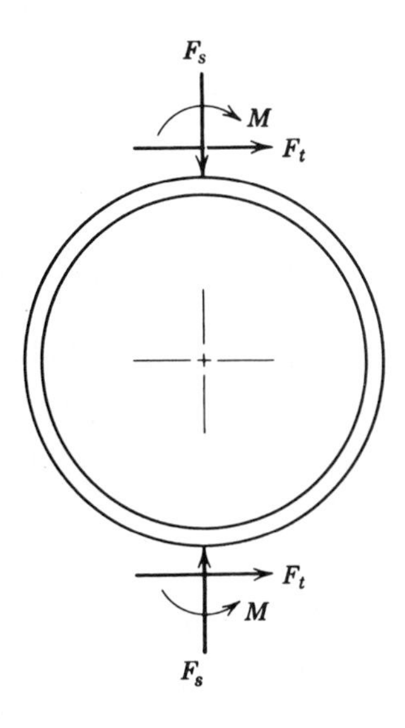

FIGURE 6.40. Resolution of gear tooth loading on outer ring.

angle ψ_i due to tangential forces F_t are given by

$$ {}_t u_i = {}_t C_i F_t \tag{6.166}$$

in which

$$ {}_t C_i = \frac{2\mathcal{R}^3}{\pi E I} \sum_{m=2}^{m=\infty} \frac{\sin \dfrac{m\pi}{2} \cos m\psi_i}{m(m^2 - 1)^2} \tag{6.167}$$

Equations (6.166) are not self-sufficient and an appropriate equilibrium equation must be used to define a rigid ring translation.

The separating forces F_s are self-equilibrating and thus do not cause a rigid ring translation. The radial deflections at angles ψ_i are given by

$$ {}_s u_i = {}_s C_i F_s \tag{6.168}$$

in which

$$ {}_s C_i = \frac{2\mathcal{R}^3}{\pi E I} \sum_{m=2}^{m=\infty} \frac{\cos \dfrac{m\pi}{2} \cos m\psi_i}{(m^2 - 1)^2} \tag{6.169}$$

Note that equations (6.169) are special cases of (6.164) in which position angle ψ_j is 90° and loads Q_j are external.

Similarly, the moment loads applied at $\psi = 90°$ are self-equilibrating.

The radial deflections are given by

$$_{\mathrm{M}}u_i = {}_{\mathrm{M}}C_i M \tag{6.170}$$

in which

$$_{\mathrm{M}}C_i = -\frac{2\mathcal{R}^2}{\pi EI} \sum_{m=2}^{m=\infty} \frac{\sin \dfrac{m\pi}{2} \cos m\psi_i}{m(m^2 - 1)} \tag{6.171}$$

To find the ring radial deflections at any angular position due to the combination of applied and resisting loads, the principle of superposition is used. Hence for the planet gear bearing, the radial deflection at any angular position ψ_i is the sum of the radial deflections due to each individual load, that is,

$$u_i = {}_{s}u_i + {}_{\mathrm{M}}u_i + {}_{t}u_i + {}_{Q_j}u_i \tag{6.172}$$

or

$$u_i = {}_{\mathrm{s}}C_i F_{\mathrm{s}} + {}_{\mathrm{M}}C_i M + {}_{\mathrm{t}}C_i F_{\mathrm{t}} + \textstyle\sum_Q C_{ij} Q_j \tag{6.173}$$

Relative Radial Approach of Rolling Elements to the Ring

A load may not be transmitted through a rolling element unless the outer ring deflects sufficiently to consume the radial clearance at the angular position occupied by the rolling element. Furthermore, because a contact deformation is caused by loading of the rolling element, the ring deflections cannot be determined without considering these contact deformations. Therefore the loading of a rolling element at angular position ψ_j depends on the relative radial clearance. The relative radial approach of the rings includes the translatory movement of the center of the outer ring relative to the initial center of that ring, which position is fixed in space. Hence for the planet gear bearing the relative radial approach at angular position ψ_i is

$$\delta_i = \delta_1 \cos \psi_i + u_i \tag{6.174}$$

From equation (6.4) the relative radial approach is related to the rolling element load as follows:

$$\begin{aligned} Q_j &= K(\delta_j - r_j)^n \qquad && \delta_j > r_j \\ Q_j &= 0 && \delta_j \le r_j \end{aligned} \tag{6.175}$$

in which r_j is the radial clearance at angular position ψ_j. Here r_j is the sum of $P_d/2$ and the condition of ring ellipticity.

Determination of Rolling Element Loads

Using the example of the planet gear bearing, the complete loading of the outer ring is shown in Fig. 6.41, which also illustrates the rigid ring translation δ_1. Combination of equations (6.173)–(6.175) yields

$$\delta_i - \delta_1 \cos \psi_i - {}_{\mathrm{s}}C_i F_{\mathrm{s}} - {}_{\mathrm{M}}C_i M - {}_{\mathrm{t}}C_i F_{\mathrm{t}} - iK \sum_{j=2}^{j=Z/2+2} {}_{Q}C_{ij}(\delta_j - r_j)^n = 0 \tag{6.176}$$

The required equilibrium equation is

$$F_{\mathrm{t}} - iK \sum_{j=2}^{j=Z/2+2} \tau_j(\delta_j - r_j)^n \cos \psi_j = 0 \tag{6.177}$$

considering symmetry about the diameter parallel to the load. In equation (6.177), $\tau_j = 0.5$ if the rolling element is located at $\psi_j = 0°$ or at $\psi_j = 180°$; otherwise $\tau_j = 1$.

Equations (6.176) and (6.177) constitute a set of simultaneous nonlinear equations which may be solved by numerical analysis. The Newton–Raphson method is recommended. Set equations (6.176) and (6.177) equal to $\mathcal{G}_i$ and $\mathcal{G}_1$, respectively. Then

$$\mathcal{G}_i + \sum_{j=1}^{j=Z/2+2} \frac{\partial \mathcal{G}_i}{\partial \delta_j} \mathcal{E}_j = 0 \tag{6.178}$$

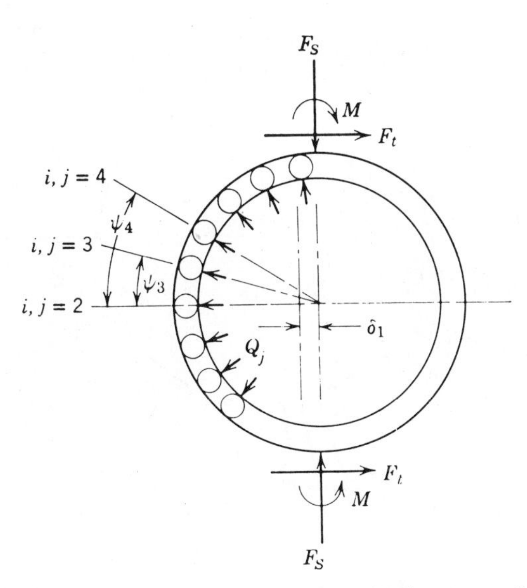

FIGURE 6.41. Total loading of outer ring in planet gear bearing.

Equations (6.178) constitute a system of simultaneous linear equations in error variables $\mathcal{E}_j$. Assuming initial values of δ_j, it is apparent that error $\mathcal{E}_j$ may be determined. Since

$$\mathcal{E}_j = \delta_j' - \delta_j^0 \tag{6.179}$$

new values of δ_j may be determined. The process is repeated until the root mean square error is sufficiently small, that is,

$$\left[\frac{1}{\frac{Z}{2} + 2} \sum_{j=1}^{j=Z/2+2} \mathcal{E}_j^2 \right]^{1/2} \approx 0 \tag{6.180}$$

It can be seen from equations (6.176) and (6.177) that

$$\begin{aligned} \frac{\partial \mathcal{G}_i}{\partial \delta_j} &= -iK_Q C_{ij}(\delta_j - r_j)^{n-1} \\ \frac{\partial \mathcal{G}_1}{\partial \delta_j} &= -iK(\delta_j - r_j)^{n-1} \cos \psi_j \\ \frac{\partial \mathcal{G}_1}{\partial \delta_1} &= 0 \\ \frac{\partial \mathcal{G}_i}{\partial \delta_1} &= -\cos \psi_i \end{aligned} \tag{6.181}$$

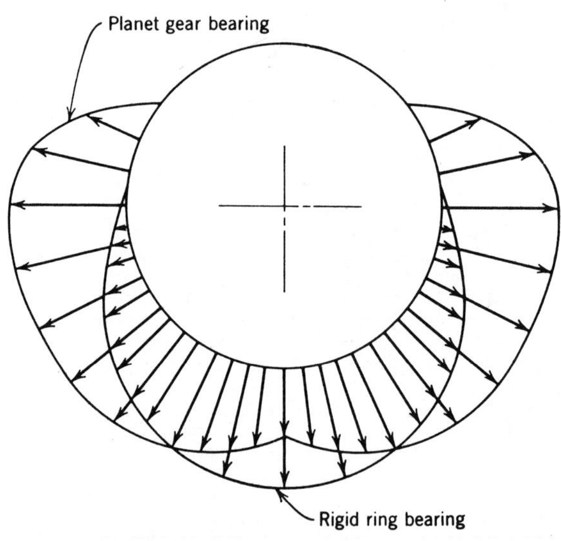

FIGURE 6.42. Comparison of load distribution for a rigid ring bearing and planet gear bearing.

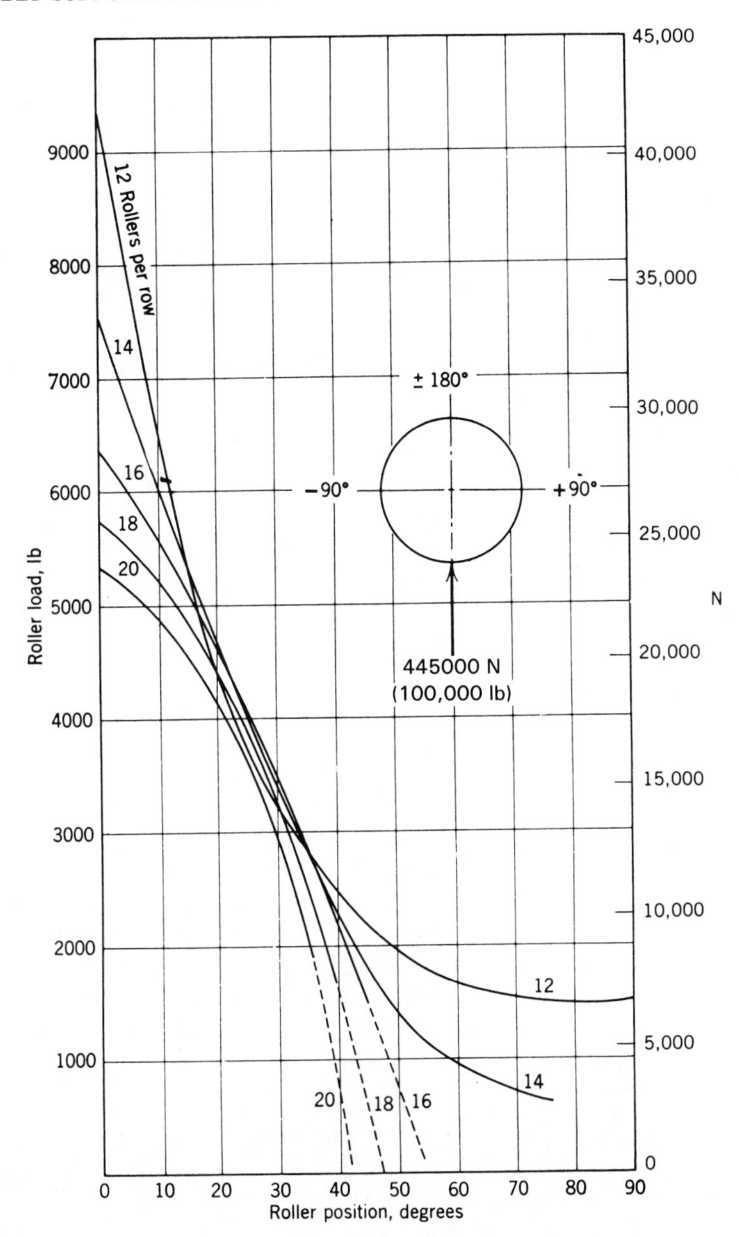

FIGURE 6.43. Roller load vs number of rollers and roller position. 445,000 N (100,000 lb) applied at 0° position, inner ring dimensions constant. Outer ring section thickness increases as the number of rollers is increased and roller diameter is subsequently decreased.

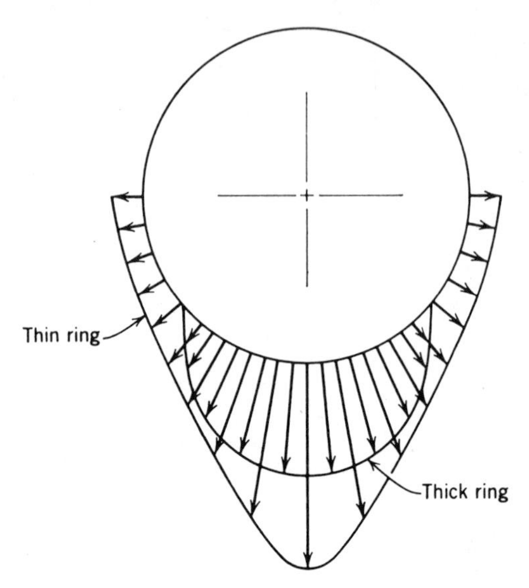

FIGURE 6.44. Comparison of load distribution of thin and thick outer ring, point-loaded back-up roll bearings.

Thus δ_j at each rolling element position can be determined and, by using (6.175) the individual rolling element loads may be calculated.

Figure 6.42 shows a typical load distribution for a planet gear bearing compared to that of a rigid ring bearing subjected to a radial load of $2F_t$.

Figure 6.43 shows typical load distributions for the back-up roll bearings of Fig. 6.35 supporting the individual line loads F_1. In this illustration the roller diameter is varied as the number of rollers is changed. Figure 6.44 compares one of these load distributions with that of the comparable rigidly supported bearing. Figure 6.45 shows typical load distributions for the back-up roll bearing of Fig. 6.35, which supports the paired line loads F_2. Figure 6.46 [6.11], which is a photoelastic study of a similarly loaded roller bearing, verifies the data of Fig. 6.45.

Finite Element Methods

In the foregoing to specify ring deflections, closed from integral analytical methods as well as influence coefficients calculated using infinite series techniques have been indicated for ring shapes, which are assumed simple both in circumference and cross section. For more complex structures, the finite element method of calculation can be used to obtain a solution whose accuracy depends only upon the fineness of the grid selected to represent the structure.

In finite element methods a function, customarily a polynomial, is chosen to define uniquely the displacement in each element (in terms of

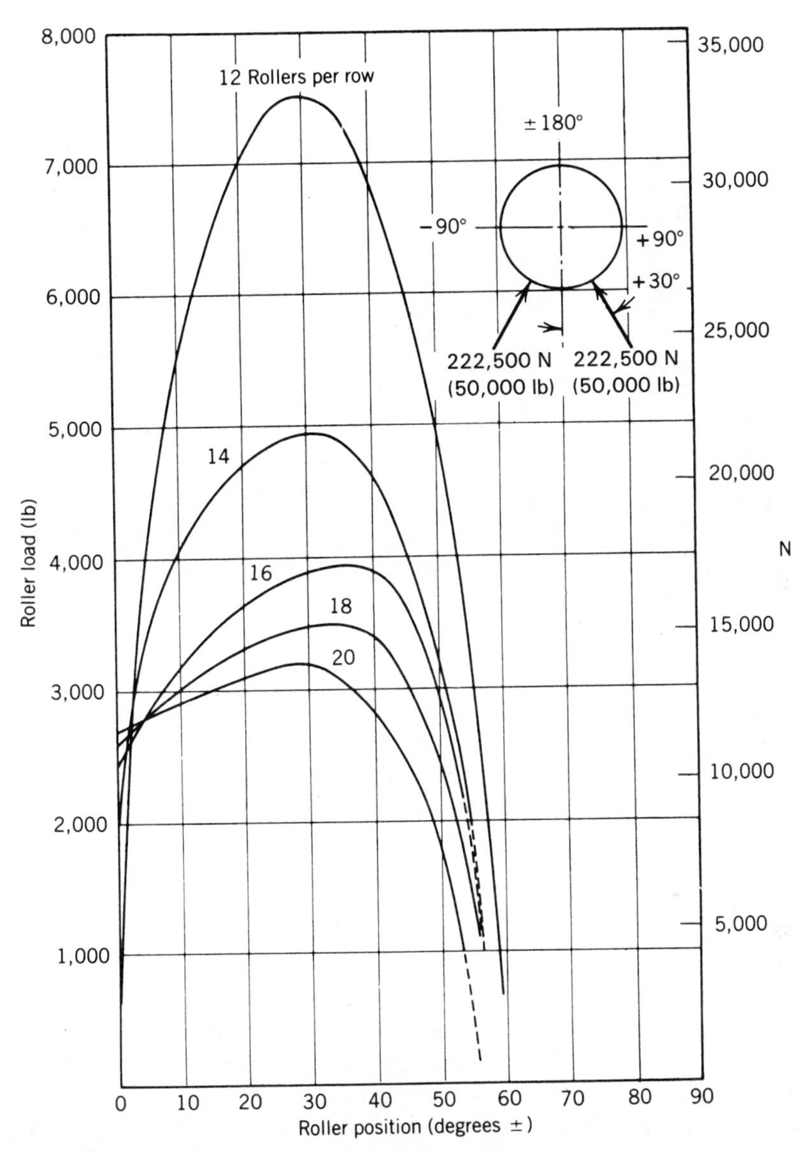

FIGURE 6.45. Roller load vs number of rollers and position. 222,500 N (50,000 lb) at $\pm 30°$, inner dimensions constant. Outer ring section thickness increases as the number of rollers is increased and roller diameter is subsequently decreased.

nodal displacements). The element stiffness matrix is obtained from equilibrium. The stiffness matrix of the complete structure is assembled, the boundary conditions are introduced, and solution of the resulting matrix equation produces the nodal displacements. Usually there is a very large quantity of unknowns; hence, a very large memory, very fast

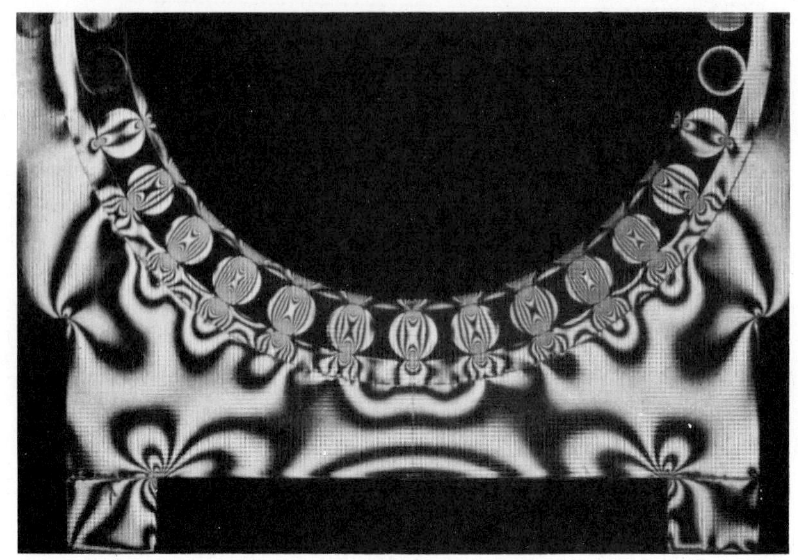

FIGURE 6.46. Photoelastic study of a roller bearing supporting loads applied at approximately $\pm 30°$ to the bearing axis.

digital computer is required to solve the displacements and load distribution accurately in a rolling bearing mounted in a flexible support. Figure 6.47 shows the grid used to analyze the combined automotive wheel hub bearing and constant velocity joint unit of Fig. 1.55.

CLOSURE

The methods developed in this chapter to calculate distribution of load among the balls and rollers of rolling bearings can be used in most bearing applications because rotational speeds are usually slow to moderate. Under these speed conditions, the effects of rolling element centrifugal forces and gyroscopic moments are negligible. At high speeds of rotation these body forces become significant, tending to alter contact angles and clearance. Thus, they can affect the static load distribution to a great extent. In Chapter 8 the effect of these parameters on high speed bearing load distribution will be evaluated.

In the foregoing discussion the effect of load distribution on the bearing deflection has been demonstrated. Further, since the contract stresses in a bearing depend on load, maximum contact stress in a bearing is also a function of load distribution. Consequently, bearing fatigue life that is governed by stress level is significantly affected by the rolling element load distribution.

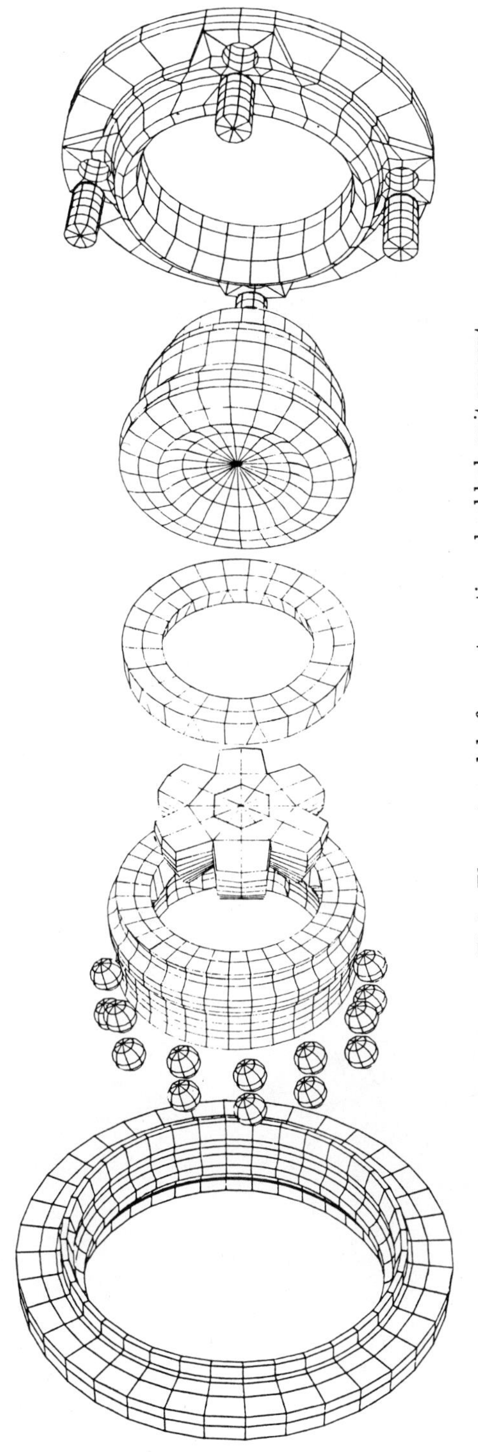

FIGURE 6.47. Finite Element model of an automotive wheel hub unit concept.

REFERENCES

6.1. R. Stribeck, "Ball Bearings for Various Loads," *Trans. ASME*, **29,** 420–463 (1907).

6.2. A. B. Jones, *Analysis of Stresses and Deflections*, New Departure Engineering Data, Bristol, Conn. (1946).

6.3. J. H. Rumbarger, "Thrust Bearings with Eccentric Loads," *Mach. Des.* (Feb. 15, 1962).

6.4. H. Sjoväll, "The Load Distribution within Ball and Roller Bearings under Given External Radial and Axial Load," *Teknisk Tidskrift*, Mek., h.9 (1933).

6.5. T. A. Harris, "The Effect of Misalignment on the Fatigue Life of Cylindrical Roller Bearings Having Crowned Rolling Members," *ASME J. Lubr. Technol.* 294–300 (April 1969).

6.6. A. B. Jones and T. A. Harris, "Analysis of a Rolling Element Idler Gear Bearing Having a Deformable Outer Race Structure," *ASME J. Basic Eng.* 273–278 (June 1963).

6.7. T. A. Harris, "Optimizing the Design of Cluster Mill Rolling Bearings," *ASLE Trans.* **7** (April 1964).

6.8. T. A. Harris and J. L. Broschard, "Analysis of an Improved Planetary Gear-Transmission Bearing," *ASME J. Basic Eng.* 457–462 (September 1964).

6.9. S. Timoshenko, *Strength of Materials, Part I*, 3rd ed., D. Van Nostrand, New York (1955).

6.10. W. Lutz, Discussion of Reference 6, presented at ASME Spring Lubrication Symposium, Miami Beach, Florida (June 5, 1962).

6.11. H. Eimer, *Aus dem Gebiet der Wälzlagertechnik*, Semesterentwurf, Technische Hochschule München (June 1964).

6.12. T. A. Harris, "The Endurance of a Thrust-loaded, Double Row, Radial Cylindrical Roller Bearing," *Wear* **18,** 429–438 (1971).

7

INTERNAL SPEEDS AND MOTIONS

LIST OF SYMBOLS

Symbol	Description	Units
a	Semimajor axis of projected contact ellipse	mm (in.)
b	Semiminor axis of projected contact ellipse	mm (in.)
d_m	Pitch diameter	mm (in.)
D	Ball or roller diameter	mm (in.)
$\mathcal{E}$	Complete elliptic integral of the second kind	
f	r/D	
h	Center of sliding	mm (in.)
M_g	Gyratory moment	N-mm (in. · lb)
n	Rotational speed	rpm
r	Raceway groove radius	mm (in.)
r'	Rolling radius	mm (in.)
R	Radius of curvature of deformed surface	mm (in.)
v	Surface velocity	mm/sec (in./sec.)

Symbol	Description	Units
x	Distance in x direction	mm (in.)
$\ddot{x}$	Acceleration in x direction	mm/sec^2 (in./sec^2)
y	Distance in y direction	mm (in.)
$\ddot{y}$	Acceleration in y direction	mm/sec^2 (in./sec^2)
z	Distance in z direction	mm (in.)
$\ddot{z}$	Acceleration in z direction	mm/sec^2 (in./sec^2)
α	Contact angle	°
β'	Angle between projection of the U axis on the $x'y'$ plane and the x' axis (Fig. 4.4)	rad
β	Angle between the W axis and z' axis (Fig. 4.4)	rad
γ'	D/d_m	
γ	$D \cos \alpha / d_m$	
κ	a/b	
ω	Rotational speed	rad/sec
θ_f	Flange angle	°

SUBSCRIPTS

f	Refers to flange
g	Refers to gyroscopic motion
i	Refers to inner raceway
m	Refers to orbital motion
o	Refers to outer raceway
R	Refers to rolling element
RE	Refers to roller end
roll	Refers to rolling motion
s	Refers to spinning motion
sl	Refers to sliding motion on flange-roller end
x	Refers to x direction (Fig. 4.4)
x'	Refers to x' direction (Fig. 4.4)
y	Refers to y direction (Fig. 4.4)
y'	Refers to y' direction (Fig. 4.4)
z	Refers to z direction (Fig. 4.4)
z'	Refers to z' direction (Fig. 4.4)

GENERAL

Ball and roller bearings are used to support various kinds of loads while permitting rotational and/or translatory motion of a shaft or slider. In this book treatment has been restricted to shaft rotation or oscillation. Unlike hydrodynamic or hydrostatic bearings, motions occurring in rolling bearings are not restricted to simple movements. For instance, in a rolling bearing mounted on a shaft that rotates at n rpm, the rolling elements orbit the bearing axis at a speed of n_m rpm, and they simultaneously revolve about their own axes at speeds of n_R rpm. Additionally, the rolling motions are accompanied by a degree of sliding that occurs in the contact areas. In ball bearings, substantial amounts of spinning motion occur simultaneously with rolling if the contact angles between balls and raceways are not zero, that is, for other than simple radial bearings. Also, gyroscopic pivotal motions occur, particularly in oil- and grease-lubricated ball bearings. In this chapter, rolling bearing internal rotational speeds and relative surface velocities, that is, sliding velocities, will be investigated and equations for their subsequent calculation will be developed.

SIMPLE ROLLING MOTION

Cage Speed

In the case of slow speed of rotation and/or an applied load of large magnitude, rolling bearings can be analyzed while neglecting dynamic effects. The resulting kinematic behavior is described in the following paragraphs.

As a general case it will be initially assumed that both inner and outer rings are rotating in a bearing having a common contact angle α (see Fig. 7.1). It is known that for a rotation about an axis,

$$v = \omega r \tag{7.1}$$

in which ω is in radians per second.

Consequently,

$$v_i = \tfrac{1}{2}\,\omega_i(d_m - D\cos\alpha)$$

or

$$v_i = \tfrac{1}{2}\,\omega_i d_m(1 - \gamma) \tag{7.2}$$

Similarly,

$$v_o = \tfrac{1}{2}\,\omega_o d_m(1 + \gamma) \tag{7.3}$$

Since

$$\omega = \frac{2\pi n}{60} \tag{7.4}$$

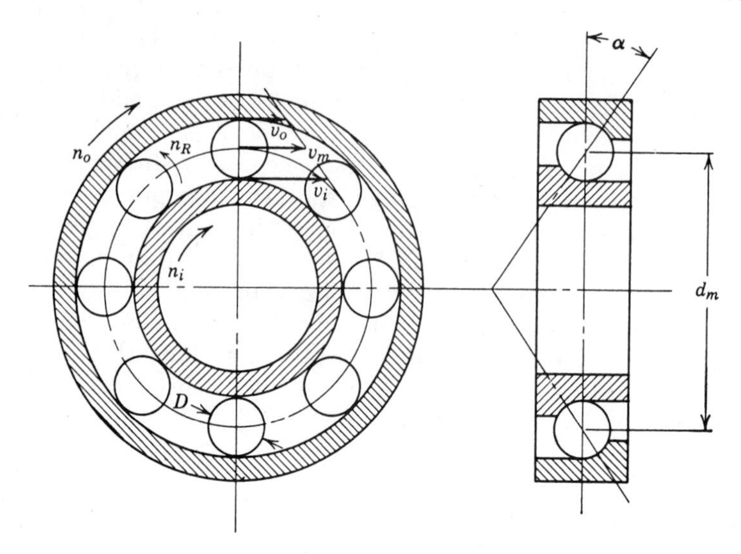

FIGURE 7.1. Rolling speeds and velocities.

in which n is in rpm, therefore

$$v_{\mathrm{i}} = \frac{\pi n_{\mathrm{i}} d_{\mathrm{m}}}{60}(1 - \gamma) \tag{7.5}$$

$$v_{\mathrm{o}} = \frac{\pi n_{\mathrm{o}} d_{\mathrm{m}}}{60}(1 + \gamma) \tag{7.6}$$

If there is no gross slip at the raceway contact, then the velocity of the cage and rolling element set is the mean of the inner and outer raceway velocities. Hence

$$v_{\mathrm{m}} = \tfrac{1}{2}(v_{\mathrm{i}} + v_{\mathrm{o}}) \tag{7.7}$$

Substituting equations (7.5) and (7.6) into (7.7) yields

$$v_{\mathrm{m}} = \frac{\pi d_{\mathrm{m}}}{120}[n_{\mathrm{i}}(1 - \gamma) + n_{\mathrm{o}}(1 + \gamma)] \tag{7.8}$$

Since

$$v_{\mathrm{m}} = \tfrac{1}{2}\,\omega_{\mathrm{m}} d_{\mathrm{m}} = \frac{\pi d_{\mathrm{m}} n_{\mathrm{m}}}{60}$$

therefore,

$$n_m = \tfrac{1}{2}[n_i(1 - \gamma) + n_o(1 + \gamma)] \tag{7.9}$$

Rolling Element Rotation

The angular speed of the cage relative to the inner raceway is

$$n_{mi} = n_m - n_i \tag{7.10}$$

Assuming no gross slip at the inner raceway–ball contact, the velocity of the ball is identical to that of the raceway at the point of contact. Hence,

$$\tfrac{1}{2}\,\omega_{mi} d_m(1 - \gamma) = \tfrac{1}{2}\,\omega_R D$$

Therefore, since n is proportional to ω and by substituting n_{mi} as in (7.10),

$$n_R = (n_m - n_i)\frac{d_m}{D}(1 - \gamma) \tag{7.11}$$

Substituting equation (7.9) for n_m yields

$$n_R = \frac{1}{2}\frac{d_m}{D}(1 - \gamma)(1 + \gamma)(n_o - n_i) \tag{7.12}$$

Considering only inner ring rotation, equations (7.9) and (7.12) become

$$n_m = \tfrac{1}{2}\,n_i(1 - \gamma) \tag{7.13}$$

$$n_R = \tfrac{1}{2}\frac{d_m}{D}\,n_i(1 - \gamma^2) \tag{7.14}$$

For a thrust bearing whose contact angle is 90°, $\cos\alpha = 0$, therefore,

$$n_m = \tfrac{1}{2}(n_i + n_o) \tag{7.15}$$

$$n_R = \frac{1}{2}\frac{d_m}{D}(n_o - n_i) \tag{7.16}$$

Example 7.1. Determine the cage and ball speeds of the 209 radial ball bearing of Example 6.1 if the shaft turns at 1800 rpm.

$$D = 12.7 \text{ mm } (0.5 \text{ in.}) \qquad \text{Ex. 2.1}$$

$$d_m = 65 \text{ mm } (2.559 \text{ in.}) \qquad \text{Ex. 2.1}$$

$$\alpha = 0^\circ \quad \text{(under radial load)}$$

$$\gamma = 0.1954 \qquad \text{Ex. 2.5}$$

$$n_m = \tfrac{1}{2}\, n_i(1 - \gamma) \qquad (7.13)$$

$$= 0.5 \times 1800(1 - 0.1954) = 724.1 \text{ rpm}$$

$$n_R = \frac{1}{2} \times \frac{d_m}{D}\, n_i(1 - \gamma^2) \qquad (7.14)$$

$$= \frac{0.5 \times 65}{12.7} \times 1800[1 - (0.1954)^2] = 4430 \text{ rpm}$$

Example 7.2. Estimate the cage speed of the 218 angular-contact ball bearing of Example 6.5 if the shaft turns at 1800 rpm.

$$D = 22.23 \text{ mm } (0.875 \text{ in.}) \qquad \text{Ex. 2.3}$$

$$d_m = 125.3 \text{ mm } (4.932 \text{ in.}) \qquad \text{Ex. 2.6}$$

$$\alpha = 41^\circ 36' \qquad \text{Ex. 6.5}$$

$$\gamma = \frac{D \cos \alpha}{d_m} \qquad (2.27)$$

$$= \frac{22.23 \cos (41^\circ 36')}{125.3} = 0.1327$$

$$n_m = \tfrac{1}{2}\, n_i(1 - \gamma) \qquad (7.13)$$

$$= \tfrac{1}{2} \times 1800(1 - 0.1327) = 780.6 \text{ rpm}$$

This estimate is satisfactory in this application because of the following:

$$F_c = 2.26 \times 10^{-11} D^3 n_m{}^2 d_m \qquad (4.41)$$

$$= 2.26 \times 10^{-11} \times (22.23)^3 (780.6)^2 \times 125.3$$

$$= 18.9 \text{ N } (4.247 \text{ lb}) \qquad \text{Ex. 6.5}$$

$$F_{\mathrm{a}} = 17{,}800 \text{ N } (4000 \text{ lb}) \qquad \text{Ex. 6.5}$$

$$Z = 16 \qquad \text{Ex. 6.5}$$

$$Q_{\mathrm{ia}} = \frac{F_{\mathrm{a}}}{Z} = \frac{17800}{16} = 1113 \text{ N/ball } (250 \text{ lb/ball})$$

Since $F_{\mathrm{a}}/Z \gg F_{\mathrm{c}}$, α_{i} $(41°36')$ is very nearly equal to α_{o}.

ROLLING AND SLIDING

Geometrical Considerations

The only conditions that can sustain *pure* rolling between two contacting surfaces are

1. Mathematical line contact under zero load
2. Line contact in which the contacting bodies are identical in length
3. Mathematical point contact under zero load

Even when the foregoing conditions are achieved it is possible to have sliding. Sliding is then a condition of overall relative movement of the rolling body over the contact area.

The motion of a rolling element with respect to the raceway consists of a rotation about the generatrix of motion. If the contact surface is a straight line in one of the principal directions, the generatrix of motion may intersect the contact surface at one point only, as in Fig. 7.2. The component ω_{R} of angular velocity ω, which acts in the plane of the contact surface, produces rolling motion. As indicated in Fig. 7.3, the component ω_{s} of angular velocity ω that acts normal to the surface causes a spinning motion about a point of pure rolling O. The instantaneous direction of sliding in the contact zone is shown in Fig. 7.4.

In ball bearings with nonzero contact angles between balls and raceways, during operation at any shaft or outer ring speed, a gyroscopic moment occurs on each loaded ball, tending to cause a sliding motion.

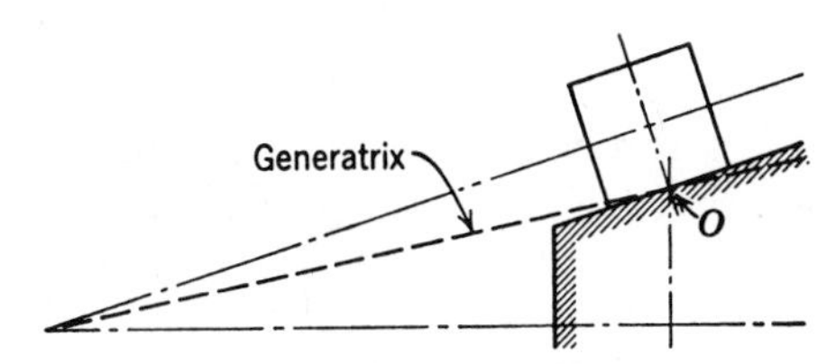

FIGURE 7.2. Roller–raceway contact; generatrix of motion pierces contact surface.

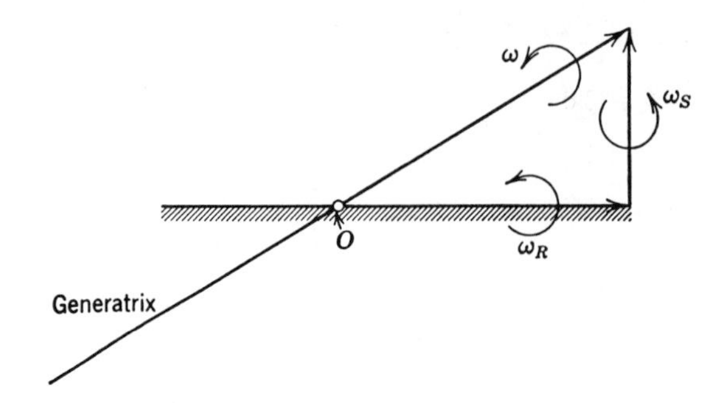

FIGURE 7.3. Resolution of angular velocities into rolling and spinning motions.

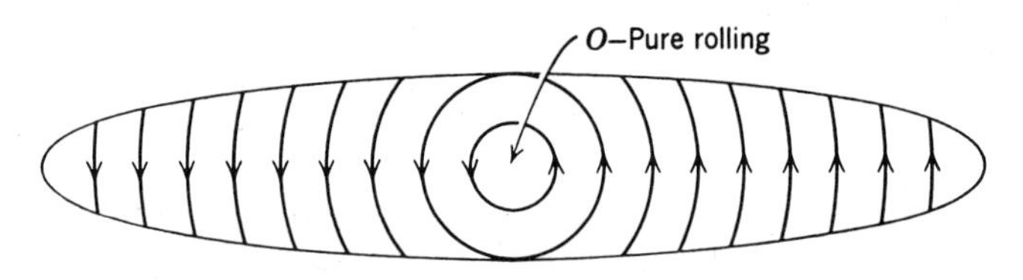

FIGURE 7.4. Contact ellipse showing sliding lines and point of pure rolling.

In most applications, because of relatively slow input speeds and/or heavy loading, such gyroscopic moments and hence motions can be neglected. In high speed applications with an oil-film lubrication between balls and raceways, such motion will occur.

The sliding velocity due to gyroscopic motion is given by (see Fig. 7.5)

$$v_{\mathrm{g}} = \tfrac{1}{2}\,\omega_{\mathrm{g}} D \tag{7.17}$$

The sliding velocities caused by gyroscopic motion and spinning of the balls are vectorially additive such that at some distance h from O they cancel each other. Thus,

$$v_{\mathrm{g}} = \omega_{\mathrm{s}} h \tag{7.18}$$

and

$$h = \frac{D}{2} \times \frac{\omega_{\mathrm{g}}}{\omega_{\mathrm{s}}} \tag{7.19}$$

The distance h defines the center of sliding about which a rotation of angular velocity ω_{s} occurs. This center of sliding (spinning) may occur within or outside of the contact surface. Figure 7.6 shows the pattern of sliding lines in the contact area for simultaneous rolling, spinning, and gyroscopic motion in a ball bearing operating under heavy load and moderate speed. Figure 7.7, which corresponds to low load and high speed

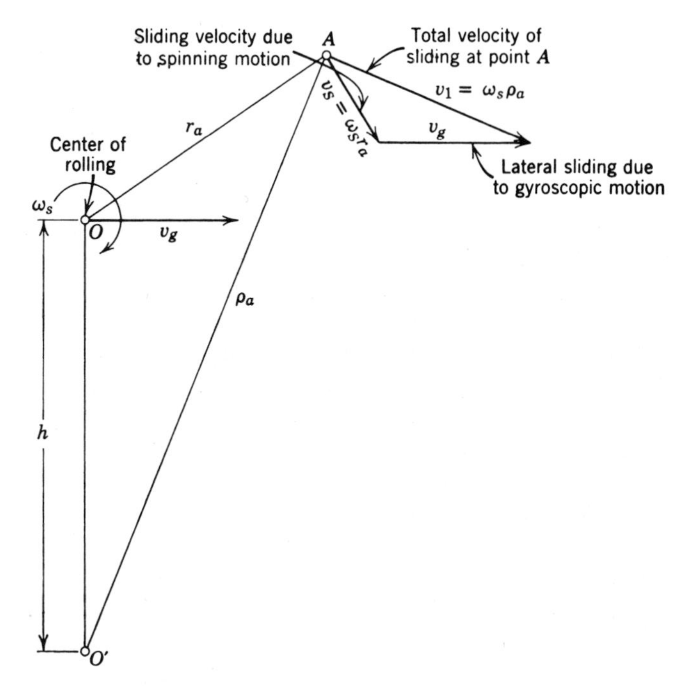

FIGURE 7.5. Velocities of sliding at arbitrary point A in contact area.

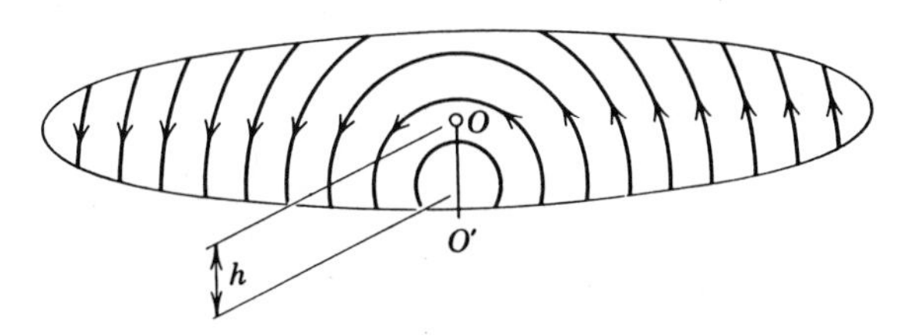

FIGURE 7.6. Sliding lines in contact area for simultaneous rolling, spinning, and gyroscopic motions—low speed operation of a ball bearing.

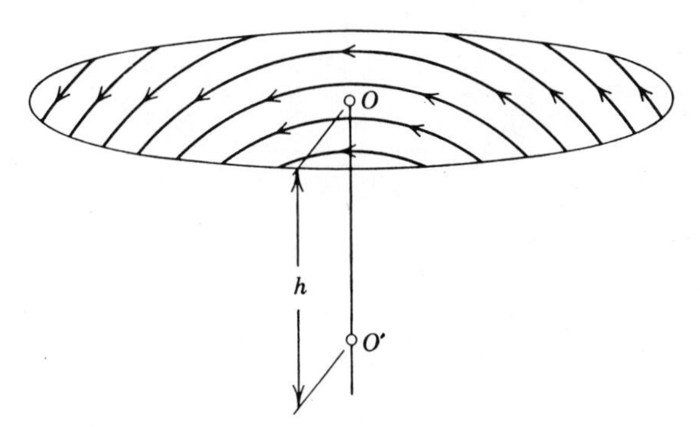

FIGURE 7.7. Sliding lines in contact area for simultaneous rolling, spinning, and gyroscopic motions—high speed operation of a ball bearing (not considering *skidding*).

conditions (however, not considering skidding*), indicates that the center of sliding is outside of the contact surface and sliding occurs over the entire contact surface. The distance h between the centers of contact and sliding is a function of the magnitude of the gyroscopic moment that can be compensated by contact surface friction forces.

Sliding and Deformation

Even when the generatrix of motion apparently lies in the plane of the contact surface, as for radial cylindrical roller bearings, sliding on the contact surface can occur when a roller is under load. In accordance with the Hertzian radius of the contact surface in the direction transverse to motion, the contact surface has a harmonic mean profile radius, which means that the contact surface is not plane, but generally curved as shown by Fig. 7.8 for a radial bearing. The generatrix of motion, being parallel to the tangent plane of the center of the contact surface, therefore pierces the contact surface at two points at which rolling occurs. Since the rigid rolling element rotates with a singular angular velocity about its axis, surface points at different radii from the axis have differ-

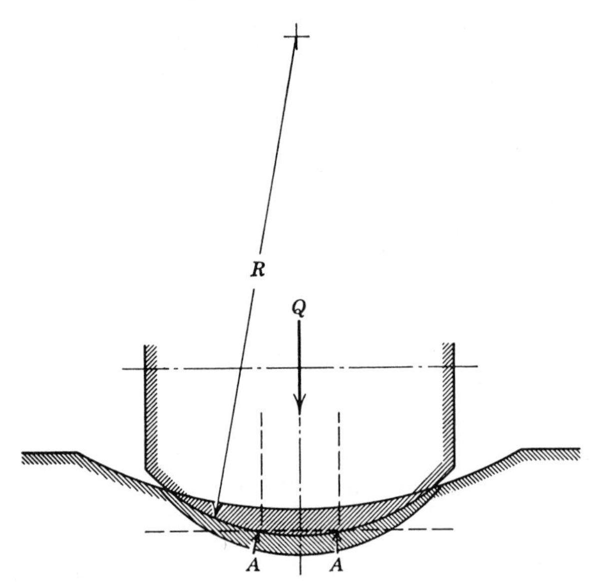

FIGURE 7.8. Roller–raceway contact showing harmonic mean radius and points of rolling A–A.

*Skidding is a very gross sliding condition occurring generally in oil-film lubricated ball and roller bearings operating under relatively light load at very high speed or rapid accelerations and decelerations. When skidding occurs, cage speed will be less than predicted by equation (7.9) for bearings with inner ring rotation.

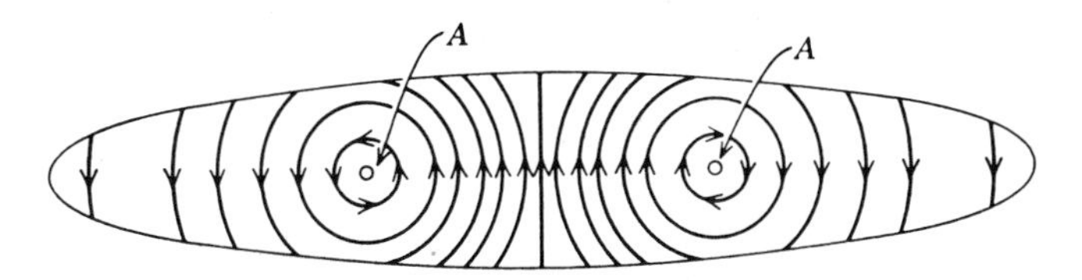

FIGURE 7.9. Sliding lines in contact area of Fig. 7.8.

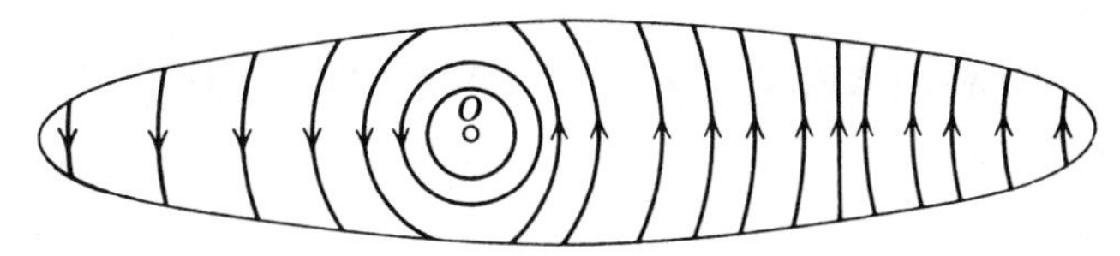

FIGURE 7.10. Sliding lines for roller–raceway contact area when load is applied; generatrix of motion pierces contact area.

ent surface velocities only two of which being symmetrically disposed about the roller geometrical center can exhibit *pure* rolling motion. In Fig. 7.9 points within area *A–A* slide backward with regard to the direction of rolling and points outside of *A–A* slide forward with respect to the direction of rolling. Figure 7.9 shows the pattern of sliding lines in the elliptical contact area.

If the generatrix of motion is angled with respect to the tangent plane at the center of the contact surface, the center of rolling is positioned unsymmetrically in the contact ellipse and, depending on the angle of the generatrix to the contact surface, one point or two points of intersection may occur at which rolling obtains. Figure 7.10 shows the sliding lines for this condition.

For a ball bearing in which rolling, spinning, and gyroscopic motions occur simultaneously, the pattern of sliding lines in the elliptical contact area is as shown in Figs. 7.11 and 7.12. More detailed information on sliding in the elliptical contact area may be found in the work by Lundberg [7.4].

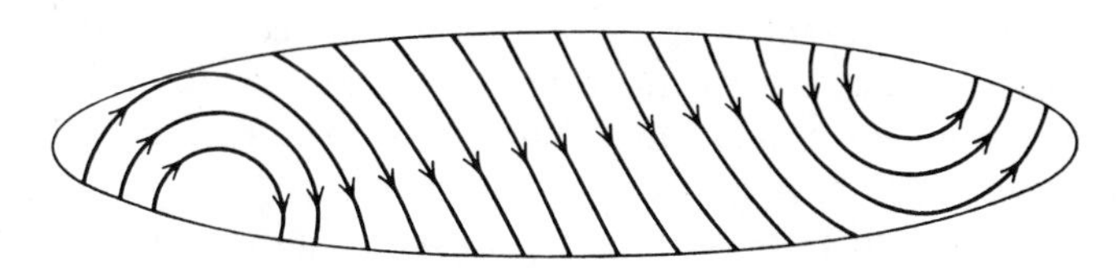

FIGURE 7.11. Sliding lines for ball–raceway contact area for simultaneous rolling, spinning, and gyroscopic motions—high load and low speed operation of an angular-contact ball bearing.

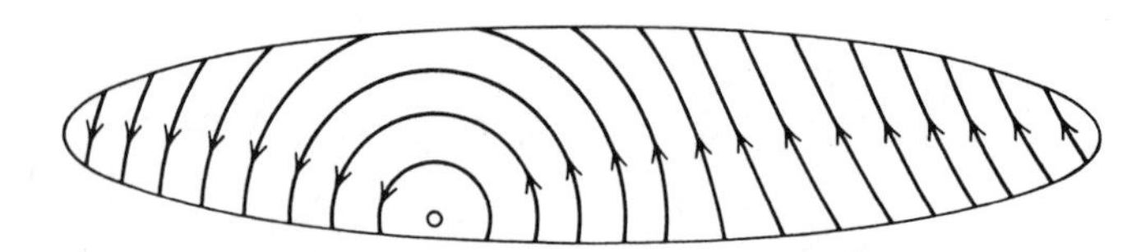

FIGURE 7.12. Sliding lines for ball–raceway contact area for simultaneous rolling, spinning, and gyroscopic motions—low load and high speed operation of an angular-contact ball bearing (not considering *skidding*).

ORBITAL, PIVOTAL, AND SPINNING MOTIONS IN BALL BEARINGS

General Motions

Figure 7.13 shows a ball contacting the outer raceway such that the normal force Q between the ball and raceway is distributed over an elliptical surface defined by projected major and minor semiaxes, a_o and b_o, respectively. The radius of curvature of the deformed pressure surface as defined by Hertz is

$$R_o = \frac{2r_oD}{2R_o + D} \tag{7.20}$$

in which r_o is the outer raceway groove curvature radius. In terms of curvature f_o:

$$R_o = \frac{2f_oD}{2f_o + 1} \tag{7.21}$$

Assume for the present purpose that the ball center is fixed in space and that the outer raceway rotates with angular speed ω_o. (The vector of ω_o is perpendicular to the plane of rotation and therefore collinear with the x axis.) Moreover, it can be seen from Fig. 4.4 that ball rotational speed ω_R has components $\omega_{x'}$ and $\omega_{z'}$ lying in the plane of the paper when $\psi = 0$.

Because of the deformation at the pressure surface defined by a_o and b_o, the radius from the ball center to the raceway contact point varies in length as the contact ellipse is traversed from $+a_o$ to $-a_o$. Therefore because of symmetry about the minor axis of the contact ellipse, pure rolling motion of the ball over the raceway occurs at most at two points. The radius at which pure rolling occurs is defined as r_o' and must be determined by methods of contact deformation analysis.

It can be seen from Figure 7.13 that the outer raceway has a component $\omega_o \cos \alpha_o$ of the angular velocity vector in a direction parallel to the

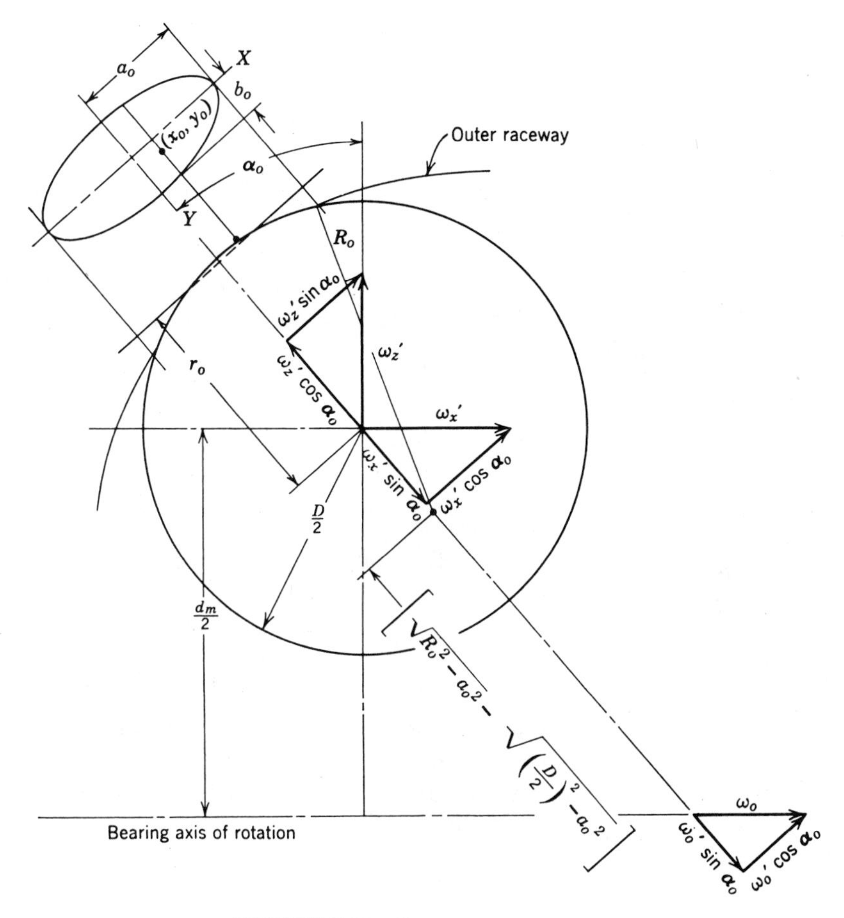

FIGURE 7.13. Outer raceway contact.

major axis of the contact ellipse. Therefore, a point (x_o, y_o) on the outer raceway has a linear velocity v_{lo} in the direction of rolling as defined below:

$$v_{lo} = -\frac{d_m \omega_o}{2} - \left\{ (R_o^2 - x_o^2)^{1/2} - (R_o^2 - a_o^2)^{1/2} + \left[\left(\frac{D}{2} \right)^2 - a_o^2 \right]^{1/2} \right\} \omega_o \cos \alpha_o \tag{7.22}$$

Similarly, the ball has angular velocity components, $\omega_{x'} \cos \alpha_o$ and $\omega_{z'} \sin \alpha_o$ of the angular velocity vector ω_R lying in the plane of the paper and parallel to the major axis of the contact ellipse. Thus, a point (x_o, y_o)

on the ball has a linear velocity v_{2o} in the direction of rolling defined as follows:

$$v_{2o} = -(\omega_{x'} \cos \alpha_o + \omega_{z'} \sin \alpha_o) \times \left\{ (R_o^2 - x_o^2)^{1/2} - (R_o^2 - a_o^2)^{1/2} + \left[\left(\frac{D}{2}\right)^2 - a_o^2 \right]^{1/2} \right\} \quad (7.23)$$

Slip or sliding of the outer raceway over the ball in the direction of rolling is determined by the difference between the linear velocities of raceway and ball. Hence

$$v_{yo} = v_{1o} - v_{2o} \quad (7.24)$$

or

$$v_{yo} = -\frac{d_m \omega_o}{2} + (\omega_{x'} \cos \alpha_o + \omega_{z'} \sin \alpha_o - \omega_o \cos \alpha_o) \times \left\{ (R_o^2 - x_o^2)^{1/2} - (R_o^2 - a_o^2)^{1/2} + \left[\left(\frac{D}{2}\right)^2 - a_o^2 \right]^{1/2} \right\} \quad (7.25)$$

Additionally, the ball angular velocity vector ω_R has a component $\omega_{y'}$ in a direction perpendicular to the plane of the paper. This component causes a slip v_{xo} in the direction transverse to the rolling, that is, in the direction of the major axis of the contact ellipse. This slip velocity is given by

$$v_{xo} = -\omega_{y'} \left\{ (R_o^2 - x_o^2)^{1/2} - (R_o^2 - a_o^2)^{1/2} + \left[\left(\frac{D}{2}\right)^2 - a_o^2 \right]^{1/2} \right\} \quad (7.26)$$

From Fig. 7.13 it can be observed that both the ball angular velocity vectors $\omega_{x'}$ and $\omega_{z'}$ and the raceway angular velocity vector ω_o have components normal to the contact area. Hence, there is a rotation about a normal to the contact area, in other words a spinning of the outer raceway relative to the ball, the net magnitude of which is given by

$$\omega_{so} = -\omega_o \sin \alpha_o + \omega_{x'} \sin \alpha_o - \omega_{z'} \cos \alpha_o \quad (7.27)$$

From Fig. 4.4 it can be determined that

$$\omega_{x'} = \omega_R \cos \beta \cos \beta' \quad (7.28)$$

$$\omega_{y'} = \omega_R \cos \beta \sin \beta' \quad (7.29)$$

$$\omega_{z'} = \omega_R \sin \beta \quad (7.30)$$

Substitution of equations (7.28) and (7.30) into (7.25), (7.26), and (7.27) yields

$$v_{yo} = -\frac{d_m \omega_o}{2} + \left\{(R_o^2 - x_o^2)^{1/2} - (R_o^2 - a_o^2)^{1/2} + \left[\left(\frac{D}{2}\right)^2 - a_o^2\right]^{1/2}\right\} \times \left(\frac{\omega_R}{\omega_o} \cos\beta \cos\beta' \cos\alpha_o + \frac{\omega_R}{\omega_o} \sin\beta \sin\alpha_o - \cos\alpha_o\right)\omega_o \quad (7.31)$$

$$v_{xo} = -\left\{(R_o^2 - x_o^2)^{1/2} - (R_o^2 - a_o^2)^{1/2} + \left[\left(\frac{D}{2}\right)^2 - a_o^2\right]^{1/2}\right\} \omega_o \left(\frac{\omega_R}{\omega_o}\right) \cos\beta \sin\beta' \quad (7.32)$$

$$\omega_{so} = \left(\frac{\omega_R}{\omega_o} \cos\beta \cos\beta' \sin\alpha_o - \frac{\omega_R}{\omega_o} \sin\beta \cos\alpha_o - \sin\alpha_o\right)\omega_o \quad (7.33)$$

Note that at the radius of rolling r_o' on the ball, the translational velocity of the ball is identical to that of the outer raceway. From Fig. 7.13, therefore,

$$\left(\frac{d_m}{2\cos\alpha_o} + r_o'\right)\omega_o \cos\alpha_o = r_o'(\omega_{x'} \cos\alpha_o + \omega_{z'} \sin\alpha_o) \quad (7.34)$$

Substituting equations (7.28) and (7.29) into (7.34) and rearranging terms yields

$$\frac{\omega_R}{\omega_o} = \frac{(d_m/2) + r_o' \cos\alpha_o}{r_o'(\cos\beta \cos\beta' \cos\alpha_o + \sin\beta \sin\alpha_o)} \quad (7.35)$$

A similar analysis may be applied to the inner raceway contact as illustrated in Fig. 7.14. The following equations can be determined:

$$v_{yi} = -\frac{d_m \omega_i}{2} - \left\{(R_i^2 - x_i^2)^{1/2} - (R_i^2 - a_i^2)^{1/2} + \left[\left(\frac{D}{2}\right)^2 - a_i^2\right]^{1/2}\right\} \times \left(\frac{\omega_R}{\omega_i} \cos\beta \cos\beta' \cos\alpha_i + \frac{\omega_R}{\omega_i} \sin\beta \sin\alpha_i - \cos\alpha_i\right)\omega_i \quad (7.36)$$

$$v_{xi} = -\left\{(R_i^2 - x_i^2)^{1/2} - (R_i^2 - a_i^2)^{1/2} + \left[\left(\frac{D}{2}\right)^2 - a_i^2\right]^{1/2}\right\} \omega_i \left(\frac{\omega_R}{\omega_i}\right) \cos\beta \sin\beta' \quad (7.37)$$

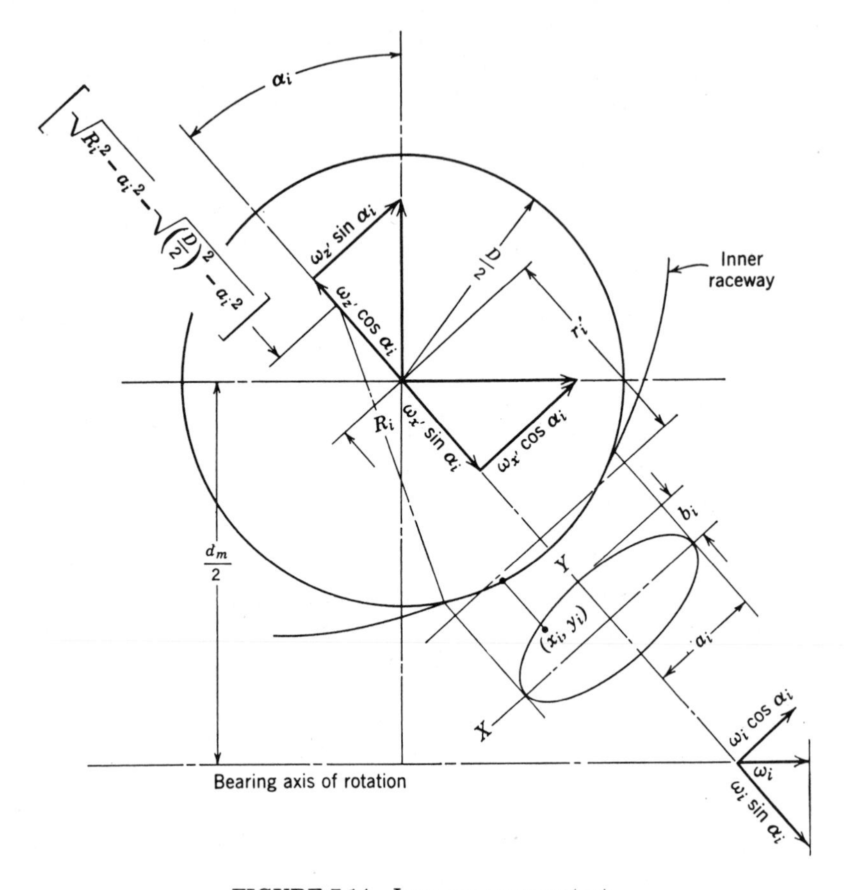

FIGURE 7.14. Inner raceway contact.

$$\omega_{si} = \left(-\frac{\omega_R}{\omega_i} \cos \beta \cos \beta' \sin \alpha_i + \frac{\omega_R}{\omega_i} \sin \beta \cos \alpha_i + \sin \alpha_i \right) \omega_i \quad (7.38)$$

$$\frac{\omega_R}{\omega_i} = \frac{-(d_m/2) + r_i' \cos \alpha_i}{r_i' (\cos \beta \cos \beta' \cos \alpha_i + \sin \beta \sin \alpha_i)} \quad (7.39)$$

If instead of the ball center being fixed in space, the outer raceway is fixed, then the ball center must orbit about the center 0 of the fixed coordinate system with an angular speed $\omega_m = -\omega_o$. Therefore the inner raceway must rotate with absolute angular speed $\omega = \omega_i + \omega_m$. By using these relationships, the relative angular speeds ω_i and ω_o can be described in terms of the absolute angular speed of the inner raceway as follows:

$$\omega_{\mathrm{i}} = \frac{\omega}{1 + \dfrac{r'_{\mathrm{o}}[(d_{\mathrm{m}}/2) - r'_{\mathrm{i}} \cos \alpha_{\mathrm{i}}](\cos \beta \cos \beta' \cos \alpha_{\mathrm{o}} + \sin \beta \sin \alpha_{\mathrm{o}})}{r'_{\mathrm{i}}[(d_{\mathrm{m}}/2) + r'_{\mathrm{o}} \cos \alpha_{\mathrm{o}}](\cos \beta \cos \beta' \cos \alpha_{\mathrm{i}} + \sin \beta \sin \alpha_{\mathrm{i}})}} \tag{7.40}$$

$$\omega_{\mathrm{o}} = \frac{-\omega}{1 + \dfrac{r'_{\mathrm{i}}[(d_{\mathrm{m}}/2) + r'_{\mathrm{o}} \cos \alpha_{\mathrm{o}}](\cos \beta \cos \beta' \cos \alpha_{\mathrm{i}} + \sin \beta \sin \alpha_{\mathrm{i}})}{r'_{\mathrm{o}}[(d_{\mathrm{m}}/2) - r'_{\mathrm{i}} \cos \alpha_{\mathrm{i}}](\cos \beta \cos \beta' \cos \alpha_{\mathrm{o}} + \sin \beta \sin \alpha_{\mathrm{o}})}} \tag{7.41}$$

Further,

$$\omega_{\mathrm{R}} = \frac{-\omega}{\dfrac{r'_{\mathrm{o}}(\cos \beta \cos \beta' \cos \alpha_{\mathrm{o}} + \sin \beta \sin \alpha_{\mathrm{o}})}{(d_{\mathrm{m}}/2) + r'_{\mathrm{o}} \cos \alpha_{\mathrm{o}}} + \dfrac{r'_{\mathrm{i}}(\cos \beta \cos \beta' \cos \alpha_{\mathrm{i}} + \sin \beta \sin \alpha_{\mathrm{i}})}{(d_{\mathrm{m}}/2) - r'_{\mathrm{i}} \cos \alpha_{\mathrm{i}}}} \tag{7.42}$$

Similarly, if the outer raceway rotates with absolute angular speed ω and the inner raceway is stationary, $\omega_{\mathrm{m}} = \omega_{\mathrm{i}}$ and $\omega = \omega_{\mathrm{m}} + \omega_{\mathrm{o}}$. Therefore,

$$\omega_{\mathrm{o}} = \frac{\omega}{1 + \dfrac{r'_{\mathrm{i}}[(d_{\mathrm{m}}/2) + r'_{\mathrm{o}} \cos \alpha_{\mathrm{o}}](\cos \beta \cos \beta' \cos \alpha_{\mathrm{i}} + \sin \beta \sin \alpha_{\mathrm{i}})}{r'_{\mathrm{o}}[(d_{\mathrm{m}}/2) - r'_{\mathrm{i}} \cos \alpha_{\mathrm{i}}](\cos \beta \cos \beta' \cos \alpha_{\mathrm{o}} + \sin \beta \sin \alpha_{\mathrm{o}})}} \tag{7.43}$$

$$\omega_{\mathrm{i}} = \frac{-\omega}{1 + \dfrac{r'_{\mathrm{o}}[(d_{\mathrm{m}}/2) - r'_{\mathrm{i}} \cos \alpha_{\mathrm{i}}](\cos \beta \cos \beta' \cos \alpha_{\mathrm{o}} + \sin \beta \sin \alpha_{\mathrm{o}})}{r'_{\mathrm{i}}[(d_{\mathrm{m}}/2) + r'_{\mathrm{o}} \cos \alpha_{\mathrm{o}}](\cos \beta \cos \beta' \cos \alpha_{\mathrm{i}} + \sin \beta \sin \alpha_{\mathrm{i}})}} \tag{7.44}$$

$$\omega_{\mathrm{R}} = \frac{\omega}{\dfrac{r'_{\mathrm{o}}(\cos \beta \cos \beta' \cos \alpha_{\mathrm{o}} + \sin \beta \sin \alpha_{\mathrm{o}})}{(d_{\mathrm{m}}/2) + r'_{\mathrm{o}} \cos \alpha_{\mathrm{o}}} + \dfrac{r'_{\mathrm{i}}(\cos \beta \cos \beta' \cos \alpha_{\mathrm{i}} + \sin \beta \sin \alpha_{\mathrm{i}})}{(d_{\mathrm{m}}/2) - r'_{\mathrm{i}} \cos \alpha_{\mathrm{i}}}} \tag{7.45}$$

Inspection of the final equations relating the relative motions of the balls and raceways reveals the following unknown quantities: r'_{o}, r'_{i}, β', β, α_{i} and α_{o}. It is apparent that analysis of the forces and moments acting on each ball will be required to evaluate the unknown quantities. As a practical matter, however, it is sometimes possible to avoid this lengthy procedure requiring digital computation by using the simplifying assumption that a ball will roll on one raceway without spinning and spin and roll simultaneously on the other raceway. The raceway on which only rolling occurs is called the "controlling" raceway. Moreover, it is also possible to assume that gyroscopic pivotal motion is negligible; some criteria for this will be discussed.

No Gyroscopic Pivotal Motion

In the event that gyroscopic rotation is prevented or is minimal then the angle β' is 0° (see Fig. 4.4). Therefore, the angular rotation $\omega_{y'}$ is zero and further

$$\omega_{x'} = \omega_R \cos \beta \tag{7.46}$$

$$\omega_{z'} = \omega_R \sin \beta \tag{7.47}$$

A second consequence of $\beta' = 0$ is that

$$\frac{\omega_R}{\omega_o} = \frac{(d_m/2) + r'_o \cos \alpha_o}{r'_o(\cos \alpha_o \cos \beta + \sin \beta \sin \alpha_o)} \tag{7.48}$$

and

$$\frac{\omega}{\omega_i} = \frac{-(d_m/2) + r'_i \cos \alpha_i}{r'_i (\cos \beta \cos \alpha_i + \sin \beta \sin \alpha_i)} \tag{7.49}$$

Spin-to-Roll Ratio

Assuming for this calculation that r_i, r_o, and $\frac{1}{2} D$ are essentially equal, the ball rolling speed relative to the outer raceway is

$$\omega_{roll} = -\omega_o \frac{d_m}{D} = -\frac{\omega_o}{\gamma'} \tag{7.50}$$

From equation (7.33) for negligible gyroscopic moment ($\beta' = 0$),

$$\omega_{so} = \omega_R \cos \beta \sin \alpha_o - \omega_R \sin \beta \cos \alpha_o - \omega_o \sin \alpha_o \tag{7.51}$$

or

$$\omega_{so} = \omega_R \sin (\alpha_o - \beta) - \omega_o \sin \alpha_o \tag{7.52}$$

Dividing by ω_{roll} according to equation (7.50) yields

$$\left(\frac{\omega_{\text{s}}}{\omega_{\text{roll}}}\right)_{\text{o}} = -\gamma'\frac{\omega_{\text{R}}}{\omega_{\text{o}}}\sin(\alpha_{\text{o}} - \beta) + \gamma' \sin \alpha_{\text{o}} \tag{7.53}$$

According to equation (7.48), replacing $2r_{\text{o}}'/d_{\text{m}}$ by γ':

$$\frac{\omega_{\text{R}}}{\omega_{\text{o}}} = \frac{1 + \gamma' \cos \alpha_{\text{o}}}{\gamma'(\cos \beta \cos \alpha_{\text{o}} + \sin \beta \sin \alpha_{\text{o}})} \tag{7.54}$$

or

$$\frac{\omega_{\text{R}}}{\omega_{\text{o}}} = \frac{1 + \gamma' \cos \alpha_{\text{o}}}{\gamma' \cos(\alpha_{\text{o}} - \beta)} \tag{7.55}$$

Therefore substitution of equation (7.55) into (7.53) yields

$$\left(\frac{\omega_{\text{s}}}{\omega_{\text{roll}}}\right)_{\text{o}} = -(1 + \gamma' \cos \alpha_{\text{o}}) \tan(\alpha_{\text{o}} - \beta) + \gamma' \sin \alpha_{\text{o}} \tag{7.56}$$

Similarly, for an inner raceway contact

$$\left(\frac{\omega_{\text{s}}}{\omega_{\text{roll}}}\right)_{\text{i}} = (1 - \gamma' \cos \alpha_{\text{i}}) \tan(\alpha_{\text{i}} - \beta) + \gamma' \sin \alpha_{\text{i}} \tag{7.57}$$

Assuming now that pure rolling occurs only at the outer raceway contact, therefore ω_{so} is 0, and substitution of equation (7.48) into (7.33) for this condition indicates that

$$\tan \beta = \frac{(d_{\text{m}} \sin \alpha_{\text{o}})/2}{(d_{\text{m}} \cos \alpha_{\text{o}})/2 + r_{\text{o}}'} \tag{7.58}$$

Since $r_{\text{o}}' \simeq \frac{1}{2}D$ and $D/d_{\text{m}} = \gamma'$, equation (7.58) becomes

$$\tan \beta = \frac{\sin \alpha_{\text{o}}}{\cos \alpha_{\text{o}} + \gamma'} \tag{7.59}$$

Raceway Control

Harris [7.5] showed that, in general, it is not possible for pure rolling, that is, without simultaneous spinning motion, to occur at either the inner or outer raceway contacts as long as the ball–raceway contact angle is nonzero. For high speed operation of relatively lightly loaded oil–

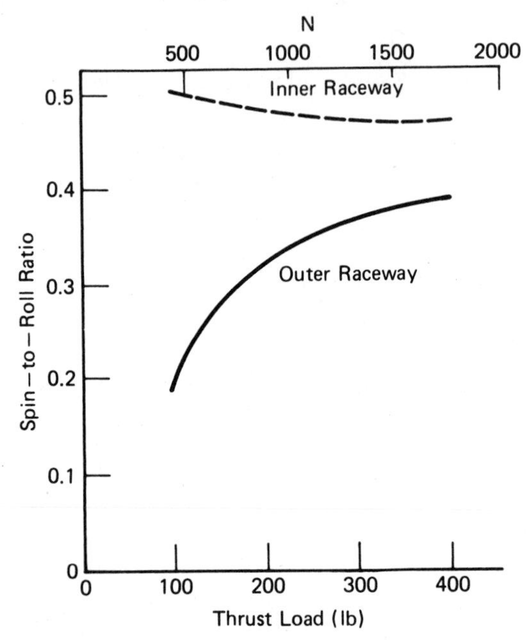

FIGURE 7.15. Spin-to-roll ratio vs thrust load for an oil–film lubricated angular-contact ball bearing.

film lubricated bearings, however, the condition of "outer raceway control" tends to be approximated. Figure 7.15 taken from reference [7.5] illustrates this condition for a high speed thrust-loaded aircraft gas turbine, angular-contact ball bearing. It must be noted that skidding also tends to occur at the same time.

Hence, for oil–film lubricated ball bearings (including grease-lubricated ball bearings), determination of actual internal speeds and motions requires a rather sophisticated mathematical analysis together with use of digital computation. Such methods require an understanding of friction and will be discussed later in this text.

For dry film-lubricated ball bearings or for ball bearings in which a constant coefficient of friction may be assumed in the ball–raceway contacts, Harris [7.6] has shown for a thrust-loaded angular-contact ball bearing that, at relatively slow speed, spinning and rolling occur simultaneously at both inner and outer ball–raceway contacts. For a given load, as speed is increased, a transition takes place in which outer raceway control is approximated; however, the outer raceway contact spin-to-roll ratio is always nonzero (see Figs. 7.16 and 7.17). It is illustrated

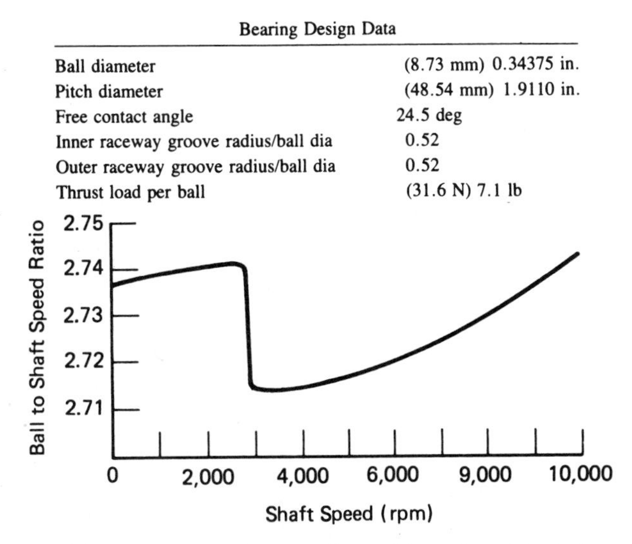

Bearing Design Data

Ball diameter	(8.73 mm) 0.34375 in.
Pitch diameter	(48.54 mm) 1.9110 in.
Free contact angle	24.5 deg
Inner raceway groove radius/ball dia	0.52
Outer raceway groove radius/ball dia	0.52
Thrust load per ball	(31.6 N) 7.1 lb

FIGURE 7.16. Ball–shaft speed ratio vs shaft speed for a thrust-loaded, angular-contact ball bearing operating with dry friction.

by Fig. 7.15–7.17 that the condition of "inner raceway control" is nonexistent; hence no equations for that condition are presented herein.

From equations (7.40) and (7.41), setting β' equal to 0 and substituting for equation (7.59), the ratio between ball and raceway angular velocities is determined:

$$\frac{\omega_R}{\omega} = \frac{\pm 1}{\left(\dfrac{\cos\alpha_o + \tan\beta\sin\alpha_o}{1 + \gamma'\cos\alpha_o} + \dfrac{\cos\alpha_i + \tan\beta\sin\alpha_i}{1 - \gamma'\cos\alpha_i}\right)\gamma'\cos\beta} \tag{7.60}$$

The upper sign pertains to outer raceway rotation and the lower sign to inner raceway rotation.

Again, using the condition of outer raceway control as established in equation (7.59), it is possible to determine the ratio of ball orbital angular velocity to raceway speed. For a rotating inner raceway $\omega_m = -\omega_o$; therefore, from equation (7.41) for β' equal to 0:

$$\frac{\omega_m}{\omega} = \frac{1}{1 + \dfrac{(1 + \gamma'\cos\alpha_o)(\cos\alpha_i + \tan\beta\sin\alpha_i)}{(1 - \gamma'\cos\alpha_i)(\cos\alpha_o + \tan\beta\sin\alpha_o)}} \tag{7.61}$$

Equation (7.61) is based on the valid assumption that $r_o \approx r_i \approx D/2$. Similarly, for a rotating outer raceway and by equation (7.44),

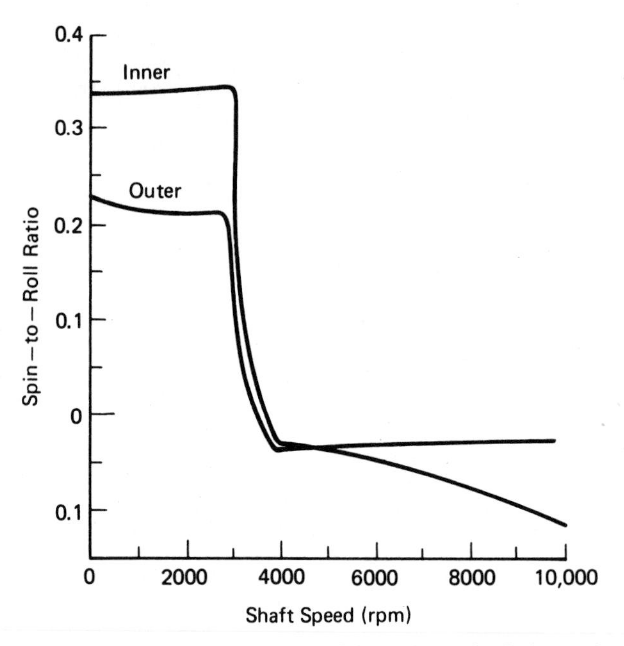

FIGURE 7.17. Spin-to-roll ratio vs shaft speed for a thrust-loaded, angular-contact ball bearing operating with dry friction.

$$\frac{\omega_m}{\omega} = \frac{1}{1 + \dfrac{(1 - \gamma' \cos \alpha_i)(\cos \alpha_o + \tan \beta \sin \alpha_o)}{(1 + \gamma' \cos \alpha_o)(\cos \alpha_i + \tan \beta \sin \alpha_i)}} \tag{7.62}$$

Substitution of equation (7.59) describing the condition of outer raceway control into equations (7.61) and (7.62) establishes the equations of the required ratio ω_m/ω. Hence, for a bearing with rotating inner raceway:

$$\frac{\omega_m}{\omega} = \frac{1 - \gamma' \cos \alpha_i}{1 + \cos (\alpha_i - \alpha_o)} \tag{7.63}$$

For a bearing with a rotating outer raceway

$$\frac{\omega_m}{\omega} = \frac{\cos (\alpha_i - \alpha_o) + \gamma' \cos \alpha_i}{1 + \cos(\alpha_i + \alpha_o)} \tag{7.64}$$

An indicated above, equations (7.59), (7.60), (7.63), and (7.64) are valid only when ball gyroscopic pivotal motion is negligible, that is, $\beta' = 0$. It was shown in Chapter 4 for ball thrust bearings that ball gyroscopic torque in inch-pounds may be calculated from

$$M_g = 4.47 \times 10^{-12} D^5 n_R n_m \sin \beta \tag{4.74}$$

Palmgren [7.1] further states that at high speeds due to better lubricant films, the coefficient of sliding friction may be as little as 0.02. To prevent sliding due to gyroscopic action:

$$0.02\, QD > M_g \tag{7.65}$$

or

$$Q > 2.24 \times 10^{-12} D^4 n_R n_m \sin \beta \tag{7.66}$$

Jones [7.2], on the other hand, mentions that a coefficient of sliding friction from 0.06 to 0.07 suffices for most ball bearing applications to prevent such sliding. In an oil–film lubricated bearing, however, it is doubtful that this motion can be prevented; rather, it can be made negligible.

Jones (7.3) has established a condition to determine whether outer raceway control is approximated in a given application. If

$$Q_o a_o \mathcal{E}_o \cos(\alpha_i - \alpha_o) > Q_i a_i \mathcal{E}_i \tag{7.67}$$

then outer raceway control may be assumed for calculation purposes. In equation (7.67), $\mathcal{E}$ is the complete elliptic integral of the second kind with modulus $\kappa = a/b$ as defined in equation (5.32). Equation (7.67), when the inequality is fulfilled, indicates only that more resistance to gyroscopic pivotal motion occurs at the ball–outer raceway contact than at the ball–inner raceway contact. It should further be apparent that in high speed ball bearings, ball centrifugal force is the significant factor in influencing a condition approximating "outer raceway control"

Example 7.3. Estimate whether outer raceway control occurs for the 218 angular-contact ball bearing of Example 7.2.

$$Q_{ia} = 1113 \text{ N (250 lb)} \qquad \text{Ex. 7.2}$$

$$\alpha_i = 41°36' \qquad \text{Ex. 6.5}$$

$$Q_i = \frac{Q_{ia}}{\sin \alpha_i} \tag{4.48}$$

$$= \frac{1113}{\sin(41°36')} = 1676 \text{ N (376.6 lb)}$$

For the purpose of this illustration, it is assumed that the values of $\Sigma\rho_i$, $F(\rho)_i$, $\Sigma\rho_o$, and $F(\rho)_o$ are nearly identical to those calculated in Example 2.6, that is,

$$\Sigma\rho_i = 0.108 \text{ mm}^{-1} \; (2.747 \text{ in.}^{-1})$$

$$F(\rho)_i = 0.9260$$

$$\Sigma\rho_o = 0.0832 \text{ mm}^{-1} \; (2.114 \text{ in.}^{-1})$$

$$F(\rho)_o = 0.9038$$

These values are accurate within 2%.

$$a_i^* = 3.50 \qquad \text{Ex. 5.1}$$

$$a_o^* = 3.10 \qquad \text{Ex. 5.1}$$

$$b_i^* = 0.430 \qquad \text{Ex. 5.1}$$

$$b_o^* = 0.455 \qquad \text{Ex. 5.1}$$

$$\kappa_i = \frac{a_i}{b_i} = \frac{a_i^*}{b_i^*} \tag{5.24}$$

$$= \frac{3.50}{0.430} = 8.14$$

$$\kappa_o = \frac{a_o^*}{b_o^*} \tag{5.24}$$

$$= \frac{3.10}{0.455} = 6.81$$

$$b_i^* = \left(\frac{2\mathcal{E}_i}{\pi\kappa_i}\right)^{1/3} \tag{5.40}$$

$$0.430 = \left(\frac{2\mathcal{E}_i}{\pi \times 8.14}\right)^{1/3}$$

$$\mathcal{E}_i = 1.017$$

$$b_o^* = \left(\frac{2\mathcal{E}_o}{\pi\kappa_o}\right)^{1/3} \tag{5.40}$$

$$0.455 = \left(\frac{2\mathcal{E}_o}{\pi \times 6.81}\right)^{1/3}$$

$$\mathcal{E}_o = 1.008$$

$$a_i = 0.0236 a_i^* \left(\frac{Q_i}{\Sigma\rho_i}\right)^{1/3} \tag{5.34}$$

$$= 0.0236 \times 3.50 \left(\frac{1676}{0.108}\right)^{1/3} = 2.056 \text{ mm } (0.08096 \text{ in.})$$

$$a_o = 0.0236 a_o^* \left(\frac{Q_o}{\Sigma\rho_o}\right)^{1/3} \tag{5.34}$$

$$= 0.0236 \times 3.10 \times \left(\frac{1676}{0.0832}\right)^{1/3} = 1.991 \text{ mm } (0.07839 \text{ in.})$$

Since $Q_i = Q_o$ (approximately) and since $\alpha_i = \alpha_o$ (approximately), $a_o \mathcal{E}_o < a_i \mathcal{E}_i$, that is,

$$a_o \mathcal{E}_o = 1.991 \times 1.008 = 2.007$$

$$a_i \mathcal{E}_i = 2.056 \times 1.017 = 2.091$$

Therefore, "outer raceway control" cannot be assumed for the purpose of calculations of speeds, and so on.

ROLLER END-FLANGE SLIDING IN ROLLER BEARINGS

Roller End-Flange Contact

Roller bearings react axial roller loads through concentrated contacts between roller ends and flange. Tapered roller bearings and spherical roller bearings (with asymmetrical rollers) require such contact to react the component of the raceway–roller contact load that acts in the roller axial direction. Some cylindrical roller bearing designs require roller end-flange contacts to react skewing-induced and/or externally applied roller axial loads. As these contacts experience sliding motions between roller ends and flange, their contribution to overall bearing frictional heat generation becomes substantial. Furthermore, there are bearing failure modes associated with roller end-flange contact such as wear and smearing of the contacting surfaces. These failure modes are related to the ability of the roller end-flange contact to support roller axial load under the prevailing speed and lubrication conditions within the contact. Both the frictional characteristics and load-carrying capability of roller end-flange contacts are highly dependent on the geometry of the contacting members. This is discussed further in Chapter 13.

Roller End-Flange Geometry

Numerous roller end and flange geometries have been used successfully in roller bearing designs. Typically, performance requirements as well as manufacturing considerations dictate the geometry incorporated into a bearing design. Most designs use either a flat (with corner radii) or sphere end roller contacting an angled flange. The angled flange surface can be described as a portion of a cone at an angle θ_f with respect to a radial plane perpendicular to the ring axis. This angle, known as the flange angle or flange layback angle, can be zero, indicating that the flange surface lies in the radial plane. Examples of cylindrical roller bearing roller end-flange geometries are shown in Fig. 7.18. The flat end roller in Fig. 7.18*a* under zero skewing conditions contacts the flange at a single point (in the vicinity of the intersection between the roller end flat and roller corner radius). As the roller skews, the point of contact travels along this intersection on the roller toward the tip of the flange, as shown in Fig. 7.19*b*. If properly designed, a sphere end roller will contact the flange on the roller end sphere surface. For no skewing the contact will be centrally positioned on the roller, as shown in Fig. 7.19*c*. As the skewing angle is increased, the contact point moves off center and toward the flange tip, as shown in Fig. 7.19*d* for a flanged inner ring. For typical designs sphere end roller contact location is less sensitive to skewing than a flat end roller contact.

The location of the roller end-flange contact has been determined analytically [7.7] for sphere end rollers contacting an angled flange. Consider the cylindrical roller bearing arrangement shown in Fig. 7.20. The flanged ring coordinate system X_{I}, Y_{I}, Z_{I} and roller coordinate system X_{i}, Y_{i}, Z_{i} are indicated. The flange contact surface is modeled as a portion of a cone with an apex at point C as shown in Fig. 7.21. The equation of this cone, expressed as a function of the x and y ring coordinates, is

$$z = [(x - C)^2 \cot^2 \theta_f - y^2]^{1/2} = f(x, y) \tag{7.68}$$

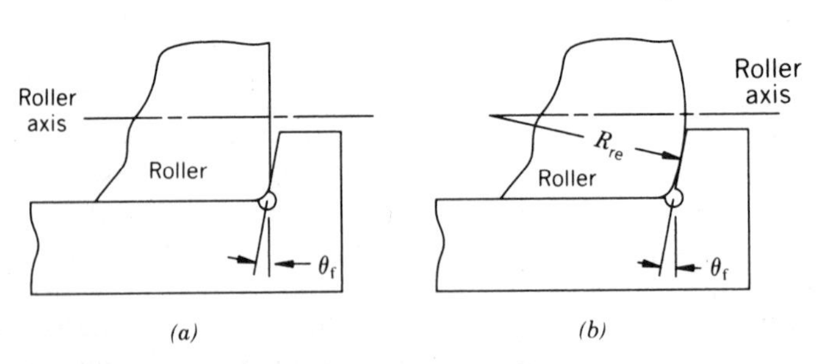

FIGURE 7.18. Cylindrical roller bearing, roller end-flange contact geometry. (*a*) Flat end roller. (*b*) Sphere end roller.

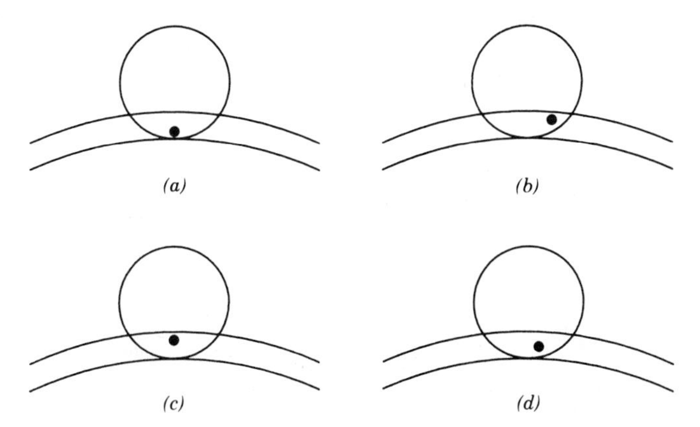

FIGURE 7.19. Cylindrical roller bearing roller end-flange contact location for flat and sphere end rollers. (*a*) Flat end roller, zero skew angle. (*b*) Flat end roller, nonzero skew angle. (*c*) Sphere end roller, zero skew angle. (*d*) Sphere end roller, nonzero skew angle.

For a point on flange surface P_x, P_y, P_z the equation of the surface normal at P can be expressed as

$$\frac{x - P_x}{\left.\dfrac{\partial f}{\partial x}\right|_{x=P_x, y=P_y}} = \frac{y - P_y}{\left.\dfrac{\partial f}{\partial y}\right|_{x=P_x, y=P_y}} = -(z - P_z) \tag{7.69}$$

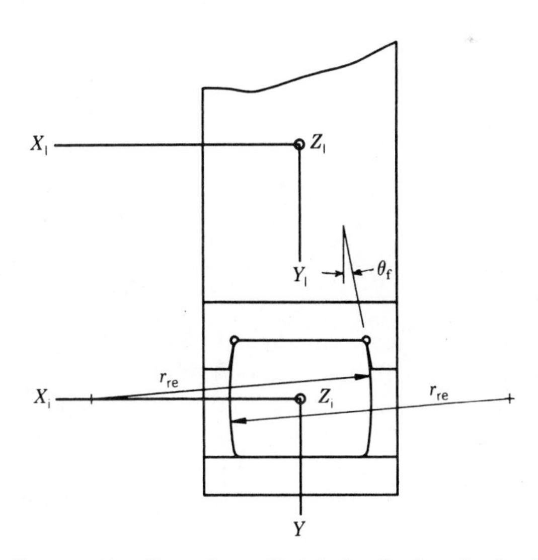

FIGURE 7.20. Cross section through a cylindrical roller bearing having a flanged inner ring.

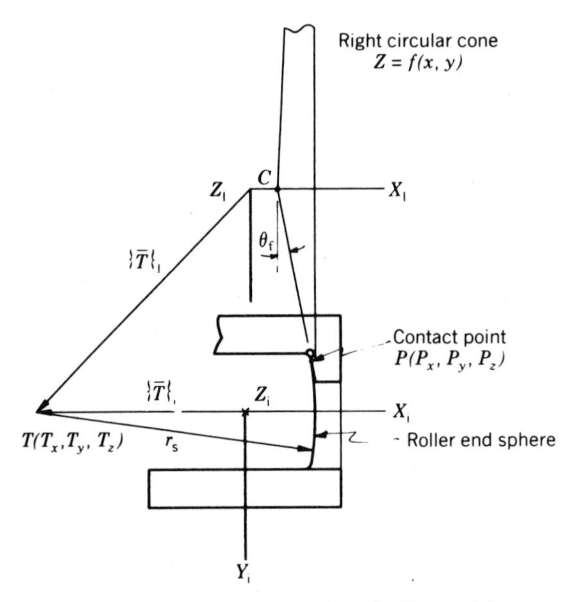

FIGURE 7.21. Coordinate system for calculation of roller end-flange contact location.

The location of the origin of the roller end sphere radius is defined as point T with coordinates (T_x, T_y, T_z) expressed in the flanged ring coordinate system. Since the resultant roller end-flange elastic contact force is normal to the end sphere surface, its line of action must pass through the sphere origin (T_x, T_y, T_z). Evaluating equations (7.68) and (7.69) at T yields the following three equations:

$$T_x - P_x = -\frac{(T_z - P_z)(P_x - C)\cot^2\theta_f}{[(P_x - C)^2\cot^2\theta_f - P_y^2]^{1/2}} \tag{7.70}$$

$$T_y - P_y = \frac{(T_z - P_z)P_y}{[(P_x - C)^2\cot^2\theta_f - P_y^2]^{1/2}} \tag{7.71}$$

$$P_z = [(P_x - C)^2\cot^2\theta_f - P_y^2]^{1/2} \tag{7.72}$$

Equations (7.70)–(7.72) contain three unknowns (P_x, P_y, P_z) and are sufficient to determine the theoretical point of contact between the roller end and flange. By introducing a fourth equation and unknown, however, namely the length of the line from points (T_x, T_y, T_z) to (P_x, P_y, P_z), the added benefit of closed-form solution is obtained. The length of a line normal to the flange surface at the point (P_x, P_y, P_z), which joins this point with the sphere origin (T_x, T_y, T_z), is given by

$$\mathfrak{D} = [(T_x - P_x)^2 + (T_y - P_y)^2 + (T_z - P_z)^2]^{1/2} \tag{7.73}$$

After algebraic reduction, $\mathfrak{D}$ is obtained from the positive root of the quadratic equation

$$\mathfrak{D} = \frac{-\mathcal{S} \pm (\mathcal{S}^2 - 4\mathcal{R}\mathfrak{I})^{1/2}}{2\mathcal{R}} \tag{7.74}$$

where values for $\mathcal{S}$, $\mathcal{R}$, and $\mathfrak{I}$ are

$$\mathcal{R} = \tan^2 \theta_f - 1$$

$$\mathcal{S} = \frac{2 \sin^2 \theta_f}{\cos \theta_f} [(T_x - C) - \tan \theta_f (T_y^2 + T_z^2)^{1/2}$$

$$\mathfrak{I} = [(T_x - C) - \tan \theta_f (T_y^2 + T_z^2)^{1/2}]$$

The coordinates $P(P_x, P_y, P_z)$ are given by the following closed-form functions of $\mathfrak{D}$:

$$P_x = T_y \tan \theta_f \left[1 + \left(\frac{T_z}{T_y}\right)^2\right]^{1/2} \left[1 - \frac{\mathfrak{D} \sin \theta_f}{(T_y^2 + T_z^2)^{1/2}}\right] + C \tag{7.75}$$

$$P_y = T_y \left[1 - \frac{\mathfrak{D} \sin \theta_f}{(T_y^2 + T_z^2)^{1/2}}\right] \tag{7.76}$$

$$P_z = T_z \left[1 - \frac{\mathfrak{D} \sin \theta_f}{(T_y^2 + T_z^2)^{1/2}}\right] \tag{7.77}$$

At a point of contact between the roller end and flange, $\mathfrak{D}$ is equal to the roller end sphere radius. Therefore, knowing the roller and flanged ring geometry as well as the coordinate location (with respect to the flanged ring coordinate system) of the roller end sphere origin, it is possible to calculate directly the theoretical roller end-flange contact location.

The foregoing analysis, although shown for a cylindrical roller bearing, is general enough to apply to any roller bearing having sphere end rollers that contact a conical flange. Tapered and spherical roller bearings of this type may be treated if the sphere radius origin is properly defined.

These equations have several notable applications since flange contact location is of interest in bearing design and performance evaluation. It is desirable to maintain contact on the flange below the flange rim (including edge break) and above the undercut at the base of the flange. To do otherwise causes loading on the flange rim (or edge of undercut) and produces higher contact stresses and less than optimum lubrication of the contact. The preceding equations may be used to determine the maximum theoretical skewing angle for a cylindrical roller bearing if the roller axial play (between flanges) is known. Also, by calculating the location of the theoretical contact point, sliding velocities between roller ends and flange can be calculated and used in an estimate of roller end-flange contact friction and heat generation.

Sliding Velocity

The kinematics of a roller end-flange contact causes sliding to occur between the contacting members. The magnitude of the sliding velocity between these surfaces substantially affects friction, heat generation, and load-carrying characteristics of a roller bearing design. The sliding velocity is represented by the difference between the two vectors defining the linear velocities of the flange and the roller end at the point of contact. A graphical representation of the roller velocity v_{ROLL} and the flange velocity v_{F} at their point of contact C is shown in Fig. 7.22. The sliding velocity vector v_{s} is shown as the difference of v_{RE} and v_{F}. When considering roller skewing motions, v_{s} will have a component in the flanged ring axial direction, albeit small in comparison to the components in the bearing radial plane. If the roller is not subjected to skewing, the contact point will lie in the plane containing the roller and flanged ring axes. The roller end-flange sliding velocity may be calculated as

$$v_{\mathrm{sl}} = v_{\mathrm{F}} - v_{\mathrm{RE}} = \omega_f R_{\mathrm{c}} - (\omega_{\mathrm{o}} R_{\mathrm{c}} + \omega_{\mathrm{R}} r_{\mathrm{c}}) \tag{7.78}$$

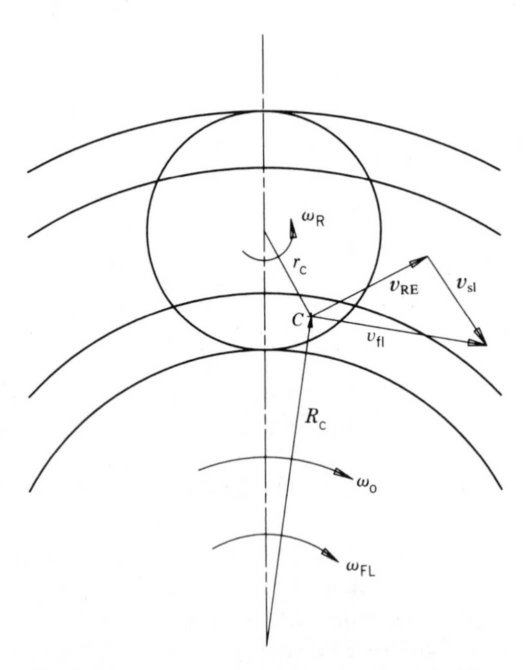

FIGURE 7.22. Roller end-flange contact velocities.

where clockwise rotations are considered positive. Varying the position of contact point C over the elastic contact area between roller end and flange allows the distribution of sliding velocity to be determined.

CLOSURE

In this chapter, methods for calculation of rolling and cage speeds in ball and roller bearings were developed for conditions of rolling and spinning motions. It will be shown in Chapter 8 how the dynamic loading that is derived from ball and roller speeds can significantly affect ball bearing contact angles, diametral clearance, and subsequently rolling element load distribution. Moreover, spinning motions that occur in ball bearings tend to alter contact area stresses, and hence they affect bearing endurance. Other quantities affected by bearing internal speeds are friction torque and frictional heat generation. It is therefore clear that accurate determinations of bearing internal speeds are necessary for analysis of rolling bearing performance.

It will be demonstrated, however, in Chapter 12 that hydrodynamic action of the lubricant in the contact areas can transform what is presumed to be only rolling motions into combinations of rolling and translatory motions. In general, this combination of rotation and translation cannot be tolerated, and bearing design and/or loading must be modified to achieve only rolling moition. In high speed ball bearings, for example, gyroscopic pivotal motions must usually be minimized. In high speed roller bearings, it is frequently necessary to alter load distribution to prevent the inner raceway from skidding past the rolling elements.

REFERENCES

7.1. A. Palmgren, *Ball and Roller Bearing Engineering*, 3rd ed., Burbank, Philadelphia, pp. 70–72 (1959).

7.2. A. B. Jones, "Ball Motion and Sliding Friction in Ball Bearings," *ASME J. Basic Eng.* **81,** 1–12 (1959).

7.3. A. B. Jones, "A General Theory for Elastically Constrained Ball and Radial Roller Bearings under Arbitrary Load and Speed Conditions," *ASME J. Basic Eng.* **82,** 309–320 (1960).

7.4. G. Lundberg, "Motions in Loaded Rolling Element Bearings," SKF unpublished report (1954).

7.5. T. A. Harris, "An Analytical Method to Predict Skidding in Thrust-Loaded, Angular-Contact Ball Bearings," *ASME J. Lubr. Technol.* **93,** 17–24 (1971).

7.6. T. A. Harris, "Ball Motion in Thrust-Loaded, Angular-contact Bearings with Coulomb Friction," *ASME J. Lubr. Technol.* **93,** 32–38 (1971).

7.7. R. Kleckner and J. Pirvics, "High Speed Cylindrical Roller Bearing Analysis—SKF Computer Program CYBEAN, Vol. 1: Analysis," SKF Report AL78PO22, NASA Contract NAS3-20068 (July 1978).

8

LOAD DISTRIBUTION IN HIGH SPEED BEARINGS

LIST OF SYMBOLS

Symbol	Description	Units
B	$f_i + f_o - 1$	
D	Ball or roller diameter	mm (in.)
d_m	Pitch diameter	mm (in.)
F_f	Friction force	N (lb)
F_c	Centrifugal force	N (lb)
f	r/D	
g	Gravitational constant	mm/sec^2 (in./sec^2)
H	Ball or roller hollowness ratio	
J	Mass moment of inertia	kg · mm^2 (in. · lb · sec^2)
K	Load-deflection constant	N/mm^x (lb/in.^{x})
l	Roller length	mm (in.)
m	Ball or roller mass	kg (lb · sec^2/in.)
M_g	Gyroscopic moment	N · mm (in. · lb)
$\mathfrak{M}$	Applied moment	N · mm (in. · lb)
P_d	Diametral clearance	mm (in.)
Q	Ball or roller load	N (lb)

Symbol	Description	Units
$\mathcal{R}$	Radius to locus of raceway groove curvature centers	mm (in.)
s	Distance between inner and outer raceway groove curvature center loci	mm (in.)
X_2	Radial projection of distance between ball center and outer raceway groove curvature center	mm (in.)
X_1	Axial projection of distance between ball center and outer raceway groove curvature center	mm (in.)
α	Contact angle	°
β	Ball attitude angle	rad,°
γ	$D \cos \alpha / d_m$	
δ	Deflection or contact deformation	mm (in.)
θ	Angular misalignment of bearing	rad,°
ψ	Angular location	rad,°
ω	Rotational speed	rad/sec
$\Delta\psi$	Angular distance between rolling elements	rad

SUBSCRIPTS

a	Refers to axial direction
i	Refers to inner raceway
j	Refers to angular position
m	Refers to cage motion and orbital motion
o	Refers to outer raceway
r	Refers to radial direction
R	Refers to rolling element
x	Refers to x direction
z	Refers to z direction

GENERAL

In high speed operation of ball and roller bearings the rolling element centrifugal forces can be significantly large compared to the forces applied to the bearing. In roller bearings this increase in loading on the outer raceway causes larger contact deformations in that member; this effect is similar to that of increasing clearance. An increase in clearance as demonstrated in Chapter 6 tends to increase maximum roller loading due to a decrease in the extent of the load zone. For relatively thin section bearings supported at only a few points on the outer ring; for example, an aircraft gas turbine mainshaft bearing, the centrifugal forces can cause bending of the outer ring thus affecting the load distribution among the rolling elements.

In high speed ball bearings, depending on the contact angles, ball gyroscopic moments and ball centrifugal forces can be of significant magnitude such that inner raceway contact angles tend to increase and outer raceway contact angles tend to decrease. This affects the deflection vs load characteristics of the bearing and thus also affects the dynamics of the ball bearing-supported rotor system.

High speed also affects the lubrication characteristics and thereby the friction in both ball and roller bearings. This will have an influence on bearing internal speeds, which in turn alters the rolling element inertial loading, that is, centrifugal forces and gyroscopic moments. It is possible, however, to determine the internal distribution of load, and hence stresses, in many high speed rolling bearing applications with sufficient accuracy while not considering the frictional loading of the rolling elements. This will be demonstrated in this chapter. The effects of friction, including skidding, on internal load distribution will be considered later.

HIGH SPEED BALL BEARINGS

To determine the load distribution in a high speed ball bearing, consider Fig. 6.19, which shows the displacements of a ball bearing due to a generalized loading system including radial, axial, and moment loads. Figure 8.1 shows the relative angular position of each ball in the bearing.

Under zero load the centers of the raceway groove curvature radii are separated by a distance BD defined by

$$BD = (f_o + f_i - 1)D \tag{2.7}$$

Under an applied static load, the distance between centers will increase by the amount of the contact deformations δ_i plus δ_o, as shown by Fig. 6.18. The line of action between centers is collinear with BD. If, however, a centrifugal force acts on the ball, then because the inner and outer raceway contact angles are dissimilar, the line of action between raceway groove curvature radii centers is not collinear with BD, but is discontinuous as indicated by Fig. 8.2. It is assumed in Fig. 8.2 that the outer raceway groove curvature center is fixed in space and the inner raceway groove curvature center moves relative to that fixed center. Moreover, the ball center shifts by virtue of the dissimilar contact angles.

The distance between the fixed outer raceway groove curvature center

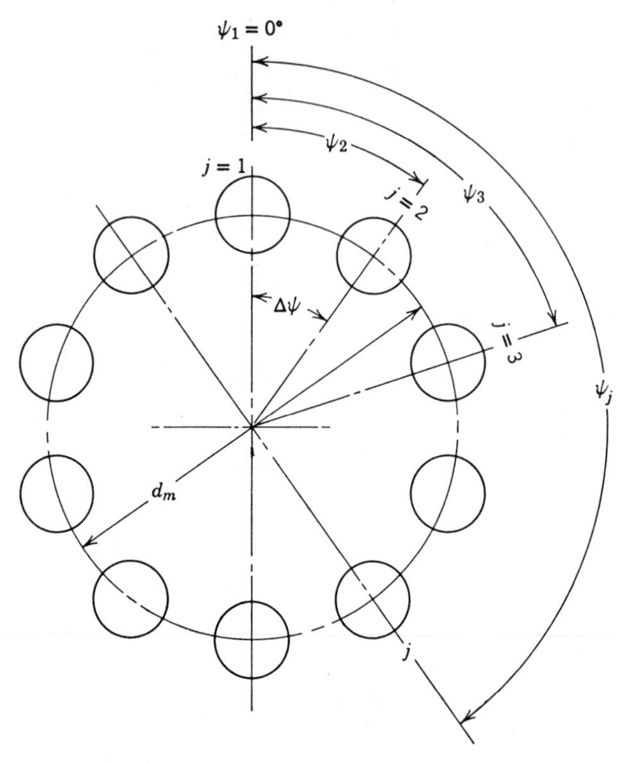

FIGURE 8.1. Angular position of rolling elements in *yz* plane (radial). $\Delta\psi = 2\pi/Z$, $\psi_j = 2\pi/Z(j-1)$.

and the final position of the ball center at any ball location j is

$$\Delta_{oj} = r_o - \frac{D}{2} + \delta_{oj} \tag{8.1}$$

Since

$$r_o = f_o D$$

$$\Delta_{oj} = (f_o - 0.5)D + \delta_{oj} \tag{8.2}$$

Similarly,

$$\Delta_{ij} = (f_i - 0.5)D + \delta_{ij} \tag{8.3}$$

δ_{oj} and δ_{ij} are the normal contact deformations at the outer and inner raceway contacts, respectively.

In accordance with the relative axial displacement of the inner and outer rings δ_a and the relative angular displacement θ, the axial distance

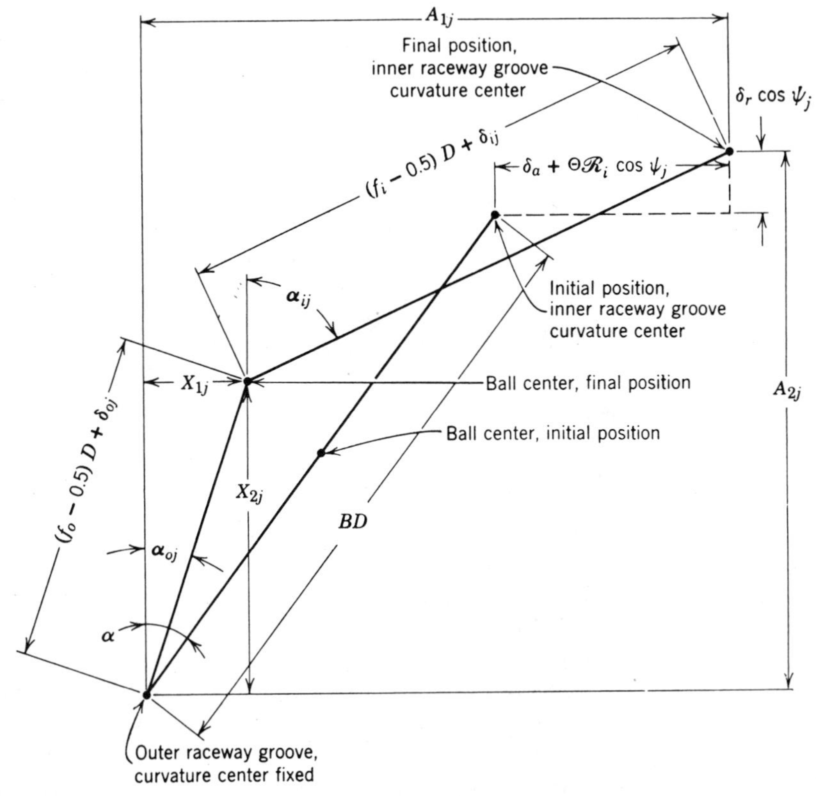

FIGURE 8.2. Positions of ball center and raceway groove curvature centers at angular position ψ, with and without applied load.

between the loci of inner and outer raceway groove curvature centers at any ball position is

$$A_{1j} = BD \sin \alpha^\circ + \delta_a + \theta \mathcal{R}_i \cos \psi_j \tag{8.4}$$

in which $\mathcal{R}_i$ is the radius of the locus of inner raceway groove curvature centers and α° is the initial contact angle prior to loading. Further, in accordance with a relative radial displacement of the ring centers δ_r, the radial displacement between the loci of the groove curvature centers at each ball location is

$$A_{2j} = BD \cos \alpha^\circ + \delta_r \cos \psi_j \tag{8.5}$$

The foregoing data are intended as an explanation of Fig. 8.2.

Jones [8.1] found it convenient to introduce new variables X_1 and X_2, as shown by Fig. 8.2. It can be seen from Fig. 8.2 that at any ball location j,

$$\cos \alpha_{oj} = \frac{X_{2j}}{(f_o - 0.5)D + \delta_{oj}} \tag{8.6}$$

$$\sin \alpha_{oj} = \frac{X_{1j}}{(f_o - 0.5)D + \delta_{oj}} \tag{8.7}$$

$$\cos \alpha_{ij} = \frac{A_{2j} - X_{2j}}{(f_i - 0.5)D + \delta_{ij}} \tag{8.8}$$

$$\sin \alpha_{ij} = \frac{A_{1j} - X_{1j}}{(f_i - 0.5)D + \delta_{ij}} \tag{8.9}$$

Using the Pythagorean theorem, it can be seen from Fig. 8.2 that

$$(A_{1j} - X_{1j})^2 + (A_{2j} - X_{2j})^2 - [(f_i - 0.5)D + \delta_{ij}]^2 = 0 \tag{8.10}$$

$$X_{1j}^2 + X_{2j}^2 - [(f_o - 0.5)D + \delta_{oj}]^2 = 0 \tag{8.11}$$

Considering the plane passing through the bearing axis and the center of a ball located at azimuth ψ_j (see Fig. 8.1), the load diagram of Fig. 8.3 obtains if noncoplanar friction forces are insignificant. If the bearing speed of rotation is sufficient that "outer raceway control" is approximated at a given ball location, then it can be assumed with little effect on calculational accuracy that the ball gyroscopic moment is resisted entirely by frictional force at the ball–outer raceway contact. Otherwise, it is safe to assume that the ball gyroscopic moment is resisted equally at the ball–inner and ball–outer raceway contacts. In Fig. 8.3, therefore, $\lambda_{ij} = 0$ ad $\lambda_{oj} = 2$ for "outer raceway control"; otherwise $\lambda_{ij} = \lambda_{oj} = 1$.

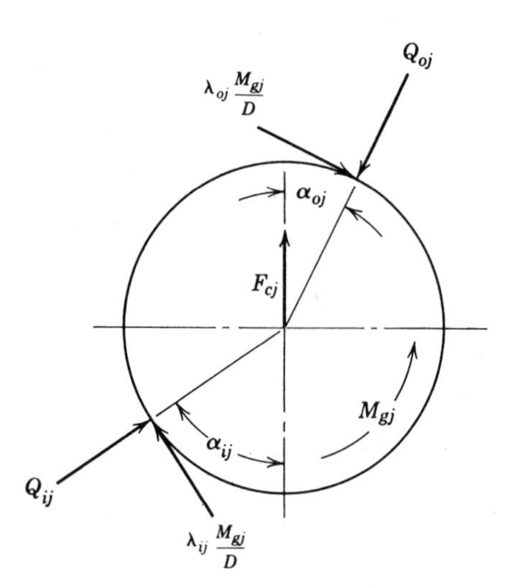

FIGURE 8.3. Ball loading at angular position ψ_j.

The normal ball loads in accordance with equation (6.4) are related to normal contact deformations as follows:

$$Q_{oj} = K_{oj}\delta_{oj}^{1.5} \tag{8.12}$$

$$Q_{ij} = K_{ij}\delta_{ij}^{1.5} \tag{8.13}$$

From Fig. 8.3 considering the equilibrium of forces in the horizontal and vertical directions:

$$Q_{ij}\sin\alpha_{ij} - Q_{oj}\sin\alpha_{oj} - \frac{M_{gj}}{D}(\lambda_{ij}\cos\alpha_{ij} - \lambda_{oj}\cos\alpha_{oj}) = 0 \tag{8.14}$$

$$Q_{ij}\cos\alpha_{ij} - Q_{oj}\cos\alpha_{oj} + \frac{M_{gj}}{D}(\lambda_{ij}\sin\alpha_{ij} - \lambda_{oj}\sin\alpha_{oj}) + F_{cj} = 0 \tag{8.15}$$

Substituting equations (8.12) and (8.13) and (8.6) to (8.9) into (8.14) and (8.15) yields

$$\frac{\dfrac{\lambda_{oj}M_{gj}X_{2j}}{D} K_{oj}\delta_{oj}^{1.5}X_{1j}}{(f_o - 0.5)D + \delta_{oj}} + \frac{K_{ij}\delta_{ij}^{1.5}(A_{1j} - X_{1j}) - \dfrac{\lambda_{ij}M_{gj}}{D}(A_{2j} - X_{2j})}{(f_i - 0.5)D + \delta_{ij}} = 0 \tag{8.16}$$

$$\frac{K_{oj}\delta_{oj}^{1.5}X_{2j} + \dfrac{\lambda_{oj}M_{gj}X_{1j}}{D}}{(f_o - 0.5)D + \delta_{oj}} - \frac{K_{ij}\delta_{ij}^{1.5}(A_{2j} - X_{2j}) + \dfrac{\lambda_{ij}M_{gj}}{D}(A_{1j} - X_{1j})}{(f_i - 0.5)D + \delta_{ij}} - F_{cj} = 0 \tag{8.17}$$

Equations (8.10), (8.11), (8.16), and (8.17) may be solved simultaneously for X_{1j}, X_{2j}, δ_{ij}, and δ_{oj} at each ball angular location once values for δ_a, δ_r, and θ are assumed. The most probable method of solution is the aforementioned Newton–Raphson method for solution of simultaneous nonlinear equations (see Chapter 6).

In equation (4.34) it was stated that the centrifugal force acting on a ball is calculated as follows:

$$F_c = \tfrac{1}{2}\, md_m \omega_m^2 \tag{4.34}$$

in which ω_m is the orbital speed of the ball. Substituting the identity $\omega_m^2 = (\omega_m/\omega)^2 \omega^2$ in equation (4.34), the following equation for centrifugal force is obtained:

$$F_{cj} = \tfrac{1}{2}\, md_m \omega^2 \left(\frac{\omega_m}{\omega}\right)_j^2 \tag{8.18}$$

in which ω is the speed of the rotating ring and ω_m is the orbital speed of the ball at angular position ψ_j. It should be apparent that because orbital speed is a function of contact angle, it is not constant for each ball location.

Moreover, it must be kept in mind that this analysis does not consider frictional forces that tend to retard ball and hence cage motion. Therefore, in a high speed bearing, it is to be expected that ω_m will be less than that predicted by equation (7.63) and greater than that predicted by equation (7.64). Unless the loading on the bearing is relatively light, however, the cage speed differential is usually insignificant in affecting the accuracy of the calculations ensuing in this chapter.

Gyroscopic moment at each ball location may be described as follows:

$$M_{gj} = J\left(\frac{\omega_R}{\omega}\right)_j \left(\frac{\omega_m}{\omega}\right)_j \omega^2 \sin\beta \tag{8.19}$$

where β is given by equation (7.59), ω_R/ω by equation (7.60), and ω_m/ω by equation (7.63) or (7.64).

Since K_{oj}, K_{ij}, and M_{gj} are functions of contact angle, equations (8.6)–(8.9) may be used to establish these values during the iteration.

To find the values of δ_r, δ_a, and θ, it remains only to establish the conditions of equilibrium applying to the entire bearing. These are

$$F_a - \sum_{j=1}^{j=Z}\left(Q_{ij}\sin\alpha_{ij} - \frac{\lambda_{ij}M_{gj}}{D}\cos\alpha_{ij}\right) = 0 \tag{8.20}$$

or

$$F_a - \sum_{j=1}^{j=Z}\left(\frac{K_{ij}(A_{1j} - X_{1j})\delta_{ij}^{1.5} - \dfrac{\lambda_{ij}M_{gj}}{D}(A_{2j} - X_{2j})}{(f_i - 0.5)D + \delta_{ij}}\right) = 0 \tag{8.21}$$

$$F_r - \sum_{j=1}^{j=Z} \left(Q_{ij} \cos \alpha_{ij} + \frac{\lambda_{ij} M_{gj}}{D} \sin \alpha_{ij} \right) \cos \psi_j = 0 \quad (8.22)$$

or

$$F_r - \sum_{j=1}^{j=Z} \left(\frac{K_{ij}(A_{2j} - X_{2j})\delta_{ij}^{1.5} + \dfrac{\lambda_{ij} M_{gj}}{D}(A_{1j} - X_{1j})}{(f_i - 0.5)D + \delta_{ij}} \right) = 0 \quad (8.23)$$

$$\mathfrak{M} - \sum_{j=1}^{j=Z} \left[\left(Q_{ij} \sin \alpha_{ij} - \frac{\lambda_{ij} M_{gj}}{D} \cos \alpha_{ij} \right) \mathcal{R}_i + \frac{\lambda_{ij} M_{gj}}{D} r_i \right] \cos \psi_j = 0 \quad (8.24)$$

or

$$\mathfrak{M} - \sum_{j=1}^{j=Z} \left[\frac{\left(K_{ij}(A_{1j} - X_{1j})\delta_{ij}^{1.5} - \dfrac{\lambda_{ij} M_{gj}}{D}(A_{2j} - X_{2j}) \right) \mathcal{R}_i}{(f_i - 0.5)D + \delta_{ij}} + \lambda_{ij} f_i M_{gj} \right]$$

$$\times \cos \psi_j = 0 \quad (8.25)$$

$$\mathcal{R}_i = \tfrac{1}{2} d_m + (f_i - 0.5)D \cos \alpha^\circ \quad (6.86)$$

Having computed values of X_{1j}, X_{2j}, δ_{ij}, and δ_{oj} at each ball position and knowing F_a, F_r, and $\mathfrak{M}$ as input conditions, the values δ_a, δ_r, and θ may be determined by equations (8.21), (8.23), and (8.25). After obtaining the primary unknown quantities δ_a, δ_r, and θ, it is then necessary to repeat the calculation of X_{1j}, X_{2j}, δ_{ij}, and δ_{oj}, and so on, until compatible values of the primary unknown quantities δ_a, δ_r, and θ are obtained.

Example 8.1. The 218 angular-contact ball bearing of Example 6.5 is to be operated through the load range of 0–44,500 N (10,000 lb) thrust and at speeds of 3000, 6000, 10,000, and 15,000 rpm. Determine such operating characteristics of the bearing as α_i, α_o, β, Q_i, Q_o, n_m, δ_a, M_g, and ω_s/ω_{roll}.

To obtain the answers to this study, clearly a computer program must be developed to solve equations (8.10), (8.11), (8.16), (8.17), and (8.21) simultaneously for each load-speed condition. In these equations, which may be solved by iterative techniques, the load-deflection "constants" K_i and K_o are functions of α_i and α_o, which are in turn functions of X_1 and X_2 according to equations (8.6)–(8.9). Similarly, F_c and M_g are functions of ω_m/ω and ω_R/ω which depend on α_i and α_o

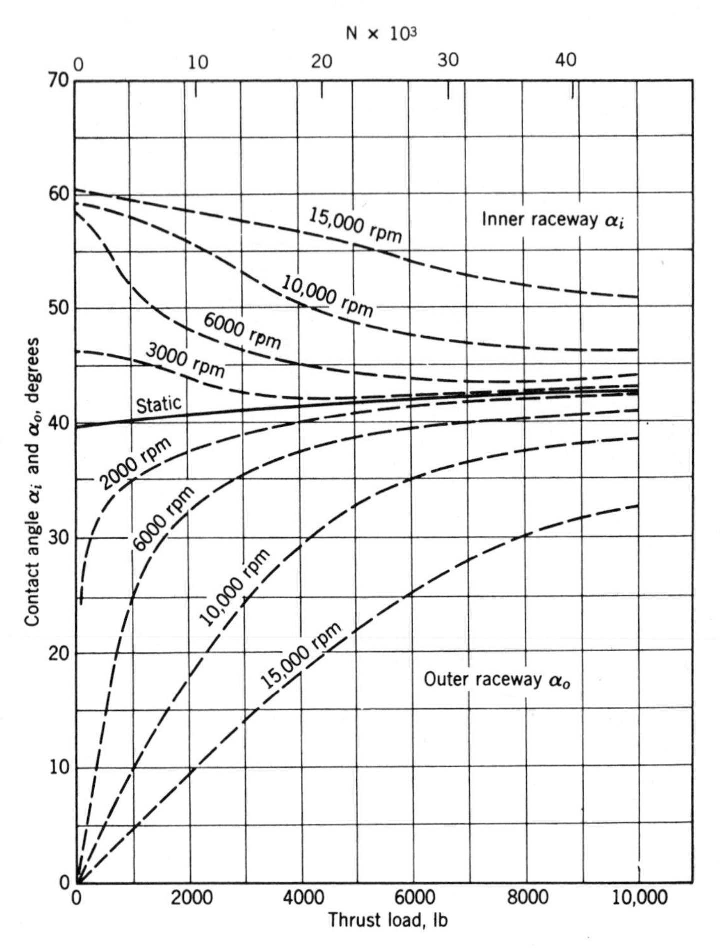

FIGURE 8.4. α_i and α_o vs thrust load. 218 angular-contact ball bearing, $\alpha^\circ = 40^\circ$.

according to equations (7.60) and (7.61). Hence the solution is not simple and care must be exercised to include all variations in the iteration. From such a computer program, the data of Figs. 8.4–8.7 were developed.

Ball Excursions

For an angular-contact ball bearing subjected only to thrust loading, the orbital travel of the balls occurs in a single radial plane, whose axial location is defined by X_{1j} in Fig. 8.2, that is, X_{1j} is the same at all azimuth angles ψ_j. For a bearing that supports combined load, that is, radial and thrust loads and perhaps also a moment load, X_{1j} is different at

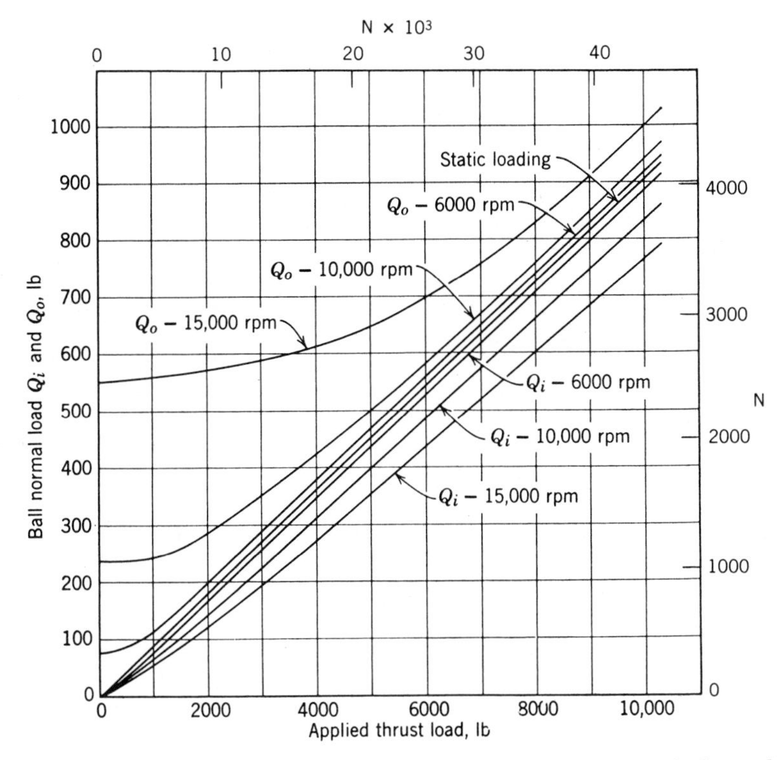

FIGURE 8.5. Ball normal loads Q_o and Q_i vs thrust load for various shaft speeds. 218 angular-contact ball bearing, $\alpha° = 40°$.

each azimuth angle ψ_j. Therefore, a ball undergoes an axial "excursion" as it orbits the shaft or housing center. Unless this excursion is accommodated by providing sufficient axial clearance between the ball and the cage pocket, the cage will experience nonuniform and possibly heavy loading in the axial direction. This can also cause a complex motion of the cage, that is, no longer simple rotation in a single plane, but rather including an out-of-plane vibrational component. Such motion together with the aforementioned loading can lead to rapid destruction and seizure of the bearing.

Under combined loading, because of the variation in the ball–raceway contact angles α_{ij} and α_{oj} as a ball orbits the bearing center, there is a tendency for the ball to advance or lag its "central" position in the cage pocket. The orbital or circumferential travel of the ball relative to the cage is, however, limited by the cage pocket. Therefore, a load occurs between the ball and the pocket in the circumferential direction. Under steady-state cage rotation, the sum of these ball–cage pocket loads in the circumferential direction is close to zero, being balanced only by fric-

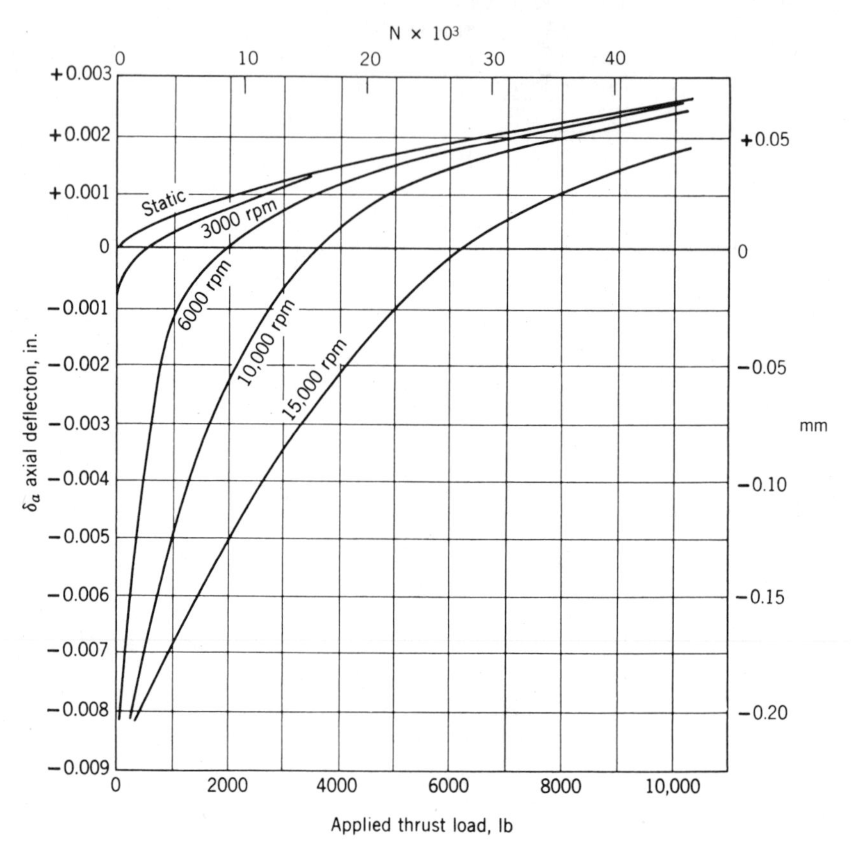

FIGURE 8.6. δ_a—axial deflection vs thrust load for various shaft speeds. 218 angular-contact ball bearing, $\alpha° = 40°$.

tional forces. Moreover, the forces and moments acting on the ball in the bearing's plane of rotation must be in balance, including acceleration or deceleration loading and frictional forces.

To achieve this condition of equilibrium, the ball speeds, including orbital speed, will be different from those calculated using the equations of Chapter 7. This condition is called *skidding*, and it will be covered in Chapter 13.

Lightweight Balls

To permit angular-contact ball bearings to operate at higher speeds, it is possible to reduce the adverse ball inertial effects by reducing the ball mass. A possible method of achieving this reduction is the use of hollow balls. For these elements

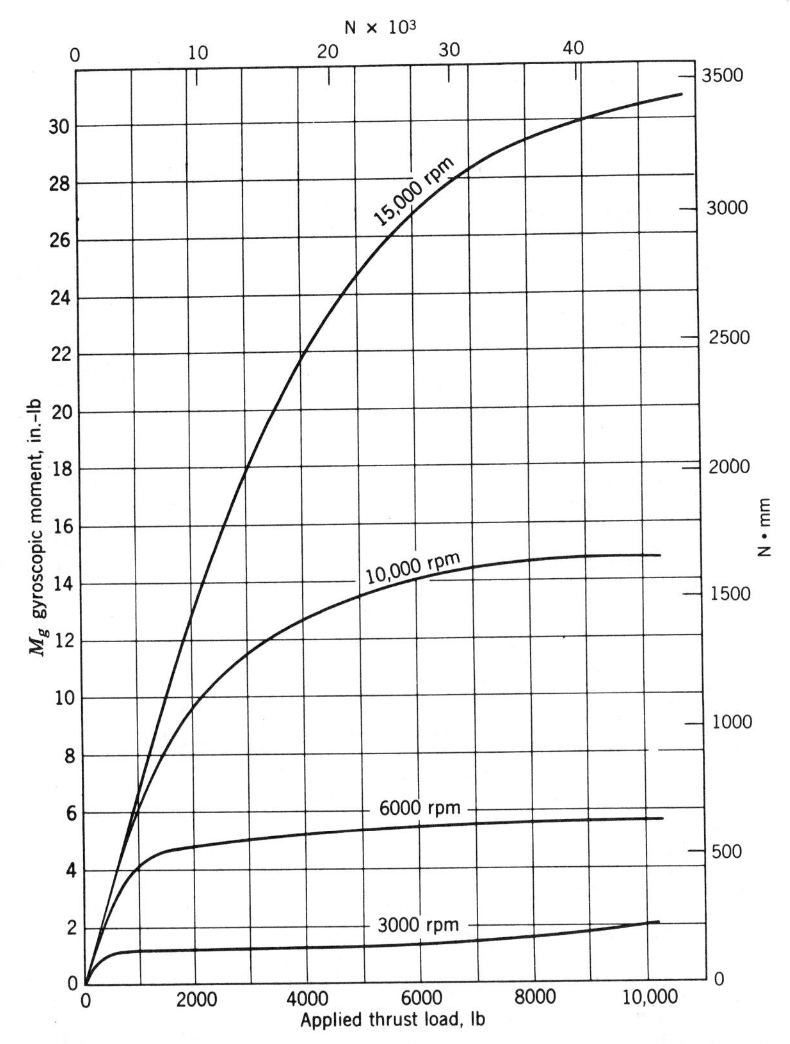

FIGURE 8.7. M_g, gyroscopic moment, vs thrust load for various shaft speeds. 218 angular-contact ball bearing, $\alpha° = 40°$.

$$F_c = \tfrac{1}{2}\, m d_m \left(\frac{\omega_m}{\omega}\right)^2 \omega^2 (1 - H^3) \tag{8.26}$$

$$M_g = J\left(\frac{\omega_R}{\omega}\right)\left(\frac{\omega_m}{\omega}\right)\omega^2 (1 - H^5) \sin\beta \tag{8.27}$$

where H is the ratio of the inside diameter of the ball to the outside. It is clear that a substantial amount of "hollowness" is required to achieve a major reduction in ball inertial loading. Figures 8.8–8.12 taken from

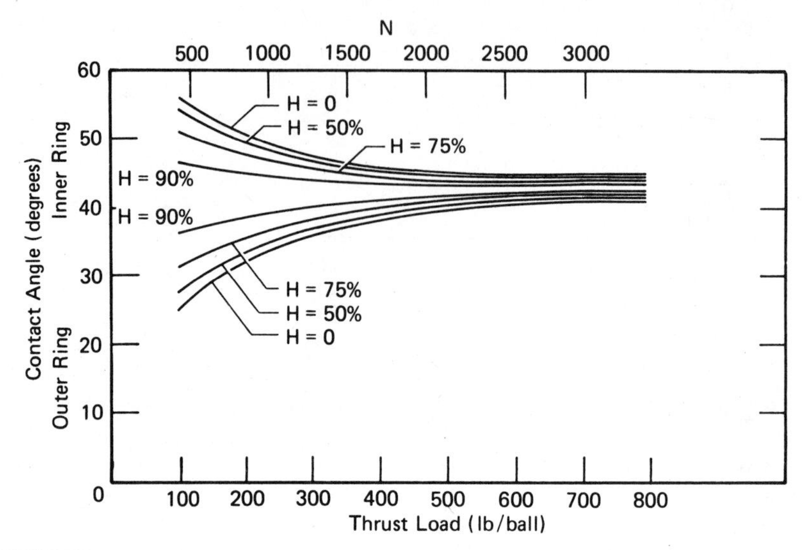

FIGURE 8.8. Contact angle vs applied thrust load, and hollowness $Nd_m = 1 \times 10^6$.

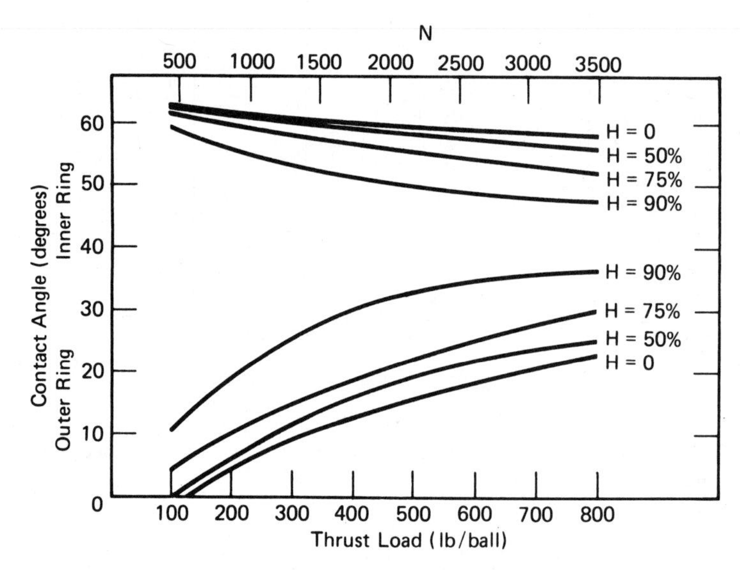

FIGURE 8.9. Contact angle vs thrust load, and hollowness, $Nd_m = 3 \times 10^6$.

reference [8.3] illustrate the effects of ball hollowness on the 218 angular-contact ball bearing used in Example 8.1.

The ability to mass produce hollow balls adequate for high speed operation in an angular-contact ball bearing has not been demonstrated. A ball must be fabricated from two hollow hemispheres and problems occur with the weld on the inside of the ball resulting in stress raisers that are significant due to the bending loads caused by the heavy concentrated surface loading. Another method to obtain lightweight balls is

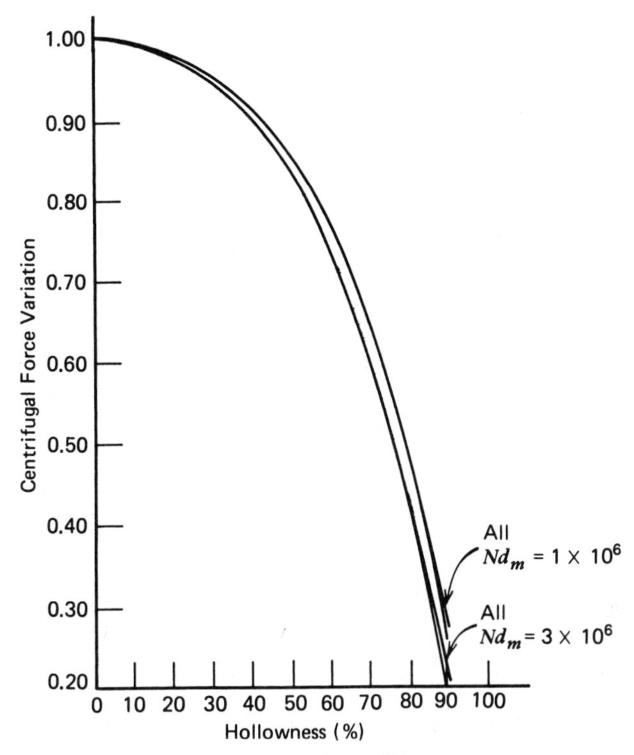

FIGURE 8.10. Centrifugal force vs hollowness.

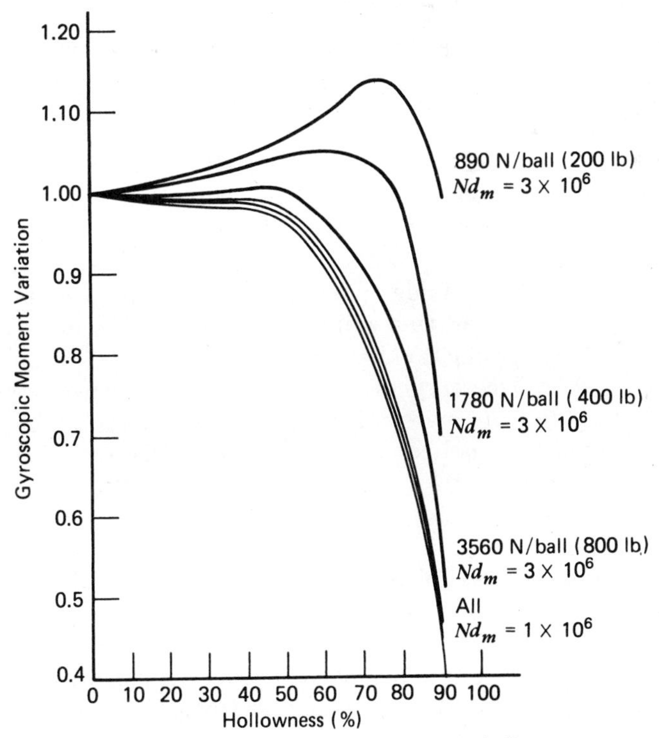

FIGURE 8.11. Gyroscopic moment vs hollowness.

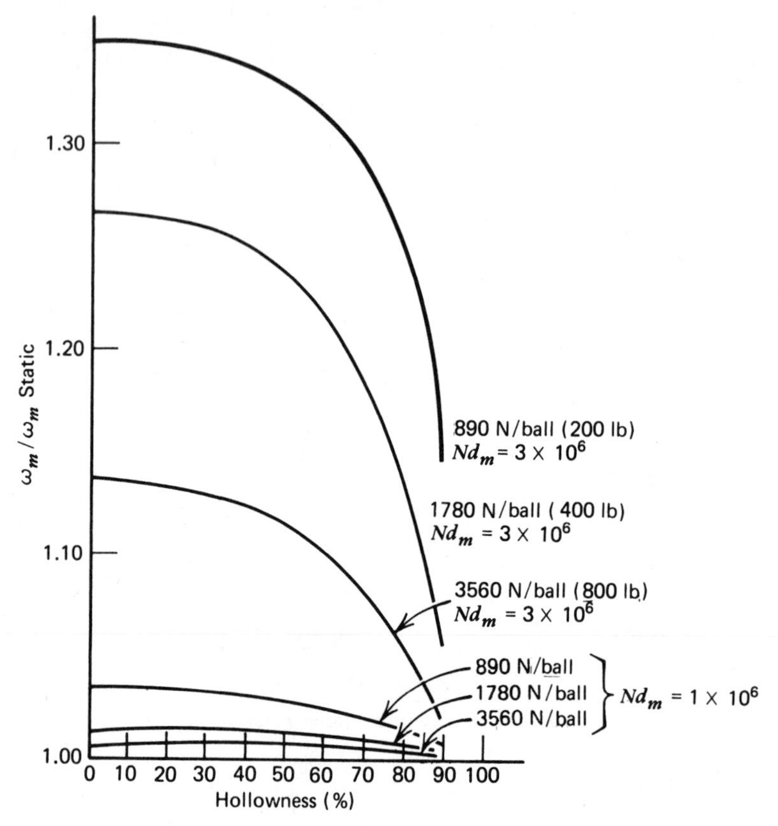

FIGURE 8.12. Cage speed ratio vs hollowness.

to make them from hot isostatically pressed (HIP) silicon nitride (Si_3N_4). This ceramic material appears excellent in compressive strength and rolling contact fatigue characteristics while exhibiting a mass density 42% of that for steel.

A problem with silicon nitride appears to be high cost per completely manufactured ball; however, this will be solved in time by more general usage in high volume production bearings (see [8.4]).

Additionally, it is necessary to establish nondestructive evaluation (NDE) means to verify the material integrity since electronic and magnetic methods used for steel components are not effective with silicon nitride.

HIGH SPEED RADIAL CYLINDRICAL ROLLER BEARINGS

Because of the high rate of heat generation accompanying relatively high friction torque, tapered roller and spherical roller bearings have not historically been employed for high speed applications. Generally, cylindrical roller bearings have been used; however, improvements in bearing internal design, accuracy of manufacture and methods of removing generated heat via circulating oil lubrication have gradually increased the allowable operating speeds for both tapered roller and spherical roller bearings. The simplest case for analytical investigation is still a radially loaded cylindrical roller bearing and this will be considered in the following discussion.

Figure 8.13 indicates the forces acting on a roller of a high speed cylindrical roller bearing subjected to a radial load F_{r}. Thus, considering equilibrium of forces,

$$Q_{\mathrm{o}j} - Q_{\mathrm{i}j} - F_{\mathrm{c}} = 0 \tag{8.28}$$

Since by equation (6.4),

$$Q = K\delta^{1.11} \tag{6.4}$$

therefore

$$K\delta_{\mathrm{o}j}^{1.11} - K\delta_{\mathrm{i}j}^{1.11} - F_{\mathrm{c}} = 0 \tag{8.29}$$

[K varies with roller length according to equation (6.9).] Since

$$\delta_{\mathrm{r}j} = \delta_{\mathrm{i}j} + \delta_{\mathrm{o}j} \tag{8.30}$$

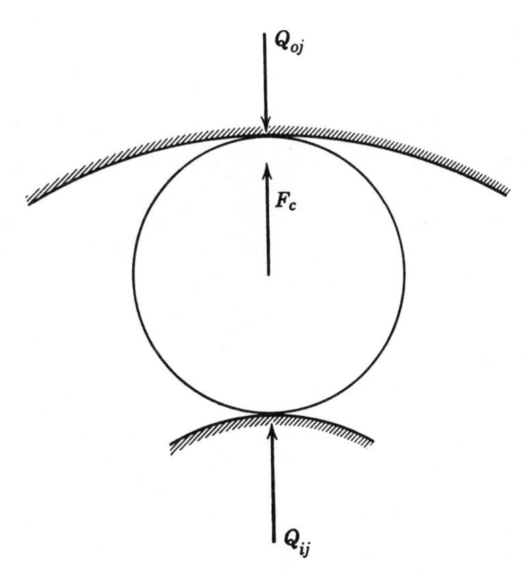

FIGURE 8.13. Roller loading at angular position ψ_j.

equation (8.29) may be rewritten as follows:

$$(\delta_{rj} - \delta_{ij})^{1.11} - \delta_{ij}^{1.11} - \frac{F_c}{K} = 0 \tag{8.31}$$

Equilibrium of forces in the direction of the applied radial load dictates that

$$F_r - \sum_{j=1}^{j=Z} Q_{ij} \cos \psi_j = 0 \tag{8.32}$$

or

$$\frac{F_r}{K} - \sum_{j=1}^{j=Z} \delta_{ij}^{1.11} \cos \psi_j = 0 \tag{8.33}$$

By considering the geometry of the loaded bearing, it can be determined that the total radial compression at any roller angular location ψ_j is

$$\delta_{rj} = \delta_r \cos \psi_j - \frac{P_d}{2} \tag{8.34}$$

Substitution of equation (8.34) into (8.31) yields

$$\left(\delta_r \cos \psi_j - \frac{P_d}{2} - \delta_{ij}\right)^{1.11} - \delta_{ij}^{1.11} - \frac{F_c}{K} = 0 \tag{8.35}$$

Equations (8.33) and (8.35) represent a system of simultaneous nonlinear equations with unknowns δ_r and δ_{ij}. As before, the Newton–Raphson method is suggested to evaluate the unknown deformations. After calculating δ_r and δ_{ij}, it is possible to calculate roller loads as follows:

$$Q_{ij} = K\delta_{ij}^{1.11} \tag{6.4}$$

$$Q_{oj} = K\delta_{ij}^{1.11} + F_c \tag{8.36}$$

Centrifugal force per roller can be calculated by using equation (4.52).

The foregoing equations apply to roller bearings with line or modified line contact. Fully crowned rollers or crowned raceways may cause point contact, in which case K_i is different from K_o and these values can be determined from equation (6.8). Information on high speed roller bearings having flexibly supported rings is given by Harris [8.2].

Example 8.2 For the 209 cylindrical roller bearing of Example 6.3 compare the load distributions at shaft speeds of 1000, 5000, 10,000, and 15,000 rpm for a radially applied load of 4450 N (1000 lb), if the bearing has no diametral clearance in the assembled condition.

$$K = 5.869 \times 10^5 \text{ N/mm}^{1.11}(4.799 \times 10^6 \text{ lb/(in.)}^{1.11}) \quad \text{Ex. 6.3}$$

$$Z = 14 \quad \text{Ex. 2.7}$$

$$P_d = 0$$

$$\Delta\psi = \frac{360°}{Z} = \frac{360}{14} = 25.71°$$

$$\frac{F_r}{K} - \sum_{j=1}^{j=Z} \delta_{ij}^{1.11} \cos\psi_j = 0 \tag{8.31}$$

$$\frac{4450}{5.869 \times 10^5} - 2\sum_{j=1}^{j=8} \tau_j \delta_{ij}^{1.11} \cos\psi_j = 0 \qquad \begin{matrix} \tau_j = 0.5 & \psi_j = 0 \\ \tau_j = 1 & \psi_j > 0 \end{matrix}$$

$$0.007581 - [0.5\delta_{i1}^{1.11} \cos(0) + \delta_{i2}^{1.11} \cos(25.71°) + \cdots \delta_{i8}^{1.11} \cos(180°)] = 0$$

$$D = 10 \text{ mm } (0.3937 \text{ in.}) \quad \text{Ex. 2.7}$$

$$d_m = 65 \text{ mm } (2.559 \text{ in.}) \quad \text{Ex. 2.7}$$

$$\gamma = \frac{d_m}{D} \cos\alpha \tag{2.27}$$

$$= \frac{65}{10} \cos(0°)$$

$$= 0.1538$$

$$n_m = \tfrac{1}{2} n_i(1 - \gamma) \tag{7.13}$$

$$= 0.5 \times n_i(1 - 0.1538)$$

$$= 0.4231 n_i$$

$$l = 9.6 \text{ mm } (0.3780 \text{ in.}) \quad \text{Ex. 2.7}$$

$$F_c = 3.39 \times 10^{-11} D^2 l d_m n_m^2 \tag{4.52}$$

$$= 3.39 \times 10^{-11}(10)^2(9.6)^2 \times 65 \times (0.4231 n_i)^2$$

$$= 3.788 \times 10^{-7} n_i^2$$

$$\left(\delta_{\mathrm{r}} \cos \psi_j - \frac{P_{\mathrm{d}}}{2} - \delta_{\mathrm{i}j}\right)^{1.11} - \delta_{\mathrm{i}j}^{1.11} - \frac{F_{\mathrm{c}}}{K} = 0 \tag{8.35}$$

$$(\delta_{\mathrm{r}} \cos \psi_j - \delta_{\mathrm{i}j})^{1.11} - \delta_{\mathrm{i}j}^{1.11} - \frac{3.788 \times 10^{-7} n_{\mathrm{i}}^2}{5.869 \times 10^5} = 0$$

This gives eight equations as follows:

$$[\delta_{\mathrm{r}} \cos (0°) - \delta_{\mathrm{i}1}]^{1.11} - \delta_{\mathrm{i}1}^{1.11} - 6.453 \times 10^{-13} n_{\mathrm{i}}^2 = 0 \tag{b}$$

$$[\delta_{\mathrm{r}} \cos (25.71°) - \delta_{\mathrm{i}2}]^{1.11} - \delta_{\mathrm{i}2}^{1.11} - 6.453 \times 10^{-13} n_{\mathrm{i}}^2 = 0 \tag{c}$$

$$\vdots$$

$$[\delta_{\mathrm{r}} \cos (180°) - \delta_{\mathrm{i}8}]^{1.11} - \delta_{\mathrm{i}8}^{1.11} - 6.453 \times 10^{-13} n_{\mathrm{i}}^2 = 0 \tag{i}$$

Equations (a)–(i) can be solved uniquely for $\delta_{\mathrm{i}1}, \ldots, \delta_{\mathrm{i}8}$ and δ_{r} for given values of n_{i}.

Figure 8.14 shows the resultant load distribution at each required speed. These loads are obtained by using equation (6.4).

$$Q_{\mathrm{i}j} = K_{\mathrm{i}j} \delta_{\mathrm{i}j}^{1.11} \tag{6.4}$$

Figure 8.15 shows the variation in δ_{r} with speed.

Hollow Rollers

As with high speed ball bearings, rollers can be made hollow to reduce roller centrifugal forces. Hollow rollers are more flexible than hollow balls, however, and greater care must be exercised to assure that accuracy of shaft location under the applied load is satisfied. Roller centrifugal force as a function of hollowness ratio D_{i}/D is given by

$$F_{\mathrm{c}} = 3.39 \times 10^{-11} D^2 l d_{\mathrm{m}} n_{\mathrm{m}}^2 (1 - H^2) \tag{8.37}$$

Figure 8.16 taken from reference [8.5] shows the effect of roller hollowness in a high speed cylindrical roller bearing on bearing radial deflection.

Figure 8.17 for the same bearing illustrates the internal load distribution.

An added criterion for evaluation in a bearing with hollow rollers is the roller bending stress. Figure 8.18 shows the effect of roller hollow-

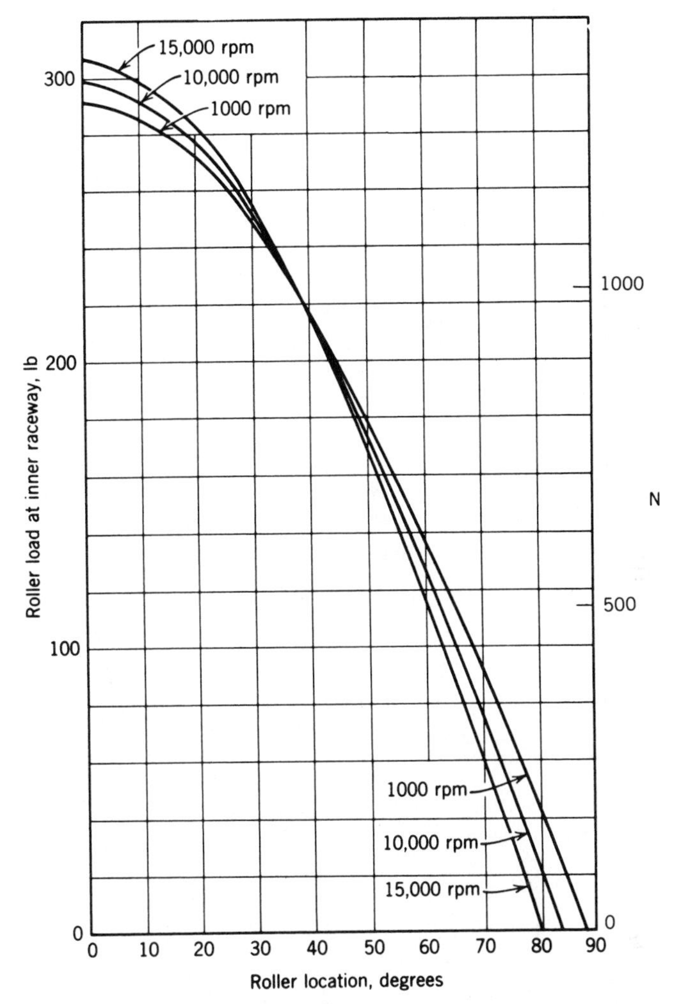

FIGURE 8.14. Roller load distribution. 209 cylindrical roller bearing, $P_d = 0$, $F_r = 4450$ N (1000 lb).

ness on maximum roller bending stress. Practical limits for roller hollowness are indicated.

Great care must be given to the smooth finishing of the inside surface of a hollow roller during manufacturing as the stress raisers that occur due to a poorly finished inside surface will reduce the allowable roller hollowness ratios still further than indicated by Fig. 8.18.

Lightweight rollers made from a ceramic material such as silicon nitride appear feasible to reduce roller centrifugal forces.

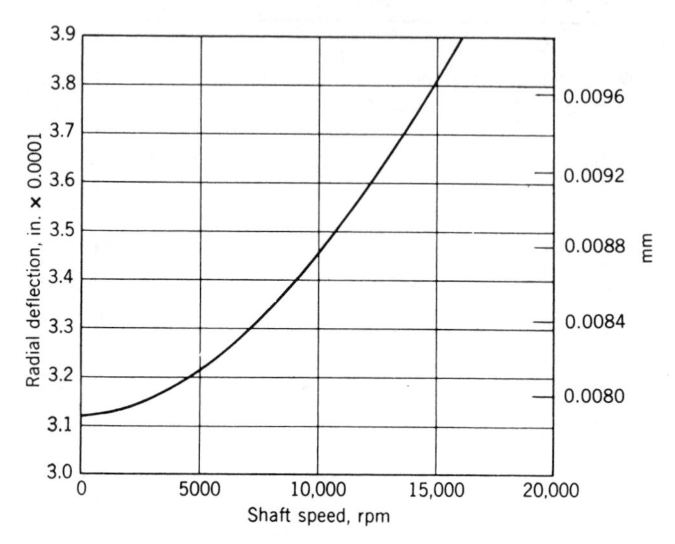

FIGURE 8.15. Radial deflection versus speed. 209 cylindrical roller bearing, $P_d = 0$, $F_r = 4450$ N (1000 lb).

Dimensions of Sample Roller Bearing

Z 21	d_m 114.3 mm (4.5 in.)
l_c 15 mm (0.59 in.)	P_d 0.0064 mm (0.00025 in.)
D 14 mm (0.55 in.)	

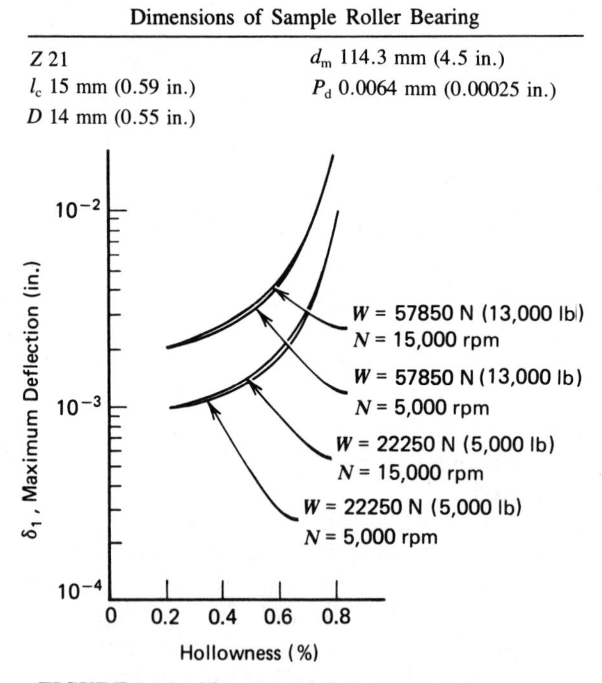

FIGURE 8.16. Maximum deflection vs hollowness.

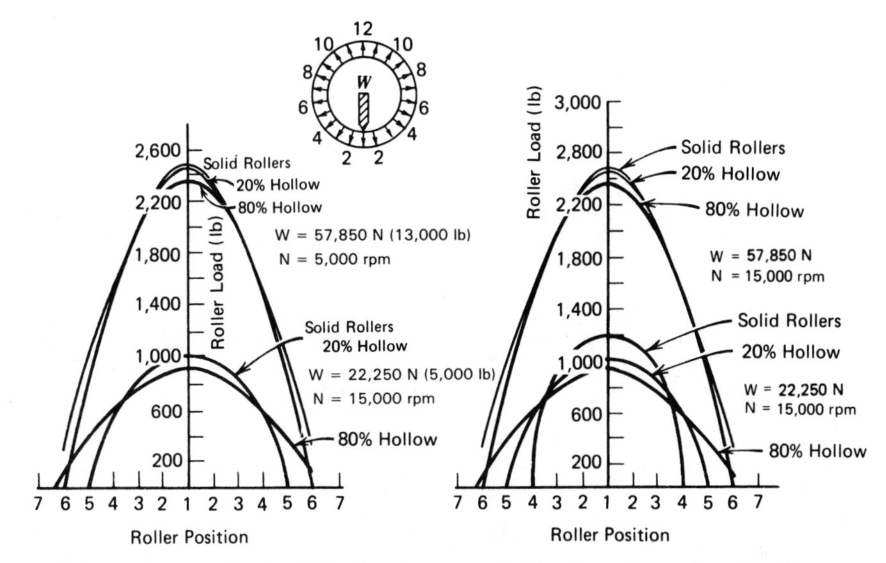

FIGURE 8.17. Roller load distribution vs applied load, shaft speed, and hollowness.

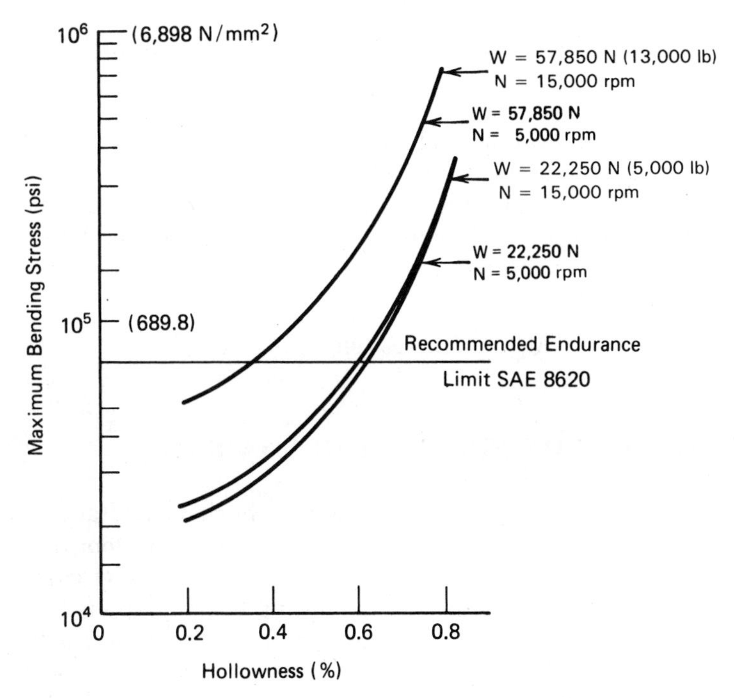

FIGURE 8.18. Maximum bending stress vs hollowness.

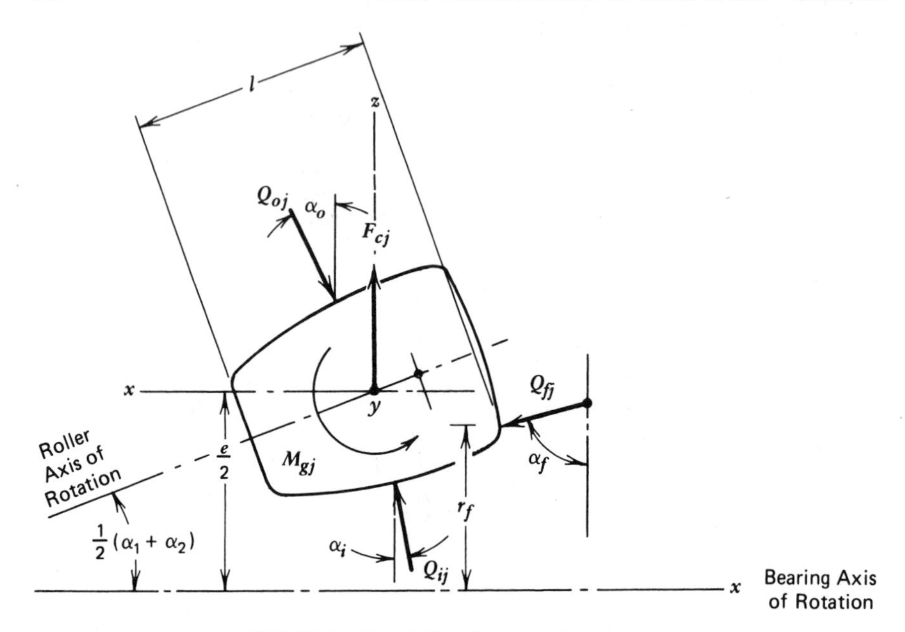

FIGURE 8.19. Roller forces and geometry.

HIGH SPEED TAPERED AND SPHERICAL ROLLER BEARINGS

Using digital computation and methods similar to those indicated in Chapter 6, the load distribution in other types of high speed roller bearings can be analyzed. Harris [8.6] indicates all of the necessary equations. The forces acting on a generalized roller are shown by Fig. 8.19. In this case, roller gyroscopic moment is given by

$$M_{gj} = J\omega_{mj}\omega_{Rj} \sin\left[\tfrac{1}{2}(\alpha_i + \alpha_o)\right] \tag{8.38}$$

FIVE DEGREES OF FREEDOM IN LOADING

In this text up until this point, all load distribution calculational methods have been limited to, at most, three degrees of freedom in loading. This has been done in the interest of simplifying the analytical methods and the understanding thereof. Every rolling bearing applied load situation can be analyzed using a system with five degrees of freedom, considering only the applied loading. Then every specialized applied loading condition, for example, simple radial load, can be analyzed using this more complex system. Reference [8.6] shows the following illustrations

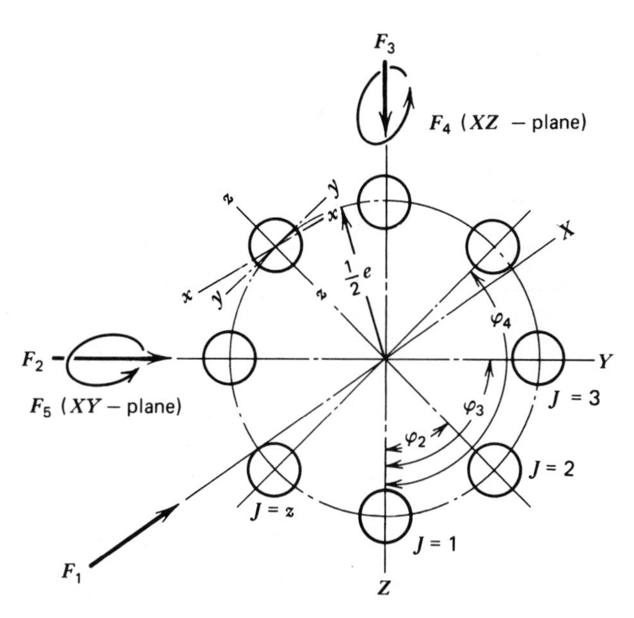

FIGURE 8.20. Bearing operating in YZ plane.

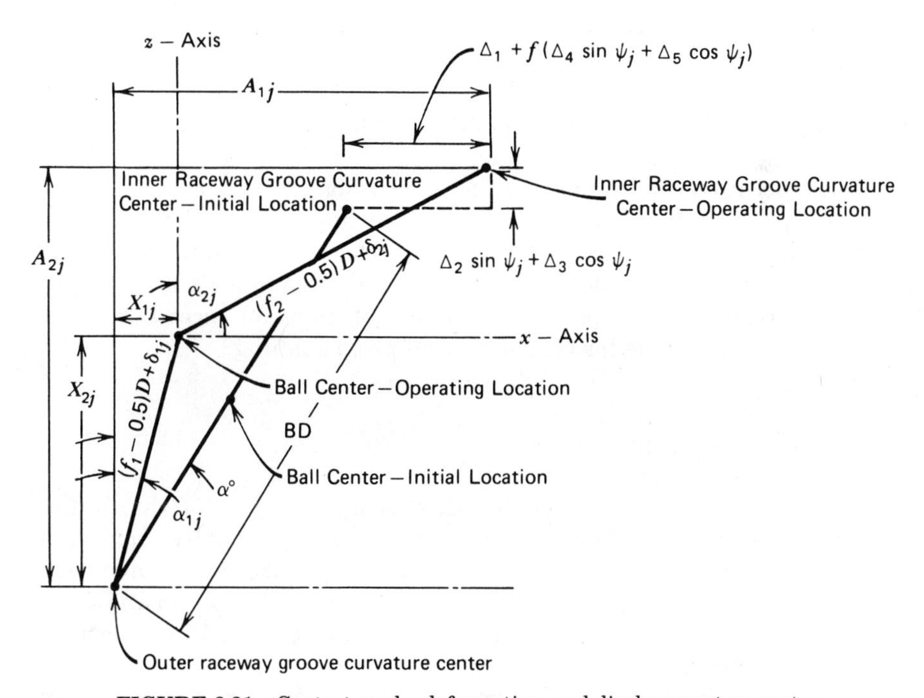

FIGURE 8.21. Contact angle, deformation, and displacement geometry.

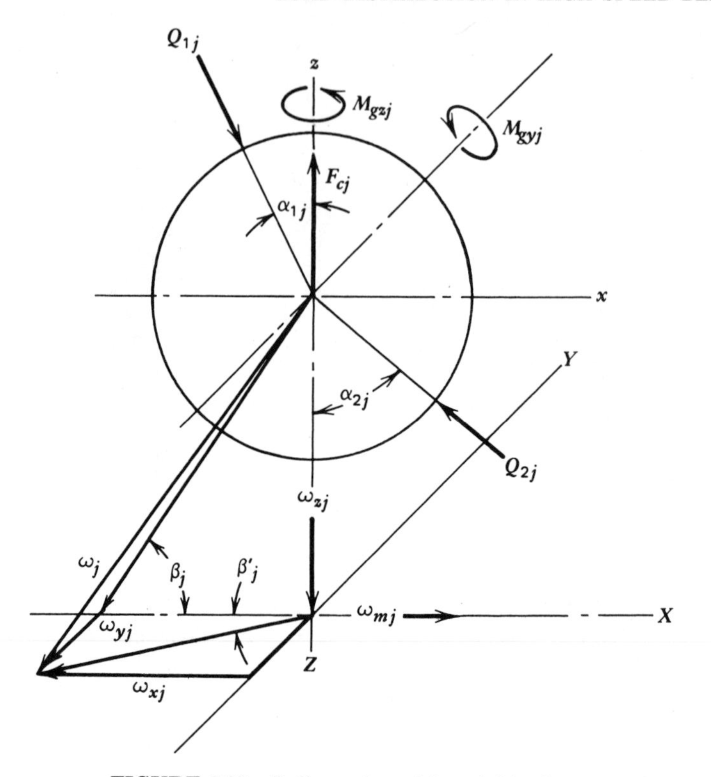

FIGURE 8.22. Ball speeds and inertial loading.

that apply to an analytical system for a ball bearing with five degrees of freedom in applied loading (see Fig. 8.20).

Note the numerical notation of applied loads, that is, $F_1, \ldots, F_5$, in lieu of F_a, F_r, and $\mathcal{M}$. Figure 8.21 shows the contact angles, deformations, and displacements for the ball–raceway contacts at azimuth ψ_j. Figure 8.22 shows the ball speed vectors and inertial loading for a ball with its center at azimuth ψ_j. Note the numerical notations for raceways; 1 = o and 2 = i. This is done for ease of digital programming.

CLOSURE

As demonstrated in the foregoing discussion, analysis of the performance of high speed roller bearings is complex and usually requires a computer to obtain numerical results. The complexity can become even greater for ball bearings. In this chapter as well as Chapters 6 and 7, for simplicity of explanation, most illustrations are confined to situations involving symmetry of loading about an axis in the radial plane of the bearing and passing through the bearing axis of rotation. The more general and complex applied loading system with five degrees of freedom is, however, discussed.

The effect of lubrication has also been neglected in this discussion. For ball bearings, it has been assumed that gyroscopic pivotal motion is minimal and can be neglected. This, of course, depends on the friction forces in the contact zones, which are affected to a great extent by lubrication. Bearing skidding is also a function of lubrication at high speeds of operation. If the bearing skids, centrifugal forces will be lower in magnitude and performance will accordingly be different.

Notwithstanding the preceding conditions, the analytical methods presented in this chapter are extremely useful in establishing optimum bearing designs for given high speed applications.

REFERENCES

8.1. A. B. Jones, "A General Theory for Elastically Constrained Ball and Radial Roller Bearings under Arbitrary Load and Speed Conditions," *ASME Trans, Journal of Basic Eng.* **82,** 309–320 (1960).

8.2. T. A. Harris, "Optimizing the Fatigue Life of Flexibly-Mounted Rolling Bearings," *Lub. Eng.* 420–428 (October, 1965).

8.3. T. A. Harris, "On the Effectiveness of Hollow Balls in High-Speed Thrust Bearings," *ASLE Transact.* **11,** 209–294 (October, 1968).

8.4. L. B. Sibley, "Silicon Nitride Bearing Elements for High Speed, High Temperature Applications," presented to NATO Advisory Group for Aerospace Research and Development, Ottawa, Canada (June, 1982).

8.5. T. A. Harris and S. F. Aaronson, "An Analytical Investigation of Cylindrical Roller Bearings Having Annular Rollers," *ASLE Preprint* No. 66LC-26 (October 18, 1966).

8.6. T. A. Harris and M. H. Mindel, "Rolling Element Bearing Dynamics," *Wear* **23,** 311–337 (1973).

9

BEARING DEFLECTION

LIST OF SYMBOLS

Symbol	Description	Units
a	Semimajor axis of the projected contact ellipse	mm (in.)
b	Semiminor axis of the projected contact ellipse	mm (in.)
d_l	Land diameter	mm (in.)
D	Ball or roller diameter	mm (in.)
F	Applied force	N (lb)
$J_r(\epsilon)$	Radial load integral	
K	Load-deflection constant	N/mm^x (lb/in.^{x})
l	Roller effective length	mm (in.)
$\mathfrak{M}$	Moment Load	N · mm (in. · lb)
Q	Rolling element load	N (lb)
Z	Number of rolling elements per row	
α	Contact angle	rad, °
γ	$D \cos \alpha / d_m$	
δ	Deflection or contact deformation	mm (in.)

Symbol	Description	Units
δ'	Deflection rate	mm/N (in./lb)
ϵ	Projection of radial load zone on bearing pitch diameter	
θ	Angle of land	rad,°
σ_{max}	Maximum contact stress	N/mm^2 (psi)
$\Sigma\rho$	Curvature sum	mm^{-1} (in.$^{-1}$)
ϕ	Angle	rad,°

SUBSCRIPTS

a	Refers to axial direction
i	Refers to inner raceway
n	Refers to direction collinear with rolling element load
o	Refers to outer raceway
p	Refers to preload condition
r	Refers to radial direction
R	Refers to ball or roller
1	Refers to bearing 1
2	Refers to bearing 2

GENERAL

In Chapter 5 a method was developed for determining the elastic contact deformation, that is, Hertzian deformation, between a raceway and rolling element. For bearings with rigidly supported rings the elastic deflection of a bearing as a unit depends on the maximum elastic contact deformation in the direction of the applied load or in the direction of interest to the designer. Because the maximum elastic contact deformation is dependent on the rolling element loads, it is necessary to analyze the load distribution occurring within the bearing prior to determination of the bearing deflection. Chapters 6 and 8 demonstrated methods for evaluating the load distribution among the rolling elements for rolling bearings subjected to static and dynamic loading, respectively.

Again, in Chapters 6 and 8 the methods for analyzing load distribution caused by generalized bearing loading (radial, axial, and moment loads applied simultaneously) utilized the variables δ_r, δ_a, and θ, which are, in fact, the principal bearing deflections. These deflections that are the subjects of this chapter may be critical in determining system stability, dynamic loading on other components, and accuracy of system operation in many applications.

DEFLECTIONS OF BEARINGS WITH RIGID RINGS

In the beginning of Chapter 6 and somewhat in Chapter 8 some simplified methods for calculating internal load distribution were discussed. Also, in those chapters methods to determine internal load distribution for complex applied loading situations were defined. The latter, which require digital computer programs to obtain solutions, generally use bearing deflections, radial, axial, and misalignment, as unknown variables. These deflections are therefore determined directly from the solution of the system of nonlinear equations. For applications with relatively simple applied loading, the methods for determining bearing deflection were not defined and these will be discussed herein.

It is possible to calculate the maximum rolling element load Q_{max} due to a combination of radial and axial loads. Q_{max} has attendant contact deformations $\delta_{o_{max}}$ and $\delta_{i_{max}}$ measured along the line of contact at the outer and inner raceways, respectively. From equation (6.4) it can be seen that

$$\delta_{o_{max}} = \left(\frac{Q_{o_{max}}}{K_o}\right)^{1/n} \tag{9.1}$$

$$\delta_{i_{max}} = \left(\frac{Q_{i_{max}}}{K_i}\right)^{1/n} \tag{9.2}$$

In equations (9.1) and (9.2), $n = 1.5$ and $n = 1.11$ for ball and roller bearings, respectively. The radial deflection of the bearing from Fig. 8.2 is therefore

$$\delta_r = [(f_i - 0.5)D + \delta_{i_{max}}] \cos \alpha_i + [(f_o - 0.5)D + \delta_{o_{max}}] \cos \alpha_o - BD \cos \alpha^\circ \tag{9.3}$$

or

$$\delta_r = (f_i - 0.5)D(\cos \alpha_i - \cos \alpha^\circ) + (f_o - 0.5)D(\cos \alpha_o - \cos \alpha^\circ) + \delta_{i_{max}} \cos \alpha_i + \delta_{o_{max}} \cos \alpha_o \tag{9.4}$$

Similarly, the axial deflection is given by

$$\delta_a = (f_i - 0.5)D(\sin \alpha_i - \sin \alpha^\circ) + (f_o - 0.5)D(\sin \alpha_o - \sin \alpha^\circ) + \delta_{i_{max}} \sin \alpha_i + \delta_{o_{max}} \sin \alpha_o \tag{9.5}$$

$Q_{o_{max}}$ includes the effect of centrifugal loading.

In lieu of the more rigorous approach to bearing deflection outlined above, Palmgren [9.1] gives a series of formulas to calculate bearing deflection for specific conditions of loading. For slow and moderate speed deep-groove and angular-contact ball bearings subjected to radial load which causes only radial deflection, that is, $\delta_a = 0$,

$$\delta_r = 4.36 \times 10^{-4} \frac{Q_{max}^{2/3}}{D^{1/3} \cos \alpha} \tag{9.6}$$

For self-aligning ball bearings,

$$\delta_r = 6.98 \times 10^{-4} \frac{Q_{max}^{2/3}}{D^{1/3} \cos \alpha} \tag{9.7}$$

For slow and moderate speed radial roller bearings with point contact at one raceway and line contact at the other,

$$\delta_r = 1.81 \times 10^{-4} \frac{Q_{max}^{3/4}}{l^{1/2} \cos \alpha} \tag{9.8}$$

For radial roller bearings with line contact at each raceway,

$$\delta_r = 7.68 \times 10^{-5} \frac{Q_{max}^{0.9}}{l^{0.8} \cos \alpha} \tag{9.9}$$

To the values given above must be added the appropriate radial clearance and any deflection due to a nonrigid housing.

The axial deflection under pure axial load, that is, $\delta_r = 0$, for angular-contact ball bearings is given by

$$\delta_a = 4.36 \times 10^{-4} \frac{Q_{max}^{2/3}}{D^{1/3} \sin \alpha} \tag{9.10}$$

For self-aligning ball bearings

$$\delta_a = 6.98 \times 10^{-4} \frac{Q_{max}^{2/3}}{D^{1/3} \sin \alpha} \tag{9.11}$$

For thrust ball bearings

$$\delta_a = 5.24 \times 10^{-4} \frac{Q_{max}^{2/3}}{D^{1/3} \sin \alpha} \tag{9.12}$$

For radial ball bearings subjected to axial load, the contact angle α must be determined prior to using equation (9.10). For roller bearings with point contact at one raceway and line contact at the other,

$$\delta_a = 1.81 \times 10^{-4} \frac{Q_{max}^{3/4}}{l^{1/2} \sin \alpha} \tag{9.13}$$

For roller bearings with line contact at each raceway,

$$\delta_a = 7.68 \times 10^{-5} \frac{Q_{max}^{0.9}}{l^{0.8} \sin \alpha} \tag{9.14}$$

Example 9.1. For the 209 cylindrical roller bearing of Example 6.4 estimate the bearing radial deflection. Compare this value with δ_{max} obtained in Example 6.3 assuming a diametral clearance of 0.0406 mm (0.0016 in.).

$$Q_{max} = 1589 \text{ N } (357.1 \text{ lb}) \qquad \text{Ex. 6.4}$$

$$l = 9.6 \text{ mm } (0.3789 \text{ in.}) \qquad \text{Ex. 2.7}$$

$$\delta_r = 7.68 \times 10^{-5} \frac{Q^{0.9}{}_{max}}{l^{0.8} \cos \alpha} \tag{9.9}$$

$$= 7.68 \times 10^{-5} \frac{(1589)^{0.9}}{(9.6)^{0.8} \cos (0°)}$$

$$= 0.00953 \text{ mm } (0.000375 \text{ in.})$$

Total shaft movement is

$$\delta_{max} = \delta_r + \frac{P_d}{2}$$

$$= 0.00953 + 0.0203$$

$$= 0.02983 \text{ mm } (0.001175 \text{ in.})$$

From Example 6.3, $\delta_{max} = 0.03251$ mm (0.00128 in.).

Example 9.2. For the 218 angular-contact ball bearing of Example 8.1, estimate the axial deflection at 44,500 N (10,000 lb) thrust load. Compare this value against the data of Fig. 8.8.

$$Z = 16 \qquad \text{Ex. 6.5}$$

$$\alpha° = 40° \qquad \text{Ex. 2.3}$$

$$D = 22.23 \text{ mm } (0.875 \text{ in.}) \qquad \text{Ex. 2.3}$$

$$Q = \frac{F_a}{Z \sin \alpha} \qquad (6.26)$$

$$= \frac{44500}{16 \times \sin (40°)}$$

$$= 4239 \text{ N } (972.8 \text{ lb})$$

$$\delta_a = 4.36 \times 10^{-4} \frac{Q_{max}^{2/3}}{D^{1/3} \sin \alpha} \qquad (9.10)$$

$$= 4.36 \times 10^{-4} \frac{(4239)^{2/3}}{(22.23)^{1/3} \sin (40°)}$$

$$= 0.064 \text{ mm } (0.00252 \text{ in.})$$

From Fig. 8.8 it can be seen that this value is a satisfactory estimate of δ_a at slow speeds. At high speed δ_a will be less than this estimate.

PRELOADING

Axial Preloading

A typical curve of ball bearing deflection vs load is shown by Fig. 9.1. It can be seen from Fig. 9.1 that as load is increased uniformly, the rate of deflection increase declines. Hence, it would be advantageous with regard to bearing deflection under load to operate above the "knee" of the

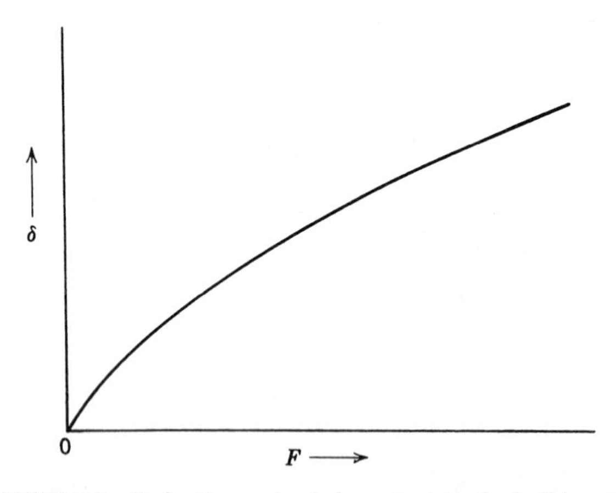

FIGURE 9.1. Deflection vs load characteristic for ball bearings.

$$\delta_1 = \delta_p + \delta_a$$

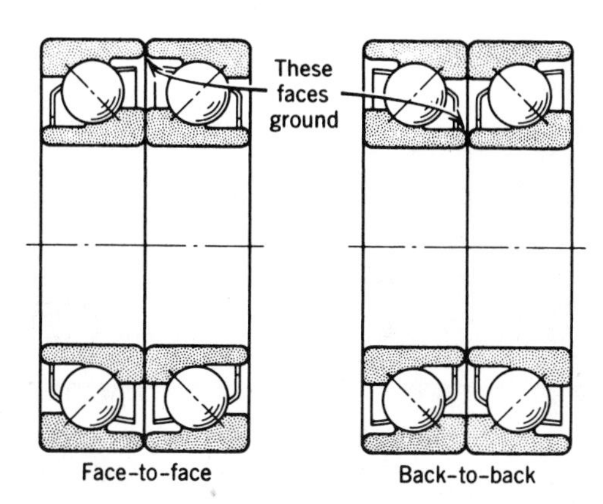

FIGURE 9.2. Duplex sets of angular-contact ball bearings.

load–deflection curve. This condition can be realized by axially preloading angular-contact ball bearings. This is usually done, as shown in Fig. 9.2, by grinding stock from opposing end faces of the bearings and then locking the bearings together on the shaft. Figure 9.3 shows preloaded bearing sets before and after the bearings are axially locked together. Figure 9.4 illustrates, graphically, the improvement in load–deflection characteristics obtained by preloading ball bearings.

Suppose two identical angular-contact ball bearings are placed back-to-back or face-to-face on a shaft, as shown in Fig. 9.5, and drawn together by a locking device. Each bearing experiences an axial deflection δ_p due to preload F_p. The shaft is thereafter subjected to thrust load F_a, as shown in Fig. 9.5, and because of the thrust load, the bearing combination undergoes an axial deflection δ_a. In this situation the total axial deflection at bearing 1 is

$$\delta_1 = \delta_p + \delta_a \tag{9.15}$$

and at bearing 2,

$$\begin{aligned} \delta_2 &= \delta_p - \delta_a \qquad \delta > \delta_a \\ \delta_2 &= 0 \qquad\qquad\; \delta_p \leq \delta_a \end{aligned} \tag{9.16}$$

The total load in the bearings is equal to the applied thrust load:

$$F_a = F_1 - F_2 \tag{9.17}$$

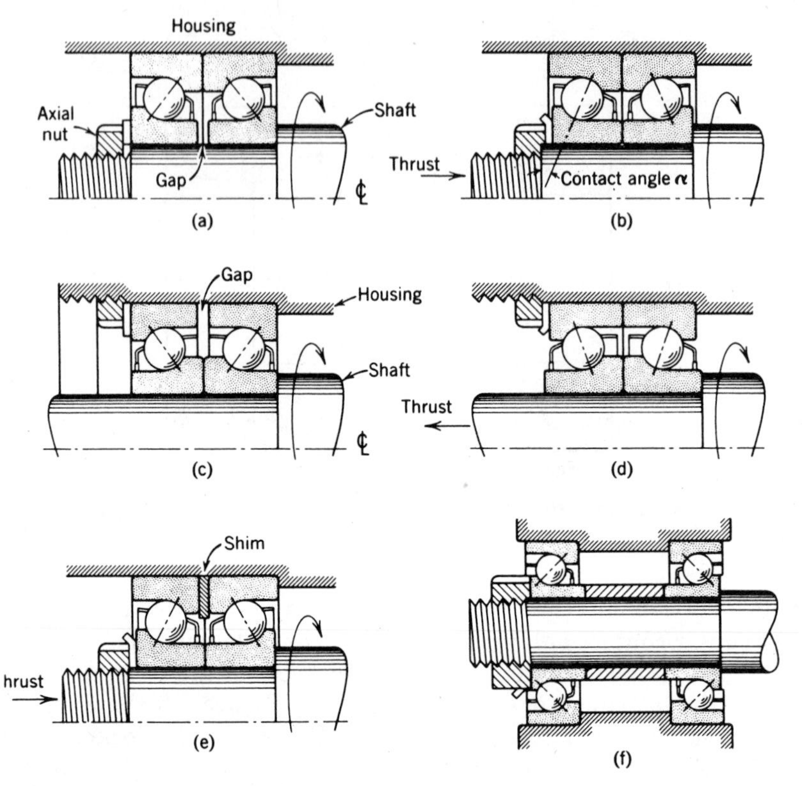

FIGURE 9.3. (*a*) Duplex set with back-to-back angular-contact ball bearings prior to axial preloading. The inner ring faces are ground to provide a specific axial gap. (*b*) Same unit as in (*a*) after tightening axial nut to remove gap. The contact angles have increased. (*c*) Face-to face angular-contact duplex set prior to preloading. In this case it is the outer ring faces that are ground to provide the required gap. (*d*) Same set as in (*c*) after tightening the axial nut. The convergent contact angles increase under preloading. (*e*) Shim between two standard-width bearings avoids need for grinding the faces of the outer rings. (*f*) Precision spacers between automatically provide proper preload by making the inner spacer slightly shorter than the outer.

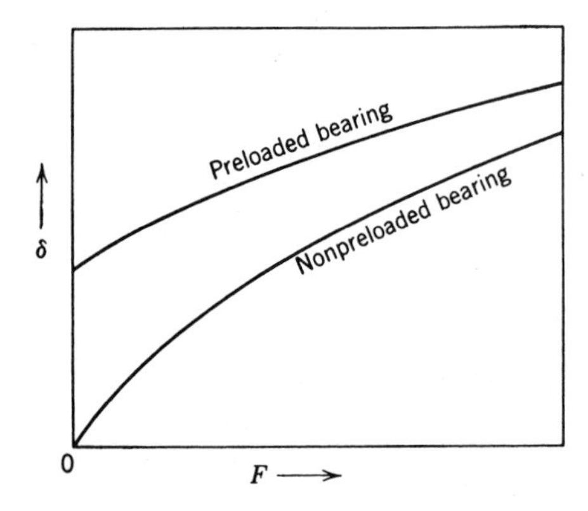

FIGURE 9.4. Deflection vs load characteristics for ball bearings. As the load increases, the rate of the increase of deflection decreases, therefore preloading (top line) tends to reduce the bearing deflection under additional loading.

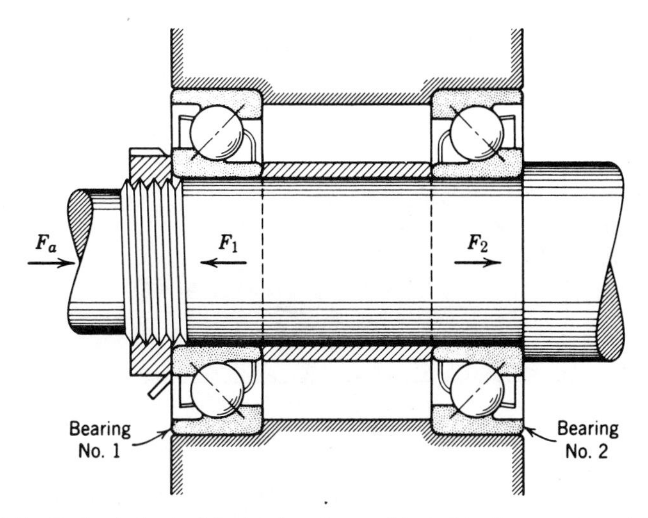

FIGURE 9.5. Preloaded set of duplex bearings subjected to F_a, an external thrust load. The computation for the resulting deflection is complicated by the fact that the preload at bearing 1 is increased by load F_a while the preload at bearing 2 is decreased.

For the purpose of this analysis consider only centric thrust load applied to the bearing; therefore, from equation (6.33),

$$\frac{F_a}{ZD^2K} = \sin\alpha_1\left(\frac{\cos\alpha^\circ}{\cos\alpha_1} - 1\right)^{1.5} - \sin\alpha_2\left(\frac{\cos\alpha^\circ}{\cos\alpha_2} - 1\right)^{1.5} \tag{9.18}$$

Combining equations (9.15) and (9.16) yields

$$\delta_1 + \delta_2 = 2\delta_p \tag{9.19}$$

Substitution of equation (9.15) for δ_1 and equation (9.16) for δ_2 in (6.36) gives

$$\sin(\alpha_1 - \alpha^\circ) + \sin(\alpha_2 - \alpha^\circ) = \frac{2\delta_p \cos\alpha^\circ}{BD} \tag{9.20}$$

Equations (9.18) and (9.20) may now be solved for α_1 and α_2. Subsequent substitution of α_1 and α_2 into equation (6.36) yields values of δ_1 and δ_2. The data pertaining to the selected preload F_p and deflection δ_p may be obtained from the following equations:

$$\frac{F_p}{ZD^2K} = \sin\alpha_p\left(\frac{\cos\alpha^\circ}{\cos\alpha_p} - 1\right)^{1.5} \tag{9.21}$$

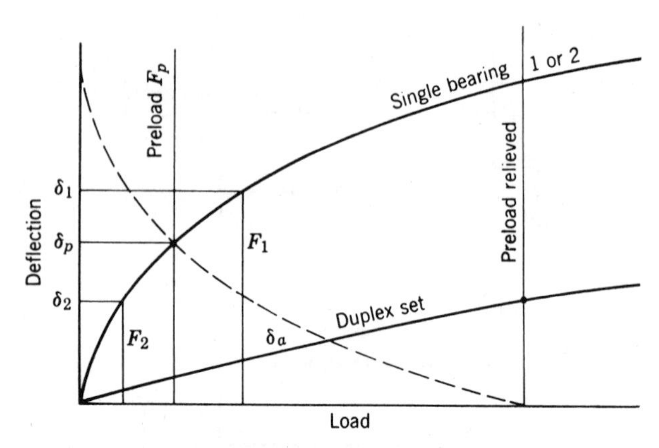

FIGURE 9.6. Deflection vs load for a preloaded duplex set of ball bearings.

$$\delta_p = \frac{BD \sin (\alpha_p - \alpha^\circ)}{\cos \alpha_p} \tag{9.22}$$

Figure 9.6 shows a typical plot of bearing deflection δ_a vs load. Note that deflection is everywhere less than that for a nonpreloaded bearing up to the load at which preload is removed. Thereafter, the unit acts as a single bearing under thrust load and assumes the same load–deflection characteristics as given by the single-bearing curve. The point at which bearing 2 loses load may be determined graphically by inverting the single-bearing load–deflection curve about the preload point. This is shown by Fig. 9.6.

Since roller bearing deflection is almost linear with respect to load, there is not as much advantage to be gained by axially preloading tapered or spherical roller bearings; hence this is not a universal practice as it is for ball bearings. Figure 9.7, however, shows tapered roller bearings axially locked together in a light preload arrangement.

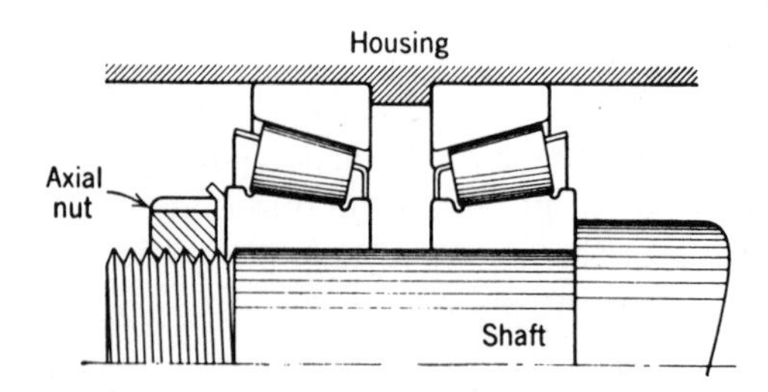

FIGURE 9.7. Lightly preloaded tapered roller bearings.

Example 9.3. A duplex pair of 218 angular-contact ball bearings are mounted back-to-back, as shown in Fig. 9.3. If the pair is preloaded to 4450 N (1000 lb), determine the axial deflection under 8900 N (2000 lb) applied thrust load. Compare these results with the static deflection data of Fig. 8.8.

$$Z = 16 \qquad \text{Ex. 6.5}$$

$$D = 22.23 \text{ mm } (0.875 \text{ in.}) \qquad \text{Ex. 2.3}$$

$$\alpha^\circ = 40^\circ \qquad \text{Ex. 2.3}$$

$$K = 896.7 \text{ N/mm}^2 \ (130{,}000 \text{ psi}) \qquad \text{Ex. 6.5}$$

From the static curve of Fig. 8.4, at 4450 N (1000 lb) $\alpha_p = 40.61^\circ$

$$B = 0.0464 \qquad \text{Ex. 2.3}$$

$$\delta_p = \frac{BD \sin(\alpha_p - \alpha^\circ)}{\cos \alpha_p} \tag{9.22}$$

$$= \frac{0.0464 \times 22.23 \sin(40.61^\circ - 40^\circ)}{\cos(40.61^\circ)}$$

$$= 0.0145 \text{ mm } (0.0005694 \text{ in.})$$

This number could have been obtained from Fig. 8.6.

$$\frac{F_a}{ZD^2K} = \sin \alpha_1 \left(\frac{\cos \alpha^\circ}{\cos \alpha_1} - 1 \right)^{1.5} - \sin \alpha_2 \left(\frac{\cos \alpha^\circ}{\cos \alpha_2} - 1 \right)^{1.5} \tag{9.18}$$

$$\frac{8900}{16 \times (22.23)^2 \times 896.7} = \sin \alpha_1 \left(\frac{\cos(40^\circ)}{\cos \alpha_1} - 1 \right)^{1.5} - \sin \alpha_2 \left(\frac{\cos(40^\circ)}{\cos \alpha_2} - 1 \right)^{1.5}$$

$$0.001255 = \sin \alpha_1 \left(\frac{0.7660}{\cos \alpha_1} - 1 \right)^{1.5} - \sin \alpha_2 \left(\frac{0.7660}{\cos \alpha_2} - 1 \right)^{1.5} \tag{a}$$

$$\frac{\sin(\alpha_1 - \alpha^\circ)}{\cos\alpha_1} + \frac{\sin(\alpha_2 - \alpha^\circ)}{\cos\alpha_2} = \frac{2\delta_p}{BD} \tag{9.20}$$

$$\frac{\sin(\alpha_1 - 40^\circ)}{\cos\alpha_1} + \frac{\sin(\alpha_2 - 40^\circ)}{\cos\alpha_2} = \frac{2 \times 0.0145}{0.0464 \times 22.23}$$

$$\frac{\sin(\alpha_1 - 40^\circ)}{\cos\alpha_1} + \frac{\sin(\alpha_2 - 40^\circ)}{\cos\alpha_2} = 0.02805 \tag{b}$$

Equations (a) and (b) can be solved simultaneously to yield α_1 and α_2 and thence δ_a. Alternatively, the static deflection curve of Fig. 8.6 can be used as follows to

1. Assume values of δ_a.
2. Create tabular values of $\delta_1 = \delta_p + \delta_a$ (9.15) and $\delta_2 = \delta_p - \delta_a$ (9.16).
3. Find F_1 and F_2 corresponding to δ_1 and δ_2, respectively.
4. Find $F_a = F_1 - F_2$ (9.17).

δ_a mm (in.)	δ_1 mm (in.)	δ_2 mm (in.)
0.0025 (0.0001)	0.0168 (0.00066)	0.0117 (0.00046)
0.0051 (0.0002)	0.0193 (0.00076)	0.0091 (0.00036)
0.0076 (0.0003)	0.0218 (0.00086)	0.0066 (0.00026)
0.0102 (0.0004)	0.0244 (0.00096)	0.0041 (0.00016)
0.0127 (0.0005)	0.0269 (0.00106)	0.0015 (0.00006)

F_a N (lb)	F_1 N (lb)	F_2 N (lb)
2,225 (500)	5,785 (1300)	3560 (800)
4,895 (1100)	7,343 (1650)	2448 (550)
6,987 (1570)	8,678 (1950)	1691 (380)
9,345 (2100)	10,240 (2300)	890 (200)
11,440 (2570)	11,790 (2650)	356 (80)

5. Plot δ_a vs F_a and find $\delta_a = 0.00968$ mm (0.000381 in.) corresponding to $F_a = 8900$ N (2000 lb).

From Fig. 8.6 at $F_a = 8900$ N (2000 lb), $\delta_a = 0.0221$ mm (0.00087 in.). Therefore, preloading of 4450 N (1000 lb) reduced δ_a by 56%.

If it is desirable to preload ball bearings that are not identical, equations (9.18) and (9.20) become

$$F = Z_1 D_1^2 K_1 \sin \alpha_1 \left(\frac{\cos \alpha_1^\circ}{\cos \alpha_1} - 1 \right)^{1.5} - Z_2 D_2^2 K_2 \times \sin \alpha_2 \left(\frac{\cos \alpha_2^\circ}{\cos \alpha_2} - 1 \right)^{1.5} \tag{9.23}$$

$$\frac{(B_1 D_1) \sin (\alpha_1 - \alpha_1^\circ)}{\cos \alpha_1} + \frac{(B_2 D_2) \sin (\alpha_2 - \alpha_2^\circ)}{\cos \alpha_2} = 2\delta_p \tag{9.24}$$

Equations (9.23) and (9.24) must be solved simultaneously for α_1 and α_2. As before, equation (6.36) yields the corresponding values of δ.

To reduce axial deflection still further, more than two bearings can be locked together axially as shown in Fig. 9.8. The disadvantages of this system are increased space, weight, and cost. More data on axial preloading are given in reference [9.2].

Radial Preloading

Radial preloading of rolling bearings is not usually used to eliminate initial large magnitude deflection as is axial preload. Instead, its purpose is generally to obtain a greater number of rolling elements under load and thus reduce the maximum rolling element load. It is also used to prevent skidding. Methods used to calculate maximum radial rolling element load are given in Chapter 6. Figure 9.9 shows various methods to radially preload roller bearings.

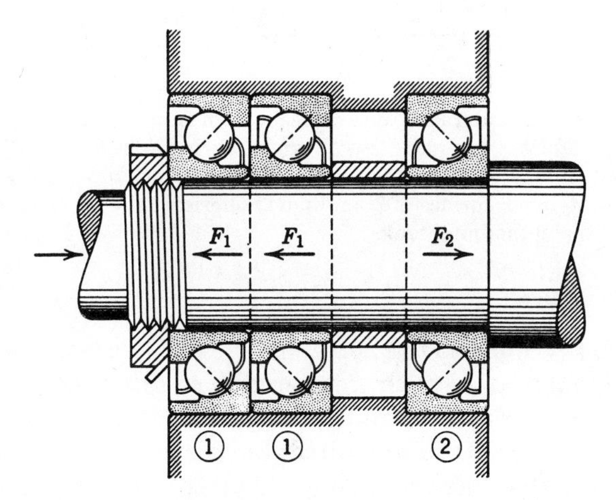

FIGURE 9.8. Triplex set of angular-contact bearings, mounted two in tandem and one opposed. This arrangement provides an even higher axial stiffness and longer bearing life than with a duplex set, but requires more space.

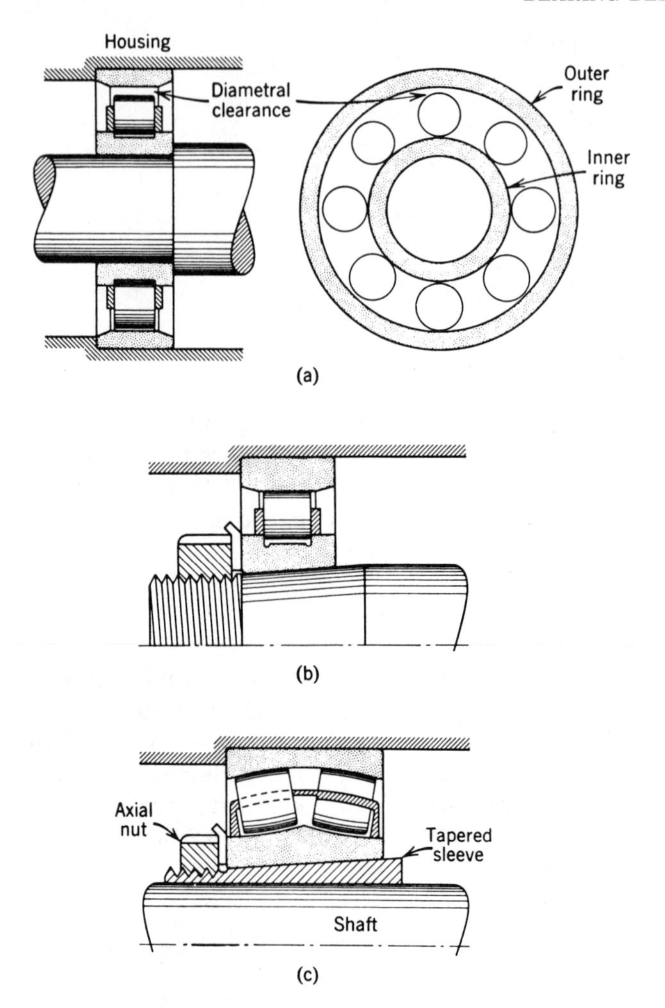

FIGURE 9.9. (*a*) Diametral (radial) clearance found in most-off-the-shelf rolling bearings. One object of preloading is to remove this clearance during assembly. (*b*) Cylindrical roller bearing mounted on tapered shaft, to expand inner ring. Such bearings are usually made with a taper on the inner surface of $\frac{1}{12}$ in./in. (*c*) Spherical roller bearing mounted on tapered sleeve to expand the inner ring.

Example 9.4. Suppose the 209 cylindrical roller bearing of Example 6.3 was manufactured with a tapered bore and was driven up a tapered shaft as in Fig. 9.9b until a negative clearance or interference of 0.00254 mm (0.0001 in.) resulted. For a radial load of 4450 N (1000 lb), determine the maximum roller load, the extent of the load zone, and the radial deflection. Compare these results with those of Example 6.3.

$$\frac{F_r}{ZK_n} = 0.001170 \qquad \text{Ex. 6.3}$$

$$F_r = ZK_n(\delta_{max} - \tfrac{1}{2} P_d)^{1.11} J_r(\epsilon) \qquad (6.22)$$

$$0.001170 = \left[\delta_{max} - \frac{(-0.00254)}{2}\right]^{1.11} J_r(\epsilon)$$

$$(\delta_{max} + 0.00127)^{1.11} J_r(\epsilon) = 0.001170 \qquad (a)$$

$$\epsilon = \frac{1}{2}\left(1 - \frac{P_d}{2\delta_{max}}\right) \qquad (6.12)$$

$$= \frac{1}{2}\left[1 - \left(\frac{-0.00254}{2\delta_{max}}\right)\right]$$

$$= 0.5 + \frac{0.000635}{\delta_{max}} \qquad (b)$$

1. Assume $\epsilon = 0.8$. From Fig. 6.2, $J_r = 0.266$.

2. $(\delta_{max} + 0.00127)^{1.11} \times 0.266 = 0.001170$ (a)

$$\delta_{max} = 0.00635 \text{ mm } (0.00025 \text{ in.})$$

3. $\epsilon = 0.5 + \dfrac{0.000635}{0.00635} = 0.6$ (b)

$0.6 \neq 0.8$

4. Assume $\epsilon = 0.6$.
From Fig. 6.2, $J_r = 0.256$.

5. $(\delta_{max} + 0.00127)^{1.11} \times 0.256 = 0.001170$ (a)

$$\delta_{max} = 0.00660 \text{ mm } (0.000260 \text{ in.})$$

6. $\epsilon = 0.5 + \dfrac{0.000635}{0.00660} = 0.596$ (b)

This answer is sufficiently close to $\epsilon = 0.6$

$$Z = 14 \qquad \text{Ex. 2.7}$$

$$F_r = ZQ_{max}J_r(\epsilon) \qquad (6.19)$$

$$4450 = 14Q_{max} \times 0.256$$

$$Q_{max} = 1242 \text{ N } (279.0 \text{ lb})$$

$$\delta_{max} = 0.00660 \text{ mm } (0.000260 \text{ in.})$$

$$\psi_l = \cos^{-1}(1 - 2\epsilon) \qquad (6.12), (6.13)$$

$$= \cos^{-1}(1 - 2 \times 0.6) = \pm 101°32'$$

Comparing results with Example 6.3

	Example 6.3	Example 9.4
P_d, mm	+0.0406 (+0.0016 in.)	−0.0025 (−0.0001 in.)
Q_{max}, N	1915 (430.3 lb)	1242 (279 lb)
δ_{max}, mm	0.032 (0.00126 in.)	0.0066 (0.00026 in.)
ψ_l°	±50°35′	±101°32′

Preloading to Achieve Isoelasticity

It is sometimes desirable that the axial and radial yield rates of the bearing and its supporting structures be as nearly identical as possible. In other words, a load in either the axial or radial direction should cause identical deflections (ideally). This necessity for *isoelasticity* in the ball bearings came with the development of the highly accurate, low drift inertial gyros for navigational systems, and for missile and space guidance systems. Such inertial gyros usually have a single degree of freedom tilt axis and are extremely sensitive to error moments about this axis.

Consider a gyro in which the spin axis (Fig. 9.10) is coincident with the x axis, the tilt axis is perpendicular to the paper at the origin O, and the center of gravity of the spin mass is acted on by a disturbing force F in the xz plane and directed at an oblique angle ϕ to the x axis, this force will tend to displace the spin mass center of gravity from 0 to 0′. If, as shown in Fig. 9.10, the displacements in the directions of the x and z

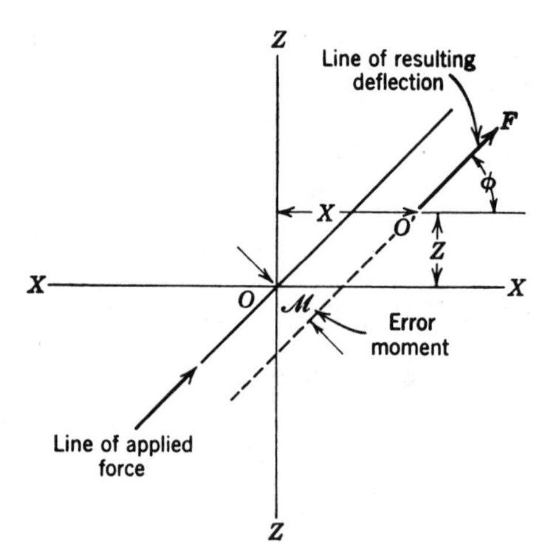

FIGURE 9.10. Effect of disturbing force F on the center of gravity of spring mass. It is frequently desirable to obtain isoelasticity in bearings in which the displacement in any direction is in line with the disturbing force.

axes are not equal, the force F will create an error moment about the tilt axis.

In terms of the axial and radial yield rates of the bearings, the error moment $\mathfrak{M}$ is

$$\mathfrak{M} = \tfrac{1}{2} F^2(\delta_z' - \delta_x') \sin 2\phi \tag{9.25}$$

where the bearing yield rates δ_z' and δ_x' are in deflection per unit of force.

To minimize $\mathfrak{M}$ and subsequent drift, δ_z' must be as nearly equal to δ_x' as possible—a requirement for pinpoint navigation or guidance. Also, from Fig. 9.10 it can be noted that improving the rigidity of the bearing, that is, decreasing δ_z' and δ_x' collectively, reduces the magnitude of the minimal error moments achieved through isoelasticity.

In most radial ball bearings, the radial rate is usually smaller than the axial rate. This is best overcome by increasing the bearing contact angle, which reduces the axial yield rate and increases radial yield rate. One-to-one ratios can be obtained by using bearings with contact angles that are 30° or higher.

At these high angles, the sensitivity of the axial-to-radial yield rate ratio to the amount of preload is quite small. It is, however, necessary to preload the bearings to maintain the desired contact angles.

LIMITING BALL BEARING THRUST LOAD

General Considerations

Most radial ball bearings can accommodate a thrust load and function properly provided that the contact stress thereby induced is not excessively high or that the ball does not override the land. The latter condition results in severe stress concentration and attendant rapid fatigue failure of the bearing. It may therefore be necessary to ascertain for a given bearing the maximum thrust load that the bearing can sustain and still function. The situation in which the balls override the land will be examined first.

Thrust Load Causing Ball to Override Land

Figure 9.11 shows an angular-contact bearing under thrust in which the balls are riding at an extreme angular location without the ring lands cutting into the balls.

From Fig. 9.11 it can be seen that the thrust load, which causes the major axis of the contact ellipse to just reach the land of the bearing, is the maximum permissible load that the bearing can accommodate with-

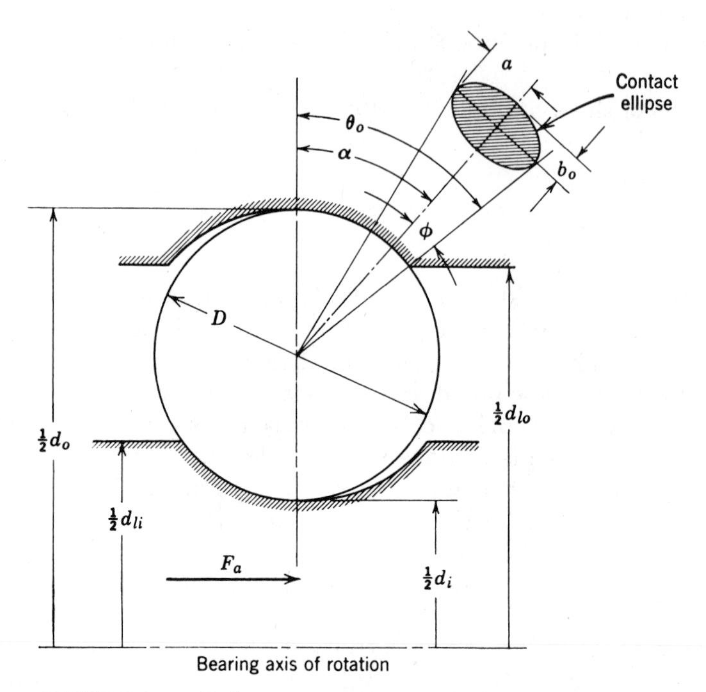

FIGURE 9.11. Ball–raceway contact under limiting thrust load.

out overriding the corresponding land. Both the inner and outer ring lands must be considered.

From Fig. 9.11 it can be determined that the angle θ_o describing the juncture of the outer ring land with the outer raceway is equal to $\alpha + \phi$ in which α is the raceway contact angle under the load necessary to cause the major axis of the contact ellipse, that is, $2a_o$, to extend to θ_o and ϕ is the one-half of the angle subtended by the chord $2a$. The angle θ_o is given approximately by

$$\theta_o = \cos^{-1}\left(1 - \frac{d_o - d_{lo}}{D}\right) \tag{9.26}$$

Since the contact deformation is small, r'_o to the midpoint of the chord $2a_o$ is approximately equal to $D/2$; therefore, $\sin\phi \approx 2a_o/D$ or

$$\sin(\theta_o - \alpha) = \frac{2a_o}{D} \tag{9.27}$$

For steel balls contacting steel raceways, the semimajor axis of the contact ellipse is given by

$$a = 0.0236a^*\left(\frac{Q}{\Sigma\rho_o}\right)^{1/3} \tag{5.34}$$

in which $\Sigma\rho_o$ is given by

$$\Sigma\rho_o = \frac{1}{D}\left(4 - \frac{1}{f_o} - \frac{2\gamma}{1+\gamma}\right) \tag{2.30}$$

and a_o^* is a function of $F(\rho)_o$ defined by

$$F(\rho)_o = \frac{\dfrac{1}{f_o} - \dfrac{2\gamma}{(1+\gamma)}}{4 - \dfrac{1}{f_o} - \dfrac{2\gamma}{(1+\gamma)}} \tag{2.31}$$

$$\gamma = \frac{D\cos\alpha}{d_m} \tag{2.27}$$

According to equation (6.26) for a thrust-loaded ball bearing,

$$Q = \frac{F_a}{Z\sin\alpha} \tag{6.26}$$

Combining equations (5.24), (2.30), (9.27), and (6.26) one obtains

$$F_{ao} = Z\sin\alpha\Sigma\rho_o\left[\frac{D\sin(\theta_o - \alpha)}{0.0472a_o^*}\right]^3 \tag{9.28}$$

In Chapter 6 equation (6.33) was developed, defining the resultant contact angle α in terms of thrust load and mounted contact angle.

$$\frac{F_a}{ZD^2K} = \sin\alpha\left(\frac{\cos\alpha^\circ}{\cos\alpha} - 1\right)^{1.5} \tag{6.33}$$

in which K is Jones' axial deflection constant, obtainable from Fig. 6.5. Combining equation (6.33) with (9.28) yields the following relationship:

$$\sin(\theta_o - \alpha) = 0.0472\,\frac{a_o^* K^{1/3}\left(\dfrac{\cos\alpha^\circ}{\cos\alpha} - 1\right)^{0.5}}{(D\Sigma\rho_o)^{1/3}} \tag{9.29}$$

This equation may be solved iteratively for α using numerical methods. Having calculated α, it is then possible to determine the limiting thrust load F_{ao} for the ball overriding the outer land from equation (6.33).

Similarly, for the inner raceway

$$\sin(\theta_i - \alpha) = 0.0472 \frac{a_i^* K^{1/3}\left(\frac{\cos\alpha^\circ}{\cos\alpha} - 1\right)^{0.5}}{(D\Sigma\rho_i)^{1/3}} \tag{9.30}$$

$$\theta_i = \cos^{-1}\left(\frac{d_{li} - d_i}{D}\right) \tag{9.31}$$

and $\Sigma\rho_i$ and $F(\rho)_i$ are determined from equations (2.28) and (2.29), respectively.

Thrust Load Causing Excessive Contact Stress

It is possible that prior to overriding of either land an excessive contact stress may occur at the inner raceway contact (or outer raceway contact for a self-aligning ball bearing). The maximum contact stress due to ball load Q is

$$\sigma_{max} = \frac{3Q}{2\pi ab} \tag{5.42}$$

in which

$$b = 0.0236 b_i^* \left(\frac{Q}{\Sigma\rho_i}\right)^{1/3} \tag{5.36}$$

Combination of equations (5.36), (5.34), (5.42), (6.33) yields

$$\left(\frac{\cos\alpha^\circ}{\cos\alpha} - 1\right)^{1/3} = \frac{1.166 \times 10^{-3} a_i^* b_i^* \sigma_{max}}{(D^2K)^{1/3}(\Sigma\rho_i)^{2/3}} \tag{9.32}$$

Assuming a value of maximum permissible contact stress σ_{max} permits a numerical solution for α; thereafter the limiting F_a may be calculated from equation (6.33). Present-day practice uses σ_{max} = 2069 N/mm^2 (300,000 psi) as a practical limit for steel ball bearings. If the balls do not override the lands, however, it is not uncommon to allow stresses to exceed 3449 N/mm^2 (500,000 psi) for short time periods.

Example 9.5. The outer ring land diameter of the 218 angular-contact ball bearing of Example 6.5 is 133.8 mm (5.269 in.). Determine the thrust load that will cause the balls to override the outer ring land.

$$d_o = 147.7 \text{ mm } (5.816 \text{ in.}) \qquad \text{Ex. 2.3}$$

$$D = 22.23 \text{ mm } (0.875 \text{ in.}) \qquad \text{Ex. 2.3}$$

$$\alpha^\circ = 40^\circ \qquad \text{Ex. 2.3}$$

$$B = 0.0464 \qquad \text{Ex. 2.3}$$

$$d_m = 125.3 \text{ mm } (4.932 \text{ in.}) \qquad \text{Ex. 2.6}$$

$$\theta_o = \cos^{-1}\left(1 - \frac{d_o - d_{lo}}{D}\right) \qquad (9.26)$$

$$= \cos^{-1}\left(1 - \frac{147.7 - 133.8}{22.23}\right) = 67^\circ 59'$$

$$\gamma = D\frac{\cos\alpha}{d_m} \qquad (2.27)$$

$$= \frac{22.23 \times \cos\alpha}{125.3} = 0.1774 \cos\alpha$$

$$f_o = f_i = 0.5232 \qquad \text{Ex. 2.3}$$

$$\Sigma\rho_o = \frac{1}{D}\left(4 - \frac{1}{f_o} - \frac{2\gamma}{1+\gamma}\right) \qquad (2.30)$$

$$= \frac{1}{22.23}\left(4 - \frac{1}{0.532} - \frac{2 \times 0.1774\cos\alpha}{1 + 0.1774\cos\alpha}\right) \qquad (2.30)$$

$$= 0.09396 - \frac{0.01596\cos\alpha}{1 + 0.1774\cos\alpha} \qquad \text{(a)}$$

$$F(\rho)_o = \frac{\dfrac{1}{f_o} - \dfrac{2\gamma}{1+\gamma}}{4 - \dfrac{1}{f_o} - \dfrac{2\gamma}{1+\gamma}} \qquad (2.31)$$

$$= \frac{1.911 - \dfrac{0.3548\cos\alpha}{1 + 0.1774\cos\alpha}}{2.089 - \dfrac{0.3548\cos\alpha}{1 + 0.1774\cos\alpha}} \qquad \text{(b)}$$

$$K = 896.7 \text{ N/mm}^2 \ (130{,}000 \text{ psi}) \qquad \text{Ex. 6.5}$$

$$\sin(\theta - \alpha) = \frac{0.0472\, a_o^* K^{1/3}\left(\dfrac{\cos\alpha^\circ}{\cos\alpha} - 1\right)^{0.5}}{(D\Sigma\rho_o)^{1/3}} \tag{9.29}$$

$$\sin(67°59' - \alpha) = \frac{0.0472\, a_o^* \times (896.7)^{1/3}\left(\dfrac{\cos 40°}{\cos\alpha} - 1\right)^{0.5}}{(D\Sigma\rho_o)^{1/3}}$$

$$\sin(67°59' - \alpha) = \frac{0.454\, a_o^*\left(\dfrac{0.7660}{\cos\alpha} - 1\right)^{0.5}}{(D\Sigma\rho_o)^{1/3}} \tag{c}$$

Using trial and error:

1. Assume $\alpha = 45°$, $\cos\alpha = 0.7071$.

2. $\Sigma\rho_o = 0.0839\ \text{mm}^{-1}\ (2.132\ \text{in.}^{-1})$ (a)
$F(\rho)_o = 0.9046$ (b)
From Fig. 5.4, $a_o^* = 3.11$

3. $$\sin(67°59' - \alpha) = \frac{0.454 \times 3.11 \times \left(\dfrac{0.7660}{0.7071} - 1\right)^{0.5}}{(22.23 \times 0.0839)^{1/3}} \tag{c}$$
$\alpha = 48°35'$

4. Assume $\alpha = 47°$, $\cos\alpha = 0.6820$

5. $\Sigma\rho_o = 0.0843\ \text{mm}^{-1}\ (2.141\ \text{in.}^{-1})$ (a)
$F(\rho)_o = 0.9050$ (b)
From Fig. 5.4, $a_o^* = 3.12$

6. $$\sin(67°59' - \alpha) = \frac{0.454 \times 3.12 \times \left(\dfrac{0.7660}{0.6820} - 1\right)^{0.5}}{(22.23 \times 0.0843)^{1/3}} \tag{c}$$
$\alpha = 44°8'$

7. Assume $\alpha = 46°$, $\cos\alpha = 0.6947$
$\Sigma\rho_o = 0.0841\ \text{mm}^{-1}\ (2.136\ \text{in.}^{-1})$ (a)
$F(\rho)_o = 0.9048$ (b)
From Fig. 5.4, $a_o^* = 3.11$

8. $$\sin(67°59' - \alpha) = \frac{0.454 \times 3.11 \times \left(\dfrac{0.7660}{0.6947} - 1\right)^{0.5}}{(22.23 \times 0.0841)^{1/3}} \tag{c}$$
$\alpha = 46°21'$

9. Assume $\alpha = 46°30'$, $\cos\alpha = 0.6884$
$\Sigma\rho_o = 0.00842\ \text{mm}^{-1}\ (2.138\ \text{in.}^{-1})$ (a)
$F(\rho)_o = 0.9049$ (b)

From Fig. 5.4, $a_o^* = 3.12$

$$\sin(67°59' - \alpha) = \frac{0.454 \times 3.12 \times \left(\dfrac{0.7660}{0.6884} - 1\right)^{0.5}}{(2.138)^{1/3}} \tag{c}$$

$$\alpha = 46°19'$$

This result is satisfactory for the purpose of this example. Use α = 46°21′

$$\frac{F_{ao}}{ZD^2K} = \sin\alpha\left(\frac{\cos\alpha°}{\cos\alpha} - 1\right)^{1.5} \tag{6.33}$$

$$\frac{F_{ao}}{16 \times (22.23)^2 \times 896.7} = \sin(46°21')\left(\frac{\cos 40°}{\cos(46°21')} - 1\right)^{1.5}$$

$$F_{ao} = 187{,}200 \text{ N } (42{,}070 \text{ lb})$$

CLOSURE

In many engineering applications bearing deflection must be known to establish the dynamic stability of the rotor system. This consideration is important in high speed systems such as aircraft gas turbines. The bearing radial deflection in this case can contribute to the system eccentricity. In other applications, such as inertial gyroscopes, radiotelescopes, and machine tools, minimization of bearing deflection under load is required to achieve system accuracy or accuracy of manufacturing. That the bearing deflection is a function of bearing internal design, dimensions, clearance, speeds, and load distribution has been indicated in the previous chapters. However, for applications in which speeds are slow and extreme accuracy is not required, the simplified equations presented in this chapter are sufficient to estimate bearing deflection.

To minimize deflection, axial or radial preloading may be employed. Care must be exercised, however, not to excessively preload rolling bearings since this can cause increased friction torque, resulting in bearing overheating and reduction in endurance.

REFERENCES

9.1. A. Palmgren, *Ball and Roller Bearing Engineering*, 3rd ed., Burbank, Philadelphia, pp. 49–51 (1959).

9.2. T. Harris, "How to Compute the Effects of Preloaded Bearings," *Prod. Eng.* 84–93 (July 19, 1965).

10

SHAFT AND BEARING SYSTEMS

LIST OF SYMBOLS

Symbol	Description	Units
a	Distance to load point from right-hand bearing	mm (in.)
A	Distance between raceway groove curvature centers	mm (in.)
D	Rolling element diameter	mm (in.)
d_m	Pitch diameter	mm (in.)
$\mathfrak{D}_o$	Outside diameter of shaft	mm (in.)
$\mathfrak{D}_i$	Inside diameter of shaft	mm (in.)
E	Modulus of elasticity	N/mm^2 (psi)
F	Bearing radial load	N (lb)
f	Raceway groove radius $\div$ D	
I	Section moment of inertia	mm^4 (in.^{4})
K	Load-deflection constant	N/mm^x (lb/in^x)
l	Distance between bearing centers	mm (in.)
$\mathfrak{M}$	Bearing moment load	N · mm (in. · lb)
P	Applied load at a	N (lb)
Q	Rolling element load	N (lb)
R	Radius from bearing centerline to raceway groove center	mm (in.)

Symbol	Description	Units
T	Applied moment load at a	N · mm (in. · lb)
w	Load per unit length	N/mm (lb/in.)
x	Distance along the shaft	mm (in.)
y	Deflection in the y direction	mm (in.)
z	Deflection in the z direction	mm (in.)
α°	Free contact angle	rad,°
γ	$D \cos \alpha / d_m$	
δ	Bearing radial deflection	mm (in.)
θ	Bearing angular misalignment	rad
$\Sigma\rho$	Curvature sum	mm^{-1} ($in.^{-1}$)
ψ	Rolling element azimuth angle	rad,°

SUBSCRIPTS

1, 2, 3	Refers to bearing location
a	Refers to axial direction
h	Refers to bearing location
j	Refers to rolling element location
y	Refers to y direction
z	Refers to z direction
xy	Refers to xy plane
xz	Refers to xz plane

SUPERSCRIPT

k	Refers to location of applied load or moment

GENERAL

The loading that a rolling bearing supports is usually transmitted to the bearing through the shaft on which the bearing is mounted. Sometimes the loading is, however, transmitted through the housing that encompasses the outer ring. In either situation, and in most engineering applications, it is sufficient to consider the bearing as simply resisting the applied load and not as an integral part of the loaded system. In some modern engineering applications of rolling bearings, such as high speed gas turbines, machine tools, and gyroscopes, the bearings often must be treated as integral to the system to be able to accurately determine shaft deflections and dynamic shaft loading as well as to ascertain the performance of the bearings. In this chapter equations for the analysis of bearing loading in both of the aforementioned situations will be developed.

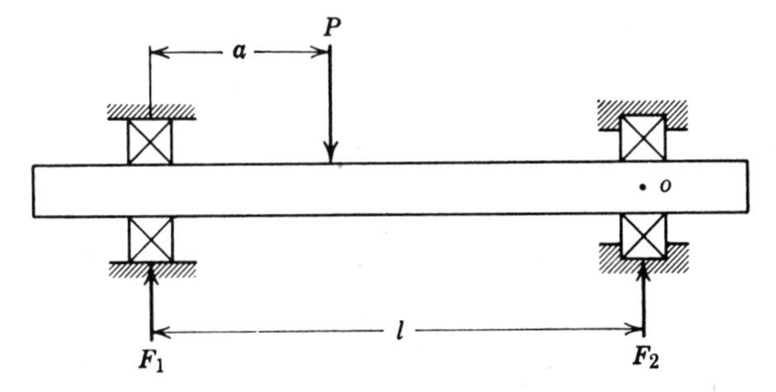

FIGURE 10.1. Simple two-bearing–shaft system.

SIMPLE SYSTEMS

The most elementary rolling bearing–shaft assembly is that shown in Fig. 10.1 in which a concentrated load is supported between two bearings. This concentrated load may be caused by a gear, pulley, piston and crank, electric motor rotor, turbine, compressor, and so on. Generally, the shaft is relatively rigid and bearing misalignment due to shaft bending is of inconsequential magnitude. Thus, the system is statically determinate, that is, the bearing load reactions F may be determined from the simple equations of static equilibrium. Hence

$$\Sigma F = 0 \tag{10.1}$$

$$F_1 + F_2 - P = 0 \tag{10.2}$$

$$\Sigma M_0 = 0 \tag{10.3}$$

$$F_1 l - P(l - a) = 0 \tag{10.4}$$

Solving equations (10.2) and (10.4) simultaneously yields

$$F_1 = \frac{P(l - a)}{l} \tag{10.5}$$

$$F_2 = \frac{Pa}{l} \tag{10.6}$$

For an overhanging load as shown in Fig. 10.2 equations (10.5) and (10.6) remain valid if distances measured to the left of the left-hand bearing support are considered negative. Equations (10.5) and (10.6) therefore become

$$F_1 = \frac{P(l \mp a)}{l} \tag{10.7}$$

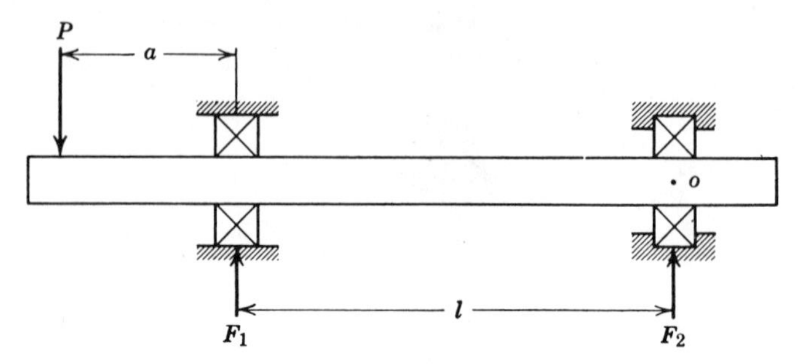

FIGURE 10.2. Simple two-bearing–shaft system, overhung load.

$$F_2 = \mp \frac{Pa}{l} \tag{10.8}$$

If a number of loads P^k act on the shaft as shown in Fig. 10.3, then the magnitudes of the bearing reactions may be obtained by the principle of superposition such that

$$F_1 = \frac{1}{l} \sum_{k=1}^{k=n} P^k(l \mp a^k) \tag{10.9}$$

$$F_2 = \frac{1}{l} \sum_{k=1}^{k=n} \pm P^k a^k) \tag{10.10}$$

Sometimes the load is distributed over a portion of the shaft as indicated in Fig. 10.4. If the loading is irregular, then it may be considered as a series of smaller loads P^k each acting at its individual distance a^k from the left-hand bearing support. Equations (10.9) and (10.10) may then be used to evaluate reactions F_1 and F_2. For a distributed load for which the load per unit length w may be represented by a continuous function as $w = \sin x$ or $w = w_0 + bx$ equations (10.9) and (10.10) become (see Fig. 10.5)

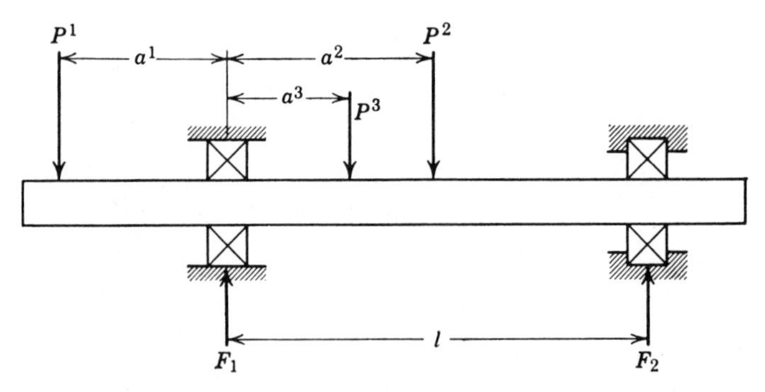

FIGURE 10.3. Simple two-bearing–shaft system, multiple loading.

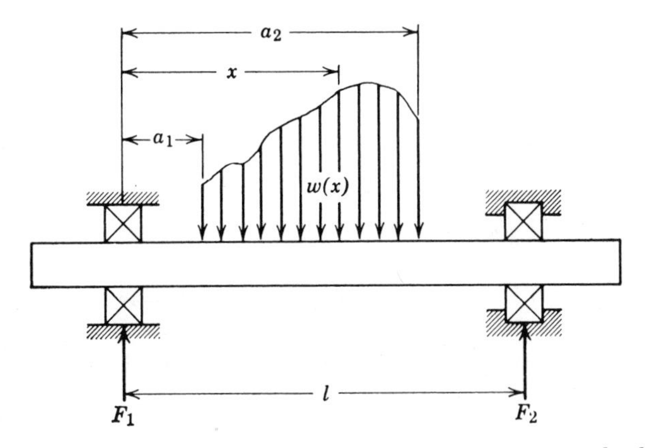

FIGURE 10.4. Simple two-bearing–shaft system, continuous loading.

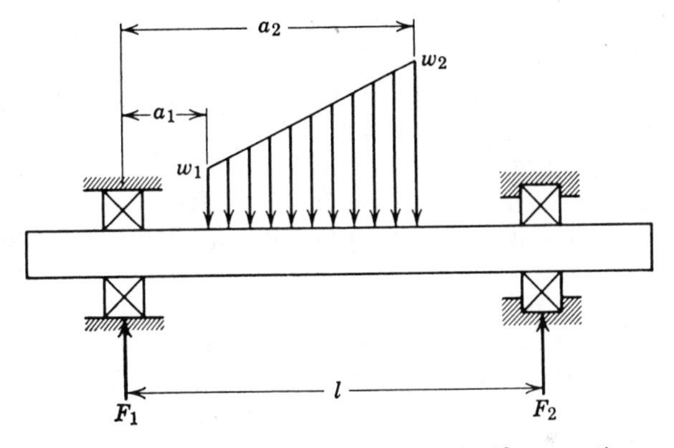

FIGURE 10.5. Simple two-bearing–shaft system, uniform continuous loading.

$$F_1 = \frac{1}{l} \int_{a_1}^{a_2} (l - x)w \, dx \tag{10.11}$$

$$F_2 = \frac{1}{l} \int_{a_1}^{a_2} wx \, dx \tag{10.12}$$

Some machine elements such as helical and bevel gears cause moment loads T^k in addition to radial loads P^k. In this case, for moment loads directed as shown in Fig. 10.6,

$$F_1 = \frac{P^k(l \mp a^k)}{l} - \frac{T^k}{l} \tag{10.13}$$

$$F_2 = \pm \frac{P^k a^k}{l} + \frac{T^k}{l} \tag{10.14}$$

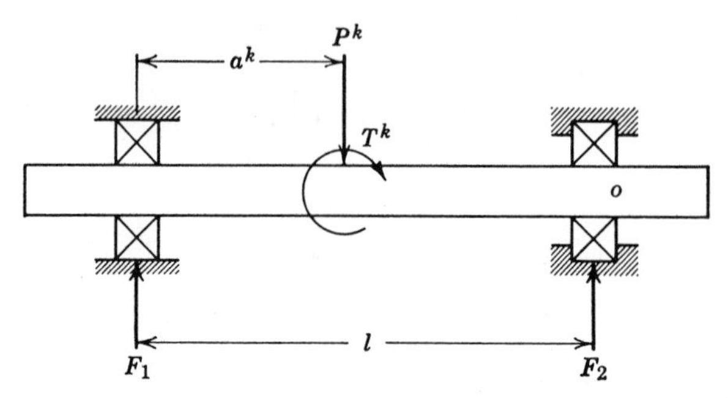

FIGURE 10.6. Simple two-bearing–shaft system, concentrated normal and moment loading.

In each of the foregoing bearing–shaft systems, it is assumed that only radial load is transmitted to the bearings. Since shafts are not infinitely rigid, it is certain that because of bending of the shaft, moments loads on non-self-aligning rolling bearings are induced to some degree. These moment loads are generally small in magnitude compared to the radial loads and are usually not considered. Later in this section, however, such loading will be evaluated. Bending of the shaft may also induce thrust load in a rolling bearing if the shaft is not free to float axially. Such induced thrust loads are usually of small magnitude relative to the radial loads.

Occasionally, a single rolling bearing may be used in a cantilever mounting as indicated in Fig. 10.7. This system is also statically determinate and yields the following reactive loading:

$$F = \sum_{k=1}^{k=n} P^k \tag{10.15}$$

$$\mathfrak{M} = \sum_{k=1}^{k=n} P^k a^k \tag{10.16}$$

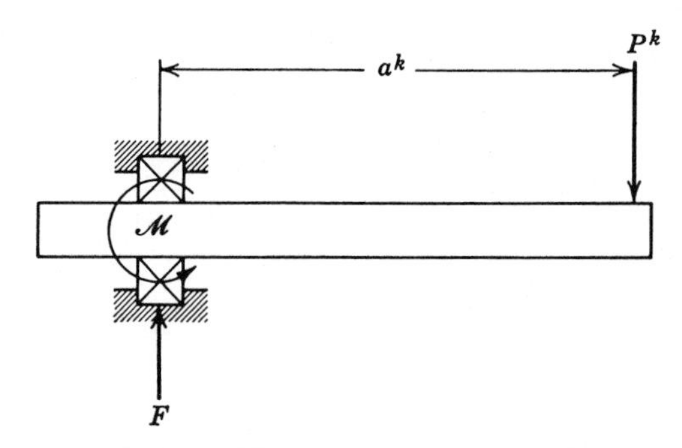

FIGURE 10.7. Cantilever shaft–bearing system.

Chapters 6 and 8 of this book detail methods of calculation of rolling element load distribution for a bearing subjected to a combination of radial and moment loading. Once the load distribution is known, the bearing fatigue life can be estimated. This is discussed in Chapter 18.

STATICALLY INDETERMINATE SYSTEMS

Two-Bearing Systems

A statically indeterminate system is one for which the equations of static equilibrium are insufficient to solve the bearing reactions. In the previous discussion it was assumed that the flexure of the shaft did not affect the bearing loading. Hence the bearings were considered similar to pinned supports that accommodate only loads normal to the shaft, that is, radial loads. This situation is substantially correct in most bearing applications; thus the previous developments concerning bearing load calculations are sufficiently accurate for most applications.

In the more general case, however, the flexure of the shaft induces moment loads. $\mathfrak{M}_h$ at the non-self-aligning bearing supports in addition to the acknowledged radial loads F_h. Thus, an apparently simple, statically determinate, shaft–bearing system such as that shown in Fig. 10.6 may be as complex as that shown in Fig. 10.8. The latter system is statically indeterminate in that there are four unknowns, that is F_1, F_2, $\mathfrak{M}_1$, and $\mathfrak{M}_2$, but only two equilibrium equations, which are as follows:

$$\Sigma F = 0$$

$$F_1 + F_2 - P = 0 \tag{10.17}$$

$$\Sigma M = 0$$

$$F_1 l - \mathfrak{M}_1 + T - P(l - a) + \mathfrak{M}_2 = 0 \tag{10.18}$$

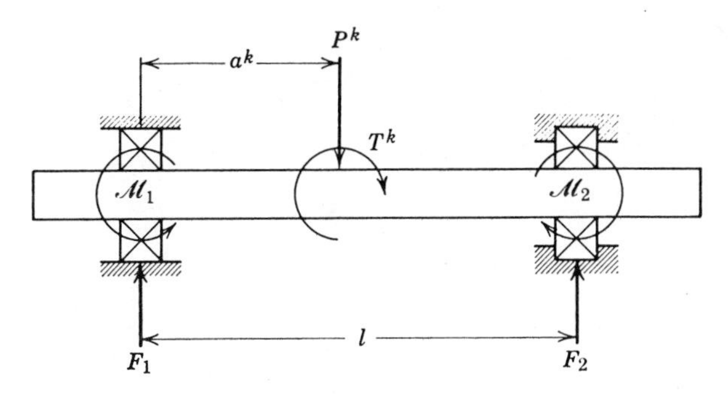

FIGURE 10.8. Statically indeterminate two-bearing–shaft system.

Considering the bending of the shaft, the bending moment at any section is given as follows:

$$EI\frac{d^2y}{dx^2} = -M \tag{10.19}$$

in which E is the modulus of elasticity, I is the shaft cross-section moment of inertia, and y is the shaft deflection at the section. For shafts having circular cross sections,

$$I = \frac{\pi}{64}(\mathfrak{D}_o^4 - \mathfrak{D}_i^4) \tag{10.20}$$

For a cross section at $0 \leq x \leq a$ illustrated in Fig. 10.9,

$$EI\frac{d^2y}{dx^2} = -F_1x + \mathfrak{M}_1 \tag{10.21}$$

Integrating equation (10.21) yields

$$EI\frac{dy}{dx} = -\frac{F_1x^2}{2} + \mathfrak{M}_1x + C_1 \tag{10.22}$$

Integrating equation (10.22) yields

$$EIy = -\frac{F_1x^3}{6} + \frac{\mathfrak{M}_1x^2}{2} + C_1x + C_2 \tag{10.23}$$

In equations (10.22) and (10.23), C_1 and C_2 are constants of integration. At $x = 0$, the shaft assumes the bearing deflection δ_{r1}. Also at $x = 0$, the shaft assumes a slope θ_1 in accordance with the resistance of the

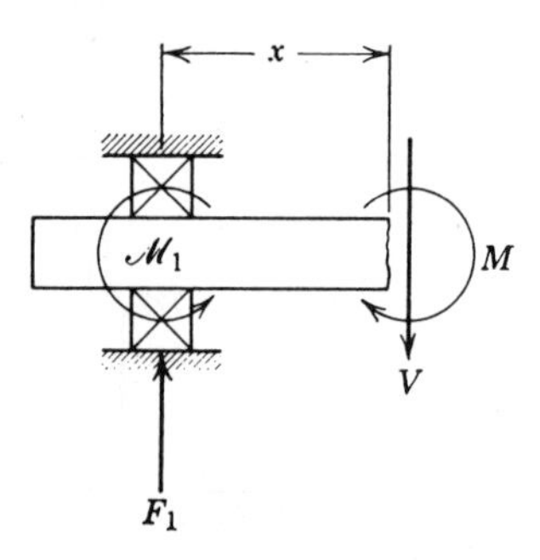

FIGURE 10.9. Statically indeterminate two-bearing–shaft system forces and moments acting on a section to the left of the load application point.

bearing to moment loading; hence

$$C_1 = EI\theta_1$$

$$C_2 = EI\delta_{r1}$$

Therefore, equations (10.22) and (10.23) become

$$EI\frac{dy}{dx} = -\frac{F_1x^2}{2} + \mathcal{M}_1x + EI\theta_1 \tag{10.24}$$

and

$$EIy = -\frac{F_1x^3}{6} + \frac{\mathcal{M}_1x^2}{2} + EI\theta_1x + EI\delta_{r1} \tag{10.25}$$

For a cross section at $a \leq x \leq l$ as shown in Fig. 10.10,

$$EI\frac{d^2y}{dx^2} = -F_1x + \mathcal{M}_1 + P(x - a) - T \tag{10.26}$$

Integrating equation (10.26) twice yields

$$EI\frac{dy}{dx} = -\frac{F_1x^2}{2} + (\mathcal{M}_1 - T)x + Px\left(\frac{x}{2} - a\right) + C_3 \tag{10.27}$$

$$EIy = -\frac{F_1x^3}{6} + (\mathcal{M}_1 - T)\frac{x^2}{2} + Px^2\left(\frac{x}{6} - \frac{a}{2}\right) + C_3x + C_4 \tag{10.28}$$

At $x = l$, the slope of the shaft is θ_2 and the deflection is δ_{r2}, therefore

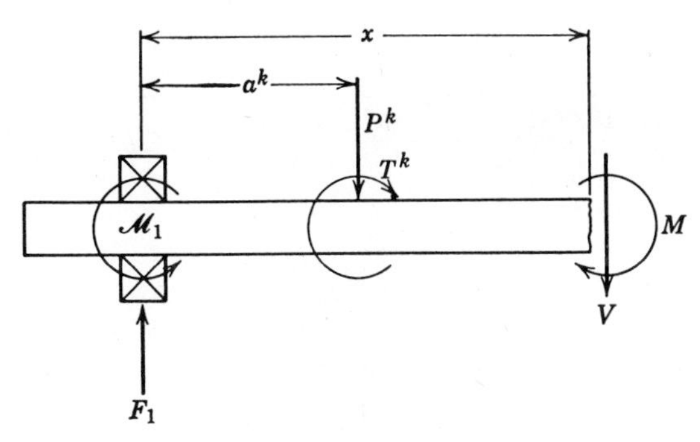

FIGURE 10.10. Statically indeterminate two-bearing–shaft system forces and moments acting on a section to the right of the load application.

$$EI\frac{dy}{dx} = \frac{F_1(l^2 - x^2)}{2} + (T - \mathfrak{M}_1)(l - x)$$
$$+ \frac{P}{2}[x(x - 2a) - l(l - 2a)] + EI\theta_2 \qquad (10.29)$$

$$EIy = -\frac{F_1}{6}[l^2(2l - 3x) + x^3] + \frac{(\mathfrak{M}_1 - T)}{2}(l - x)^2$$
$$+ \frac{P}{6}[x^2(x - 3a) - l^2(3x + 3a - 2l) + 6xla]$$
$$+ EI[\delta_{r2} - \theta_2(l - x)] \qquad (10.30)$$

At $x = a$, singular conditions of slope and deflection occur. Therefore at $x = a$, equations (10.24) and (10.29) are equivalent as are equations (10.25) and (10.30). Solving the resultant simultaneous equations yields

$$F_1 = \frac{P(l - a)^2(l + 2a)}{l^3} - \frac{6Ta(l - a)}{l^3}$$
$$- \frac{6EI}{l^2}\left[\theta_1 + \theta_2 + \frac{2(\delta_{r1} - \delta_{r2})}{l}\right] \qquad (10.31)$$

$$\mathfrak{M}_1 = \frac{Pa(l - a)^2}{l^2} + \frac{T(l - a)(l - 3a)}{l^2}$$
$$- \frac{2EI}{l}\left[2\theta_1 + \theta_2 + \frac{3(\delta_{r1} - \delta_{r2})}{l}\right] \qquad (10.32)$$

Substituting equations (10.31) and (10.32) into (10.17) and (10.18) yields

$$F_2 = \frac{Pa^2(3l - 2a)}{l^3} + \frac{6Ta(l - a)}{l^3} + \frac{6EI}{l^2}\left[\theta_1 + \theta_2 + \frac{2(\delta_{r1} - \delta_{r2})}{l}\right] \qquad (10.33)$$

$$\mathfrak{M}_2 = \frac{Pa^2(l - a)}{l^2} + \frac{Ta(2l - 3a)}{l^2} + \frac{2EI}{l}\left[\theta_1 + 2\theta_2 + \frac{3(\delta_{r1} - \delta_{r2})}{l}\right] \qquad (10.34)$$

In equations (10.31)–(10.34), slope θ_1 and δ_{r1} are considered positive and the signs of θ_2 and δ_{r2} may be determined from equations (10.29) and (10.30). The relative magnitudes of P and T and their directions will determine the sense of the shaft slopes at the bearings. To determine the reactions it is necessary to develop equations relating bearing misalignment angles θ_h to the misaligning moments $\mathfrak{M}_h$ and bearing radial deflections δ_{rh} to loads F_h. This may be done by using the data of Chapters 6 and 8.

In their most simple form, in which the bearings are considered as axially free pinned supports, equations (10.31) and (10.33) are identical to (10.13) and (10.14). This format is obtained by setting $\mathfrak{M}_h$ and δ_{rh} equal to zero and solving equations (10.32) and (10.34) simultaneously for θ_1 and θ_2. Substitution of these values into equations (10.31) and (10.33) produces the resultant equations. If the shaft is very flexible and the bearings are rigid with regard to misalignment, then θ_1 and θ_2 are approximately zero. This substitution into equations (10.31)–(10.34) yields the classical solution for a beam with both ends built in. The various types of two-bearing support may be examined by using equations (10.31)–(10.34). If more than one load and/or torque is applied between the supports, then by the principle of superposition

$$F_1 = \frac{1}{l^3}\sum_{k=1}^{k=n} P^k(l - a^k)^2(l + 2a^k) - \frac{6}{l^3}\sum_{k=1}^{k=n} T^k a^k(l - a^k) - \frac{6EI}{l^2}\left[\theta_1 + \theta_2 + \frac{2(\delta_{r1} - \delta_{r2})}{l}\right] \quad (10.35)$$

$$\mathfrak{M}_1 = \frac{1}{l^2}\sum_{k=1}^{k=n} P^k a^k(l - a^k)^2 + \frac{1}{l^2}\sum_{k=1}^{k=n} T^k(l - a^k)(l - 3a^k) - \frac{2EI}{l}\left[2\theta_1 + \theta_2 + \frac{3(\delta_{r1} - \delta_{r2})}{l}\right] \quad (10.36)$$

$$F_2 = \frac{1}{l^3}\sum_{k=1}^{k=n} P^k(a^k)^2(3l - 2a^k) + \frac{6}{l^3}\sum_{k=1}^{k=n} T^k a^k(l - a^k) + \frac{6EI}{l^2}\left[\theta_1 + \theta_2 + \frac{2(\delta_{r1} - \delta_{r2})}{l}\right] \quad (10.37)$$

$$\mathfrak{M}_2 = \frac{1}{l^2}\sum_{k=1}^{k=n} P^k(a^k)^2(l - a^k) + \frac{1}{l^2}\sum_{k=1}^{k=n} T^k a^k(2l - 3a^k) + \frac{2EI}{l}\left[\theta_1 + 2\theta_2 + \frac{3(\delta_{r1} - \delta_{r2})}{l}\right] \quad (10.38)$$

Example 10.1. A pair of 209 radial ball bearings mounted on a hollow steel shaft having a 45 mm (1.772 in.). o.d. and 40.56 mm (1.597 in.) i.d. support a 13,350 N (3000 lb) radial load acting on the shaft midway between the bearings. The span between the bearing centers is 254 mm (10 in.). If each of the bearings is mounted according to the fits of Example 3.1 and their dimensions are as given by Example 2.1, determine the bearing load distribution, radial deflection, moment loading, and angle of misalignment.

$$I = \frac{\pi}{64} (\mathfrak{D}_o^4 - \mathfrak{D}_i^4) \qquad (10.20)$$

$$= \frac{3.1416}{64} [(45)^4 - (40.56)^4]$$

$$= 6.855 \times 10^4 \text{ mm}^4 \ (0.1647 \text{ in.}^4)$$

Bearings 1 and 2 are identically loaded because load is applied at mid-span, therefore, $F_1 = F_2 = 6675$ N (1500 lb).

$$\mathfrak{M} = \frac{Pa(l - a)^2}{l^2} - \frac{2EI}{l}\left[2\theta_1 + \theta_2 + \frac{3(\delta_{r1} - \delta_{r2})}{l}\right] \qquad (10.32)$$

But, $\delta_{r1} = \delta_{r2}$ and $\theta_2 = -\theta_1$. Hence

$$\mathfrak{M} = \frac{Pa(l - a)^2}{l^2} - \frac{2EI\theta}{l}$$

$$\mathfrak{M} = \frac{6675 \times 127 \times (254 - 127)^2}{(254)^2} - \frac{2 \times 2.069 \times 10^5 \times 6.855 \times 10^4\theta}{254}$$

or (a)

$$\mathfrak{M} = 2.119 \times 10^5 - 1.115 \times 10^8\theta$$

$f = 0.52$	Ex. 2.2
$\Delta\psi = 40°$	Ex. 6.1
$A = 0.508$ mm (0.02 in.)	Ex. 3.1
$\alpha° = 7°30'$	Ex. 3.1
$D = 12.7$ mm (0.5 in.)	Ex. 2.1
$d_m = 65$ mm (2.559 in.)	Ex. 2.1
	Ex. 2.1

$$\mathcal{R}_i = \tfrac{1}{2} d_m + (r_i - 0.5D) \cos \alpha° \qquad (6.86)$$

Since $f = r/D$,

$$\mathcal{R}_i = \tfrac{1}{2} d_m + (f_i - 0.5)D \cos \alpha°$$

$$= \frac{65}{2} + (0.52 - 0.5) \times 12.7 \cos (7°30')$$

$$= 32.75 \text{ mm } (1.289 \text{ in.})$$

$$F_r - K_n A^{1.5} \sum_{\psi=0}^{\psi=\pm\pi} \frac{\{[(\sin \alpha^\circ + \bar{\delta}_a + \mathcal{R}_i \bar{\theta} \cos \psi)^2 + (\cos \alpha^\circ + \bar{\delta}_r \cos \psi)^2]^{1/2} - 1\}^{1.5} \times (\cos \alpha^\circ + \bar{\delta}_r \cos \psi) \cos \psi}{[(\sin \alpha^\circ + \bar{\delta}_a + \mathcal{R}_i \bar{\theta} \cos \psi)^2 + (\cos \alpha^\circ + \bar{\delta}_r \cos \psi)^2]^{1/2}} = 0 \tag{6.105}$$

$$6675 - 0.3621\, K_n \sum_{\psi=0}^{\psi=\pm\pi} \frac{\{[(0.1305 + 32.75\bar{\theta} \cos \psi)^2 + (0.9914 + \bar{\delta}_r \cos \psi)^2]^{1/2} - 1\}^{1.5} \times (0.9914 + \bar{\delta}_r \cos \psi) \cos \psi}{[(0.1305 + 32.75\bar{\theta} \cos \psi)^2 + (0.9914 + \bar{\delta}_r \cos \psi)^2]^{1/2}} = 0 \tag{b}$$

Note that $\delta_a = 0$, therefore, $\bar{\delta}_a = 0$.

$$\mathfrak{M} - \tfrac{1}{2}\, d_m K_n A^{1.5} \sum_{\psi=0}^{\psi=\pm\pi} \frac{\{[(\sin \alpha^\circ + \bar{\delta}_a + \mathcal{R}_i \bar{\theta} \cos \psi)^2 + (\cos \alpha^\circ + \bar{\delta}_r \cos \psi)^2]^{1/2} - 1\}^{1.5} \times (\sin \alpha^\circ + \bar{\delta}_a + \mathcal{R}_i \bar{\theta} \cos \psi) \cos \psi}{[(\sin \alpha^\circ + \bar{\delta}_a + \mathcal{R}_i \bar{\theta} \cos \psi)^2 + (\cos \alpha^\circ + \bar{\delta}_r \cos \psi)^2]^{1/2}} = 0 \tag{6.106}$$

$$\mathfrak{M} - 11.77\, K_n \sum_{\psi=0}^{\psi=\pm\pi} \frac{\{[(0.1305 + 32.75\bar{\theta} \cos \psi)^2 + (0.9914 + \bar{\delta}_r \cos \psi)^2]^{1/2} - 1\}^{1.5} \times (0.1395 + 32.75\bar{\theta} \cos \psi) \cos \psi}{[(0.1305 + 32.75\bar{\theta} \cos \psi)^2 + (0.9914 + \bar{\delta}_r \cos \psi)^2]^{1/2}} = 0 \tag{c}$$

In the preceding equations (b) and (c), K_n is determined as follows:

$$K_n = \left(\frac{1}{(1/K_i)^{0.667} + (1/K_0)^{0.667}} \right)^{1.5} \tag{6.6}$$

in which

$$K_i = 2.15 \times 10^5\, \Sigma\rho_i^{-1/2} (\delta_i^*)^{-3/2} \tag{6.8}$$

and

$$\Sigma\rho_i = \frac{1}{D} \left(4 - \frac{1}{f_i} + \frac{2\gamma}{1-\gamma} \right) \tag{2.28}$$

To determine δ_i^* one needs $F(\rho)_i$

$$F(\rho)_i = \frac{\dfrac{1}{f_i} + \dfrac{2\gamma}{1-\gamma}}{4 - \dfrac{1}{f_i} + \dfrac{2\gamma}{1-\gamma}} \tag{2.29}$$

$$\gamma = \frac{D \cos \alpha}{d_m} \tag{2.27}$$

A similar procedure is followed to determine K_o. It can be seen that K_i and K_o depend on the contact angle. If this is significantly different at each ball location, that is, at $\psi = 0°, 40°, 80°, 120°$, and $160°$, K_n is different at each location and must be included in the summation terms of equations (b) and (c). (Example 6.1 indicates the method of calculation of K_n when α is constant.)

In equations (b) and (c),

$$\bar{\delta}_r = \frac{\delta_r}{A} = \frac{\delta_r}{1.969} = 0.508\ \delta_r \tag{6.92}$$

$$\bar{\theta} = \frac{\theta}{A} = \frac{\theta}{1.969} = 0.508\ \theta \tag{6.93}$$

Equations (a)–(c) constitute a set of simultaneous nonlinear equations in unknowns δ_r, θ, and $\mathfrak{M}$. By using the method of equations (6.107) and (6.108), the equations may be solved. A computer is required. The answers are accordingly

$$\delta_r = 0.02073 \text{ mm } (0.000816 \text{ in.})$$

$$\theta = 0°11'$$

$$\mathfrak{M} = 6.804 \times 10^4 \text{ N} \cdot \text{mm } (602 \text{ in.} \cdot \text{lb})$$

The corresponding load distribution and contact angles are

ψ	α	Q(N)
0	18°36′	3540 (795.5 lb)
40°	16°18′	2213 (497.2 lb)
80°	10°3′	194 (43.6 lb)
120°	0	0
160°	0	0

Additionally, Figs. 10.11–10.13 show δ_r, θ, and $\mathfrak{M}$ for various combinations of shaft hollowness and span between bearings.

Three-Bearing Systems

Nonrigid Shafts. Whether or not moment loads are induced in the bearings, a three-bearing support system constitutes a statically indeterminate system. In general, the three-bearing system is illustrated by Fig. 10.14*a*. When the shaft is relatively flexible, the analysis of the foregoing section may be applied. The system of Fig. 10.14*a* may be

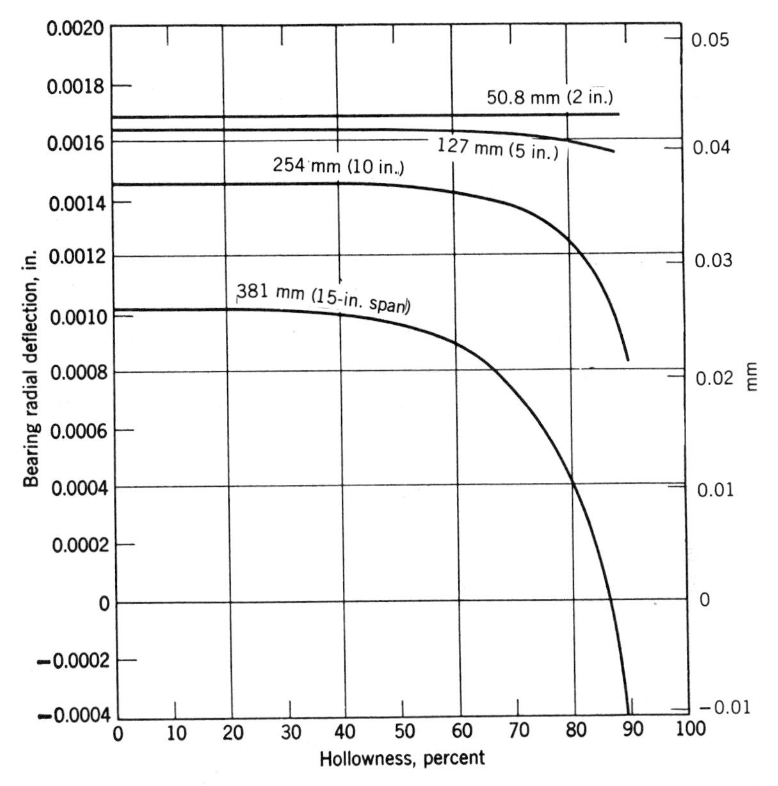

FIGURE 10.11. Bearing radial deflection versus shaft hollowness and span, 209 radial ball bearing, 13,350 N (3000 lb) at midspan.

reduced to the two systems of Fig. 10.14*b* provided that

$$F_2' + F_2'' = F_2 \tag{10.39}$$

$$\mathfrak{M}_2' - \mathfrak{M}_2'' = \mathfrak{M}_2 \tag{10.40}$$

Hence from equations (10.35)–(10.38),

$$F_1 = \frac{1}{l_1^3} \sum_{k=1}^{k=n} P_1^k (l_1 - a_1^k)^2 (l_1 + 2a_1^k) - \frac{6}{l_1^3} \sum_{k=1}^{k=n} T_1^k a_1^k (l_1 - a_1^k) - \frac{6EI_1}{l_1^2} \left[\theta_1 + \theta_2 + \frac{2(\delta_{r1} - \delta_{r2})}{l_1} \right] \tag{10.41}$$

$$\mathfrak{M}_1 = \frac{1}{l_1^2} \sum_{k=1}^{k=n} P_1^k a_1^k (l_1 - a_1^k)^2 + \frac{1}{l_1^2} \sum_{k=1}^{k=n} T_1^k (l_1 - a_1^k)(l_1 - 3a_1^k) - \frac{2EI_1}{l_1} \left[2\theta_1 + \theta_2 + \frac{3(\delta_{r1} - \delta_{r2})}{l_1} \right] \tag{10.42}$$

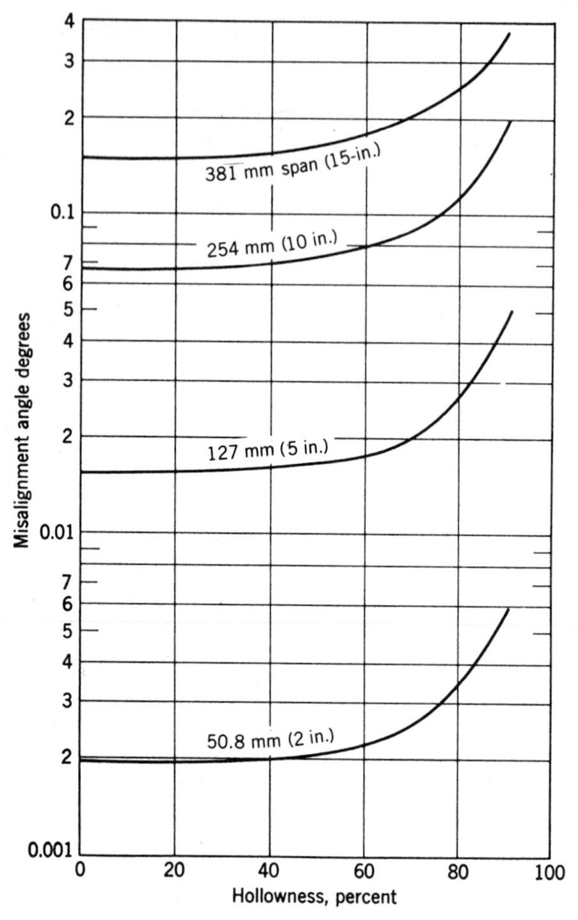

FIGURE 10.12. Bearing misalignment angle vs shaft hollowness and span, 209 radial ball bearing, 13,350 N (3000 lb) at midspan.

$$F_2 = \frac{1}{l_1^3} \sum_{k=1}^{k=n} P_1^k (a_1^k)^2 (3l_1 - 2a_1^k) + \frac{1}{l_2^3} \sum_{k=1}^{k=m} P_2^k (l_2 - a_2^k)(l_2 + 2a_2^k)$$
$$+ \frac{6}{l_1^3} \sum_{k=1}^{k=n} T_1^k a_1^k (l_1 - a_1^k) - \frac{6}{l_2^3} \sum_{k=1}^{k=m} T_2^k a_2^k (l_2 - a_2^k)$$
$$+ 6E \left[\frac{I_1}{l_1^2} (\theta_1 + \theta_2) - \frac{I_2}{l_2^2} (\theta_2 + \theta_3) \right]$$
$$+ 12E \left[\frac{I_1}{l_1^3} (\delta_{r1} - \delta_{r2}) - \frac{I_2}{l_2^3} (\delta_{r2} - \delta_{r3}) \right] \quad (10.43)$$

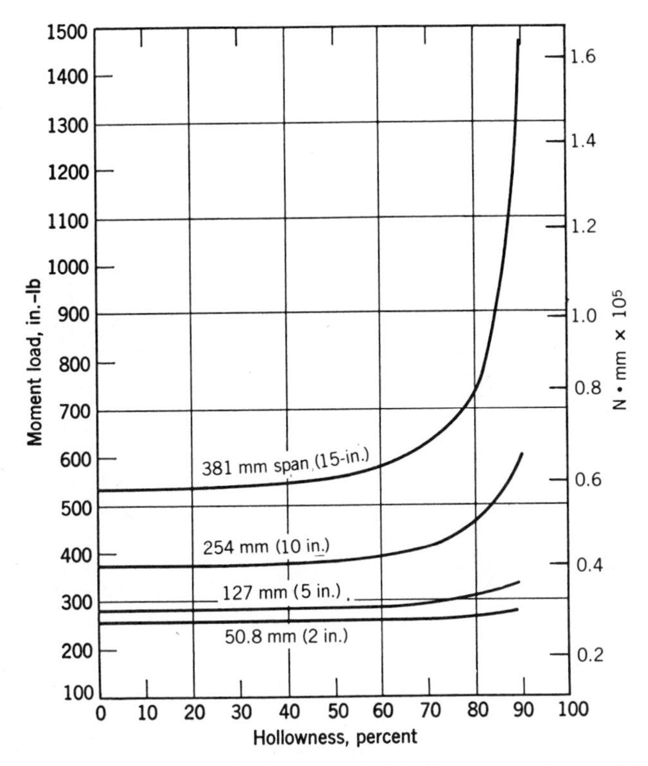

FIGURE 10.13. Bearing moment load versus shaft hollowness and span, 209 radial ball bearing 13,350 N (3000 lb) at midspan.

$$\begin{aligned}\mathfrak{M}_2 = &\frac{1}{l_1^2}\sum_{k=1}^{k=n} P_1^k(a_1^k)^2(l_1 - a_1^k) - \frac{1}{l_2^2}\sum_{k=1}^{k=m} P_2^k a_2^k(l_2 - a_2^k)^2 \\ &+ \frac{1}{l_1^2}\sum_{k=1}^{k=n} T_1^k a_1^k(2l_1 - 3a_1^k) - \frac{1}{l_2^2}\sum_{k=1}^{k=m} T_2^k(l_2 - a_2^k)(l_2 - 3a_2^k) \\ &+ 2E\left[\frac{I_1}{l_1}(\theta_1 + 2\theta_2) + \frac{I_2}{l_2}(2\theta_2 + \theta_3)\right] \\ &+ 6E\left[\frac{I_1}{l_1^2}(\delta_{r1} - \delta_{r2}) + \frac{I_2}{l_2^2}(\delta_{r2} - \delta_{r3})\right]\end{aligned} \tag{10.44}$$

$$\begin{aligned}F_3 = &\frac{1}{l_2^3}\sum_{k=1}^{k=m} P_2^k(a_2^k)^2(3l_2 - 2a_2^k) + \frac{6}{l_2^3}\sum_{k=1}^{k=m} T_2^k a_2^k(l_2 - a_2^k) \\ &+ \frac{6EI_2}{l_2^2}\left[\theta_2 + \theta_3 + \frac{2(\delta_{r2} - \delta_{r3})}{l_2}\right]\end{aligned} \tag{10.45}$$

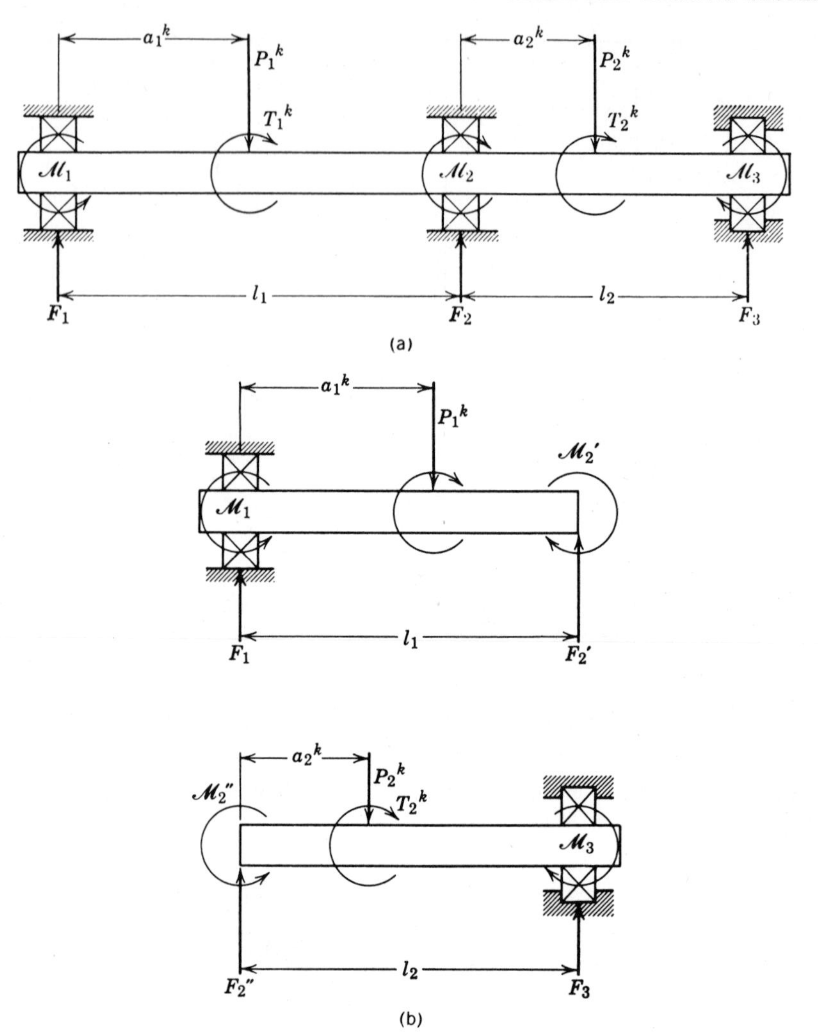

FIGURE 10.14. (*a*) Three-bearing–shaft system. (*b*) Equivalent two-bearing–shaft systems.

$$\mathfrak{M}_3 = \frac{1}{l_2^2} \sum_{k=1}^{k=m} P_2^k (a_2^k)^2 (l_2 - a_2^k) + \frac{1}{l_2^2} \sum_{k=1}^{k=m} T_2^k a_2^k (2l_2 - 3a_2^k)$$
$$+ \frac{2EI_2}{l_2} \left[\theta_2 + 2\theta_3 + \frac{3(\delta_{r2} - \delta_{r3})}{l_2} \right] \tag{10.46}$$

An example of the utility of the generalized equations (10.41)–(10.46) is the simple system illustrated by Fig. 10.15. for that system it is as-

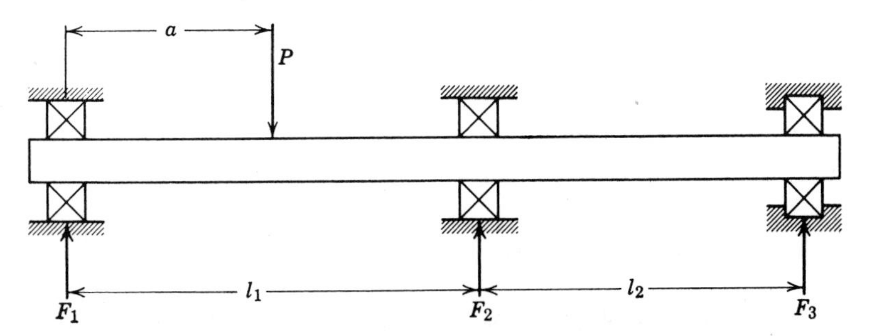

FIGURE 10.15. Simple three-bearing–shaft system.

sumed that moment loads are zero and that the differences between bearing radial deflections are negligibly small. Hence, equations (10.41)–(10.46) become

$$F_1 = \frac{P(l_1 - a)^2(l_1 + 2a)}{l_1^3} - \frac{6EI}{l_1^2}(\theta_1 + \theta_2) \tag{10.47}$$

$$2\theta_1 + \theta_2 = \frac{Pa(l_1 - a)^2}{2EIl_1} \tag{10.48}$$

$$F_2 = \frac{Pa^2(3l_1 - 2a)}{l_1^3} + 6EI\left[\frac{(\theta_1 + \theta_2)}{l_1^2} - \frac{(\theta_2 + \theta_3)}{l_2^2}\right] \tag{10.49}$$

$$\frac{(\theta_1 + 2\theta_2)}{l_1} + \frac{(2\theta_2 + \theta_3)}{l_2} = -\frac{Pa^2(l_1 - a)}{2EIl_1^3} \tag{10.50}$$

$$F_3 = \frac{6EI(\theta_2 + \theta_3)}{l_2^2} \tag{10.51}$$

$$\theta_2 + 2\theta_3 = 0 \tag{10.52}$$

Equations (10.48), (10.50), and (10.52) can be solved for θ_1, θ_2, and θ_3. Subsequent substitution of these values into (10.47), (10.49), and (10.51) yields the following result:

$$F_1 = \frac{P(l_1 - a)[2l_1(l_1 + l_2) - a(l_1 + a)]}{2l_1^2(l_1 + l_2)} \tag{10.53}$$

$$F_2 = \frac{Pa[(l_1 - l_2)^2 - a^2 - l_2^2]}{2l_1^2 l_2} \tag{10.54}$$

$$F_3 = \frac{-Pa(l_1^2 - a^2)}{2l_1 l_2(l_1 + l_2)} \tag{10.55}$$

Rigid Shafts. When the distances between bearings are small or the shaft is otherwise very stiff, the bearing radial deflections determine the load distribution among the bearings. From Fig. 10.16 it can be seen that by considering similar triangles

$$\frac{\delta_{r1} - \delta_{r2}}{l_1} = \frac{\delta_{r2} - \delta_{r3}}{l_2} \tag{10.56}$$

This identical relationship can be obtained from equations (10.41)–(10.46) by setting I to an infinitely large value. From equation (6.4),

$$Q = K\delta^n \tag{6.4}$$

For a bearing with rigid rings, the maximum rolling element load is directly proportional to the applied load and the maximum rolling element deflection determined the bearing deflection; therefore,

$$F = K\delta_r^n \tag{10.57}$$

Rearranging equation (10.57),

$$\delta_r = \left(\frac{F}{K}\right)^{1/n} \tag{10.58}$$

Substitution of equation (10.58) into (10.56) yields

$$\left(\frac{F_1}{K_1}\right)^{1/n} - \left(\frac{F_2}{K_2}\right)^{1/n} = \frac{l_1}{l_2}\left[\left(\frac{F_2}{K_2}\right)^{1/n} - \left(\frac{F_3}{K_3}\right)^{1/n}\right] \tag{10.59}$$

Equation (10.59) is valid for bearings that support a radial load only. More complex relationships are required in the presence of simultaneous thrust and/or moment loading. Equation (10.59) can be solved simultaneously with the equilibrium equations to yield values of F_1, F_2, and F_3.

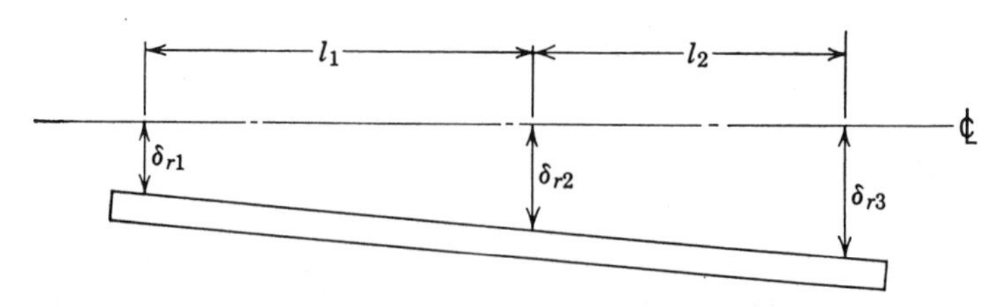

FIGURE 10.16. Deflection of three-bearing-shaft system due to rigid shaft.

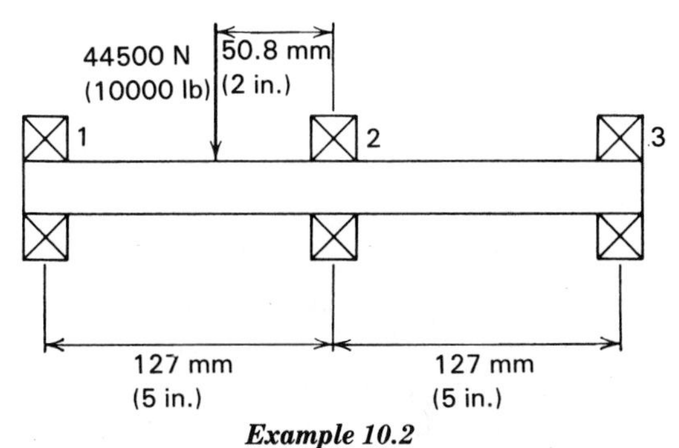

Example 10.2

Example 10.2. Three identical ball bearings are mounted on a shaft and located 127 mm (5 in.) apart. A load of 44,500 N (10,000 lb) is supported 50.8 mm (2 in.) from the centrally located bearing. Determine the radial load on each bearing if the shaft may be considered rigid.

$$\Sigma F_r = 0$$

$$F_1 + F_2 + F_3 - 10{,}000 = 0 \tag{a}$$

$$\Sigma M_3 = 0$$

$$F_1 \times 254 + F_2 \times 127 - 44500 \times 177.8 = 0$$

$$254F_1 + 127F_2 - 7.912 \times 10^6 = 0 \tag{b}$$

$$\left(\frac{F_1}{K_1}\right)^{1/1.5} - \left(\frac{F_2}{K_2}\right)^{1/1.5} = \frac{l_1}{l_2}\left[\left(\frac{F_2}{K_2}\right)^{1/1.5} - \left(\frac{F_3}{K_3}\right)^{1/1.5}\right] \tag{10.59}$$

But $K_1 = K_2 = K_3$ and $l_1 = l_2$. Therefore,

$$F_1^{0.67} - 2F_2^{0.67} + F_3^{0.67} = 0 \tag{c}$$

Solving equations (a)–(c) simultaneously yields

$$F_1 = 24{,}030 \text{ N (5400 lb)}$$

$$F_2 = 14{,}240 \text{ N (3200 lb)}$$

$$F_3 = 6230 \text{ N (1400 lb)}$$

Multiple-Bearing Systems

Equations (10.41)–(10.46) inclusive may be used to determine the bearing reactions in a multiple bearing system such as that shown in Fig.

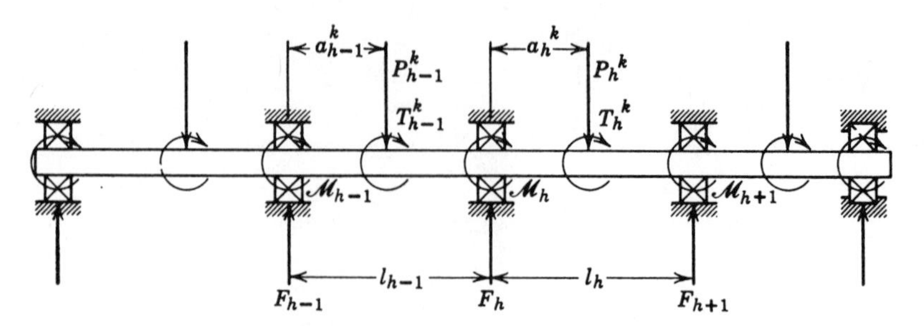

FIGURE 10.17. Multiple bearing-shaft system.

10.17 having a flexible shaft. It is evident that the reaction at any bearing support location h is a function of the loading existing at and in between the bearing supports located at $h - 1$ and $h + 1$. Therefore, from equations (10.41)–(10.46), the reactive loads at each support location h are given as follows:

$$
\begin{aligned}
F_h = {} & \frac{1}{l_{h-1}^3} \sum_{k=1}^{k=p} P_{h-1}^k (a_{h-1}^k)^2 (3l_{h-1} - 2a_{h-1}) \\
& + \frac{1}{l_h^3} \sum_{k=1}^{k=q} P_h^k (l_h - a_h^k)^2 (l_h + 2a_h^k) \\
& + \frac{6}{l_{h-1}^3} \sum_{k=1}^{k=r} T_{h-1}^k a_{h-1}^k (l_{h-1} - a_{h-1}^k) - \frac{6}{l_h^3} \sum_{k=1}^{k=s} T_h^k a_h^k (l_h - a_h^k) \\
& + 6E \left\{ \frac{I_{h-1}}{l_{h-1}^2} \left[\theta_{h-1} + \theta_h + \frac{2}{l_{h-1}} (\delta_{r,h-1} - \delta_{r,h}) \right] \right. \\
& \left. - \frac{I_h}{l_h^2} \left[\theta_h + \theta_{h+1} + \frac{2}{l_h} (\delta_{r,h} - \delta_{r,h+1}) \right] \right\}
\end{aligned}
\tag{10.60}
$$

$$
\begin{aligned}
\mathfrak{M}_h = {} & \frac{1}{l_{h-1}^2} \sum_{k=1}^{k=p} P_{h-1}^k (a_{h-1}^k)^2 (l_{h-1} - a_{h-1}^k) - \frac{1}{l_h^2} \sum_{k=1}^{k=q} P_h^k a_h^k (l_h - a_h^k)^2 \\
& + \frac{1}{l_{h-1}^2} \sum_{k=1}^{k=r} T_{h-1}^k a_{h-1}^k (2l_{h-1} - 3a_{h-1}^k) \\
& - \frac{1}{l_h^2} \sum_{k=1}^{k=s} T_h^k (l_h - a_h^k)(l_h - 3a_h^k) \\
& + 2E \left\{ \frac{I_{h-1}}{l_{h-1}} \left[\theta_{h-1} + 2\theta_h + \frac{3}{l_{h-1}} (\delta_{r,h-1} - \delta_{r,h}) \right] \right. \\
& \left. + \frac{I_h}{l_h} \left[2\theta_h + \theta_{h+1} + \frac{3}{l_h} (\delta_{r,h} - \delta_{r,h+1}) \right] \right\}
\end{aligned}
\tag{10.61}
$$

For a shaft-bearing system of n supports, that is, $h = n$, equations (10.60) and (10.61) represent a system of $2n$ equations. In the most elementary case, all bearings are considered as being sufficiently self-aligning that all $\mathfrak{M}_h$ equal zero; furthermore, all δ_{rh} are considered negligible compared to shaft deflection. Equations (10.60) and (10.61) thereby degenerate to the familiar equation of "three moments."

It is evident that the solution of equations (10.60) and (10.61) to obtain bearing reactions $\mathfrak{M}_h$ and F_h depends on relationships between radial load and radial deflection and moment load and misalignment angle for each radial bearing in the system. These relationships have been defined in Chapters 6 and 8. Thus, for a very sophisticated solution to a shaft-bearing problem as illustrated in Fig. 10.18 one could consider a shaft having two degrees of freedom with regard to bending, that is, deflection in two of three principal directions, supported by bearings h and accommodating loads k. At each bearing location h, one must establish the following relationships:

$$\delta_{y,h} = f_1(F_{x,h}, F_{y,h}, F_{z,h}, \mathfrak{M}_{xy,h}, \mathfrak{M}_{xz,h}) \tag{10.62}$$

$$\delta_{z,h} = f_2(F_{x,h}, F_{y,h}, F_{z,h}, \mathfrak{M}_{xy,h}, \mathfrak{M}_{xz,h}) \tag{10.63}$$

$$\theta_{xy,h} = f_3(F_{x,h}, F_{y,h}, F_{z,h}, \mathfrak{M}_{xy,h}, \mathfrak{M}_{xz,h}) \tag{10.64}$$

$$\theta_{xz,h} = f_4(F_{x,h}, F_{y,h}, F_{z,h}, \mathfrak{M}_{xy,h}, \mathfrak{M}_{xz,h}) \tag{10.65}$$

To accommodate the movement of the shaft in two principal directions, the following expressions will replace equations (8.4) and (8.5) for each

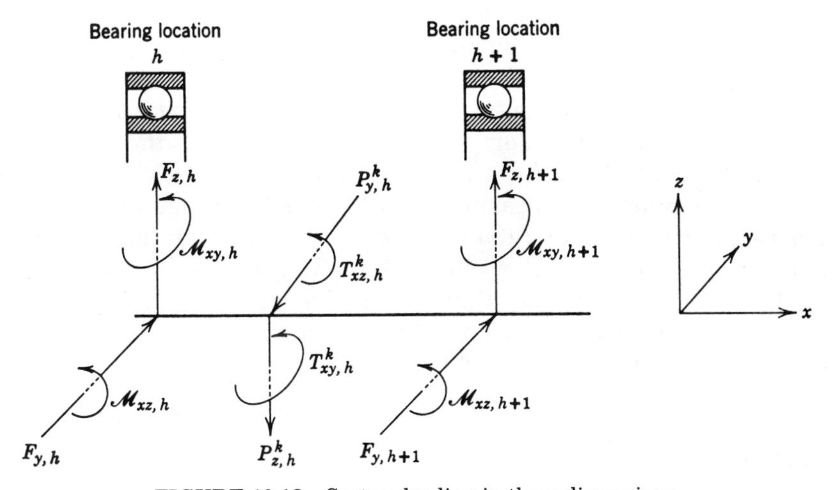

FIGURE 10.18. System loading in three dimensions.

ball bearing (see Jones [10.1]):

$$s_{xj} = BD \sin \alpha^{\circ} + \delta_x + \theta_{xz} \mathcal{R}_i \sin \psi_j + \theta_{xy} \mathcal{R}_i \cos \psi_j \qquad (10.66)$$

$$s_{zj} = BD \cos \alpha^{\circ} + \delta_y \sin \psi_j + \delta_z \cos \psi_j \qquad (10.67)$$

In equations (10.66) and (10.67) it will be recalled that j refers to rolling element location within a given bearing, and δ_x is axial deflection.

CLOSURE

For most rolling bearing applications it is usually sufficient to consider the shaft and housings as rigid structures. Thus, in simple two-bearing systems the determination of bearing loads is a matter of simple solution of static equilibrium equations. For three- or multiple-bearing systems, the simple radial deflection formulas for rolling bearings must be introduced to solve the statically indeterminate bearing loads.

As demonstrated by Example 10.1, however, when the shaft is considerably hollow, for example, 75% or more, and/or the span between bearing supports is sufficiently great, the shaft bending characteristics cannot be separated from the bearing deflection characteristics with the expectation of accurately ascertaining the bearing loads or the overall system deflection characteristics. Indeed, the bearings may be stiffer than might be anticipated by the simplified deflection formulas or even stiffer than a more elegant solution that employs accurate evaluation of load distribution might predict for the assumed loading. Ultimately, the penalty for increased stiffness will be paid in bearing endurance since the improved stiffness is obtained at the expense of induced moment loading.

It is of interest to note that the accurate solution of bearing loading in integral bearing–shaft–housing systems depends on the use of a computer. For example, a high speed shaft in three ball bearings each of which has 10 rolling elements, the shaft being loaded such as to cause each bearing to have five degrees of freedoms, requires the solution of 142 simultaneous nonlinear equations. More than likely, the system would include some roller bearings instead of ball bearings; however, the roller complement is likely to exceed 20 per row and considering inner and outer assembly flexibility, the problem may be even more complex. It is therefore very fortunate that bearing loading on most bearing systems can be analyzed by using equations of static equilibrium, at worst coupled with simple equations of bearing deflection.

REFERENCE

10.1. A. B. Jones, "A General Theory for Elastically Constrained Ball and Radial Roller Bearings under Arbitrary Load and Speed Conditions," *ASME J. Basic Eng.* **82**, 309–320 (1960).

11

ROTOR DYNAMICS AND CRITICAL SPEEDS

LIST OF SYMBOLS

Symbol	Description	Units
B	Bearing total curvature = $(f_o + f_i - 1)$	
$[B]$	Damping matrix	
C	Damping coefficient	N-sec/mm (lb-sec/in.)
D	Ball diameter	mm (in.)
e	Eccentricity	mm (in.)
F	Applied force	N (lb)
F_0	Total amplitude of harmonic forcing function	N (lb)
I	Identity matrix	
J_G	Mass moment of inertia	N-mm-sec^2 (in.-lb.-sec^2)
K	Bearing deflection constant (x = 1.5 for ball bearings, 1.1 for roller bearings)	N/mm^x (lb/in.^{x})
$[K]$	Stiffness matrix	
M	Large body mass (rotating unbalance case)	kg (lb-sec^2/in.)

Symbol	Description	Units
m	Mass	kg (lb-sec^2/in.)
P_d	Bearing internal diametral clearance	mm (in.)
Q	Ball normal load	N (lb)
r	Radius of gyration	mm (in.)
S	Stiffness	N/mm (lb/in.)
t	Time	sec
x	Displacement (lateral)	mm (in.)
X_0	Displacement amplitude	mm (in.)
Z	Number of rolling elements	
α	Contact angle (loaded bearing)	°, rad
α°	Contact angle (unloaded bearing)	°, rad
δ	Elastic deflection	mm (in.)
θ	Angular displacement or inclination	°, rad
Θ_0	Angular displacement amplitude	°, rad
ϕ	Phase angle	°, rad
λ	Eigenvalue	
ζ	Damping factor	
ω	Frequency	rad/sec
ω_n	Natural or critical frequency	rad/sec
ν	Frequency of vibration	

SUBSCRIPTS

1, 2	Refers to first or second natural frequency
a	Refers to axial direction
c	Refers to critical
i	Refers to ith frequency or ith value
ir	Refers to inner raceway
o	Refers to amplitude value
or	Refers to outer raceway
r	Refers to radial direction
x	Refers to coordinate direction
y	Refers to coordinate direction
z	Refers to coordinate direction

GENERAL

In 1948 DenHartog [11.1] published the first text to address the mathematical principles of vibrating motion in mechanical systems. These principles have been built upon and extended over the years to provide vital tools to machine designers. With the advent of digital computer techniques, sophisticated design tools have been created that provide the ability to predict critical speeds and rotor behavior in high-speed, shaft-bearing systems. The specific topic of rotor-bearing dynamics rose in sophistication and importance to such a degree that in 1965 the U.S. Air Force sponsored the construction of a 10-part design guide on the subject [11.2]. The guide was later updated in 1978 [11.3].

The majority of this book deals with the technologies of rolling bearing design ranging through materials, performance prediction, lubrication, and fatigue life. This chapter deals with the one characteristic of rolling bearings that directly influences shaft-bearing dynamic behavior—stiffness. The subject of bearing-rotor systems interaction will be addressed in a threefold manner. First, the basics of mechanical vibration to form a foundation for understanding the analytics involved in rotor dynamics are presented. Second, the concept of bearing stiffness and the nature of its behavior in the rotor system environment are considered. Finally, a brief overview of rotor dynamic analysis is given.

DAMPED FORCED VIBRATIONS

The basic principles of flexible rotor dynamics stem from the mathematical representation of damped forced vibrations [11.4]. Figure 11.1*a* shows the system as a mass with viscous damping being forced by a harmonic

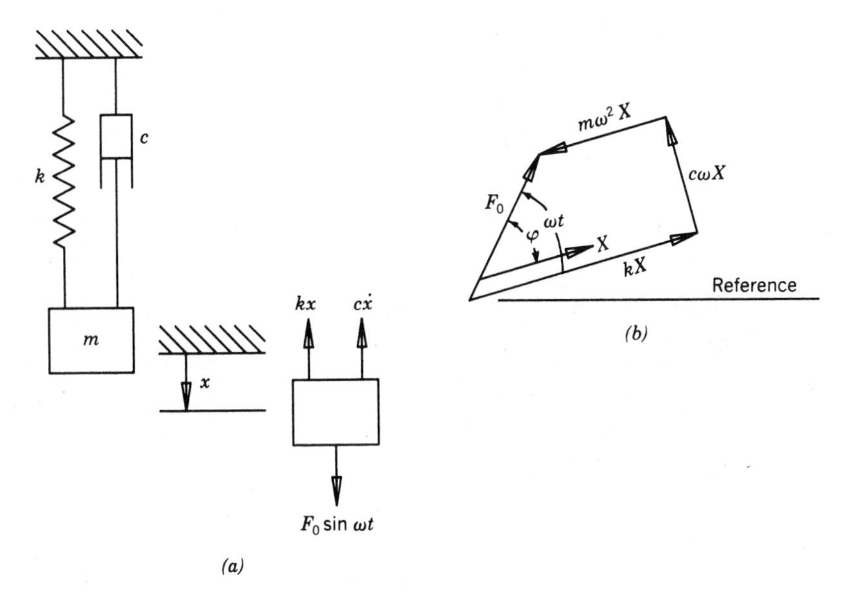

FIGURE 11.1. (*a*) Free-body diagram of a viscously damped system with a harmonic forcing function. (*b*) Vector relationship of (*a*).

exciting force F_0. The equation of motion that follows from the free-body diagram is

$$m\ddot{x} + c\dot{x} + kx = F_0 \sin \omega t \tag{11.1}$$

The solution of this differential equation is characterized by two parts: the complementary function and the particular integral. The particular solution is of the form

$$x = X \sin (\omega t - \phi) \tag{11.2}$$

where X is the amplitude of oscillation and ϕ is the phase of the displacement with respect to the exciting force. Vectorily, this is shown in Fig. 11.1*b*. From the vector diagram it is seen that

$$X = \frac{F_0}{[(k - m\omega^2)^2 + (c\omega)^2]^{1/2}} \tag{11.3}$$

and

$$\phi = \tan^{-1}\left(\frac{c\omega}{k - m\omega^2}\right) \tag{11.4}$$

Dividing by the stiffness k allows the expressions to be made dimensionless. Thus

$$X = \frac{F_0/k}{[(1 - m\omega^2/k)^2 + (c\omega/\mathrm{k})^2]^{1/2}} \tag{11.5}$$

and

$$\tan \phi = \frac{c\omega/k}{1 - m\omega^2/k} \tag{11.6}$$

The following quantities are now defined:

$$\begin{aligned}
\omega_\mathrm{n} &= (k/m)^2 = \text{natural frequency} \\
C_\mathrm{c} &= 2m\omega_\mathrm{n} = \text{critical damping parameter} \\
\zeta &= C/C_\mathrm{c} = \text{damping factor} \\
c\omega/k &= (C/C_\mathrm{c})(C_\mathrm{c}\omega/k) = (2\zeta\omega/\omega_\mathrm{n})
\end{aligned}$$

Thus, equations (11.5) and (11.6) can be rewritten as

$$\frac{Xk}{F_0} = \frac{1}{\{[1 - (\omega/\omega_n)^2]^2 + (2\zeta\omega/\omega_\mathrm{n})^2\}^{1/2}} \tag{11.7}$$

and

$$\tan \phi = \frac{2\zeta\omega/\omega_n}{1 - (\omega/\omega_n)^2} \tag{11.8}$$

Equations (11.7) and (11.8) are plotted in Fig. 11.2 to show the relationship between frequency (ω/ω_n) and damping (ζ) on amplitude and phase. The curves show the impact of damping factor on amplitude and phase, particularly in the region of resonance.

Constructing vector diagrams of the system such as Fig. 11.3 yields a better understanding of the effects of frequency. For values of ω/ω_n that are much less than 1 (Fig. 11.3*a*), both inertial and damping forces are small, resulting in a small phase angle ϕ and an impressed force nearly equal to the spring force. Figure 11.3*b* shows the situation for $\omega/\omega_n = 1$. Here the phase angle is 90°, and the inertial and spring forces balance each other, allowing the impressed force to overcome the damping force. At large values of ω/ω_n (Fig. 11.3*c*) the phase angle is approaching 180°, and the impressed force is stretched out to compensate for a large inertial force.

In summary, the differential equation of motion and its complete solution are written

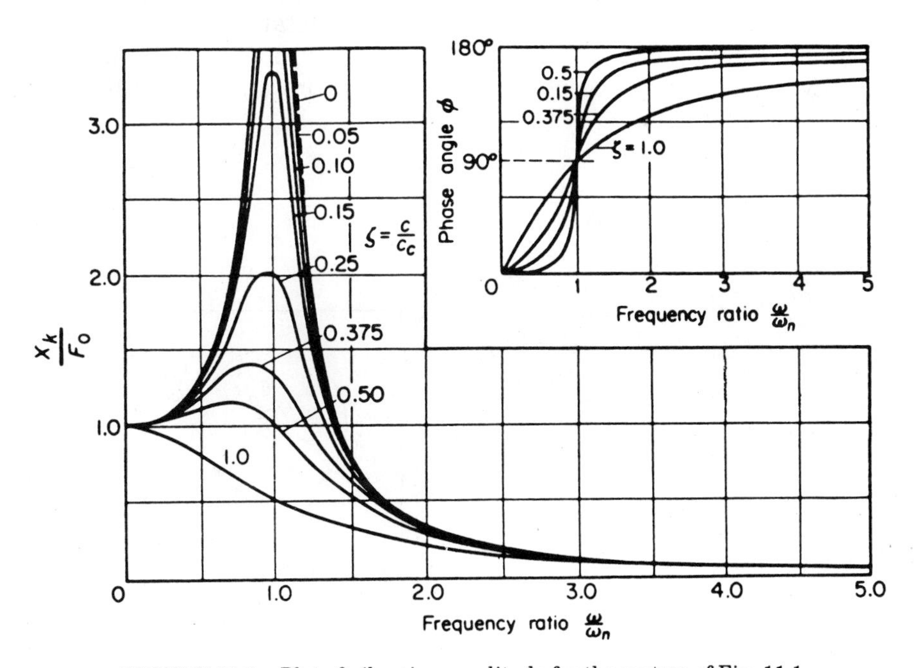

FIGURE 11.2. Plot of vibration amplitude for the system of Fig. 11.1.

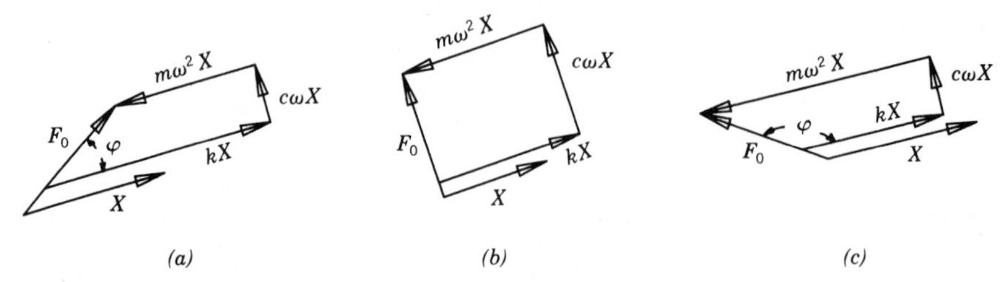

FIGURE 11.3. Vector relationships for forced damped vibration systems. (a) $\omega/\omega_n \ll 1$, (b) $\omega/\omega_n = 1$, (c) $\omega/\omega_n \gg 1$.

$$\ddot{x} + 2\zeta\omega_n\dot{x} + \omega_n^2 x = \frac{F_0}{m}\sin\omega t \tag{11.9}$$

$$x(t) = \left(\frac{F_0}{k}\right)\frac{\sin(\omega t - \phi)}{\{[1 - (\omega/\omega_n)^2]^2 + (2\zeta\omega/\omega_n)^2\}^{1/2}} + X_1 e^{-\zeta\omega_n t}\sin[(1 - \zeta^2)^{1/2}\,\omega_n t + \phi_1] \tag{11.10}$$

The most common source of harmonic excitation in rotor-bearing systems is rotating unbalance. A simple rotating unbalance system with a spring and damper support can be represented as shown in Fig. 11.4. The unbalance can be represented by a mass rotating at some radius represented by an eccentricity e. If the mass is rotating with angular velocity ω and x is the displacement of the nonrotating mass $M - m$, then the displacement of m is

$$x_m = x + e\sin\omega t \tag{11.11}$$

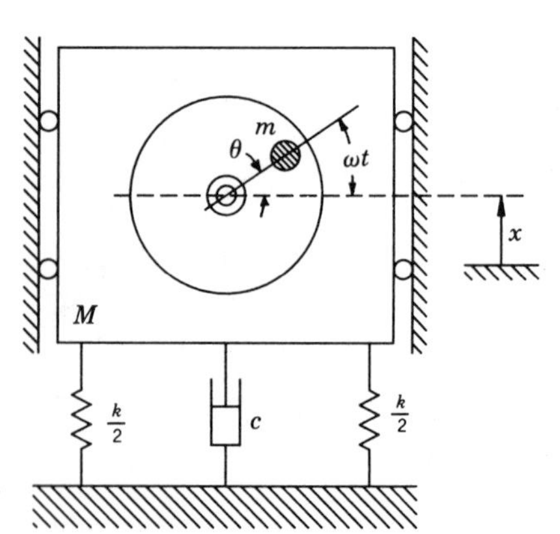

FIGURE 11.4. Schematic of a simple rotating unbalance system.

The differential equation of motion is thus

$$(M - m)\,\ddot{x} + m\,\frac{d^2}{dt^2}(x + e \sin \omega t) = -kx - c\dot{x} \qquad (11.12)$$

which rearranges to

$$M\ddot{x} + c\dot{x} + kx = me\omega^2 \sin \omega t \qquad (11.13)$$

Equation (11.13) is the same as equation (11.1), where F_0 is represented by $me\omega^2$.

It follows that the nondimensional form of the solutions is

$$\frac{Mx}{me} = \frac{(\omega/\omega_n)^2}{\{[1 - (\omega/\omega_n)^2]^2 + (2\zeta\omega/\omega_n)^2\}^{1/2}} \qquad (11.14)$$

and

$$\tan \phi = \frac{2\zeta\omega/\omega_n}{1 - (\omega/\omega_n)^2} \qquad (11.15)$$

The solution is presented graphically in Fig. 11.5.

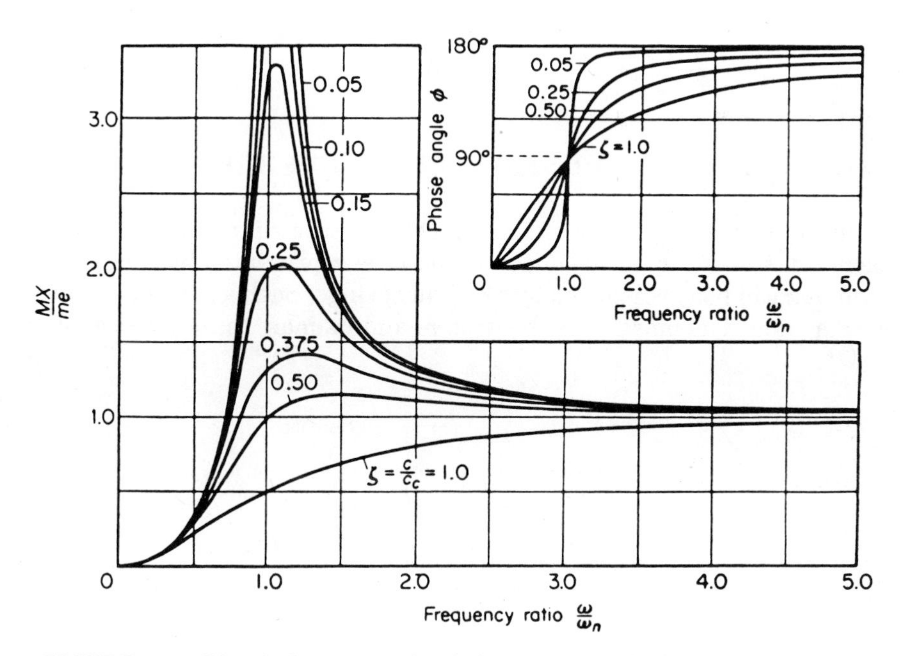

FIGURE 11.5. Plot of vibration amplitude for a system with a rotating unbalance.

Example 11.1. Assume a system as shown schematically in Fig. 11.4. A combination of rotating unbalances causes a resonant amplitude of 1.2 mm (0.0472 in.). Upon considerably increasing the speed of rotation of the unbalanced masses, the resonant amplitude seemed to level out at 0.16 mm (0.0063 in.). Calculate the apparent damping factor.

From equation (11.14) the resonant amplitude is

$$X = \frac{me/M}{2\zeta}$$

$$1.2 = \frac{me/M}{2\zeta}$$

When the ratio of ω to ω_n is very large, the equation becomes

$$X = \frac{me}{M}$$

$$0.16 = \frac{me}{M}$$

Solving both equations simultaneously yields

$$1.2 = \frac{0.16}{2\zeta}$$

$$\zeta = 0.0666$$

COUPLED VIBRATORY MOTION (RIGID SHAFT)

Consider a free-body diagram that can be constructed to represent a rigid beam on elastic supports [11.5]. It would be represented by a large unsymmetrical body supported by unequal springs. Such a system is shown in Fig. 11.6. The mass of the body is m, and its mass center is located at

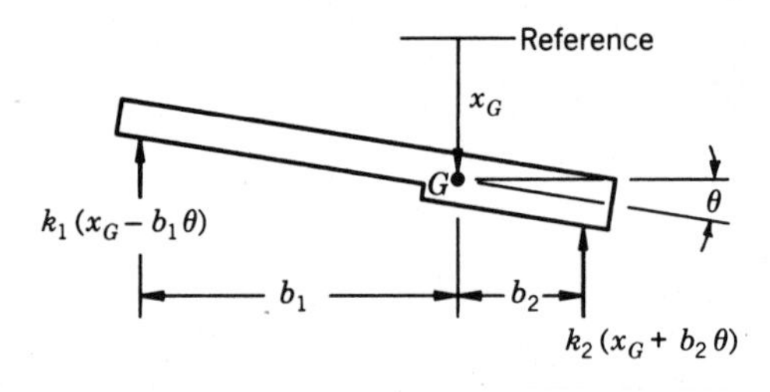

FIGURE 11.6. Free-body diagram of a coupled system (rigid beam).

G. To define the motion of the body, a coordinate x will be used to specify linear position and a coordinate θ to specify angular position. Both will be measured from some reference position or neutral plane. Because of this coordinate mix, the system is said to be coupled. The nature of the coupling can only be determined by analysis and depends on the coordinates selected in defining the equations of motion.

For instance, if x_G were measured as the linear displacement of the mass from its mass center G, then the equations of motion become

$$m\ddot{x}_G + (k_1 + k_2)\,x_G + (k_2 b_2 - k_1 b_1)\,\theta = 0$$

$$J_G\ddot{\theta} + (k_2 b_2 - k_1 b_1)\,x_G + (k_1 b_1^2 + k_2 b_2^2)\,\theta = 0 \tag{11.16}$$

These equations are said to be "statically coupled" through the coupling coefficient $k_2 b_2 - k_1 b_1$. This system model closely resembles the rotor-bearing system situation. Assuming motions to be harmonic,

$$x = X_0 \sin(\omega t + \phi) \qquad \theta = \Theta_0 \sin(\omega t + \phi) \tag{11.17}$$

Solving for amplitude gives

$$(k_1 + k_2)\,X_0 + (k_2 b_2 - k_1 b_1)\,\Theta_0 = m\omega^2 X_0$$

$$(k_2 b_2 - k_1 b_1)\,X_0 + (k_1 b_1^2 + k_2 b_2^2)\,\Theta_0 = J_G \omega^2 \Theta_0 \tag{11.18}$$

The following simplifications are now made:

$$(k_1 + k_2)/m = a^* \qquad (k_2 b_2 - k_1 b_1)/m = b^* \qquad (k_1 b_1^2 + k_2 b_2^2)/m = c^*$$

where $J_G = mr^2$, r being the radius of gyration with respect to G. Equations (11.18) now become

$$a^* X_0 + b^* \Theta_0 = \omega^2 X_0 \qquad \frac{b^* X_0}{r^2} + \frac{c^* \Theta_0}{r^2} = \omega^2 \Theta_0 \tag{11.19}$$

Setting the determinant equation to zero and solving for the principal frequencies yields

$$\omega_{1,2}^2 = \frac{1}{2}\left\{\frac{c^*}{r^2} + a^* \pm \left[\left(\frac{c^*}{r^2} - a^*\right)^2 + \left(\frac{2b^*}{r}\right)^2\right]^{1/2}\right\} \tag{11.20}$$

Example 11.2. Consider a rigid beam system, as depicted by Fig. 11.7a, with the following data: $b_1 = 400$ mm (15.75 in.), $k_1 = 100$ N/mm (570.8 lb/in.), $b_2 = 200$ mm (7.874 in.), $k_2 = 70$ N/mm (399.6 lb/in.). Assume that the beam mass is 5.1813 kg (0.0296 lb-sec^2/in.) and

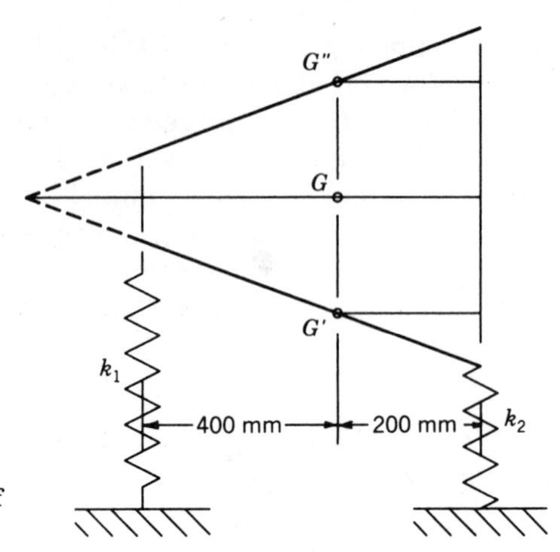

FIGURE 11.7. Free-body diagram of system for Example 11.2.

that the radius of gyration is 300 mm (11.81 in.). Determine the natural frequencies.

$$a^* = \frac{k_1 + k_2}{m} = \frac{100 + 70}{5.181} = 32.8 \text{ rad}^2/\text{sec}^2$$

$$b^* = \frac{k_2 b_2 - k_1 b_1}{m} = \frac{70 \times 200 - 100 \times 400}{5.181}$$

$$= -5018 \text{ mm/sec}^2 \ (-197.6 \text{ in./sec}^2)$$

$$c^* = \frac{k_1 b_1^2 + k_2 b_2^2}{m} = \frac{100 \times 400^2 + 70 \times 200^2}{5.181}$$

$$= 3.628 \times 10^6 \text{ mm}^2/\text{sec}^2 \ (1.429 \times 10^5 \text{ in.}^2/\text{sec}^2)$$

$$\omega_{1,2}^2 = \frac{1}{2}\left\{\frac{c^*}{r^2} + a^* \pm \left[\left(\frac{c^*}{r^2} - a^*\right)^2 + \left(\frac{2b^*}{r}\right)^2\right]^{1/2}\right\}$$

$$= \tfrac{1}{2}\{40.3 + 32.8 \pm [56.48 + 1119]^{1/2}\}$$

$$= 53.7,\ 19.4 \text{ rad}^2/\text{sec}^2$$

Thus, $\omega_1 = 7.33$ rad/sec and $\omega_2 = 4.4$ rad/sec.

In matrix notation the amplitude equations (11.18) are

$$\begin{bmatrix} k_1 + k_2 & k_2 b_2 - k_1 b_1 \\ k_2 b_2 - k_1 b_1 & k_1 b_1^2 + k_2 b_2^2 \end{bmatrix} \begin{bmatrix} X_0 \\ \Theta_0 \end{bmatrix} = \omega^2 \begin{bmatrix} m & 0 \\ 0 & J_G \end{bmatrix} \begin{bmatrix} X_0 \\ \Theta_0 \end{bmatrix} \tag{11.21}$$

This presents the coupling term more visibly as the off-diagonal elements of the spring matrix. Further, if the spring moments are equal, then $k_2 b_2 - k_1 b_1 = 0$ and the solutions become

$$\omega_1 = \left(\frac{k_1 + k_2}{m}\right)^{1/2}, \qquad \omega_2 = \left(\frac{k_1 b_1^2 + k_2 b_2^2}{J_G}\right)^{1/2} \tag{11.22}$$

These solutions represent the translation (x) and rocking (θ) natural frequencies. If $k_2 b_2 - k_1 b_1 \neq 0$, then the frequencies are mixed. That is, the first frequency is translation with some rocking, and the other frequency is predominantly rocking with some translation.

Associated with each eigenvector solution or natural frequency is a natural mode shape. The mode shape is defined mathematically as the ratio of the displacement and rotation amplitudes at the mass center. For the uncoupled case ($k_2 b_2 - k_1 b_1 = 0$) they are

$$X_\omega = \frac{b^*}{\omega_1^2 - a^*}, \qquad \Theta_\omega = \frac{\omega_2^2 r^2 - c^*}{b} \tag{11.23}$$

Thus, when the system is decoupled, the behavior is governed by two distinct single-degree-of-freedom systems. Conversely, it can be shown that, at the extreme of the coupled case, the system behaves as a single-degree-of-freedom system. This occurs when

$$\left(\frac{c^*}{r^2} - a^*\right)^2 + \left(\frac{2b^*}{r}\right)^2 = 0 \tag{11.24}$$

This means

$$\frac{c^*}{r^2} = a^* \qquad b^* = 0 \tag{11.25}$$

This condition can only be satisfied when

$$\frac{k_1}{k_2} = \left(\frac{r}{b_1}\right)^2 \tag{11.26}$$

and

$$k_1 b_1 = k_2 b_2 \tag{11.27}$$

This gives

$$\omega_1^2 = \omega_2^2 = \frac{1}{2}\left(\frac{k_1 + k_2}{m} + \frac{k_1 b_1^2 + k_2 b_2^2}{J_G}\right) \tag{11.28}$$

MULTI-DEGREE-OF-FREEDOM SYSTEM (FLEXIBLE SHAFT)

A series of n masses connected by springs is shown in Fig. 11.8 with its free-body diagram. This represents the general case of a multi-degree-of-freedom system. The differential equations of motion are thus written as

$$
\begin{aligned}
m_1\ddot{x}_1 &= -k_1x_1 + k_2(x_2 - x_1)\\
m_2\ddot{x}_2 &= -k_2(x_2 - x_1) + k_3(x_3 - x_2)\\
&\vdots\\
m_i\ddot{x}_i &= -k_i(x_i - x_{i-1}) + \mathrm{k}_{i+1}(x_{i+1} - x_i)\\
&\vdots\\
m_n\ddot{x}_n &= -k_n(x_n - x_{n-1}) - k_{n+1}x_n
\end{aligned}
\tag{11.29}
$$

Rearranged and in matrix form they become

$$
\begin{bmatrix} m_1 & 0 & 0 & \cdots & 0\\ 0 & m_2 & 0 & \cdots & 0\\ 0 & 0 & m_3 & \cdots & 0\\ \vdots & \cdots & & & \\ 0 & \cdots & & & m_n \end{bmatrix}
\begin{Bmatrix} \ddot{x}_1\\ \ddot{x}_2\\ \ddot{x}_3\\ \vdots\\ \ddot{x}_n \end{Bmatrix}
+ \begin{bmatrix} k + k_2 & -k_2 & 0 & \cdots & 0\\ -k_2 & k_2 + k_3 & -k_3 & \cdots & 0\\ 0 & -k_3 & & \cdots & 0\\ \vdots & \vdots & \vdots & \cdots & k_{n+1} + k_n \end{bmatrix}
\times \begin{Bmatrix} X_1\\ X_2\\ X_3\\ \vdots\\ X_n \end{Bmatrix} = \begin{Bmatrix} 0\\ \vdots\\ 0 \end{Bmatrix}
\tag{11.30}
$$

This is a particular case of the general equation

$$
[m]\{\ddot{x}\} + [k]\{x\} = 0 \tag{11.31}
$$

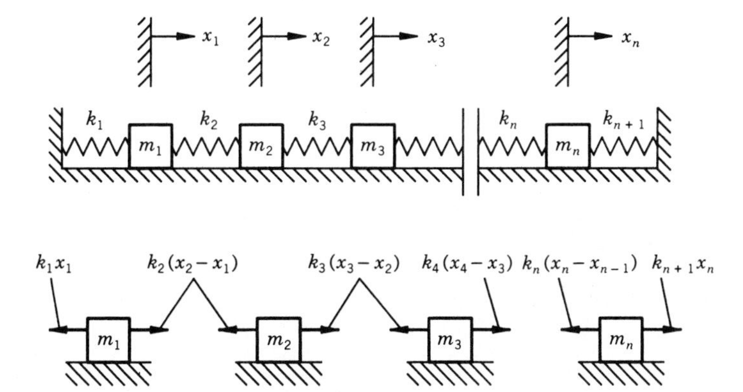

FIGURE 11.8. Free-body schematic of a multi-degree-of-freedom system.

where the square matrices $[m]$ and $[k]$ are the mass matrix and stiffness matrix, respectively, and are given by

$$[m] = \begin{bmatrix} m_{11} & m_{12} & m_{13} & \cdots & m_{1n} \\ m_{21} & m_{22} & m_{23} & \cdots & m_{2n} \\ \vdots & & & & \\ m_{n1} & n_{n2} & m_{23} & \cdots & m_{nn} \end{bmatrix} \tag{11.32}$$

and

$$[k] = \begin{bmatrix} k_{11} & k_{12} & k_{13} & \cdots & k_{1n} \\ k_{21} & k_{22} & k_{23} & \cdots & k_{2n} \\ \vdots & & & & \\ k_{n1} & k_{n2} & k_{n3} & \cdots & k_{nn} \end{bmatrix} \tag{11.33}$$

If equation (11.31) is multiplied by $[m]^{-1}$, the inverse of the mass matrix, then is

$$[I]\,\{\ddot{x}\} + [A]\,\{x\} = 0 \tag{11.34}$$

where $[I]$ is a unit matrix and $[A]$ is a dynamic matrix.

For simplicity, the bracket and brace notation is dropped and harmonic motion is assumed, giving

$$[A - \lambda I]\,\{X\} = 0 \tag{11.35}$$

where $\ddot{x} = -\lambda X$ with $\lambda = \omega^2$. λ is known as the direct frequency factor. The determinant of equation (11.35),

$$|A - \lambda I| = 0 \tag{11.36}$$

is the characteristic equation of the system. The roots of this characteristic equation (eigenvalues) represent the natural frequencies of the system:

$$\lambda_i = \omega_i^2 \tag{11.37}$$

The eigenvectors can now be determined by substituting λ_i into equation (11.35) and solving for the corresponding mode shapes X_i. For an n-degree-of-freedom system there are n eigenvalues and n eigenvectors.

It is also possible to solve this system by methods using the adjoint matrix (see any reference on matrix algebra). Let

$$B = A - \lambda I \tag{11.38}$$

Then by definition the inverse is

$$B^{-1} = \left(\frac{1}{|B|}\right) \text{adj } B \tag{11.39}$$

Multiplying $|B|$ by B yields

$$|B|\, I = B \text{ adj } B \tag{11.40}$$

or, resubstituting for B,

$$|A - I|\, I = [\mathrm{A} - \lambda \mathrm{I}] \text{ adj } [A - \lambda I] \tag{11.41}$$

If the eigenvalue is introduced into the equation ($I = \lambda_i$), then the determinant is zero and

$$[0] = [A - \lambda_i I] \text{ adj } [A - \lambda_i I] \tag{11.42}$$

From equation (11.35), for some ith mode,

$$[A - \lambda_i I]\, [X_i] = 0 \tag{11.43}$$

It must follow that the adjoint matrix adj $[A - \lambda_i I]$ must be a column matrix of the eigenvectors X_i (multiplied by some arbitrary constant).

Example 11.3. Considering the generalized system of Fig. 11.9, perform a complete eigenvalue/eigenvector solution [11.4].

The equations of motion in matrix form are

$$\begin{bmatrix} m & 0 \\ 0 & 2m \end{bmatrix} \begin{Bmatrix} \ddot{x}_1 \\ \ddot{x}_2 \end{Bmatrix} + \begin{bmatrix} 2k & -k \\ -k & 2k \end{bmatrix} \begin{Bmatrix} x_1 \\ x_2 \end{Bmatrix} = \begin{Bmatrix} 0 \\ 0 \end{Bmatrix} \tag{11.44}$$

Assuming harmonic motion, substituting $\lambda = \omega^2$, and multiplying by the inverse of the mass matrix yield

$$\begin{bmatrix} 2k/m - \lambda & -k/m \\ -\frac{1}{2}\,\mathrm{k}/m & k/m - \lambda \end{bmatrix} \begin{Bmatrix} x_1 \\ x_2 \end{Bmatrix} = 0 \tag{11.45}$$

The characteristic equation is thus

$$\lambda^2 - 3\left(\frac{k}{m}\right)\lambda + \frac{3}{2}\left(\frac{k}{m}\right)^2 = 0 \tag{11.46}$$

from which the eigenvalues follow as

$$\lambda_1 = 0.634\left(\frac{k}{m}\right)^{1/2} \qquad \lambda_2 = 2.366\left(\frac{k}{m}\right)^{1/2} \tag{11.47}$$

Returning to equation (11.45) and substituting in the eigenvalues, the adjoint matrix is obtained:

$$\operatorname{adj}\,[A - \lambda_i I] = \begin{bmatrix} k/m - \lambda_i & k/m \\ k/2m & 2k/m - \lambda_i \end{bmatrix} \tag{11.48}$$

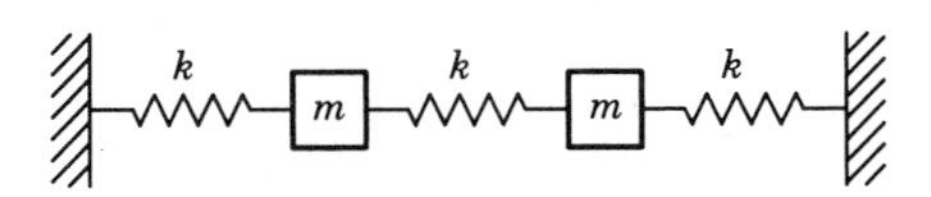

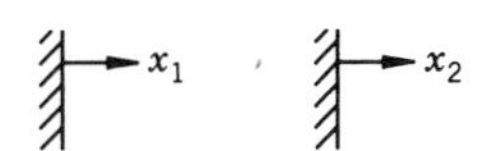

FIGURE 11.9. System for Example 11.3.

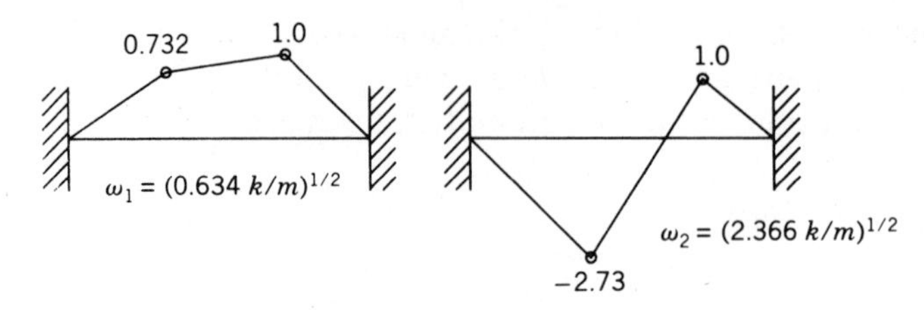

FIGURE 11.10. Mode results of Example 11.3.

Substituting $\lambda_1 = 0.634\,(k/m)^{1/2}$ in (11.48) gives

$$\begin{Bmatrix} X_1 \\ X_2 \end{Bmatrix} = \begin{bmatrix} 0.366 & 1.000 \\ 0.500 & 1.366 \end{bmatrix} \left(\frac{k}{m}\right)^{1/2}$$

Normalizing each column to unity results in the first eigenvector

$$\begin{Bmatrix} X_1 \\ X_2 \end{Bmatrix} = \begin{bmatrix} 0.732 & 0.732 \\ 1.000 & 1.000 \end{bmatrix} \left(\frac{k}{m}\right)^{1/2}$$

or

$$X_1 = \begin{Bmatrix} 0.732 \\ 1.000 \end{Bmatrix} \left(\frac{k}{m}\right)^{1/2}$$

Likewise, it can be shown that

$$X_2 = \begin{Bmatrix} -2.73 \\ 1.00 \end{Bmatrix} \left(\frac{k}{m}\right)^{1/2}$$

The two normal modes are shown schematically in Fig. 11.10.

BEARING STIFFNESS

A common design parameter for oil film bearings is stiffness, which is defined as follows:

$$\mathcal{S} = \frac{dF}{d\delta} \tag{11.49}$$

In other words, stiffness $\mathcal{S}$ is the inverse of the rate of change of bearing deflection with applied load. To find the stiffness of a particular rolling bearing under a given type of loading, one must first determine the load distribution among the rollers and then relate the maximum rolling element load to the applied load. This is simply done for single-row radial ball bearings with nominal clearance for which

$$Q_{\max} = \frac{5F_r}{Z \cos \alpha} \tag{6.24}$$

Substituting for $Q_{\max}$ in equation (9.6) yields

$$\delta_r = \frac{1.976 \times 10^{-6} F_r^{2/3}}{Z^{2/3} D^{1/3} \cos^{5/3} \alpha} \tag{11.50}$$

From equation (11.50) according to (11.49), the stiffness of the bearing is

$$\mathcal{S} = 6.916 \times 10^{-4} Z D^{1/2} \cos^{5/2} \alpha \delta_r^{1/2} \tag{11.51}$$

From equation (11.51) it is apparent that the stiffness of a ball bearing is a nonlinear relationship because it is dependent on the square root of radial deflection. In this respect a ball bearing is unlike a simple spring for which deflection is linear with respect to load. Roller bearings react in similar fashion except that load varies with deflection to the 1.11 power and stiffness varies directly with deflection to the 0.11 power.

Another relatively simple case is that of an angular-contact ball bearing subjected to simple axial load. The load is distributed equally among the balls. Hence

$$Q = \frac{F_a}{Z \sin \alpha} \tag{6.26}$$

where α is the contact angle in the loaded condition; α° is the contact angle in the unloaded condition, given by

$$\cos \alpha^\circ = 1 - \frac{P_d}{2BD} \tag{2.9}$$

in which P_d is the mounted internal diametral clearance. Referring to Fig. 6.4, it is seen that a thrust load F_a applied to the inner ring results in an axial deflection δ_a.

From Fig. 6.4 it was shown that

$$\delta_n = BD \left(\frac{\cos \alpha^\circ}{\cos \alpha} - 1 \right) \tag{6.28}$$

and since

$$Q = K_n \delta_n^{1.5} \tag{6.2}$$

equation (6.29) was obtained:

$$Q = K_n \,(BD)^{1.5} \left(\frac{\cos \alpha^\circ}{\cos \alpha} - 1 \right)^{1.5} \tag{6.29}$$

Substituting (6.26) into (6.29) yielded

$$\frac{F_{\mathrm{a}}}{ZK_n (BD)^{1.5}} = \sin \alpha \left(\frac{\cos \alpha^\circ}{\cos \alpha} - 1 \right)^{1.5} \tag{6.30}$$

and

$$\delta_{\mathrm{a}} = \frac{BD \sin (\alpha - \alpha^\circ)}{\cos \alpha} \tag{6.36}$$

Computing bearing stiffness involves solving equation (6.30) for bearing equilibrium by iteration. In Chapter 6, it was shown that K_n is also a function of α and therefore must be determined by iteration schemes as well. In Chapter 6, however, it was further shown that for a thrust-loaded, angular-contact ball bearing K_n can be replaced by $KD^{0.5}/B^{1.5}$, where K is determined by Fig. 6.5. Hence (6.30) became

$$\frac{F_{\mathrm{a}}}{ZD^2K} = \sin \alpha \left(\frac{\cos \alpha^\circ}{\cos \alpha} - 1 \right)^{1.5} \tag{6.33}$$

Since stiffness

$$\mathcal{S} = \frac{dF_{\mathrm{a}}}{d\delta_{\mathrm{a}}} = \frac{dF_{\mathrm{a}}}{d\alpha} \frac{d\alpha}{d\delta_{\mathrm{a}}}$$

$$\mathcal{S} = \frac{ZDK}{B \cos \alpha^\circ} \left(\frac{\cos \alpha^\circ}{\cos \alpha} - 1 \right)^{1/2} [\cos^3 \alpha \,(\cos \alpha^\circ - \cos \alpha) + 1.5 \sin^2 \alpha] \tag{11.52}$$

Example 11.4. A 218 angular-contact ball bearing is loaded axially with 17,800 N (4000 lb). Determine the axial stiffness.

For a 218 angular-contact ball bearing, $Z = 16$, $D = 22.23$ mm (0.875 in.), $B = 0.0464$, $\alpha^\circ = 40°$. From Fig. 6.5, $K = 896.7$ N/mm^2 (130,000 psi).

The new contact angle can be determined from equation (6.30) by interation:

$$\frac{F_a}{ZD^2K} = \sin\alpha\left(\frac{\cos\alpha^\circ}{\cos\alpha} - 1\right)^{3/2} \quad (6.33)$$

$$\frac{17{,}800}{(16)\,(22.23)^2\,(896.7)} = \sin\alpha\left(\frac{\cos 40^\circ}{\cos\alpha} - 1\right)^{3/2}$$

If K remains unchanged, iteration yields $\alpha = 41.36^\circ$.

$$\mathcal{S} = \frac{ZDK}{B\cos\alpha^\circ}\left(\frac{\cos\alpha^\circ}{\cos\alpha} - 1\right)^{1/2}\left[\cos^3\alpha\,(\cos\alpha^\circ - \cos\alpha) + 1.5\sin^2\alpha\right]$$

$$= \frac{16 \times 22.23 \times 896.7}{0.0464\cos 40^\circ}\left(\frac{\cos 40^\circ}{\cos 41.36^\circ} - 1\right)^{1/2} \quad (11.52)$$

$$\times\left[(\cos^3 41.36^\circ)(\cos 40^\circ - \cos 41.36^\circ) + 1.5\sin^2 41.36^\circ\right]$$

$$= 8.522 \times 10^5\ \text{N/mm}\ (4.864 \times 10^6\ \text{lb/in.})$$

Bearing stiffness behavior can be extrapolated to the whole bearing. Thus, the bearing locations in a rotor-bearing system can be modeled as spring supports. Remembering rigid beam theory, it is seen that there is a need to analytically represent a bearing location with both lateral and angular stiffnesses. In the analytical model the reaction force and the reaction moment of each bearing are felt by the rotor through a single location on the rotor axis [11.6]. Schematically this is illustrated in Fig. 11.11.

A typical linear spring restraining the lateral displacement and a torsional spring opposing the rotation are attached to the same point on the rotor axis. A complete description of the characteristics of the support

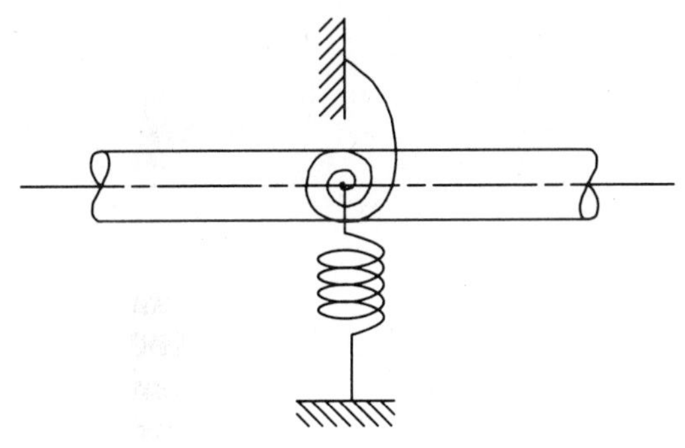

FIGURE 11.11. Basic model concept for bearing stiffness.

bearings, however, involves much more than the specification of the two spring constants [11.6] for the following reasons:

The lateral motion of the rotor axis is concerned with two displacement components and two inclination components.
The restraining characteristics may include cross-coupling among various displacement/inclination coordinates.
The restraining force/moment may not be temporally in phase with the displacement/inclination.
The restraining characteristics of the bearing may depend on the rotor speed or the vibration frequency or both.
Bearing pedestal compliance might not be negligible.

To accommodate the above concerns, the support bearing characteristics are described analytically as a 4-degree-of-freedom impedance matrix:

$$[R_N] = -[Z_N] \times [W_N] \tag{11.53}$$

where $[W_N]$ is a column vector containing elements that are the two lateral displacements (δ_x, δ_y) and the two lateral inclinations (θ_x, θ_y) of the rotor axis at some bearing location N. With a right-hand Cartesian coordinate system as shown in Fig. 11.12, directions can be defined. From the figure the z axis is coincident with the spin vector of the rotor or the rotor axis. This results in the x axis and y axis being perpendicular and in the direction of radially applied external static loads. Intuitively, δ_x, δ_y and θ_x, θ_y are, respectively, the displacements and rotations at the appropriate axis.

Returning to equation (11.53), it is noted that $[Z_N]$ is a complex 4 × 4 matrix that represents the stiffness and damping coefficients at the bearing support. According to common notation this is

$$[Z_N] = [K_N] + i\nu[B_N] \tag{11.54}$$

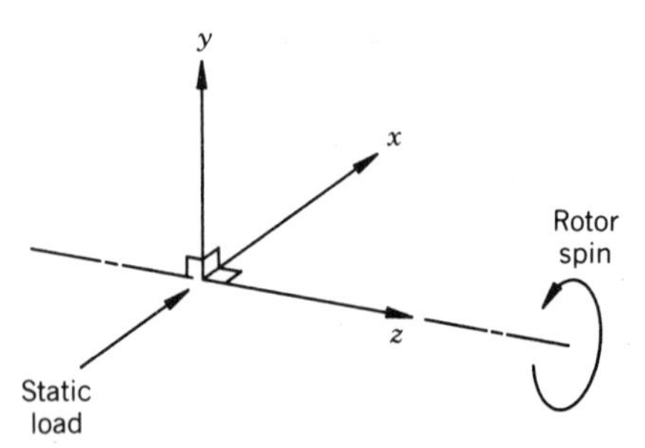

FIGURE 11.12. Bearing-shaft coordinate system.

where $[K_N]$ is the stiffness matrix and $[B_N]$ is the damping matrix. ν is the frequency of vibration. Most commonly, lateral, linear, and angular displacements do not interact with each other, so the nonvanishing portions of $[K_N]$ and $[B_N]$ are separate 2×2 matrices. Thus

$$[K_N] = \begin{bmatrix} [K_N]_{\text{linear}} & 0 & 0 \\ 0 & 0 \; 0 & 0 \\ 0 & 0 \; [K_N]_{\text{angular}} & \end{bmatrix} \tag{11.55}$$

$$[B_N] = \begin{bmatrix} [B_N]_{\text{linear}} & 0 & 0 \\ 0 & 0 \; 0 & 0 \\ 0 & 0 \; [B_N]_{\text{angular}} & \end{bmatrix} \tag{11.56}$$

Here the four 2×2 matrices characterize the support and are constructed as follows:

$$[K]_{\text{linear}} = \begin{bmatrix} K_{xx} & K_{xy} \\ K_{yx} & K_{yy} \end{bmatrix}_{\text{linear}} \tag{11.57}$$

$$[B]_{\text{linear}} = \begin{bmatrix} B_{xx} & B_{xy} \\ B_{yx} & B_{yy} \end{bmatrix}_{\text{linear}} \tag{11.58}$$

$$[K]_{\text{angular}} = \begin{bmatrix} K_{\theta_x\theta_x} & K_{\theta_x\theta_y} \\ K_{\theta_y\theta_x} & K_{\theta_y\theta_y} \end{bmatrix}_{\text{angular}} \tag{11.59}$$

$$[B]_{\text{angular}} = \begin{bmatrix} B_{\theta_x\theta_x} & B_{\theta_x\theta_y} \\ B_{\theta_y\theta_x} & B_{\theta_y\theta_y} \end{bmatrix}_{\text{angular}} \tag{11.60}$$

Figure 11.13 depicts an angular-contact ball bearing arranged with an orthogonal *xyz* coordinate system. The outer ring is assumed fixed in space while the inner ring is permitted to displace with respect to the coordinate system.

For complex loading such as combined radial, axial, and moment loading, the location of the inner ring is defined by the three linear displacements δ_x, δ_y, δ_z and the two angular displacements θ_x, θ_y. The components of the 2×2 stiffness matrices of equations (11.57) and (11.59) then be-

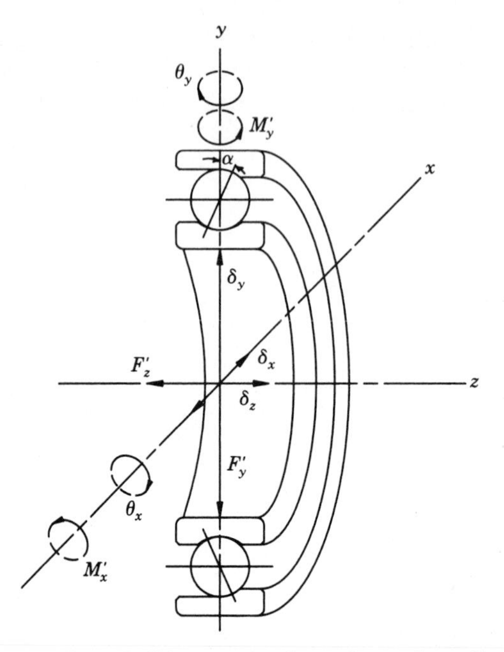

FIGURE 11.13. Basic ball bearing configuration and coordinate system.

come

$$[K]_{\text{linear}} = \begin{vmatrix} \dfrac{\partial F_x}{\partial \delta_x} & \dfrac{\partial F_x}{\partial \delta_y} \\ \dfrac{\partial F_y}{\partial \delta_x} & \dfrac{\partial F_y}{\partial \delta_y} \end{vmatrix} \tag{11.61}$$

$$[K]_{\text{angular}} = \begin{vmatrix} \dfrac{\partial M_x}{\partial \theta_x} & \dfrac{\partial M_x}{\partial \theta_y} \\ \dfrac{\partial M_y}{\partial \theta_x} & \dfrac{\partial M_y}{\partial \theta_y} \end{vmatrix} \tag{11.62}$$

The axial stiffness components are not presented, given that the analysis is commonly lateral. The relationship between the individual rolling element as well as total bearing stiffness and external applied load agrees with Hertzian contact theory and is nonlinear. Figure 11.14 illustrates a one-dimensional load displacement curve.

Since the theories upon which rotor dynamic analyses are built are linear, the bearing stiffnesses are linearized, as shown in Fig. 11.14, by

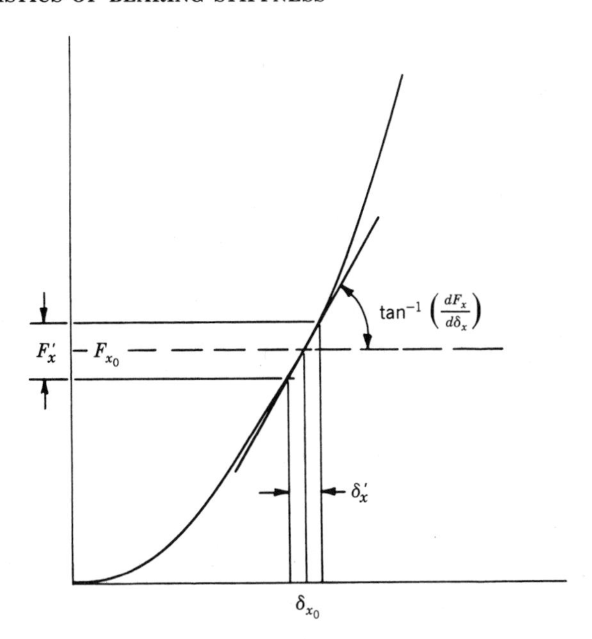

FIGURE 11.14. Linearization method for calculating bearing stiffness.

taking a tangent to the load-deflection function near the point of static equilibrium.

CHARACTERISTICS OF BEARING STIFFNESS

Speed Effects

As already developed, the deflection of a rolling element bearing under load is due to the contact compliances at both the inner and outer raceways. At higher operating speeds centrifugal force acting on a rolling element increases the load at the outer raceway contact, which causes, for a thrust-loaded, angular-contact ball bearing, a load decrease at the inner raceway contact. See Fig. 11.15.

Owing to the nonlinear nature of the load-deflection characteristic and the load-speed behavior described before, the stiffness of a rolling element bearing is speed-dependent. As speed increases, there is an additional effect that occurs simultaneously with the contact load effects already described; the outer raceway contact angle decreases and outer raceway contact stiffness increases while the inner sees an opposite effect.

The radial stiffness components can be formulated in terms of the individual contact stiffnesses as follows:

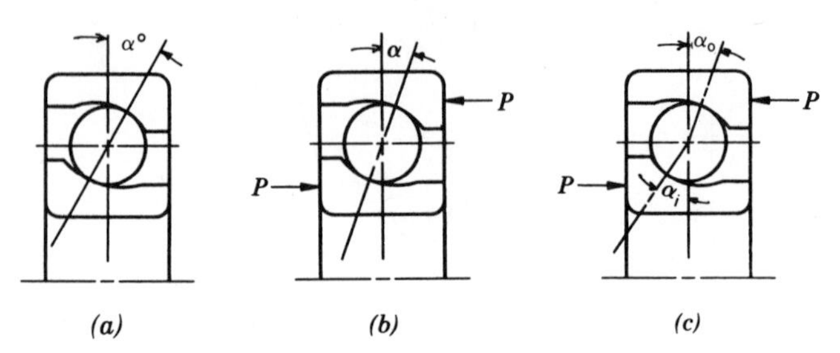

FIGURE 11.15. Angular-contact ball bearing contact angle change due to load and speed. (*a*) Free contact angle (no load). (*b*) Contact angle under load. (*c*) Contact angles under load and high speed.

$$k_{\mathrm{or}} = k_{\mathrm{o}} \cos \alpha_{\mathrm{o}} \tag{11.63}$$

$$k_{\mathrm{ir}} = k_{\mathrm{i}} \cos \alpha_{\mathrm{i}} \tag{11.64}$$

Combining as springs in series yields

$$k_{\mathrm{r}} = \frac{k_{\mathrm{i}} \cos \alpha_{\mathrm{i}}}{1 + (k_{\mathrm{i}}/k_{\mathrm{o}}) \cos \alpha_{\mathrm{i}}/\cos \alpha_{\mathrm{o}}}. \tag{11.65}$$

As a result of the kinematic changes, the increase in α_{i} causes a decrease in the numerator $k_{\mathrm{i}} \cos \alpha_{\mathrm{i}}$. Since this is the dominant term, k_{r} undergoes

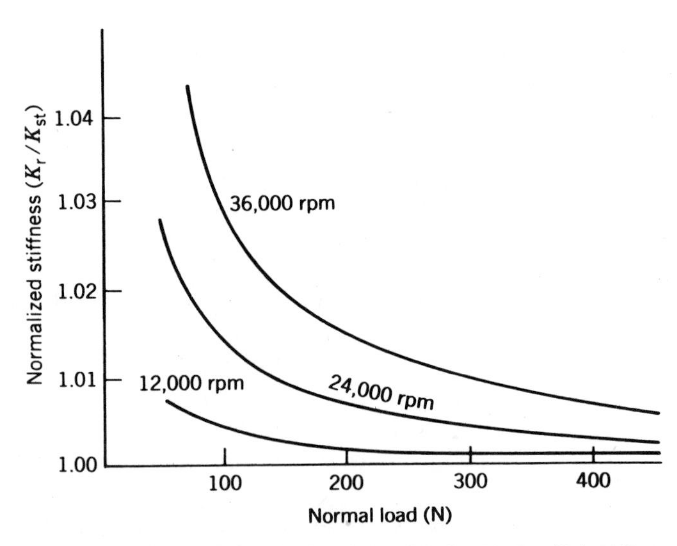

FIGURE 11.16. Effect of load and speed on bearing radial stiffness.

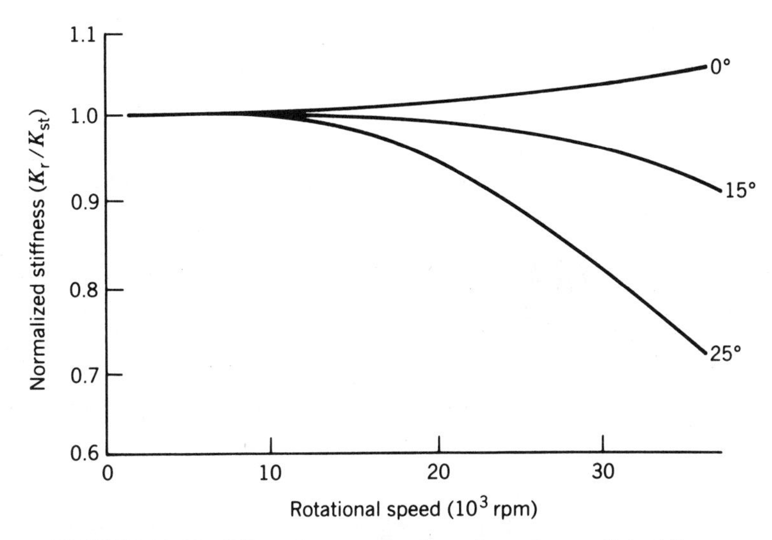

FIGURE 11.17. Effect of contact angle and speed on radial stiffness.

a decrease with increasing speed. Figure 11.17 illustrates the combined effect of load and speed on bearing radial stiffness. Plotted against normal load is a normalized radial stiffness value that is the ratio of radial stiffness at load and speed to the radial stiffness at static conditions.

Contact Angle Effects

From the analytical development of Section 11.6, the interdependence of load and stiffness with contact angle and deflection was established. From these developments it can be shown that

$$k_r = \frac{\partial Q}{\partial \delta_r} = \frac{3}{2} K^{2/3} Q^{1/3} \cos^2 \alpha \tag{11.66}$$

If speed effects are neglected, increasing the contact angle will decrease the bearing radial stiffness; see Fig. 11.18, where, for any given speed, stiffness decreases with increased contact angle. Further, the effect is more dramatic at higher speed.

Preload Effects

A well-documented effect in radial bearings is the effect upon load distribution of internal clearance. Of particular interest to the stiffness characteristics of bearings is the condition of preload or negative internal clearance. Figure 11.18 illustrates the preloading effect on load dis-

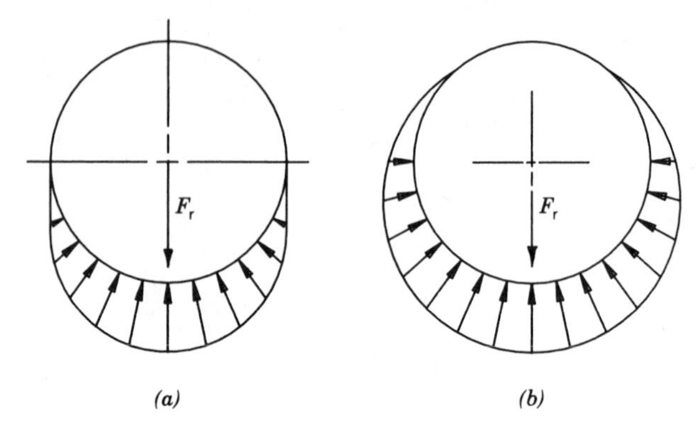

FIGURE 11.18. Load distribution in a radially loaded bearing. (*a*) Zero clearance. (*b*) Negative clearance (preload).

tribution in a radially loaded bearing. As preload is increased (clearance reduced or more negative), the number of elements sharing a given radial load F_r is increased. This lowers the individual contact load and thereby decreases stiffness. Eventually, a preload level will be reached that will create a 100% load zone, thereby creating an increase in stiffness with further preloading.

The combined load-preload effect on bearing radial stiffness warrants further explanation [11.6]. In general, the radial stiffness versus radial load curve for an angular-contact bearing is composed of three distinct behavior regions. Figure 11.19 illustrates this for a typical angular-contact ball bearing with heavy preload. In the region of light radial load

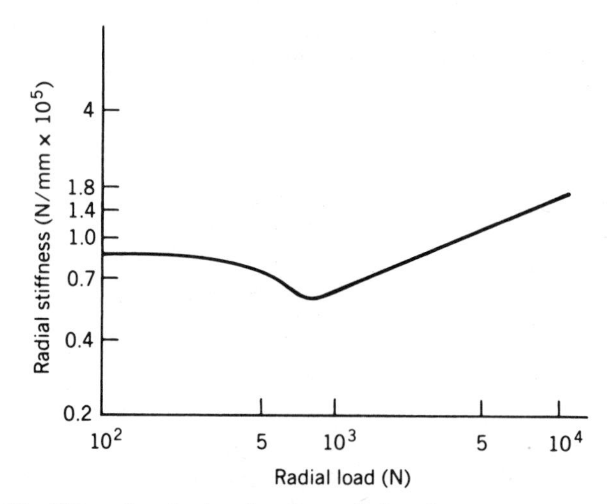

FIGURE 11.19. Effect of preload on bearing radial stiffness (angular-contact ball bearing—heavy preload).

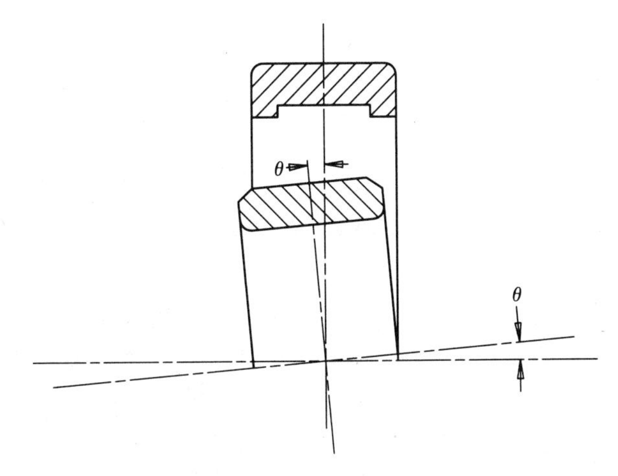

FIGURE 11.20. Misalignment of bearing rings.

the stiffness is nearly constant. The reason is that in this region, the preload is dominant and the radial load is not sufficient to effect a change. In the middle region or region of medium radial load, there is a slight drop in stiffness to some minimum value. Here the radial load is nearly equal to, and then surpasses, the preload. This causes greater load sharing among the elements, therefore reducing total stiffness. The third region shows the effect of the radial load dominating the system, resulting in a linearly increasing stiffness.

Shaft Bending Effects

The interactive nature of the shaft-bearing system is terms of its mechanical behavior dictates that shaft deflections influence bearing deflection and, hence, stiffness. One such shaft deflection is angular deflection or misalignment, which results from the bearing resistance to applied moments. Figure 11.20 illustrates a cylindrical roller bearing experiencing shaft misalignment. It is obvious that shaft bending (misalignment) will influence the load distribution. The general trend is that as shaft stiffness decreases, bending increases, resulting in increased bearing stiffness.

ROTOR DYNAMICS ANALYSIS

Critical Speed

The major objective of rotor dynamics analysis is to allow development of rotating machinery that will be free from vibrational problems detrimental to its performance. This is generally a two-step process consisting of a critical speed analysis and a synchronous response analysis. Critical speed analysis is the process of determining the natural frequencies of a rotor-bearing system and identifying the mode shapes associated with them. This must be done to insure that the machine operating speed is located at a frequency safely spaced from any undamped natural frequencies. If allowed to operate at or near a natural frequency, a machine may enter an unstable and destructive vibration situation.

Synchronous Response

Synchronous response analysis allows the further examination of the rotor-bearing system behavior as a function of operating speed. It relates rotor displacement and bearing loads to operating frequency. Both of the foregoing analytical techniques are accomplished by computer programs constructed using the mathematical principles reviewed in this chapter. In general, the programs mathematically model the shaft-bearing system as a series of rigidly connected, multispan beams supported on springs. Division of the rotor into segments allows for the proper modeling of variations, such as cross-sectional thickness, material, location of gears or discs, and location of bearing supports.

Figure 11.21 is an example of a shaft-bearing system model. Section mass can be either distributed across the element or lumped at the ends of the element. The problem is then solved by a transfer matrix technique. The transfer matrix is derived directly from the differential equation describing the dynamic behavior of each beam segment.

Mode Shapes

Associated with each identified natural frequency (critical speed) is a normalized mode shape. An estimation of the relative displacement of each shaft station is computed. Figure 11.22 illustrates two common bending modes associated with systems involving stiff bearings. These are the shapes with which the system would vibrate if excited at the

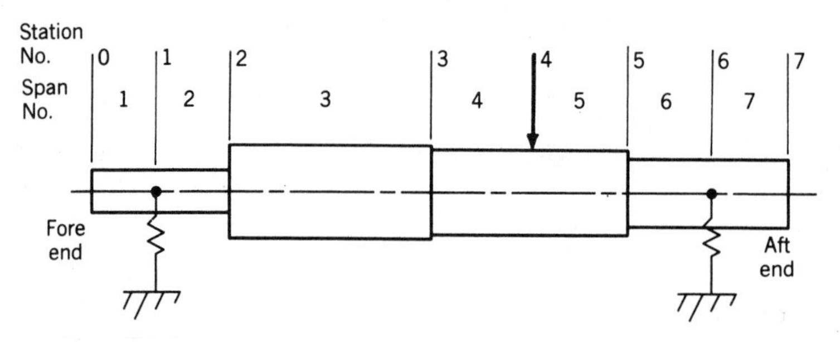

FIGURE 11.21. Model representation of a shaft-bearing system.

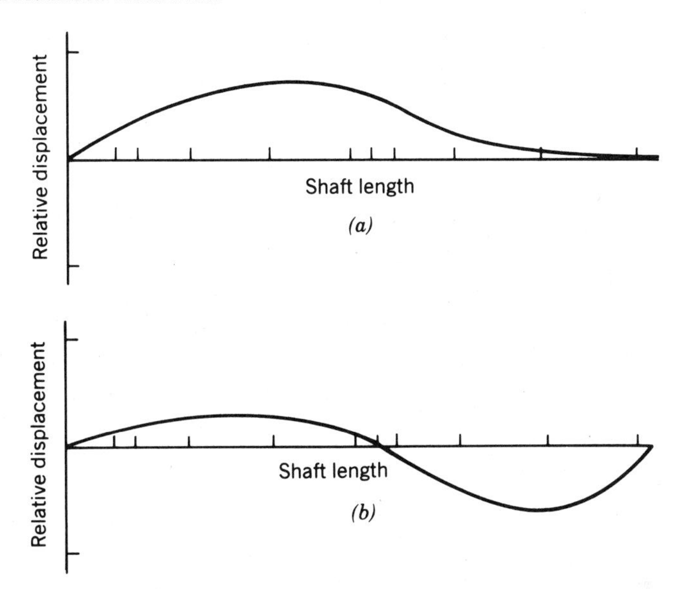

FIGURE 11.22. Typical mode shape results for a flexible rotor-bearing system. (*a*) First bending mode, ω_{c1}. (*b*) Second bending mode, ω_{c2}.

corresponding critical speed. Rotor-bearing systems can be excited into vibration modes by any number of periodic forces. The most common cause of periodic forces in rotating machinery is mass unbalance. Mass unbalance may be a result of inadequate balancing techniques, shaft bending due to gravity, debris deposits, or unstable rotor structures.

Analyzing the effects of unbalance forces on shaft vibration and bearing loads is done by synchronous response analysis. By solving the differential equation of motion for each shaft segment, with the harmonic driving forces being represented by the unbalance mass, vibration amplitude can be computed. Then by sweeping through a range of frequencies (operating speeds) synchronous response plots for each shaft station can be constructed. Figure 11.23 illustrates the typical results of a synchronous response. They identify the location of the critical operating speeds and allow the determination of safe operating frequencies based on severity of vibration amplitude (Fig. 11.23*a*) and bearing load (Fig. 11.22*b*).

Shaft Whirl

Rotating shaft-bearing systems have a natural tendency to form a first bending mode at certain speeds. The shaft motion under these conditions is complex and is referred to as whirling. Whirling is defined as the rotation of the plane made by the bent shaft and the line of centers of the

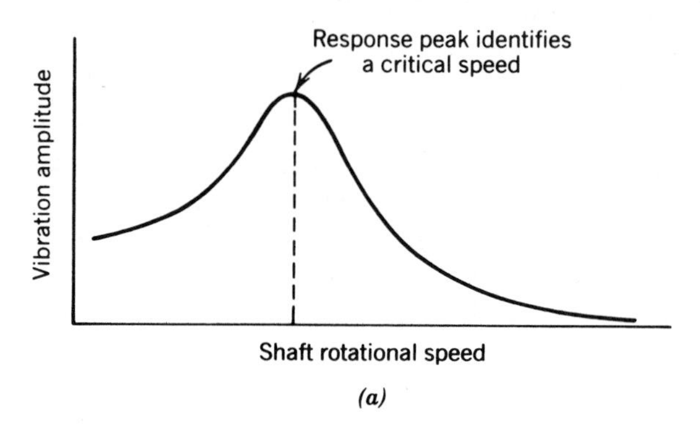

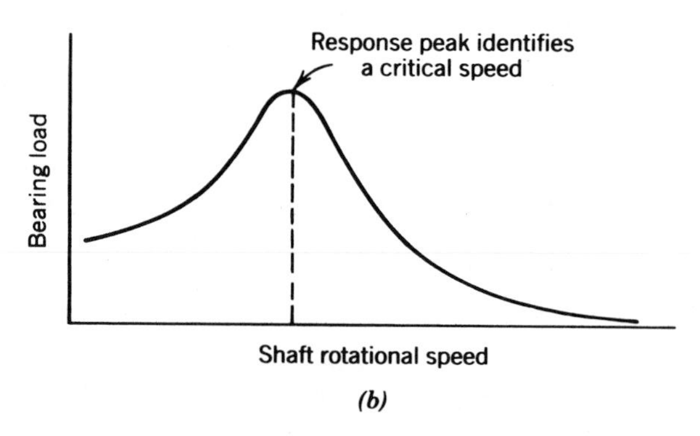

FIGURE 11.23. Typical synchronous response plots for a flexible rotor-bearing system. (*a*) Shaft station vibration amplitude. (*b*) Bearing station load magnitude.

bearings. Whirling is most prevalent in systems with little or no bearing damping, such as systems involving rolling bearings. The shaft whirl may take place in the same direction (natural precession) as that of the rotating shaft or in the opposite direction (reverse precession). In addition, the whirling speed may be equal to (synchronous) or different from (asynchronous) the shaft rotational speed. In general, whirl is a self-excited force. Whirling in lightly damped systems may be the result of internal friction. Studies into this phenomenon [11.7] have pointed out the importance of the use of flexible foundations for reducing the whirl threshold for high-speed systems with rolling bearing supports and thereby achieving rotor stability.

CLOSURE

Understanding the mechanical interactions between the various components (shaft, bearing, seals, impellers, etc.) of rotating machinery is critical to their successful operations. Without incorporating this understanding in the design stages of new machinery, catastrophic failures of shafting and/or bearings could occur. Since digital computers have made complex and repetitive calculations quite simple, engineers and designers have been applying the foundations of mechanical vibrations into useful design tools. Rotor dynamics and the analysis of rotor-bearing systems is an outgrowth of their efforts.

REFERENCES

11.1. J. DenHartog, *Mechanical Vibrations*, 4th ed., McGraw-Hill, New York (1956).

11.2. Air Force Aero Propulsion Laboratory Report AFAPL-TR-65-45, Parts I–X, "Rotor Bearing Dynamics Design Technology," (June 1965).

11.3. Air Force Aero Propulsion Laboratory Report AFAPL-TR-78-6, Parts I–IV, "Rotor Bearing Dynamics Design Technology," (February 1978).

11.4. W. Thomson, *Theory of Vibration with Applications*, 2nd ed., Prentice-Hall, Englewood Cliffs, NJ (1972).

11.5. R. Vierck, *Vibration Analysis*, Harper & Row, New York (1979).

11.6. A. Jones and J. McGrew, "Rotor Bearing Dynamics Design Guide, Part II, Ball Bearings," AFFAPL-TR-78-6, Part II (February 1978).

11.7. E. Gunter, "Dynamic Stability of Rotor-Bearing Systems," NASA SP-113, U.S. Government Printing Office, Washington, DC (1966).

12

MECHANICS OF LUBRICATION IN CONCENTRATED CONTACTS

LIST OF SYMBOLS

Symbol	Description	Units
a	Semimajor axis of elliptical contact area	mm (in.)
A	Viscosity–temperature calculation constants	
b	Semiwidth of rectangular contact area, semiminor axis of elliptical contact area	mm (in.)
C	Lubrication regime and film thickness calculation constants	
D	Roller or ball diameter	mm (in.)
d_m	Pitch diameter of bearing	mm (in.)
E	Modulus of elasticity	N/mm^2 (psi)
E'	$E/(1 - \xi^2)$	N/mm^2 (psi)
f	r/D for ball bearing	
F	Force	N (lb)
F_c	Centrifugal force	N (lb)

Symbol	Description	Units
$\overline{F}$	$F/lE'\mathcal{R}$	
g	Gravitational constant	mm/sec^2 (in./sec^2)
$\mathcal{G}$	$\lambda E'$	
G	Shear modulus	N/mm^2 (psi)
h	Film thickness	mm (in.)
h^0	Minimum film thickness	mm (in.)
H	$h/\mathcal{R}$	
I	Viscous stress integral	
J	Polar moment of inertia per unit length	N · sec^2 (lb · sec^2)
$\overline{J}$	$J/E'\mathcal{R}_i$	mm · sec^2 (in. · sec^2)
k_b	Lubricant thermal conductivity	W/m · °C (Btu/hr · in. · °F)
l	Roller effective length	mm (in.)
L	Factor for calculating film thickness reduction due to thermal effects	
M	Moment	N · mm (in. · lb)
n	Speed	rpm
p	Pressure	N/mm^2 (psi)
Q	Force acting on roller or ball	N (lb)
$\overline{Q}$	$Q/lE'\mathcal{R}$	
R	Cylinder radius	mm (in.)
$\mathcal{R}$	Equivalent radius	mm (in.)
s	rms surface finish (height)	mm (in.)
SSU	Saybolt universal viscosity	sec
t	Time	sec
T	Lubricant temperature	°C, °K (°F, °R)
u	Fluid velocity	mm/sec (in./sec)
U	Entrainment velocity ($U_1 + U_2$)	mm/sec (in./sec)
$\overline{U}$	$\eta_0 U/2E'\mathcal{R}$	
v	Fluid velocity, displacement in y direction	mm/sec, mm (in./sec, in.)
V	Sliding velocity ($U_1 - U_2$)	mm/sec (in./sec)
$\overline{V}$	$\eta_0 V/E'\mathcal{R}$	
w	Deformation in z direction	mm (in.)
y	Distance in y direction	mm (in.)
z	Distance in z direction	mm (in.)
β'	Coefficient for calculating viscosity as a function of temperature	
γ	$D \cos \alpha/d_m$	

Symbol	Description	Units
$\dot{\gamma}$	Lubricant shear rate	sec^{-1}
Γ	Surface roughness orientation parameter	
Δ	Surface roughness parameter	mm (in.)
ϵ	Strain	mm/mm (in./in.)
η	Lubricant viscosity	cp (lb · sec/in.^{2})
η_b	Base oil viscosity (grease)	cp (lb · sec/in.^{2})
η_{eff}	Effective viscosity (grease)	cp (lb · sec/in.^{2})
η_0	Fluid viscosity at atmospheric pressure	cp (lb · sec/in.^{2})
κ	Ellipticity ratio a/b	
λ	Pressure coefficient of viscosity	mm^2/N (in.^{2}/lb)
Λ	Lubricant film parameter	
ν_b	Kinematic viscosity	stokes (cm^2/sec)
ξ	Poisson's ratio	
ρ	Weight density	g/mm^3 (lb/in.^{3})
σ	Normal stress	N/mm^2 (psi)
τ	Shear stress	N/mm^2 (psi)
θ	Angle	rad
$\overline{Y}$	Factor to calculate φ_{TS}	
φ	Film thickness reduction factor	
Φ	Factor to calculate φ_S	
ψ	Angular location of roller	rad
ω	Rotational speed	rad/sec

SUBSCRIPTS

b	Refers to entrance to contact zone
e	Refers to exit from contact zone
G	Refers to grease
i	Refers to inner raceway film
j	Refers to roller location
m	Refers to orbital motion
NN	Refers to non-Newtonian lubricant
o	Refers to outer raceway film
R	Refers to roller
S	Refers to lubricant starvation
SF	Refers to surface roughness (finish)
T	Refers to temperature
TS	Refers to temperature and lubricant starvation
x	Refers to x direction, that is, transverse to rolling
y	Refers to y direction, that is, direction of rolling

Symbol	Description	Units
	SUBSCRIPTS	
z	Refers to z direction	
μ	Refers to rotating raceway	
ν	Refers to nonrotating raceway	
0	Refers to minimum lubricant film	
1, 2	Refers to contacting bodies	

GENERAL

It has been accepted for a long time that ball and roller bearings require fluid lubrication if they are to perform satisfactorily for long periods of time. Although modern rolling bearings in extreme temperature, pressure, and vacuum environment aerospace applications have been adequately protected by dry film lubricants, such as molybdenum disulfide among many others, these bearings have not been subjected to severe demands regarding heavy load and longevity of operation without fatigue. It is further recognized that in the absence of a high temperature environment only a small amount of lubricant is required for excellent performance. Thus many rolling bearings can be packed with greases containing only small amounts of oil and then be mechanically sealed to retain the lubricant. Such rolling bearings usually perform their required functions for indefinitely long periods of time. It is also known that bearings that are lubricated with excessive quantities of oil or grease tend to overheat and "burn" up.

The mechanism of the lubrication of rolling elements operating in concentrated contact with a raceway was not established mathematically until the late 1940s; it was not proven experimentally until the early 1960s. This is to be compared with the existence of hydrodynamic lubrication in journal bearings, which knowledge was established by Reynolds in the 1880s. It is known, for instance, that a fluid film completely separates the bearing surface from the journal or slider surface in a properly designed bearing. Moreover, the lubricant can be oil, water, gas, or some other fluid that exhibits adequate viscous properties for the intended application. In rolling bearings, however, it was only relatively recently established that fluid films could, in fact, separate rolling surfaces subjected to extremely high pressures in the zones of contact. Today, the existence of lubricating fluid films in rolling bearings is substantiated; in many successful applications, these films are effective in completely separating the rolling surfaces. In this chapter, methods will be presented for the calculation of the thickness of lubricating films in rolling bearing applications.

HYDRODYNAMIC LUBRICATION

Reynolds' Equation

Because it appeared possible that lubricant films of significant proportions do occur in the contact zones between rolling elements and raceways under certain conditions of load and speed, several investigators have examined the hydrodynamic action of lubricants on rolling bearings according to classical hydrodynamic theory. Martin [12.1] presented a solution for rigid rolling cylinders as early as 1916. In 1959, Osterle [12.2] considered the hydrodynamic lubrication of a roller bearing assembly.

It is of interest at this stage to examine the mechanism of hydrodynamic lubrication at least in two dimensions. Accordingly, consider an infinitely long roller that is rolling on an infinite plane and is lubricated by an incompressible isoviscous Newtonian fluid having viscosity η. By reference to a Newtonian fluid, it is asserted that the fluid shear stress τ at any point obeys the relationship

$$\tau = -\eta \frac{\partial u}{\partial z} \tag{12.1}$$

in which $\partial u/\partial z$ is the local fluid velocity gradient in the z direction (see Fig. 12.1). Because the fluid is viscous, fluid inertia forces are small compared to the viscous fluid forces. Hence a particle of fluid is subjected only to fluid pressure and shear stresses as shown in Fig. 12.2

Noting the stresses of Fig. 12.2 and recognizing that static equilibrium exists, the sum of the forces in any direction must equal zero. Therefore,

$$\Sigma F_y = 0$$

$$p\,dz - \left(p + \frac{\partial p}{\partial y}\,dy\right) dz + \tau\,dy - \left(\tau + \frac{\partial \tau}{\partial z}\,dz\right) dy = 0$$

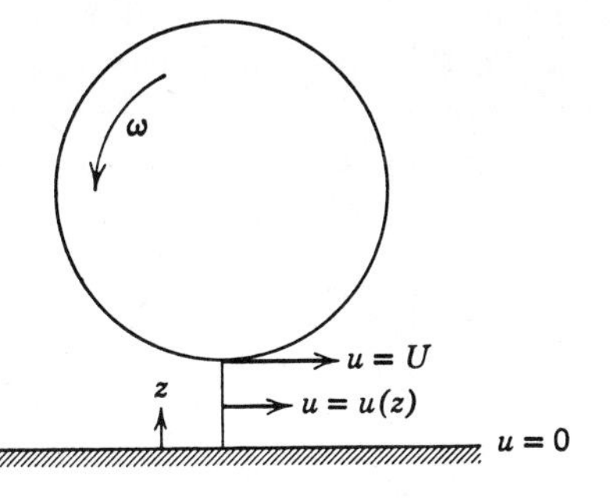

FIGURE 12.1. Cylinder rolling on a plane with lubricant between cylinder and plane.

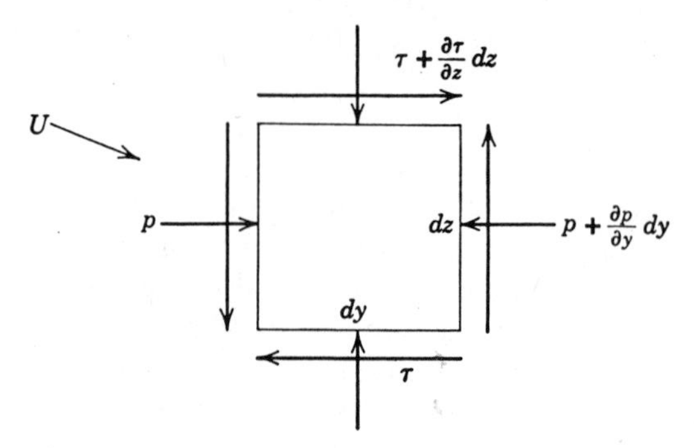

FIGURE 12.2. Stresses on a fluid particle in a two-dimensional flow field.

and

$$\frac{\partial p}{\partial y} = -\frac{\partial \tau}{\partial z} \tag{12.2}$$

Differentiating equation (12.1) once with respect to z yields

$$\frac{\partial \tau}{\partial z} = -\eta \frac{\partial^2 u}{\partial z^2} \tag{12.3}$$

Substituting equation (12.3) into (12.2) one obtains

$$\frac{\partial p}{\partial y} = \eta \frac{\partial^2 u}{\partial z^2} \tag{12.4}$$

Assuming for the moment that $\partial p/\partial y$ is constant, equation (12.4) may be integrated twice with respect to z. This procedure gives the following expression for local fluid velocity u:

$$u = \frac{1}{2\eta} \frac{\partial p}{\partial y} z^2 + c_1 z + c_2 \tag{12.5}$$

The velocity U may be ascribed to the fluid adjacent to the plane that translates relative to a roller. At a point on the opposing surface, it is proper to assume that $u = 0$, that is, at $z = 0$, $u = U$ and at $z = h$, $u = 0$. Substituting these boundary conditions into equation (12.5), it can be determined that

$$u = \frac{1}{2\eta} \frac{\partial p}{\partial y} z(z - h) + U\left(1 - \frac{z}{h}\right) \tag{12.6}$$

in which h is the film thickness.

Considering the fluid velocities surrounding the fluid particle as shown by Fig. 12.3, one can apply the law of continuity of flow in steady state, that is, mass influx equals mass efflux. Hence, since density is constant for an incompressible fluid

$$u\,dz - \left(u + \frac{\partial u}{\partial y}\,dy\right) dz + v\,dy - \left(v + \frac{\partial v}{\partial z}\,dz\right) dy = 0 \qquad (12.7)$$

Therefore,

$$\frac{\partial u}{\partial y} = -\frac{\partial v}{\partial z} \qquad (12.8)$$

Differentiating equation (12.6) with respect to y and equating this to equation (12.8) yields

$$\frac{\partial v}{\partial z} = -\frac{\partial}{\partial y}\left[\frac{1}{2\eta}\frac{\partial p}{\partial y}\,z(z-h) + U\left(1 - \frac{z}{h}\right)\right] \qquad (12.9)$$

Integrating equation (12.9) with respect to z,

$$\int \frac{\partial v}{\partial z}\,dz = -\int_0^h dv = 0 = \int_0^h \frac{\partial}{\partial y}\left[\frac{1}{2\eta}\frac{\partial p}{\partial y}\,z(z-h) + U\left(1 - \frac{z}{h}\right)\right] dz \qquad (12.10)$$

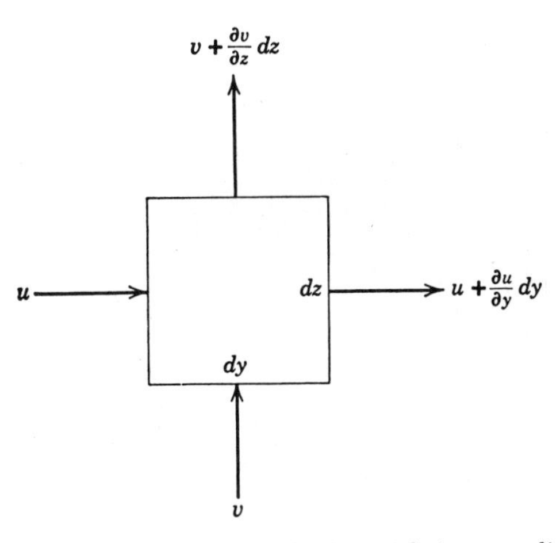

FIGURE 12.3. Velocities associated with a fluid particle in a two-dimensional flow field.

and

$$\frac{\partial}{\partial y}\left(h^3 \frac{\partial p}{\partial y}\right) = 6\eta U \frac{\partial h}{\partial y} \tag{12.11}$$

Equation (12.11) is commonly called the Reynolds equation in two dimensions.

Film Thickness

To solve the Reynolds equation, it is only necessary to evaluate film thickness as a distinct function of y, that is, $h = h(y)$. For a cylindrical roller near a plane as shown by Fig. 12.4, it can be seen that

$$h = h^0 + \frac{y^2}{2R} \tag{12.12}$$

in which h^0 is the minimum film thickness.

Substituting equation (12.12) into (12.11) gives

$$\frac{\partial}{\partial y}\left[\left(h^0 + \frac{y^2}{2R}\right)^3 \frac{\partial p}{\partial y}\right] = \frac{6\eta U y}{R} \tag{12.13}$$

Equation (12.13) varies only in y; hence

$$\frac{d}{dy}\left[\left(h^0 + \frac{y^2}{2R}\right)^3 \frac{dp}{dy}\right] = \frac{6\eta U y}{R} \tag{12.14}$$

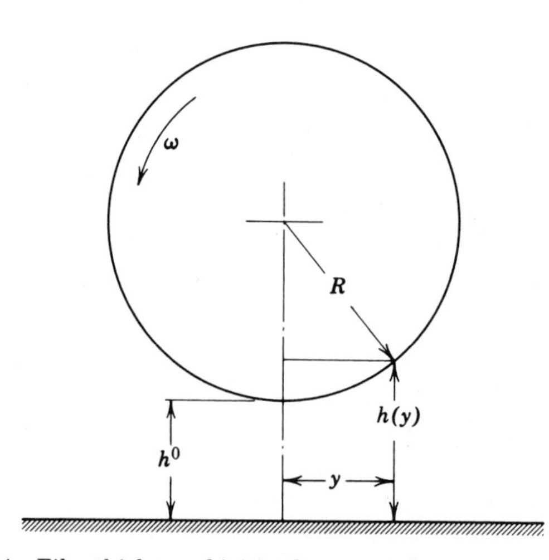

FIGURE 12.4. Film thickness $h(y)$ in the contact between a roller and plane.

Pressure Distribution

The integration of equation (12.14) to determine $p(y)$ is rather intricate. This equation was integrated by Sternlicht et al. [12.3] as follows:

$$p = 6\eta U \frac{(2Rh^0)^{1/2}}{(h^0)^2}\left[\frac{\beta}{2} + \frac{\sin 2\beta}{4} - \frac{\pi}{4} - K\left(\frac{3\beta}{8} - \frac{3\pi}{16} + \frac{\sin 2\beta}{4} + \frac{\sin 4\beta}{32}\right)\right] \tag{12.15}$$

and

$$p_{\max} = 0.76\eta U (2R)^{1/2} (h^0)^{-3/2} \tag{12.16}$$

In equation (12.15), $\tan \beta = y/(2Rh^0)^{1/2}$ and K is a constant depending on the boundary condition. If both surfaces are construed to be rotating cylinders, then

$$U = U_1 + U_2 \tag{12.17}$$

in which subscripts 1 and 2 refer to the respective cylinders. Moreover, an equivalent cylinder radius $\mathcal{R}$ is defined as

$$\mathcal{R} = (R_1^{-1} + R_2^{-1})^{-1} \tag{12.18}$$

Note that for an outer raceway R_2^{-1} is negative.

To find the load which will be carried,

$$Q' = \int p(y)\, dy \tag{12.19}$$

in which Q' is the supported load per unit length of the cylinder.

Unfortunately, for all the complexity of the foregoing analysis, the film thicknesses indicated by subsequent calculations are far smaller than the composite surface roughness achievable on rolling bearings. Thus, it would appear that hydrodynamic lubrication by an isoviscous fluid alone cannot explain the existence of a satisfactory fluid film in the contact zone between rolling element and raceway. It would not be proper to state that the foregoing analysis is worthless since it does adequately represent a limiting case which can be achieved at light loads.

Pressure and Friction Forces

Dowson and Higginson [12.4] used a similar analysis to evaluate the friction forces that occur in a lightly loaded roller bearing. They reasoned

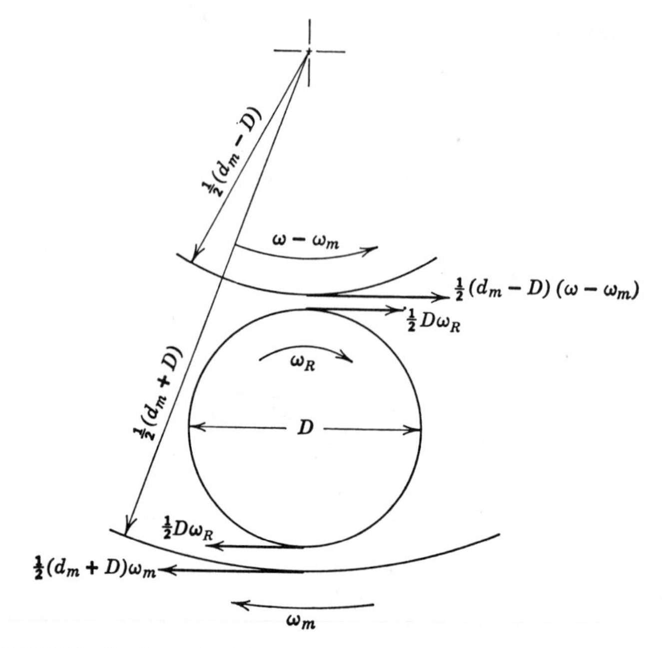

FIGURE 12.5. Surface velocities and rotational speeds of raceways and roller.

that since the width of the contact surface is extremely small compared to the length, the assumption of negligible end leakage could be supported and the Reynolds equation in two dimensions was adequate. Considering a roller between inner and outer raceways, they used the model demonstrated by Figs. 12.5 and 12.6. In Fig. 12.5, ω is the rotational speed of the inner raceway, ω_R is the speed of the roller about its own axis, and ω_m is the orbital speed of the roller. In Fig. 12.6, F_i is the viscous friction force at the inner raceway contact, F_o is the viscous friction force at the outer raceway contact, Q_{yi} and Q_{yo} are the hydrodynamic pressure forces acting on the roller and Q_{zi} and Q_{zo} are the normal forces acting on the roller. All forces are based on a roller of unit length. It should be remembered that Q_{zi} may be determined from a static load distribution analysis as specified in Chapter 6.

If one refers to an equivalent cylinder as shown by Fig. 12.7, the forces acting on the cylinder can be determined as follows:

$$Q_{yo} = 0 \tag{12.20}$$

$$Q_{yh} = A\eta(U_o + U_h)\left(\frac{\Re}{h^0}\right)^{1/2} \tag{12.21}$$

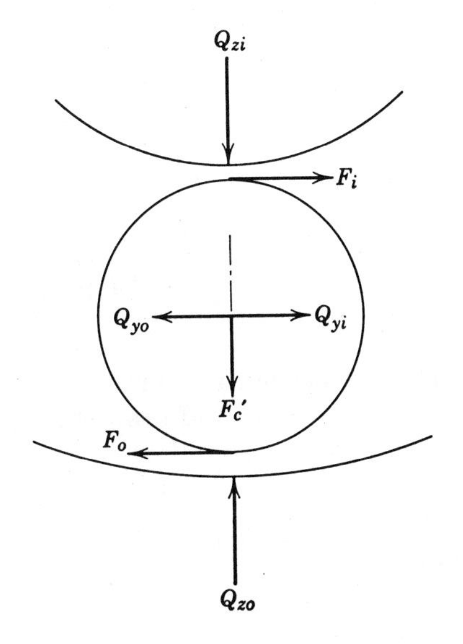

FIGURE 12.6. Forces acting on a roller in a bearing.

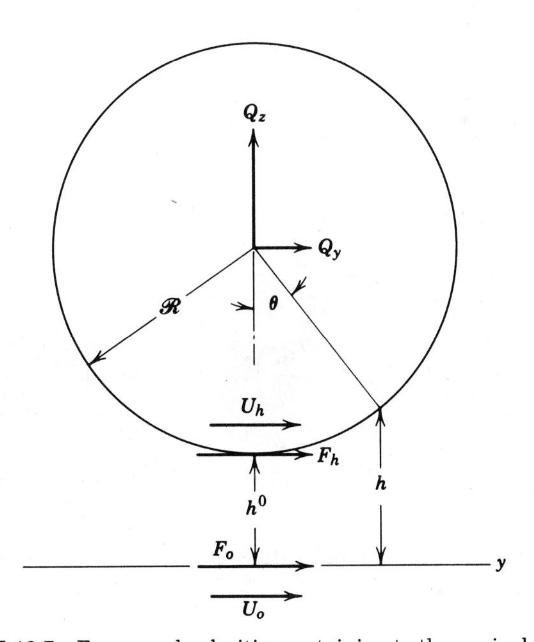

FIGURE 12.7. Forces and velocities pertaining to the equivalent roller.

$$Q_z = B\eta(U_o + U_h)\left(\frac{\mathcal{R}}{h^0}\right)^{1/2} \tag{12.22}$$

$$F_o = -\frac{Q_{yh}}{2} - C\eta(U_o + U_h)\left(\frac{\mathcal{R}}{h^0}\right)^{1/2} \tag{12.23}$$

$$F_h = -\frac{Q_{yh}}{2} + C\eta(U_o + U_h)\left(\frac{\mathcal{R}}{h^0}\right)^{1/2} \tag{12.24}$$

As stated by Dowson and Higginson [12.4] the values of the constants A, B, and C are determined from the boundary conditions and are dependent upon $\mathcal{R}/h^0$. The boundary conditions assumed are as follows:

1. Pressure generation commences at $\theta = -\pi/2$.
2. Pressure generation terminates in the divergent film where $p = 0$ and $dp/dy = 0$.

For values of $10^2 < \mathcal{R}/h^0 < 10^6$, A was set equal to 4.5, $B = 2.45$, and $C = 3.48$.

Referring once again to the model of Fig. 12.6, the following force components were estimated:

$$Q_{yi} = 4.5(1 - \gamma)\eta U_i \left(\frac{\mathcal{R}}{h^0}\right)_i^{1/2} \tag{12.25}$$

$$Q_{yo} = 4.5(1 + \gamma)\eta U_o \left(\frac{\mathcal{R}}{h^0}\right)_o^{1/2} \tag{12.26}$$

$$Q_{zi} = 2.45\eta U_i \left(\frac{\mathcal{R}}{h^0}\right)_i^{1/2} \tag{12.27}$$

$$Q_{zo} = 2.45\eta U_o \left(\frac{\mathcal{R}}{h^0}\right)_o^{1/2} \tag{12.28}$$

$$F_i = -2.25\eta U_i \left(\frac{\mathcal{R}}{h^0}\right)_i^{1/2} + 3.48\eta V_i \left(\frac{\mathcal{R}}{h^0}\right)_i^{1/2} \tag{12.29}$$

$$F_o = -2.25\eta U_o \left(\frac{\mathcal{R}}{h^0}\right)_o^{1/2} + 3.48\eta V_o \left(\frac{\mathcal{R}}{h^0}\right)_o^{1/2} \tag{12.30}$$

In equations (12.25)–(12.30), $\mathcal{R}$ is the equivalent roller radius as defined by equation (12.18). Thus

$$\mathcal{R}_i = \frac{D}{2}(1 - \gamma) \tag{12.31}$$

$$\mathcal{R}_o = \frac{D}{2}(1 + \gamma) \tag{12.32}$$

Also U is the entrainment velocity, which is equal to

$$U_i = \frac{d_m}{2}[(1 - \gamma)(\omega - \omega_m) + \gamma\omega_R] \tag{12.33}$$

$$U_o = \frac{d_m}{2}[(1 + \gamma)\omega_m + \gamma\omega_R] \tag{12.34}$$

V is the sliding velocity which is equal to

$$V_i = \frac{d_m}{2}[(1 - \gamma)(\omega - \omega_m) - \gamma\omega_R] \tag{12.35}$$

$$V_o = \frac{d_m}{2}[(1 + \gamma)\omega_m - \gamma\omega_R] \tag{12.36}$$

Considering the forces acting on the roller of Fig. 12.6, one may apply the conditions of static equilibrium such that

$$\Sigma F_y = 0$$

$$Q_{yi} + F_i - Q_{yo} - F_o = 0 \tag{12.37}$$

$$\Sigma F_z = 0$$

$$Q_{zo} - Q_{zi} - F_c' = 0 \tag{12.38}$$

$$\Sigma M_o = 0$$

$$F_i + F_o = 0 \tag{12.39}$$

It is recognized that centrifugal force per unit length is given by

$$F_c' = \frac{m\omega_m^2 d_m}{2l} \tag{12.40}$$

Equations (12.37)–(12.39) can be solved for ω_R, ω_m, and $(h^0)_o$, since Q_{zi} is known, and Q_{zo} is dependent on Q_{zi} and ω. It is apparent from equation (12.27) that

$$(h^0)_i^{1/2} = \frac{2.45\,\eta U_i \mathcal{R}_i^{1/2}}{Q_{zi}} \tag{12.41}$$

Thus, in the foregoing analysis, minimum film thickness varies inversely with load.

VISCOSITY VARIATION WITH PRESSURE

According to the sample calculations of compressive contact stress in Chapter 5, the normal pressure between contacting rolling bodies is likely to be of the magnitude of 689.8 N/mm^2 (100,000 psi) and higher. In normal fluid film bearing applications pressures of this magnitude do not exist and, consequently, it is usually assumed that viscosity is virtually unaffected by pressure. Figure 12.8 shows some experimental data on viscosity variation with pressure for different bearing lubricants. It is to be noted that fluid viscosity is an exponential function of pressure such that between the contacting surfaces in a loaded rolling bearing assembly, viscosity is likely to be 10,000 times its base value at zero pressure.

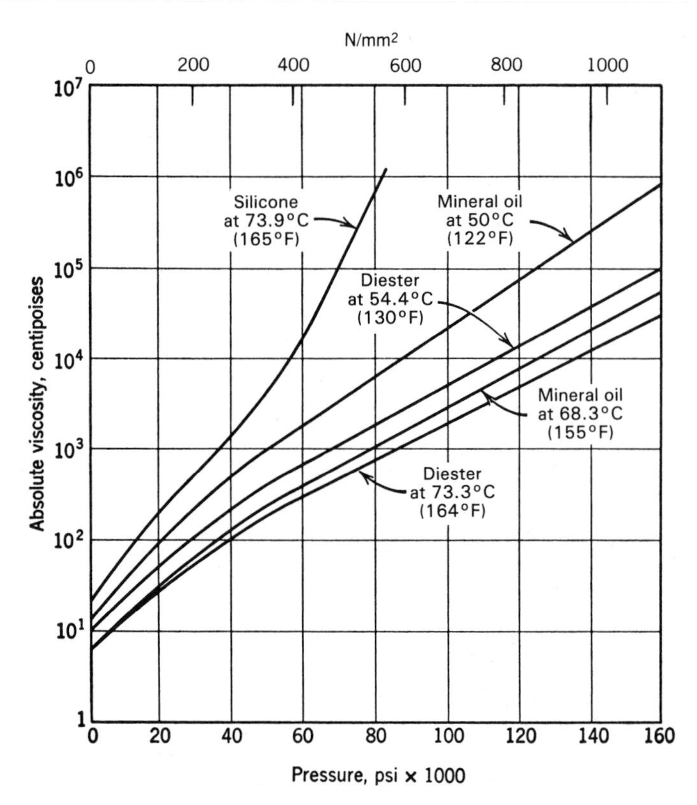

FIGURE 12.8. Pressure viscosity of lubricants (ASME data [12.15]).

In using this viscosity variation with pressure, scientists have approximated the curve as follows:

$$\eta = \eta_0 e^{\lambda p} \tag{12.42}$$

An expression for λ for mineral oils is obtained from the data of reference [12.15]:

$$\lambda = 7.74 \times 10^{-4} \left(\frac{\nu_0}{10^4}\right)^{0.163} \tag{12.43}$$

Hence equation (12.4) becomes

$$\frac{\partial p}{\partial y} e^{-\lambda p} = \eta_0 \frac{\pi^2 u}{\partial z^2} \tag{12.44}$$

Kapitsa [12.5] conducted an analysis similar to that of Osterle [12.2], except that the pressure variation of viscosity was considered. Others who obtained similar results are Sternlicht et al. [12.3]. They stated the following formulation for pressure distribution in the direction of rolling:

$$p = -\frac{1}{\lambda} \ln \left\{ 1 - \frac{6\lambda\eta_0 U}{(h^0)^2} (2\mathcal{R}h^0)^{1/2} \times \left[\frac{\beta}{2} + \frac{\sin 2\beta}{4} - \frac{\pi}{4} - K\left(\frac{3\beta}{8} - \frac{3\pi}{16} + \frac{\sin 2\beta}{4} + \frac{\sin 4\beta}{32}\right)\right]\right\} \tag{12.45}$$

in which $\tan \beta = y/(2\mathcal{R}h^0)^{1/2}$ and K is a constant depending on the boundary conditions. Kapitsa [12.5] developed the following formulation for film thickness:

$$h^0 = [6U\lambda\eta_0(2\mathcal{R}^{1/2})\, p']^{2/3} \tag{12.46}$$

in which p' is defined:

$$p' = \int \frac{Y^2 - m_1^2}{(Y^2 + 1)^3}\, dY \tag{12.47}$$

$$Y = \frac{y}{(2\mathcal{R}h^0)^{1/2}} \tag{12.48}$$

and

$$m_1 = l/(2\mathcal{R}h^0)^{1/2} \tag{12.49}$$

Although analyses of film thickness utilizing variation of fluid viscosity with pressure represent an improvement over those that do not consider this effect, film thicknesses so determined are still too thin to be of any consequence in explaining rolling bearing lubrication.

ISOTHERMAL ELASTOHYDRODYNAMIC LUBRICATION

Deformation of Contact Surfaces

Because of the fluid pressures present between contacting rolling bodies causing the increases in viscosity noted on Fig. 12.8, it is certain that the rolling surfaces deform appreciably in proportion to the thickness of a fluid film between the surfaces. The combination of the deformable surface with the hydrodynamic lubricating action constitutes the "elastohydrodynamic" problem. The solution of this problem established the first feasible analytical means of estimating the thickness of fluid films, the local pressures, and the tractive forces that occur in rolling bearings.

Dowson and Higginson [12.7] for the model of Fig. 12.7 used the following formulation for film thickness at any point in the contact:

$$h = h^0 + \frac{y^2}{2R_1} + \frac{y^2}{2R_2} + w_1 + w_2 \tag{12.50}$$

Solid displacements w are calculated for a semiinfinite solid in a condition of plane strain. Since the width of the loaded zone is extremely small compared to the dimensions of the contacting bodies, an approximation that $w_1 = w_2$ is valid. Hence for the equivalent cylinder radius

$$\mathcal{R} = (R_i^{-1} + R_o^{-1})^{-1} \tag{12.18}$$

the film thickness is given by

$$h = h^0 + \frac{y^2}{2\mathcal{R}} + w \tag{12.51}$$

To solve the plane strain problem, the following stress function was assumed:

$$\Phi = -\frac{Q}{\pi} y \tan^{-1} \frac{y}{z} \tag{12.52}$$

Using this stress function, the stresses due to a narrow strip of pressure over the width ds in the y direction are determined as follows:

$$\sigma_y = -\frac{2y^2zp\,ds}{\pi(y^2 + z^2)^2} \tag{12.53}$$

$$\sigma_z = -\frac{2z^3p\,ds}{\pi(y^2 + z^2)^2} \tag{12.54}$$

$$\tau_{yz} = -\frac{2yz^2p\,ds}{\pi(y^2 + z^2)^2} \tag{12.55}$$

σ_y and σ_z are normal stresses and τ_{yz} is the shear stress. By Hooke's law, the strains are given by

$$\epsilon_y = \frac{(1 - \xi^2)\sigma_y}{E} - \frac{\xi(1 + \xi)\sigma_z}{E} \tag{12.56}$$

$$\epsilon_z = \frac{(1 - \xi^2)\sigma_z}{E} - \frac{\xi(1 + \xi)\sigma_y}{E} \tag{12.57}$$

$$\epsilon_{yz} = \frac{2(1 + \xi)\tau_{yz}}{E} = \frac{\tau_{yz}}{G} \tag{12.58}$$

in which G is the shear modulus of elasticity and ξ is Poisson's ratio. In plain strain

$$\epsilon_y = \frac{\partial v}{\partial y}, \qquad \epsilon_z = \frac{\partial w}{\partial z}, \qquad \text{and} \qquad \epsilon_{yz} = \frac{\partial v}{\partial z} + \frac{\partial w}{\partial y}$$

Using these relationships, and equations (12.53)–(12.58), it can be established that at the surface, that is, at $z = 0$:

$$w = -\frac{2(1 - \xi^2)}{\pi E}\int_{S_1}^{S_2} p \ln(y - S)\,dS + \text{constant} \tag{12.59}$$

To solve for w, Dowson et al. [12.7] divided the pressure curve into segments and represented the pressure thereunder by

$$p = \zeta_1 + \zeta_2 S + \zeta_3 S^2 \tag{12.60}$$

in which ζ_1, ζ_2, and ζ_3 are constants for that segment. Using p in this form, equation (12.59) can be integrated to obtain surface deformation. This procedure, of course, is used for an assumed pressure distribution.

To obtain h^0, the Reynolds equation is used in accordance with the pressure variation of viscosity.

$$\frac{d}{dy}\left(h^3 e^{-\lambda p}\frac{dp}{dy}\right) = -6\eta_0 U\frac{dh}{dy} \tag{12.61}$$

Performing the indicated differentiation and rearranging yields:

$$h^3 e^{-\lambda p}\left[\frac{d^2p}{dy^2} - \lambda\left(\frac{dp}{dy}\right)^2\right] + \frac{dh}{dy}\left(6u\eta_0 + 3h^2 e^{-\lambda p}\frac{dp}{dy}\right) = 0 \quad (12.62)$$

At the inlet and at the outlet of the contact:

$$\frac{d^2p}{dy^2} - \lambda\left(\frac{dp}{dy}\right)^2 = 0 \quad (12.63)$$

such that equation (12.62) becomes

$$\frac{dh}{dy}\left(6U\eta_0 + 3h^2 e^{-\lambda p}\frac{dp}{dy}\right) = 0 \quad (12.64)$$

At the outlet end of the pressure curve $dh/dy = 0$. This condition applies to the point of minimum film thickness. At the inlet, equation (12.64) is solved by

$$\frac{dp}{dy} = -\frac{2\eta_0 e^{\lambda p} U}{h^2} \quad (12.65)$$

Thus, if viscosity and speed are known, the value of h for the point at which equation (12.64) is satisfied in the inlet region can be evaluated for a given pressure curve. Solving (12.65) for h_b (at inlet) gives

$$h_b = \left[-\frac{2\eta_0 e^{\lambda p} U}{(dp/dy)_b}\right]^{1/2} \quad (12.66)$$

Once h_b has been determined, then the entire film shape can be estimated by using the integrated form of the Reynolds equation, that is,

$$\frac{dp}{dy} = -6\eta_0 e^{\lambda p} U\left(\frac{1}{h^2} - \frac{h_e}{h^2}\right) \quad (12.67)$$

Substitution of dp/dy from equation (12.65) for the point at which $h = h_b$ determines that $h_e = 2h_b/3$. At other positions y, film thickness h may be determined from the following cubic equation developed from (12.67)

$$\frac{dp/dy}{6\eta_0 e^{\lambda p} U} h^3 + h - h_e = 0 \quad (12.68)$$

At the point of maximum pressure $dp/dy = 0$ and equation (12.62) becomes

$$\frac{dh}{dy} = -\frac{h^3}{6\eta_0 e^{\lambda p} U} \frac{d^2 p}{dy^2} \tag{12.69}$$

In cases of most interest the pressure curve is predominantly Hertzian such that

$$p = p_0 \left[1 - \left(\frac{y}{b} \right)^2 \right]^{1/2} \tag{12.70}$$

in which p_0 is the maximum pressure and b is the semiwidth of the contact zone.

Thus at $y = 0$, $p = p_0$ and equation (12.69) becomes

$$\frac{dh}{dy} = \frac{p_0 h^3}{3\eta_0 e^{\lambda p_0} U b^2} \tag{12.71}$$

Consequently, if h is small (as it must be in a rolling bearing under load) and viscosity is high (as it will become because of high pressure), dh/dy is very small and the film is essentially of uniform thickness. This result is shown by Dowson et al. [12.7], and also by Grubin [12.8].

Pressure and Stress Distribution

In a later presentation Dowson et al. [12.9] indicated that dimensionless film thickness $H = h/\mathcal{R}$ could be expressed as follows:

$$H = f(\overline{Q}_z, \overline{U}, G) \tag{12.72}$$

in which

$$\overline{Q}_z = \frac{Q_z}{lE'\mathcal{R}} \tag{12.73}$$

$$\overline{U} = \frac{\eta_0 U}{2E'\mathcal{R}} \tag{12.74}$$

$$\mathcal{G} = \lambda E' \tag{12.75}$$

$$E' = \frac{E}{1 - \xi^2} \tag{12.76}$$

In reference [12.9], Dowson et al. presented the results shown by Figs. 12.9 and 12.10 for $\mathcal{G} = 2500$ and 5000. $\mathcal{G} = 2500$ corresponds to bronze rollers lubricated by mineral oil and $\mathcal{G} = 5000$ corresponds to steel rollers lubricated by mineral oil. The load $\overline{Q}_z = 0.00003$ corresponds to ap-

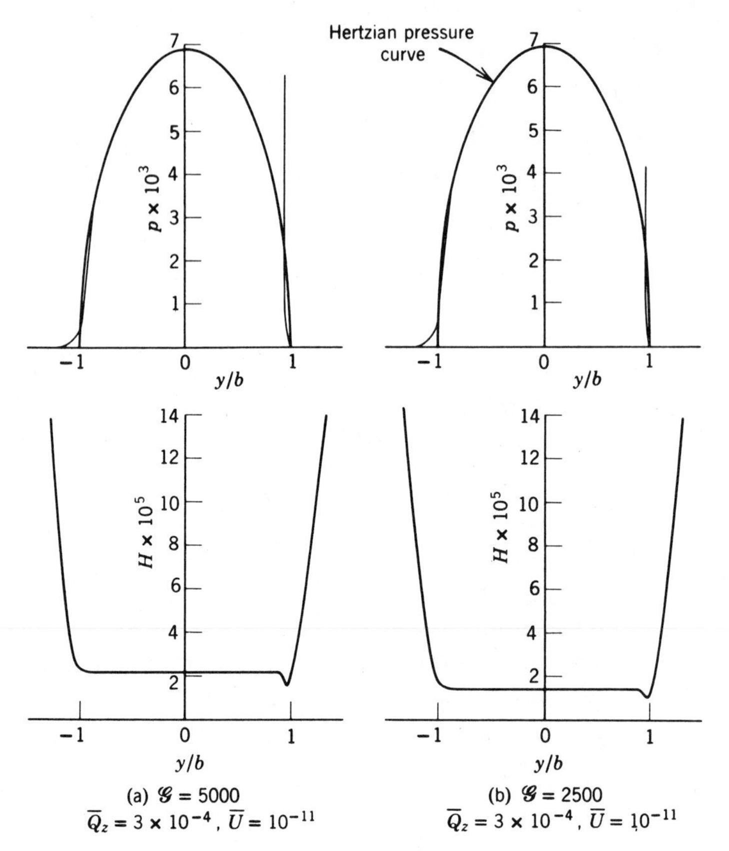

FIGURE 12.9. Pressure distribution and film thickness for high load conditions (reprinted from [12.9] by permission of the Institution of Mechanical Engineers).

proximately 483 N/mm^2 (70,000 psi) and $\overline{Q}_z = 0.0003$ corresponds to 1380 N/mm^2 (200,000 psi) approximately. Dimensionless speed $\overline{U} = 10^{-11}$ corresponds to surface velocities on the order of 1524 mm/sec (5 ft/sec) for an equivalent roller radius of 25.4 mm (1 in.) operating in mineral oil.

Note from Figs. 12.9 and 12.10 that the departure from the Hertzian pressure distribution is less significant as load increases. The second pressure peak at the outlet end of the contact corresponds to a local decrease in the film thickness at that point. Otherwise, the film is essentially of uniform thickness. The latter condition was confirmed by tests conducted by Sibley and Orcutt [12.10].

Additionally, Dowson et al. [12.9] demonstrated the effect of the distorted pressure distribution on maximum subsurface shear stress. Figure 12.11 shows contours of $\tau_{yz_{\max}}/p_{\max}$. Note that the shear stress increases in the vicinity of the second pressure peak and tends toward the surface. This condition was indicated in Chapter 5.

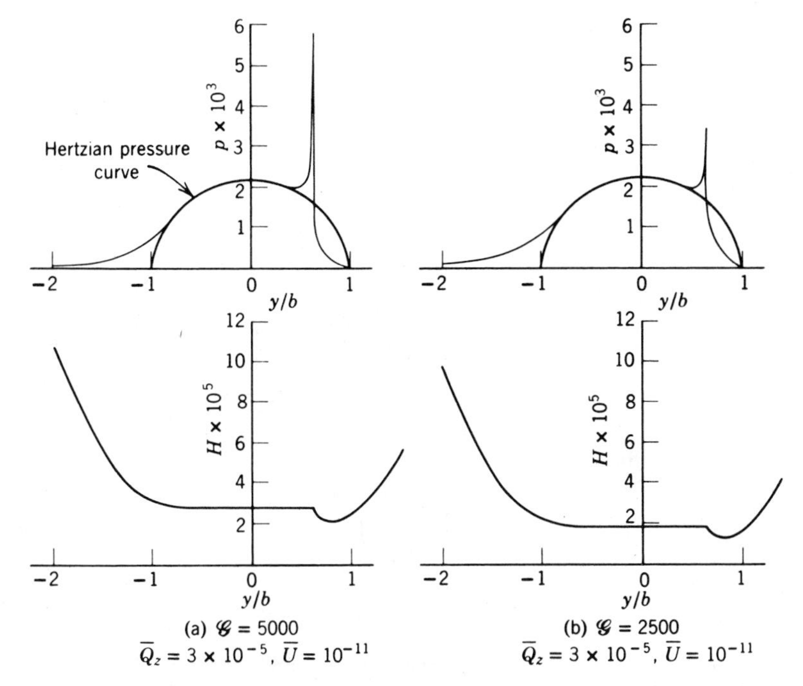

FIGURE 12.10. Pressure distribution and film thickness for light load conditions (reprinted from [12.9] by permission of the Institution of Mechanical Engineers).

Lubricant Film Thickness

Grubin [12.8] developed a formula for minimum film thickness in line contact, that is, the thickness of the lubricant film between the protuberance at the trailing edge of the contact on the equivalent roller surface and the opposing surface of the relative flat. The Grubin formula is based on the assumption that the rolling surfaces deform as if dry contact occurs and is given in dimensionless format.

$$H^0 = \frac{1.95(\mathcal{G}\overline{U})^{0.727}}{\overline{Q}_z^{0.091}} \tag{12.77}$$

where $H^0 = h^0/\mathcal{R}_y$.

Based upon analytical studies and experimental results, Dowson et al. [12.11] established the following formula to calculate the minimum film thickness:

$$H^0 = \frac{2.65\,\overline{U}^{0.7}\,\mathcal{G}^{0.54}}{\overline{Q}_z^{0.13}} \tag{12.78}$$

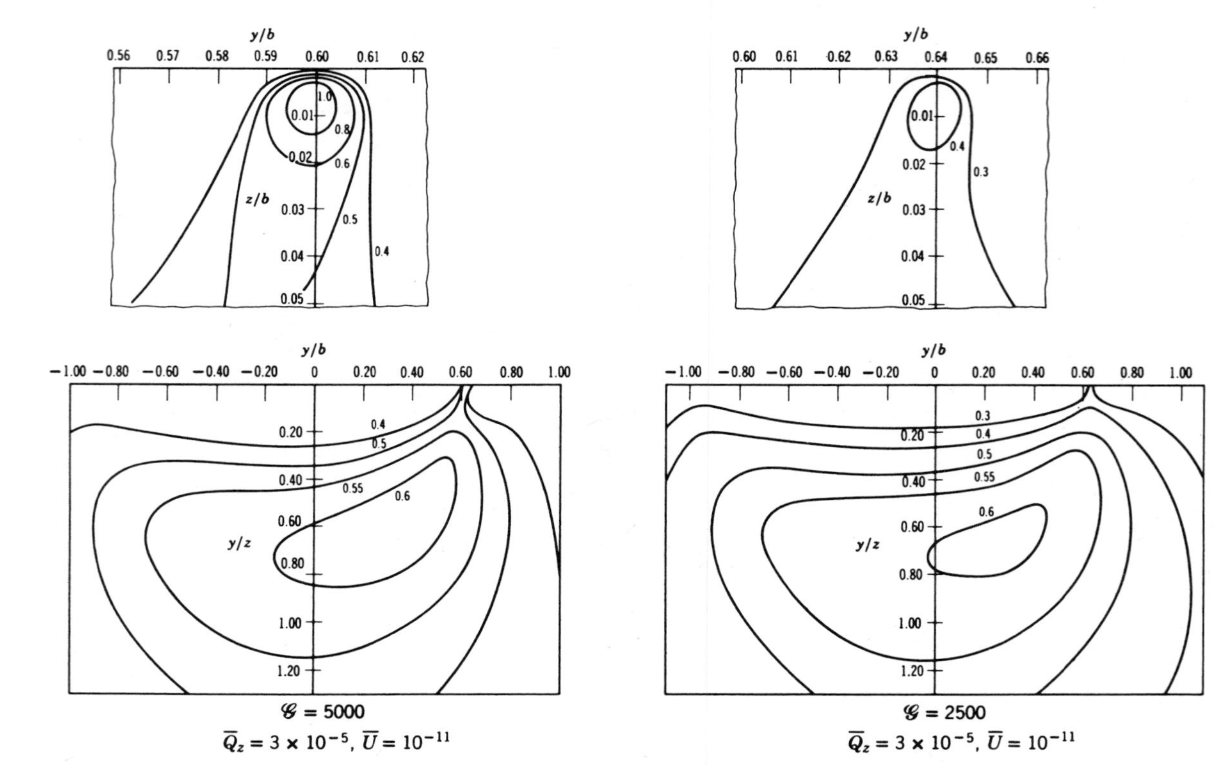

$\mathcal{G} = 5000$

$\overline{Q}_z = 3 \times 10^{-5}$, $\overline{U} = 10^{-11}$

$\mathcal{G} = 2500$

$\overline{Q}_z = 3 \times 10^{-5}$, $\overline{U} = 10^{-11}$

FIGURE 12.11. Contours of maximum shear stress amplitude—maximum Hertz pressure (reprinted from [12.9] by permission of the Institution of Mechanical Engineers).

The salient feature of both equations is the relatively large dependency of film thickness on speed and lubricant viscosity and the comparative insensitivity to load. Testing conducted by Sibley et al. [12.10] using radiation techniques seemed to confirm the Grubin equation; however, the agreement between the Dowson and Grubin formulas is apparent. Today, the Dowson equation is recommended as representative of line contact lubrication conditions.

The foregoing equations describe the minimum lubricant film thickness. The film thickness at the center of the contact is approximated by

$$H_c = \tfrac{4}{3} H^0 \tag{12.79}$$

Archard and Kirk [12.12] described the minimum film thickness between two spheres as

$$H^0 = \frac{0.84(\mathcal{G}\overline{U})^{0.741}}{\overline{Q}_z^{0.074}} \tag{12.80}$$

It is interesting to note once again the relative insensitivity to load. Subsequently, the Hamrock and Dowson [12.20] formula for minimum film thickness in point contact was developed:

$$H^0 = \frac{3.63\overline{U}^{0.68}\mathcal{G}^{0.49}(1 - e^{-0.68\kappa})}{\overline{Q}_z^{0.073}} \tag{12.81}$$

where $\overline{Q}_z$ for point contact is given by

$$\overline{Q}_z = \frac{Q}{E'\mathcal{R}^2} \tag{12.82}$$

Sometimes for elliptical point contact, an equivalent line contact load is considered as follows:

$$\overline{Q}_{ez} = \frac{3Q}{4E'\mathcal{R}_y a} \tag{12.83}$$

In equation (12.81), κ is the ellipticity ratio a/b. The central lubricant film thickness for point contact is given by

$$H_c = \frac{2.69\overline{U}^{0.67}\mathcal{G}^{0.53}(1 - 0.61e^{-0.73\kappa})}{\overline{Q}_z^{0.067}} \tag{12.84}$$

Example 12.1. The 209 cylindrical roller bearing of Example 8.2 is lubricated by a naphthenic oil having a kinematic viscosity of 100

SSU at operating temperature. Considering that the bearing supports a 4450 N (1000 lb) radial load at a shaft speed of 10,000 rpm estimate the minimum lubricant film thickness.

$$Q_{max} = 1335 \text{ N (300 lb)} \qquad \text{Fig. 8.12}$$

$$D = 10 \text{ mm (0.3937 in.)} \qquad \text{Ex. 2.7}$$

$$l = 9.6 \text{ mm (0.378 in.)} \qquad \text{Ex. 2.7}$$

Marks [12.21] gives the following formula for kinematic viscosity in stokes (cm^2/sec):

$$\nu_b = 2.26 \times 10^{-3}\text{SSU} - \frac{1.95}{\text{SSU}} \tag{12.85}$$

$$\nu_b = 0.207 \text{ cm}^2\text{/sec (stokes)}$$

$$\lambda = 0.1122 \left(\frac{\nu_0}{10^4}\right)^{0.163} \tag{12.43}$$

$$= 0.1122 \left(\frac{0.207}{10^4}\right)^{0.163}$$

$$= 0.01934 \text{ mm}^2\text{/N } (1.33 \times 10^{-4} \text{ in.}^2\text{/lb}) \tag{12.76}$$

$$E' = \frac{E}{1 - \xi^2}$$

$$= \frac{206{,}900}{1 - (0.3)^2} = 227{,}400 \text{ N/mm}^2 \ (33 \times 10^6 \text{ lb/in.}^2)$$

$$\mathcal{G} = \lambda E' \tag{12.75}$$

$$= 0.01934 \times 227{,}400 = 4398$$

$$\eta_b = \nu_b \rho g$$

$$= 0.207 \times 0.86 \frac{\text{g}}{\text{cm}^3} \times \frac{1 \text{ kg}}{10^3 \text{ g}} \times \frac{1 \text{ cm}}{10 \text{ mm}} \times \frac{1 \text{ m}}{10^3 \text{ mm}}$$

$$= 1.780 \times 10^{-8} \frac{\text{N} \cdot \text{sec}}{\text{mm}^2} \left(2.59 \times 10^{-6} \frac{\text{lb} \cdot \text{sec}}{\text{in.}^2}\right)$$

$$\mathcal{R}_i = \frac{D}{2}(1 - \gamma) \tag{12.31}$$

$$= \frac{10\,(1 - 0.1538)}{2} = 4.231 \text{ mm (0.167 in.)}$$

$$n_m = 0.4231 n_i$$

$$\omega_i - \omega_m = (n_i - n_m) \times \frac{2\pi}{60}$$ Ex. 8.2

$$= n_i(1 - 0.4231) \times \frac{2\pi}{60} = \frac{1.154\pi}{60} n_i$$

$$n_R = \tfrac{1}{2} n_i(1 - \gamma^2) \times \frac{d_m}{D} \tag{7.14}$$

$$= \frac{n_i}{2\gamma}(1 - \gamma^2) = \frac{\pi}{60}\left(\frac{1}{\gamma} - \gamma\right) n_i$$

$$= \frac{\pi}{60}\left(\frac{1}{0.1538} - 0.1538\right) n_i = \frac{6.348\pi}{60} n_i$$

$$U_i = \frac{d_m}{2}[(1 - \gamma)(\omega_i - \omega_m) + \gamma\omega_R] \tag{12.33}$$

$$= \frac{65}{2}[(1 - 0.1538) \times 1.154 + 0.1538 \times 6.348] \frac{10000\pi}{60}$$

$$= 3.324 \times 10^4 \text{ mm/sec } (1308 \text{ in./sec})$$

$$\overline{U}_i = \frac{\eta_0 U_i}{2E'\mathcal{R}_i} \tag{12.74}$$

$$= \frac{1.780 \times 10^{-8} \times 3.324 \times 10^4}{2 \times 2.274 \times 10^5 \times 4.231} = 3.075 \times 10^{-10}$$

$$\overline{Q}_{zi} = \frac{Q_{zi}}{lE'\mathcal{R}_y} \tag{12.73}$$

$$= \frac{1335}{9.6 \times 2.274 \times 10^5 \times 4.231} = 1.44 \times 10^{-4}$$

$$H_i^0 = \frac{2.65\,\overline{U}_i^{0.7}\mathcal{G}^{0.54}}{\overline{Q}_{zi}^{0.13}} \tag{12.78}$$

$$= \frac{2.65 \times (3.075 \times 10^{-10})^{0.7} \times (4398)^{0.54}}{(1.44 \times 10^{-4})^{0.13}} = 1.704 \times 10^{-4}$$

$$h_i^0 = H_i^0 \mathcal{R}_i$$

$$= 1.704 \times 10^{-4} \times 4.231$$

$$= 0.000721 \text{ mm} = 0.721\ \mu\text{m } (28.4 \times 10^{-6} \text{ in.})$$

The average or "plateau" film thickness is

$$h_{ci} = \tfrac{4}{3} h^0$$

$$= 0.000961 \text{ mm} = 0.961\ \mu\text{m } (37.9 \times 10^{-6} \text{ in.})$$

Example 12.2. Estimate the minimum lubricant film thickness assuming the bearing of Example 12.1 is lubricated by a mineral oil having a kinematic viscosity of 40 SSU at operating temperature.

$$\nu_b = 2.26 \times 10^{-3}\ \text{SSU} - \frac{19.5}{\text{SSU}} \tag{12.85}$$

$$= 2.26 \times 10^{-3} \times 32 - \frac{1.95}{32} = 0.01138\ \text{cm}^2/\text{sec}$$

$$\lambda = 0.1122 \left(\frac{\nu_b}{10^4}\right)^{0.163} \tag{12.43}$$

$$= 0.1122 \left(\frac{0.01138}{10^4}\right)^{0.163}$$

$$= 0.01206\ \text{mm}^2/\text{N}\ (0.8316 \times 10^{-4}\ \text{in.}^2/\text{lb})$$

$$h^0 = 7.209 \times 10^{-4} \left(\frac{\lambda_{40\,\text{SSU}}}{\lambda_{100\,\text{SSU}}}\right)^{0.54} \left(\frac{\nu_{b-40\,\text{SSU}}}{\nu_{b-100\,\text{SSU}}}\right)^{0.7}$$

$$= 7.209 \times 10^{-4} \left(\frac{0.01206}{0.01934}\right)^{0.54} \left(\frac{0.01138}{0.207}\right)^{0.7}$$

$$= 0.0000733\ \text{mm} = 0.0733\ \mu\text{m}\ (2.89 \times 10^{-6}\ \text{in.})$$

THERMAL EFFECTS

At high bearing operating speeds, some of the frictional heat generated in the concentrated contacts is dissipated in the lubricant momentarily residing in the inlet zone of the contacts. This effect, examined first by Cheng [12.22], tends to increase the temperature of the lubricant in the contacts. Vogels [12.23] gives the following expression for viscosity:

$$\eta_b = A_1 e^{\beta'/(T_b + A_2)} \tag{12.86}$$

in which T_b is in °C and A_1, A_2, and β' are three parameters to be defined for each lubricant. Three temperature–viscosity data points are required to determine A_1, A_2, and β' as follows:

$$A_1 = \eta_1 e^{-\beta'/(T_1 + A_2)} \tag{12.87}$$

$$A_2 = \frac{A_3 T_1 - T_3}{1 - A_3} \tag{12.88}$$

$$\beta' = \frac{(T_2 + A_2)(T_1 + A_2)}{T_2 - T_1} \ln \frac{\eta_1}{\eta_2} \tag{12.89}$$

$$A_3 = \frac{T_3 - T_2}{T_2 - T_1} \times \frac{\ln(\eta_1/\eta_2)}{\ln(\eta_2/\eta_3)} \tag{12.90}$$

If only two temperature–viscosity data points are known and A_2 can be fixed to 273, equation (12.86) can be simplified to

$$\eta_b = \eta_{ref} e^{\beta(1/T_b - 1/T_{ref})} \tag{12.91}$$

where T is now in °K and η_{ref} is the absolute viscosity at reference temperature, that is, T_{ref}. Since T_{ref} is generally at room temperature and since T_b is usually at higher than room temperature, equation (12.91) generally takes the form

$$\eta_b = \eta_{ref} e^{-A_4 \beta} \tag{12.92}$$

showing that as temperature increases, lubricant viscosity decreases.

In accordance with the foregoing, it is clear that the lubricant film thickness will reduce as a result of temperature increases in the contacts. For line contact Wilson [12.24] provides the following reduction factor to be applied to the central lubricant film thickness:

$$\varphi_T = \frac{1}{1 + 0.39L^{0.548}} \tag{12.93}$$

where

$$L = -\frac{U^2}{4k_b} \times \left(\frac{d\eta}{dT}\right)_b \tag{12.94}$$

and from equation (12.93),

$$\frac{d\eta}{dT} = \frac{-\beta\eta_b}{T^2} \tag{12.95}$$

Goksem et al. [12.25] provide a more complex expression for the thermal reduction factor as follows:

$$\phi_T = \frac{(11.81\, \mathcal{G}\overline{U}/\overline{Q}_z^{1.5})^{0.00736L^{0.234}}}{1 + 0.303L^{0.606}} \tag{12.96}$$

For applications where lubricant starvation is not a factor, equation (12.93) will suffice; however, as will be discussed subsequently in more detail, the simple combination of thermal reduction and lubricant starvation factors is not correct and equation (12.96) is then recommended as the thermal reduction factor.

Example 12.3. Consider that the lubricant used in the 209 cylindrical roller bearing of Example 12.1 is SAE 10W achieving its viscosity of 100 SSU at 54.4°C (100°F). According to [12.21], SAE 10W has a viscosity of 45 SSU at 98.9°C (210°F). What is the average lubricant film thickness of Example 12.1 considering thermal effects?

$$\nu_{b54.4°C} = 0.207 \text{ cm}^2/\text{sec} \qquad \text{Ex. 12.1}$$

$$\nu_{b98.9°C} = 2.26 \times 10^{-3} \text{ SSU} - \frac{1.95}{\text{SSU}} \qquad (12.85)$$

$$= 2.26 \times 10^{-3} \times 45 - \frac{1.95}{45} = 0.0584 \text{ cm}^2/\text{sec}$$

$$54.4°C = 327.6°K$$

$$98.9°C = 372.1°K$$

$$\beta' = \frac{(T_2 + A_2)(T_1 + A_2)}{(T_2 - T_1)} \ln \frac{\eta_1}{\eta_2} \qquad (12.89)$$

Assume $A_2 = 273.2°C$. Since $\nu = \eta/\rho$ and since in the temperature range considered oil density may be assumed constant, equation (12.89) becomes for T in °K,

$$\beta' = \frac{T_2 \times T_1}{(T_2 - T_1)} \ln \frac{\eta_1}{\eta_2}$$

$$= \frac{372.1 \times 327.6}{(372.1 - 327.6)} \ln \frac{0.207}{0.0584} = 3466°K$$

$$\eta_b = 1.780 \times 10^{-8} \text{ lb} \cdot \text{sec/in.}^2 \qquad \text{Ex. 12.1}$$

$$\frac{d\eta}{dT} = \frac{-\beta \eta_b}{T^2} \qquad (12.95)$$

$$= \frac{-3466 \times 0.01780}{(327.6)^2} = -5.749 \times 10^{-4} \text{ N} \cdot \text{sec/m}^2 °K$$

$$U = 33.22 \text{ m/sec (1308 in./sec)} \qquad \text{Ex. 12.1}$$

According to McAdams [12.34], the thermal conductivity of a mineral oil in the temperature range of this example is 0.1385 W/m · °C (0.08 Btu/hr · ft · °F).

$$L = \frac{-U^2}{4k_b} \left(\frac{d\eta}{dT}\right)_b \qquad (12.94)$$

$$= \frac{-(33.22)^2}{4 \times 0.1385} \times -5.749 \times 10^{-4} = 1.145$$

$$\varphi_{\mathrm{T}} = \frac{1}{1 + 0.39L^{0.548}} \tag{12.93}$$

$$= \frac{1}{1 + 0.39(1.145)^{0.548}} = 0.7042$$

$$h_{\mathrm{ci}} = 0.961\ \mu\mathrm{m}\ (37.9 \times 10^{-6}\ \mathrm{in.}) \qquad \text{Ex. 12.1}$$

$$h'_{\mathrm{ci}} = \varphi_{\mathrm{T}} h_{\mathrm{ci}}$$

$$= 0.7042 \times 9.61 \times 10^{-4}$$

$$= 0.677\ \mu\mathrm{m}\ (26.7 \times 10^{-6}\ \mathrm{in.})$$

STARVATION OF LUBRICANT

The basic formulas for calculation of lubricant film thickness assume an adequate supply of lubricant to the contact zones. The condition in which the volume of lubricant on the surfaces entering the contact is insufficient to develop a full lubricant film is called *starvation*. Factors to determine the reduction of the apparent lubricant film thickness have been developed as functions of the distance of the lubricant meniscus in the inlet zone from the center of the contact. As yet, no definitive equations have been developed to accurately calculate the aforementioned distance; therefore, the meniscus distance has to be determined experimentally. Figure 12.12 illustrates the concept of meniscus distance. References [12.26–12.29] give further detail about this concept.

In consideration of the meniscus distance problem, a condition of *zero reverse flow* is defined. Under this condition, the minimum velocity of the point situated at the meniscus distance from the contact center is, by definition, zero. If the meniscus distance is greater, the latter point will have a negative velocity, that is, *reverse flow*. The zero reverse flow

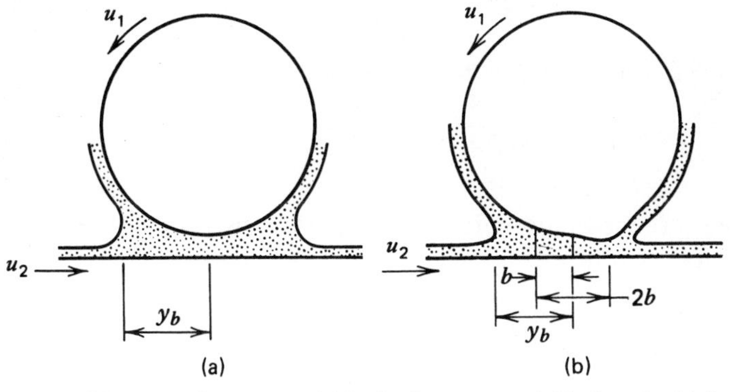

FIGURE 12.12. Meniscus distance in (*a*) hydrodynamic and (*b*) elastohydrodynamic lubrication.

condition is therefore a quasistable situation because no lubricant is lost to the contact owing to reverse flow. In the case of a minimum quantity of lubricant supplied, for example, oil mist or grease lubrication, the lubricant film thickness reduction factor owing to starvation effects, according to references [12.25] and [12.28], lies between 0.71 (in pure rolling) and 0.46 (in pure sliding). Castle et al. [12.28] give the following equation for line contact:

$$\varphi_s = 1 - \exp\left[-1.347\,\Phi^{(0.69\Phi^{0.13})}\right] \tag{12.97}$$

where

$$\Phi = \frac{\dfrac{y_b}{b} - 1}{\left[2\left(\dfrac{\mathcal{R}_y}{b}\right)^2 H_c\right]^{2/3}} \tag{12.98}$$

It is clear that Φ is zero if the meniscus distance should equal b and in that case $\varphi_s = 0$. Accordingly, the accurate estimation of the meniscus distance is necessary to the effective employment of a lubricant starvation factor. In the absence of this value, the condition of zero reverse flow provides a practical limitation and a starvation factor of $\varphi_s = 0.70$.

Thermal effects on lubricant film formation under conditions approaching lubricant starvation are extremely significant owing to the absence of excess lubricant to help dissipate frictional heat generation in the contacts. Accordingly, the lubricant film reduction factors for thermal effects and starvation are not multiplicative and a combined factor is required. Goksem et al. [12.25] derived the following expression for elastohydrodynamic line contact:

$$\varphi_{TS} = \varphi_T\left(1 - \frac{1}{(4.6 + 1.15L^{0.6})^{(0.67\overline{Q}_z\overline{Y}/\varphi_T H_c)^{(0.52/(1+0.001L))}}}\right) \tag{12.99}$$

where L is given by equation (12.94) and

$$Y = y_b(y_b^2 - 1)^{1/2} - \ln\left[y_b + (y_b^2 - 1)^{1/2}\right] \tag{12.100}$$

For the zero reverse flow condition, the combined reduction factor for the central lubricant film thickness is

$$\varphi_{TS} = \varphi_T\left(1 - \frac{1}{(4.6 + 1.15L^{0.6})^{(0.6345/\varphi_T)^{(0.52/(1+0.001L))}}}\right) \tag{12.101}$$

For point contact, equations (12.99)–(12.101) can be used in conjunction with (12.83) for equivalent line contact loading.

Example 12.4. Considering the 209 cylindrical roller bearing of Examples 12.1 and 12.3, what would be the average lubricant film thickness if oil mist lubrication were used instead of fully flooded lubrication?

$$L = 1.148 \qquad \text{Ex. 12.3}$$

$$\varphi_{\mathrm{T}} = 0.704 \qquad \text{Ex. 12.3}$$

For thermal plus starvation effects, use

$$\varphi_{\mathrm{TS}} = \varphi_{\mathrm{T}}\left(1 - \frac{1}{(4.6 + 1.15L^{0.6})^{(0.6345/\varphi_{\mathrm{T}})^{(0.52/(1+0.001L))}}}\right) \tag{12.101}$$

$$= 0.704\left(1 - \frac{1}{[4.6 + 1.15 \times (1.148)^{0.6}]^{(0.6345/0.704)^{(0.52/(1+0.001\times 1.148))}}}\right)$$

$$= 0.572$$

$$h'_{\mathrm{ci}} = \varphi_{\mathrm{TS}} h_{\mathrm{ci}}$$

$$= 0.572 \times 9.61 \times 10^{-4} = 0.550\ \mu\mathrm{m}\ (21.7 \times 10^{-6}\ \mathrm{in.})$$

SURFACE ROUGHNESS

Depending upon the thickness of the lubricant film relative to the roughnesses of the "contacting" surfaces, the direction of the roughness pattern can affect the film-building capability of the lubricant. In other words, if surface roughness has a pattern wherein the microgrooves are transverse to the direction of motion then the lubricant films in the contact zones will tend to be a thick as possible. Conversely, if the "lay" of the roughness is parallel to the direction of motion, then the lubricant films will be diminished as compared to the formulas stated in previous sections. Pater and Cheng [12.30] developed a correction factor φ_{SF} shown by Fig. 12.13 as a function of a correlation number Γ and film thickness parameter Λ. Figure 12.14 shows a schematic diagram of the surface roughness orientation.

The following definitions pertain to Fig. 12.13:

$$\Lambda = \frac{h_{\mathrm{c}}}{s} \tag{12.102}$$

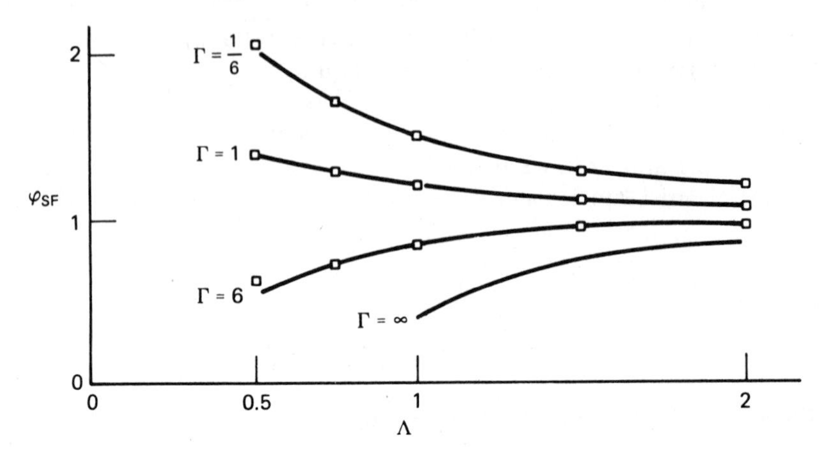

FIGURE 12.13. Lubricant film factor due to surface roughness orientation (from [12.30]).

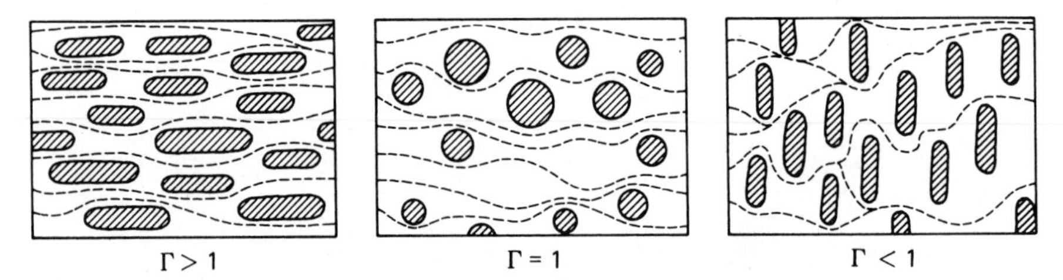

FIGURE 12.14. Typical contact area for longitudinally oriented ($\Gamma < 1$), isotropic ($\Gamma = 1$), and transversely oriented ($\Gamma < 1$) surfaces.

$$s = (s_1^2 + s_2^2)^{1/2} \tag{12.103}$$

where s_1 is the rms surface roughness of contact body 1, and so on.

$$\Gamma = \frac{\Delta_y}{\Delta_x} \tag{12.104}$$

where Δ_y and Δ_x are "correlation" lengths pertaining to distances between "hills" and "valleys" on a surface in the directions transverse and parallel to motion, respectively.

NON-NEWTONIAN LUBRICANT

A non-Newtonian lubricant is one that does not behave according to the formula

$$\tau = -\eta \frac{\partial u}{\partial y} \tag{12.1}$$

Sasaki et al. [12.6], Bell [12.17], and Smith [12.18] investigated effects of non-Newtonian lubricant behavior on the elastohydrodynamic model. Subsequently, it has been established that the non-Newtonian characteristics of lubricants tend to cause decreases in viscosity at high lubricant shear rates and therefore reductions in film thicknesses are to be expected. Gecim and Winer [12.31] considered a limiting shear stress concept with regard to film thickness and generated the following film thickness reduction factor.

$$\varphi_{NN} = 1 - \exp\{-[0.988(\tau_l/0.01p)^{0.5412}]\} \tag{12.105}$$

Typical values of the ratio τ_l/p are between 0.03 and 0.10, giving φ_{NN} between 0.83 and 0.97. Apparently, therefore, the effect of non-Newtonian lubrication is not substantial even considering the conservative assumption of a limiting lubricant shear stress.

GREASE LUBRICATION

When grease is used as a lubricant, a common tactic in calculating the lubricant film thickness is to use the properties of the base oil of the grease, ignoring the role of the thickener. There is evidence, however, that greases form films thicker than their base oils [12.36–12.39]. Kauzlarich and Greenwood [12.40] have developed an expression for the thickness of the film formed by greases in line contact under a Herschel–Bulkley constitutive law in which shear stress τ and strain rate $\dot{\gamma}$ are related by the equation

$$\tau = \tau_y + \alpha\dot{\gamma}^\beta \tag{12.106}$$

where τ_y is the yield stress, and α and β are considered physical properties of the grease.

For a Newtonian fluid

$$\tau = \eta\dot{\gamma} \tag{12.107}$$

where η is the viscosity.

The effective viscosity under a Herschel–Bulkley law is thus found by equating τ from equations (12.106) and (12.107) so that

$$\eta_{eff} = \frac{\tau_y + \alpha\dot{\gamma}^\beta}{\dot{\gamma}} \tag{12.108}$$

In this form it is seen that for $\beta > 1$, η_{eff} increases indefinitely with shear rate, and for $\beta < 1$, η_{eff} approaches zero as $\dot{\gamma}$ increases. Palacios and Palacios [12.41] argue that it is more reasonable to assume that at high shear rates greases will behave like their base oils. They accordingly propose a modification of the Herschel–Bulkley law to the form

$$\tau = \tau_y + \alpha\dot{\gamma}^{\beta} + \eta_b\dot{\gamma} \tag{12.109}$$

where η_b is the base oil viscosity. In this form, provided $\beta < 1$, $\eta_{eff} \rightarrow \eta_b$ as $\dot{\gamma} \rightarrow \infty$. Values of τ_y, α, β, and η_b are given in [12.41] for three greases from 35–80°C (95–176°F).

Since viscosity appears raised to the power 0.67 in equation (12.84), Palacios and Palacios propose that the film thickness of a grease h_G and the film thickness of the base stock h_b will be in the proportion:

$$\frac{h_G}{h_b} = (\eta_{eff}/\eta_b)^{0.67} \tag{12.110}$$

They propose that this evaluation be made at $\dot{\gamma} = 0.68u/h_G$, which requires iteration to determine h_G. A suggested approach is to calculate h_b from equation (12.84), then determine $\dot{\gamma} = 0.68u/h_b$, and then h_G from equation (12.110). The shear rate $\dot{\gamma}$ is then recalculated using h_G, and the process is repeated until convergence occurs. The analysis applies to line contact but should be valid for long elliptical contacts with a/b in the range 8–10.

LUBRICATION REGIMES

Three regimes of lubrication have been discussed with regard to the concentrated contacts occurring in ball and roller bearings—isoviscous hydrodynamic (IHD), piezoviscous hydrodynamic (PHD), and elastohydrodynamic (EHD). Dowson et al. [12.4] used Fig. 12.15 to define these regimes for line contact in terms of dimensionless quantities $\overline{U}$, $\overline{Q}_z$, and H.

Markho et al. [12.32] established a parameter, called C_1 herein, for a fixed value of $\mathcal{G}$, which parameter is used to define the lubrication regime. Dalmaz [12.34] subsequently established equation (12.111) to cover all practical values of $\mathcal{G}$.

$$C_1 = \log_{10}\left[1.5 \times 10^6\left(\frac{\mathcal{G}}{5000}\right)^2 \frac{\overline{Q}_z^3}{\overline{\overline{U}}}\right] \tag{12.111}$$

Table 12.1 shows the relationship of parameter C_1 to the operating lubrication regimes.

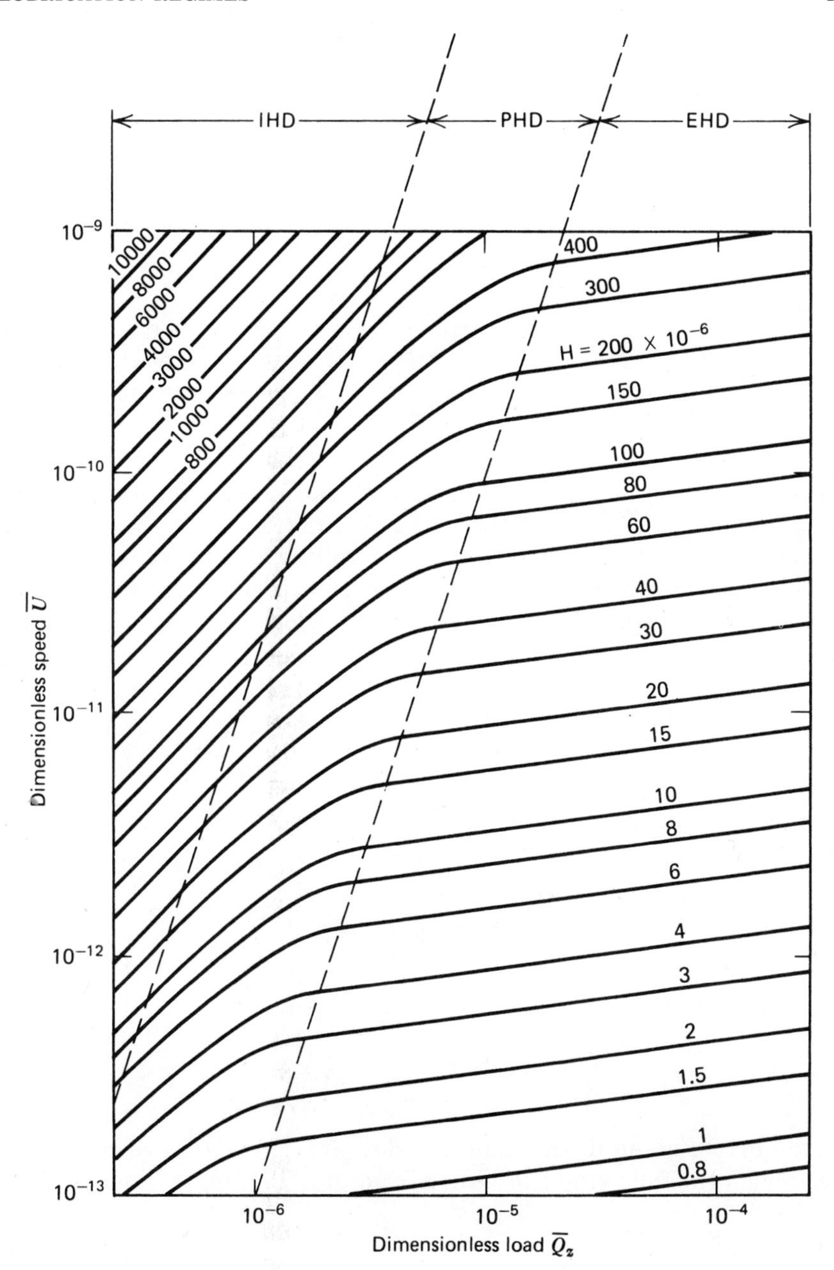

FIGURE 12.15. Film thickness vs speed and load for a line contact (from [12.4]).

TABLE 12.1. Lubrication Regimes

Parameter Limits	Lubrication Regime	Characteristics
$C_1 \leq -1$	IHD	Low contact pressure, no significant surface deformation
$-1 < C_1 < 1$	PHD	No significant surface deformation, lubricant viscosity increases with pressure
$C_1 \geq 1$	EHD	Surface deformation and lubricant viscosity increase with pressure

For calculation of the lubricant film thicknesses in rolling element-raceway contacts, only the PHD and EHD regimes need to be considered. For calculations associated with the cage–rolling element contacts, probably consideration of the hydrodynamic regime is sufficient. In this case, Martin [12.1] gives the following equation for film thickness in line contact:

$$H = 4.9 \frac{\overline{U}}{\overline{Q}_z} \tag{12.112}$$

For point contact, Brewe et al. [12.33] give

$$H = \left[\frac{\dfrac{\overline{Q}_z}{\overline{U}} \left(1 + \dfrac{2\,\mathcal{R}_x}{3\,\mathcal{R}_y}\right)}{\left(128 \dfrac{\mathcal{R}_y}{\mathcal{R}_x}\right)^{1/2} \left[0.131 \tan^{-1}\left(\dfrac{\mathcal{R}_y}{2\,\mathcal{R}_x}\right) + 1.163\right]} + 2.6511 \right]^{-2} \tag{12.113}$$

For the PHD regime in line contacts, data from [12.32] have been used to establish the following expression for minimum film thickness:

$$H = 10^{C_4} \times \left(\frac{\mathcal{G}}{5000}\right)^{0.35(1 + C_1)} \tag{12.114}$$

where

$$C_2 = \log_{10}(618\overline{U}^{0.6617}) \tag{12.115}$$

$$C_3 = \log^{10}(1.285\overline{U}^{0.0025}) \tag{12.116}$$

and C_1 is given by equation (12.111). In equation (12.114), C_4 is given by

$$C_4 = C_2 + C_1 C_3 (C_1^2 - 3) - 0.094 C_1 (C_1^2 - 0.77 C_1 - 1) \quad (12.117)$$

Dalmaz [12.34] also developed numerical results for point contact film thicknesses in the PHD regime; however, no analytical relationship is currently available.

CLOSURE

In the foregoing discussion it has been demonstrated analytically that a lubricant film can serve to separate the rolling elements from the contacting raceways. Moreover, the fluid friction forces developed in the contact zones between rolling elements and raceways can significantly alter the bearing's mode of operation. It should be apparent that it is desirable from a standpoint of preventing increased stresses caused by metal-to-metal contact that the minimum film thickness should be sufficient to completely separate the rolling surfaces. The effect of film thickness on bearing endurance will be discussed in Chapter 18.

A substantial amount of analytical and experimental research during the 1960s–1980s has contributed greatly to the understanding of the lubrication mechanics of concentrated contacts in rolling bearings. Perhaps the original work by Grubin [12.8] will prove to be as significant as that conducted by Reynolds during the 1880s. It is the feeling of the author that the full impact of Grubin's revelation has yet to be felt in the design of rolling bearings. More effort is needed to define the friction forces between contacting rolling bodies, and further the effect of those forces on rolling bearing endurance.

Besides acting to separate rolling surfaces, the lubricant is frequently used as a medium to dissipate the heat generated by bearing friction as well as to remove heat that would otherwise be transferred to the bearing from surroundings at elevated temperatures. This topic will be discussed in Chapter 15.

REFERENCES

12.1. H. M. Martin, "Lubrication of Gear Teeth," *Engineering* **102,** 199 (1916).

12.2. J. F. Osterle, "On the Hydrodynamic Lubrication of Roller Bearings," *Wear* **2,** 195 (1959).

12.3. B. Sternlicht, P. Lewis, and P. Flynn, "Theory of Lubrication and Failure of Rolling Contacts," *ASME J. Basic Eng.* 213–226 (1961).

12.4. D. Dowson and G. Higginson, "Theory of Roller Bearing Lubrication and Deformation," *Proc. Inst. Mech. Eng.* **117** (1963).

12.5. P. L. Kapitsa, "Hydrodynamic Theory of Lubrication during Rolling," *USSR J. Tech. Phys.* **25,** Sec. 4, 747–762 (1955).

12.6. T. Sasaki, H. Mori, and N. Okino, "Fluid Lubrication Theory of Roller Bearings, Parts I and II," *ASME J. Basic Eng.* **166,** 175 (1963).

12.7. D. Dowson and G. Higginson, "A Numerical Solution to the ElastoHydrodynamic Problem." *J. Mech. Eng. Sci.* **1**(1), 6 (1959).

12.8. A. Grubin, "Fundamentals of the Hydrodynamic Theory of Lubrication of Heavily Loaded Cylindrical Surfaces," *Investigation of the Contact Machine Components*, Kh. F. Ketova, ed. Translation of Russian Book No. 30, Chapter 2, Central Scientific Institute of Technology and Mechanical Engineering, Moscow (1949).

12.9. D. Dowson and G. Higginson, "The Effect of Material Properties on the Lubrication of Elastic Rollers," *J. Mech. Eng. Sci.* **2**(3) (1960).

12.10. L. Sibley and F. Orcutt, "Elasto-hydrodynamic Lubrication of Rolling Contact Surfaces," *ASLE Trans.* **4,** 234–249 (1961).

12.11. D. Dowson and G. Higginson, *Proceedings of Institution of Mechanical Engineers*, Vol. 182, Part 3A, London, 151–167 (1968).

12.12. G. Archard and M. Kirk, "Lubrication at Point Contacts," *Proc. R. Soc. Ser.* **A 261,** 532–550 (1961).

12.13. A. Crook, "The Lubrication of Rollers," *Philos. Trans. R. Soc. London, Ser.* **A 250,** 387–409 (1958).

12.14. J. Bell, J. Kannel, and C. Allen, "The Rheological Behavior of the Lubricant in the Contact Zone of a Rolling Contact System," *ASME Paper No. 63-LUB-8*, ASME-ASLE Lubrication Conference (October, 1963).

12.15. ASME Research Committee on Lubrication "Pressure-Viscosity Report—Vol. 11," ASME (1953).

12.16. Belsky, Kamenshine, Lankert, McCool, Tataiah, Tschirschnitz, Walker, and Waltrich, "A Basic Study of Sliding Contacts in Rolling Bearings," SKF Research Report AL65L032 (May 7, 1965).

12.17. J. Bell, "Lubrication of Rolling Surfaces by a Ree-Eyring Fluid," *ASLE Trans.* **5,** 160–171 (1963).

12.18. F. Smith, "Rolling Contact Lubrication—The Application of Elastohydrodynamic Theory," ASME Spring Lubrication Symposium, Paper No. 64-Lubs-2 (April 1964).

12.19. C. Roelands, *Correlational Aspects of the Viscosity-Temperature-Pressure Relationship of Lubricating Oils*, Druk V.R.B., Groningen, Netherlands (1966).

12.20. B. Hamrock and D. Dowson, "Isothermal Elastohydrodynamic Lubrication of Point Contacts—Part III—Fully Flooded Results," *ASME J. Lubr. Technol.* **99,** 264–276 (1977).

12.21. T. Baumeister and L. Marks, *Standard Handbook for Mechanical Engineers*, 7th Ed., McGraw-Hill, New York (1958).

12.22. H. Cheng, "A Numerical Solution to the Elastohydrodynamic Film Thickness in an Elliptical Contact," *ASME J. Lubr. Technol.* **92,** 155–162 (1970).

12.23. H. Vogels, "Das Temperaturabhängigkeitsgesetz der Viscosität von Flüssigkeiten," *Phys. Z.* **22,** 645–646 (1921).

12.24. A. Wilson, "An Experimental Thermal Correction for Predicted Oil Film Thickness in Elastohydrodynamic Contacts," Sixth Leeds-Lyon Symposium on Tribology, Lyon (1979).

12.25. P. G. Goksem and R. A. Hargreaves, "The Effect of Viscous Shear Heating in Both Film Thickness and Rolling Traction in an EHL Line Contact—Part II: Starved Condition," *ASME J. Lubr. Technol.* **100,** 353–358 (1978).

12.26. D. Dowson, "Inlet Boundary Conditions," Leeds-Lyon Symposium (1974).

12.27. P. Wolveridge, K. Baglin, and J. Archard, "The Starved Lubrication of Cylinders in Line Contact," *Proc. Inst. Mech. Eng.* **185,** 1159–1169 (1970–71).

12.28. P. Castle and D. Dowson, "A Theoretical Analysis of the Starved Elastohydrodynamic Lubrication Problem for Cylinders in Line Contacts," *Elastohydrodynamic Lubrication Symposium, Proc. Inst. Mech. Eng.* **131,** 131–137 (1972).

12.29. B. Hamrock and D. Dowson, "Isothermal Elastohydrodynamic Lubrication of Point Contact—Part IV: Starvation Results," *ASME J. Lub. Technol.* **99,** 15–23 (1977).

12.30. N. Pater and H. Cheng, "Effect of Surface Roughness Orientation on the Central Film Thickness in EHD Contacts," *Proc. Fifth Leeds-Lyon Symp.*, pp. 15–21 (1978).

12.31. B. Gecim and W. Winer, "A Film Thickness Analysis for Line Contacts under Pure Rolling Conditions with a Non-Newtonian Rheological Model," ASME Paper 80 C2/LUB 26 (August 8, 1980).

12.32. P. Markho and D. Clegg, "Reflections on Some Aspects of Lubrication of Concentrated Line Contacts," *ASME J. Lubr. Technol.* **101,** 528–531 (1979).

12.33. D. Brewe and B. Hamrock, "Analysis of Starvation on Hydrodynamic Lubrication in Non-Conforming Contacts," ASME Paper 81-LUB-52 (1981).

12.34. G. Dalmaz, "Le Film Mince Visquex dans les Contacts Hertziens en Regimes Hydrodynamique et Elastohydrodynamique," Docteur d'Etat Es Sciences Thesis, I.N.S.A. Lyon (1979).

12.35. W. McAdams, *Heat Transmission*, 3rd ed., McGraw-Hill, New York, 455 (1954).

12.36. A. Wilson, "The Relative Thickness of Grease and Oil Films in Rolling Bearings," *Proc. I. Mech. Eng.* **193,** 185–192 (1979).

12.37. H. Muennich and H. Gloeckner, "Elastohydrodynamic Lubrication of Grease-Lubricated Rolling Bearings," *ASLE Trans.* **23,** 45–52 (1980).

12.38. J. Palacios, A. Cameron, and L. Arizmendi, "Film Thickness of Grease in Rolling Contacts," *ASLE Trans.* **24,** 474–478 (1981).

12.39. J. Palacios, "Elastohydrodynamic Films in Mixed Lubrication: An Experimental Investigation," *Wear* **89,** 303–312 (1983).

12.40. J. Kauzlarich and J. Greenwood, "Elastohydrodynamic Lubrication with Herschel-Bulkley Model Greases," *ASLE Trans.* **15,** 269–277 (1972).

12.41. J. Palacios and M. Palacios, "Rheological Properties of Greases in EHD Contacts," *Tribol. Int.* **17,** 167–171 (1984).

13

FRICTION IN ROLLING BEARINGS

LIST OF SYMBOLS

Symbol	Description	Units
a	Semimajor axis of projected contact ellipse	mm (in.)
b	Semiminor axis of projected contact ellipse	mm (in.)
C_s	Basic static capacity	N (lb)
c_v	Viscous drag coefficient	
d	Diameter	mm (in.)
d_m	Pitch diameter	mm (in.)
D	Roller or ball diameter	mm (in.)
$\mathcal{E}$	Complete elliptic integral of second kind	
F, f	Force, friction force	N (lb)
F_c	Centrifugal force	N (lb)
g	Gravitational constant	mm/sec^2 (in./sec^2)
h	Distance between center of contact ellipse and center of spinning	mm (in.)

Symbol	Description	Units
I	Mass moment of inertia	$kg \cdot mm^2$ ($in. \cdot lb \cdot sec^2$)
l	Effective roller length	mm (in.)
M	Moment	$N \cdot mm$ ($in. \cdot lb$)
M_g	Gyroscopic moment	$N \cdot mm$ ($in. \cdot lb$)
M_f	Bearing friction torque due to flange load	$N \cdot mm$ ($in. \cdot lb$)
M_l	Bearing friction torque due to load	$N \cdot mm$ ($in. \cdot lb$)
M_v	Bearing friction torque due to lubricant	$N \cdot mm$ ($in. \cdot lb$)
m	Mass	kg ($lb \cdot sec^2/in.$)
n	Bearing rotational speed	rpm
Q	Roller or ball load	N (lb)
q	Load per unit length or x'/a	N/mm (lb/in.)
R	Radius of curvature of contact surface	mm (in.)
S	Surface area	mm^2 ($in.^2$)
t	y'/b	
T	Rolling line location on x' axis	mm (in.)
T	Cage torque	N-mm ($in. \cdot lb$)
u	Surface velocity	mm/sec (in./sec.)
v	Surface velocity	mm/sec (in./sec.)
w	Width of laminum	mm (in.)
w_{CR}	Width of cage rail	mm (in.)
$\mathcal{W}$	Lubricant flow rate through bearing	cm^3/min (gal/min.)
x	Distance in the x direction	mm (in.)
y	Distance in the y direction	mm (in.)
z	Distance in the z-direction	mm (in.)
α	Contact angle	rad, °
γ	$D \cos \alpha / d_m$	
η	Lubricant viscosity	cp ($lb \cdot sec/in.^2$)
θ	Angle	rad
κ	Ellipticity parameter	
μ	Coefficient of friction	
ν_0	Kinematic viscosity	centistokes
ρ	Radius	mm (in.)
ξ	Lubricant effective density	g/mm^3 ($lb/in.^3$)
ξ_b	Lubricant density	g/mm^3 ($lb/in.^3$)
σ	Normal stress	N/mm^2 (psi)
τ	Shear stress	N/mm^2 (psi)
ϕ	Angle	rad
ψ	Azimuth angle	rad, °

Symbol	Description	Units
ω	Rotational speed	rad/sec
Ω	Ring rotational speed	rad/sec

SUBSCRIPTS

CG	Refers to cage
CL	Refers to cage land
CP	Refers to cage pocket
CR	Refers to cage rail
drag	Refers to viscous friction on cage
g	Refers to gyroscopic motion
i	Refers to inner raceway
n	Refers to outer or inner raceway, o or i
m	Refers to orbital motion
o	Refers to outer raceway
R	Refers to rolling motion
S	Refers to spinning motion
v	Refers to viscous friction on rolling element
x	Refers to x direction
x'	Refers to x' direction
y	Refers to y direction
y'	Refers to y' direction
z	Refers to z direction
z'	Refers to z' direction
λ	Refers to laminum

GENERAL

It is universally recognized that friction due to rolling of nonlubricated surfaces over each other is considerably less than the dry friction encountered by sliding the identical surfaces over each other. Notwithstanding the motions of the contacting elements in rolling bearings are more complex than is indicated by pure rolling, rolling bearings exhibit considerably less friction than most fluid film or sleeve bearings of comparable size and load-carrying ability. A notable exception to the foregoing generalization is, of course, the hydrostatic gas bearing; however, such a bearing is not self-sustaining, as is a rolling bearing, and it requires a complex gas supply system.

Friction of any magnitude represents an energy loss and causes a retardation of motion. Hence friction in a rolling bearing is witnessed as a temperature increase and may be measured as a retarding torque.

The sources of friction in rolling bearings are manifold, the principal sources being as follows:

1. Elastic hysteresis in rolling
2. Sliding in rolling element–raceway contacts due to geometry of contacting surfaces
3. Sliding due to deformation of contacting elements
4. Sliding between the cage and rolling elements and, for a land-riding cage, sliding between the cage and bearing rings
5. Viscous drag of the lubricant on the rolling elements and cage
6. Sliding between roller ends and inner and/or outer ring flanges
7. Seal friction

These sources of friction are discussed in the following section.

SOURCES OF FRICTION

Elastic Hysteresis in Rolling

As a rolling element under compressive load travels over a raceway, the material in the forward portion of the contact surface, that is, in the direction of rolling, will undergo a compression while the material in the rear of the contact is being relieved of stress. It is recognized that as load is increasing, a given stress corresponds to a smaller deflection than when load is decreasing (see Fig. 13.1). The area between the curves of Fig. 13.1 is called the hysteresis loop and represents an energy loss. (This is readily determined if one substitutes force times a constant for stress and deformation times a constant for strain.) Generally, the energy loss or friction due to elastic hysteresis is small compared to other types of friction occurring in rolling bearings. Drutowski [13.1] has verified this by experimenting with balls rolling between flat plates. Coefficients of rolling friction as low as 0.0001 can be determined from the reference [13.1] data for 12.7mm (0.5 in.) diameter chrome steel balls rolling on chrome steel plates under normal loads of about 356 N (80 lb).

Drutowski [13.2] has demonstrated the apparent linear dependence of rolling friction on the volume of significantly stressed material. In both references [13.1] and [13.2] Drutowski further demonstrates the dependence of elastic hysteresis on the material under stress and on the specific load in the contact area.

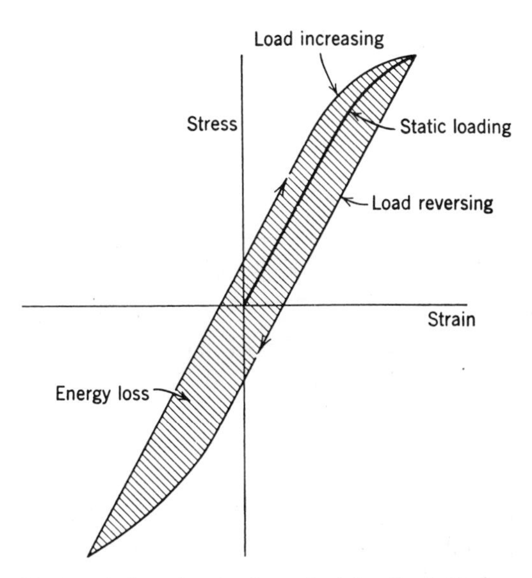

FIGURE 13.1. Hysteresis loop for elastic material subjected to reversing stresses.

Rolling and Deformation

Nominally, the balls or rollers in a rolling bearing are subjected to loads that are perpendicular to the tangent plane at each contact surface. Because of these normal loads the rolling elements and raceways are deformed at each contact, producing, according to Hertz, a radius of curvature of the contacting surface equal to the harmonic mean of the radii of the contacting bodies. Hence for a roller of diameter D, bearing on a cylindrical raceway of diameter d_i, the radius of curvature of the contact surface is

$$R = \frac{d_i D}{d_i + D} \tag{13.1}$$

Because of the deformation indicated above and because of the rolling motion of the roller over the raceway, which requires a tangential force to overcome rolling resistance, raceway material is squeezed up to form a bulge in the forward portion of the contact, as shown in Fig. 13.2. A subsequent depression is formed in the rear of the contact area. Thus, an additional tangential force is required to overcome the resisting force of the bulge.

Sliding Friction in Rolling Element–Raceway Contacts

Macrosliding due to Rolling Motion. In Chapter 7, it was demonstrated that sliding occurs in most ball and roller bearings simply due to the

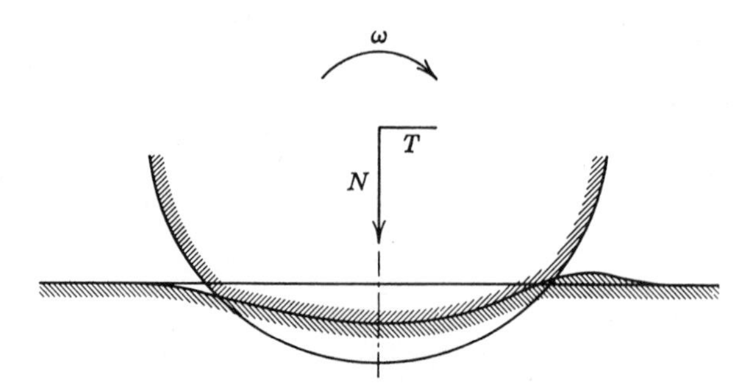

FIGURE 13.2. Roller–raceway contact showing bulge due to tangential forces.

macro or basic internal geometry of the bearing. Theoretically, if a radial cylindrical roller bearing had rollers and raceways of exactly the same length, if the rollers were very accurately guided by frictionless flanges, and if the bearing operated with zero misalignment, then sliding in the roller–raceway contacts would be avoided. In the practical situation, however, rollers and/or raceways are crowned to avoid "edge loading," and under applied load the contact surface is curved in the plane passing through the bearing axis of rotation and the center of "rolling" contact. Since pure rolling is defined by instant centers at which no relative motion of the contacting elements occurs, that is, the surfaces have the same velocities at such points, then even in a radial cylindrical roller bearing, only two points of pure rolling can exist on the major axis of each contact surface. At all other points, sliding must occur. In fact, the major source of friction in rolling bearings is sliding.

Most rolling bearings are lubricated by a viscous medium such as oil, provided either directly as a liquid or indirectly exuded by a grease. Some rolling bearings are lubricated by less viscous fluids and some by dry lubricants such as molybdenum disulfide (MoS_2). In the former cases, the coefficient of sliding friction in the contact areas, that is, the ratio of the shear force caused by sliding to the normal force pressing the surfaces together, is generally significantly lower than with "dry" film lubrication. It was indicated in Chapter 12 that a Newtonian fluid is one that exhibits the relationship

$$\tau = \eta \frac{\partial u}{\partial z} \tag{13.1}$$

Various researchers [13.13, 13.14] have investigated traction in rolling/sliding contacts and found that equation (13.1) pertains only to a situation involving a relatively low slide-to-roll ratio, for example, as shown in Fig. 13.3, less than 0.003. This however depends upon the amount of normal pressure in the contact zone. Note that traction coefficient is syn-

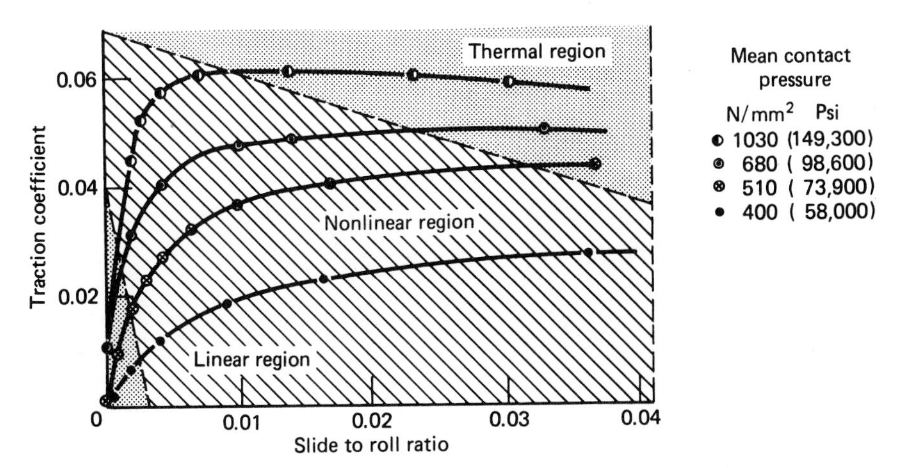

FIGURE 13.3. Typical curves of traction measured on disk machine in line contact.

onymous with friction coefficient. For calculation purposes, the curves of Fig. 13.3 can be approximated by polynomial expressions or approximated by a linear portion and a constant portion. In the linear portion, surface velocities can be calculated using the methods of Chapter 7 and lubricant film thicknesses, by the methods of Chapter 12.

Macrosliding due to Gyroscopic Action. In Chapter 7 for angular-contact ball bearings, ball motions induced by gyroscopic moments were discussed. This motion occasions pure sliding in directions colinear with the major axes of the ball–raceway elliptical areas of contact. Jones [13.4] considered that gyroscopic motion can be prevented if the friction coefficient is sufficiently great, for example, as stated in Chapter 7, 0.06–0.07. Figure 13.3 pertaining to lubrication with a viscous fluid indicates such a high friction coefficient is not probable at normal contact pressures when a sufficiently thick lubricant film separates the "contacting" surfaces. This is especially true for modern ball bearings, which have extremely smooth ball and raceway surfaces. As shown in Chapter 12, the lubricant film thickness generated depends heavily upon the speed of bearing operation. Upon the relationship of the lubricant film thickness to the composite roughness of the contacting surfaces depends the magnitude of the friction coefficient. Obviously, if the "separation" of the surfaces is insufficient, metal-to-metal contact can occur, which even with boundary lubrication will result in a higher friction coefficient than that occurring with complete separation of the surfaces by an elastohydrodynamic lubricant film. Therefore, when a lubricant film is generated that is sufficiently thick to "separate" the surfaces in "rolling" contact, gyroscopic ball motion will occur in angular-contact ball bearings and thrust ball bearings.

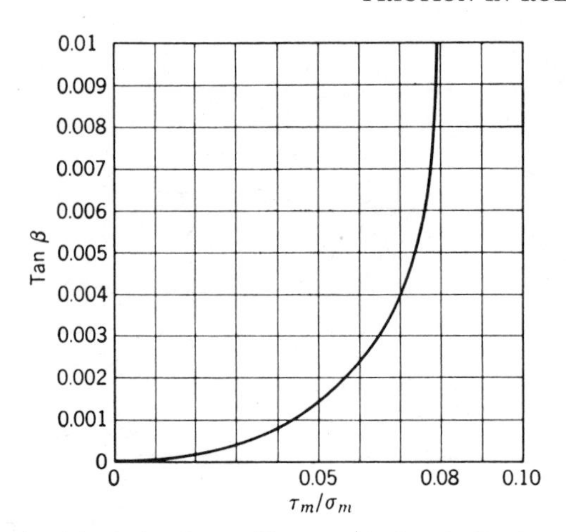

FIGURE 13.4. Angle of deviation from rolling motion for a ball subjected to a tangential load perpendicular to the direction of rolling.

Palmgren [13.11] called the gyroscopic motion *creep* and in experiments he found that if the tangential force attitude was perpendicular to the direction of rolling, the relationship of the angle β by which the motion of a ball deviates from the direction of rolling can be shown to be a function of the ratio of the mean tangential stress to the mean normal stress. Figure 13.4 shows for lubricated surfaces that creep becomes infinite as $2\tau_m/\sigma_m$ approaches 0.08. Palmgren further deduced as a consequence of creep that a ball can never remain rolling between surfaces that form an angle to each other, regardless of the minuteness of the angle. The ball, while rolling, always seeks surfaces that are parallel.

Microslip. Reynolds [13.5] first referred to microslip when in his experiments involving the rolling of an elastically stiff cylinder on rubber, he observed that since the rubber stretched in the contact zone, the cylinder rolled forward a distance less than its circumference in one complete revolution about its axis. The classical demonstration of the microslip or creep phenomenon was developed in two dimensions by Poritsky [13.6]. He considered the action of a locomotive driving wheel as shown in Fig. 13.5. The normal load between the cylinders was assumed to generate a parabolic stress distribution over the contact surface. Superimposed on the Hertzian stress distribution was a tangential stress on the contact surface, as shown in Fig. 13.5. Using this motion Poritsky demonstrated the existence of a "locked" region over which no slip occurs and a slip region of relative movement in a contact area over which it has been historically assumed that only rolling had occurred. Cain [13.7] further determined that in rolling the "locked" region coincided with the leading edge of the contact area, as shown in Fig. 13.6. In general, the "locked

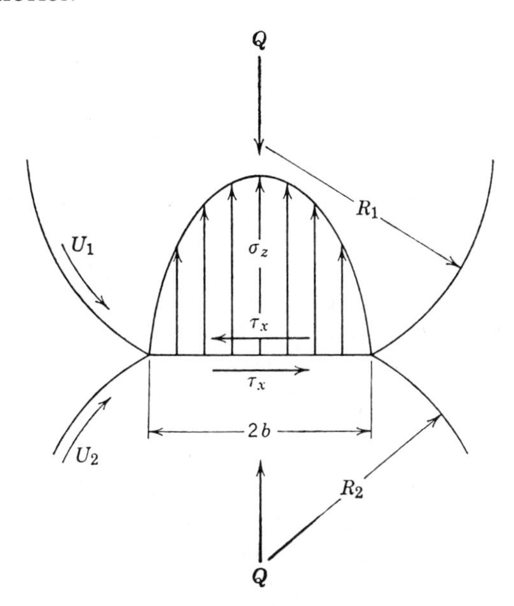

FIGURE 13.5. Rolling under action of surface tangential stress (reprinted from [13.9] by permission of American Elsevier Publishing Company).

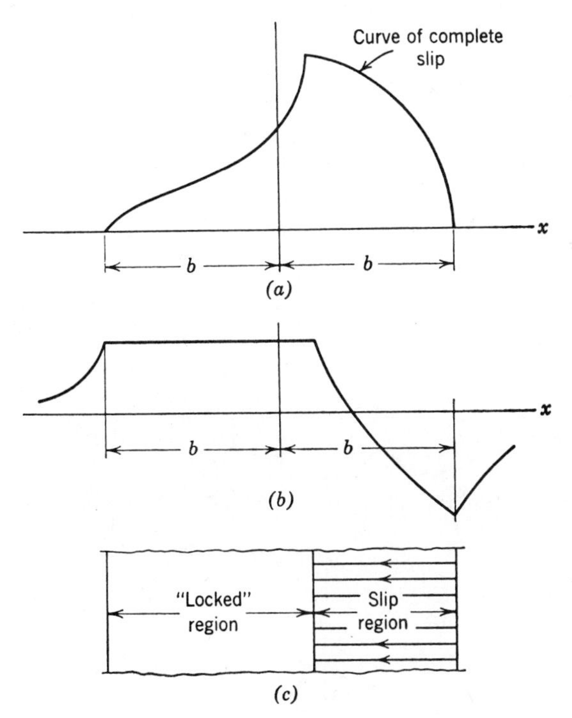

FIGURE 13.6. (*a*) Surface tangential transactions; (*b*) surface strains; (*c*) region of traction and microslip (reprinted from [13.9] by permission of American Elsevier Publishing Company).

region" phenomenon can occur only when the friction coefficient is very high as between unlubricated surfaces.

Heathcote "slip" is very similar to that which occurs because of rolling element–raceway deformation. Heathcote [13.8] determined that a hard ball "rolling" in a closely conforming groove can roll without sliding on two narrow bands only. Ultimately, Heathcote obtained a formula for the "rolling" friction in this situation. Heathcote's analysis takes no account of the ability of the surfaces to elastically deform and accommodate the difference in surface velocities by differential expansion. Johnson [13.9] expanded on the Heathcote analysis by slicing the elliptical contact area into differential slabs of area, as shown in Fig. 13.7, and thereafter applying the Poritsky analysis in two dimensions to each slab. Generally, Johnson's analysis using tangential elastic compliance demonstrates a lower coefficient of friction than does the Heathcote analysis, which assumes sliding rather than microslip. Figure 13.8 shows the "locked" and slip regions that obtain within the contact ellipse. Greenwood and Tabor [13.10] evaluated the rolling resistance due to elastic hysteresis. It is of interest to indicate that the frictional resistance due to elastic hysteresis as determined by Greenwood and Tabor is generally less than that due to sliding if normal load is sufficiently large.

Viscous Drag

Owing to its orbital speed, each ball or roller must overcome a viscous drag force imposed by the lubricant within the bearing cavity. It can be assumed that drag caused by a gaseous atmosphere is insignificant; how-

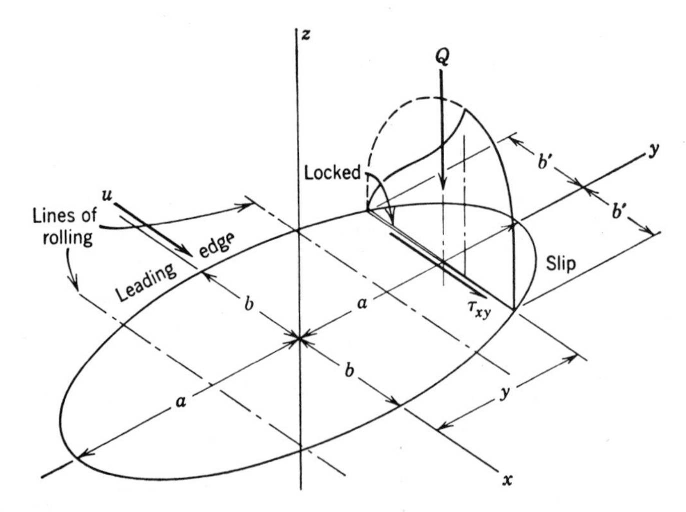

FIGURE 13.7. Ball–raceway contact ellipse showing "locked" region and microslip region–radial ball bearing (reprinted from [13.9] by permission of American Elsevier Publishing Company).

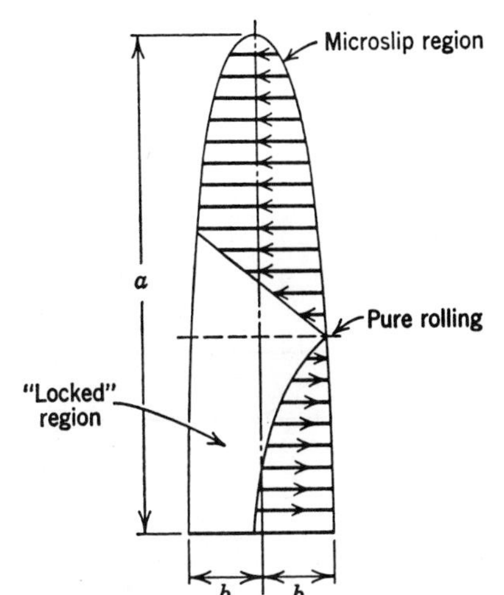

FIGURE 13.8. Semiellipse of contact showing sliding lines and rolling point (reprinted from [13.9] by permission of American Elsevier Publishing Company).

ever, the lubricant viscous drag depends upon the quantity of the lubricant dispersed in the bearing cavity. Hence, the effective fluid within the cavity is a gas–lubricant mixture having an effective viscosity and an effective specific gravity. The viscous drag force acting on a ball as indicated in [13.15] can be approximated by

$$F_v = \frac{\pi \xi c_v D^2 (d_{\mathrm{m}} \omega_{\mathrm{m}})^{1.95}}{32g} \tag{13.2}$$

where ξ is the weight of lubricant in the bearing cavity divided by the free volume within the bearing boundary dimensions. Similarly, for an orbiting roller

$$F_v = \frac{\xi l D c_v (d_{\mathrm{m}} \omega_{\mathrm{m}})^{1.95}}{16g} \tag{13.3}$$

The drag coefficients c_{v} in equations (13.2) and (13.3) can be obtained from reference [13.16] among others.

Sliding between the Cage and Bearing Rings

Three basic cage types are used in ball and roller bearings: (1) ball riding (BR) or roller riding (RR), (2) inner ring land riding (IRLR), and (3) outer ring land riding (ORLR). These are illustrated schematically in Fig. 13.9.

BR and RR cages are usually of relatively inexpensive manufacture and are usually not used in critical applications. The choice of an IRLR or ORLR cage depends largely upon the application and designer preference. An IRLR cage is driven by a force between the cage rail and inner ring land as well as by the rolling elements. ORLR cage speed is

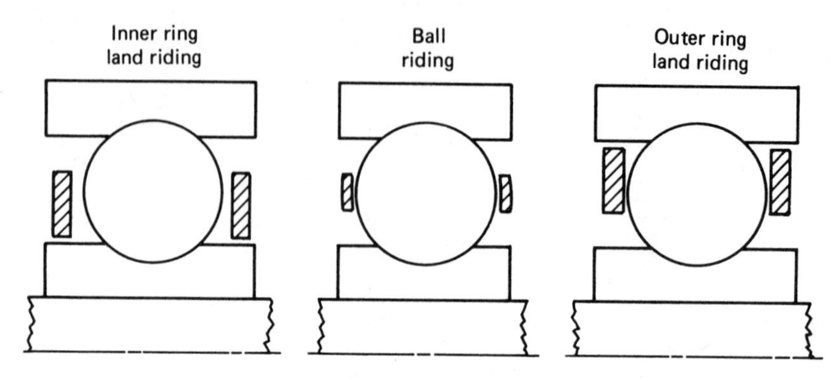

FIGURE 13.9. Cage types.

retarded by cage rail/outer ring land drag force. The magnitude of the drag or drive force between the cage rail and ring land depends upon the resultant of the cage/rolling element loading, the eccentricity of the cage axis of rotation and the speed of the cage relative to the ring on which it is piloted. If the cage rail/ring land normal force is substantial, hydrodynamic short bearing theory [13.17] might be used to establish the friction force F_{CL}. For a properly balanced cage and a very small resultant cage/rolling element load, Petroff's law can be applied; for example,

$$F_{CL} = \frac{\eta \pi \omega_{CR} c_n d_{CR}(\omega_c - \omega_n)}{1 - (d_1/d_2)} \qquad \begin{aligned} c_o &= 1 \\ c_i &= -1 \end{aligned} \tag{13.4}$$

where d_2 is the larger of the cage rail and ring land diameters and d_1 is the smaller.

Sliding between Rolling Elements and Cage Pockets

At any given azimuth location, there is generally a normal force acting between the rolling element and its cage pocket. This force can be positive or negative depending upon whether the rolling element is driving the cage or vice versa. It is also possible for a rolling element to be *free* in the pocket with no normal force exerted; however, this situation will be of less usual occurrence. Insofar as rotation of the rolling element about its own axes is concerned, the cage is stationary. Therefore, pure sliding occurs between rolling elements and cage pockets. The amount of friction that occurs thereby depends on the rolling element–cage normal loading, lubricant properties, rolling element speeds, and cage pocket geometry. The last variable is substantial in variety. Generally, application of simplified elastohydrodynamic theory should suffice to analyze the friction forces.

Sliding between Roller Ends and Ring Flanges

In a tapered roller bearing and in a spherical roller bearing having asymmetrical rollers, concentrated contacts always occur between the roller ends and the inner (or outer) ring flange owing to a force component that drives the rollers against the flange. Also, in a radial cylindrical roller bearing, which can support thrust load in addition to the predominant radial load by virtue of having flanges on both inner and outer rings (see Fig. 1.30), sliding occurs simultaneously between the roller ends and both inner and outer rings. In these cases, the geometries of the flanges and roller ends are extremely influential in determining the sliding friction between those contacting elements.

The most general case for roller end-flange contact occurs, as shown by Fig. 13.10, in a spherical roller thrust bearing. The different types of contact are illustrated in Table 13.1 for rollers having sphere ends.

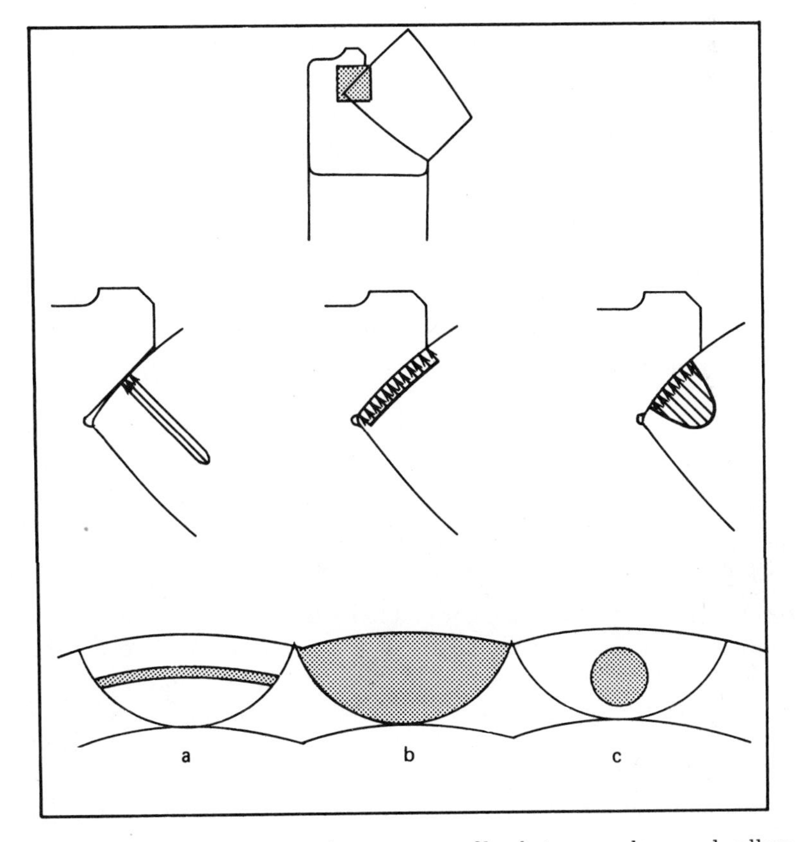

FIGURE 13.10. Contact types and pressure profiles between sphere end rollers and flanges in a spherical roller thrust bearing.

TABLE 13.1. Roller End–Flange Contact vs Geometry[a]

	Flange Geometry	Type of Contact
a	Portion of a cone	Line
b	Portion of sphere, $R_f = R_{re}$	Entire surface
c	Portion of sphere, $R_f > R_{re}$	Point

[a] R_f is the flange surface radius of curvature; R_{re} is the roller end radius of curvature.

Rydell [13.36] indicates that optimal frictional characteristics are achieved with point contacts between roller ends and flanges. Additionally Brown et al. [13.33] studied roller end wear criteria for high speed cylindrical roller bearings. They found that increasing roller corner radius runout tends to increase wear. Increasing roller end clearance and l/D ratio also tend toward increased roller wear, but, are of lesser consequence than roller corner radius runout.

Seals

An integral seal on a ball or roller bearing generally consists of an elastomer partially encased in a steel or plastic carrier. This is shown in Fig. 1.7.

The elastomeric sealing element bears either on a ring "land" or on a special recess in a ring. In either case, the seal friction normally substantially exceeds the sum total of all other sources of friction in the bearing unit. The technology of seal friction depends frequently on the specific mechanical structure of the seal and on the elastomeric properties. See Chapter 17 for some information on integral seals.

FRICTION FORCES AND MOMENTS IN ROLLING ELEMENT–RACEWAY CONTACTS

Ball Bearings

The sliding that occurs in the contact area has been discussed only qualitatively insofar as determination of friction forces is concerned. The analysis performed in Chapter 8 to evaluate the normal load on each ball and the contact angles took no account of friction forces in the contact other than to recognize the necessity to balance the gyroscopic moments which occur in angular-contact and thrust ball bearings. Of the many components that constitute the frictional resistance to motion in a ball–raceway contact, sliding is the most significant. It is further possible for the purpose of analysis to utilize a coefficient of friction even though the latter is a variable. Coefficient of friction in this section will be handled as a constant defined by

$$\mu = \frac{\tau}{\sigma} \tag{13.5}$$

where τ is surface shear stress and σ is the normal stress. Jones [13.4] first utilized the methods developed in this section.

In the ball–raceway elliptical contact area of a ball bearing consider a differential area of dS as shown by Fig. 13.11. The normal stress on the differential area is given by equation (5.43):

$$\sigma = \frac{3Q}{2\pi ab}\left[1 - \left(\frac{x}{a}\right)^2 - \left(\frac{y}{b}\right)^2\right]^{1/2} \tag{5.43}$$

In accordance with a sliding friction coefficient of friction μ, the differential friction force at dS is given by

$$dF = \frac{3\mu Q}{2\pi ab}\left[1 - \left(\frac{x}{a}\right)^2 - \left(\frac{y}{b}\right)^2\right]^{1/2} dS \tag{13.6}$$

The friction force of equation (13.6) has a component in the y direction $dF_y = dF \cos\phi$; therefore the total friction force in the y direction due to sliding is

$$F_y = \frac{3\mu Q}{2\pi ab}\int_{-a}^{+a}\int_{-b[1-(x/a)^2]^{1/2}}^{+b[1-(x/a)^2)]^{1/2}}\left[1 - \left(\frac{x}{a}\right)^2 - \left(\frac{y}{b}\right)^2\right]^{1/2}\cos\phi \, dy \, dx \tag{13.7}$$

Similarly, the friction force in the x direction is

$$F_x = \frac{3\mu Q}{2\pi ab}\int_{-a}^{+a}\int_{-b[1-(x/a)^2]^{1/2}}^{+b[1-(x/a)^2]^{1/2}}\left[1 - \left(\frac{x}{a}\right)^2 - \left(\frac{y}{b}\right)^2\right]^{1/2}\sin\phi \, dy \, dx \tag{13.8}$$

Since the differential friction force dF does not necessarily act at right angles to a radius drawn from the geometrical center of the contact ellipse, the moment of dF about the center of the contact ellipse is

$$dM_s = \rho\cos(\phi - \theta)\, dF \tag{13.9}$$

or

$$dM_s = (x^2 + y^2)^{1/2}\cos(\phi - \theta)\, dF \tag{13.10}$$

in which

$$\theta = \tan^{-1}\frac{y}{x} \tag{13.11}$$

The total frictional moment about the center of the contact ellipse is, therefore,

$$M_s = \frac{3\mu Q}{2\pi ab} \int_{-a}^{+a} \int_{-b[1-(x/a)^2]^{1/2}}^{+b[1-(x/a)^2]^{1/2}} (x^2 + y^2)^{1/2} \times \left[1 - \left(\frac{x}{a}\right)^2 - \left(\frac{y}{b}\right)^2\right]^{1/2} \cos(\phi - \theta)\, dy\, dx \tag{13.12}$$

Additionally, the moment of dF about the y' axis is (see Figs. 4.4, 7.13, 7.14, and 13.11)

$$dM_{y'} = \left\{(R^2 - x^2)^{1/2} - (R^2 - a^2)^{1/2} + \left[\left(\frac{D}{2}\right)^2 - a^2\right]^{1/2}\right\} \sin\phi\, dF \tag{13.13}$$

Integration of equation (13.13) over the entire contact ellipse yields

$$M_{y'} = \frac{3\mu Q}{2\pi ab} \int_{-a}^{+a} \int_{-b[1-(x/a)^2]^{1/2}}^{+b[1-(x/a)^2]^{1/2}} \left[1 - \left(\frac{x}{a}\right)^2 - \left(\frac{y}{b}\right)^2\right]^{1/2} \times \left\{(R^2 - x^2)^{1/2} - (R^2 - a^2)^{1/2} + \left[\left(\frac{D}{2}\right)^2 - a^2\right]^{1/2}\right\} \sin\phi\, dy\, dx \tag{13.14}$$

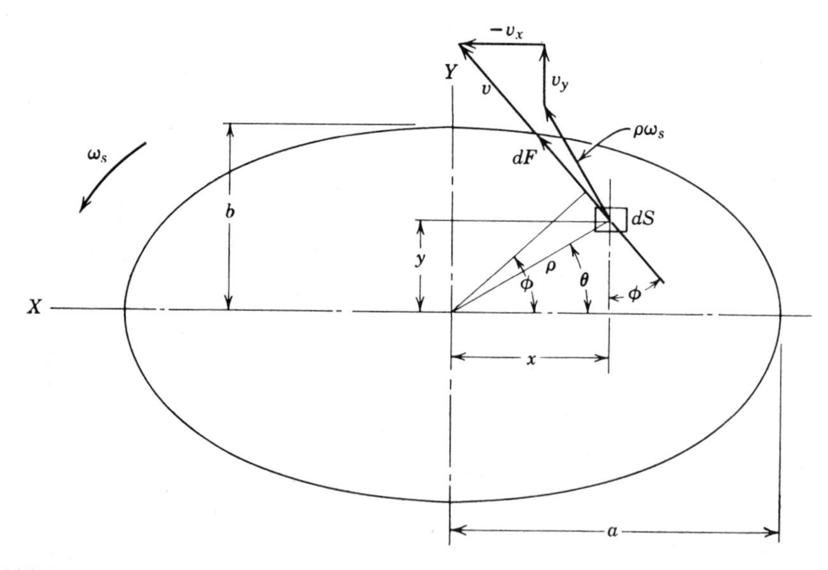

FIGURE 13.11. Friction force and sliding velocities acting on area dS of the elliptical contact surface.

Similarly, the frictional moment about an axis through the ball center perpendicular to the line defining the contact angle which line lies in the $x'\,z'$ plane of Fig. 4.4 is given by

$$M_R = \frac{3\mu Q}{2\pi ab} \int_{-a}^{+a} \int_{-b[1-(x/a)^2]^{1/2}}^{+b[1-(x/a)^2]^{1/2}} \left[1 - \left(\frac{x}{a}\right)^2 - \left(\frac{y}{b}\right)^2\right]^{1/2}$$
$$\times \left\{(R^2 - x^2)^{1/2} - (R^2 - a^2)^{1/2} + \left[\left(\frac{D}{2}\right)^2 - a^2\right]^{1/2}\right\} \cos\phi \, dy \, dx \tag{13.15}$$

Referring once again to Fig. 13.11, there are associated with area dS sliding velocities v_y and v_x according to equations (7.31) and (7.32) and (7.36) and (7.37) for the outer and inner raceway contacts, respectively. Also, there is associated with each contact a spinning speed ω_s according to equations (7.33) and (7.38). These velocities determine the angle ϕ (see Fig. 13.11) such that

$$\phi = \tan^{-1} \frac{\rho\omega_s \sin\theta - v_x}{\rho\omega_s \cos\theta + v_y} \tag{13.16}$$

Therefore,

$$\phi = \tan^{-1} \frac{y\omega_s - v_x}{x\omega_s + v_y} \tag{13.17}$$

The moments acting on a ball, both gyroscopic and frictional, are shown in Fig. 13.12. $M_{y'}$ and $M_{z'}$ may be calculated from equations (4.35) and (4.36), respectively. The summation of the moments in each direction must equal zero; therefore,

$$-M_{Ro} \sin\alpha_o + M_{so} \cos\alpha_o + M_{z'} + M_{Ri} \sin\alpha_i - M_{si} \cos\alpha_i = 0 \tag{13.18}$$

$$-M_{Ro} \cos\alpha_o - M_{so} \sin\alpha_o + M_{Ri} \cos\alpha_i + M_{si} \sin\alpha_i = 0 \tag{13.19}$$

$$M_{gy'} - M_{yo} - M_{yi} = 0 \tag{13.20}$$

The forces acting on a ball can be disposed as in Fig. 13.13. $F_{z'}$ is the ball centrifugal force defined by equation (4.34). F_y and F_x are defined by equations (13.7) and (13.8), respectively. From Fig. 13.13 it can be seen that equation (13.20) becomes

$$\tfrac{1}{2} D(F_{xi} + F_{xo}) - M = 0 \tag{13.21}$$

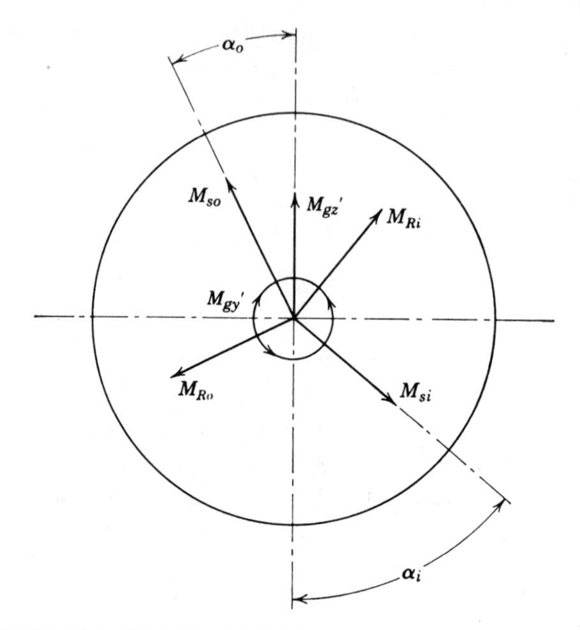

FIGURE 13.12. Gyroscopic and frictional moments acting on a ball.

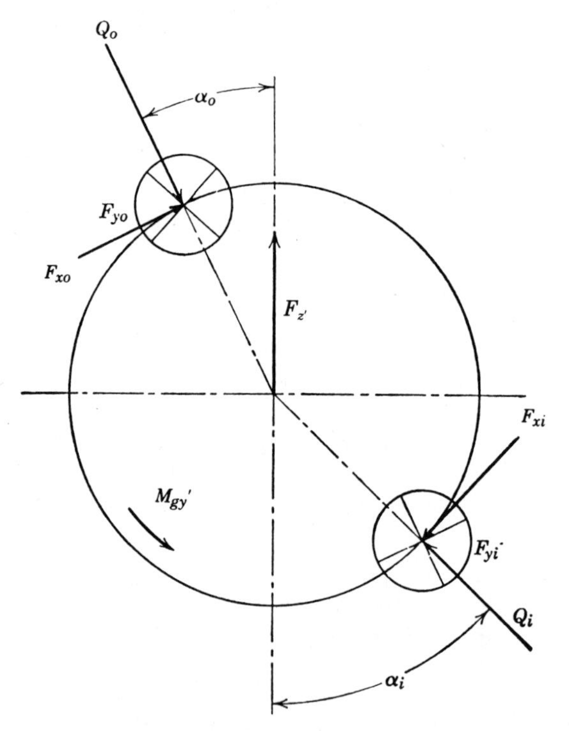

FIGURE 13.13. Centrifugal, normal and frictional forces acting on a ball. *Note:* F_{yo} and F_{yi} act normal to the plane of the paper.

and

$$F_{yi} + F_{yo} = 0 \tag{13.22}$$

Note also that equations (13.18) and (13.19) can be combined to yield

$$-M_{Ro}(\sin\alpha_o + \cos\alpha_o) + M_{so}(\cos\alpha_o - \sin\alpha_o)$$
$$+ M_{Ri}(\sin\alpha_i + \cos\alpha_i) - M_{si}(\cos\alpha_i - \sin\alpha_i) + M_{z'} = 0 \tag{13.23}$$

Simplifying assumptions may be made at this point for relatively slow speed bearings such that ball gyroscopic moment is negligible and that outer raceway control is approximated. Although the latter is not necessarily true of slow speed bearings, the result of calculations using these assumptions will permit the investigator to obtain a qualitative idea of the sliding zones in the ball–raceway contacts and an order of magnitude idea of friction in the contacts. Moreover, Q_o, Q_i, α_o, and α_i may be determined by methods of Chapters 6 or 8. Therefore, to calculate the frictional forces and moments in the contact area, one needs only to determine the radii of rolling r'_o and r'_i.

In Chapter 7 it was demonstrated that pure rolling can occur at most at two points in the contact area. If spinning is absent at a raceway contact, then all points on lines parallel to the direction of rolling and passing through the aforementioned points of *pure* rolling roll without sliding. The sliding velocities v_{yo} or v_{yi} are defined by equations (7.25) and (7.31), respectively; The distribution of sliding velocity on the contact surface is illustrated by Fig. 13.14. As in Fig. 13.14 the lines of pure

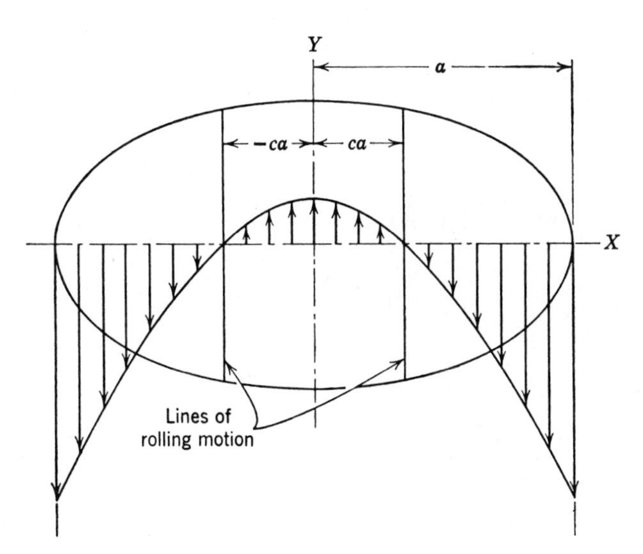

FIGURE 13.14. Distribution of sliding velocity on the elliptical contact surface for negligible gyroscopic motion and zero spin.

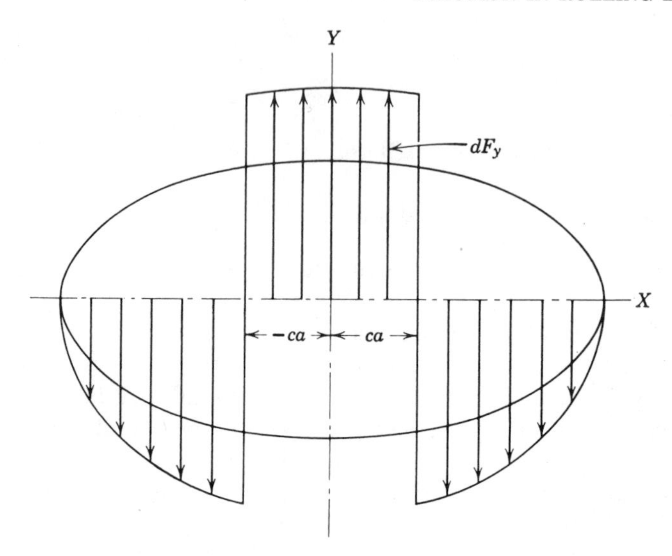

FIGURE 13.15. Distribution of sliding friction forces dF_y on the elliptical contact surface.

rolling lie at $x = \pm\, ca$. Then the frictional forces of sliding are distributed as in Fig. 13.15. Using equation (13.6) to describe the differential frictional force dF, it can be seen that the net sliding frictional force in the direction of rolling at a raceway contact is

$$F_y = \pm \frac{3\mu Q}{2\pi ab}\left\{\int_0^{ca}\int_{-b[1-(x/a)^2]^{1/2}}^{+b[1-(x/a)^2]^{1/2}}\left[1-\left(\frac{x}{a}\right)^2-\left(\frac{y}{b}\right)^2\right]^{1/2} dy\, dx - \int_{ca}^{a}\int_{-b[1-(x/a)^2]^{1/2}}^{+b[1-(x/a)^2]^{1/2}}\left[1-\left(\frac{x}{a}\right)^2-\left(\frac{y}{b}\right)^2\right]^{1/2} dy\, dx\right\} \tag{13.24}$$

Performing the integration of equation (13.24) yields

$$F_y = \pm\mu Q(3c - c^3 - 1) \tag{13.25}$$

Thus, for a given value F_y obtainable from equation (13.7), the value c may be established. Referring to Fig. 7.13 or 7.14, it can be seen that the radius of rolling is given by

$$r' = [R^2 - (ca)^2]^{1/2} - (R^2 - a^2)^{1/2} + \left[\left(\frac{D}{2}\right)^2 - a^2\right]^{1/2} \tag{13.26}$$

or

$$r' = R\left\{\left[1 - \left(\frac{ca}{R}\right)^2\right]^{1/2} - \left[1 - \left(\frac{a}{R}\right)^2\right]^{1/2} + \left[\left(\frac{D}{2R}\right) - \left(\frac{a}{R}\right)^2\right]^{1/2}\right\} \quad (13.27)$$

The rolling moment dM_R about the U axis through the center of the ball as determined from Fig. 7.13 or 7.14 is

$$dM_R = \left\{(R^2 - x^2)^{1/2} - (R^2 - a^2)^{1/2} + \left[\left(\frac{D}{2}\right)^2 - a^2\right]^{1/2}\right\} dF \quad (13.28)$$

Rearranging equation (13.28) and converting to integral form yields

$$\begin{aligned} M_R = \pm \frac{3\mu QR}{\pi ab} & \\ \times \Bigg(\int_0^{ca} \int_{-b[1-(x/a)^2]^{1/2}}^{+b[1-(x/a)^2]^{1/2}} & \left\{\left[1 - \left(\frac{x}{R}\right)^2\right]^{1/2} - \left[1 - \left(\frac{a}{R}\right)^2\right]^{1/2} \right. \\ & \left. + \left[\left(\frac{D}{2R}\right)^2 - \left(\frac{a}{R}\right)^2\right]\right\}^{1/2} \left[1 - \left(\frac{x}{a}\right)^2 - \left(\frac{y}{b}\right)^2\right]^{1/2} dy\, dx \\ - \int_{ca}^{a} \int_{-b[1-(x/a)^2]^{1/2}}^{+b[1-(x/a)^2]^{1/2}} & \left\{\left[1 - \left(\frac{x}{R}\right)^2\right]^{1/2} - \left[1 - \left(\frac{a}{R}\right)^2\right]^{1/2} \right. \\ & \left. + \left[\left(\frac{D}{2R}\right)^2 - \left(\frac{a}{R}\right)^2\right]^{1/2}\right\} \left[1 - \left(\frac{x}{a}\right)^2 - \left(\frac{y}{b}\right)^2\right]^{1/2} dy\, dx \Bigg) \end{aligned} \quad (13.29)$$

Performing the indicated integration and rearranging yields

$$\begin{aligned} M_R = \pm \frac{3\mu QR}{4 \sin \Gamma_1} & \\ \times \Bigg(& \sin 2\Gamma_2 - \tfrac{1}{2} \sin 2\Gamma_1 - \frac{(\sin 4\Gamma_1 - 2 \sin 4\Gamma_2)}{16 \sin^2 \Gamma_1} \\ & + (\Gamma_1 - 2\Gamma_2)\left(\frac{1}{4 \sin^2 \Gamma_1} - 1\right) \\ & - \tfrac{2}{3}\left\{\sin 2\Gamma_1 - 2 \sin \Gamma_1 \left[\left(\frac{D}{2R}\right)^2 - \sin^2 \Gamma_1\right]\right\}(3c - c^3 - 1)\Bigg) \end{aligned} \quad (13.30)$$

in which

$$\sin \Gamma_1 = \frac{a}{R} \tag{13.31}$$

$$\sin \Gamma_2 = \frac{ca}{R} \tag{13.32}$$

It is now possible to calculate r'_o and r'_i. The steps are as follows:

1. Assume $r'_o = r'_i = r$ and calculate centrifugal force F_c from equation (8.18); ω_m/ω is determined from equation (7.63) or (7.64). It is recognized in the calculation of F_c and ω_m/ω that pitch diameter is a variable defined as follows (see Fig. 8.2):

$$d_{mj} = d_m + 2\{[(f_o - 0.5)D + \delta_{oj}] \cos \alpha_{oj} - (f_o - 0.5)D \cos \alpha^\circ\} \tag{13.33}$$

wherein δ_{oj} is obtained from equation (8.12).

2. Calculate F_{yi} from equation (13.7), M_{si} from (13.12), and M_{Ri} from (13.15). The angle ϕ is calculated by using equation (13.17). v_{xi} and v_{yi} are determined by using equations (7.36) and (7.37) for $\beta' = 0$. Numerical integrations are generally necessary to complete these calculations.

3. Having determined F_{yi}, calculate $F_{yo} = -F_{yi}$ by equation (13.22). Thereafter, calculate c by using equation (13.25). The upper sign of equation (13.25) is used if v_{yo} at $x = 0$ is positive.

4. Having determined c, calculate M_{so} from equation (13.23) for $M'_z = 0$.

5. The final condition to be satisfied is that the input torque at each ball location must equal the output torque. Hence,

$$M_{Ro} \frac{\left(\frac{d_m}{2} + r'_o \cos \alpha_o\right)}{r'_o} + M_{so} \sin \alpha_o + M_{Ri} \frac{\left(\frac{d_m}{2} - r'_i \cos \alpha_i\right)}{r'_i} - M_{si} \sin \alpha_i = 0 \tag{13.34}$$

If equation (13.34) is not satisfied, a new value of c_i, that is, r'_i, is assumed and the process is repeated until equation (13.34) is satisfied.

If the motion of a raceway relative to the ball was merely a spinning about the normal to the center of the contact area, all other relative surface velocities being reduced to zero, the magnitude of the spinning moment as determined from equation (13.12) for $\phi = \theta$ is given by

$$M_s = \frac{3\mu Q a \mathcal{E}}{8} \tag{13.35}$$

in which $\mathcal{E}$ is the complete elliptic integral of the second kind with modulus $[1 - (b/a)^2]^{1/2}$. For the condition of outer raceway control M_{so} as calculated from equation (13.23) for rolling and spinning is less than M_s as calculated from (13.35) for the outer raceway contact with only spinning motion.

Example 13.1. For the 218 angular-contact ball bearing of Example 8.1, estimate the friction torque due to spinning about the axis normal to the inner raceway contact area for the condition of 22,250 N (5000 lb) thrust load and 10,000 rpm shaft speed. Assume a coefficient of friction equal to 0.03.

$$\alpha_i = 48.8° \qquad \text{Fig. 8.4}$$

$$D = 22.23 \text{ mm } (0.875 \text{ in.}) \qquad \text{Ex. 2.3}$$

$$d_m = 125.3 \text{ mm } (4.932 \text{ in.}) \qquad \text{Ex. 2.6}$$

$$Q_i = 1788 \text{ N } (401.7 \text{ lb}) \qquad \text{Fig. 8.6}$$

$$f_i = 0.5232 \qquad \text{Ex. 2.3}$$

$$\gamma_i = \frac{D \cos \alpha_i}{d_m} \tag{2.27}$$

$$= \frac{22.23 \cos (48.8°)}{125.3} = 0.1169$$

$$\Sigma\rho_i = \frac{1}{D}\left(4 - \frac{1}{f_i} + \frac{2\gamma_i}{1 - \gamma_i}\right) \tag{2.28}$$

$$= \frac{1}{22.23}\left(4 - \frac{1}{0.5232} + \frac{2 \times 0.1169}{1 - 0.1169}\right)$$

$$= 0.1058 \text{ mm}^{-1} (2.690 \text{ in.}^{-1})$$

$$F(\rho)_i = \frac{1/f_i + (2\gamma_i)/(1 - \gamma_i)}{4 - 1/f_i + 2\gamma_i/(1 - \gamma_i)} \tag{2.29}$$

$$= \frac{1/0.5232 + (2 \times 0.1169)/(1 - 0.1169)}{4 - 1/0.5232 + (2 \times 0.1169)/(1 - 0.1169)}$$

$$= 0.9244$$

From Fig. 5.4, $a_i^* = 3.47$; $b_i^* = 0.433$

$$a_i = 0.0236 a_i^* \left(\frac{Q_i}{\Sigma\rho_i}\right)^{1/3} \tag{5.34}$$

$$= 0.0236 \times 3.47 \times \left(\frac{1788}{0.1058}\right)^{1/3} = 2.101 \text{ mm } (0.0827 \text{ in.})$$

$$\kappa_i = \frac{a_i^*}{b_i^*}$$

$$= \frac{3.47}{0.433} = 8.01$$

$$b_i^* = \left(\frac{2\mathcal{E}_i}{\pi\kappa_i}\right)^{1/3} \tag{5.40}$$

$$0.433 = \left(\frac{2\mathcal{E}_i}{3.1416 \times 8.01}\right)^{1/3}$$

$$\mathcal{E}_i = 1.022$$

$$M_{si} = \frac{3\mu Q_i a_i \mathcal{E}_i}{8}$$

$$= \frac{3 \times 0.03 \times 1788 \times 2.101 \times 1.022}{8} \tag{13.35}$$

$$= 43.19 \text{ N} \cdot \text{mm } (0.382 \text{ in.} \cdot \text{lb})$$

Thus far, the solution of the friction force and moment equilibrium equations has assumed that outer raceway control was approximated. A more general solution was achieved by Harris [13.18] for a thrust-loaded angular contact ball bearing operating with Coulomb friction in the ball–raceway contacts. In this case, the forces and moments actirg on a ball are shown in Fig. 13.16.

Gyroscopic motion about the axis y' is assumed negligible and the contact ellipse is divided into two or three sliding zones as shown in Fig. 13.17.

Now for the raceway contacts as shown in Fig. 13.17,

$$F_{y'n} = 2\mu a_n b_n c_n \left(\int_{-1}^{T_{n1}} \int_0^{\sqrt{1-q^2}} p_n \, dt \, dq - \int_{T_{n1}}^{T_{n2}} \int_0^{\sqrt{1-q^2}} p_n \, dt \, dq + \int_{T_{n2}}^{1} \int_0^{\sqrt{1-q^2}} p_n \, dt \, dq \right) \tag{13.36}$$

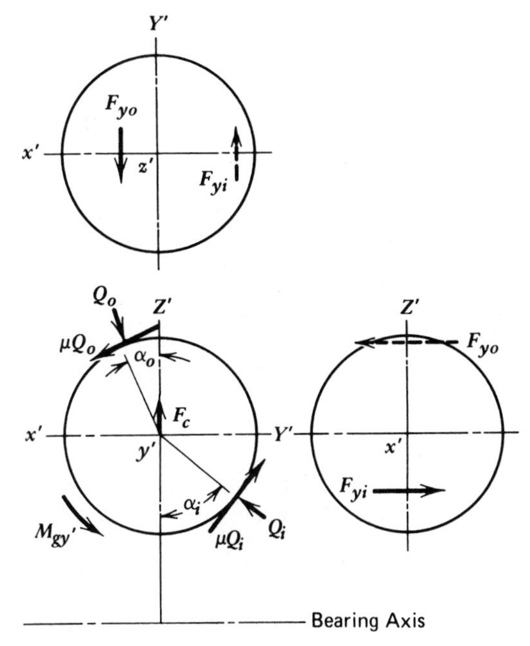

FIGURE 13.16. Forces and moments acting on a ball.

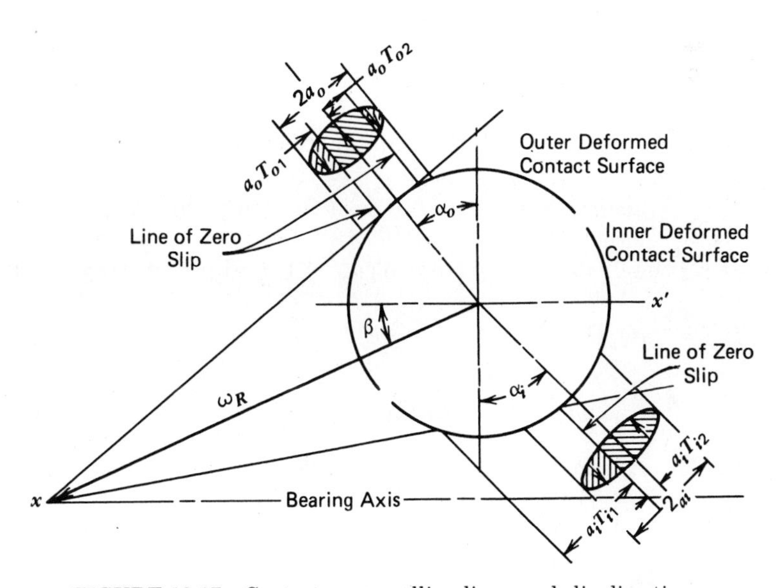

FIGURE 13.17. Contact areas, rolling lines, and slip directions.

where $q = x'/a_n$, $t = y'/b_n$, T_{n1}, and T_{n2} define rolling lines, n refers to inner or outer ball–raceway contact, that is, $n = \mathrm{o}$ or $n = \mathrm{i}$; and σ_n the pressure at any point in the contact ellipse is given by

$$\sigma_n = \frac{3Q_n}{2\pi a_n b_n}(1 - q^2 - t^2)^{1/2} \tag{13.37}$$

Substituting equation (13.37) into (13.36) and integrating yields

$$F_{y'n} = 3\mu Q_n c_n \left[\frac{2}{3} + \sum_{k=1}^{k=2} c_k T_{nk}\left(1 - \frac{T_{nk}^2}{3}\right)\right] \qquad \begin{array}{l} n = \mathrm{o,i} \\ c_\mathrm{o} = 1; \quad c_\mathrm{i} = -1 \\ c_1 = 1; \quad c_2 = -1 \end{array} \tag{13.38}$$

Using Figures 7.13 and 7.14 to define the radii r_n from the ball center to points on the inner and outer ball–raceway contact areas, the equations from frictional moments are

$$\begin{aligned} M_{x'n} = 2\mu a_n b_n c_n \Bigg[& \int_{-1}^{T_{n1}} \int_0^{\sqrt{1-q^2}} \sigma_n r_n \cos(\alpha_n + \theta_n)\, dt\, dq \\ & - \int_{T_{n1}}^{T_{n2}} \int_0^{\sqrt{1-q^2}} \sigma_n r_n \cos(\alpha_n + \theta_n)\, dt\, dq \\ & + \int_{T_{n2}}^{1} \int_0^{\sqrt{1-q^2}} \sigma_n r_n \cos(\alpha_n + \theta_n)\, dt\, dq \Bigg] \\ & n = \mathrm{o,\ i} \\ & c_\mathrm{o} = 1; \quad c_\mathrm{i} = -1 \end{aligned} \tag{13.39}$$

where $\sin\theta_n = x'/r_n$. Using the trigonometric identity

$$\cos(\alpha_n + \theta_n) = \cos\alpha_n \cos\theta_n - \sin\alpha_n \sin\theta_n \tag{13.40}$$

recognizing the θ_n is small giving $\cos\theta_n \to 1$, and integrating yields

$$\begin{aligned} M_{x'n} = {} & 3\mu Q_n D c_n \\ & \times \Bigg\{ \tfrac{2}{3}\cos\alpha_n + \sum_{k=1}^{k=2} c_k T_{nk} \\ & \times \left[\left(1 - \frac{T_{nk}^2}{3}\right)\cos\alpha_n - \frac{a_n T_{nk}}{D}\left(1 - \frac{T_{nk}^2}{2}\right)\sin\alpha_n\right]\Bigg\} \\ & n = \mathrm{o,\ i}; \quad c_\mathrm{o} = 1; \quad c_\mathrm{i} = -1 \\ & k = 1, 2; \quad c_1 = 1; \quad c_2 = -1 \end{aligned} \tag{13.41}$$

Similarly,

$$
\begin{aligned}
M_{z'n} &= 3\mu Q_n D c_n \\
&\times \left\{ \tfrac{2}{3} \sin \alpha_n + \sum_{k=1}^{k=2} c_k T_{nk} \right. \\
&\left. \times \left[\left(1 - \frac{T_{nk}^2}{3}\right) \sin \alpha_n + \frac{a T_{nk}}{D} \left(1 - \frac{T_{nk}^2}{2}\right) \cos \alpha_n \right] \right\} \\
n &= \mathrm{o}, \mathrm{i}; \quad c_\mathrm{o} = 1; \quad c_\mathrm{i} = -1 \\
k &= 1, 2; \quad c_1 = 1; \quad c_2 = -1
\end{aligned} \tag{13.42}
$$

Using Fig. 13.16 it can be established that four conditions of force and moment equilibrium about the x', y', and z' axes must be satisfied together with four ball position equations determined in Chapter 8. These eight equations must be solved for two position variables, two contact deformations, bearing axial deflection, and speeds ω_m, $\omega_{x'}$, and $\omega_{z'}$.

Thus, there are eight equations and eight unknowns; but, the rolling lines T_{nk}, of which there are three as shown in Fig. 13.17, are functions of speed ω_m, $\omega_{x'}$, and $\omega_{z'}$. To establish the required relationship, the major axes of the deformed contact surfaces as shown by Figs. 7.13 and 7.14 are considered arcs of great circles defined by

$$(x_n' - X_n)^2 + (z_n' - Z)^2 - (\zeta_n D)^2 = 0 \tag{13.43}$$

where $\zeta = 2f/(2f + 1)$ and $f = r/D$. From Figs. 7.13 and 7.14, it can be determined that the offset of the ball center from the circle center is given by the coordinates

$$X = \frac{D}{2} [(4\zeta_n^2 - k_n^2)^{1/2} - (1 - k_n^2)^{1/2}] \sin \alpha_n \tag{13.44}$$

$$Z = \frac{D}{2} [(4\zeta_n^2 - k_n^2)^{1/2}] \cos \alpha_n \tag{13.45}$$

where $k_n = 2a_n/D$. Zero sliding velocity is determined from the equations

$$(\Omega_\mathrm{o} - \omega_\mathrm{m}) \left(\frac{d_\mathrm{m}}{2} + z'\right) + \omega_{x'} z' + \omega_{z'} x' = 0 \tag{13.46}$$

$$(\omega_\mathrm{m} - \Omega_\mathrm{i}) \left(\frac{d_\mathrm{m}}{2} - z'\right) + \omega_{x'} z' + \omega_{z'} x' = 0 \tag{13.47}$$

Equations (13.43), (13.46), and (13.47) can be solved simultaneously to yield x_{nk}', z_{nk}' locations at which zero sliding velocity occurs on the deformed surface circle. It can be shown that

$$T_{nk} = \frac{1}{a_n}(x'^2_{nk} + z'^2_{nk})^{1/2} \sin\left[\frac{\pi}{2} - \alpha_n - \tan^{-1}\left(\frac{z'_{nk}}{x'_{nk}}\right)\right] \qquad k = 1, 2 \qquad (13.48)$$

Using the foregoing method Harris [13.18] was able to prove the impossibility of an "inner raceway control" situation, even with bearings operating with "dry film" lubrication. Moreover, a speed transition point seems to occur in a thrust-loaded angular-contact ball bearing at which a radical shift of the ball speed pitch angle β must occur to achieve load equilibrium in the bearing (see Figs. 7.16, 7.17, 13.17, 13.18, and 13.19)

Additionally, Table 13.2 shows the corresponding locations of rolling lines in the inner and outer contact ellipses for this example.

Roller Bearings

A similar approach may be applied to roller bearings having point contact at each raceway. Usually, however, roller bearings, are designed to operate in the line contact or modified line contact regime (see Chapter 5) in which the area of contact is essentially rectangular, it generally being an ellipse truncated at each end of the major axis (see Fig. 5.18). In this case the major sliding forces on the contact surface are essentially parallel to the direction of rolling and are principally due to the deformation of the surface. Thus, the sliding forces acting on the contact sur-

Bearing Design Data	
Ball diameter	8.731 mm (0.34375 in.)
Pitch diameter	48.54 mm (1.9110 in.)
Free contact angle	24.5°
Inner raceway grove radius/ball diameter	0.52
Outer raceway groove radius/ball diameter	0.52
Thrust load per ball	31.6 N (7.1 lb)

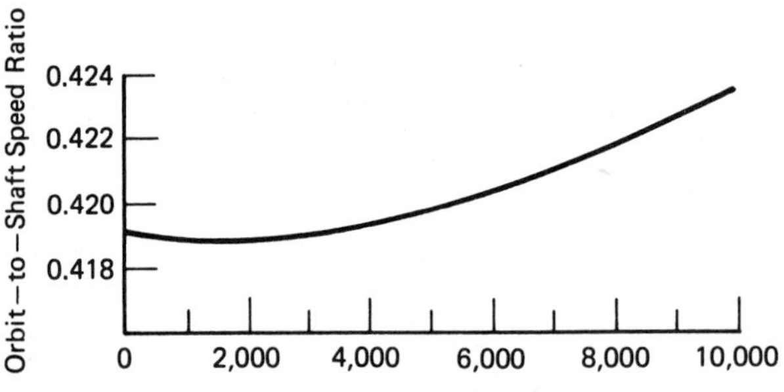

FIGURE 13.18. Orbit/shaft speed ratio vs shaft speed.

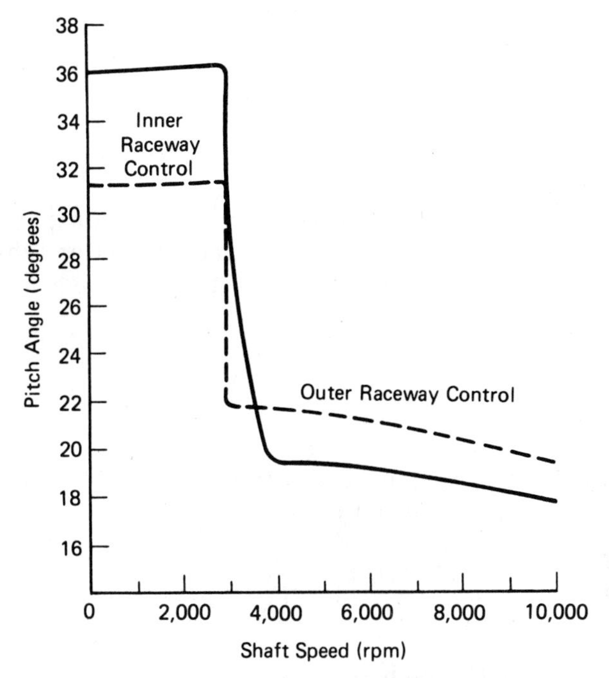

FIGURE 13.19. Ball speed vector pitch angle vs shaft speed.

TABLE 13.2. Locations of Lines of Zero Slip in Contact Ellipses

Shaft Speed	Outer Raceway T_1	Outer Raceway T_2	Inner Raceway T_1	Inner Raceway T_2
1000	0.0001	—	−0.00605	0.92123
1500	0.00183	—	−0.00672	0.92376
2000	0.00129	—	−0.00537	0.93140
2500	0.00047	—	−0.00353	0.94272
3000	—	0.02975	0.02995	—
3500	—	−0.00156	—	−0.00190
4000	−0.95339	0.00156	—	0.00052
4500	−0.93237	0.00376	—	0.00064
5000	−0.91449	0.00627	—	0.00077
5500	−0.89730	0.01055	—	−0.00039

faces of a loaded roller bearing are usually less complex than for ball bearings.

Dynamic loading of roller bearings does not generally affect contact angles, and hence geometry of the contacting surfaces is virtually identical to that occurring under static loading. Because of the relatively slow speeds of operation necessitated when contact angle differs from zero degrees, gyroscopic moments are negligible. In any event, gyroscopic moments of any magnitude do not substantially alter normal motion of the rolling elements. In this analysis therefore, the sliding on the contact surface of a properly designed roller bearing will be assumed to be a function only of the radius of the deformed contact surface in a direction transverse to rolling.

To perform the analysis, it is assumed that the contact area between roller and either raceway is substantially rectangular and that the normal stress at any distance from the center of the rectangle is adequately defined by

$$\sigma = \frac{2Q}{\pi l b}\left[1 - \left(\frac{y}{b}\right)^2\right]^{1/2} \tag{5.45}$$

Thus, the differential friction force acting at any distance x from the center of the rectangle is given by

$$dF_y = \frac{2\mu Q}{\pi l b}\left[1 - \left(\frac{y}{b}\right)^2\right]^{1/2} dy\, dx \tag{13.49}$$

Integrating equation (13.49) between $y = \pm b$ yields

$$dF_y = \frac{\mu Q}{l}\, dx \tag{13.50}$$

Referring to Fig. 13.20, it can be determined that the differential frictional moment in the direction of rolling at either raceway is given by

$$dM_{\mathrm{R}} = \left[(R^2 - x^2)^{1/2} - \left(R - \frac{D}{2}\right)\right] dF \tag{13.51}$$

or

$$dM_{\mathrm{R}} = \frac{2\mu Q}{\pi l b}\left[1 - \left(\frac{y}{b}\right)^2\right]^{1/2}\left[(R^2 - x^2)^{1/2} - \left(R - \frac{D}{2}\right)\right] dy\, dx \tag{13.52}$$

in which R is the radius of curvature of the deformed surface. Integrating equation (13.52) with respect to y between limits $y = \pm b$ yields

$$dM_{\mathrm{R}} = \frac{\mu Q}{l}\left[(R^2 - x^2)^{1/2} - \left(R - \frac{D}{2}\right)\right] dx \tag{13.53}$$

Because of the curvature of the deformed surface, *pure* rolling exists at most at two points $x = \pm (cl)/2$ on the deformed surface; the radius of

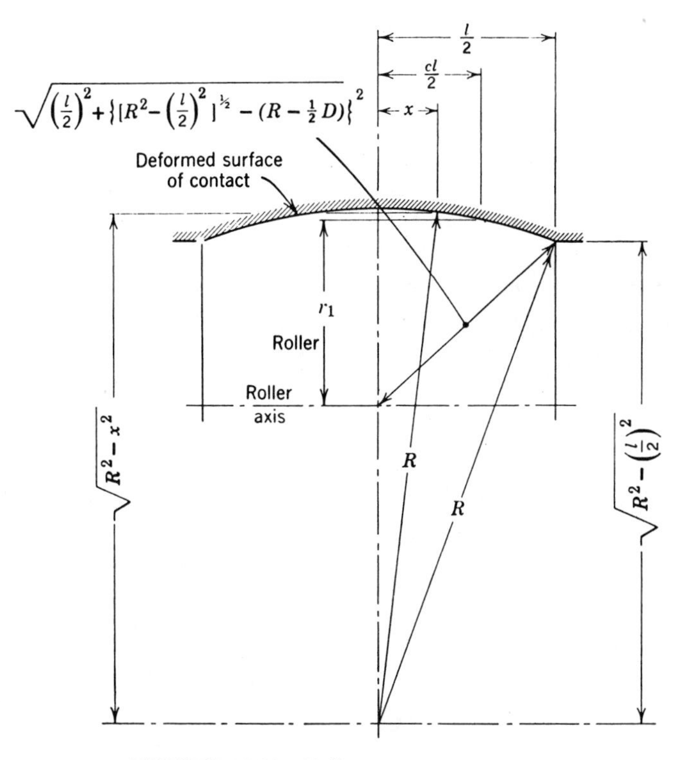

FIGURE 13.20. Roller–raceway contact.

rolling measured from the roller axis of rotation is r'. Thus

$$F_y = \frac{2\mu Q}{l}\left(\int_0^{(cl)/2} dx - \int_{(cl)/2}^{l/2} dx\right) \tag{13.54}$$

or

$$F_y = \mu Q(2c - 1) \tag{13.55}$$

Also

$$M_R = \frac{2\mu Q}{l}\left\{\int_0^{(cl)/2}\left[(R^2 - x^2)^{1/2} - \left(R - \frac{D}{2}\right)\right] dx - \int_{(cl)/2}^{l/2}\left[(R^2 - x^2)^{1/2} - \left(R - \frac{D}{2}\right)\right] dx\right\} \tag{13.56}$$

or

$$M_R = \mu Q \left\{ \frac{R^2}{l} \left(4 \sin^{-1} \frac{cl}{2R} - \sin^{-1} \frac{l}{2R} \right) + (1 - 2c) \left(R - \frac{D}{2} \right) + cR \left[1 - \left(\frac{cl}{2R} \right)^2 \right]^{1/2} - \frac{R}{2} \left[1 - \left(\frac{2R}{l} \right)^2 \right]^{1/2} \right\} \tag{13.57}$$

Considering the equilibrium of forces acting on the roller at the inner and outer raceway contacts (see Fig. 13.21), $F_{yo} = -F_{yi}$. Therefore, from (13.55) assuming $\mu_o = \mu_i$:

$$c_o + c_i = 1 \tag{13.58}$$

Furthermore, since in uniform rolling motion the sum of the torques at the outer and inner raceway contacts is equal to zero, therefore

$$M_{Ro} \frac{\left(\frac{d_m}{2} + r'_o \right)}{r'_o} + M_{Ri} \frac{\left(\frac{d_m}{2} - r'_i \right)}{r'_i} = 0 \tag{13.59}$$

From Fig. 13.20, it can be seen that the roller radius of rolling is

$$r' = \left[R^2 - \left(\frac{cl}{2} \right)^2 \right]^{1/2} - \left(r - \frac{D}{2} \right) \tag{13.60}$$

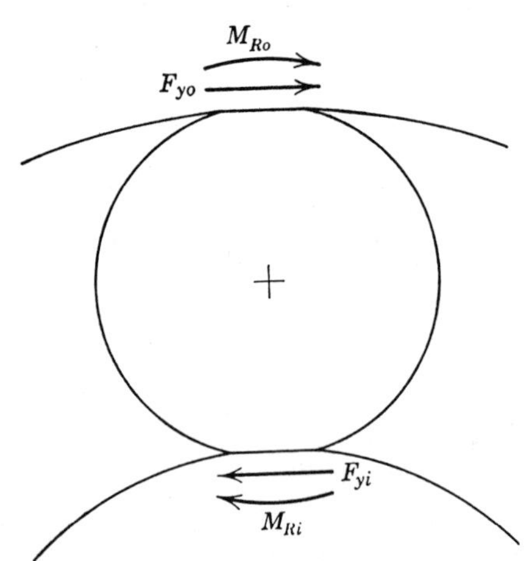

FIGURE 13.21. Friction forces and moments acting on a roller.

Hence, assuming $\mu_o = \mu_i$, from equations (13.57), (13.59), and (13.60):

$$\left\{\frac{R_o^2}{l}\left(2\sin^{-1}\frac{c_o l}{2R_o} - \sin^{-1}\frac{l}{2R_o}\right) + (1-2c_o)\left(r_o - \frac{D}{2}\right)\right.$$
$$\left. + c_o R_o\left[1-\left(\frac{c_o l}{2R_o}\right)^2\right]^{1/2} - \frac{R_o}{2}\left[1-\left(\frac{l}{2R_o}\right)^2\right]^{1/2}\right\}$$
$$\times\left\{1 + \frac{d_m}{2\left\{\left[R_o^2 - \left(\frac{c_o l}{2}\right)^2\right]^{1/2} - \left(R_o - \frac{D}{2}\right)\right\}}\right\}$$
$$-\left\{\frac{R_i^2}{l}\left(2\sin^{-1}\frac{c_i l}{2R_i} - \sin^{-1}\frac{l}{2R_i}\right) + (1-2c_i)\left(R_i - \frac{D}{2}\right)\right.$$
$$\left. + c_i R_i\left[1-\left(\frac{c_i l}{2R_i}\right)^2\right]^{1/2} - \frac{R_i}{2}\left[1-\left(\frac{l}{2R_i}\right)^2\right]^{1/2}\right\}$$
$$\times\left\{\frac{d_m}{2\left[\left[R_i^2 - \left(\frac{c_i l}{2}\right)^2\right]^{1/2} - \left(R_i - \frac{D}{2}\right)\right]} - 1\right\} = 0 \tag{13.61}$$

Equations (13.58) and (13.61) can be solved simultaneously for c_o and c_i. Note that if R_o and R_i, the radii of curvature of the outer and inner contact surfaces respectively are infinite, the foregoing analysis does not apply. In this case sliding on the contact surfaces is obviated and only rolling occurs.

Having determined c_o and c_i, one may revert to equation (13.55) to determine the net sliding forces F_{yo} and F_{yi}. Similarly, M_{Ro} and M_{Ri} may be calculated from equation (13.57).

SKIDDING AND CAGE FORCES

In all of the analytical development regarding rolling element and cage speeds so far, at least one location could be found in each of the rolling element–raceway contact areas that was an instant center; that is, at that location no relative motion (sliding) occurs between the contacting surfaces. If during bearing operation, no instant center can be found in either the inner or outer raceway contacts, particularly at the azimuth location of the most heavily loaded rolling element, then skidding is said to occur. Skidding is therefore gross sliding of a contact surface relative to the opposing surface. Skidding results in surface shear stresses of significant magnitudes in the contact areas. If the lubricant film generated by the relative motion of the rolling element–raceway surfaces is insufficient to completely separate the surfaces, surface damage called *smearing* will occur. An example of smearing is shown by Fig. 13.22. Tallian [13.19] defines smearing as a severe type of wear characterized by metal tightly bonded to the surface in locations into which it has been transferred from remote locations of the same or opposing surfaces and the transferred metal is present in sufficient volume to connect more than one distinct asperity contact. When the number of asperity contacts connected is small, it is called *microsmearing*. When the number of such contacts is large enough to be seen with the unaided eye, this is called *gross* or *macroscopic* smearing.

If possible, skidding is to be avoided in any application since at the very least it results in increased friction and heat generation even if smearing does not occur. Skidding can occur in high speed operation of liquid-lubricated ball and roller bearings. Rolling element centrifugal forces in such applications tend to cause higher normal load at the outer raceway–rolling element contact as compared to the inner raceway–rolling element contact at any azimuth location. Therefore, the balance of the friction forces and moments acting on a rolling element requires a higher coefficient of friction at the inner raceway contact to compensate for the lower normal contact load thereat. It was shown in Chapter 12 that the lubricant film thickness generated in a fluid film-lubricated rolling element–raceway contact depends upon the velocities of the surfaces in contact. Moreover, considering as a simplistic case Newtonian lubrication, the surface shear stress is a direct function of the sliding velocity of the surfaces and an inverse function of the lubricant film thickness. Hence, considering equations (13.1) and (13.5), the coefficient of friction in the contact is a function of sliding speed, which is greatest at the inner raceway contacts. Generally, skidding can be minimized by increasing the applied load on the bearing, thus decreasing the relative magnitude of the rolling element centrifugal force to the contact load at the most heavily loaded rolling element. As will be seen in Chapter 18, this remedy will tend to reduce fatigue endurance. Therefore, a compromise between the degree of skidding allowed and bearing endurance must be accepted. Of course, by making the contacting surfaces extremely smooth, the effectiveness of the lubricant film thicknesses is improved, and skidding is more tolerable.

Notwithstanding, skidding is generally a high speed phenomenon caused by a difference between inner and outer raceway–rolling element loading; it is also aggravated by any rolling element or cage loading that tends to retard motion. The most significant of such loadings is the viscous drag of the lubricant in the bearing cavity on the rolling elements. Therefore, a high speed bearing operating submerged in lubricant will skid more than the same bearing operating in mist-type lubrication. In this case another compromise is required because, in a high speed appli-

FIGURE 13.22. Raceway surface smearing damage caused by skidding. (*a*) magnification × 100; (*b*) magnification; × 500.

cation, a copious supply of lubricant is generally used to carry away the frictional heat generated by the bearing. Rolling element–cage friction and cage–bearing ring friction as well as cage–lubricant friction also affect skidding.

Skidding in Ball Bearings

One of the most important applications with regard to skidding is the mainshaft angular-contact ball bearing in aircraft gas turbines. This bearing is predominantly thrust loaded, and it is therefore only necessary to divide the thrust load uniformly among the bearing balls to determine the applied load.

The ball loading is shown by Fig. 13.23 for the coordinate system and ball speeds of Fig. 4.4.

The sliding velocities in the y' and x' directions are given by equations (7.31), (7.32), 7.36), and (7.37). The fluid entrainment velocities are given by

$$u_n = \frac{D}{4}\left[\frac{\omega_n}{\gamma} + (c_n\omega_n + \omega_{x'})\,\varphi_n \cos(\alpha_n + \theta_n) + \omega_{z'}\varphi_n \sin(\alpha_n + \theta_n)\right] \quad n = \text{o, i} \tag{13.62}$$

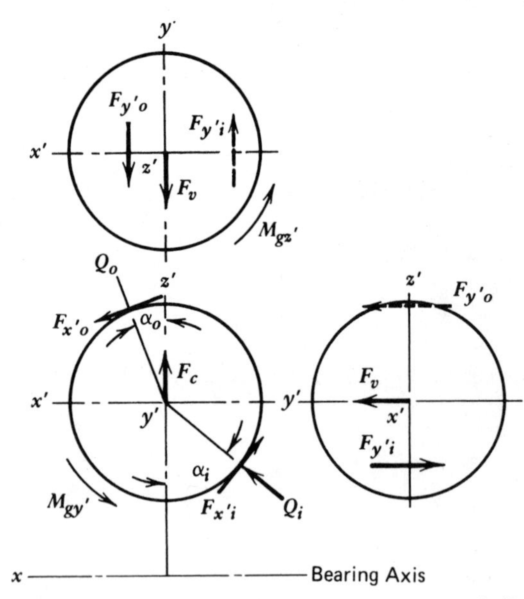

FIGURE 13.23. Forces and moments acting on a ball.

where $\omega_n = c_n(\omega_m - \Omega_n)$, $c_o = 1$, $c_i = -1$, $\theta = \sin^{-1}(x'_n/r_n)$, $\varphi_n = \tan^{-1}(X/Z)$, and X and Z are given by equations (13.44) and (13.45). From equation (13.1), it can be seen that at every point along the x' axis of the contact ellipses

$$\tau_{y'n} = \frac{\eta v_{y'n}}{h_n} \tag{13.63}$$

This relationship holds true as long as the degree of sliding is relatively low. At increased sliding velocities, as shown by Fig. 13.3, the coefficient of friction may be assumed constant and $\tau_{y'n} = \mu p_n$, where p_n is given by equation (13.37).

Values of lubricant film thickness h_n can be determined from methods given in Chapter 12 using equation (13.62). Thus, for any assumed contact loads, the frictional shear stresses can be numerically evaluated at every point in the contact areas. The frictional forces in the contact areas are given by

$$F_{y'n} = a_n b_n \int_{-1}^{1} \int_{-\sqrt{1-q^2}}^{\sqrt{1-q^2}} \tau_{y'n}\, dt\, dq \qquad n = \text{o, i} \tag{13.64}$$

$$F_{x'n} = a_n b_n \int_{-1}^{1} \int_{-\sqrt{1-t^2}}^{\sqrt{1-t^2}} \tau_{x'n}\, dq\, dt \qquad n = \text{o, i} \tag{13.65}$$

The moments due to shear stresses in the contact areas are given by

$$M_{x'n} = a_n b_n \int_{-1}^{1} \int_{-\sqrt{1-q^2}}^{\sqrt{1-q^2}} \tau_{y'n} r_n \cos(\alpha_n + \theta_n)\, dt\, dq \tag{13.66}$$

$$M_{z'n} = a_n b_n \int_{-1}^{1} \int_{-\sqrt{1-q^2}}^{\sqrt{1-q^2}} \tau_{y'n} r_n \sin(\alpha_n + \theta)\, dt\, dq \tag{13.67}$$

$$M_{x'n} = a_n b_n \int_{-1}^{1} \int_{-\sqrt{1-t^2}}^{\sqrt{1-t^2}} \tau_{x'n} r_n\, dq\, dt \tag{13.68}$$

where $r_n = \frac{1}{2} D\varphi_n$. Hence the equations of force and moment equilibrium are

$$Q_o \sin\alpha_o + F_{x'o}\cos\alpha_o - \frac{F_a}{Z} = 0 \tag{13.69}$$

$$\sum_{n=o}^{n=i} c_n(Q_n \cos\alpha_n - F_{x'n}\sin\alpha_n) - F_c = 0 \qquad \begin{matrix} n = \text{o, i} \\ c_o = 1;\ c_i = -1 \end{matrix} \tag{13.70}$$

$$\sum_{n=o}^{n=i} c_n(Q_n \sin \alpha_n + F_{x'n} \cos \alpha_n) = 0 \qquad (13.71)$$

$$\sum_{n=o}^{n=i} c_n F_{y'n} + F_v = 0 \qquad (13.72)$$

$$\sum_{n=o}^{n=i} M_{x'n} = 0 \qquad (13.73)$$

$$\sum_{n=o}^{n=i} M_{y'n} - M_{gy'} = 0 \qquad (13.74)$$

$$\sum_{n=o}^{n=i} M_{z'n} - M_{gz'} = 0 \qquad (13.75)$$

where

$$M_{gy'} = J\omega_{\mathrm{m}}\omega_{y'} \qquad (13.76)$$

$$M_{gz'} = J\omega_{\mathrm{m}}\omega_{z'} \qquad (13.77)$$

and J is the polar moment of inertia. F_v in equation (13.72) is determined from equation (13.2). A ball-riding cage with negligible friction in the ball pockets is assumed. Since only a simple thrust load is assumed, cage speed is identical to ball orbital speed ω_{m}. The unknowns in equations (13.69)–(13.75) are inner and outer raceway-ball contact deformations, ball contact angles or position variables, bearing axial deflection, and ball speeds, $\omega_{x'}$, $\omega_{y'}$, $\omega_{z'}$, and ω_{m}. Hence, there are nine unknowns and seven equations. The remaining two equations pertaining to ball position are obtained from Chapter 8. The solution of the equations requires the use of a computer. These equations were first solved by Harris [13.20] using the simplifying assumption of an isothermal Newtonian lubricant, adequately supplied to the ball–raceway contacts.

Figures 13.24 and 13.25 show the comparison of the analytical results with the experimental data of Shevchenko and Bolan [13.21] and Poplawski and Mauriello [13.22]. Note the deviations from the outer raceway control approximation.

Parker [13.23] established an empirical formula to estimate the percentage of the bearing "free space" occupied by fluid lubricant. Using Parker's formula it is possible to calculate the effective fluid density ξ in equation (13.3) and hence F_v in equation (13.72). The effective density so determined is given by equation (13.78).

$$\xi = 10^5 \frac{\xi_b \mathscr{W}^{0.37}}{n d_{\mathrm{m}}^{1.7}} \qquad (13.78)$$

This equation was developed from ball bearing tests.

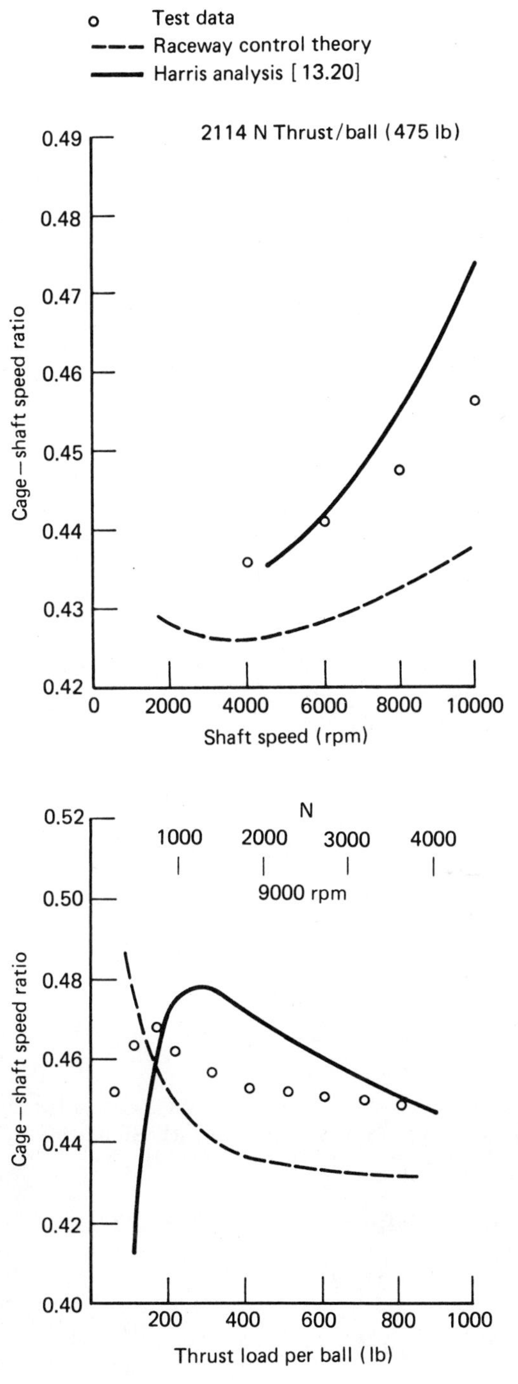

FIGURE 13.24. Experimental data of Shevchenko and Bolan [13.21] for an angular-contact ball bearing with three 28.58 mm (1.125 in.) balls.

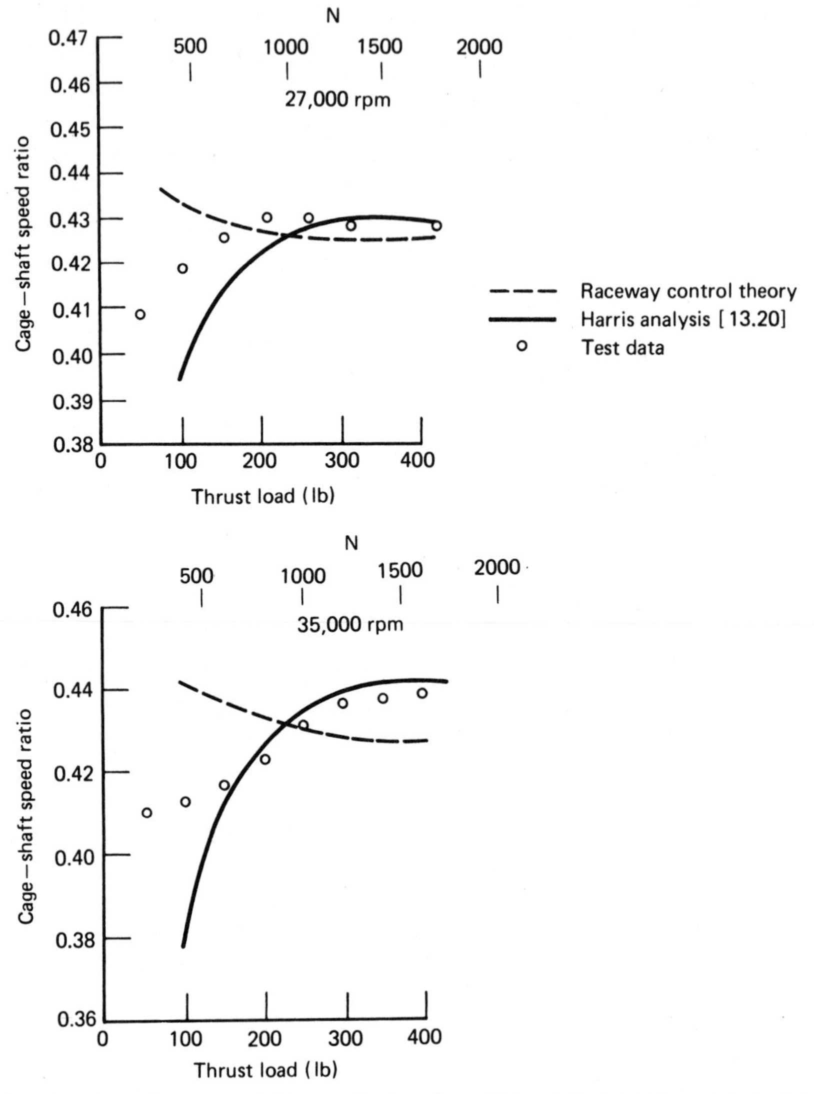

FIGURE 13.25. Experimental data of Poplawski and Mauriello [13.22] vs analytical data of Harris [13.20] for a 35 mm · 62 mm angular-contact ball bearing.

Skidding in Radial Cylindrical Roller Bearings

Skidding is a particular problem in cylindrical roller bearings used to support the mainshaft in aircraft gas turbines. These bearings, which are used principally for location are very lightly loaded while operating at high speeds. Harris [13.15] indicates the method to predict skidding in this application. Considering the roller-raceway contacts to be divided into laminae as in Chapter 6, the sliding velocity at a given "slice" is given by

$$v_{\lambda nj} = \tfrac{1}{2}\left\{\left[d_{\mathrm{m}} + c_n\left(D_\lambda + \tfrac{2}{3}\,\delta_{\lambda nj}\right)\right]\omega_{nj} - \left(D_\lambda - \tfrac{1}{3}\,\delta_{\lambda nj}\right)\right\} \qquad \begin{matrix} n = \mathrm{o, i} \\ c_{\mathrm{o}} = 1;\ c_{\mathrm{i}} = -1 \\ j = 1 - z \end{matrix} \tag{13.79}$$

where D_λ is the equivalent roller diameter at laminum λ. It is assumed in equation (13.79) that one-third of the elastic deformation occurs in the roller and two-thirds in the raceway. Furthermore, owing to assumed zero clearance between the roller and cage pocket, roller orbital speed is constrained to equal cage speed, hence, $\omega_{\mathrm{o}j} = \omega_{\mathrm{m}} - \Omega_{\mathrm{o}}$ and $\omega_{\mathrm{i}j} = \Omega_{\mathrm{i}} - \omega_{\mathrm{m}}$. Fluid entrainment velocities are given by equations (12.33) and (12.34) and the lubricant film thickness by equation (12.76). Thus, for isothermal, Newtonian lubrication the contact surface shear stresses can be determined from

$$\tau_{\lambda nj} = \frac{\eta_{\lambda nj} v_{\lambda nj}}{h_{nj}} \tag{13.80}$$

Again, if the sliding velocities in a contact are sufficiently great, equation (13.81) can be used in conjunction with a constant coefficient of friction (see Fig. 13.3) and

$$\tau_{\lambda nj} = \mu_n \sigma_{\lambda nj} \tag{13.81}$$

From equation (5.45),

$$\tau_{\lambda nj} = \frac{2q_{\lambda nj}(1 - t^2)}{\pi b_{nj}} \tag{13.82}$$

where $t = y'/b$ and $q = Q/l$. Contact friction force is then given by

$$F_{nj} = 2w_n \sum_{\lambda=1}^{\lambda=k} b_{\lambda nj} \int_0^1 \tau_{\lambda nj}\, dt \tag{13.83}$$

Figure 13.26 shows the forces and moment acting on a roller in a radially loaded cylindrical roller bearing with negligible roller end–flange friction. From Fig. 13.26, the equilibrium equations (13.84) and (13.85) obtain

$$\sum_{n=\mathrm{o}}^{n=\mathrm{i}} c_n Q_{nj} - F_{cj} = 0 \qquad \begin{matrix} n = \mathrm{o, i} \\ c_{\mathrm{o}} = 1;\ c_{\mathrm{i}} = -1 \end{matrix} \tag{13.84}$$

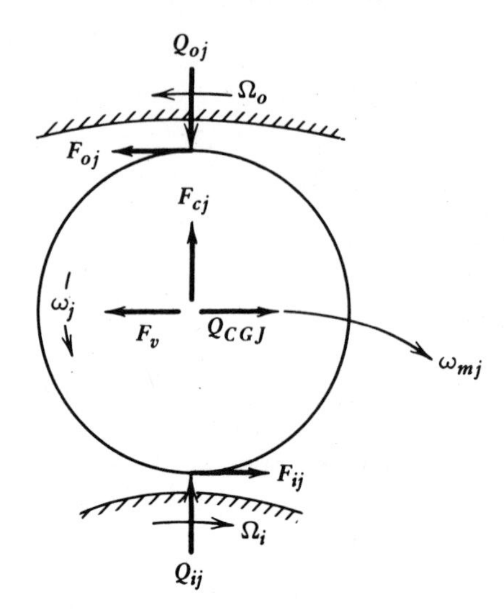

FIGURE 13.26. Forces acting on a roller.

where F_{cj} is given by equation (4.52), and

$$\sum_{n=0}^{n=1} c_n F_{nj} + F_v - Q_{CGj} = 0 \tag{13.85}$$

where the viscous drag force is given by equation (13.3). Note that if there is clearance between the roller and the cage webbing, then the roller is free to orbit at other than cage speed and equation (13.85) is nonzero, being equal to the inertial load, $\frac{1}{2}\, m d_{\mathrm{m}} \omega_{\mathrm{m}j}\, d\omega_{\mathrm{m}j}/d\psi$. The frictional moments about the roller axis due to shear stresses are given by

$$M_{nj} = w_n \sum_{\lambda=1}^{\lambda=k} b_{\lambda nj} D_\lambda \int_0^1 \tau_{\lambda nj}\, dt \tag{13.86}$$

and

$$\sum_{n=\mathrm{o}}^{n=\mathrm{i}} M_{nj} - \tfrac{1}{2}\, \mu_{\mathrm{c}} D Q_{\mathrm{CG}j} = J\,\omega_{\mathrm{m}} \frac{d\omega_j}{d\psi} \tag{13.87}$$

Finally, the radial equilibrium equation for the bearing is

$$\sum_{j=1}^{j=Z} Q_{\mathrm{i}j} - F_{\mathrm{r}} = 0 \tag{13.88}$$

and if the bearing operates at constant speed, the sum of the moments on the cage in the circumferential direction must equate to zero, or

$$d_m \sum_{j=1}^{j=Z} Q_{CGj} \pm D_{CR} F_{CL} = 0 \tag{13.89}$$

where F_{CL} is given by equation (13.14).

As in Chapter 6, normal loads Q_{nj} can be written in terms of contact deformations, and bearing radial deflection can be related to contact deformations and radial clearance. Accordingly, equations (13.84), (13.85), (13.87), (13.88), and (13.89), a set of $3Z + 2$ equations, can be solved for δ_r, δ_{ij}, ω_m, ω_j, and Q_{CGj}. Reference [13.15] gives the general solution for all types of roller bearings (and ball bearings); that is, for five degrees of freedom in applied bearing loading, freedom for each roller (and ball) to orbit at a speed other than cage speed (ω_{mj} instead of ω_m), and any shape of raceway and/or roller.

Harris [13.24] using a simpler form of the analysis, considering only

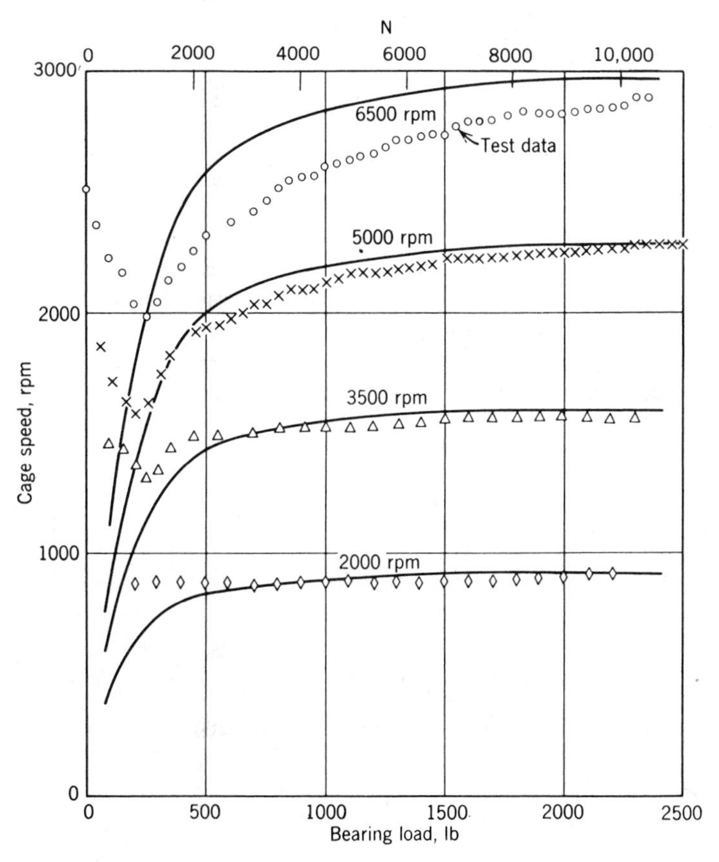

FIGURE 13.27. Cage speed vs load and inner ring speed for cylindrical roller bearing, lubricant-diester type according to MIL-L-7808 Specification. $Z = 36$ rollers, $l = 20$ mm (0.787 in.), $D = 19$ mm (0.551 in.), $d_m = 183$ mm (7.204 in.), $P_d = 0.0635$ mm (0.0025 in).

isothermal lubrication conditions and neglecting viscous drag on the rollers, nevertheless managed to demonstrate the adequacy of the analytical method. Figure 13.27, taken from reference [13.24], compares analytical data against experimental data. Figures 13.28 and 13.29 show the effects of load, speed, and number of rollers on skidding as determined by the cage speed slip. Figure 13.30 shows the relative lack of effect on skidding of the type of lubricant, considering only normally acceptable lubricants for the application.

Several aircraft engine manufacturers assemble their bearings in an "out-of-round" outer raceway to achieve the load distribution of Fig. 13.31 as a means of minimizing skidding. This artificial loading of the bearing increases the maximum roller load and doubles the number of the rollers so loaded. Figure 13.32 taken from reference [13.24] illustrates the effect on skidding of an out-of-round outer raceway. Another

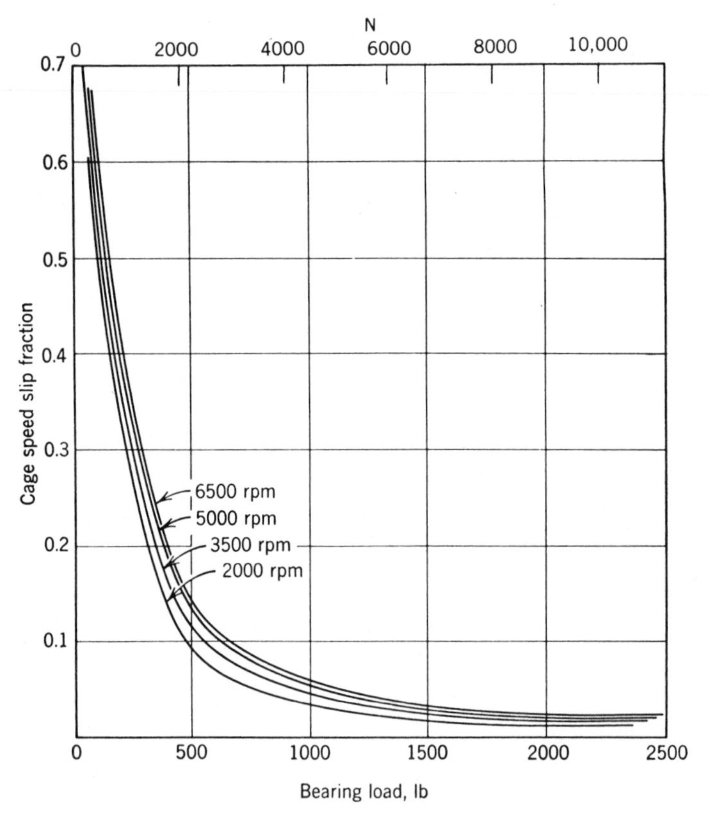

FIGURE 13.28. Cage speed slip fraction vs. load and inner ring speed, lubricant-diester type according to MIL-L-7808 Specification. $Z = 36$, $i = 1$, $l = 20$ mm (0.787 in.), $D = 14$ mm (0.551 in.), $d_m = 183$ mm (7.204 in.), $P_d = 0.0635$ mm (0.0025 in.).

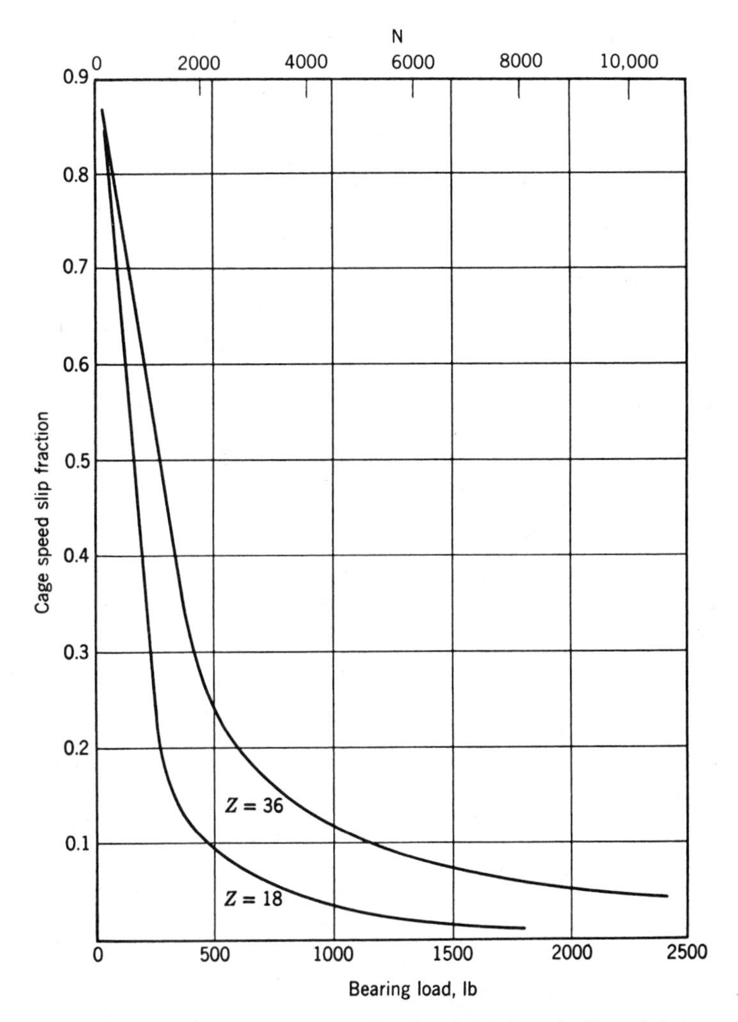

FIGURE 13.29. Cage speed slip fraction vs load and number of rollers, lubricant-diester type according to MIL-L-7808 Specification. $i = 1$, $l = 20$ mm (0.787 in.), $D = 14$ mm (0.551 in.), $P_d = 0.0635$ mm (0.0025 in.), $n = 6500$ rpm.

method to minimize skidding is to use a few, for example, three, equally spaced hollow rollers that provide an interference fit with the raceways under zero radial load and static conditions. Figure 13.33, taken from reference [13.25], illustrates such an assembly, while Fig. 13.34 indicates the effectiveness to minimize skidding.

Figure 13.35, taken from the commentary to reference [13.25], confirms the adequacy of the analytical method by showing a high degree of skidding for 2–4 90% hollow rollers tested in a 207 cylindrical roller bearing.

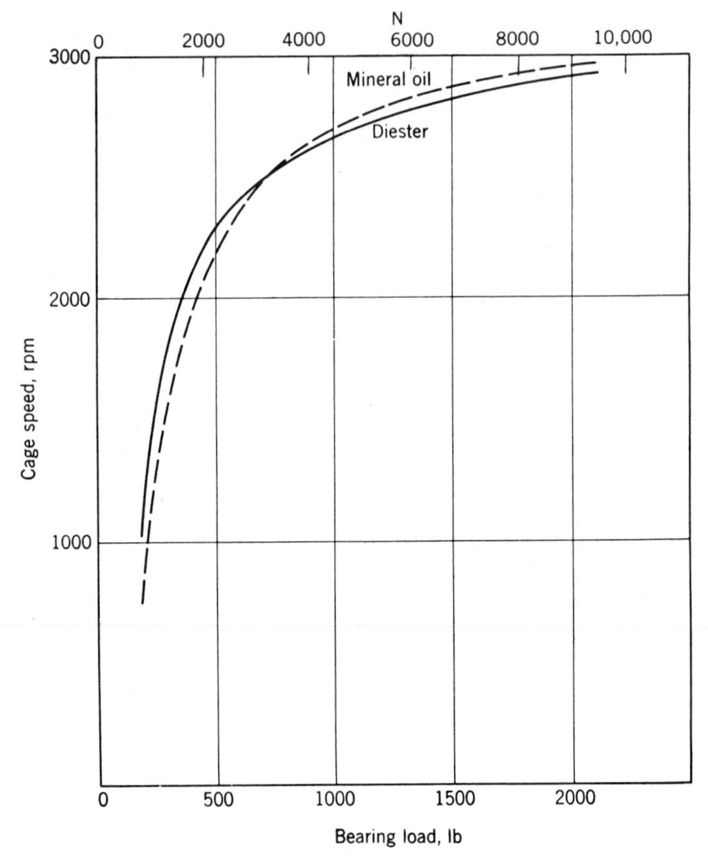

FIGURE 13.30. Cage speed vs bearing load for different lubricants.

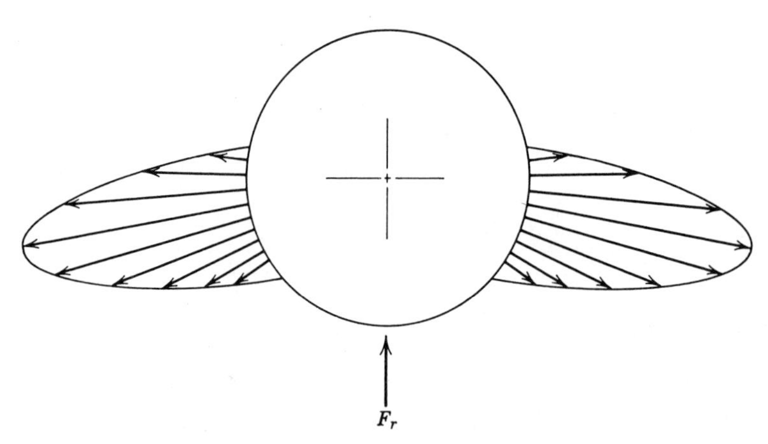

FIGURE 13.31. Distribution of load among the rollers of a bearing having an out-of-round outer ring and subjected to radial load F_r.

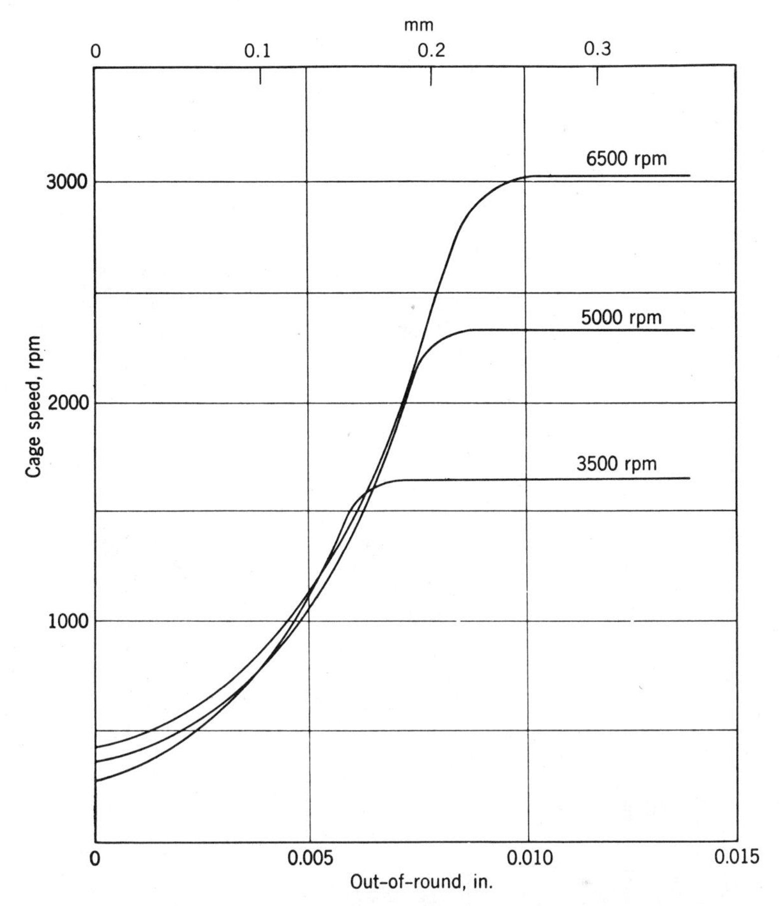

FIGURE 13.32. Cage speed vs out-of-round and inner ring speed. Lubricant-diester type according to MIl-L-7808 Specification. $Z = 36$, $i = 1$, $l = 20$ mm (0787 in.), $D = 14$ mm (0.551 in.), $d_m = 183$ mm (7.204 in.), $P_d = 0.0635$ mm (0.0025 in.), $F_r = 222.5$ N (50 lb).

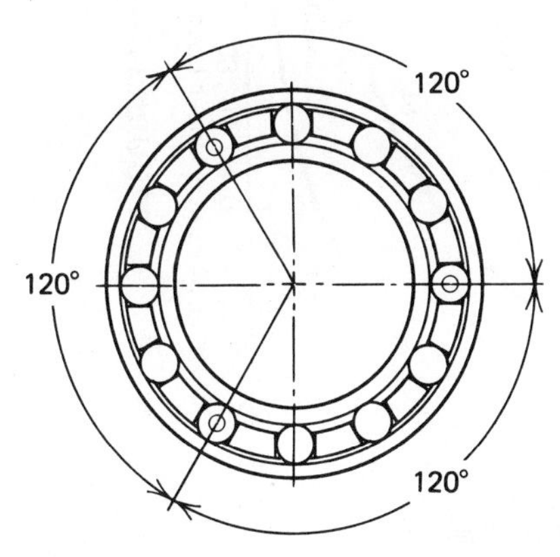

FIGURE 13.33. Cylindrical roller bearing having three preloaded annular rollers.

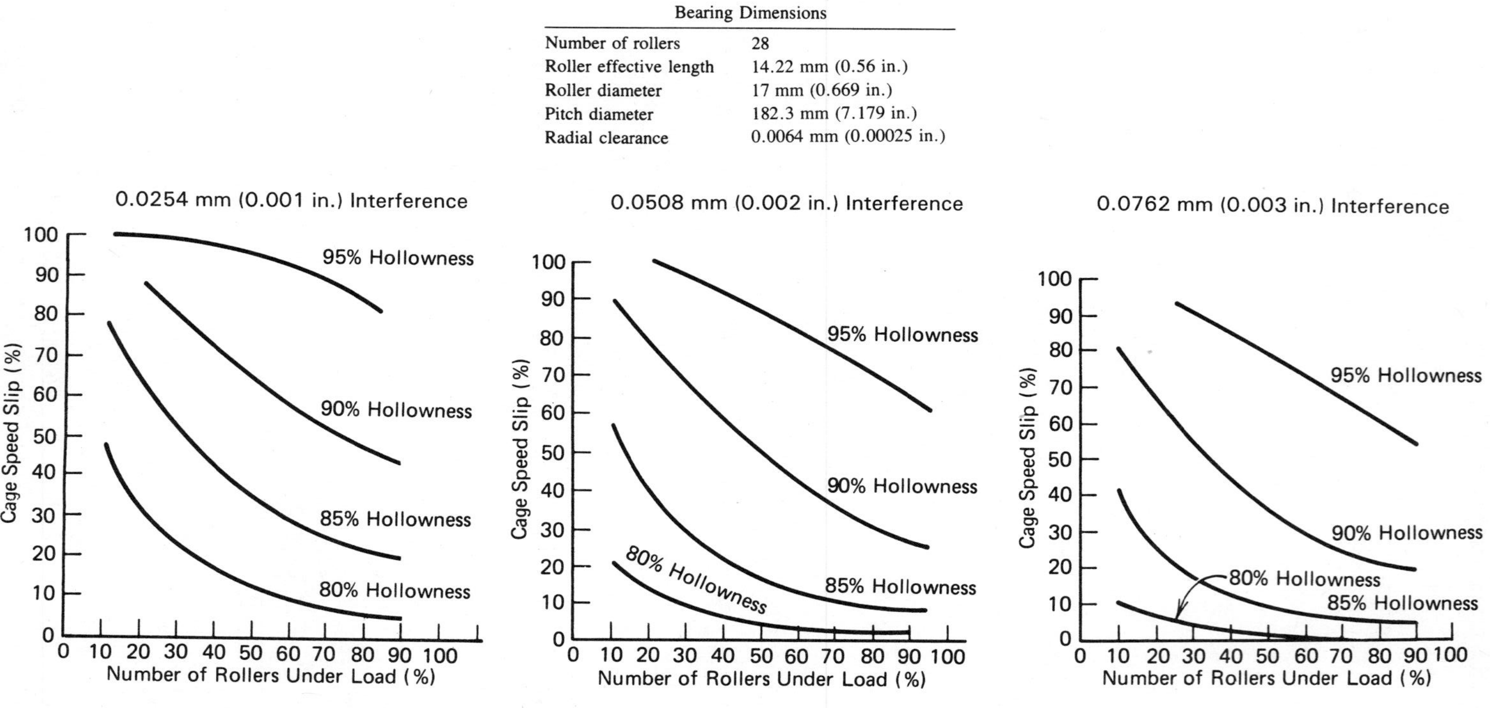

Bearing Dimensions	
Number of rollers	28
Roller effective length	14.22 mm (0.56 in.)
Roller diameter	17 mm (0.669 in.)
Pitch diameter	182.3 mm (7.179 in.)
Radial clearance	0.0064 mm (0.00025 in.)

FIGURE 13.34. Skidding in cylindrical roller bearings having spaced preloaded hollow rollers.

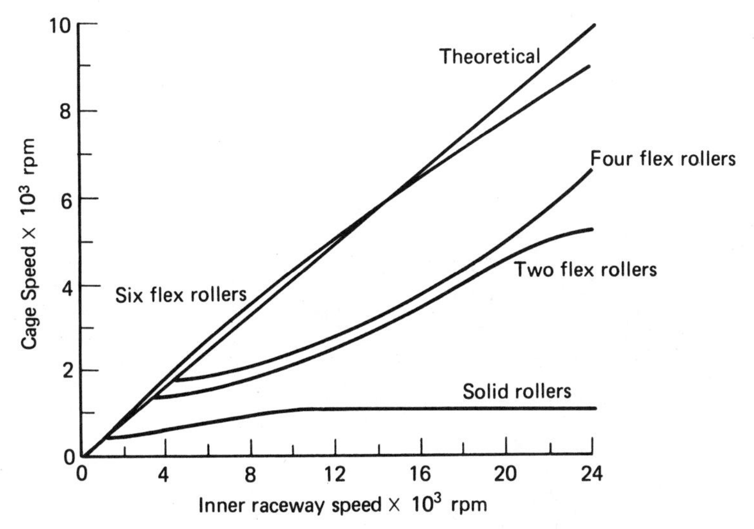

FIGURE 13.35. Cage speed vs inner raceway speed: 207 roller bearing, $F_r = 0$, $P_d = -0.061$ mm ($-$ 0.0024 in.), 90% hollow rollers, lubricant MIL-L-6085A at 0.85 kg/min.

CAGE MOTIONS AND FORCES

Influence of Speed

With respect to rolling element bearing performance, cage design has become more important in recent years. Cage problems tend to be encountered more frequently as bearing rotational speeds increase. In instrument ball bearings undesirable torque variations have been traced to cage dynamic instabilities. In the development of solid-lubricated bearings for high-speed, high-temperature gas turbine engines, the cage remains a major concern.

A key to successful cage design is a detailed analysis of the forces acting on the cage and the motions it undergoes. Both steady-state and dynamic formulations of varying complexity have been developed.

Forces Acting on the Cage

The primary forces acting on the cage are due to the interactions between the rolling element and cage pocket (F_{CP}) and the cage rail and the piloting land (F_{CL}). As Fig. 13.36 shows, a roller can contact the cage on either side of the pocket, depending on whether the cage is driving the roller, or vice versa. The direction of the cage pocket friction force (F_{CP}) depends on which side of the pocket contact occurs. For an inner land riding cage a friction torque (T_{CL}) in the direction of cage rotation

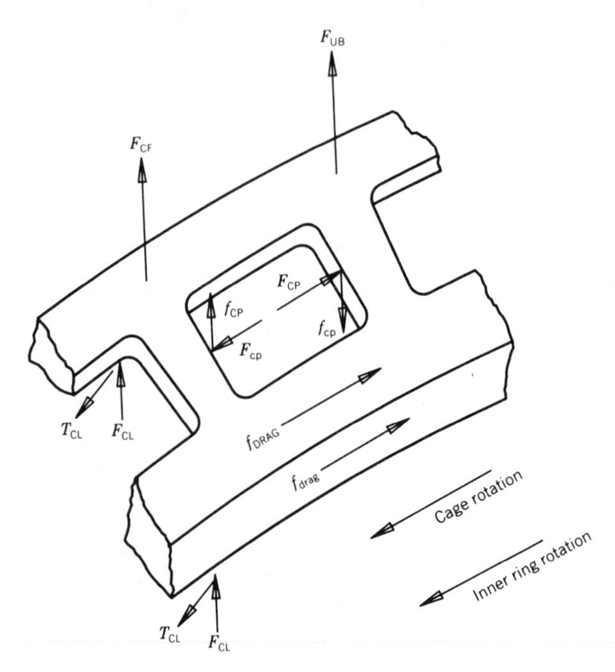

FIGURE 13.36. Cage forces.

develops at the cage-land contact. For an outer land riding cage a friction torque tending to retard cage rotation develops at the cage-land contact.

A lubricant viscous drag force (f_{DRAG}) develops on the cage surfaces resisting motion of the cage. Centrifugal body forces (shown as F_{CF}) due to cage rotation make the cage expand uniformly outward radially and induce tensile hoop stresses in the cage rails. An unbalanced force (F_{UB}), the magnitude of which depends on how accurately the cage is balanced, acts radially outward.

Hydrodynamic short bearing theory can be used to model the cage-land interaction as indicated in [13.26]. The contact between the rolling element and cage pocket can be hydrodynamic, elastohydrodynamic (EHD), or elastic in nature, depending on the proximity of the two bodies and the magnitude of the rolling element forces. In most cases the rolling element–cage interaction forces are small enough that hydrodynamic lubrication considerations prevail.

Steady-State Conditions

In a previous section it was demonstrated that adequate analytical means exist to predict skidding in ball and roller bearings in any fluid-lubricated application. All of the calculations, even for the least complex

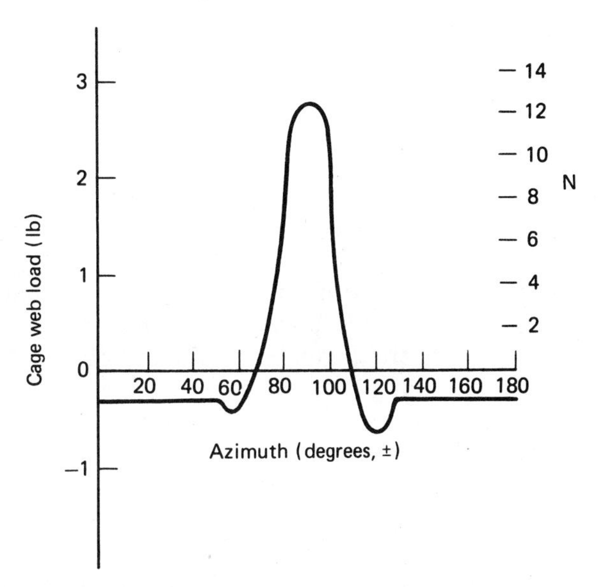

FIGURE 13.37. Cage-to-roller load vs azimuth for a gas turbine main shaft cylindrical roller bearing. Thirty 12 mm × 12 mm rollers on a 152.4 mm (6-in.) pitch diameter. Roller i.d./o.d. = 0.6, outer ring out-of-roundness = 0.254 mm (0.01 in.), radiai load = 445 N (100 lb), shaft speed = 25,000 rpm.

application, require the use of a computer. As a spin-off from the skidding analysis, rolling element-cage forces are determined. For an out-of-round outer raceway cylindrical roller bearing under radial load, Fig. 13.37, from reference [13.27], illustrates cage web loading for steady-state, centric cage rotation.

Whereas the analysis of [13.26] considered only centric rotation in the radial plane, Kleckner and Pirvics [13.28] used three degrees of freedom in the radial plane; that is, the cage rotational speed and two radial displacements locating the cage center in the plane of rotation. The corresponding cage equilibrium equations are

$$\sum_{j=1}^{Z} [(-F_{CPj}) \cos \psi_j - (f_{CPj}) \sin \psi_j] - W_y = 0 \qquad (13.90)$$

$$\sum_{j=1}^{Z} [(F_{CPj}) \sin \psi_j - (f_{CPj}) \cos \psi_j] - W_z = 0 \qquad (13.91)$$

$$\tfrac{1}{2} d_m \sum_{j=1}^{Z} (F_{CPj}) \pm T_{CL} = 0 \qquad (13.92)$$

where W_y, W_z = components of F_{CL} in the y and z directions
F_{CPj} = cage pocket normal force for the jth rolling element
f_{CPj} = cage pocket friction force for the jth rolling element

The cage coordinate system is shown in Fig. 13.38.

Equations (13.90) and (13.91) represent equilibrium of cage forces in the radial plane of motion. The summation of the cage pocket normal forces and friction forces equilibrate the cage-land normal force. Equation (13.92) establishes torque equilibrium for the cage about its axis of rotation. The cage pocket normal forces are assumed to act at the bearing pitch circle. The sign of the cage-land friction torque T_{CL} depends on whether the cage is inner ring land-riding or outer ring land-riding. In the formulation of [13.37] each roller is allowed to have different rotational and orbital speeds.

Dynamic Conditions

Rolling element bearing cages are subjected to transient motions and forces due to accelerations caused by contact with rolling elements, rings, and eccentric rotation. In some applications, notably with very high speed or rapid acceleration, these transient cage effects may be of sufficient magnitude to warrant evaluation. The steady-state analytical approaches discussed do not address the time-dependent behavior of rolling element bearing cages. Several researchers have developed analytical models for transient cage response [13.26, 13.29–13.33]. Due to the complexity of the calculations involved, these models must be implemented in programs operating in large-scale, very high-speed computers to be of practical value.

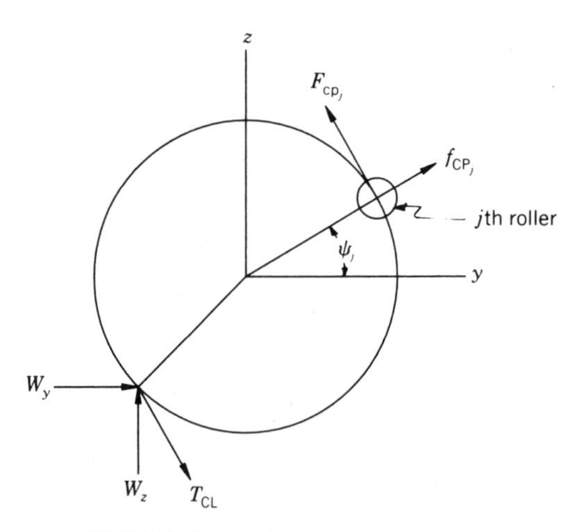

FIGURE 13.38. Cage coordinate system.

In general, the cage is treated as a rigid body subjected to a complex system of forces. These forces may include the following:

1. Impact and frictional forces at the cage–rolling element interface
2. Normal and frictional forces at the cage–land surface (if land-guided cage)
3. Cage mass unbalance force
4. Gravitational force
5. Cage inertial forces
6. Others (i.e., lubricant drag on the cage and lubricant churning forces)

Forces 1 and 2 are intermittent. For example the cage might or might not be in contact with a given rolling element or guide flange at a given time, depending on the relative position of the bodies in question. Frictional forces can be modeled as hydrodynamic, EHD, or dry friction, depending on the nature of the lubricant, contact load, and geometry. Both elastic and inelastic impact models appear in the literature. General equations of motion for the cage may be written. The Euler equations describing cage rotation about its center of mass (in Cartesian coordinates) are as follows:

$$I_x \dot{\omega}_x - (I_y - I_z)\,\omega_y \omega_z = M_x \tag{13.93}$$

$$I_y \dot{\omega}_y - (I_z - I_x)\,\omega_z \omega_x = M_y \tag{13.94}$$

$$I_z \dot{\omega}_z - (I_x - I_y)\,\omega_x \omega_y = M_z \tag{13.95}$$

where I_x, I_y, I_z are the cage principal moments of inertia, and ω_x, ω_y, ω_z are the angular velocities of the cage about the inertial x, y, z axes. The total moment about each axis is denoted by M_x, M_y, and M_z, respectively. The equations of motion for translation of the cage center of mass in the inertial reference frame are

$$m\ddot{r}_x = F_x \tag{13.96}$$

$$m\ddot{r}_y = F_y \tag{13.97}$$

$$m\ddot{r}_z = F_z \tag{13.98}$$

where m is cage mass, r_x, r_y, r_z describe the position of the cage center of mass, and F_x, F_y, F_z are the net force components acting on the cage.

Once cage force and moment components are determined, accelerations can be computed. Numerical integration of the equations of motion (with respect to discrete time increments) will yield cage translational velocity, rotational velocity, and displacement vectors. In some approaches [13.26, 13.30] the cage dynamics model is solved in conjunction with roller and ring equations of motion. Other researchers have devised less cumbersome approaches by limiting the cage to in-plane motion [13.29] or by considering simplified dynamic models for the rolling elements [13.31].

Meeks and Ng [13.31] developed a cage dynamics model for ball bearings, which treats both ball- and ring land-guided cages. This model considers six cage degrees of freedom and inelastic contact between balls and cage and between cage and rings. This model was used to perform a cage design optimization study for a solid-lubricated, gas turbine engine bearing [13.32].

The results of the study indicated that ball-cage pocket forces and wear are significantly affected by the combination of cage-land and ball-pocket clearances. Using the analytical model to identify more suitable clearance values improved experimental cage performance. Figures 13.39 and 13.40 contain typical output data from the cage dynamics analysis.

In Fig. 13.39 the cage center of mass motion is plotted versus time for X and Y (radial plane) directions. The time scale relates to approximately five shaft revolutions at a shaft speed of 40,000 rpm. Figure 13.40 shows plots of ball–cage pocket normal force for two representative pockets positioned approximately 90° apart.

In addition to the works of Meeks [13.32], Mauriello et al. [13.37] succeeded in measuring ball-to-cage loading in a ball bearing subjected to combined radial and thrust loading. They observed impact loading between balls and cage to be a significant factor on high speed bearing cage design.

ROLLER SKEWING

Thus far in this section, rollers have been assumed to run "true" in cylindrical, spherical, and tapered roller bearings. In fact, due to slightly imperfect geometry there is an inevitable tendency for unbalance of frictional loading between the roller-inner raceway and roller–outer raceway contacts, and thus a tendency for rollers to skew. Additionally, in a misaligned radial cylindrical roller bearing, as indicated schematically in Fig. 6.23, rollers are "squeezed" at one end and thereby forced against the "guide" flange. The latter causes a roller end-flange frictional force and hence a roller skewing moment that must be substantially resisted by the cage. In tapered roller bearings, even without misalignment, the rollers are forced against the large end flange and skewing moments occur. The thrust load applied to radial cylindrical roller bearings, as discussed in Chapter 6 , results in a roller skewing moment that is aug-

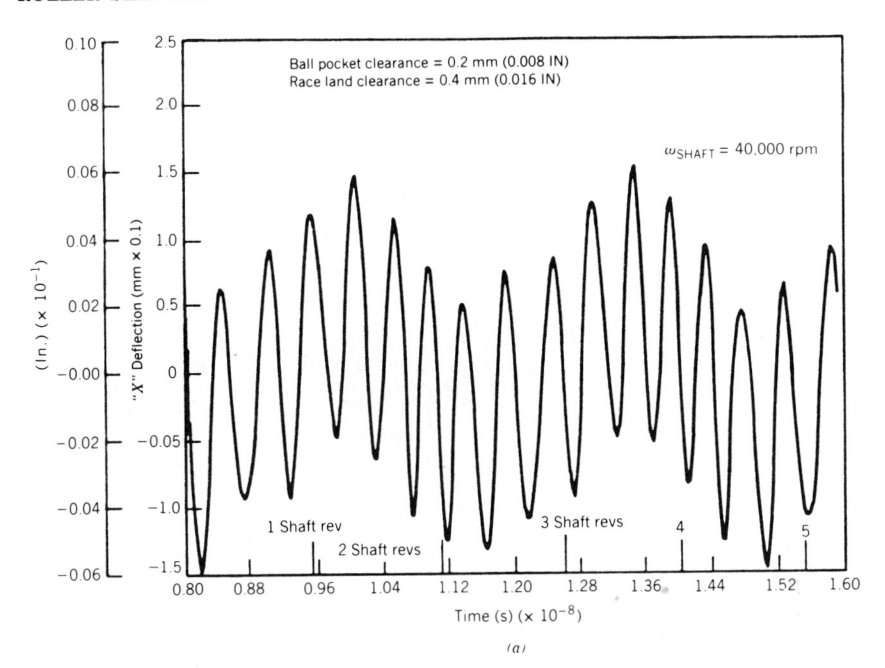

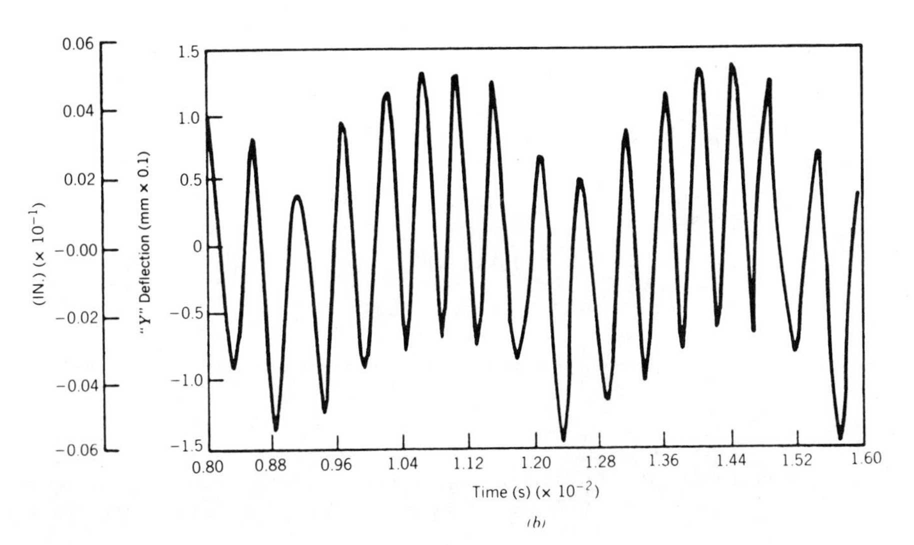

FIGURE 13.39. Calculated cage motion versus time. (*a*) Prediction of cage motion, X vs time. (*b*) Prediction of cage motion, Y vs time (from [13.32]).

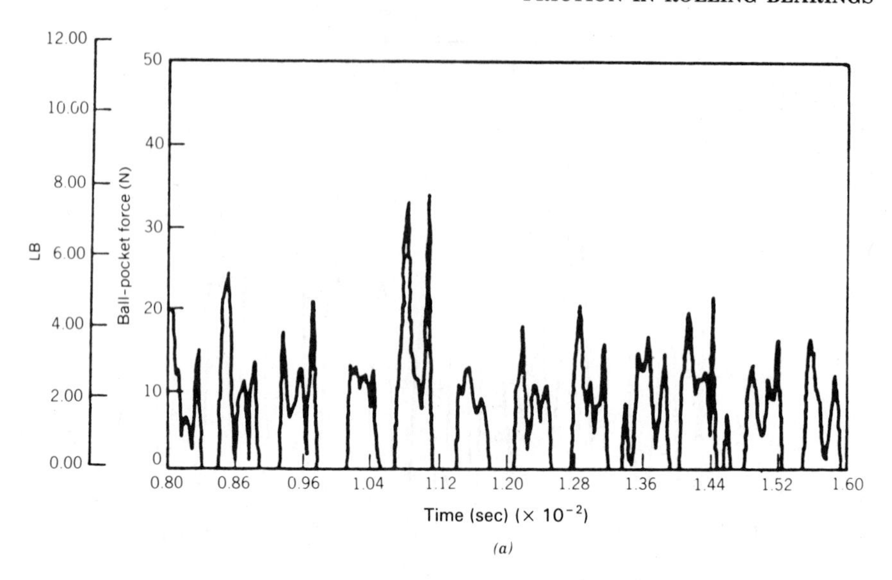

(a)

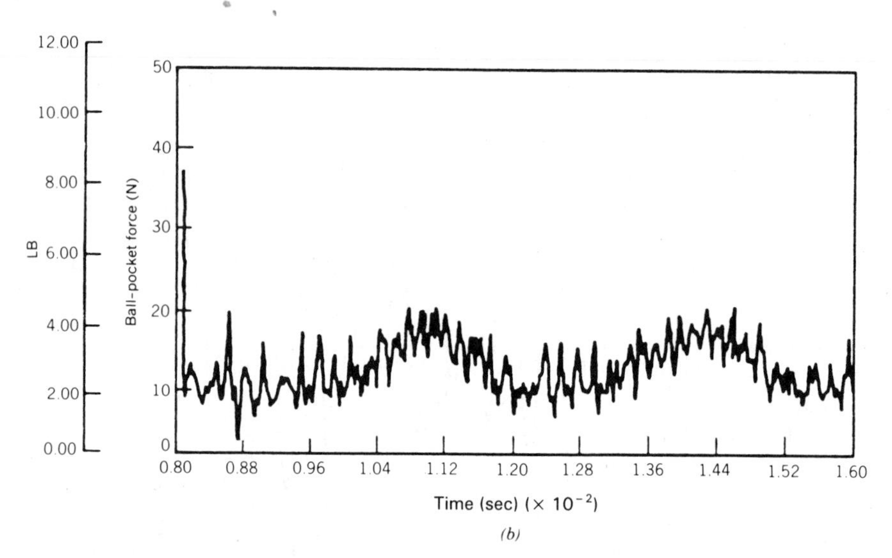

(b)

FIGURE 13.40. Calculated ball-pocket force vs time. (*a*) Prediction of cage ball-pocket force vs time (pocket No. 1). (*b*) Prediction of cage ball-pocket force vs time (pocket No.4) (from [13.32]).

mented by unbalance of raceway–roller friction forces, as indicated in Fig. 13.41.

In most cases roller skewing is detrimental to roller bearing operation because it causes increased friction torque and frictional heat generation as well as necessitating a cage strong enough to resist the roller moment loading.

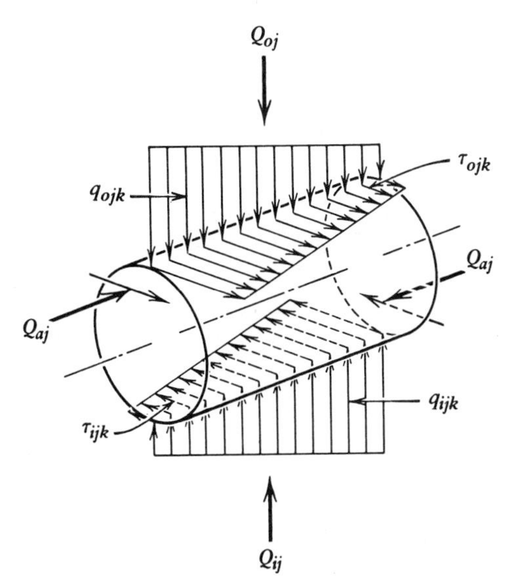

FIGURE 13.41. Normal, axial, and frictional loading of a roller at azimuth ψ_j in a radial cylindrical roller bearing subjected to radial and thrust applied loading.

Equilibrium Roller Skewing Angle

The notion that rollers skew until skewing moment equilibrium is achieved has implications beyond those of roller end-flange load determination. In spherical roller bearings with symmetric roller profiles, proper management of roller skewing can reduce frictional losses and corresponding friction torque. Early spherical roller bearing designs employing asymmetrical roller profiles, because of their close osculations and primary skewing guidance from cage and flange contacts, exhibit greater friction than current bearings with symmetrical roller designs. The temperature rise associated with friction is the factor that limits performance in many applications. Designing the bearings so that skewing equilibrium is provided by raceway guidance alone lowers losses and increases load-carrying capacity. Kellstrom [13.34, 13.35] investigated skewing equilibrium in spherical roller bearings considering the complex changes in roller force and moment balance caused by roller tilting and skewing in the presence of friction.

Any rolling element that contacts a raceway along a curved contact surface will undergo sliding in the contact. For an unskewed roller there will be at most two points along each contact where the sliding velocity is zero. These zero sliding points form the generatrices of a theoretical "rolling" cone, which represents the contact surface on which pure kinematic rolling would occur for a given roller orientation. At all other points along the contact, sliding is present in the direction of rolling or opposite to it, depending on whether the roller radius is greater or less

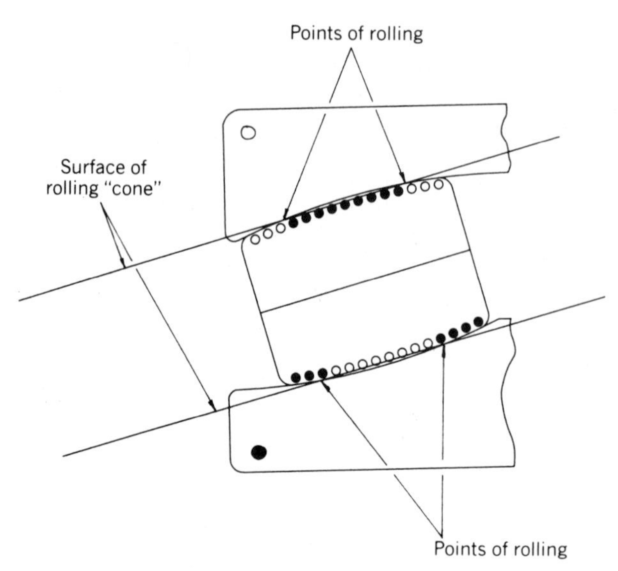

FIGURE 13.42. Spherical roller bearing, symmetrical roller-tangential friction force directions. Motion and force direction: ○ out of page; ● into page.

than the radius to the theoretical rolling cone. This situation is illustrated in Fig. 13.42.

Friction forces or tractions due to sliding will be oriented to oppose the direction of sliding on the roller. In the absence of tangential roller forces from cage or flange contacts, the roller-raceway traction forces in each contact must sum to zero. Additionally, the sum of the inner and outer raceway contact skewing moments must equal zero. These two conditions will determine the position of the rolling points along the contacts and thus the theoretical rolling cone. These conditions are met at the equilibrium skewing angle. If the moments tend to restore the roller to the equilibrium skewing angle when it is disturbed, the equilibrium skewing angle is said to be stable.

As a roller skews relative to its contacting raceway a sliding component is generated in the roller axial direction and traction forces are developed that oppose axial sliding. These traction forces may be beneficial in that, if suitably oriented, they help to carry the axial bearing load, as indicated in Fig. 13.43.

Those skewing angles that produce axial tractions opposing the applied axial load and reducing the roller contact load required to react the applied axial load are termed positive (Fig. 13.43*a*). Conversely, those skewing angles producing axial tractions that add to the applied axial load are termed negative (Fig. 13.43*b*). For a positive skewing roller the normal contact loading is reduced, and an improvement in contact fatigue life is achieved.

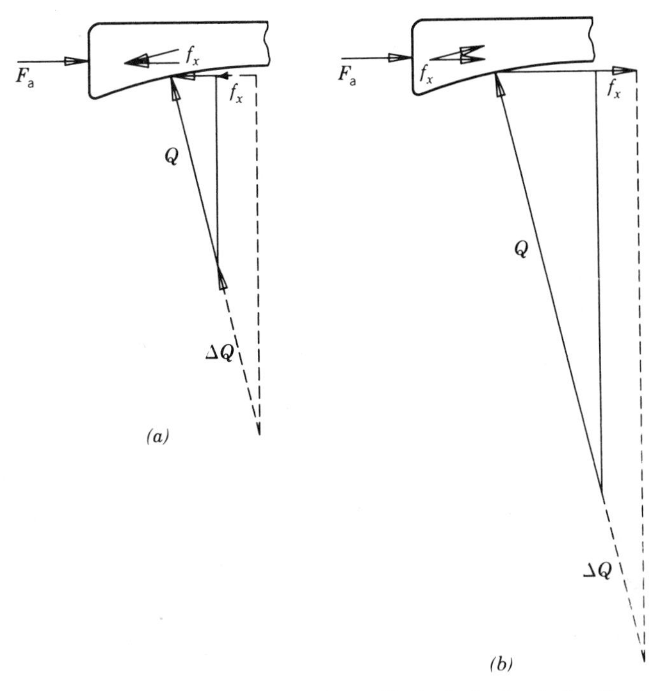

FIGURE 13.43. Forces on outer raceway of axially loaded spherical roller bearing with positive and negative skewing. (*a*) Positive skewing angle. (*b*) Negative skewing angle.

The axial traction forces acting on the roller also produce a second effect. These forces, acting in different directions on the inner and on outer ring contacts, create a moment about the roller and cause it to tilt. The tilting motion repositions the inner and outer ring contact load distributions with respect to the theoretical points of rolling and distribution of sliding velocity. Detailed evaluations [13.34, 13.35] of this behavior have shown that skewing in excess of the equilibrium skewing angle generates a net skewing moment opposing the increasing skewing motion. A roller that skews less than the equilibrium skewing angle will generate a net skewing moment tending to increase the skew angle. This set of interactions explains the existence of stable equilibrium skewing angles.

To apply this concept to the design of spherical roller bearings, specific design geometries over a wide range of operating conditions must be evaluated. There are tradeoffs involved between minimizing friction losses and maximizing contact fatigue life. Some designs may exhibit unstable skewing control in certain operating regimes or stable skewing equilibrium and require impractically large skewing angles. Computer programs that predict spherical roller bearing performance contribute to more accurate evaluations. See Fig. 13.44, which shows the possible

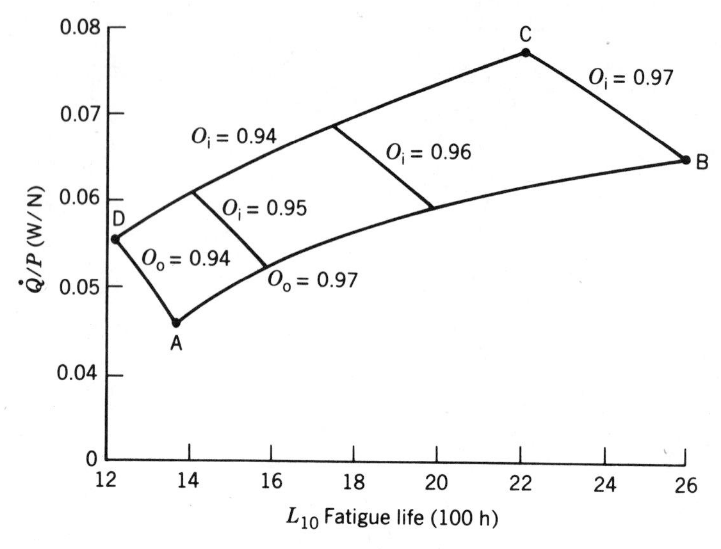

FIGURE 13.44. Study of frictional power loss vs calculated fatigue life of spherical roller bearing with equilibrium skewing control. Q/P = ratio of bearing power loss to applied load. O_o = outer raceway osculation. O_i = inner raceway osculation.

tradeoffs between frictional power loss and calculated fatigue life for a bearing design using skewing control. Results are shown for several values of outer and inner raceway osculation.

BEARING FRICTION TORQUE

Torque Due to Applied Load

Exclusive of an analytical approach to determine bearing friction torque, Palmgren [13.11] empirically evaluated bearing friction torque due to all mechanical friction phenomena with the exception of friction owing to the quantity of lubricant contained within the bearing boundary dimensions; that is, within the bearing cavity. Data were compiled on each basic bearing type. Palmgren [13.11] gave the following equation to describe this torque:

$$M_1 = f_1 F_\beta d_m \tag{13.99}$$

in which f_1 is a factor depending upon bearing design and relative bearing load. For ball bearings,

$$f_1 = z(F_s/C_s)^y \tag{13.100}$$

TABLE 13.3. Values of z and y

Ball Bearing Type	Nominal Contact Angle	z	y
Radial deep groove	0°	0.0002–0.0004[a]	0.55
Angular-contact	30–40°	0.001	0.33
Thrust	90°	0.0008	0.33
Double-row, self-aligning	10°	0.0003	0.40

[a]Lower values pertain to light series bearings; higher values pertain to heavy series bearings.

in which F_s is static equivalent load and C_s basic static load rating (these terms are explained in Chapter 21 covering plastic deformation and static capacity). Table 13.3 gives appropriate values of z and y. Values of C_s are generally given in manufacturers' catalogs along with data to enable calculation of F_s. The internal designs of roller bearings have changed both from macrogeometrical and microgeometrical bases since the publication by Palmgren [13.11]. Therefore, Table 13.4 as updated according to data from [13.38] gives empirical values of f_1 for roller bearings. For modern design, double-row, radial spherical roller bearings, SKF [13.38] uses the formula:

$$M_1 = f_1 F^a d^b \tag{13.101}$$

in which constant f_1 and exponents a and b depend upon the specific bearing series. As the internal design of these bearings is specific to SKF, the catalog [13.38] should be consulted to obtain the required values of f_1, a, and b.

F_β in equation (13.99) depends on the magnitude and direction of the applied load. It may be expressed in equation form as follows for radial

TABLE 13.4. f_1 for Roller Bearings

Roller Bearing Type	f_1
Radial cylindrical with cage	0.0002–0.0004[a]
Radial cylindrical, full complement	0.00055
Tapered	0.0004
Radial needle	0.002
Thrust cylindrical	0.0015
Thrust needle	0.0015
Thrust spherical	0.00023–0.0005[a]

[a]Lower values pertain to light series bearings; higher values pertain to heavy series bearings.

ball bearings:

$$F_\beta = 0.9F_a \operatorname{ctn} \alpha - 0.1F_r \quad (13.102)$$

or

$$F_\beta = F_r \quad (13.102)$$

Of equations (13.102), the one yielding the larger value of F_β is used. For deep groove ball bearings, with nominal contact angle 0°, the first equation can be approximated by

$$F_\beta = 3F_a - 0.1F_r \quad (13.103)$$

For radial roller bearings,

$$\begin{aligned} F_\beta &= 0.8F_a \operatorname{ctn} \alpha \\ F_\beta &= F_r \end{aligned} \quad (13.104)$$

Again, the larger value of F_β is used. For thrust bearings, either ball or roller, $F_\beta = F_a$.

These values of torque as calculated from equation (13.99) appear to be reasonably accurate for bearings operating under reasonable load and relatively slow speed conditions. (Harris [13.12] used these data successfully in the thermal evaluation of a submarine propeller shaft thrust bearing assembly. Other minor analyses have been performed with seemingly good results.)

Viscous Friction Torque

Complex methods for calculating viscous friction forces in lubricated rolling bearings were indicated in Chapter 12, which dealt with elastohydrodynamic lubrication. In lieu of those methods to estimate friction torque, a simpler, empirical method was developed to cover standard bearing types.

For bearings that again operate at moderate speeds and under not-excessive load, Palmgren [13.11] determined empirically that viscous friction torque can be expressed as follows:

$$M_v = 10^{-7} f_o (\nu_o n)^{2/3} d_m^3 \qquad \nu_o n \geq 2000 \quad (13.105)$$

$$M_v = 160 \times 10^{-7} f_o d_m^3 \qquad \nu_o n \leq 2000 \quad (13.106)$$

in which ν_o is given in centistokes and n in revolutions per minute. In equations (13.105) and (13.106), f_o is a factor depending upon type of bearing and method of lubrication. Table 13.5 as updated in [13.38] gives

TABLE 13.5. Values of f_0 vs Bearing Type and Lubrication

Bearing Type	Type of Lubrication			
	Grease	Oil Mist	Oil Bath	Oil Bath (vertical shaft) or Oil Jet
Deep groove ball[a]	0.7–2[b]	1	2	3–4[b]
Self-aligning ball[c]	1.5–2[b]	0.7–1[b]	1.5–2[b]	3–4[b]
Thrust ball	5.5	0.8	1.5	3
Angular-contact ball[a]	2	1.7	3.3	6.6
Cylindrical roller				
with cage[a]	0.6–1[b]	1.5–2.8[b]	2.2–4[b]	2.2–4[b,d]
full complement	5–10[b]	—	5	—
Spherical roller[c]	3.5–7[b]	1.7–3.5[b]	3.5–7[b]	7–14[b]
Tapered roller[a]	6	3	6	8–10[b,d]
Needle roller	12	6	12	24
Thrust cylindrical roller	9	—	3.5	8
Thrust spherical roller	—	—	2.5–5[b]	5–10[b]
Thrust needle roller	14	—	5	11

[a]Use 2 × f_0 value for paired bearings or double row bearings.
[b]Lower values are for light series bearings; higher values are for heavy series bearings.
[c]Double row bearings only.
[d]For oil bath lubrication and vertical shaft, use 2 × f_0.

values of f_0 for various types of bearings subjected to different conditions of lubrication. Equations (13.105) and (13.106) are valid for oils having a specific gravity of approximately 0.9. Palmgren [13.11] gave a more complete formula for oils of different densities. For grease-lubricated bearings, kinematic viscosity ν_0 refers to the oil within the grease, and the equation is approximately valid shortly after the addition of lubricant.

Radial cylindrical roller bearings with flanges on both inner and outer rings can carry thrust load in addition to the normal radial load. In this case, the rollers are loaded against one flange on each ring. The bearing friction torque due to the roller end motions against properly designed and manufactured flanges is given by

$$M_f = f_f F_a d_m \tag{13.107}$$

Values of f_f are given in Table 13.6 when $F_a/F_r \leq 0.4$ and the lubricant is sufficiently viscous.

Total Friction Torque

A reasonable estimate of the friction torque of a given rolling bearing under moderate load and speed conditions is the sum of the *load* friction

TABLE 13.6. Values of f_f for Radial Cylindrical Roller Bearings

Bearing Type	Type of Lubrication	
	Grease	Oil
With cage, optimum design	0.0003	0.002
With cage, other designs	0.009	0.006
Full complement, single row	0.006	0.003
Full complement, double row	0.015	0.009

torque, *viscous* friction torque, and roller end-flange friction torque, if any, that is,

$$M = M_1 + M_v + M_f \tag{13.108}$$

Since M_1 and M_v are based on empirical formulas, the effect of rolling element-cage pocket sliding friction is included.

For high speed ball bearings for which friction due to spinning motions becomes important, the equations previously given should enable a calculation of friction torque. This torque should be added to that of equation (13.108). It must be remembered also that equation (13.108) does not account for friction torque due to seals, which friction torque in most instances far exceeds the friction torque of the bearing alone.

Example 13.2. Estimate the total friction torque for a 209 cylindrical roller bearing rotating at 10,000 rpm and supporting a radial load of 4450 N (1000 lb). The bearing is lubricated by a mineral oil bath, the oil having a kinematic viscosity of 20 centistokes.

$$d_m = 65 \text{ mm } (2.559 \text{ in.}) \qquad \text{Ex. 2.7}$$

$$D = 10 \text{ mm } (0.3937 \text{ in.}) \qquad \text{Ex. 2.7}$$

$$\gamma = 0.1538 \qquad \text{Ex. 2.7}$$

$$Z = 14 \qquad \text{Ex. 2.7}$$

$$l = 9.6 \text{ mm } (0.378 \text{ in.}) \qquad \text{Ex. 2.7}$$

From Table 13.4, assume $f_1 = 0.0003$ for a medium series bearing having a cage.

$$M_1 = f_1 F_\beta d_m \tag{13.99}$$

$$= 0.0003 \times 4450 \times 65 = 86.78 \text{ N} \cdot \text{mm } (0.7677 \text{ in.} \cdot \text{lb})$$

$$\nu_o n = 20 \times 10{,}000 = 200{,}000$$

$$M_v = 10^{-7} f_o (\nu_o n)^{2/3} d_m^3 \tag{13.105}$$

For oil bath lubrication from Table 13.5, assume $f_o = 3$ for a medium series bearing,

$$M_v = 10^{-7} \times 3 \times (200{,}000)^{2/3} \times (65)^3$$

$$= 281.8 \text{ N} \cdot \text{mm} \ (2.493 \text{ in.} \cdot \text{lb})$$

$$M = M_1 + M_v + M_f \qquad (13.108)$$

$$= 86.8 + 281.8 + 0 = 368.6 \text{ N} \cdot \text{mm} \ (3.261 \text{ in.} \cdot \text{lb})$$

Example 13.3. Estimate the rolling friction torque and viscous friction torque of the 218 angular-contact ball bearing operating at a shaft speed of 10,000 rpm and a thrust load of 22,250 N (5000 lb). The bearing is jet lubricated by a highly refined mineral oil having a kinematic viscosity of 5 centistokes at operating temperature.

$$d_m = 125.3 \text{ mm} \ (4.932 \text{ in.}) \qquad \text{Ex. 2.6}$$

$$D = 22.23 \text{ mm} \ (0.875 \text{ in.}) \qquad \text{Ex. 2.3}$$

$$\alpha = 40° \ (\text{nominal}) \qquad \text{Ex. 2.3}$$

$$\gamma = 0.1359 \qquad \text{Ex. 2.6}$$

$$f = 0.5232 \qquad \text{Ex. 2.3}$$

$$Z = 16 \qquad \text{Ex. 2.5}$$

$$C_s = \varphi_s i Z D^2 \cos \alpha \qquad (21.8)$$

From Table 21.2 at $\gamma = 0.1359$, $\varphi_s = 15.48$

$$C_s = 15.48 \times 1 \times 16 \times (22.23)^2 \cos 40°$$

$$= 93{,}760 \text{ N} \ (21{,}070 \text{ lb})$$

$$F_s = X_s F_r + Y_s F_a \qquad (21.15)$$

From Table 21.2, $X_s = 0.5$; $Y_s = 0.26$ for $\alpha = 40°$,

$$F_s = 0.5 \times 0 + 0.26 \times 22{,}250$$

$$= 5785 \text{ N} \ (1300 \text{ lb})$$

$$f_1 = z(F_s/C_s)^y \qquad (13.100)$$

From Table 13.3, $z = 0.001$; $y = 0.33$ for $\alpha = 40°$

$$f_1 = 0.001 \ (5{,}785/93{,}760)^{0.33}$$

$$= 0.0003988$$

$$F_\beta = 0.9F_a \text{ ctn } \alpha^\circ - 0.1F_r \tag{13.102}$$

$$= 0.9 \times 22{,}250 \text{ ctn } 40^\circ - 0.1 \times 0$$

$$= 23{,}860 \text{ N } (5363 \text{ lb})$$

$$M_1 = f_1 F_\beta d_m \tag{13.99}$$

$$= 0.0003988 \times 23{,}860 \times 125.3$$

$$= 1192 \text{ N} \cdot \text{mm } (10.55 \text{ in.} \cdot \text{lb})$$

From Table 13.5, $f_o = 6.6$ for oil jet lubrication

$$\nu_o n = 5 \times 10{,}000 = 50{,}000$$

$$M_v = 10^{-7} f_o (\nu_o n)^{2/3} d_m^3 \tag{13.105}$$

$$= 10^{-7} \times 6.6 \times (50{,}000)^{2/3} \times (125.3)^3$$

$$= 1762 \text{ N} \cdot \text{mm } (15.59 \text{ in.} \cdot \text{lb})$$

$$M = M_1 + M_v + M_f \tag{13.108}$$

$$= 1192 + 1762 + 0$$

$$= 2954 \text{ N} \cdot \text{mm } (26.13 \text{ in.} \cdot \text{lb})$$

CLOSURE

Rolling bearing are sometimes called *antifriction* bearings to emphasize the small amount of frictional power consumed during their operation. Notwithstanding this commercially acceptable terminology, it has been shown in this chapter that the rolling process does involve frictional power losses from various sources. Recent basic research had done much to define the mechanics of rolling friction, and for certain ideal conditions of rolling, estimates of rolling friction torque can be made. The operation of industrial rolling bearings that employ curved raceways, cages, and seals is, however, far from ideal in that sources of frictional power loss other than rolling are present in the bearings. Therefore, although it is important to understand the mechanics of rolling friction, empirical data are usually required to define friction torque of rolling bearing assemblies. These empirical data are presented in the previous section.

Rolling bearing friction is manifested in the form of temperature rises in the rolling bearing structure and lubricant unless effective heat removal methods are employed or naturally occur. When excessive temperature level occurs, the rolling bearing steel suffers loss in its ability to resist rolling surface fatigue and the lubricant undergoes deterioration such that it is ineffective. Subsequently, rapid bearing failure may be anticipated. Bearing thermal analysis and methods of heat removal are discussed further in Chapter 15.

Rolling bearing friction also tends to retard motion. In sensitive control systems such as those employing instrument bearings, torque due to bearing friction can significantly affect rotor speed.

REFERENCES

13.1. R. Drutowski, "Energy Losses of Balls Rolling on Plates," *Friction and Wear*, Elsevier, Amsterdam, 16–35 (1959).

13.2. R. Drutowski, "Linear Dependence of Rolling Friction on Stressed Volume," *Rolling Contact Phenomena*, Elsevier, Amsterdam (1962).

13.3. G. Lundberg, "Motions in Loaded Rolling Element Bearings," SKF unpublished report (1954).

13.4. A. Jones, "Ball Motion and Sliding Friction in Ball Bearings," *ASME J. Basic Eng.* 1–12 (March 1959).

13.5. O. Reynolds, *Philos. Trans. R. Soc. London* **166,** 155 (1875).

13.6. H. Poritsky, *J. Appl. Mech.* **72,** 191 (1950).

13.7. B. S. Cain, *J. Appl. Mech.* **72,** 465 (1950).

13.8. H. Heathcote, *Proc. Inst. Automobile Eng. London* **15,** 569 (1921).

13.9. K. Johnson, "Tangential Tractions and Micro-slip," *Rolling Contact Phenomena*, Elsevier, Amsterdam, 6–28 (1962).

13.10. J. Greenwood and D. Tabor, *Proc. Phys. Soc. London* **71,** 989 (1958).

13.11. A. Palmgren, *Ball and Roller Bearing Eng.* 3rd ed., Burbank, Philadelphia, 34–41 (1959).

13.12. T. Harris, "Prediction of Temperature in a Rolling Contact Bearing Assembly," *Lubr. Eng.* 145–150 (April 1964).

13.13. E. Trachman and H. Cheng, "Thermal and Non-Newtonian Effects on Traction in Elastohydrodynamic Contacts," *Proceedings of the Second Symposium on Elastohydrodynamic Lubrication*, Institute of Mechanical Engineers, London, 142–148 (1972).

13.14. K. Johnson and J. Tevaarwerk, "Shear Behaviour of Elastohydrodynamic Oil Films," *Proc. R. Soc. London Ser. A* **356,** 215–236 (1977).

13.15. T. Harris, "Rolling Element Bearing Dynamics," *Wear* **23,** 311–337 (1973).

13.16. V. Streeter, *Fluid Mechanics*, McGraw-Hill, New York, 313–314 (1951).

13.17. E. Bisson and W. Anderson, *Advanced Bearing Technology*, NASA SP-38 (1964).

13.18. T. Harris, "Ball Motion in Thrust-Loaded, Angular-Contact Bearings with Coulomb Friction," *ASME J. Lubr. Tech.* **93,** 32–38 (1971).

13.19. T. Tallian, G. Baile, H. Dalal, and O. Gustafson, *Rolling Bearing Damage Atlas*, SKF Industries, Inc., King of Prussia, Penn., 119–143 (1974).

13.20. T. Harris, "An Analytical Method to Predict Skidding in Thrust-Loaded, Angular-Contact Ball Bearings," *ASME J. Lubr. Tech.* **93,** 17–24 (1971).

13.21. R. Shevchenko and P. Bolan, "Visual Study of Ball Motion in a High Speed Thrust Bearing," SAE Paper No. 37 (January 14–18, 1957).

13.22. J. Poplawski and J. Mauriello, "Skidding in Lightly Loaded, High Speed, Ball Thrust Bearings," ASME Paper 69-LUBS-20 (1969).

13.23. R. Parker, "Comparison of Predicted and Experimental Thermal Performance of Angular-Contact Ball Bearings," NASA Tech. Paper 2275 (1984).

13.24. T. Harris, "An Analytical Method to Predict Skidding in High Speed Roller Bearings," *ASLE Trans.* **9,** 229–241 (1966).

13.25. T. Harris and S. Aaronson, "An Analytical Investigation of Skidding in a High Speed, Cylindrical Roller Bearing Having Circumferentially Spaced, Preloaded Annular Rollers," *Lubr. Eng.* 30–34 (January 1968).

13.26. C. Walters, "The Dynamics of Ball Bearings," *ASME J. Lub. Tech.* **93**(1), 1–10 (Jan. 1971).

13.27. F. Wellons and T. Harris, "Bearing Design Considerations," *Interdisciplinary Approach to the Lubrication of Concentrated Contacts*, NASA SP-237, 529–549 (1970).

13.28. R. Kleckner and J. Pirvics, "High Speed Cylindrical Roller Bearing Analysis—SKF Computer Program CYBEAN, Vol. I: Analysis," SKF Report AL78P022, submitted to NASA–Lewis Research Center under contract NAS3-20068 (July 1978).

13.29. J. Kannel and S. Bupara, "A Simplified Model of Cage Motion in Angular-Contact Bearings Operating in the EHD Lubrication Regime," *ASME J. Lub. Tech.* **100,** 395–403 (July 1978).

13.30. P. Gupta, "Dynamics of Rolling Element Bearings—Part I–IV Cylindrical Roller Bearing Analysis," *ASME J. Lub. Tech.* **101,** 293–326 (1979).

13.31 C. Meeks and K. Ng, "The Dynamics of Ball Separators in Ball Bearings—Part I: Analysis," ASLE Paper No. 84-AM-6C-2 (May 1984).

13.32. C. Meeks, "The Dynamics of Ball Separators in Ball Bearings—Part II: Results of Optimization Study," ASLE Preprint No. 84-AM-6C-3 (May 1984).

13.33. P. Brown, L. Dobek, F. Hsing, and J. Miner, "Mainshaft High Speed Cylindrical Roller Bearings for Gas Turbine Engines," U.S. Navy Contract N00140-76-C-0383, Interim Report FR-8615 (April 1977).

13.34. M. Kellstrom and E. Blomqvist, U.S. Patent 3,990,753, "Roller Bearings Comprising Rollers with Positive Skew Angle."

13.35. M. Kellstrom, "Rolling Contact Guidance of Rollers in Spherical Roller Bearings," ASME Paper 79-LUB-23, ASME/ASLE Lubrication Conference (October 1979).

13.36. B. Rydell, "New Spherical Roller Thrust Bearings, the E. Design," *Ball Bear. J. (SKF)*, No. 202, 1–7 (1980).

13.37. J. Mauriello, N. Lagasse, A. Jones, and W. Murray, "Rolling Element Bearing Retainer Analysis," USA AMRDL Technical Report 72–45 (November 1973).

13.38. SKF, General Catalogue 4000E (April 1989).

14

MICROCONTACT PHENOMENA

LIST OF SYMBOLS

Symbol	Description	Units
a	Semimajor axis of contact ellipse	mm (in.)
A_c	True average contact area	mm^2 (in.^{2})
A_0	Apparent area	mm^2 (in.^{2})
b	Semiminor axis of contact ellipse	mm (in.)
d	Separation of mean plane of summits and smooth plane	mm (in.)
D	$\eta u/Gb$, Deborah number	
D_{SUM}	Summit density	mm^{-2} (in.$^{-2}$)
E_1, E_2	Elastic moduli of bodies 1 and 2	N/mm^2 (psi)
E'	Reduced elastic modulus	N/mm^2 (psi)
$F_0(\ )$, $F_1(\ )$, $F_{3/2}(\ )$	Tabular functions for the Greenwood-Williamson model	
F	Traction force	N (lb)
F_x, F_y	Components of traction force	N (lb)

Symbol	Description	Units
$\overline{G}$	Average elastic modulus	N/mm^2 (psi)
h	Film thickness	mm (in.)
m_0	Zero-order spectral moment, $\equiv R_q^2 \equiv \sigma^2$	μm^2 ($\mu in.^2$)
m_2	Second-order spectral moment	(nondimensional)
m_4	Fourth-order spectral moment	μm^{-2} ($\mu in.^{-2}$)
n	Contact density	mm^{-2} ($in.^{-2}$)
n_p	Plastic contact density	mm^{-2} ($in.^{-2}$)
p_0	Maximum pressure	N/mm^2 ($lb/in.^2$)
P	Applied load	N (lb)
Q_a	Asperity supported load	N (lb)
Q_f	Fluid supported load	N (lb)
r_{1x}, r_{2x}	Radii of bodies 1 and 2 in rolling direction	mm (in.)
R	Summit sphere radius	μm ($\mu in.$)
R_x	Equivalent radius in rolling direction	mm (in.)
R_q	Root mean square (rms) value of surface profile	μm ($\mu in.$)
u_1, u_2	Surface velocities of bodies 1 and 2	m/sec (in./sec)
u	Entrainment velocity	m/sec (in./sec)
w	Deflection of summit	μm ($\mu in.$)
w_p	Variable governing plastic asperity density	μm ($\mu in.$)
x	Coordinate in rolling direction	m (in.)
y	Coordinate in direction transverse to rolling	mm (in.)
Y	Yield strength in simple tension, dimensionless coordinate y/a	N/mm^2 (psi)
z_s	Summit height relative to summit mean plane	mm (in.)
$\bar{z}_s$	Distance between surface and summit mean plane	mm (in.)
$z(x)$	Surface profile	mm (in.)
α	Pressure-viscosity coefficient	mm^2/N ($in.^2/lb$)
β	Temperature-viscosity coefficient	$°C^{-1}$ ($°F^{-1}$)
$\dot{\gamma}$	Shear rate	sec^{-1}
Δu	Sliding velocity in rolling direction	m/sec (in./sec)

Symbol	Description	Units
Δv	Sliding velocity transverse to rolling direction	m/sec (in./sec)
δ_x	Shift in fluid pressure distribution	mm (in.)
η	Absolute viscosity	N-sec/m^2 (lb-sec/in.^{2})
Λ	Lubricant film parameter, h/σ	
μ_a	Coulomb friction coefficient	
μ_f	Fluid friction coefficient	
ν_1, ν_2	Poisson's ratio for bodies 1 and 2	
σ	Composite surface roughness	mm (in.)
σ_1, σ_2	Rms profile roughness for bodies 1 and 2	mm (in.)
τ	Shear stress	N/mm^2 (psi)
τ_c	Limiting shear stress	N/mm^2 (psi)
τ_e	Equivalent shear stress	N/mm^2 (psi)
τ_{max}	Maximum shear stress	N/mm^2 (psi)
τ_y	Yield strength	N/mm^2 (psi)
ϕ	Constant in Herschel-Bulkley law	
$\phi(\ )$	Gaussian probability density function	mm^{-1} (in.$^{-1}$)
ψ	Auxiliary function in traction model	
ω	Spinning speed	sec^{-1}

GENERAL

In its full complexity, a concentrated contact cannot be represented in an analytical expression. The combined action of an applied load and kinematic constraints produces some combination of rolling, sliding, and spinning motions. These motions act to draw lubricant into the contact where, its properties altered by the pressure and temperature that vary throughout the contact region, it forms a film that serves to separate the contacting bodies to an extent depending on both the microgeometry of the bodies, and the properties of the lubricant. When the separating film is small relative to the composite surface roughness, a myriad of microcontacts of highly irregular shapes forms within the macrocontact, causing pressure, temperature, and film thickness perturbations on a microscale. Moreover, these microcontacts may deform plastically as well as elastically with the result that the microgeometry varies with time.

Sliding and spinning motions on the macrocontact act to shear the separating lubricant film and, if separation is only partial, to drag the microcontacts across each other. A tangential force is produced from these combined effects. This tangential or traction force alters the stress distribution in the solids and is a critical factor in determining fatigue life. The magnitude of the fluid contribution to the traction depends on the lubricant properties under the locally variable pressure and temperature and shear rates that prevail in the macrocontact. The contribution to the traction caused by the sliding microcontacts will depend on the local film conditions or the nature of the surface boundary films that result from oxidation and additives present in the lubricant.

Given the complexity of the interfacial conditions, attempts to model the contact of real surfaces have necessarily been crude and not likely to give an accurate absolute prediction of such key quantities as traction force, number of microcontacts, number of plastic microcontacts, film thickness, and so on. The chief value of such models is that they are useful for comparing the relative severities of various macrogeometries. Armed with such a descriptive model, a designer may evaluate which of the possible design options offers the best hope for satisfactory performance.

This chapter describes an approach that synthesizes state-of-the-art models for lubricant film thickness, asperity load sharing, and fluid traction into a practicable, analytical description of a real contact.

ANALYTICAL PROCEDURE

To determine the effect of microgeometry, that is, surface asperities, on the normal stresses and frictional stresses in a rolling element-raceway contact, McCool [14.1] established a method using the lubricant film thickness equation of Dowson and Hamrock [14.2] [equation (12.84)], the Greenwood–Williamson [14.3] microcontact model and the Tevaarwerk–Johnson [14.4] fluid traction model. The steps used in the calculation are the following:

1. The lubricant film thickness in the plateau region of the macrocontact is calculated using equation (12.84) for isothermal conditions. Lubricant heating in the inlet to the contact and also possible starvation effects are accommodated using methods discussed in Chapter 12.
2. Using the Greenwood–Williamson (GW) microcontact model, the mean load supported by the contacting asperities is determined considering the roughness mean planes of the contacting bodies are separated by the plateau lubricant film thickness. Addition-

ally parameters for judging the severity of the interfacial condition are calculated; for example:

(i) The "true" contact area as a fraction of the macrocontact area.
(ii) The number of microcontacts per unit area, that is, density of microcontacts in the macrocontact.
(iii) The number of contacts per unit area for which the maximum subsurface shear stress exceeds the material yield strength in shear, that is, the density of plastic contacts.

3. The load supported by the lubricant film is determined by subtracting the load supported by the asperities.
4. Using the Tevaarwerk–Johnson [14.4] fluid traction relationship, the fluid friction coefficient is calculated.
5. Finally, the total traction force in the macrocontact is determined by the following equation:

$$F = \mu_f Q_f + \mu_a Q_a \tag{14.1}$$

McCool [14.1] developed a computer program TRIBOS to perform the calculations indicated above. In the sections that follow the various elements of the calculations are discussed in detail.

MICROCONTACTS

"Rough" Surfaces

In calculating the lubricant film thickness in Chapter 12, it is assumed that the surfaces are perfectly smooth. The assumption is now made that when the surfaces are rough the lubricant film thickness, calculated as if the surfaces were smooth, separates the mean planes of the rough surfaces, as shown in Fig. 14.1.

The surfaces fluctuate randomly about their mean planes in accordance with a probability distribution. The root-mean-square (rms) value of this distribution is denoted σ_1 for the upper surface and σ_2 for the lower

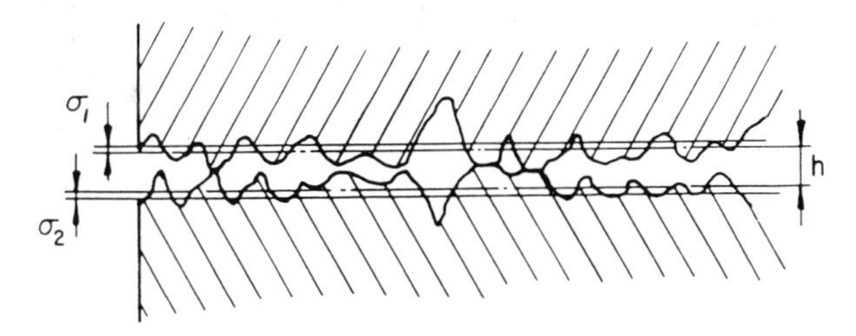

FIGURE 14.1. Asperity contacts through partial oil film.

surface. When the combined surface fluctuations at a given position exceed the gap h due to the lubricant film, a microcontact occurs. At the microcontacts the surfaces deform elastically and possibly plastically. The aggregate of the microcontact areas is generally a small fraction ($<5\%$) of the nominal area of contact.

A microcontact model uses surface microgeometry data to predict, at a minimum, the density of microcontacts, the real area of contact, and the elastically supported mean load. One of the earliest and simplest microcontact models is that of Greenwood and Williamson (GW) [14.3]. Generalizations of this model applicable to isotropic surfaces have been developed by Bush et al. [14.5] and by O'Callaghan and Cameron [14.6]. Bush et al. [14.7] also treated a strongly anisotropic surface. One of the most comprehensive models yet developed is ASPERSIM [14.8], which requires a nine-parameter microgeometry description and accounts for anisotropic as well as isotropic surfaces. A comparison of various microcontact models conducted by McCool [14.9] has shown that the GW model, despite its simplicity, compares favorably with the other models. Because it is much easier to implement than the other models, the GW model is the microcontact model recommended here.

GW Model

For the contact of real surfaces Greenwood and Williamson [14.3] developed one of the first models that specifically accounted for the random nature of interfacial phenomena. The model applies to the contact of two flat elastic planes, one rough and the other smooth. It is readily adapted to the case of two rough surfaces as discussed further below. In the GW model the rough surface is presumed to be covered with local high spots or asperities whose summits are spherical. The summits are presumed to have the same radius R, but randomly variable heights, and to be uniformly distributed over the rough surface with a known density D_{SUM} of summits/unit area.

The mean height of summits lies above the mean height of the surface as a whole by the amount $\bar{z}_s$, indicated in Fig. 14.2. The summit heights

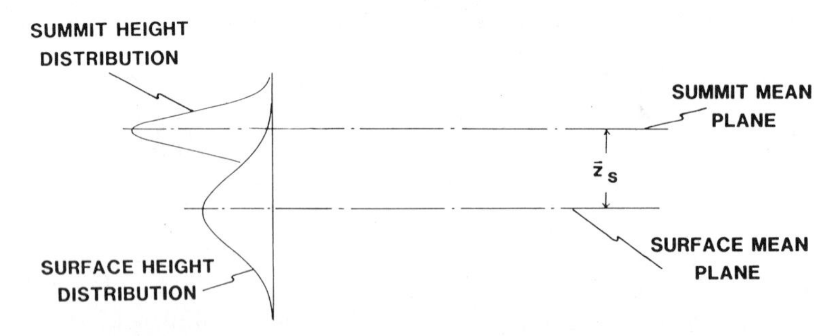

FIGURE 14.2. Surface and summit mean planes and distributions.

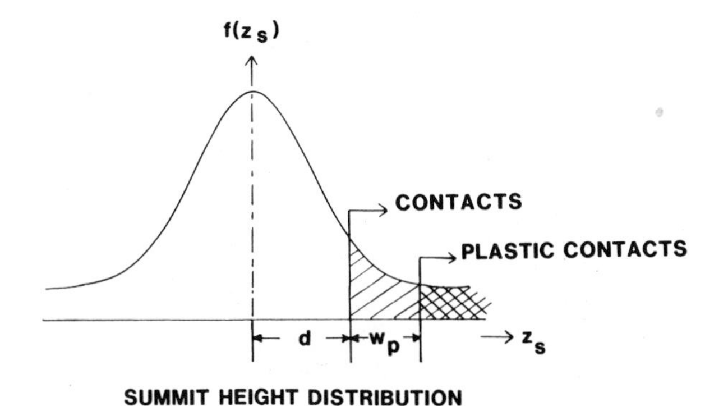

FIGURE 14.3. Distribution of summit heights.

z_s are assumed to follow a Gaussian probability law with a standard deviation σ_s. Figure 14.3 shows the assumed form for the summit height distribution or probability density function (pdf) $f(z_s)$. It is symmetrical about the mean summit height. The probability that a summit has a height, measured relative to the summit mean plane in the interval (z_s, $z_s + dz_s$) is expressed in terms of the pdf as $f(z_s)\,dz_s$. The probability that a randomly selected summit has a height in excess of some value d is the area under the pdf to the right of d. The equation of the pdf is

$$f(z_s) = \frac{e^{-(z_s/2\sigma_s)^2}}{\sigma_s\sqrt{2\pi}} \tag{14.2}$$

So the probability that a randomly selected summit has height in excess of d is

$$P[z_s > d] = \int_d^\infty f(z_s)\,ds \tag{14.3}$$

This integration must be performed numerically. Fortunately, however, the calculation can be related to tabulated areas under the standard normal curve for which the mean is 0 and the standard deviation is 1.0.

Using the standard normal density function $\phi(x)$, the probability that a summit has a height greater than d above the summit mean plane is calculated.

$$P[z_s > d] = \int_{d/\sigma_s}^\infty \phi(x)\,dx = F_0\left(\frac{d}{\sigma_s}\right) \tag{14.4}$$

where $F_0(t)$ is the area under the standard normal curve to the right of the value t. Values $F_0(t)$ for t ranging from 1.0 to 4.0, are given in column 2 of Table 14.1.

TABLE 14.1. Functions for the Greenwood-Williamson model

t	$F_0(t)$	$F_1(t)$	$F_{3/2}(t)$
0.0	0.5000	0.3989	0.4299
0.1	0.4602	0.3509	0.3715
0.2	0.4207	0.3069	0.3191
0.3	0.3821	0.2668	0.2725
0.4	0.3446	0.2304	0.2313
0.5	0.3085	0.1978	0.1951
0.6	0.2743	0.1687	0.1636
0.7	0.2420	0.1429	0.1363
0.8	0.2119	0.1202	0.1127
0.9	0.1841	0.1004	0.9267×10^{-1}
1.0	0.1587	0.8332×10^{-1}	0.7567×10^{-1}
1.1	0.1357	0.6862×10^{-1}	0.6132×10^{-1}
1.2	0.1151	0.5610×10^{-1}	0.4935×10^{-1}
1.3	0.9680×10^{-1}	0.4553×10^{-1}	0.3944×10^{-1}
1.4	0.8076×10^{-1}	0.3667×10^{-1}	0.3129×10^{-1}
1.5	0.6681×10^{-1}	0.2930×10^{-1}	0.2463×10^{-1}
1.6	0.5480×10^{-1}	0.2324×10^{-1}	0.1925×10^{-1}
1.7	0.4457×10^{-1}	0.1829×10^{-1}	0.1493×10^{-1}
1.8	0.3583×10^{-1}	0.1428×10^{-1}	0.1149×10^{-1}
1.9	0.2872×10^{-1}	0.1105×10^{-1}	0.8773×10^{-2}
2.0	0.2275×10^{-1}	0.8490×10^{-2}	0.6646×10^{-2}
2.1	0.01786	6.468×10^{-3}	0.4995×10^{-2}
2.2	0.01390	4.887×10^{-3}	0.3724×10^{-2}
2.3	0.01072	3.662×10^{-3}	0.2754×10^{-2}
2.4	0.8198×10^{-2}	2.720×10^{-3}	0.2020×10^{-2}
2.5	0.6210×10^{-2}	2.004×10^{-3}	0.1469×10^{-2}
2.6	0.4661×10^{-2}	1.464×10^{-3}	0.1060×10^{-2}
2.7	0.3467×10^{-2}	1.060×10^{-3}	0.7587×10^{-3}
2.8	0.2555×10^{-2}	7.611×10^{-4}	0.5380×10^{-3}
2.9	0.1866×10^{-2}	5.417×10^{-4}	0.3784×10^{-3}
3.0	0.1350×10^{-2}	3.822×10^{-4}	0.2639×10^{-3}
3.2	6.871×10^{-4}	1.852×10^{-4}	0.1251×10^{-3}
3.4	3.369×10^{-4}	8.666×10^{-5}	0.5724×10^{-4}
3.6	1.591×10^{-4}	3.911×10^{-5}	0.2529×10^{-4}
3.8	7.235×10^{-5}	1.702×10^{-5}	0.1079×10^{-4}
4.0	3.167×10^{-5}	7.145×10^{-6}	0.4438×10^{-5}

It is assumed that when large flat surfaces are pressed together, their mean planes remain parallel. Thus, if a rough surface and a smooth surface are pressed against each other until the summit mean plane of the rough surface and the mean plane of the smooth surface are separated by an amount d, the probability that a randomly selected summit will be a microcontact is

$$P[\text{summit is a contact}] = P[z_s > d] = F_0\,(d/\sigma_s) \tag{14.5}$$

Since the number of summits per unit area is D_{SUM}, the average expected number of contacts in any unit area is

$$n = D_{SUM} F_0\,(d/\sigma_s) \tag{14.6}$$

Given that a summit is in contact because its height z_s exceeds d, the summit must deflect by the amount $w = z_s - d$, as shown in Fig. 14.4.

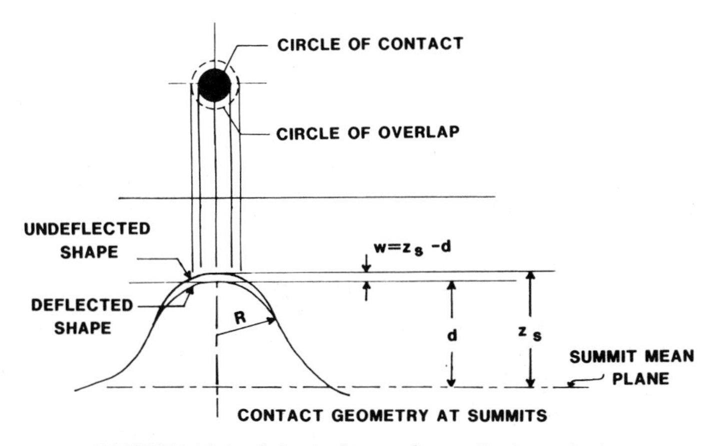

FIGURE 14.4. Spherical capped asperity in contact.

For notational simplicity the subscript on z_s is henceforth deleted. For a sphere of radius R elastically deflecting by the amount w, the Hertzian solution gives the contact area

$$A = \pi R w = \pi R \,(z - d) = \pi a^2 \qquad z > d \tag{14.7}$$

where a = contact radius.

The corresponding asperity load is

$$P = \tfrac{4}{3} E' R^{1/2} w^{3/2} = \tfrac{4}{3} \mathrm{E'R}^{1/2} \,(z - d)^{3/2} \qquad z > d \tag{14.8}$$

where $E' = [(1 - \nu_1^2)/E_1 + (1 - \nu_2^2)/E_2]^{-1}$ and E_i, ν_i $(i = 1, 2)$ are Young's moduli and Poisson's ratios for the two bodies. The maximum Hertzian pressure in the microcontact is

$$p_0 = 1.5 \frac{P}{A} = \frac{2E' w^{1/2}}{\pi R^{1/2}} = \left(\frac{2E'}{\pi R^{1/2}} \right) (z - d)^{1/2} \tag{14.9}$$

Both A and P are functions of the random variable z. The average or expected value of functions of random variables are obtained by integrating the function and the probability density of the random variable over the space of possible values of the random variable. The expected summit contact area is thus

$$A = \int_d^{\infty} \pi R \,(z - d)\, f(z)\, dz \tag{14.10}$$

which transforms to

$$A = \pi R\sigma_s \int_{d/\sigma_s}^{\infty} \left(x - \frac{d}{\sigma_s}\right) \phi_x \, dx = \pi R\sigma_s F_1 \left(\frac{d}{\sigma_s}\right) \tag{14.11}$$

where

$$F_1 (t) = \int_t^{\infty} (x - t) \, \phi_x \, dx \tag{14.12}$$

$F_1(t)$ is also given in Table 14.1.

The expected total contact area as a fraction of the apparent area is obtained as the product of the average asperity contact area contributed by a single randomly selected summit and the density of summits. Thus, the ratio of contact to apparent area, A_c/A_0, is

$$\frac{A_c}{A_0} = \pi R\sigma_s D_{\text{SUM}} F_1 \left(\frac{d}{\sigma_s}\right) \tag{14.13}$$

By the same argument the total load per unit area supported by asperities is

$$\frac{P}{A_0} = \frac{4}{3} E' R^{1/2} \sigma_s^{3/2} D_{\text{SUM}} F_{3/2} \left(\frac{d}{\sigma_s}\right) \tag{14.14}$$

where

$$F_{3/2} (t) = \int_t^{\infty} (x - t)^{3/2} \, \phi (x) \, dx \tag{14.15}$$

$F_{3/2}(t)$ is also given in Table 14.1.

Plastic Contacts

A contacting summit will experience some degree of plastic flow when the maximum shear stress exceeds half the yield stress in simple tension. In the contact of a sphere and a flat, the maximum shear stress is related to the maximum Hertzian stress p_0 by

$$\tau_{\max} = 0.31 p_0 \tag{14.16}$$

Thus, some degree of plastic deformation is present at a contact if $\tau_{\max} > Y/2$. Using the expression for p_0 [(eq. (14.9)] gives

$$\frac{0.31 \cdot 2E'(z-d)^{1/2}}{\pi R^{1/2}} > \frac{Y}{2} \tag{14.17}$$

or

$$z - d > 6.4R\left(\frac{Y}{E'}\right)^2 \equiv w_p \tag{14.18}$$

$$z > d + w_p \tag{14.19}$$

Thus, any summit whose height exceeds $d + w_p$ will have some degree of plastic deformation. The probability of a plastic summit is given by the shaded area in Fig. 14.3 to the right of $d + w_p$. The expected number of plastic contacts per unit area becomes

$$n_p = D_{SUM} F_0 \left(\frac{d}{\sigma_s} + w_p^*\right) \tag{14.20}$$

where

$$w_p^* \equiv \frac{w_p}{\sigma_s} \equiv 6.4\left(\frac{R}{\sigma_s}\right)\left(\frac{Y}{E'}\right)^2 \tag{14.21}$$

For fixed d/σ_s the degree of plastic asperity interaction is determined by the value of w_p^*: the higher w_p^*, the fewer plastic contacts. Accordingly, GW use the inverse, $1/w_p^*$, as a measure of the plasticity of an interface. For a given nominal pressure P/A_0, d/σ_s is found by solving equation (14.14), assuming that most of the load is elastically supported.

Application of the GW Model to a Lubricated Contact of Two Rough Surfaces

To use the GW model for a lubricated contact, (1) the height d relative to the mean plane of the summit heights to h, the thickness of the lubricant film that separates the two surfaces, must be determined, and (2) the values of the GW parameters R, D_{SUM}, and σ_s must be established. For (1) the first step is to compute the composite rms value of the two "rough" surfaces as

$$\sigma = (\sigma_1^2 + \sigma_2^2)^{1/2} \tag{14.22}$$

When the mean plane of a rough surface with this rms value is held at a height h above a smooth plane, the rms value of the gap width is the same as shown in Fig. 14.3, where both surfaces are rough. It is in this sense that the surface contact of two rough surfaces may be translated into the equivalent contact of a rough surface and a smooth surface. As shown in Fig. 14.2, the summit and surface mean planes are separated by an amount $\bar{z}_s$.

For an isotropic surface with normally distributed height fluctuations, the value of $\bar{z}_s$ has been found by Bush et al. [14.10] to be

$$\bar{z}_s = \frac{4\sigma}{\sqrt{\pi\alpha}} \tag{14.23}$$

The quantity α, known as the bandwidth parameter, is defined by

$$\alpha = \frac{m_0 m_4}{m_2^2} \tag{14.24}$$

where m_0, m_2, and m_4 are known as the zeroth, second, and fourth spectral moments of a profile. They are equivalent to the mean square height, slope, and second derivative of a profile in an arbitrary direction; that is

$$m_0 = E\,(z^2) = \sigma^2 \tag{14.25}$$

$$m_2 = E\left[\left(\frac{dz}{dx}\right)^2\right] \tag{14.26}$$

$$m_4 = E\left[\left(\frac{d^2z}{dx^2}\right)^2\right] \tag{14.27}$$

where $z(x)$ is a profile in an arbitrary direction x, $E[\ ]$ denotes statistical expectation, and m_0 is simply the mean square surface height. The square root of m_0 or root mean square (rms) is sometimes referred to as σ or R_q and forms part of the usual output of a stylus measuring device. Some of the newer profile measuring devices also give the rms slope, which is the same as $(m_2)^{1/2}$ converted from radians to degrees. No commercial equipment is yet available to measure m_4. Measurements of m_4 made so far have used custom computer processing of the signal output of profile measurement equipment.

Bush et al. [14.10] also show that the variance σ_s^2 of the surface summit height distribution is related to σ^2, the variance of the composite surfaces, by

$$\sigma_s^2 = \left(1 - \frac{0.8968}{\alpha}\right)\sigma^2 \tag{14.28}$$

A summit located a distance d from the summit height mean plane is at a distance $h = d + \bar{z}_s$ from the surface mean plane. Thus,

$$d = h - \bar{z}_s \tag{14.29}$$

Using equation (14.23) for $\bar{z}_s$ and equation (14.28) for σ_s gives

$$\frac{d}{\sigma_s} = \frac{h/\sigma - 4/\sqrt{\pi\alpha}}{(1 - 0.8968/\alpha)^{1/2}} \tag{14.30}$$

Equation (14.30) shows that d/σ_s is linearly related to h/σ. The ratio h/σ is also referred to as the lubricant film parameter Λ. When $\Lambda > 3$, contacts are few and the surfaces may be considered to be well lubricated.

For a specified or calculated value of Λ, d/σ_s is computed from equation (14.30) for use in the GW model. For an isotropic surface the two parameters D_{SUM} and R, the average radius of the spherical caps of asperities, may be expressed as (Nayak [14.11]):

$$D_{SUM} = \frac{m_4}{6\pi m_2 \sqrt{3}} \tag{14.31}$$

$$R = \tfrac{3}{8}\sqrt{\frac{\pi}{m_4}} \tag{14.32}$$

ASPERITY- AND FLUID-SUPPORTED LOAD AND TRACTION

Computational Method

For a specified contact with semiaxes a and b, under a load P, with plateau lubricant film thickness h and given values of m_0, m_2, and m_4, the asperity load Q_a is determined by first computing P/A_o from equation (14.14) and using

$$Q_a = \pi ab \left(\frac{P}{A_o}\right) \tag{14.33}$$

The fluid-supported load is then

$$Q_f = P - Q_a \tag{14.34}$$

If $Q_a > P$, the implication is that the lubricant film thickness is larger than computed under smooth surface theory. In this case, equation (14.14) could be solved iteratively until $Q_a = P$.

With a fluid friction coefficient μ_f and an asperity Coulomb friction coefficient μ_a, the traction force in the absence of spinning is

$$F = \mu_f Q_f + \mu_a Q_a \tag{14.1}$$

The force F and its two components act in the direction of the sliding velocity vector. The effective traction coefficient is

$$T = \frac{F}{P} \tag{14.35}$$

For an anisotropic surface, the value of m_2 will vary with the direction in which the profile is taken on the surface. The maximum and minimum values occur in two orthogonal "principal" directions. Sayles and Thomas [14.12] recommend the use of an equivalent isotropic surface for which m_2 is computed as the harmonic mean of the m_2 values found along the principal directions. The value of m_4 is similarly taken as the harmonic mean of the m_4 values in these two directions.

Example 14.1. An isotropic surface has roughness parameters $\sigma^2 = m_0 = 0.0625\ \mu m^2$, $m_2 = 0.0018$, and $m_4 = 1.04 \times 10^{-4}\ \mu\text{m}^{-2}$. Calculate the summit density D_{SUM}, the height of the summit mean plane above the surface mean plane, the mean summit radius R, and the standard deviation σ_s of the summit height distribution.

From equation (14.31) the summit density is

$$D_{\text{SUM}} = \frac{m_4}{6\pi m_2 \sqrt{3}} = \frac{1.04 \times 10^{-4}}{1.8 \times 10^{-3} \times 32.65} \tag{14.31}$$

$$= 1.77 \times 10^{-3}\ \mu\text{m}^{-2}\ (1.142\ \mu\text{in.}^{-2})$$

The separation of the surface and summit mean plane is, by equations (14.23) and (14.24),

$$\bar{z}_s = \frac{4\sigma}{\sqrt{\pi\alpha}} = 4\left(\frac{m_0}{\pi\alpha}\right)^{1/2} \tag{14.23}$$

$$\alpha = \frac{m_0 m_4}{m_2^2} = \frac{0.0625 \times 1.04 \times 10^{-4}}{(1.8 \times 10^{-3})^2} \tag{14.24}$$

$$= 2.006$$

$$\bar{z}_s = 4\left(\frac{0.0625}{2.006\pi}\right)^{1/2}$$

$$= 0.399\ \mu\text{m}\ (1.571 \times 10^{-5}\ \text{m.})$$

The mean summit tip radius is, from equation (14.32),

$$R = \frac{3}{8}\left(\frac{\pi}{m_4}\right)^{1/2} = \frac{3}{8}\left(\frac{\pi}{1.04 \times 10^{-4}}\right)^{1/2} \tag{14.32}$$
$$= 65.2\ \mu\text{m}\ (2.567 \times 10^{-3}\ \text{in.})$$

The standard deviation of the summit height distribution is calculated from equation (14.28) to be

$$\sigma_s = \left[\left(1 - \frac{0.8968}{\alpha}\right) m_0\right]^{1/2} \tag{14.28}$$
$$= \left[\left(1 - \frac{0.8968}{2.006}\right) 0.0625\right]^{1/2}$$
$$= 0.186\ \mu\text{m}\ (7.323 \times 10^{-6}\ \text{in.})$$

Let a steel surface having these roughness characteristics make rolling contact with a smooth plane forming an EHD lubricated contact for which the plateau lubricant film thickness, computed from equation (12.84) and adjusted for starvation and inlet heating, is $h = 0.5\ \mu$m. Using the GW microcontact model, calculate the nominal pressure P/A_o, the relative contact area A_c/A_o, the mean real pressure P/A_c, the contact density n, and, for a tensile yield strength of 2070 N/mm^2, the plastic contact density n_p.

The computed film parameter $\Lambda = 0.5/(0.0625)^{1/2} = 2.0$. From equation (14.30),

$$\frac{d}{\sigma_s} = \frac{h/\sigma - 4/(\pi\alpha)^{1/2}}{(1 - 0.8968/\alpha)^{1/2}} \tag{14.30}$$
$$= \frac{2 - 4/(2\pi)^{1/2}}{(1 - 0.8968/2.006)^{1/2}}$$
$$= 0.544$$

Interpolating in Table 14.1 gives

$$F_0\,(0.544) = 0.2935$$
$$F_1\,(0.544) = 0.1850$$
$$F_{3/2}\,(0.544) = 0.1812$$

The nominal pressure is calculated from equation (14.14) with $E' = 1.14 \times 10^5$ N/mm^2 (16.53×10^6 psi) for steel:

$$\frac{P}{A_o} = \frac{4}{3} E' R^{1/2} \sigma_s^{3/2} D_{SUM} F_{3/2}\left(\frac{d}{\sigma_s}\right) \tag{14.14}$$

$$= \frac{4}{3} \times 1.14 \times 10^5 \,(0.0652)^{1/2}\,(0.186 \times 10^{-3})^{3/2} \times 1770 \times 0.1812$$

$$= 31.6 \text{ N/mm}^2 \text{ (4581 psi)}$$

From equation (14.13) the ratio of mean real contact area A_c to nominal contact area A_o is

$$\frac{A_c}{A_o} = \pi R \sigma_s D_{SUM} F_1\left(\frac{d}{\sigma_s}\right) \tag{14.13}$$

$$= \pi \times 0.0652 \times 0.186 \times 10^{-3} \times 1770 \times 0.185$$

$$= 0.0125$$

The actual contact area thus averages only 1.25% of the nominal contact area. The mean actual pressure P/A_c is

$$\frac{P}{A_c} = \frac{P/A_o}{A_c/A_o}$$

$$= \frac{31.6}{0.0125} = 2528 \text{ N/mm}^2 \text{ } (3.665 \times 10^5 \text{ psi})$$

From equation (14.6) the contact density n is

$$n = D_{SUM} F_0\left(\frac{d}{\sigma_s}\right) \tag{14.6}$$

$$= 1770 \times 0.2935 = 519 \text{ contacts/mm}^2 \text{ } (3.35 \times 10^5/\text{in}^2.)$$

From equation (14.21),

$$w_p^* = 6.4\left(\frac{R}{\sigma_s}\right)\left(\frac{Y}{E'}\right)^2 \tag{14.21}$$

$$= 6.4 \times \frac{0.0652}{0.186 \times 10^{-3}}\left(\frac{2070}{1.14 \times 10^5}\right)^2$$

$$= 0.740.$$

From equation (14.20),

$$n_p = D_{SUM} F_0 \left(\frac{d}{\sigma_s} + w_p^* \right) \tag{14.20}$$

$$= 1770\ F_0\ (0.544 + 0.740)$$

$$= 1770\ F_0\ (1.284)$$

Interpolating in Table 14.1 gives F_0 (1.284) = 0.100 and n_p = 177/ mm^2 (114,000/in.^{2}).

If the macrocontact is elliptical with semiaxes a = 3 mm (0.01181 in.) and b = 0.33 mm (0.01299 in.) under a load of P = 3500 N (786.5 lb), the mean asperity-supported load is

$$Q_a = \pi ab \left(\frac{P}{A_o} \right) = \pi \times 3 \times 0.33 \times 31.6$$

$$= 98.3 \text{ N } (22.1 \text{ lb})$$

The fluid-supported load is

$$Q_f = P - Q_a = 3500 - 98.3 = 3402 \text{ N } (764.5 \text{ lb})$$

FLUID TRACTION MODEL

The Idealized Contact

Predicting the traction transmitted by the fluid in an EHD lubricated contact is difficult because the fluid properties that obtain under high pressures, shear rates, and transit times characteristic of most EHD lubricated contacts could differ substantially from the values measured under usual laboratory conditions. Tevaarwerk and Johnson (TJ) [14.4] proposed a model for predicting the fluid traction in EHD lubricated contacts of smooth surfaces that overcomes this difficulty. Their model uses two numerical values measured from an actual traction test as input, and it has been shown to lead to predictions that agree well with measurements made with crowned discs.

Figure 14.5 shows the idealized contact to which the model is applied. Figure 14.5*a* represents the contact cross section. The upper surface is moving with a velocity u_1, the lower surface with a velocity u_2 relative to the contact. The film thickness is designated h and is presumed to be constant across the contact zone; that is, the constriction that forms at the rear and sides of an EHD lubricated contact is neglected. For $u_1 > u_2$ the fluid film is progressively strained in traversing the contact, as shown in Fig. 14.5*a*.

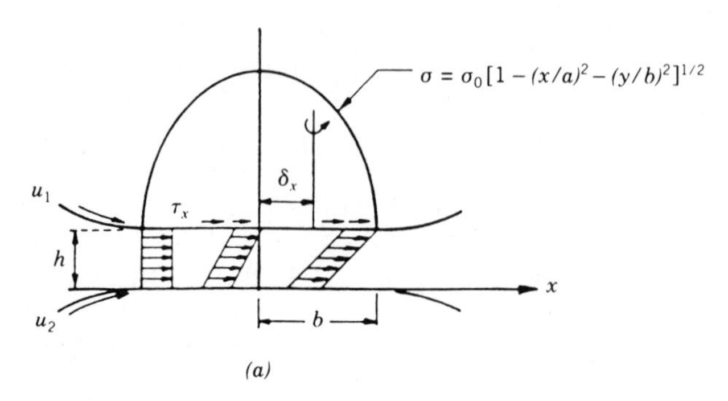

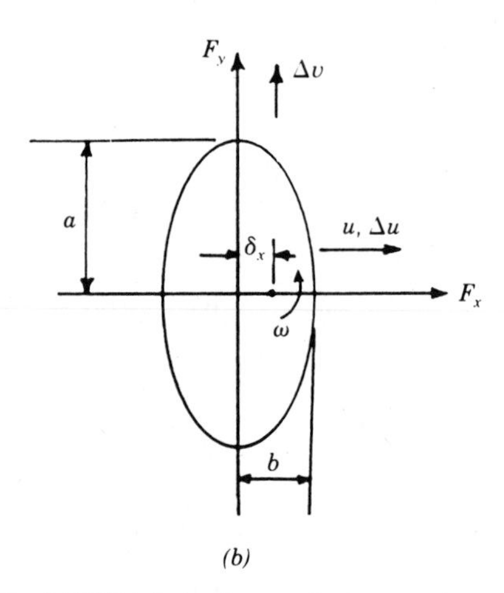

FIGURE 14.5. Idealized EHD lubrication contact geometry and kinematics. (*a*) Oil film shearing. (*b*) Contact geometry.

The entrainment velocity u—the velocity with which lubricant is swept into the contact—is

$$u = \tfrac{1}{2}(u_1 + u_2). \tag{14.36}$$

The sliding velocity Δu is

$$\Delta u = u_1 - u_2. \tag{14.37}$$

The variation of sliding across the contact ellipse due to the curved interfacial geometry is not explicitly considered, and thus the model strictly applies to the case where gross sliding of the contact is large relative to the microslip. Accounting for microslip requires a slice-by-slice approach within a numerical treatment. Such a treatment could employ the TJ model on a slice-by-slice basis. A coordinate system is established with the x direction coincident with the direction of u and Δu and the y direction lying in the plane of the contact ellipse. The z direction is coincident with the line of action of the applied force P. The pressure distribution is taken to be the same as in dry Hertzian contact:

$$p(x, y) = p_0 \left[1 - \left(\frac{x}{b}\right)^2 - \left(\frac{y}{a}\right)^2 \right]^{1/2} \tag{14.38}$$

where p_0 is the maximum Hertzian pressure [see equation (5.43)].

The redistribution of Hertzian pressure by the lubricant film is considered to the first order by treating the resultant force normal to the contact as being displaced forward by an amount δ_x along the x axis, which can be calculated by the formula in [14.4]

$$\delta_x = 4.25\, b(g_1)^{0.022}\, (g_2)^{-0.35} \left(\frac{a}{b}\right)^{0.91} \tag{14.39}$$

where g_1 and g_2 are the following dimensionless variables:

$$g_1 = \frac{\alpha p^3}{\eta_0 u^2 R_x^4} \qquad g_2 = \frac{p^4}{\eta_0^3 u^3 E' R_x^5}. \tag{14.40}$$

In addition to the sliding velocity in the x direction there may be transverse sliding in the y direction in the amount Δv and spinning with angular velocity ω about a "spin pole" located at the load center.

Constitutive Equation

At each coordinate position (x, y) within the contact ellipse, the sliding and/or spinning induces a fluid shearing stress τ, with components τ_x and τ_y in the directions of the x and y axes. The relation between τ_x, τ_y, and the spinning and sliding velocities is expressed by a "constitutive" law. When applied to concentrated EHD lubricated contacts, the TJ constitutive law reduces to the pair of simultaneous differential equations:

$$\frac{u}{Gb}\frac{d\tau_x}{dx} + \frac{\tau_x}{\tau_e} F(\tau_e) = \frac{\Delta u - \omega y}{h}$$

$$\frac{u}{Gb}\frac{d\tau_y}{dx} + \frac{\tau_y}{\tau_e} F(\tau_e) = \frac{\Delta v - \omega(x - \delta_x)}{h} \tag{14.41}$$

where G = local elastic shear modulus of the fluid
$\tau_e = (\tau_x^2 + \tau_y^2)^{1/2}$ = local equivalent shear stress

$F(\tau_e)$ is a nonlinear function that, if suitably chosen, can represent many forms of nonlinear viscous behavior.

The boundary conditions are that τ_x and τ_y vanish on the inlet border of the contact ellipse; in other words $\tau_x = \tau_y = 0$ when $X = -(1 - Y)^{1/2}$, where $X = x/b$ and $Y = y/a$.

TJ suggest that a useful choice for $F(\tau_e)$ is

$$F(\tau_e) = 0 \qquad \tau_e < \tau_c,$$

$$F(\tau_e) = \frac{\left\{\tau_x\left(\dfrac{\Delta u - \omega y}{h}\right) + \tau_y\left(\dfrac{\Delta v + \omega(x - \delta_x)}{h}\right)\right\}}{\tau_c}, \qquad \tau_e = \tau_c \tag{14.42}$$

where τ_c is a locally variable critical shear stress. They assume that τ_c varies in proportion to the Hertzian pressure

$$\tau_c = 1.5\,\bar{\tau}_c(1 - X^2 - Y^2)^{1/2} \tag{14.43}$$

where $\bar{\tau}_c$ is the average value of τ_c within the contact.

This choice of $F(\tau_e)$ represents the case where the fluid deforms elastically at low strain rates and plastically with a stress independent of the strain rate for high values of the strain rate. TJ show experimental evidence that this model for $F(\tau_e)$ leads to reasonable predictions when the Deborah number (D) is greater than 1000. The Deborah number is defined as

$$D = (\eta_0 e^{\alpha\bar{p}})\,\frac{u}{Gb} \tag{14.44}$$

where η_0 is the ambient absolute viscosity of the inlet oil and $\bar{p}$ is the average pressure $= P/\pi ab$, P being the load applied to the contact. D represents the ratio of the relaxation time (η/G) to the transit time through the contact (u/a). TJ suggest that local variation in G within the contact may be neglected and that G may be taken as constant at its average value $\overline{G}$.

Solution of Constitutive Equations

For specified values of h, u, Δu, Δv, ω, a, b, $\bar{\tau}_c$, and $\overline{G}$, equations (14.38) and (14.39) must be solved numerically to determine τ_x and τ_y as functions of the x and y coordinates within the contact ellipse. Following this evaluation, the components F_x and F_y of the fluid traction force are calculated by integrating the shear stresses over the contact ellipse.

$$F_x = \iint \tau_x \, dx \, dy \qquad F_y = \iint \tau_y \, dx \, dy \tag{14.45}$$

The spinning torque T is

$$T = \iint (x\tau_y - y\tau_x) \, dx \, dy \tag{14.46}$$

Closed-form solutions were developed in [14.4] when $\Delta v = \omega = 0$—that is, slip in the x direction only—and $\omega = 0$—that is, slip in the x and y directions but no spinning.

$$F_x = \bar{\tau}_c ab\psi \qquad \Delta v = \omega = 0 \qquad F_y = T = 0 \tag{14.47}$$

where

$$\psi = \frac{\pi}{2} - \sin^{-1}\left(\frac{1 - S^2}{1 + S^2}\right) + \frac{2S}{1 + S^2} \tag{14.48}$$

and

$$S = \frac{2}{3}\frac{b\bar{G}}{\bar{\tau}_c h} \times \frac{\Delta u}{u} \tag{14.49}$$

For the combined action of slip and side slip—that is, for $\omega = 0$—the spinning torque $T = 0$ and F_x and F_y are given by

$$F_x = \frac{\bar{\tau}_c ab\psi}{[1 + (\Delta v/\Delta u)^2]^{1/2}} \qquad F_y = \frac{\bar{\tau}_c ab\psi}{[1 + (\Delta u/\Delta v)^2]^{1/2}} \tag{14.50}$$

where ψ is evaluated with S redefined as

$$S = \frac{2b\bar{G}}{3\bar{\tau}_c hu}[(\Delta u)^2 + (\Delta v)^2]^{1/2} \tag{14.51}$$

The general case of slip, side slip, and spinning requires numerical integration.

Determining $\bar{G}$ and $\bar{\tau}_c$

In a simple traction test the velocity u and load P are fixed, and the traction force F_x is measured as a function of Δu. Under the TJ fluid traction model, a plot of the traction coefficient F_x/P as a function of the ratio $\Delta u/u$, slip to entrainment velocity, will have the general appear-

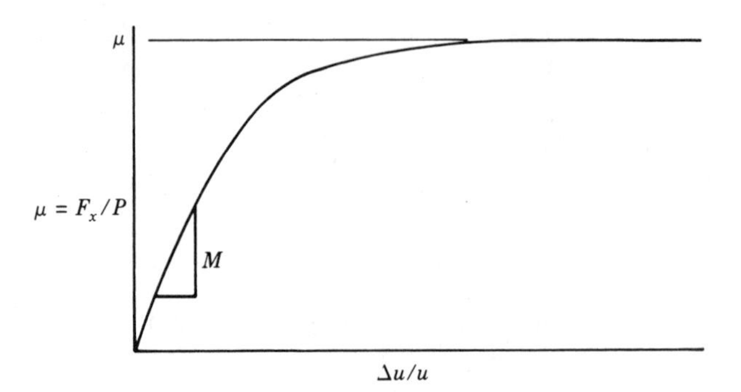

FIGURE 14.6. Typical traction curve.

ance of Fig. 14.6. The curve increases linearly with a slope m at low slip values and reaches an asymptotic value μ at high slip rates.

From equation (14.45) for small $\Delta u/u$ (or S) values, F_x can be shown to vary as

$$F_x = 4\bar{\tau}_c abS \tag{14.52}$$

or, using equation (14.51), as

$$F_x = \frac{8\bar{G}}{3} \times \frac{b^2 a}{h} \times \frac{\Delta u}{u} \tag{14.53}$$

The ratio $F_x/P/(\Delta u/u) = m$ is then

$$m = \frac{8}{3}\left(\frac{b^2 a}{P}\right) h\bar{G} \tag{14.54}$$

Thus, in terms of the measured slope m from a simple traction test, $\bar{G}$ may be calculated as

$$\bar{G} = \frac{3}{8} P \left(\frac{h}{b^2 a}\right) m \tag{14.55}$$

For large values of S the quantity $\psi(S)$ in equation (14.48) converges to π and F_x approaches the constant value

$$F_x = \pi\bar{\tau}_c ab \tag{14.56}$$

so

$$\mu = \frac{F_x}{P} = \frac{\pi\bar{\tau}_c ab}{P} \tag{14.57}$$

In terms of the maximum traction coefficient measured in a simple traction test, the value of $\bar{\tau}_c$ may be calculated as

$$\bar{\tau}_c = \frac{P\mu}{\pi ab} \tag{14.58}$$

Example 14.2. For the contact given, the applied load P was 3500 N (786.5 lb) and the rolling velocity u was 150 m/sec (5906 in./sec). The fluid-supported portion of this load was 3400 N (764 lb). Assume that with smooth surfaces and an applied load of 3400 N (764 lb), a simple traction curve (sliding only in x direction) run using this same fluid at the same ambient temperature exhibited a slope $m = 15$ and a maximum value $\mu_{\mathrm{max}} = 0.03$ (see Fig. 14.6). Calculate the total traction force due to fluid and asperity traction when the sliding velocity is $\Delta u = 0.2$ in the x direction and $\Delta v = 0.2$ in the y direction and the Coulomb friction coefficient for asperity contact is assumed to be $\mu_a = 0.10$.

From equation (14.55),

$$\begin{aligned} \overline{G} &= \frac{3}{8} P \left(\frac{h}{b^2 a} \right) m \qquad (14.55) \\ &= \frac{3}{8} \times \frac{3400 \times 0.5 \times 10^{-3} \times 15}{(0.33)^2 \times 3} \\ &= 29.2 \text{ N/mm}^2 \text{ (4233 psi)} \end{aligned}$$

From equation (14.58),

$$\begin{aligned} \bar{\tau}_c &= \frac{P\mu}{\pi ab} = \frac{3400 \times 0.03}{\pi \times 3 \times 0.33} \qquad (14.58) \\ &= 32.5 \text{ N/mm}^2 \text{ (4712 psi)} \end{aligned}$$

From equation (14.51)

$$\begin{aligned} S &= \frac{2b\overline{G}}{3\bar{\tau}_c hu} [(\Delta u)^2 + (\Delta v)^2]^{1/2} \qquad (14.51) \\ &= \frac{2}{3} \times \frac{0.33 \times 29.2}{32.5 \times 0.5 \times 10^{-3} \times 150} [(0.2)^2 + (0.2)^2]^{1/2} \\ &= 0.745 \end{aligned}$$

From equation (14.48),

$$\psi = \frac{\pi}{2} - \sin^{-1}\left(\frac{1 - S^2}{1 + S^2}\right) + \frac{2S}{1 + S^2} \tag{14.48}$$

$$= \frac{\pi}{2} - \sin^{-1}\left(\frac{1 - (0.745)^2}{1 + (0.745)^2}\right) + \frac{2 \times 0.745}{1 + (0.745)^2}$$

$$= 2.239$$

From equation (14.50),

$$F_x = \frac{\bar{\tau}_c ab\psi}{[1 + (\Delta u/\Delta v)^2]^{1/2}} \tag{14.50}$$

$$= \frac{32.5 \times 3 \times 0.33 \times 2.239}{[1 + (0.2/0.2)^2]^{1/2}}$$

$$= 50.94 \text{ N } (11.45 \text{ lb})$$

and

$$F_y = 50.94 \text{ N } (11.45 \text{ lb})$$

The resultant fluid traction is thus

$$F_f = (F_x^2 + F_y^2)^{1/2} = 72.04 \text{ N } (16.18 \text{ lb})$$

The asperity load is $Q_a \approx 100$ N (22.5 lb) so the asperity component of the traction force is

$$F_a = \mu_a Q_a = 0.10 \times 100 = 10 \text{ N } (2.25 \text{ lb})$$

The total traction force is thus

$$F = F_f + F_a = 72.8 + 10 = 82.8 \text{ N } (18.6 \text{ lb})$$

The composite traction coefficient is

$$\mu = F/P = 82.8/3500 = 0.023$$

CLOSURE

This chapter contains an approach to predicting key performance-related parameters descriptive of real EHD lubricated contacts, including contact density, true contact area, plastic contact density, fluid and asperity load sharing, and the relative contributions of the fluid and asperities to total traction. It offers a rational way to assess on a comparative basis the joint effect of fluid, roughness, and material variables on the tribological performance of alternative contact designs.

REFERENCES

14.1. J. McCool, "TRIBOS: A Performance Evaluation Tool for Traction Transmitting Partial EHD Contacts," *ASLE Trans.* **29**(3), 432–440 (July 1986).

14.2. B. Hamrock and D. Dowson, "Isothermal Elastohydrodynamic Lubrication of Point Contacts, Part IV, Starvation Results," *J. Lub. Tech.* **99**(1), 15–23 (Jan. 1977).

14.3. J. Greenwood, and J. Williamson, "Contact of Nominally Flat Surfaces," *Proc. R. Soc. London* **A295,** 300–319 (1966).

14.4. J. Tevaarwerk, and K. Johnson, "The Influence of Fluid Rheology on the Performance of Traction Drives," ASME Paper No. 78-LUB-10 (October, 1978).

14.5. A. Bush, R. Gibson, and T. Thomas, "The Elastic Contact of a Rough Surface," *Wear* **35,** 87–111 (1975).

14.6. M. O'Callaghan and M. Cameron, "Static Contact Under Load Between Nominally Flat Surfaces," *Wear* **36,** 79–97 (1976).

14.7. A. Bush, R. Gibson, and G. Keogh, "Strongly Anisotropic Rough Surfaces," ASME Paper 78-LUB-16 (1978).

14.8. J. McCool and S. Gassel, "The Contact of Two Surfaces Having Anisotropic Roughness Geometry," ASLE Special Publication (SP-7), pp. 29–38 (1981).

14.9. J. McCool, "Comparison of Models for the Contact of Rough Surfaces," *Wear* **107,** 37–60 (1986).

14.10. A. Bush, R. Gibson, and G. Keogh, "The Limit of Elastic Deformation in the Contact of Rough Surfaces," *Mech. Res. Comm.* **3,** 169–174 (1976).

14.11. P. Nayak, "Random Process Model of Rough Surfaces," *Trans. ASME, J. Lub. Tech.* **93F,** 398–407 (1971).

14.12. R. Sayles and T. Thomas, "Thermal Conductance of a Rough Elastic Contact," *Appl. Energy* **2,** 249–267 (1976).

15

ROLLING BEARING TEMPERATURES

LIST OF SYMBOLS

Symbol	Description	Units
$\mathfrak{D}$	diameter	m (ft)
$\mathcal{E}$	thermal emissivity	
F	temperature coefficient	W-sec/°C (Btu/°F)
H	heat flow	W (Btu/hr)
h	film coefficient of heat transfer	$W/m^2 \cdot °C$ ($Btu/hr \cdot ft^2 \cdot °F$)
J	conversion factor, 1.0003×10^3 N · mm = 1 W · sec	
k	thermal conductivity	W/m · °C (Btu/hr · ft · °F)
$\mathcal{L}$	length of heat conduction path	m (ft)
M	friction torque	N · mm (in. · lb)
n	rotational speed	rpm
P_r	Prandtl Number	
q	error function	
$\mathcal{R}$	radius	m (ft)

Symbol	Description	Units
S	area normal to heat flow	m^2 (ft^2)
T	temperature	°C, °K (°F, °R)
u_s	fluid velocity	m/sec (ft/sec)
v	velocity	m/sec (ft/sec)
$\mathcal{W}$	width	m (ft)
x	distance in x direction	m (ft)
ϵ	error	
ν	fluid viscosity	m^2/sec (ft^2/sec)
ω	rotational speed	rad/sec

SUBSCRIPTS

a	refers to air or ambient condition
c	refers to heat conduction
f	refers to friction
j	refers to rolling element position
o	refers to oil
r	refers to heat radiation
s	refers to spinning motion
v	refers to heat convection
1	refers to temperature node
2	refers to temperature node, and so on

GENERAL

The temperature level at which a rolling bearing operates is a function of many variables among which the following are predominant:

1. bearing load
2. bearing speed
3. bearing friction torque
4. lubricant type and viscosity
5. bearing mounting and/or housing design
6. environment of operation

In the steady-state operation of a rolling bearing, as for any other machine element, whatever heat is generated internally is dissipated. Therefore, the steady-state temperature level of one bearing system compared to another system using identical sizes and number of bearings is a measure of the relative ability of that system's efficiency of heat dissipation.

Of course, if the rate of heat dissipation is less than the heat generation rate, then an unsteady state exists and the system temperature will rise until lubricant distress occurs, with ultimate bearing failure. The temperature level at which this occurs is determined largely by the type of lubricant and the bearing material. This dissertation is limited to the steady-state thermal operation of rolling bearings since this is a common concern of bearing users regarding satisfactory operation.

Most rolling bearing applications perform at temperature levels that are relatively cool and therefore do not require any special consideration regarding thermal adequacy. This is due to either one of the following conditions:

1. The bearing heat generation rate is low because of light load and/or relatively slow speed.
2. The ability to remove heat from the bearing is sufficient because of location of the bearing assembly in a moving air stream or because of adequate heat conduction through adjacent metal.

Some applications occur in certain adverse environmental conditions such that it is certain that external cooling is required. A rapid determination of the bearing cooling requirements may then suffice to establish the cooling capability that must be applied to the lubricating fluid. In other applications it is not obvious whether external cooling is required, and it may be economically advantageous to establish analytically the thermal conditions of bearing operation.

HEAT GENERATION

Although rolling bearings have been called *antifriction* bearings, nevertheless, they exhibit a small amount of friction during rotation. This should be evident since if friction were not present, the rolling elements would slip on the rotating ring rather than roll.

Friction in a rolling bearing as in most other mechanisms represents a wasteful power dissipation manifested in the form of heat generation. This frictional power must be effectively removed or an unsatisfactory temperature condition will obtain in the bearing.

Having determined the bearing friction torque by methods of Chapter 13, one can obtain the bearing frictional power loss in watts from the following equation:

$$H_f = 1.047 \times 10^{-4}\ nM \tag{15.1}$$

For a high speed ball bearing, the local heat generation rate due to sliding at each contact is given by

$$H_{fj} = \frac{1}{J} \int_{S_j} v_k \, dF_{kj} \tag{15.2}$$

where J is a constant converting Newton-meters per second to watts, and v_k is the sliding velocity component in the direction of differential friction force dF_k. The rate of heat generation at a raceway contact due to spinning alone is given by

$$H_{fj} = \frac{1}{J} \int_{S_j} \omega_{sj} \, dM_{sj} \tag{15.3}$$

Equations (15.2) and (15.3) must be evaluated separately for inner and outer raceway contacts at each rolling element location.

Having established values for the rates of heat generation due to friction torque, it is then possible to determine the bearing temperature structure by heat transfer analysis.

Example 15.1. For the 218 angular-contact ball bearing of Examples 8.1 and 13.1, estimate the rate of heat generation due to spinning motion at the inner raceway contacts for 22,250-N (5000-lb) thrust load at 10,000 rpm shaft speed.

$Z = 16$ Ex. 6.5

$\beta = 28.5°$ Fig. 8.5

$\alpha_i = 48.8°$ Fig. 8.4

$\alpha_o = 33.3°$ Fig. 8.4

$D = 22.23$ mm (0.875 in.) Ex. 2.3

$d_m = 125.3$ mm (4.932 in.) Ex. 2.3

$n_m = 4498$ rpm Fig. 8.7

$$\gamma' = \frac{D}{d_m}$$

$$= \frac{22.23}{125.3} = 0.1774$$

$$\left(\frac{\omega_s}{\omega_{roll}}\right)_i = (1 - \gamma' \cos \alpha_i) \tan (\alpha_i - \beta) + \gamma' \sin \alpha_i \tag{7.63}$$

$$= (1 - 0.1774 \cos 48.8°) \tan (48.8° - 28.5°)$$

$$+ 0.1774 \sin 48.8° = 0.4602$$

$$\omega_o = -\omega_m \text{ for inner raceway rotation}$$

$$\omega_{roll} = -\frac{\omega_o}{\gamma'} = \frac{\omega_m}{\gamma'} \tag{7.56}$$

or

$$\frac{\omega_m}{\omega_{roll}} = \frac{1}{\gamma'}$$

$$\frac{\omega_s}{\omega_m} = \frac{0.4602}{\gamma'}$$

$$= \frac{0.4602}{0.1774} = 2.594$$

$$\frac{\omega_s}{\omega_m} = \frac{n_s}{n_m}$$

$$n_s = 2.594 \times 4498 = 11{,}670 \text{ rpm}$$

$$M_{si} = 43.19 \text{ N} \cdot \text{mm } (0.382 \text{ in.} \cdot \text{lb}) \qquad \text{Ex. 13.1}$$

$$H_{fj} = 1.047 \times 10^{-4} \text{ n } M_{sij} \tag{15.1}$$

$$= 1.05 \times 10^{-4} \times 11670 \times 43.19 = 52.92 \text{ W } (180.6 \text{ Btu/hr})$$

Since $Z = 16$,

$$H_f = 16 \times H_{fj}$$

$$= 16 \times 52.92 = 846.7 \text{ W } (2889 \text{ Btu/hr})$$

$$M_l + M_v = 2860 \text{ N} \cdot \text{mm } (25.27 \text{ in.} \cdot \text{lb}) \qquad \text{Ex. 13.3}$$

The total heat generation rate is calculated as follows:

$$H_{l+v} = 1.047 \times 10^{-4} \, nM_{l+v} \tag{15.1}$$

$$= 1.05 \times 10^{-4} \times 10{,}000 \times 2860 = 3003 \text{ W } (10248 \text{ Btu/hr})$$

$$H_{tot} = 3003 + 847 = 3850 \text{ W } (13140 \text{ Btu/hr})$$

Figure 15.1 shows heat generation rates for this bearing for a load range of 44,500 N (0–10,000 lb) and a shaft speed of 10,000 rpm.

HEAT TRANSFER

Modes of Heat Transfer

There exist three fundamental modes for the transfer of heat between masses having different temperature levels. These are conduction of heat within solid structures, convection of heat from solid structures to fluids

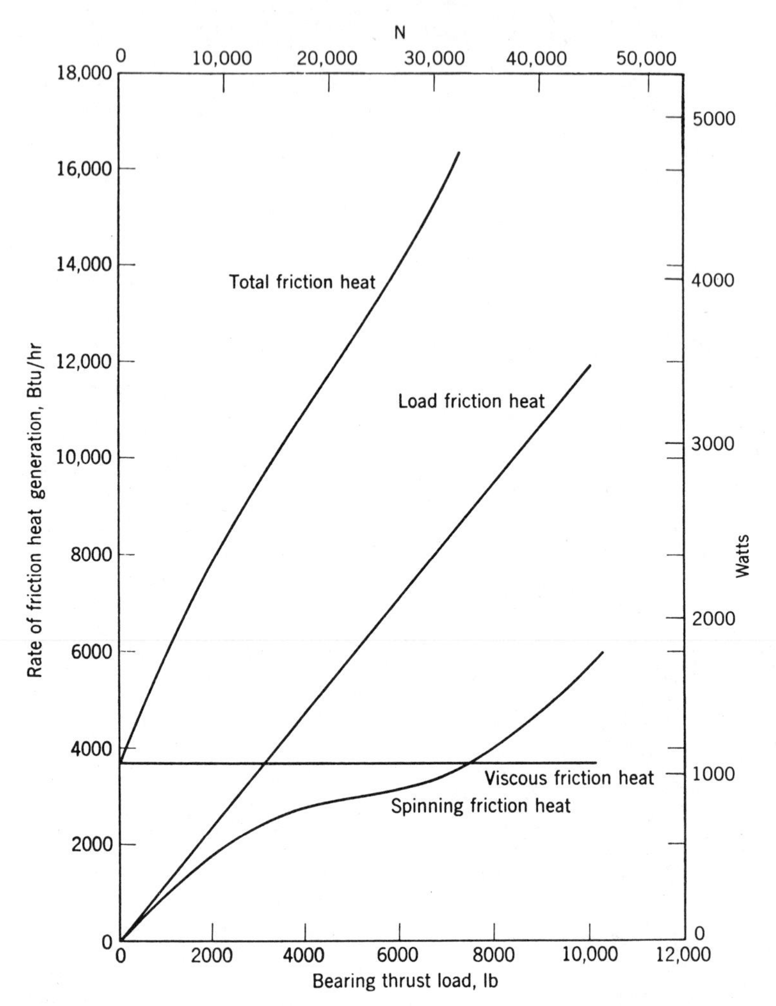

FIGURE 15.1. Friction heat generation vs load; 218 angular-contact ball bearing, 10,000 rpm, 5 centistokes oil, jet lubrication.

in motion (or apparently at rest), and radiation of heat between masses separated by space. Although other modes exist, such as radiation to gases and conduction within fluids, their effects are minor for most bearing applications and may usually be neglected.

Heat Conduction

Heat conduction, which is the simplest form of heat transfer, may be described for the purpose of this discussion as a linear function of the difference in temperature level within a solid structure, that is,

$$H_c = \frac{kS}{\mathcal{L}}(T_1 - T_2) \qquad (15.4)$$

The quantity S in equation (15.4) is the area normal to the flow of heat between two points and $\mathcal{L}$ is the distance between the same two points. The thermal conductivity k is a function of the material and temperature level; however, the latter variation is generally minor for structural solids and will be neglected here. For heat conduction in a radial direction within a cylindrical structure such as a bearing inner or outer ring, the following equation is useful:

$$H_c = \frac{2\pi k \mathcal{W}(T_i - T_o)}{\ln(\mathcal{R}_o/\mathcal{R}_i)} \qquad (15.5)$$

In equation (15.5), $\mathcal{W}$ is the width of the annular structure and $\mathcal{R}_o$ and $\mathcal{R}_i$ are the inner and outer radii defining the limits of the structure through which heat flow occurs. If $\mathcal{R}_i = 0$, an arithmetic mean area is used and the equation assumes the form of equation (15.4).

Heat Convection

Heat convection is the most difficult form of heat transfer to estimate quantitatively. It occurs within the bearing housing as heat is transferred to the lubricant from the bearing and from the lubricant to other structures within the housing as well as to the inside walls of the housing. It also occurs between the outside of the housing and the environmental fluid—generally air, but possibly oil, water, another gas, or a working fluid medium.

Heat convection from a surface may generally be described as follows:

$$H_v = h_v S(T_1 - T_2) \qquad (15.6)$$

in which h_v, the film coefficient of heat transfer, is a function of surface and fluid temperatures, fluid thermal conductivity, fluid velocity adjacent to the surface, surface dimensions and attitude, fluid viscosity, and density. It can be seen that many of the foregoing properties are temperature dependent. Therefore, heat convection is not a linear function of temperature unless fluid properties can be considered reasonably stable over a finite temperature range.

Heat convection within the housing is most difficult to describe, and a rough approximation will be used for the heat transfer film coefficient. Since oil is used as a lubricant and viscosity is high, laminar flow is assumed. Eckert [15.1] states for a plate in a laminar flow field:

$$h_v = 0.0332kP_r^{1/3}\left(\frac{u_s}{\nu_o x}\right)^{1/2} \tag{15.7}$$

The use of equation (15.7) taking u_s equal to bearing cage surface velocity and x equal to bearing pitch diameter seems to yield workable values for h_v, considering heat transfer from the bearing to the oil that contacts the bearing. For heat transfer from the housing inside surface to the oil, taking u_s equal to one-third cage velocity and x equal to housing diameter yields adequate results. In equation (15.7), ν_o represents kinematic viscosity and P_r the Prandtl number of the oil.

If cooling coils are submerged in the oil sump, it is best that they be aligned parallel to the shaft so that a laminar cross flow obtains. In this case Eckert [15.1] shows for a cylinder in cross flow, the outside heat transfer film coefficient may be approximated by

$$h_v = 0.06\frac{k_o}{\mathfrak{D}}\left(\frac{u_s\mathfrak{D}}{\nu_o}\right)^{1/2} \tag{15.8}$$

in which $\mathfrak{D}$ is the outside diameter of the tube and k_o is the thermal conductivity of the oil. It is recommended that u_s be taken as approximately one-fourth of the bearing inner ring surface velocity.

The foregoing approximations for film coefficient are necessarily crude. If greater accuracy is required, Reference [15.1] indicates more refined methods for obtaining the film coefficient. In lieu of a more elegant analysis, the values yielded by equations (15.7) and (15.8), and (15.9) and (15.10) that follow, should suffice for general engineering purposes.

In quiescent air, heat transfer by convection from the housing external surface may be approximated by using an outside film coefficient in accordance with equation (15.9) (see Jakob and Hawkins [15.2]):

$$h_v = 2.3 \times 10^{-5}(T - T_a)^{0.25} \tag{15.9}$$

For forced flow of air of velocity u_s over the housing, Reference [15.1] yields:

$$h_v = 0.03\frac{k_a}{\mathfrak{D}}\left(\frac{u_s\mathfrak{D}_h}{\nu_a}\right)^{0.57} \tag{15.10}$$

in which $\mathfrak{D}_h$ is the approximate housing diameter. Palmgren [15.3] gives the following formula to approximate the external area of a bearing housing or pillow block:

$$S = \pi\mathfrak{D}_h(\mathcal{W}_h + \tfrac{1}{2}\mathfrak{D}_h) \tag{15.11}$$

in which $\mathfrak{D}_h$ is the maximum diameter of the pillow block and $\mathcal{W}_h$ is the width.

Heat Radiation

The remaining mode of heat transfer to be considered is radiation from the housing external surface to surrounding structures. For a small structure in a large enclosure Reference [15.2] gives

$$H_r = 5.73\ \mathcal{E}S\left[\left(\frac{T}{100}\right)^4 - \left(\frac{T_a}{100}\right)^4\right] \tag{15.12}$$

in which temperature is in degrees Kelvin (absolute). Equation (15.12) being nonlinear in temperature is sometimes written in the following form:

$$H_r = h_r S(T - T_a) \tag{15.13}$$

in which

$$h_r = 5.73 \times 10^{-8}\mathcal{E}(T + T_a)(T^2 + T_a^2) \tag{15.14}$$

Equations (15.13) and (15.14) are useful for hand calculation in which problem T and T_a are not significantly different. Upon assuming a temperature T for the surface, the pseudofilm coefficient of radiation h_r may be calculated. Of course, if the final calculated value of T is significantly different from that assumed, then the entire calculation must be repeated. Actually, the same consideration is true for calculation of h_v for the oil film. Since k_o and ν_o are dependent upon temperature, the assumed temperature must be reasonably close to the final calculated temperature. How close is dictated by the actual variation of those properties with oil temperature.

ANALYSIS OF HEAT FLOW

Systems of Equations

Because of the discontinuities of the structures that comprise a rolling bearing assembly, classical methods of heat transfer analysis cannot be applied to obtain a solution describing the system temperatures. By classical methods is meant the description of the system in terms of differential equations and the analytical solution of these equations. Instead, methods of finite difference as demonstrated by Dusinberre [15.4] must be applied to obtain a mathematical solution.

For finite difference methods applied to steady-state heat transfer, various points or nodes are selected throughout the system to be analyzed. At each of these points, temperature is determined. In steady-state heat transfer, heat influx to any point equals heat efflux; therefore, the sum of all heat flowing toward a temperature node is equal to zero. Figure 15.2 is a heat flow diagram at a temperature node, demonstrating that the nodal temperature is affected by the temperatures of each of the four indicated surrounding nodes. (Although the system depicted by Fig. 15.2 shows only four surrounding nodes, this is purely by choice of grid and the number of nodes may be greater or smaller.) Since the sum of the heat flows is zero, therefore

$$H_{1-0} + H_{2-0} + H_{3-0} + H_{4-0} = 0 \tag{15.15}$$

For this example, it is assumed that heat flow occurs only by conduction and that the grid is nonsymmetrical, making all areas S and lengths of flow path different. Furthermore, the material is assumed nonisotropic so that thermal conductivity is different for all flow paths. Substitution of equation (15.4) into (15.15) therefore yields

$$\frac{k_1 S_1}{\mathcal{L}_1}(T_1 - T_0) + \frac{k_2 S_2}{\mathcal{L}_2}(T_2 - T_0) + \frac{k_3 S_3}{\mathcal{L}_3}(T_3 - T_0) + \frac{k_4 S_4}{\mathcal{L}_4}(T_4 - T_0) = 0 \tag{15.16}$$

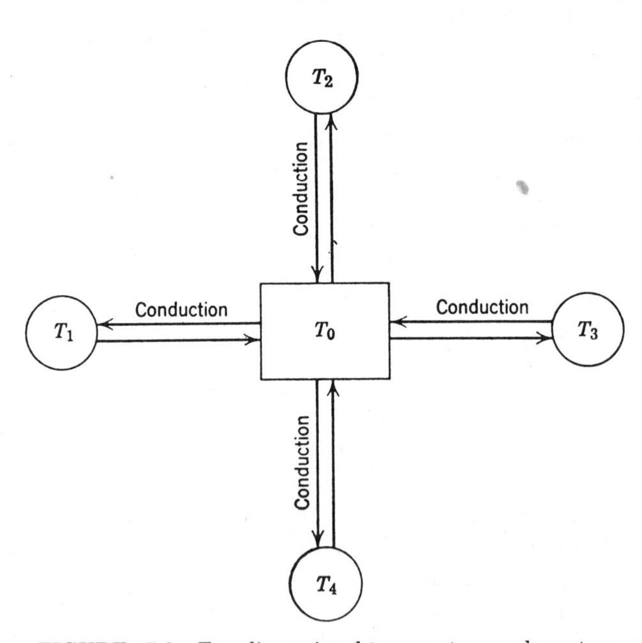

FIGURE 15.2. Two-dimensional temperature node system.

By rearranging terms, one obtains

$$\frac{k_1S_1}{\mathcal{L}_1}T_1 + \frac{k_2S_2}{\mathcal{L}_2}T_2 + \frac{k_3S_3}{\mathcal{L}_3}T_3 + \frac{k_4S_4}{\mathcal{L}_4}T_4 - \sum_{i=1}^{i=4}\frac{k_iS_i}{\mathcal{L}_i}T_o = 0 \qquad (15.17)$$

or

$$F_1T_1 + F_2T_2 + F_3T_3 + F_4T_4 - \sum_{i=1}^{i=4} F_iT_o = 0 \qquad (15.18)$$

Dividing by ΣF_i yields

$$\frac{F_1}{\Sigma F_i}T_1 + \frac{F_2}{\Sigma F_i}T_2 + \frac{F_3}{\Sigma F_i}T_3 + \frac{F_4}{\Sigma F_i}T_4 - T_0 = 0 \qquad (15.19)$$

More concisely, equation (15.19) may be written

$$\phi_iT_i = 0 \qquad (15.20)$$

in which the ϕ_i are influence coefficients of temperature equal to $F_i/\Sigma F_i$. If the material were isotropic and a symmetrical grid was chosen, then $\phi_0 = 1$ and the other $\phi_i = 0.25$.

In the foregoing example, only heat conduction was illustrated. If, however, heat flow between points 4 and 0 was by convection, then according to equation (15.5), $F_4 = h_{v4}S_4$. For a multinodal system, a series of equations similar to (15.19) may be written. If the equations are linear in temperature T, they may be solved by classical methods for the solution of simultaneous linear equations or by numerical methods (see Reference [15.5]).

The system may include heat generation and be further complicated, however, by nonlinear terms caused by heat radiation and free convection. Consider the example schematically illustrated by Fig. 15.3. In that illustration, heat is generated at point 0, dissipated by free convection and radiation between points 1 and 0 and dissipated by conduction be-

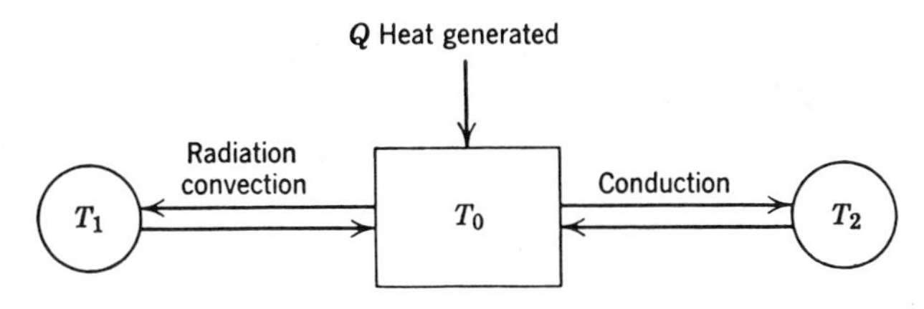

FIGURE 15.3. Convective, radiation, and conductive heat transfer system.

tween points 2 and 0. Thus,

$$H_{f0} + H_{1-0,v} + H_{1-0,r} + H_{2-0} = 0 \tag{15.21}$$

The use of equations (15.4), (15.6), (15.9), and (15.12) gives

$$H_{f0} + 2.3 \times 10^{-5} S_1 (T_1 - T_0)^{1.25} + 5.73 \times 10^{-8} \epsilon S_1 (T_1^4 - T_0^4) + \frac{K_2 S_2}{\mathfrak{L}_2}(T_2 - T_0) = 0 \tag{15.22}$$

or

$$H_{f0} + F_{1v}(T_1 - T_0)^{1.25} + F_{1r}(T_1^4 - T_0^4) + F_2(T_2 - T_0) = 0 \tag{15.23}$$

Solution of Equations

A system of nonlinear equations similar to equation (15.23) is difficult to solve by direct numerical methods of iteration or relaxation. Therefore, the Newton-Raphson method [15.5] is recommended for solution.

The Newton-Raphson method states that for a series of nonlinear functions q_i of variables T_j:

$$q_i + \sum \frac{\partial q_i}{\partial T_j} \epsilon_j = 0 \tag{15.24}$$

Equations (15.24) represent a system of simultaneous linear equations which may be solved for ϵ_j (error on T_j).

Then, the new estimate of T_j is

$$T_j' = T_j(0) + \epsilon_j \tag{15.25}$$

and new values q_i may be determined. The process is continued until the functions q_i are virtually zero. With a system of nonlinear equations similar to equation (15.23), such equations must be linearized according to equation (15.24). Thus, let equation (15.23) be rewritten as follows:

$$H_{f0} + F_{1v}(T_1 - T_0)^{1.25} + F_{1r}(T_1^4 - T_0^4) + F_2(T_2 - T_0) = q_0 \tag{15.26}$$

Now

$$\frac{\partial q_0}{\partial T_0} = -1.25 F_{1v}(T_1 - T_0)^{0.25} - 4F_{1r}T_0^3 + F_2$$

$$\frac{\partial q_0}{\partial T_1} = 1.25 F_{1v}(T_1 - T_0)^{0.25} + 4F_{1r}T_1^3$$

$$\frac{\partial q_0}{\partial T_2} = F_2 \tag{15.27}$$

Substitution of equations (15.26) and (15.27) into (15.24) yields one equation in variables ϵ_0, ϵ_1, and ϵ_2.

The system of nonlinear equations is solved for T_0, T_1, and T_2 when $\sqrt{\frac{1}{3}\,\Sigma e_j^2}$ is sufficiently small, say, less than one-tenth of a degree temperature.

Example 15.2. A 23072 double-row radial spherical roller bearing has a 444.5 mm (17.5-in.) pitch diameter and is mounted in the pillow block shown by Fig. 15.4. The bearing is operated at a shaft speed of 350 rpm while it supports a 489,500 (110,000-lb) radial load. The bearing is lubricated by 100-SSU (20-centistokes) oil in an oil bath at operating temperature. The pillow block that houses the bearing is situated in an atmosphere of quiescent air at a temperature of 48.9°C (120°F). Estimate the bearing and sump oil temperatures. From Table 13.4, use $f_1 = 0.0005$,

$$F_\beta = 489{,}500 \text{ (110,000 lb)}$$

$$M_1 = f_1 F_\beta d_m \tag{13.106}$$

$$= 0.005 \times 489{,}500 \times 444.5 = 108.8 \text{ N} \cdot \text{m (962.5 in.} \cdot \text{lb)}$$

From Table 13.5, use $f_0 = 6$ for oil bath lubrication:

$$\nu_o n = 20 \times 350 = 7000$$

$$M_v = 9.79 \times 10^{-8} \times f_0 (\nu_o n)^{2/3} d_m^3 \tag{13.104}$$

$$= 9.79 \times 10^{-8} \times 6 \times (7000)^{2/3} \times (444.5)^3$$

$$= 18.88 \text{ N} \cdot \text{m (167.0 in.} \cdot \text{lb)}$$

$$M = M_1 + M_v \tag{13.107}$$

$$= 108.8 + 18.9 = 127.7 \text{ N} \cdot \text{m (1130 in.} \cdot \text{lb)}$$

$$H_f = 1.047 \times 10^{-4}\, nM \tag{15.1}$$

$$= 1.047 \times 10^{-4} \times 350 \times 127{,}700$$

$$= 4680 \text{ W (15970 Btu/hr)}$$

Since this problem is for illustrative purposes only, it has been designed to be as simple as possible such that all equations and methods of solution may be demonstrated. Therefore, the following conditions will be assumed:

1. Nine temperature nodes are sufficient to describe the system shown by Fig. 15.4.

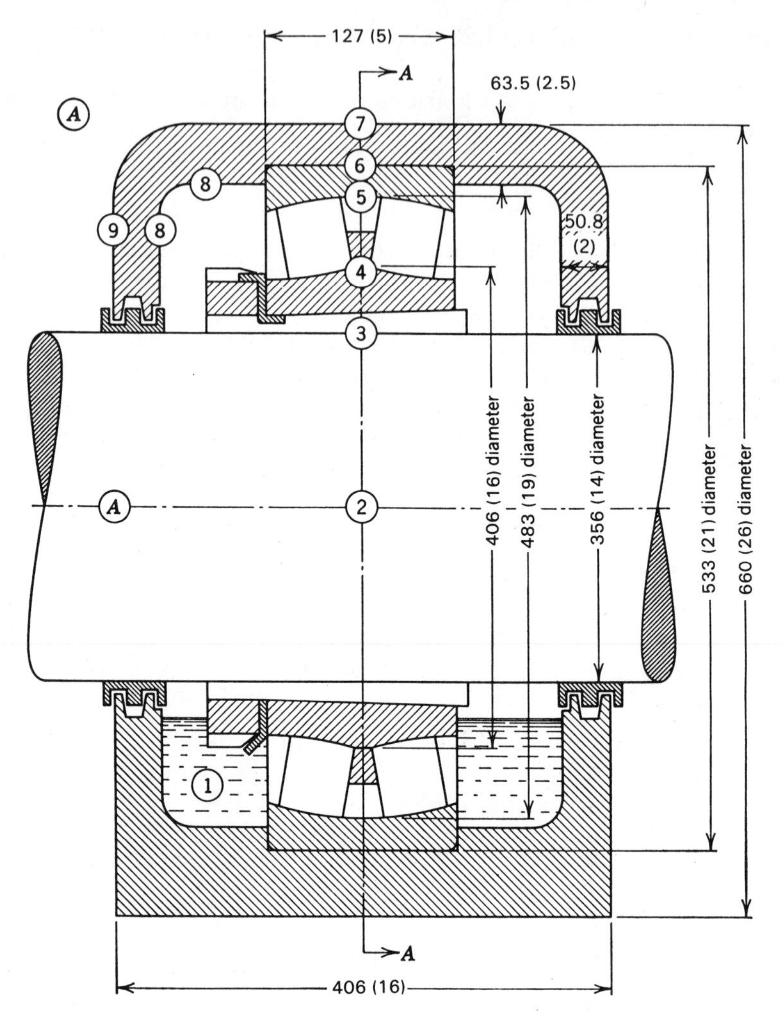

FIGURE 15.4. Temperature node selection.

2. The inside of the housing is coated with oil and may be described by a single temperature.
3. The inner ring raceway may be described by a single temperature.
4. The outer ring raceway may be described by a single temperature.
5. The housing is symmetrical about the shaft centerline and vertical section A-A. Thus, heat transfer in the circumferential direction need not be considered.
6. The sump oil may be considered at a single temperature.
7. The shaft ends at the extremities of the housing are at ambient temperature.

TABLE 15.1. Heat Transfer System[a]

Temperature Node	A	1	2	3	4	5	6	7	8	9
1	—	—	—	Convection (15.6), (15.7)	Convection (15.6), (15.7)	Convection (15.6), (15.7)	—	—	Convection (15.6), (15.7)	—
2	Conduction (15.4)	—	—	Conduction (15.5)	—	—	—	—	—	—
3	—	Convection (15.6), (15.7)	Conduction (15.5)	—	Conduction (15.5)	—	—	—	—	—
4	—	Convection (15.6), (15.7)	—	Conduction (15.5)	Heated Generated (15.1), (15.2), (15.3)	—	—	—	—	—
5	—	Convection (15.6), (15.7)	—	—	—	Heated Generated (15.1), (15.2), (15.3)	Conduction (15.5)	—	—	—
6	—	—	—	—	—	Conduction (15.5)	—	Conduction (15.5)	Conduction (15.4)	—
7	Convection (15.6), (15.9) Radiation (15.12)	—	—	—	—	—	Conduction (15.5)	—	Conduction (15.5)	Conduction (15.4)
8	—	Convection (15.6), (15.7)	—	—	—	—	Conduction (15.4)	Conduction (15.5)	—	Conduction (15.4)
9	Convection (15.6), (15.9) Radiation (15.12)	—	—	—	—	—	—	Conduction (15.4)	Conduction (15.4)	—
—										

[a] Numbers in parentheses refer to equations used to calculate heat flow, film coefficients, and heat generation rate.

Considering the temperature nodes indicated in Fig. 15.4, the heat transfer system is that indicated by Table 15.1. Table 15.1 also indicates which equations are used to determine heat flow, film coefficient of heat transfer, and rate of heat generation. The heat flow areas and lengths of flow path are obtained from the dimensions of Fig. 15.4, considering the location of each temperature node.

Based on Table 15.1 and Fig. 15.4, a set of nine simultaneous, nonlinear equations with unknown variables T_1–T_9 can be developed. Since this system is nonlinear in temperature because of free convection and radiation from the housing to ambient, the Newton-Raphson method of equations (15.24) will be used to obtain a solution. The final values of temperature are shown in the proper location in Fig. 15.5.

The system chosen for evaluation was necessarily simple. A more realistic system would consider variation of bearing temperature in a circumferential direction also. For this case, viscous torque may be

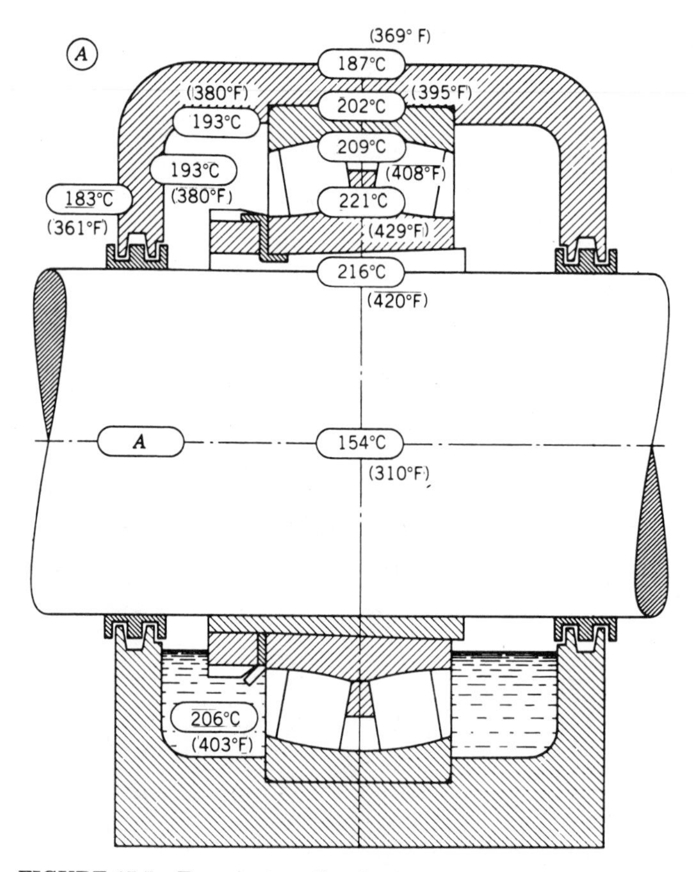

FIGURE 15.5. Temperature distribution, natural convection of air.

considered constant with respect to angular position; however, load torque varies as the individual rolling element load on the stationary ring but may be considered invariant with respect to angular position on the rotating ring. A three-dimensional analysis such as that indicated by load torque variation on the stationary ring should, however, show little variation in temperature around the bearing rings so that a two-dimensional system should suffice for most engineering applications. Of course, if temperatures of structures surrounding or abutting the housing are significantly different, then a three-dimensional study is required. A three-dimensional study will require a computer.

It is not intended that the results of the foregoing method of analysis will be of extreme accuracy, but only that accuracy will be sufficient to determine the approximate thermal level of operation such that corrective measures may be taken in the event excessive steady-state operating temperatures are indicated. Moreover, in the event that cooling of the assembly is required, the same methods may be used to evaluate the adequacy of the cooling system.

Generally, the more temperature nodes selected or the finer the grid, the more accurate will be the analysis.

HIGH TEMPERATURE CONSIDERATIONS

Special Lubricants and Steels

Having established the operating temperatures in a rolling bearing assembly while using a conventional mineral oil lubricant and lubrication system, and having estimated that the bearing and/or lubricant temperatures are excessive, it then becomes necessary to redesign the system to either reduce the operating temperatures or make the assembly compatible with the temperature level. Of the two alternatives, the former is safest when considering prolonged duration of operation of the assembly; however, when shorter finite lubricant life and/or bearing life are acceptable, it may be expeditious and even economical to simply accommodate the increased temperature level by using special lubricants and/or bearing steels. The later approach is effective when space and weight limitations preclude the use of external cooling systems and is further necessitated in many applications in which the bearing is not the prime source of heat, such as in aircraft gas turbine engines.

Heat Removal

For situations in which the bearing is the prime source of heat and in which the ambient conditions surrounding the housing do not permit an adequate rate of heat removal, simply placing the housing in a moving air stream may be sufficient to reduce operating temperatures. For instance, placing the housing of Fig. 15.4 in a fanned air stream of 15.2 m/sec (50 ft/sec) velocity will create a heat transfer coefficient of about 2.386 $\times 10^{-5}$ W/mm$^2 \cdot$ °C (4.2 Btu/hr $\cdot$ ft$^2 \cdot$ °F) on the outside surface of the housing, which is approximately 4 times that for the natural convection system and gives a maximum bearing temperature of 168 °C (334 °F) as opposed to 221 °C (429 °F) obtained for free convection. Figure 15.6 shows the remaining system temperatures. Thus, in this case, the system temperatures can be significantly reduced if a fan is used to circulate air over the housing.

If a fan is used, increased heat transfer from the housing to the air stream may be effected by placing fins on the housing. This increases the effective area for heat transfer from the housing. Consider that the external area of the housing of Fig. 15.4 is doubled by use of fins. Using

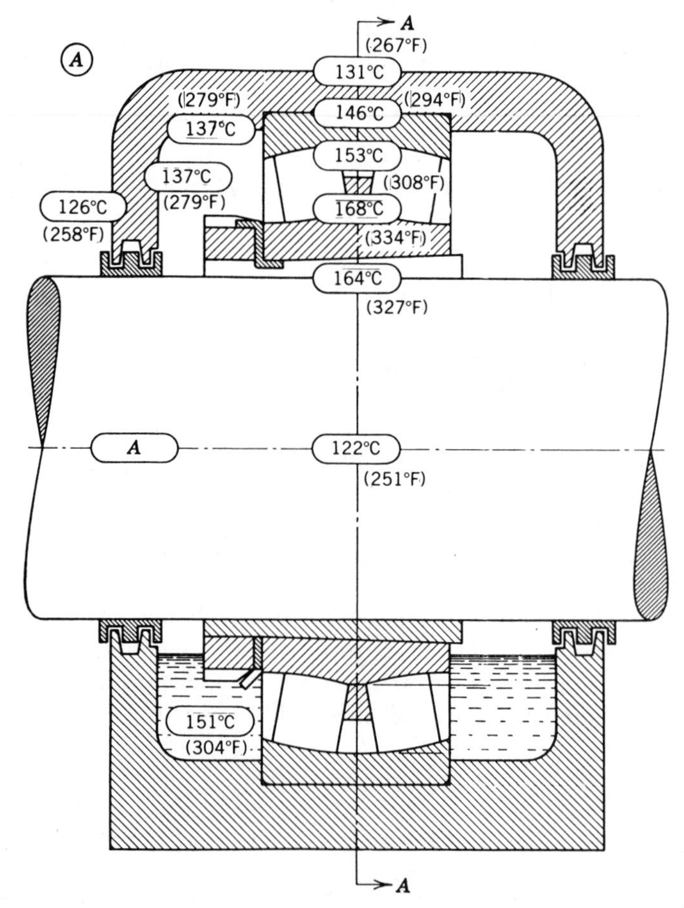

FIGURE 15.6. Temperature distribution, outside air velocity 15.2 m/sec (50 ft/sec).

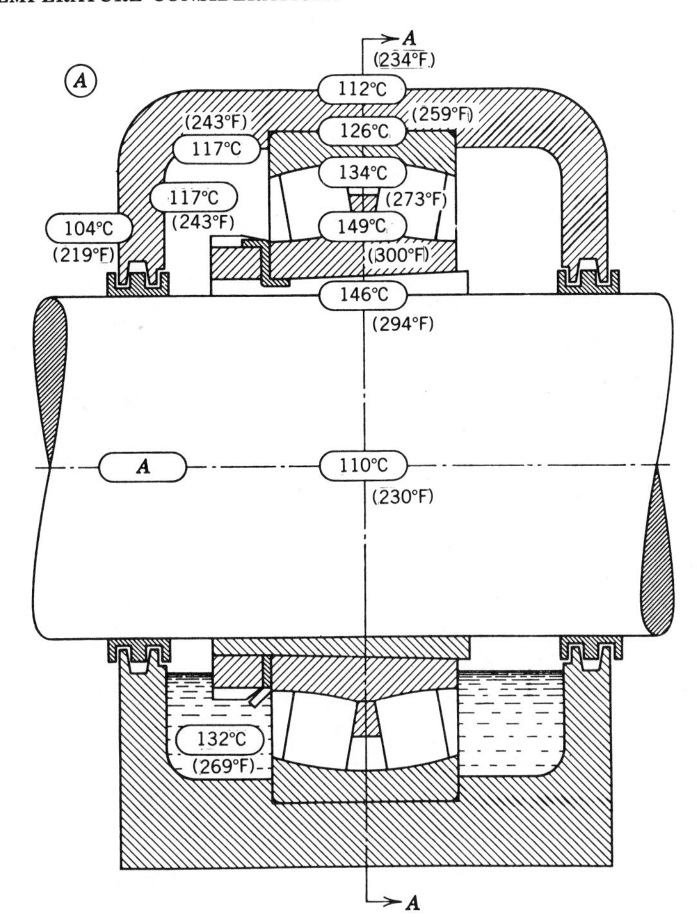

FIGURE 15.7. Temperature distribution, air velocity 15.2 m/sec (50 ft/sec) finned housing.

the film coefficient of 2.386×10^{-5} W/mm^2 · °C (4.2 Btu/hr · ft^2 · °F) for the moving air stream and 2 times the external housing area yields a maximum bearing temperature of 149°C (300°F). Figure 15.7 shows other system temperatures.

When the bearing is not the prime source of heat, cooling of the housing will generally not suffice to maintain the bearing and lubricant cool. For example, consider a shaft temperature of 260°C (500°F) instead of 48.9°C (120°F). With the aforementioned moving air system in operation, the maximum bearing temperature for the reference system is 196°C (385°F). (Figure 15.8 shows other system temperatures.) Thus, it is necessary to cool the lubricant and permit the lubricant to cool the bearing. The most effective way of accomplishing this is to pass the oil through an external heat exchanger and direct jets of the cooled oil on

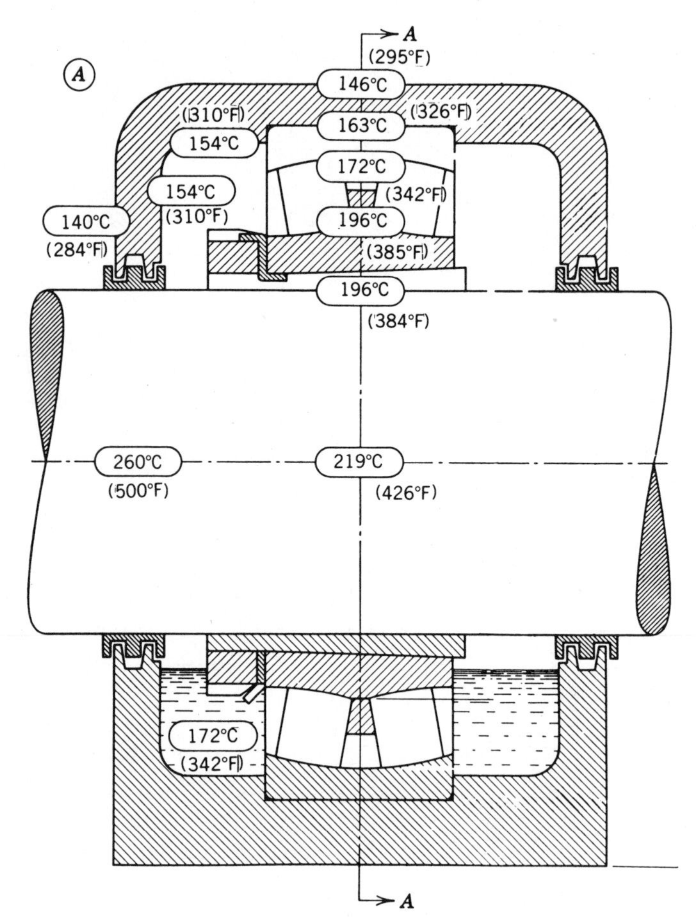

FIGURE 15.8. Temperature distribution, outside air velocity 15.2 m/sec (50 ft/sec).

the bearing. To save space when a supply of moving coolant is readily available, it may be possible to place heat exchanger coils directly in the sump. The cooled lubricant is then circulated by bearing rotation. The latter method is not quite as efficient thermally as jet cooling although bearing friction torque and heat generation may be less by not resorting to jet lubrication and the attendant churning of excess oil. The adequacy of either system of cooling may be demonstrated approximately by assuming that oil temperature is maintained at an average of 60 °C (140 °F) with shaft temperature of 260 °C (500 °F) as above and ambient temperature of 49 °C (120 °F) in quiescent air. Maximum bearing temperature is thereby suppressed to 93 °C (200 °F), which would appear to be a satisfactory operating level. (See Fig. 15.9 for other temperatures.)

It was intended to demonstrate in the foregoing discussion that it is possible to estimate with a reasonable degree of accuracy the tempera-

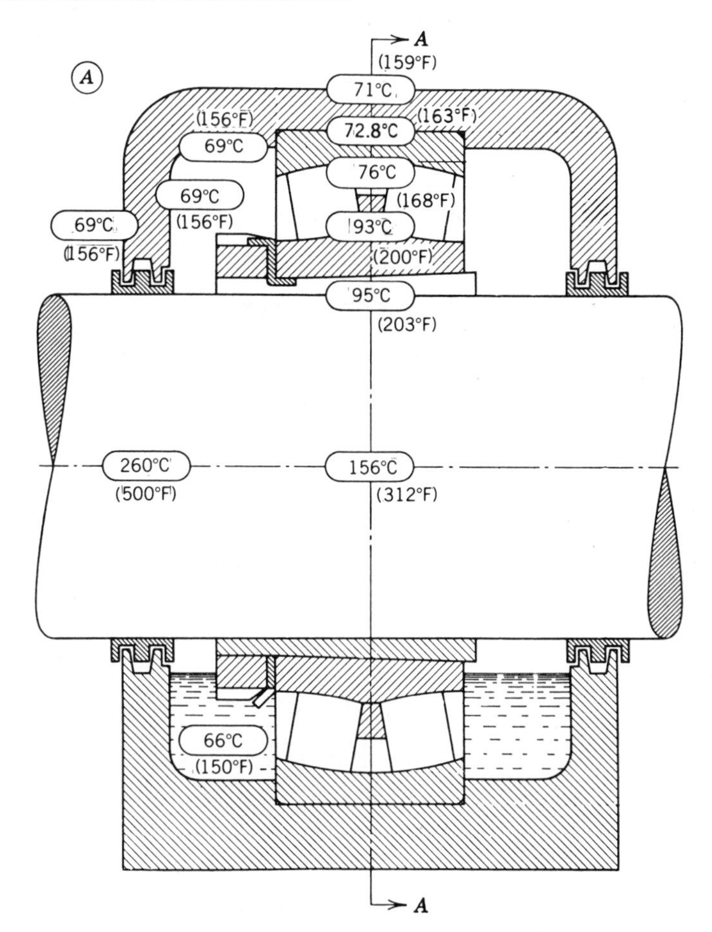

FIGURE 15.9. Temperature distribution, oil is cooled to 66°C (150°F) average temperature.

tures occurring in an oil-lubricated rolling bearing assembly. Furthermore, if the bearing and oil temperatures so calculated are excessive, it is possible to determine the type and degree of cooling capability required to maintain a satisfactory temperature level.

Several researchers have applied the foregoing methods to effectively predict temperatures in rolling bearing applications. Initially, Harris [15.6, 15.7] applied the method to relatively slow speed spherical roller bearings. Subsequently, these methods have been successfully applied to both high speed ball and roller bearings [15.8–15.10].

Good agreement with experimentally measured temperatures has been reported [15.12] using the steady-state temperature calculation option of SHABERTH, a computer program to analyze the thermomechanical performance of shaft-rolling bearing systems. Figure 15.10 shows a

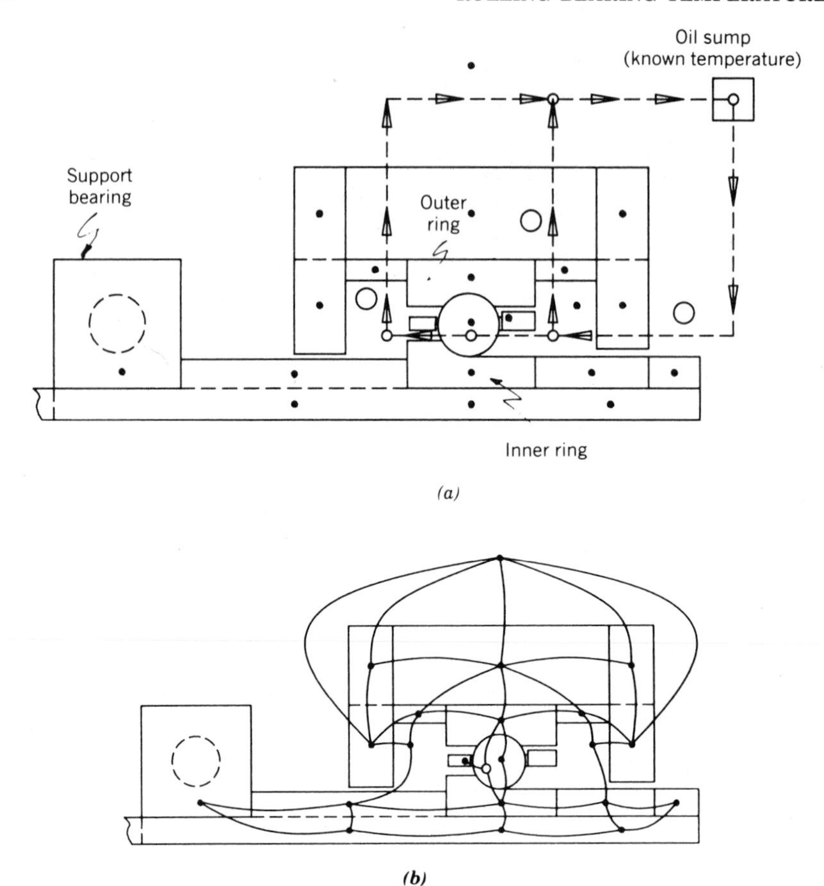

FIGURE 15.10. Bearing system nodal network and heat flow paths for steady-state thermal analysis. (*a*) Metal, air, and lubricant temperature nodes: ● metal or air node; ○ lubricant node; · → lubricant flow path. (*b*) Conduction and convection heat flow paths (from [15.12]).

nodal network model and the associated heat flow paths for a 35-mm-bore ball bearing. Figure 15.11 shows the agreement achieved between calculated and experimentally measured temperatures. It must be pointed out, however, that construction of a thermal model that accurately models a bearing often requires a considerable amount of effort and heat transfer expertise.

CLOSURE

The temperature level at which a rolling bearing operates dictates the type and amount of lubricant required as well as the materials from

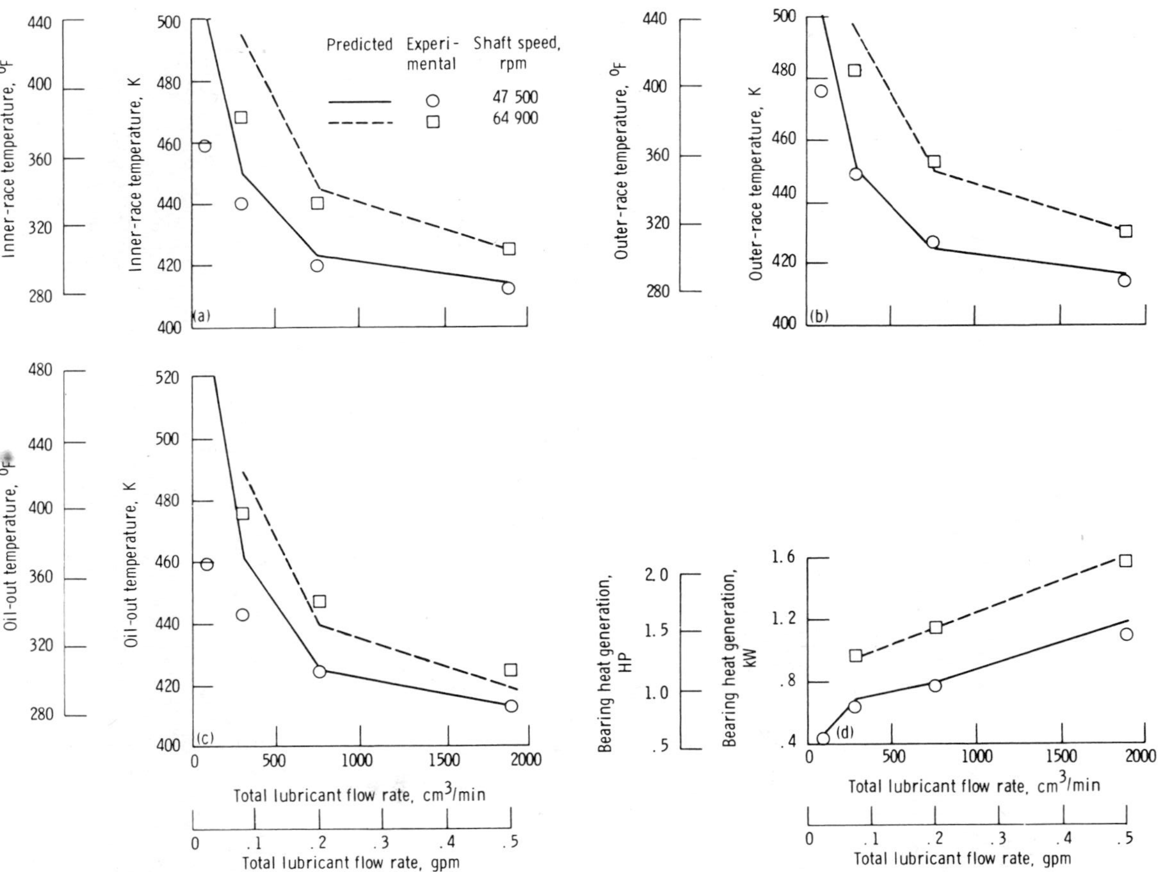

FIGURE 15.11. Comparison of predicted and experimental temperatures using SHABERTH. (*a*) Inner raceway temperature. (*b*) Outer raceway temperature. (*c*) Oil-out temperature. (*d*) Bearing heat generation (from [15.12]).

which the bearing components may be fabricated. In some applications the environment in which the bearing operates establishes the temperature level, whereas in other applications the bearing is the prime source of heat. In either case, depending on the bearing materials and the endurance required of the bearing, it may be necessary to cool the bearing using the lubricant as a coolant.

General rules cannot be formulated to determine temperature level for a given bearing operating under a given load at a given speed. The environment in which the bearing operates is generally different for each specialized application. Using the friction torque formulas of Chapter 13 to establish the rate of bearing heat generation in conjunction with the heat transfer methods presented in this chapter, however, it is possible to estimate the bearing system temperatures with an adequate degree of accuracy.

REFERENCES

15.1. E. Eckert, *Introduction to the Transfer of Heat and Mass*, McGraw-Hill, New York (1950).

15.2. M. Jakob and G. Hawkins, *Elements of Heat Transfer and Insulation*, 2nd ed., Wiley, New York (1950).

15.3. A. Palmgren, *Ball and Roller Bearing Engineering*, 3rd ed., Burbank, Philadelphia (1959).

15.4. G. Dusinberre, *Numerical Methods in Heat Transfer*, McGraw-Hill, New York (1949).

15.5. G. Korn and T. Korn, *Mathematical Handbook for Scientists and Engineers*, McGraw-Hill, New York (1961).

15.6. T. Harris, "Prediction of Temperature in a Rolling Bearing Assembly," *Lubr. Eng.*, 145–150 (April 1964).

15.7. T. Harris, "How to Predict Temperature Increases in Rolling Bearings," *Prod. Eng.*, 89–98 (December 9, 1963).

15.8. J. Pirvics and R. Kleckner, "Prediction of Ball and Roller Bearing Thermal and Kinematic Performance by Computer Analysis," *Advanced Power Transmission Technology*, NASA Conference Publication 2210, 185–201 (1982).

15.9. H. Coe, "Predicted and Experimental Performance of Large-Bore High-Speed Ball and Roller Bearings," *Advanced Power Transmission Technology*, NASA Conference Publication 2210, 203–220 (1982).

15.10. R. Kleckner and G. Dyba, "High Speed Spherical Roller Bearing Analysis and Comparison with Experimental Performance," *Advanced Power Transmission Technology*, NASA Conference Publication 2210, 239–252 (1982).

15.11. W. Crecelius, "User's Manual for SKF Computer Program SHABERTH, Steady State and Transient Thermal Analysis of a Shaft Bearing System Including Ball, Cylindrical, and Tapered Roller Bearings," SKF Report AL77P015, submitted to U.S. Army Ballistic Research Laboratory (February 1978).

15.12. R. Parker, "Comparison of Predicted and Experimental Thermal Performance of Angular-Contact Ball Bearings," NASA Technical Paper 2275 (February 1984).

16

BEARING STRUCTURAL MATERIALS

GENERAL

The functional performance and endurance of a "dimensionally perfect" bearing with ideal internal geometries and surfaces, correct mounting, and preferential operating conditions is significantly influenced by the characteristics of its materials. Major criteria to be considered for satisfactory bearing performance include material selection and processing with resultant physical properties. This chapter contains brief descriptions of various bearing steel analyses, melting practices, manufacturing process variables, and the influence of these factors on the physical and metallurgical properties with respect to bearing performance. It also contains discussions concerning metallic and nonmetallic materials used for cages, seals, and shields.

ROLLING BEARING STEELS

Types of Steels for Rolling Components

Rolling bearing steels, from their inception, were selected on the basis of hardenability, fatigue strength, wear resistance, and toughness.

American Iron & Steel Institute (AISI) 52100 steel, an alloy machinable in its annealed condition and exhibiting high hardness in the heat-treated state, was introduced around 1900 and is still the most-used steel for ball bearing plus most roller bearing applications. For large bearing sizes, particularly with respect to cross sectional thickness, modifications to this basic analysis incorporating silicon, manganese, and molybdenum were introduced. Carburizing steel came into being when the tapered roller bearing was introduced. Over the years more demanding product requirements promoted the introduction of high-speed steels and stainless steels for high-temperature operating conditions and corrosion resistance.

TABLE 16.1. Chemical Composition of Through-Hardening Bearing Steels

	Composition (%)				
Grade[a]	C	Mn	Si	Cr	Mo
ASTM[b]-A295 (52100)	0.98	0.25	0.15	1.30	—
ISO[c] grade 1,683/XVII	1.10	0.45	0.35	1.60	0.10
ASTM-A295 (51100)	0.98	0.25	0.15	0.90	—
DIN[d] 105 Cr4	1.10	0.45	0.35	1.15	0.10
ASTM-A295 (50100)	0.98	0.25	0.15	0.40	—
DIN 105 Cr2	1.10	0.45	0.35	0.60	0.10
ASTM-A295 (5195)	0.90	0.75	0.15	0.70	—
	1.03	1.00	0.35	0.90	0.10
ASTM-A295 (K19526)	0.89	0.50	0.15	0.40	—
	1.01	0.80	0.35	0.60	0.10
ASTM-A295 (1570)	0.65	0.80	0.15	—	—
	0.75	1.10	0.35	—	0.10
ASTM-A295 (5160)	0.56	0.75	0.15	0.70	—
	0.64	1.00	0.35	0.90	0.10
ASTM-A485 grade 1	0.95	0.95	0.45	0.90	—
ISO Grade 2,683/XVII	1.05	1.25	0.75	1.20	0.10
ASTM-A485 grade 2	0.85	1.40	0.50	1.40	—
	1.00	1.70	0.80	1.80	0.10
ASTM-A485 grade 3	0.95	0.65	0.15	1.10	0.20
	1.10	0.90	0.35	1.50	0.30
ASTM-A485 grade 4	0.95	1.05	0.15	1.10	0.45
	1.10	1.35	0.35	1.50	0.60
DIN 100 Cr Mo 6	0.92	0.25	0.25	1.65	0.30
ISO-grade 4,683/XVII	1.02	0.40	0.40	1.95	0.40

[a]Phosphorus and sulfur limitation for each alloy is 0.025% maximum (each element).
[b]American Society for Testing Materials [16.1].
[c]International Standards Organization.
[d]West German Standards Organization.

Through-Hardening Steels

The largest tonnage of bearing steels currently produced is the category of through-hardening steels. Table 16.1 lists common grade designations and respective chemical compositions for this family of alloys.

Through-hardening steels are classified as hypereutectoid-type steels when containing greater than 0.8% carbon by weight and essentially containing less than 5% by weight of the total alloying elements. Assuming satisfactory material availability, the bearing producer selects the appropriate grade of steel based on bearing size, geometry, dimensional characteristics, specific product performance requirements, and manufacturing methods and associated costs.

Case-Hardening Steels

Table 16.2 outlines the grade designations and corresponding chemical compositions of the common carburizing steels.

TABLE 16.2. Chemical Composition of Carburizing Bearing Steels

	Composition (%)					
Grade[a]	C	Mn	Si	Ni	Cr	Mo
SAE[b] 4118	0.18	0.70	0.15	—	0.40	0.08
	0.23	0.90	0.35	—	0.60	0.15
SAE 8620, ISO 12	0.18	0.70	0.15	0.40	0.40	0.15
DIN 20 NiCrMo2	0.23	0.90	0.35	0.70	0.60	0.25
SAE 5120	0.17	0.70	0.15	—	0.70	—
AFNOR[c] 18C3	0.22	0.90	0.35	—	0.90	—
SAE 4720, ISO 13	0.17	0.50	0.15	0.90	0.35	0.15
	0.22	0.70	0.35	1.20	0.55	0.25
SAE 4620	0.17	0.45	0.15	1.65	—	0.20
	0.22	0.65	0.35	2.00	—	0.30
SAE 4320, ISO 14	0.17	0.45	0.15	1.65	0.40	0.20
	0.22	0.65	0.35	2.00	0.60	0.30
SAE E9310	0.08	0.45	0.15	3.00	1.00	0.08
	0.13	0.65	0.35	3.50	1.40	0.15
SAE E3310	0.08	0.45	0.15	3.25	1.40	—
	0.13	0.60	0.35	3.75	1.75	—
KRUPP	0.10	0.45	0.15	3.75	1.35	—
	0.15	0.65	0.35	4.25	1.75	—

[a]Grades are listed in ASTM A534 [16.2]; phosphorus and sulfur limitation for each alloy is 0.025% maximum (each element).
[b]Society of Automotive Engineers.
[c]French Standard.

TABLE 16.3. Chemical Composition of Special Bearing Steels

	Typical Composition (%)							
Grade	C	Mn	Si	Cr	Ni	V	Mo	W
M50	0.80	0.25	0.25	4.00	0.10	1.00	4.25	—
BG-42	1.15	0.50	0.30	14.50	—	1.20	4.00	—
440-C	1.10	1.00	1.00	17.00	—	—	0.75	—
CBS-600	0.20	0.60	1.00	1.45	—	—	1.00	—
CBS-1000	0.15	3.00	0.50	1.05	3.00	0.35	4.50	—
VASCO X-2	0.22	0.30	0.90	5.00	—	0.45	1.40	1.35
M50-NIL	0.15	0.15	0.18	4.00	3.50	1.00	4.00	—

These grades are essentially very high alloy, through-hardening steels; however, M50-NIL is a surface-hardening steel.

These steels are classified as hypoeutectoid steels; their carbon contents are generally below 0.80%. Carburizing steels are alloyed with nickel, chromium, molybdenum, and manganese to increase hardenability. The higher hardenability grades are used in applications requiring ring components of heavier cross section. Carbon is diffused into the surface layer of the machined components to approximately 0.65–1.10% carbon during the heat treatment operation to achieve surface hardnesses comparable to those attained with through-hardening grades of steel.

Steels for Special Bearings

Demands for good bearing performance under hostile operating conditions have consistently increased. The aerospace industry, in particular, demands products capable of operating at increased temperatures, higher speeds, and greater loads. Other challenging applications include performing in a corrosive atmosphere, under cryogenic temperatures, and in a hard vacuum. Reducing the adverse effect on rolling contact fatigue life, due to undesirable nonmetallic inclusions, in these critical applications required changes in steel melting practice. VIM-VAR (vacuum induction melted–vacuum arc remelted) alloys, such as M50 and BG42, were developed to provide the high reliability required. Table 16.3 lists the compositions of these two grades along with many of their counterparts.

STEEL MANUFACTURE

Melting Methods

During the past 30 years high-quality bearing steels have been melted by the electric arc furnace melting process. During the oxidizing period in the furnace cycle, impurities such as phosphorus and some sulfur are removed from the steel. Further refining removes dissolved oxides and other impurities that might negatively affect the steel's performance. Unfortunately, this furnace practice by itself does not remove undesirable gases absorbed by the molten steel during melting and entrapped during solidification. Although vacuum ladle degassing was patented in 1943 [16.3], it was not until the 1960s through the 1970s that cost-effective vacuum degassing production facilities were used to principally remove oxygen and hydrogen and further improve material quality. These processes provide bearing steels with the good machinability, hardenability, and homogeneity required for manufacturing economy and good product performance.

The advent of high-performance, aircraft gas turbine engines and the corresponding demand for premium-quality bearing steels led to the development of sophisticated induction and consumable electrode vacuum melting techniques. Electroslag remelting capability was developed in concert with these two vacuum processes. These special melting techniques have significantly influenced the development of higher-alloy tool steels for elevated temperature and corrosion-resistant bearing applications.

Raw Materials

The demand upon the steel industry to provide higher-quality alloy steels promoted the development of the cold charge in the basic lined, electric arc furnace. These furnaces let high-alloy scrap and lower-grade alloy scrap be mixed with plain carbon scrap; economical operation and product quality are achieved through proper selection and weight control of these materials. Knowing the exact chemistry of the scrap charge reduces the consumption of more costly alloy additions to the melt and minimizes introduction of undesirable "tramp" alloying elements. An abundance of certain trace metallic elements, occurring unintentionally in an AISI 52100 steel, has shown negative correlation on fatigue life in ball bearings [16.4].

The furnace melter selects scrap that, when melted, contains a lower percentage of the specified alloying elements for the specific grade analysis and permits final adjustment with selected alloying and carbon additions. The furnace charge must not contain elements that are not to be part of the final heat chemistry. Raw material selection for a furnace charge is also judged by physical size and weight. Light scrap takes up furnace volume. Heavy scrap reduces protection to furnace walls and roof during meltdown. A proper balance and distribution of scrap must be maintained to enable proper thermal and electrical distribution during the initial stage of meltdown. Variation from these parameters can produce "off-analysis" heat. Careful attention must also be given to slag-making materials, such as lime, silica, and fluorspar, to assure consistent quality and slag performance.

Basic Electric Furnace Process

The basic electric arc furnace (Fig. 16.1) [16.5] is circular, lined with heat-resistant brick, and contains three electrodes in a removable roof. The charge is blended to provide efficient melting, the electrodes are lowered, and arcing begins. A layer of complex slag is produced that covers the molten surface layer and absorbs impurities from the steel. During this oxidizing period a carbon boil occurs, which produces gases from the molten bath. This "complex" slag is then replaced with a "reducing" slag to decrease oxygen levels.

Because the molten metal in the refining cycle is less active than in the oxidizing cycle, furnaces may be equipped with inductive stirrers. These stirrers generate a magnetic field that imparts a circulatory motion to the molten bath, enhancing both temperature control and homogeneity of chemical composition throughout the melt. When the chemical composition of the molten steel has been adjusted to the desired range and the proper pouring temperature has been reached, the "heat" is "tapped" or removed from the furnace. Material not to be vacuum treated is then ready for "teeming" into ingot molds.

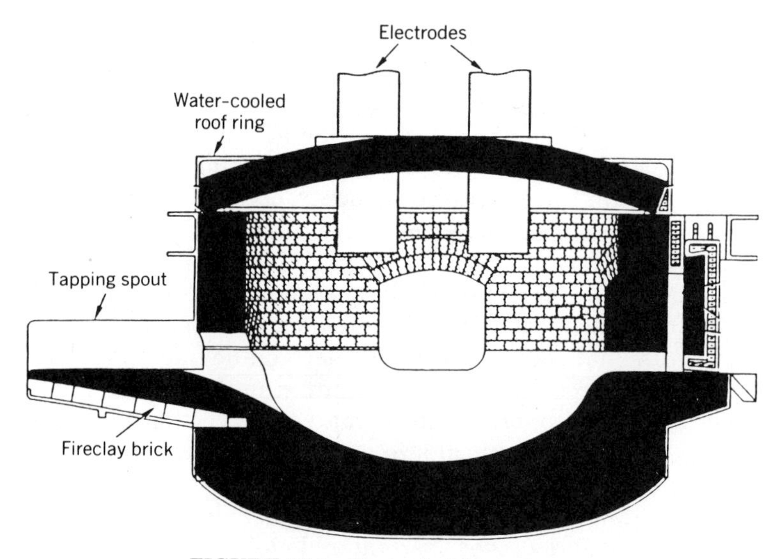

FIGURE 16.1. Basic electric arc furnace.

Vacuum Degassing of Steel

Material quality produced in the basic electric arc furnace process can be improved through vacuum degassing by various methods, including the ladle, stream, D-H (Dortmund-Hörder), and R-H (Ruhrstahl-Heraeus) processes. In conjunction with these refining practices, inert gas shrouding may be incorporated with both bottom pouring and uphill teeming. These methods economically reduce undesirable gases and remove nonmetallic inclusions from the molten steel.

Ladle Degassing. The ladle degassing process requires a ladle of molten metal to be placed in a sealed, evacuated chamber. Gases resulting from pressure differentials cause turbulence and a moderate stirring action in the molten bath. Ladle degassing units may be equipped with induction stirrers and/or injection devices for bubbling inert gases such as argon or helium through the molten metal to further agitate the bath. See Fig. 16.2 [16.6]. With this technique the quantity of metal exposed to the vacuum is greatly increased, which allows small alloy additions to be made to the melt.

Stream Degassing. The molten metal in stream degassing is poured from an intermediate or "pony" ladle, as illustrated in Fig. 16.3 [16.7], into the vacuum chamber containing an empty ladle. As the molten metal enters the vacuum chamber, the stream explodes into fine spray

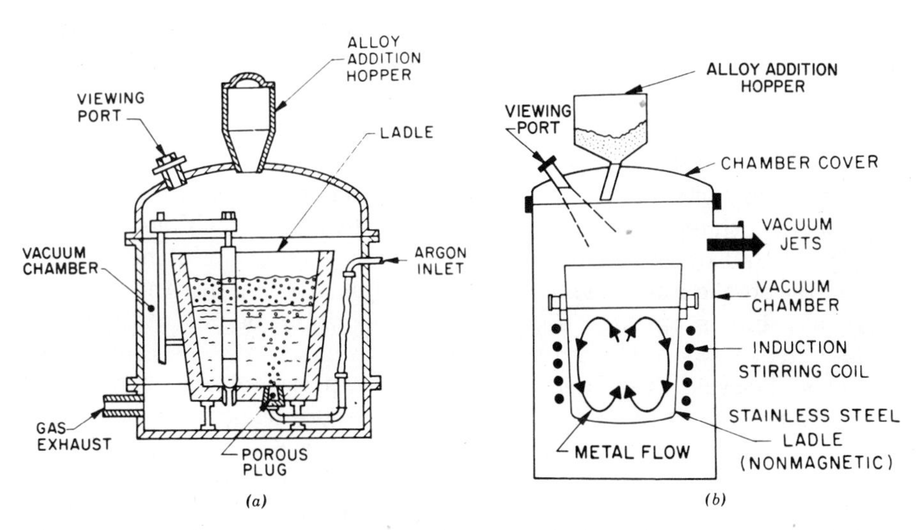

FIGURE 16.2. Schematic arrangement of equipment for ladle degassing. (*a*) Gas stirring. (*b*) Induction stirring.

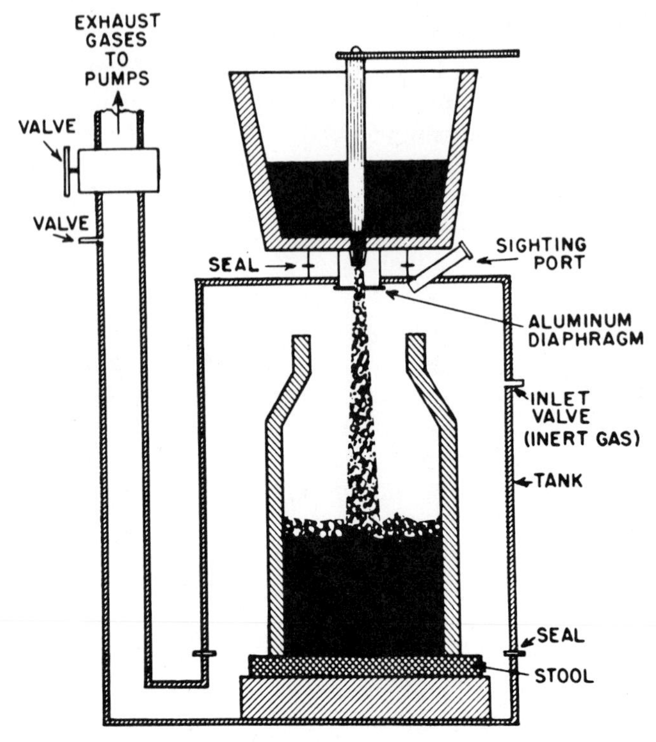

FIGURE 16.3. Schematic illustration of stream degassing method.

or droplets that fall into a second, or teeming, ladle. When this ladle is full, the vacuum seal is broken and the molten metal is poured into the ingot molds.

D-H (Dortmund-Hörder) Process. The D-H or Dortmund-Hörder process is a cyclic type of vacuum operation. This practice necessitates positioning the furnace ladle under a refractory-lined vessel, which moves alternately up and down. As the vacuum vessel descends into the molten metal, the steel enters the nozzle, or snorkel, is lifted up into the vacuum chamber, and is agitated vigorously as degassing begins. As the degassed steel flows back into the ladle, it is mixed with the remaining steel and the cycle is repeated. A total cycle requires approximately 20 sec and is repeated 40 to 60 times during the entire degassing operation. Heat losses occurring during the cycle are compensated by a graphite, electrically resistant, heating element positioned in the upper part of the vacuum chamber. Vacuum-sealed hoppers allow alloys to be added during the operation. When the vacuum degassing cycle is completed, the vacuum vessel is purged with an inert gas so that the accumulated flammable gas in the vessel will not be ignited before the nozzle is raised above the surface level of the liquid steel and the ladle is removed.

R-H (Ruhrstahl-Heraeus) Process. The R-H (Ruhrstahl-Heraeus) vacuum chamber (Fig. 16.4) [16.8] straddles the ladle containing the molten metal. Two vertical legs extending downward from the base of the vacuum chamber are submerged just below the surface level of the metal in the holding ladle. The vacuum chamber is evacuated, and a pressure imbalance is created by flowing inert gas into one leg extension. The resulting pumping action circulates the molten metal between the holding ladle and the vacuum chamber. Outgassing of the molten metal occurs as it enters the vacuum chamber. The circulation process continues until the specified gas content is obtained.

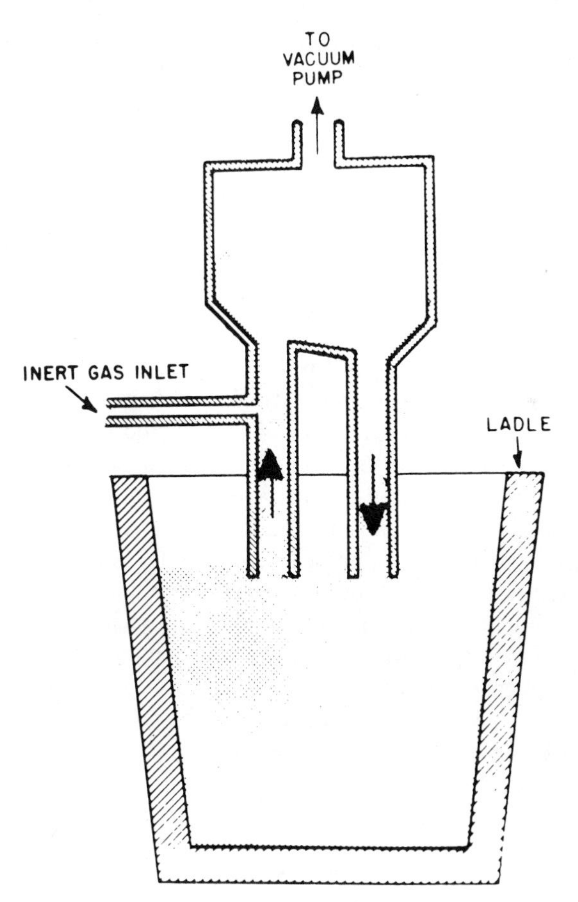

FIGURE 16.4. Principle of operation of the R-H (Ruhrstahl-Haraeus) degassing process.

Ladle Furnace

Ladle metallurgy of molten steel is conducted in a ladle furnace outside the normal constraints of the initial electric arc melting furnace. The ladle furnace is equipped with independent electrodes for bath temperature control and electromagnetic stirrers for bath circulation. Therefore there is no need to superheat steel in the electric arc furnace to compensate for subsequent temperature drops experienced in standard ladle degassing practices as previously described.

Lance injection permits powdered alloys to be inserted deep into the ladle. Argon is used as a carrier gas for these powders; the resulting bubbling action helps disperse particles uniformly throughout the molten bath. Lance injection combined with wire feeding provides inclusion shape control, reduction of sulfur content, and improvement of fluidity, chemical homogeneity, and overall microcleanliness.

Ladle furnace technology permits very rapid meltdown of scrap in the electric arc furnace and improved refining capability in a subsequent ladle furnace operation. This system generates improved product quality with a correspondingly improved economy in steel melting.

SKF M-R Process

Consistent demands for cleaner bearing steels have resulted in the SKF melting and refining process, termed the SKF M-R process. This method employs a twin-shell, electric arc melting furnace in parallel with a SKF-ASEA ladle-refining unit.* The twin-shell furnace has two vessels with oxy-fuel burners and two roofs—that is, one contains graphite electrodes, and the other has no electrodes. Melting occurs in one furnace while charging and preheating of scrap can be carried out in the other shell. In the melting furnace carbon and phosphorus contents are adjusted to values below the final maximum limits. The furnace is then tapped, and the ladle of molten metal is transferred to the ASEA ladle-refining furnace containing an independent electrode roof. The equipment provides many metallurgical options, including vacuum degassing, desulfurizing, deoxidizing, and adjusting the chemical composition of the molten steel. The additional ability to induction stir under conditions of close temperature control permits precipitation deoxidation with aluminum, resulting in steels with very low contents of oxygen and nonmetallic inclusions. The sequence of steps in the melting and refining process is illustrated in Fig. 16.5 [16.9].

*This method was developed by SKF Steel AB in conjunction with ASEA in Sweden. Subsequently, SKF Steel AB formed a joint venture with OVAKO OY of Finland. The resulting company OVAKO Steel AB employs the M-R process.

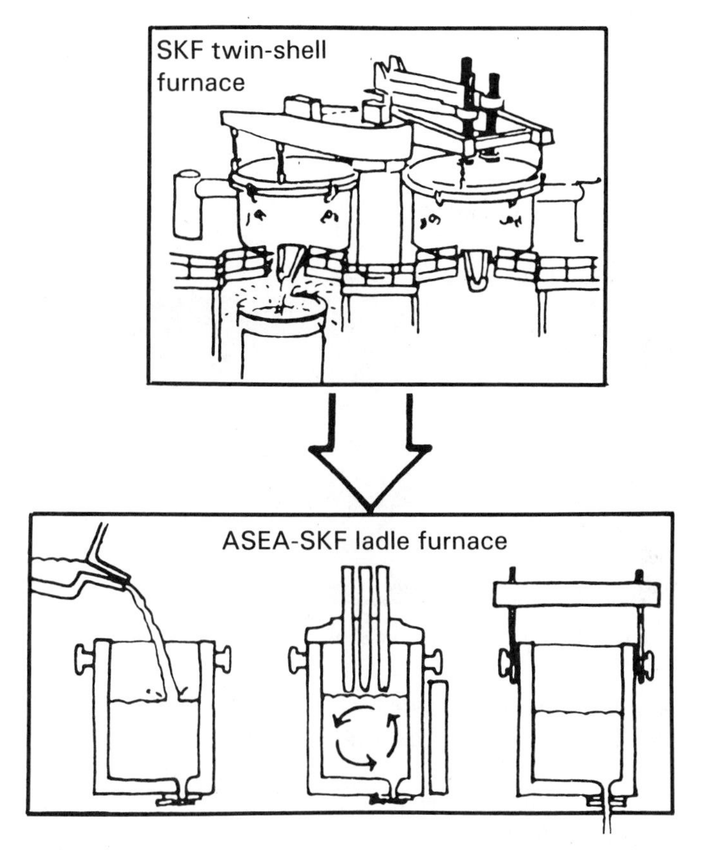

FIGURE 16.5. SKF MR steel process.

Methods for Producing Ultrahigh-Purity Steel

Vacuum Induction Melting. Yet more sophisticated steel melting processes were introduced during the 1960s for producing ultrahigh-purity steels, also called "clean" bearing steels, which are essentially free from deleterious nonmetallic-type inclusions.

In vacuum induction melting, selected scrap material containing few impurities and comparable in chemical composition to the alloy grade being melted is charged into a small electrical induction furnace. The furnace (Fig. 16.6) [16.10] is encapsulated in a large vacuum chamber containing sealed hoppers strategically located for adding required alloys.

Outgassing of the melt occurs early in a very rapid meltdown and refining period. After the melting cycle, furnace tilting and pouring the molten metal into ingot molds take place. The molds are automatically manipulated into and out of the pouring position while still within the

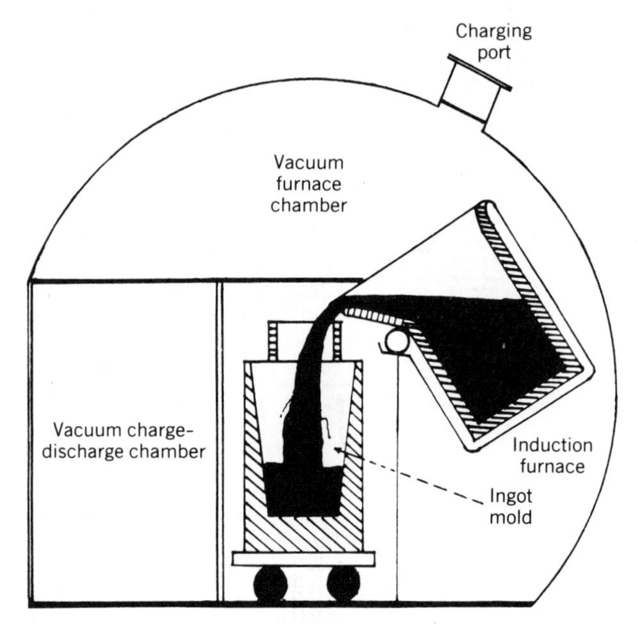

FIGURE 16.6. Schematic view of induction vacuum melting facility.

vacuum-sealed chamber. This vacuum induction melting furnace process was among the first vacuum processing methods employed in manufacturing premium aircraft quality bearing steels. One of its primary functions today is to provide electrodes used to produce ultrahigh-purity, vacuum arc remelted steels.

Vacuum Arc Remelting. Vacuum technology for bearing steel alloy production, as described in the foregoing sections on vacuum degassing and induction vacuum melting, provided a way to reduce the gas content and nonmetallic inclusions in steel. Steel electrodes, melted in furnaces using vacuum technology can be remelted by still more sophisticated techniques, such as the consumable electrode vacuum melt practice, to provide material for bearings requiring the utmost reliability. This process, illustrated in Fig. 16.7, involves inserting an electrode of the desired chemical composition into a water-cooled, copper mold in which a vacuum is created.

An electrical arc is struck between the bottom face of the electrode and a base plate of the same alloy composition. As the electrode is consumed under extremely high vacuum conditions, it is automatically lowered and the voltage is controlled to maintain constant melting parameters. Because the solidification pattern is controlled, the remelted product is essentially free from center porosity and ingot segregation. The product has improved mechanical properties, particularly in the transverse direction. Aircraft bearing material specifications for critical applications now specify the VIM-VAR steel melting practice.

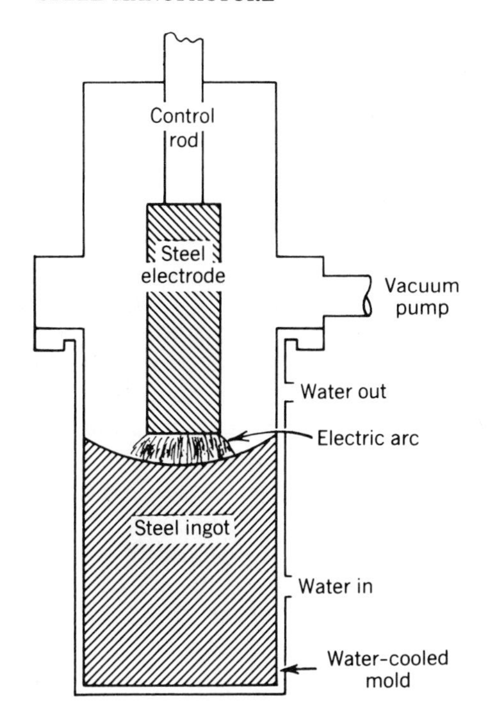

FIGURE 16.7. Vacuum arc remelting furnace.

Electroslag Refining. The electroslag refining process (ESR) is very similar to that of the consumable electrode vacuum melting process except that a liquid slag bath positioned at the base of the electrode provides the electrical resistance required for melting. The slag bath is either introduced into the furnace chamber in the molten state or provided as a powder slag that will quickly melt upon striking an arc between the electrode and base plate. Refinement of the molten steel occurs as steel droplets pass through the slag bath. Control of slag composition permits removal of sulfur and oxygen and undesirable impurities. The resulting ingot solidification pattern reduces porosity, minimizes segregation, and provides for improved physical properties in the transverse and longitudinal directions.

Steel Products

Product Forms. Rolling bearings cover a broad spectrum of sizes with ring components varying in both cross-sectional configurations and material grades. Balls or rollers vary in size and shape to accommodate their mating ring components. Generally, these components are manufactured from forgings, tubing, bar stock, or wire. The bearing manufacturer orders from the steel producer the form and condition of the raw material best suited for the selected method of processing and that will meet bearing performance plus customer requirements.

Regardless of the melting practice employed, the resulting ingots are stripped from the molds, homogenized in soaking pits, rolled into blooms or billets, and subsequently conditioned to remove surface defects. Billets are reheated and hot-worked into bars, tubes, forgings, or rolled rings. Additional cold-working operations convert the hot-rolled tubes and bars into cold-reduced tubes, bars, and cold-drawn wire. As an ingot passes back and forth between the rolls on a blooming mill, its cast structure is broken up and refined. Continued hot-working operations elongate and break up nonmetallic inclusions and alloy segregation. Hot mechanical working operations also permit plastic deformation of the material into desired shape or form. The subsequent cold-working of material results in induced stresses and improved machinability. Cold-working also produces changes in the mechanical properties, improved surface finish, and closer tolerances. Dimensional tolerances and eccentricities of cold-worked tubes are far superior to those of hot-rolled tubes.

Most mill product forms require thermal treatment at the mill before or after final finishing so that the forms are ready for machining or forming. The thermal treatment may include annealing, normalizing, or stress relieving. The product is then straightened, if necessary, inspected, and readied for transport.

Product Inspection. The steel producer performs two major functions to satisfy customer product quality requirements. First, an inspection plan is implemented to ensure that during the various manufacturing stages the product meets the specified quality limits. Second, the product is tested in its final form to ensure that it conforms to predetermined internal and external standards. Nondestructive testing methods, including hot acid etching, magnetic particle, eddy current, ultrasonic, and hydrostatic pressure tests, in conjunction with standard dimensional testing of the product are successfully used in statistical process control (SPC) and final inspection programs.

Numerous industrial, military, aerospace, or other independent customer specifications usually form part of the purchaser's requirement for bearing quality steels. These standards necessitate testing to satisfy heat characteristics, including chemical analyses, hardenability, macroetch and microinclusion ratings. Special product requirements include surface defects, microstructure, hardness, and dimensional tests.

Steel Metallurgical Characteristics

Quality Requirements. Cleanliness, segregation, and microstructure are steel product characteristics that influence bearing manufacture and subsequent product performance. The cleanliness of steel pertains to nonmetallic inclusions entrapped during ingot solidification that cannot be subsequently removed. Segregation pertains to an undesirable nonuniform distribution of alloying elements. Microstructure of the mill product relates directly to its suitability for machining and/or forming by the bearing manufacturer. The producer therefore incorporates tests early in the manufacture of the product and before shipment to be sure that pertinent mill and customer quality requirements are met.

Cleanliness. Steel quality with respect to nonmetallic inclusions depends on the initial raw material charge, selection of the melting furnace type/practice, and control of the entire process, including teeming and conditioning of the ingot molds. Exogenous inclusions result from erosion or breakdown of furnace refractory material or from other dirt particles outside the melt that are entrapped during tapping and teeming. Indigenous inclusions are products of deoxidation occurring within the melt. Inclusions less than 0.5 mm (0.020 in.) long are considered microinclusions; larger ones are considered macroinclusions. Nonmetallic inclusions are classified by their composition and morphology as sulfides, aluminates, silicates, and oxides. Occasionally, nitrides are included in rating steel cleanliness.

It is well established through testing that nonmetallic inclusions are detrimental to rolling contact fatigue life. The data further indicate that hard, brittle-type inclusions, based on their size, shape, and distribution, are more detrimental than soft deformable-type particles such as sulfides. By encapsulating harder, nonmetallic particles and forming a cocoon or cushion around them so that they do not become points of stress concentration under cyclic loading, sulfides act beneficially. Sulfide inclusions can also enhance machinability by performing as a lubricant. Because aluminates and silicates have sharp corners and are very brittle, they can act as stress raisers and initiate early fatigue failures. Globular oxides are inclusions formed with such elements as calcium. These particles are very hard and brittle and are considered to be most detrimental to machinability and rolling contact fatigue life.

Many methods have been devised for detecting or quantitatively defining nonmetallic inclusions. One of the most popular involves microscopic examination at 100X magnification of polished specimens of a predetermined size from specific ingot locations and comparing the worst field observed against standard photographs weighted on a numbered rating system [16.1]. Material specifications containing this method of evaluation stipulate the testing frequency and the acceptance or rejection limits. Industry standards also cite product acceptance tests such as fracture, magnaflux stepdown (AMS 2300/AMS 2301), and visual rating–counting indications on machined bar surfaces to determine material quality with respect to nonmetallic inclusion content.

Premium-quality aircraft bearing steel with respect to detection of nonmetallic inclusions is automatically checked on a 100% bar product basis by employing both eddy current and ultrasonic test procedures.

Segregation. Nonuniform solidification rate of molten metal within an ingot mold might lead to segregation of alloying constituents. Because of the rapid freezing rate occurring in the mold-ingot interface, this outermost portion or shell will solidify first and form columnar crystals. At the centermost portion of the ingot, which cools at a much slower rate, the grains are equiaxed. Because solidification is not spontaneous throughout the ingot, the molten metal that freezes last will also become richer in alloying elements such as manganese, phosphorus, and sulfur, because the elements have inherently lower melting points.

Segregation is only slightly improved by thermal treatment and hot-working. Precautions must be taken in the melting and pouring practices, such as incorporating special molds designed to control rates of freezing and thereby prevent or minimize segregation during solidification.

Macroetching of properly prepared billet or bar slices, employing dilute, hot hydrochloric acid solution, which preferentially attacks these numerous alloying constituents, is used to reveal material segregation.

Structure. Both macroscopic and microscopic methods of inspection are used to evaluate steel structure. Discs cut from the ends of bars or billets for macroscopic examination are prepared according to industry standards, acid etched, and examined with the unaided eye or under magnification generally not exceeding 10X. Although numerous etching reagents are available, the generally recommended and accepted solution is hot dilute, hydrochloric (muriatic) acid.

In addition to the detection of alloy segregates, material may be microstructurally evaluated for other objectionable characteristics such as "pipe," porosity, blow holes, decarburization, excessive inclusions, cracks, and banding. The aerospace industry uses hot acid etching and forging inspection to ensure conformance of grain flow patterns to a previously type-tested product.

Microscopic examination of steels involves a more detailed study of the structure at magnifications generally between 100X and 1000X. Numerous reagents are available to help identify and rate specific microconstituents. The steel producers' manufacturing processes incorporating thermal cycles influence the resulting microstructure of the finished mill product. Because microstructures are a reflection of material physical properties, microstructural ratings are contained within specification or purchase order requirements. Bars and tubes fed into automatic single or multiple spindle machines for turning into ring components must be in a soft annealed condition. Carbides should be uniform in size and well distributed throughout a ferrite matrix. Tool life may be expected to increase as the sizes of the "spheroidal" carbides increase. Conversely, presence of "lamellar" carbide will adversely affect machinability and tool life.

When choosing the optimum microstructure for maximum machinability of low-carbon, carburizing steel, a "blocky" microstructure of ferrite and pearlite, should be selected. A very soft annealed structure in these low-carbon grades, such as AISI 8620, is appropriate for cold-forming operations but is considered gummy and unsatisfactory in machining operations. Each material grade in its finished mill product form must exhibit the proper microstructure and hardness so that it can be economically converted to its designated configuration.

EFFECTS OF PROCESSING METHODS ON STEEL COMPONENTS

Many of the mechanical properties of finished bearing components are developed by manufacturing methods that dictate the form and condition for raw material. Generally, raw material is produced by either hot- or cold-reduction processes and furnished as tubing, bars, wire, and forgings. Cold-reduction for producing bars, tubing, balls, and rollers of AISI 52100 will lower both the austenite transformation temperature during heating for hardening and the martensite start temperature upon cooling [16.12]. The resulting fracture grain size of a ring produced from a cold-reduced tube will be finer than that of an identical ring from a hot-rolled annealed tube. Although the volume change for the hot-rolled and cold-reduced components is the same, the ring from the cold-reduced tube will have a smaller diameter after heat treatment.

Bars and tubes are elongated during manufacture and display directional properties; that is, the mechanical properties are different in the longitudinal direction compared to the transverse direction. Forging the bar into ring components provides a more homogeneous product. Ring rolling might provide beneficial grain flow conforming to the rolling contact surface. Bearing endurance tests demonstrate that end grain is detrimental to rolling contact fatigue life [16.13].

Raw material intended for machining operations before heat treatment should be received in a readily machinable condition. The material should have sufficient stock to render an "as-machined" component free from carburization, decarburization, and/or other surface defects.

HEAT TREATMENT OF STEEL

Basic Principles

Heat treatment of bearing steel components necessitates heating and cooling under controlled atmospheric conditions to impart desired material characteristics and properties such as hardness, a diffused high-carbon surface layer, high fracture toughness or ductility, high tensile strength, improved machinability, proper grain size, or reduced stress state. Specific thermal cycles that produce these material characteristics are called annealing, normalizing, hardening, carburizing, tempering, and stress relieving. Selective thermal cycles provide distinctive microstructures such as bainite, martensite, austenite, ferrite, and pearlite.

Iron and carbon are the basic constituents in bearing steels along with specific amounts of manganese, silicon, or other alloying elements such as chromium, nickel, molybdenum, vanadium, or tungsten. Bearing steels have a distribution in carbon content from 0.08% minimum (AISI 3310) to 1.10% maximum (AISI 52100). Beginning with ingot solidification, bearing steels take on a crystalline structure. These crystals are composed of atoms placed at fixed locations within a unit cell. Spacings remain constant at fixed temperature. Although there are 14 different space lattice types, the bearing metallurgist deals primarily with only three: body-centered cubic (bcc), face-centered cubic (fcc), and body-centered tetragonal (bct). See Fig. 16.8.

These types of three-dimensional cells have different physical and mechanical properties because of differences in atomic spacings; they also have a different solubility for atoms of other alloying elements. An atom of one or more such alloying elements residing in the high-carbon bearing steel may be substituted for an iron atom. Elements with very small atomic radius, such as carbon, which is about one eighth the size of iron, can be placed in the interstitial spaces in the lattice.

Pure iron has a bcc structure at room temperature and an fcc structure within a specific elevated temperature range. The temperature upon heating or cooling, at which the atoms shift from one unit-cell type to another, is called a transformation temperature. These alterations can be observed in the time-temperature cooling curve for iron, as shown by Fig. 16.9.

Pure iron is bcc below 912°C (1673°F) and fcc above. When carbon is added to iron, the transformation temperature is lowered and extended over a broader temperature range. This information is displayed in the iron-carbon phase diagram in Fig. 16.10.

Because bearing steels rarely exceed 1.1% carbon and their heat treatment does not exceed a metal temperature of 1302°C (2375°F) [for T-1 (C–0.70, Cr–4.00, W–1.00, V–18.00)], only a section of the iron-carbon phase diagram, in Fig. 16.11, will be required for further discussion.

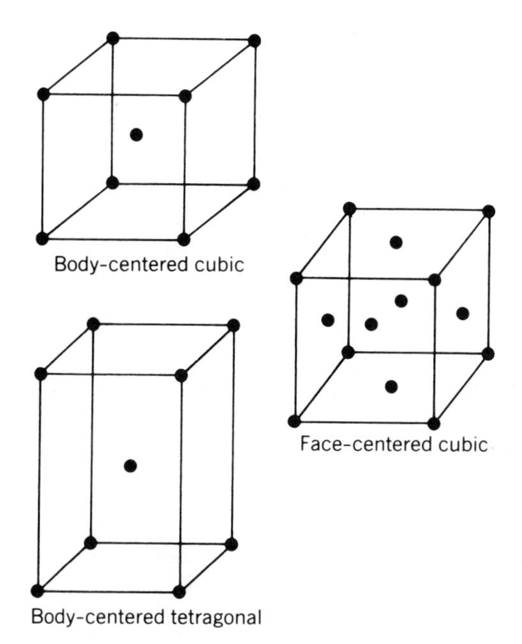

FIGURE 16.8. Crystal structures of steel.

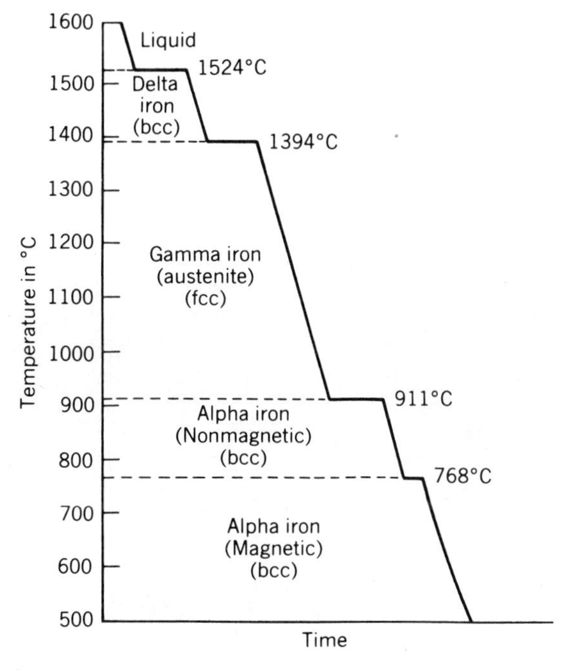

FIGURE 16.9. Time-temperature cooling curve for pure iron.

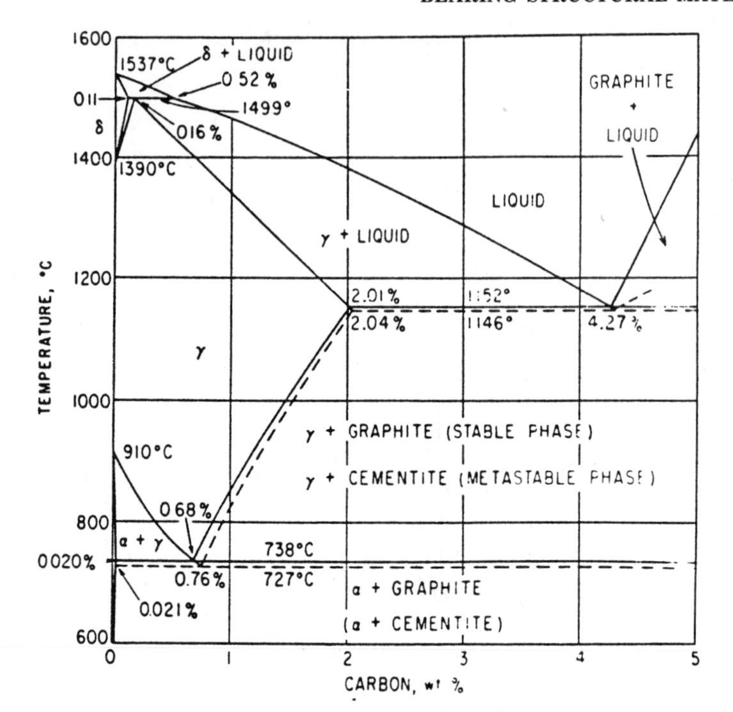

FIGURE 16.10. Iron-carbon phase diagram.

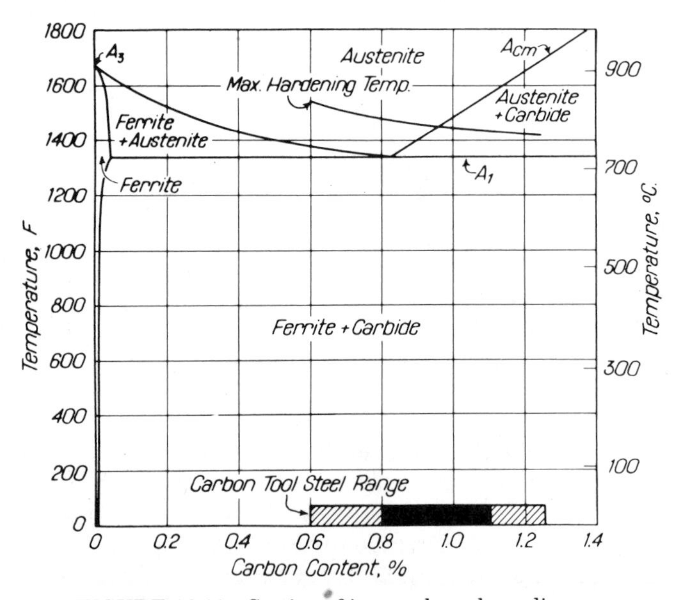

FIGURE 16.11. Section of iron-carbon phase diagram.

Carbon is dissolvable in molten iron just as salt is dissolvable in water. It is this action occurring in solid solution that enables alteration of mechanical properties of steel. High-carbon-chromium bearing steels, as received from the steel producer, are generally in a soft, spheroidized annealed condition suitable for machining. The microstructure consists of spheroidal carbide particles in a ferritic matrix. This mixture of ferrite and carbide that exists at room temperature transforms to austenite at approximately 727 °C (1340 °F). The austenite is capable of dissolving far larger quantities of carbon than that contained within the ferrite. By altering the cooling rate from the austenitizing temperature, the distribution of the resulting ferrite and carbide can be modified, thus giving a wide variation in resultant material properties.

Based on carbon content, steel can be put into three categories: eutectoid, hypoeutectoid, and hypereutectoid. Eutectoid steels are those containing 0.80% carbon, which upon heating above 727 °C (1340 °F) become 100% austenite. This composition upon cooling from the austenitic range to approximately 727 °C (1340 °F) simultaneously forms ferrite and cementite. This product is termed pearlite, and it will revert to austenite if it is reheated to slightly above 727 °C (1340 °F).

Hypoeutectoid steels are those containing less than 0.80% carbon. The iron-carbon diagram indicates that for a 0.40% carbon steel approximately 843 °C (1550 °F) is required to dissolve all the carbon into the austenite. Under conditions of slow cooling, ferrite separates from the austenite until the mixture reaches 727 °C (1340 °F). At this point the remaining austenite, containing 0.80% carbon, transforms into pearlite. The resulting microstructure is a mixture of ferrite and pearlite. The pearlite will dissolve into solid solution when it is reheated to approximately 727 °C (1340 °F). At temperatures above 727 °C (1340 °F) the ferrite will dissolve into austenite.

The iron-carbon diagram indicates the existing phases when very slow heating and cooling rates are enacted.

Time-Temperature Transformation Curve

The time-temperature transformation diagram is an isothermal transformation diagram. Steel will transform when cooled rapidly from the austenitizing temperature to a lower temperature than the minimum at which the austenite is stable. Diagrams for various grades of steel have been developed at specific austenitizing temperatures to depict the time required for the austenite to begin to transform and to be completely transformed at any constant temperature studied. Figure 16.12 [16.14] shows an isothermal time-temperature transformation (TTT) diagram for a typical high-carbon steel (AISI 52100). The shape and the position of the curves change with increased alloy content, grain size of the austenite, and austenitizing temperature. Figure 16.13 [16.15] depicts the TTT diagram for a typical alloy steel (AISI 4337).

Continuous Cooling Transformation Curves

A eutectoid steel, upon slow cooling, will transform to pearlite at approximately 727°C (1340°F). If this same steel specimen is quenched into a liquid medium controlled at a temperature just below 727°C (1340°F), a coarse pearlite structure will result. As the temperature of the holding medium is lowered, however, the diffusion of carbon atoms is decreased and the lattice spacing of ferrite and cementite is reduced, thus producing a pearlitic microstructure. These microstructures indicate that the formation of pearlite is a nucleation and growth process. At still lower temperatures carbon atoms move more slowly, and the resulting transformation product is bainite, which consists of ferrite needles containing a fine dispersion of cementite. Under still further cooling, the transformation product martensite is formed, which consists

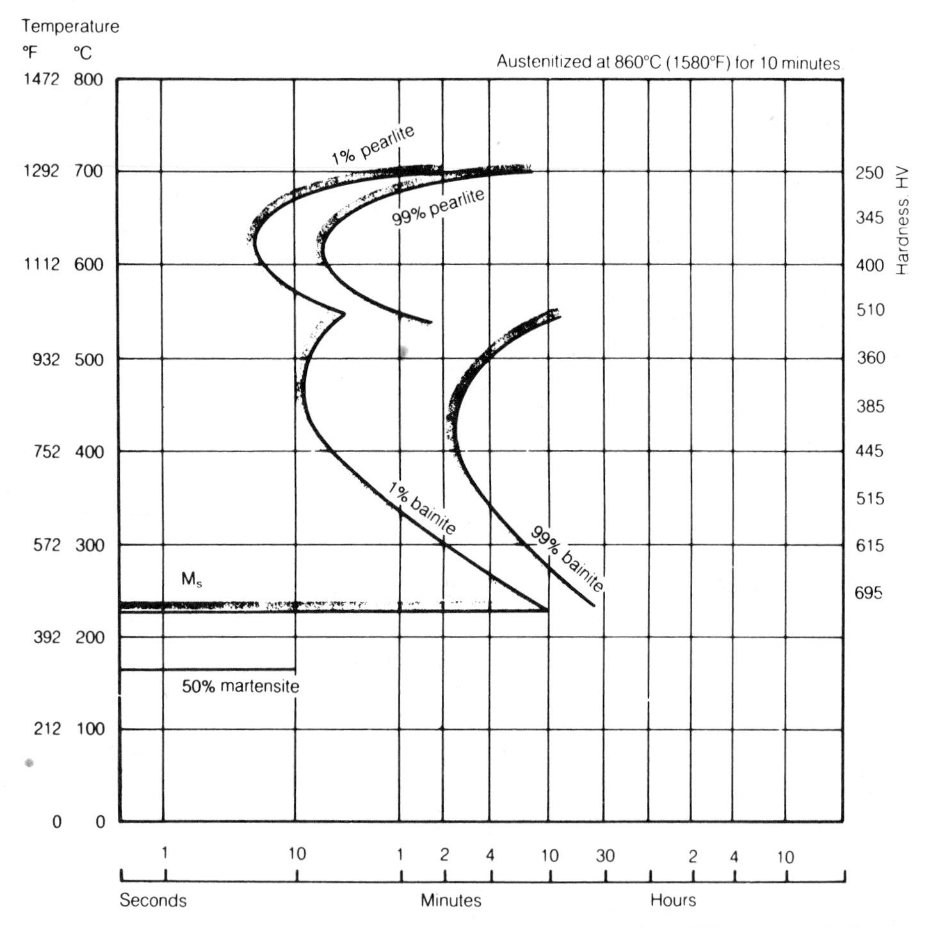

FIGURE 16.12. Time-temperature transformation diagram for AISI 52100 steel (from [16.14]).

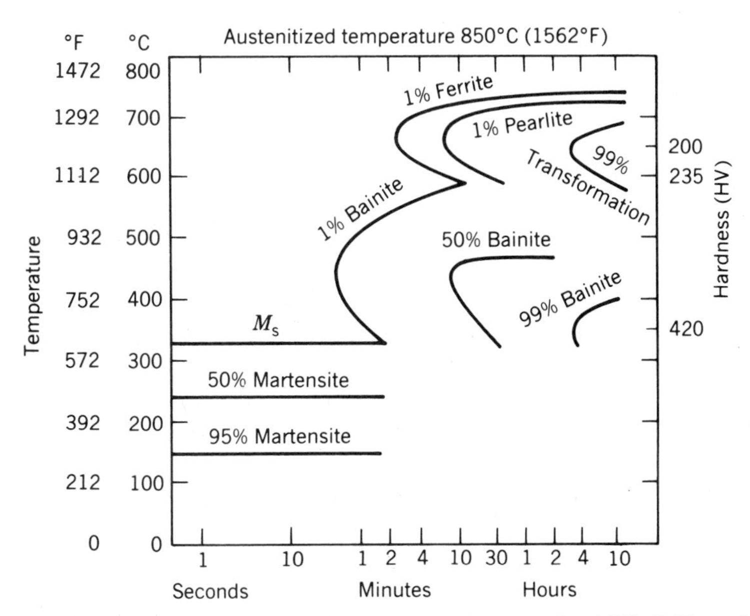

FIGURE 16.13. Time-temperature transformation diagram for AISI 4337 steel (from [16.15]).

of a very fine, needlelike structure. Martensite forms athermally involving a shear mechanism in the microstructure; it is not a product of isothermal transformation. The quenching must be done very rapidly into a medium such as molten salt or oil at a controlled temperature to prevent the austenite from converting to a soft transformation product such as pearlite.

The TTT diagrams reveal the microstructures that form at a single constant temperature; however, steel heat treatment uses rapid cooling, and transformation occurs over a range of temperatures. Continuous cooling transformation (CCT) curves have been developed to explain the resulting transformations. Figure 16.14 is a cooling transformation diagram for AISI 52100 steel.

An AISI 52100 steel Jominy test bar 25.4 mm (1 in.) in diameter by 76.2 or 101.6 mm (3 or 4 in.) long may be used to explain the value of the CCT diagram. The piece is austenitized at 843°C (1550°F) and, while held vertically, is sprayed with a stream of water on the lower end face. The cooling rate then varies from the quenched surface to the extreme opposite end, which cools much slower. Microstructures can then be correlated with various cooling rates occurring along the length of the bar.

Hardenability

Hardness should not be confused with hardenability. Hardness is resistance to penetration and is normally measured by an indenter of fixed

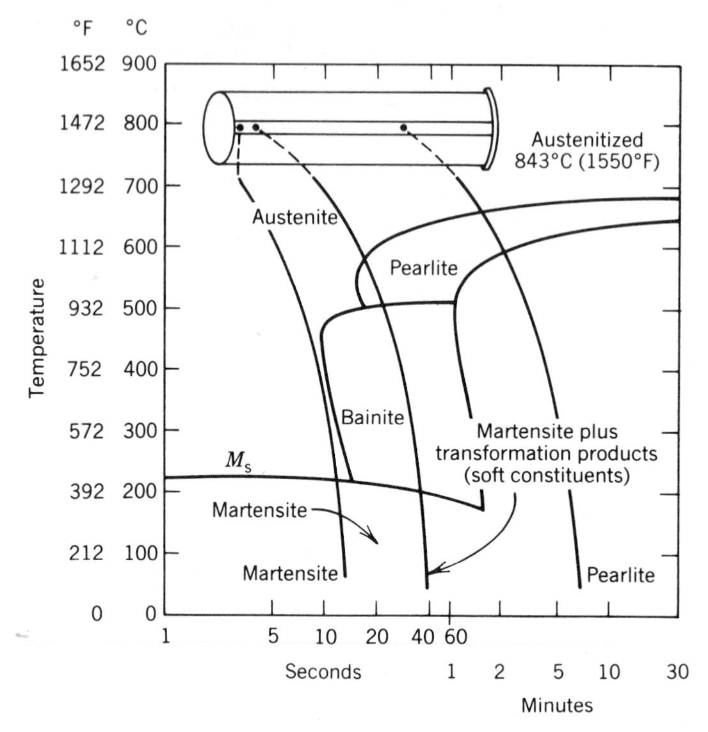

FIGURE 16.14. Continuous cooling transformation diagram for AISI 52100 steel.

geometry applied under static load in a direction perpendicular to the material surface being tested. Hardenability pertains to the depth of hardness achievable in an alloy. The alloying elements in the steel, as witnessed by the movement of the isothermal transformation curves to the right on the TTT diagram, permit additional cooling time from the austenitizing temperature to the point of martensite transformation. This positive effect of alloy additions to steel readily explains the need for the numerous modifications of the basic AISI 52100 for varying section thicknesses of the bearing components.

Hardenability is also influenced by the effect of grain size and the degree of hot-working. The hardenability of a coarse-grained steel is much greater than that of a fine-grained steel. Hot-working of material into progressively smaller bar sizes correspondingly reduces the hardenability spread found in ingots and blooms by reducing the segregation of carbon and other alloying elements normally experienced during ingot solidification. Higher austenitizing temperatures and longer soaking times at temperatures that promote grain coarsening also enhance hardenability by permitting more carbon to go into solid solution.

Because hardenability is a measure of the depth of hardness achieved under perfect heat treatment parameters, it is possible to quench bar products varying in diameter and to measure the resulting cross-sectional hardness patterns to determine hardenability. Grossman [16.16] defined the ideal critical size of a bar processed in this manner to be one in which the core hardens in an "ideal quench" to 50% martensite and fully to 100% martensite at the surface. Ideal quench is one in which the surface of the heated test specimen instantaneously reaches the quench medium temperature. Quenching identical specimens into media of less severity reduces the extent of hardening. Under these conditions Grossman defined the smaller bar diameter that hardened to 50% martensite in the core, as the "actual critical diameter." This variance led to the development of severity quench curves (H-value) relating to both the ideal critical diameter and the actual critical diameter.

The Jominy end-quench hardenability test is standardized [16.17] with respect to specimen geometry, apparatus, water temperature, and flow rate such that all results can be rated on a comparison basis. The hardness ratings at 1.6-mm (0.0625-in.) intervals when plotted against specimen length provide a curve indicative of the hardenability of an alloy. End-quench hardenability data normally incorporate both the maximum and minimum hardenability limits anticipated under specific heat treatment parameters [16.18].

Hardening Methods

Heat treatment practices used for bearing components are either through-hardening or surface-hardening. Heat treatments for the through-hardening martensitic grades are substantially comparable to one or another in that they necessitate heating (to an austenitizing temperature), quenching, washing, and tempering. Time-temperature parameters, primarily based on weight and cross-sectional thickness of the part being processed, have been established for the various through-hardening bearing alloys.

Ring components, particularly of large diameter and thin section thickness, require elaborate means of handling to minimize physical damage. In furnace construction, precautions are taken to avoid mass loading and excessive weight, which could adversely influence the geometry of the parts during heating or quenching.

Furnace manufacturers generally use natural gas or electricity as their heat source for equipment. Arrangements for protective atmospheres are normally provided to minimize carburization or decarburization of the high-carbon-chromium steel parts during processing. Furnaces of comparable construction and processing capability are also selected for the heat treatment of carburizing grades of steel except that the atmosphere is controlled to provide the carbon potential necessary for carbon to diffuse into the steel.

Precise, uniform furnace temperatures are maintained and controlled, providing exact reproducibility of processing cycles. Adequate quenching facilities are provided for salt, oils, water, or synthetic-type quenchants. Temperature control, agitation, and fixtures are used independently or in combination to reduce distortion in the heat-treated components.

Induction heating, often using synthetic-type quenchants, can be used for automated heat treatment of special bearing components. This can be a selective-type heat treatment in which only the rolling contact surface is hardened.

After hardening, the parts are washed to remove all quenchant residue before tempering. Tempering furnaces are generally electric or gas fired. Parts may be either batch loaded or automatically transported through these units.

Many types of furnaces are being used to process bearing components: for example, roller hearths, rotary drum, rotary hearth, shaker hearth, batch/pit type, conveyor belt/cast link, and pusher tray. In addition, automated salt lines using programmable hoists are in operation for steels requiring austenitizing temperatures of 802–1302°C (1475–2375°F).

Through-Hardening, High-Carbon-Chromium Bearing Steels

General Treatment. The high hardness and high strength required for through-hardening bearing steels are achieved by first austenitizing at a temperature sufficiently high to provide carbon solution and then cooling sufficiently fast into the bainite or martensite temperature ranges to avoid the formation of undesirable soft constituents. Heat treatment of these steels generally involves temperatures of approximately 802–871°C (1475–1600°F), uniform soaking, and quenching into a medium of salt, water, or synthetic oil controlled between approximately 27 and 230°C (80–445°F). The resulting "as-quenched" hardness range for martensite-hardened components is normally Rockwell C (R_C) 63–67; for bainite-hardened components, 57–62. Although bainite-hardened components do not require subsequent thermal treatment, martensite-hardened components are tempered.

Martensite. The martensite start (MS) temperature is lowered as the austenitizing temperature and the time at temperature are increased, permitting more carbon to go into solid solution. Correspondingly, the tendency exists for more austenite to be retained during the martensite transformation. The morphology of the resultant martensite also depends on the dissolved carbon content: high amounts of dissolved carbon are associated with plate martensite formation, and low amounts promote a tendency to form lath martensite.

High austenitizing temperatures also have the tendency to coarsen the material grain size. This condition is evident by both visual and low-power magnification of fracture surfaces. Properly heat-treated, high-carbon-chromium grades of steel show a fine, "silkylike" appearance on fracture faces.

After quenching, components are washed and tempered to relieve stresses and improve toughness. Tempering at temperatures at or slightly above the MS point will also transform retained austenite to bainite. The penalty for tempering at higher temperatures is loss of hardness, which can adversely affect load-carrying capacity and endurance of the bearing component. Components of lower hardness are also more prone to handling and functional surface damage than their harder counterparts are.

Marquenching. Quenching into a low-temperature medium [49–82°C (120–180°F)] can produce thermal shock and nonuniform phase transformation stresses. Components with nonuniform cross sections and/or sharp corners can warp or fracture. Transformation stresses may be reduced by quenching the part into a hot oil or hot salt medium controlled at a temperature between 177 and 218°C (350–425°F), in the uppermost portion of the martensite transformation range. Temperature equalization throughout the cross section of the component permits uniform phase transformation to progress during subsequent air cooling to room temperature. Although the as-quenched hardness is normally R_C 63–65, tempering cycles for martempered parts are similar to those used in straight martensite hardening operations.

Bainite. Bainite hardening is an "austempering"-type heat treatment in which the component is quenched from the austenitizing temperature to a temperature slightly above the MS temperature, which is the lower bainite transformation zone. Molten salt baths between 220 and 230°C (425–450°F) are normally used for this type of heat treatment. Water can be added to the quench bath to achieve the critical quench rate, thus avoiding the formation of undesirable soft constituents. Bainite-hardening grades of steel are again selected on the basis of component cross-sectional area. The higher the hardenability is, the greater the permissible cross-sectional area or thickness of a given component. As alloy content increases, the "nose" and "knee" of the transformation curve are pushed further to the right, which lengthens the time for bainite transformation to begin.

These alloys normally require 4 hours or more for complete transformation to bainite. Hardness values of 57 to 63 R_C are achieved on components processed in this manner. Subsequent tempering is not required. Quenching into molten salts and holding at these temperatures significantly reduce stresses induced due to thermal shock and phase transformation.

Bainite hardening produces components with small compressive surface stresses, in contrast to martensite hardening, which produces small tensile stresses in the as-quenched surface layers. A bainite microstructure is coarser, with a more "feathery" needle than that produced in straight martensite hardening.

Surface Hardening

Methods. Surface hardening is done by altering the chemical composition of the base material—for example, by carburizing or carbonitriding—or by selectively heat treating the surface layer of a given high-carbon bearing steel component. Induction and flame-hardening practices are used to fabricate production bearings. Laser beam and electron beam processes are also possible, depending on the hardness depth required.

Surface hardening of bearing steels produces well-defined depths of high surface hardness and wear characteristics. High residual compressive stresses, present in the surface layer, enhance rolling and bending fatigue resistance. The surface layer is supported by a softer and tougher core which tends to retard crack propagation.

Carburizing. The carburizing source or medium (gas, liquid, or solid) supplies carbon for absorption and diffusion into the steel. The same precautions followed for through-hardening furnace operations are followed in carburizing to minimize handling damage, reduce part distortion, and provide process economy. The normal carburizing temperature range is 899–982°C (1650–1800°F) with the carbon diffusion rate increasing with temperature. Therefore, it is easier to control narrow case depth ranges at the lower carburizing temperatures.

Based on the alloy steel being processed, time, temperature, and atmospheric composition determine the resulting carbon gradient. The resulting carbon content affects the hardness, amount of retained austenite, and microstructure of the carburized case. The hardness profile and compressive stress field depend on the carbon profile.

Although the practice of quenching directly from the carburizing furnace is used to heat treat bearing components, it is general practice to reharden carburized components to develop both case and core properties and, at the same time, to employ fixture-quenching devices to reduce part distortion.

Based on the grade of steel being carburized, the carbon potential of the furnace atmosphere must be adjusted so that large carbides and/or a carbide network are not formed. Those alloying elements such as chromium which lower the eutectoid carbon content are most likely to form globular carbides. Carbon can be further precipitated to the grain boundaries if the steel is then slow cooled before quenching. These grain boundary carbides and/or the carbide network can reduce mechanical properties.

Choosing the bearing material not only involves considering proper surface hardness and microstructure, but it also must incorporate core properties to prevent case crushing. Resistance to case crushing is generally provided by increasing subsurface strength. Therefore, a material with a section thickness and hardenability that will provide a core hardness R_C 30 to 45 is selected. Carburizing grades should be fine grained to minimize sensitivity to grain growth at high carburizing temperatures.

Direct quenching from the carburizing furnace has the advantage that one can obtain a case microstructure free of soft constituents, such as bainite, while using a leaner alloy steel. This heat treatment practice offers less part distortion than is experienced in reheating and quenching, particularly if the temperature is lowered to 816–843°C (1500–1550°F) before quenching. Adversely, this practice can produce parts with too much retained austenite and possible microcracking. The excess austenite in the case could permit plastic deformation of components under heavy load; microcracking could provide initiation points for fatigue. Microcracking can be minimized by keeping the carbon content in the as-carburized component lower than eutectoid level. Reheating at the lower austenitizing temperature and quenching tend to reduce microcracking.

Gas carburizing is common to the roller bearing industry because gas flow rates and carbon potential of the atmosphere may be accurately controlled. Gases present in furnace atmospheres include carbon dioxide, carbon monoxide, water vapor, methane, nitrogen, and hydrogen.

Over a period of time at a predetermined temperature, the specified case depth is established. This effective case depth (ECD) is generally defined as the perpendicular distance from the surface to the farthest point where the hardness drops to R_C 50. Normal ECDs for bearing components range between 0.5 and 5 mm (0.020–0.200 in.) with a surface carbon content between 0.75 and 1.00%.

Carburized components are tempered after quenching to increase their toughness. Cold-treating might be introduced to transform retained austenite to martensite. Additional tempering is then required.

Carbonitriding. Carbonitriding is a modified gas-carburizing process. Because of the health hazards and ecological problems in disposing of cyanide salts, the preferred method is to use a gaseous atmosphere. At an elevated temperature an atmosphere is generated having a given carbon potential to which ammonia is added. Nitrogen and carbon are diffused into the steel forming the hard, wear-resistant case. Because these hard carbonitrided cases are generally shallow in nature, ranging from approximately 0.07 to 0.75 mm (0.003–0.03 in.), produced at furnace temperatures ranging from 788 to 843°C (1450–1550°F), the case-core interface is easily differentiated. These same beneficial shallow case characteristics can also be achieved in components requiring excessively heavy case depths. In this instance the parts are generally carburized to the heavy case depths and then reheated in a carbonitriding atmosphere.

Ammonia added to the carburizing atmosphere dissociates to form nascent nitrogen at the work surface. The combination of the carbon and nitrogen being absorbed into the surface layer of the steel lowers the critical cooling rate of the steel; that is, the hardenability of the steel is significantly increased by the nitrogen. This characteristic permits lower-cost materials, such as AISI 1010 and 1020, to be processed to the desired high hardness by oil quenching and thus minimizes distortion during heat treatment.

All parameters being constant, the carbonitrided component will evidence a more uniform case depth than that produced by carburizing. Because nitrogen lowers the transformation temperature, carbonitrided components will have more retained austenite than carburized components of the same carbon content. These high levels of austenite may be reduced by increasing the carbonitriding temperature, controlling the surface concentrations to approximately 0.70–0.85% carbon, keeping the ammonia content at a minimum during processing, and introducing a diffusion cycle before quenching.

The presence of nitrogen in the carbonitrided case also enhances resistance to tempering. Carbonitrided components are tempered in the 190–205°C (375–400°F) range to increase toughness and maintain a minimum hardness of R_C 58.

Induction Heating. Induction heating is a means for rapidly bringing the surface layer of a high-carbon–low-alloy bearing steel component into the austenitic temperature range, from which it can be quenched directly to martensite. Induction heating is accomplished by passing an alternating current through a work coil or inductor. A concentrated magnetic field is then induced within the coil. This magnetic field will in turn induce an electrical potential in a part placed within the coil. Since a part represents a closed circuit, the induced potential establishes an electrical current within the part. Heating of the part is then the result of the material's resistance to the flow of induced current.

Power generating equipment is selected according to frequency requirements. Motor generators have historically been used to provide medium-frequency ranges from 1–10 kHz and to provide deep, hardened surface layers. These units are currently being replaced by solid-state inverters using silicon-controlled-rectifier (SCR) switching devices. Radio-frequency generators provide frequencies ranging from 100 to 500 kHz for very shallow case depth requirements.

The chief factors influencing the success of the induction-heating operation are frequency selection, power density, heating time, and coupling distance:

Frequency selection—the size of the part and the depth of heating desired dictate the frequency requirement.

Power density—the watts available per square mm of inductor surface influence the depth to which a part can be surface hardened.

Heating time—the heating time required to bring the part to temperature is a critical factor with respect to overheating and resulting case depth.

Coupling distance—the coupling distance is defined as the distance between the coil and the part surface.

Quenching of induction-hardened components is generally accomplished by either a spray or immersion method. Spray quenching involves a pressure deposition of the quenchant onto the component by a series of holes machined into the inductor or by a separate quench ring. The immersion method necessitates dropping the part out of the inductor into an agitated quench bath. The required physical and metallurgical properties in high-carbon-chromium bearing steel can be achieved by using a synthetic quench in lieu of water and/or oil. Concentrations may be adjusted to provide maximum quenchability while minimizing the tendency for cracking.

All surface-hardened components require tempering after a quenching. Although the case depth may be similar to those achieved by carburizing, a steeper hardness gradient exists in the case-core transition zone.

Properly induction-hardened AISI 52100 steel bearing components will generally achieve hardness values of R_C 65–67 as quenched. If the part before heat treatment is in the annealed condition, the microstructure of the as-hardened surface zone will consist of fine spheroidal carbide and a matrix of untempered martensite. When it is examined for fractures, a fine grain size can be seen.

Flame-Hardening. Flame-hardening is used primarily in the heat treatment of very large rings of high-carbon–low-alloy steel components more than 1 m (approximately 3 ft) in diameter. A combustible gas is mixed with oxygen to fire a cluster of burners directed to selectively heat the ring component as it rotates at a fixed rate through the impinging flame. The depth of the heat-affected zone is a function of the dwell time of the part at the heat source. The rotating part, upon reaching the proper austenitic temperature, is water quenched. The core material, being in the unaffected heat zone, remains in the annealed condition. Subsequent tempering is mandatory to relieve stresses and increase the ductility of the as-quenched component.

Flame-hardening is not a capital intensive process from an equipment standpoint. It is very versatile for selective hardening and rapidly adaptable for changing ring sizes with varying cross-sectional configurations and thicknesses.

The progressive zone heating method means an overlap will occur after 360° are completed. The resulting overtempering effect in the heat sink zone will result in a spot of lower hardness. Precautions must be taken to minimize the thermal and transformation stresses at this overlap point to prevent cracking.

Thermal Treatment for Structural Stability

Knowledge of size and shape changes of rolling bearing components occurring in heat treatment is critical to subsequent manufacturing operations and to the components' functional suitability. Basic high-carbon–low-alloy steel with a bcc structure expands rapidly (Fig. 16.15) as it is heated to approximately 727°C (1340°F) due to the coefficient of thermal expansion. At this critical temperature the material, as previously stated, undergoes a phase transformation to an fcc structure (i.e., austenite), resulting in component shrinkage. The specific volume of austenite is less than that of ferrite. If the material is heated to still higher temperatures within the austenite range, it continues to increase in volume due to the coefficient of thermal expansion. Conversely, when rapidly cooled, the material shrinks to the martensite transformation temperature. The martensite, formed as the part continues to contract while cooling to room temperature, is a bct structure. The resulting increase in volume, due to this transformation occurring at such low temperatures, stresses the material. Because it is virtually impossible in production heat treatment to achieve complete transformation from the austenitic (fcc) structure to the untempered martensite (bct), varying amounts of austenite, depending on the severity of the quench, are retained in the as-quenched microstructure. Components must be thermally treated to reduce the residual stresses and to provide required structural stability.

Dimensional changes occurring in bearing steels essentially depend on precipitation of fine carbide from martensite and decomposition or transformation of retained austenite. Because changes can also be induced during bearing operation, due to the temperature or stress environment, the manufacturer must select the appropriate heat treatment to provide required stability. Tempering of high-carbon-chromium steels generally occurs in the range 66–260°C (150–500°F). At these temperatures fine carbide is precipitated, and the tempered martensite remains essentially bct with some shrinkage. Tempering in the range 205–288°C (400–550°F) results in a time-temperature-dependent decomposition of retained austenite to bainite and volume increase. Loss of hardness at high temperatures is prevented by tempering below 260°C (500°F).

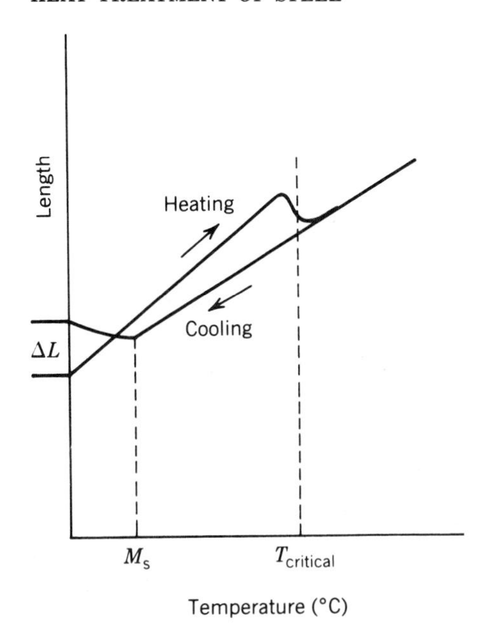

FIGURE 16.15. Volume changes during heating and quenching (hardening) of high-carbon bearing steel.

The annealed microstructures of high-speed steels, providing maximum machinability, contain numerous hard metallic carbides, such as tungsten, molybdenum, vanadium, or chromium, imbedded in a soft ferritic matrix. Unlike the high-carbon-chromium steels, temperatures far above the critical temperature must be attained to dissolve the desired amount of these hard carbide particles. Carbide precipitation is avoided by rapidly cooling the steel from the austenitizing temperature into the martensite transformation temperature range. After further cooling to room temperature, the structure normally contains 20–30% retained austenite. Heating to temperatures required for tempering high-carbon-chromium steel produces only slight tempering of the martensite. Between 427 and 593°C (800–1100°F), "secondary hardening" occurs; that is, the austenite is conditioned and subsequently transforms to martensite upon cooling back through the MS temperature transformation range. Multiple tempering at these high temperatures is required to complete the transformation of austenite to martensite and to precipitate very fine alloy carbides, which are responsible for the secondary hardening phenomenon and provide for the high-temperature hardness retention characteristic of high-speed steels.

Subzero treatments are often used after the initial quench or intermittently between tempering cycles to complete austenite-to-martensite transformation upon cooling. However, because cold-treatment sets up high internal stresses in the as-quenched components, it is generally recommended that cold-treating be practiced only after the first tempering cycle.

Corrosion-resistant steels, for example, AISI 440C and BG42 (AMS 5749), are generally heat-treated incorporating deep freezing immediately after rapid cooling from the austenitizing temperature. AISI 440C may be subsequently multiple tempered at approximately 149°C (300°F) or 316°C (600°F), depending on the product hardness requirements. Because of its alloy composition BG42 is heat-treated according to standard practices for high-speed steels—that is, multiple tempering at 524°C (975°F) incorporating refrigeration cycles.

Retained austenite, present in the case microstructure of case-carburized steels, is a relatively soft constituent providing some tolerance for stress concentrations arising from inclusions, handling damage, and surface roughness. Case properties are preserved by generally tempering bearing components from 135 to 196°C (285–385°F). The core is stable at normal bearing operating temperatures.

Mechanical Properties Affected by Heat Treatment

Elasticity. The elastic properties of rolling bearing steel are not significantly affected by heat treatment. Hence the modulus of elasticity at normal temperatures is 202 kN/mm^2 (29.3 $\times$ 10^6 psi) for both through-hardened and case-hardened steels.

The limit of elastic behavior—that is, the stress under maximum uniaxial loading giving insignificant plastic deformation or permanent set—is described for rolling bearing steels by a 0.2% offset yield strength—that is, 0.2% remaining plastic strain. Figure 16.16 illustrates that strength properties tend to decline as transformation temperature increases for a given rolling bearing steel composition.

Ultimate Strength. Ultimate strength is the stress at which the sample breaks in the uniaxial test described before; it is significantly affected by heat treatment. For through-hardened AISI 52100 ultimate strength for martensitic steel generally lies between 2900 and 3500 N/mm^2 (420–510 ksi). For the best case-hardening bearing steels, for example, AISI 8620, ultimate strength approximates 2600 N/mm^2 (380 ksi).

Fatigue Strength. Fatigue strength is determined in a cyclic push-pull or reversed bending test as the maximum stress that can be endured with no failure before accumulation of 10 million cycles. These data depend strongly on heat treatment, surface finish and treatment, test conditions, and so on. Accordingly, it is difficult to generalize and no numerical values are given herein. It is best to test the individual steel.

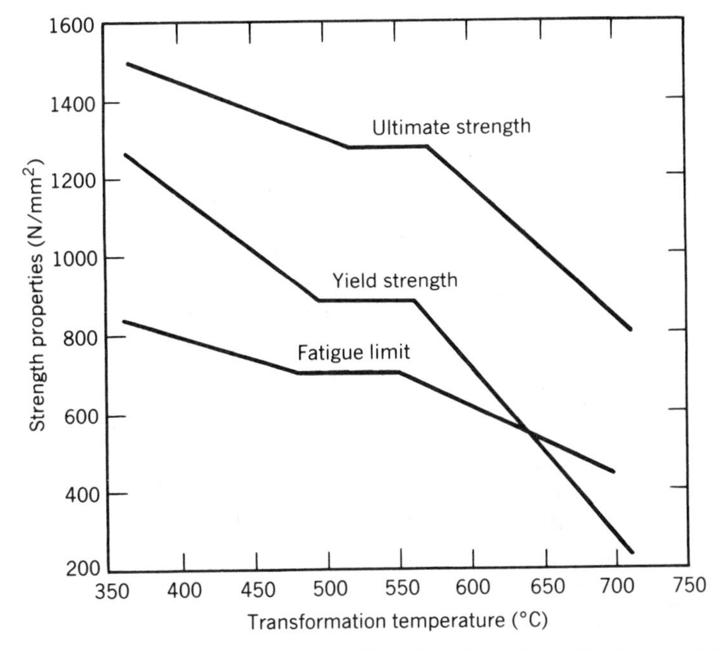

FIGURE 16.16. Properties of a 0.8% C steel vs. transformation temperature.

Toughness. Two test methods are used to determine toughness of bearing steels: the fracture toughness test and the impact test. In the first, a plain stress value K_{IC} is measured in N/mm^2-m$^{1/2}$; this is the stress related to the defect size that can be tolerated without incipient structural failure. For martensitic AISI 52100, K_{IC} falls between 15 and 22, depending on heat treatment. A slight increase in K_{IC} occurs as temperature increases. Case-hardening steel tends to have greater fracture toughness than through-hardening steel. A K_{IC} value of 60 is not uncommon for surface-hardened steel.

The second test—the impact of a hammer blow of defined energy on a sample—measures the energy absorbed in breaking the sample. For martensite-hardened AISI 52100 this is only 4.5 J (3.3 ft-lb) compared to 172 J (127 ft-lb) for the soft annealed material.

Hardness. The manner in which carbon is distributed in steel dictates the resulting hardness and mechanical properties. Although carbon makes by far the greatest contribution to hardness, increasing the alloy content also increases hardness.

Hardness, a material's resistance to penetration and, hence, wear, can be measured by static or dynamic methods. Static testing involves applying a load through a penetrator of defined geometry. Depending on the type of hardness tester employed, either the depth of penetration or the size of the indentation becomes the measurement of the material hardness. See Fig. 16.17.

Dynamic testing involves bouncing a diamond-tipped hammer from a specified height onto the surface of the test specimen. The resultant rebound height is a measure of material hardness. The scleroscope is the only piece of equipment based on the dynamic test principle.

Hardness attainable is a function of carbon content, as shown by Fig. 16.18. In general, as hardness is increased, toughness decreases for a given alloy steel.

Residual Stress. Stresses induced in a component through fabrication or thermal treatments are totally eliminated upon uniform heating and soaking in the austenite temperature range. Quenching of the component can once again generate tremendous internal stresses in the part. Through-hardening of the martensitic high-carbon steels may produce surface tensile stresses that can produce part distortion or even cracking. Surface-hardening heat treatments, including carburizing, carbonitriding, induction or flame-hardening, generally produce parts showing surface compressive stresses. Regardless of the heating method selected for austenitizing, subsequent thermal cycles with or without subzero treatment can appreciably alter the established stress state in the as-quenched parts.

The stresses induced in a through-hardened component during quenching are principally the result of temperature variances and nonuniform phase transformations. Bearing rings, essentially being thin hoops of varying cross-sectional thickness, are prone to both size and shape changes. Fixture quenching, employed to retain the components' as-machined dimensional characteristics, may hamper quench medium flow and induce additional nonuniform stress distribution in the part because of mechanical restraints that do not adapt to size or shape changes. The machined undercuts, grooves, filling slots, oil holes, and flanges having sharp corners and recesses provide additional focal points for stress concentration.

The high-carbon-chromium bearing steels under recommended austenitizing temperatures have an MS temperature range of approximately 204–232°C (400–450°F). Increasing the carbon content and various alloying elements in this family of steels will tend to depress the MS temperature. Coarse-grained materials will also have a lower MS point than fine-grained materials of the same chemistry. Therefore, austenitizing at very high temperatures will reduce the MS point into the range where the material is less able to adjust plastically. The higher

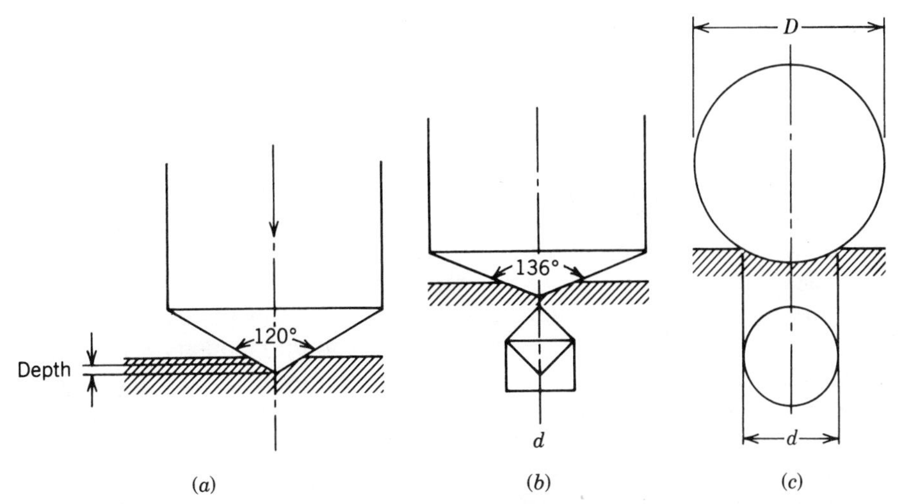

FIGURE 16.17. Hardness tests. (*a*) Rockwell R_C; indentation body: diamond cone; load 150 kgf (including 10-kgf preload), indentation depth for R_C 63: 74 μm, hardness testing range: R_C 20–67. (*b*) Vickers; indentation body: diamond pyramid 136°, indentation depth: $\frac{1}{7}d$—for V 782 $\approx$ R_C: 22 μm, hardness testing range: up to V 2000. (*c*) Brinell; Indentation body: hardened steel ball (*D*), hard metal ball (*D*), Indentation depth: $\sim\frac{1}{5}d$, Hardness testing range: up to 400 B ($\sim$42.5 R_C), up to 600 B ($\sim$57 R_C)

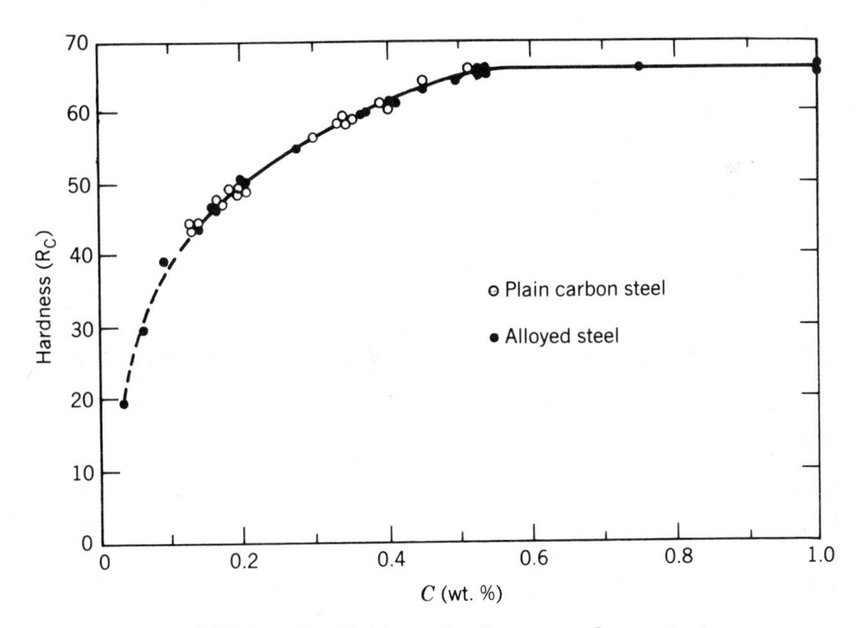

FIGURE 16.18. Maximum hardness vs carbon content.

austenitizing temperature will also result in higher thermal gradients occurring in the parts during quenching. Subzero treatments that permit the austenite-to-martensite transformation to be further completed could cause high stress levels, which can crack the parts. Bainitic heat treatment or martempering will appreciably reduce transformation stresses during quenching.

It is standard practice in the rolling bearing industry to cool martensite-hardened parts to room temperature from the quenching bath, wash the parts, and subject them to a tempering cycle. A low tempering temperature for a long time is equivalent to a high temperature for a short time from the standpoint of reducing the residual component stresses. This sequence of operation may be interrupted by a subzero treatment following the washing operation to permit the completion of the austenite-to-martensite transformation. Parts processed in this manner are very prone to cracking because of the resulting high residual stress state. These parts must be tempered as soon as they warm to room temperature.

Surface-hardening heat treatments, by a diffusion process altering the composition of the material or by the rapid heating of a selected surface area of a homogeneous steel, are developed and controlled to provide surface compressive stresses with normal counterbalancing tensile stresses in the core. Induction surface hardening of an appropriate material to the proper case depth results in the maximum compressive stress being located at the case-core transition zone. The magnitude of this compressive stress in a surface-hardened high-carbon steel alloy will normally be less than that produced in a carburized part at its point of maximum compressive stress, which is at the approximate midpoint of the total case depth. This point corresponds to the carbon content of approximately 0.50%.

Tempering of as-quenched, surface-hardened components with or without support of subzero treatment will generally reduce the retained austenite level and modestly alter the level of compressive stresses.

ROLLING CONTACT FATIGUE: MODES AND CAUSES

Failure Modes

Contact fatigue failure in rolling bearings occurs when local material stresses exceed the local fatigue limit; cracks are initiated and then propagated. Even if the stresses induced by cyclic loading between rolling elements and raceways are generally below the fatigue limit, additional stresses can be caused locally by material inhomogeneities and defects acting as stress concentrators. These stresses are superimposed on those arising due to normal bearing operation. Inhomogeneities occurring from the steelmaking process are distributed throughout the entire material: for example, slag inclusions and pores. They can also be generated by the manufacturing process, where they are mainly limited to surface zones: for example, ring marks, scratches, and grinding burns. The different types of material inhomogeneities and defects are the reason that bearings fatigue in two observable modes. In one mode, failure is caused by a stress-raising inhomogeneity in the subsurface region where the normal stresses induced by cyclic loading are maximum. In the other mode, failure is initiated by an inhomogeneity or defect in the surface, thus increasing the surface stresses. Tallian et al. [16.19] provide a compendium of results concerning rolling bearing failure modes and causes.

Subsurface Failures

Inhomogeneities. Material inhomogeneities that can lead to subsurface-initiated failures are macro- and microinclusions, pores, and bandings. The stresses surrounding these inhomogeneities depend on the nature, size, distribution, shape, and interface between the inhomogeneity and the base material.

Macroinclusions. Macroinclusions are impurities originating from the refractory material because of their large size [above 0.5 mm (0.02 in.)], high brittleness, very low cold deformability, and irregular shape. They always cause early failure when situated in the surface or subsurface zones.

Microinclusions. All oxide-type slag microinclusions are brittle and practically impossible to deform. They occur either in stringers or in globular shape. Above 40 μm (0.0016 in.) diameter for the globular shape their negative influence on life is great; see Fig. 23.13. High tensile stresses can occur in the matrix surrounding the inclusions, owing to the difference in thermal contraction between matrix and particle during quenching. They are called tessellated stresses. A relationship exists between the size and size distribution of oxides, on the one hand, and the total oxygen content of the steel, on the other hand; the higher the oxygen content, the bigger the maximum size and the total number of oxides.

Sulfide Inclusions. Sulfide inclusions are relatively soft and easily cold-deformed. Therefore, high stress peaks surrounding sulfide inclusions can be partially reduced by plastic deformation. In some cases sulfides surround the oxides and, by plastic deformation, close the voids so that the oxides become less dangerous. The allowable size and distribution of sulfides are greater than those of oxides.

Pores. Pores are generated by gas bubbles enclosed during steel solidification and often occur as clusters. If their surfaces are not oxidized, they are closed during warm- and cold-processing of the steel. In the oxidized condition of the surface they have a negative influence on fatigue life and, regarding size and shape, can be treated similarly to globular-type slag inclusions.

Banding. High local differences in steel chemical composition (segregations) caused by improper solidification conditions can lead to different component microstructures after hardening. The high stresses between microstructural zones, originating from the different MS points, and the reduced fatigue limits, can lead to early fatigue failures.

Surface-Initiated Failures

In addition to the foregoing defects, other material defects in the surfaces, such as high-tempered or rehardened grinding burns, decarburized layers, and marks can lead to surface-initiated failure.

Grinding Burns. Grinding burns occur under improper grinding conditions of hardened components. The material is then locally tempered, yielding insufficient hardness, or it is rehardened. The heat-affected zone varies between a few microns and a few hundred microns. In the latter case tensile stresses up to 1200 N/mm^2 (175 ksi) have been measured.

Decarburization. Decarburization is the result of heat treatment in an oxygen rich atmosphere. The carbon content in the surface is thereby depleted. The different MS points in the base material and the surface layer give rise to residual tensile stresses.

Marks and Indentations. Marks and indentations occur because of incorrect handling of components during manufacture or mounting. During bearing operation the normally induced stresses can add to the residual tensile stress caused around oxides, decarburized zones, bandings, or areas of grinding burns. The resulting stresses can get so high that fatigue can be initiated. The voids around marks, indentations, and similar defects are discontinuities in the surface material from which fatigue can also begin.

MATERIALS FOR SPECIAL BEARINGS

For most rolling bearing applications, the through-hardening steel AISI 52100 and case-hardening steels described in Table 16.2 are sufficient to provide good performance characteristics such as fatigue and wear resistance, appropriate fracture toughness, and consistently reliable mechanical properties. The advent of the aircraft piston engine, however, created the demand for long-lived endurance at higher operating temperatures. This demand was met by using the tool steels M1, M2, M10, and T1. These steels lost their prominence in the 1950s with the introduction of vacuum-melted M50 to meet the needs of aircraft gas turbine engine bearings. In many applications, particularly instrument ball bearings, corrosion resistance became important and this requirement was met by using AISI 440C and BG42 stainless steels generally at the sacrifice of fatigue endurance when compared to bearings fabricated from AISI 52100. Because of light applied loading, however, fatigue endurance is not a major consideration in such applications. The chemical compositions of some of the foregoing steels are given in Table 16.3.

The exploration of space and the continuing development of the aircraft gas turbine engine provided the demand for yet increased development of *exotic* materials. Examples are sapphire for balls, precipitation-hardening stainless steels, and nickel-based superalloys. Additionally, the nuclear power industry created the need for cobalt alloys such as L-605, Stellite-3, and Stellite-6. Powder metal-forming techniques have now provided the means to create *steels* of differential properties; for example, extremely hard, corrosion-resistant surfaces combined with tough, high strength substrates.

The requirement for aircraft gas turbine engine mainshaft ball and roller bearings to operate at ever-increasing speeds initiated the search for a relatively high temperature capability, fracture-tough steel. Because of the bearing ring hoop stresses caused by ring centrifugal stresses and rolling element centrifugal forces at high speeds, fatigue spalls under such conditions can lead to fracture of rings fabricated from through-hardening steels such as M50. Thus, the operating speeds of aircraft gas turbine engines were limited to approximately 2.4 million dN (bearing bore in mm $\times$ shaft speed in rpm). With the development of M50-Nil, a case-hardening derivative of M50 whose chemical composition is shown in Table 16.3, this limitation has been overcome. Figure 16.19 from Spitzer [16.24] shows the effect of the higher fracture toughness of M50-Nil on speed capability assuming various levels of induced surface defects.

The need for bearings to operate at ultrahigh temperatures has trig-

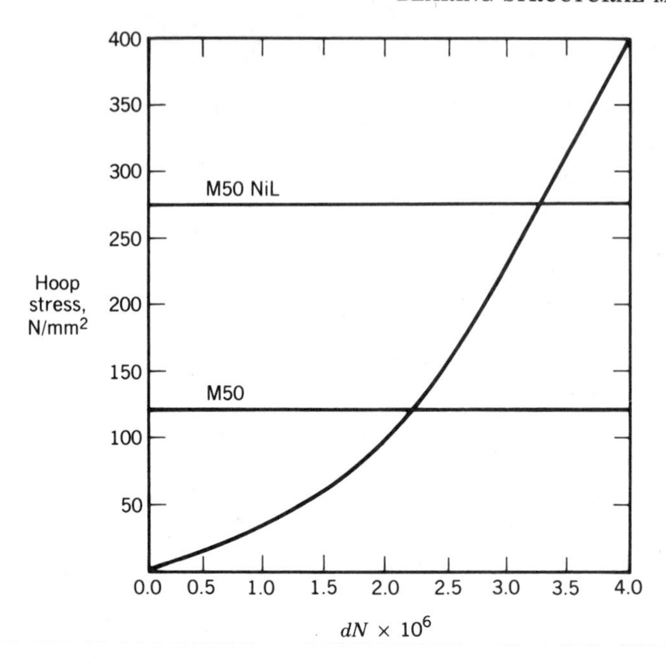

FIGURE 16.19. Comparison of M50 vs M50-Nil steel-ring hoop stress vs bearing *dN* (from [16.24]).

FIGURE 16.20. Silicon nitride begins life as a powder before a series of processes transforms it into a highly engineered bearing component (SKF photograph).

gered the development of cemented carbides and ceramics as rolling bearing materials. Materials such as titanium carbide, tungsten carbide, silicon carbide, sialon, and particularly silicon nitride are being investigated. At elevated temperatures, these materials retain hardness, have corrosion resistance, and provide some unique properties, some of which are advantageous, such as low specific gravity for silicon nitride. Conversely, other properties of these materials, such as extremely high elastic modulus and low thermal coefficient of expansion for silicon nitride as compared to steel, create significant bearing design problems that must be overcome if these materials are to succeed for use in rolling bearing structural components, particularly rings. Ceramic materials such as silicon nitride, as illustrated by Fig. 16.20, commence life as

TABLE 16.4. Properties of Special Bearing Structural Materials

Material	Rockwell Hardness C (room temp)	Max Useful Temperature °C (°F)	Specific Gravity	Elastic Modulus N/mm^2 $\times 10^3$ (psi $\times 10^6$)	Poisson's Ratio	Coefficient of Thermal Expansion 10^{-6}/°C (10^{-6}/°F)
440C stainless steel	62	260(500)	7.8	200(29)	0.28	10.1(100°C) (5.61)
M50 tool steel	64	320(600)	7.6	190(28)	0.28	12.3(300°C) (6.83)
M2 tool steel	66	480(900)	7.6	190(28)	0.28	12.3(300°C) (6.83)
T5 tool steel	65	560(1050)	8.8	190(28)	0.28	11.3 (6.28)
T15 tool steel	67	590(1100)	8.2	190(28)	0.28	11.9 (6.61)
Titanium carbide cermet	67	800(1470)	6.3	390(57)	0.23	10.7 (5.94)
Tungsten carbide	78	815(1500)	14.0	533(77.3)	0.24	5.9 (3.28)
Silicon nitride	78	1200(2200)	3.2	310(48)	0.26	2.9 (1.61)
Silicon carbide	90	1200(2200)	3.2	410(59)	0.25	5.0 (2.78)
Sialon 201	78	1300(2372)	3.3	288(42)	0.23	3.0 (1.67)

powders that after a series of processes, principal among which is hot isostatic pressing, are transformed into highly engineered bearing components.

Table 16.4 excerpted from Pallini [16.25] gives significant mechanical properties and allowable operating temperatures for several of the materials described above. Considering the density and elastic properties of silicon nitride, the 218 angular-contact ball bearing of Example 8.1 was reexamined as a hybrid bearing; that is, having ceramic balls and steel rings and as a completely ceramic bearing. Figures 16.21–16.23 indicate some relative performance characteristics for the various bearing types (all steel, hybrid, and all silicon nitride components) considering an applied thrust load of 22,250 N (4000 lb). These illustrations assume the bearings have identical internal geometries; it would clearly be appropriate to optimize the bearing internal design to take full advantage of the material properties. Additionally, Fig. 16.24 indicates the frictional properties of silicon nitride when used with various types of lubricants, both liquid and *dry*. It is to be noted that the friction coefficient of this material is highly dependent upon the type of lubricant used, temperature of operation, and operating environment. It is further to be noted that while the compressive strength of silicon nitride is excellent, the tensile strength is only 30% that of M50 steel. The fracture toughness is also only a small percentage of that of M50, let alone M50-Nil. Furthermore, although in rolling contact under heavy load the material tends to fail by surface fatigue and even tends to have longer fatigue life

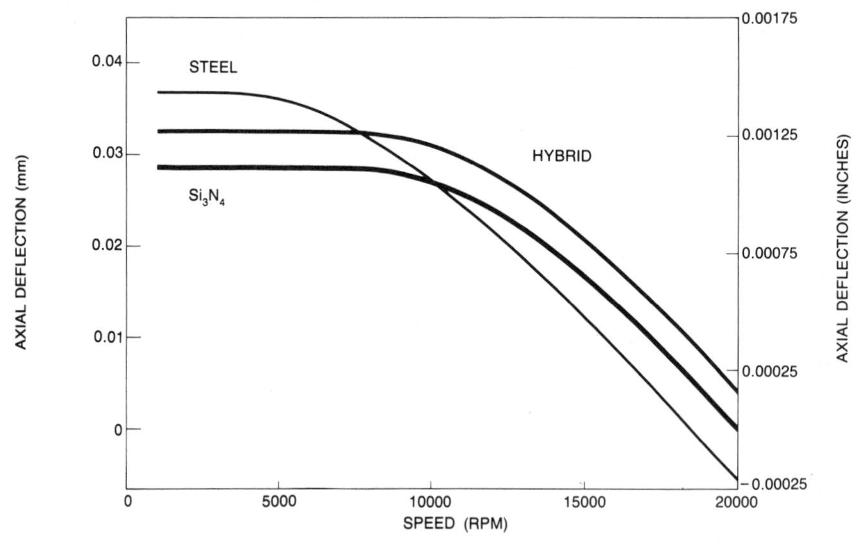

FIGURE 16.21. Axial deflection vs shaft speed for the 218 angular-contact ball bearing of Example 8.1 for the bearing having (1) steel rings and balls, (2) steel rings-silicon nitride balls, and (3) silicon nitride rings and balls supporting a 22,250 N (4000 lb) thrust load.

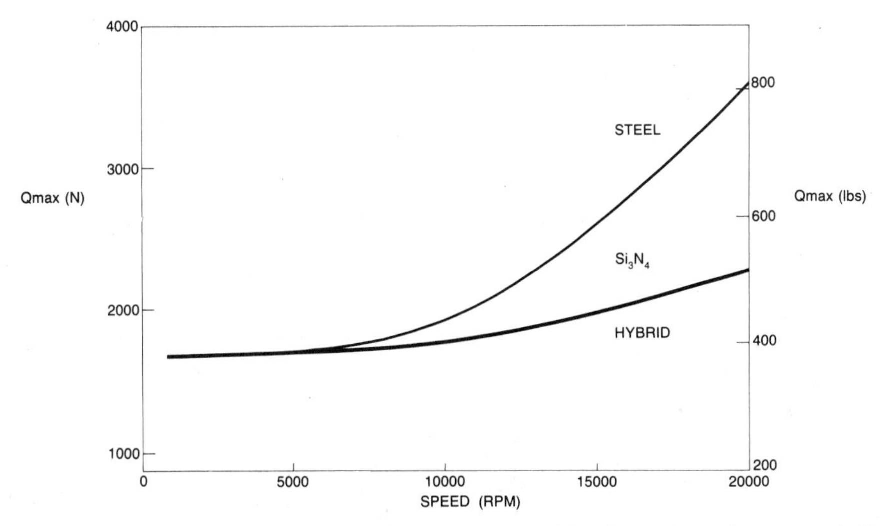

FIGURE 16.22. Outer raceway-ball load vs shaft speed for the 218 angular-contact ball bearing of Example 8.1 for the bearing having (1) steel rings and balls, (2) steel rings-silicon nitride balls, and (3) silicon nitride rings and balls supporting a 22,250 N (4000 lb) thrust load.

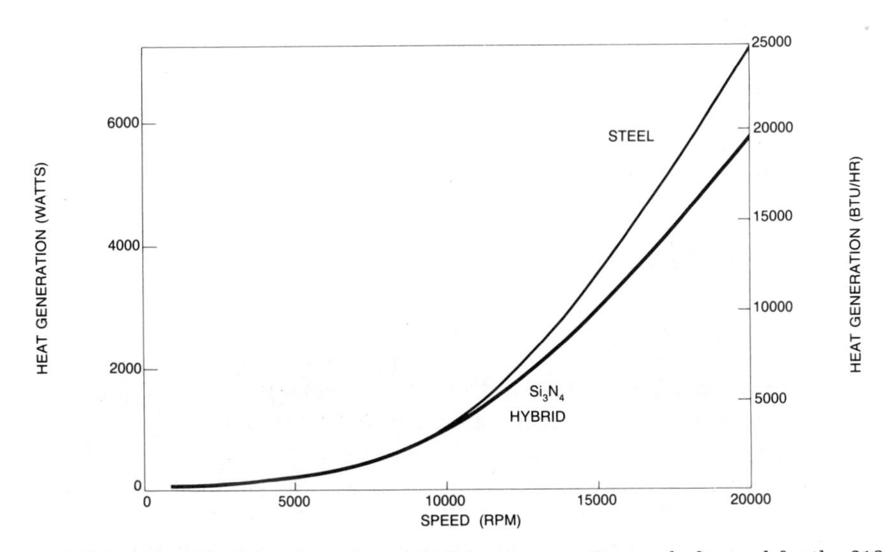

FIGURE 16.23. Total bearing raceway-ball heat generation vs shaft speed for the 218 angular-contact ball bearing of Example 8.1 for the bearing having (1) steel rings and balls, (2) steel rings and silicon nitride balls, and (3) silicon nitride rings and balls supporting a 22,250 N (4000 lb) thrust load.

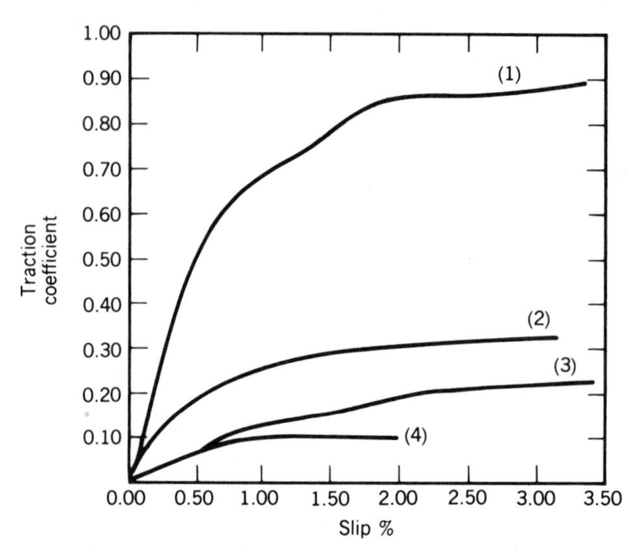

FIGURE 16.24. Traction coefficient vs percentage slip in the contact between two bodies of hot pressed silicon nitride. Contact stress is 2068 N/mm^2 (300,000 psi); nominal speed is 3800 mm/sec (150 in./sec) (from [16.25]).

Operating conditions

(1) 25°C (77°F)—dry contact
(2) 370°C (698°F)—graphite lubrication
(3) 538°C (1000°F)—graphite lubrication
(4) 25°C (77°F)—oil lubrication

than steel, any disruption of the surface can lead to rapid crumbling of the surface under continued operation. Means for failure detection is therefore an important consideration in life-critical applications.

CAGE MATERIALS

Material Types

The generally stated function of the rolling bearing cage is to maintain the rolling elements at properly spaced intervals for assembly purposes. It is sometimes inferred that, in normal bearing operation, the cage is not necessary; rather it "goes along for the ride"; that is, it is not a highly stressed component requiring the strength of the accompanying ring components. There are more exceptions to this statement than examples. For example, mainshaft and accessory aircraft gas turbine engine bearings require AISI 4340 steel (AMS 6414 or AMS 6415) cages supplied in the hardness range of R_C 28–35. These cages are also silver plated (AMS 2410 or AMS 2412) to provide corrosion resistance and added lubricity. In many bearing applications not only do the rolling elements contact the cage pockets, but the cages themselves are either inner ring or outer ring land "riding."

Although cages are manufactured from many types of material, including aluminum, S-Monel, graphite, nylon, and cast iron, the major bearing product lines use brass or steel. In ball bearings principally, but also in some roller bearings, plastics are starting to replace these metals. See Chapter 13.

Low-Carbon Steel

Plain low-carbon strip steel, suitable for cold-forming (0.1 to 0.23% C) is used in the bulk fabrication of pressed, two-piece, or finger-retention-type steel cages. Two-piece cages are joined by mechanical lock joints, rivets, or welds. The material has a tensile strength of 300–400 N/mm^2 (44–58 ksi). The AISI 4340 machined steel cages previously mentioned for aircraft applications have approximately 0.4% C for increased strength. Additionally, low-carbon steel tubes and forgings are used to make cages for bearing applications that need unique features for lubrication or greater material strength. Many steel cages are surface hardened or phosphate coated to provide improved wear characteristics.

Brass

Brass cages are generally manufactured from continuously cast rounds, centrifugally cast cylinders, sand castings, or sheet metal/plate. Because high tensile strength, 300–380 N/mm^2 (44–55 ksi), alpha brass has poor machinability but is readily capable of deep drawing (ductility increasing with increasing zinc content up to 38%), it is used to cold-form one-piece cages. When zinc is increased from 38 to 46%, a mixture of alpha and beta phases is formed. Ultimate strength is higher with the higher amounts of zinc. Adding phosphorus and/or aluminum provides alloys that can be centrifugally cast, readily machined, drilled, or broached for making ball and roller bearings. Other nonferrous brass alloys may be centrifugally cast, but they are hot-worked by upsetting or ring rolling to meet specific product requirements. Cage blanks may also be produced by extruding the centrifugally cast billet.

Bronze

Silicon-iron-bronze (Cu: 91.5%, Zn: 3.5%, Si: 3.25% Mn: 1% and Fe: 1.20%) is an alloy recommended for ball and roller bearing cages operating at temperatures up to 316°C (600°F). The as-cast billet material must be extensively hot-worked and extruded to promote optimum material properties.

Polymeric Cage Materials

Advantages and Disadvantages. The use of polymer, particularly nylon (polyamide) 6,6, as a cage material is widespread in many rolling bearing applications. Polymeric cages have the following advantages over metallic cages in both production and operation:

1. Processing of polymeric materials often allows one-step fabrication of complex designs, thus eliminating the machining operations necessary to produce a comparable metallic retainer and saving money.
2. Plastic cages tend to be free from the debris that accompanies the production of metallic cages. The increased cleanliness contributes to reduced bearing noise.
3. Polymers are more flexible than metals. This is advantageous in cage assembly and in bearing operation under some difficult loading conditions.
4. Favorable physical properties of polymeric materials lead to cage performance advantages in many applications; for example, low density (reduced cage weight), good chemical resistance, low friction and damping properties for low torque and quiet running.

The primary disadvantage of polymer usage is the deterioration of initial properties of the material due to temperature, lubricant, and environmental exposure. Polymer deterioration causes loss of strength and flexibility, which is important to the cage function during bearing operation. Bearing rotation causes centrifugal forces to act on the cage, which deform it radially. Misalignment of the inner and outer rings can cause large stresses on the cage during bearing operation. Hence, loss of cage strength can lead to failure. Therefore polymeric material candidates must be evaluated for the rate and degree of deterioration under conditions of extreme temperature, lubricant exposure, and other environmental factors.

Rolling Bearing Plastic Cages. Some examples of polymeric cage designs are shown in Fig. 16.25. Properties of cage polymeric materials are

- Low coefficient of thermal expansion
- Good physical property retention, especially strength and flexibility, throughout the temperature range of operation
- Compatibility with lubricants and environmental factors
- Development of a suitable cage design to minimize friction and provide proper lubrication

This list indicates essential differences between polymeric and metallic cage materials. Lubricant compatibility is rarely a factor, and loss of

FIGURE 16.25. Polymeric cage designs. (*a*) Snap cage for ball bearing (nylon 6,6). (*b*) Cage for cylindrical roller bearing (nylon 6,6). (*c*) Cage for high-angular-contact bearing (nylon 6,6). (*d*) Phenolic cage for precision ball bearing.

physical properties does not occur within bearing operation temperatures with metals. Cage design depends on the specific polymer used in a more intimate fashion than when steel or brass is used.

Polymeric Types for Cages. Fabric-reinforced phenolic resin cages have been used for many years in high-speed bearing applications where decreased cage mass is a benefit. The low density of the material, approximately 15% that of steel, results in a low cage mass. The centrifugal force on a phenolic cage is consequently only 15% of the force acting on a steel cage. At high speeds centrifugal force causes a cage to spread radially. The low-density cage therefore offers better dimensional stability at high speeds. Use of a phenolic resin, however, is limited to bearing applications that do not exceed temperatures of 100°C (212°F) continuous and 120°C (248°F) peak. Another disadvantage with the phenolic resin is the necessity of machining operations to obtain the final shape. Other resins, discussed in the following paragraphs, can be injection molded into a final shape directly, thus reducing process cost. Resins of this type, particularly nylon 6,6, have replaced phenolic in many rolling bearing applications.

The nylon 6,6 (polyamide 6,6) resin is the most widely used plastic for bearing cages. It provides a low material price, desirable physical properties, and low processing costs in one product. The material is constructed of aliphatic linkages connected by amide linkages to form a polymer of molecular weight between 25,000 and 40,000. Nylon 6,6 is synthesized from carbon hexamethylenediamine and adipic acid, both of which have six carbons, hence the 6,6 designation.

$$H_2N(CH_2)_6NH_2 + HOOC(CH_2)_4COOH \xrightarrow{\text{heat}} -[NH(CH_2)_6NHOC(CH_2)_4CO]_x- + H_2O$$

The material is semicrystalline and thermoplastic. It possesses many desirable properties for cage applications: strength, toughness, abrasion resistance, chemical resistance, and impact resistance. The resin is somewhat hygroscopic (to 3%), and absorbed water causes dimensional changes that must be considered during cage design.

Product modifications containing additives are abundant for nylon 6,6. The variations provide improved physical properties, environmental inertness, and improved processing characteristics. Being thermoplastic, it is an injection-moldable resin allowing direct production of complex cage shapes with obvious cost advantages. In general, resin compatibility with lubricants is very good. Cages formed from this resin exhibit a high degree of flexibility, which allows easy assembly and operation under misalignment of the inner and outer bearing rings. Glass fiber reinforcement is often used with the resin at levels of 25% fill. The glass fiber gives better retention of strength and toughness at high temperatures, but with loss of flexibility.

Rolling bearings selected from manufacturers' catalogs are designed to operate in wide varieties of applications. Therefore, the strength/toughness properties afforded to nylon 6,6 cages by glass-fiber reinforcement are required for bearing series employing such cage material. Figure 16.26 from [16.27] illustrates the endurance capability of 25% glass-fiber-filled nylon 6,6 as a function of operating temperature. In Fig. 16.26, the "black band" indicates the spread determined with various lubri-

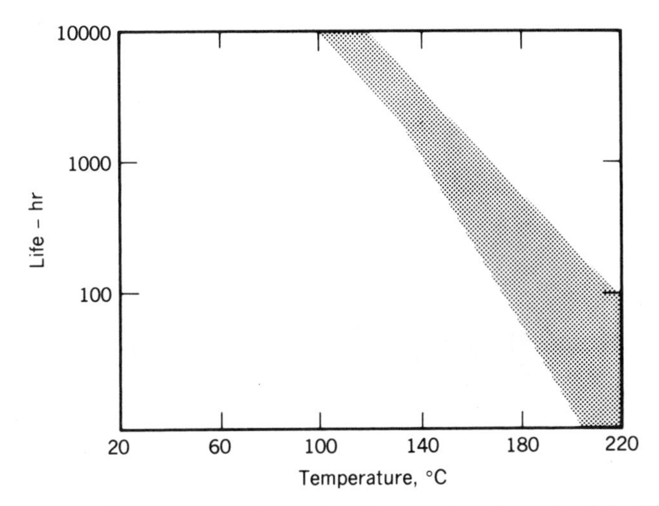

FIGURE 16.26. Life expectancy vs operating temperature for nylon 6,6 with 25% glass-fiber-fill (from [16.27]).

cants. The lower edge of the band is applicable for *aggressive* lubricants such as transmission oils (with EP additives), while the upper edge pertains to *mild* lubricants such as motor oils and normal greases. Table 16.5 from [16.27] indicates the strength, thermal, chemical, and structural properties of this material in the dry and conditioned states. The conditioned state is that in which some water has been absorbed. By comparison of Fig. 16.26 with Table 16.5, it can be seen that the *permissible operating temperature* of 120°C (250°F) corresponds to a probable endurance of approximately 5000 to 10,000 hr depending upon lubricant type. This refers to continuous operation at 120°C (250°F); operation at lesser temperatures will extend satisfactory cage performance for greater duration.

High Temperature Polymers

A variety of high-temperature resins with and without glass-fiber fill have been evaluated for use as cage materials. Included in the list are polybutylene terephthalate (PBT), polyethylene terephthalate (PET), polyethersulfone (PES), polyamideimide (PAI), and polyether-etherketone (PEEK). To date of these materials only PES and PEEK have demonstrated sufficient promise as high temperature bearing cage materials; these materials are discussed in further detail below.

Polyethersulfone is a high-temperature thermoplastic material with good strength, toughness, and impact behavior for cage applications. The resin consists of diaryl sulfone groups linked together by ether groups. The structure is wholly aromatic, providing the basis for excellent high-

TABLE 16.5. Properties of Glass-Fiber Filled[a] Thermoplastic Polymers for Cages

Property	PA[b] 6,6 dry	PA 6,6 Conditioned[c]	PES	PEEK
Tensile strength[d] N/mm^2 (psi)	160 (23,200)	110 (15,900)	150 (21,700)	130 (18,800)
Yield stress[d]	4	5	2.3	4.4
Bending strength[d] N/mm^2 (psi)	270 (39,100)	230 (33,300)	210 (30,400)	240 (34,800)
Impact resistance[d] kJ/m^2 (lb/in.)	30 (171)	50 (286)	30 (171)	30 (171)
Coefficient of thermal expansion $10^{-5}/°C$ ($10^{-5}/°F$)	2–3 (1.1–1.7)	2–3 (1.1–1.7)	2–6 (1.1–3.3)	2–3 (1.1–1.7)
Specific gravity	1.3	1.3	1.51	1.44
Operating temperature, max. °C (°F)	120 (248)	120 (248)	170 (338)	250 (482)
Operating temperature, min. °C (°F)	−60 (−76)	−60 (−76)	−100 (−148)	−70 (−94)
Resistance to grease	Good	Good	Satisfactory	Very good
Resistance to oil	Very good	Very good	Good	Very good

[a] PA 6,6 has 25% glass-fiber-fill; PES and PEEK 20%.
[b] Nylon is a polyamide (PA).
[c] Conditioned refers to increased flexibility owing to absorption of a small amount of water.
[d] Strength properties determined at 20°C (68°F).

temperature properties. Being thermoplastic, it is processible using conventional molding equipment. This allows direct part production; that is, without subsequent machining or finishing. In lubricant-temperature exposure tests the resin has performed well to 170°C (338°F). The material is suitable for applications using petroleum and silicone lubricants; however, there are some problems with polymer degradation after exposure to ester-based lubricants and greases. The properties of PES are also shown in Table 16.5; it can be seen that PES is not as strong as nylon 6,6. When it is desired to use a "snap-in" type assembly of balls or rollers in a one-piece cage as illustrated in Fig. 16.25, this somewhat lesser strength can result in crack formation during assembly of the bearing.

Polyether-etherketone is a wholly aromatic thermoplastic that shows excellent physical properties to 250°C (482°F). It is particularly good for cage applications because of its abrasion resistance, fatigue strength, and toughness. It is a crystalline material and can be injection molded. Lubricant compatibility tests show excellent performance to 200°C (392°F) and above. Tests also indicate antiwear performance equal to or better than nylon 6,6. Table 16.5 compares the properties of PEEK with those of PES and nylon 6,6. The only known drawback to the extensive use of PEEK as a bearing cage material at this time is cost. This currently restricts its use to specialized applications.

SEAL MATERIALS

Function, Description, and Illustration

To prevent lubricant loss and contaminant ingress, manufacturers provide bearings with sealing. The effectiveness of the sealing has a critical effect on bearing endurance. When choosing a sealing arrangement for a bearing application, rotational speed at the sealing surface, seal friction and resultant temperature rise, type of lubricant, available volume, environmental contaminants, misalignment, and cost must all be considered.

A bearing can be protected by an integral *seal* consisting of an elastomeric ring with a metallic support ring, the elastomer riding on an inner ring surface (see Fig. 17.14), or by a stamped *shield* of mild steel staked into the outer ring and approaching the inner ring closely but not in intimate contact with it (see Fig. 17.13).

Shields cost less and do not increase torque for the bearing in operation. This design is useful for excluding gross particulate contamination (150 μm). Often used with greased bearings, it is used in bearings lubricated by liquids that must pass through the bearing. The seal configuration is more expensive because of design and materials. Depending on seal lip design, it adds to bearing friction torque to a greater or lesser extent. Seals are used in greased bearings when moisture and all contamination must be excluded. They are also the best choice for minimizing grease purging.

Elastomeric Seal Materials

Because of the prevalence of elastomeric seals in rolling bearings, a variety of materials has been developed to meet the requirements of differing applications. Important properties of elastomeric seal materials include lubricant compatibility, high- and low-temperature performance, wear resistance, and frictional characteristics.

Table 16.6 summarizes physical properties, and Table 16.7 lists general application guidelines.

In the following discussion of elastomeric types, it is important to note that compounding variations starting with a particular elastomer type can lead to products of distinct properties. The general inputs to a formulated compound may be taken as follows:

Elastomer—basic polymer that determines the ranges of final product properties

Curing agents, activators, accelerators—determine degree and rate of elastomeric vulcanization (cross-linking)

Plasticizers—improve flexibility characteristics and serve as processing aids

Antioxidants—improve antifatigue and antioxidation properties of product

"Nitrile" rubber represents the most widely used elastomer for bearing seals. This material, consisting of copolymers of butadiene and acrylonitrile, is also known as Buna N and NBR. Varying the ratio of butadiene to acrylonitrile has a major effect on the final product properties.

The general polymer reaction can be repesented as

$$\underset{\text{Butadiene}}{CH_2{=}CH{-}CH{=}CH_2} + \underset{\text{Acrylonitrile}}{CH_2{=}\underset{\displaystyle CN}{\underset{|}{C}H}} \rightarrow$$

$$\underset{\text{Copolymer}}{-[CH_2{-}CH{=}CH{-}CH_2{-}CH_2{-}\underset{\displaystyle CN}{\underset{|}{C}H}]-}$$

Nitrile rubbers are commercially available with a range of acrylonitrile contents from 20 to 50% and containing a variety of antioxidants. Particular polymer selection will depend on lubricant low-temperature requirements and thermal resistance required.

The nitrile rubber seal is used in many standard bearing application areas. Material cost is low compared to other elastomers. The material is injection-moldable, which allows one-step processing of complex lip shapes. Lubricant compatibility with petroleum-based lubricants is good for high acrylonitrile versions. This elastomer is suitable for applications to 100°C (212°F) and is therefore not indicated for high-temperature bearing applications.

"Polyacrylic" elastomers have been used in bearing applications. The acrylic polymer is generally based on ethyl acrylate and/or butyl acrylate, usually with an acrylonitrile comonomer present. As with nitrile rubbers, the higher the percentage of acrylonitrile present, the better the lubricant resistance. However, higher acrylonitrile levels degrade low-temperature properties of these rubbers.

These materials are able to withstand operating temperatures up to 150°C (302°F) and, if properly formulated, show very good resistance to mineral oils and extreme pressure (EP) lubricant additives.

Negative features of this material are poor water resistance, substandard strength and wear resistance for most seal applications, and high cost. Although no longer used for high temperature applications, it still is used when a low sealing force is required.

Silicon rubbers are used as seal materials in some high-temperature and food-contacting bearing applications. Silicon rubbers have a backbone structure made up of silicon-oxygen linkages, which give excellent thermal resistance. A typical polymer is

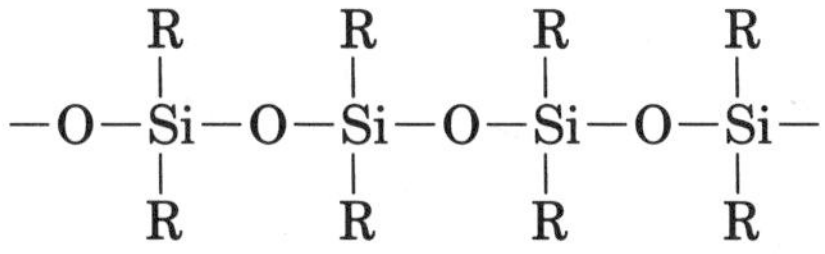

$$
\begin{array}{ccccccccc}
 & R & & R & & R & & R & \\
 & | & & | & & | & & | & \\
-O- & Si & -O- & Si & -O- & Si & -O- & Si & - \\
 & | & & | & & | & & | & \\
 & R & & R & & R & & R &
\end{array}
$$

The silicon polymer is modified by introducing different side groups, R, into the structure in varying amounts. Typical organic substitutes are methyl, phenyl, and vinyl. If $R{=}CH_3$, the polymer is dimethylpolysiloxane.

Advantages of silicon rubber seal use are high-temperature performance to 180°C (356°F) and good low-temperature flexibility to −60°C (−76°F). The material is nontoxic and inert; hence it is chosen for food, beverage, and medical applications. It is stable with regard to the effects of repeated high temperature. Their excellent low-temperature flexibility makes these elastomers useful for very low-temperature applications where sealing is required.

Silicon rubbers are very expensive compared to nitrile rubbers. Lubricant resistance and mechanical strength are poor for most seal applications. On the whole, silicon elastomers have limited usefulness.

Fluoroelastomers have become increasingly popular as seal materials

TABLE 16.6. Physical Properties of Seal Elastomers[a]

	Fluoroelastomers			Nitrile Rubber				
Elastomer	Standard	Peroxide-cured	Fluoro-Silicone	Standard	Heat Resistant	Hydro-genated	Polyacrylate	Silicone
Material designation (ASTM D1418)	FKM	FKM	FVMQ	NBR	NBR		ACM	VMQ, PVMQ
Material designation (ASTM D-2000/ SAE J-200)	HK	HK	FK	BF, BG, BK, CH	CH, CK	DH, DK	DF, DH	FC, FE, GE
Mechanical properties								
Hardness range (Durometer)	60–95	60–95	60–80	40–90	40–90	40–95	40–80	40–85
Tensile strength	B	B	C	A	A	A	C	C
Resilience (73°F)	C	C	C	B	B	B	C	C-A
Tear strength	C	C	D	B	B	A	D	D
Abrasion resistance	B	B	D	A	A	A	C	D
Brittle point (°F)	−40	−40	−85	−40	−40	−30	−40	−90 to −180
Adhesion to metal	C	C	D	A	A	A	BC	C

Electrical properties	B	B	A	C	C	C	C	A
Resistance to								
Gas permeability	A	A	D	B	B	B	B	D
Ozone	A	A	A	D	D	A	A	A
Weather	A	A	A	D	D	A	A	A
Water	A	A	A	A	A	A	D	B
Steam	B	A	C	C	C	B	NR	C
Synthetic lubricants (diester)	A	A	A	B	B	B	D	NR
Lubricating oils	A	A	A	A	A	A	A	C
Aliphatic hydrocarbons	A	A	A	A	A	A	B	NR
Aromatic hydrocarbons	A	A	A	B	B	B	D	NR
Acids	A	A	B	B	B	B	D	C
Bases	A	A	B	B	B	B	D	A

[a] A = excellent, B = good, C = fair, D = use with caution, NR = not recommended.
(from reference [16.29]).

TABLE 16.7. Application Guidelines for Seal Elastomers

	Fluoroelastomers			Nitrile Rubber				
Elastomer	Standard	Peroxide-cured	Fluoro-Silicone	Standard	Heat Resistant	Hydro-genated	Poly-acrylate	Silicone
Material designation (ASTM D1418)	FKM	FKM	FVMQ	NBR	NBR		ACM	VMQ, PVMQ
Material designation (ASTM D-2000/ SAE J-200)	HK	HK	FK	BF, BG, BK, CH	CH, CK	DH, DK	DF, DH	FC, FE, GE
Temperature service range (°F)	−40 to 450	−40 to 450	−80 to 400	−40 to 225	−40 to 250	−40 to 300	40 to 325	−80 to 450
Advantages	Excellent heat resistance	Excellent heat resistance	Good low-temperature flexibility	Low cost	Fair heat resistance	Good heat resistance	Moderate heat resistance	Excellent heat resistance
	Excellent resistance to fluids & additives	Excellent resistance to fluids & additives	Good fluid resistance	Good mechanical strength	Good mechanical strength	Good mechanical strength	Good fluid resistance	Excellent low-temperature flexibility
		Excellent steam resistance	Good heat resistance	Good fluid resistance	Good fluid resistance	Good fluid resistance	Resistance to EP additives	
						Good EP additive resistance		

Disadvantages	High cost	High cost	High cost	Limited heat resistance	Limited heat resistance	High Cost	Poor mechanical strength	Poor fluid resistance
	Poor hot mechanical strength	Poor hot mechanical strength	Poor mechanical strength	Attacked by EP additives	Attacked by EP additives		Poor abrasion	Poor mechanical strength
	Poor abrasion	Poor abrasion	Difficult to process				Poor water & steam resistance	
	Poor water, steam & amine resistance		Difficult to bond to n					

(from reference [16.29])

because of excellent high-temperature and lubricant-compatibility characteristics. A typical polymer of this class is the copolymer of vinylidine fluoride and hexafluoropropylene, which an be represented as

$$-CH_2-CF_2-CH_2-CF_2-CH_2-\underset{}{\overset{CF_3}{\overset{|}{C}}}F-$$

Materials of this general type have become common for bearing seal applications at temperatures exceeding 130°C (266°F). Suitably compound fluoroelastomers show good wear resistance and water resistance for bearing seal applications. As would be expected, material cost is very high compared to nitrile rubbers.

SURFACE TREATMENTS FOR BEARING COMPONENTS

Coatings in General

Several coatings exist to improve surface characteristics of bearing or bearing components without affecting the gross properties of the bearing material. Within the realm of standard bearing applications, coatings are used to provide wear resistance, initial lubrication, sliding characteristics and cosmetic improvements. In addition, bearings operating in extreme environments of temperature, wear, or corrosivity can be specially treated.

Phosphate Coating

Zinc and manganese phosphate coatings are applied to finished bearings and components to provide

Increased corrosion protection by providing a porous base for preservative oils

Initial lubrication during bearing run-in by preventing metal-to-metal contacts and providing a lubricant reservoir

Prepared surfaces for other surface coatings—that is, MoS_2

Parts are immersed in acidic solutions of metal phosphates at temperature. This produces a conversion coating integrally bonded to the bearing surface. The coated surface is now nonmetallic and nonconductive. The zinc phosphate process gives a finer structure, which may be preferred cosmetically. The manganese phosphate process gives a heavier structure that is generally preferred for wear resistance and lubricant retention. Phosphating in itself does not provide for substantial improvements in rust protection. It is only when a suitable preservative is employed that full benefits are obtained.

Black Oxide

Black oxide conversions have been used on bearings and components for

Cosmetic uniformity appearance to components
Lubrication during run-in
Rust protection during extended storage

Black oxide is a generic term referring to the formation of a mixture of iron oxides on a steel surface. An advantage of the process is that no dimensional change results from the process, so tolerances can be maintained after treatment.

A common approach to obtain this coating consists of treating a steel component in a highly oxidizing bath. Because the chemical process results in dissolution of surface iron, close process control is necessary to prevent objectionable surface damage. The black color is obtained from the presence of Fe_3O_4.

Plating Processes

Both electroplating and electroless disposition have long been employed in the rolling bearing industry to provide wear-resistant coatings for cages. In response to aircraft bearing requirements, silver plating over a nickel- or copper-struck cage is commonplace. In this case the strike metal provides an oxygen barrier to the base metal to prevent corrosion. The silver plating offers reduced friction. Cadmium, tin, and chrome plating are also used for certain bearings and accessories.

Extreme Environment Coatings

Several coating techniques and materials somewhat reduce sliding friction and markedly improve wear and corrosion resistance in extreme environments. The techniques include physical vapor deposition (PVD), chemical vapor deposition (CVD), and special process electroplating. Coating materials include titanium nitride (TiN), titanium carbide (TiC), and hard chromium. Some of these process-coating material combinations have demonstrated excellent performance on rolling contact surfaces.

Chemical Vapor Deposition

In chemical vapor deposition of TiC, titanium tetrachloride is vaporized and allowed to react with the substrate at high temperatures in the presence of hydrogen and methane gas. Typically, CVD processing needs temperatures of 850–1050°C (1562–1922°C). Although these temperatures will promote diffusion with the substrate, the processing temperatures exceed the tempering temperatures of bearing steels requiring heat treating after coating. Postcoating heat treatment may cause dimensional distortion. This post treatment of the CVD coating diminishes the attractiveness of this process for bearing components.

Physical Vapor Deposition

The principal advantage of PVD over CVD is that substrate temperatures below 550°C (1022°F) are used. High bond strength with PVD is achieved with ion bombardment of the substrate surface. Consequently, postcoating heat treatment of high-speed steels is not required. Therefore there has been considerable interest in applying PVD coatings to bearing surfaces, with TiN being a usual coating. Initial work indicates excellent bonding with bearing steel and compatibility with a high contact stress environment.

Special Process Chromium Electrodeposition

Super chrome-plating techniques have been developed that produce coatings free of the surface cracks that characterize conventional hard chrome deposits. Increased corrosion resistance is reported for the coated bearing steel; the coating does not negatively affect the rolling contact fatigue life. Substrate temperature is below 66°C (151°F) during plating, and the coating, with a reported hardness of R_C 70 deforms plastically rather than cracking when overloaded.

CLOSURE

An operating rolling bearing is a system containing rings, raceways, rolling elements, cage, lubricant, seals, and ring support. In general, ball and roller bearings selected from listings in manufacturers' catalogs must be able to satisfy broad ranges of operating conditions. Accordingly, the materials used must be universal in their applicability. Through-hardened AISI 52100 steel, nylon 6,6, lithium-based greases, and so on, are among the materials that have met the test of universality for many years. Moreover, these materials as indicated in this chapter have undergone significant improvement, particularly in the past few decades.

For special applications involving extra heavy applied loading, very high speeds, high temperatures, very low temperatures, severe ambient environment, and combinations of these, the bearing system materials must be carefully matched to each other to achieve the desired operational longevity. In an aircraft gas turbine engine mainshaft bearing for example, it is insufficient that the M50 or M50-Nil bearing rings provide

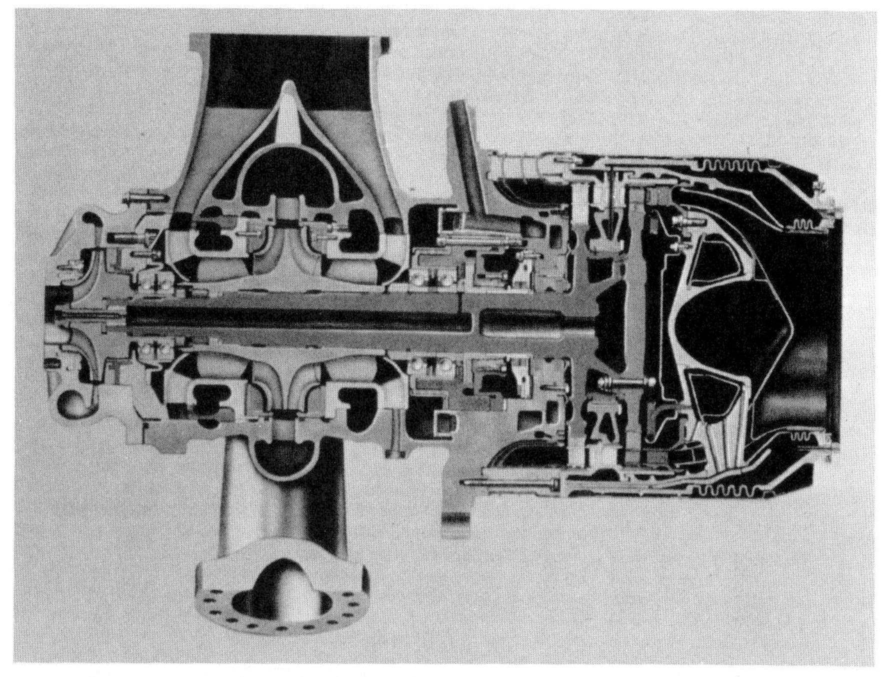

FIGURE 16.27. Section through high-pressure liquid oxygen fuel pump for space shuttle main engines (from [16.28]).

long-term operating capability at engine operating temperatures and speeds; rather, the bearing cage materials and lubricant must also survive for the same operating period. Therefore cages for such applications are generally fabricated from tough steel and are silver plated; nylon cages are precluded by the elevated operating temperatures and possibly by incompatibility with the lubricant. The upper limit of bearing operating temperature is established by the lubricant; in most cases this is a synthetic oil according to United States military specification MIL-L-23699 or MIL-L-7808.

An example of an extreme operating condition is the liquid oxygen (LOX) turbopump for the space shuttle main engine as shown in Fig. 16.27. In this application, the bearings must rotate at very high speed (30,000 rpm) while being *lubricated* by LOX. The LOX vaporizes in the confines of the bearing, and the bearing tends to *burn up* and wear notwithstanding the initial cryogenic temperature [−150°C (−302°F)] of the LOX. To achieve sufficient duration of satisfactory operation, the ball bearing cage has been fabricated from Armalon, a woven fiberglass-reinforced PTFE material [16.28] that *lubricates* by transfer of PTFE film from the cage pockets to the balls. The bearing rings are fabricated from vacuum-melted AISI 440C stainless steel. Target duration for bearing operation is only a few hours.

For most bearing applications, currently available materials as described in this chapter are sufficient to attain the bearing performance required. Considering the capabilities of modern bearing steels, bearing service life tends to be limited by sealing capabilities at normal and somewhat elevated operating temperatures and by lubricating capabilities at very high and very low temperatures. Continuing material development is necessary to meet the ever-increasing demands of special rolling bearing applications.

REFERENCES

16.1. American Society for Testing and Materials, Std. A295-84, "High Carbon Ball and Roller Bearing Steels;" Std. A485-79, "High Hardenability Bearing Steels," American Society for Testing and Materials, Philadelphia.

16.2. American Society for Testing and Materials, Std. A534-79, "Carburizing Steels for Anti-Friction Bearings," American Society for Testing and Materials, Philadelphia.

16.3. C. Finkl, "Degassing—Then and Now," *Iron and Steelmaker*, 26–32 (December 1981).

16.4. T. Morrison, T. Tallian, H. Walp, and G. Baile, "The Effect of Material Variables on the Fatigue Life of AISI 52100 Steel Ball Bearings," *ASLE Trans.* **5,** 347–364 (1962).

16.5. *Making, Shaping and Treating of Steel*, 9th ed. United States Steel Corporation, p. 551 (1971).

16.6. *Making, Shaping and Treating of Steel*, 9th ed., United States Steel Corporation, pp. 596–597 (1971).

16.7. *Making, Shaping and Treating of Steel*, 9th ed., United States Steel Corporation, p. 594 (1971).

16.8. *Making, Shaping and Treating of Steel*, 9th ed., United States Steel Corporation, p. 598 (1971)

16.9. J. Akesson and T. Lund, "Rolling Bearing Steelmaking at SKF Steel", Technical Report 7 (1984).

16.10. *Making, Shaping and Treating of Steel*, 9th ed., United States Steel Corporation, p. 580 (1971).

16.11. American Society for Testing and Materials, Std. E45-81, "Standard Practice for Determining the Inclusion Content of Steel," American Society for Testing and Materials, Philadelphia.

16.12. J. Beswick, "Effect of Prior Cold Work on the Martensite Transformation in SAE 52100," *Metall. Trans. A*, **15A**, 299–305 (1984).

16.13. R. Butler, H. Bear, and T. Carter, "Effect of Fiber Orientation on Ball Failures Under Rolling-Contact Contact," NASA TN 3933 (1975).

16.14. *The Black Book*, SKF Steel, p. 194 (1984).

16.15. *The Black Book*, SKF Steel, p. 151 (1984).

16.16. M. Grossman, *Principles of Heat Treatment*, American Society for Metals (1962).

16.17. American Society for Testing and Materials, Std. A255-67, "End-Quench Test for Hardenability of Steel," American Society for Testing and Materials, Philadelphia (1979).

16.18. *Atlas of Isothermal Transformation and Cooling Transformation Diagrams*, American Society for Metals (1977).

16.19. T. Tallian, G. Baile, H. Dalal, and O. Gustafsson, *Rolling Bearing Damage Atlas*, SKF Industries, Inc. (1974).

16.20. G. Winspiar, Ed., *The Vanderbilt Rubber Handbook*, R. T. Vanderbilt Co., Inc., New York (1968).

16.21. *Modern Plastics Encyclopedia*, McGraw-Hill, New York (1985–1986).

16.22. *Metal Finishing Guidebook and Directory 85*, Metals and Plastics Publication, Inc. Hackensack, N.J. (1985).

16.23. A. Graham, *Electroplating Engineering*, 3rd ed., Van Nostrand Reinhold, New York (1971).

16.24. R. Spitzer, "New Case-Hardening Steel Provides Greater Fracture Toughness," *SKF Ball Bearing J.* No. 234, 6–11 (July 1989).

16.25. R. Pallini, "Turbine Engine Bearings for Ultra-High Temperatures," *SKF Ball Bearing J.* No. 234, 12–15 (July 1989).

16.26. A. Olschewski, "High Temperature Cage Plastics," *SKF Ball Bearing J.* No. 228, 13–16 (November 1986).

16.27. H. Lankamp, "Materials for Plastic Cages in Rolling Bearings," *SKF Ball Bearing J.* No. 227, 14–18 (August 1986).

16.28. R. Maurer and L. Wedeven, "Material Selection for Space Shuttle Fuel Pumps, *SKF Ball Bearing J.* No. 226, 2–9 (April 1986).

16.29. Delta Rubber Company "Elastomer Selection Guide."

17

LUBRICANTS AND LUBRICATION TECHNIQUES

LIST OF SYMBOLS

Symbol	Description
Ba	Barium
C	Carbon
Ca	Calcium
F	Fluorine
H	Hydrogen
Li	Lithium
O	Oxygen
P	Phosphorus
R, R′, R″	Reaction group
S	Sulfur
VI	Viscosity index
W	Tungsten

GENERAL

The primary function of a lubricant is to lubricate the rolling and sliding contacts of a bearing to enhance its performance through the prevention of wear. This can be accomplished through various lubricating mechanisms such as hydrodynamic lubrication, elastohydrodynamic (EHD) lubrication, and boundary lubrication. The rolling/sliding contacts of concern are those between rolling element and raceway, rolling element and cage (separator), cage and supporting ring surface, and roller end and ring guide flanges.

In addition to wear prevention the lubricant performs many other vital functions. The lubricant can minimize the frictional power loss of the bearing. It can act as a heat transfer medium to remove heat from the bearing. It can redistribute the heat energy within the bearing to minimize geometrical effects due to differential thermal expansions. It can protect the precision surfaces of the bearing components from corrosion. It can remove wear debris from the roller contact paths. It can prevent extraneous dirt from entering the roller contact paths, and it can provide a damping medium for separator dynamic motions.

No single lubricant or class of lubricants can satisfy all these requirements for bearing operating conditions from cryogenic to ultrahigh temperatures, from very slow to ultrahigh speeds, and from benign to highly reactive operating environments. As for most engineering tasks, a compromise is generally exercised between performance and economic constraints. The economic constraints involve not only the cost of the lubricant and the method of application but also its impact on the life cycle cost of the mechanical system.

Cost and performance decisions are frequently complicated because many other components of a mechanical system also need lubrication or cooling, and they might dominate the selection process. For example, an automobile gearbox typically comprises gears, a ring synchronizer, rolling bearings of several types operating in very different load and speed regimes, plain bearings, clutches, and splines.

TYPES OF LUBRICANTS

Selection Criteria

The selection of lubricants is based on their flow properties and chemical properties in connection with lubrication. Additional considerations, which sometimes may be of overriding importance, are associated with operating temperature, environment, and the transport or retention properties of the lubricant in the bearing.

Liquid Lubricants

Liquid lubricants are usually mineral oils; that is, fluids produced from petroleum base stocks. They have a wide range of molecular constituents and chain lengths giving rise to a large variation in flow properties and chemical performances. These lubricants are generally additive enhanced for both viscous and chemical performance improvement.

Synthetic hydrocarbon fluids are manufactured from petroleum-based materials. They are synthesized with both narrowly limited and specifically chosen molecular compounds to provide the most favorable properties for lubrication purposes. Other "synthetic" fluids that can have unique properties are manufactured from non-petroleum-based oils. These include polyglycols, phosphate esters, dibasic acid esters, silicone fluids, silicate esters, and fluorinated polyethers.

Greases

Greases have two major constituents: an oil phase and a thickener system that physically retains the oil by capillary action. The thickener is normally composed of a material with very long twisted and/or contorted molecules that both physically interlock and provide the necessarily large surface area to retain the oil. The resultant material behaves as a soft solid, capable of bleeding oil at controlled rates to meet the consumption demands of the bearing.

Polymeric Lubricants

Polymeric lubricants are related to greases in that these materials consist of an oil phase and a retaining matrix. They differ in one crucial point: the matrix is a true solid sponge that retains its physical shape and location in the bearing. Lubrication functions are provided by the oil alone after it has bled from the sponge. The oil content can be made higher than in greases, and a greater quantity can be installed in the void space within the bearing. This greater oil volume portends longer bearing life before all fluid is consumed by oxidation, evaporation, or leakage. The latter is particularly significant for vertical axis bearing applications.

Solid Lubricants

Solid lubricants are substituted for liquid lubricants when extreme environments such as high temperature or vacuum make liquid lubricants or greases impractical. Solid lubricants, unless melted, do not utilize the mechanism of hydrodynamic or EHD lubrication. Their performance is less predictable, and there is generally much greater heat generation due to friction. Solid lubricants perform as boundary lubricants consisting of thin layers that provide lower shear strength than the bearing materials. Solid lubricants can consist of layered structures that shear easily or nonlayered structures that deform plastically at relatively low temperatures. Graphites and molybdenum disulfide (MoS_2) are common examples of materials with layered structures. Fluorides such as calcium fluoride (CaF_2) are nonlayered materials that perform well at or near their melting temperatures.

LUBRICATION METHODS

Oil Sump or Bath

Decisions in connection with the selection of lubricants must parallel decisions in connection with the supply of the lubricant to the bearing for maintaining conditions that will prevent rapid deterioration of the lubricant and bearing. An oil sump applicable to horizontal, inclined, and vertical axis arrangements provides a small pool of oil contained in contact with the bearing, as in Fig. 17.1.

The liquid level in the stationary condition is arranged to just reach the lower portion of the rolling elements. Experience has shown that higher levels lead to excessive lubricant churning and resultant excessive temperature. This churning in turn can cause premature lubricant oxidation and subsequent bearing failure. Lower liquid levels threaten oil starvation at operating speeds where windage can redistribute the oil and cut off communication with the working surfaces. Maintenance of proper level is thus very important and provision of a "sight" glass is recommended.

Oil bath systems are used at low-to-moderate speeds where grease is ruled out by short relubrication intervals, hot environments, or where purging of grease could cause problems. Heat dissipation is somewhat

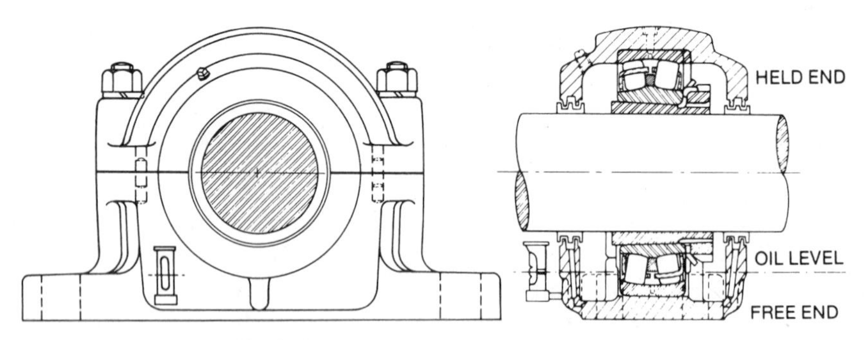

FIGURE 17.1. Pillow block with oil sump.

better than for a greased bearing due to fluid circulation, offering improved performance under conditions of heavy load where contact friction losses are greater than the lubricant churning losses. This method is often used when conditions warrant a specially formulated oil not available as a grease. A cooling coil is sometimes used to extend the applicable temperature range of the oil bath. This usually takes the form of a water-circulating loop or, in some recent applications, the fitting of one or more heat pipes.

Wick-feed and oil-ring methods of raising oil from a sump to feed the bearing are not generally used with rolling bearings, but, occasionally shaft motion is used to drive a viscous pump for oil elevation, thus reducing the sensitivity of the system to oil level. A disc dipping into the sump drags oil up a narrow groove in the housing to a scraper blade or stop that deflects the oil to a drilled passage leading to the bearing. A major limitation of all sump systems is the lack of filtration or debris entrapment. Fitting a magnetic drain plug is advantageous for controlling ferrous particles, but otherwise sump systems are only suitable for clean conditions.

Circulating Oil Systems

As the speeds and loads on a bearing are increased, the need for deliberate means of cooling also increases. The simple use of a reservoir and a pump to supply a lubricant flow increases the heat dissipation capabilities significantly. Pressure feed permits the introduction of appropriate heat exchange arrangements. Not only can excess heat be removed, but heat can be added to assure flow under extremely cold start-ups. Some systems are equipped with thermostatically controlled valves to keep the oil in an optimum viscosity range.

Equally important, a circulating system can be fitted with a filtration system to remove the inevitable wear particles and extraneous debris. The mechanisms of debris-induced wear and the effects of even microscale indentations on the EHD lubrication processes and the consequent reductions in fatigue life are discussed in Chapters 23 and 24. Finer filtration is being introduced in existing circulating systems with beneficial effects; however, increased pressure drops, space, weight, cost, and reliability have to be considered.

Circulating systems are used exclusively in critical high-performance applications, of which the main shaft support bearings of an aircraft gas turbine engine constitute prime examples. Subjected to heavy thrusts at near limiting speeds, the angular-contact ball bearings generate considerable frictional heat, particularly at the inner raceway contacts. This heat must be removed effectively together with leakage heat conducted to the bearing cavity from the surrounding engine components.

Heavily loaded bearings running at moderate speeds can be supplied with oil jets aimed at the rolling elements. At higher speeds, bearing windage deflects the jets, and lubrication and cooling become ineffective. This problem can be avoided by routing the oil to pickup scoops on the shaft with centrifugal force taking the oil via drilled passages to the inner ring, as shown in Fig. 17.2. Much of the flow passes through axial slots in the bore of the inner ring, removing heat as it does so. Only a small portion of the lubricant is metered to the rolling contacts through grooves between the inner ring halves. Separate drilled holes may be used to supply the cage lands.

Adequate space should be provided on both sides of the bearing to facilitate lubricant drainage. Often, space is at a premium, so a system of baffles can be substituted to shield the lubricant from the windage, permitting it to be scavenged without severe churning. When the lubricant pump is activated at the same time as the main machinery, these baffles act as a dam and retain a small pool of lubricant in the bottom of the bearing to provide lubrication at start-up until the circulating flow becomes established.

Hydrocarbon-based fluids are satisfactory for circulating systems operating at temperatures to about 274°C (525°F), where oxidation becomes a problem. Use of an inert cover gas to exclude oxygen extends

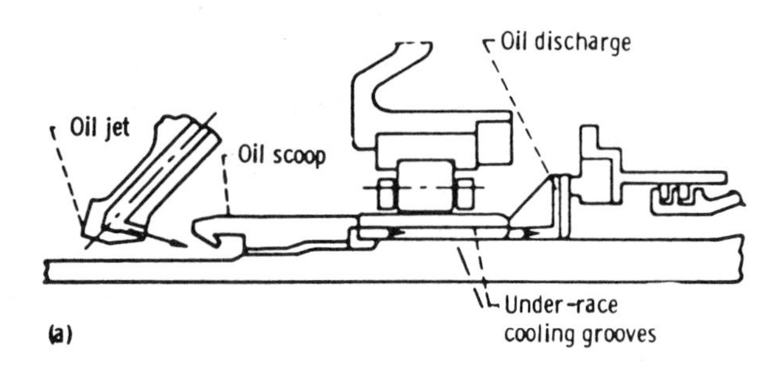

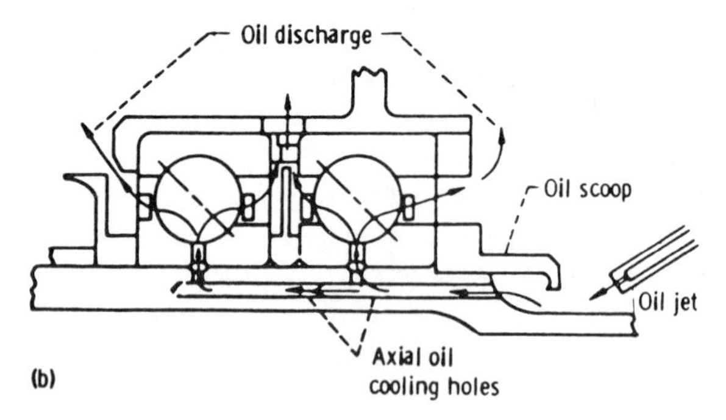

FIGURE 17.2. Under-raceway lubricating systems for mainshaft bearings in an aircraft gas turbine engine. (*a*) Cylindrical roller bearing. (*b*) Ball bearings.

the working range to about 449°C (840°F) where thermal degradation becomes a limiting factor. Beyond 449°C (840°F), fluorocarbon-based fluids are serviceable, but conspicuously lack the lubrication properties of the hydrocarbon lubricants they replace. At this time they cannot reach the temperature limits inherent in the tool steels from which modern high-temperature performance bearings are made.

Once-Through Systems

A separate class of lubrication arrangements can be used when minimum bearing friction is essential at moderate-to-high speeds and where loads are sufficiently low that heat removal is not a major concern. Lubricant is delivered to the bearing as a fine spray or an air-entrained mist in just sufficient quantities to maintain the necessary lubricant films in the contacts. Lubricant churning is virtually eliminated, and the volume of lubricant is so small that it can be discarded after a single passage through the bearing. Scavenging, cooling, and storage facilities are unnecessary. The one-time exposure to high shear stresses and/or temperatures relaxes the oxidation and the stability requirements of the fluid to some extent. The necessity for satisfactory air quality in the workplace requires that the exhaust droplets be reclassified and lubricant collected before discharge.

Recent work has shown that the spray does not even need to be continuous. Trace injection of minute quantities of lubricant at intervals of up to one hour is sufficient to keep precision spindle assemblies running at friction torque levels unobtainable by any other method.

Grease Application

In the majority of rolling bearing applications grease lubrication can be employed. In ease of application, grease has some advantages compared to oil in that it is easily retained, and it also helps seal the bearing operating surfaces from particulate contaminants and moisture. Grease lubrication is, however, in general restricted to relatively slower speed applications owing to reduced capability for frictional heat dissipation as compared to oil; hence, limiting speeds as shown in bearing catalogs are less for grease lubrication than for oil lubrication. Moreover, care must be exercised when charging a bearing with grease. Too much grease will cause a rapid temperature rise and potential bearing seizure. Therefore, while the bearing *free space* may be filled with grease, the surrounding space in the housing in general should only be partially (30–50%) filled. For very slow speed operation, to provide maximum corrosion protection, the housing may be completely filled.

If the service life of the grease used to lubricate the bearing is less than the expected bearing life, the bearing needs to be relubricated prior to lubricant deterioration. Relubrication intervals are dependent upon bearing type, size, speed, operating temperature, grease type, and the ambient conditions associated with the application. As operating conditions become more severe, particularly in terms of frictional heat generation and operating temperature, the bearing must be relubricated more frequently. Some manufacturers specify relubrication intervals for their catalog bearings; for example, reference [17.1]. Such recommendations, given in the form of charts, are specific to the manufacturer's bearing internal designs and are generally based on good quality, lithium soap-based greases (see "Grease Lubricants") operating at temperatures not exceeding 70°C (158°F). It is interesting to note that for every 15°C (27°F) above 70°C (158°F) relubrication intervals must be halved. Bearings operating at temperatures lower than 70°C (158°F) tend to require relubrication less often; however, the lower operating temperature limit of the grease may not be exceeded [−30°C (−22°F) for a lithium-based grease]. Also, bearings operating on vertical shafts need to be relubricated approximately twice as often as bearings on horizontal shafts. (Relubrication interval charts are generally based on the latter application.) It is presumed that in no case is the grease upper operating temperature limit exceeded; this limit is 110°C (230°F) for a lithium-based grease.

Because relubrication intervals depend on specific bearing internal design features such as rolling element proportions, working surface finishes and cage configuration, they are different for each manufacturer even for basic bearing sizes. Therefore no such charts are given in this text; they may be found in manufacturers catalogs. Of course, the methods given in Chapter 15 may be used to estimate the operating bearing grease temperature in a given application, and the grease manufacturer's recommendations for replenishment may be employed.

If the relubrication interval is greater than 6 months, then all of the used grease should be removed from the bearing arrangement and replaced with new grease. If the interval is less than 6 months, then an incremental grease charge may be added according to the bearing manufacturer's recommendation. The 6 month limit is understood to be a rough guideline.

Relubrication interval requirements vary significantly according to the types of grease used and even where apparently similar greases are used. For small ball bearings, the relubrication interval is often longer than the life of the bearing application, and relubrication is not normally required. Ball bearings fitted with seals that are lubricated for life are often used in such cases. Where there is definite risk of contamination, the recommended relubrication intervals should be reduced. This also applies to applications in which the grease is required to seal against moisture.

Solid Lubrication Application

In applications where conventional lubricants are not appropriate, thin solid films of soft or hard materials are applied to bearing surfaces to reduce friction and enhance wear resistance of contacting surfaces. There are many methods of applying solid lubricants, each of which provides varying degrees of success with respect to adhesion to the substrate, thickness, and uniformity of coverage.

Resin-bonded solid lubricants are very commonly used. These materials usually consist of a lubricating solid and a bonding agent. The lubricating solid may be a single material or a mixture of several materials. It can be applied in a thin film by spraying or dipping. Depending on the binding agent used, it may be a heat-cured or air-cured material. Heat-cured materials are generally superior to the air-dried materials. Metal surfaces are usually pretreated prior to application. Pretreatment may be chemical or mechanical; the latter tends to increase the surface area, which gives the binder greater holding power.

The application of solid lubricants frequently relies on the transfer of thin solid films from one contacting surface to another. The interaction of rolling elements with a solid-lubricated or impregnated separator transfers thin solid films to the rolling elements, which in turn are transferred to the rolling contact raceways. When the rubbing action against the solid lubricant occurs with sufficient load, the solid lubricant will compact itself into the existing surface imperfections. This burnishing action provides little control over film thickness and uniformity of coverage.

Much greater control of solid lubricant film thickness, composition, and adhesion can be obtained by using various electrically assisted thin-film deposition techniques. These include ion plating, activated reactive evaporation, dc and rf sputtering, magnetron sputtering, arc coating, and coating with high-current plasma discharge. Coatings of virtually all of the soft metals and hard materials and a number of nonequilibrium materials can be produced with one or another of the electrically assisted, film deposition techniques. When vacuum techniques of deposition are used, the vapor of the solid lubricant species being deposited can be reacted with process gases to produce various synthesized compounds.

Application of Polymeric Lubricants

Two approaches are used to apply polymeric lubricants. The first approach is to make a suitably shaped part from a porous material, either by molding or machining, and to place it in the bearing in one or more pieces. A vacuum impregnation process then charges the material with lubricant. The need to insert the porous structure governs the amount of bearing free space that can be used. Often rivets or other fasteners must be used to hold the pieces in place, further reducing the volume available.

The second method entails the formation of a lubricant-saturated rigid gel in the bearing itself by filling the bearing with the fluid mixture and using a curing or polymerization step to effect a transition to a solid structure. Essentially all the void space in the bearing can be used. Only a very few polymers have been identified that will function in this manner. Further, there appears to be a tradeoff between bleeding, shrinkage, strength, and temperature limit characteristics.

LIQUID LUBRICANTS

Advantages Over Other Lubricants

As compared to any other lubricant, in particular grease, a liquid lubricant provides the following advantages:

1. It is easier to drain and refill, a particular advantage for applications requiring short relubricating intervals.
2. The lubricant quantity is more accurately controlled.
3. It is suitable for lubricating multiple sites in a complex system.
4. It lends itself to higher temperature applications when used with a circulating system.

Guidelines for Use

In most applications pure petroleum oils are satisfactory as lubricants. They must be free from contamination that might cause wear in the bearing, and should show high resistance to oxidation, gumming, and deterioration by evaporation. The oil must not promote corrosion of any parts of the bearing during standing or operation.

The friction torque in a liquid-lubricated bearing is a function of the bearing design, the load imposed, the viscosity and quantity of the lubricant, and the speed of operation. Only enough lubricant is needed to form a thin film over the contacting surfaces. Friction torque will increase with larger quantities and with increased viscosity of the lubricant.

Energy loss in a bearing depends on the product of torque and speed. It is dissipated as heat, causing increased temperature of the bearing and its mounting structures. The temperature rise will always cause a decreased viscosity of the oil and, consequently, a decrease in friction torque from initial values. The overall heat balance of the bearing and mounting structures will determine the steady-state operating conditions.

It is not possible to give definite lubricant recommendations for all bearing applications. A bearing operating throughout a wide temperature range requires a lubricant with high viscosity index—that is, having the least variation with temperature. Very low starting temperatures necessitate a lubricant with a sufficiently low pouring point to enable the bearing to rotate freely on start-up. For specialized bearing applications involving unusual conditions, the recommendation of the bearing or lubricant manufacturer should be followed.

Mineral Oils

Mineral oil is a generic term referring to fluids produced from petroleum oils. Chemically, these fluids consist of paraffinic, naphthenic, and aromatic groups combined into many molecules. See Fig. 17.3. Also present in crude stocks are trace amounts of molecules containing sulfur, oxygen, or nitrogen. Elementally, the composition of petroleum oils is quite constant: 83–87% carbon, 11–14% hydrogen, and the remainder sulfur, nitrogen, and oxygen. The molecular makeup of the fluid is very complex and depends on its source.

For the purposes of lubricant production, crude petroleum oils are characterized by the type of hydrocarbon distillates obtained. For this method it is common to speak of paraffinic, mixed, and naphthenic crude oils. Aromatics are generally a minor component. Depending on the source, the crude petroleum may consist of gasoline and light solvents, or it may consist of heavy asphalts. Modern distillation, refining, and blending techniques allow the production of a wide range of oil types from a given crude stock; however, some crude stocks are more desirable for lubricant formulation.

With respect to lubricant properties, a few generalizations can be made. Paraffinic base crudes have the viscosity-temperature characteristics for lubrication. Usually such crudes are low in asphalt and trace materials. The earliest commercial crude, Pennsylvania, was of this type. Naphthenic-based crudes do not contain paraffin waxes, so they are bet-

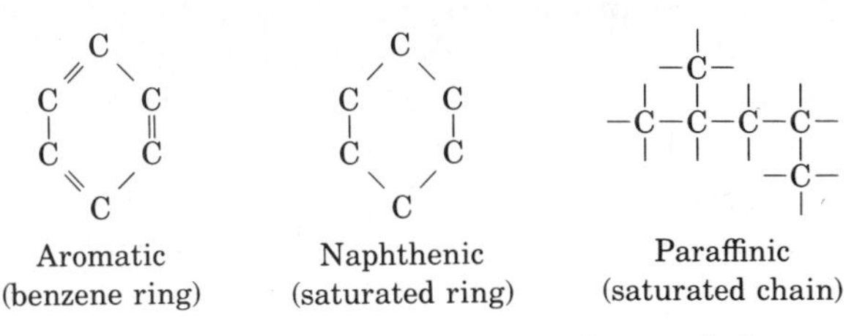

FIGURE 17.3. Chemical structures of mineral oils.

ter suited for certain low-temperature application. Naphthenic oils also have lower flash points and are more volatile than comparable paraffinic oils.

The two most common lubricants used for industrial rolling bearing applications may be described as rust and oxidation (R & O) inhibited oils and extreme pressure (EP) oils. The R & O oils are often used when bearings and gears share a common lubricant reservoir. These products may incorporate antifoam and antiwear agents. They are suitable for light-to-moderate loadings and for temperatures from −20 to 120°C (−4 to 248°F).

Extreme pressure oils usually encompass the additive package of R & O oil with an additional EP additive. The EP additive essentially generates a lubricating surface to prevent metal-to-metal contact. Two approaches exist in formulating EP additives. The first employs an active sulfur, chlorine, or phosphorous compound to generate sacrificial surfaces on the bearing itself. These surfaces will shear rather than weld upon contact. The second approach uses a suspended solid lubricant to impose between two otherwise contacting surfaces.

Extreme pressure oils are used where bearing (or associated gear) loadings are high or where shock loadings may be present. The normal temperature range for such lubricants is −20 to 120°C (−4 to 248°F). Some precautions are necessary when using EP oils of either type. EP solids will reduce internal clearances that can cause failure in certain bearings. These solids might also be lost in close filtration processes. EP sulfur-chlorine-phosphorous compounds might be corrosive to bronze cages and accessory items.

Synthetic Hydrocarbons

Synthetic hydrocarbons are manufactured petroleum fluids. Being synthesized products, the particular compounds present can be both narrowly limited and specifically chosen. This allows production of a petroleum fluid with the most favorable properties for lubrication purposes. One commercially important type is the polyalphaolefin fluids, which have been widely used as turbine lubricants, as hydraulic fluids, and in grease formulations.

These fluids show improved thermal and oxidation stability over refined petroleum oils, allowing higher temperature performance for lubricants compounded from them. These materials also exhibit inherently high viscosity indexes, leading to better viscosity retention at elevated temperatures. Other properties showing improvement include flash point, pouring point, and low volatility. Although synthetic, the materials are compatible with petroleum products because of the compositions involved.

Viscosity of Lubricants

The most important property of a lubricating oil is viscosity. Defined as the resistance to flow, viscosity physically is the factor of proportionality between shearing stress and the rate of shearing. As described in an earlier section, increased viscosity relates to the increased ability of a fluid to separate microsurfaces under pressure, the fundamental process of lubrication. For bearing applications viscosity is usually measured kinematically per ASTM specification D-445. This method measures the passage time required under the force of gravity for a specified volume of liquid to pass through a calibrated capillary tube.

A related concept of importance is viscosity index (VI), which is an arbitrary number indicating the effect of temperature on the kinematic viscosity for a fluid. The higher the VI for an oil, the smaller the viscosity change will be with temperature. For typical paraffinic base stocks VI is 85–95. Polymer additions may be made to petroleum base stocks to obtain VI of 190 or more. The shearing stability of these additives is questionable, and VI generally deteriorates with time. Many synthetic base stocks have VIs far in excess of mineral oils, as Table 17.1 shows. The method of calculating VI from measured viscosities is described in ASTM specification D-567.

Selection of Proper Viscosity for Petroleum Oil Lubricants

Figures 17.4 and 17.5 can be used to derive a minimum acceptable viscosity for an application. Figure 17.4 indicates the minimum required viscosity as a function of bearing size and rotational speed for a petroleum-based lubricant. The viscosity of a lubricating oil decreases with increasing temperature. Therefore, the viscosity at the operating temperature rather than the viscosity at the standardized reference temperature of 40°C (104°F) must be used. Figure 17.5 can be used to determine the actual viscosity at the operating temperature if the viscosity grade (VG) of the lubricant is known.

Example 17.1. A bearing having a pitch diameter of 65 mm (2.559 in.) operates at a speed of 2000 rpm. As shown in Fig. 17.4, the intersection of d_m = 65 mm with the oblique line representing 2000 rpm yields a minimum required kinematic viscosity of 13 mm^2/sec (0.02 in^2/sec), assuming that the operating temperature is 80°C (176°F); in Fig. 17.5 the intersection between 80°C (176°F) and the required viscosity of 13 mm^2/sec (0.02 in.^{2}/sec) is between the oblique lines for VG46 and VG68. Therefore, a lubricant with minimum viscosity grade VG46 should be used; that is, a minimum lubricant viscosity of 46 mm^2/sec (0.07 in.^{2}/sec) at standard reference temperature of 40°C

TABLE 17.1. Typical Properties of Lubricant Base Stocks[a]

	Density (g/cm^3)	Viscosity (40°C CST)	Viscosity Index	Flash Point (°C)	Pour Point (°C)	Volatility	Oxidation Resistance	Lubricity	Thermal Stability	Application Range
Mineral oils						5	5	5	5	Standard lubricants either as oil or grease base.
Paraffinic	0.881	95	100	210	−7					
Naphthenic	0.894	70	65	180	−18					
Mixed base	0.884	80	99	218	−12					
VI improved	0.831	33	242	127	−40					
Polyalphaolefin	0.853	32	135	227	−54	4	4	5	4	
Esters						3–5	3–7	3–6	4–7	Jet turbine lubricants, hydraulic fluids, heat transfer products. Used as bases for low-volatility, low-viscosity grease. Phosphate esters fire resistant.
Dibasic	0.945	14	152	232	< −60	3	5	5	5	
Polyol	0.971	60	132	275	−54					
Tricresyl phosphate	1.160	37	−65	235	−23	3	3	3	7	
Polyglycol ether	0.984	36	150	210	−46	3–5	7	4	5	
Silicate	0.909	6.5	150	188	< −60	3	7	6	4	
Silicones						1	1	7	1–5	High-temperature/low-volatility applications either as oil or grease base. Lightly loaded bearings. Excellent thermal stability.
Dimethyl	0.968	100	400	> 300	< −60					
Phenyl methyl	0.990	75	350	260	< −60					
Chlorophenyl	1.050	55	160	288	< −60					
Perfluoromethyl	1.230	44	158	> 300	< −60					
Phenyl ether						1	1–3	5–7	1	
Low viscosity	1.180	75	−20	263	9.5					
High viscosity	1.210	355	−74	343	4.4					
Perfluoroalky Ether	1.910	320	138	—	−32	3–7	1	5–7	1–3	Extreme temperature fluid. Used in very low volatility applications.

[a] For volatility, oxidation resistance, lubricity, and thermal Stability, 5 is taken as characteristic of highly refined mineral oil. Numbers > 5 reflect superior performance; numbers > 5 reflect inferior performance to mineral oil with respect to lubrication properties.

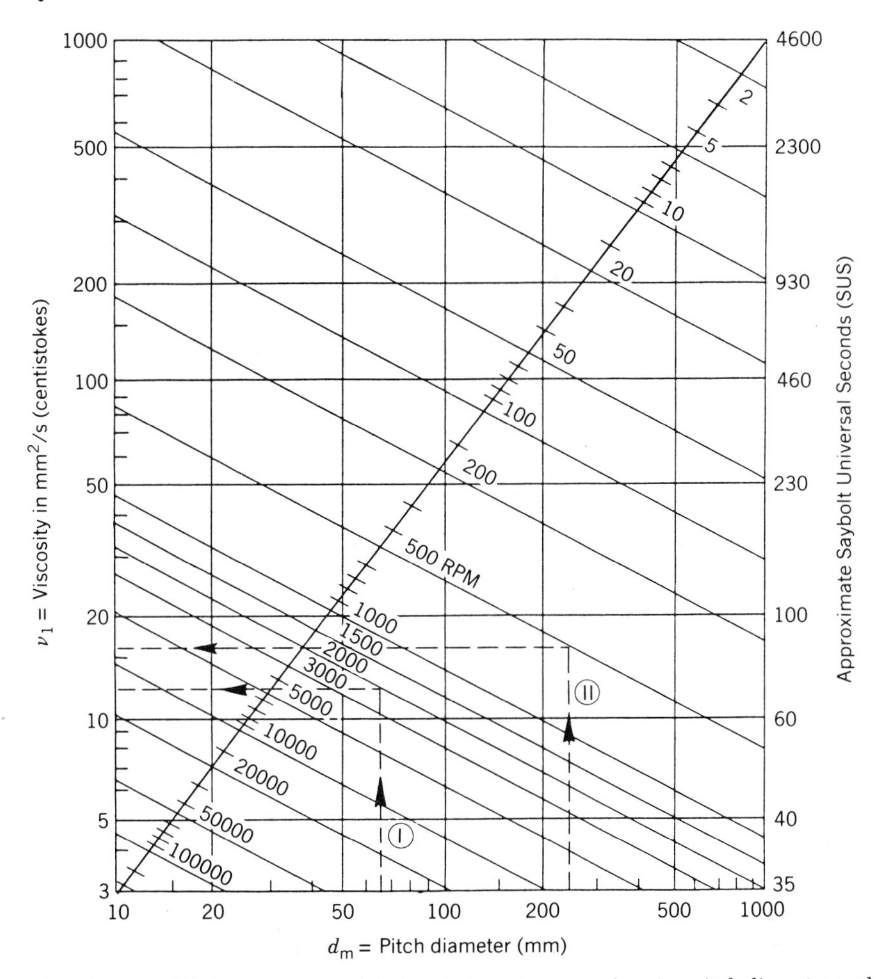

FIGURE 17.4. Minimum required lubricant viscosity versus bearing pitch diameter and speed. d_m = (bearing bore + bearing O.D.) ÷ 2, ν_1 = required lubricant viscosity for adequate lubrication at the opening temperature.

(104°F). When determining operating temperature, it must be kept in mind that oil temperature is usually 3–11°C (5–20°F) higher than bearing housing temperature.

If a lubricant with higher than required viscosity is selected, an improvement in bearing fatigue life can be expected; however, since increased viscosity raises the bearing operating temperature, there is frequently a practical limit to the lubrication improvement that can be obtained by this means. For exceptionally low or high speeds, for critical loading conditions, or for unusual lubrication conditions, the bearing manufacturer should be consulted.

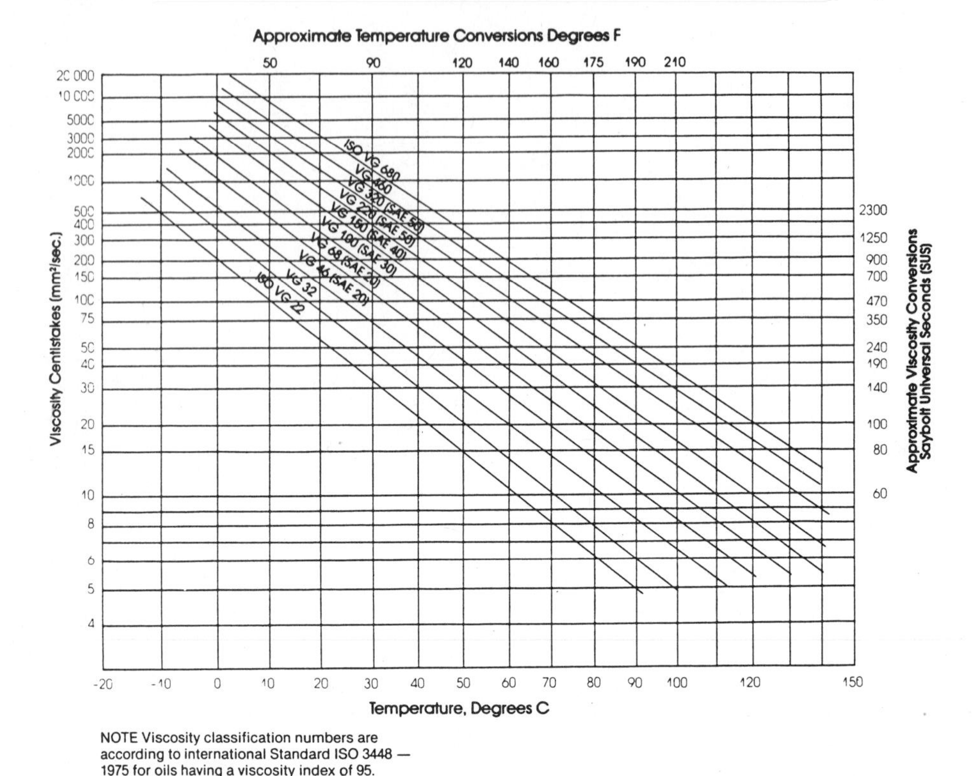

NOTE Viscosity classification numbers are according to international Standard ISO 3448 — 1975 for oils having a viscosity index of 95. Approximate equivalent SAE viscosity grades are shown in parentheses.

FIGURE 17.5. Viscosity–temperature chart.

Types and Properties of Nonpetroleum Oils

Many types of "synthetic" fluids have been developed in response to lubrication requirements not adequately addressed by petroleum oils. These areas include extreme temperature, fire resistance, low volatility, and high viscosity index.

Table 17.1 lists some typical properties of various lubricant-base stocks and indicates application areas for finished products of each type. As for petroleum oils, many additive chemistries have been developed to provide property enhancement.

Using synthetic lubricants requires a thorough understanding of the application requirements involved. The favorable properties shown by some synthetics are obtained only with unsuitable performance characteristics in such areas as load-carrying ability and high-speed operation. Also, many very high-temperature fluids, principally developed for military applications, have short service lives compared to commercial requirements.

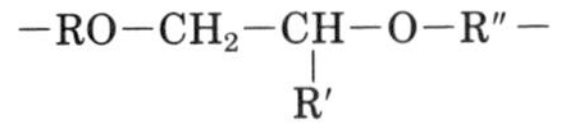

$$-RO-CH_2-\underset{\displaystyle R'}{\underset{|}{CH}}-O-R''-$$

FIGURE 17.6. Chemical structure of a polyglycol.

"Polyglycols" are often used as synthetic lubricant bases in water-emulsion fluids. These products are linear polymers of the general formula shown by Fig. 17.6: R, R′, R″ are alkyl groups, and R′ may be hydrogen.

This class of fluids includes glycols, polyethers, and polyalkylene glycols. Properties of the class include excellent hydrolytic stability, high viscosity index, and low volatility. The most prevalent usage is as a component of fire-resistant hydraulic fluids.

"Phosphate esters" (tertiary) have properties that make them useful as lubricants. They are generally represented as shown in Fig. 17.7: R, R′ and R″ are organic groups.

Phosphate esters have poor hydrolytic stability and low viscosity index. Their outstanding characteristic is fire resistance, and as such these fluids are often used as hydraulic fluids in high-temperature applications.

Dibasic acid esters represent a family of synthetic base stocks widely used in aircraft gas turbine engine applications and as a basis for low-volatility lubricants. They are synthesized by reacting aliphatic dicarboxylic acids (adepic to sebacic) with primary branch alcohols (butyl to octyl). Some are available from such natural sources as castor beans and animal tallow. Characteristic properties of these fluids are low volatility and high viscosity index. Polyol esters formed by linking dibasic acids through a polyglycol center have been found suitable as high film strength lubricants. Blends of dibasic esters, complex esters with suitable antiwear additives, VI improvers, and antioxidates are used to form the current generation of aircraft gas turbine engine lubricants. Generally, these products show excellent viscosity-temperature relationships, good low-temperature properties, and acceptable hydrolysis resistance. Elastomeric seals used with these materials must be chosen carefully because they chemically attack standard rubbers.

Silicone fluids (organosiloxanes) exhibit outstanding viscosity retention with temperature and are functional in conditions of extreme heat and cold. These fluids are the basis for many high-temperature, 204°C

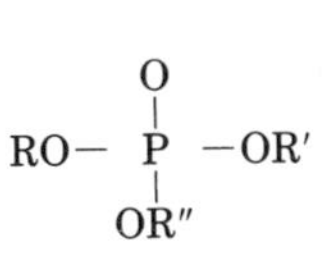

$$RO-\overset{\displaystyle O}{\overset{|}{\underset{\displaystyle OR''}{\underset{|}{P}}}}-OR'$$

FIGURE 17.7. Chemical structure of phosphate esters.

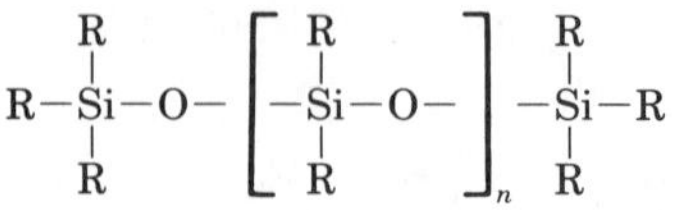

$$R-\overset{\displaystyle R}{\underset{\displaystyle R}{\overset{|}{\underset{|}{Si}}}}-O-\left[-\overset{\displaystyle R}{\underset{\displaystyle R}{\overset{|}{\underset{|}{Si}}}}-O-\right]_n-\overset{\displaystyle R}{\underset{\displaystyle R}{\overset{|}{\underset{|}{Si}}}}-R$$

FIGURE 17.8. Chemical structure of a silicone fluid.

(400°F), lubricants. The general structure of silicone fluids may be shown as in Fig. 17.8: R represents methyl, phenyl, or some other organic group.

In addition to favorable viscosity-temperature characteristics, both thermal and oxidation resistance are excellent. As a family, these fluids also exhibit low volatility and good hydrolytic stability. These materials are inert towards most elastomers and polymers as long as very high temperatures are avoided. Oxygen exposure with high-temperature use, however, can result in gelation and loss of fluidity. The lubrication properties of the base oils are not impressive compared to other lubricating fluids. Typical applications of these materials as lubricants are electric motors, brake fluids, oven preheater fans, plastic bearings, and electrical insulating fluids.

Silicate esters represent a mating of the previous two lubricant fluid types. As a class, these fluids possess good thermal stability and low volatility. These materials are used in high-temperature hydraulic fluids and low-volatility greases.

Fluorinated polyethers as a class represent the highest-temperature lubricating fluids commercially available. Although distinct chemical versions are marketed, all of these fluids are fully fluorinated and completely free of hydrogen. This structural characteristic makes them inert to most chemical reactions, nonflammable, and extremely oxidation resistant. Products from these oils show very low volatility and excellent resistance to radiation-induced polymerization. The products are essentially insoluble in common solvents, acids, and bases. Density is approximately double that of petroleum oils. Products of this chemical family are used to lubricate rolling bearings at extremely high temperatures—that is, 204–260°C (400–500°F). Other application areas are in high vacuums, corrosive environments, and oxygen-handling systems. The cost of these lubricants is very high.

GREASE LUBRICANTS

General Conditions of Use

Grease is a thickened oil that allows localization of the lubricant to areas of contact in the bearing. Rolling bearing greases are a suspension of fluid dispersed into a soap or nonsoap thickener, with the addition of a variety of performance additives.

Grease provides lubricant by bleeding; that is, when the moving parts of a bearing come in contact with grease, a small quantity of thickened oil will adhere to the bearing surfaces. The oil is gradually degraded by oxidation or lost by evaporation, centrifugal force, and so on; in time the oil in the grease near the bearing will be depleted.

Several differing viewpoints currently exist concerning the mechanism of grease operation. Until recently, grease was looked upon as merely a sponge holding oil near the working contacts. As these contacts consumed oil by way of evaporation and oxidation, a replenishment flow maintained an equilibrium as long as the supply lasted. Research using optical EHD and microflow lubrication assessment techniques has shown that the thickener phase plays rather complex roles in both the development of a separating film between the surfaces and in the modulation of the replenishing flows. The manner in which the thickener controls oil outflow, reabsorbs fluid thrown from the contacts, and acts as a trap for debris is little understood at this time. The mechanism is not steady state, but is characterized by a series of identifiable events.

Greases offer the following advantages compared to fluid lubricants:

1. Maintenance is reduced because there is no oil level to maintain. New lubricant needs to be added less frequently.
2. Lubricant in proper quantity is confined to the housing. Design of enclosures can therefore be simplified.
3. Freedom from leakage can be accomplished, avoiding contamination of products in food, textile and chemical industries.
4. The efficiency of labyrinth "seals" is improved, and better sealing is offered for the bearing in general.
5. The friction torque and temperature rise are generally more favorable.

Grease Property Definitions

Bomb Oxidation. The procedures described in the following are available from the American Society for Testing and Materials (ASTM). The determination of the resistance of lubricating greases to oxidation when stored under static conditions for a long time is described by ASTM specification D-942. A sample is oxidized in a "bomb" heated to 99°C (210°F) and filled with oxygen at 0.76 N/mm^2 (110 psi). Pressure is observed and recorded at stated intervals. The degree of oxidation after a given period of time is determined by the corresponding decrease in oxygen pressure.

Dropping Point. Dropping point is the temperature at which a grease becomes a liquid and is sometimes referred to as the melting point. The test is performed per ASTM specification D-566.

Evaporation Loss. The method of determining evaporation loss is described by ASTM specification D-972. The sample in an evaporation cell is placed in a bath maintained at the desired test temperature [usually 99–149°C (210–300°F)]. Heated air is passed over the cell surface for 22 hr. The evaporation loss is calculated from the sample weight loss.

Flash Point. Flash point is the lowest temperature at which an oil gives off inflammable vapor by evaporation, per ASTM specification D-566.

Low-Temperature Torque. Low-temperature torque is the extent to which a low-temperature grease retards the rotation of a slow-speed ball bearing when subjected to subzero temperature. The method of testing is described by ASTM specification D-1478.

Oil Separation. This is the tendency of lubricating grease to separate oil during storage in both conventional and cratered containers, as described by ASTM specification D-1742; the sample is determined by supporting on a 74-μm sieve subjected to 0.0017 N/mm^2 (0.25 psi) air pressure for 24 hr at 25°C (77°F). Any oil seepage drains into a beaker and is weighed.

Penetration. The penetration is determined at 25°C (77°F) by releasing a cone assembly from a penetrometer and allowing the cone to drop into the grease for 5 sec. The greater is the penetration, the softer is the grease. Worked penetrations are determined immediately after working the sample for 60 strokes in a standard grease worker. Prolonged penetrations are performed after 100,000 strokes in a standard grease worker. A common grease characteristic is described by NLGI (National Lubricating Grease Institute) grade assigned, as shown in Table 17.2. Most rolling bearing applications employ an NLGI 1, 2, or 3 grade grease.

TABLE 17.2. NLGI Penetration Grades

NLGI Grade	Penetration (60 Strokes)
000	445–475
00	400–430
0	355–385
1	310–340
2	265–295
3	220–250
4	175–205
5	130–160
6	85–115

Pour Point. Pour point is the lowest temperature at which an oil will pour or flow. The pour point is measured under the conditions in ASTM specification D-97. The pour point together with measured low-temperature viscosities gives an indication of the low-temperature serviceability of an oil.

Viscosity, Viscosity Index. The values of viscosity and VI generally refer to the base oil values of these properties as discussed in "Liquid Lubricants".

Water Washout Resistance. Water washout resistance is the resistance of a lubricating grease to washout by water from a bearing tested at 38°C (100°F) and 79.5°C (145°F), as described by ASTM specification D-1264.

Types of Grease Thickeners

Thickener composition is critical to grease performance, particularly with respect to temperature capability, water-resistance, and oil-bleeding characteristics. Thickeners are divided into two broad classes: soaps and nonsoaps. Soap refers to a compound of a fatty acid and a metal. Common metals for soaps include aluminum, barium, calcium, lithium, and sodium. The great majority of commercial greases are soap type, with lithium being the most widely used.

Lithium soap greases—Lithium soaps are divided into two types: 12-hydroxysterate and complex. The latter material is derived from organic acid components and permits higher temperature performance. The upper operating temperature limit of the usual lithium-based grease is approximately 110°C (230°F). For a lithium complex-based grease the upper temperature limit is extended to 140°C (284°F). Conversely, the lower operating temperature limits are −30°C (−22°F) and −20°C (−4°F), respectively. High-quality lithium soap greases of both types have excellent service histories in rolling bearings and have been used extensively in prelubricated; that is, sealed and *greased-for-life* applications. Lithium-based products have found acceptance as multipurpose greases and have no serious deficiencies except in severe temperature or loading extremes.

Calcium soap greases—The oldest of the metallic soap types, calcium-based greases, has undergone several important technical changes. In the first formulations, substantial water (0.5–1.5%) was needed to stabilize the finished product. Loss of water destroys grease consistency; as such, grease upper temperature operating limit is only

60°C (140°F). [Correspondingly, the lower operating temperature is only −10°C (14°F).] Regardless of temperature, evaporation of water occurs, requiring frequent relubrication of the bearing. Alternately, the ability of the grease to entrain water is of some advantage; such greases have been widely used in food processing plants, water pumps, and wet applications in general. Today, this type of formulation has been made obsolete by newer products with better temperature performance.

The latest development in calcium-thickened greases is the calcium complex-based grease. Herein the soap is modified by adding an acetate, and a substantially different product results having upper and lower operating temperature limits of 130°C (266°F) and −20°C (−4°F), respectively. Performance of these greases in rolling bearings is sometimes less than optimum. Although high temperature and *EP* (extreme pressure) characteristics have been exhibited, there are some problems with excessive grease thickening in use, causing an eventual loss of lubrication to the bearing.

Sodium soap greases—Sodium soap greases were developed to provide an increase in the limited temperature capability of early calcium soap greases. An inherent problem with this thickener is poor water-washout resistance; however, small amounts of water are emulsified into the grease pack, which helps to protect metal surfaces from rusting. The upper operating temperature limit for such greases is only 80°C (176°F). The lower operating temperature limit is −30°C (−22°F). Sodium-base greases have been superceded by more water-resistant products in applications such as electric motors and front wheel bearings. Sodium complex-base greases have subsequently been developed having upper and lower operating temperature limits of 140°C (284°F) and −20°C (−4°F), respectively.

Aluminum complex greases—Aluminum stearate greases are seldom used in rolling bearings, but aluminum complex-base greases are being used more often. Greases formed from the complex soap perform favorably on water-resistance tests; however, the upper operating temperature limit is somewhat low at 110°C (230°F) compared to other types of high-quality greases. The lower operating temperature limit is satisfactory at −30°C (−22°F). These greases find use in rolling mills and food-processing plants.

Non-soap-base greases—Organic thickeners, including ureas, amides, and dyes, are used to provide higher temperature capability than is available with metallic soap thickeners. Improved oxidation stability over metallic soaps occurs because these materials do not catalyze base oil oxidation. Dropping points for greases of these types are generally above 260°C (500°F) with generally good low temperature properties. The most popular of these thickeners is polyurea, which is extensively used in high-temperature ball bearing greases for the electric motor industry. The recommended upper operating temperature limit for polyurea-base grease is 140°C (284°F); the lower temperature limit is −30°C (−22°F).

Inorganic thickeners include various clays such as bentonite. Greases made from a clay base do not have a melting point, so service temperature depends on the oxidation and thermal resistance of the base oil. These greases find use in special military and aerospace applications requiring very high temperature performance for short intervals, for example, greater than 170°C (338°F). On the other hand, the recommended upper temperature limit for continuous operation is only 130°C (266°F); the lower temperature limit is −30°C (−22°F).

Grease Compatibility. Mixing greases of differing thickeners and/or base oils can produce an incompatibility and loss of lubrication with eventual bearing failure. When differing thickeners are mixed—that is, soap and nonsoap or differing soap types—dramatic changes in consistency can result, leading to a grease either too stiff to lubricate properly or too fluid to remain in the bearing cavity. Mixing greases of differing base oils—that is, petroleum and silicone oils—can produce a two-component fluid phase that will not provide a continuous lubrication medium. Early failures can be expected under these conditions. The best practice to follow is to not mix lubricants but rather purge bearing cavities and supply lines with new lubricant until previous product cannot be detected before starting operation.

POLYMERIC LUBRICANTS

A polymeric lubricant uses a matrix or spongelike material that retains its physical shape and location in the bearing. Lubrication functions are provided by the oil alone after it has bled from the sponge. Ultrahigh molecular weight polyethylene forms a pack with generally good performance properties, but it is temperature limited to about 100°C (212°F), precluding its use in many applications within the temperature capability of standard rolling bearings. Some higher temperature materials, such as polymethylpentene, form excellent porous structures but are relatively expensive and suffer from excessive shrinkage. Fillers and blowing agents, tools of the plastics industry, interfere with the oil-flow behavior, and contribute little in this situation. New solutions must be developed.

Despite its temperature limitations, a polyethylene-based material,

FIGURE 17.9. Polymer-lubricated rolling bearing.

designated W64 by SKF, has achieved some notable successes in the solution of everyday bearing lubrication problems. Fig. 17.9 shows a W64-filled bearing. Perhaps the most notable success is where a bearing must operate under severe acceleration conditions, most typical of which are found in planetary systems. Bearing rotational speed about its own axis may be moderate, but the centrifuging action due to the planetary motion is strong enough to throw conventional greases out of the bearing despite the presence of seals. When polymerically lubricated bearings are substituted, life improvements of two orders of magnitude are not uncommon. Such situations occur in cablemaking, tire-cord winding, and textile mill applications.

Another major market for polymer lubricants is food processing. Food machinery must be cleaned frequently, often daily, using steam, caustic, or sulfamic acid solutions. These degreasing fluids tend to remove lubricant from the bearings, and it is standard practice to follow every cleaning procedure with a relubrication sequence. Polymer lubricants have proven to be highly resistant to washout by such cleansing methods, hence the need for regreasing is reduced. The reservoir effect of polymer lubricants has been exploited to a degree in bearings normally lubricated by a circulating oil system where there can be a delay in the oil reaching a critical location. The same effect has been used to provide a backup in case the oil supply system should fail.

The high occupancy ratio of the void space by the polymer minimizes the opportunity for the bearing to "breathe" as temperatures change. Corrosion due to internal moisture condensation is therefore reduced. Since all ferrous surfaces are very close to the pack, conditions are optimum for using vapor phase corrosion-control additives in the formulations.

Despite these advantages polymeric lubricants have some specific drawbacks. There tends to be considerable physical contact between the pack and the moving surfaces of the bearing. This leads to increased frictional torque, which produces more heat in the bearing. In conjunction with thermal insulating properties of the polymer and its inherently limited temperature tolerance, the speed capability is reduced. Moreover, compared to grease, the solid polymer is relatively incapable of entrapping wear debris and dirt particles.

SOLID LUBRICANTS

Solid lubricants are used where conventional lubricants are not suitable. Extreme environment conditions frequently make solid lubricants a preferred choice of lubrication. Solid lubricants can survive temperatures well above the decomposition temperatures of oils. They can also be used in chemically reactive environments. The disadvantages of solid lubricants are (1) high coefficient of friction, (2) inability to act as a coolant, (3) finite wear life, (4) difficult replenishment, and (5) little damping effect for controlling vibrational instabilities of rolling elements and separator components.

Many common solid lubricants, such as graphite and molybdenum disulfide (MoS_2), are layered lattice compounds that shear easily along preferred planes of their structure. MoS_2 has weak van der Waals forces between sulfur bonds, giving the material a characteristic relatively low coefficient of friction. MoS_2 oxidizes at approximately 399°C (750°F) in air; the oxides can be abrasive.

The low friction associated with graphite depends on intercalation with gases, liquids, or other substances. For example, the presence of absorbed water in graphite imparts good lubricating qualities. Thus, pure graphite has deficiencies as a lubricant except when used in an environment containing contaminants such as gases and water vapor. With proper additives graphite can be effective up to 649°C (1200°F).

Tungsten disulfide (WS_2) is similar to MoS_2 in that it is a type of layered lattice solid lubricant. It does not need absorbable vapors to develop low-shear-strength characteristics.

Other "solid" lubricating materials are solid at bulk temperatures of the bearing but melt from frictional heating at points of local contact, giving rise to a low-shear-strength film. This melting may be very localized and of very short duration. Soft oxides, such as lead monoxide (PbO), are relatively nonabrasive and have relatively low friction coefficients, especially at high temperatures where their shear strengths are reduced. At these temperatures deformation occurs by plastic flow rather than by brittle fracture. Melted oxides can form a glaze on the surface. This glaze can increase or decrease friction, depending on the "viscosity" of the glaze within the contact region. Stable fluorides such as lithium fluoride (LiF_2), calcium fluoride (CaF_2), and barium fluoride (BaF_2) also lubricate well at high temperatures but over a broader range than lead oxides.

SEALS

Functions of Seals

Once the general method of lubricating a bearing has been determined, the question of a suitable sealing method frequently needs to be addressed. A seal has two basic tasks. It must keep lubricant where it belongs and keep contaminating materials from the bearing and its lubricant. This separation must be accomplished between surfaces in relative motion, usually a shaft or bearing inner ring and a housing. The seal must not only accommodate rotary motion, but it must also accommodate eccentricities due to run-outs, bearing clearance, misalignments, and deflections. The selection of a seal design depends on the category of lubricant employed (grease, oil, or solid). Also, the amount and nature of the contaminant that must be kept out needs to be assessed. Speed, friction, wear, ease of replacement, and economics govern the final choice. Bearings run under a great variety of conditions, so it is necessary to judge which seal type will be sufficiently effective in each particular circumstance.

Seals with Greased Bearings

Grease is the simplest lubricant to seal. The fluidity of the oil has been deliberately reduced by blending with the thickener. The stiff nature of grease means that it requires little in the way of constraint yet at the same time readily plugs small spaces. Given a suitably small gap, grease can form layers on the opposing moving surfaces, effectively closing the gap completely. This principle is used in labyrinth-type designs.

Because it has an oily consistency, dirt or dust particles that penetrate a seal are caught by the grease and prevented from entering the bearing. The wicking type of oil delivery to the bearing means that the particles are permanently kept out of circulation unless the grease becomes stirred in some way.

Some seals make physical contact between the surfaces. A film of grease provides the necessary continuous supply of lubricant to establish hydrodynamic separation with its attendant low friction and wear.

Seals with Oil-Lubricated Bearings

Oils are more difficult to seal than greases. They will flow through the smallest gaps if there is any hydraulic head. Either the possibility of a head developing must be prevented by the cavity design, or running gaps must be eliminated by use of a movable seal lip. Oils are excellent as dirt traps, but they lack the ability to keep the dirt out of circulation unless backed up by a filter system. Dirt can only be kept out by positive gap elimination.

Seals for Solid-Lubricated Bearings

Only the powder and reactive gas forms of solid lubricants pose sealing challenges. Even then, since they are essentially once-through systems, some leakage is tolerable unless the material is toxic. Furthermore sealing of such lubricants can pose grave difficulties, particularly since they are almost exclusively used at extremely high temperatures at which gas dynamic behavior increases. Zero gap conditions are then necessary with extra provision made to prevent the powder or soot-type exhaust products from compacting and causing separation.

Types of Seals

Labyrinth Seals. Labyrinth seals consist of an intricate series of narrow passages that protect well against dirt intrusion. An example is shown in Fig. 17.10. This type is suitable for use in pillow blocks or other assemblies where the outer stationary structure is separable. The inner part is free to float on the shaft so that it can position itself relative to

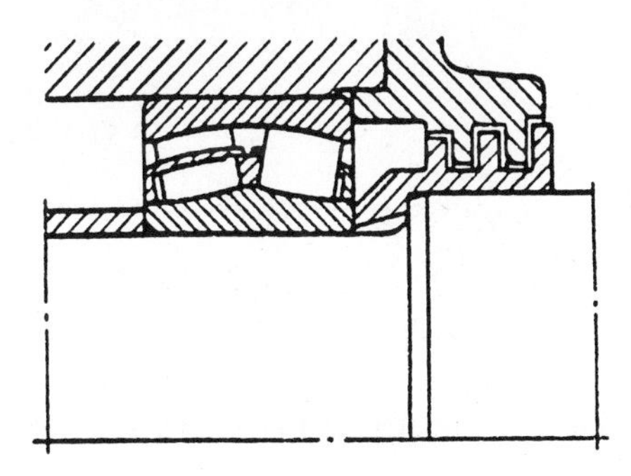

FIGURE 17.10. Bearing housing with labyrinth seal.

the fixed sections. The mechanism of sealing is complex, being associated with turbulent flow fluid mechanics. It is reasonably effective with liquids, greases, and gases, provided that there is no continuous static head across the assembly.

It is normal practice to add grease to the labyrinth, making the gaps even smaller than can be achieved mechanically due to tolerance stack-ups. Dirt has virtually no chance of penetrating such a system without becoming ensnared in the grease. A further advantage accrues at re-greasing. Spent lubricant can purge readily through the labyrinth and flush the trapped debris with it.

The relatively moving parts are separated by a finite gap, so wear, in the absence of large bridging particles of dirt, is essentially nonexistent. Likewise, frictional losses are extremely low. The number of convolutions of the labyrinth passage can be increased with the severity of the dirt exclusion requirements. Separate flingers and trash guards or cutters may be added on the outboard side to deal with wet or fibrous contaminants that could damage or penetrate the labyrinth. Figure 17.11

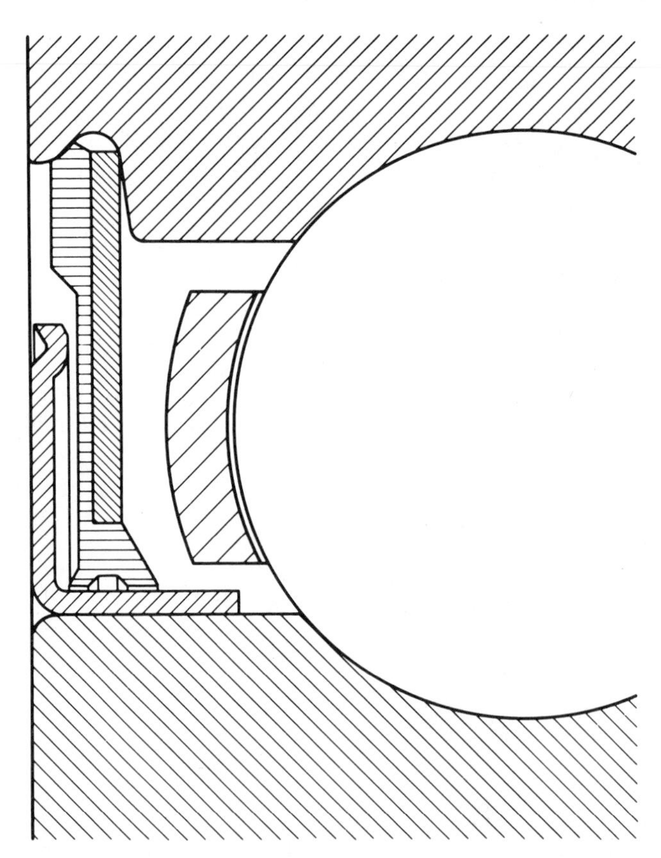

FIGURE 17.11. Deep-groove ball bearing assembly with integral labyrinth seal and flinger ring. (SKF photograph.)

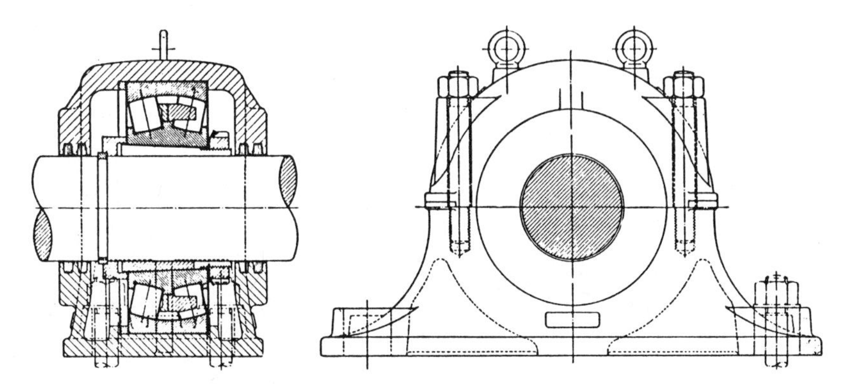

FIGURE 17.12. Bearing pillow block with felt seals.

shows a ball bearing with an integral labryinth seal and outboard flinger ring.

Felt seals.—Semicircular pieces of felt, pressed into trapezoidal section grooves in the housing, lightly contact the shaft surface, as shown in Fig. 17.12. Inexpensive and simple to install and replace, the grease-laden felts keep dirt out of the enclosure; however, the dirt entrapped in the felt fibers can cause serious shaft surface wear. Also, the felt can become compacted, eventually leaving an air gap. Friction is often high and difficult to control. For these reasons, felt seals, though once popular, are not currently in significant use.

Shields—As Fig. 17.13 shows, a shield takes up very little axial space and can usually be accommodated within the standard boundary dimensions of the bearing. The near knife edge standing just clear of the ring land is, in effect, a single-stage labyrinth seal. Effective enough to keep all but the most fluid greases in the bearing, the shield can be considered as a modest dirt excluder, suitable for use in most workplace environments. Under harsher conditions it must be backed up with extra guards. Special greases or acceptance of leakage and reduced lubricant life are necessary when shielded bearings are used in vertical axis applications. The absence of contact friction permits these bearings to be used at the highest speed allowed by the mode of lubrication and type of lubricant.

Elastomeric lip seals—The narrow gap between a shield and an inner ring groove or chamfer can be closed by a carefully designed section of elastomer (nitrile rubber for general purposes). Figure 17.14 illustrates a typical configuration. The flexible material makes rubbing contact with the ring and establishes a barrier to the outward flow of lubricant or the ingress of contaminants. When the bearing is in motion, the elastomer must slide over the metal surface, and

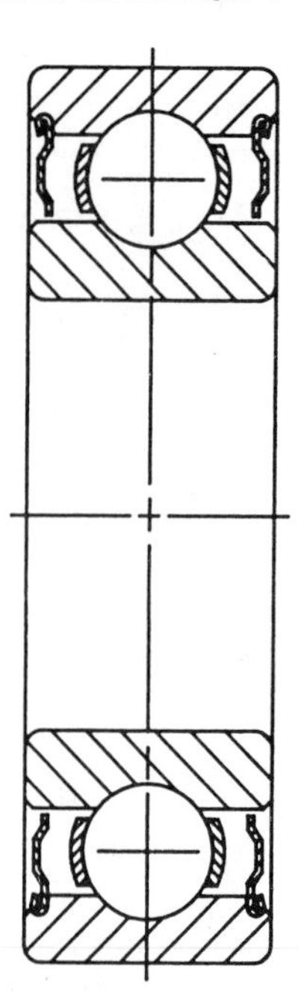

FIGURE 17.13. Radial ball bearing with shields.

a frictional drag is produced, which even for a well-designed seal, is generally greater than the frictional torque of the bearing. Often more important is the seal *breakaway* torque, which can be several times the running torque. Considerable research has been devoted to finding both elastomers and seal designs that achieve a suitable balance between sealing efficacy, lip or ring wear, and frictional torque.

The lip of the seal must bear on the ring with sufficient pressure to follow the relative motions of the running surface caused by eccentricities and roundness errors. This pressure is achieved by a slight interference fit, producing a dilation of the seal. The spring rate of the lip governs the speed at which the lip can respond to the running errors without a gap being formed through which fluid can pass. Higher bearing speeds demand better running accuracies. Spring rate is regulated by the elastic properties of the seal material and the design of the bending section.

At first glance, even though a lubricant is present, no hydrodynamic lift would be expected on the lip, due to the axial symmetry. Recent work has established, both theoretically and experimentally, that a very thin, stable dynamic film persists over much of the operating regime. The mechanism of sealing is a complex one involving the elastomeric lip, the counterface, and the grease, or at least the oil phase of the grease. Figure 17.15 shows the seal of Fig. 17.14 composed of a molded annulus of polymer bonded to a thin steel disc.

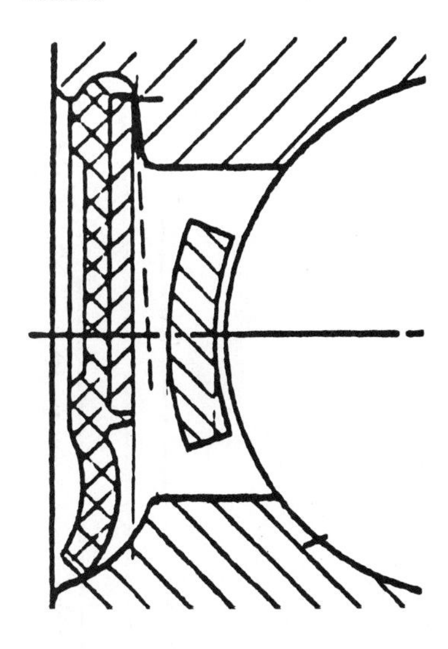

FIGURE 17.14. Radial ball bearing with integral single lip seal.

The disc provides mechanical support against minor pressure differences that can occur across the seal and also assures a slight compression of the polymer against the outer ring recess, thus creating a fluid-tight static seal at that point. The inside diameter of the disc, in conjunction with the waisted section of the molding, defines the flexure point of the lip itself and is located so that the deflected lip bears against the counterface groove with suitable pressure and at a predetermined angle. This angle of contact produces appropriate convergence and divergence on either side of the contact, which helps the sealing function appreciably. The lip pressure induced by the interference between the seal and its counterface is sufficient to prevent fluid leakage under static conditions.

To function adequately, the elastomer must exhibit specific properties. Beyond compatibility with the common types of lubricating oils and swell that can be accommodated by the seal configuration, it must survive the frictional heating at the lip and heat from the bearing or its environment without hardening, cracking, or otherwise aging. To survive start-up and the presence of dirt, it must have wear and abrasion resistance. Care must be taken when forming the elastomer and its fillers that the final cured product does not promote corrosion of the counterface under humid conditions. The range of candidate materials, their chemical structures, and physical properties are discussed in Chapter 16.

Lip seals require the presence of lubricant, for if allowed to run dry, wear and failure are usually rapid. The grease charge for the bearing must be positioned to wet the seals upon assembly. In most cases the grease volume is sufficient to require a period of working when the bearing first operates. This is followed by channeling, and the formation of

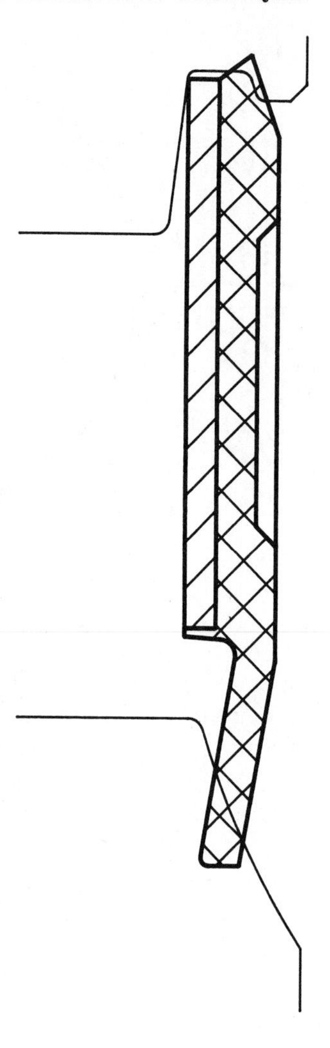

FIGURE 17.15. Lip seal construction showing interference with bearing inner ring seal groove and retention in outer ring groove.

grease packs against the inside surfaces of the seals. Operational sufficiency is then assured.

In operation, EHD lubrication develops under the lip, which progressively lowers the seal torque by changing the friction from Coulomb to viscous shear. Wear is thereby greatly reduced. Currently, there are two schools of thought concerning lubricant film formation. One approach ascribes the film to asperities on the seal lip, producing localized hydrodynamic pressure fluctuations, as illustrated in Fig. 17.16. Cavitation downstream from each asperity limits the negative pressure remaining to separate the surfaces. The countervailing view invokes the viscoelastic properties of the seal material and the inability of the elastomer to

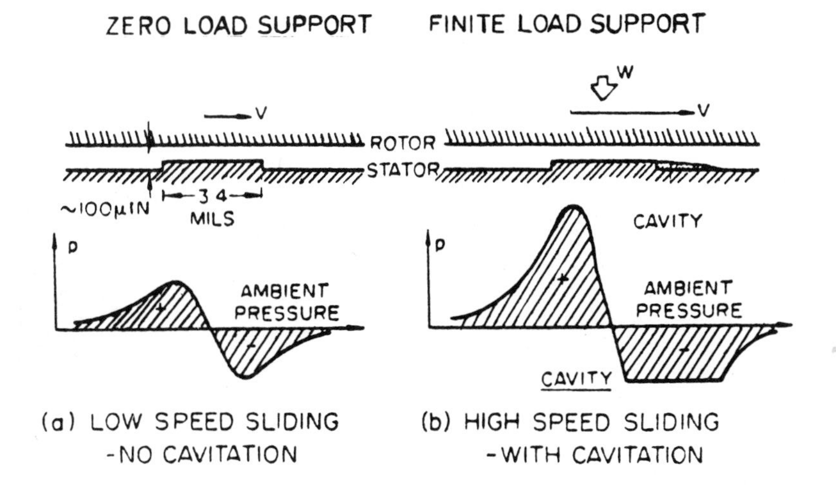

FIGURE 17.16. Inducing hydrodynamic separation of sealing surfaces by asperities.

follow precisely the radial motions of the counterface produced by eccentricity and out-of-roundness.

Both of these mechanisms may be valid and function simultaneously and essentially independently of one another. The first is governed by the microgeometry of the lip as modified by wear, abrasion scratches, thermal and installation distortions, and possibly inhomogeneities in elastomer properties. The second is a by-product of manufacturing process characteristics and nonrotary displacements of the inner ring.

Seal torque arises from four sources: adhesion between asperities, abrasion, viscous shearing of the film, and hysteresis in the elastomer. The last two are strongly influenced by temperature and so tend to be self-limiting; otherwise they all depend not only on the application but on the detail installation itself. Methods for exact prediction of torque and operating temperature have not yet been devised. In seemingly identical conditions one sealed bearing frequently runs cooler than another, or one will leak slightly and another will not. Much work needs to be done to predict seal performance in a given application.

The primary task of the single lip seal is to contain grease. It can exclude moderate dust as found in typical home or commercial atmospheres, and it finds a great many suitable applications. Some dusts, such as from wood sanders or lint accumulating on the bearings in textile machinery, have the ability to wick considerable amounts of oil through the lip film, which shortens bearing life. In these situations and

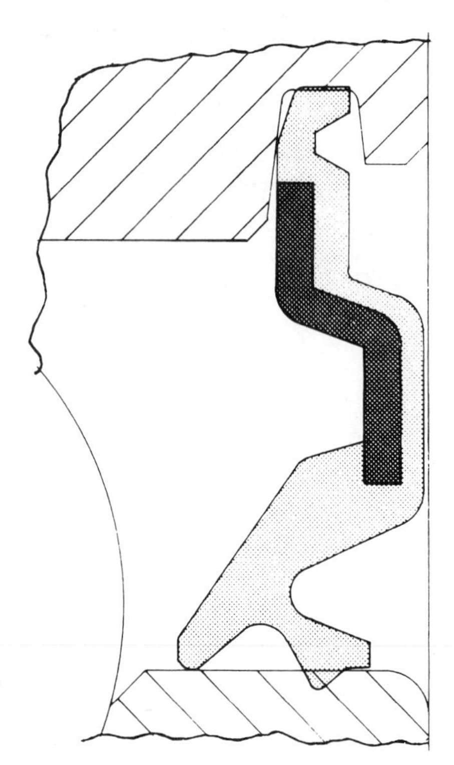

FIGURE 17.17. Double-lip seal. Inside lip is for fluid retention; outside lip is for dust exclusion.

where there is heavy exposure to dirt, particularly waterborne dirt, such as in automotive uses, additional protection in the form of dust lips and flingers should be provided. Figure 17.17 shows an example of a double-lip seal.

Garter Seals. Similar in many respects to the lip seal, the garter seal uses a hoop spring or garter to apply an essentially constant inward pressure on the lip. As shown in Fig. 17.18, the arrangement requires more axial space than is available in a bearing of standard envelope dimensions. Either extra wide rings must be used, or the seal must be fitted as a separate entity in the assembly.

The spring-induced pressure gives a very positive sealing effect and is used to contain oil rather than grease, for two reasons: Oil can be thrown or pumped by an operating bearing with considerable velocity, sufficient to cause leakage through a lip seal, and the lip itself requires a generous supply of oil for lubrication and removal of frictional heat. Relieved of the need to provide the closing force, the elastomer section can be designed to hinge freely so that relatively large amplitudes of shaft eccentricity can be accommodated. The strictly radial nature of the spring force precludes the use of anything other than a cylindrical counterface surface. Axial floating of the shaft is therefore accommodated well.

FIGURE 17.18. Garter seal X-section showing retaining spring.

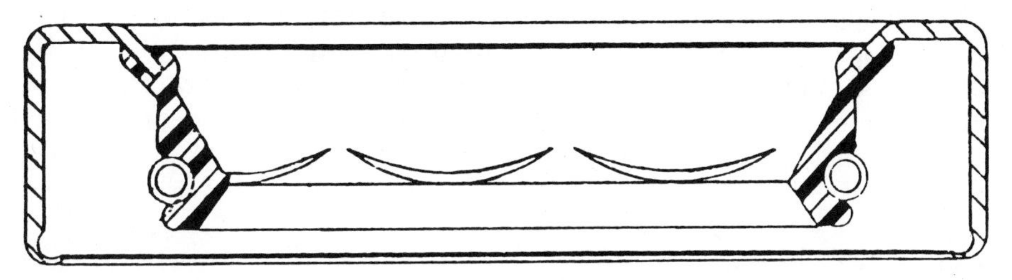

FIGURE 17.19. Radial seal with "quarter moon" projections moulded on the lip to develop a hydrodynamic lubricant film during operation.

The design lends itself to molding, and artificial asperities and other film generating devices can readily be formed in the elastomer. Figure 17.19 shows an example of a helical rib pattern intended not only to enhance the oil film thickness but to act as screw pump to minimize leakage.

Ferrofluidic Seals. Magnetic fluids are a recent introduction to the arsenal of tools available to the sealing engineer. A ferrolubricant is basically a dispersion of very fine particles of ferrite in oil. The particles are typically 100 Å in diameter and are coated with a molecular dispersing agent to prevent coalescence. Brownian movement inhibits sedimentation. The result is a lubricant that responds to magnetic fields.

Fig. 17.20 shows a two-stage seal, each stage composed of several gaps across which is suspended a ferrofluid film. This type of seal has proved very effective where bearing and shaft systems penetrate a vacuum enclosure and in computer disc drive spindle assemblies where absolutely clean internal conditions must be maintained.

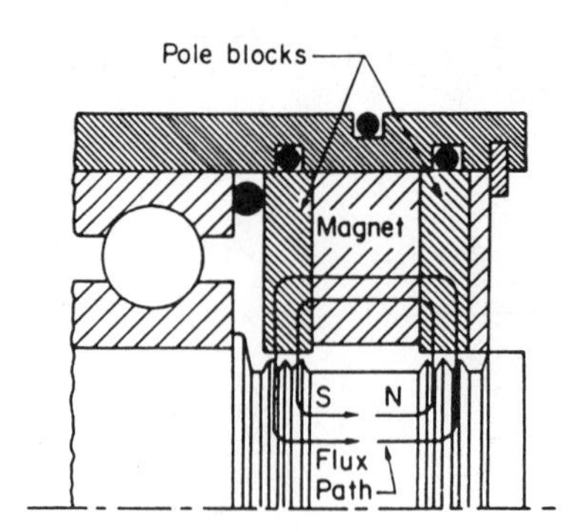

FIGURE 17.20. Ferromagnetic fluid shaft seal.

Its ability to withstand pressure gradients to 0.345 N/mm^2 (50 psi) (by multistaging) and accept high eccentricities with 100% fluid tightness makes the ferrofluidic seal essentially unique. Two things prevent greater application. The ferrite increases the apparent viscosity of the fluid, and viscous heating limits the speed capability. The greatest drawback is the need to introduce magnets into the system. Tramp iron is attracted to the seals unless considerable conventional sealing is applied outboard, negating much of the seal's advantages.

CLOSURE

Following the design and manufacture of a rolling element bearing, the technology associated with creating and maintaining the internal environment of the bearing during its operation is the single most important factor connected with its performance and life. This environment is intimately associated with the lubricant selected, its means of application, and the method of sealing. In this chapter, a brief overview has been given to each of these important considerations. No attempt has been made to provide an exhaustive study of lubricant types, means of lubrication, or means of sealing. It remains for the reader to explore each of these topics to the depth required by the individual application.

REFERENCES

17.1. SKF, *General Catalogue 4000E* (April 1989).

18

ROLLING BEARING FATIGUE

LIST OF SYMBOLS

Symbol	Description	Units
A	Material factor for ball bearings, constant	
$\mathcal{A}$	Life adjustment factor	
a	Semimajor axis of projected contact ellipse	mm (in.)
a^*	Dimensionless semimajor axis	
B	Material factor for roller bearings with line contact	
b	Semiminor axis of projected contact ellipse	mm (in.)
b^*	Dimensionless semiminor axis	
b_m	Rating factor for contemporary material	
C	Basic dynamic capacity of a bearing raceway or entire bearing	N (lb)
c	Exponent on τ_0	
d	Diameter	mm (in.)

Symbol	Description	Units
d_m	Pitch diameter	mm (in.)
D	Ball or roller diameter	mm (in.)
E	Modulus of elasticity	N/mm^2 (psi)
e	Weibull slope	
$\mathfrak{F}$	Probability of failure	
F_r	Applied radial load	N (lb)
F_a	Applied axial load	N (lb)
F_e	Equivalent applied load	N (lb)
f	r/D	
f_m	Material factor	
g_c	Factor combining the basic dynamic capacities of the separate bearing raceways	
h	Exponent on z_0	
i	Number of rows	
J_1	Factor relating mean load on a rotating raceway to Q_{max}	
J_2	Factor relating mean load on a nonrotating raceway to Q_{max}	
J_r	Radial load integral	
J_a	Axial load integral	
K	A constant	
L	Fatigue life	revolutions $\times 10^6$
L_{10}	The fatigue life that 90% of a group of bearings will endure	revolutions $\times 10^6$
L'	Adjusted fatigue life	revolutions $\times 10^6$
L_{50}	The fatigue life that 50% of a group of bearings will endure	revolutions $\times 10^6$
L_s	Fatigue life corresponding to probability of survival $\mathcal{S}$	revolutions $\times 10^6$
l	Effective roller length	mm (in.)
$\mathfrak{L}$	Length of rolling path	mm (in.)
N	Number of revolutions	revolutions
n	Rotational speed	rpm
$\mathfrak{N}$	Number of bearings in a group	
n_{mi}	Orbital speed of rolling elements relative to inner raceway	rpm
Q	Ball or roller load	N (lb)
Q_c	Basic dynamic capacity of a raceway contact	N (lb)
Q_e	Equivalent rolling element load	N (lb)
R	Roller contour radius	mm (in.)
r	Raceway groove radius	mm (in.)
$\mathcal{S}$	Probability of suvival	

Symbol	Description	Units
T	τ_0/σ_{max}	
u	Number of stress cycles per revolution	
$\mathcal{V}$	Volume under stress	mm^3 ($in.^3$)
V	Rotation factor	
v	$J_2(0.5)/J_1(0.5)$	
X	Radial load factor	
Y	Axial load factor	
$\mathcal{Y}_e$	Standardized excess fatigue life	
Z	Number of rolling elements per row	
z_0	Depth of maximum reversing shear stress	mm (in.)
α	Contact angle	rad,°
γ	$D \cos \alpha/d_m$	
ϵ	Factor describing load distribution	
ζ	z_0/b	
η	Capacity reduction factor	
Λ	Lubricant film parameter	
λ	Reduction factor to account for edge loading and nonuniform stress distribution on the rolling elements	
ν	Reduction factor used in conjunction with a load-life exponent $n = 10/3$	
σ	Normal stress	N/mm^2 (psi)
τ_0	Maximum orthogonal subsurface shear stress	N/mm^2 (psi)
ϕ	Amplitude of oscillation	rad,°
ψ	Position angle of rolling element	rad,°
ψ_l	Limiting position angle	rad,°
ω_s	Spinning speed	rad/sec
ω_{roll}	Rolling speed	rad/sec
$\Sigma\rho$	Curvature sum	mm^{-1} ($in.^{-1}$)
$F(\rho)$	Curvature difference	

SUBSCRIPTS

a	Refers to axial direction
c	Refers to a single contact
e	Refers to an equivalent load
i	Refers to inner raceway
j	Refers to a rolling element location
l	Refers to line contact
μ	Refers to a rotating raceway
ν	Refers to nonrotating raceway
o	Refers to the outer raceway

Symbol	Description	Units
	SUBSCRIPTS	
r	Refers to the radial direction	
s	Refers to probability of survival $\mathcal{S}$	
R	Refers to rolling element	
I	Refers to body I	
II	Refers to body II	

GENERAL

It has been considered that if a rolling bearing in service is properly lubricated, properly aligned, kept free of abrasives, moisture, and corrosive reagents, and properly loaded, then all causes of damage are eliminated save one, material fatigue. Historically, rolling bearing theory postulated that no rotating bearing can give unlimited service, because of the probability of fatigue of the surfaces in rolling contact. As indicated in Chapter 5, the stresses repeatedly acting on these surfaces can be extremely high as compared to other stresses acting on engineering structures. In the latter situation, some steels appear to have an endurance limit, as shown in Fig. 18.1. This endurance limit is a level of cyclically applied, reversing stress, which, if not exceeded, the structure will accommodate without fatigue failure. The endurance limit for structural fatigue has been established by rotating beam and/or torsional testing of simple bars for various materials.

In Chapter 23, the concept of a fatigue endurance limit for rolling bearings will be discussed in detail as well as the correlation of struc-

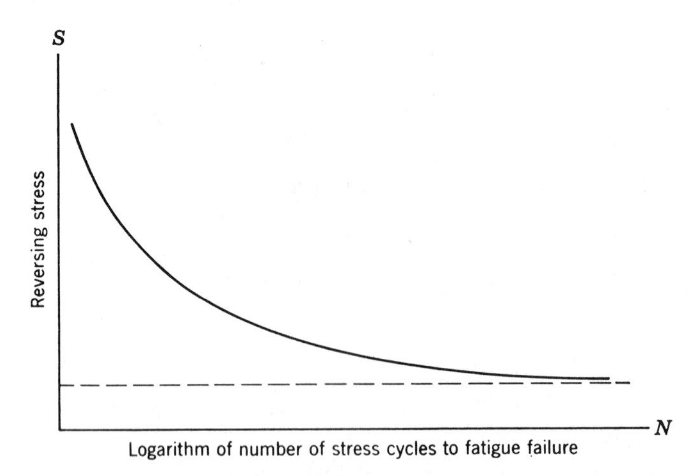

FIGURE 18.1. *S–N* curve for mild steel.

tural fatigue with rolling contact fatigue. In this Chapter, the concept of rolling contact fatigue and its association with bearing load and life ratings is covered.

Rolling contact fatigue is manifested as a flaking off of metallic particles from the surface of the raceways and/or rolling elements. For well lubricated, properly manufactured bearings, this flaking usually commences as a crack below the surface and is propagated to the surface, eventually forming a pit or spall in the load-carrying surface. Lundberg et al. [18.1] postulated that it is the maximum orthogonal shear stress τ_0 of Chapter 5 that initiates the crack and that this shear stress occurs at depth z_0 below the surface. Figure 18.2 is a photograph of a typical fatigue failure in a ball bearing raceway. Figure 18.3, taken from reference [18.2], indicates the typical depth in a spalled area.

Not all researchers accept the maximum orthogonal shear stress as the significant stress initiating failure. Another criterion is the Von Mises distortion energy theory, which yields a scalar "stress" level of similar magnitude to the double amplitude; that is, $2\tau_0$, of the maximum orthogonal shear stress. Moreover, the subsurface depth at which this value is a maximum is approximately 50% greater than for τ_0. According to reference [5.16], this greater depth for failure initiation appears to be verified.

Lundberg et al. [18.1] postulated that fatigue cracking commences at weak points below the surface of the material. Hence, changing the chemical composition, metallurgical structure, and homogeneity of the steel can significantly affect the fatigue characteristics of a bearing, all other factors remaining the same. In referring to weak points, one does not include macroscopic slag inclusions, which cause imperfect steel for bearing fabrication and hence premature failure. Rather microscopic inclusions and metallurgical dislocations that are undetectable except by laboratory methods are possibly the weak points in question. Figure 18.4, taken from reference [18.2], shows a fracture failure at weak points developed during rolling. This type of experimental study tends to confirm the Lundberg–Palmgren theory insofar as failure that initiates at weak points. That the weak points are those at a specified depth below the

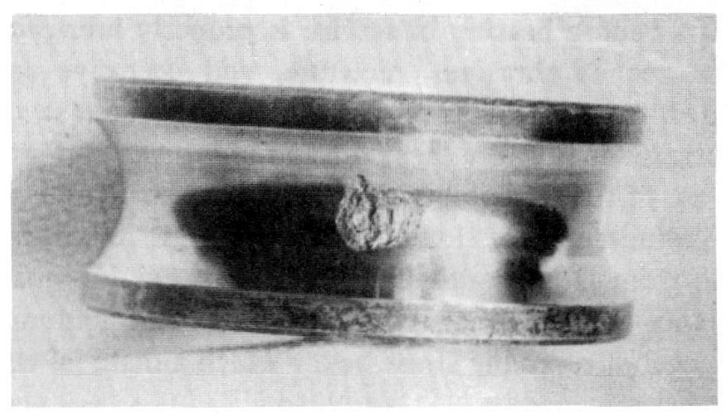

FIGURE 18.2. Rolling bearing fatigue failure.

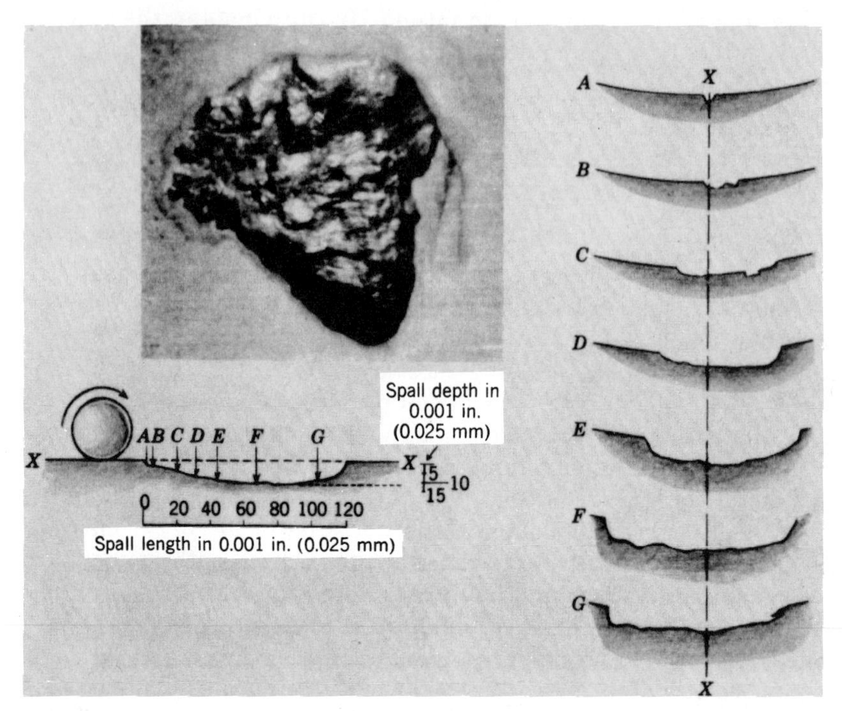

FIGURE 18.3. Characteristics of a fatigue-spalled area. Photographs of a typical fatigue spall showing sections cut through spall.

rolling contact surface, rather than at other depths or even at the surface, will be discussed later.

FATIGUE LIFE DISPERSION

Even if a population of apparently identical rolling bearings is subjected to identical load, speed, lubrication, and environmental conditions, all the bearings do not exhibit the same life in fatigue. Instead the bearings fail according to a dispersion such as that presented in Fig. 18.5. Figure 18.5 indicates that the number of revolutions a bearing may accomplish with 100% probability of survival, that is, $\mathcal{S} = 1$, in fatigue is zero. Alternatively, the probability of any bearing in the population having infinite endurance is zero. For this model, fatigue is assumed to occur when the first crack or spall is observed on a load-carrying surface. It is apparent, owing to the time required for a crack to propagate from the subsurface depth of initiation to the surface, that a practial fatigue life of zero is not possible. This will be discussed in greater depth later; how-

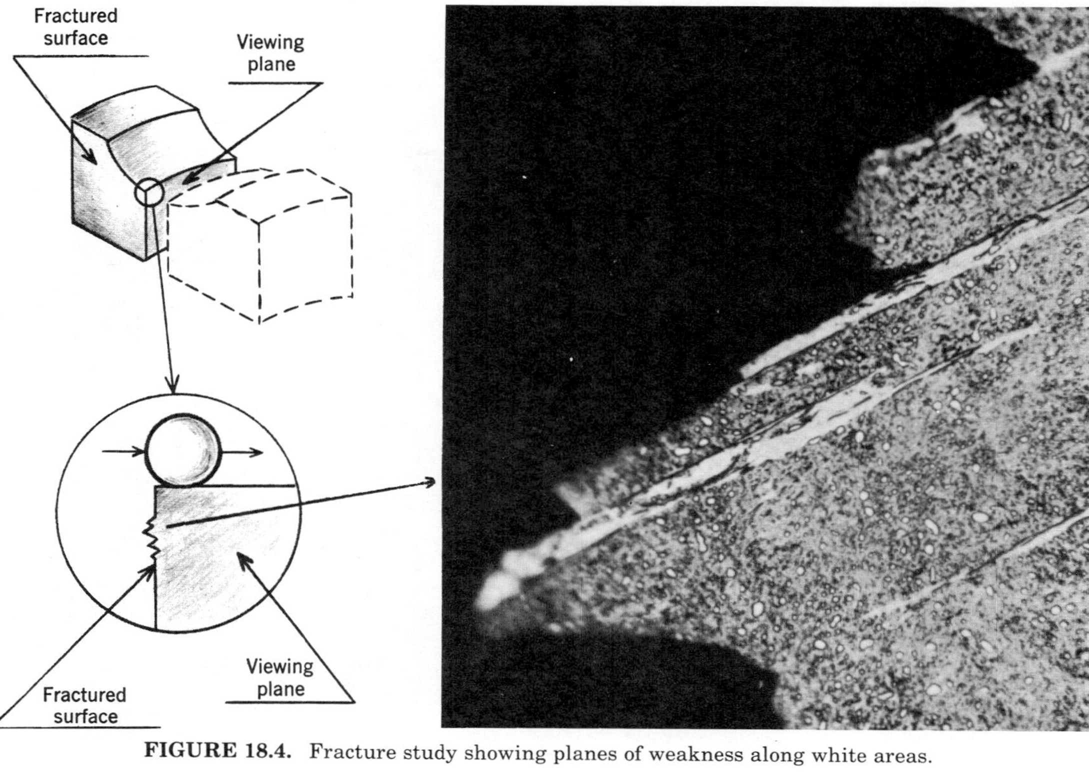

FIGURE 18.4. Fracture study showing planes of weakness along white areas.

ever, for the purpose of discussing the general concept of bearing fatigue life, Fig. 18.5 is appropriate.

Since such a life dispersion exists, bearing manufacturers have chosen to use one or two points (or both) on the curve to describe bearing endurance. These are

1. L_{10} the fatigue life that 90% of the bearing population will endure.
2. L_{50} the median life, that is, the life that 50% of the bearing population will endure.

In Fig. 18.5, $L_{50} = 5L_{10}$ approximately. This relationship is based on fatigue endurance data for all types of bearings tested and is a good rule of thumb when more exact information is unavailable.

The probability of survival $\mathcal{S}$ is described as follows:

$$\mathcal{S} = \frac{\mathfrak{N}_s}{\mathfrak{N}} \tag{18.1}$$

in which $\mathfrak{N}_s$ is the number of bearings that have successively endured L_s revolutions of operation and $\mathfrak{N}$ is the total number of bearings under

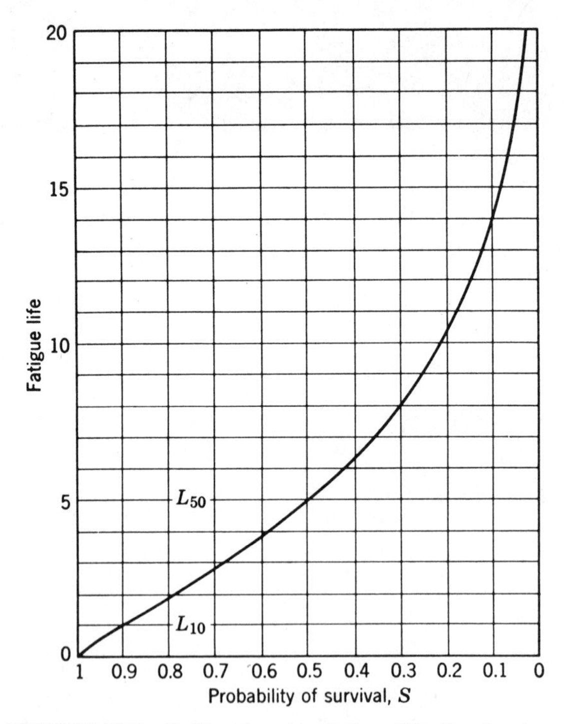

FIGURE 18.5. Rolling bearing fatigue life distribution.

test. Thus if 100 bearings are being tested and 12 bearings have failed in fatigue at L_{12} revolutions, the probability of survival of the remaining bearings is $\mathcal{S} = 0.88$. Conversely, a probability of failure may be defined as follows:

$$\mathfrak{F} = 1 - \mathcal{S} \tag{18.2}$$

Bearing manufacturers almost universally refer to a "rating life" as a measure of the fatigue endurance of a given bearing operation under given load conditions. This *rating life* is the estimated L_{10} fatigue life of a large population of such bearings operating under the specified loading. In fact, it is not possible to ascribe a given fatigue life to a solitary bearing application. One may however refer to the reliability of the bearing. Thus, if for a given application using a given bearing, a bearing manufacturer will estimate a rating life, the manufacturer is, in effect, stating that the bearing will survive the rating life (L_{10} revolutions) with 90% reliability. Reliability is therefore synonymous with probability of survival.

Fatigue life is generally stated in millions of revolutions. As an alternative it may be and frequently is given in hours of successful operation at a given speed.

An interesting aspect of bearing fatigue is the life of multirow bearings. As an example of this effect, Fig. 18.6 shows actual endurance data of a group of single-row bearings superimposed on the dispersion curve of Fig. 18.5. Next consider that the test bearings are randomly grouped in pairs. The fatigue life of each pair is evidently the least life of the pair if one considers a pair is essentially a double-row bearing. Note from Fig. 18.6 that the life dispersion curve of the paired bearings falls below the curve for the single bearings. Thus, the life of a double-row bearing subjected to the same specified loading as a single-row bearing of identical design is less than the life of a single-row bearing. Hence in the fatigue of rolling bearings, the product law of probability [18.3] is in effect.

When one considers the postulated cause of surface fatigue, the physical truth of this rule becomes apparent. If fatigue failure is, indeed, a function of the number of weak points in a highly stressed region, then as the region increases in volume, the number of weak points increases and the probability of failure increases although the specific loading is unaltered. This phenomenon is further explained by Weibull [18.4, 18.5].

WEIBULL DISTRIBUTION

In a statistical approach to the static failure of brittle engineering materials, Weibull [18.5] determined that the ultimate strength of a material cannot be expressed by a single numerical value and that a statis-

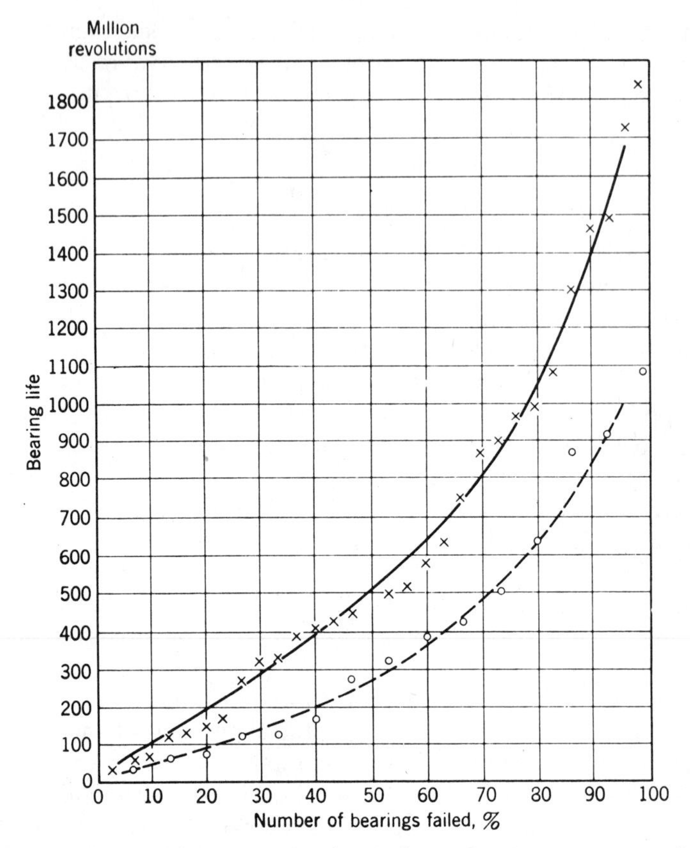

FIGURE 18.6. Fatigue life comparison of a single-row bearing to a two-row bearing. A group of single-row bearings programmed for fatigue testing, was numbered by random selection, no. 1–30, inclusive. The resultant lives, plotted individually, give the upper curve. The lower curve results if bearings nos. 1 and 2, nos. 3 and 4, nos. 5 and 6, and so on were considered two-row bearings and the shorter life of the two plotted as the life of a two-row bearing.

tical distribution was required for this purpose. The application of the calculus of probability led to the fundamental law of the Weibull theory:

$$\ln (1 - \mathfrak{F}) = -\int_{\mathcal{V}} n(\sigma)\, dv \tag{18.3}$$

Equation (18.3) describes the probability of rupture $\mathfrak{F}$ due to a given distribution of stress σ over volume $\mathcal{V}$ in which $n(\sigma)$ is a material characteristic. Weibull's principal contribution is the determination that structural failure is a function of the volume under stress. The theory is based on the assumption that the initial crack leads to a break. In the fatigue of rolling bearings, experience has demonstrated that many cracks are formed below the surface that do not propagate to the surface. Thus Weibull's theory is not directly applicable to rolling bearings. Lundberg et al. [18.1] theorized that consideration ought to be given to the fact that the probability of the occurrence of a fatigue break should be a function of the depth z_0 below the load-carrying surface at which the most severe shear stress occurs. The Weibull theory and rolling bearing statistical methods are discussed in greater detail in Chapter 20.

According to Lundberg et al. [18.1] let $\Gamma(n)$ be a function that describes the condition of material at depth z after n loadings. Therefore $d\Gamma(n)$ is the change in that condition after a small number of dn subsequent loadings. The probability that a crack will occur in the volume element $\Delta\mathcal{V}$ at depth z for that change in condition is given by

$$\mathfrak{F}(n) = g[\Gamma(n)]\, d\Gamma(n)\Delta\mathcal{V} \tag{18.4}$$

Thus, the probability of failure is assumed to be proportional to the condition of the stressed material, the change in the condition of the stressed material, and the stressed volume. The magnitude of the stressed volume is evidently a measure of the number of weak points under stress.

In accordance with equation (18.4), $\mathcal{S}(n) = 1 - \mathfrak{F}(n)$ is the probability that the material will endure at least n cycles of loading. The probability that the material will survive at least $n + dn$ loadings is the product of the probabilities that it will survive n load cycles and that the material will endure the change of condition $d\Gamma(n)$. In equation format, this is

$$\Delta\mathcal{S}(n + dn) = \Delta\mathcal{S}(n)\{1 - g[\Gamma(n)]\, d\Gamma(n)\Delta\mathcal{V}\} \tag{18.5}$$

Rearranging equation (18.5) and taking the limit as dn approaches zero yields

$$\frac{1}{\Delta\mathcal{S}(n)} \frac{d\Delta\mathcal{S}(n)}{dn} = -g[\Gamma(n)] \frac{d\Gamma(n)}{dn} \Delta\mathcal{V} \tag{18.6}$$

Integrating equation (18.6) between 0 and N and recognizing that $\Delta\mathcal{S}(0) = 1$ gives

$$\ln \frac{1}{\Delta\mathcal{S}} = \Delta\mathcal{V} \int_0^N g[\Gamma(n)] \frac{d\Gamma(n)}{dn}\, dn \tag{18.7}$$

or

$$\ln \frac{1}{\Delta\mathcal{S}(N)} = G[\Gamma(n)]\Delta\mathcal{V} \tag{18.8}$$

By the product law of probability, it is known that the probability that $\mathcal{S}(N)$ the entire volume $\mathcal{V}$ will endure is

$$\mathcal{S}(N) = \Delta_1\mathcal{S}(N) \times \Delta_2\mathcal{S}(N) \cdots \tag{18.9}$$

Combining equations (18.8) and (18.9) and taking the limit as $\Delta\mathcal{V}$ approaches zero yields

$$\ln \frac{1}{\mathcal{S}(n)} = \int_{\mathcal{V}} G[\Gamma(n)]\, d\mathcal{V} \tag{18.10}$$

Equation (18.10) is similar in form to Weibull's function equation (18.3) except that $G[\Gamma(n)]$ includes the effect of depth z on failure. Alternatively, (18.10) could be written as follows:

$$\ln \frac{1}{\mathcal{S}} = f(\tau_0, N, z_0)\mathcal{V} \tag{18.11}$$

in which τ_0 is the maximum orthogonal shear stress, z_0 is the depth below the load-carrying surface at which this shear stress occurs, and N is the number of stress cycles survived with probability $\mathcal{S}$. It can be seen here that τ_0 and z_0 could be replaced by another stress–depth relationship.

Lundberg et al. [18.1] empirically determined the following relationship, which they felt adequately matched their test results:

$$f(\tau_0, N, z_0) \sim \tau_0^c N^e z_0^{-h} \tag{18.12}$$

Furthermore, the assumption was made that the stressed volume was effectively bounded by the width $2a$ of the contact ellipse, the depth z_0, and the length $\mathcal{L}$ of the path, that is,

$$\mathcal{V} \sim a z_0 \mathcal{L} \tag{18.13}$$

Substituting equations (18.12) and (18.13) into (18.11) gives

$$\ln \frac{1}{\mathcal{S}} \sim \tau_0^c z_0^{1-h} a\mathcal{L}N^e \tag{18.14}$$

Today it is known that a lubricant film fully separates the rolling elements from the raceways in an accurately manufactured bearing that is properly lubricated. In this situation, the surface shear stress in a rolling contact is negligible. Considering the operating conditions and the bearings used by Lundberg and Palmgren in the 1940s to develop their theory, it is probable that surface shear stesses of magnitudes greater than zero occurred in the rolling element–raceway contacts. It has been shown by many researchers that, if a surface shear stress occurs in addition to the normal stress, the depth at which the maximum subsurface shear stress occurs will be closer to the surface than z_0. Hence, the use of z_0 in equations (18.12)–(18.14), must be questioned considering the Lundberg–Palmgren test bearings and probable test conditions. Moreover, if z_0 is in question, then the use of a in the stressed volume relationship must be reconsidered. This particular problem is covered in Chapter 23.

If the number of stress cycles N equals uL, in which u is the number of stress cycles per revolution and L is the life in revolutions, then

$$\ln \frac{1}{\mathcal{S}} \sim \tau_0^c z_0^{1-h} a \mathcal{L} u^e L^e \tag{18.15}$$

More simply, for a given bearing under a given load,

$$\ln \frac{1}{\mathcal{S}} = A L_s^e \tag{18.16}$$

or

$$\ln \ln \frac{1}{\mathcal{S}} = e \ln L_s + \ln A \tag{18.17}$$

Equation (18.17) defines what is called a Weibull distribution of rolling bearing fatigue life. The exponent e is called the Weibull slope.

It has been found experimentally that between the L_7 and L_{60} lives of the bearing life distribution, the Weibull distribution fits the test data extremely well (see Tallian [18.6]). From equation (18.17) it can be seen that $\ln \ln 1/\mathcal{S}$ vs $\ln L$ plots as a straight line. Figure 18.7 shows a Weibull plot of bearing test data. It should be evident from the foregoing discussion and Fig. 18.7 that the Weibull slope e is a measure of bearing fatigue life dispersion. From equation (18.17) it can be determined that the Weibull slope for a given test group is given by

$$e = \frac{\ln \dfrac{\ln (1/\mathcal{S}_1)}{\ln (1/\mathcal{S}_2)}}{\ln \dfrac{L_1}{L_2}} \tag{18.18}$$

in which $(L_1, \mathcal{S}_1)$ and $(L_2, \mathcal{S}_2)$ are any two points on the best straight line passing through the test data. This best straight line may be accurately determined from a given set of endurance test data by using methods of

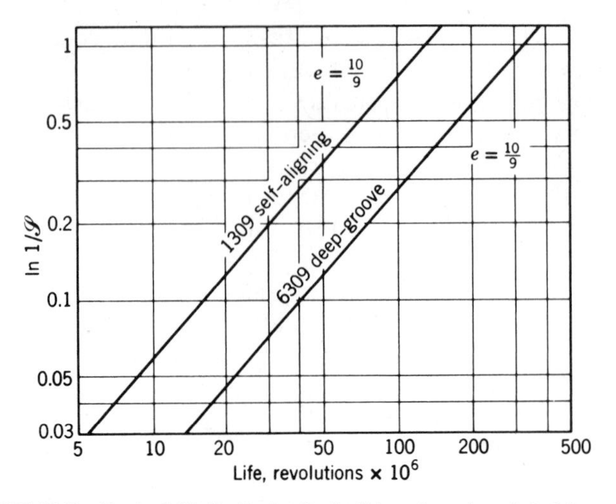

FIGURE 18.7. Typical Weibull plot for ball bearings (reprinted from [18.1]).

extreme value statistics as described by Lieblein [18.7]. According to Lundberg et al. [18.1, 18.8], $e = 10/9$ for ball bearings and $e = 9/8$ for roller bearings. These values are based on actual bearing endurance data for bearings fabricated from through-hardened AISI 52100 steel. Palmgren [18.9] states that for commonly used bearing steels, e is in the range 1.1–1.5. For modern, ultraclean, vacuum-remelted steels, values of e in the range 0.8 to 1.0 have been found. The lower value of e indicates greater dispersion of fatigue life.

At $L = L_{10}$, $\mathcal{S} = 0.9$. Setting these values into equation (18.17) gives

$$\ln \ln \frac{1}{0.9} = e \ln L_{10} + \ln A \tag{18.19}$$

Eliminating A between equations (18.17) and (18.19) yields

$$\ln \frac{1}{\mathcal{S}} = \left(\frac{L_{\mathcal{S}}}{L_{10}}\right)^e \ln \frac{1}{0.9} \tag{18.20}$$

or

$$\ln \frac{1}{\mathcal{S}} = 0.1053 \left(\frac{L_{\mathcal{S}}}{L_{10}}\right)^e \tag{18.21}$$

Equation (18.21) enables the estimation of $L_{\mathcal{S}}$, the bearing fatigue life at reliability $\mathcal{S}$ (probability of survival) once the Weibull slope e and "rating life" have been determined for a given application. The equation is valid between $\mathcal{S} = 0.93$ and $\mathcal{S} = 0.40$—a range that is useful for most bearing applications.

Example 18.1. A 209 radial ball bearing in a certain application yields a fatigue life of 100 million revolutions with 90% reliability. What fatigue life would be consistent with a reliability of 95%?

$$\ln \frac{1}{\mathcal{S}} = 0.1053 \left(\frac{L_5}{L_{10}} \right)^e \tag{18.21}$$

$$\ln \frac{1}{0.95} = 0.1053 \left(\frac{L_5}{100 \times 10^6} \right)^{10/9}$$

$$L_5 = 52.2 \times 10^6 \text{ revolutions}$$

Example 18.2. Of a group of 100 ball bearings on a given application 30 have failed in fatigue. Estimate the L_{10} life which may be expected of the remaining bearings.

At the moment 70 bearings remain, the relative number of surviving bearings is

$$\mathcal{S}_a = \frac{70}{100} = 0.70$$

The corresponding consumed life is obtained from

$$\ln \frac{1}{\mathcal{S}_a} = 0.1053 \left(\frac{L_a}{L_{10}} \right)^{10/9} \tag{18.21}$$

$$\ln \frac{1}{0.70} = 0.1053 \left(\frac{L_a}{L_{10}} \right)^{10/9}$$

$$L_a = 3.00 L_{10}$$

After additional L_{10} life of the surviving 70 bearings has been attained, the number of surviving bearings is 0.9 × 70 or 63. The relative number of surviving bearings is 0.63. The corresponding consumed total life is given by

$$\ln \frac{1}{\mathcal{S}_b} = 0.1053 \left(\frac{L_b}{L_{10}} \right)^{10/9} \tag{18.21}$$

$$\ln \frac{1}{0.63} = 0.1053 \left(\frac{L_b}{L_{10}} \right)^{10/9}$$

$$L_b = 3.79 L_{10}$$

$$L'_{10} = L_b - L_a$$

$$= 3.79 L_{10} - 3.00 L_{10}$$

$$= 0.79 L_{10}$$

Example 18.3. A group of ball bearings has an L_{10} life of 5000 hr in a given application. The bearings have been operated for 10,000 hr and some have failed. Estimate the amount of additional L_{10} life that can be expected from the remaining bearings.

The relative number of bearings attaining or exceeding life L_a is $\mathcal{S}_a$ and

$$\ln \frac{1}{\mathcal{S}_a} = 0.1053 \left(\frac{L_a}{L_{10}} \right)^e \tag{18.21}$$

After the additional L_{10} life is attained, the relative number of bearings remaining is $\mathcal{S}_b = 0.9\mathcal{S}_a$ corresponding to life L_b.

$$\ln \frac{1}{\mathcal{S}_b} = 0.1053 \left(\frac{L_b}{L_{10}} \right)^e \tag{18.21}$$

since $\mathcal{S}_b = 0.9\mathcal{S}_a$,

$$\ln \frac{1}{\mathcal{S}_b} = \ln \frac{1}{0.9} + \ln \frac{1}{\mathcal{S}_a}$$

Therefore,

$$\ln \frac{1}{\mathcal{S}_a} + 0.1053 = \left(\frac{L_b}{L_{10}} \right)^e \times 0.1053$$

By subtraction

$$0.1053 = 0.1053 \frac{(L_b^e - L_a^e)}{L_{10}^e}$$

or

$$L_b = (L_{10}^e + L_a^e)^{1/e}$$

The additional L_{10} life is given by

$$L'_{10} = L_b - L_a$$

or

$$\begin{aligned} L'_{10} &= (L_{10}^e + L_a^e)^{1/e} - L_a \\ &= [(5000)^{1.11} + (10{,}000)^{1.11}]^{0.9} - 10{,}000 = 4100 \text{ hr} \end{aligned}$$

DYNAMIC CAPACITY AND LIFE OF A ROLLING CONTACT

In equation (5.65) it was established that at a point contact

$$\frac{2\tau_0}{\sigma_{\max}} = \frac{(2t - 1)^{1/3}}{t(t + 1)} \tag{5.65}$$

More simply,

$$\tau_0 = T\sigma_{\max} \tag{18.22}$$

in which T is a function of the contact surface dimensions, that is, b/a (see Fig. 5.14). From equation (5.42) the maximum compressive stress within the contact ellipse is

$$\sigma_{\max} = \frac{3Q}{2\pi ab} \tag{5.42}$$

Furthermore, from equations (5.33) and (5.35) a and b are

$$a = a^* \left(\frac{3E_0 Q}{\Sigma\rho} \right)^{1/3} \tag{5.33}$$

$$b = b^* \left(\frac{3E_0 Q}{\Sigma\rho} \right)^{1/3} \tag{5.35}$$

in which

$$E_0 = \left[\frac{(1 - \xi_{\mathrm{I}}^2)}{E_{\mathrm{I}}} + \frac{(1 - \xi_{\mathrm{II}}^2)}{E_{\mathrm{II}}} \right] \tag{18.23}$$

By equation (5.52),

$$z_0 = \zeta b$$

in which ζ is a function of b/a per equation (5.66) and Fig. (5.14).

Substituting equations (5.42) and (5.35) into (18.15) yields

$$\ln \frac{1}{\mathcal{S}} \sim \frac{T^c a \mathcal{L}}{(\zeta b)^{h-1}} \left(\frac{Q}{ab} \right)^c u^e L^e \tag{18.24}$$

Letting d equal the raceway diameter, then $\mathcal{L} = \pi d$ and

$$\ln \frac{1}{\mathcal{S}} \sim \frac{T^c u^e L^e d}{\zeta^{h-1}} \left(\frac{1}{b} \right)^{c+h-1} \left(\frac{1}{a} \right)^{c-1} Q^c \tag{18.25}$$

Rearranging equation (18.25)

$$\ln\frac{1}{\mathcal{S}} \sim \frac{T^c u^e L^e d}{\zeta^{h-1}}\left(\frac{Q}{ab^2}\right)^{(c+h-1)/2} \times \left(\frac{1}{a}\right)^{(c-h-1)/2} Q^{(c-h+1)/2} \quad (18.26)$$

From equations (5.33) and (5.35),

$$\frac{Q}{ab^2} = \frac{E_0\Sigma\rho}{3a^*(b^*)^2} \quad (18.27)$$

Creating the identity

$$D^{(c+h-1)/2}D^{(c-h-1)/2}\left(\frac{1}{D^2}\right)^{(c-h+1)/2} D^{2-h} = 1 \quad (18.28)$$

and substituting equations (18.27) and (18.28) into (18.26) yields

$$\ln\frac{1}{\mathcal{S}} \sim \frac{T^c}{\zeta^{h-1}}\left[\frac{E_0 D\Sigma\rho}{3a^*(b^*)^2}\right]^{(c+h-1)/2}\left(\frac{D}{a}\right)^{(c-h-1)/2}\left(\frac{Q}{D^2}\right)^{(c-h+1)/2} dD^{2-h}u^e L^e \quad (18.29)$$

Substituting equation (5.33) for the semimajor axis a in point contact into equation (18.29):

$$\ln\frac{1}{\mathcal{S}} \sim \frac{T^c}{\zeta^{h-1}}\left[\frac{E_0 D\Sigma\rho}{3a^*(b^*)^2}\right]^{(c+h-1)/2}\left[\frac{D}{a^*}\left(\frac{E_0\Sigma\rho}{3Q}\right)^{1/3}\right]^{(c-h-1)/2} \times \left(\frac{Q}{D^2}\right)^{(c-h+1)/2} dD^{2-h}u^e L^e \quad (18.30)$$

Rearranging equation (18.30) gives

$$\ln\frac{1}{\mathcal{S}} \sim \frac{T^c dD^{2-h}u^e L^e}{\zeta^{h-1}(a^*)^{c-1}(b^*)^{c+h-1}}\left(\frac{E_0 D\Sigma\rho}{3}\right)^{(2c+h-2)/3}\left(\frac{Q}{D^2}\right)^{(c-h+2)/3} \quad (18.31)$$

Equation (18.31) can be further rearranged. Recognizing that the probability of survival $\mathcal{S}$ is a constant for any given bearing application,

$$\left(\frac{Q}{D^2}\right)^{(c-h+2)/3} L^e \sim \left[\frac{T^c dD^{2-h}u^e}{\zeta^{h-1}(a^*)^{c-1}(b^*)^{c+h-1}}\left(\frac{E_0 D\Sigma\rho}{3}\right)^{(2c+h-2)/3}\right]^{-1} \quad (18.32)$$

Letting $T = T_1$ and $\zeta = \zeta_1$ when $b/a = 1$, then

$$\left(\frac{Q}{D^2}\right)^{(c-h+2)/3} L^e \sim \left[\left(\frac{T}{T_1}\right)^c \left(\frac{\zeta_1}{\zeta}\right)^{h-1} \frac{(D\Sigma\rho)^{(2c+h-2)/3}}{(a^*)^{c-1}(b^*)^{c+h-1}} \frac{d}{D} u^e\right]^{-1} D^{-(3-h)} \tag{18.33}$$

Further rearrangement yields

$$QL^{(3e)/(c-h+2)} = A_1 \Phi D^{(2c+h-5)/(c-h+2)} \tag{18.34}$$

in which A_1 is a material constant and

$$\Phi = \left[\left(\frac{T}{T_1}\right)^c \left(\frac{\zeta_1}{\zeta}\right)^{h-1} \frac{(D\Sigma\rho)^{(2c+h-2)/3}}{(a^*)^{c-1}(b^*)^{c+h-1}} \times \frac{d}{D} u^e\right]^{-3/(c-h+2)} \tag{18.35}$$

For a given probability of survival the basic dynamic capacity of a rolling element–raceway contact is defined as that load which the contact will endure for one million revolutions of a bearing ring. Hence, Q_c the basic dynamic capacity of a contact is

$$Q_c = A_1 \Phi D^{(2c+h-5)/(c-h+2)} \tag{18.36}$$

For a bearing of given dimensions, by equating equation (18.34) to (18.36), one obtains

$$QL^{(3e)/(c-h+2)} = Q_c \tag{18.37}$$

or

$$L = \left(\frac{Q_c}{Q}\right)^{(c-h+2)/(3e)} \tag{18.38}$$

Thus for an applied load Q and a basic dynamic capacity Q_c (of a contact), the fatigue life in millions of revolutions may be calculated.

Endurance tests of ball bearings [18.1] have shown the load-life exponent to be very close to 3. Figure 18.8 is a typical plot of fatigue life vs load for a ball bearing. The adequacy of the value of 3 has also been substantiated through statistical analysis by the U.S. National Bureau of Standards [18.11]. Equation (18.38) thereby becomes

$$L = \left(\frac{Q_c}{Q}\right)^3 \tag{18.39}$$

This equation is also accurate for roller bearings having point contact.

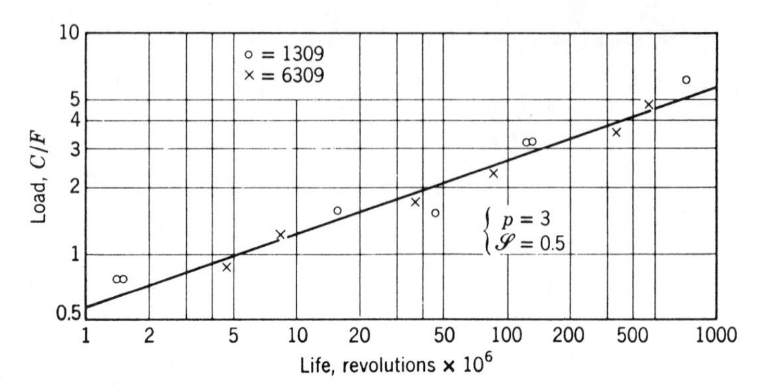

FIGURE 18.8. Load vs life for ball bearings (reprinted from [18.1]).

Since $e = 10/9$ for point contact, therefore

$$c - h = 8 \tag{18.40}$$

Evaluating the endurance test data of approximately 1500 bearings, Lundberg et al. [18.1] determined that $c = 31/3$ and $h = 7/3$. Substituting the values for c and h into equations (18.35) and (18.36), respectively, gives

$$\Phi = \left(\frac{T_1}{T}\right)^{3.1}\left(\frac{\zeta}{\zeta_1}\right)^{0.4}\frac{(a^*)^{2.8}(b^*)^{3.5}}{(D\Sigma\rho)^{2.1}}\left(\frac{D}{d}\right)^{0.3}u^{-1/3} \tag{18.41}$$

$$Q_c = A_1\Phi D^{1.8} \tag{18.42}$$

Recall that for a roller–raceway point contact in a roller bearing,

$$F(\rho) = \frac{\dfrac{2}{D} - \dfrac{1}{R} \pm \dfrac{2\gamma}{D(1 \mp \gamma)} + \dfrac{1}{r}}{\Sigma\rho} \tag{2.38, 2.40}$$

Therefore,

$$\frac{D}{2}\Sigma\rho F(\rho) = 1 - \frac{D}{2R} \pm \frac{\gamma}{1 \mp \gamma} + \frac{D}{2r} \tag{18.43}$$

Also, from equation (2.37),

$$\frac{D}{2}\Sigma\rho = 1 + \frac{D}{2R} \pm \frac{\gamma}{1 \mp \gamma} - \frac{D}{2r} \tag{18.44}$$

Adding equations (18.43) and (18.44) gives

$$[1 + F(\rho)]\frac{D}{2}\Sigma\rho = \frac{2}{1 \mp \gamma} \tag{18.45}$$

Subtracting equation (18.43) from (18.44) yields

$$[1 - F(\rho)]\frac{D}{2}\Sigma\rho = D\left(\frac{1}{R} - \frac{1}{r}\right) \tag{18.46}$$

From equation (18.45),

$$D\Sigma\rho = \frac{4}{[1 + F(\rho)](1 \mp \gamma)} \tag{18.47}$$

At this point in the analysis define Ω as follows:

$$\Omega = \frac{1 - F(\rho)}{1 + F(\rho)} = \frac{D}{2R}\frac{r - R}{r}(1 \mp \gamma) \tag{18.48}$$

Let

$$\Omega_1 = [1 + F(\rho)]^{2.1}\left(\frac{T}{T_1}\right)^{3.1}\left(\frac{\zeta_1}{\zeta}\right)^{0.4}(a^*)^{2.8}(b^*)^{3.5} \tag{18.49}$$

Also recognize that d in (18.41) is given by

$$d = d_{\mathrm{m}}(1 \mp \gamma) \tag{18.50}$$

Therefore, substituting equations (18.49) and (18.50) into (18.41) yields

$$\Phi = \frac{\Omega_1}{[1 + F(\rho)]^{2.1}} \times \frac{1}{(D \times \Sigma\rho)^{2.1}} \times \left[\frac{D}{d_{\mathrm{m}}(1 \mp \gamma)}\right]^{0.3} u^{-1/3} \tag{18.51}$$

Lundberg et al. [18.1] determined that within the range corresponding to ball and roller bearings, Ω_1 very nearly is given by

$$\Omega_1 = 1.3\Omega^{-0.41} \tag{18.52}$$

Figure 18.9 from reference [18.1] establishes the validity of this assumption. Substituting (18.52) and (18.47) into (18.51) gives

$$\Phi = 0.0706\left[\frac{2R}{D} \times \frac{r}{(r - R)}\right]^{0.41}(1 \mp \gamma)^{1.39}\left(\frac{D}{d_{\mathrm{m}}}\right)^{0.3} u^{-1/3} \tag{18.53}$$

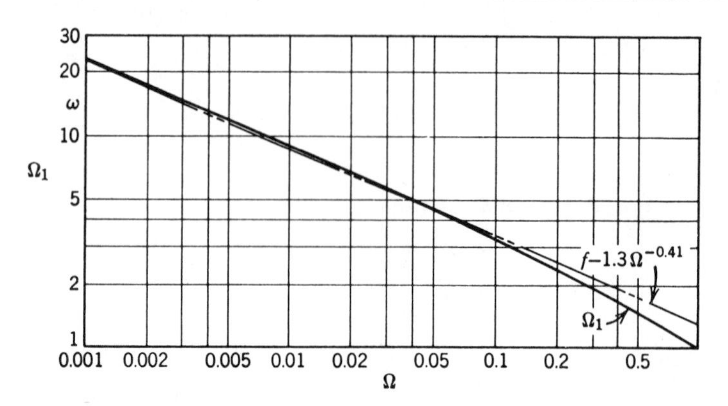

FIGURE 18.9. Ω_1 vs Ω for point-contact ball and roller bearings (reprinted from [18.1].

The number of stress cycles u per revolution is the number of rolling elements which pass a given point (under load) on the raceway of one ring while the other ring has turned through one complete revolution. Hence from Chapter 7 the number of rolling elements passing a point on the inner ring per unit time is

$$u_i = Z\frac{n_{mi}}{n}$$

$$= 0.5Z(1+\gamma) \tag{18.54}$$

For the outer ring,

$$u_o = 0.5Z(1-\gamma) \tag{18.55}$$

or

$$u = 0.5Z(1 \pm \gamma) \tag{18.56}$$

in which the upper sign refers to the inner ring and the lower sign refers to the outer ring.

Substitution of equation (18.56) into (18.53) gives the following expression for Φ:

$$\Phi = 0.089\left(\frac{2R}{D}\times\frac{r}{r-R}\right)^{0.41}\frac{(1\mp\gamma)^{1.39}}{(1\pm\gamma)^{1/3}}\left(\frac{D}{d_m}\right)^{0.3}Z^{-1/3} \tag{18.57}$$

Combining equation (18.57) with (18.42) yields an equation for Q_c, the basic dynamic capacity of a point contact, in terms of the bearing design parameters:

$$Q_c = A\left(\frac{2R}{D} \times \frac{r}{(r-R)}\right)^{0.41} \frac{(1 \mp \gamma)^{1.39}}{(1 \pm \gamma)^{1/3}} \left(\frac{\gamma}{\cos \alpha}\right)^{0.3} D^{1.8} Z^{-1/3} \quad (18.58)$$

Test data of Lundberg et al. [18.1] resulted in an average value $A = 98.1$ in mm · N units (7450 in in. · lb units) for bearings fabricated from 52100 steel through-hardened to Rockwell C = 61.7–64.5. This value strictly pertains to the steel quality and manufacturing accuracies achievable at that time, that is, up to approximately 1960. Subsequent improvements in steel-making and manufacturing processes have resulted in significant increases in this ball bearing material factor. This situation will be discussed in detail later in the chapter.

Line Contact

Equation (18.29) is equally valid for line contact. It can be shown for line contact that as b/a approaches zero, $(a^*)(b^*)^2$ approaches the limit $2/\pi$. Therefore, the following expression can be written for line contact:

$$\ln \frac{1}{\mathcal{S}} \sim \frac{T^c}{\zeta^{h-1}} \left(\frac{\pi E_0 D \Sigma\rho}{6}\right)^{(c+h-1)/2} \left(\frac{4D}{3l}\right)^{(c-h-1)/2} \left(\frac{Q}{D^2}\right)^{(c-h+1)/2} dD^{2-h} u^e L^e \quad (18.59)$$

In a manner similar to that used for point contact, it can be developed that

$$Q_c = B_1 \Psi D^{(c+h-3)/(c-h+1)} l^{(c-h-1)/(c-h+1)} \quad (18.60)$$

in which

$$B_1 = \left(\frac{3}{4}\right)^{(c-h-1)/(c-h+1)} \left(\frac{\pi}{2}\right)^{-(c+h-1)/(c-h+1)} \left(\frac{T_1}{T_0}\right)^{(2c)/(c-h+1)} \times \left(\frac{\zeta_0}{\zeta_1}\right)^{2(h-1)/(c-h+1)} \left(\frac{E_0}{3}\right)^{(c-h-1)/[3(c-h+1)]} A^{(2c-2h+4)/(3c-3h+3)} \quad (18.61)$$

$$\Psi = \left[(D\Sigma\rho)^{(c+h-1)/2} \frac{d}{D} u^e\right]^{-2/(c-h+1)} \quad (18.62)$$

It can be further established that

$$\Psi = 0.513 \frac{(1 \mp \gamma)^{29/27}}{(1 \pm \gamma)^{1/4}} \left(\frac{D}{d_m}\right)^{2/9} Z^{-1/4} \quad (18.63)$$

and

$$Q_c = B \frac{(1 \mp \gamma)^{29/27}}{(1 \pm \gamma)^{1/4}} \left(\frac{\gamma}{\cos \alpha} \right)^{2/9} D^{29/27} l^{7/9} Z^{-1/4} \tag{18.64}$$

in which $B = 552$ in mm · N units (49,500 in in. · lb units) for bearings fabricated from through-hardened 52100 steel. As for ball bearings, the material factor for roller bearings has undergone substantial increase since the investigations of Lundberg and Palmgren. This situation will be covered in detail later in the chapter.

For line contact it was determined that

$$L = \left(\frac{Q_c}{Q} \right)^4 \tag{18.65}$$

and further, from Lundberg et al. [18.8], that

$$\frac{c - h + 1}{2e} = 4 \tag{18.66}$$

Since $e = 9/8$ for line contact, from (18.66)

$$c - h = 8$$

which is identical for point contact, establishing that c and h are material constants.

Some roller bearings have fully crowned rollers such that "edge loading" does not occur under the probable maximum loads, that is, modified line contact occurs under such loads. Under lighter loading, however, point contact occurs. For such a condition, equation (18.64) should yield the same capacity value as equation (18.58). Unfortunately, this is a deficiency in the original Lundberg–Palmgren theory owing to the calculational tools then available. This situation can be rectified for the sake of continuity by utilizing the exponent $\frac{20}{81}$ in lieu of $\frac{1}{4}$ (and $-\frac{20}{81}$ in lieu of $-\frac{1}{4}$) in equation (18.64). Also, the value of constant B becomes 488 in mm · N units (43800 in in. · lb units) for roller bearings fabricated from through-hardened 52100 steel. Again, this material factor strictly pertains to the roller bearings of the Lundberg–Palmgren era.

FATIGUE LIFE OF A ROLLING BEARING

Point Contact, Radial Bearings

According to the foregoing analysis, the fatigue life of a rolling element–raceway point contact subjected to normal load Q may be estimated by

$$L = \left(\frac{Q_c}{Q}\right)^3 \tag{18.39}$$

in which L is in millions of revolutions and

$$Q_c = 98.1\left(\frac{2R}{D}\frac{r}{r-R}\right)^{0.41}\frac{(1 \mp \gamma)^{1.39}}{(1 \pm \gamma)^{1/3}}\left(\frac{\gamma}{\cos\alpha}\right)^{0.3} D^{1.8}Z^{-1/3} \tag{18.58}$$

For ball bearings this equation becomes

$$Q_c = 98.1^*\left(\frac{2f}{2f-1}\right)^{0.41}\frac{(1 \mp \gamma)^{1.39}}{(1 \pm \gamma)^{1/3}}\left(\frac{\gamma}{\cos\alpha}\right)^{0.3} D^{1.8}Z^{-1/3} \tag{18.67}$$

in which the upper signs refer to the inner raceway contact and the lower signs refer to the outer raceway contact. Since stress is usually higher at the inner raceway contact than at the outer raceway contact, failure generally occurs on the inner raceway first. This is not necessarily true for self-aligning ball bearings for which stress is high on the outer raceway, it being a portion of a sphere.

A rolling bearing consists of a plurality of contacts. For instance, a point on the inner raceway of a bearing with inner ring rotation may experience a load cycle as shown in Fig. 18.10. Although the maximum load and hence maximum stress is significant in causing failure, the statistical nature of fatigue failure requires that the load history be considered. Lundberg et al. [18.1] determined empirically that a cubic mean load fits the test data very well for point contact. Hence for a ring which rotates relative to a load,

$$Q_{e\mu} = \left(\frac{1}{Z}\sum_{j=1}^{j=z} Q_j^3\right)^{1/3} \tag{18.68}$$

In the terms of the angular disposition of the rolling element,

$$Q_{e\mu} = \left(\frac{1}{2\pi}\int_0^{2\pi} Q_\psi^3\, d\psi\right)^{1/3} \tag{18.69}$$

*Palmgren recommended reducing this constant to 93.2 (7080) for single-row ball bearings and to 88.2 (6700) for double-row, deep-groove ball bearings to account for inaccuracies in raceway groove form owing to the manufacturing processes at that time. Subsequent improvements in the steel quality and in the manufacturing accuracies have seen the material factor increase significantly for groove-type bearings. This increase is accommodated by a factor b_m that augments the above-indicated material factors; this is discussed in detail later in this chapter.

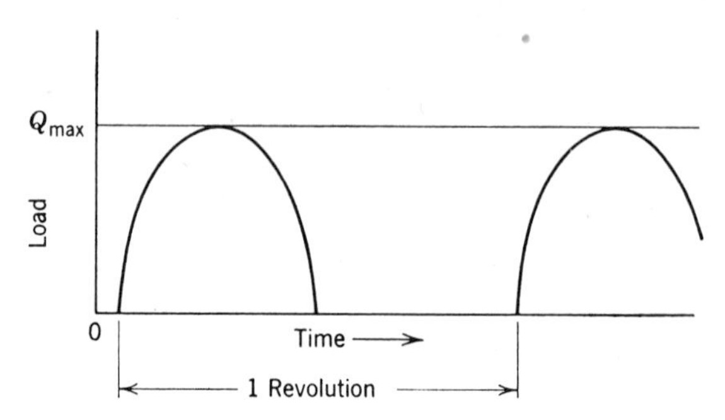

FIGURE 18.10. Typical load cycle for a point on inner raceway of a radial bearing.

The fatigue life of a rotating raceway is therefore calculated as follows:

$$L_{\mu} = \left(\frac{Q_{c\mu}}{Q_{e\mu}} \right)^3 \tag{18.70}$$

Each point on a raceway that is stationary relative to the applied load is subjected to virtually a constant stress amplitude. Only the space between rolling elements causes the amplitude to fluctuate with time. From equation (18.31) it can be determined that the probability of survival of any given contact point on the nonrotating raceway is given by

$$\ln \frac{1}{\mathbb{S}_{\nu j}} \sim Q_j^{(c-h+2)/3} L_j^e \tag{18.71}$$

According to the product law of probability, the probability of failure of the ring is the product of the probability of failure of the individual parts; hence since $3e = c - h + 2$,

$$\ln \frac{1}{\mathbb{S}} \sim L^e \int_0^{2\pi} Q_{\psi}^{3e} \, d\psi = L^e Q_{e\nu}^{3e} \tag{18.72}$$

in which $Q_{e\nu}$ is defined as follows:

$$Q_{e\nu} = \left(\frac{1}{2\pi} \int_0^{2\pi} Q_{\psi}^{3e} \, d\psi \right)^{1/3e} = \left(\frac{1}{2\pi} \int_0^{2\pi} Q_{\psi}^{10/3} \, d\psi \right)^{0.3} \tag{18.73}$$

In discrete numerical format, equation (18.73) becomes

$$Q_{e\nu} = \left(\frac{1}{Z} \sum_{j=1}^{j=z} Q_j^{10/3} \right)^{0.3} \tag{18.74}$$

From equations (18.74) and (18.39), the fatigue life of a nonrotating ring may be calculated by

$$L_\nu = \left(\frac{Q_{c\nu}}{Q_{e\nu}}\right)^3 \tag{18.75}$$

To determine the life of an entire bearing the lives of the rotating and nonrotating (inner and outer or vice versa) raceways must be statistically combined according to the product law. The probability of survival of the rotating raceway is given by

$$\ln \frac{1}{\mathcal{S}_\mu} = K_\mu L_\mu^e \tag{18.76}$$

Similarly for the nonrotating raceway

$$\ln \frac{1}{\mathcal{S}_\nu} = K_\nu L_\nu^e \tag{18.77}$$

and for the entire bearing

$$\ln \frac{1}{\mathcal{S}} = (K_\mu + K_\nu) L^e \tag{18.78}$$

Since $\mathcal{S}_\mu = \mathcal{S}_\nu = \mathcal{S}$, the combination of equations (18.76)–(18.78) yields

$$L = (L_\mu^{-e} + L_\nu^{-e})^{-1/e} \tag{18.79}$$

Since $e = 10/9$ for point contact equation (18.79) becomes

$$L = (L_\mu^{-1.11} + L_\nu^{-1.11})^{-0.9} \tag{18.80}$$

Based on the preceding development, it is possible to calculate a rolling bearing fatigue life in point contact if the normal load is known at each rolling element position. These data may be calculated by methods established in Chapters 6 and 8.

It is seen that the bearing lives determined according to the methods given above are based on subsurface-initiated fatigue failure of the raceways. Ball failure was not considered apparently because it was not frequently observed in the Lundberg-Palmgren fatigue life test data. It was rationalized that, because a ball could change rotational axes readily, the entire ball surface was subjected to stress, spreading the stress cycles over greater volume consequently reducing the probability of ball fatigue failure prior to raceway fatigue failure. Some researchers have since observed that in most applications, each ball tends to seek a single axis of rotation irrespective of original orientation prior to bearing operation. This observation tends to negate the Lundberg-Palmgren assumption. It is perhaps more correct to assume that Lundberg and Palmgren did not observe significant numbers of ball-fatigue failures because at that time the ability to manufacture accurate geometry balls of good metallurgical parameters exceeded that for the corresponding raceways. The ability to accurately manufacture raceways of good quality steel has consistently improved since that era, and for many ball bearings of current manufacture the incidence of ball-fatigue failure in lieu of raceway-fatigue failure, particularly in heavily loaded bearing applications, is frequently observed. Obviously, the accuracy of ball manufacture, and ball materials and processing has also improved; however, the gap has narrowed significantly. It is now clear that bearing-fatigue life based upon ball-fatigue failure also needs to be considered.

Example 18.4. The 209 radial ball bearing of Example 6.1 is operated at a shaft speed of 1800 rpm. Estimate the L_{10} life of the bearing.

$$Z = 9 \qquad \text{Ex. 2.1}$$

$$\gamma = 0.1954 \qquad \text{Ex. 2.5}$$

$$D = 12.7 \text{ mm } (0.5 \text{ in.}) \qquad \text{Ex. 2.1}$$

$$f = 0.52 \qquad \text{Ex. 2.2}$$

$$Q_{ci} = 93.2\left(\frac{2f_i}{2f_i - 1}\right)^{0.41} \frac{(1-\gamma)^{1.39}}{(1+\gamma)^{1/3}}\gamma^{0.3}D^{1.8}Z^{-1/3} \tag{18.67}$$

$$= 93.2\left(\frac{2 \times 0.52}{2 \times 0.52 - 1)}\right)^{0.41} \frac{(1 - 0.1954)^{1.39}}{(1 + 0.1954)^{1/3}}$$

$$\times (0.1954)^{0.3}(12.7)^{1.8}(9)^{-1/3}$$

$$= 7058 \text{ N } (1586 \text{ lb})$$

$$Q_{co} = 93.2\left(\frac{2f_o}{2f_o - 1}\right)^{0.41} \frac{(1+\gamma)^{1.39}}{(1-\gamma)^{1/3}}\gamma^{0.3}D^{1.8}Z^{-1/3} \tag{18.67}$$

$$= 93.2\left(\frac{2 \times 0.52}{2 \times 0.52 - 1}\right)^{0.41} \frac{(1 + 0.1954)^{1.39}}{(1 - 0.1954)^{1/3}}$$

$$\times (0.1954)^{0.3}(12.7)^{1.8}(9)^{-1/3}$$

$$= 13{,}970 \text{ N } (3140 \text{ lb})$$

From Example 6.1,

ψ	Q_ψ
0°	4536 N (1019 lb)
40°	2842 N (638.6 lb)
80°	58 N (13.0 lb)
120°	0
160°	0

Since the inner ring rotates,

$$Q_{\mathrm{ei}} = \left(\frac{1}{Z}\sum_{j=1}^{j=z} Q_j^3\right)^{1/3} \tag{18.68}$$

$$= \{\tfrac{1}{9}[(4536)^3 + 2 \times (2842)^3 + 2 \times (58)^3 + 0 + 0]\}^{1/3}$$

$$= 2475 \text{ N } (556.3 \text{ lb})$$

$$L_{\mathrm{i}} = \left(\frac{Q_{\mathrm{ci}}}{Q_{\mathrm{ei}}}\right)^3 \tag{18.70}$$

$$= \left(\frac{7058}{2475}\right)^3 = 23.2 \text{ million revolutions}$$

$$Q_{\mathrm{eo}} = \left(\frac{1}{Z}\sum_{j=1}^{j=z} Q_j^{10/3}\right)^{0.3} \tag{18.74}$$

$$= \{\tfrac{1}{9}[(4536)^{10/3} + 2 \times (2842)^{10/3} + 2 \times (58)^{10/3} + 0 + 0]\}^{0.3}$$

$$= 2605 \text{ N } (585.3 \text{ lb})$$

$$L_{\mathrm{o}} = \left(\frac{Q_{\mathrm{co}}}{Q_{\mathrm{eo}}}\right)^3 = \left(\frac{13970}{2605}\right)^3 \tag{18.75}$$

$$= 154.4 \text{ million revolutions}$$

$$L = (L_{\mathrm{i}}^{-1.11} + L_{\mathrm{o}}^{-1.11})^{-0.9} \tag{18.79}$$

$$= [(23.2)^{-1.11} + (154.4)^{-1.11}]^{-0.9} \times 10^6 = 20.9 \times 10^6 \text{ revolutions}$$

or

$$L = \frac{20.9 \times 10^6}{60 \times 1800} = 194^* \text{ hr}$$

In lieu of the foregoing rigorous approach to the calculation of bearing fatigue life, an approximate method has been developed by Lundberg et al. [18.1] for bearings having rigidly supported rings and operating at moderate speeds. It was developed in Chapter 6 that

*This L_{10} fatigue life was calculated according to the basic Lundberg–Palmgren theory and is based upon the standard bearing materials and manufacturing processes employed until approximately 1960. To be able to compare the numerical example results with the graphical data of Lundberg–Palmgren as well as with graphical data generated to demonstrate the effects of nonstandard load distributions, all numerical examples in this and the next two sections are calculated using the basic Lundberg–Palmgren theory. Later in the chapter the effects on fatigue life of subsequent improvements in materials and manufacturing processes are discussed in detail.

$$Q_\psi = Q_{\max}\left[1 - \frac{1}{2\epsilon}(1 - \cos\psi)\right]^n \tag{6.15}$$

and $n = 1.5$ for point contact. This equation may be substituted into equation (18.69) for $Q_{e\mu}$ to yield

$$Q_{e\mu} = Q_{\max}\left\{\frac{1}{2\pi}\int_{-\psi_1}^{+\psi_1}\left[1 - \frac{1}{2\epsilon}(1 - \cos\psi)\right]^{4.5} d\psi\right\}^{1/3} \tag{18.81}$$

or

$$Q_{e\mu} = Q_{\max}J_1 \tag{18.82}$$

Similarly for the nonrotating ring,

$$Q_{e\nu} = Q_{\max}\left\{\frac{1}{2\pi}\int_{-\psi_1}^{+\psi_1}\left[1 - \frac{1}{2\epsilon}(1 - \cos\psi)\right]^{5} d\psi\right\}^{0.3} \tag{18.83}$$

or

$$Q_{e\nu} = Q_{\max}J_2 \tag{18.84}$$

Table 18.1 gives values of J_1 and J_2 for point contact and various values of ϵ.

TABLE 18.1. J_1 and J_2 for Point Contact

Single-Row Bearings			Double-Row Bearings			
ϵ	J_1	J_2	ϵ_{I}	ϵ_{II}	J_1	J_2
0	0	0				
0.1	0.4275	0.4608	0.5	0.5	0.6925	0.7233
0.2	0.4806	0.5100	0.6	0.4	0.5983	0.6231
0.3	0.5150	0.5427	0.7	0.3	0.5986	0.6215
0.4	0.5411	0.5673	0.8	0.2	0.6105	0.6331
0.5	0.5625	0.5875	0.9	0.1	0.6248	0.6453
0.6	0.5808	0.6045	1.0	0	0.6372	0.6566
0.7	0.5970	0.6196				
0.8	0.6104	0.6330				
0.9	0.6248	0.6453				
1.0	0.6372	0.6566				
1.25	0.6652	0.6821				
1.67	0.7064	0.7190				
2.5	0.7707	0.7777				
5	0.8675	0.8693				
∞	1	1				

Again referring to Chapter 6, equation (6.66) states for a radial bearing

$$F_r = ZQ_{max}J_r \cos \alpha \tag{6.66}$$

Setting $F_r = C_\mu$, the basic dynamic capacity of the rotating ring (relative to the applied load), and substituting for Q_{max} according to equation (18.82) gives

$$C_\mu = Q_{c\mu}Z \cos \alpha \frac{J_r}{J_1} \tag{18.85}$$

Basic dynamic capacity is defined here as that radial load that 90% of a group of apparently identical bearing rings will survive for one million revolutions. Tables 6.1 and 6.4 give values of J_r.

Similarly, for the nonrotating ring

$$C_\nu = Q_{c\nu}Z \cos \alpha \frac{J_r}{J_2} \tag{18.86}$$

At $\epsilon = 0.5$, which is a nominal value for radial rolling bearings,

$$C_\mu = 0.407Q_{c\mu}Z \cos \alpha \tag{18.87}$$

$$C_\nu = 0.389Q_{c\nu}Z \cos \alpha \tag{18.88}$$

Again, the product law of probability is introduced to relate bearing fatigue life of the components. From equation (18.31) it can be established that

$$\ln \frac{1}{\mathcal{S}_\mu} = K_\mu F^{3e} = K_\mu C_\mu^{10/3} \tag{18.89}$$

Similarly,

$$\ln \frac{1}{\mathcal{S}_\nu} = K_\nu C_\nu^{10/3} \tag{18.90}$$

$$\ln \frac{1}{\mathcal{S}} = (K_\mu + K_\nu)C^{10/3} \tag{18.91}$$

Combining equations (18.89)–(18.91) determines

$$C = (C_\mu^{-10/3} + C_\nu^{-10/3})^{-0.3} \tag{18.92}$$

in which C is the basic dynamic capacity of the bearing. Rearrangement of equation (18.92) gives

$$C = C_\mu \left[1 + \left(\frac{C_\mu}{C_\nu} \right)^{10/3} \right]^{-0.3} = g_c C_\mu \tag{18.93}$$

A similar approach may be taken toward calculation of the effect of a plurality of rows of rolling elements. Consider that a bearing with point contact has two identical rows of rolling elements, each row being loaded identically. Then for each row the basic dynamic capacity is C_1 and the basic dynamic capacity of the bearing is C. From equation (18.93).

$$C = 2C_1(1 + 1)^{-0.3} = 2^{0.7} C_1 = 1.625 C_1$$

Hence, a two-row bearing does not have twice the basic dynamic capacity of a single-row bearing because of the statistical nature of fatigue failure.

In general, for a bearing with point contact having a plurality of rows i of rolling elements,

$$C = i^{0.7} C_k \tag{18.94}$$

in which C_k is the basic dynamic capacity of one row. Equations (18.85) and (18.86) can now be rewritten as follows:

$$C_\mu = Q_{c\mu} i^{0.7} Z \cos\alpha \frac{J_r}{J_1} \tag{18.95}$$

or

$$C_\mu = 0.407 Q_{c\mu} i^{0.7} Z \cos\alpha \qquad (\epsilon = 0.5) \tag{18.96}$$

$$C_\nu = Q_{c\nu} i^{0.7} Z \cos\alpha \frac{J_r}{J_2} \tag{18.97}$$

$$C_\nu = 0.389 Q_{c\nu} i^{0.7} Z \cos\alpha \qquad (\epsilon = 0.5) \tag{18.98}$$

Substitution of Q_c from equation (18.58) into (18.95) gives the following expression for basic dynamic capacity of a rotating ring:

$$C_\mu = 98.1 \left(\frac{2R}{D} \frac{r}{r - R} \right)^{0.41} \frac{(1 \mp \gamma)^{1.39}}{(1 \pm \gamma)^{1/3}} \gamma^{0.3} (i \cos\alpha)^{0.7} Z^{2/3} D^{1.8} \frac{J_r}{J_1} \tag{18.99}$$

$$C_\mu = 39.9 \left(\frac{2R}{D} \frac{r}{r - R} \right)^{0.41} \frac{(1 \mp \gamma)^{1.39}}{(1 \pm \gamma)^{1/3}} \gamma^{0.3} (i \cos\alpha)^{0.7} Z^{2/3} D^{1.8} \qquad (\epsilon = 0.5) \tag{18.100}$$

For the nonrotating ring,

$$C_\nu = 98.1\left(\frac{2R}{D}\frac{r}{r-R}\right)^{0.41}\frac{(1\mp\gamma)^{1.39}}{(1\pm\gamma)^{1/3}}\gamma^{0.3}(i\cos\alpha)^{0.7}Z^{2/3}D^{1.8}\frac{J_r}{J_2} \quad (18.101)$$

$$C_\nu = 38.2\left(\frac{2R}{D}\frac{r}{r-R}\right)^{0.41}\frac{(1\mp\gamma)^{1.39}}{(1\pm\gamma)^{1/3}}\gamma^{0.3}(i\cos\alpha)^{0.7}Z^{2/3}D^{1.8} \quad (\epsilon = 0.5)$$

$$(18.102)$$

According to equation (18.93) the basic dynamic capacity of the bearing assembly is as follows for $\epsilon = 0.5$:

$$C = f_c(i\cos\alpha)^{0.7}Z^{2/3}D^{1.8\,*} \quad (18.103)$$

in which

$$f_c = 39.9\left\{1+\left[1.04\left(\frac{1\mp\gamma}{1\pm\gamma}\right)^{1.72}\left(\frac{r_\mu}{r_\nu}\times\frac{(2r_\nu-D)}{(2r_\mu-D)}\right)^{0.41}\right]^{10/3}\right\}^{-0.3} \times\frac{\gamma^{0.3}(1\mp\gamma)^{1.39}}{(1\pm\gamma)^{1/3}}\left(\frac{2r_\mu}{2r_\mu-D}\right)^{0.41} \quad (18.104)$$

Generally, it is the inner raceway that rotates relative to the load and therefore

$$f_c = 39.9\left\{1+\left[1.04\left(\frac{1-\gamma}{1+\gamma}\right)^{1.72}\left(\frac{r_i}{r_o}\times\frac{r_o-D}{r_i-D}\right)^{0.41}\right]^{10/3}\right\}^{-0.3} \times\frac{\gamma^{0.3}(1-\gamma)^{1.39}}{(1+\gamma)^{1/3}}\left(\frac{2r_i}{2r_i-D}\right)^{0.41} \quad (18.105)$$

For ball bearings, equation (18.105) becomes

$$f_c = 39.9^{\dagger}\left\{1+\left[1.04\left(\frac{1-\gamma}{1+\gamma}\right)^{1.72}\left(\frac{f_i}{f_o}\times\frac{2f_o-1}{2f_i-1}\right)^{0.41}\right]^{10/3}\right\}^{-0.3} \times\frac{\gamma^{0.3}(1-\gamma)^{1.39}}{(1+\gamma)^{1/3}}\left(\frac{2f_i}{2f_i-1}\right)^{0.41} \quad (18.106)$$

*ANSI [18.10] recommends using D raised to the 1.4 power in lieu of 1.8 for bearings having balls of diameter greater than 25.4 mm (1 in.).
†According to Palmgren [18.9] this factor can be as low as 37.9 (2880) for single-row ball bearings and 35.9 (2730) for double-row deep-groove ball bearings to account for manufacturing inaccuracies.

Equation (18.103) in conjunction with (18.106) is generally considered valid for ball bearings whose rings and balls are fabricated from AISI 52100 steel heat treated at least to Rockwell C 58 hardness throughout. If the hardness of the bearing steel is less than Rockwell C 58, a reduction in the bearing basic dynamic capacity according to the following formula may be used:

$$C' = C\left(\frac{RC}{58}\right)^{3.6} \tag{18.107}$$

in which RC is the Rockwell C scale hardness.

By using equations (18.103) and (18.106) the basic dynamic capacity* of a radially loaded bearing may be calculated. The pertinent L_{10} fatigue life formula is given below:

$$L = \left(\frac{C}{F_e}\right)^{3} \tag{18.108}$$

in which F_e is an equivalent radial load which will cause the same L_{10} fatigue life as the applied load.

From equation (6.66) it can be seen that

$$Q_{max} = \frac{F_r}{Z \cos \alpha J_r} \tag{6.66}$$

in which F_r is an applied radial load and Q_{max} is the maximum rolling element load. For a rotating ring, from equation (18.82) $Q_{e\mu} = Q_{max} J_1$; therefore,

$$Q_{e\mu} = \frac{F}{Z \cos \alpha} \times \frac{J_1}{J_r} \tag{18.109}$$

in which $Q_{e\mu}$ is the mean equivalent rolling element load in a combined loading defined by J_r. At $\epsilon = 0.5$ [see Chapter 6 and equations (18.82) and (18.84)] loading is ideal and purely radial; therefore,

*The term *basic dynamic capacity* was created by Lundberg and Palmgren [18.1]. ANSI [18.10, 18.12] uses the term *basic load rating* and ISO [18.13] using *basic dynamic load rating.* These terms are interchangeable.

$$Q_{e\mu} = \frac{F_{e\mu}}{Z \cos \alpha} \times \frac{J_1(0.5)}{J_r(0.5)} \tag{18.110}$$

in which $F_{e\mu}$ is the equivalent radial load.

Similarly, for a nonrotating ring

$$Q_{e\nu} = \frac{F_{e\nu}}{Z \cos \alpha} \times \frac{J_2(0.5)}{J_r(0.5)} \tag{18.111}$$

The fatigue life of the rotating ring may be described by

$$\ln \frac{1}{\mathcal{S}_\mu} = \left(\frac{F_{e\mu}}{C_\mu}\right)^{3.33} L_\mu^{1.11} \ln \frac{1}{0.9} \tag{18.112}$$

[see equations (18.20) and (18.89)–(18.91)]. Similarly, for the nonrotating ring,

$$\ln \frac{1}{\mathcal{S}_\nu} = \left(\frac{F_{e\nu}}{C_\nu}\right)^{3.33} L_\nu^{1.11} \ln \frac{1}{0.9} \tag{18.113}$$

For the bearing,

$$\ln \frac{1}{\mathcal{S}} = \left(\frac{F_e}{C}\right)^{3.33} L^{1.11} \ln \frac{1}{0.9} \tag{18.114}$$

Combining equations (18.112)–(18.114) yields

$$\left(\frac{F_e}{C}\right)^{3.33} = \left(\frac{F_{e\mu}}{C_\mu}\right)^{3.33} + \left(\frac{F_{e\nu}}{C_\nu}\right)^{3.33} \tag{18.115}$$

Equation (18.109) to (18.110) gives

$$F_{e\mu} = \frac{J_r(0.5)J_1}{J_1(0.5)J_r} F_r \tag{18.116}$$

Similarly,

$$F_{e\nu} = \frac{J_r(0.5)J_2}{J_2(0.5)J_r} F_r \tag{18.117}$$

Substituting for $F_{e\mu}$ and $F_{e\nu}$ in equation (18.115) yields the following expression for equivalent radial load:

$$F_e = \left[\left(\frac{C}{C_\mu} \times \frac{J_r(0.5)}{J_1(0.5)} \times \frac{J_1}{J_r}\right)^{3.33} + \left(\frac{C}{C_\nu} \times \frac{J_r(0.5)}{J_2(0.5)} \times \frac{J_2}{J_r}\right)^{3.33}\right]^{0.3} F_r \tag{18.118}$$

In terms of an axial load F_a applied to a radial bearing:

$$Q_{max} = \frac{F_a}{Z \sin \alpha J_a} \tag{18.119}$$

(see Chapter 6 for evaluation of J_a).

In a manner similar to that developed for a radial load F_r,

$$Q_{e\mu} = \frac{F_a}{Z \sin \alpha} \times \frac{J_1}{J_a} \tag{18.120}$$

$$Q_{e\nu} = \frac{F_a}{Z \sin \alpha} \times \frac{J_2}{J_a} \tag{18.121}$$

Combining equations (18.110) and (18.120) yields

$$F_{e\mu} = \left[\frac{J_r(0.5)}{J_1(0.5)} \times \frac{J_1}{J_a} \tan \alpha\right] F_a \tag{18.122}$$

Similarly, equations from (18.111) and (18.121),

$$F_{e\nu} = \left[\frac{J_r(0.5)}{J_2(0.5)} \times \frac{J_2}{J_a} \tan \alpha\right] F_a \tag{18.123}$$

Substituting equations (18.122) and (18.123) for $F_{e\mu}$ and $F_{e\nu}$, respectively, in equation (18.115) gives

$$F_e = \left\{\left[\frac{C}{C_\mu} \times \frac{J_1}{J_1(0.5)}\right]^{3.33} + \left[\frac{C}{C_\nu} \times \frac{J_2}{J_2(0.5)}\right]^{3.33}\right\}^{0.3} \frac{J_r(0.5)}{J_a \tan \alpha} F_a \tag{18.124}$$

In equations (18.118) and (18.124), for inner ring rotation, that is, with load stationary relative to the outer ring, $C_\mu = C_i$ and $C_\nu = C_o$. For pure radial displacement of the bearing rings ($\epsilon = 0.5$); therefore,

$$F_e = \left[\left(\frac{C}{C_i}\right)^{3.33} + \left(\frac{C}{C_o}\right)^{3.33}\right]^{0.3} F_r \tag{18.125}$$

For outer ring rotation, that is, with the inner ring stationary relative to load, $C_\mu = vC_o$ and $C_\nu = C_i/v$, in which $v = J_2(0.5)/J_1(0.5)$. For this case in pure radial load,

$$F_e = VF_r \tag{18.126}$$

in which

$$V = \left[\left(\frac{C}{vC_o}\right)^{3.33} + \left(\frac{vC}{C_i}\right)^{3.33}\right]^{0.3} \tag{18.127}$$

The factor V, which is a rotation factor, can be rearranged as follows

$$V = v\left[\frac{1 + \left(\dfrac{C_i}{v^2 C_o}\right)^{3.33}}{1 + \left(\dfrac{C_i}{C_o}\right)^{3.33}}\right]^{0.3} \tag{18.128}$$

When C_i/C_o approaches 0, then $V = v = 1.04$ for point contact. In the other extreme, when C_i/C_o becomes infinitely large, $V = 1/v = 0.962$ for point contact. Figure 18.11 shows the variation of V with C_i/C_o for outer ring rotation. For most applications $V = 1$ is sufficiently accurate.

Both ANSI [18.10] and ISO [18.13] neglect the rotation factor and simply recommend the following equation for equivalent radial load:

$$F_e = XF_r + YF_a \tag{18.129}$$

Values of X and Y are given in Table 18.2 for radial ball bearings.

Point Contact Thrust Bearings

For a bearing loaded in pure thrust every rolling element is loaded equally as follows:

$$Q = \frac{F_a}{Z \sin \alpha} \tag{6.26}$$

For both the rotating and stationary raceways, the mean equivalent rolling element load is simply Q as defined by equation (6.26). Setting $F_a = C_a$, therefore,

$$C_{a\mu} = Q_{c\mu} Z \sin \alpha \tag{18.130}$$

$$C_{a\nu} = Q_{c\nu} Z \sin \alpha \tag{18.131}$$

TABLE 18.2. Values of X and Y for Radial Ball Bearings

Bearing Type	$\frac{F_a}{C_o}$	$\frac{F_a}{iZD^2}$ Units (N · mm)	$\frac{F_a}{iZD^2}$ Units (lb · in.)	Single-Row Bearings $\frac{F_a}{F_r} > e$ X	Single-Row Bearings $\frac{F_a}{F_r} > e$ Y	Double Row Bearings $\frac{F_a}{F_r} \leq e$ X	Double Row Bearings $\frac{F_a}{F_r} \leq e$ Y	Double Row Bearings $\frac{F_a}{F_r} > e$ X	Double Row Bearings $\frac{F_a}{F_r} > e$ Y	e
Radial-contact groove ball bearings	0.014	0.172	25		2.30				2.30	0.19
	0.028	0.345	50		1.99				1.99	0.22
	0.056	0.689	100		1.71				1.71	0.26
	0.084	1.03	150	0.56	1.56	1	0	0.56	1.55	0.28
	0.11	1.38	200		1.45				1.45	0.30
	0.17	2.07	300		1.31				1.31	0.34
	0.28	3.45	500		1.15				1.15	0.38
	0.42	5.17	750		1.04				1.04	0.42
	0.56	6.89	1000		1.00				1.00	0.44
Angular-contact groove ball bearings with contact angle	0.014	0.172	25	For this type use the X, Y, and e values applicable to single-row radial contact bearings			2.78		3.74	0.23
	0.028	0.345	50				2.40		3.23	0.26
	0.056	0.689	100				2.07		2.78	0.30
	0.085	1.03	150			1	1.87	0.78	2.52	0.34
5°	0.11	1.38	200				1.75		2.36	0.36
	0.17	2.07	300				1.58		2.13	0.40
	0.28	3.45	500				1.39		1.87	0.45
	0.42	5.17	750				1.26		1.69	0.50
	0.56	6.89	1000				1.21		1.63	0.52

10°	0.014	0.172	25		1.88		2.18		3.06	0.29
	0.029	0.345	50		1.71		1.98		2.78	0.32
	0.057	0.689	100		1.52		1.76		2.47	0.36
	0.086	1.03	150		1.41		1.63		2.20	0.38
	0.11	1.38	200	0.46	1.34	1	1.55	0.75	2.18	0.40
	0.17	2.07	300		1.23		1.42		2.00	0.44
	0.29	3.45	500		1.10		1.27		1.79	0.49
	0.43	5.17	750		1.01		1.17		1.64	0.54
	0.57	6.89	1000		1.00		1.16		1.63	0.54
15°	0.015	0.172	25		1.47		1.65		2.39	0.38
	0.029	0.345	50		1.40		1.57		2.28	0.40
	0.058	0.689	100		1.30		1.46		2.11	0.43
	0.087	1.03	150		1.23		1.38		2.00	0.46
	0.12	1.38	200	0.44	1.19	1	1.34	0.72	1.93	0.47
	0.17	2.07	300		1.12		1.26		1.82	0.50
	0.29	3.45	500		1.02		1.14		1.66	0.55
	0.44	5.17	750		1.00		1.12		1.63	0.56
	0.58	6.89	1000		1.00		1.12		1.63	0.56
20°				0.43	1.00	1	1.09	0.70	1.63	0.57
25°				0.41	0.87	1	0.92	0.67	1.41	0.68
30°				0.39	0.76	1	0.78	0.63	1.24	0.80
35°				0.37	0.66	1	0.66	0.60	1.07	0.95
40°				0.35	0.57	1	0.55	0.57	0.98	1.14
Self-aligning ball bearings				0.40	0.4 ctn α	1	0.42 ctn α	0.65	0.65 ctn α	1.5 tan α

[a] Two similar single-row angular-contact ball bearings mounted face-to-face or back-to-back are considered as one double-row angular-contact bearing.
[b] Values of X, Y, and e for a load or contact angle other than shown are obtained by linear interpolation.
[c] Values of X, Y, and e do not apply to filling slot bearings for applications in which ball–raceway contact areas project substantially into the filling slot under load.
[d] For single-row bearings when $F_a/F_r \leq e$, use $X = 1$, $Y = 0$.

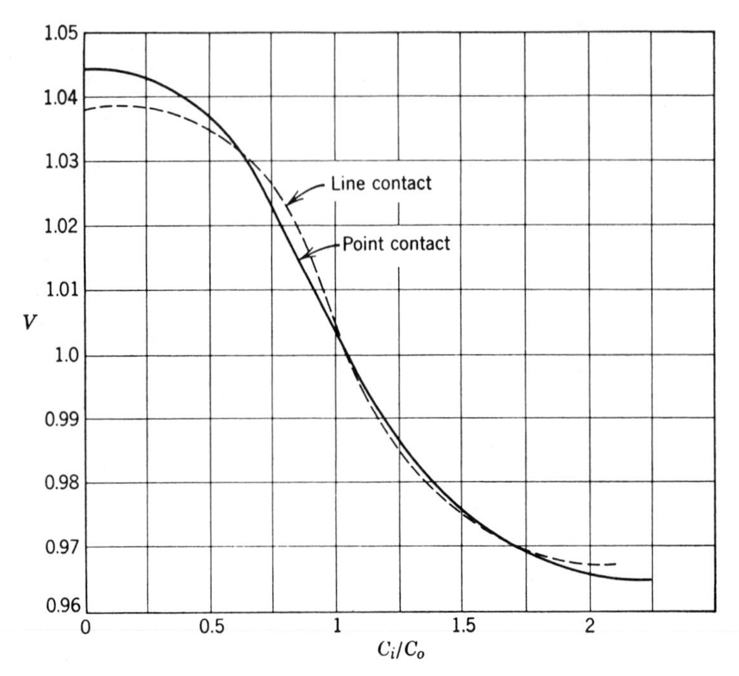

FIGURE 18.11. Rotation factor V vs C_i/C_o.

Hence, by equation (18.58),

$$C_{a\mu} = 98.1\left(\frac{2R}{D} \times \frac{r_\mu}{r_\mu - R}\right)^{0.41} \frac{(1 \mp \gamma)^{1.39}}{(1 \pm \gamma)^{0.33}}\, \gamma^{0.3}(\cos \alpha)^{0.7} Z^{0.67} \tan \alpha D^{1.8} \tag{18.132}$$

$$C_{a\nu} = 98.1\left(\frac{2R}{D} \times \frac{r_\nu}{r_\nu - R}\right)^{0.41} \frac{(1 \mp \gamma)^{1.39}}{(1 \pm \gamma)^{0.33}}\, \gamma^{0.3}(\cos \alpha)^{0.7} Z^{0.67} \tan \alpha D^{1.8} \tag{18.133}$$

In equations (18.132) and (18.133) the upper signs refer to an inner raceway and the lower signs to an outer raceway. The basic dynamic capacity of an entire thrust bearing assembly is given by

$$C_a = 98.1\left\{1 + \left[\left(\frac{1 \mp \gamma}{1 \pm \gamma}\right)^{1.72} \left(\frac{r_\mu}{r_\nu} \times \frac{2r_\nu - D}{2r_\mu - D}\right)^{0.41}\right]^{3.33}\right\}^{-0.3}$$

$$\times \left(\frac{2r_\mu}{2r_\mu - D}\right)^{0.41} \frac{\gamma^{0.3}(1 \mp \gamma)^{1.39}}{(1 \pm \gamma)^{0.33}} (\cos \alpha)^{0.7} \tan \alpha Z^{0.67} D^{1.8} \tag{18.134}$$

For ball bearings with inner ring rotation equation (18.134) becomes

$$C_a = 98.1\left\{1 + \left[\left(\frac{1-\gamma}{1+\gamma}\right)^{1.72}\left(\frac{f_i}{f_o} \times \frac{2f_o - 1}{2f_i - 1}\right)^{0.41}\right]^{3.33}\right\}^{-0.3}$$
$$\times \left(\frac{2f_i}{2f_i - 1}\right)^{0.41} \frac{\gamma^{0.3}(1-\gamma)^{1.39}}{(1+\gamma)^{0.33}} (\cos\alpha)^{0.7} \tan\alpha Z^{0.67} D^{1.8} \quad (18.135)$$

Lundberg et al. [18.8] recommended a reduction in the material constant to accommodate inaccuracies in manufacturing that cause unequal internal load distribution. Hence, equation (18.135) becomes

$$C_a = 88.2^*(1 - 0.33\sin\alpha)$$
$$\times \left\{1 + \left[\left(\frac{1-\gamma}{1+\gamma}\right)^{1.72}\left[\frac{f_i}{f_o} \times \frac{(2f_o - 1)}{(2f_i - 1)}\right]^{0.41}\right]^{3.33}\right\}^{-0.3}$$
$$\times \left(\frac{2f_i}{2f_i - 1}\right)^{0.41} \frac{\gamma^{0.3}(1-\gamma)^{1.39}}{(1+\gamma)^{0.33}} (\cos\alpha)^{0.7} \tan\alpha Z^{0.67} D^{1.8} \quad (18.136)$$

In (18.136) as recommended by Palmgren [18.9] the term $(1 - 0.33 \sin \alpha)$ accounts for reduction in C_a caused by added friction due to spinning (presumably). The following is the formula for basic dynamic thrust capacity:

$$C_a = f_c(\cos\alpha)^{0.7} \tan\alpha Z^{2/3} D^{1.8\dagger} \quad (18.137)$$

from which it is apparent that (approximately)

$$f_c = 88.2(1 - 0.33\sin\alpha)$$
$$\times \left\{1 + \left[\left(\frac{1-\gamma}{1+\gamma}\right)^{1.72} \times \left(\frac{f_i}{f_o} \times \frac{2f_o - 1}{2f_i - 1}\right)^{0.41}\right]^{3.33}\right\}^{-0.3}$$
$$\times \frac{\gamma^{0.3}(1-\gamma)^{1.39}}{(1+\gamma)^{0.33}} \left(\frac{2f_i}{2f_i - 1}\right)^{0.41} \quad (18.138)$$

For thrust bearings with a 90° contact angle

$$C_a = f_c Z^{2/3} D^{1.8} \quad (18.139)$$

*This value can be as high as 93.2 (7080) for thrust-loaded angular-contact ball bearings.
†ANSI [18.10] recommends using D raised to the 1.4 power in lieu of 1.8 for bearings having balls of diameter greater than 25.4 mm (1 in.).

TABLE 18.3. X and Y Factors for Ball Thrust Bearings

	Single Direction Bearings $F_a/F_r > e$		Double Direction Bearings[a]				
			$F_a/F_r \leq e$		$F_a/F_r > e$		
Bearing Type	X	Y	X	Y	X	Y	e
Thrust ball bearings with contact angle[b]							
$\alpha = 45°$	0.66	1	1.18	0.59	0.66	1	1.25
$\alpha = 60°$	0.92	1	1.90	0.54	0.92	1	2.17
$\alpha = 75°$	1.66	1	3.89	0.52	1.66	1	4.67

[a] Double direction bearings are presumed to be symmetrical.
[b] For $\alpha = 90°$: $F_r = 0$ and $Y = 1$.

in which (approximately)

$$f_c = 59.1\left[1 + \left(\frac{f_i}{f_o} \times \frac{2f_o - 1}{2f_i - 1}\right)^{1.36}\right]^{-0.3} \gamma^{0.3}\left(\frac{2f_i}{2f_i - 1}\right)^{0.41} \tag{18.140}$$

For thrust bearings having i rows of balls in which Z_k is the number of rolling elements per row and C_{ak} is the basic dynamic capacity per row, the basic dynamic capacity C_a of the bearing may be determined as follows:

$$C_a = \sum_{k=1}^{k=i} Z_k \left[\sum_{k=1}^{k=i} \left(\frac{Z_k}{C_{ak}}\right)^{3.33}\right]^{-0.3} \tag{18.141}$$

As for radial bearings, the L_{10} life of a thrust bearing is given by

$$L = \left(\frac{C_a}{F_{ea}}\right)^3 \tag{18.142}$$

in which F_{ea} is the equivalent axial load. As before,

$$F_{ea} = XF_r + YF_a$$

X and Y as recommended by ANSI are given by Table 18.3.

Example 18.5. The 218 angular-contact ball bearing of Example 8.1 is operated at 10,000 rpm under a 22,250 N (5000 lb) thrust load. Estimate the L_{10} fatigue life of the bearing for inner raceway rotation.

$$Z = 16 \qquad \text{Ex. 6.5}$$

$$D = 22.23 \text{ mm } (0.875 \text{ in.}) \qquad \text{Ex. 2.3}$$

$$d_m = 125.3 \text{ mm } (4.932 \text{ in.}) \qquad \text{Ex. 2.6}$$

$$f_i = f_o = 0.5232 \qquad \text{Ex. 2.3}$$

$$\alpha_i = 48.8° \qquad \text{Ex. 8.4}$$

$$\alpha_o = 33.3° \qquad \text{Ex. 8.4}$$

$$Q_i = 1788 \text{ N } (401.7 \text{ lb}) \qquad \text{Ex. 8.6}$$

$$Q_o = 2241 \text{ N } (503.7 \text{ lb}) \qquad \text{Ex. 8.6}$$

$$\gamma_i = \frac{D \cos \alpha_i}{d_m} \tag{2.27}$$

$$= \frac{22.23 \cos (48.8°)}{125.3} = 0.1169$$

$$Q_{ci} = 93.2^*\left(\frac{2f_i}{2f_i - 1}\right)^{0.41} \frac{(1 - \gamma_i)^{1.39}}{(1 + \gamma_i)^{1/3}} \left(\frac{\gamma_i}{\cos \alpha_i}\right)^{0.3} D^{1.8} Z^{-1/3} \tag{18.67}$$

$$= 93.2\left(\frac{2 \times 0.5232}{2 \times 0.5232 - 1}\right)^{0.41} \frac{(1 - 0.1169)^{1.39}}{(1 + 0.1169)^{1/3}}$$

$$\times \left(\frac{0.1169}{\cos 48.8°}\right)^{0.3} (22.23)^{1.8}(16)^{-1/3}$$

$$= 17{,}040 \text{ N } (3830 \text{ lb})$$

According to Fig. 8.5 at 10,000 rpm and 22,250-N (5000-lb) thrust load, this bearing operation approximates "outer raceway control." To account for spinning, the inner raceway capacity is reduced by a factor of $(1 - 0.33 \times \sin \alpha_i)$.

$$Q'_{ci} = Q_{ci}(1 - 0.33 \sin \alpha_i)$$

$$= 17{,}040[1 - 0.33 \sin (48.8°)] = 12{,}810 \text{ N } (2879 \text{ lb})$$

$$L_i = \left(\frac{Q_{ci}}{Q_{ei}}\right)^3 \tag{18.70}$$

*Strictly speaking an angular-contact ball bearing with $\alpha° < 40°$ is classified as a radial bearing and would be rated by using $f_m = 93.2$.

Under pure thrust $Q_{ei} = Q_i$.

$$L_i = \left(\frac{12{,}810}{1788}\right)^3 = 368 \text{ million revolutions}$$

$$\gamma_o = \frac{D \cos \alpha_o}{d_m} \tag{2.27}$$

$$= \frac{22.23 \cos (33.3°)}{125.3} = 0.1483$$

$$Q_c = 93.2\left(\frac{2f_o}{2f_o - 1}\right)^{0.41} \frac{(1 + \gamma_o)^{1/3}}{(1 - \gamma_o)^{1/3}} \left(\frac{\gamma_o}{\cos \alpha_o}\right)^{0.3} D^{1.8} Z^{-1/3} \tag{18.67}$$

$$= 93.2\left(\frac{2 \times 0.5232}{2 \times 0.5232 - 1}\right)^{0.41} \frac{(1 + 0.1483)^{1.39}}{(1 - 0.1483)^{1/3}} \left[\frac{0.1483}{\cos (33.3°)}\right]^{0.3}$$

$$\times (22.23)^{1.8}(16)^{-1/3}$$

$$= 26{,}880 \text{ N } (6040 \text{ lb})$$

$$L_o = \left(\frac{Q_{co}}{Q_{eo}}\right)^3 \tag{18.75}$$

$$= \left(\frac{26880}{2241}\right)^3 = 1724 \text{ million revolutions}$$

$$L = (L_i^{-1.11} + L_o^{-1.11})^{-0.9} \tag{18.79}$$

$$= [(368)^{-1.11} + (1724)^{-1.11}]^{-0.9} \times 10^6$$

$$= 315 \times 10^6 \text{ revolutions}$$

or

$$L = \frac{315 \times 10^6}{10{,}000 \times 60} = 525 \text{ hr}$$

Line Contact Radial Bearings

The L_{10} fatigue life of a roller–raceway line contact subjected to normal load Q may be estimated by

$$L = \left(\frac{Q_c}{Q}\right)^4 \tag{18.65}$$

in which L is in millions of revolutions and

$$Q_c = 552 \frac{(1 \mp \gamma)^{29/27}}{(1 \pm \gamma)^{1/4}} \left(\frac{\gamma}{\cos \alpha}\right)^{2/9} D^{29/27} l^{7/9} Z^{-1/4} \tag{18.64}$$

The upper signs refer to an inner raceway contact and the lower signs refer to an outer raceway contact.

To account for stress concentrations due to edge loading of rollers and noncentered roller loads, Lundberg et al. [18.8] introduced a reduction factor λ such that

$$Q_c = 552\lambda \frac{(1 \mp \gamma)^{29/27}}{(1 \pm \gamma)^{1/4}} \left(\frac{\gamma}{\cos\alpha}\right)^{2/9} D^{29/27} l^{7/9} Z^{-1/4} \tag{18.143}$$

Based on their test results, the schedule of Table 18.4 for λ_i and λ_o was developed. Variation in λ for line contact is probably due to method of roller guiding, for example, in some bearings rollers are guided by flanges that are integral with a bearing ring; other bearings employ roller guiding cages.

In lieu of a cubic mean roller load for a raceway contact, a quartic mean will be used such that

$$Q_{e\mu} = \left(\frac{1}{Z}\sum_{j=1}^{j=Z} Q_j^4\right)^{1/4} = \left(\frac{1}{2\pi}\int_0^{2\pi} Q_\psi^4 \, d\psi\right)^{1/4} \tag{18.144}$$

The difference between a cubic mean load and a quartic mean load is substantially negligible. The fatigue life of the rotating raceway is

$$L_\mu = \left(\frac{Q_{c\mu}}{Q_{e\mu}}\right)^4 \tag{18.145}$$

As with point-contact bearings, the equivalent loading of a nonrotating raceway is given by

$$Q_{e\nu} = \left(\frac{1}{Z}\sum_{j=1}^{j=Z} Q_j^{4e}\right)^{1/4e} = \left(\frac{1}{2\pi}\int_0^{2\pi} Q_\psi^{4.5} \, d\psi\right)^{1/4.5} \tag{18.146}$$

The life of the stationary raceway is

$$L_\nu = \left(\frac{Q_{c\nu}}{Q_{e\nu}}\right)^4 \tag{18.147}$$

As with point-contact bearings, the life of a roller bearing having line contact is calculated from

$$L = (L_\mu^{-9.8} + L_\nu^{-9.8})^{-8/9} \tag{18.148}$$

Thus, if each roller load has been determined by methods of Chapters 6 and 8, the fatigue life of the bearing may be estimated by using equations (18.144) and (18.148).

TABLE 18.4. Values of λ_i and λ_o

Contact	Inner Raceway	Outer Raceway
Line contact	0.41–0.56	0.38–0.6
Modified line contact	0.6–0.8	0.6–0.8

Example 18.6. Assuming modified line contact, estimate the L_{10} fatigue life of the 209 cylindrical roller bearing of Example 6.3. Assume also inner raceway rotation.

$$Z = 14 \qquad \text{Ex. 2.7}$$

$$D = 10 \text{ mm } (0.3937 \text{ in.}) \qquad \text{Ex. 2.7}$$

$$d_m = 65 \text{ mm } (2.559 \text{ in.}) \qquad \text{Ex. 2.7}$$

$$l = 9.6 \text{ mm } (0.378 \text{ in.}) \qquad \text{Ex. 2.7}$$

Ex. 6.3

ψ	Q_ψ
0	1915 N (430.3 lb)
25.71°	1348 N (302.9 lb)
51.42°	0
⋮	⋮
180°	0

$$\gamma = D\frac{\cos\alpha}{d_m} \tag{2.27}$$

$$= \frac{10}{65} = 0.1538$$

From Table 18.4, use $\lambda_i = \lambda_o = 0.61$ (see also Table 18.10),

$$Q_{ci} = 552\lambda_i \frac{(1-\gamma)^{29/27}}{(1+\gamma)^{1/4}} \gamma^{2/9} D^{29/27} l^{7/9} Z^{-1/4} \tag{18.143}$$

$$= 552 \times 0.61$$

$$\times \frac{(1-0.1538)^{29/27}}{(1+0.1538)^{1/4}} (0.1538)^{2/9}(10)^{29/27}(9.6)^{7/9}(14)^{-1/4}$$

$$= 6381 \text{ N } (1434 \text{ lb})$$

$$Q_{ei} = \left(\frac{1}{Z}\sum_{j=1}^{j=Z} Q_j^4\right)^{1/4} \tag{18.144}$$

$$= \{\tfrac{1}{14}[(1915)^4 + 2(1348)^4 + 0 + \cdots 0]\}^{1/4} = 1095 \text{ N } (246 \text{ lb})$$

$$L_i = \left(\frac{Q_{ci}}{Q_{ei}}\right)^4 \tag{18.145}$$

$$= \left(\frac{6381}{1095}\right)^4 = 1155 \text{ million revolutions}$$

$$Q_{co} = 552\lambda_o \frac{(1+\gamma)^{29/27}}{(1-\gamma)^{1/4}} \gamma^{2/9} D^{29/27} l^{7/9} Z^{-1/4} \tag{18.143}$$

$$= 552 \times 0.61$$

$$\times \frac{(1 + 0.1538)^{29/27}}{(1 - 0.1538)^{1/4}} (0.1538)^{2/9}(10)^{29/27}(9.6)^{7/9}(14)^{-1/4}$$

$$= 9621 \text{ N } (2162 \text{ lb})$$

$$Q_{eo} = \left(\frac{1}{Z}\sum_{j=1}^{j=Z} Q_j^{4.5}\right)^{1/4.5} \tag{18.146}$$

$$= \{\tfrac{1}{14}[(1915)^{4.5} + 2(1348)^{4.5} + 0 + \cdots 0]\}^{1/4.5}$$

$$= 1148 \text{ N } (258 \text{ lb})$$

$$L_o = \left(\frac{Q_{co}}{Q_{eo}}\right)^4 \tag{18.147}$$

$$= \left(\frac{9621}{1148}\right)^4 = 4937 \text{ million revolutions}$$

$$L = (L_i^{-9/8} + L_o^{-9/8})^{-8/9} \tag{18.148}$$

$$= [(1155)^{-9/8} + (4937)^{-9/8}]^{-8/9} \times 10^6 = 9.85 \times 10^8 \text{ revolutions}$$

To simplify the rigorous method of calculating bearing fatigue life just outlined, an approximate method was developed by Lundberg et al. [18.1, 18.8] for roller bearings having rigid rings and moderate speeds. In a manner similar to point contact bearings,

$$Q_{e\mu} = Q_{\max} J_1 \tag{18.149}$$

$$J_1 = \left\{\frac{1}{2\pi}\int_{-\psi_l}^{+\psi_l}\left[1 - \frac{1}{2\epsilon}(1 - \cos\psi)\right]^{4.4} d\psi\right\}^{1/4} \tag{18.150}$$

TABLE 18.5. J_1 and J_2 for Line Contact

Single Row			Double Row			
ϵ	J_1	J_2	ϵ_I	ϵ_{II}	J_1	J_2
0	0	0				
0.1	0.5287	0.5633	0.5	0.5	0.7577	0.7867
0.2	0.5772	0.6073	0.6	0.4	0.6807	0.7044
0.3	0.6079	0.6359	0.7	0.3	0.6806	0.7032
0.4	0.6309	0.6571	0.8	0.2	0.6907	0.7127
0.5	0.6495	0.6744	0.9	0.1	0.7028	0.7229
0.6	0.6653	0.6888				
0.7	0.6792	0.7015				
0.8	0.6906	0.7127				
0.9	0.7028	0.7229				
1	0.7132	0.7323				
1.25	0.7366	0.7532				
1.67	0.7705	0.7832				
2.5	0.8216	0.8301				
5	0.8989	0.9014				
∞	1	1				

$$Q_{e\nu} = Q_{\max} J_2 \tag{18.151}$$

$$J_2 = \left\{ \frac{1}{2\pi} \int_{-\psi_l}^{+\psi_l} \left[1 - \frac{1}{2\epsilon}(1 - \cos\psi) \right]^{4.95} d\psi \right\}^{2/9} \tag{18.152}$$

Table 18.5 gives values of J_1 and J_2 as functions of ϵ.

As for point-contact bearings, equations (6.66), (18.85), and (18.86) are equally valid for radial roller bearings in line contact. Therefore at $\epsilon = 0.5$,

$$C_\mu = 0.377 i^{7/9} Q_\mu Z \cos\alpha \tag{18.153}$$

$$C_\nu = 0.363 i^{7/9} Q_\nu Z \cos\alpha \tag{18.154}$$

According to the product law of probability,

$$C = C_\mu \left[1 + \left(\frac{C_\mu}{C_\nu} \right)^{9/2} \right]^{2/9} = g_c C_\mu \tag{18.155}$$

The reduction factor λ accounting for edge loading may be applied to the entire bearing assembly. For line contact at one raceway and point contact at the other $\lambda = 0.54$ if a symmetrical pressure distribution similar to that shown by Fig. 5.17*b* is attained along the roller length. Figure 18.12, taken from reference [18.8] shows the fit obtained to the test data while using $\lambda = 0.54$. Table 18.6 is a schedule for λ for bearing assemblies.

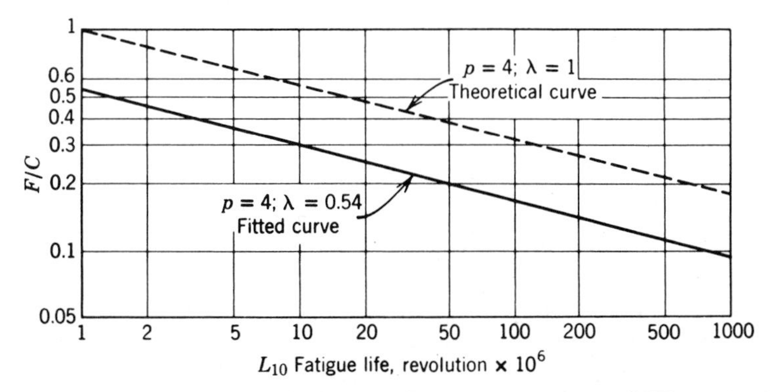

FIGURE 18.12. L_{10} vs F/C for roller bearing. Test points are for an SKF 21309 spherical roller bearing.

TABLE 18.6. Reduction Factor λ

Contact Condition	λ Range
Line contact at both raceways	0.4–0.5
Line contact at one raceway Point contact at other raceway	0.5–0.6
Modified line contact	0.6–0.8

Using the reduction factor, λ, the resulting expression for basic dynamic capacity of a radial roller bearing is

$$C = 207\lambda\left\{1 + \left[1.04\left(\frac{1 \mp \gamma}{1 \pm \gamma}\right)^{143/108}\right]^{9/2}\right\}^{-2/9} \frac{\gamma^{2/9}(1 \mp \gamma)^{29/27}}{(1 \pm \gamma)^{1/4}} \times (il\cos\alpha)^{7/9} Z^{3/4} D^{29/27} \tag{18.156}$$

In most bearing applications the inner raceway rotates and

$$C = 207\lambda\left\{1 + \left[1.04\left(\frac{1 - \gamma}{1 + \gamma}\right)^{143/108}\right]^{9/2}\right\}^{-2/9} \frac{\gamma^{2/9}(1 - \gamma)^{29/27}}{(1 + \gamma)^{1/4}} \times (il\cos\alpha)^{7/9} Z^{3/4} D^{29/27} \tag{18.157}$$

As for point-contact bearings an equivalent radial load can be developed and

$$F_e = \left\{\left[\frac{C}{C_i} \times \frac{J_r(0.5)J_1}{J_1(0.5)J_r}\right]^{9/2} + \left[\frac{CJ_r(0.5)J_2}{C_oJ_2(0.5)J_r}\right]^{9/2}\right\}^{2/9} F_r + \left\{\left[\frac{CJ_1}{C_iJ_1(0.5)}\right]^{9/2} + \left[\frac{CJ_2}{C_oJ_2(0.5)}\right]^{9/2}\right\}^{2/9} \frac{J_r(0.5)}{J_a \tan\alpha} F_a \tag{18.158}$$

The rotation factor V is given by

$$V = \upsilon \left[\frac{1 + \left(\frac{C_i}{\upsilon^2 C_o} \right)^{9/2}}{1 + \left(\frac{C_i}{C_o} \right)^{9/2}} \right]^{2/9} \tag{18.159}$$

in which $\upsilon = J_2(0.5)/J_1(0.5)$. Figure 18.11 shows the variation of V with C_i/C_o for both point and line contact.

ANSI [18.12] gives the same formula for equivalent radial load for radial roller bearings as for radial ball bearings. (Rotation factor V is once again neglected.)

$$F_e = XF_r + YF_a \tag{18.129}$$

X and Y for spherical self-aligning and tapered roller bearings are given in Table 18.7.

The life of a roller bearing in line contact is given by

$$L = \left(\frac{C}{F_e} \right)^4 \tag{18.160}$$

Line-Contact Thrust Bearings

For thrust bearings, Lundberg et al. [18.8] introduced the reduction factor η, in addition to λ, to account for variations in raceway groove dimensions which may cause a roller from experiencing the theoretical uniform loading:

$$Q = \frac{F_a}{Z \sin \alpha} \tag{6.26}$$

According to reference [18.8], for thrust roller bearings

$$\eta = 1 - 0.15 \sin \alpha \tag{18.161}$$

TABLE 18.7. *X* and *Y* for Radial Roller Bearings

	$F_a/F_r \leq 1.5 \tan \alpha$		$F_a/F_r > 1.5 \tan \alpha$	
	X	Y	X	Y
Single-row bearing	1	0	0.4	0.4 ctn α
Double-row bearing	1	0.45 ctn α	0.67	0.67 ctn α

Considering the capacity reductions due to λ and η, for thrust roller bearings in line contact, the following equations may be used for thrust bearings in which $\alpha \neq 90°$:

$$C_{ak} = 552\lambda\eta\gamma^{2/9}\frac{(1 \mp \gamma)^{29/27}}{(1 \pm \gamma)^{1/4}}(l\cos\alpha)^{7/9}\tan\alpha Z^{3/4}D^{29/27} \qquad (18.162)$$

In equation (18.162) the upper signs refer to the inner raceway, that is, $k = i$; the lower signs refer to the outer raceway, that is, $k = \text{o}$.

For thrust roller bearings in which $\alpha = 90°$;

$$C_{ai} = C_{ao} = 469\lambda\gamma^{2/9}l^{7/9}D^{29/27}Z^{3/4} \qquad (18.163)$$

Equations (18.162) and (18.163) may be substituted into (18.155) to obtain the basic dynamic capacity of a bearing row in thrust loading. Equation (18.164) may be used to determine the basic dynamic capacity in thrust loading for a thrust roller bearing having i rows and Z_i rollers in each row:

$$C_a = \sum_{k=1}^{k=i} Z_k \left[\sum_{k=1}^{k=i}\left(\frac{Z_k}{C_{ak}}\right)^{9/2}\right]^{-2/9} \qquad (18.164)$$

The fatigue life of a thrust roller bearing can be calculated by the following equation:

$$L = \left(\frac{C_a}{F_{ea}}\right)^4 \qquad (18.165)$$

According to ANSI [18.13] the equivalent thrust load may be estimated by

$$F_{ea} = XF_r + YF_a \qquad (18.166)$$

Table 18.8 gives values of X and Y.

TABLE 18.8. *X* and *Y* for Thrust Roller Bearings

Bearing Type	Contact Angle	Loading	X	Y
Single direction	$\alpha < 90°$	$F_a/F_r \leq 1.5\tan\alpha$	0	1
	$\alpha = 90°$	$F_r = 0$	0	1
	$\alpha < 90°$	$F_a/F_r > 1.5\tan\alpha$	$\tan\alpha$	1
Double direction	$\alpha < 90°$	$F_a/F_r \leq 1.5\tan\alpha$	$1.5\tan\alpha$	0.67
	$\alpha < 90°$	$F_a/F_r > 1.5\tan\alpha$	$\tan\alpha$	1

Radial Roller Bearings with Point and Line Contact

If a roller bearing contains rollers and raceways having straight profiles, then line contact obtains at each contact and the formulations of the preceding two sections are valid. If, however, the rollers have a curved profile (crowned; see Fig. 1.33) of smaller radius than one or both of the conforming raceway profiles or if one or both raceways have a convex profile and the rollers have straight profiles, then depending on the angular position of a roller and its roller load, one of the contact conditions in Table 18.9 will occur.

Of the contact conditions shown in Table 18.9, optimum roller bearing design for any given application is generally achieved when the most heavily loaded roller is in modified line contact. As stated in Chapter 5 this condition produces the most nearly uniform stress distribution along the roller profile, and edge loading is precluded. It is also stated in Chapter 5 that a logarithmic profile roller can produce an even better load distribution over a wider range of loading; however, this roller profile tends to be special. The more usual profile is that of the partially crowned roller. It should be apparent that the optimum crown radii or osculations necessary to obtain modified line contact can only be evaluated for a given bearing after the loading has been established. Series of bearings, however, are often optimized by basing the crown radii or osculations on an estimated load, for example, $0.5C$ or $0.25C$, in which C is the basic dynamic capacity. Depending on the applied loads, bearings in such a series may operate anywhere from point to line contact at the most heavily loaded roller.

Because it is desirable to use one rating method for a given roller bearing, and because in any given roller bearing application it is possible to have combinations of line and point contact, Lundberg and Palmgren [18.8] estimated the fatigue life should be calculated from

$$L = \left(\frac{C}{F_e}\right)^{10/3} \tag{18.167}$$

TABLE 18.9. Roller–Raceway Contact

Condition	Inner Raceway	a_i[a]	Outer Raceway	a_o[b]	Load
1	Line	$2a_i > 1.5l$	Line	$2a_o > 1.5l$	Heavy
2	Line	$2a_i > 1.5l$	Point	$2a_o < l$	Moderate
3	Point	$2a_i < l$	Line	$2a_o > 1.5l$	Moderate
4	Modified line	$l \le 2a_i \le 1.5l$	Modified line	$l \le 2a_o \le 1.5l$	Moderate
5	Point	$2a_i < l$	Point	$2a_o < l$	Light

[a] a_i is the semimajor axis of inner raceway contact ellipse.
[b] a_o is the semimajor axis of outer raceway contact ellipse.

Note that $3 < 10/3 < 4$. In equation

$$C = \nu C_l \tag{18.168}$$

C_l is the basic dynamic capacity in line contact as calculated by equations (18.157) or (18.163).

If both inner and outer raceway contacts are line contacts and $\lambda = 0.45$ to account for edge loading and nonuniform stress distribution, curve 1 of Fig. 18.13 shows the variation of load with life by using equation (18.160) and the fourth power slope. Assuming $\nu = 1.36$ and using equation (18.167), curve 2 illustrates the approximation to curve 1. The shaded area shows the error which occurs when using the approximation. Points A on Fig. 18.13 represent 5% error.

If outer and inner raceway contacts are point contacts for loads arbitrarily less than $0.21C$ ($L = 100$ million revolutions), then for $\lambda = 0.65$ curve 1 of Fig. 18.14 shows the load-life variation of the bearing. Note the inverse slope of the curve decreases from 4 to 3 at $L = 100$ million revolutions. Curve 2 of Fig. 18.14 shows the fit obtained while using equation (18.167) and $\nu = 1.20$.

Lastly, if one raceway contact is line contact and the other is point contact, curve 1 of Fig. 18.15 shows load-life variation for $\lambda = 0.54$. Transformation from point to line contact is arbitrarily assumed to occur at $L = 100$ million revolutions. Curve 2 of Fig. 18.15 shows the fit obtained while using equation (18.167) and $\nu = 1.26$.

In Fig. 18.13, using equation (18.167), fatigue lives between 150 and 1500 million revolutions have a calculational error less than 5%. Simi-

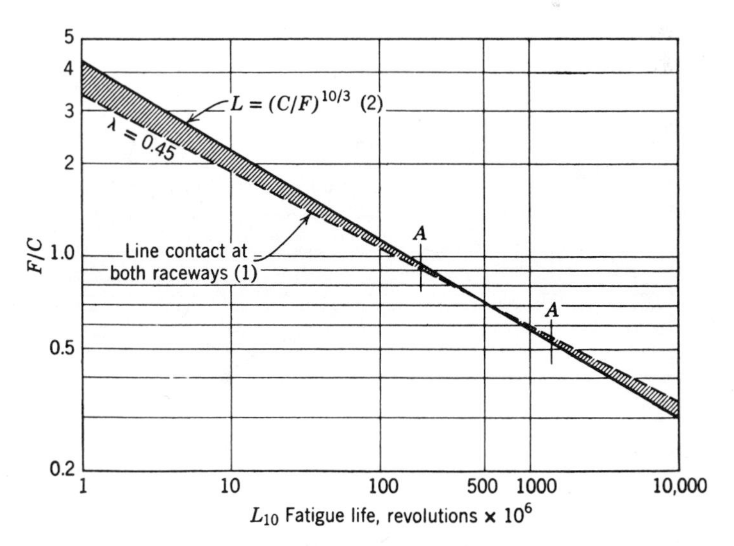

FIGURE 18.13. L_{10} vs F/C for line contact at both raceways.

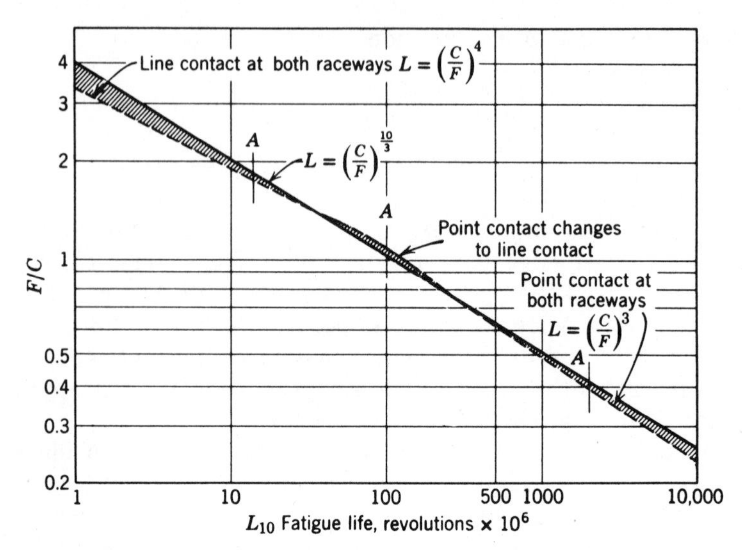

FIGURE 18.14. L_{10} vs F/C for point or line contact at both raceways.

larly, in Fig. 18.14 lives between 15 and 2000 million revolutions have less than 5% calculational error, and in Fig. 18.15 lives between 40 and 10,000 million revolutions have less than 5% calculational error. Since the foregoing ranges represent probable regions of roller bearing operation, Lundberg et al. [18.8] considered that equation (18.167) was a satisfactory approximation by which to estimate fatigue life of roller bearings. Accordingly, the data in Table 18.10 were developed.

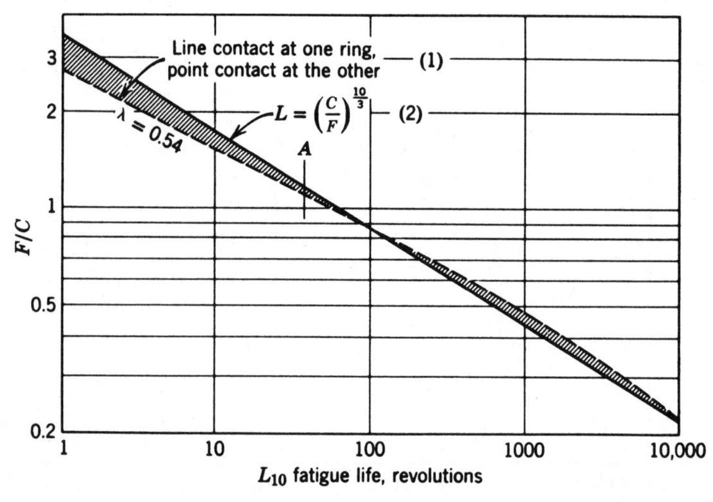

FIGURE 18.15. L_{10} vs F/C for combination of line and point contact (reprinted from [18.8]).

TABLE 18.10. ν and λ for Roller Bearings

Condition	λ	ν	$\lambda\nu$
Modified line contact in cylindrical roller bearings	0.61	1.36	0.83
Modified line contact in spherical and tapered roller bearings	0.57	1.36	0.78
Combination line and point contact (inner and outer raceways)	0.54	1.26	0.68
Line Contact	0.45	1.36	0.61

Equation (18.156) becomes

$$C = 207\lambda\nu\left\{1 + \left[1.04\left(\frac{1 \mp \gamma}{1 \pm \gamma}\right)^{143/108}\right]^{9/2}\right\}^{-2/9} \frac{\gamma^{2/9}(1 \mp \gamma)^{29/27}}{(1 \pm \gamma)^{1/4}} \cdot (il \cos \alpha)^{7/9} Z^{3/4} D^{29/27} \qquad (18.169)$$

Thrust Roller Bearing with Point and Line Contact

For thrust roller bearings operating with a combination point and line contact at the raceways, equations (18.162) and (18.163) become

$$C_{ak} = 552\lambda\eta\nu\gamma^{2/9} \frac{(1 \mp \gamma)^{29/27}}{(1 \pm \gamma)^{1/4}} (l \cos \alpha)^{7/9} \tan \alpha Z^{3/4} D^{29/27} \qquad \alpha \neq 90° \qquad (18.170)$$

$$C_{ai} = C_{ao} = 469\lambda\nu\gamma^{2/9} l^{7/9} Z^{3/4} D^{29/27} \qquad \alpha = 90° \qquad (18.171)$$

EFFECT OF STEEL COMPOSITION AND PROCESSING ON FATIGUE LIFE

All of the equations pertaining to basic dynamic capacity of a raceway contact Q_c, or bearing raceway C_i or C_o, or of an entire bearing C developed thus far are based on bearings fabricated from AISI 52100 steel hardened at least to 58 Rockwell C. This is air-melted, air-cast steel whose chemical composition is shown in Table 16.1. Moreover, the equations are based upon steel processing methods and manufacturing methods that existed at the time the original load rating standards were created. Equation (18.107) gives a recommended reduction in basic dynamic capacity if the steel is not as hard as 58 Rockwell C. Additionally, if the bearing material hardness is greater than 58 Rockwell C as occasioned by a special heat treatment, the bearing fatigue life can tend to be greater than that predicted by the standard life rating formulas. As shown in Fig. 18.16 from reference [18.14], this effect is diminished if the bearing operates at elevated temperature.

Developments in the processing of rolling bearing steel have been continuous since 1960, the date corresponding to the introduction of carbon vacuum-degassed (CVD) steel in the United States. Improvements in melting practices have yielded bearing steels of significantly increased fatigue endurance capability. For instance, AISI 52100 steel melted in a vacuum has fewer impurities and therefore tends to yield increased rolling bearing fatigue life. Walp et al. [18.15] determined that the presence of traces of metallic impurities such as aluminum, copper, and vanadium is detrimental to fatigue life. Remelting of vacuum-melted steel while using a consumable electrode furnace tends to produce a more homogeneous steel, which inherently has a better resistance to fatigue. Some rolling bearing steels afford increased corrosion resistance in certain applications however at the expense of potentially decreased fatigue endurance. In Chapter 16, the various bearing steels and steel manufacturing processes are discussed in detail. When rolling bearings are fabricated of special steels, the expected increased or decreased fatigue lives may be estimated using life adjustment factors discussed later in this chapter.

Koistinen [18.16] demonstrated a two-stage heat-treating process that forms a surface layer in compression conducive to increased fatigue en-

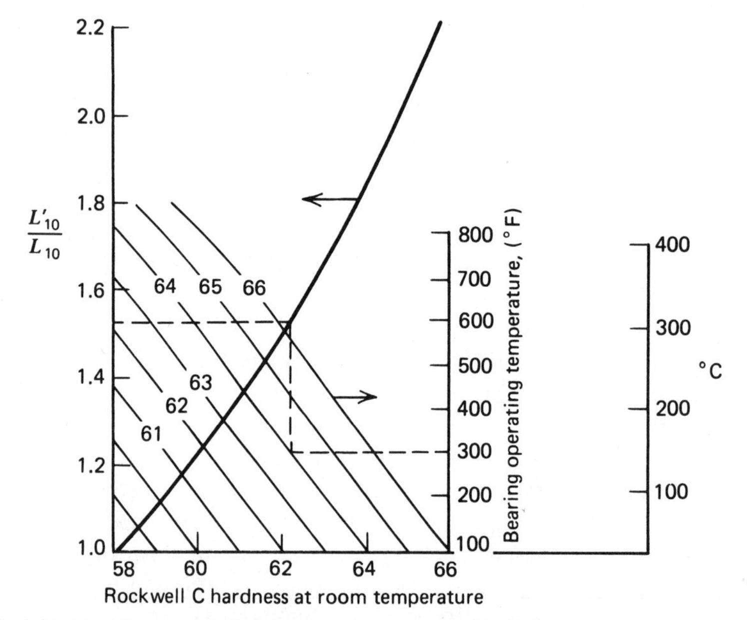

FIGURE 18.16. Nomograph for determining relative life at bearing operating temperature as a function of room temperature hardness.

durance. The rationale for the increased endurance is discussed in Chapter 23. Additionally, for AISI 52100 and M50 bearing steels, ring and rolling element forming processes that tend to eliminate end grain in the bearing rolling contact surfaces have been used to manufacture rolling bearings of increased endurance capability.

The endurance characteristics of case-hardened bearing steels must also be considered. These steels, covered in detail in Chapter 16, have the advantage of a tough, fracture-resistant core as well as a fatigue-resistant, hard surface layer (case). Since the information in Chapter 5 on subsurface shear stress indicates the critical stress occurs close to the raceway or rolling element surface, case-hardened steels are appropriate for rolling bearings. In fact, the fracture-resistant core is essential in many applications. Timken [18.17] presented endurance data for tapered roller bearing fabricated from case-carburized steel; Harris [18.18] demonstrated that the Timken data tended to confirm the load rating method developed by Lundberg and Palmgren [18.8].

One may question the value of the data developed in this chapter concerning L_{10} fatigue life and basic dynamic capacity since these data are based only upon air-melted AISI 52100 steel of a type no longer acceptable in modern industrial practice. The answer is threefold: (1) it is important to understand the origin of the load rating standards in current worldwide use in order to use them effectively and accurately, (2) a comparison may be conducted between the endurance characteristics of similar and dissimilar bearings of different manufacturers on the basis of geometry alone, and (3) the equations may be used to optimize rolling bearing design for any given application.

LOAD RATING STANDARDS

To accommodate the improvements in bearing geometrical accuracies afforded by modern manufacturing methods and the improvements in the modern basic rolling bearing steel chemistries and metallurgies, both ANSI and ISO have included in the formulas for basic load rating* "a rating factor for contemporary, normally used material and manufacturing quality." ISO [18.13] uses the b_m factor directly in the equations for basic dynamic radial load rating* and basic dynamic axial load rating*; for example, equation (18.103) for radial ball bearings becomes:

$$C = b_m f_c (i \cos \alpha)^{0.7} Z^{2/3} D^{1.8}\dagger \tag{18.172}$$

*The terms basic loading rating, basic dynamic radial load rating (or basic dynamic axial load rating), and basic dynamic capacity may be used interchangeably. The last term was that created by Lundberg and Palmgren.

†ANSI [18.10] recommends using D raised to the 1.4 power in lieu of 1.8 for bearings having balls of diameter greater than 25.4 mm (1 in.). In this case, for metric units calculation of basic load rating, f_{cm} values must be multiplied by 3.647; that is, $f_{cm} = 3.647 \times f_{cm}$ (tabular value).

TABLE 18.11. Rating Factor for Contemporary Bearing Material

Bearing Type	b_m Rating Factor
Radial ball (except filling slot and self-aligning types)	1.3
Angular-contact ball	1.3
Filling slot ball	1.1
Thrust ball	1.1
Radial spherical roller	1.15
Radial cylindrical roller	1.1
Radial tapered roller	1.1
Radial needle roller with machined rings	1.1
Drawn cup needle roller	1.0
Thrust tapered roller	1.1
Thrust spherical roller	1.15
Thrust cylindrical roller	1.0
Thrust needle roller	1.0

ISO [18.13] gives values of the factor b_m which may be applied to the formulas for basic dynamic capacity for each of the various executions of ball and roller bearings as deemed appropriate. Table 18.11 summarizes these b_m values.

The ANSI load rating standards [18.10 and 18.12] have incorporated the b_m factors into f_{cm} factors which are simply $f_{cm} = b_m \times f_c$. Tables 18.12 and 18.13 give the f_{cm} values to be used in the standard basic load rating equation (18.173) for radial ball bearings.

$$C = f_{cm}(i \cos \alpha)^{0.7} Z^{2/3} D^{1.8*} \qquad (18.173)$$

Similarly, for thrust ball bearings, Tables 18.14 and 18.15 give the f_{cm} values to be used in the standard basic load rating equations (18.174) and (18.175).

$$C_a = f_{cm}(\cos \alpha)^{0.7} \tan \alpha Z^{2/3} D^{1.8*} \qquad \alpha \neq 90° \qquad (18.174)$$

$$C_a = f_{cm} Z^{2/3} D^{1.8*} \qquad \alpha = 90° \qquad (18.175)$$

For radial roller bearings, Tables 18.16 and 18.17 provide f_{cm} values for use in the standard basic load rating equation (18.176).

$$C = f_{cm}(il \cos \alpha)^{7/9} Z^{3/4} D^{29/27} \qquad (18.176)$$

*ANSI [18.10] recommends using D raised to the 1.4 power in lieu of 1.8 for bearings having balls of diameter greater than 25.4 mm (1 in.). In this case, for metric units calculation of basic load rating, f_{cm} values must be multiplied by 3.647; that is, $f_{cm} = 3.647 \times f_{cm}$ (tabular value).

TABLE 18.12. Metric Values for f_{cm} for Radial- and Angular-Contact Ball Bearings[a]

$\frac{D \cos \alpha}{d_m}$	Single-Row Radial-Contact Groove Ball Bearings; Single and Double-Row Angular-Contact Groove Ball Bearings; Insert Bearings[b]	Filling Slot Ball Bearings	Double-Row Radial-Contact Groove Ball Bearings	Single-Row and Double-Row Self-Aligning Ball Bearings	Single-Row Radial-Contact Separable Ball Bearings
0.01	37.83	32.01	35.75	12.87	12.22
0.02	46.54	39.38	44.07	16.12	15.21
0.03	52.39	44.33	49.66	18.59	17.42
0.04	56.94	48.18	53.95	20.67	19.37
0.05	60.71	51.37	57.46	22.49	21.06
0.06	63.83	54.01	60.45	24.18	22.62
0.07	66.43	56.21	62.92	25.87	24.05
0.08	68.64	58.08	65.00	27.43	25.35
0.09	70.59	59.73	66.82	28.99	26.78
0.10	72.15	61.05	68.38	30.42	27.95
0.11	73.58	62.26	69.68	31.85	29.25
0.12	74.75	63.25	70.85	33.28	30.42
0.13	75.66	64.02	71.76	34.58	31.72
0.14	76.44	64.68	72.41	36.01	32.89
0.15	77.09	65.23	72.93	37.31	34.06
0.16	77.48	65.56	73.45	38.61	35.23
0.17	77.74	65.78	73.71	39.91	36.27
0.18	77.87	65.89	73.84	41.21	37.44
0.19	78.00	66.00	73.84	42.38	38.48
0.20	77.87	65.89	73.84	43.55	39.65

(continued)

TABLE 18.12. *(Continued)*

$\frac{D \cos \alpha}{d_m}$	Single-Row Radial-Contact Groove Ball Bearings; Single and Double-Row Angular-Contact Groove Ball Bearings; Insert Bearings[b]	Filling Slot Ball Bearings	Double-Row Radial-Contact Groove Ball Bearings	Single-Row and Double-Row Self-Aligning Ball Bearings	Single-Row Radial-Contact Separable Ball Bearings
0.21	77.74	65.78	73.58	44.72	40.69
0.22	77.48	65.56	73.45	45.76	41.73
0.23	77.09	65.23	73.06	46.93	43.77
0.24	76.70	64.90	72.67	47.84	43.80
0.25	76.18	64.46	72.15	48.75	44.72
0.26	75.66	64.02	71.63	49.66	45.76
0.27	75.01	63.47	70.98	50.44	46.67
0.28	74.23	62.81	70.33	51.22	47.58
0.29	73.58	62.26	69.68	51.87	48.36
0.30	72.80	61.60	68.90	52.39	49.14
0.31	71.89	60.83	68.12	52.78	49.92
0.32	70.98	60.06	67.34	53.17	50.57
0.33	70.07	59.29	66.43	53.43	51.22
0.34	69.16	58.52	65.52	53.56	51.74
0.35	68.12	57.64	64.61	53.69	52.13
0.36	67.21	56.87	63.57	53.69	52.52
0.37	66.17	55.99	62.66	53.56	52.91
0.38	65.00	55.00	61.62	53.30	53.04
0.39	63.96	54.12	60.58	52.91	53.17
0.40	62.92	53.24	59.54	52.52	53.17

[a] Use to obtain C in pounds when D and d_m are given in inches. Values of f_{cm} for intermediate values of $D \cos \alpha / d_m$ are obtained by linear interpolation.

[b] Insert bearings are not in accordance with [18.13].

TABLE 18.13. Inch Values for f_{cm} for Radial- and Angular-Contact Ball Bearings[a]

$\frac{D \cos \alpha}{d_m}$	Single-Row Radial-Contact Groove Ball Bearings; Single and Double-Row Angular-Contact Groove Ball Bearings; Insert Bearings[b]	Filling Slot Ball Bearings	Double-Row Radial-Contact Groove Ball Bearings	Single-Row and Double-Row Self-Aligning Ball Bearings	Single-Row Radial-Contact Separable Ball Bearings
0.01	2875	2433	2717	978	929
0.02	3537	2993	3349	1225	1156
0.03	3982	3369	3774	1413	1324
0.04	4327	3662	4100	1571	1472
0.05	4614	3904	4367	1709	1601
0.06	4851	4105	4594	1838	1719
0.07	5049	4272	4782	1966	1828
0.08	5217	4414	4940	2085	1927
0.09	5365	4539	5078	2203	2035
0.10	5483	4640	5197	2312	2124
0.11	5592	4732	5296	2421	2223
0.12	5681	4807	5385	2529	2312
0.13	5750	4866	5454	2628	2411
0.14	5809	4916	5503	2737	2500
0.15	5859	4983	5543	2836	2589
0.16	5888	4993	5582	2934	2677
0.17	5908	4999	5602	3033	2757
0.18	5918	5008	5612	3132	2845
0.19	5928	5016	5612	3221	2924
0.20	5918	5008	5612	3310	3013

(continued)

TABLE 18.13. *(Continued)*

$\frac{D \cos \alpha}{d_m}$	Single-Row Radial-Contact Groove Ball Bearings; Single and Double-Row Angular-Contact Groove Ball Bearings; Insert Bearings[b]	Filling Slot Ball Bearings	Double-Row Radial-Contact Groove Ball Bearings	Single-Row and Double-Row Self-Aligning Ball Bearings	Single-Row Radial-Contact Separable Ball Bearings
0.21	5908	4999	5592	3399	3092
0.22	5888	4983	5582	3478	3171
0.23	5859	4957	5553	3567	3251
0.24	5829	4932	5523	3636	3330
0.25	5790	4899	5510	3705	3399
0.26	5750	4866	5444	3774	3478
0.27	5701	4824	5394	3833	3547
0.28	5641	4774	5345	3893	3616
0.29	5592	4732	5296	3942	3675
0.30	5528	4678	5236	3982	3735
0.31	5464	4623	5177	4011	3794
0.32	5394	4565	5118	4041	3843
0.33	5325	4506	5049	4061	3893
0.34	5256	4448	4980	4071	3932
0.35	5177	4381	4910	4080	3962
0.36	5108	4322	4831	4080	3992
0.37	5029	4255	4762	4071	4021
0.38	4940	4180	4683	4051	4031
0.39	4861	4113	4604	4021	4041
0.40	4782	4046	4525	3992	4041

[a] Use to obtain C in pounds when D and d_m are given in inches. Values of f_{cm} for intermediate values of $D \cos \alpha / d_m$ are obtained by linear interpolation.

[b] Insert bearings are not in accordance with [18.13].

TABLE 18.14. Metric Values for f_{cm} for Thrust Ball Bearings[a]

$\frac{D^b}{d_m}$	f_{cm}	$\frac{D \cos \alpha^b}{d_m}$	f_{cm}		
	$\alpha = 90°$		$\alpha = 45°$[c]	$\alpha = 60°$	$\alpha = 75°$
0.01	47.71	0.01	54.73	50.96	48.49
0.02	58.76	0.02	67.21	62.53	59.67
0.03	66.43	0.03	75.66	70.46	67.21
0.04	72.41	0.04	82.29	76.57	72.93
0.05	77.35	0.05	87.49	81.38	77.61
0.06	81.77	0.06	91.91	85.54	81.51
0.07	85.54	0.07	95.55	88.92	84.76
0.08	89.05	0.08	98.67	91.78	87.49
0.09	92.30	0.09	101.40	94.38	89.96
0.10	95.29	0.10	103.61	96.46	91.91
0.11	98.02	0.11	105.43	98.15	—
0.12	100.62	0.12	106.99	99.58	—
0.13	103.09	0.13	108.29	100.75	—
0.14	105.43	0.14	109.33	101.79	—
0.15	107.51	0.15	110.11	102.44	—
0.16	109.72	0.16	110.63	102.96	—
0.17	111.67	0.17	111.02	103.35	—
0.18	113.62	0.18	111.15	103.48	—
0.19	115.44	0.19	111.15	103.48	—
0.20	117.26	0.20	111.02	103.35	—
0.21	118.95	0.21	110.76	—	—
0.22	120.64	0.22	110.37	—	—
0.23	122.33	0.23	109.85	—	—
0.24	123.89	0.24	109.20	—	—
0.25	125.45	0.25	108.42	—	—
0.26	126.88	0.26	107.64	—	—
0.27	128.31	0.27	106.60	—	—
0.28	129.74	0.28	105.69	—	—
0.29	131.04	0.29	104.52	—	—
0.30	132.47	0.30	103.48	—	—
0.31	133.77	—	—	—	—
0.32	135.07	—	—	—	—
0.33	136.24	—	—	—	—
0.34	137.54	—	—	—	—
0.35	138.71	—	—	—	—

[a]Use to obtain C in newtons when D and d_m are given in mm.
[b]Values of f_{cm} for D/d_m or $D \cos \alpha/d_m$ and/or angles other than those shown are obtained by linear interpolation.
[c]For thrust bearings $\alpha > 45°$. Values for $\alpha = 45°$ permit interpolation of values for α between 45° and 60°.

TABLE 18.15. Inch Values for f_{cm} for Thrust Ball Bearings[a]

$\frac{D}{d_m}$[b]	$\frac{f_{cm}}{\alpha = 90°}$	$\frac{D \cos \alpha}{d_m}$[b]	f_{cm}		
			$\alpha = 45°$[c]	$\alpha = 60°$	$\alpha = 75°$
0.01	3626	0.01	4159	3873	3685
0.02	4466	0.02	5108	4752	4535
0.03	5049	0.03	5750	5355	5108
0.04	5503	0.04	6254	5819	5543
0.05	5879	0.05	6649	6185	5898
0.06	6215	0.06	6985	6501	6195
0.07	6501	0.07	7262	6758	6442
0.08	6768	0.08	7499	6975	6649
0.09	7015	0.09	7706	7173	6837
0.10	7242	0.10	7874	7331	6985
0.11	7450	0.11	8013	7459	—
0.12	7647	0.12	8131	7568	—
0.13	7835	0.13	8230	7657	—
0.14	8013	0.14	8309	7736	—
0.15	8171	0.15	8368	7785	—
0.16	8339	0.16	8408	7825	—
0.17	8487	0.17	8438	7855	—
0.18	8635	0.18	8447	7864	—
0.19	8773	0.19	8447	7864	—
0.20	8912	0.20	8438	7855	—
0.21	9040	0.21	8418	—	—
0.22	9169	0.22	8388	—	—
0.23	9297	0.23	8349	—	—
0.24	9416	0.24	8299	—	—
0.25	9534	0.25	8240	—	—
0.26	9643	0.26	8181	—	—
0.27	9752	0.27	8102	—	—
0.28	9860	0.28	8034	—	—
0.29	9959	0.29	7944	—	—
0.30	10068	0.30	7864	—	—
0.31	10167	—	—	—	—
0.32	10265	—	—	—	—
0.33	10354	—	—	—	—
0.34	10453	—	—	—	—
0.35	10542	—	—	—	—

[a]Use to obtain C in pounds when D and d_m are given in inches.
[b]Values of f_{cm} for D/d_m or $D \cos \alpha/d_m$ and/or angles other than those shown are obtained by linear interpolation.
[c]For thrust bearings $\alpha > 45°$. Values for $\alpha = 45°$ permit interpolation of values for α between 45° and 60°.

TABLE 18.16. Metric Values for f_{cm} for Radial Roller Bearings[a]

$\frac{D \cos \alpha^b}{d_m}$	Cylindrical Roller Bearings, Tapered Roller Bearings, and Needle Roller Bearings with Machined Rings	Drawn Cup Needle Roller Bearings	Spherical Roller Bearings
0.01	57.310	52.100	59.915
0.02	66.880	60.800	69.920
0.03	73.150	66.500	76.475
0.04	77.770	70.700	81.305
0.05	81.510	74.100	85.215
0.06	84.590	76.900	88.435
0.07	87.120	79.200	91.080
0.08	89.210	81.100	93.265
0.09	91.080	82.800	95.220
0.10	92.620	84.200	96.830
0.11	93.830	85.300	98.095
0.12	95.040	86.400	99.360
0.13	95.810	87.100	100.165
0.14	96.470	87.700	100.855
0.15	97.020	88.200	101.430
0.16	97.350	88.500	101.775
0.17	97.570	88.700	102.005
0.18	97.680	88.800	102.120
0.19	97.680	88.800	102.120
0.20	97.570	88.700	102.005
0.21	97.350	88.500	101.775
0.22	97.020	88.200	101.430
0.23	96.580	87.800	100.970
0.24	96.250	87.500	100.625
0.25	95.590	86.900	99.935
0.26	95.040	86.400	99.360
0.27	94.380	85.800	98.670
0.28	93.720	85.200	97.980
0.29	92.840	84.400	97.060
0.30	92.070	83.700	96.255
0.31	91.300	83.000	95.450
0.32	90.420	82.200	94.530
0.33	89.430	81.300	93.495
0.34	88.440	80.400	92.460
0.35	87.450	79.500	91.425
0.36	86.460	78.600	90.390
0.37	85.360	77.600	89.240

(continued)

TABLE 18.16. *(Continued)*

$\frac{D \cos \alpha^b}{d_m}$	Cylindrical Roller Bearings, Tapered Roller Bearings, and Needle Roller Bearings with Machined Rings	Drawn Cup Needle Roller Bearings	Spherical Roller Bearings
0.38	84.370	76.700	88.205
0.39	83.270	75.700	87.055
0.40	82.060	74.600	85.790
0.41	80.960	73.600	84.640
0.42	79.750	72.500	83.375
0.43	78.540	71.400	82.110
0.44	77.330	70.300	80.845
0.45	76.120	69.200	79.580
0.46	74.910	68.100	78.315
0.47	73.700	67.000	77.050
0.48	72.380	65.800	75.670
0.49	71.060	64.600	74.290
0.50	69.850	63.500	73.025

[a] Use to obtain C in newtons when D and d_m are given in mm.
[b] Values of f_{cm} for intermediate values of $D \cos \alpha / d_m$ are obtained by linear interpolation.

TABLE 18.17. Inch Values for f_{cm} for Radial Roller Bearings[a]

$\frac{D \cos \alpha^b}{d_m}$	Cylindrical Roller Bearings, Tapered Roller Bearings, and Needle Roller Bearings with Machined Rings	Drawn Cup Needle Roller Bearings	Spherical Roller Bearings
0.01	5149	4681	5383
0.02	6009	5463	6282
0.03	6573	5975	6871
0.04	6987	6352	7305
0.05	7324	6658	7657
0.06	7600	6909	7945
0.07	7828	7116	8183
0.08	8016	7287	8380
0.09	8184	7440	8556
0.10	8322	7565	8700
0.11	8431	7665	8814
0.12	8539	7763	8927
0.13	8609	7826	9000
0.14	8668	7880	9062
0.15	8718	7925	9114

TABLE 18.17. (*Continued*)

$\frac{D \cos \alpha}{d_m}$ [b]	Cylindrical Roller Bearings, Tapered Roller Bearings, and Needle Roller Bearings with Machined Rings	Drawn Cup Needle Roller Bearings	Spherical Roller Bearings
0.16	8747	7952	9145
0.17	8767	7970	9166
0.18	8778	7979	9176
0.19	8778	7979	9176
0.20	8767	7970	9166
0.21	8747	7952	9145
0.22	8718	7925	9114
0.23	8678	7889	9073
0.24	8648	7862	9041
0.25	8589	7808	8980
0.26	8539	7763	8927
0.27	8480	7709	8865
0.28	8421	7655	8803
0.29	8342	7584	8721
0.30	8273	7521	8649
0.31	8204	7458	8577
0.32	8125	7386	8494
0.33	8036	7305	8401
0.34	7946	7224	8308
0.35	7857	7143	8214
0.36	7768	7062	8121
0.37	7669	6972	8018
0.38	7580	6891	7925
0.39	7482	6802	7822
0.40	7373	6703	7708
0.41	7274	6613	7605
0.42	7165	6514	7491
0.43	7057	6415	7377
0.44	6948	6316	7263
0.45	6840	6218	7151
0.46	6731	6119	7037
0.47	6622	6020	6923
0.48	6503	5912	6799
0.49	6384	5804	6675
0.50	6276	5705	6561

[a] Use to obtain C in pounds when D and d_m are given in inches.
[b] Values of f_{cm} for intermediate values of $D \cos \alpha / d_m$ are obtained by linear interpolation.

And for thrust roller bearings of various executions, Tables 18.18 through 18.23 give f_{cm} values to be used in the standard basic load rating equations (18.177) and (18.178).

$$C_a = f_{cm}(l \cos \alpha)^{7/9} \tan \alpha Z^{3/4} D^{29/27} \qquad \alpha \neq 90° \tag{18.177}$$

$$C_a = f_{cm} l^{7/9} Z^{3/4} D^{29/27} \qquad \alpha = 90° \tag{18.178}$$

With regard to the use of the standard load rating equations and tabular data on f_{cm}, X and Y factors to estimate rolling bearing fatigue life endurance, certain limitations should be observed:

1. Load ratings pertain only to bearings fabricated from properly hardened, good quality steel.
2. Rating life calculations assume the bearing inner and outer rings are rigidly supported.
3. Rating life calculations assume the bearing inner and outer ring axes are properly aligned.
4. Rating life calculations assume the bearing has only a nominal internal clearance during operation.
5. With regard to ball bearings, the raceway groove radii must fall within $0.52 \leq f/D \leq 0.53$.
6. With regard to roller bearings, the load ratings pertain only to bearings manufactured to achieve optimized contact. This involves good roller guidance by flanges or cage as well as optimum roller and/or raceway profile.
7. For both ball and roller bearings, no stress concentrations may occur due to loading conditions. In a ball bearing this condition can be caused if the applied thrust load is sufficient to cause the balls to override the land edges.

Example 18.7.* Estimate the L_{10} life of the 209 radial ball bearing of Example 6.1 according to the ANSI method. Use a shaft speed of 1800 rpm as in Example 18.4.

$$Z = 9 \qquad \text{Ex. 2.1}$$

$$D = 12.7 \text{ mm } (0.5 \text{ in.}) \qquad \text{Ex. 2.1}$$

$$\gamma = 0.1954 \qquad \text{Ex. 2.5}$$

*Example 18.7 continues on page 739.

TABLE 18.18. Metric Values for f_{cm} for Tapered Roller Thrust Bearings[a]

$\frac{D}{d_m}$[b]	$\alpha = 90°$	$\frac{D \cos \alpha}{d_m}$[b]	$\alpha = 50°$[c]	$\alpha = 65°$[d]	$\alpha = 80°$[e]
0.01	115.94	0.01	120.67	117.81	116.16
0.02	135.19	0.02	140.58	137.17	135.30
0.03	147.95	0.03	153.45	149.82	147.73
0.04	157.74	0.04	163.13	159.17	157.08
0.05	165.77	0.05	170.72	166.65	164.34
0.06	172.59	0.06	176.99	172.70	170.39
0.07	178.64	0.07	182.16	177.76	175.34
0.08	183.92	0.08	186.45	182.05	179.52
0.09	188.87	0.09	190.08	185.57	183.04
0.10	193.27	0.10	193.05	188.54	185.90
0.11	197.45	0.11	195.58	190.96	188.32
0.12	201.30	0.12	197.67	192.94	190.30
0.13	204.93	0.13	199.21	194.48	191.84
0.14	208.34	0.14	200.53	195.69	193.05
0.15	211.53	0.15	201.41	196.68	193.93
0.16	214.61	0.16	202.07	197.23	—
0.17	217.47	0.17	202.40	197.56	—
0.18	220.33	0.18	202.51	197.67	—
0.19	222.97	0.19	202.40	197.56	—
0.20	225.50	0.20	202.07	197.23	—
0.21	227.92	0.21	201.52	—	—
0.22	230.34	0.22	200.86	—	—
0.23	232.65	0.23	199.98	—	—
0.24	234.85	0.24	198.99	—	—
0.25	236.94	0.25	197.78	—	—
0.26	239.03	0.26	196.57	—	—
0.27	241.01	—	—	—	—
0.28	242.99	—	—	—	—
0.29	244.97	—	—	—	—
0.30	246.73	—	—	—	—

[a] Use to obtain C_a in newtons when D and d_m are given in mm.
[b] Values of f_{cm} for intermediate values of D/d_m or $D \cos \alpha / d_m$ are obtained by linear interpolation.
[c] Applicable for $45° < \alpha < 60°$.
[d] Applicable for $60° \leq \alpha < 75°$.
[e] Applicable for $75° \leq \alpha < 90°$.

TABLE 18.19. Inch Values for f_{cm} for Tapered Roller Thrust Bearings[a]

$\frac{D}{d_m}$[b]	$\alpha = 90°$	$\frac{D \cos \alpha}{d_m}$[b]	$\alpha = 50°$[c]	$\alpha = 65°$[d]	$\alpha = 80°$[e]
0.01	10400	0.01	10824	10568	10420
0.02	12127	0.02	12610	12304	12136
0.03	13271	0.03	13764	13439	13251
0.04	14149	0.04	14633	14278	14090
0.05	14870	0.05	15314	14949	14741
0.06	15481	0.06	15876	15491	15284
0.07	16024	0.07	16340	15945	15728
0.08	16498	0.08	16725	16330	16103
0.09	16942	0.09	17050	16646	16419
0.10	17336	0.10	17317	16912	16675
0.11	17711	0.11	17544	17129	16892
0.12	18057	0.12	17731	17307	17070
0.13	18382	0.13	17869	17445	17208
0.14	18688	0.14	17988	17553	17317
0.15	18974	0.15	18066	17642	17396
0.16	19251	0.16	18126	17692	—
0.17	19507	0.17	18155	17721	—
0.18	19764	0.18	18165	17731	—
0.19	20000	0.19	18155	17721	—
0.20	20227	0.20	18126	17692	—
0.21	20444	0.21	18076	—	—
0.22	20661	0.22	18017	—	—
0.23	20869	0.23	17938	—	—
0.24	21066	0.24	17849	—	—
0.25	21254	0.25	17741	—	—
0.26	21441	0.26	17632	—	—
0.27	21619	—	—	—	—
0.28	21796	—	—	—	—
0.29	21974	—	—	—	—
0.30	22132	—	—	—	—

[a] Use to obtain C_a in pounds when D and d_m are given in inches.
[b] Values of f_{cm} for intermediate values of D/d_m or $D \cos \alpha/d_m$ are obtained by linear interpolation.
[c] Applicable for $45° < \alpha < 60°$.
[d] Applicable for $60° \le \alpha < 75°$.
[e] Applicable for $75° \le \alpha < 90°$.

TABLE 18.20. Metric Values for f_{cm} for Cylindrical Roller Thrust Bearings and Needle Roller Thrust Bearings[a]

$\frac{D}{d_m}$ [b]	$\alpha = 90°$	$\frac{D \cos \alpha}{d_m}$ [b]	$\alpha = 50°$ [c]	$\alpha = 65°$ [d]	$\alpha = 80°$ [e]
0.01	105.4	0.01	109.7	107.1	105.6
0.02	122.9	0.02	127.8	124.7	123.0
0.03	134.5	0.03	139.5	136.2	134.3
0.04	143.4	0.04	148.3	144.7	142.8
0.05	150.7	0.05	155.2	151.5	149.4
0.06	156.9	0.06	160.9	157.0	154.9
0.07	162.4	0.07	165.6	161.6	159.4
0.08	167.2	0.08	169.5	165.5	163.2
0.09	171.7	0.09	172.8	168.7	166.4
0.10	175.7	0.10	175.5	171.4	169.0
0.11	179.5	0.11	177.8	173.6	171.2
0.12	183.0	0.12	179.7	175.4	173.0
0.13	186.3	0.13	181.1	176.8	174.4
0.14	189.4	0.14	182.3	177.9	175.5
0.15	192.3	0.15	183.1	178.8	176.3
0.16	195.1	0.16	183.7	179.3	—
0.17	197.7	0.17	184.0	179.6	—
0.18	200.3	0.18	184.1	179.7	—
0.19	202.7	0.19	184.0	179.6	—
0.20	205.0	0.20	183.7	179.3	—
0.21	207.2	0.21	183.2	—	—
0.22	209.4	0.22	182.6	—	—
0.23	211.5	0.23	181.8	—	—
0.24	213.5	0.24	180.9	—	—
0.25	215.4	0.25	179.8	—	—
0.26	217.3	0.26	178.7	—	—
0.27	219.1	—	—	—	—
0.28	220.9	—	—	—	—
0.29	222.7	—	—	—	—
0.30	224.3	—	—	—	—

[a] Use to obtain C_a in newtons when D and d_m are given in mm.

[b] Values of f_{cm} for intermediate values of D/d_m or $D \cos \alpha / d_m$ are obtained by linear interpolation.

[c] Applicable for $45° < \alpha < 60°$.

[d] Applicable for $60° \leq \alpha < 75°$.

[e] Applicable for $75° \leq \alpha < 90°$.

TABLE 18.21. Inch Values for f_{cm} for Cylindrical Roller Thrust Bearings and Needle Roller Thrust Bearings[a]

$\frac{D}{d_m}$[b]	$\alpha = 90°$	$\frac{D \cos \alpha}{d_m}$[b]	$\alpha = 50°$[c]	$\alpha = 65°$[d]	$\alpha = 80°$[e]
0.01	9454	0.01	9840	9607	9472
0.02	11024	0.02	11464	11186	11033
0.03	12065	0.03	12513	12217	12047
0.04	12863	0.04	13303	12980	12809
0.05	13518	0.05	13921	13590	13401
0.06	14074	0.06	14433	14083	13895
0.07	14567	0.07	14854	14496	14298
0.08	14998	0.08	15204	14845	14639
0.09	15401	0.09	15500	15132	14926
0.10	15760	0.10	15742	15375	15159
0.11	16101	0.11	15949	15572	15357
0.12	16415	0.12	16119	15733	15518
0.13	16711	0.13	16245	15859	15644
0.14	16989	0.14	16352	15958	15742
0.15	17249	0.15	16424	16038	15814
0.16	17500	0.16	16478	16083	—
0.17	17734	0.17	16505	16110	—
0.18	17967	0.18	16514	16119	—
0.19	18182	0.19	16505	16110	—
0.20	18389	0.20	16478	16083	—
0.21	18586	0.21	16433	—	—
0.22	18783	0.22	16379	—	—
0.23	18972	0.23	16307	—	—
0.24	19151	0.24	16227	—	—
0.25	19321	0.25	16128	—	—
0.26	19492	0.26	16029	—	—
0.27	19653	—	—	—	—
0.28	19815	—	—	—	—
0.29	19976	—	—	—	—
0.30	20120	—	—	—	—

[a] Use to obtain C_a in pounds when D and d_m are given in inches.
[b] Values of f_{cm} for intermediate value of D/d_m or $D \cos \alpha / d_m$ are obtained by linear interpolation.
[c] Applicable for $45° < \alpha < 60°$.
[d] Applicable for $60° \leq \alpha < 75°$.
[e] Applicable for $75° \leq \alpha < 90°$.

TABLE 18.22. Metric Values for f_{cm} for Spherical Roller Thrust Bearings[a]

$\frac{D}{d_m}$[b]	$\alpha = 90°$	$\frac{D \cos \alpha}{d_m}$[b]	$\alpha = 50°$[c]	$\alpha = 65°$[d]	$\alpha = 80°$[e]
0.01	121.210	0.01	126.155	123.165	121.440
0.02	141.335	0.02	146.970	143.405	141.450
0.03	154.675	0.03	160.425	156.630	154.445
0.04	164.910	0.04	170.545	166.405	164.220
0.05	173.305	0.05	178.480	174.225	171.810
0.06	180.435	0.06	185.035	180.550	178.135
0.07	186.760	0.07	190.440	185.840	183.310
0.08	192.280	0.08	194.925	190.325	187.680
0.09	197.455	0.09	198.720	194.005	191.360
0.10	202.055	0.10	201.825	197.110	194.350
0.11	206.425	0.11	204.470	199.640	196.880
0.12	210.450	0.12	206.655	201.710	198.950
0.13	214.245	0.13	208.265	203.320	200.560
0.14	217.810	0.14	209.645	204.585	201.825
0.15	221.145	0.15	210.565	205.620	202.745
0.16	224.365	0.16	211.255	206.195	—
0.17	227.355	0.17	211.600	206.540	—
0.18	230.345	0.18	211.715	206.655	—
0.19	233.105	0.19	211.600	206.540	—
0.20	235.750	0.20	211.255	206.195	—
0.21	238.280	0.21	210.680	—	—
0.22	240.810	0.22	209.990	—	—
0.23	243.225	0.23	209.070	—	—
0.24	245.525	0.24	208.035	—	—
0.25	247.710	0.25	206.770	—	—
0.26	249.895	0.26	205.505	—	—
0.27	251.965	—	—	—	—
0.28	254.035	—	—	—	—
0.29	256.105	—	—	—	—
0.30	257.945	—	—	—	—

[a] Use to obtain C_a in newtons when D and d_m are given in mm.
[b] Values of f_{cm} for intermediate values of D/d_m or $D \cos \alpha / d_m$ are obtained by linear interpolation.
[c] Applicable for $45° < \alpha < 60°$.
[d] Applicable for $60° \leq \alpha < 75°$.
[e] Applicable for $75° \leq \alpha < 90°$.

TABLE 18.23. Inch Values for f_{cm} for Spherical Roller Thrust Bearings[a]

$\frac{D}{d_m}$[b]	$\alpha = 90°$	$\frac{D \cos \alpha}{d_m}$[b]	$\alpha = 50°$[c]	$\alpha = 65°$[d]	$\alpha = 80°$[e]
0.01	10873	0.01	11316	11048	10893
0.02	12678	0.02	13183	12863	12688
0.03	13874	0.03	14390	14050	13854
0.04	14792	0.04	15298	14927	14731
0.05	15545	0.05	16010	15628	15411
0.06	16185	0.06	16598	16195	15979
0.07	16752	0.07	17082	16670	16443
0.08	17248	0.08	17485	17072	16835
0.09	17712	0.09	17825	17402	17165
0.10	18124	0.10	18104	17681	17433
0.11	18516	0.11	18341	17908	17660
0.12	18877	0.12	18537	18093	17846
0.13	19218	0.13	18681	18238	17990
0.14	19538	0.14	18774	18351	18104
0.15	19837	0.15	18888	18444	18186
0.16	20126	0.16	18950	18496	—
0.17	20394	0.17	18981	18527	—
0.18	20662	0.18	18991	18537	—
0.19	20910	0.19	18981	18527	—
0.20	21148	0.20	18950	18496	—
0.21	21374	0.21	18898	—	—
0.22	21601	0.22	18836	—	—
0.23	21817	0.23	18754	—	—
0.24	22024	0.24	18661	—	—
0.25	22220	0.25	18547	—	—
0.26	22416	0.26	18434	—	—
0.27	22601	—	—	—	—
0.28	22787	—	—	—	—
0.29	22973	—	—	—	—
0.30	23138	—	—	—	—

[a] Use to obtain C_a in pounds when D and d_m are given in inches.

[b] Values of f_{cm} for intermediate values of D/d_m or $D \cos \alpha/d_m$ are obtained by linear interpolation.

[c] Applicable for $45° < \alpha < 60°$.

[d] Applicable for $60° \le \alpha < 75°$.

[e] Applicable for $75° \le \alpha < 90°$.

From Table 18.12, for $\gamma = 0.1954$, $f_{cm} = 77.93$

$$C = f_{cm}(i \cos \alpha)^{0.7} Z^{2/3} D^{1.8} \tag{18.173}$$

$$= 77.93 \times (9)^{2/3} (12.7)^{1.8} = 32{,}710 \text{ N (7351 lb)}$$

$$F_e = F_r = 8900 \text{ N (2000 lb)} \qquad \text{Ex. 6.1}$$

$$L = (C/F_e)^3 \tag{18.108}$$

$$= (32{,}710/8900)^3 = 49.64 \text{ million revolutions}$$

$$L = 49.64 \times 10^6/(60 \times 1800) = 459.6 \text{ hr}$$

To compare this result with that of Example 18.4, it is necessary to multiply the result of that example by b_m^3. From Table 18.11, b_m = 1.3. Accordingly, the comparable L_{10} life of Example 18.4 is 2.197 × 194 = 426.2 hr. It is evident that the ANSI-estimated life is not as accurate as that calculated in Example 18.4. In this particular case, however, the difference is not significant.

Example 18.8. Estimate the L_{10} fatigue life of the 209 cylindrical roller bearing of Example 6.3 according to ANSI methods. Radial loading is 4450 N (1000 lb).

$$Z = 14 \qquad \text{Ex. 2.7}$$

$$i = 1$$

$$D = 10 \text{ mm (0.3987 in.)} \qquad \text{Ex. 2.7}$$

$$l = 9.6 \text{ mm (0.378 in.)} \qquad \text{Ex. 2.7}$$

$$\gamma = 0.1538 \qquad \text{Ex. 13.7}$$

Using Table 18.16, $f_{cm} = 97.15$

$$C = f_{cm}(il \cos \alpha)^{7/9} Z^{3/4} D^{29/27} \tag{18.176}$$

$$= 97.15 \times (1 \times 9.6)^{7/9} (14)^{3/4} (10)^{29/27} = 48{,}430 \text{ N (10,880 lb)}$$

$$L = (C/F_e)^{10/3} \tag{18.167}$$

$$= (48{,}430/4450)^{10/3} = 2856 \text{ million revolutions}$$

To compare the result of this ANSI method calculation with that of Example 18.6 it is necessary to introduce b_m factor into the latter set of calculations. From Table 18.11, b_m is 1.1 for cylindrical roller bearings. Hence, the bearing L_{10} life would be $(1.1)^4 \times 985 \times 10^6$ or 1442 $\times 10^6$ revolutions. It is apparent that the ANSI method which does not account for the precise load distribution gives only an approximate life estimate and in this case tends to overstate the bearing fatigue endurance capability.

Example 18.9. The 22317 two-row spherical roller bearing of Example 6.8 experiences an outer raceway rotation of 900 rpm. Estimate the L_{10} life according to ANSI methods.

$$\alpha = 12° \qquad \text{Ex. 2.9}$$

$$Z = 14 \qquad \text{Ex. 2.8}$$

$$l = 20.71 \text{ mm } (0.8154 \text{ in.}) \qquad \text{Ex. 2.8}$$

$$D = 25 \text{ mm } (0.9843 \text{ in.}) \qquad \text{Ex. 2.8}$$

$$\gamma = 0.1810 \qquad \text{Ex. 2.10}$$

From Table 18.16, $f_{cm} = 97.68$

$$C = f_{cm}(il \cos \alpha)^{7/9} Z^{3/4} D^{29/27} \qquad (18.176)$$

$$= 97.68 \times (2 \times 20.71 \times \cos 12°)^{7/9} (14)^{3/4} (25)^{29/27}$$

$$= 399{,}300 \text{ N } (89{,}720 \text{ lb})$$

$$F_a = 22{,}250 \text{ N } (5000 \text{ lb}) \qquad \text{Ex. 6.8}$$

$$F_r = 89{,}000 \text{ N } (20{,}000 \text{ lb}) \qquad \text{Ex. 6.8}$$

$$F_a/F_r = 22{,}250/89{,}000 = 0.25$$

$$1.5 \tan \alpha = 1.5 \tan 12° = 0.3189$$

Since $F_a/F_r < 1.5 \tan \alpha$, from Table 18.7 use $X = 1$ and $Y = 0.45 \text{ ctn } \alpha$

$$Y = 0.45 \text{ ctn } 12° = 2.117$$

$$F_e = XF_r + YF_a \qquad (18.129)$$

$$= 136{,}100 \text{ N } (30{,}950 \text{ lb})$$

$$L = (C/F_a)^{10/3} \qquad (18.160)$$

$$= (399{,}300/136{,}000)^{10/3} = 36.15 \text{ million revolutions}$$

or

$$L = 36.15 \times 10^6/(60 \times 900) = 669.5 \text{ hr}$$

THE EFFECT OF LOAD DISTRIBUTION ON FATIGUE LIFE

Clearance

In most rolling bearing applications, fatigue life may be calculated using ANSI formulas [18.10, 18.12] for basic load rating, equivalent load, and load-life variation. The formulas are valid for rolling bearing applications in which only contact deformations are significant, that is, for bearings having nonflexible (rigidly supported rings). The formulas are further valid only for bearings having essentially zero diametral clearance.

The fatigue life of a rolling bearing is strongly dependent on the maximum rolling element load Q_{max}. Hence, if Q_{max} is significantly increased, fatigue life is significantly decreased. Any factor, therefore, that affects Q_{max} affects bearing fatigue life. One such factor is diametral clearance. In Chapter 6 the effect of clearance on load distribution in radial bearings was examined. Figure 18.17, from reference [18.19], which gives the L_{10} life reduction factor as determined by ϵ, the extent of rolling element loading, was developed by using the load distribution data of Chapter 6 in accordance with the cubic and quartic mean load formulas presented in this chapter. An increase in maximum rolling element load for rigid ring bearings is accompanied by a decrease in the number of rollers loaded. This decrease in load zone, however, has less effect on the mean effective rolling element load than does the increase in the peak rolling element load. Figure 6.3 shows the variation of load

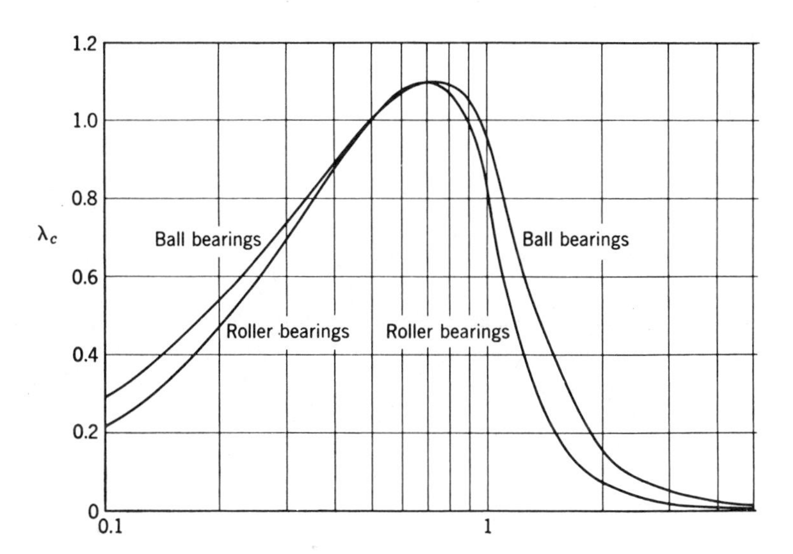

FIGURE 18.17. Fatigue life reduction factor λ_c, based on bearing diametral clearance.

distribution among the rolling elements for different amounts of diametral clearance.

Example 18.10. The 209 radial ball bearing of Example 6.1 has 0.0150 mm (0.0006 in.) diametral clearance that according to that example gives an ϵ value of 0.434 for a radial load of 8900 N (2000 lb). The ANSI load rating method is based on $\epsilon = 0.5$, that is, zero clearance. What L_{10} life reduction may be expected considering the ANSI rating life?

From Fig. 18.17,

$$\text{at } \epsilon = 0.434, \qquad \lambda_c = 0.93$$

$$\text{at } \epsilon = 0.5, \qquad \lambda_c = 1$$

$$\frac{L_{10}}{L_{10}(\text{ANSI})} = 0.93$$

Consequently, a 7% reduction in rating life is required.

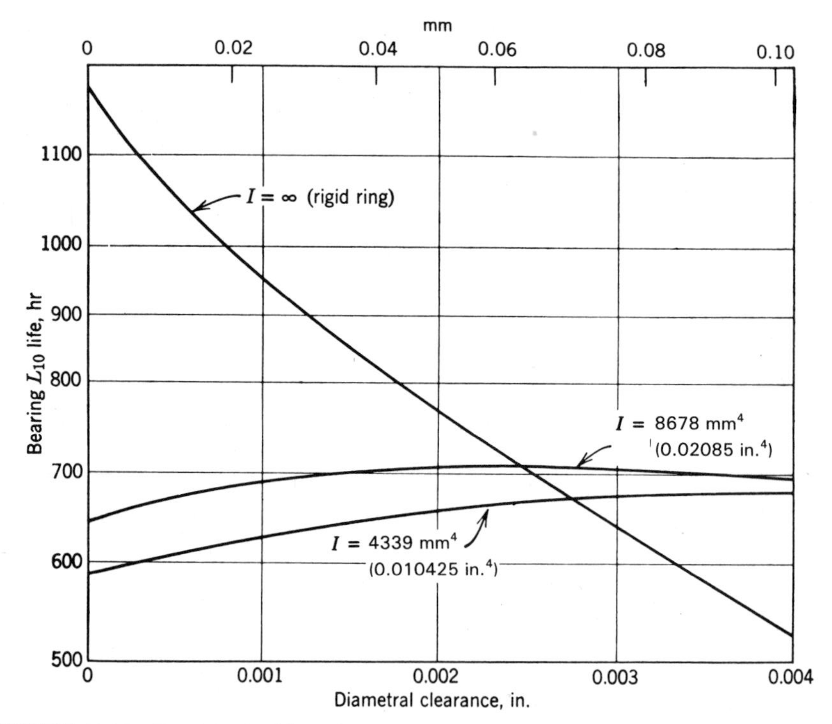

FIGURE 18.18. Bearing L_{10} life vs diametral clearance as a function of geared outer ring section moment of inertia.

Flexibly Supported Bearings

If one or both rings of a rolling bearing bend under the applied loads such as in a planet gear application [18.20, 18.21] or other aircraft bearing applications in which ring and housing cross sections are optimized for aircraft weight reduction, then load distribution may be considerably different from that of a rigid ring bearing. Depending on the flexibility of the ring and bearing clearance, it may be possible for a flexible ring bearing to yield superior endurance characteristics when compared to a rigid ring bearing. Figure 18.18, taken from reference [18.20] shows the variation of bearing fatigue life with outer ring section and clearance for a planet gear bearing as shown by Fig. 6.25. The load distribution obtained is illustrated by Fig. 6.34.

When the bearing rings are flexibly supported, it may be possible to alter bearing design and obtain increased fatigue life. Harris and Broschard [18.21] applied clearance selectively at the planet gear bearing maximum load positions. This was accomplished by making the bearing inner ring elliptical. Figure 18.19 demonstrates the variation of fatigue life with diametral clearance and out-of-round. Out-of-round is the difference between major and minor axes of the ring. In reference [18.22] Harris also demonstrated that rolling bearing ring dimensions can be optimized to obtain maximum fatigue life. Figures 18.20–18.27 taken from reference [18.22] demonstrate for a cylindrical roller bearing [Z =

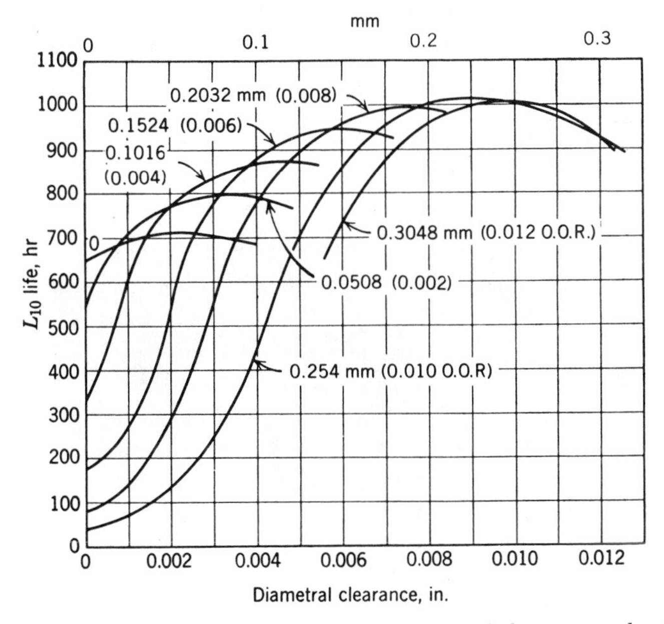

FIGURE 18.19. Bearing L_{10} life as a function of diametral clearance and out-of-round.

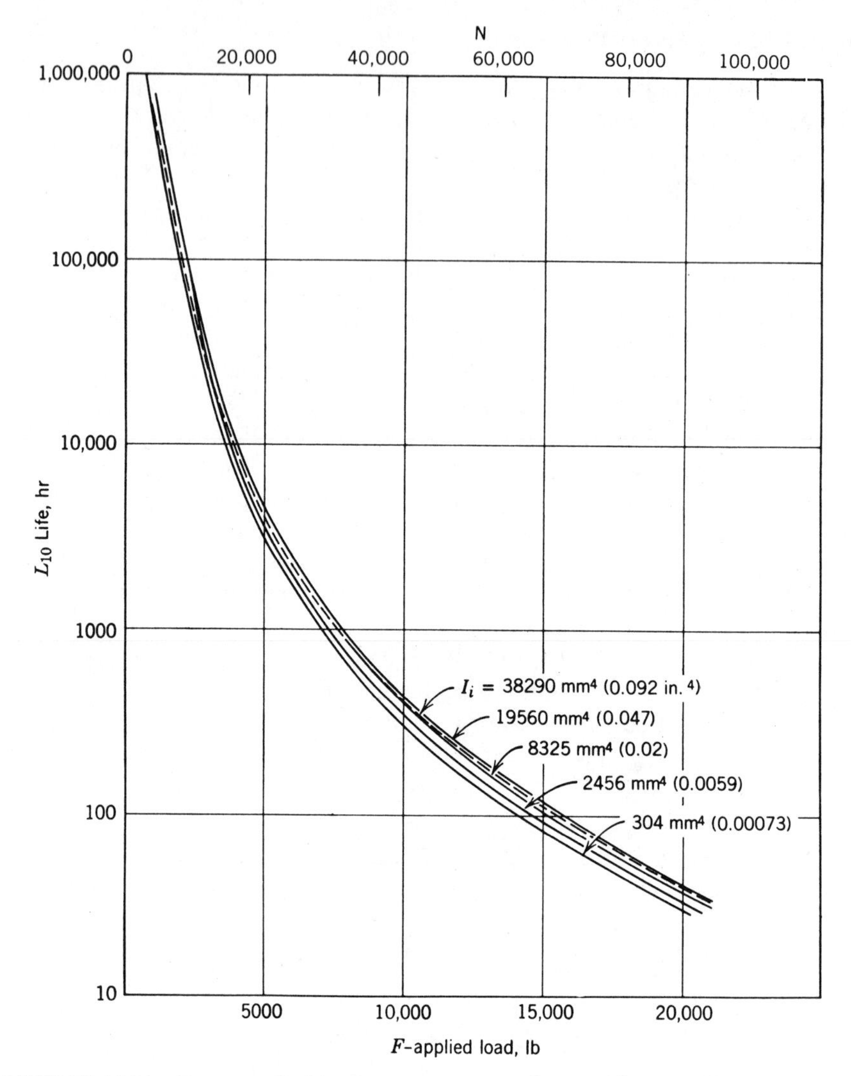

FIGURE 18.20. L_{10} vs applied load; I_o = 41,620 mm^4 (0.1 in.^{4}), c = 0.0254 mm (0.001 in.), n = 4000 rpm.

34, l = 15.24 mm (0.6 in.), D = 17.02 mm (0.67 in.), d_m = 193.04 mm (7.6 in.)] in an aircraft gas turbine mainshaft application that L_{10}* life can be optimized by proper ring design for various conditions of load, speed, clearance, and housing design.

*Note that L_{10} lives shown in Fig. 18.18–18.27 were not calculated using the b_m factors of Table 18.11. Therefore, they should be used for comparison purposes only.

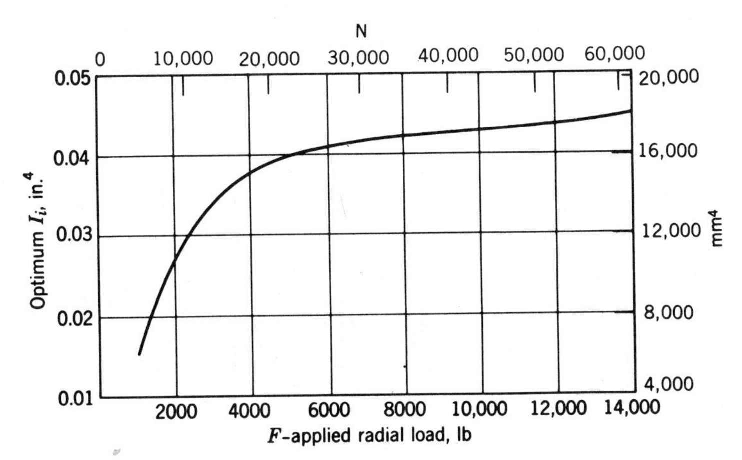

FIGURE 18.21. Optimum I_i vs applied radial load; n = 4000 rpm, c = 0.0254 mm (0.001 in.), I_o = 41,620, mm^4 (0.1 in.^{4}).

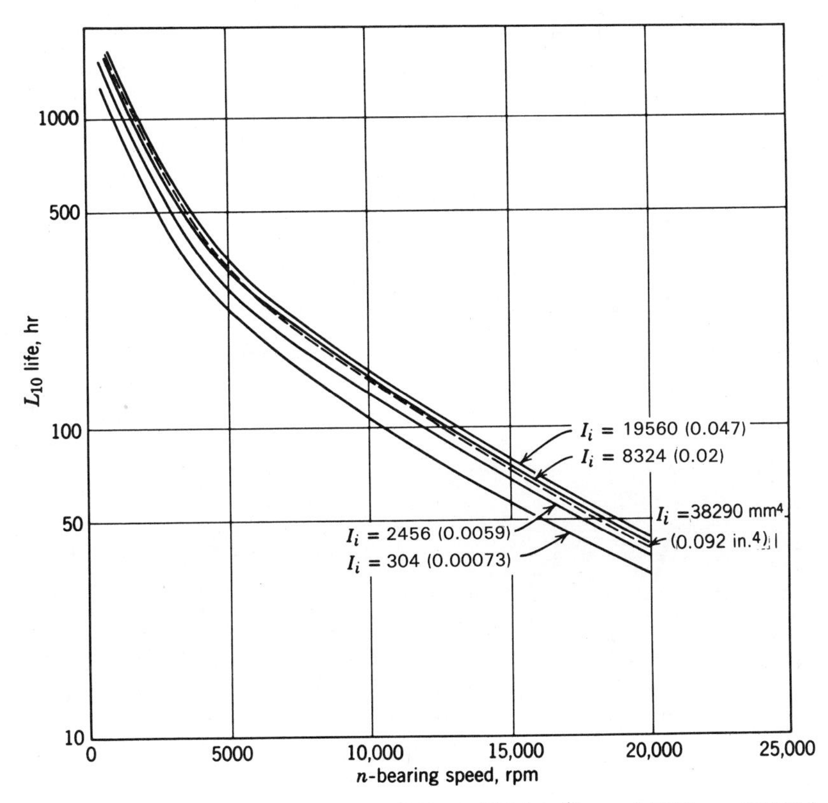

FIGURE 18.22. L_{10} life vs speed; I_o = 41,620 mm^4 (0.1 in.^{4}), c = 0.0254 mm (0.001 in.), F = 44,500 N (10,000 lb).

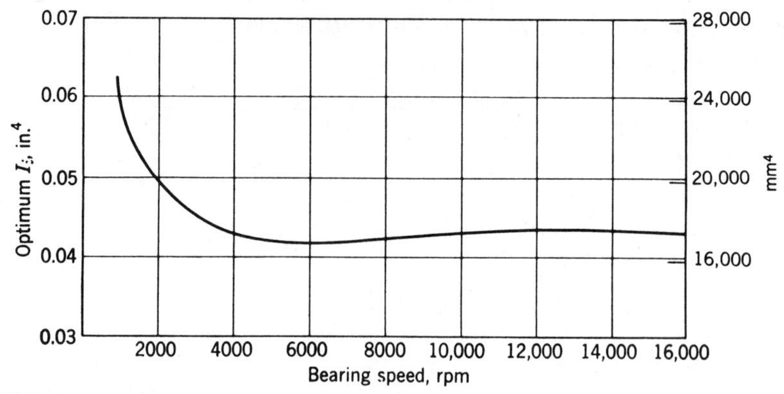

FIGURE 18.23. Optimum I_i vs bearing speed; F = 44,500 N (10,000 lb), c = 0.0254 mm (0.001 in.), I_o = 41,620 mm^4 (0.1 in.^{4}).

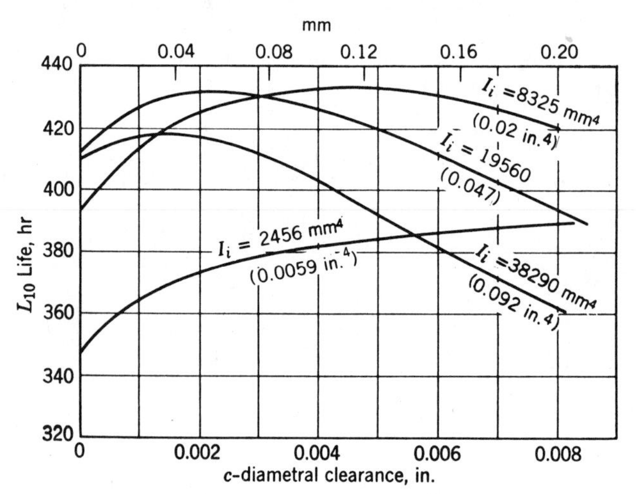

FIGURE 18.24. L_{10} life vs clearance; I_o = 41,620 mm^4 (0.1 in.^{4}), n = 4000 rpm, F = 44,500 N (10,000 lb).

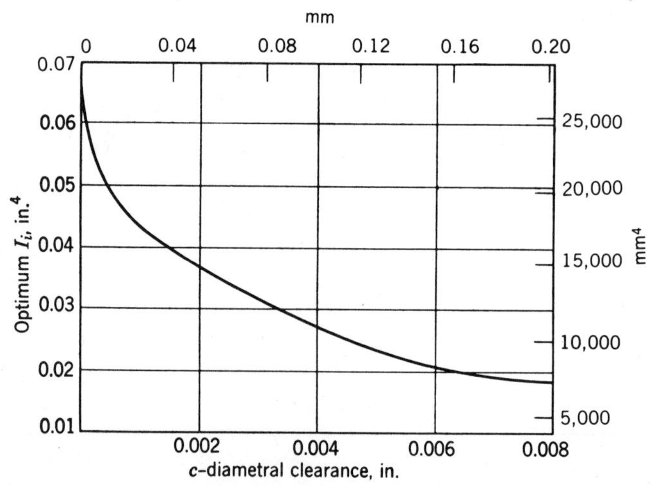

FIGURE 18.25. Optimum I_i vs clearance, n = 4000 rpm, F = 44,500 N (10,000 lb), I = 41,620 mm^4 (0.1 in.^{4}).

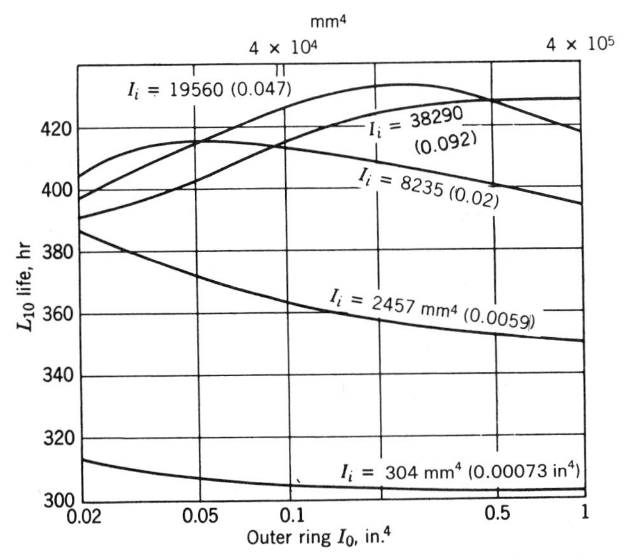

FIGURE 18.26. L_{10} life vs I_o; c = 0.0254 mm (0.001 in.), n = 4000 rpm, F = 44,500 N (10,000 lb).

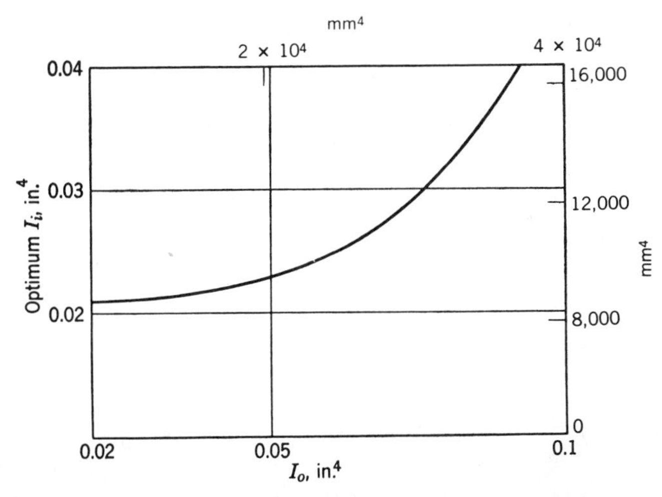

FIGURE 18.27. Optimum I_i vs I_o; n = 4000 rpm, c = 0.0254 mm (0.001 in.), F = 44,500 N (10,000 lb).

High Speed Operation

Operation at high speeds, as shown in Chapter 8, affects the bearing load distribution due to the increased magnitude of rolling element centrifugal forces and gyroscopic moments. The ANSI standard method of calculating fatigue life does not account for these body forces and moments and other subsequent effects such as changes in ball bearing contact angles, hence, the deviation in fatigue life from that calculated according

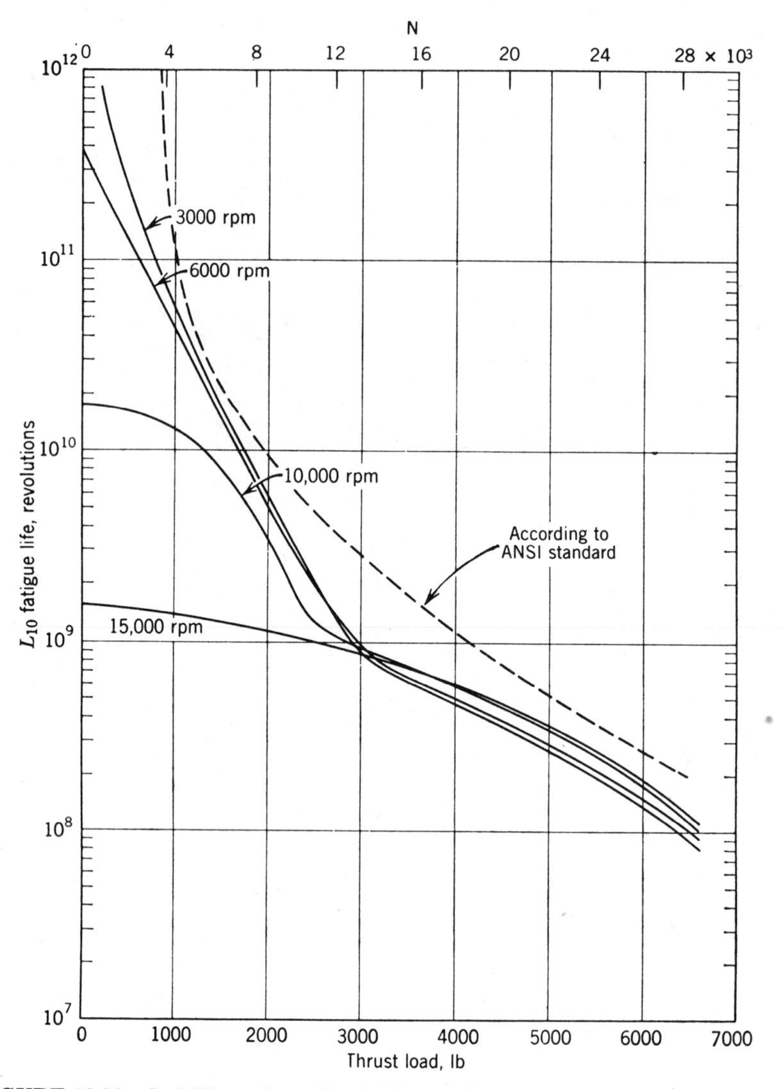

FIGURE 18.28. L_{10}* life vs thrust load; 218 angular-contact ball bearing, $\alpha° = 40°$.

*Note that the b_m factor was not included in the L_{10} life calculations. The graphical data should be used for comparison purposes only.

to the ANSI method can be considerable. In Chapter 8 methods were developed to calculate load distribution in high speed ball and roller bearings. Methods for using these load distributions in the estimation of fatigue life have been given in this chapter. Figure 18.28 demonstrates the variation of L_{10} fatigue life with load and speed for the 218 angular-contact ball bearing of Example 8.1. Note that the data shown in Fig. 18.28 do not consider the effect of skidding, which results in a reduction

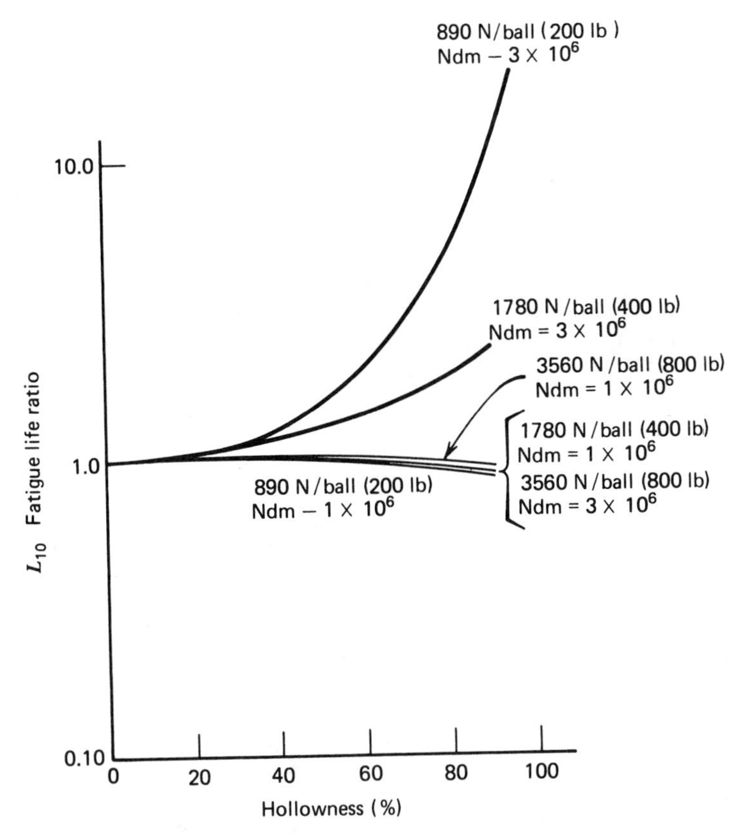

FIGURE 18.29. Fatigue life ratio vs ball hollowness.

in ball orbital speed and hence reduced ball centrifugal and gyroscopic loading. This in turn tends to result in an increase in fatigue life; however, depending on the thicknesses of the lubricant films separating the balls from the raceways, sliding in the ball–raceway contacts, with its potential deleterious effect on fatigue endurance, may more than eliminate the beneficial effect of reduced inertial loading.

Fig. 18.29 taken from reference [18.23] shows the effect on L_{10} life of reducing ball inertial loading by the use of hollow balls. Figure 18.30 shows L_{10} life vs speed for the 209 cylindrical roller bearing of Example 8.2. Figure 18.29 also does not include the effects of skidding.

Misalignment

Misalignment in nonaligning rolling bearings tends to distort the load distribution and thus alter fatigue life. In Chapter 6, methods were developed to determine the misalignment angle in ball and roller bearings as a function of the applied moment. In ball bearings, the load distribution from ball to ball is altered by misalignment; in roller bearings,

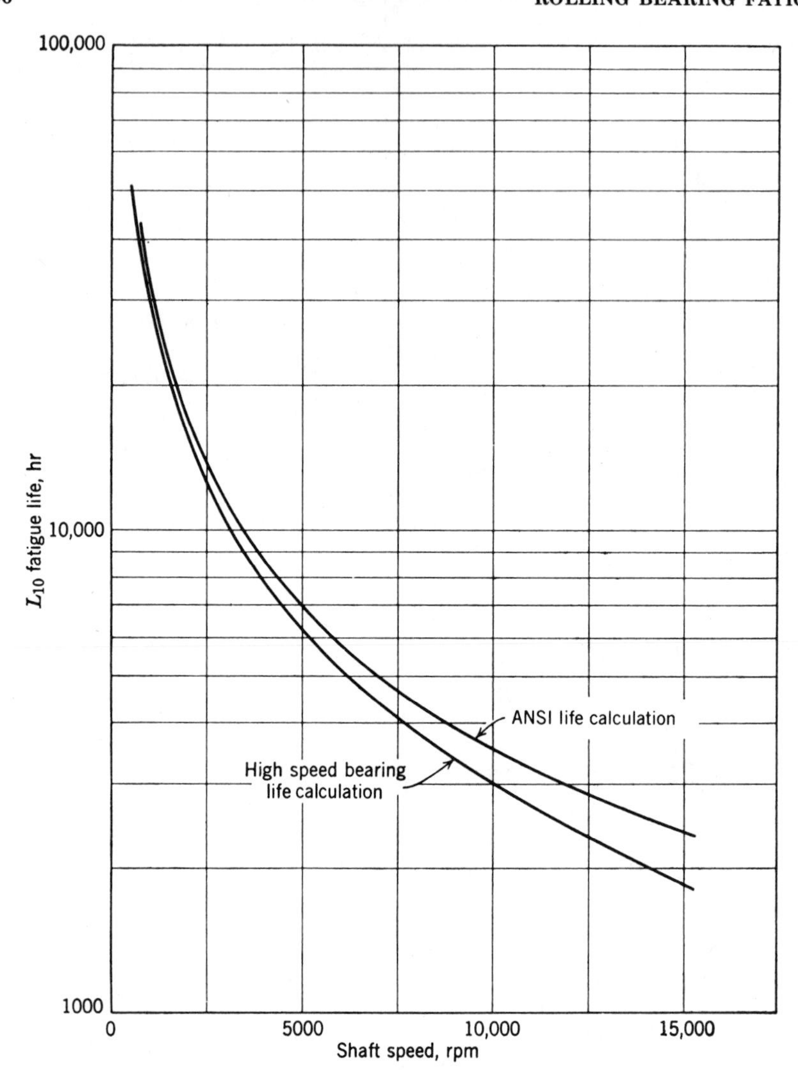

FIGURE 18.30. L_{10}* life vs speed; 209 cylindrical roller bearing, $F_1 = 44{,}500$ N (10,000 lb), $P_d = 0$.

*Note that the b_m factor was not included in the L_{10} life calculations. The graphical data should be used for comparison purposes only.

however, the distribution of roller load per unit length becomes skewed, as shown by Fig. 6.25. This variable load per unit length q is given by equation (6.115).

Since the statistical analysis of the last section pertained only to a uniform distribution of load per unit length along the roller length, the methods developed in this chapter for calculating fatigue life in line contact roller bearings must be modified if misalignment is to be considered.

According to equation (18.59), it can be seen that for a line contact

$$\ln \frac{1}{\mathcal{S}} = G_1 q^{(c-h+1)/2} L^e l \tag{18.179}$$

in which G_1 is a constant. Therefore, considering as in Chapter 6 that the roller is made up of laminae each of length dx:

$$\ln \frac{1}{\mathcal{S}_x} = G_1 q_x^{(c-h+1)/2} L^e \, dx \tag{18.180}$$

According to the product law of probability, the probability of survival of the entire raceway is the product of the probabilities of survival of the individual disks. Hence

$$\mathcal{S} = \mathcal{S}_{x1} \times \mathcal{S}_{x2} \cdots \mathcal{S}_{xn} \tag{18.181}$$

Inverting and taking logarithms of both sides of the equation gives

$$\ln \frac{1}{\mathcal{S}} = \ln \frac{1}{\mathcal{S}_{x1}} + \ln \frac{1}{\mathcal{S}_{x2}} + \cdots \ln \frac{1}{\mathcal{S}_{xn}} \tag{18.182}$$

Assuming that each disk survives with fatigue life L and substituting equation (18.180) into (18.182) yields

$$\ln \frac{1}{\mathcal{S}} = G_1 L^e \sum_{j=1}^{j=n} q_j^{(c-h+1)/2} \, dx \tag{18.183}$$

In integral format, equation (18.183) becomes

$$\ln \frac{1}{\mathcal{S}} = G_1 L^e \int_0^l q^{(c-h+1)/2} \, dx \tag{18.184}$$

For roller bearings with line contact, Lundberg et al. [18.1] indicate that $(c - h + 1)/2$ is equal to 9/2. Further, according to Lundberg et al. [18.8], $e = 9/8$; therefore,

$$\ln \frac{1}{\mathcal{S}} = G_1 L^{9/8} \int_0^l q^{9/2} \, dx \tag{18.185}$$

In terms of an equivalent load per unit length q_e and roller effective length l, bearing life L may be expressed as follows:

$$\ln \frac{1}{\mathcal{S}} = G_1 L^{9/8} q_e^{9/2} l \tag{18.186}$$

Equating equation (18.185) to (18.186) yields the following expression for equivalent roller load per unit length:

$$q_e = \left(\frac{1}{l} \int_0^l q^{9/2}\, dx \right)^{2/9} \tag{18.187}$$

If it is assumed that the bearing is mounted properly and that essentially zero misalignment occurs after the load is applied, then

$$q_e = \frac{Q}{l} \tag{18.188}$$

In finite difference format, the equation for roller equivalent load is

$$Q_{ej} = m^{7/9} w \left(\sum_{k=1}^{k=m} q_{jk}^{9/2} \right)^{2/9} \tag{18.189}$$

where m is the number of increments of width w into which the roller equivalent length is divided for calculational purposes. The fatigue life of the raceways can now be calculated using equations (18.143)–(18.147) and the fatigue life of the entire bearing using equation (18.148). Figure 18.31 taken from reference [18.24] shows the effect of misalignment on a 309 cylindrical roller bearing as a function of roller crowning and applied load.

Table 18.24 taken from reference [18.14] indicates, based on experience data in bearing manufacturers' catalogs, maximum acceptable misalignments for the various rolling bearing types.

It should be apparent that misalignment can lead very quickly to *edge loading* in the roller–raceway contacts and edge loading of significant magnitude will rapidly diminish bearing fatigue life. In Chapter 5, references were cited indicating that the magnitude of edge loading can be calculated for any roller–raceway contact profile. The calculations require the use of a computer, and occupy substantial computer time even for a single roller–raceway contact condition. An approximation can, however, be used in the *laminated* contact calculation introduced in Chapter 6 for roller bearing load distribution calculations. Accordingly, the roller equivalent load given by equation (18.182) can be augmented by correction factors applied to the loading at the end laminae.

$$Q_{ej} = \left[\frac{1}{mw} \left(f_e q_{j1}^{9/2} D + w \sum_{k=1}^{k=m} q_{jk}^{9/2} + f_e q_{jm}^{9/2} D \right) \right]^{2/9} \tag{18.189}$$

A conservative value for the "edge correction factor" f_e is 5.4.

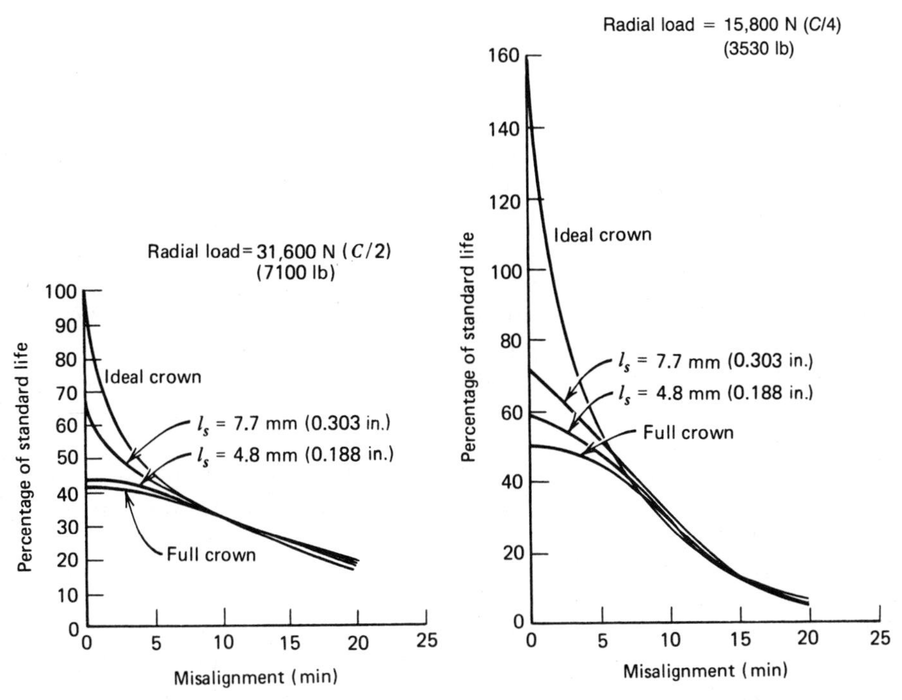

FIGURE 18.31. Life vs misalignment for a 309 cylindrical roller bearing as a function of crowning and applied load.

TABLE 18.24. Limitations on Rolling Bearing Misalignment

Bearing Type	Misalignment Angle (min)	Misalignment Angle (rad)
Cylindrical roller	3–4	0.001
Tapered roller	3–4	0.001
Spherical roller	30	0.0087
Deep groove ball	12–16	0.0035–0.0047

EFFECT OF LUBRICATION ON FATIGUE LIFE

In Chapter 12, it was indicated that if a rolling bearing is adequately designed and lubricated, the rolling surfaces can be separated by a lubricant film. Endurance testing of rolling bearings as shown by Tallian et al. [18.25] and Skurka [18.26] has demonstrated the considerable effect of lubricant film thickness on fatigue life. In Chapter 12 methods for estimating this lubricant film thickness were given. It was also demonstrated that lubricant film thickness is sensitive to bearing speed of operation and lubricant viscous properties and, moreover, the film thickness is virtually insensitive to load.

The test results reported both in references [18.25] and [18.26] showed that at high speeds of operation, a considerable improvement in fatigue life occurs. Moreover, a similar effect can be achieved by using a sufficiently viscous lubricant at slower speeds. The effectiveness of the lubricant film thickness generated depends upon its magnitude relative to the surface topographies of the contacting elements, that is, rolling elements and raceways. In other words, a bearing with very smooth raceway and rolling element surfaces requires less of a lubricant film than does a bearing with relatively rough surfaces (see Fig. 18.32).

Thus far, the relationship of lubricant film thickness to surface roughness is signified in rolling bearing literature by Λ, which utilizes the simple rms value of the roughnesses of the surfaces of the contacting bodies. Tallian [18.27] among many other researchers has introduced the use of asperity slopes as well as height of asperity peaks. Chapter 14 covering microcontact phenomena provides additional, more complex means to evaluate the effect of a "rough" surface on contact and hence bearing lubrication and performance. In the absence of more elegant means however, using Λ, Harris [18.28] indicated the effect of lubrication on bearing fatigue life in Fig. 18.33. According to reference [18.28], if Λ

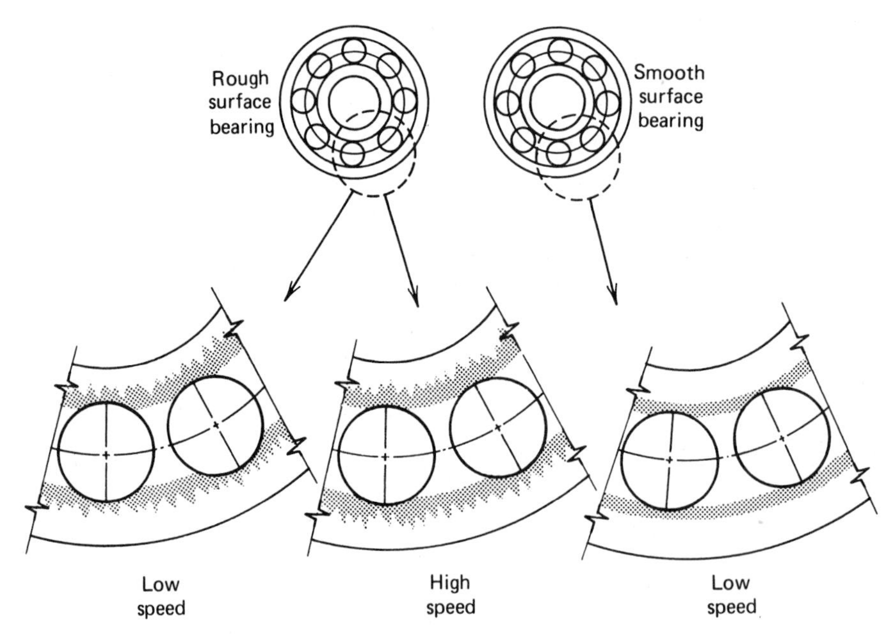

FIGURE 18.32. Illustration of the effect of surface roughness on the lubricant film thickness required to prevent metal-to-metal contact.

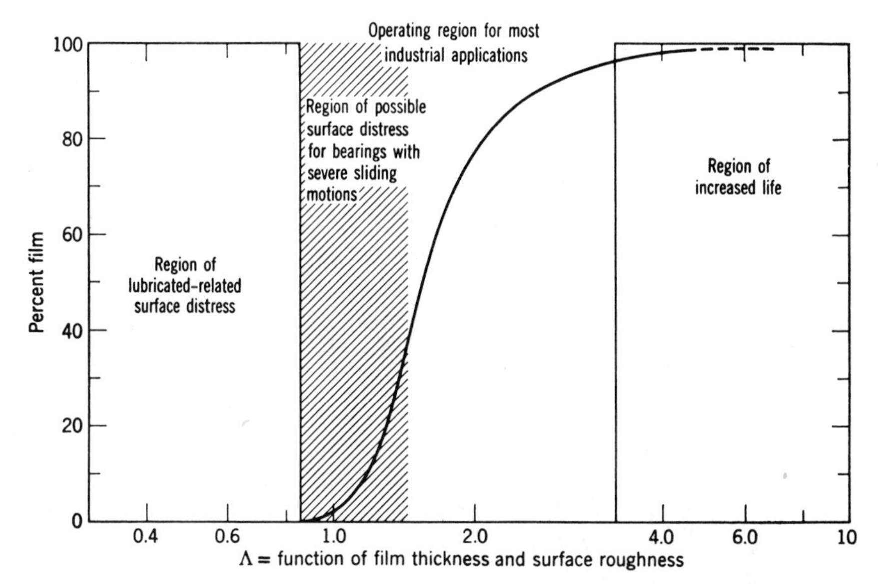

FIGURE 18.33. Percent film vs Λ.

is numerically equal to or greater than 4, fatigue life can be expected to exceed L_{10} estimates calculated according to ANSI methods by at least 100%. On the other hand, if Λ is less than unity, the bearing will probably not attain the calculated L_{10} estimates because of surface distress such as smearing or pulling which can lead to rapid fatigue failure of the rolling surfaces. Fig. 18.33 shows the various operating regions just described. In Fig. 18.33 the ordinate, that is, percent film, is a measure of the time percentage during which the "contacting" surfaces are fully separated by an oil or lubricant film.

Tallian [18.29] showed a more definitive estimate of rolling bearing fatigue life vs Λ as did Skurka [18.26]. Zaretsky et al. [18.14] show the combination of the foregoing in Fig. 18.34, recommending the use of the mean curve. Recent experimental data indicate that for $\Lambda > 4$, the L/L_{10} ratios given by Fig. 18.34 are substantially greater for accurately manufactured, bearings lubricated by minimally contaminated oil. See Chapter 23.

Schouten [18.30] developed a microtransducer to accurately measure the pressure distribution in an oil-lubricated line contact in the direction of rolling. Subsequently, it was shown in [18.31] that the *edge stress* in a line contact is substantially reduced if an adequate lubricant film separates the *contacting* rolling bodies. Thus, in this situation, the lubricant film tends to permit an increase in fatigue life by reducing the magnitude of normal stress at the end(s) of heavily loaded contacts.

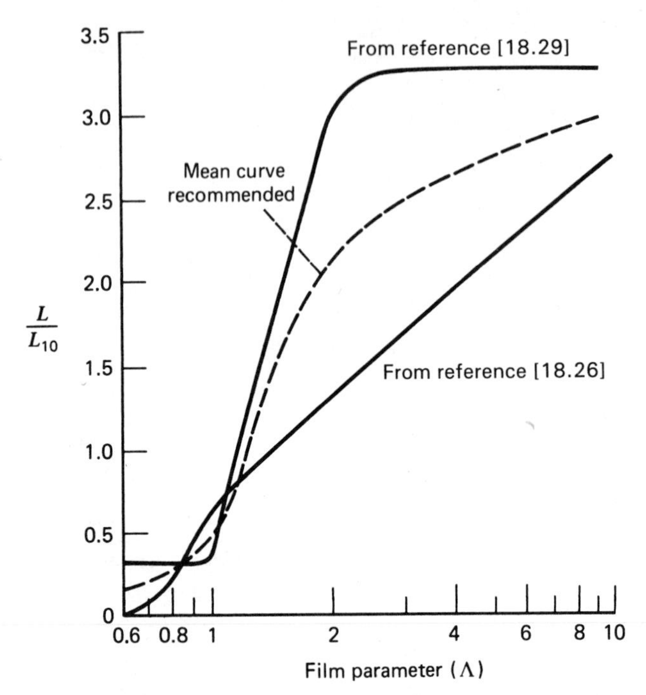

FIGURE 18.34. Lubrication-life factor vs Λ.

Example 18.11. In the 209 cylindrical roller bearing of Example 12.1, the rollers have surface finishes of 0.102 μm (4-μin.) rms and the raceways have surface finishes of 0.203 μm (8-μin.) rms. Determine the effect of lubrication on fatigue life.

$$h^\circ = 0.731 \ \mu\text{m} \ (28.4 \ \mu\text{in.}) \ (\text{naphthenic oil}) \qquad \text{Ex.12.1}$$

$$SF = (SF_R^2 + SF_i^2)^{1/2}$$

$$= [(0.102)^2 + (0.203)^2]^{1/2}$$

$$= 0.227 \ \mu\text{m} \ (8.94 \ \mu\text{in.})$$

$$\Lambda = \frac{h^\circ}{SF}$$

$$= \frac{0.726}{0.227} = 3.2$$

According to Fig. 18.34, the expected fatigue life would be between 1.7 and 3.3 L_{10} or a mean value of 2.5L_{10}.

Example 18.12. Estimate the effect of lubrication on the life of the 209 cylindrical roller bearing of Example 12.2.

$$h^\circ = 0.0734\ \mu\text{m}\ (2.89\ \mu\text{in.}) \qquad \text{Ex.12.2}$$

$$\Lambda = \frac{h^\circ}{SF}$$

$$= \frac{0.0734}{0.227} = 0.32$$

From Fig. 18.34 it can be seen that a reduction in L_{10} fatigue life will most likely be experienced. If any degree of gross sliding occurs in the roller–raceway contacts, the fatigue life reduction may be very severe.

RELIABILITY AND DEVIATIONS FROM THE WEIBULL DISTRIBUTION

Based upon an analysis of the endurance data pertaining to more than 2500 test bearings, Tallian [18.6] confirmed that a Weibull distribution fits the test data in the most-used cumulative failure probability region, that is, between $\mathfrak{F} = 0.07$ and $\mathfrak{F} = 0.60$. Everywhere outside of this region, fatigue life is greater than that predicted by a Weibull distribution.

Generally, bearing users are not concerned about fatigue lives in excess of the median life, that is, L_{50}; therefore, the upper failure region, which deviates from the Weibull distribution, is not of interest here. However, rolling bearing users are extremely interested in the failure region between $\mathfrak{F} = 0$ and $\mathfrak{F} = 0.10$. Aerospace applications demand rolling bearings having better than 90% reliability ($\mathfrak{F} = 0.1$, $\mathcal{S} = 0.9$). In fact, reliability of better than 99% is not an uncommon requirement. Automobile manufacturers are on record with multi year unconditional warranties on parts including bearings and hence would like bearings of greater reliability. Considering the improved bearing steels which are available, it is possible to manufacture bearings having increased fatigue life. It may therefore be necessary to specify the fatigue of a given rolling bearing application in terms of L_1, L_5, and so on, instead of L_{10}.

Figure 18.35 taken from Tallian's paper [18.6], shows the deviation from the Weibull distribution for probability of failure $\mathfrak{F}$ less than 10%. Note that below approximately a standardized bearing life of 0.004, that is, below

$$\mathcal{Y} = \left(\frac{L}{L_{50}}\right)^e \ln 2 = 0.004$$

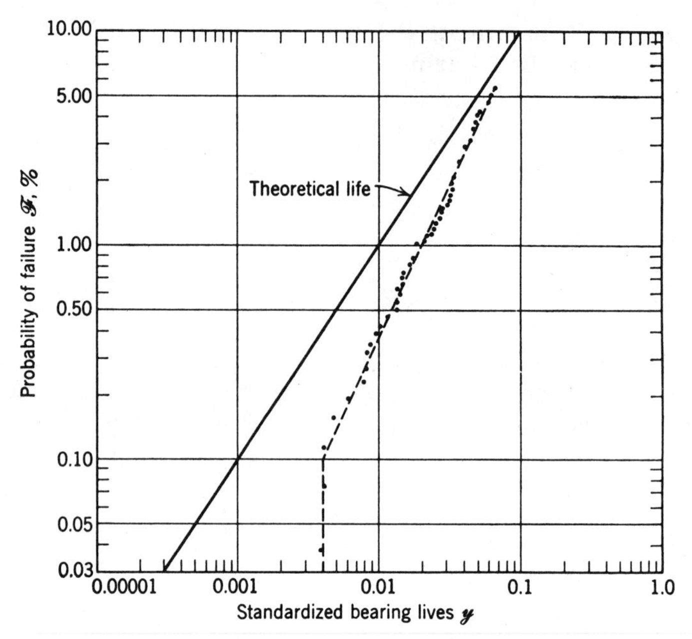

FIGURE 18.35. Life distribution in the early failure region.

there is apparently no decrease in fatigue life. Figure 18.36, which is a similar plot shown by Harris [18.32] on semilogarithmic coordinates, demonstrates more dramatically the significance of Tallian's research. It is apparent that "no-failure" fatigue life may be predicted with an attendant reliability of 100% ($\mathcal{S} = 1.0$, $\mathcal{F} = 0$). According to Tallian [18.6], the no-failure fatigue life may be approximated by

$$L_{NF} \cong 0.05 L_{10} \tag{18.190}$$

Tallian [18.6] reasoned that bearing fatigue life may be separated into two discrete phases:

1. The time between commencement of rotation and initiation of the crack below the surface, that is, L_a.
2. The time necessary for the crack to propagate to the surface, that is, L_b.

Hence, the fatigue life measured represents the sum of the two durations:

$$L = L_a + L_b \tag{18.191}$$

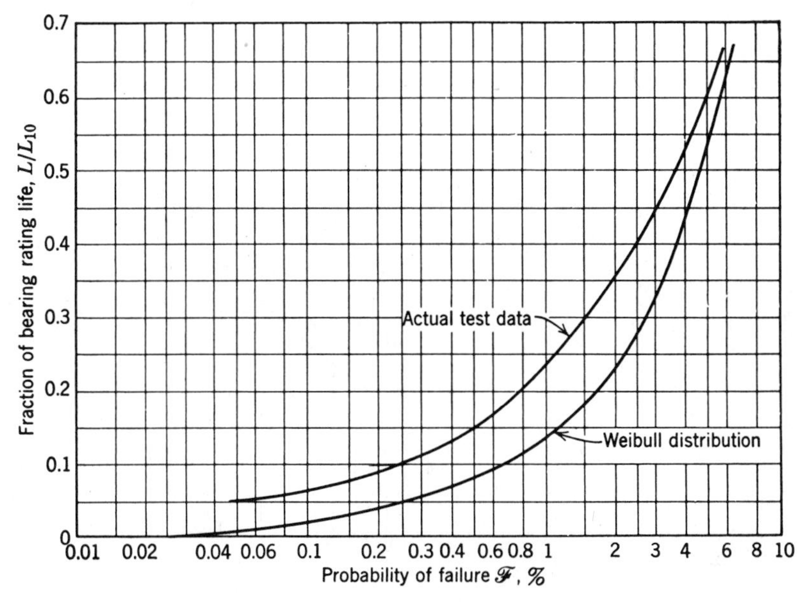

FIGURE 18.36. Fraction L_{10} life vs probability of failure.

L_b may be called an excess life since L_a is the fatigue life predicted by Weibull theory. If a crack is initiated when rotation has just commenced, L_b is very much greater than L_a and L is not a valid Weibull life measurement. If a crack is initiated a considerable time after rotation has started, then L_b is insignificant compared to L_a and L is a reasonably accurate Weibull life estimator. Thus, deviation of early failures from the Weibull distribution is explained. In equation format, one may determine the fatigue life at reliabilities other than $\mathcal{S} = 0.9$ as follows:

$$\frac{L}{L_{10}} = \left[\frac{\ln \frac{1}{\mathcal{S}} + \mathcal{Y}_e}{\ln (1/0.9)}\right]^{1/e} \tag{18.192}$$

in which $\mathcal{Y}_e$ is the standardized excess life. Tallian [18.6] gives the schedule for $\mathcal{Y}_e$ presented in Table 18.25.

TABLE 18.25. Excess Standardized Life $\mathcal{Y}_e$

Probability of Survival (%)	Standardized Theoretical Life	Standardized Life
$\mathcal{S} \geq 99.9$	$\mathcal{Y} \leq 0.001$	$\mathcal{Y} + \mathcal{Y}_e = 0.004$
$99.9 \geq \mathcal{S} \geq 95$	$0.001 < \mathcal{Y} < 0.05$	$\ln(\mathcal{Y} + \mathcal{Y}_e) = 0.690 \ln(0.328\ \mathcal{Y})$
$95 > \mathcal{S} > 40$	$0.05 < \mathcal{Y} < 0.6$	$\mathcal{Y}_e = 0.013$

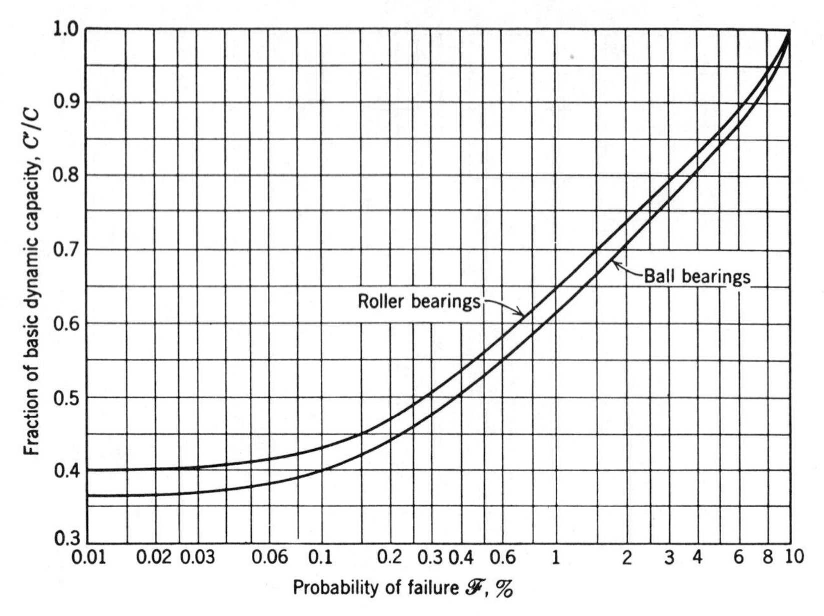

FIGURE 18.37. Reduction in basic dynamic capacity required for increased reliability.

Increase in reliability may be assigned to a given bearing by reduction of its basic dynamic capacity. Harris [18.32] developed Fig. 18.37, which determines the reduction in basic dynamic capacity required to achieve reliability higher than 90%.

Example 18.13. In example 18.6, the L_{10} fatigue life of the 218 angular-contact ball bearing that supports a 22,250 N (5000 lb) thrust load while the shaft rotates at 10,000 rpm was estimated at 525 hr according to the basic Lundberg–Palmgren formulas. Consider that the bearing is fabricated fron contemporary 52100 steel and according to contemporary manufacturing methods; estimate the fatigue life that may be attained with 99% reliability.

$$\mathfrak{F} = 100 - 99 = 1\%$$

From Table 18.11, $b_{\text{m}} = 1.3$, therefore

$$Q_{\text{c}} = b_{\text{m}} \times Q_{\text{c}}$$

$$= 1.3Q_{\text{c}}$$

$$L_{10} = (Q_{\text{c}}/Q_{\text{e}})^3 \qquad (18.70, 18.75)$$

Therefore,

$$L_{10} = (1.3Q_c/Q_e)^3$$

or

$$L'_{10} = 2.20\,L_{10}$$

From Fig. 18.36,

$$L/L_{10} = 0.23$$

Therefore,

$$L_1 = 0.23 \times 2.20 \times 525 = 265 \text{ hr}$$

Example 18.14. The 209 radial ball bearing of Example 18.7 achieved an estimated L_{10} life of 460 hr. Assuming contemporary steel and manufacturing processes, the bearing has a basic load rating C = 32,710 N (7350 lb). To achieve the same life with 95% reliability, what is the basic load rating of the required ball bearing?

$$\mathfrak{F} = 100 - 95 = 5\%$$

From Fig. 18.37, $C'/C = 0.845$

$$C' = 32{,}710/0.845 = 38{,}710 \text{ N (8699 lb)}$$

COMBINING LIFE ADJUSTMENT FACTORS

Both the ANSI [18.10, 18.12] and ISO [18.13] have recommended the combination of the life adjustment factors for reliability ($\mathcal{A}_1$), material ($\mathcal{A}_2$), and operating conditions ($\mathcal{A}_3$) as follows:

$$L'_{10} = \mathcal{A}_1 \mathcal{A}_2 \mathcal{A}_3 \left(\frac{C}{F_e}\right)^p \tag{18.193}$$

where $p = 3$ for ball bearings and $p = 10/3$ for roller bearings. The standards advise that indiscriminate application of the life adjustment factors in equation (18.193) may lead to serious overestimation of bearing endurance since fatigue life is only one criterion for bearing selection. Moreover, care must be exercised to select bearings of sufficient size for the application. Undersizing of shaft and housing structures by using bearings that appear adequate from a fatigue life standpoint could lead to misalignment and fitting problems, which can invalidate the standard formulas. Furthermore, it may not be assumed that when lubrication is marginal, that is, $\mathcal{A}_3$ is less than 1, such deficiency can be overcome by using an improved steel, that is $\mathcal{A}_2$ greater than 1. Accuracy of manufacture of the rolling contact surfaces also has a significant influence so that material cannot be taken as a singular effect. Reduction of $\mathcal{A}_3$ values should be considered where the lubricant viscosity is less than 13 centistokes (mm^2/sec)(70 SSU) for ball bearings or 20 centistokes (100 SSU) for roller bearings at the operating temperature and/or where rotational speed is exceptionally low, for example, n(rpm) $\times$ d_m(mm) is less than 10,000. When $\mathcal{A}_3$ is less than 1, values of $\mathcal{A}_2$ greater than 1 should therefore normally not be used.

For rolling bearings properly fabricated from contemporary, good quality AISI 52100 steel, SKF [18.33] has combined the $\mathcal{A}_2$ and $\mathcal{A}_3$ factors into an $\mathcal{A}_{23}$ factor dependent upon the adequacy of lubrication as determined by the relative viscosity κ. This recognizes the interdependency of the material and lubrication factors, particularly under conditions of marginal lubrication. Figure 18.38 from [18.33] gives the value of $\mathcal{A}_{23}$ versus κ. κ is defined as the ratio of the actual operating lubricant vis-

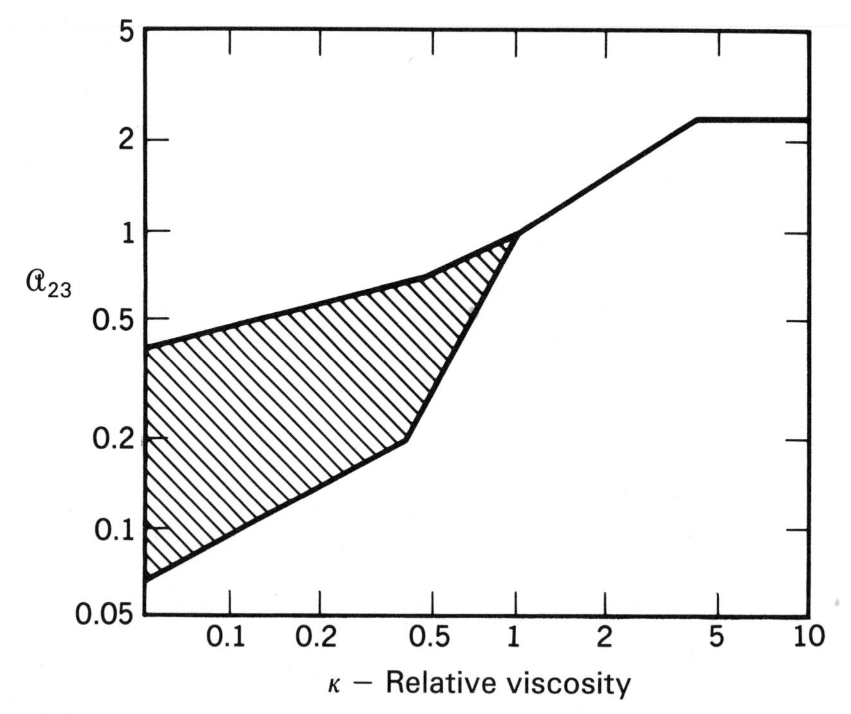

FIGURE 18.38. Material-lubrication factor as a function of lubricant relative viscosity (from [18.33]). In the shaded area, the higher values may be obtained if lubricants containing lead-based additives of the EP type are used. If, however, the EP additives are based on compounds aggressive to bearing steels, even the lower values may not obtain in the shaded zone.

TABLE 18.26. Life Adjustment Factor for Reliability

Reliability (%)	$\mathcal{A}_1$
90	1
95	0.62
96	0.53
97	0.44
98	0.33
99	0.21

cosity to viscosity associated with just adequate lubrication, that is, $\Lambda = 1$.

The life adjustment factors for reliability ($\mathcal{A}_1$) recommended by the standards are given by Table 18.26. These $\mathcal{A}_1$ values are slightly more conservative than those shown in Fig. 18.37 at high reliability requirements; the "no-failure" life consideration is not addressed. Reference [18.14] gives some insight into values for the $\mathcal{A}_2$ and $\mathcal{A}_3$ life adjustment factors.

EFFECT OF SLIDING ON FATIGUE LIFE

The Lundberg–Palmgren [18.1, 18.8] method for evaluation of the fatigue life of rolling bearings is based on the premise that rolling bearings so analyzed are properly lubricated and properly operated. *Properly lubricated* refers to an adequate supply of good mineral oil lubricant under a bearing temperature not sufficiently elevated to cause deterioration of the lubrication. *Properly operated* refers to proper alignment, adequate loading to preclude skidding, dirt and moisture free atmosphere, and so on.

In has been shown in Chapter 7 that even when a rolling bearing operates properly, sliding occurs within the contact areas because of deformations of surfaces. Also as shown in Chapter 13, microslip occurs.

Since to an extent the Lundberg–Palmgren [18.1, 18.8] method is based on empirical data, this type of sliding may be assumed to be accounted for in the equations for fatigue life. Sliding that occurs because of skidding is not, however, accounted for in the Lundberg–Palmgren [18.1] equations since this is not a condition attendant to good rolling bearing operation and thus was not analyzed.

It is recognized that sliding on the contact surfaces tends to cause higher subsurface stresses because of the tangential stress component (in addition to normal stress) on the surface. Furthermore, the increased subsurface stress level tends to decrease fatigue life. In Chapter 23, a method is defined to more accurately predict the effect of surface shear stress due to sliding on fatigue endurance. Previously, however, Jones [18.34] gave the following relationship for dynamic capacity reduction of a point contact caused by spinning:

$$Q'_{cj} = \left[1 - \frac{b}{a}\left(\frac{\omega_{sj}}{\omega_{rollj}}\right)\right]^{\phi} Q_{cj} \tag{18.194}$$

In equation (18.194), $(\omega_{sj}/\omega_{rollj})$ is the absolute value of the spin-to-roll ratio at raceway contact j. Values of the spin-to-roll ratio may be calculated from equations (7.56) and 7.57). In equation (18.138), the reduction in capacity caused by spinning was approximated by the factor $(1 - 0.33 \sin \alpha)$.

FATIGUE LIFE DUE TO A FLUCTUATING LOAD

To determine the fatigue life of a rolling bearing subjected to a time-variant load, one must determine a mean effective bearing load F_m such that

$$L = \left(\frac{C}{F_m}\right)^p \tag{18.195}$$

in which $p = 3$ for point-contact bearings and $p = 4$ for line-contact bearings. Actually, there is little difference in F_m calculated for $p = 3$ as compared to F_m calculated for $p = 4$; therefore, a cubic mean effective load is frequently used for roller bearings with little error.

Consider a bearing subjected to a load F_1 for N_1 revolutions and F_2 for N_2 revolutions. The calculated fatigue lives corresponding to these loads are

$$L_1 = \left(\frac{C}{F_1}\right)^p \tag{18.196}$$

and

$$L_2 = \left(\frac{C}{F_2}\right)^p \tag{18.197}$$

The load F_1, however, only acts for N_1 revolutions and F_2 only acts for N_2 revolutions. Because of loads F_1 and F_2, therefore, N_1/L_1 and N_2/L_2, respectively, are the consumed fractions of the bearing's ability to resist fatigue and

$$\frac{N_1}{L_1} + \frac{N_2}{L_2} = 1 \tag{18.198}$$

Substituting equations (18.196) and (18.197) into (18.198) yields

$$\frac{F_1^p N_1}{C^p} + \frac{F_2^p N_2}{C^p} = 1 \tag{18.199}$$

Dividing by L, the total bearing fatigue life, and substituting for L according to equation (18.195) yields

$$\frac{F_1^p N_1}{L} + \frac{F_2^p N_2}{L} = F_{\mathrm{m}}^p \tag{18.200}$$

Hence,

$$F_{\mathrm{m}} = \left(\frac{F_1^p N_1 + F_2^p N_2}{L}\right)^{1/p} \tag{18.201}$$

It is readily apparent that $L = N_1 + N_2$. Thus, in general for k loads each operating for N_k revolutions, the mean effective load is given by

$$F_{\mathrm{m}} = \left(\frac{\Sigma F_k^p N_k}{N}\right)^{1/p} \tag{18.202}$$

in which N is the total number of revolutions in one load cycle. In integral format, equation (18.202) becomes

$$F_{\mathrm{m}} = \left(\frac{1}{N}\int_0^N F^p \, dN\right)^{1/p} \tag{18.203}$$

For a cyclic load of period τ,

$$F_{\mathrm{m}} = \left(\frac{1}{\tau}\int_0^\tau F_t^p \, dt\right)^{1/p} \tag{18.204}$$

in which F_t is a defined function of time.

Example 18.15. The 22317 two-row spherical roller bearing of Example 18.9 was shown to have a basic load rating of 399,300 N (89,720 lb). Assuming the bearing experiences the following repetitive loading cycle while operating at 900 rpm shaft speed, estimate the L_{10} fatigue life according to standard (ANSI, ISO) methods.

Condition	Time (min)	F_r (N)	F_r (lb)	F_a (N)	F_a (lb)
1	20	89,000	(20,000)	22,250	(5000)
2	30	44,500	(10,000)	0	
3	10	22,250	(5000)	22,250	(5000)

Condition 1 $\quad F_{e1} = 136{,}100$ N (30,950 lb) $\quad$ Ex. 18.9

Condition 2 $\quad F_{e2} = 44{,}500$ N (10,000 lb)

Condition 3 $\quad F_a/F_r = 22{,}250/22{,}250 = 1$

$1.5 \tan \alpha = 0.3189$ $\quad$ Ex. 18.9

$F_a/F_r > 1.5 \tan \alpha$

From Table 18.7, $X = 0.67 \quad Y = 0.67 \text{ ctn } \alpha$

$$\alpha = 12° \qquad \text{Ex. 2.9}$$

$$Y = 0.67 \text{ ctn } (12°) = 3.152$$

$$F_{e3} = XF_r + YF_a \tag{18.165}$$

$$= 0.67 \times 22{,}250 + 3.152 \times 22{,}250$$

$$= 85{,}040 \text{ N } (19{,}100 \text{ lb})$$

$$F_m = (\Sigma F_k^{10/3} N_k/N)^{3/10} \tag{18.202}$$

$$= \left\{\frac{20(136{,}100)^{10/3} + 30(44{,}500)^{10/3} + 10(85{,}040)^{10/3}}{20 + 30 + 10}\right\}^{3/10}$$

$$= 101{,}700 \text{ N } (22{,}860 \text{ lb})$$

$$L = (C/F_m)^{10/3} \tag{18.195}$$

$$= (399{,}300/101{,}700)^{10/3}$$

$$= 95.48 \text{ million revolutions}$$

$$L = 95.48 \times 10^6/(60 \times 900)$$

$$= 1768 \text{ hr}$$

Some special cases of fluctuating load may now be defined. Palmgren [18.9] states that for a bearing load that varies nearly linearly between F_{min} and F_{max} as shown in Fig. 18.39, the following approximation is valid:

$$F_m = \tfrac{1}{3}F_{min} + \tfrac{2}{3}F_{max} \tag{18.205}$$

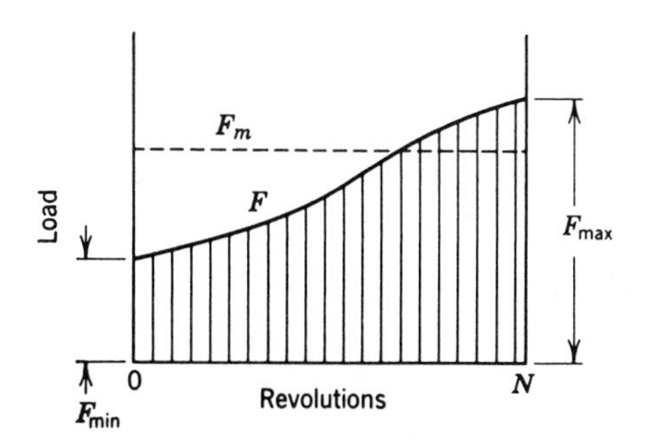

FIGURE 18.39. Load vs time; nearly linear.

If the load variation is truly linear, then

$$F_m = F_{min} + \left(\frac{F_{max} - F_{min}}{\tau}\right) \times t \tag{18.206}$$

and

$$F_m = F_{min}\left\{\int_0^1 \left[1 + \left(\frac{F_{max}}{F_{min}} - 1\right)z\right]^p dz\right\}^{1/p} \tag{18.207}$$

in which z is a dummy variable.

A bearing load may be composed of a steady load F_1 upon which a sinusoidally varying load of amplitude F_3 in phase with F_1 is superimposed as shown in Fig. 18.40. In this case,

$$F_t = F_1 + F_3 \cos \omega t \tag{18.208}$$

in which ω is the circular frequency in radians per second. Thus,

$$F_m = F_1\left[\frac{1}{\tau}\int_0^\tau \left(1 + \frac{F_3}{F_1 \cos \omega t}\right)^p dt\right]^{1/p} \tag{18.209}$$

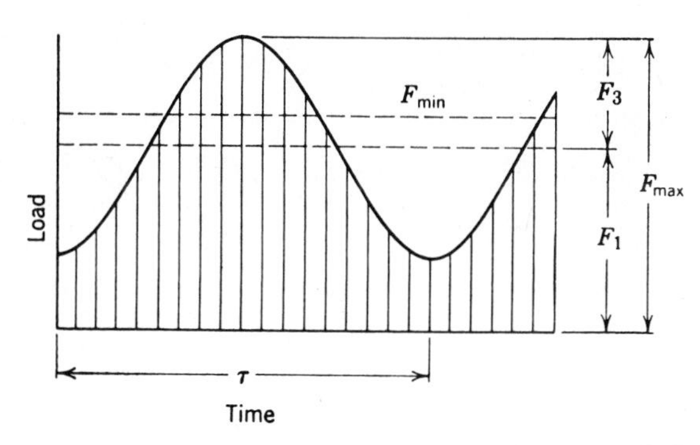

FIGURE 18.40. Sinusoidal load F_3 superimposed on a steady load F_1.

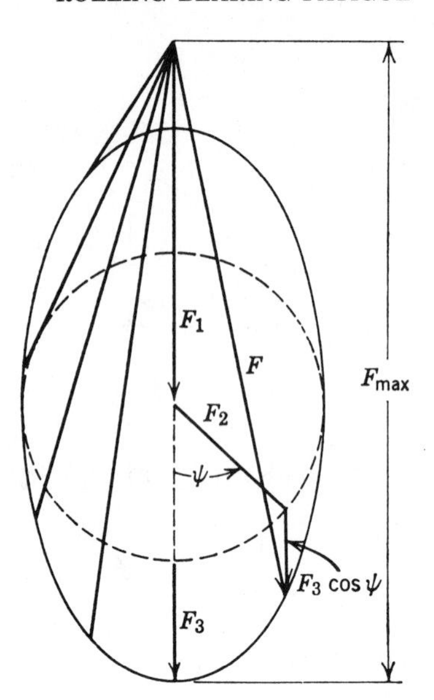

FIGURE 18.41. Load vector diagram for generalizing bearing loading consisting of steady load F_1, rotating load F_2, and sinusoidal load F_3 in phase with F_1.

A more general case of loading is that of a steady load F_1, a rotating load F_2, and a sinusoidally varying load of amplitude F_3 (in phase with F_1) simultaneously applied to a rolling bearing. Figure 18.41 demonstrates this form of loading. Maximum bearing load occurs when F_1, F_2, and F_3 assume the same line of action. Steady loads F_1 are caused by the weight of machine elements on a shaft and also by their imposed loading such as gear or belt loads. Rotating loads F_2 are caused by unbalance in spinning mechanisms, either intentional or not. Sinusoidal loads F_3 are caused by inertial forces of reciprocating machinery. For this general loading, one may simply state that

$$F_{\mathrm{m}} = \phi_{\mathrm{m}}(F_1 + F_2 + F_3) \tag{18.210}$$

in which

$$\phi_{\mathrm{m}} = \frac{\left\{\dfrac{1}{2\pi} \displaystyle\int_0^{2\pi} ([F_1 + (F_2 + F_3) \cos \psi]^2 + (F_2 \sin \psi)^2)^{p/2} \, d\psi \right]^{1/p}}{F_1 + F_2 + F_3} \tag{18.211}$$

SKF has developed a series of curves depicting this relationship in terms of the relative magnitudes of F_1, F_2, and F_3. Figure 18.42a applies to

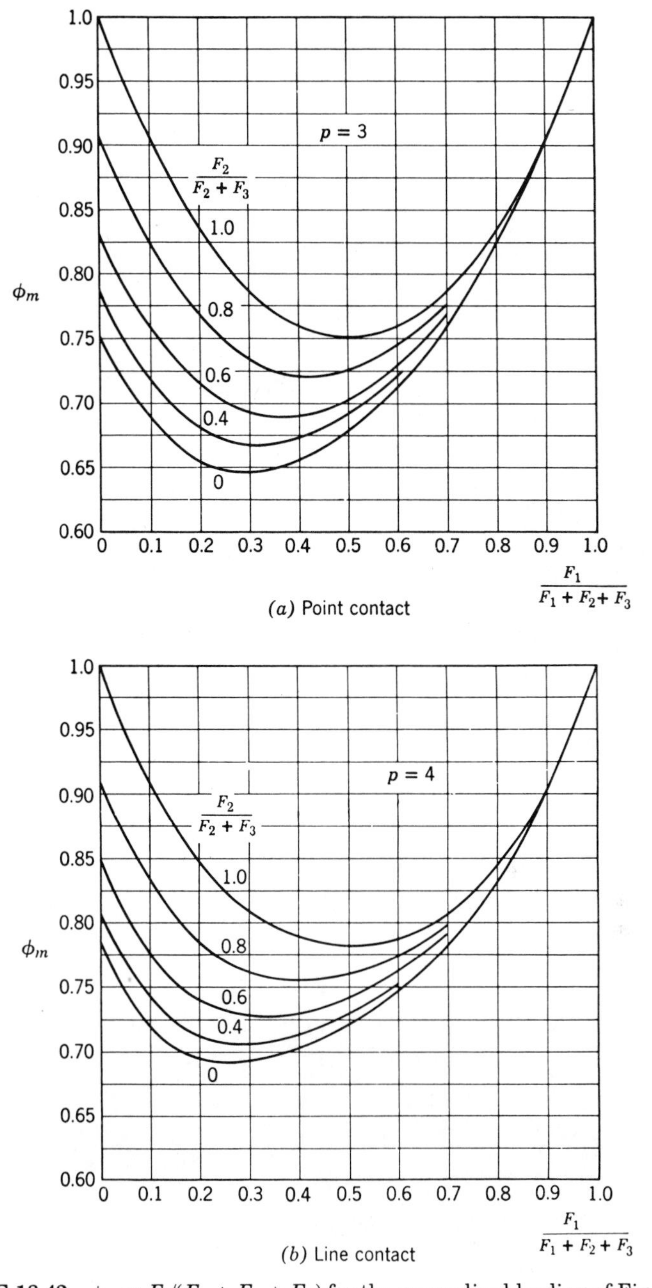

FIGURE 18.42. ϕ_m vs $F_1/(F_1 + F_2 + F_3)$ for the generalized loading of Fig. 18.41.

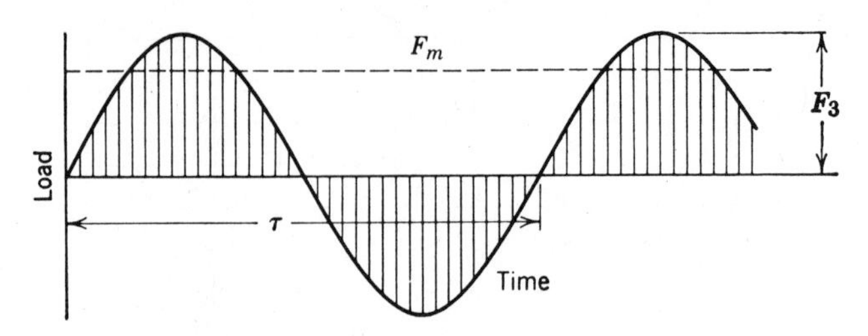

FIGURE 18.43. Sinusoidal bearing load.

point contact, that is, $p = 3$, and Fig. 18.42*b* applies to line contact, $p = 4$. Note that when $F_2 = 0$, the loading of Fig. 18.43 occurs. Thus, the lowest curves of Figs. 18.42 refer to that loading situation. If F_1 and F_2 are individually absent, that is, for the sinusoidal bearing loading demonstrated by Fig. 18.43, then according to Fig. 18.42, $\phi_m = 0.75$ for point contact and $\phi_m = 0.79$ for line contact.

Figure 18.44 demonstrates a bearing loading in which F_3, the sinusoidal load, acts 90° out-of-phase to the steady load F_1. Figures 18.45*a* and 18.45*b* yield values of ϕ_m for this type of loading for point and line contact, respectively. When steady load F_1 is absent and maximum F_3 occurs 90° out-of-phase with F_2, then Fig. 18.46 illustrates the bearing loading with time, and Fig. 18.47 gives values of ϕ_m for point and line contact.

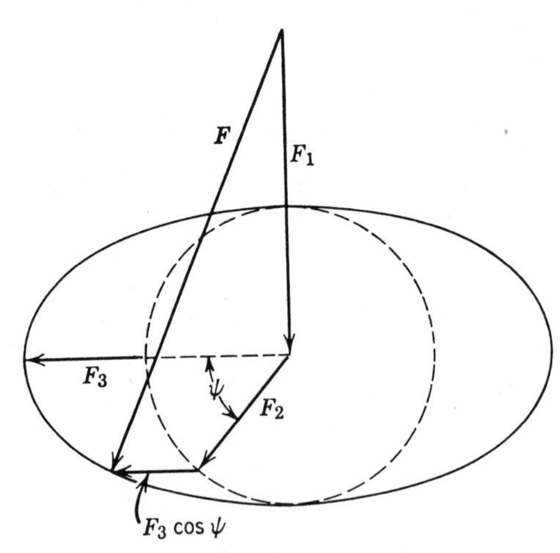

FIGURE 18.44. Load vector diagram for generalized bearing loading consisting of steady load F_1, rotating load F_2, and sinusoidal load F_3, 90° out-of-phase with F_1.

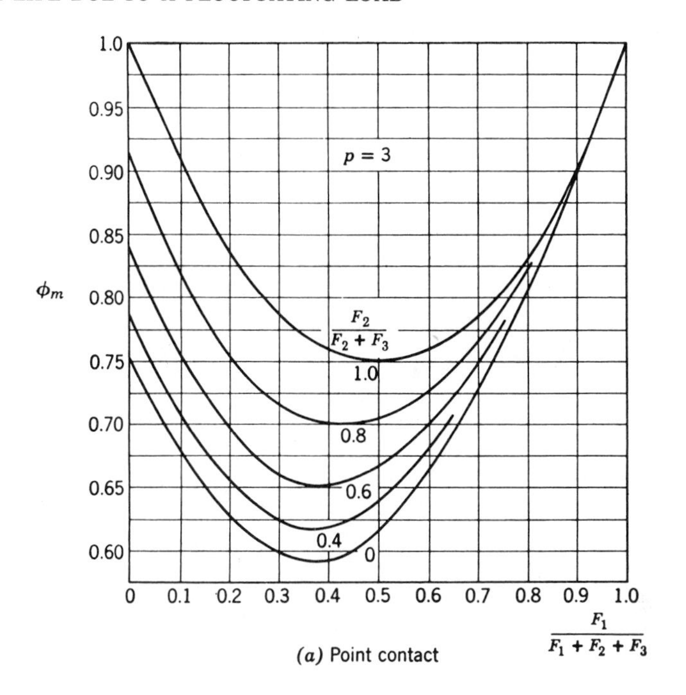

(a) Point contact

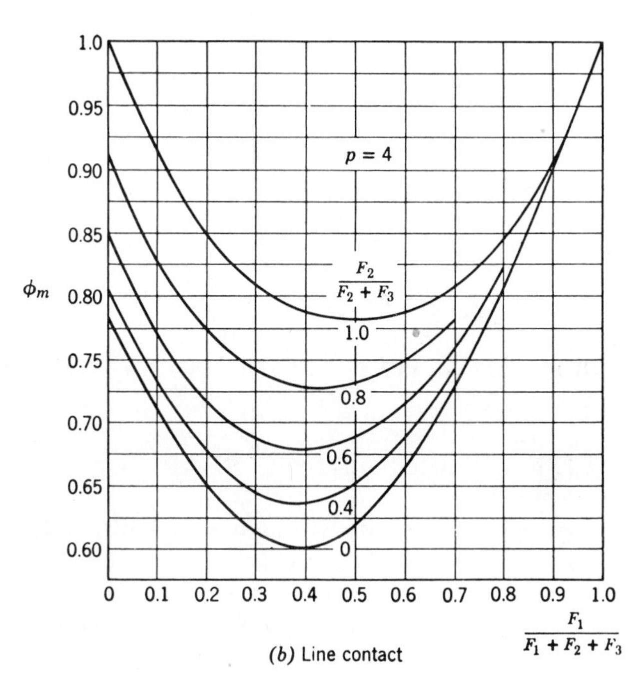

(b) Line contact

FIGURE 18.45. ϕ_m vs $F_1/(F_1 + F_2 + F_3)$ for the generalized loading of Fig. 18.44.

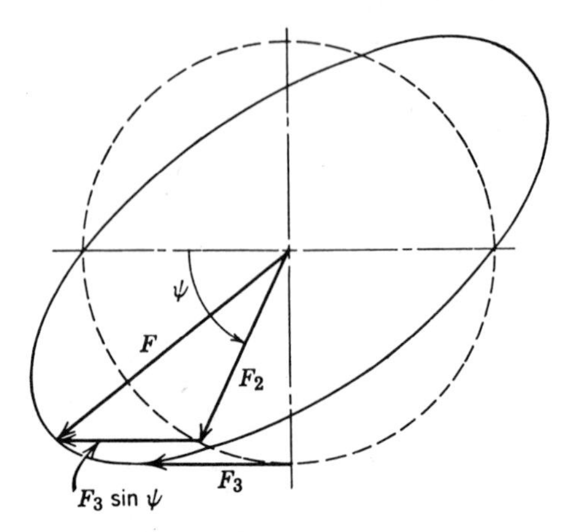

FIGURE 18.46. Load vector diagram for bearing loading consisting of a rotating load F_2 and a sinusoidal load F_3. Maximum F_3 occurs 90° out-of-phase with F_2.

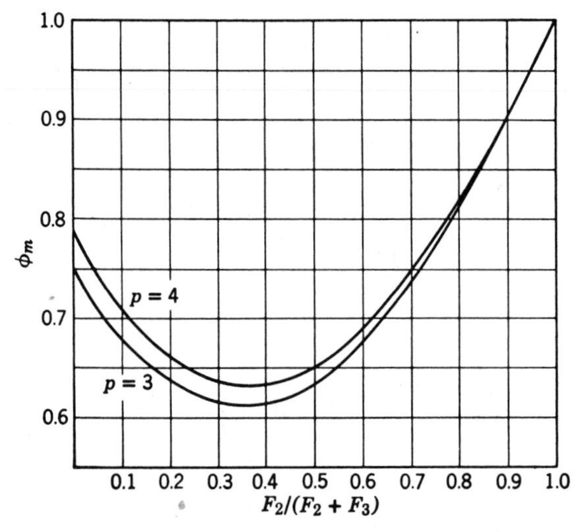

FIGURE 18.47. ϕ_m vs $F_2/(F_2 + F_3)$ for the loading of Fig. 18.46.

FATIGUE LIFE OF OSCILLATING BEARINGS

Oscillating bearings do not turn through complete revolutions. If one refers to the frequency of oscillation as n cycles per minute, then an oscillating bearing operating at frequency n will have a longer fatigue life than the same bearing rotating at n rpm under the same load. This is due to a lesser mean equivalent load per rolling element. One oscillation (cycle) refers to the motion of the bearing from one extreme position to the other, and return. The time to traverse the angular amplitude

ϕ of oscillation starting from 0° position is one-fourth of a cycle, or $\tau/4$, in which τ is the period of the oscillation.

To determine the fatigue life of an oscillating bearing, the applied load may be converted to an equivalent load for a rotating bearing, thus accounting for the decreased number of stress cycles. For a given bearing design having a given fatigue life, equation (18.31) yields the following relationship for point contact:

$$Q^{(c-h+2)/3}u^e = \mathcal{J} \tag{18.212}$$

in which $\mathcal{J}$ is a constant, u is the number of stress cycles per revolution, and e is the Weibull slope. It was shown in equations (18.37) and (18.38) that

$$\frac{c - h + 2}{3e} = 3 = p \tag{18.213}$$

Therefore

$$Q^p u = \mathcal{J} \tag{18.214}$$

Using the subscript RE for the equivalent rotating bearing, O for an oscillating bearing, and R for a rotating bearing, from equation (18.214),

$$Q_{\mathrm{RE}} = \left(\frac{u_{\mathrm{O}}}{u_{\mathrm{R}}}\right)^{1/p} Q_{\mathrm{O}} \tag{18.215}$$

The length of arc stressed during one complete revolution of a rotating bearing is $2\pi r$, in which r is the raceway radius. The length of arc stressed during one complete oscillation is $4\phi r$ in which ϕ is the amplitude of oscillation in radians. It is apparent that

$$\frac{u_{\mathrm{R}}}{u_{\mathrm{O}}} = \frac{\pi}{2\phi} \tag{18.216}$$

Thus,

$$Q_{\mathrm{RE}} = \left(\frac{2\phi}{\pi}\right)^{1/p} Q_{\mathrm{O}} \tag{18.217}$$

In this example Q is the load of a contact yielding a fatigue life of L million revolutions. Therefore, for an oscillating bearing

$$L = \left(\frac{C}{F_{\mathrm{RE}}}\right)^p \tag{18.218}$$

in which

$$F_{\mathrm{RE}} = \left(\frac{2\phi}{\pi}\right)^{1/p} F \tag{18.219}$$

For ϕ in degrees,

$$F_{\mathrm{RE}} = \left(\frac{\phi}{90}\right)^{1/p} F \tag{18.220}$$

The foregoing development is valid for point contact in which $p = 3$; the equations are similar for line contact except that $p = 4$.

If $\phi/90^\circ$ is less than $1/Z$, in which Z is the number of rolling elements per row, then a strong possibility exists that indentation of the raceways will occur. In this situation, surface fatigue may not be a valid criterion of failure in view of the vibration which can obtain.

CLOSURE

The rolling bearing industry was among the first to use fatigue life as a design criterion. As a result the space-age term, *reliability*, which is synonymous with probability of survival, is familiar to rolling bearing users and manufacturers. The concepts of L_{10} or rating life and L_{50} or median life are and have been for a long time used as yardsticks of bearing performance. By means of the ANSI or ISO load rating standards, the rolling bearing industry has presented relatively uncomplicated methods to evaluate the rating life. These standards enjoy almost universal acceptance and they can be applied to compare the adequacy of diverse bearing types from different manufacturers for use in most engineering applications.

In certain applications, however, the simple use of ANSI formulas to rate rolling bearing performance can lead to inadequate estimates of fatigue life. These applications include those involving high speed, flexible inner and outer ring support structures, extremely slow speed, and unusual loading conditions. For these situations, methods to estimate fatigue life have been presented in this chapter. These methods rely on the same basic fatigue life analysis as the ANSI standards; however, they frequently require a more complete analysis of bearing contact angles, load distribution, speeds, lubrication, and so on. In these analyses the use of a computer is usually necessary.

The ANSI standards and the more complicated approach terminate with the calculation of L_{10} life based on a bearing fabricated from a good quality hardened bearing steel. Through intensive research the industry has developed and continues to develop new and superior bearing steels with much longer L_{10} lives. Improvement in rolling bearing manufacturing methods to produce more accurate bearings has also succeeded in yielding bearings of increased endurance. Therefore, in lieu of L_{10} life, it is becoming more and more feasible to design based on L_5, L_1, or even "no failure" life, that is, L_0. In fact, the aerospace industry usually demands this type of bearing performance.

The Lundberg–Palmgren theory, on which standard rolling bearing load ratings and fatigue life predictions are based, has served industry well for more than 30 years. Recent endurance test investigations, however, have shown the need for improvement. Consideration of the pertinent failure stress, stressed volume of material, rate of application of stress, effect of surface shear stresses, endurance limit and so on is required. Chapter 23 details an extension of the Lundberg–Palmgren theory that permits the accurate consideration of these effects.

REFERENCES

18.1. G. Lundberg and A. Palmgren, "Dynamic Capacity of Rolling Bearings," *Acta Polytech. Mech. Eng. Ser.* **1,** R.S.A.E.E., No. 3, 7 (1947).

18.2. T. Martin, S. Borgese, and A. Eberhardt, "Microstructural Alterations of Rolling Bearing Steel Undergoing Cyclic Stressing," *ASME Preprint 65-WA/CF-4*, Winter Annual Meeting, Chicago (November 1965).

18.3. P. Hoel, *Introduction to Mathematical Statistics*, 2nd ed., Wiley (1954).

18.4. W. Weibull, "A Statistical Representation of Fatigue Failure in Solids," *Acta Polytech. Mech. Eng. Ser.* **1,** R.S.A.E.E., No. 9, 49 (1949).

18.5. W. Weibull, "A Statistical Theory of the Strength of Materials," *Proc. R. Swedish Inst. Eng. Res.*, No. 151, Stockholm (1939).

18.6. T. Tallian, "Weibull Distribution of Rolling Contact Fatigue Life and Deviations Therefrom," *ASLE Trans.* **5**(1) (April 1962).

18.7. J. Lieblein, "A New Method of Analyzing Extreme-Value Data," Technical Note 3053, NACA (January 1954).

18.8. G. Lundberg and A. Palmgren, "Dynamic Capacity of Roller Bearings," *Acta Polytech. Mech. Eng. Ser.* **2,** R.S.A.E.E., No. 4, 96 (1952).

18.9. A. Palmgren, *Ball and Roller Bearing Engineering*, 3rd ed., Burbank, Philadelphia (1959).

18.10. American National Standards Institute, *American National Standard (ANSI/ AFBMA) Std 9-1990*, "Load Ratings and Fatigue Life for Ball Bearings."

18.11. J. Lieblein and M. Zelen, "Statistical Investigation of Fatigue Life of Ball Bearings," National Bureau of Standards, Report No. 3996 (March 28, 1955).

18.12. American National Standards Institute, *American National Standard (ANSI/ AFBMA) Std 11-1990*, "Load Ratings and Fatigue Life for Roller Bearings."

18.13. International Standards Organization, *International Standard ISO 281/1*, "Rolling Bearings—Dynamic Load Ratings and Rating Life—Part 1; Calculation Methods" (1991).

18.14. E. Zaretsky, E. Bamberger, T. Harris, W. Kacmarsky, C. Moyer, R. Parker, and J. Sherlock, *Life Adjustment Factors for Ball and Roller Bearings*, ASME Engineering Design Guide (1971).

18.15. H. Walp, T. Morrison, T. Tallian, and G. Baile, "The Effect of Material Variables on the Fatigue Life of AISI 52100 Steel Ball Bearings," *ASLE Trans.* **5**(2) (1962).

18.16. D. Koistinen, "Heat Treated Steel Article," U.S. Patent No. 3117041 (January 7, 1964).

18.17. C. Moyer and R. McKelvey, "A Rating Formula for Tapered Roller Bearings," SAE National Farm, Construction and Industrial Machinery Meeting, Milwaukee (September 12, 1962).

18.18. T. Harris, Discussion on [18.17], SAE National Farm, Construction and Industrial Machinery Meeting, Milwaukee (September 12, 1962).

18.19. T. Harris, "How to Compute the Effects of Preloaded Bearings," *Prod. Eng.* 84–93 (July 19, 1965).

18.20. A. Jones and T. Harris, "Analysis of Rolling Element Idler Gear Bearing Having a Deformable Outer Race Structure," *ASME J. Basic Eng.* 273–277 (June 1963).

18.21. T. Harris and J. Broschard, "Analysis of an Improved Planetary Gear Transmission Bearing," *ASME J. Basic Eng.* 457–462 (September 1964).

18.22. T. Harris, "Optimizing the Fatigue Life of Flexibly Mounted, Rolling Bearings," *Lubrication Eng.* 420–428 (October 1965).

18.23. T. Harris, "On the Effectiveness of Hollow Balls in High Speed Thrust Bearings," *ASLE Trans.* **11,** 290–294 (1968).

18.24. T. Harris, "The Effect of Misalignment on the Fatigue Life of Cylindrical Roller Bearings Having Crowned Rolling Members," *ASME J. Lubr. Tech.* 294–300 (April 1969).

18.25. T. Tallian, L. Sibley, and R. Valori, "Elastohydrodynamic Film Effects on the Load-Life Behavior of Rolling Contacts," *ASME Paper 65-LUBS-11*, ASME Spring Lubrication Symposium, New York (June 8, 1965).

18.26. J. Skurka, "Elastohydrodynamic Lubrication of Roller Bearings," *ASME Paper 69-LUB-18* (1969).

18.27. T. Tallian, "Theory of Partial Elastohydrodynamic Contacts," *Wear* **21,** 49–101 (1972).

18.28. T. Harris, "The Endurance of Modern Rolling Bearings," *AGMA Paper 269.01*, American Gear Manufacturers' Ass'n Roller Bearing Symposium, Chicago (October 26, 1964).

18.29. T. Tallian, "On Competing Failure Modes in Rolling Contact," *ASLE Trans.* **10**(4) 418–439 (October 1967).

18.30. M. Schouten, "*Einfluss Elastohydrodynamischer Schmierung auf Reibung, Verschliess und Lebendauer von Getrieben*," Doctoral Dissertation, Technische Hogeschool Eindhoven (1973).

18.31. M. Schouten, *Lebensduur van Overbrengingen*, TH Eindhoven, (November 10, 1976).

18.32. T. Harris, "Predicting Bearing Reliability," *Mach. Des.* 129–132 (January 3, 1963).

18.33. J. Wuttkowski and E. Ioannides, "The New Life Theory and Its Practical Consequences," *SKF Ball Bearing J., Special Issue* 6–11, (April 1989).

18.34. A. Jones, "A General Theory for Elastically Constrained Ball and Radial Roller Bearings under Arbitrary Load and Speed Conditions," *ASME J. Basic Eng.* 309–320 (January 1960).

19

ENDURANCE TESTING

LIST OF SYMBOLS

Symbol	Description	Units
$\mathcal{A}_n$	Life adjustment factors in rating life formula $n = 1$ for reliability $n = 2$ for materials $n = 3$ for operating conditions $n = 23$ for combined material and lubrication conditions	
C	Basic dynamic capacity	N (lb)
F_e	Equivalent applied load	N (lb)
h	Lubricant film thickness	μm (μin.)
Λ	Lubricant film parameter	
κ	Ratio of actual lubricant viscosity to minimum calculated viscosity	
σ	Rms surface roughness	μm (μin.)

GENERAL

In service even a rolling contact bearing that is well lubricated, properly aligned, and adequately protected from the effects of abrasives or moisture can fail from rolling contact fatigue if the loading is sufficiently high. It is important for the user to be able to predict reliably the length of service that can be achieved from a bearing in a specific application. As indicated in Fig. 1.3, the ability to make these types of predictions is hampered since rolling contact fatigue is probabilistic, similar to human life and the service life of light bulbs. Therefore, the life of any one bearing operating in a specific environment can differ significantly from that of an apparently identical unit. This distribution of rolling bearing fatigue data is illustrated in Fig. 18.5.

Over the years, statistical procedures have been established for the analysis of bearing fatigue data. A "rating" life system has been defined that uses two specific points on the failure distribution curve to compare the fatigue endurance of bearing designs. These points are the L_{10} life, or the life that 90% of the bearings can be expected to survive, and the L_{50} life, which 50% of the bearings can be expected to exceed. Concurrently, testing procedures have been developed to measure bearing fatigue life (endurance). As shown in Chapter 18, these experimental techniques have been utilized to collect a quantity of fatigue data which allowed the derivation of theoretical life prediction formulas based on a calculated bearing basic dynamic capacity, the equivalent radial load applied to the bearing, and a number of environmental and manufacturing-related factors. This basic concept is routinely used to select bearings that will yield the desired performances in specific applications.

Bearing endurance tests are used to evaluate bearing materials, new heat treatment processes, and improved forming or surface finishing techniques. The spread in experimental fatigue data and the limitations of the statistical analysis techniques require that many bearings be tested for a long time to obtain valid estimates of bearing life. Testing costs are directly related to the manufacturing costs of the test specimens and the length of the test process; therefore, conducting a life test series on full bearing assemblies is expensive. Effort has thus been expended to develop life testing techniques using simple partial bearings or single rolling contacts. To date, this approach has not been entirely successful since the results obtained from these element-type tests have not as yet been able to be reliably extrapolated to actual bearings. Nevertheless, these techniques have merit as initial screening tools as long as the limitations of the process are recognized.

This chapter discusses the concepts and philosophies of conducting endurance tests on complete rolling bearing assemblies and on elemental rolling contact configurations.

THEORETICAL BASIS OF LIFE TESTING

The ability to use life test data collected on a particular bearing type and size under a specific set of operating conditions to predict general bearing performance requires a systematic relationship between applied load and life. The basic form of this relationship was originally defined by Lundberg and Palmgren [19.1] in 1947. Recently, the mathematical form of this relationship has undergone modification, as discussed in detail in Chapter 23. For the purposes herein, however, the life formulation currently defined by the Anti-Friction Bearing Manufacturers Association (AFBMA) and the International Standards Organization (ISO) will be considered. This relationship is

$$L' = \mathcal{a}_1\mathcal{a}_2\mathcal{a}_3 \left(\frac{C}{F_e}\right)^p \tag{18.193}$$

The form of the load-life relationship for ball bearings was graphically illustrated in Fig. 18.8 as a straight line on a log-log chart. This formula provides the means to use experimental life data collected under one set of test conditions to establish projected bearing performance under a wide range of conditions.

The time to initiation of a fatigue-originated spalling failure on a rolling contact in a typical application is about 10 to 15 years. It is therefore obvious that any practical laboratory test sequence must be conducted under accelerated conditions if the necessary data are to be accumulated within a reasonable time. Several potential methods exist for test acceleration. Rolling contact failure modes exist, however, that compete in individual bearings to produce the final failure [19.2]. Care must be taken to ensure that the method of test acceleration does not change the natural failure mode. In an endurance test—that is, a test series conducted to establish the experimental life of a lot of bearings for comparison with theoretical lives and/or other test data—this means that the primary failure mode must remain fatigue related. Generally, there have been two methods of test acceleration used in most endurance testing: increasing the applied load levels and/or increasing the operating speed.

The experimental results obtained under increased load levels can be rather easily extrapolated to other conditions by using the basic load-life formula. Thus, this is probably the most widely used method of life test acceleration. It is important to note that consistency must be maintained with the basic assumptions used in the derivation of the life formulas. All of the stresses generated at and below the bearing raceways must remain within the elastic regime. Exceeding elastic stress levels will produce modifications to the load-life relationship as reported in [19.3] and illustrated in Fig. 19.1.

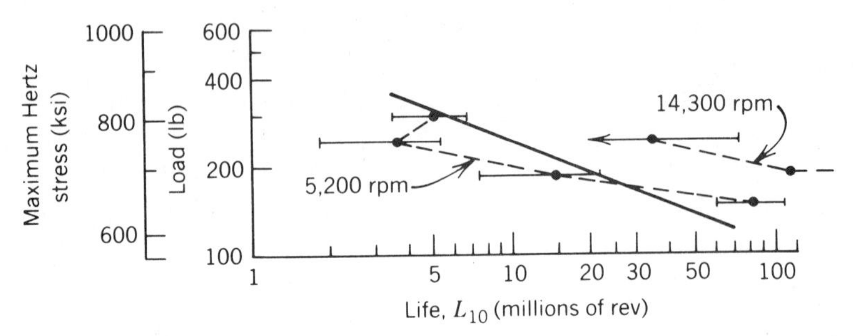

FIGURE 19.1. Life relationship deviation due to overloading. Theoretical (———). Experimental (– – –).

Intuitively, it is surprising that such extreme loads can produce apparent life increases of significant magnitude over theoretical predictions. This phenomenon is, however, readily explained by the plastic deformations produced in the raceways, creating a conforming contact pattern that then significantly reduces the existing stress levels. Testing in this regime produces results that are inconsistent with operating practice and cannot be reliably extrapolated. The practical load limit for testing acceleration is usually considered to be that load that produces a maximum Hertz stress of approximately 3300 N/mm^2 (475 ksi).

Some cases require special consideration in general life testing application. One involves testing self-aligning ball bearings. The spherical geometry of the outer ring raceway needed to provide the self-aligning capability produces circular point contacts between balls and raceway with very high stress levels. As loads are increased, these contacts develop plastic stresses more rapidly than is considered by the dynamic capacity calculation. It has been shown in past experiments [19.4] that the load should be no greater than $C/F_e = 8.0$ to prevent substantial plastic deformations in these bearings. Similarly, it would be anticipated that other types of bearings having nonstandard internal geometries could also experience significant plastic deformations at lower than anticipated loads. This situation must be evaluated before initiating an endurance test on specially designed bearings.

Roller bearings generally have contact ellipses that are much longer than they are wide; that is, they have line rather than point contacts. Roller and raceway geometries must be profiled in the axial direction to ensure that stress concentrations do not occur at the edges of these contacts and thus produce substantial plastic deformations there. The profiles used in standard bearings are often insufficient for the heavy loads used in an accelerated life test series. Therefore care must be taken to ensure that life reduction effects due to edge loading are not experienced during the testing process.

Even when edge stresses do not occur, roller bearings exhibit behavior under high loads that does not coincide with the stated load-life relationship. Life test series conducted on cylindrical roller bearings [19.5] have shown that under loads with C/F_e less than about 4.5, the load-life relationship of these bearings conforms to a power law function with an exponent closer to 4 rather than the stated 10/3 value. Modification factors presented in [19.5] allow the normal catalog capacities to be converted for use in the latter relationship to yield a more reliable estimate of the rating life. Also, extra care must be used when calculating the life-modifying effects of the lubricant film. Experimental indications are that roller bearings generate elastohydrodynamic (EHD) lubricant films thicker than those generally calculated. These thicker films will then produce greater than expected life enhancements. To insure the proper interpretation of life test data, these effects must be established before extrapolating the data to sets of operating conditions containing different speeds or lubricants.

The second usual method of accelerating life testing is by increasing operating speed. Operating a rolling bearing assembly faster produces a more rapid accumulation of stress cycles, but not necessarily a shorter test time. The increased operating speed also produces thicker EHD lubricant films, which then enhance bearing endurance (i.e., increase the adjusted rating life). The life enhancement effect may overshadow the increased stressing rate, and test times will be increased.

As the operating speeds are increased to even higher speed levels, other life-confounding effects can come into play. Under these operating conditions ball or roller centrifugal loads are increased, which causes increased outer raceway loading, increased clearance, and more concentrated inner raceway loading that reduce assembly life, as shown by Fig. 18.28.

The magnitude of this effect increases rapidly, since centrifugal force varies with the square of the speed. Bearing size also has a major effect, producing variations in both rolling element mass and radius of rotation. These effects are particularly important in angular-contact ball bearings, where the centrifugal forces alter the outer ring contact angle, producing even more significant life alteration effects. A parameter often used to express the severity of bearing speed conditions is dN, the product of the bearing bore diameter expressed in millimeters and the rotational speed of the bearing in revolutions per minute. It is normal to consider high-speed bearings as those with dN values of 1 million or more. For bearing operation under high-speed conditions, sophisticated analytical techniques are required to reliably calculate bearing rating lives for comparison with collected life data.

Using speed to accelerate bearing test programs has its limitations. Functional speed limitations exist in standard bearing designs because of the stamped metal or molded plastic cage designs, which are inadequate for high-speed operation. Excessive heat generation rates may occur at the raceway contacts, which have been designed primarily for maximum load-carrying capabilities at lower speeds, and component precision may be altered due to the dynamic loading of high-speed operation. System operating effects can also produce significant life effects on high-speed bearings. Some effects are insufficient cooling or the inadequate distribution of the cooling medium, creating thermal gradients in the bearings that affect internal clearances and geometries. Higher operating speeds will also produce higher bearing operating temperature levels. The lubricants used then must be capable of sustaining extended exposure to these high temperatures without suffering degradation. The conduct of high-speed life tests requires extra care to ensure that the failures obtained are fatigue related and not precipitated by some speed-originated performance malfunction.

In Chapter 12 it was shown that rolling contacts in ball and roller bearings operate in the regime of elastohydrodynamic (EHD) lubrication and that the thickness of the lubricant film generated by EHD lubrication is a strong function of bearing internal speeds, increasing as these speeds are increased. Furthermore in Chapter 18 it was shown that the fatigue life of a rolling bearing is a function of the thicknesses of the lubricant films it generates compared to the *roughnesses* of the rolling contact surfaces. Bearing fatigue life tends to increase as the surfaces in rolling contact are effectively separated by lubricant films. The adequacy of lubrication is currently expressed as Λ, the ratio of the lubricant film thickness to the composite rms roughness of the *contacting surfaces*. Values of Λ in excess of 3 tend to yield extended bearing fatigue life. To ensure an adequate lubricant film, a sufficient lubricant supply must be available to preclude lubricant *starvation* (see Chapter 12). Lubricant starvation is a particular consideration when performing endurance testing with grease lubrication at high speeds.

In Chapters 18 and 23, the means to quantify the lubrication-associated effect of speed on bearing endurance have been demonstrated. Particularly in the regime of marginal lubrication, the effect is complex owing to the interactions of rolling component surface finishes and chemistry, lubricant chemical and mechanical properties, lubrication adequacy, contaminant types, and contamination levels. Considering adequate lubricant supply and minimal contamination, increasing testing speed to the point that complete separation of rolling contact surfaces is achieved has tended to extend test duration appreciably, thus thwarting the desired acceleration of endurance testing. Testing at speeds slow enough to cause operation in the marginal lubrication regime can achieve shortened testing time; however, the above-indicated side effects must be considered in the evaluation of test results.

PRACTICAL TESTING CONSIDERATIONS

As indicated earlier, an individual bearing can fail for several reasons. The results of an endurance test series, however, are only meaningful when the test bearings fail by fatigue-related mechanisms. The experimenter must control the test process to ensure that this happens. Some of the other failure modes that can be experienced are discussed in detail in [19.2]. The following paragraphs deal with a few specific failure types that can affect the conduct of a life test sequence.

Earlier the importance of bearing lubrication was discussed from the standpoint of EHD lubricant film generation. There are, however, other lubrication-related effects that can affect the outcome of the test series. The first is lubricant contamination. The overall thickness of the lubricant film developed at the contacts in an operating bearing is about 0.2–0.5 μm (10–20 μin.). Solid particles larger than the film can be mechanically trapped in the contact region and damage the rolling element surfaces, leading to substantially foreshortened endurances, as indicated by Sayles and MacPherson [19.6].

Thus, filtration of the fluid lubricant used to the desired level is necessary to ensure meaningful test results. The desired level is determined by the application, which the testing purports to approximate. If this degree of filtration is not provided, effects of contamination must be considered when evaluating test results.

The moisture content in the lubricant is another important consideration. It has long been known that quantities of free water in the oil have a detrimental effect on the life of bearings through the surface deterioration produced by corrosion. Recent studies [19.7], however, have shown that water levels as low as 50 to 100 parts per million (ppm) could also have a detrimental effect even with no evidence of corrosion on the surfaces. A working limit of 40 ppm appears necessary to eliminate these life reduction effects. Filter manufacturers are currently developing hardware for this purpose, and it has performed satisfactorily in a laboratory environment. The actual long-term capabilities of these systems in service are not now known, however. Also some lubricant producers now market commercial oils that contain water absorption additives. Again the long-term effects of using these types of fluids need to be established. Thus moisture control in test lubrication systems continues to be a concern, and the effect of moisture needs to be considered during the evaluation of life test results.

The chemical composition of the test lubricant also requires consideration. Most commercial lubricants contain a number of producer proprietary additives developed for specific purposes: for example, to provide antiwear properties, to achieve extreme pressure and/or thermal stability, and to provide boundary lubrication in cases of marginal lubricant films. These additives can also affect the endurance of rolling bearings, either immediately or after experiencing time-related degradation. Care must be taken to ensure that the additives included in the test lubricant will not suffer excessive deterioration as a result of accelerated life test conditions. Also for consistency of results and comparing life test groups, it is good practice to utilize one standard test lubricant from a particular producer for the conduct of all general life tests.

The statistical nature of rolling contact fatigue requires many test samples to obtain a reasonable estimate of life. A bearing life test sequence thus needs a long time. A major job of the experimentalist is to ensure the consistency of the applied test conditions throughout the entire test period. This process is not simple because subtle changes can occur during the test period. Such changes might be overlooked until their effects become major. At that time it is often too late to salvage the collected data, and the test must be redone under better controls.

For example, the stability of the additive packages in a test lubricant can be a source of changing test conditions. Some lubricants have been known to suffer additive depletion after an extended period of operation. The degradation of the additive package can alter the EHD lubrication conditions in the rolling contacts, altering bearing life. Generally, the normal chemical tests used to evaluate lubricants do not determine the conditions of the additive content. Therefore if a lubricant is used for endurance testing over a long time, a sample of the fluid should be returned to the producer at regular intervals, say annually, for a detailed evaluation of its condition.

Adequate temperature controls must also be employed during the test. The thickness of the EHD lubricant film is sensitive to the contact temperature. Most test machines are located in standard industrial environments where rather wide fluctuations in ambient temperature are experienced over a period of a year. In addition, the heat generation rates of individual bearings can vary as a result of the combined effects of normal manufacturing tolerances. Both of these conditions produce variations in operating temperature levels in a lot of bearings and affect the validity of the life data. A means must be provided to monitor and control the operating temperature level of each bearing to achieve a degree of consistency. A tolerance level of $\pm 3C$ is normally considered adequate for the endurance test process.

The deterioration of the condition of the mounting hardware used with the bearings is another area requiring constant monitoring. The heavy loads used for life testing require heavy interference fits between the bearing inner rings and shafts. Repeated mounting and dismounting of bearings can produce damage to the shaft surface, which in turn can alter the geometry of a mounted ring. The shaft surface and the bore of the housing are also subject to deterioration from fretting corrosion.

Fretting corrosion results from the oxidation of the fine wear particles generated by the vibratory abrasion of the surfaces, which is accelerated by the heavy endurance test loading. This mechanism can also produce significant variations in the geometry of the mounting surfaces, which can alter the internal bearing geometry. Such changes can have a major effect in reducing bearing test life.

The detection of bearing failure is also a major consideration in a life test series. The fatigue theory considers failure as the initiation of the first crack in the bulk material. Obviously there is no way to detect this occurrence in practice. To be detectable the crack must propagate to the surface and produce a spall of sufficient magnitude to produce a marked effect on an operating parameter of the bearing: for example, noise, vibration, and/or temperature. Techniques exist for detecting failures in application systems. The ability of these systems to detect early signs of failure varies with the complexity of the test system, the type of bearing under evaluation, and other test conditions. Currently no single system exists that can consistently provide the failure discrimination necessary for all types of bearing life tests. It is then necessary to select a system that will repeatedly terminate machine operation with a consistent minimal degree of damage.

The rate of failure propagation is therefore important. If the degree of damage at test termination is consistent among test elements, the only variation between the experimental and theoretical lives is the lag in failure detection. In standard through-hardened bearing steels the failure propagation rate is quite rapid under endurance test conditions, and this is not a major factor, considering the typical dispersion of endurance test data and the degree of confidence obtained from statistical analysis. This may not, however, be the case with other experimental materials or with surface-hardened steels or steels produced by experimental techniques. Care must be used when evaluating these latter results and particularly when comparing the experimental lives with those obtained from standard steel lots.

The ultimate means of ensuring that an endurance test series was adequately controlled is the conduct of a post-test analysis. This detailed examination of all the tested bearings uses high-magnification optical inspection, higher-magnification scanning electron microscopy, metallurgical and dimensional examinations, and chemical evaluations as required. The characteristics of the failures are examined to establish their origins and the residual surface conditions are evaluated for indications of extraneous effects that may have influenced the bearing life. This technique allows the experimenter to ensure that the data are indeed valid. A comprehensive atlas containing typical photographs of numerous bearing failure modes was published [19.8] to provide guidance for these types of determinations.

FIGURE 19.2. Typical fatigue spalling failure (from ref. [19.8]).

FIGURE 19.3. Bearing failure precipitated by misalignment (ref. [19.8]).

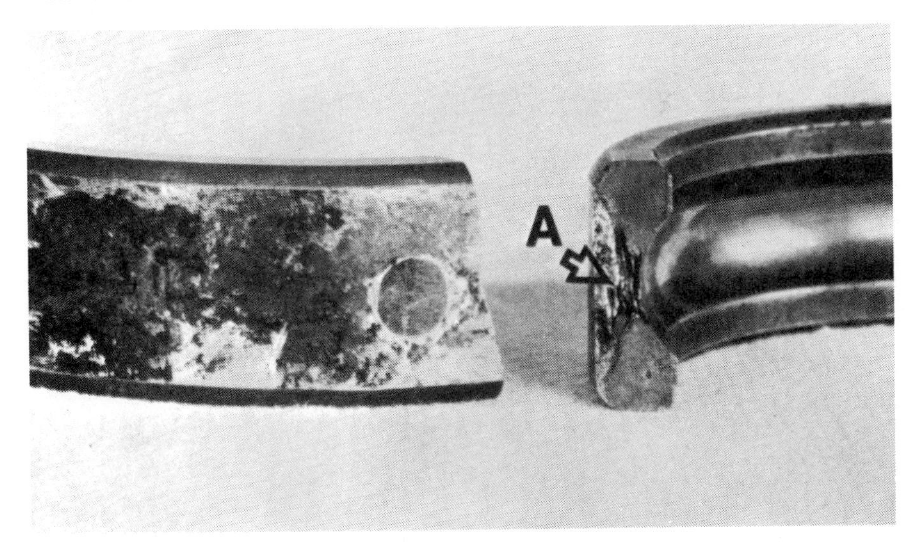

FIGURE 19.4. Spall precipitated by fretting corrosion crack (ref. [19.8]).

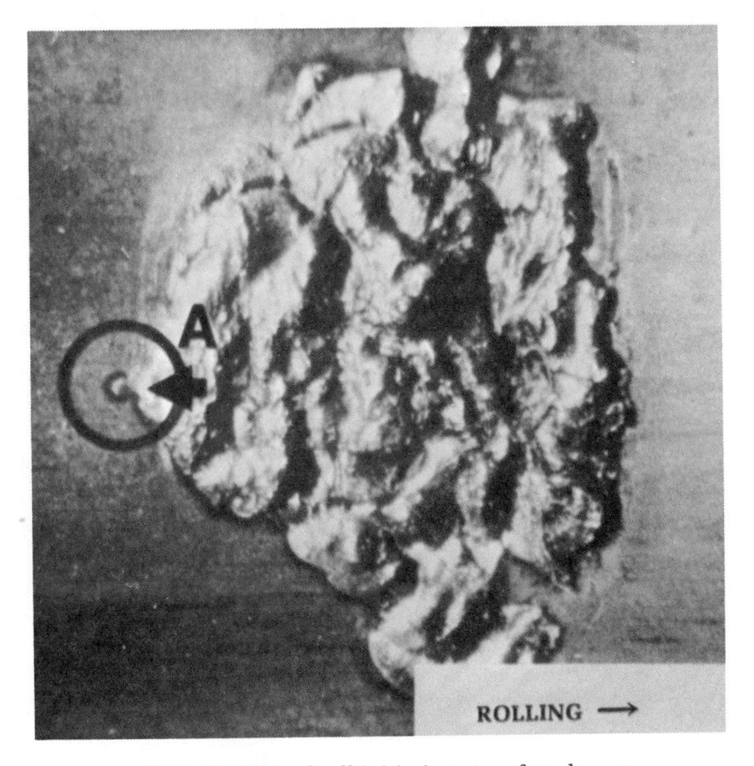

FIGURE 19.5. Spall initiating at surface damage.

FIGURE 19.6. Failure resulting from thermal imbalance (ref. [19.8]).

The post-test analysis is, by definition, after the fact. To provide control throughout the test series and to eliminate all questionable areas, the experimenter should conduct a preliminary study whenever a bearing is removed from the test machine. In this portion of the investigation each bearing is examined optically at magnifications up to 30× for indications of improper or out-of-control test parameters. Examples of the types of indications that can be observed are given in Figs. 19.2–19.6.

Figure 19.2 illustrates the appearance of a typical fatigue-originated spall on a ball bearing raceway. Figure 19.3 contains a spalling failure on the raceway of a roller bearing that resulted from bearing misalignment, and Fig. 19.4 contains a spalling failure on the outer ring of a ball bearing produced by fretting corrosion on the outer diameter. Figure 19.5 illustrates a more subtle form of test alteration, where the spalling failure originated from the presence of a debris dent on the surface. Figure 19.6 gives an example of a totally different failure mode produced by the loss of internal bearing clearance due to thermal unbalance of the system.

The last four failures are not valid fatigue spalls and indicate the need to correct the test methods. Furthermore, these data points would need to be eliminated from the failure data to obtain a valid estimate of the experimental bearing life.

TEST SAMPLES

Specific requirements have been established for a test sample to be used in an endurance test sequence. The statistical techniques used to evaluate the failure data require that the bearings be statistically similar assemblies. Therefore the individual components must be manufactured in the same processing lot from one heat of material. Generally, it is considered prudent to manufacture the total bearing assembly in this manner; however, when highly experimental materials or processes are considered, this is often not cost effective or even possible. In those cases the inner ring, the most critical element in a bearing assembly from a fatigue point of view, can be used as the test element, with the other components being manufactured from standard material. The effects of failures occurring on the other parts can be eliminated during analysis of the test data. There is some risk in this approach because it is possible that too many failures could occur on these nontest parts, rendering it impossible to calculate an accurate life estimate for the material under evaluation. In the cases cited, however, this risk is small because an initial result indicating the superior performance of an experimental process is usually sufficient to justify continued developmental effort even without a firm numerical estimate. In any case, additional life tests would be required to establish the magnitude of the expected lot-to-lot variation before adopting a new material or implementing a new manufacturing process.

The number of bearings to be tested and the test strategy to be employed must also be carefully considered. Statistical analysis provides a numerical estimate of the value of the experimental life enclosed by upper- and lower-bound estimates at specified confidence levels. The precision of the experimental life estimate can be defined by the ratio of these upper and lower confidence limits, and the experimental aim is to minimize this spread. The magnitude of the confidence interval decreases as the size of the test lot is increased; the cost of conducting the test series also increases with test lot size. Therefore, the degree of precision required in the test result should be established during the planning stages to define the size of the test lot to achieve the required result.

The test strategy employed also affects testing precision. The classical method of performing endurance tests is to use one large group running each individual bearing to failure. This process is time consuming, but it provides the best experimental estimates of both the L_{10} and L_{50} lives. Primary interest is, however, in the magnitude of the experimental L_{10}, so considerable time savings can be achieved by curtailing the test runs after a finite operating period equal to at least three times the achieved experimental L_{10} life. Recently it has been shown that additional savings in test time accompanied by increases in test precision can be obtained by using a sudden death test strategy [19.9]. In this test approach the original test lot is subdivided into smaller groups of equal sizes. Each subgroup is then run as a unit until one bearing fails, at which time the testing of that subgroup is terminated. Figure 19.7 illustrates the effect of both lot size and test strategy on the precision of life test estimates obtainable from an endurance testing series.

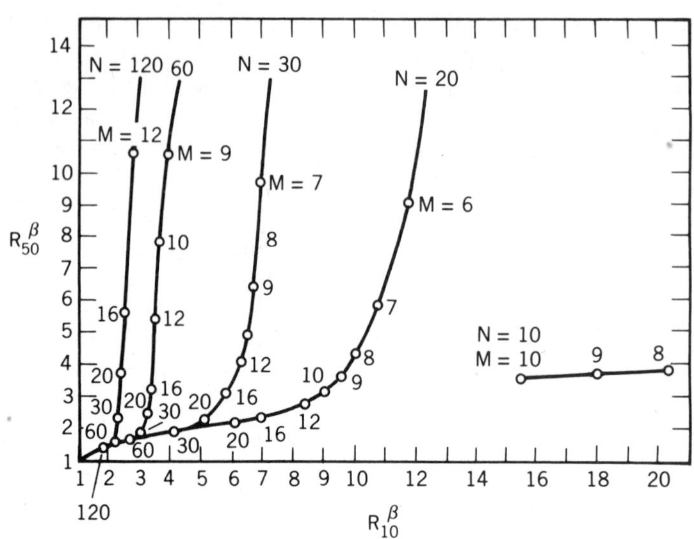

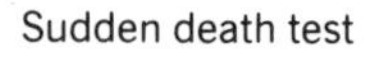

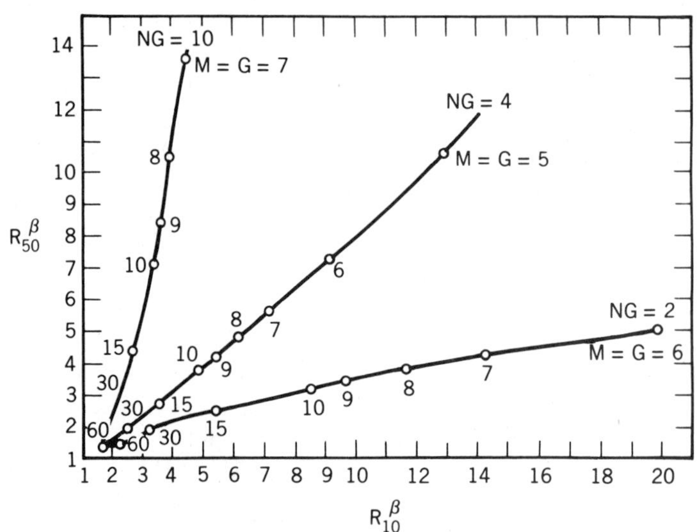

N = size of test series (total number of bearings)
M = number of bearing failures
G = number of groups
NG = group size (all groups are run to one failure)
β = true Weibull slope

FIGURE 19.7. Effect of lot size on confidence of life test results. N = size of test series (total number of bearings). M = number of bearing failures. G = number of groups. NG = group size (all groups are run to one failure). β = true Weibull slope (from ref. [19.9]).

To provide an accurate life estimate for the variable under evaluation, the experimenter must be sure that the test bearings are free from material and manufacturing defects and that all parts conform to established dimensional and form tolerances. Although this is an obvious requirement, it is not always easy to attain. Experimental materials might respond quite differently to standard manufacturing processes, or they could require unique processing steps that are not yet totally defined. Experimental manufacturing processes require additional verification, or their use might produce unexpected variations in related metallurgical or dimensional parameters. Therefore, adequate test control is achieved by detailed pretest auditing of the test parts to supplement the standard in-process evaluations. Tables 19.1 and 19.2 contain lists of

TABLE 19.1. Typical Metallurgical Audit Parameters

100% Nondestructive Tests: Ring Raceways Only

Magnaflux for near-surface materials defects
Etch inspection for surface processing defects

Sample Destructive Tests: All Components

Microhardness to 0.1 mm (0.004 in.) depth below raceway surfaces
Microstructure to 0.3 mm (0.012 in.) depth below raceway surfaces
Retained austenite levels
Fracture grain size
Inclusions ratings

TABLE 19.2. Typical Dimensional Audit Parameters

100% Assembled Bearings

Radial looseness
Average and peak vibration levels

Statistical Sample of Ring Grooves

Diameter and waviness
Radius and form
Cross-groove surface texture

Statistical Sample of Balls

Diameter and out-of-round
Set size variation
Waviness
Surface texture

those metallurgical and dimensional parameters considered mandatory in a typical pretest audit, as well as an indication of the number of samples that need to be checked in each case.

These lists are not complete. Other parameters could be evaluated beneficially if time and money permit.

TEST RIG DESIGN CONSIDERATIONS

Some specific characteristics are desired in an endurance test system to achieve the control requirements of a life test series. An individual test run uses a long time, so the test machine must be capable of running unattended without experiencing variation in the applied test parameters, such as load(s), speed, lubrication conditions, and operating temperature. The basic test system components that could also be subject to fatigue, such as load support bearings, shafts, and load linkages, should be many times stronger than the test bearings so that test runs can be completed with the fewest interruptions from extraneous causes. The assembly of the test machine should have a minor influence on the experimental test conditions to minimize variations between individual test runs. For example, the alignment of the test bearing relative to the shaft should be automatically ensured by the assembly of the test housing. If not, a simple direct means of monitoring and adjusting this parameter must be provided. Again, since a test series requires multiple setups, easy assembly and disassembly of the test system is desirable to minimize turnaround time and manpower requirements for test bearing changes. In addition, the test system must be easy to maintain and should be capable of operating reliably and efficiently for years to ensure the long-term compatibility of test results. Basically, design simplicity is a key ingredient in meeting all of these demands. A more comprehensive discussion of the design philosophy of life test rigs has been published in [19.10]. Figure 19.8 illustrates some of the typical endurance test rig configurations discussed there.

The application of these design concepts to actual endurance test systems will be briefly addressed. Figure 19.9 is the schematic of an SKF rig for testing 35–50-mm-bore ball and roller bearings under radial, axial, or combined loads. Figure 19.10 is an actual photograph.

Operating speed may be varied within limits to achieve a given test condition and bearing lubrication can be provided by grease, sump oil, circulating oil, or air-oil mist.

Practical life test rig designs will vary, depending on the type of bearing to be tested and its normal operating mode. For example, Fig. 19.11 illustrates a four-bearing test rig concept used in the life testing of tapered roller bearings [19.11]. In this instance, while testing is conducted under an externally applied radial load, the bearing also sees an inter-

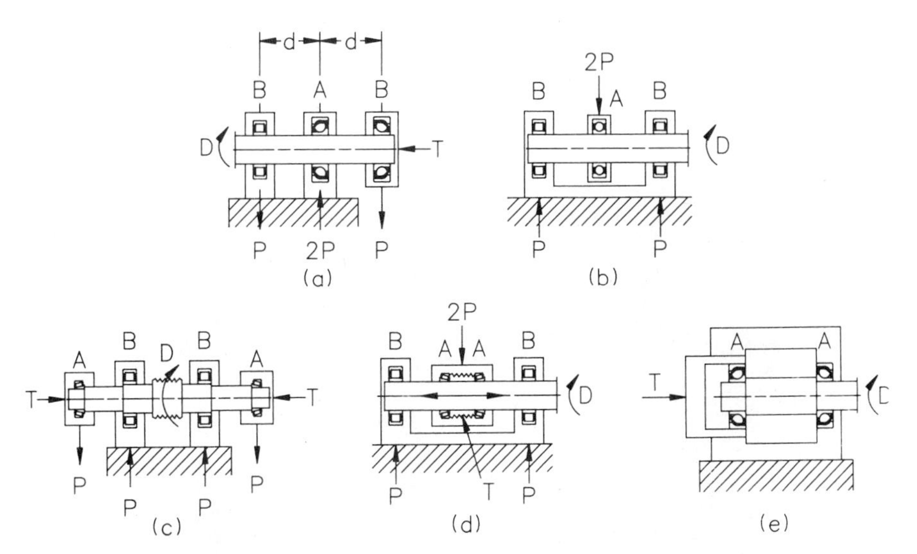

FIGURE 19.8. Typical test rig configuration. A = test bearing, B = load bearing, T = thrust load, P = radial load, D = drive (from ref. [19.8]).

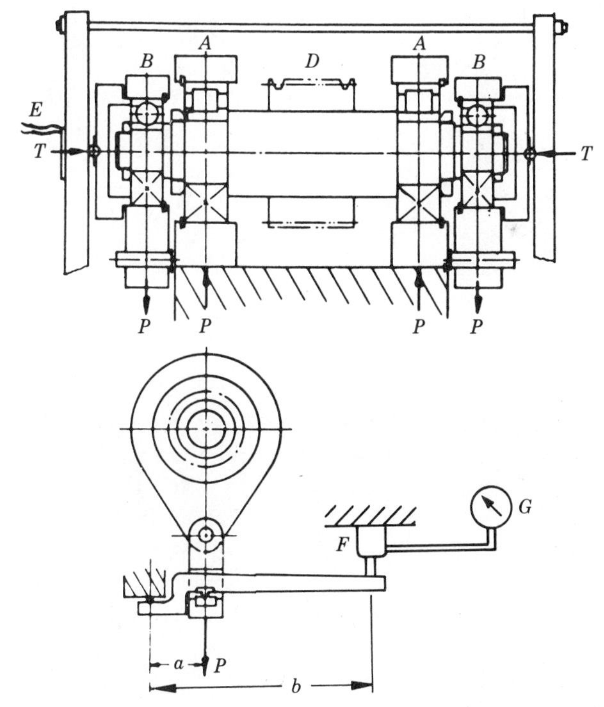

FIGURE 19.9. Schematic layout of endurance test system. A = load bearing, B = test bearing, P = radial load, T = thrust load, D = drive, E = strain gage, F = hydraulic cylinder, G = pressure gage (from ref. [19.11]).

FIGURE 19.10. R2 tester.

nally induced axial load. The size of this latter load is a function of the magnitude of the applied radial load, the fixed axial locations of the bearing cups and cones in the test housing, and the basic internal design of the test bearings. Figure 19.12 shows a rig using this design concept.

Tests are often conducted to define the life of bearings as used in specific applications. They are frequently called life or endurance tests, but, more correctly, they are extended duration performance tests. The same basic test practices and rig considerations are required for these tests, but some modifications of philosophy are required to simulate the major operating parameters of the application while achieving realistic test acceleration. An example of this type of tester is the SKF-developed "A-frame" tester shown in Fig. 19.13, which is used for evaluating automotive wheel hub assemblies.

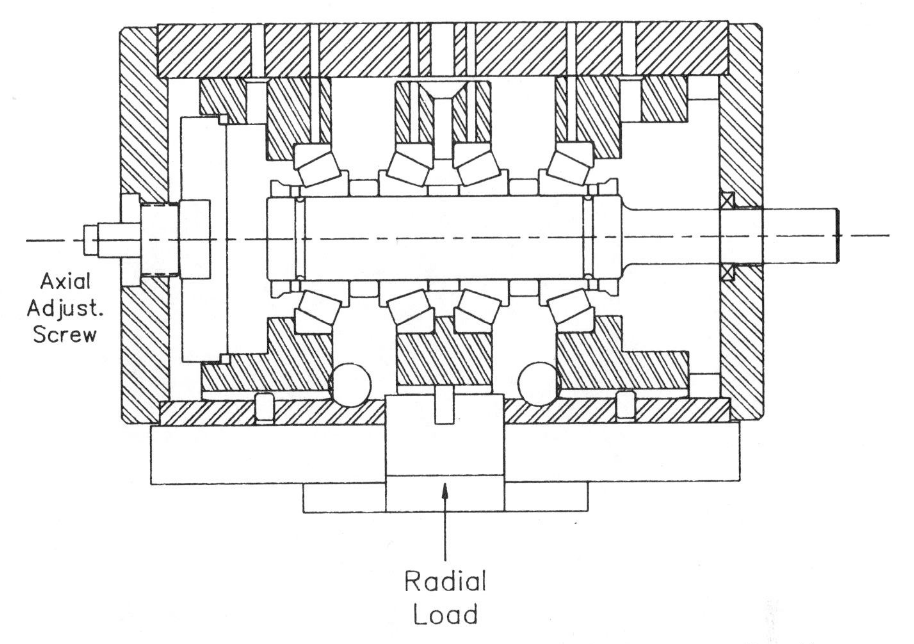

FIGURE 19.11. Schematic layout of tapered roller bearing test configuration.

FIGURE 19.12. Tapered roller bearing tester.

FIGURE 19.13. A-frame hub unit tester.

This tester simulates an automotive wheel bearing environment by using actual car mounting hardware, combined radial and axial loads applied at the tire periphery to produce moment loads on the bearing assembly, grease lubrication, and forced-air cooling. Dynamic wheel loading cycles equivalent to those produced by vehicle lateral loading are applied cyclically to simulate a critical driving sequence. Testing is conducted in the sudden death mode so that hub unit life in this simulated environment can be calculated from standard life test statistics. This test provides a way to compare the relative performance of automotive wheel support designs using life data generated under conditions similar to those of actual applications.

ELEMENT TESTING

Because numerous test samples are required to obtain a usable experimental life estimate, conducting an endurance test series on full-scale bearings is expensive. The identification of simpler, less costly, life testing methods has therefore been a longstanding goal. The use of elemental test configurations would appear to offer a solution to this need. In this approach, a test specimen having a simplified geometry (e.g., flat washer, rod, or ball) is used, and rolling contact is developed at a few test locations. In the ultimate case life test data that can be extrapolated to a real bearing situation are obtained at a fraction of the calendar time and cost. Unfortunately, this has not yet proven to be the case. The simplified geometries of the test configurations usually produce very small contact areas that are easily overstressed in the attempt to accelerate the testing process. These contacts are also difficult to lubricate adequately and are subject to accelerated rates of surface deterioration. These factors have inhibited the direct extrapolation of the data to actual bearing situations. Nevertheless, element test techniques have proven to be useful for the ranking of material performance in initial screening test sequences or in adverse environments, such as extremely high or low temperature, oxidizing atmospheres, and vacuum. Therefore, some discussion of these techniques is warranted. Caution must always be used, however, when using these techniques because the precision of the ranking process is open to question. Performance reversals have been seen in screening test results when the materials have been retested in actual bearings.

The oldest and perhaps most widely used element test configuration is the rolling four-ball machine developed in the early 1950s [19.12]. This system uses four 12.7-mm (0.5-in.) diameter balls to simulate an angular-contact ball bearing operating with a vertical axis under a pure thrust load. One ball is the primary test element serving as the inner ring of the bearing assembly. It is supported in pyramid fashion on the remaining three balls, which rotate freely in a conforming cup at a predetermined contact angle. A modification of this test method, the rolling five-ball tester, was subsequently developed at NASA Lewis Research Laboratories [19.13], and it uses four balls in the intermediate position. This latter system, illustrated in Fig. 19.14 has been used to generate an extensive amount of life test data on standard and experimental bearing materials.

Another widely used element test system is the RC (rolling contact) tester developed at General Electric [19.14] (Fig. 19.15). The test element in this configuration is a 4.76-mm (0.1875-in.) rod rotating under load between two 95.25-mm (3.75-in.) diameter disks. The rod can be axially repositioned to achieve a number of rolling contact tracks on a single

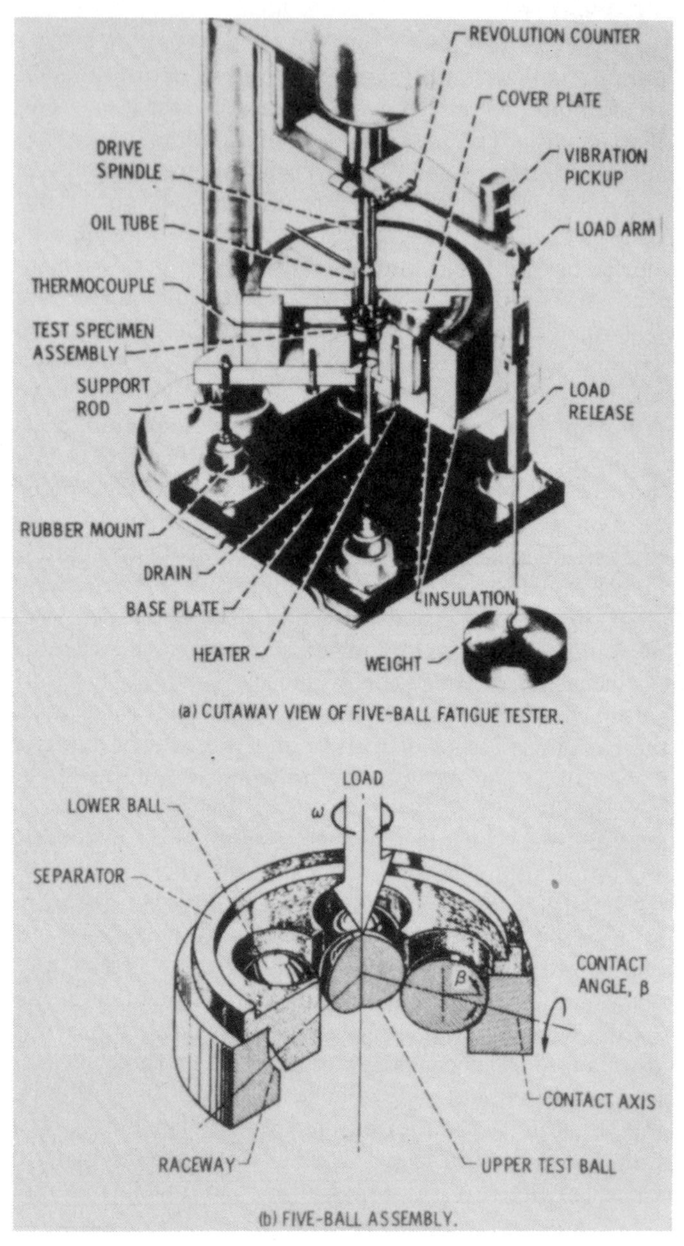

FIGURE 19.14. NASA rolling five-ball test system.

FIGURE 19.15. GE/polymet rolling contact disc machine.

test specimen. Unfortunately, this configuration is not as cost effective as it first appears. Stress concentrations will occur at the edges of the rod contact unless the discs are profiled in the axial direction. This significantly increases the cost of manufacturing the discs. During operation, fatigue failures on the rod also tend to damage the disc surfaces, requiring that these be refinished at regular intervals.

An interesting modification of this element test concept has been recently developed to eliminate the need for the expensive test discs [19.15]. In this version, illustrated in Fig. 19.16, three standard balls supported in standard tapered roller bearing outer rings serve the function of the discs and increase the number of test contacts. This approach appears to offer significant advantages, but only a limited amount of test experience has been obtained to date. It is thus too early to establish the overall efficacy of this test method.

One other element test configuration warranting consideration is the single-ball tester developed at Pratt and Whitney Aircraft for evaluating balls used in aircraft gas turbine engine bearings [19.16]. This system, shown in Fig. 19.17, tests a 25.4 mm (approximately 1 in.) diameter ball

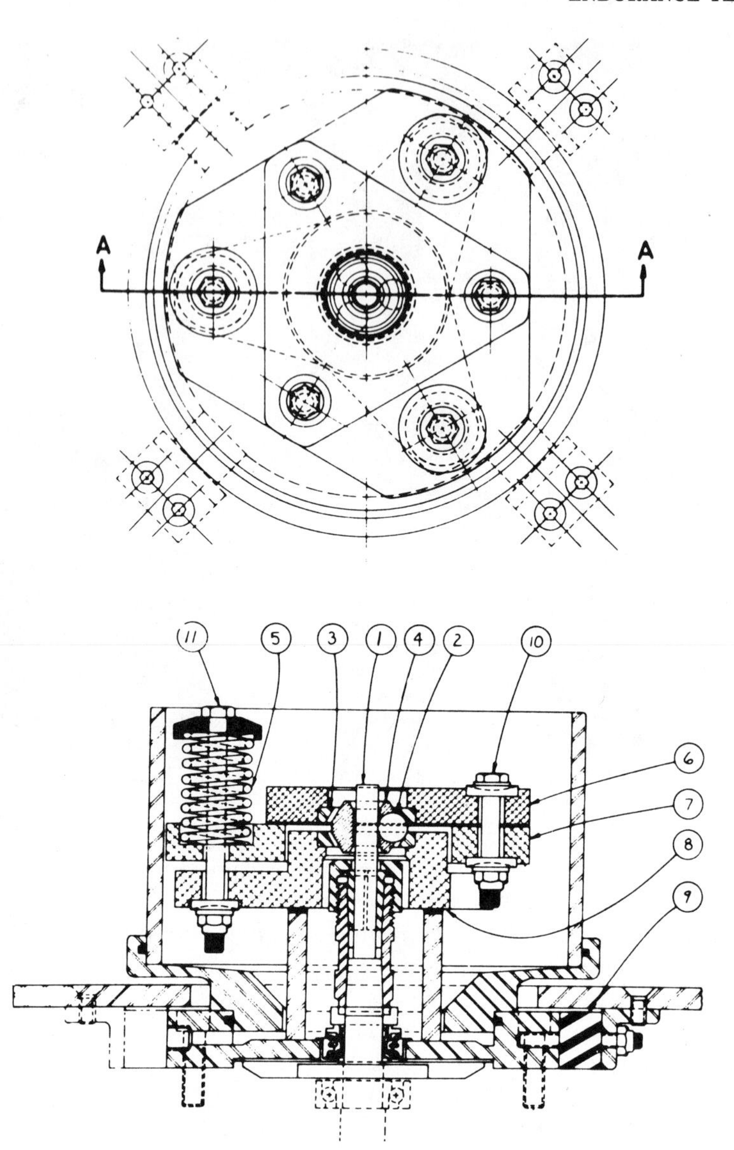

1. Specimen
2. Ball
3. Tapered bearing cup
4. Ball retainer
5. Compression spring
6. Upper cup housing
7. Spring retainer plate
8. Lower cup housing
9. Shock mount
10. Load application bolt
11. Spring calibration bolt

FIGURE 19.16. Ball-rod rolling contact fatigue tester.

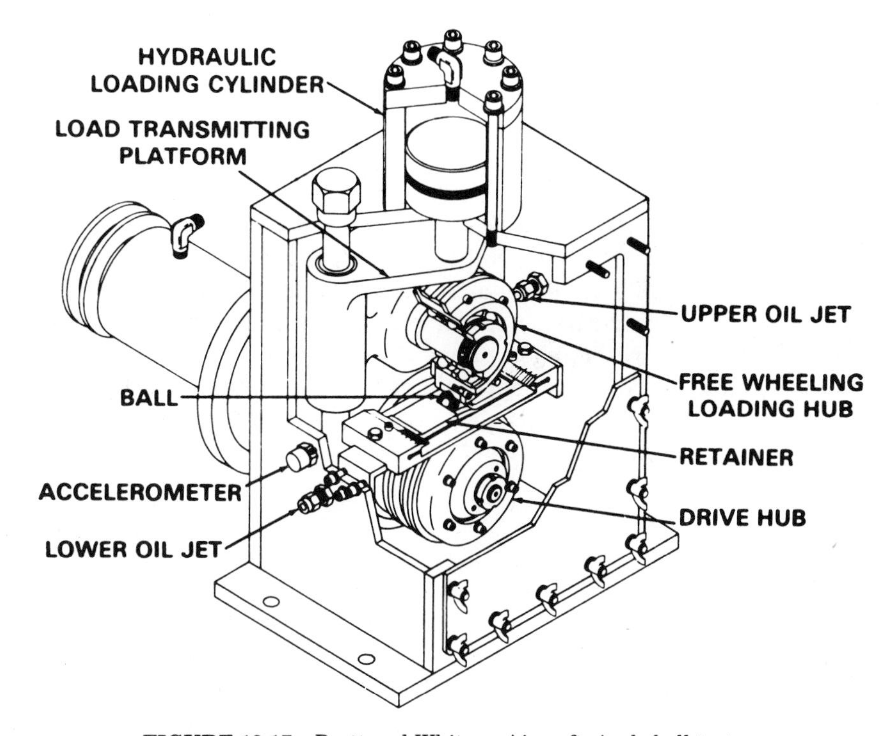

FIGURE 19.17. Pratt and Whitney Aircraft single-ball tester.

in two V-ring raceways with lubrication provided to simulate the application. It is reported that the correlation is very good between the results obtained on this rig and those achieved with balls in full-scale angular-contact bearings.

CLOSURE

In Chapter 18, it was demonstrated that although ball and roller bearing fatigue life rating and endurance formulas are founded in theory, they are semiempirical relationships requiring the establishment of various constants to enable their use. These constants, which depend upon the bearing raceway and rolling element materials, can be established only by appropriate testing. Because of the probabalistic nature of rolling bearing fatigue endurance, testing procedures necessarily require bearing and/or material populations of sufficient size to render the test results meaningful. Sample sizing effects are discussed in detail in Chapter 20.

At the present time, to establish sufficiently accurate rating formula constants, it appears necessary to test complete bearings or at least the most fatigue-vulnerable ring raceway. Since rolling element and raceway surface forming and finishing methods are influential in the fatigue endurance of the bearing, and since these methods vary according to component type and size, it is difficult to justify basing the fatigue endurance constants on rolling element, that is, ball or roller, fatigue test results. The same situation currently exists with regard to other types of fatigue testing, for example, rotating beam, torsion bar, and "push-pull." Nevertheless, improved rolling contact fatigue relationships discussed in Chapter 23 give reasonable hope that this can be accomplished in the not too distant future. Accordingly, the cost of endurance testing could be reduced by resorting to element-type testing in lieu of complete bearing testing. At the very least, element testing might be used to effectively screen various materials, material processing methods, and methods of finishing rolling element surfaces.

REFERENCES

19.1. G. Lundberg and A. Palmgren, "Dynamic Capacity of Rolling Bearings," *Acta Polytech. Mech. Eng. Ser. 1, R.S.A.E.E.* No. 3, 7 (1947).

19.2. T. Tallian, "On Competing Failure Modes in Rolling Contact," *ASLE Trans.* **10,** 418–439 (1967).

19.3. R. Valori, T. Tallian, and L. Sibley, "Elastohydrodynamic Film Effects on the Load Life behavior of Rolling Contacts," ASME Publication 65-LUBS-11 (1965).

19.4. G. Johnston, T. Andersson, E. Van Amerongen, and A. Voskamp, "Experience of Element and Full Bearing Testing Over Several Years," *Rolling Contact Fatigue Testing of Bearing Steels*, ASTM STP 771, J. J. Hoo, Ed., American Society for Testing Materials, Philadelphia (1982).

19.5. G. Lundberg and A. Palmgren, "Dynamic Capacity of Roller Bearings," *Acta Polytech. Mech. Eng. Ser. 2, R.S.A.E.E.*, No. 4, 96 (1952).

19.6. R. Sayles and P. MacPherson, "Influence of Wear Debris on Rolling Contact Fatigue," *Rolling Contact of Bearing Steels*, ASTM STP 771, J. J. Hoo, Ed., American Society for Testing Materials, Philadelphia (1982).

19.7. E. Fitch, *An Encyclopedia of Fluid Contamination Control*, Fluid Power Research Center, Oklahoma State University (1980).

19.8. T. Tallian, G. Baile, H. Dalal, and O. Gustafsson, *Rolling Bearing Damage, A Morphological Atlas*, SKF Industries, Inc., Philadelphia (1974).

19.9. T. Andersson, "Endurance Testing in Theory," *Ball Bearing J.* **217,** 14–23 (1983).

19.10. G. Sebok and U. Rimrott, "Design of Rolling Element Endurance Testers," ASME Publication 69-DE-24 (1969).

19.11. R. Hacker, "Trials and Tribulations of Fatigue Testing of Bearings," SAE Technical Paper Series 831372 (1983).

19.12. F. Barwell and D. Scott, *Engineering* **182**(4713), 9–12 (1956).

19.13. E. Zaretsky, R. Parker, and W. Anderson, "NASA Five-Ball Tester—Over 20 years of Research," *Rolling Contact Fatigue Testing of Bearing Steels*, ASTM STP 771, J. J. Hoo, Ed., American Society for Testing Materials, Philadelphia (1982).

19.14. E. Bamberger and J. Clark, "Development and Application of the Rolling Contact Fatigue Test Rig," *Rolling Contact Fatigue Testing of Bearing Steels*, ASTM STP 771, J. J. Hoo, Ed., American Society for Testing Materials, Philadelphia (1982).

19.15. D. Glover, "A Ball-Rod Rolling Contact Fatigue Tester," *Rolling Contact Fatigue Testing of Bearing Steels*, ASTM STP 771, J. J. Hoo, Ed., American Society for Testing Materials, Philadelphia (1982).

19.16. P. Brown, Jr., G. Bogardus, R. Dayton, and D. Schulze, "Evaluation of Powder-Processed Metals for Turbine Engine Ball Bearings," *Rolling Contact Fatigue Testing of Bearing Steels*, ASTM STP 771, J. J. Hoo, Ed., American Society for Testing Materials, Philadelphia (1982).

20

STATISTICAL METHODS TO ANALYZE ENDURANCE

LIST OF SYMBOLS

Symbol	Description
$f(x)$	Probability density function
$F(x)$	Cumulative distribution function
h	Hazard rate
H	Cumulative hazard rate
i	Failure order number
k	Number of samples
k_p	$-\ln(1-p)$
l	Number of subgroups in a sudden death test
m	Sample size of a sudden death subgroup
n	Sample size
p	A probability value
q	A probability value
$q(l, m, p)$	A pivotal function used for sudden death test analysis
r	Number of failures in a censored sample
R	Ratio of upper to lower confidence limits for β

Symbol	Description
$R_{0.50}$	Median ratio of upper to lower confidence limits for $x_{0.10}$
$t_1(r, n, k)$	Pivotal function for testing for differences among k estimates of $x_{0.10}$
$u(r, n, p)$	Pivotal function for setting confidence limits on x_p
$u_1(r, n, p, k)$	Pivotal function for setting confidence limits on x_p using k data samples
$v(r, n)$	Pivotal function for setting confidence limits on β
$v_1(r, n, k)$	Pivotal function for setting confidence limits on β using k data samples
$w(r, n, k)$	Pivotal function for testing whether k Weibull populations have a common β
x	A random variable
x_p	The pth percentile of the distribution of the random variable x
β	The Weibull shape parameter
η	The Weibull scale parameter

GENERAL

Many statistical distributions have been used to describe the random variability of the life of manufactured products. Such choices can be variously justified. For example, if a product has a reservoir of a substance that is used up at a uniform rate through the product's life, and if the initial supply of the substance varies from item to item according to a normal (Gaussian) distribution, then the product life will be normally distributed. Correspondingly, if the initial amount of the substance follows a gamma distribution, item life will be gamma distributed.

The Weibull distribution is a popular product life model generally justified by its property of describing, under fairly general circumstances, the way that the smallest values in large samples vary among sets of large samples. Thus if item life is determined by the smallest life among many potential failure sites, it is reasonable to expect that life will vary from item to item according to a Weibull distribution.

Another property that makes the Weibull distribution a reasonable choice for some products is that it can account for a steadily increasing failure rate characteristic of wear-out failures, a steadily decreasing failure rate characteristic of a product that benefits from "burn-in," or a constant failure rate typical of products that fail due to the occurrence of a random shock.

The two-parameter Weibull distribution was adopted by Lundberg and Palmgren [20.1] to describe bearing fatigue life on the strength of the excellence of the empirical fit to bearing fatigue life data. As will be described in Chapter 23, because of improvements in both the material and the methods of finishing bearings, it has been found recently that there is, under moderate load, a finite probability that a bearing will endure for an indefinitely long period. The Weibull model cannot describe this aspect of fatigue life. Nonetheless, under the relatively high loads common in fatigue testing practice, the Weibull model will closely approximate the observed fatigue life behavior of rolling bearings.

What follows is an account of the principal features of the Weibull distribution that are of interest to bearing engineers and an explanation of the methods of inferring the Weibull parameter values from life test data acquired under various experimental schemes.

THE TWO-PARAMETER WEIBULL DISTRIBUTION

Probability Functions

When it is said that a random variable, for example, bearing life, follows the two-parameter Weibull distribution, it is implied that the probability that an observed value of that variable is less than some arbitrary value x can be expressed as

$$\text{Prob(life} < x) = F(x) = 1 - e^{-(x/\eta)^\beta} \qquad x, \eta, \beta > 0 \tag{20.1}$$

The function $F(x)$ is known as the cumulative distribution function (CDF). The constants η and β are the scale parameter and the shape parameter, respectively. The function $F(x)$ may be thought of as the area under a curve $f(x)$ between 0 and the arbitrary value x. This curve is known as the probability density function (pdf) and has the form

$$f(x) = \frac{x^{\beta-1}}{\eta^\beta} e^{-(x/\eta)^\beta} \tag{20.2}$$

Figure 20.1 is a plot of the Weibull pdf for various values of β. Note that a wide diversity of distribution forms are encompassed by the Weibull family, depending on the value of β. For $\beta = 1.0$ the Weibull distribution reduces to the exponential distribution. For β in the range 3.0–3.5 the Weibull distribution is nearly symmetrical and approximates the normal pdf. The ability to assume such a range of shapes accounts for the extraordinary applicability of the Weibull distribution to many types of data.

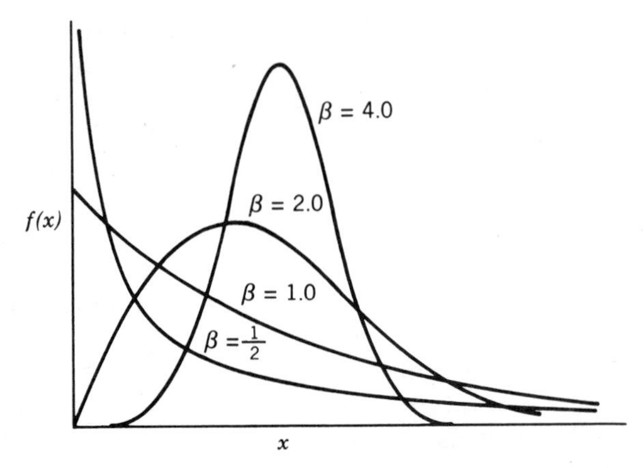

FIGURE 20.1. The two-parameter Weibull distribution for various values of the shape parameter β.

Mean Time Between Failures

The average or expected value of a random variable is a useful measure of its "central tendency"; it is a single numerical value that can be considered to typify the random variable. It is defined as

$$E(x) = \int_0^\infty xf(x)\,dx = \int_0^\infty x\left(\frac{x^{\beta-1}}{\eta^\beta}\right)e^{-(x/\eta)^\beta}\,dx \tag{20.3}$$

The value of the integral in (20.3) is

$$E(x) = \eta\Gamma\left(\frac{1}{\beta} + 1\right) \tag{20.4}$$

$\Gamma(\ )$ is the widely tabulated gamma function. Table 20.1 gives values of $\Gamma(1/\beta + 1)$ for β ranging from 1.0 to 5.0.

In reliability theory $E(x)$ is known as the mean time between failures and is commonly referred to as MTBF. It represents the average time between the failures of two consecutively run bearings—that is, the time between the failure of a bearing and the failure of its replacement. It does *not* represent the mean time between consecutive failures in a group of simultaneously running bearings. For this latter situation, provided $\beta \neq 1.0$, MTBF will vary with the failure order number. For example, the mean time between the first and second failures in samples of size 20 is different from the mean time between the 19th and 20th failures.

The scatter of a random variable is often characterized by a quantity

TABLE 20.1. Values of $\Gamma(1/\beta + 1)$ and $\Gamma(2/\beta + 1) - \Gamma^2(1/\beta + 1)$

β	$\Gamma(1/\beta + 1)$	$\Gamma(2/\beta + 1) - \Gamma^2(1/\beta + 1)$
1.0	1.0000	1.0000
1.1	0.9649	0.7714
1.2	0.9407	0.6197
1.3	0.9236	0.5133
1.4	0.9114	0.4351
1.5	0.9027	0.3757
1.6	0.8966	0.3292
1.7	0.8922	0.2919
1.8	0.8893	0.2614
1.9	0.8874	0.2360
2.0	0.8862	0.2146
2.5	0.8873	0.1441
3.0	0.8930	0.1053
3.5	0.8997	0.0811
4.0	0.9064	0.0647
5.0	0.9182	0.0442

known as variance, defined as the average or expected value of the squared deviation of the variable from its expected value; that is,

$$\sigma^2 = \int (x - E(x))^2 f(x)\, dx$$
$$= \int \left(x - \eta\Gamma\left(\frac{1}{\beta} + 1\right)\right)^2 \frac{x^{\beta - 1}}{\eta^\beta} e^{-(x/\eta)^\beta}\, dx \tag{20.5}$$

The value of this integral is

$$\sigma^2 = \eta^2\left[\Gamma\left(\frac{2}{\beta} + 1\right) - \Gamma^2\left(\frac{1}{\beta} + 1\right)\right] \tag{20.6}$$

Values of the quantity $[\Gamma(2/\beta + 1) - \Gamma^2(1/\beta + 1)]$ are listed in Table 20.1 for β ranging from 1.0–5.0.

The units of σ^2 are the square of the units in which life is measured, $(\text{hr})^2$. The square root of σ^2 is often preferred as a measure of scatter because it is expressed in the same units as the random variable itself. It is called the standard deviation. Neither the variance nor the standard deviation are cited much for the Weibull distribution; it is more usual to convey the magnitude of scatter by citing the values of a low percentile and a high percentile.

Percentiles

Equation (20.1) gives the probability that the observed value of a Weibull random variable is less than an arbitrary value. The inverse problem is to find the value of x, say x_p, for which the probability is a specified value p such that the life will not exceed it. The term x_p is defined implicitly as

$$F(x_p) = 1 - e^{-(x_p/\eta)^\beta} = p \tag{20.7}$$

The solution of equation (20.7) is

$$x_p = \eta\left[\ln\left(\frac{1}{1-p}\right)\right]^{1/\beta} \tag{20.8}$$

An important special case in rolling bearing engineering is the 10th percentile $x_{0.10}$, since it is historically customary that bearings are rated by the value of their tenth percentile life. In the literature $x_{0.10}$ is referred to as L_{10}. For consistency with the statistical literature on the Weibull distribution, $x_{0.10}$ is used in this part of the chapter. It is expressible as

$$x_{0.10} = \eta\left[\ln\left(\frac{1}{1-0.10}\right)\right]^{1/\beta} = \eta(0.1054)^{1/\beta} \tag{20.9}$$

Also of some interest is the median life, $x_{0.50}$:

$$x_{0.50} = \eta\left[\ln\left(\frac{1}{1-0.50}\right)\right]^{1/\beta} = \eta(0.6931)^{1/\beta} \tag{20.10}$$

Equation (20.8) shows that the ratio of two percentiles, say x_p and x_q, is

$$\frac{x_q}{x_p} = \left[\frac{\ln(1-q)^{-1}}{\ln(1-p)^{-1}}\right]^{1/\beta} \tag{20.11}$$

Thus

$$\frac{x_{0.50}}{x_{0.10}} = \left(\frac{0.6931}{0.1054}\right)^{1/\beta} = (6.579)^{1/\beta}$$

For $\beta = 1.1$, therefore, $x_{0.50} = 5.54$. This supports the rule, often quoted in the rolling bearing industry, that L_{50} ($x_{0.50}$) is roughly five times L_{10} ($x_{0.10}$).

Graphical Representation of the Weibull Distribution

From equation (20.1) the probability that a bearing survives a life x, denoted $S(x)$, is given by

$$S(x) = 1 - F(x) = e^{-(x/\eta)^{\beta}} \tag{20.12}$$

Taking natural logarithms twice of both sides of equation (20.12), leads to

$$\ln \ln \left(\frac{1}{S}\right) = \beta[\ln (x) - \ln (\eta)] \tag{20.13}$$

The right side is a linear function of ln (x). On special graph paper, called Weibull probability paper, for which the ordinate is ruled proportionately to ln [ln $(1/S)$] and the abscissa is logarithmically scaled, the values of S versus the associated values of x plot as a straight line. If in the design of the paper the same cycle lengths are used for the logarithmic scale on both coordinate axes, the slope of the straight line representation will be numerically equal to β. In any case, the Weibull shape parameter or Weibull slope will be related to the slope of the straight line representation, and in some designs of probability paper an auxiliary scale is provided to relate the shape parameter and the slope.

Figure 20.2 is a piece of Weibull probability paper on which the dis-

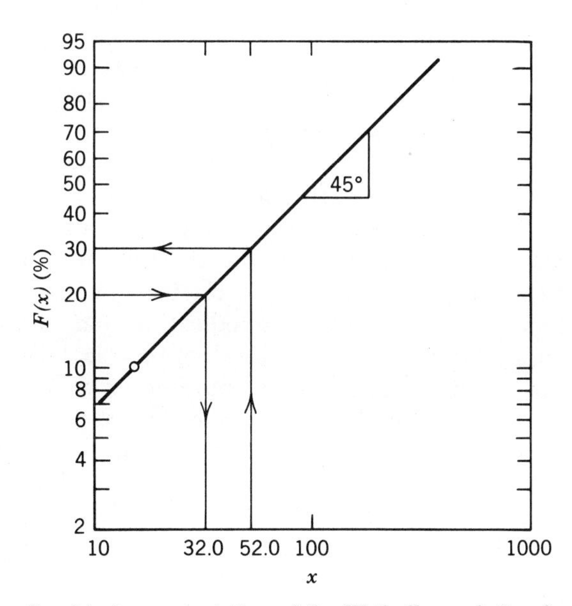

FIGURE 20.2. Graphical representation of the Weibull population for which $\beta = 1.0$, $x_{0.10} = 15.0$.

tribution with $\beta = 1.0$ and $x_{0.10} = 15.0$ is represented. It was constructed by passing a 45° line through the point corresponding to the failure probability value $F = 0.1$ ($S = 0.9$) and the life value $x_{0.10} = 15.0$. From this plot may be read the 20th percentile as the abscissa value at which a horizontal line at the ordinate value $F = 0.2$ intersects the straight line. Within graphical accuracy, $x_{0.20} = 32.0$. Inversely, the probability of failing prior to the life $x = 52.0$ is read to be roughly 30%. Representing a Weibull population on probability paper thus offers a graphical alternative to the use of equations (20.1) and (20.8) for the computation of probabilities and percentiles. The graphical approach is sufficiently accurate for most purposes. The primary use of probability paper is not, however, for representing known Weibull distributions but for estimating the Weibull parameters from life test results.

ESTIMATION IN SINGLE SAMPLES

Application of the Weibull Distribution

Up to this point it has been assumed that the Weibull parameters are known, and additionally required quantities such as probabilities, percentiles, expected values, variances, and standard deviations have been calculated in terms of these known parameters. This is a common situation in bearing application engineering, in which, given a catalog calculation of $x_{0.10}$ (L_{10}) and the standard Weibull slope of $\beta = 1.1$, it is required to compute the median life or the MTBF, and so on. In developmental work involving new variables such as materials, lubricants, or finishing processes, the focus is on determining the effect of these factors on the Weibull parameters. Accordingly, a sample of bearings modified from standard in some way is subjected to testing under standardized conditions of load and speed until some or all fail. When all fail, the sample is said to be uncensored. In a censored sample some bearings are removed from test prior to failure. Given the lives to failure or to test suspension for the unfailed bearings, the aim is to deduce the underlying Weibull parameters. This process is called estimation, because it is recognized that, since life is a random variable, identical samples will result in different test lives. The Weibull parameter values estimated in any single sample must themselves therefore be regarded as observed values of random variables that will vary from sample to sample according to a probability distribution known as the sampling distribution of the estimate. The scatter in the sampling distribution will decrease with sample size. Sample size thus affects the degree of precision with which the parameters are determined by a life test. The precision is expressed by an uncertainty or confidence interval within which the parameter value is likely to lie. An estimation procedure that results in the computation of a confidence interval is called interval estimation. A procedure that results in a single numerical value for the parameter is called point estimation. Point estimates in themselves are virtually useless, since without some qualification there is no way of judging how precise they are.

Accordingly, an analytical technique is given in the sequel for computing interval estimates of Weibull parameters. It is recommended that this technique be supplemented, however, with a point estimate obtained graphically. The graphical approach to estimation gives a synoptic view of the entire distribution and offers the opportunity to detect anomalies in the data that could easily be overlooked if reliance is placed entirely on an analytical technique.

Point Estimation in Single Samples: Graphical Methods

Assume that a sample of n bearings is tested until all fail. The ordered times to failure are denoted $x_1 < x_2 < \cdots x_n$. Consider the CDF of the Weibull population from which the sample was drawn. If this function were known, it would follow that the lives x_i and the values $F(x_i)$, $i = 1, \cdots, n$, would plot as a straight line on Weibull probability paper. It has been shown that even though the function $F(x)$ is not known, nonetheless $F(x_i)$ will vary in repeated samples according to a known pdf. The mean or expected value of $F(x_i)$ has been shown to equal $i/(n + 1)$. The median value of $F(x_i)$, also known as the median rank, has been shown by Johnson [20.2] to be approximately $(i - 0.3)/(n + 0.4)$. The procedure then is to plot the mean or median value of $F(x_i)$ against x_i for $i = 1, 2, \cdots, n$. The tradition in the bearing industry is to use the median rather than the mean as a plotting position choice, but the difference is small compared to the sampling variability.

Table 20.2 lists the ordered lives at failure for a sample of size $n =$

TABLE 20.2. Random Uncensored Sample Size of $n = 10$

Failure Order No. (i)	Life	Median Rank	$\frac{i - 0.3}{n + 0.4}$
1	14.01	0.06697	0.06731
2	15.38	0.16226	0.16346
3	20.94	0.25857	0.25962
4	29.44	0.35510	0.35577
5	31.15	0.45169	0.45192
6	36.72	0.54831	0.54808
7	40.32	0.64490	0.64423
8	48.61	0.74142	0.74038
9	56.42	0.83774	0.83654
10	56.97	0.93303	0.93269

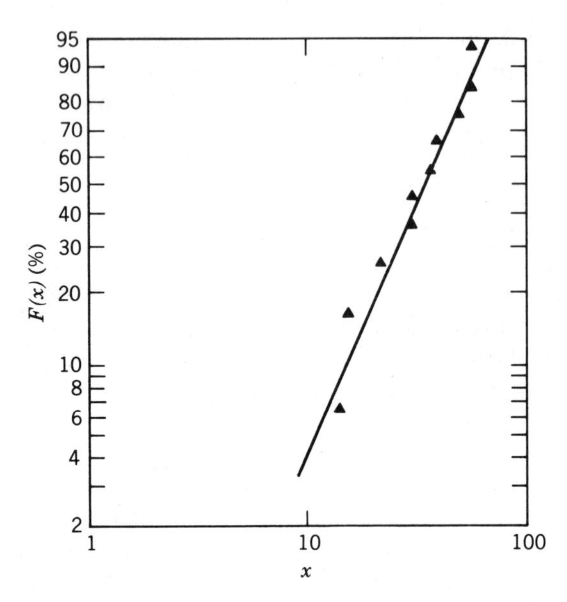

FIGURE 20.3. Probability plot for uncensored random sample for size $n = 10$.

10, along with the actual and approximated values of the median ranks. Hence the approximation is adequate within the limits of graphical approximation. The median ranks are shown plotted against the lives in Fig. 20.3.

The straight line fitted to the plotted points represents the graphical estimate of the entire $F(x)$ curve. Estimates of the percentiles of interest are then read from the fitted straight line. For example, to within graphical accuracy the $x_{0.10}$ value is estimated as 15.3. The Weibull shape parameter, estimated simply as the measured slope of the straight line, is roughly 2.2.

The same graphical approach applies for right-censored data in which the censored observations achieve a longer running time than the failures do. The full sample size n is used to compute the plotting positions, but only the failures are plotted. When there is *mixed* censoring—that is, there are suspended tests among the failures—the plotting positions are no longer computable by the method given since the suspensions cause ambiguity in determining the order numbers of the failures. Several alternative approaches are available for this situation, with generally negligible differences among them. The method known as hazard plotting due to Nelson [20.3], is recommended because it is easy to use. Column 1 of Table 20.3 gives the lives of failure or test suspension in a sample of size $n = 10$. Of the 10 bearings, $r = 4$ have failed, and the lives at failure are marked with an "F" in Table 20.3. Similarly the lives at test suspension are indicated by "S." The lives in column 1 are in

TABLE 20.3. Computation of Plotting Positions for Hazard Plot

Life	Reverse Rank	Hazard (h)	Cumulative Hazard (H)	$F = 1 - e^{-H}$
0.569 S	10	—	—	—
8.910 F	9	0.1111	0.1111	0.1052
21.410 S	8	—	—	—
21.960 F	7	0.1429	0.2540	0.2243
32.620 S	6	—	—	—
39.290 F	5	0.2000	0.4540	0.3649
42.990 S	4	—	—	—
50.400 F	3	0.3333	0.7873	0.5449
53.270 S	2	—	—	—
102.600 S	1	—	—	—

ascending order of time on test irrespective of whether the bearing failed. Column 2, termed the reverse rank by Nelson [20.3], assigns the value n to the lowest time on test, the value $n - 1$ to the next lowest, and so on. Column 3, called the hazard, is the reciprocal of the reverse rank, but is computed only for the failed bearings. Column 4 is the cumulative hazard and contains for each failure the sum of the hazard values in column 3 for that failure and each failure that occurred at an earlier running time. Thus for the second failure the cumulative hazard is $0.2540 = 0.1111 + 0.1429$. The cumulative hazard can then be plotted directly against life on probability paper that has been designed with an extra "hazard" scale. If paper of this type is unavailable, it is only necessary to compute an estimate of the plotting position applicable to ordinary probability paper by transforming the cumulative hazard H to $F = 1 - \exp(-H)$. This computation is shown in column 5 of Table 20.3. Figure 20.4 shows the resultant plot. Note that as in the right censored case only the failures are plotted. The suspended tests have played a role, however, in determining the plotting positions for the failures.

Point Estimation in Single Samples: Method of Maximum Likelihood

The method of maximum likelihood is a general approach to the estimation of the parameters of probability distributions. The central idea is to estimate the parameters as the values for which the observed test sample would most likely have occurred.

Consider an uncensored sample of size n. The "likelihood" is the product of the probability density functions $f(x) = x^{\beta - 1}/\eta^{\beta} \exp[-(x/\eta)^{\beta}]$ evaluated at each observed life value. The maximum likelihood estimates of η and β are the values that maximize this product. For censored samples

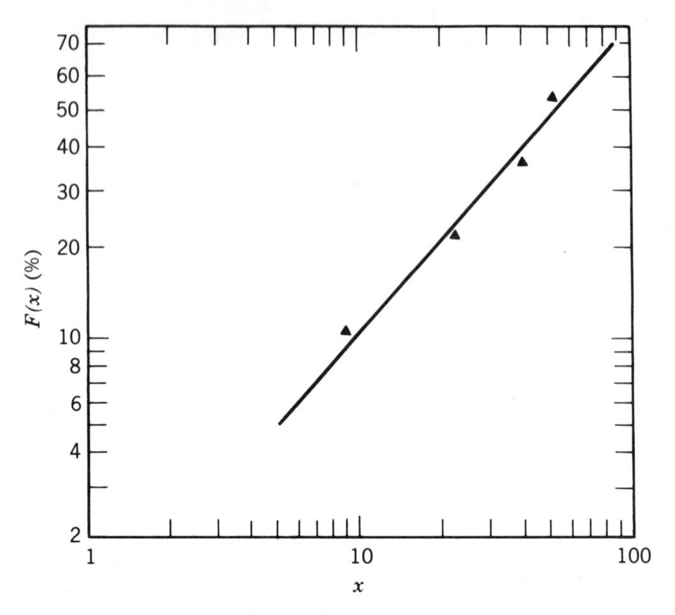

FIGURE 20.4. Probability plot for mixed censoring. Plotting positions are calculated based on cumulative hazard.

with $r < n$ failures the likelihood function contains, in lieu of the density function $f(x)$, the term $1 - F(x) = \exp[-(x/\eta)^{\beta}]$ evaluated at each suspended life value. It can be shown that the maximum likelihood (ML) estimate of β, denoted by a caret (ˆ), is the solution of the following nonlinear equation:

$$\frac{1}{\hat{\beta}} + \frac{\sum_{i=1}^{i=r} \ln x_i}{r} - \frac{\sum_{i=1}^{i=n} x_i^{\hat{\beta}} \ln x_i}{\sum_{i=1}^{i=n} x_i^{\hat{\beta}}} = 0 \tag{20.14}$$

This equation has only a single positive solution according to [20.4]. That solution is readily found by the Newton-Raphson method, although in highly censored cases the initial guess used to start the method might need to be modified to avoid convergence to a negative value for β.

Having determined β from equation (20.14), the ML estimate of η is obtained:

$$\hat{\eta} = \left(\sum_{i=1}^{i=n} \frac{x_i^{\hat{\beta}}}{r} \right)^{1/\hat{\beta}} \tag{20.15}$$

The ML estimate of a general percentile is

$$\hat{x}_p = \hat{\eta} k_p^{1/\hat{\beta}} \tag{20.16}$$

where k_p is defined as

$$k_p = -\ln (1 - p) \tag{20.17}$$

Confidence limits can be set if the censoring mode corresponds to the suspension of testing when the rth earliest failure occurs. This type of censoring is customarily referred to as type II censoring, as contrasted to type I censoring in which testing is suspended at a predetermined running time. In type I censoring the number of failures r is a random variable. In type II censoring the number of failures is predetermined by the experimenter.

The basis for confidence intervals for β is that the random function $v(r, n) = \hat{\beta}/\beta$ follows a sampling distribution that depends on the sample size n and censoring number r, not on the underlying values of β or η. Functions with this property are known as pivotal functions. The sampling distribution of $v(r, n)$ cannot be found analytically, but may be determined empirically to whatever precision is needed by Monte Carlo sampling. In the Monte Carlo method repeated samples are drawn by computer simulation from a Weibull distribution having arbitrary parameter values, say $\beta = 1.0$ and $\eta = 1.0$. The ML estimate $\hat{\beta}$ is formed for each sample and divided by the underlying population value of β to yield a value of $v(r, n)$. With typically 10,000 such values the percentiles may be computed from the sorted set and their average equated to the expected value of the distribution.

Denoting the 5th and 95th percentiles of $v(r, n)$ as $v_{0.05}(r, n)$ and $v_{0.95}(r, n)$ leads to the following 90% confidence interval for β:

$$\frac{\hat{\beta}}{v_{0.95}(r, n)} < \beta < \frac{\hat{\beta}}{v_{0.05}(r, n)} \tag{20.18}$$

The raw maximum likelihood estimates of the Weibull parameters are biased; that is, both the average and the median of the β estimates in an indefinitely large number of samples will differ somewhat from the true β value for the population from which the samples were drawn. It is possible to correct the raw ML estimate so that either its average or its median will coincide with the underlying population value of β. Because the distribution of $v(r, n)$ is not symmetrical, it is necessary to choose whether the adjusted estimator should be median or mean unbiased. Median unbiasedness is recommended because then the ML point estimate will have the reasonable property that it is just as likely to be larger

TABLE 20.4. 5th, 50th, and 95th Percentiles of $v(r, n)$ and $u(r, n, 0.10)$

		$v(r, n)$			$u(r, n, 0.10)$				
r	n	0.05	0.50	0.95	0.05	0.50	0.95	R	$R^{\beta}_{0.50}$
3	5	0.6351	1.6510	6.7596	−1.2672	0.8483	9.9607	10.6	8.98
5	5	0.6795	1.2346	2.8146	−1.1422	0.4465	4.4453	4.14	92.4
3	10	0.6208	1.7223	7.6478	−1.4304	0.4313	7.0208	12.3	135
5	10	0.6482	1.3117	3.2791	−0.9571	0.3737	3.7698	5.06	36.7
10	10	0.7361	1.1031	1.8363	−0.8794	0.2125	2.1304	2.49	15.3
5	15	0.6430	1.3321	3.3937	−0.9223	0.2435	2.9446	5.28	18.2
10	15	0.7130	1.1269	1.9428	−0.6184	0.1933	1.9477	2.72	9.75
15	15	0.7715	1.0679	1.5634	−0.7648	0.1393	1.5091	2.03	8.41
5	20	0.6432	1.3353	3.5078	−0.9601	0.1482	2.5445	5.45	13.8
10	20	0.7047	1.1328	1.9913	−0.7274	0.1604	1.7473	2.83	8.89
15	20	0.7459	1.0754	1.6327	−0.7055	0.1215	1.4431	2.19	7.37
20	20	0.7949	1.0476	1.4454	−0.6740	0.0958	1.2262	1.82	6.13
5	30	0.6430	1.3475	3.4437	−1.1306	0.0176	1.6920	5.36	8.12
10	30	0.6996	1.1309	2.0236	−0.6348	0.0931	1.3891	2.89	5.99
15	30	0.7410	1.0819	1.6771	−0.6038	0.0928	1.2152	2.26	5.39
20	30	0.7662	1.0569	1.5182	−0.5955	0.0790	1.1130	1.98	5.04
30	30	0.8259	1.0290	1.3353	−0.5672	0.0536	0.9147	1.62	4.22

than the underlying true value as to be smaller. It is shown in [20.4] that the median unbiased estimate of β, denoted by writing $\hat{}'$ above the symbol, is expressible as

$$\hat{\beta}' = \hat{\beta}/v_{0.50}(r, n) \tag{20.19}$$

Table 20.4 gives values of $v_{0.05}(r, n)$, $v_{0.50}(r, n)$, and $v_{0.95}(r, n)$ for n from 5–30 and various values of r.

Correcting the bias of the estimate and setting confidence limits for a general percentile x_p depends on the pivotality of the random function $u(r, n, p) = \hat{\beta} \ln (\hat{x}_p/x_p)$. Given percentiles of $u(r, n, 0.10)$ determined by Monte Carlo sampling, a 90% confidence interval on $x_{0.10}$ can be set up:

$$\hat{x}_{0.10} e^{-u_{0.95}(r,n,0.10)/\hat{\beta}} < x_{0.10} < \hat{x}_{0.10} e^{-u_{0.05}(r,n,0.10)/\hat{\beta}} \tag{20.20}$$

A median unbiased estimate of $x_{0.10}$ can be computed as

$$\hat{x}'_{0.10} = \hat{x}_{0.10} e^{-u_{0.50}(r,n,0.10)/\hat{\beta}} \tag{20.21}$$

Values of the 5th, 50th, and 95th percentiles of $u(r, n, 0.10)$ are also given in Table 20.4.

Numerical Example

Example 20.1. Using the ML method, calculate median unbiased point estimates and 90% confidence limits for β and $x_{0.10}$ for the uncensored sample of size $n = 10$ listed in Table 20.2.

Using a computer program that solves equation (20.14) for β and equation (20.16) for $x_{0.10}$, it is found that the raw ML estimates are $\hat{\beta} = 2.58$ and $x_{0.10} = 16.55$. The following applicable values are taken from Table 20.4.

$$v_{0.05}(10, 10) = 0.736, \qquad u_{0.05}(10, 10, 0.10) = -0.879,$$

$$v_{0.50}(10, 10) = 1.103, \qquad u_{0.50}(10, 10, 0.10) = 0.213,$$

$$v_{0.95}(10, 10) = 1.836, \qquad u_{0.95}(10, 10, 0.10) = 2.130.$$

Equation (20.18) gives

$$\frac{\hat{\beta}}{v_{0.95}(r, n)} < \beta < \frac{\hat{\beta}}{v_{0.05}(r, n)} \tag{20.18}$$

$$\frac{2.58}{1.836} < \beta < \frac{2.58}{0.736}$$

$$1.41 < \beta < 3.51$$

From equation (20.19) the median unbiased estimate of β is

$$\hat{\beta}' = \frac{\hat{\beta}}{v_{0.50}(r, n)} = \frac{2.58}{1.103} = 2.15 \tag{20.19}$$

From equation (20.20), 90% confidence limits for $x_{0.10}$ are computed as

$$\hat{x}_{0.10} \exp\left(\frac{-u_{0.95}(r, n, 0.10)}{\hat{\beta}}\right) < x_{0.10} < \hat{x}_{0.10} \exp\left(\frac{-u_{0.05}(r, n, 0.10)}{\hat{\beta}}\right) \tag{20.20}$$

$$16.6\left\{\exp\left[-\left(\frac{2.130}{2.58}\right)\right]\right\} < x_{0.10} < 16.6\left\{\exp\left[-\left(\frac{-0.879}{2.58}\right)\right]\right\}$$

$$7.25 < x_{0.10} < 23.3.$$

The median unbiased estimate of $x_{0.10}$ is

$$\hat{x}'_{0.10} = \hat{x}_{0.10} \exp\left(\frac{-u_{0.50}(r, n, 0.10)}{\hat{\beta}}\right) \tag{20.21}$$

$$= 16.6\left\{\exp\left[-\left(\frac{0.213}{2.58}\right)\right]\right\} = 15.3$$

The median unbiased estimates of $x_{0.10}$ and β agree closely with the values estimated graphically.

Sudden Death Tests

A popular test strategy in the bearing industry is the "sudden death" test. In sudden death testing a test sample of size n is divided into l subgroups each of size m ($n = lm$). When the first failure occurs in each subgroup, testing is suspended on that subgroup. When the test is over, there are l failures, the first failures in each of the l subgroups. To estimate β, substitute these first failures directly in equation (20.14). Confidence limits for β are then computed from equation (20.18) with $r = n = l$. That is, the first failures are treated as members of an uncensored sample whose size is equal to the number of subgroups. Table 20.5 gives percentiles of $v(l, l)$ for $l = 2$–6. The value of $\hat{x}_{0.10}$, determined by using the sample of first failures and equation (20.16), is denoted $\hat{x}_{0.10s}$. The ML estimate applicable to the complete sample is then computed as in [20.5]:

$$\hat{x}_{0.10} = \hat{x}_{0.10s} m^{1/\hat{\beta}} \tag{20.22}$$

90% confidence limits for $x_{0.10}$ may be computed as

$$\hat{x}_{0.10} e^{-q_{0.95}(l,m,0.10)/\hat{\beta}} < x_{0.10} < \hat{x}_{0.10} e^{-q_{0.05}(l,m,0.10)/\hat{\beta}} \tag{20.23}$$

A median unbiased estimate of $x_{0.10}$ is computed from

$$\hat{x}'_{0.10} = \hat{x}_{0.10} e^{-q_{0.50}/\hat{\beta}} \tag{20.24}$$

Table 20.5 gives values of the percentiles of the random function $q(l, m, p)$ required for these calculations.

Example 20.2. A sudden death test conducted with $l = 3$ subgroups, each of size $m = 10$, yielded the following values for the subgroup first failures: 4.72, 6.64, and 14.17. Using these lifetimes as input to a computer program that solves equation (20.14) yields the shape parameter estimate $\hat{\beta} = 2.27$. The estimated value of $x_{0.10}$ based on these three failure lives is $\hat{x}_{0.10s} = 3.59$. From equation (20.22) the estimate of $x_{0.10}$ for the parent population is

$$\hat{x}_{0.10} = \hat{x}_{0.10s} m^{1/\hat{\beta}} = 3.59\,(10)^{1/2.27} = 9.9. \tag{20.22}$$

The following values are taken from Table 20.5:

$$q_{0.05}(3, 10) = -2.91 \quad q_{0.50}(3, 10) = -0.133 \quad q_{0.95}(3, 10) = 1.83$$

$$v_{0.05}(3, 3) = 0.650 \quad v_{0.50}(3, 3) = 1.53 \quad v_{0.95}(3, 3) = 5.71$$

TABLE 20.5. Percentile Values for Analyzing Sudden Death Tests

No. of Subgroups l	Subgroup Size m	$q_{0.05}$	$q_{0.50}$	$q_{0.95}$	$v_{0.05}$	$v_{0.50}$	$v_{0.95}$	R	$R^{\beta}_{0.50}$
2	5	−16.8	0.128	6.99	0.654	2.15	22.8	34.9	64000
2	10	−29.7	−0.334	3.16	0.654	2.15	22.8	34.9	4240000
3	5	−1.79	0.163	3.79	0.650	1.53	5.71	8.78	37.7
3	10	−2.91	−0.133	1.83	0.650	1.53	5.71	8.78	22.0
4	5	−1.19	0.099	2.50	0.653	1.32	3.73	5.71	16.1
5	2	−1.02	0.302	3.25	0.668	1.23	2.83	4.24	18.9
5	3	−0.980	0.208	2.63	0.668	1.23	2.83	4.24	13.2
5	4	−0.954	0.140	2.22	0.668	1.23	2.83	4.24	13.2
5	6	−0.997	0.045	1.62	0.668	1.23	2.83	4.24	9.97

The median unbiased estimate of β is

$$\hat{\beta}' = \frac{\hat{\beta}}{v_{0.50}(l, l)} = \frac{2.27}{1.53} = 1.48 \tag{20.19}$$

90% confidence limits for β are

$$\frac{\hat{\beta}}{v_{0.95}(l, l)} < \beta < \frac{\hat{\beta}}{v_{0.05}(l, l)} \tag{20.18}$$

$$\frac{2.27}{5.71} < \beta < \frac{2.27}{0.650}$$

$$0.398 < \beta < 3.50$$

The median unbiased estimator of $x_{0.10}$ is

$$\hat{x}'_{0.10} = \hat{x}_{0.10} \exp\left(\frac{-q_{0.50}}{\hat{\beta}}\right) = 9.9 \exp\left(\frac{-0.133}{2.27}\right) = 9.34 \tag{20.24}$$

90% confidence limits for $x_{0.10}$ are

$$\hat{x}_{0.10} \exp\left(\frac{-q_{0.95}(l, m, 0.10)}{\hat{\beta}}\right) < x_{0.10} < \hat{x}_{0.10} \exp\left(\frac{-q_{0.05}(l, m, 0.10)}{\hat{\beta}}\right) \tag{20.23}$$

$$9.9 \exp\left(\frac{-1.83}{2.27}\right) < x_{0.10} < 9.9 \exp\left(\frac{2.91}{2.27}\right)$$

$$4.43 < x_{0.10} < 35.6.$$

Precision of Estimation: Sample Size Selection

A confidence interval reflects the uncertainty in the value of the estimated parameter due to the finite size of the life test sample. As the sample size increases, the two ends of the confidence interval approach each other; that is, the ratio of the upper to lower ends of the confidence interval approaches 1.0. For finite sample sizes this ratio was suggested in [20.6] as a useful measure of the precision of estimation. From equation (20.18) the confidence limit ratio R for β estimation is

$$R = \frac{v_{0.95}(r, n)}{v_{0.05}(r, n)} \tag{20.25}$$

Values of R for various n and r are given in Table 20.4 for conventional tests and Table 20.5 for sudden death tests. Note that for a given sample size n, the precision improves (R decreases) as the number of failures r increases.

For $x_{0.10}$ the ratio of the upper to lower confidence limits contains the random variable $\hat{\beta}$. The approach taken in [20.6] in this case is to use as a precision measure the *median* value of this ratio, denoted $R_{0.50}$. The expression for this median ratio contains the unknown value of the true shape parameter β. For planning purposes one may use an historical value such as 1.1 or, alternatively, the value $R_{0.50}^{\beta}$ as the precision measure. Values of $R_{0.50}^{\beta}$ are given in Table 20.4 for conventional testing and Table 20.5 for sudden death testing.

ESTIMATION IN SETS OF WEIBULL DATA

Methods

Very often an experimental study of bearing fatigue life will include the testing of several samples, differing from each other with respect to the level of some qualitative factor under study. A qualitative factor is distinct from a quantitative factor, such as temperature or load, which can be assigned a numerical value. Examples of qualitative factors include lubricants, cage designs, or bearing materials.

It was shown in [20.7] that more precise estimates can be made if the data in the samples making up the complete investigation are analyzed as a set. This is possible if it can be assumed that the samples are drawn from Weibull populations, which, although they might differ in their scale parameter values, nonetheless have a common value of β.

Applicable tabular values for carrying out the analyses presuppose that the sample size n and the number of failures r are the same for each sample in the set, so henceforth this is assumed to be the case. It is thus assumed that k groups of size n have been tested until the rth failure occurred in each group. The first step is to determine whether it is plausible that the groups have a common value of β. This is done by analyzing each group individually to determine the values of $x_{0.10}$ and β. The largest and smallest of the k β estimates are then determined, and their ratio formed. If the β values did differ among the groups, this ratio would tend to be large. Table 20.6 gives the value of the 90th percentile of the ratio $w = \hat{\beta}_{max}/\hat{\beta}_{min}$ for various r, n, and k. These values were determined by Monte Carlo sampling from k Weibull populations that *did* have a common value of β. Thus, values of the ratio of largest to smallest shape parameter estimates exceeding those in Table 20.6 will occur only 10% of the time if the groups do have a common value of β. These values may be used as the critical values in deciding whether a common β assumption is justified.

TABLE 20.6. Critical Values $w_{0.90}(r, n, k)$ for Testing Homogeneity of k Weibull Shape Parameter Estimates

n	r	$k = 2$	$k = 3$	$k = 4$	$k = 5$	$k = 10$
5	3	5.45	8.73	11.0	13.4	22.2
5	5	2.77	3.59	4.23	4.69	6.56
10	3	6.04	9.93	12.5	15.4	26.8
10	5	3.21	4.35	5.16	5.69	7.98
10	10	1.87	2.23	2.47	2.61	3.16
15	5	3.20	4.48	5.28	5.90	8.39
15	10	2.02	2.44	2.69	2.90	3.56
15	15	1.65	1.87	2.05	2.16	2.45
20	5	3.31	4.54	5.28	6.24	9.05
20	10	2.11	2.50	2.80	3.00	3.69
20	15	1.72	1.96	2.15	2.30	2.70
20	20	1.52	1.70	1.80	1.90	2.14
30	5	3.28	4.47	5.38	6.11	8.95
30	10	2.11	2.54	2.90	3.10	3.88
30	15	1.78	2.05	2.27	2.40	2.82
30	20	1.61	1.82	1.95	2.06	2.40
30	30	1.41	1.53	1.61	1.67	1.84

Having determined that the common β assumption is reasonable, this common β value can be estimated using the data in each group, by solving the nonlinear equation

$$\frac{1}{\hat{\beta}_1} + \frac{1}{rk}\sum_{i=1}^{i=k}\sum_{j=1}^{j=r} \ln x_{i(j)} - \sum_{i=1}^{i=k} \frac{\sum_{j=1}^{j=n} x_{i(j)}^{\hat{\beta}_1} \ln x_{i(j)}}{k \sum_{j=1}^{j=n} x_{i(j)}^{\hat{\beta}_1}} = 0 \qquad (20.26)$$

where $\hat{\beta}_1$ denotes the ML estimate of the common β value and $x_{i(j)}$ denotes the jth ordered failure time within the ith group. Confidence limits for β may be set analogously to equation (20.17) as follows:

$$\frac{\hat{\beta}_1}{(v_1)_{0.95}} < \beta < \frac{\hat{\beta}_1}{(v_1)_{0.05}} \qquad (20.27)$$

where $v_1(r, n, k) = \hat{\beta}_1/\beta$. A median unbiased estimate of β may be computed from

$$\hat{\beta}' = \frac{\hat{\beta}_1}{(v_1)_{0.50}} \qquad (20.28)$$

Table 20.7 gives percentiles of $v_1(r, n, k)$ needed for setting 90% confidence limits and for bias correction for various values of n, r, and k. The

TABLE 20.7. Percentiles of Functions Needed for Analyzing Sets of Weibull Data

			v_1			u_1			t_1	
n	r	k	0.05	0.50	0.95	0.05	0.50	0.95	0.90	0.95
5	3	2	0.7618	1.505	3.833	−1.175	0.6861	5.325	2.833	3.752
5	3	3	0.8431	1.471	3.022	−1.135	0.6111	3.864	3.294	4.192
5	3	4	0.8915	1.462	2.673	−1.131	0.6124	3.417	3.605	4.342
5	3	5	0.9331	1.446	2.485	−1.112	0.6216	3.073	3.733	4.512
5	3	10	1.038	1.430	2.030	−1.084	0.5889	2.439	4.158	4.707
5	5	2	0.7773	1.195	2.056	−0.9715	0.3875	2.750	1.441	1.802
5	5	3	0.8265	1.184	1.816	−0.8715	0.3473	2.255	1.742	2.111
5	5	4	0.8609	1.178	1.703	−0.8321	0.3410	1.980	1.950	2.285
5	5	5	0.8888	1.177	1.621	−0.7907	0.3470	1.810	2.082	2.392
5	5	10	0.9545	1.167	1.453	−0.7314	0.3336	1.471	2.409	2.702
10	5	2	0.7677	1.256	2.326	−0.8691	0.2856	2.424	1.576	1.999
10	5	3	0.8281	1.242	2.039	−0.8526	0.2727	1.959	1.900	2.296
10	5	4	0.8679	1.234	1.877	−0.8303	0.2761	1.729	2.084	2.426
10	5	5	0.8904	1.230	1.768	−0.7987	0.2794	1.587	2.209	2.547
10	5	10	0.9669	1.219	1.575	−0.7994	0.2733	1.361	2.549	2.823
10	10	2	0.8130	1.087	1.516	−0.7161	0.1587	1.444	0.8509	1.040
10	10	3	0.8508	1.082	1.414	−0.6590	0.1460	1.204	1.062	1.231
10	10	4	0.8794	1.077	1.355	−0.6117	0.1448	1.066	1.171	1.330
10	10	5	0.8939	1.076	1.319	−0.5883	0.1519	0.9899	1.256	1.406
10	10	10	0.9405	1.073	1.235	−0.5766	0.1450	0.8384	1.468	1.598
15	5	2	0.768	1.269	2.414	−0.879	0.203	1.976	1.569	2.02
15	5	3	0.822	1.256	2.080	−0.876	0.190	1.645	1.899	2.308
15	5	4	0.857	1.246	1.927	−0.848	0.193	1.474	2.110	2.475
15	5	5	0.891	1.239	1.815	−0.864	0.189	1.377	2.206	2.548
15	10	2	0.799	1.106	1.611	−0.668	0.157	1.352	0.873	1.080
15	10	3	0.845	1.100	1.489	−0.613	0.145	1.152	1.080	1.254
15	10	4	0.866	1.097	1.412	−0.586	0.143	1.033	1.200	1.381
15	10	5	0.888	1.094	1.375	−0.573	0.138	0.987	1.289	1.457
15	15	2	0.836	1.056	1.369	−0.615	0.108	...	0.671	0.817
15	15	3	0.871	1.052	1.302	−0.559	0.098	0.904	0.827	0.949
15	15	4	0.890	1.049	1.258	−0.515	0.097	0.805	0.920	1.036
15	15	5	0.902	1.048	1.230	−0.497	0.095	0.772	0.987	1.101
20	5	2	0.761	1.278	2.428	−0.937	0.119	1.660	1.575	2.020
20	5	3	0.818	1.267	2.078	−0.912	0.127	1.403	1.922	2.348
20	5	4	0.864	1.251	1.916	−0.918	0.118	1.299	2.121	2.486
20	5	5	0.892	1.246	1.836	−0.937	0.119	1.226	2.233	2.580
20	10	2	0.796	1.109	1.650	−0.621	0.125	1.231	0.871	1.072
20	10	3	0.836	1.104	1.507	−0.587	0.125	1.035	1.098	1.282
20	10	4	0.866	1.100	1.446	−0.573	0.115	0.953	1.213	1.384
20	10	5	0.885	1.099	1.402	−0.550	0.117	0.915	1.304	1.475
20	15	2	0.826	1.064	1.425	−0.569	0.0982	1.013	0.672	0.809
20	15	3	0.857	1.062	1.337	−0.531	0.0988	0.874	0.830	0.964
20	15	4	0.882	1.058	1.289	−0.497	0.0914	0.800	0.922	1.057
20	15	5	0.896	1.059	1.263	−0.472	0.0955	0.743	0.999	1.120
20	20	2	0.855	1.040	1.300	−0.531	0.0723	0.874	0.566	0.678
20	20	3	0.881	1.037	1.242	−0.486	0.0746	0.7422	0.702	0.803
20	20	4	0.898	1.037	1.208	−0.461	0.0715	0.665	0.785	0.883
20	20	5	0.911	1.036	1.187	−0.433	0.0714	0.623	0.836	0.941
25	5	2	0.764	1.287	2.427	−1.002	0.0776	1.407	1.560	1.200
25	5	3	0.822	1.264	2.102	−0.996	0.0786	1.246	1.910	2.308
25	5	4	0.866	1.256	1.952	−0.977	0.0740	1.144	2.114	2.469
25	5	5	0.894	1.251	1.846	−0.986	0.0719	1.098	2.231	2.562
25	10	2	0.791	1.117	1.652	−0.605	0.107	1.099	0.871	1.061
25	10	3	0.832	1.107	1.527	−0.586	0.0957	0.979	1.091	1.278
25	10	4	0.864	1.103	1.456	−0.575	0.0987	0.887	1.219	1.387
25	10	5	0.885	1.100	1.401	−0.557	0.0946	0.838	1.289	1.466

Table 20.7. *(Continued)*

			v_1			u_1			t_1	
n	r	k	0.05	0.50	0.95	0.05	0.50	0.95	0.90	0.95
25	15	2	0.817	1.069	1.447	−0.551	0.0969	0.974	0.673	0.827
25	15	3	0.855	1.063	1.358	−0.501	0.0830	0.849	0.835	0.972
25	15	4	0.878	1.061	1.308	−0.477	0.0847	0.753	0.932	1.072
25	15	5	0.893	1.061	1.277	−0.458	0.0843	0.710	0.996	1.118
25	20	2	0.841	1.045	1.336	−0.519	0.0762	0.857	0.567	0.682
25	20	3	0.874	1.043	1.268	−0.469	0.0710	0.745	0.701	0.811
25	20	4	0.892	1.041	1.236	−0.435	0.0692	0.661	0.785	0.882
25	20	5	0.905	1.041	1.208	−0.426	0.0681	0.618	0.836	0.931
25	25	2	0.864	1.032	1.255	−0.494	0.0629	0.752	0.497	0.601
25	25	3	0.890	1.030	1.204	−0.445	0.0607	0.655	0.615	0.710
25	25	4	0.905	1.030	1.178	−0.409	0.0541	0.581	0.686	0.779
25	25	5	0.916	1.028	1.051	−0.387	0.0570	0.539	0.736	0.814
30	5	2	0.761	1.285	2.436	−1.066	0.0177	1.289	1.591	2.060
30	5	3	0.827	1.267	2.120	−1.070	0.0342	1.151	1.938	2.367
30	5	4	0.871	1.256	1.963	−1.065	0.0331	1.049	2.148	2.515
30	5	5	0.895	1.256	1.838	−1.062	0.0275	1.019	2.256	2.607
30	10	2	0.789	1.117	1.674	−0.587	0.0874	1.020	0.880	1.095
30	10	3	0.831	1.109	1.528	−0.576	0.0804	0.896	1.099	1.303
30	10	4	0.861	1.108	1.454	−0.570	0.0819	0.828	1.228	1.407
30	10	5	0.883	1.104	1.410	−0.570	0.0795	0.787	1.313	1.492
30	15	2	0.813	1.071	1.458	−0.513	0.0823	0.886	0.669	0.813
30	15	3	0.815	1.068	1.365	−0.583	0.0767	0.776	0.832	0.965
30	15	4	0.875	1.065	1.319	−0.470	0.0734	0.721	0.940	1.067
30	15	5	0.893	1.066	1.288	−0.460	0.0768	0.692	1.004	1.136
30	20	2	0.833	1.049	1.349	−0.490	0.0737	1.083	0.560	0.679
30	20	3	0.871	1.045	1.280	−0.442	0.0680	0.704	0.698	0.796
30	20	4	0.889	1.045	1.243	−0.430	0.0664	0.629	0.781	0.884
30	20	5	0.903	1.045	1.219	−0.410	0.0701	0.621	0.842	0.942
30	25	2	0.853	1.037	1.280	−0.481	0.0572	0.882	0.500	0.595
30	25	3	0.882	1.035	1.226	−0.425	0.0593	0.627	0.609	0.699
30	25	4	0.899	1.033	1.195	−0.396	0.0575	0.583	0.683	0.771
30	25	5	0.913	1.033	1.175	−0.380	0.0599	0.549	0.736	0.816
30	30	2	0.873	1.026	1.224	−0.458	0.0530	0.668	0.448	0.534
30	30	3	0.895	1.025	1.184	−0.406	0.0513	0.574	0.552	0.635
30	30	4	0.911	1.024	1.156	−0.387	0.0467	0.529	0.618	0.698
30	30	5	0.921	1.024	1.142	−0.368	0.0493	0.496	0.665	0.740

scale parameter for the ith group may be reestimated with β_1 as follows:

$$\hat{\eta}_i = \left(\frac{\sum x_{i(j)}^{\hat{\beta}_1}}{r} \right)^{1/\hat{\beta}_1} \tag{20.29}$$

The value of x_{pi} may be estimated from

$$\hat{x}_{pi} = \hat{\eta}_i k^{1/\hat{\beta}_1} \tag{20.30}$$

Confidence limits for $x_{0.10}$ may be computed as follows:

$$\hat{x}_{0.10} e^{-(u_1)_{0.95}/\hat{\beta}_1} < x_{0.10} < \hat{x}_{0.10} e^{-(u_1)_{0.05}/\hat{\beta}_1} \tag{20.31}$$

where $u_1 = \hat{\beta}_1 \ln(\hat{x}_{0.10}/x_{0.10})$ is the k sample generalization of $u(r, n, 0.10)$.

The median unbiased estimate of $x_{0.10}$ may be computed as

$$\hat{x}'_{0.10} = \hat{x}_{0.10} e^{-(u_1)_{0.50}/\hat{\beta}_1} \tag{20.32}$$

Now that $x_{0.10}$ has been estimated for each group using the ML estimate $\hat{\beta}_1$ of the common shape parameter, the next question of interest is whether these $x_{0.10}$ values differ significantly. That is, are the apparent differences among the $x_{0.10}$ estimates real, or could they be due to chance? To test whether the underlying true $x_{0.10}$ values could all be equal, the magnitude of variation that could occur in the estimated values due to chance alone must be assessed. This can be done by using the random function $t_1(r, n, k)$ defined by

$$t_1(r, n, k) = \hat{\beta}_1 \ln \left(\frac{(\hat{x}_{0.10})_{\max}}{(\hat{x}_{0.10})_{\min}} \right) \tag{20.33}$$

where $(\hat{x}_{0.10})_{\max}$ and $(\hat{x}_{0.10})_{\min}$ are the largest and smallest values of $\hat{x}_{0.10}$ calculated among the k samples. The 90th or 95th percentiles of $t_1(r, n, k)$ may be used to assess the observed difference in the $x_{0.10}$ values. Any two samples, say sample i and sample j, for which values the quantity $\hat{\beta}_1 \ln [(\hat{x}_{0.10})_i/(\hat{x}_{0.10})_j]$ exceeds $(t_1)_{0.90}$, may be declared to differ from each other at the 10% level of significance. Correspondingly, using the 95th percentile of $t_1(r, n, k)$ results in a 5% significance level test for the equality of the $x_{0.10}$ values.

Example 20.3. Ten bearings are fabricated from each of five different steel alloys. Uncensored life tests are performed. The raw ML estimates of β and $x_{0.10}$ for each group are given below:

Group No.	1	2	3	4	5
$\hat{\beta}$	2.59	2.32	3.13	1.94	3.65
$\hat{x}_{0.10}$	5.06	2.60	4.72	3.49	8.83

To test whether a common β assumption is plausible, the ratio of the largest to the smallest shape parameter estimates is computed:

$$w = \frac{\hat{\beta}_{\max}}{\hat{\beta}_{\min}} = \frac{3.65}{1.94} = 1.88.$$

From Table 20.7, with $n = r = 10$ and $k = 5$, it is found that $v_{0.90} = 2.61$. Since $1.88 < 2.61$, the data are consistent with the assumption of a constant β.

Having established the plausibility of a common β, its ML estimate using equation (20.26) is calculated and $\hat{\beta}_1 = 2.48$.

Recomputing the values of $\hat{x}_{0.10i}$ by equation (20.30), the confidence limits by equation (20.31), and the median unbiased estimates by equation (20.32) gives

Group No.	1	2	3	4	5
Raw ML estimate $\hat{x}_{0.10}$	4.84	2.81	3.80	4.87	6.34
Median unbiased estimate $\hat{x}'_{0.10}$	4.55	2.65	3.57	4.58	5.97
Lower confidence limit	3.25	1.89	2.55	3.27	4.25
Upper confidence limit	6.13	3.56	4.82	6.18	8.04

From Table 20.7 the 90th percentile of $t_1(10, 10, 5)$ is 1.26. The smallest ratio of raw $\hat{x}_{0.10}$ values that will be significant, denoted SSR (shortest significant ratio) is defined implicitly by

$$1.26 = \hat{\beta}_1 \ln(\mathrm{SSR}) = 2.480 \ln(\mathrm{SSR}).$$

Solving for SSR gives

$$\mathrm{SSR} = \exp\left(\frac{1.26}{2.48}\right) = 1.66.$$

Thus groups for which the ratio of raw ML estimates of $x_{0.10}$ exceeds 1.66 may be declared different. This would include Groups 5 and 3, because their $x_{0.10}$ estimates are in the ratio 6.34/3.80 = 1.67 > 1.66. Note that the confidence limits on $x_{0.10}$ overlap for these two groups. In general, if the confidence limits do not overlap, the groups differ. However the groups may differ as in the present case even though the confidence limits do overlap.

CLOSURE

Chapter 19 explored the reasons for, and concepts and methods of, endurance testing of ball and roller bearings and components. In this chapter the means to relate such test results to applications of standard and special bearing products were covered. The applicability of the Weibull distribution to such test data was demonstrated. It is further shown that, having specified a Weibull population, for example, by a catalog calculation of theoretical life, it is possible to calculate other characteristics that may occasionally be of interest, such as the mean time between failures.

The methodology for forming a graphical estimate of a Weibull population using a censored or an uncensored data sample was described and illustrated.

Examples were given for using the method of maximum likelihood for the point and interval estimation of the Weibull parameters from either type II censored or sudden death test samples.

Finally, the procedure was given for analyzing Weibull samples in sets. This procedure gives more precise estimated parameters because it extracts information from the entire set of data.

REFERENCES

20.1. G. Lundberg and A. Palmgren, "Dynamic Capacity of Rolling Bearings," *Acta Polytech. Mech. Eng. Ser. 1, R.S.A.E.E.*, No. 3, 7 (1947).

20.2. L. Johnson, *Theory and Technique of Variation Research*, Elsevier, New York (1970).

20.3. W. Nelson, "Theory and Application of Hazard Plotting for Censored Failure Data," *Technometrics* **14,** 945–966 (1972).

20.4. J. McCool, "Inference on Weibull Percentiles and Shape Parameter for Maximum Likelihood Estimates," *IEEE Trans. Reliab.* **R-19,** 2–9 (1970).

20.5. J. McCool, "Analysis of Sudden Death Tests of Bearing Endurance," *ASLE Trans.* **17,** 8–13 (1974).

20.6. J. McCool, "Censored Sample Size Selection for Life Tests," *Proceedings 1973 Annual Reliability and Maintainability Symposium*, IEEE Cat. No. 73CHO714-64 (1973).

20.7. J. McCool, "Analysis of Sets of Two-Parameter Weibull Data Arising in Rolling Contact Endurance Testing," *Rolling Contact Fatigue Testing of Bearing Steels*, ASTM STP 771, J. J. C. Hoo, Ed., American Society for Testing and Materials, Philadelphia, pp. 293–319 (1982).

21

PERMANENT DEFORMATION

LIST OF SYMBOLS

Symbol	Description	Units
C_s	Basic static load rating	N (lb)
d_m	Pitch diameter	mm (in.)
D	Ball or roller diameter	mm (in.)
F	Load	N (lb)
f	r/D	
FS	Factor of safety	
HV	Vickers hardness	
i	Number of rows	
l	Roller effective length	mm (in.)
P_d	Radial clearance	mm (in.)
Q	Rolling element load	N (lb)
r	Groove curvature radius	mm (in.)
R	Roller contour radius	mm (in.)
X_s	Radial load factor	
Y_s	Axial load factor	
Z	Number of rolling elements per row	

Symbol	Description	Units
α	Contact angle	°
γ	$D \cos \alpha / d_m$	
δ_s	Permanent deformation	mm (in.)
η	Hardness reduction factor	
ρ	Curvature	mm^{-1} ($in.^{-1}$)
σ	Yield or limit stress	N/mm^2 (psi)
φ_s	Load rating factor	

SUBSCRIPTS

a	Refers to axial direction
i	Refers to inner raceway
ip	Refers to incipient plastic flow of material
o	Refers to outer raceway
r	Refers to radial direction
s	Refers to static loading
SD	Refers to shakedown
yield	Refers to yield stress

GENERAL

Many structural materials exhibit a strain limit under load beyond which full recovery of the original elemental dimensions is not possible when the load is removed. Bearing steel loaded in compression behaves in a similar manner. Thus when a loaded ball is pressed on a bearing raceway, an indentation may remain in the raceway and the ball may exhibit a "flat" spot after load is removed. These permanent deformations, if they are sufficiently large, can cause excessive vibration and possibly stress concentrations of considerable magnitude.

CALCULATION OF PERMANENT DEFORMATION

In practice, permanent deformations of small magnitude occur even under light loads. Figure 21.1 taken from reference [21.4] shows a very large magnification of the contacting rolling element surfaces in a typical ball bearing both in the direction of rolling motion and transverse to that direction.

Figure 21.2 also from reference [21.4] shows an isometric view of a ground surface having spatial properties similar to honed and lapped raceway surfaces. Noting the occurrence of "peaks and valleys" even with a finely finished surface, it is apparent that prior to distributing a load between rolling element and raceway over the entire contact area

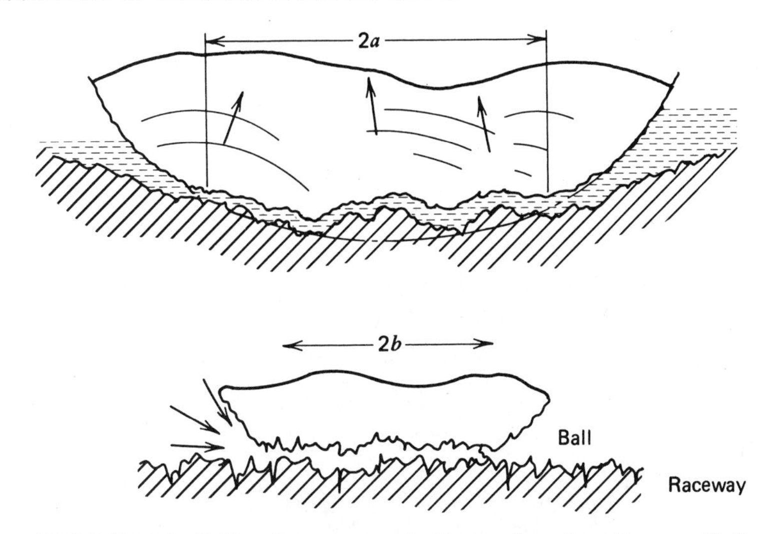

FIGURE 21.1. Ball and raceway contacting surfaces (greatly magnified).

thus giving an average compressive stress $\sigma = Q/A$, the load is distributed only over the smaller area of contacting peaks, giving a much larger stress than σ. Thus, it is probable that the compressive yield strength is exceeded locally and both surfaces are somewhat flattened and polished in operation. According to Palmgren [21.1] this flattening has little effect on bearing operation because of the extremely small magnitude of deformation. It may be detected by a slight change in reflection of light from the surface.

It was shown in Chapter 5 that the relative approach of two solid steel bodies loaded elastically in point contact is given by

$$\delta = 2.79 \times 10^{-4} \delta^* Q^{2/3} \sum \rho^{1/3} \tag{5.38}$$

in which δ^* is a constant depending on the shapes of the contacting surfaces. As the load between the surfaces is increased, deformation gradually departs from that depicted by equation (5.38) and becomes larger for any given load (see Fig. 21.3). The point of departure is the bulk compressive yield strength. Based on empirical data for bearing quality steel hardened between 63.5 and 65.5 Rockwell C, Palmgren [21.1] developed the following formula to describe permanent deformation in point contact:

$$\delta_s = 1.3 \times 10^{-7} \frac{Q^2}{D} (\rho_{I1} + \rho_{II1})(\rho_{I2} + \rho_{II2}) \tag{21.1}$$

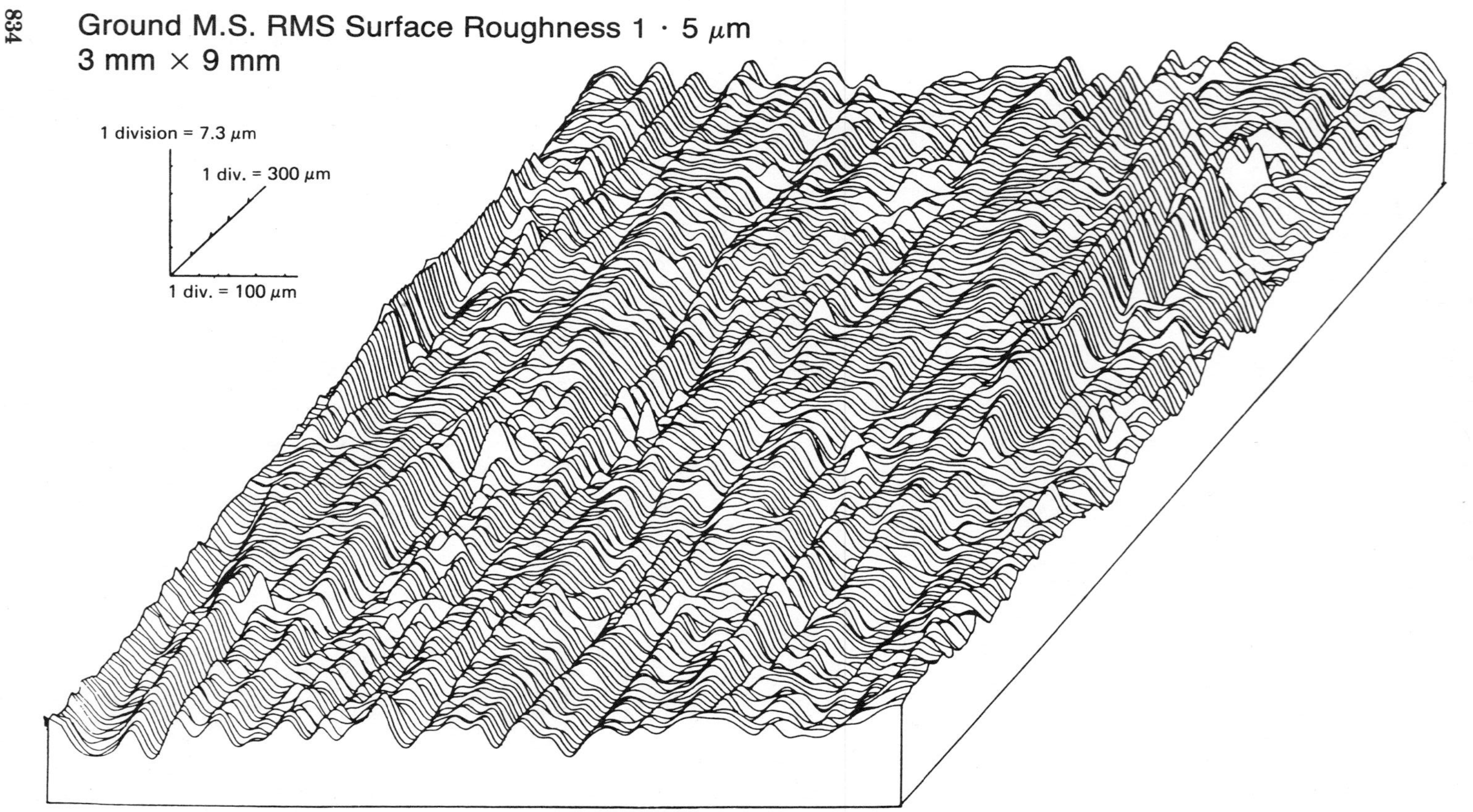

FIGURE 21.2. Isometric view of a typical honed and lapped surface.

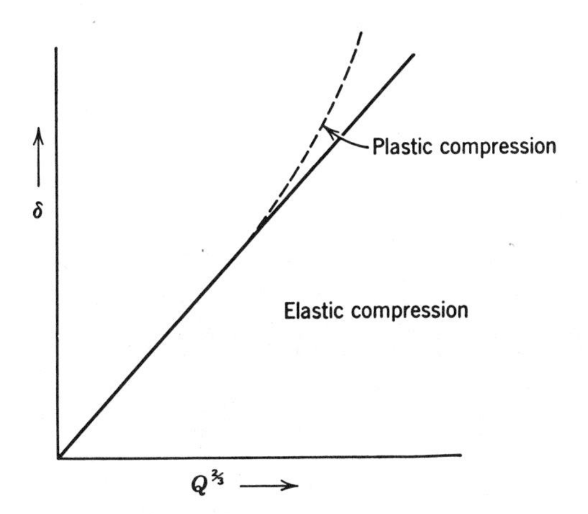

FIGURE 21.3. Deflection vs load in point contact.

in which $\rho_{\mathrm{I}1}$ is the curvature of body I in plane 1, and so on. For ball–raceway contact, equation (21.1) is

$$\delta_s = 5.25 \times 10^{-7} \frac{Q^2}{D^3}\left[1 \pm \frac{\gamma}{(1 \mp \gamma)}\right]\left(1 - \frac{1}{2f}\right) \tag{21.2}$$

in which the upper signs refer to the inner raceway contact and the lower signs refer to the outer raceway contact. For roller–raceway point contact the following equation obtains:

$$\delta_s = 2.52 \times 10^{-7} \left(\frac{Q}{D}\right)^2 \left[1 \pm \frac{\gamma}{(1 \mp \gamma)}\right]\left(\frac{1}{R} - \frac{1}{r}\right) \tag{21.3}$$

in which R is the roller contour radius and r is the groove radius. The foregoing formulas are valid for permanent deformation in the vicinity of the compressive elastic limit (yield point) of the steel.

Example 21.1. For the 209 radial ball bearing of Example 6.1 estimate the maximum permanent deformation at the inner raceway. Compare this value to the maximum elastic deformation.

$$D = 12.7 \text{ mm } (0.5 \text{ in.}) \qquad \text{Ex. 2.1}$$

$$f_i = 0.52 \qquad \text{Ex. 2.2}$$

$$Q_{max} = 4536 \text{ N } (1019 \text{ lb}) \qquad \text{Ex. 6.1}$$

$$\delta_{max} = 0.0604 \text{ mm } (0.00238 \text{ in.}) \qquad \text{Ex. 6.1}$$

$$P_d = 0.0150 \text{ mm } (0.0006 \text{ in.}) \qquad \text{Ex. 2.1}$$

$$\gamma = 0.1954 \qquad \text{Ex. 2.5}$$

$$\delta_s = 5.25 \times 10^{-7} \frac{Q^2}{D^3}\left[1 + \frac{\gamma}{(1-\gamma)}\right]\left(1 - \frac{1}{2f}\right) \tag{21.2}$$

$$= 5.25 \times 10^{-7} \frac{(4536)^2}{(12.7)^3}\left(1 + \frac{0.1954}{1 - 0.1954}\right)\left(1 - \frac{1}{2 \times 0.52}\right)$$

$$= 2.521 \times 10^{-4} \text{ mm } (9.93 \times 10^{-6} \text{ in.})$$

On the other hand the elastic deformation δ_{i0} at $\psi = 0°$ is

$$\delta_{i0} = \delta_{max} - \frac{P_d}{2}$$

$$= 0.0604 - \frac{0.0150}{2} = 0.0454 \text{ mm } (0.00179 \text{ in.})$$

Thus $\delta_{i0} \gg \delta_s$.

For line contact between roller and raceway, the following formula may be used to ascertain permanent deformation with the same restrictions as earlier:

$$\delta_s = \frac{6.03 \times 10^{-11}}{D^2}\left[\frac{Q}{l}\left(\frac{1}{1 \mp \gamma}\right)^{1/2}\right]^3 \tag{21.4}$$

According to Lundberg et al. [21.6], the deformation predicted by equation (21.4) occurs at the end(s) of a line contact when the raceway length tends to exceed the roller effective length. The corresponding deformation in the center of the contact is $\delta_s/6.2$ according to reference [21.6]. Palmgren [21.1] stated that of the total permanent deformation, approximately two-thirds occur in the ring and one-third in the rolling element.

Palmgren's data were based on indentation tests carried out in the 1940s, and the data were dependent on the measurement devices available then. Later, some of these tests were repeated using modern measurement devices. The following conclusions were reached:

1. The amount of total permanent indentation occurring due to an applied load Q between a ball and a raceway appears to be less than that given by equation (21.1).
2. The amount of permanent deformation that occurs in the ball surface is virtually equal to that occurring in the raceway, when balls have not been work hardened.

Accordingly, it can be stated that permanent deformations calculated according to equations (21.1–21.4) will tend to be greater than will actually occur in modern ball and roller bearings of good quality steel and with relatively smooth surface finishes.

STATIC LOAD RATING OF BEARINGS

As indicated earlier, some degree of permanent deformation is unavoidable in loaded rolling bearings. Moreover, experience has demonstrated that rolling bearings do not generally fracture under normal operating loads. Experience further has shown that permanent deformations have little effect on the operation of the bearing if the magnitude at any given contact point is limited to a maximum of $0.0001D$. If the deformations become much larger, the cavities formed in the raceways cause the bearing to vibrate and become noisier although bearing friction does not appear to increase significantly. Bearing operation is usually not impaired in any other manner; however, indentations together with conditions of marginal lubrication can lead to surface-initiated fatigue.

The basic static load rating of a rolling bearing is defined as that load applied to a nonrotating bearing that will result in permanent deformation of $0.0001D$ at the weaker of the inner or outer raceway contacts occurring at the position of the maximum loaded rolling element. In other words in equations (21.2)–(21.4), $\delta_s/D = 0.0001$ at $Q = Q_{max}$. This concept of an allowable amount of permanent deformation consistent with smooth minimal vibration and noise operation of a rolling bearing continues to be the basis of the ISO standard [21.5] and ANSI standards [21.2, 21.3]. In the latest revision of the ISO standard [21.5], it is stated that contact stresses at the center of contact at the maximum loaded rolling element as shown in Table 21.1 yield permanent deformations of $0.0001D$ for the bearing types indicated. The ANSI standards [21.2, 21.3] use the same criteria.

For most radial ball bearing and roller bearing applications the maximum loaded rolling element load according to Chapter 6 may be ap-

TABLE 21.1. Contact Stress That Causes $0.0001D$ Permanent Deformation

	Contact Stress	
Bearing Type	N/mm^2	(psi)
Self-aligning ball bearing	4600	(667,000)
Other ball bearings	4200	(609,000)
Roller bearings	4000	(580,000)

proximated by

$$Q_{\max} = \frac{5F_r}{iZ\cos\alpha} \tag{6.24}$$

in which i is the number of rows of rolling elements. Setting $F_r = C_s$ yields

$$C_s = 0.2iZQ_{\max}\cos\alpha \tag{21.5}$$

Considering the stress criterion, equations (5.25), (5.34), and (5.36) may be used to determine $Q_{\max}$ corresponding to 4200 N/mm^2 (609,000 psi) for standard radial ball bearings. Substituting for $Q_{\max}$ in equation (21.5) yields the equation

$$C_s = \frac{23.8\,iZD^2(a_i^* b_i^*)^3 \cos\alpha}{\left[4 - \frac{1}{f_i} + \left(\frac{2\gamma}{1-\gamma}\right)\right]^2} \tag{21.6}$$

if the maximum stress occurs at the inner raceway and

$$C_s = \frac{23.8\,iZD^2(a_o^* b_o^*)^3 \cos\alpha}{\left[4 - \frac{1}{f_o} - \left(\frac{2\gamma}{1+\gamma}\right)\right]^2} \tag{21.7}$$

if the maximum stress occurs at the outer raceway. Reference [21.5] reduces these equations to

$$C_s = \varphi_s iZD^2 \cos\alpha \tag{21.8}$$

where values of φ_s are given in Table 21.2 for standard ball bearings.

The corresponding formula for radial roller bearings as taken from reference [21.5] is

$$C_s = 44(1-\gamma)ilZD\cos\alpha \tag{21.9}$$

For thrust bearings,

$$Q_{\max} = \frac{F_a}{iZ\sin\alpha} \tag{6.26}$$

TABLE 21.2. Values of φ_s for Ball Bearings[a]

γ	Radial and Angular-Contact Groove Type		Radial Self-Aligning		Thrust	
	Metric[b]	Inch[c]	Metric[b]	Inch[c]	Metric[b]	Inch[c]
0.00	14.7	2140	1.9	284	61.6	8950
0.01	14.9	2180	2.0	290	60.8	8820
0.02	15.1	2220	2.0	297	59.9	8680
0.03	15.3	2270	2.1	301	59.1	8540
0.04	15.5	2300	2.1	307	58.3	8430
0.05	15.7	2350	2.1	313	57.5	8320
0.06	15.9	2400	2.2	319	56.7	8210
0.07	16.0	2430	2.2	325	55.9	8100
0.08	16.2	2480	2.3	332	55.1	7990
0.09	16.4	2440	2.3	338	54.3	7870
0.10	16.4	2410	2.4	344	53.5	7790
0.11	16.1	2370	2.4	351	52.7	7710
0.12	15.9	2340	2.4	357	51.9	7630
0.13	15.6	2290	2.5	363	51.2	7500
0.14	15.4	2260	2.5	370	50.4	7390
0.15	15.2	2220	2.6	376	49.0	7270
0.16	14.9	2190	2.6	382	48.8	7150
0.17	14.7	2140	2.7	389	48.0	7030
0.18	14.4	2110	2.7	397	47.3	6910
0.19	14.2	2070	2.8	403	46.5	6780
0.20	14.0	2040	2.8	409	45.7	6670
0.21	13.7	2000	2.8	417	44.9	6540
0.22	13.5	1960	2.9	423	44.2	6420
0.23	13.2	1920	2.9	430	43.5	6300
0.24	13.0	1890	3.0	438	42.7	6200
0.25	12.8	1850	3.0	446	41.9	6110
0.26	12.5	1820	3.1	452	41.2	6010
0.27	12.3	1780	3.1	459	40.5	5880
0.28	12.1	1750	3.2	467	39.7	5760
0.29	11.8	1730	3.2	473	39.0	5660
0.30	11.6	1690	3.3	481	38.2	5570
0.31	11.4	1670	3.3	488	37.5	5490
0.32	11.2	1630	3.4	496	36.8	5370
0.33	10.9	1600	3.4	503	36.0	5244

TABLE 21.2. (*Continued*)

	Radial and Angular-Contact Groove Type		Radial Self-Aligning		Thrust	
γ	Metric[b]	Inch[c]	Metric[b]	Inch[c]	Metric[b]	Inch[c]
0.34	10.7	1560	3.5	511	35.3	5120
0.35	10.5	1530	3.5	519	34.6	5040
0.36	10.3	1490	3.6	526		
0.37	10.0	1460	3.6	534		
0.38	9.8	1440	3.7	541		
0.39	9.6	1400	3.8	549		
0.40	9.4	1370	3.8	558		

[a]Based on modulus of elasticity = 2.07×10^5 N/mm^2 (30×10^6 psi), Poisson's ratio = 0.3.
[b]Use to obtain C_s in newtons when D is given in millimeters.
[c]Use to obtain C_s in pounds when D is given in inches.

Setting $F_a = C_{sa}$ yields

$$C_{sa} = iZQ_{max} \sin \alpha \tag{21.10}$$

Correspondingly, the standard stress criterion formula is

$$C_{sa} = \varphi_s Z D^2 \sin \alpha \tag{21.11}$$

where φ_s is given by Table 21.2.

For thrust roller bearings with line contact,

$$C_{sa} = 220(1 - \gamma) ZlD \sin \alpha \tag{21.12}$$

When hardness of the surfaces is less than specified lower limit of validity, a correction factor may be applied directly to the basic static capacity such that

$$C_s' = \eta_s C_s \tag{21.13}$$

in which

$$\eta_s = \eta_1 \left(\frac{HV}{800} \right)^2 \leq 1 \tag{21.14}$$

and HV is the Vickers hardness. A graph of Vickers hardness versus Rockwell C hardness is shown in Fig. 21.4. Equation (21.14) was developed empirically by SKF. The values of η_1 depend on type of contact and are given by Table 21.3. η_s has a maximum value of unity.

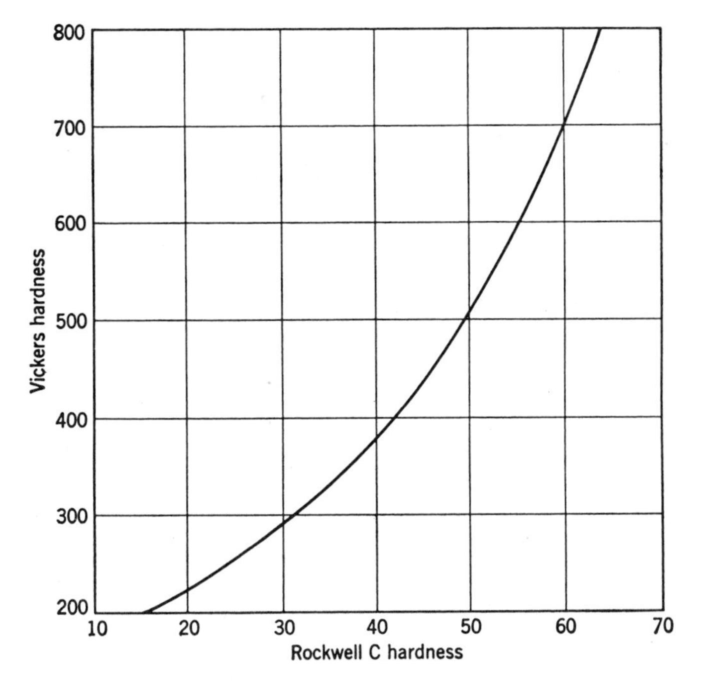

FIGURE 21.4. Vickers hardness vs Rockwell C hardness.

TABLE 21.3. Values of η_1

η_1	Type of Contact
1	Ball on plane (self-aligning ball bearings)
1.5	Ball on groove
2	Roller on roller (radial roller bearings)
2.5	Roller on plane

STATIC EQUIVALENT LOAD

To compare the load on a nonrotating bearing with the basic static capacity, it is necessary to determine the static equivalent load, that is, the pure radial or pure thrust load—whichever is appropriate—that would cause the same total permanent deformation at the most heavily loaded contact as the applied combined load. A theoretical calculation of this load may be made in accordance with methods of Chapter 6.

In lieu of the more rigorous approach, for bearings subjected to combined radial and thrust loads the static equivalent load may be calculated as follows:

$$F_s = X_s F_r + Y_s F_a \tag{21.15}$$

TABLE 21.4. Values of X_s and Y_s for Radial Ball Bearings

	Single-Row Bearings		Double-Row Bearings	
Bearing Type	X_s	Y_s[b]	X_s	Y_s[b]
Radial contact groove ball bearing[a, c]	0.6	0.5	0.6	0.5
Angular-contact groove ball bearings				
$\alpha = 15°$	0.5	0.47	1	0.94
$\alpha = 20°$	0.5	0.42	1	0.84
$\alpha = 25°$	0.5	0.38	1	0.76
$\alpha = 30°$	0.5	0.33	1	0.66
$\alpha = 35°$	0.5	0.29	1	0.58
$\alpha = 40°$	0.5	0.26	1	0.52
Self-aligning ball bearings	0.5	0.22 ctn α	1	0.44 ctn α

[a] P_o is always $\geq F_r$.
[b] Values of Y_o for intermediate contact angles are obtained by linear interpolation.
[c] Permissible maximum value of F_a/C_o depends on the bearing design (groove depth and internal clearance.)

If F_r is greater than F_s as calculated in equation (21.15), use F_s equal to F_r. Table 21.4 taken from reference [21.2] gives values of X_s and Y_s for radial ball bearings.

Data in Table 21.4 pertain to bearings having a groove curvature not greater than 53% of the ball diameter. Double-row bearings are presumed to be symmetrical. Face-to-face and back-to-back mounted angular-contact ball bearings are similar to double-row bearings; tandem mounted bearings are similar to single bearings.

For radial roller bearings the values of Table 21.5, taken from reference [21.3] apply.

For thrust bearings the static equivalent load is given by

$$F_{sa} = F_a + 2.3F_r \tan \alpha \tag{21.16}$$

When F_r is greater than 0.44 F_a ctn α, the accuracy of equation (21.16) diminishes and the theoretical approach according to Chapter 6 is warranted.

FRACTURE OF BEARING COMPONENTS

It is generally considered that the load which will fracture bearing rolling elements or raceways is greater than $8C_s$, (see reference [21.1]).

TABLE 21.5. Values of X_s and Y_s for Radial Roller Bearings[a]

	Single-Row Bearings[b]		Double-Row Bearings	
Bearing Type	X_s	Y_s	X_s	Y_s
Self-aligning and tapered roller bearings	0.5	0.22 ctn α	1	0.44 ctn α
$\alpha \neq 0°$				

[a]The ability of radial roller bearings with $\alpha = 0°$ to support axial loads varies considerably with bearing design and execution. The bearing user should therefore consult the bearing manufacturer for recommendations regarding the evaluation of equivalent load in cases where bearings with $\alpha = 0°$ are subjected to axial load.
[b]F_s is always $\geqq F_r$. The static equivalent radial load for radial roller bearings with $\alpha = 0°$, and subjected to radial load only is $F_s = F_r$.

PERMISSIBLE STATIC LOAD

It is known that the maximum load on a rotating bearing may be permitted to exceed the basic load rating provided this load acts continuously through several revolutions of bearing rotation. In this manner the permanent deformations that occur are uniformly distributed over the raceways and rolling elements, and the bearing retains satisfactory operation. If, on the other hand, the load is of short duration, unevenly distributed deformations may develop even when the bearing is rotating at the instant shock occurs. For this situation it is necessary to use a bearing whose basic static load rating exceeds the maximum applied load. When the load is of longer duration, the basic static load rating may be exceeded without impairing operation of the bearing.

According to the type of bearing service a factor of safety may be applied to the basic static load rating. Therefore the allowable load is given by

$$F_s = \frac{C_s}{FS} \tag{21.17}$$

Table 21.6 gives satisfactory values of FS for various types of service.

TABLE 21.6. Factor of Safety for Static Loading

Factor of Safety	Service
$FS \geq 0.5$	Smooth shock-free operation
$FS \geq 1$	Ordinary service
$FS \geq 2$	Sudden shocks and high requirements for smooth running

Example 21.2. The 218 angular-contact ball bearing is subjected to simultaneously applied radial and thrust loads of 17,800 N (4000 lb) each. Using the ISO standard, estimate the factor of safety for the basic static load rating of the bearing.

$$f = 0.52 \qquad \text{Ex. 2.3}$$

$$Z = 16 \qquad \text{Ex. 6.5}$$

$$D = 22.23 \text{ mm } (0.875 \text{ in.}) \qquad \text{Ex. 2.3}$$

$$\alpha = 40° \qquad \text{Ex. 2.3}$$

$$d_m = 125.3 \text{ mm } (4.932 \text{ in.}) \qquad \text{Ex. 2.6}$$

$$\gamma = \frac{D \cos \alpha}{d_m} \qquad \text{Ex. 2.6}$$

$$= \frac{22.23 \cos (40°)}{125.3} = 0.1358 \qquad (2.27)$$

$$C_s = \varphi_s iZD^2 \cos \alpha \qquad (21.8)$$

From Table 21.2 at $\gamma = 0.1358$, $\varphi_s = 15.48$

$$C_s = 15.48 \times 1 \times 16 \times (22.23)^2 \cos (40°)$$

$$= 93{,}760 \text{ N } (21{,}070 \text{ lb})$$

From Table 21.4,

$$X_s = 0.5, \qquad Y = 0.26$$

$$F_s = X_s F_r + Y_s F_a \qquad (21.15)$$

$$= 0.5 \times 17{,}800 + 0.26 \times 17{,}800 = 13{,}530 \text{ N } (3040 \text{ lb})$$

Therefore, use $F_s = 17{,}800$ N (4000 lb)

$$F_s = \frac{C_s}{FS}$$

$$FS = \frac{93{,}760}{17{,}800} = 5.3$$

CLOSURE

Smoothness of operation is an important consideration for modern ball and roller bearings. Interruptions in the rolling path such as caused by permanent deformations result in increased friction, noise, and vibration. Chapter 25 discusses the noise and vibration phenomenon in substantial detail. In this chapter, the discussion centered on bearing static load ratings, which, if not exceeded while the bearing was not rotating, would preclude permanent deformations of significant magnitude. The ratings were based on a maximum allowable permanent deformation of $0.0001D$. Subsequently, it was determined that for various types of ball and roller bearings, this deformation could be related to a value of rolling element–raceway contact stress. In accordance with this stress, basic static load ratings are developed for each rolling bearing type and size.

Generally, a load of magnitude equal to the basic static load rating cannot be continuously applied to the bearing with the expectation of obtaining satisfactory endurance characteristics. Rather, the basic static load rating is based on a sudden overload or, at most, one of short duration compared to the normal loading during continuous operation. Exceptions to this rule are bearings that undergo infrequent operation of short duration, for example, bearings on doors of missile silos or dam gate bearings. For these and simpler applications, bearing design may be based on basic static load rating rather than on endurance of fatigue. In Chapter 22, the concept of a shakedown limit stress is discussed, which pertains to raceway subsurface microstructural alterations that occur during rotation while the current static load ratings are based upon damage during nonrotation. Because of relatively slow speeds of rotation and infrequent operation, neither vibration nor surface fatigue may be as significant in such applications as excessive plastic flow of subsurface material. The bearings could thus be sized to eliminate or minimize such plastic flow and ultimately bearing failure.

REFERENCES

21.1. A. Palmgren, *Ball and Roller Bearing Engineering*, 3rd ed., Burbank, Philadelphia (1959).

21.2. American National Standard, *ANSI/AFBMA Std 9-1990*, "Load Ratings and Fatigue Life for Ball Bearings."

21.3. American National Standard, *ANSI/AFBMA Std 11-1990*, "Load Ratings and Fatigue Life for Roller Bearings."

21.4. R. S. Sayles and S. Y. Poon, "Surface Topography and Rolling Element Vibration," *Precis. Eng.* 137–144 (1981).

21.5. International Standard ISO 76, "Rolling Bearings—Static Load Ratings" (1989).

21.6. G. Lundberg, A. Palmgren, and E. Bratt, "Statiska Bärförmågan hos Kullager och Rullager," *Kullagertidningen* **3** (1943).

22

MATERIAL RESPONSE TO ROLLING CONTACT

LIST OF SYMBOLS

Symbol	Description	Units
θ	Angle of incidence of x-rays to diffracting planes	degrees
ψ	Angle between specimen normal and normal to diffracting planes	degrees
d	Spacing of crystallographic planes	Å
λ	X-ray wavelength	Å

GENERAL

Since 1945 much work has been devoted to characterizing alterations of the microstructure, hardness, and residual stress distribution occurring in rolling bearing components during bearing operation. The results of this work provide a technological foundation that serves as a basis for continuing analytical and experimental investigations into the fundamental mechanisms involved in bearing failure.

Materials-oriented investigations have been primarily phenomenological in nature; that is, attempts have been made to carefully control test conditions and to correlate material response characteristics with test conditions and bearing performance. Subsequent analyses of the metallurgical mechanisms involved are prerequisites for establishing reliable analytical models for predicting performance.

This chapter reviews the microstructural alterations resulting from rolling contact and the attendant changes in residual stress state. Observed effects of residual and applied bulk stresses, combined with rolling contact stress, on bearing performance are discussed in terms of fatigue life, failure mode, and dimensional stability.

MICROSTRUCTURES OF ROLLING BEARING STEELS

The microstructure of martensitically hardened and tempered AISI 52100 steel is shown in Fig. 22.1. The microstructure consists primarily of plate martensite [22.1], with 5–8 vol.% of $(Fe,Cr)_3C$ type carbides [22.2], and up to 20 vol.% retained austenite, depending on austenitizing and tempering conditions. Tempered hardness is generally R_c 58–64. The lower values of retained austenite content and hardness are characteristically associated with higher tempering temperatures.

The hardened and tempered microstructures of the near-surface (case)

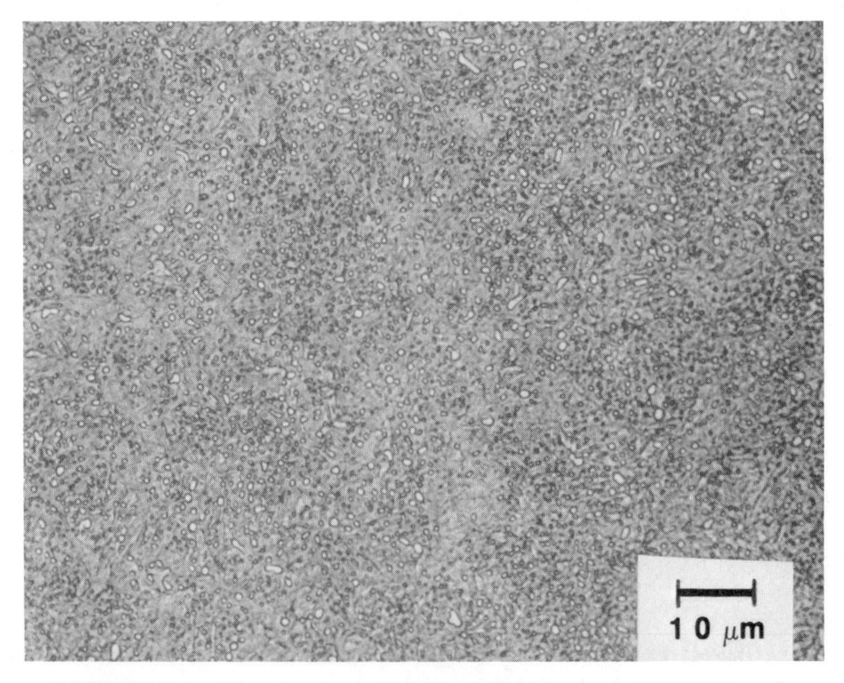

FIGURE 22.1. Microstructure of hardened and tempered AISI 52100 steel.

and subsurface (core) regions of carburized bearing steel (e.g., AISI 8620), are shown in Fig. 2.2. The case microstructure is predominately martensitic (plate type). The volume fractions of carbide and retained austenite vary widely, depending on the carburizing conditions, steel analysis, and tempering procedures used.

The microstructure of the core region—that is, the subsurface material unaffected by carburizing—is also primarily martensitic, but, due to the low carbon content (gradually 0.2 wt.% or less), is of the lath type rather than plate martensite. Characterization of these two martensite morphologies is detailed in [22.1] and [22.3].

With the relatively rare exceptions of single-stress-cycle bearing failure (e.g., fracture from heavy impact loading and wear), all material-related bearing failures involve accumulation of irreversible plastic deformation during cyclic stressing. The latter is the classical definition of metallurgical fatigue. The variety of bearing failure manifestations arises from the conditions that promote fatigue initiations and the manner in which the operating conditions sustain propagation of the fatigue mechanisms. Fatigue failure may be initiated by an exogenous inclusion in the bearing steel, by mechanical disruption of the integrity of the rolling contact surface (e.g., debris dents or scratches), or by a corrosion pit, to name just a few. The subsequent bearing failure, which by definition precludes continued functionality, results from progression of fatigue damage from the initiating cause.

Specifications for microstructural characteristics, hardness, depth of hardening, amount of retained austenite, and so forth, to a large extent have been based upon bearing performance experiences in the field and the results of bearing endurance tests conducted under controlled operating conditions. Several decades of collective experience provide today's phenomenologically based compendium of materials knowledge concerning requirements for and responses to rolling contact stressing. Using this experience to provide bearing steel quality heat treatment, manufacturing procedures, and operating conditions that minimize the likelihood of experienced early-failure modes has permitted the study of material response to long-term rolling contact stressing. The objectives of such studies have been oriented toward gaining insight into the fundamental mechanisms involved in rolling contact fatigue and to assess fatigue life potential under "ideal" operating conditions. Alterations of the microstructure that have been observed under these conditions are described in the next section.

MICROSTRUCTURAL ALTERATIONS DUE TO ROLLING CONTACT

Marked alteration of the near-surface microstructure of endurance-tested bearing inner rings has been reported in the literature since 1946 [22.4–

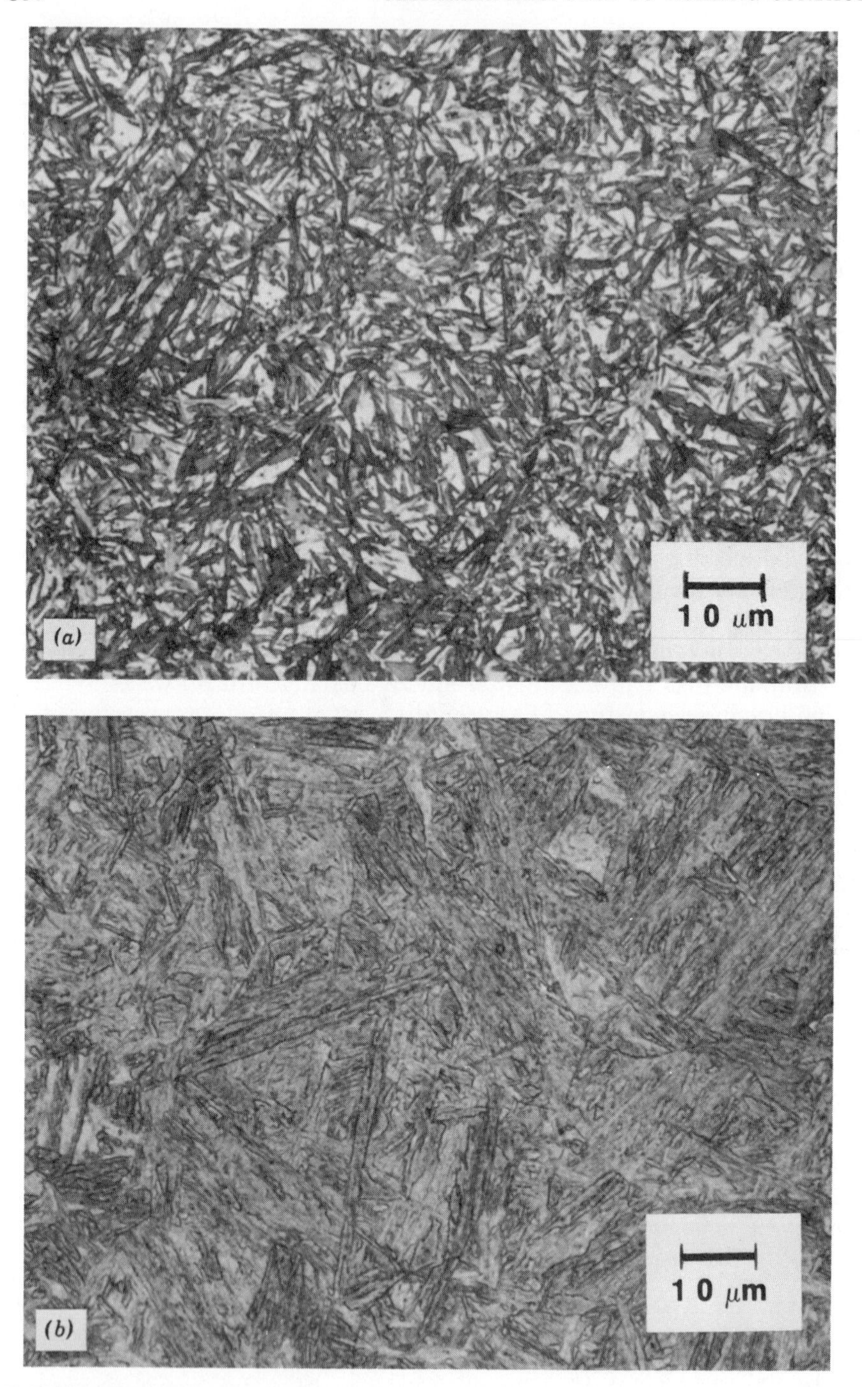

FIGURE 22.2. Microstructure of carburized, hardened, and tempered AISI 8620 steel. (*a*) Case. (*b*) Core.

22.13]. The alterations are principally characterized by differences in the etching response of the microstructure in the region just beneath the raceway surface (see Fig. 22.3), and are most heavily concentrated at a depth corresponding to the maximum unidirectional shear stress associated with the Hertzian stress field in the contact [22.5, 22.6].

Three aspects of microstructural alterations have been described [22.7, 22.8] and chronologically characterized [22.8]: the dark etching region (DER), DER + 30° bands, and DER + 30° bands + 80° bands. The structural changes obtained as a function of stress level and number of inner ring revolutions are diagrammed in Fig. 22.4. Optical micrographs of the structural alterations, in parallel sections, are shown in Fig. 22.5.

The first alteration is the formation of the DER. Transmission electron microscopy identified the DER as consisting of a ferritic phase, containing an inhomogeneously distributed excess carbon content (equivalent to that of the initial martensite) mixed with residual parent martensite. A stress-induced process of martensitic decay is indicated [22.8]. The second manifestation of altered microstructure [22.8, 22.9] is the formation of white etching, disc-shape regions of ferrite, about 0.1–0.5 μm (40–200 μin.) thick, and inclined at an angle of approximately 30° to the raceway circumferential tangent. These regions are sandwiched between carbide rich layers. The third feature, initially reported

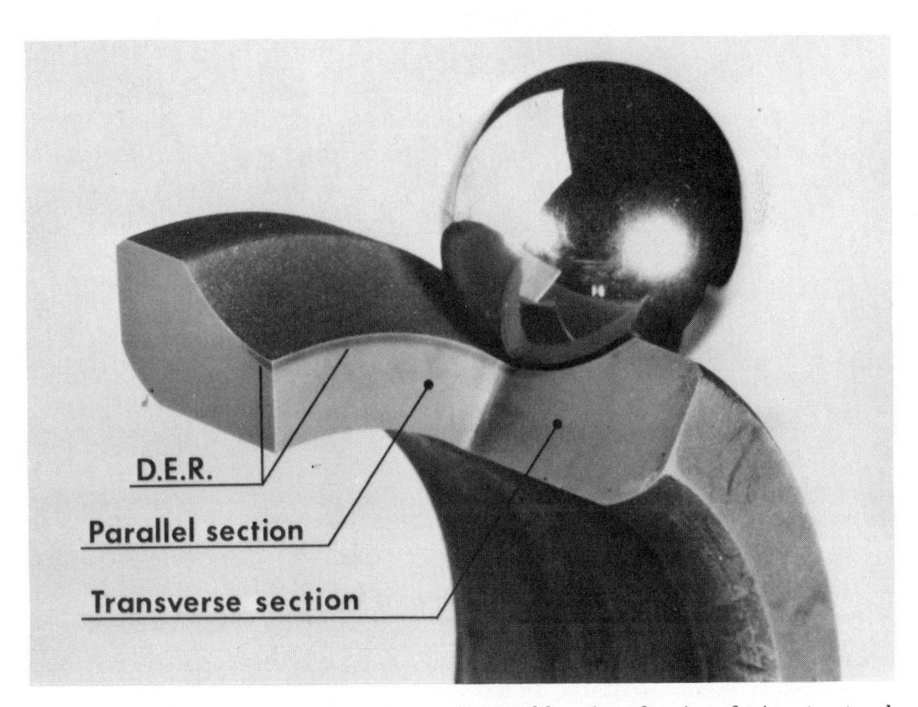

FIGURE 22.3. Orientation of viewing sections and location of region of microstructural alterations in a 309 deep-groove ball bearing inner ring (from [22.8]).

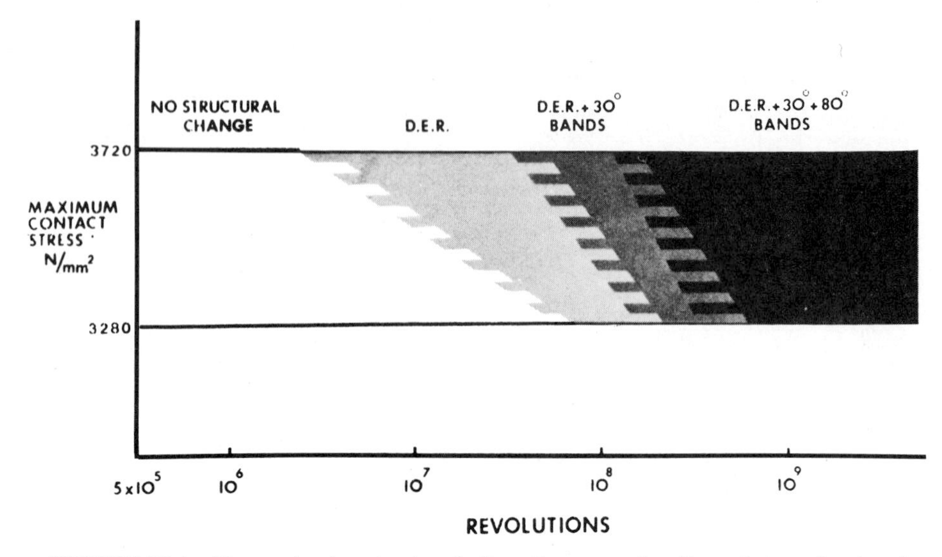

FIGURE 22.4. Observed microstructural alterations as a function of stress level and number of inner ring revolutions (from [22.8]).

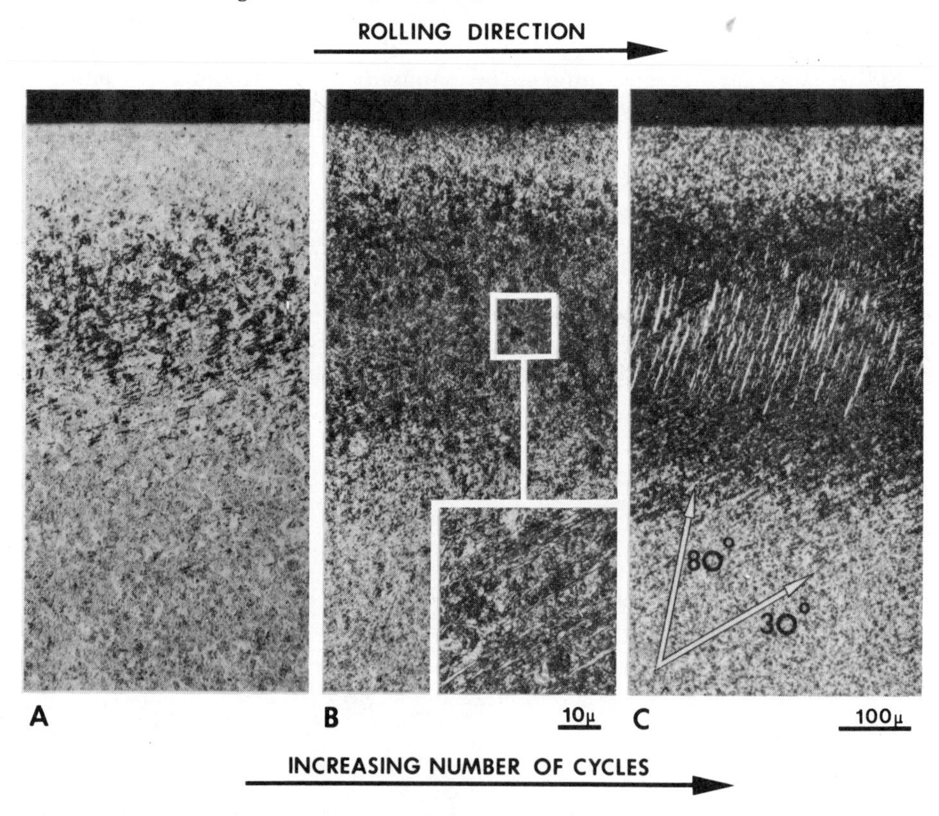

FIGURE 22.5. Optical micrographs of structural changes in 309 deep-groove ball bearing inner rings (parallel section). (*a*) DER in early stage. (*b*) Fully developed DER and 30° bands. (*c*) DER, 30° bands, and 80° bands. AISI 52100 steel (from [22.8]).

in [22.7], is a second set of white etching bands, considerably larger than the 30° bands and inclined at 80° to the raceway tangent in parallel sections. These disc-shaped regions are about 10 μm (0.0004 in.) thick and consist of severely plastically deformed ferrite [22.8].

As shown in Fig. 22.5, both the 30° and 80° bands incline toward the surface in the direction of rolling element motion. Reversing the direction of rotation reverses the orientation of the bands. The characteristic angular orientations of these white etching bands have not been satisfactorily explained. Crystallographic texturing in the near-surface region of ball bearing inner rings after high-stress operation has, however, been observed by Voskamp [22.6]. Angular characteristics of the texturing are reported to be consistent with white etching band orientations, suggesting that the bands have a crystallographically determined nature.

Moreover, the crystal alignment appears to accentuate the plane of least resistance to fracture; that is, the cube planes of body-centered-cubic (bcc) Fe become aligned parallel to the raceway surface with the so-called [110] direction parallel to the direction of rolling.

Hardness has been reported to increase slightly in the early stages of testing and then to decrease markedly in the region associated with microstructural alteration [22.8, 22.13]. The number of stress cycles required to produce the sequence of events leading to microstructural alteration may be significantly reduced by increasing the temperature of the test specimen [22.13]. Localized changes in retained austenite content and residual stress level are also associated with these microstructural alterations. See next section.

Another manifestation of microstructural alteration found in bearing components that have experienced rolling contact service is commonly called a "butterfly," beause of winglike emanations from a "body" composed of a nonmetallic inclusion. An example of a butterfly is shown in Fig. 22.6.

Characteristically, the butterfly wings are oriented at an angle of 40–45° to the raceway tangent, when viewed parallel to the raceway [22.14–22.16], and appear white, in contrast to the surrounding matrix of tempered martensite, after nital etching. Butterflies can occur at depths well below the maximum shear stress, and they are characteristically associated with microcracks running along the smooth edges of the wings.

Butterflies form around oxide, silicate, and titanium nitride particles but not in conjunction with manganese sulfide or carbide particles [22.15–22.17]. Wing development depends on stress level and the number of stress cycles [22.16].

A comprehensive characterization of the microstructural features of butterfly wings [22.18] concluded that they consist of a dispersion of ultrafine-grained ferrite and carbide, very similar in nature and formation mode to the 30° and 80° white etching bands described earlier. Further,

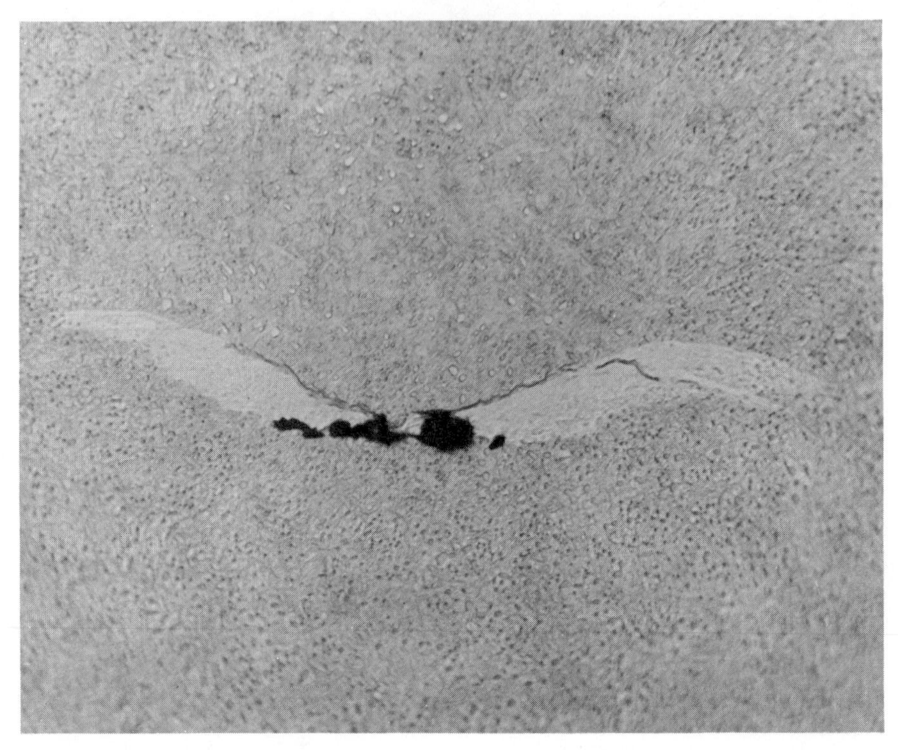

FIGURE 22.6. Optical micrograph of white etching regions in association with a subsurface, nonmetallic inclusion.

the wings are probably initiated by preexisting cracks associated with nonmetallic inclusion bodies [22.14–22.16, 22.18]. Subsequent crack and wing growth proceed together.

Although they are striking manifestations of high-stress, high-cycle, rolling contact, neither white etching bands nor butterflies have been identified as failure-initiating characteristics. They are due to stress and age.

RESIDUAL STRESSES IN ROLLING BEARING COMPONENTS

Sources of Residual Stresses

Residual stress is that which remains in a material when all externally applied forces are removed. Residual stresses arise in an object from any process that produces a nonuniform change in shape or volume. These stresses may be induced mechanically, thermally, chemically, or by combinations of these processes [22.19]. If a relatively thin sheet of malleable material, such as copper, is repeatedly struck with a hammer, the thickness of the sheet is reduced, and the length and width are correspondingly increased, preserving constancy of volume. If the same number of equally intensive hammer blows were uniformly delivered to the surface of a copper block several inches thick, the depth of penetration of plastic deformation would be relatively shallow with respect to the block thickness. The deformed surface layer would be restrained from lateral expansion by the bulk of subsurface material, which experienced less deformation. Consequently, the heavily deformed surface material would be like an elastically compressed spring, prevented from expanding to its unloaded dimensions by its association with elastically extended subsurface material. The resulting residual stress profile is one in which the surface region is in residual compression and the subsurface region is in a balancing residual tension. This example is a literal description of the shot-peening process, wherein a surface is bombarded with pellets of steel or glass. A highly desirable residual-stress pattern is established for components that experience high, cyclic tensile stresses at the surface during service. The magnitude of tensile stress experienced by the component during service is functionally reduced by the amount of residual compressive stress, thereby providing significantly increased fatigue lives for parts such as springs and shafts.

The shot-peening example illustrates the essential characteristics of a surface in which residual stress has been induced:

1. Nonuniformity of plastic deformation—that is, being near-surface only—encourages the surface material to expand laterally.
2. Subsurface material, which experienced less plastic deformation, is elastically strained (in tension, in this example) as it restrains expansion of the surface material, thereby inducing residual compressive stress in the surface region.
3. The resulting state of residual stress is a reflection of the elastic components of strain in the surface and subsurface regions, which are in equilibrium, providing a balanced tensile-compressive system.

Heat treatment, such as is used for hardening rolling bearing components, can exert very significant influence over the state of residual stress. Depending on the steel analysis, austenitizing temperature, quenching severity, component geometry, section thickness, and so forth, heat treatment can provide either residual compressive stress or residual tensile stress in the surface of the hardened component [22.19, 22.20]. Temperature gradients are established from the surface to the center of a part during quenching in a hardening treatment. Differential thermal contraction associated with these gradients provides for nonuniform plastic deformation, giving rise to residual stresses. Additionally, volumetric changes associated with the phase transformations taking place during heat treatment of steel occur at different times during quenching at the part surface and interior due to the thermal gradients established. These sequential volumetric changes, combined with differential thermal contractions, are responsible for the residual-stress state in a hardened steel component. The sequence and relative magnitudes of these contributing factors determine the stress magnitude and whether the surface is in residual tension or compression.

Grinding of a hardened steel component to finished dimensions also affects the residual surface stress. Generally, neglecting the effects of abusive grinding practices that generate excessive heat and produce microstructure alterations, it is found that the residual-stress effects associated with grinding are confined to material within the first 50 microns (0.002 in.) of the surface. Good grinding practice, as applied to bearing rings, produces circumferential residual compressive stress in a shallow surface layer. Grinding also involves some plastic deformation of the surface, producing residual compression as described earlier.

The residual-stress state in a finished bearing ring is therefore a function of heat treatment and grinding. If properly ground, the residual stress in a through-hardened bearing ring will be 0 to slightly compressive. The subsurface residual stress conditions will be determined by the prior heat treatment.

Measuring Residual Stress

The most widely used method to precisely determine residual stress in crystalline materials is x-ray diffraction. X-ray diffraction equipment and techniques are well developed and described in the literature [22.19, 22.21, 22.22].

All metals, being crystalline solids, consist of atoms arranged in planes precisely positioned in terms of interplanar distances. In a single crystal of a metal the orientation of these planes of atoms is consistent everywhere within the crystal. Most metallic objects of interest here are not single crystals but polycrystalline; that is, they consist of many crystals. Each crystal or grain in the microstructure of a polycrystalline metal is delineated from its neighbors by mismatch in the orientation of the crystallographic planes. The region of mismatch or disorder between neighboring grains is called the grain boundary.

In the unstressed condition the distances between crystallographic planes assume equilibrium values. If elastically stressed in tension (i.e., a tensile stress component perpendicular to the planes), the interplanar distance is increased. In compression the distance is decreased. Consequently, if the equilibrium interplanar distance, the stressed interplanar distance, and the orientation of the planes with respect to the stress axis are known, the elastic strain conditions are defined. Multiplying the strain by the elastic modulus for the material being studied provides the value of residual stress. X-ray diffraction is used to measure interplanar distances. The technique is therefore used to measure elastic strain, from which the associated residual stress is calculated.

The relationship stating the conditions that must be met for x-ray diffraction to occur was first formulated by Bragg [22.23] in 1912 and is known as Bragg's law:

$$\lambda = 2d \sin \theta \tag{22.1}$$

where λ is the wavelength of the x-rays used, d is the interplanar spacing, and θ is the angle of incidence of the x-ray beam to the diffracting planes. What Bragg's law states is that for a given x-ray wavelength λ and interplanar spacing d, there is an angle of incidence θ such that the x-rays penetrating the specimen surface will experience constructive interference and emerge from the surface at an angle θ to the planes of spacing d. With appropriate detection equipment the emergent (diffracted) x-rays can be detected, and the precise diffraction angle can be determined.

The orientations of a polycrystalline specimen and a particular family of crystallographics planes (in randomly oriented grains) to the x-ray beam in residual-stress determination are shown in Fig. 22.7.

Initially, the specimen is oriented such that the normal to the specimen surface (A) and the normal to the diffracting planes (B) are coincident (Fig. 22.7*a*); that is, the diffracting planes are parallel to the specimen surface. The diffraction angle, θ is determined for these planes.

The specimen is then rotated to an orientation shown in Fig. 22.7*b*. In this orientation the normal to the diffracting planes makes an angle ψ with the normal to the specimen surface. These planes, being at a nonzero angle to residual stress acting in the direction parallel to the specimen surface, are elastically separated from their equilibrium spacing d. Residual compressive stress gives a smaller value of d. According to equation (22.1), for fixed λ a decrease in d requires an increase in the value of $\sin \theta$—that is, a larger diffraction angle θ. Conversely, residual tensile stress gives a larger value of d, corresponding to a smaller θ. Therefore, comparing the θ values obtained with the two specimen orientations—that is, Fig. 22.7*a*,*b*—will show an increase or decrease, indicating residual compression or residual tension, respectively. The magnitude of the change in θ is related to the magnitude of residual stress by a calculated stress factor [22.19, 22.21, 22.22].

Values of residual stress as functions of depth below the surface (i.e., residual stress profiles) are obtained by successive material removal and x-ray residual-stress determinations. Material removal is most appro-

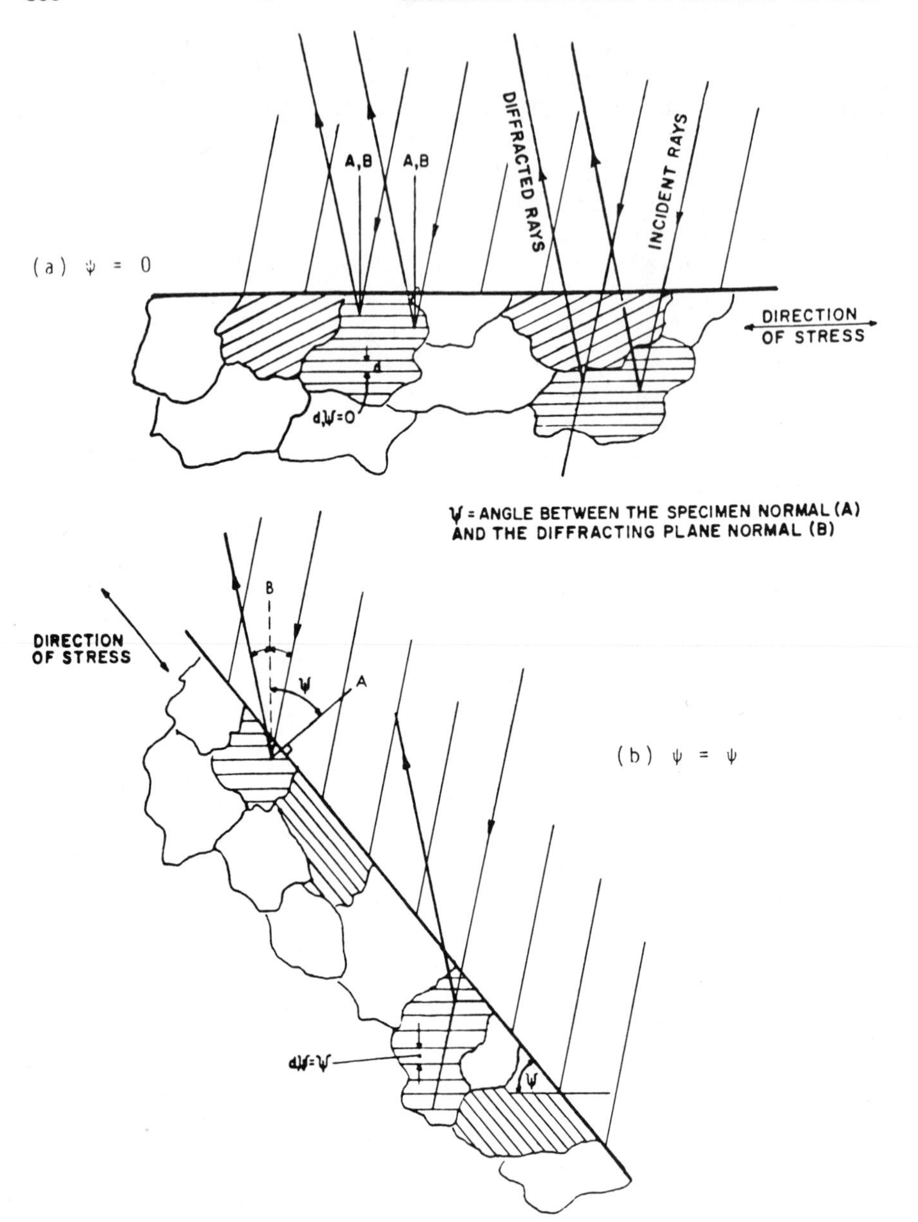

FIGURE 22.7. Orientations of specimen surface, diffracting planes, and x-ray beam for residual-stress determination via x-ray diffraction.

priately performed by electrochemical means. Mechanical material removal will invariably introduce alterations to the preexisting stress profile.

Alteration of Residual Stress Due to Rolling Contact

Associated with the microstructural alterations resulting from rolling contact stressing, significant changes in residual stress and retained austenite content have been reported [22.6, 22.10, 22.11, 22.13, 22.24, 22.25]. The forms of the changes in tangential residual stress and retained austenite content profiles are illustrated in Fig. 22.8.

The low and high load values correspond to the two levels of maxi-

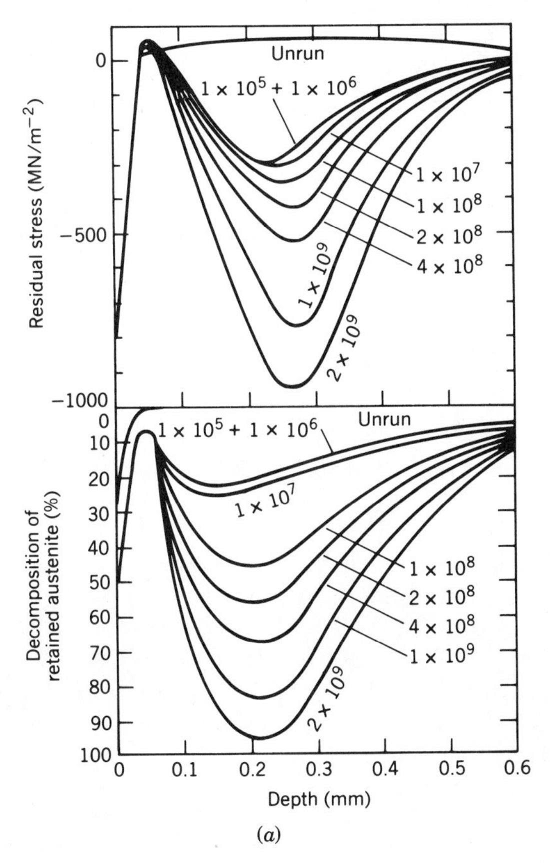

FIGURE 22.8. Residual stress and percent retained austenite decomposition vs depth below raceway surface for various numbers of inner ring revolutions. Bearing: 6309 deep-groove ball bearing. Material: AISI 52100, heat-treated hardness of R_C 64. (*a*) Maximum contact stress: 3280 N/mm^2; depth of maximum orthogonal shear stress: 0.19 mm; depth of maximum unidirectional shear stress: 0.30 mm. *continued on next page.*

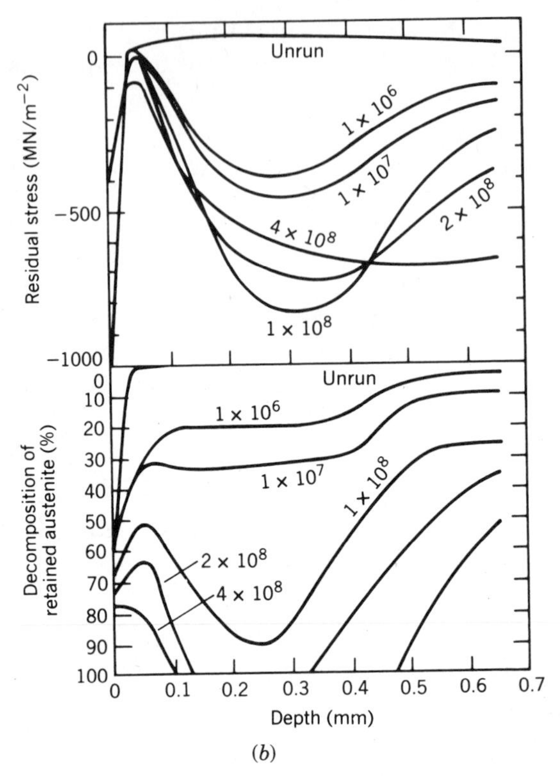

(*b*)

FIGURE 22.8. (*b*) Maximum contact stress: 3720 N/mm^2; depth of maximum orthogonal shear stress: 0.21 mm; depth of maximum unidirectional shear stress: 0.33 mm (from [22.24]).

mum contact stress indicated in Fig. 22.4, depicting the number of revolutions of which the various microstructural alterations occur. Comparison of Figs. 22.4 and 22.8 indicates that significant changes in residual-stress profile and retained austenite content precede any observable alterations in microstructure. The residual-stress data of Fig. 22.8 show peak values at increasing depths corresponding to increasing numbers of stress cycles. A similar form is indicated for decomposition of retained austenite, with peak effect depths being slightly less than for residual stress. The data in Fig. 22.8 for the high maximum-contact stress indicate more rapid rates of change for both residual stress and retained austenite content.

The slight differences in the depths at which the peak values occur in residual stress and retained austenite decomposition imply correlation with the maximum unidirectional shear stress and maximum orthogonal shear stress, respectively. Earlier work [22.13] supports the correlation of peak residual-stress values with the maximum unidirectional shear stress. There appears to be no direct relationship between retained austenite decomposition and the generation of residual compressive stress, nor any indication of which, if either, of these processes triggers microstructural alterations [22.24].

Effects of Residual Stress on Rolling Contact Fatigue Life

Experimental studies on the endurance life of rolling bearing components have indicated a positive effect of residual compressive stress over a "zero" stress state [22.25–22.28, 22.32, 22.33]. Another investigation [22.29] indicated that in addition to the positive effect of compressive stress superimposed on Hertzian contact stresses, there was a definite negative effect of superimposed tensile stress. In [22.30] the negative effect of tensile stress was demonstrated, but it was also concluded that there was no advantage of high residual compressive stress over a zero stress state superimposed on the Hertzian stress field. Another study revealed that bearings tested with inner ring raceways at two levels of residual compressive stress showed no significant difference in life [22.31].

The variety of methods used to induce the residual stresses in components for rolling contact fatigue testing included prestressing of the inner rings by running them in bearings at a load higher than the subsequent test load (thereby inducing subsurface residual compression as described previously) [22.25, 22.27, 22.30], bulk loading of test elements by shrink-fitting rings on a shaft or press-fitting into a housing [22.29], and altering the chemistry of the surface during heat treatment to provide residual compression in the quenched and tempered surface [22.31, 22.32].

As discussed earlier, the subsurface residual compressive stress generated in a bearing ring during high-contact-stress operation is accompanied by changes in hardness, microstructure, and crystallographic texture. Therefore, prestressing by high-stress operation before testing could introduce significant factors to rolling contact fatigue life and to residual compressive stress.

Bulk loading of rolling contact test specimens by heavy interference-fitting on shafts or in housings provides stress profiles across the specimen section that are quite different from self-contained, balanced, residual-stress profiles within a freestanding component. Although such a test scheme might accurately indicate performance trends for bearing applications in which bulk ring loading is similarly experienced, it is not clear that such interact with a Hertzian stress field in the same manner as a true residual state of stress.

Alteration of surface chemistry by infusion of nitrogen or carbon to provide residual surface compressive stress also changes the microstructural characteristics, the mechanical properties of the surface region, and, perhaps, physical properties such as friction coefficient. Consequently, resolution of the separate influence of residual stress is obscured.

As indicated previously, Voskamp [22.6] determined that realigning crystals caused by "overrolling," causes a plane of "weakness" to form below the raceway aligned parallel to the raceway. Moreover, he postulated that residual tensile stresses created by such overrolling tend to cause fatigue failure to occur with propagation beneath and parallel to the raceway. Voskamp [22.6] further shows photographic evidence of such failures in very highly loaded bearings: for example, 5200 N/mm^2 (750 ksi) maximum contact stress.

In any event, regardless of direct or indirect association with rolling contact fatigue life, there is general agreement that a residual or applied bulk compressive stress is a more desirable situation in a rolling bearing component than is a residual or applied bulk tensile stress. Quantitative relationships between the magnitude and sign of residual stress and bearing life have not yet been established.

Shakedown

Recently, the phenomenon of shakedown occurring in the material of bearing rings that has been subjected to the stresses of normal operating loads and speeds has been recognized. The shakedown phenomenon can be described as a self-stabilization of the material under a cyclically applied load of such magnitude that the material yield stress has been exceeded. Thus, a permanent change has occurred in the material below the rolling contact surfaces. Plastic flow of material has occurred in the structure, generally within a limited region. As a result of the plastic deformations produced during one load cycle, residual stresses occur after the load is removed, keeping the material in equilibrium. During the next load cycle, the residual stresses act together with the stresses caused by the externally applied load. If the load is not too heavy, the amount of plastic flow is less than during the previous cycle. If the load causes stresses in excess of the shakedown limit, however, the plastic flow continues and, in fact, spreads until failure occurs.

In Chapter 21, static capacities and loading were based on permanent deformations occurring in nonrotating bearings. Subsequently during bearing rotation, the indentations or rolling contact surface deformations were considered to impair bearing endurance and/or cause undue vibration. The permanent deformations occurring during the shakedown process are the result of rolling contact during normal bearing rotation, and they do not eventually impair the rolling contact surface unless the shakedown limit has been exceeded. It is conceivable therefore that a bearing "static capacity" criterion could be based on the shakedown limit.

It is possible to apply the distortion energy yield criterion; that is, the von Mises yield criterion (see reference [22.42]) to bearing steel. Experimental investigations have indicated a variation of the von Mises yield limit with heat treatment parameters. Yhland [22.43] states that for normally heat-treated through-hardened carbon chromium steel; that is, AISI 52100, as measured in tension tests, the von Mises yield limit stress is in the region of 1800–2000 N/mm^2 (260,000–290,000 psi). Rydholm [22.44] developed a method to calculate the shakedown limit considering the foregoing yield criterion. For a well-lubricated bearing operating in line contact, the shakedown limit is approximately 2.31 times the yield stress in simple tension. For a point contact bearing, the shakedown limit is approximately $2.77\sigma_{\text{yield}}$. It is further possible to evaluate a situation

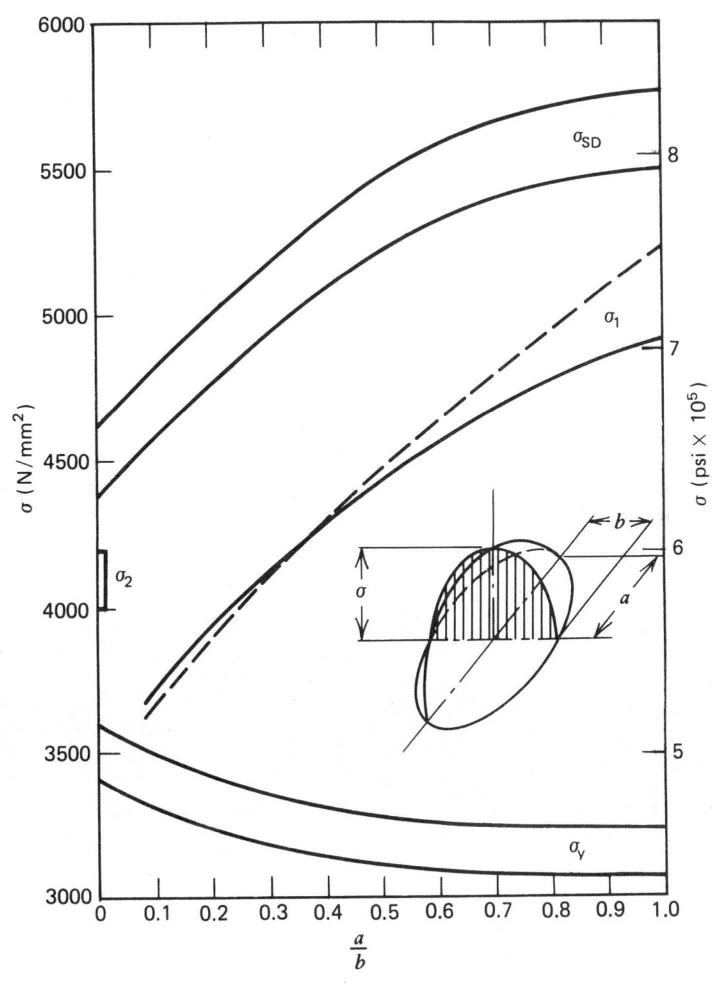

FIGURE 22.9. Comparison of stress limits—shake-down, incipient yield, and 0.0001*D* permanent deformation vs. *b*/*a* (from [22.43]).

in which no plastic flow occurs; that is, loading is relatively light and subsurface stresses are therefore low. In this case, for a well-lubricated bearing operating in line contact the "incipient plastic flow" limit is $1.79\sigma_{\text{yield}}$; for point contact bearings the limit is $1.56\sigma_{\text{yield}}$. Figure 22.9 taken from reference [22.43] shows an interesting comparison of the stresses obtaining for shakedown limit, incipient plastic flow limit, and a permanent deformation of $0.0001D$.

From Fig. 22.9, it is seen that the shakedown limit, that is, stress, is significantly greater than the $0.0001D$ permanent deformation stress.

EFFECTS OF BULK STRESSES ON MATERIAL RESPONSE TO ROLLING CONTACT

Ring Fracture

There is relatively little information in the open literature on the effects of bulk bearing ring stresses, imposed by mounting and centrifugal effects, on bearing operation, and failure manifestations. However, significant consequences are associated with them, involving both dimensional instability and catastrophic fracture of bearing rings.

For many decades a peculiar failure manifestation has been observed in bearings installed with heavy interference fit between the inner ring bore and shaft. Most commonly experienced with heavily loaded through-hardened roller bearings, the failure is characterized by through-section fracture of the inner ring on an axial plane. A typical fracture surface is shown in Fig. 22.10.

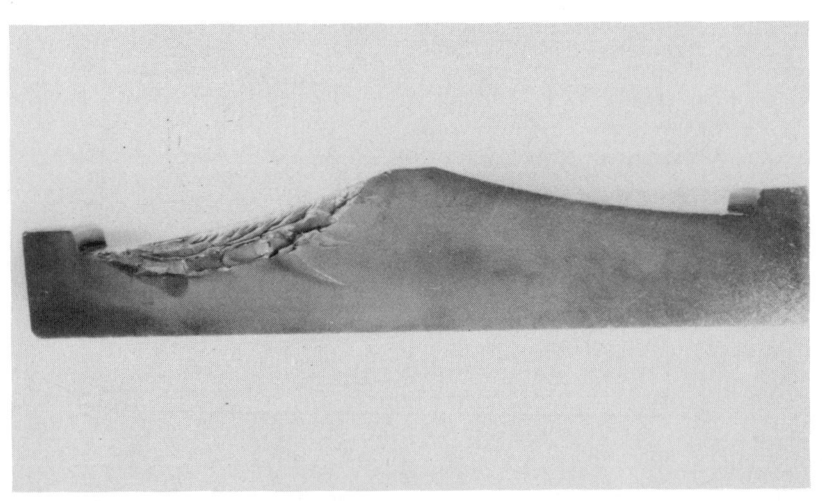

FIGURE 22.10. Fracture surface of a spherical roller bearing inner ring that failed due to excessive shaft fit (about half actual size).

The fracture surface is predominately planar and characteristically exhibits a semicircular region of stable fatigue crack propagation originating at a raceway surface. The remaining portion of the fracture surface is characteristic of unstable crack propagation—that is, rapid fracture. Spalling might or might not be associated with the axially oriented crack on the raceway surface. Such failures are almost exclusively experienced with through-hardened bearing rings. The traditional remedies for this failure experience have been to reduce the interference fit (if this can be done without permitting relative motion between the inner ring bore and shaft surface during operation) or to use a case-hardened inner ring.

This type of failure assumed greater significance when it was observed in aircraft gas turbine engine bearings. This observation prompted the first published analysis of the mechanisms involved [22.33–22.35]. The described characteristics of the failures are identical to those indicated here. The circumferential tensile stresses associated with inner ring failures were induced centrifugally (due to high-speed rotation) rather than from heavy shaft fit, but the magnitude of 172–207 N/mm^2 (25–30 ksi) agrees closely with hoop stress values calculated for fit-induced ring fractures. The failure scenario outlined in [22.35] describes the role of circumferential tensile stress of relatively modest magnitude in producing bearing ring fracture.

In classical subsurface-initiated fatigue failure, a crack initiates below the surface at a stress raiser such as a nonmetallic inclusion or carbide cluster. The crack propagates radially outward toward the surface. It also propagates radially inward but, in the absence of circumferential tensile stress, does not reach significant depth. During continued bearing operation, this crack participates in the formation of a spall. In the presence of circumferential tensile stress of sufficient magnitude [172 N/mm^2 (25 ksi) or greater], radially inward stable crack propagation continues to the point at which the critical crack size is reached. The critical crack size is defined by the magnitude of the circumferential tensile stress and the plane strain fracture toughness of the bearing steel. When the critical crack size is reached, a rapid through-section fracture occurs. Rapid fracture occurs on a plane perpendicular to the circumferential tensile stress.

Carburized materials provide both residual compressive stress in the high-hardness surface region and fracture toughness that increases with decreasing hardness from the surface to the core region. Centrifugally or fit-induced circumferential tensile stress should be offset to some degree by residual circumferential compression. The increased fracture toughness accommodates larger crack size. The individual contributions from these sources to the successful use of carburized inner rings in heavy interference fit applications have not been described in the open literature.

In [22.29], rolling contact fatigue experiments were performed with through-hardened inner ring specimens containing 1000 N/mm^2 (145 ksi) circumferential tensile stress. From the foregoing discussion it is not surprising that the inner ring failed by axial fracture with only minute indications of fatigue crack propagation (the critical crack size at this level of circumferential tension is estimated to be about 0.127 mm (0.005 in.). Perhaps of greater significance is the comparison of the running times to fracture for the high-tensile-stressed rings to running times accumulated by the "zero-stress" baseline. Ring fractures were experienced in 20 to 30 hr in the stressed ring tests, whereas baseline rings ran for 440 to 960 hr with no failures. This indicates that significant life reduction could be associated with stress conditions that promote ring fracture (e.g., bulk tensile loading).

The results of bearing tests performed with circumferentially tensile-stressed inner rings [345 N/mm^2 (50 ksi)] indicated both very early failure and radial crack propagation in through-hardened AISI M50 bearing steel [22.36]. Similarly tested bearings, with inner rings made of a carburizing version of AISI M50 (i.e., M50 Nil), completed substantially longer test times with no indications of failure. The residual compressive stress in the carburized case is cited as the reason for improved performance.

Dimensional Instability

Dimensional instability of bearing components in service, and particularly growth of bearing inner rings, is a known problem. For many years it was common knowledge that dimensional instability of hardened and tempered steel was due to retained austenite being transformed to martensite or bainite, depending on the thermal exposure conditions. Volume changes associated with the transformation of retained austenite, however, have not always been of sufficient magnitude to account for observed dimensional changes. Additionally, substantial dimensional changes of hardened and tempered bearing steel components have been reported with no reduction in retained austenite content [22.37, 22.38].

Overhaul statistics from U.S. railroads indicate that inner ring growth of carburized railway axle bearings is the major cause of bearing rejection at overhaul [22.39]. Metallurgical phase transformations [22.39] and accumulation of microplastic deformation [22.40] have been cited as causes. A detailed investigation of retained austenite and residual-stress profiles in railway journal bearings that exhibited inner ring bore growth, however, showed no alteration of either retained austenite content or residual-stress profile [22.37]; instead, ring growth correlated with retained austenite content and bearing operating temperature. Rings with higher retained austenite content exhibited more bore growth. Increased operating temperature produced increased growth. The absence of any indications of microstructural alteration or residual-stress buildup lead to the conclusion that a creep mechanism is involved in producing the observed change in inner ring dimensions, as opposed to metallurgical phase change or microplastic deformation.

Data in [22.38] are consistent with the findings in [22.37]. A cleverly designed fixture was used to determine the dimensional stability of hardened AISI 52100 steel containing 0–15% retained austenite. Testing was performed from −34 to 74°C (−30–165°F) with constant applied tensile stress levels of 0, 69, and 138 N/mm^2 (10 and 20 ksi). Test times up to 1700 hr were used. The results indicated that dimensional change (length increase) increased with increasing amounts of retained austenite. More dramatically, for a given austenite content and test temperature (which was well below the specimen tempering temperature), large increases in specimen length were associated with the application of relatively modest tensile stress. See Fig. 22.11. Also, no reduction in retained austenite content was detected, even after tests resulting in significant length extensions. A creep mechanism similar to that proposed in [22.36] could be indicated. Additionally, these data imply that assessment of growth potential via unstressed thermal exposure tests may provide an underestimate for an application such as a heavy interference-fitted bearing ring. Further, increasing interference fit to compensate for such an underestimate may result in more rapid loss of fit. See Fig. 22.11. Since this can be done without decomposition or transformation of the initial retained

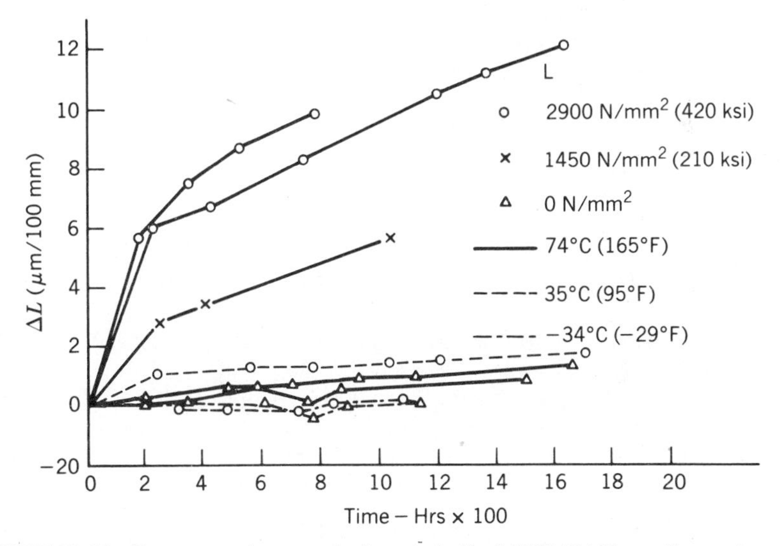

FIGURE 22.11. Permanent increase in length (ΔL) of AISI 52100 tensile specimens at various stress levels and temperatures. Hardness (R_C 64) and retained austenite content (15%) remained unchanged in specimens tested at 2900 N/mm^2 and 74°C (165°F) (from [22.38]).

austenite content, the potential for continued growth is preserved. Clearly this would not be the case if growth was experienced at the expense of retained austenite content. This could be a significant consideration when making judgments pertaining to such matters as bearing refurbishment.

Retained austenite content remains a primary consideration in terms of bearing ring dimensional stability, with definite correlations between growth potential and initial austenite content. The mechanisms by which dimensional change is effected, however, are not understood. Published work [22.41] indicates desirable contributions of retained austenite to rolling bearing performance. Consequently, it may be technologically imprudent to totally ignore possible performance advantages in pursuit of austenite-free dimensional stability.

CLOSURE

Because of the extremely high rolling element–raceway contact stresses that occur during operation of many ball and roller bearings, the microstructure of the bearing material, that is, steel, undergoes significant change. This phenomenon has been investigated for many years, and yet it has not thus far been possible to relate such microstructural changes to the imminence of rolling contact fatigue. It is known that not only do the applied stresses cause such microstructural alterations, but also bearing operating temperatures well below steel tempering temperatures significantly affect the rate and amount of microstructural change. One type of microstructural change, that is, shakedown, appears to be related to the yield strength of the steel. A shakedown limit might be established as one bearing operating criterion. Investigation in this important area of material science is continuing.

REFERENCES

22.1. J. van de Sanden, "Martensite Morphology of Low Alloy Commercial Steels," *Practical Metallogr.* **17,** 238–248 (1980).

22.2. J. Rescalvo, "Fracture and Fatigue Crack Growth in 52100, M50 and 18-4-1 Bearing Steels," Ph.D. thesis, Department of Materials Science and Engineering, Massachusetts Institute of Technology, (June, 1979).

22.3. G. Krauss, *Principles of Heat Treatment of Steel,* American Society for Metals, pp. 61–75, (1980).

22.4. A. Jones, "Metallurgical Observations of Ball Bearing Fatigue Phenomena," *Proc. ASTM* **46,** 1 (1946).

22.5. T. Tallian, "On Competing Failure Modes in Rolling Contact," *ASLE Trans.* **10,** 418–439 (1967).

22.6. A. Voskamp, "Material Response to Rolling Contact Loading," presented at the ASME/ASLE Lubrication Conference, San Diego, Calif., Oct. 22–24, 1984. ASME Preprint 84-TRIB-2

22.7. T. Lund, *Jernkontorets Ann.* **153,** 337 (1969).

22.8. H. Swahn, P. Becker, and O. Vingsbo, "Martensite Decay During Rolling Contact Fatigue in Ball Bearings," *Metallurgical Trans. A* **7A,** 1099–1110 (Aug. 1976).

22.9. J. Martin, S. Borgese, and D. Eberhardt, "Microstructural Alterations of Rolling Bearing Steel Undergoing Cyclic Stressing," *Trans. ASME* **59,** 555–567 (Sept. 1966).

22.10. A. Gentile, E. Jordan, and A. Martin, "Phase Transformations in High-Carbon, High Hardness Steels Under Contact Loads," *Trans. AIME* **233,** 1085–1093 (June 1965).

22.11. J. Bush, W. Grube, and G. Robinson, "Microstructural and Residual Stress Changes in Hardened Steel Due to Rolling Contact," *Trans. ASM* **54,** 390–412 (1961).

22.12. M. Kuroda, *Trans. Jpn. Soc. Mech. Eng.* **26,** 1256–1270 (1960).

22.13. H. Muro, and N. Tsushima, "Microstructural, Microhardness and Residual Stress Changes Due to Rolling Contact," *Wear* **15,** 309–330 (1970).

22.14. H. Styri, *Proc. ASTM* **51,** 682–700 (1951).

22.15. R. Tricot, J. Monnot, and L. Luansi, *Metals Eng. Quart.* **12,** 39–42 (1972).

22.16. W. Littmann and R. Widner, "Propagation of Contact Fatigue from Surface and Subsurface Origins," *J. Basic Eng.* **88,** 624–636 (1966).

22.17. L. Uhrus, "Clean Steel," The Iron and Steel Institute, London, pp. 104–109 (1963).

22.18. P. Becker, "Microstructural Changes Around Non-Metallic Inclusions Caused by Rolling-Contact Fatigue of Ball-Bearing Steels," *Metals Technology*, 234–243 (June 1981).

22.19. "Residual Stress Measurement by X-Ray Diffraction," SAE J784a, 2nd ed. Society for Automotive Engineers, New York (1971).

22.20. D. Koistinen, "The Distribution of Residual Stresses in Carburized Cases and Their Origins," *Trans. ASM* **50,** 227–238 (1958).

22.21. B. Cullity, *Elements of X-ray Diffraction*, Addison-Wesley, Reading: Mass. (1959).

22.22. C. Gazzara, "The Measurement of Residual Stress with X-ray Diffraction," Rept. AD-A130 614, Army Material & Mechanics Res. Center (May 1983).

22.23. W. Bragg, "The Diffraction of Short Electromagnetic Waves by a Crystal," *Proc. Camb. Phil. Soc.* **17,** 43 (1912).

22.24. A. Voskamp, R. Osterlund, P. Becker, and O. Vingsbo, "Gradual Changes in Residual Stress and Microstructure during Contact Fatigue in Ball Bearings," *Metals Technology*, 14–21 (Jan. 1980).

22.25. E. Zaretsky, R. Parker, and W. Anderson, "A Study of Residual Stress Induced during Rolling," *J. Lub. Tech.* **91,** 314–319 (1969).

22.26. R. Scott, R. Kepple, and M. Miller, "The Effect of Processing-Induced Near-Surface Residual Stress on Ball Bearing Fatigue," *Rolling Contact Phenomena*, J. B. Bidwell, Ed., Elsevier, pp. 301–316 (1962).

22.27. E. Zaretsky, R. Parker, W. Anderson, and S. Miller, "Effect of Component Differential Hardness on Residual Stress and Rolling-Contact Fatigue," NASA TND-2664 (1965).

22.28. C. Foord, C. Hingley, and A. Cameron, "Pitting of Steel Under Varying Speeds and Combined Stresses," *J. Lub. Tech.* **91,** 282–290 (1969).

22.29. R. Kepple and R. Mattson, "Rolling Element Fatigue and Macroresidual Stress," *J. Lub. Tech.* **92,** 76–82 (1970).

22.30. W. Littmann, Discussion to Reference 30, *J. Lub. Tech.* **92,** 81 (1970).

22.31. C. Stickels and A. Janotik, "Controlling Residual Stresses in 52100 Bearing Steel by Heat Treatment," *Metal Progress*, 34–40 (Sept. 1981).

22.32. D. Koistinen, "The Generation of Residual Compressive Stresses in the Surface Layers of Through-Hardening Steel Components by Heat Treatment," *Trans. ASM* **57,** 581–588 (1964).

22.33. J. Clark, "Fracture Failure Modes in Lightweight Bearings," *AIAA, Journal Aircraft* **12,** No. 4 (1975).

22.34. E. Bamberger, E. Zaretsky, and H. Signer, "Endurance and Failure Characteristics of Main-Shaft Jet Engine Bearings at 3×10^6 DN," *ASME Trans. J. Lub. Tech.* **95,** No. 4 (1976).

22.35. E. Bamberger, "Materials for Rolling Element Bearings," presented at ASME-ASLE International Lubrication Conference, San Francisco, Calif. (August 1980).

22.36. J. Clark, "Fracture Tough Bearings for High Stress Applications," presented at AIAA/SAE/ASME/ASEE 21st Joint Propulsion Conference, Monterey, Calif. (July 8–10, 1985).

22.37. A. Voskamp and B. Schalk, "Ring Growth in Case Hardened Railway Journal Roller Bearings," presented at the 2nd International Heavy Haul Railway Conference, Pueblo, Colo. (September, 1982).

22.38. E. Mikus, T. Hughel, J. Gerty, and A. Knudsen, "The Dimensional Stability of a Precision Ball Bearing Material," *Trans. ASM* **52,** 307–315 (1960).

22.39. J. McGrew, A. Krawler, and G. Moyar, "Reliability of Railroad Roller Bearings," *ASME Trans. J. Lub. Tech.* **99,** 30–40 (1977).

22.40. R. Steel, Discussion to Reference 39, *ASME Trans. J. Lub. Tech.* **99,** 39 (1977).

22.41. J. Seehan and M. Howes, "The Effect of Case Carbon Content and Heat Treatment on the Pitting Fatigue of 8620 Steel," SAE Conf. Congress, No. 720268 (January 1972).

22.42. M. F. Spotts, *Design of Machine Elements*, 3rd ed., Prentice-Hall, Englewood Cliffs, pp. 85–87 (1961).

22.43. E. Yhland, "Static Load-Carrying Capacity," *Ball Bear. Journal 211*, 1–8 (1982).

22.44. G. Rydholm, *On Inequalities and Shakedown in Contact Problems*, Linköping Studies in Science and Technology, Dissertations, No. 61, Linköping (1981).

23

FATIGUE: LIMITING STRESS EFFECTS

LIST OF SYMBOLS

Symbol	Description	Units
$\overline{A}$	Overall material constant	
$\mathcal{A}_1$	Reliability factor	
$\mathcal{A}_{SKF}$	SKF life factor	
a	Semimajor axis of contact ellipse	mm (in.)
b	Semiminor axis of contact ellipse	mm (in.)
C	Bearing basic load rating	N (lb)
L_{10}	L_{10} fatigue life	rev. $\times 10^6$
L_{50}	L_{50} fatigue life	rev. $\times 10^6$
L_{naa}	SKF fatigue life	rev. $\times 10^6$
N	Number of stress cycles	
P	Bearing equivalent load	N (lb)
P_u	Bearing fatigue load limit	N (lb)
P_o	Maximum pressure in Hertzian contact	N/mm^2 (psi)
r_r	Raceway radius	mm (in.)
$\mathcal{S}$	Probability of survival	
T	Stress-related fatigue criterion	N/mm^2 (psi)

Symbol	Description	Units
T_1	Stress-related fatigue limit	N/mm^2 (psi)
U	Rolling velocity	mm/sec (in./sec)
V	Stressed volume	mm^3 (in.^{3})
x	Distance in x (axial) direction	mm (in.)
y	Distance in y (rolling) direction	mm (in.)
z	Distance in z (depth) direction	mm (in.)
z_o	Depth to maximum orthogonal shear stress	mm (in.)
z'	Stress-weighted average depth to volume element at risk to fatigue	mm (in.)
ν	Lubricant kinematic viscosity at operating conditions	cs (in.^{2}/sec)
ν_1	Lubricant kinematic viscosity for $\mathfrak{A}_{23} = 1$	cs (in.^{2}/sec)
κ	Relative viscosity, ν/ν_1	
σ	Normal stress	N/mm^2 (psi)
τ_o	Maximum orthogonal shear stress	N/mm^2 (psi)

GENERAL

According to the U.S. National and the International Standards for calculating load ratings and life of ball and roller bearings (ANSI [23.1, 23.2] and ISO [23.3]), even considering ideal operating conditions, the fatigue life of rolling bearings in any application is finite. The standards pertain to rolling bearings of specific conventional design, properly manufactured from good quality steel, and are based on work by Lundberg and Palmgren [23.4, 23.5] conducted during the 1930s and 1940s. The load rating and life calculation methods developed in the foregoing references were representative of the manufacturing methods, materials, and lubricating methods of that time. The mechanics of lubrication of concentrated contacts—that is, elastohydrodynamic (EHD) lubrication as initially proposed by Grubin [23.6] and confirmed by Sibley and Orcutt [23.7]—were unknown to Lundberg and Palmgren. During the 1960s and 1970s, manufacturers' catalogs and the standards, respectively, acknowledged the lubrication phenomena as well as continually improving materials by inclusion of life adjustment factors applied to fatigue lives determined according to the original standard calculating methods. Additionally, the ASME developed guidelines [23.8] for values of life adjustment factors.

The Lundberg and Palmgren formulas represented a significant development in rolling bearing technology; however, it was not possible to correlate the fatigue of bearing surfaces in rolling contact so calculated to structural fatigue in spite of the desire and need to do so. Nor was it possible to correlate rolling contact fatigue in bearings to fatigue of elemental surfaces in rolling contact. Moreover, materials tested for structural fatigue have typically exhibited a fatigue limit as shown by curves similar to Fig. 18.1.

According to Fig. 18.1, for cyclic loading less than the fatigue limit, fatigue, for all practical purposes, does not occur. On the contrary, rolling bearing applications according to the standard methods of calculations were characterized by a finite fatigue life in any application. Innumerable modern rolling bearing applications, however, have defied this limitation. Recent data (e.g., [23.9] for bearings of standard design, accurately manufactured from high-quality steel—that is, having minimal impurities and homogeneous chemical and metallurgical structures) have demonstrated that infinite fatigue life is a practical consideration in some rolling bearing applications. Since the Lundberg and Palmgren formulas did not address the concept of a possible infinite fatigue life and did not relate to structural fatigue, an improvement in these formulas beyond application of some empirical life adjustment factors was required.

The Lundberg and Palmgren theory considers that a fatigue crack begins at a point below the surface in rolling contact, at which point a large-magnitude orthogonal shear stress coincides with a weak point in the material. Such weak points are assumed to be randomly distributed throughout the material. As demonstrated in Chapter 5, the orthogonal shear stress results from a concentrated load applied normal to the surfaces in contact, giving a Hertzian surface stress distribution similar to that of Fig. 5.6 for point contact. Figure 5.13 shows the orthogonal shear stress distribution in the subsurface material.

Knowledge of pressure distributions in EHD lubricated contacts has demonstrated that such pressure distributions can be substantially different from the pure Hertzian distribution indicated in Fig. 5.6. Figure 12.10 shows just how different an EHD lubrication pressure distribution can be compared to the Hertzian distribution.

Moreover, if the surfaces are nonideal—that is, not smooth but having perturbations on the smooth surface—then concepts of micro-EHD lubrication, as discussed in Chapter 14, obtain. Additionally, in their analysis Lundgren and Palmgren did not accommodate surface shear stresses, which can substantially alter the subsurface shear stresses, as indicated by Fig. 5.17.

In Fig. 5.17 the subsurface stress determined is that from the distortion energy theory of Von Mises; a similar situation would occur considering subsurface orthogonal shear stresses. There is a difference of opinion concerning which subsurface stress effectively causes rolling contact fatigue. The depths below the surface at which maximum orthogonal shear stress and maximum Von Mises stress occur are slightly different, the latter occurring at a depth approximately 50% deeper than the former. Whichever stress is considered most detrimental, the effect of surface shear stress is to bring the maximum subsurface shear stress toward the surface. The maximum of the subsurface shear stress is estimated to occur on the surface when the applied surface shear stress is approximately 30% of the applied normal stress. In general, shear stresses of this magnitude do not occur over the entire concentrated contact area in an effectively EHD lubricated contact. Such stresses could occur in micro-EHD lubricated contacts existing within the overall contact area. When the maximum subsurface shear stress occurs at the surface, the possibility of surface-initiated fatigue, as compared to subsurface-initiated fatigue, occurs. Tallian [23.10] considers competing modes of fatigue failure—that is, surface initiated and subsurface initiated. Rigorous mathematical solution requires the consideration of failure at any point in the material from the surface into the subsurface, consistent with the applied stresses both normal and tangential to the surface. Clearly the Lundberg and Palmgren theory did not cover this generalized stress situation.

The basic equation stated by Lundberg and Palmgren is

$$\ln \frac{1}{\mathcal{S}} \propto \frac{N^{\mathrm{e}} \tau_{\mathrm{o}}^{\mathrm{c}} V}{z_{\mathrm{o}}^{\mathrm{h}}} \tag{23.1}$$

In this equation τ_o is the maximum orthogonal shear stress, z_o is the depth at which it occurs, V is the stressed volume, N is the number of stress cycles, and $\mathcal{S}$ is the probability of survival of the stressed volume. From Fig. 23.1 the stressed volume of Lundberg and Palmgren is proportional to the product of the major axis of the contact ellipse, the depth

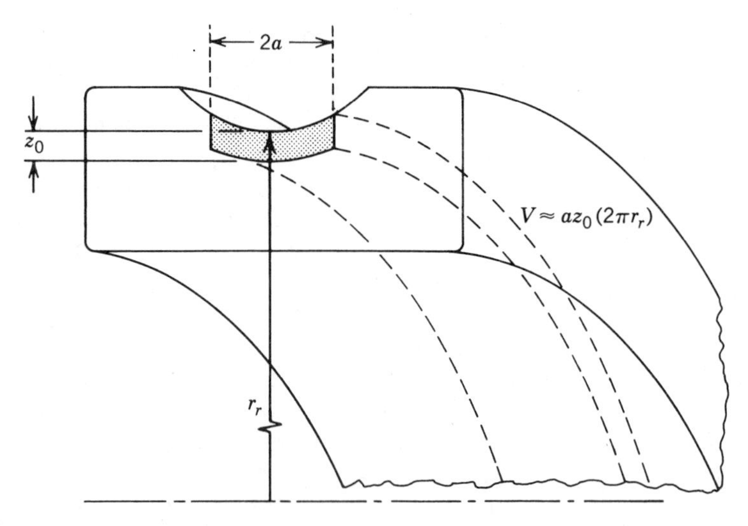

FIGURE 23.1. Volume at risk to fatigue in rolling contact according to Lundberg and Palmgren.

to maximum shear stress, and the circumference of the raceway contact. This proportionality is valid for geometrically perfect, contacting surfaces between which only normal stresses occur. If significant surface shear stresses occur, the tendency toward surface-initiated fatigue is ignored.

The Lundberg and Palmgren theory also does not account for the bearing operating temperature and its effect on material properties, let alone account for the effect of temperature on lubrication and hence on surface shear stresses. Furthermore, the theory does not consider the rate at which energy is absorbed by the surfaces in rolling contact. Bearing speeds are used simply to convert predicted fatigue lives in millions of revolutions to time values. Nor are hoop stresses induced by ring fitting on shafts or in housings or by centrifugal loading accommodated. Finally, the development of microstructural alterations and residual stresses below the raceways, induced by rolling contact, as indicated by Voskamp [23.11] must be considered.

A MODIFIED FATIGUE LIFE FORMULA

Considering all of the foregoing, Ioannides [23.12] developed the basic equation

$$\ln\left(\frac{1}{\Delta S_i}\right) = \mathrm{F}(N, T_i - T_{li})\,\Delta V_i \tag{23.2}$$

In this formula a fatigue crack is presumed incapable of initiation until a stress-related fatigue criterion T_i exceeds a threshold value of the criterion T_{li} at a given elemental volume ΔV_i. It is evident that the threshold criterion T_{li} corresponds to a fatigue limit. Consistent with the Lundberg and Palmgren theory the stress-related fatigue criterion would be the orthogonal shear stress amplitude; however, it is possible to use other criteria, such as the Von Mises critical shear stress or even the maximum subsurface shear stress if either is subsequently proved to be more appropriate. Voskamp [23.11] has also proposed consideration of a subsurface tensile stress for this criterion. In equation (23.2), in lieu of the entire stress volume used by Lundberg and Palmgren as discussed in the preceding section, only the elemental volume in which T_i exceeds T_{li} is considered at risk. See Fig. 23.2.

Therefore the probability of survival in equation (23.2) is a differential value—that is, ΔS_i. The probability of component survival is determined according to the product law of probability, and after considerable mathematical manipulation equation (23.3), which corresponds to the form of the Lundberg and Palmgren relationship (23.1), is obtained.

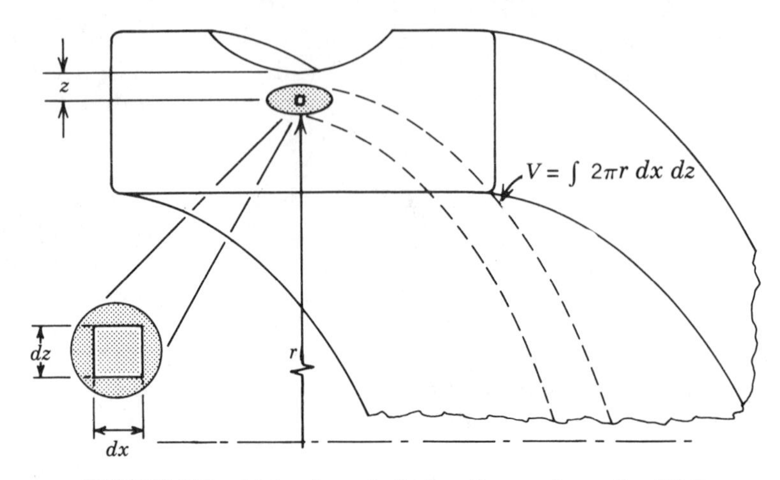

FIGURE 23.2. Risk volume in fatigue theory of equation (23.2).

$$\ln\left(\frac{1}{\mathcal{S}}\right) \approx \overline{A}N^{\mathrm{e}} \int_{V_{\mathrm{R}}} \frac{(T - T_l)^{\mathrm{c}}}{z'^{\mathrm{h}}}\, dV \tag{23.3}$$

in which $\overline{A}$ is a constant pertaining to the overall material and z' is a stress-weighted average depth to the volume at risk to fatigue. When T_l is 0, equation (23.3) reduces to equation (23.1) if it is assumed $T = \tau_{\mathrm{o}}$.

A considerable amount of development is required, starting with equation (23.3), to obtain a simplified load-life equation (e.g., similar to the current formula $L = \mathcal{A}_1\mathcal{A}_2\mathcal{A}_3(C/F_{\mathrm{e}})^3$ for ball bearings) for general use in manufacturers' catalogs or for inclusion in national and international standards. This entails the development of new endurance test data under extremely well-controlled test conditions. This effort is being accomplished, as indicated by [23.9]. Moreover, correlation with endurance test data collected during previous years is essential. It is apparent, however, that a fatigue life formula developed from (23.3) would yield a bearing load-life relationship as indicated by Fig. 23.3.

It is also clear from Fig. 23.3 that at very high loads and/or low fatigue load limit that the current standard formula is a special case of the new developing formula. Based on the relative inferiority of the rolling bearings tested by Lundberg and Palmgren to develop their formulae, it was not possible for them to observe a fatigue limit.

In [23.12], equation (23.3) was applied to fatigue data for rotating beams in bending, beams in torsion, and flat beams in reversed bending. Figures 23.4 and 23.5 illustrate how well equation (23.3) can be used to fit structural fatigue test data. Accordingly, it is anticipated that equation (23.3) is sufficient to describe metal, especially steel, fatigue in components subjected to any type of cyclical loading. To fit the equation to the test data of Figs. 23.4 and 23.5, it was only necessary to establish a

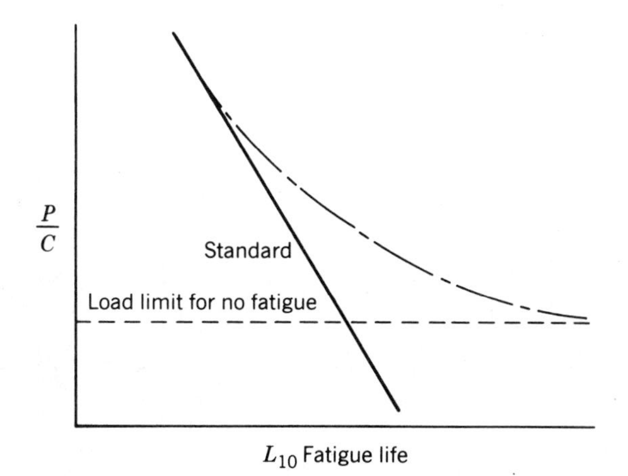

FIGURE 23.3. Load-life relationship with fatigue load limit.

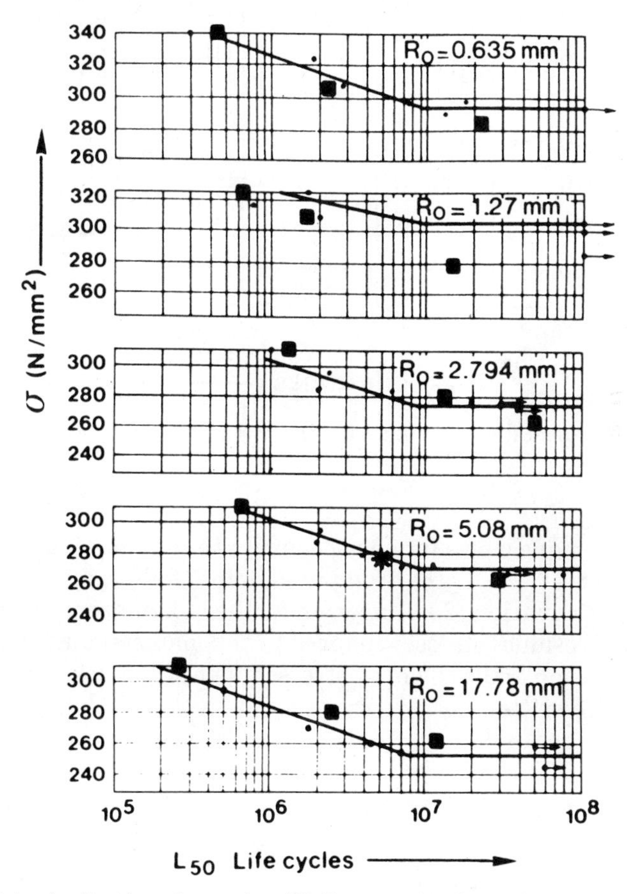

FIGURE 23.4. Application of equation (23.3) to rotating beam fatigue test data (from [23.12]).

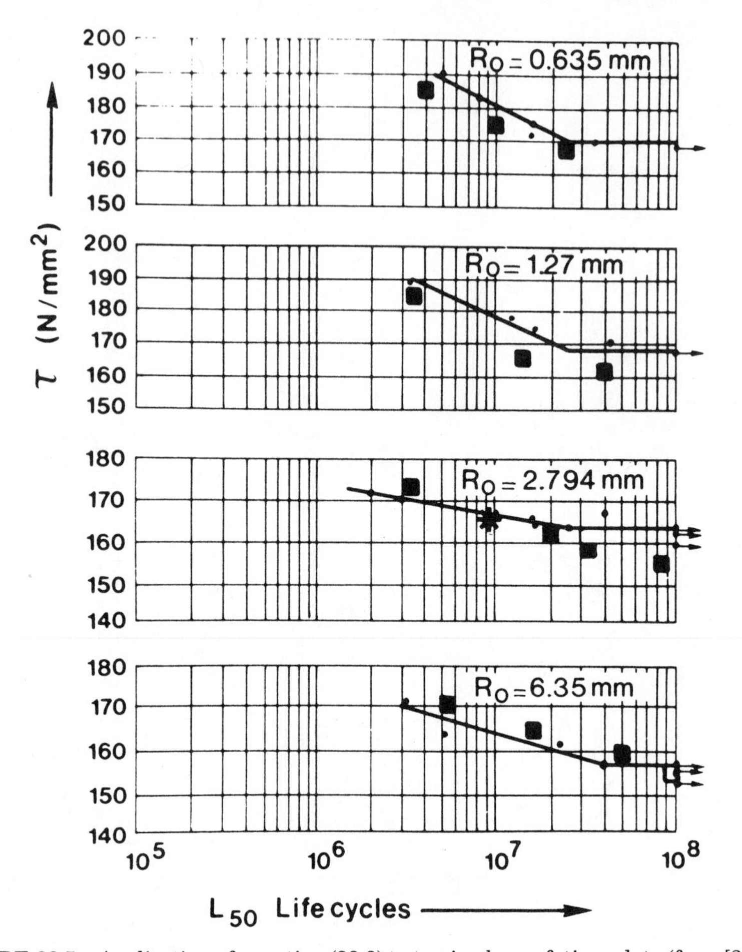

FIGURE 23.5. Application of equation (23.3) to torsion beam fatigue data (from [23.12]).

single point on one curve for one specimen; that point is identified by an asterisk. Thereafter all other computed points followed.

The use of equation (23.3) may be further extended to accommodate the effects of residual stresses induced in the material during heat treatment or operation and/or by hoop stresses. The stress portion of the equation may be rewritten as

$$T - T_l = T - (T_l - T_r - T_h) \tag{23.4}$$

where T_h and T_r are the consistent stresses resolved from the induced hoop and residual stresses, respectively. Positive values of T_h and T_r

refer to tensile hoop and residual stresses, respectively. Thus it can be seen that tensile ring stresses tend to reduce rolling contact fatigue life, whereas compressive stresses tend to increase fatigue life. Such considerations can be included in more rigorous analyses than those afforded by standard and catalog-type calculational methods.

EFFECT OF MATERIAL QUALITY ON FATIGUE LIFE

In the time of Lundberg and Palmgrem the basic electric arc (BEA) furnace was primarily used for producing the AISI 52100 high-chromium–high-carbon steel for through-hardened ball and roller bearings. As indicated by Figs. 23.6 and 23.7, it was not uncommon for such steel to have high oxygen content (in the form of oxides)—for example, 35 parts per million (ppm) and substantial amounts of macroinclusions. See Chapter 16.

During the 1970s the acid open hearth (AOH) furnace was introduced, which improved cleanliness; that is, oxygen content reduced to 20 ppm and macroinclusions reduced over a period of time to virtually nil. By 1982 the SKF M-R process, as shown by Fig. 23.6, permitted the reduction of oxygen to 10 ppm while maintaining a low level of macroinclusions. As indicated by Fig. 23.8, the decrease of oxygen content from 35 ppm to 10 ppm can afford a 10-fold improvement in L_{10} fatigue life.

Although not indicating the same level of life improvement (e.g., only three- to fourfold), Tsushima and Kashimura [23.14] nevertheless demonstrated increase of rolling contact fatigue life with reduction of oxygen. The amount of oxygen affects values of $\bar{A}$ and T_l in equation (23.3), implying increased $\bar{A}$ and T_l for steel with reduced amounts of oxygen.

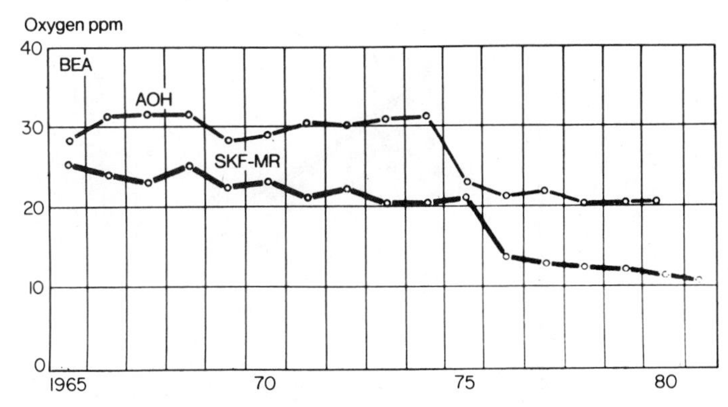

FIGURE 23.6. Oxygen content improvement of through-hardened AISI 52100 (SKF M-R) steel from 1965–1982 (from [23.13]).

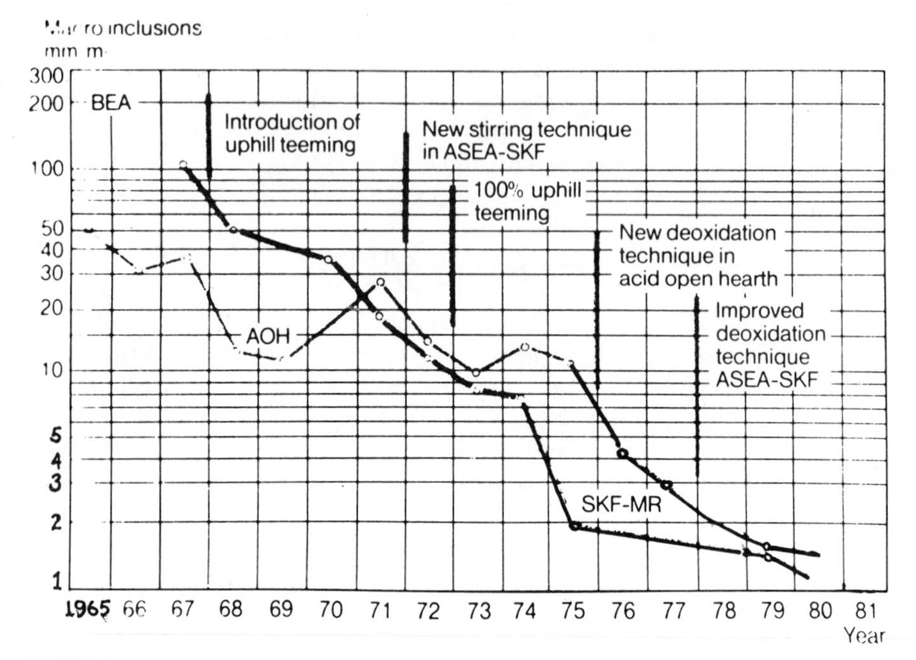

FIGURE 23.7. Reduction of macroinclusions in through-hardened AISI 52100 (SKF M-R) steel from 1965–1982 (from [23.13]).

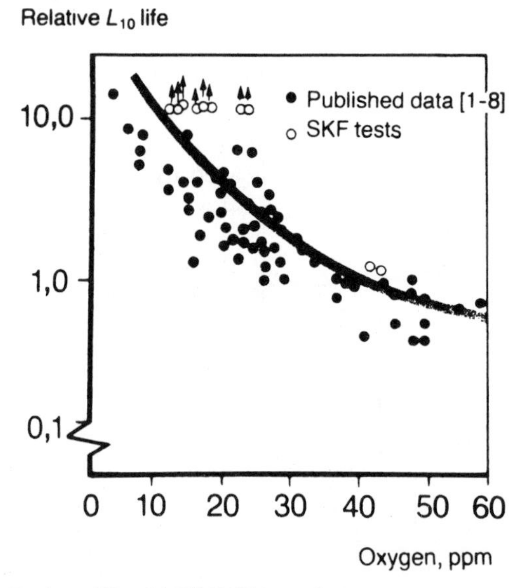

FIGURE 23.8. Fatigue life of AISI 52100 steel versus oxygen content (from [23.13]).

EFFECT OF PARTICULATE CONTAMINANTS ON FATIGUE LIFE

It has long been known that excessive particulate contamination, such as gear wear material, alumina, silica, and so on, in the lubricant will severely shorten rolling bearing fatigue life. The standards [23.1–23.3] and manufacturers' catalogs contain warning statements about this. Even small amounts of particulate contamination have significantly limiting effects on fatigue life. Sayles and MacPherson [23.15] demonstrated this phenomenon by endurance testing rolling elements with varying degrees of lubricant filtration: for example, from 40 μm (0.0016 in.) down to 1 μm (0.00004 in.) or possibly less. Figure 23.9 shows that a significant improvement in fatigue life is achieved by improving lubricant filtration—that is, removing "large" particulate matter from the lubricant.

According to Fig. 23.9, there was little difference in performance of bearings lubricated with 3-μm (0.00012-in.) or finer filtration; hence there appears to be a limit to filter effectiveness. The data of [23.15] tended to be confirmed by Tanaka et al. [23.16], who, by using sealed ball bearings in an automotive gearbox, managed to increase fatigue life severalfold, compared to that of "open" (no seals or shields) bearings in the same application.

Figure 23.10 is a photograph of dents incurred under the operating conditions indicated in the Sayles and MacPherson [23.15] investigation with 40-μm (0.0016-in.) filtration. The dents are approximately 10–30 μm (0.0004–0.00012 in.) long and about 2 μm (0.00008 in.) deep.

Webster et al. [23.17], using a numerical contact model coupled to a subsurface finite element analysis, examined the state of stress within a bearing raceway in the presence of debris-initiated dents. They indicated the process of denting by wear debris generates not only shallow depressions but also raised edges on the peripheries of the dents. Moreover, the latter are rounded, indicating a probable plastic deformation process. Nevertheless, a significant probability of fatigue failure is associated with dent shoulders, as indicated by Fig. 23.11.

The analytical work of [23.17], using the Ioannides [23.12] fatigue life formula, seems to confirm the experimental work of Sayles and MacPherson [23.15], indicating fatigue lives of bearings tested under 40-μm (0.0016-in.) filtration to be about seven times less than those tested under 3-μm (0.00012-in.) or finer filtration. Hence, for bearings with "rough" surfaces caused by random debris denting, the concepts of micro-EHD lubrication are significant in determining fatigue life.

BEARING FATIGUE LIFE CALCULATION

It is apparent from equation (23.3) that its most accurate use requires the volumetric integration of the subsurface stress fields in each of the

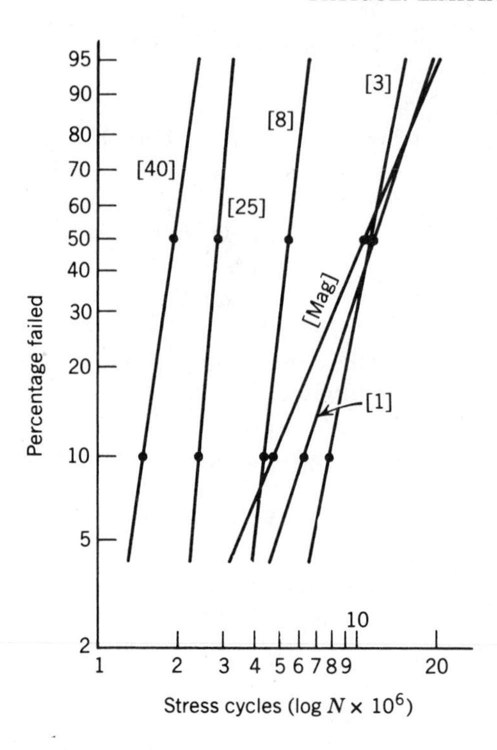

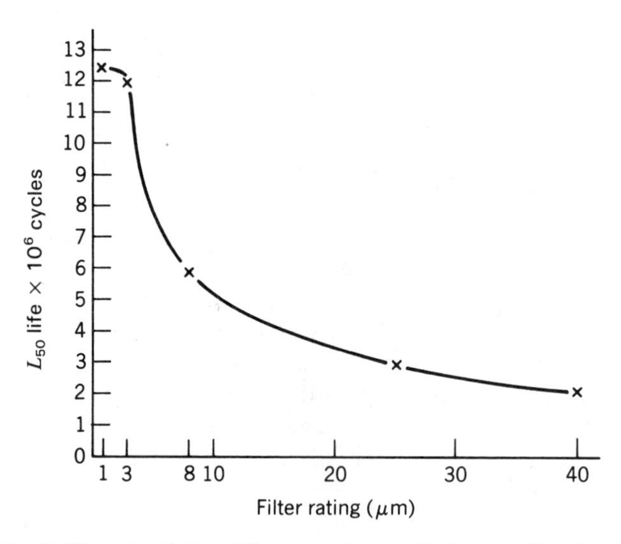

FIGURE 23.9. Ball bearing fatigue life versus degree of lubricant filtration (from [23.15]).

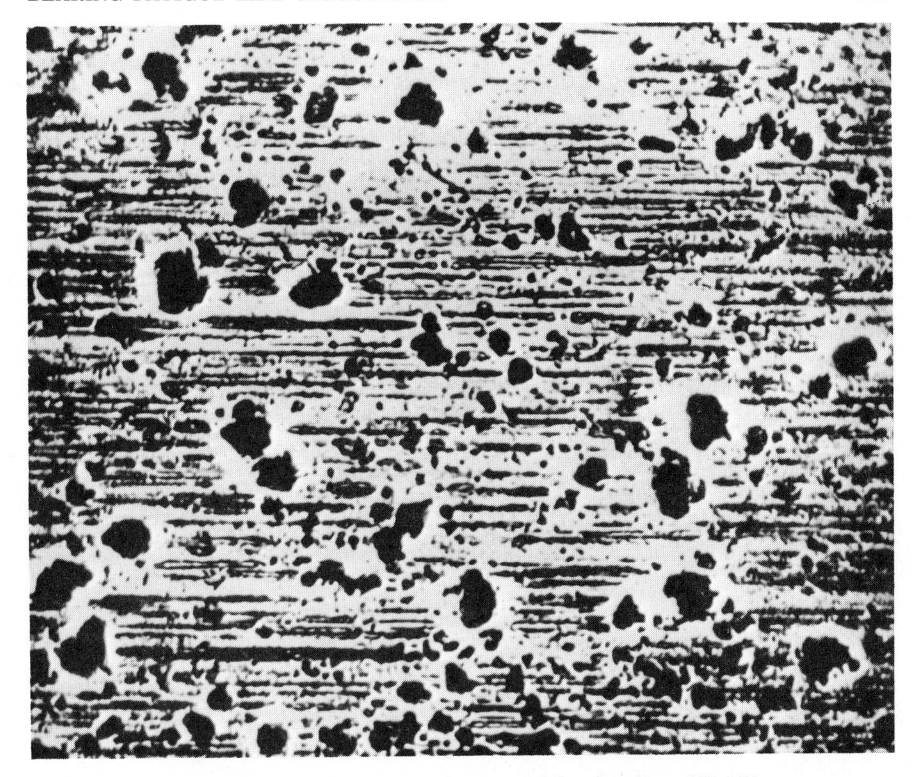

FIGURE 23.10. Contamination-caused denting (from [23.15]).

rolling element-raceway contacts. Moreover, only those elemental volumes wherein the equivalent stress exceeds the fatigue failure stress limit are subject to fatigue. Also, the precise surface stress profile may be considered in determining the subsurface stress fields. This was demonstrated by Webster et al. [23.17] who indicated how dents caused by solid contaminants in the lubricant affect fatigue endurance. The type of fatigue life calculation indicated above currently requires a substantial amount of mainframe computer time to determine the fatigue endurance of a rolling bearing for one application condition of load, speed, temperature, lubricant, contamination level, and so on. Approximations may, of course, be made, which, while providing less accurate results than the foregoing scientific method, provide nonetheless sufficiently accurate engineering representations of bearing fatigue lives in given applications. These latter calculations will produce more accurate predictions of fatigue lives than are possible to achieve using the current standard methods [23.1–23.3]. SKF [23.18] has provided such a method for use with bearing series contained in the SKF catalog [23.19]. The method strictly pertains to those bearings that are mounted on rela-

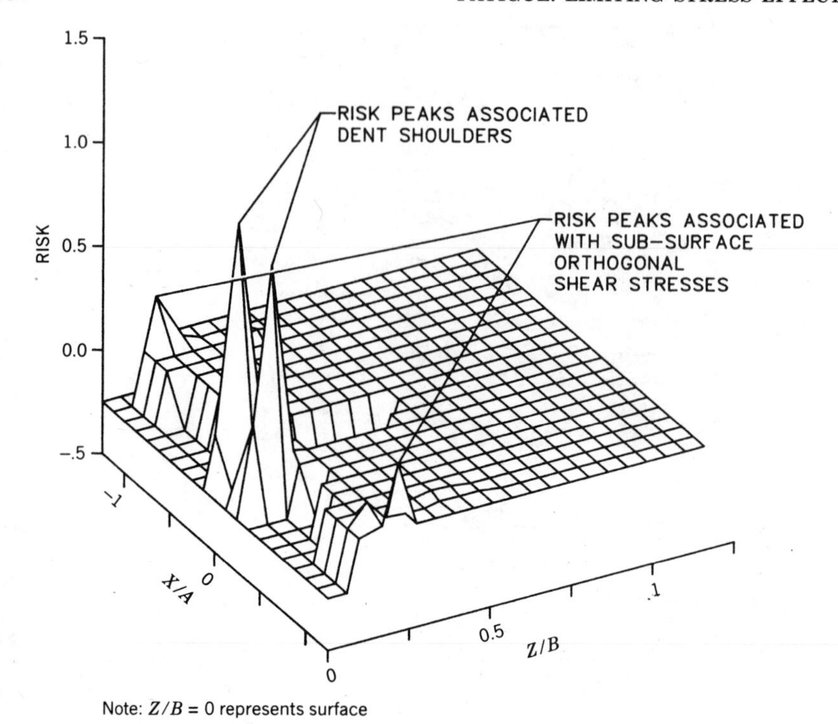

FIGURE 23.11. Plot showing relative risk of failure throughout raceway subsurface including effect of dent shoulders (from [23.17]).

tively rigid shafts and in rigid housings, which operate at speeds within the limits specified in [23.19] and in general meet all the operating conditions specified in the engineering data section of the catalog. Moreover, the method thus far strictly applies only to SKF bearings since the empirical parameters necessary to use the method have been determined only for the bearings represented in [23.19].

The SKF method is based on the following equation:

$$L_{\text{naa}} = \mathcal{A}_1 \mathcal{A}_{\text{SKF}} L_{10} \tag{23.5}$$

where L_{naa} is the adjusted rating life to the new fatigue life theory, $\mathcal{A}_1$ is the life adjustment factor for reliability (see Chapter 18), and $\mathcal{A}_{\text{SKF}}$ is the life adjustment factor based on the new life theory. The last factor represents very complex interrelationships of several bearing properties including, but not limited to, material strength and cleanliness, manufacturing accuracy of bearing components, microgeometries of "contacting" surfaces, lubricant type, operating temperatures, and contaminant type and level. These properties and conditions are all included in Fig. 23.12 taken from [23.18]. Considering the foregoing statement Fig. 23.12

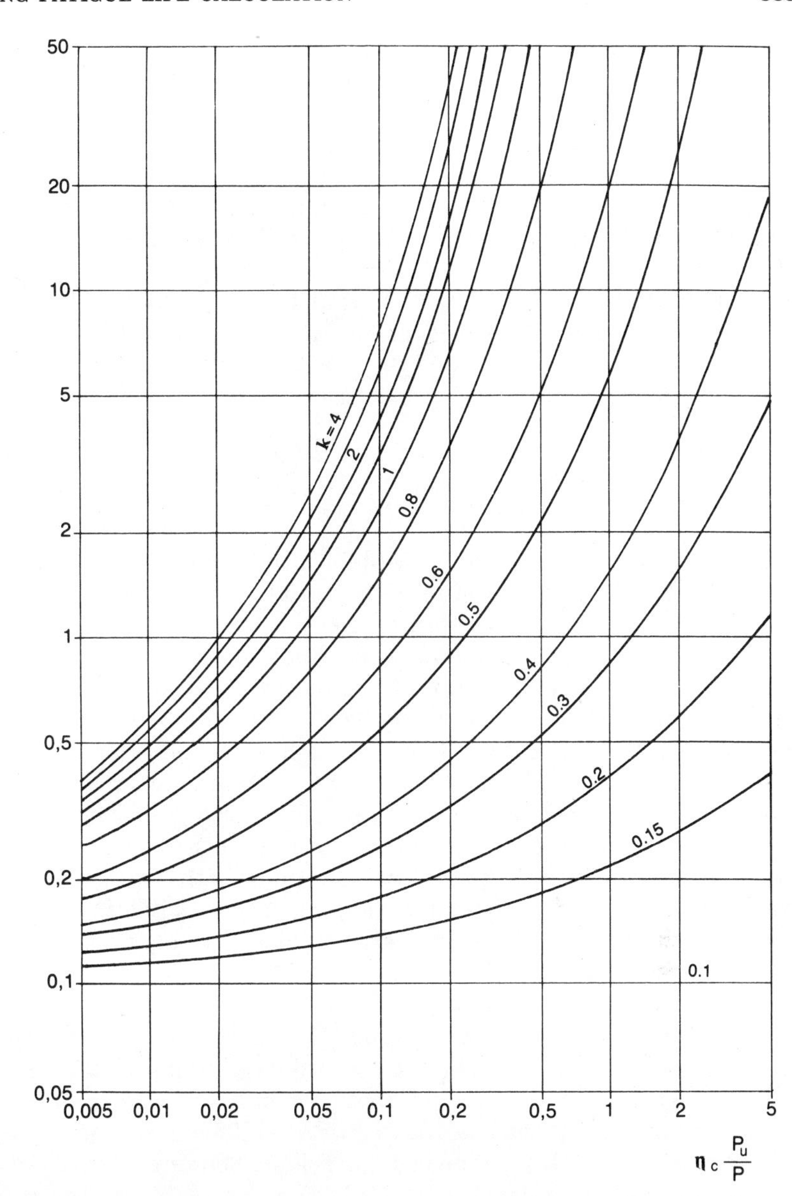

FIGURE 23.12. Fatigue life factor for $\mathfrak{a}_{SKF}$ radial ball bearings (from [23.18]).

is valid only for the radial ball bearings produced by SKF as listed in [23.19]. (Reference [23.19] contains charts similar to that of Fig. 23.12 for other types of bearings identified in [23.19].)

To use Fig. 23.12 to obtain $\mathfrak{a}_{SKF}$, it is necessary to obtain the fatigue load limit P_u for the bearing whose endurance is being evaluated. P_u

represents the load below which fatigue will not occur in the bearing. As seen in Fig. 23.12, the factor η_c is used as the abscissa. η_c is the adjustment factor for degree of contamination. Its value ranges from 1 for contaminants whose particle size is on the order of the lubricant film thickness to zero for heavy contamination, that is, contaminants having large size, very hard particulate matter. A table is given in [23.18] with recommended values for η_c; effectiveness of sealing can determine the value of η_c.

It is also noticed from Fig. 23.12 that to obtain $\mathcal{a}_{SKF}$, it is necessary to know κ, the relative viscosity in the application.

$$\kappa = \frac{\nu}{\nu_1} \tag{23.6}$$

where ν is the actual lubricant kinematic viscosity in the bearing and ν_1 is the kinematic viscosity required for adequate lubrication (see Fig. 17.4). Since both ν and ν_1 are to be determined at the operating temperature, dynamic viscosities can also be used to determine κ.

To illustrate the impact on predicted bearing endurance of the new life theory as embodied in equations (23.3) and (23.5), some numerical examples are developed herein. In this instance, the 209 radial ball bearing considered in numerical examples 6.1, 18.4, 18.5, and 18.13 will be reexamined. It is important, however, to have a valid comparison, that is, it must be recognized that the basic load ratings specified in the catalogs, for example, [23.19], have been increased to accommodate improvements in bearing steel and bearing manufacturing. Moreover, it is necessary to consider bearing operation under both heavy and light applied loading to gain a good understanding of the application of a fatigue load limit. The following numerical examples have been constructed accommodating these considerations.

Example 23.1. Using the fatigue life theory according to equation (23.3) and the calculation method according to equation (23.5) and Fig. 23.12, estimate the L_{10} life of the 209 radial ball bearing of Example 18.13 assuming (1) the bearing is an SKF 6209 ball bearing according to the catalog [23.19], (2) the bearing is just adequately lubricated, and (3) the bearing is properly protected from ingress of foreign material, that is, only negligible debris is contained in the bearing and lubricant.

From Example 18.13, L_{10} = 460 hr and applied radial load is 8900 N (2000 lb). From the SKF Catalog [23.19], P_u = 915 N (205.6 lb). For very clean operating conditions, η_c = 1. Using Fig. 23.12, P = 8900 N (2000 lb)

$$\eta_c P_u/P = 1 \times 915/8900 = 0.103$$

For adequate lubrication, $\kappa = 1$. From Fig. 23.12, $\mathcal{a}_{SKF} = 2.4$.

$$
\begin{aligned}
L_{naa} &= \mathcal{a}_1 \mathcal{a}_{SKF} L_{10} \qquad (23.5) \\
&= 1 \times 2.4 \times 460 = 1104 \text{ hr}
\end{aligned}
$$

Example 23.2. Assuming that the 209 ball bearing of Example 23.1 is not properly protected from contamination and that hard particle contaminants are present in the lubricant, what L_{10} life might be expected?

For heavy contamination $\eta_c = 0$ and $\eta_c P_u/P = 0$. For adequate lubrication $\kappa = 1$. From Fig. 23.12, $\mathcal{a}_{SKF} = 0.366$.

$$
\begin{aligned}
L_{naa} &= \mathcal{a}_1 \mathcal{a}_{SKF} L_{10} \\
&= 1 \times 0.366 \times 460 = 168 \text{ hr}
\end{aligned}
$$

Example 23.3. Assuming that the 209 radial ball bearing of Example 23.1 is subjected to a radial load of 4450 N (1000 lb) while the shaft rotates at 1800 rpm, what L_{10} life may be anticipated?

$$\eta_c P_u/P = 1 \times 915/4450 = 0.206$$

From Fig. 23.12, at $\kappa = 1$, $\mathcal{a}_{SKF} = 33.3$.

$$
\begin{aligned}
L_{naa} &= \mathcal{a}_1 \mathcal{a}_{SKF} \, (C/F_e)^3 \qquad (23.5) \\
&= 1 \times 33.3 \times (32{,}710/4450)^3 \\
&= 1.225 \times 10^{11} \text{ revolutions} \\
&= 1.135 \times 10^6 \text{ hr} \\
&= 129.6 \text{ years of continuous operation}
\end{aligned}
$$

Therefore under light load, for example, $C/8$, extremely long lives can be achieved when contamination is eliminated or at least minimized.

CLOSURE

The Lundberg and Palmgren theory to predict fatigue life was a significant advancement in the state-of-the-art of ball and roller bearings affecting the internal design and external design and dimensions for 40 years. The EHD lubrication theory, initially developed by Grubin and further advanced by scores of researchers, initially affected bearing microgeometry, but later, because of the possibility of increased endurance together with improved materials, resulted in "downsizing" of ball and roller bearings. The new fatigue life theory described in this chapter carries the development to the next plateau by substantially increasing understanding of the significance of material quality and concentrated contact surface integrity. It is now apparent that for a material that is clean, homogeneous, and free from particulate contaminants in the contact zones between rolling elements and raceways, it is probable that many applications need not rely on bearing selection and sizing according to the life-limiting rolling contact fatigue criterion. This should result in optimization of bearing design and selection in many common applications. Of course, high-quality material must be used and significant contamination must be eliminated.

Since the publication of the new theory in 1984, additional analytical and experimental effort has been on-going to both refine the methods of application and verify capability to evaluate specialized bearing applications. In this regard Ioannides and Pareti [23.20] demonstrated the ability to predict the fatigue endurance of rollers with various degrees of crowning. Also, in [23.21], Ioannides et al evaluated the effects of hoop stress and fatigue limit on cylindrical roller bearing inner ring endurance. In [23.22], Ioannides et al used the new theory to predict the extremely long L_{10} fatigue life experienced for the main thrust-carrying ball bearing of a high performance gas turbine engine. The life prediction was accurately at least two orders of magnitude greater than that afforded by the standard method. If such prediction method would have been employed in the initial bearing design phase, the bearing macrogeometry could have been substantially affected. Conversely, in [23.23] Harris et al applied the theory to explain foreshortened fatigue life caused by difficult lubrication conditions even though the bearing design and steel used should have provided superior endurance according to standard methods of life prediction.

The application of the new fatigue life theory has only just commenced. As with the introduction of the Lundberg-Palmgren theory, it will require a substantial period of time before the full effects of the new theory are realized on bearing design and application.

REFERENCES

23.1. American National Standard (ANSI/AFBMA), Std. 9-1990, "Load Ratings and Fatigue Life for Ball Bearings."

23.2. American National Standard (ANSI/AFBMA), Std. 11-1990, "Load Ratings and Fatigue Life for Roller Bearings."

23.3. International Standard ISO 281/1, "Rolling Bearings-Dynamic Load Ratings and Rating Life-Part 1; Calculation Methods (1977-03-15).

23.4. G. Lundberg and A. Palmgren, "Dynamic Capacity of Rolling Bearings," *Acta Polytech. Mech. Eng. Ser. 1, R.S.A.E.E.*, No. 3, 7 (1947).

23.5. G. Lundberg and A. Palmgren, "Dynamic Capacity of Roller Bearings," *Acta Polytech. Mech. Eng. Ser. 2, R.S.A.E.E.*, No. 4, (1952).

23.6. A. Grubin, "Fundamentals of the Hydrodynamic Theory of Lubrication of Heavily Loaded Cylindrical Surfaces," *Investigation of Contact Machine Components*, Kh.F. Ketova, Ed., Translation of Russian Book No. 30, Chapter 2, Central Scientific Institute of Technology and Mechanical Engineering, Moscow (1949).

23.7. L. Sibley and F. Orcutt, "Elastohydrodynamic Lubrication of Rolling Contact Surfaces," *ASLE Trans.* **4,** 234–249 (1961).

23.8. E. Zaretsky, E. Bamberger, T. Harris, W. Kacmarsky, C. Moyer, R. Parker, and J. Sherlock, *Life Adjustment Factors for Ball and Roller Bearings*, ASME Engineering Design Guide (1971).

23.9. T. Anderson, "Endurance Testing in Theory," *Ball Bearing J.,* **217,** 14–23 (1983).

23.10. T. Tallian, "On Competing Failure Modes in Rolling Contact," *ASLE Trans.* **10,** 418–439 (1967).

23.11. A. Voskamp, "Material Response to Rolling Contact Loading," ASME Paper 84-Trib-2.

23.12. E. Ioannides and T. Harris, "A New Fatigue Life Model for Rolling Bearings," *ASME Trans., J. Tribol.* **107,** 367–378 (July 1985).

23.13. J. Akesson and T. Lund, "SKF Rolling Bearing Steels-Properties and Processes," *Ball Bearing J.* **217,** 32–44 (1983).

23.14. N. Tsushima and H. Kashimura, "Improvement of Rolling Contact Steels," SAE Technical Paper 841123.

23.15. R. Sayles and P. MacPherson, "Influence of Wear Debris on Rolling Contact Fatigue," ASTM Special Technical Publication 771, J. Hoo, Ed., 255–274, (1982).

23.16. A. Tanaka, K. Furumura, and T. Ohkuna, "Highly Extended Life of Transmission Bearings of 'Sealed-Clean' Concept," SAE Technical Paper 830570.

23.17. M. Webster, E. Ioannides, and R. Sayles, "The Effect of Topographical Defects on the Contact Stress and Fatigue Life in Rolling Element Bearings," *Proc. 12th Leeds-Lyon Symposium on Tribology*, 207–226 (1986).

23.18. J. Wuttkowski and E. Ioannides, "The New Life Theory and Its Practical Consequences," *Ball Bearing J. Special Issue*, 6–11 (April 1989).

23.19. SKF, *General Catalogue 4000E* (April 1989).

23.20. E. Ioannides and G. Pareti, "Fatigue Life Predictions in Line Contacts With and Without Edge Stresses," *Proc. Inst. Mech Eng., Int. Conf. Fatigue of Engineering Materials and Structures*, 211–218 (September 1986).

23.21. E. Ioannides, B. Jacobson, and J. Tripp, "Prediction of Rolling Bearing Life under Practical Operating Conditions," *Tribological Design of Machine Elements*, D. Dowson et al Ed., Elsevier, 181–187 (1989).

23.22. E. Ioannides, T. Harris and M. Ragen, "Endurance of Aircraft Gas Turbine Mainshaft Ball Bearings—Analysis Using Improved Fatigue Life Theory 1: Application to a Long Life Bearing," *ASME Trans. J. Tribol.*, 304–308, (April 1990).

23.23. T. Harris, E. Ioannides, M. Ragen and H. Tam, "Endurance of Aircraft Gas Turbine Mainshaft Ball Bearings—Analysis Using Improved Fatigue Life Theory 2: Application to a Bearing Operating Under Difficult Lubrication Conditions", *ASME Trans, J. Tribol.*, 309–316, (April 1990).

24

WEAR

LIST OF SYMBOLS

Symbol	Description	Units
F	Shear force	N (lb.)
h	Mean plane separation, lubrication film thickness	μm (μin.)
N	Normal force	N (lb.)
R	Radius of surface	mm (in.)
T	Total temperature between contacting surfaces	°C (°F)
T_b	Bulk temperature of a component	°C (°F)
T_f	Flash temperature generated during a tribological encounter	°C (°F)
u	Surface velocity	mm/sec (in./sec)
σ	Combined surface roughness	μm (μin.)
Λ	Film thickness/surface roughness ratio	
	SUBSCRIPTS	
1,2	Refer to individual surfaces	

GENERAL

The forces transmitted in a bearing give rise to stresses of varying magnitudes between surfaces in both rolling and sliding motion. As a result of repeated loads in concentrated contacts, changes occur in the contact surfaces and in the regions below the surfaces. These changes cause surface deterioration or wear. Wear is the loss or displacement of material from a surface. Material loss may be loose debris. Material displacement may occur by local plastic deformation or the transfer of material from one location to another by adhesion. When wear has progressed to the degree that it threatens the essential function of the bearing, the bearing is considered to have failed. Through experience and detailed failure analysis, the bearing engineer recognizes distinct classes of failure. They are listed in Table 24.1.

These failures are defined without presupposing the exact mechanism by which they occur. They are defined in engineering terms based on a description of observations. The observations and their classifications reflect the remaining evidence of a complicated sequence of events involving many physical and chemical processes that preceded it, including those in the manufacturing of the original surfaces. Associated with the physical and chemical interactions on the surfaces are several mechanistic wear processes. They are listed in Table 24.2.

Wear prevention is accomplished by forming lubricating films by hydrodynamic lubrication, elastohydrodynamic (EHD) lubrication, and boundary lubrication. During the surface life of a bearing, the lubrica-

TABLE 24.1. Bearing Failure Classification Due to Wear

Mild mechanical wear
Adhesive wear
Smearing
Corrosive (tribochemical) wear
Plastic flow
Surface indentation
Abrasive wear
Surface distress
Pitting
Fatigue spalling

TABLE 24.2 Mechanistic Wear Processes

Adhesion	Plastic flow
Chemical reaction	Fatigue

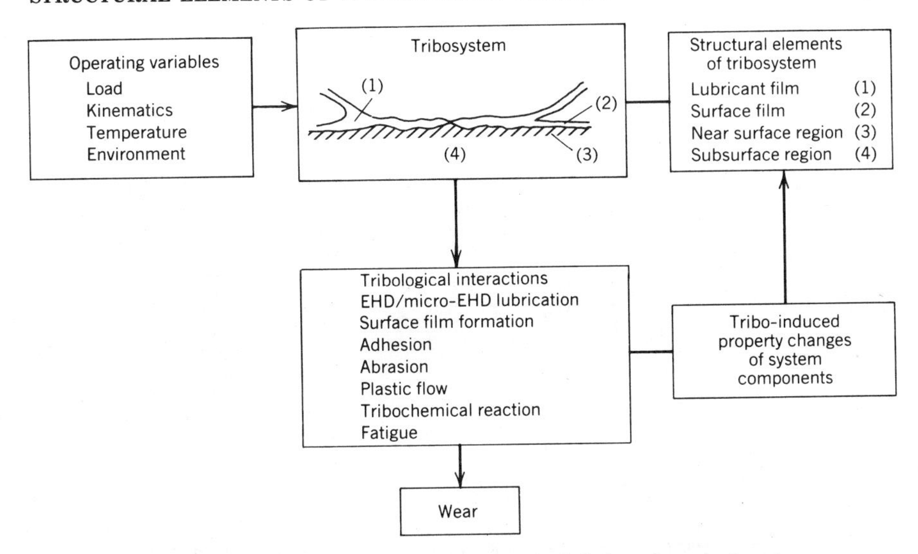

FIGURE 24.1. Tribological interactions of a lubricated contact system.

tion and wear processes are interactive and competitive. The topic of wear cannot be divorced from the topic of lubrication, and it is essential that individual contact areas within a bearing be considered as a tribological system. The tribological interactions of a system are described schematically in Fig. 24.1 similar to the description in [24.1].

Numerous technical options exist for improving wear performance through materials, lubricant base stocks, additives, finishing processes, and surface modification technologies. The correct bearing design for a particular application is derived from the synergistic assembly of many tribological contributions.

STRUCTURAL ELEMENTS OF A LUBRICATED CONTACT

Load-carrying capacity is derived from the integral strength of four general regions of a lubricated contact. Figure 24.2 shows these four regions.

The *EHD/micro-EHD lubricated region* is created by the generation of an EHD lubricant film, which on a global scale is derived from the hydrodynamic pressure generated in the inlet region of the contact; on a local scale it is derived from the micro-EHD lubrication action associated with the local topography of the surfaces. The EHD/micro-EHD region is typically less than 1 μm (40 μin.) thick.

The *surface film region* contains the outer layers of the surface, which consist of surface oxides, adsorbed films, and chemical reaction films derived from the lubricant and its additives. It is usually less than 1 μm (40 μin.) thick.

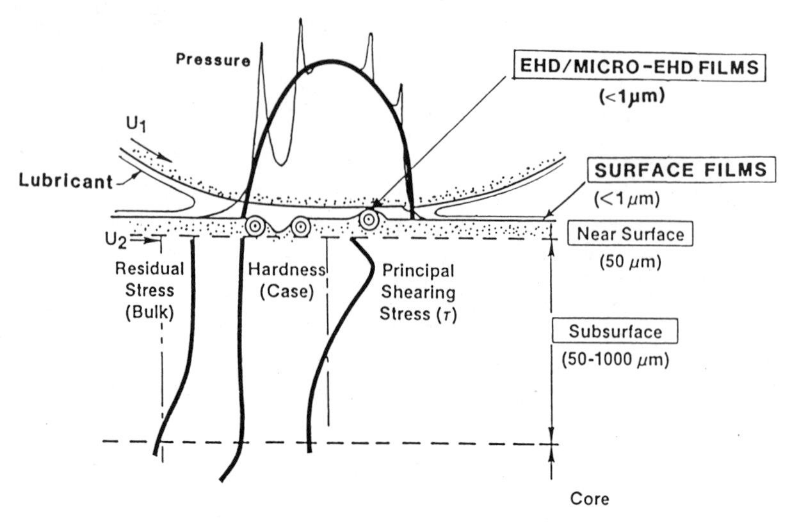

FIGURE 24.2. Structural elements of a lubricated contact.

The *near-surface region* contains the inner layers of the surface, including a finely structured and highly worked Bielby layer as well as other deformed layers. These deformed layers, which have a different microstructure than the material below them, may arise from surface preparation techniques, such as grinding and honing, or they may be induced during operation (e.g., run-in). Hardness and residual stress can vary significantly, and they could be substantially different from the material below. The near-surface region may extend to 50 μm (0.002 in.) below the surface.

For concentrated contacts a *subsurface region* can be defined, which may be 50–1000 μm (0.002–0.04 in.) below the surface. This region is not significantly affected by the mechanical processes that produce the surface or the asperity-induced changes that occur during operation. Its microstructure and hardness may still be different from the bulk material below it, and significant residual stresses might still be present. These stresses and microstructures, however, are the result of macroprocesses such as heat treating, surface hardening, and forging. For typical Hertzian contact pressures the maximum shear stress is located within the subsurface region. In other words, the detrimental global contact stresses are communicated to the subsurface region where fatigue begins. This fatigue is called subsurface-initiated.

Between the near-surface region and the subsurface region is a "quiescent zone" that resides below the surface where the local asperity and surface defect stresses are not significant and the stress field from the macroscopic Hertzian contact stress is not yet appreciable. This zone is quiescent from the point of view of stress and the accumulation of plastic flow and fatigue damage. The existence of the quiescent zone is important with regard to rolling contact fatigue. It inhibits the propagation of cracks between the stress field in the near-surface region and the stress field in the subsurface region.

With regard to rolling contact fatigue, major material improvements have been made that reduce the risk of subsurface-initiated fatigue. The risk of surface-initiated fatigue now seems to be a more dominant factor.

TRIBOLOGICAL PROCESSES ASSOCIATED WITH WEAR

Global and Local Processes

The lubricated contact system can be characterized by global processes associated with the lubricated contact as a whole, and by local processes associated with local features of the system—that is, those derived from the topography of the surface, the microstructure of the underlying materials, or the presence of wear debris. An inherent practical problem with the control or prediction of wear is that the failure process of wear is initiated generally on a local level, but it is influenced greatly by the interaction of both global and local processes. Some of the global and local processes are definable on reasonably good scientific grounds; however, their interactions in a real system seem to be the important quantity that is missing. This section discusses some of the important tribological processes in connection with wear.

Lubrication Processes

EHD Lubrication. The formation of an EHD lubricant film contributes to wear reduction by reducing the local stresses between the surfaces and by creating a lubricant film easy to shear. The global pressure and elastic shape are very similar to the Hertzian condition for dry contact, giving rise to three reasonably well-defined regions shown schematically in Fig. 24.3.

The formation of an EHD lubricant film is derived from the hydrodynamic pressure generated in the inlet region. EHD lubrication is on excellent quantitative grounds, which allows the oil film thickness to be predicted from the viscous properties of the lubricant, the geometry of the contact system, and the operating conditions. This has proven to be a very useful design tool for predicting the lubrication regime for various operating conditions; however, it is not sufficient to predict wear. This is partly because EHD lubrication is primarily an inlet phenomenon; that is, its major action occurs in a region displaced from the Hertzian region where the more local events involved in wear initiation take place. The severity of these local events can be significantly influenced by the

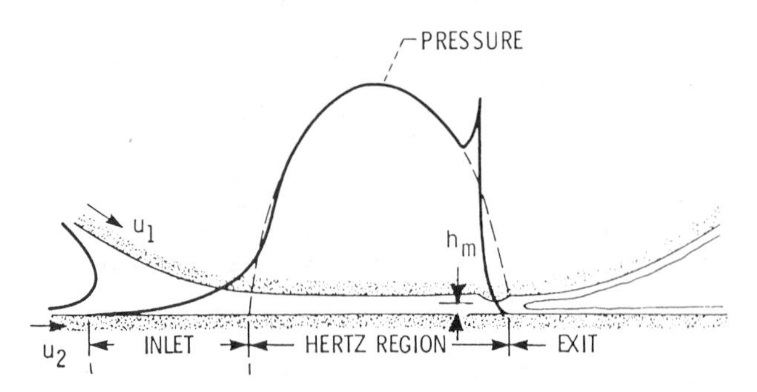

FIGURE 24.3. EHD lubrication.

EHD lubrication process by the thickness of the EHD lubricant film. It determines what may be called the "degree of asperity interaction."

Thus, the EHD lubrication process is viewed as a quantitative foundation upon which a predictive capability of wear can be established by incorporating several less quantifiable processes.

Surface Temperature. Surface temperature is not a lubrication process but a key link between lubrication and wear because it significantly influences the viscous properties of the lubricant that control the thickness of the EHD lubricant film, and it is a major driving force in the formation of chemical reaction films. Additionally, it determines the rate of lubricant degradation and influences the strength of surface films as well as the flow properties of the material in the near-surface region. Consequently, it is not surprising that the total temperature level is a frequently used criterion for failure.

From a simplistic point of view the total temperature (T) is the sum of a bulk temperature (T_b) of the bearing component and the flash temperature (T_f) associated with the instantaneous temperature rise derived from the friction within the lubricated contact. Flash temperature may arise from the traction of the lubricant film as well as from the energy dissipated from the adhesion, plastic flow of surface films, and deformation of the material within the near-surface region. The global magnitude of T_f can be predicted if simplifying assumptions about the coefficient of friction and convection heat transfer are made.

Micro-EHD Lubrication. The lambda ratio, $\Lambda = h/\sigma$, a most useful engineering quantity, is the ratio of the EHD lubricant film thickness h to the average combined roughness height σ of the interacting surfaces. It is a simple way of describing the degree of asperity interaction. Thus, when $\Lambda > 3$, fatigue life is much greater than for lower Λ, because local asperity stresses have been significantly reduced. Its connection with surface-initiated fatigue seems to be more obvious than failure modes associated with wear. The latter failure modes generally appear at low Λ (e.g., $\Lambda < 1$) where, unfortunately, it loses much of its meaning.

When σ is on the same order of magnitude as h, the surface topography becomes intimately involved in the lubrication process itself in the form of micro-EHD lubrication. This comes about from a global standpoint where the orientation of the topographical features can influence the average film thickness. It also comes about from a local standpoint through the generation of micro-EHD lubrication pressures associated with topographical features, as shown in Fig. 24.4.

Recent measurements [24.2] imply that the energy dissipation due to friction becomes concentrated at specific local topographical sites, giving rise to local stresses and temperatures even without physical contact.

Boundary Lubrication. It is well known that surface films are important to boundary lubrication because they prevent adhesion and provide a film that is easy to shear. These films may be in the form of oxides, adsorbed films from surfactants, and chemical reaction films from other additives. These surface films are schematically shown in Fig. 24.5.

Their reactions and interactions are complex. Most studies on the subject have focused on the chemical identification or phenomenological effect of surface films, but little is known about the mechanism of pro-

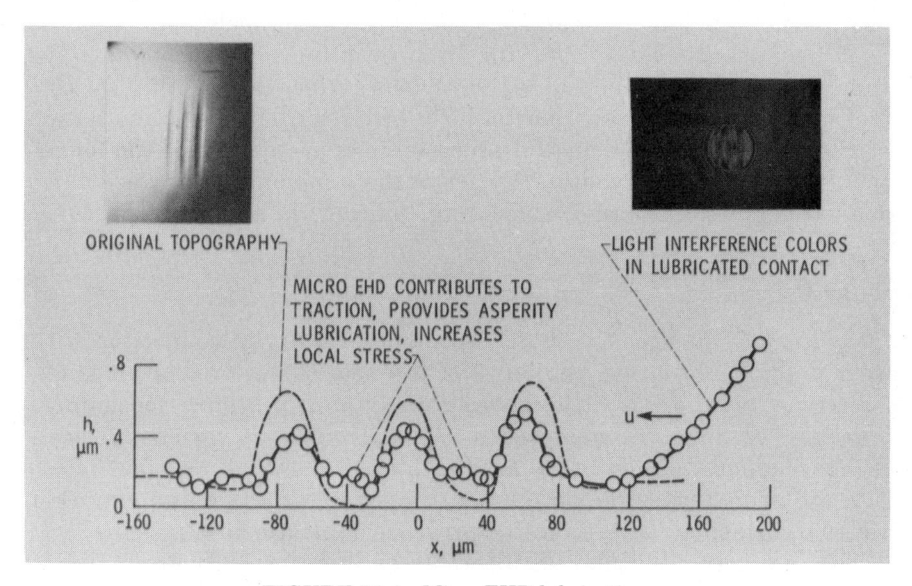

FIGURE 24.4. Micro-EHD lubrication.

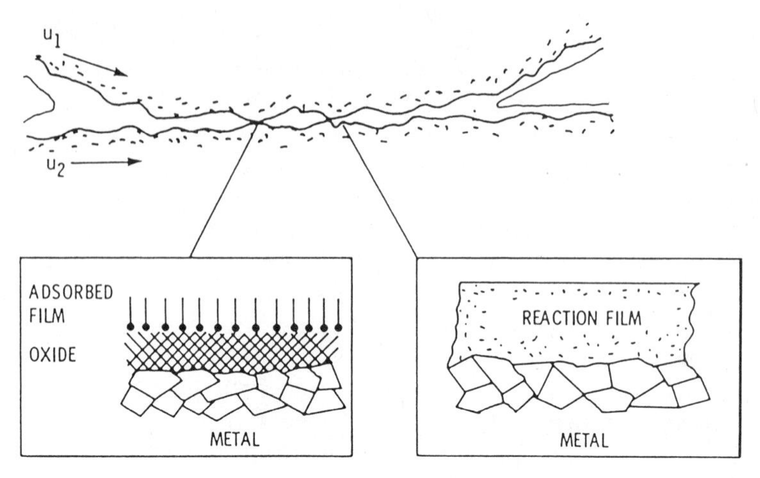

FIGURE 24.5. Surface films.

tection, the means of removal, or the rate of reformation. At high temperatures the oxidation of the base fluid can contribute to surface film formation. There have been many studies on the catalytic effect of metals on the bulk oxidation of fluids. Similar oxidative processes can occur under the thermal stress environment in the contact region where intermediate oxidation species can react with the surface or organo-metallic material. These reactions can influence boundary lubrication in several ways, such as by corrosive wear, by competition with other additives, or by forming polymeric material—that is, a friction polymer.

The contribution of surface films in preventing wear is complex. The time and spatial distribution of the various surface films within the contact seems to be important, particularly with regard to the accumulation of material (including debris of all sorts) in depressions and the formation of films at asperity sites. In view of the complexity of surface films, one wonders what the real lubricating "juice" is in a real system.

Wear Processes

Interactions. Perhaps the most important quantity in connection with wear is the deformation attributes of the near-surface region. It is unfortunate that there is little understanding of near-surface mechanical properties or the attributes needed to complement the various lubricating mechanisms to improve wear resistance. To maintain surface integrity, the near-surface region must prevent microfracture and maintain a viable surface finish even in the presence of plastic flow.

The interaction of tribological processes in an asperity encounter is shown schematically in Fig. 24.6.

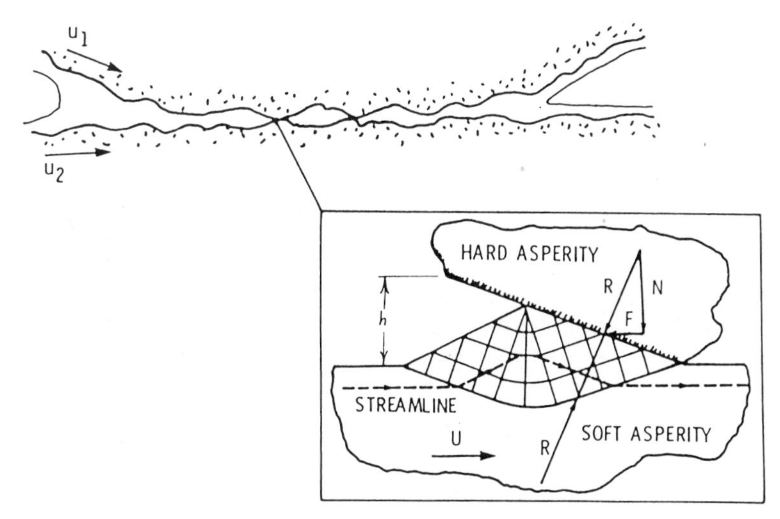

FIGURE 24.6. Surface plastic flow.

The severity of interaction is reflected in the normal load N, which is influenced by the thickness of the EHD lubricant film h. The shear force F is influenced by the various surface films and micro-EHD lubricant films, along with the flow properties in the near-surface region. The exact mechanism whereby shear stress is applied to the near-surface region is not known. This could come about through metal-to-metal adhesion but it is also possible to have sufficient compressive and shear stresses applied locally through a thin lubricant film.

In any case, the severity of interaction is important to the initiation and propagation of wear. It will determine whether the result is (1) a benign elastic encounter, (2) a further accumulation of plastic fracture sites that can lead to the generation of wear particles (e.g., microspalling, mild mechanical wear, and delamination), (3) oxidative or corrosive wear, or (4) the advancing of adhesive transfer, which can lead to smearing. The wear processes associated with these events are discussed in connection with a description of commonly recognized wear modes.

Wear can be defined in terms of four "mechanistic wear processes": adhesion, plastic deformation, fatigue, and chemical reactions [24.3]. The recognized wear modes in bearing technology, such as "smearing" or "pitting," are not singly connected to these wear processes but are associated with the interaction of the processes both simultaneously and sequentially. The importance of connecting the wear process with the commonly accepted failure mode is associated with engineering decisions required to overcome wear problems through lubrication, material selection, design, or allowable operating conditions.

Adhesion. Under high normal and tangential stresses, boundary films trapped between contacting asperities can be stretched until they rupture, which allows the formation of metal-to-metal contact on an atomic scale and gives rise to strong adhesive or welded junctions.

With relative motion between the contacting asperities, junction growth occurs by plastic deformation. Fracture ultimately takes place, and it can occur at a location other than the original interface, resulting in material transfer from surface to surface. The formation and rupture of adhesive junctions is accompanied by very high local temperatures that can form reaction films on the newly formed surface and change the mechanical properties of the underlying material. "Adhesive wear" occurs when the adhesive transfer of material is the important or controlling mechanism.

Smearing. Smearing is adhesive wear on a large scale, which occurs between rolling element and raceway when sliding is substantial. The severe plastic deformation that accompanies smearing is shown in Fig. 24.7.

Smearing is sometimes called scuffing or galling. The precise mechanism of smearing is not well understood but it does involve the gross failure of the surface and is accompanied by an increase in friction and contact temperature. A current view of smearing [24.4] is that under conditions yet to be defined it is a gradual breakdown in the lubrication of interacting asperities, the nature of which may be boundary, micro-EHD, or a mixture of the two. Although the final smearing mode may represent the gross breakdown of various lubricating films and the near-surface region, it may be triggered by the deterioration in surface topography as a result of adhesive wear or local plastic flow.

Chemical Reaction. The mechanistic wear processes associated with fatigue and plastic flow are the result of material deformation caused by stress. A significant part of wear and its control involves chemical reaction processes with the environment. The environment is defined as that portion of the contact system that is not an intrinsic part of the surfaces. The environment includes the surrounding atmosphere as well as the lubricating films.

Pure chemical reactions should be distinguished from "tribochemical" reactions, which are a consequence of the tribological interactions between the contacting surfaces. "Corrosion" results from reaction of the surface with the ambient environment under the prevailing ambient conditions; tribochemical behavior is activated by mechanical interaction of the contacting surfaces. Corrosion often occurs on bearing components because of improper handling or storage resulting from the absence or removal of a protective film. An example of corrosion is shown in Fig. 24.8.

Preventing adhesive wear is done by forming tribochemical films.

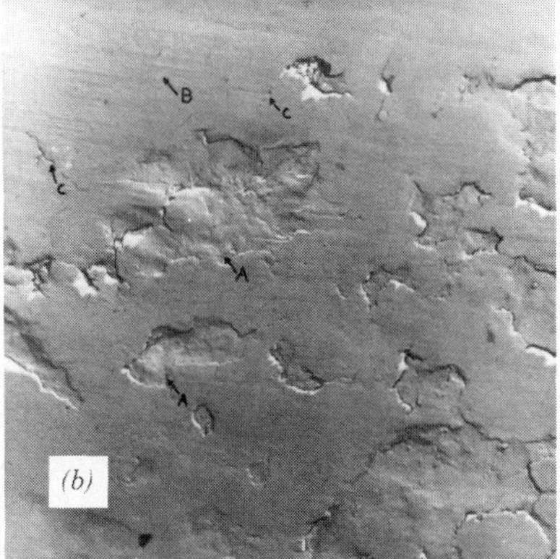

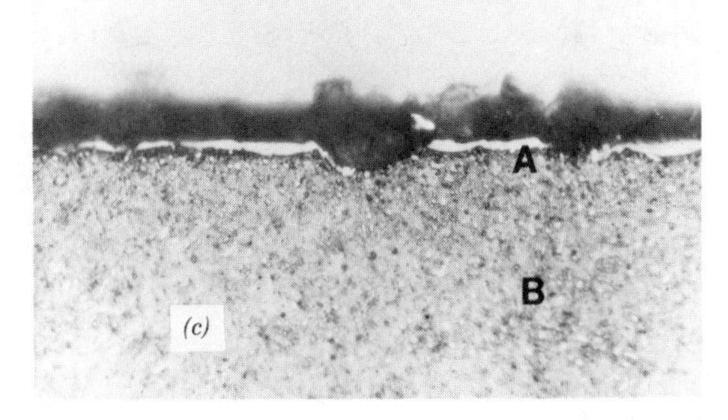

FIGURE 24.7. Smearing. (*a*) Smeared material crossing preexisting finishing marks of a honed roller. (*b*) Replica electron micrograph showing adhesive wear [A], original machining marks [B], and microcracks [C]. (*c*) Etched metallographic cross section through smeared area showing white etching bands at surface attributed to rehardening as a result of overheating (from [24.13]).

These films may be formed from oxygen in the atmosphere or from antiwear or extreme pressure additives in the lubricant. "Tribochemical wear" generally involves a continuous process of surface film formation and removal. The formation process involves chemical reaction or adsorption of chemical species on the surface. The removal process results from mechanically induced crack formation and abrasion of the reaction products in the contact. The process introduces "clean," that is, activated, local areas where new tribochemical films can be formed and subsequently removed. The tribochemical process introduces thermal and mechanical activation of the near-surface region, which can cause (1) greater chemical reactivity as a result of increased asperity temperature and (2) changes in the microstructure and mechanical properties of the near-surface layer due to high local temperatures and mechanical working.

FIGURE 24.8. Corrosion resulting from reaction of the surface with the environment. Roller subsequently run in a bearing leaving a multitude of dark-bottomed pits, a condition that subsequently creates surface-originated spalling (from [24.13]).

Under favorable operating conditions tribochemical reactions may be associated with "mild wear." Mild wear is associated with low wear rates and smooth surfaces frequently characterized by oxidation of the surfaces and subsequently removed—that is, oxidative wear [24.5, 24.6]. Unfavorable operating conditions can produce "severe wear," where the surfaces are extensively disturbed and may be characterized by extensive adhesion and plastic flow rather than oxidative wear. Severe wear can be prevented by increasing the rate of chemical reactions to form protective surface films at the same rate as clean activated local areas are generated. In this way a balance can be obtained between adhesive wear and "chemical wear." "Corrosive wear" is a term used when chemical wear dominates the adhesive wear mode by a wide margin. The rate of chemical wear is controlled by additive composition and concentration. An optimum additive formulation is achieved when there is a balance between adhesive and chemical wear for a given degree of contact severity [24.7]. This balance between adhesive and chemical wear is shown schematically in Fig. 24.9.

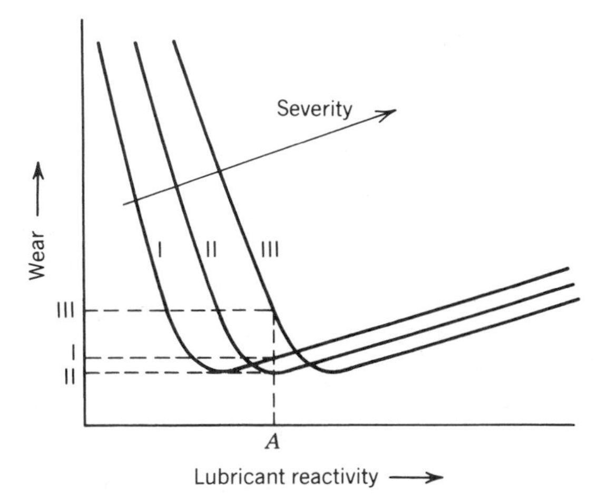

FIGURE 24.9. Adhesive/corrosive wear balance (from [24.7]).

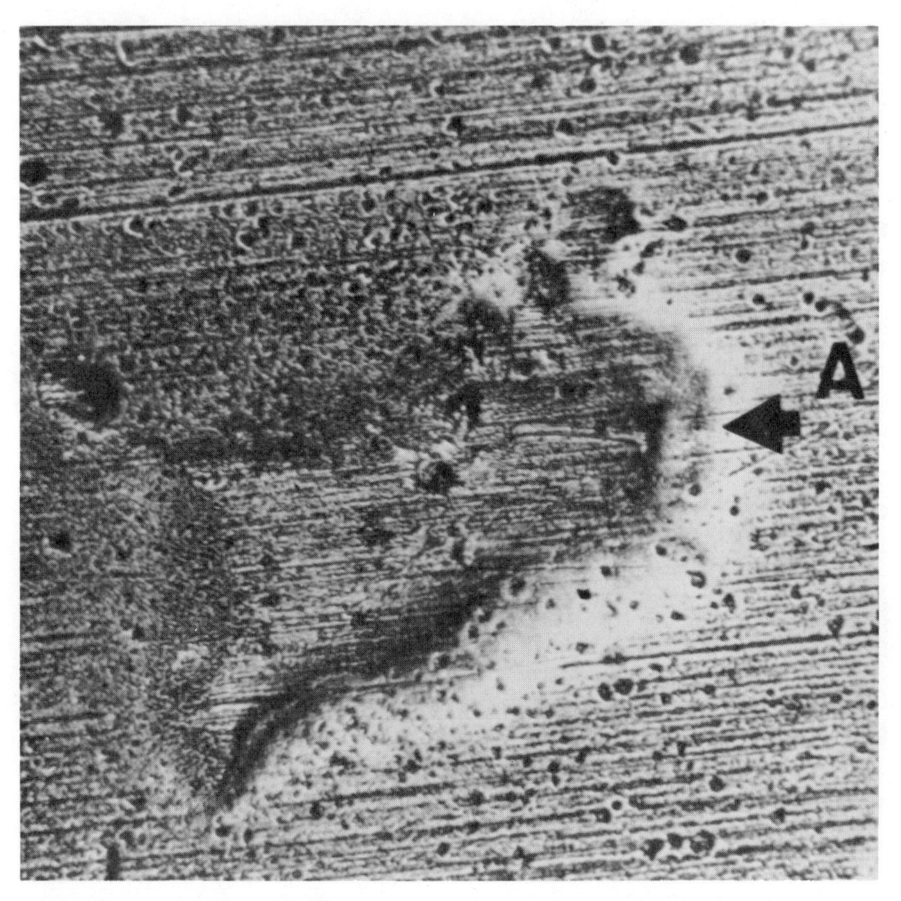

FIGURE 24.10. Debris dent showing local plastic deformation (surface distress) at the dent shoulder [A] as a result of substantial rolling over (from [24.13]).

Plastic Deformation. Depending on geometry, relative hardness, and load, the shape of a contacting surface can be permanently deformed, on both a macroscopic and a microscopic scale, as a result of plastic deformation. On a macroscopic scale the overload of rolling elements under static conditions can cause "Brinell marks" or distort the entire rolling track under operating conditions. A much more destructive macroscopic plastic deformation process occurs when the thermal balance between the rolling element bearing components becomes unstable because more heat is generated than is removed. A thermal runaway can cause the bearing materials to soften and flow plastically until the entire bearing geometry has been destroyed.

Almost all wear processes involve plastic flow on a microscopic scale. The plastic deformation that occurs from overrunning of hard particles, such as contaminants and wear debris, is "denting." Figure 24.10 is an example.

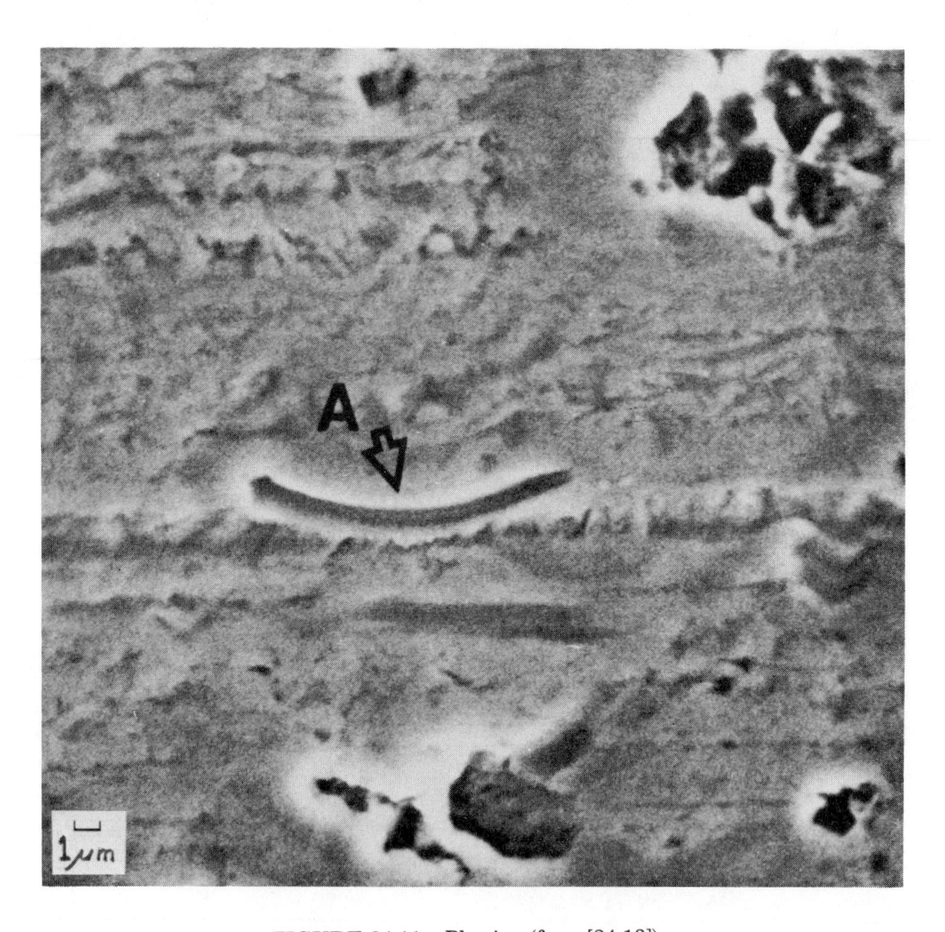

FIGURE 24.11. Plowing (from [24.13]).

FIGURE 24.12. Abrasive wear in the pockets of a brass cage (from [24.13]).

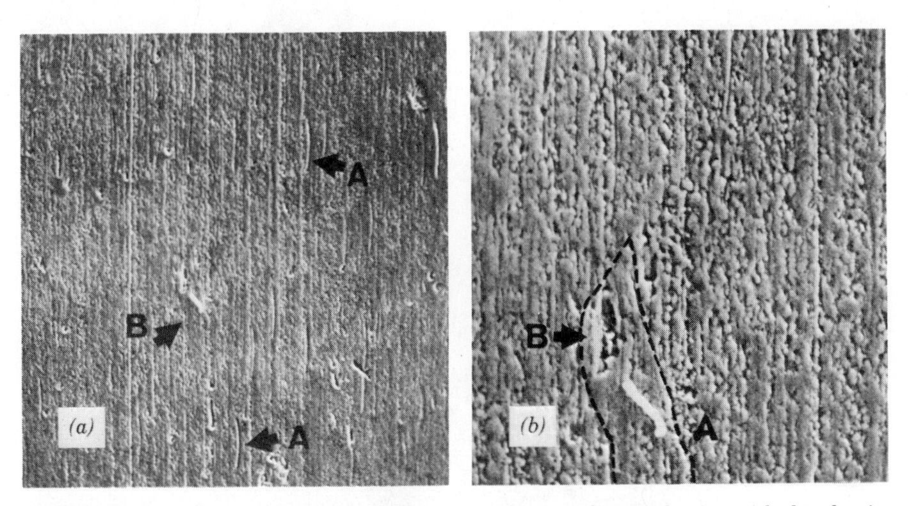

FIGURE 24.13. Inception of surface distress. (*a*) Surface distress begins with the plastic burnishing of asperity ridges [B]. (*b*) Higher magnification of (*a*). The small pit (probably preexisting) is surrounded by an area that has partially been burnished away (the slanting white mark is an artifact) (from [24.13]).

"Plowing" occurs when there is displacement of material by a hard particle under the presence of sliding or combined rolling/sliding conditions. See Fig. 24.11.

"Abrasive wear" occurs when the plastic deformation leads to material removal and wear debris. The interaction of hard rolling elements with softer separator materials often leads to abrasive wear, as shown in Fig. 24.12.

General plastic deformation of asperities and ridges on rolling contact surfaces is generally referred to as "surface distress," or at least the initial stages of surface distress. The final stages of surface distress involve the loss of material through microfracture and pitting. Figures 24.13 and 24.14 are examples of surface distress.

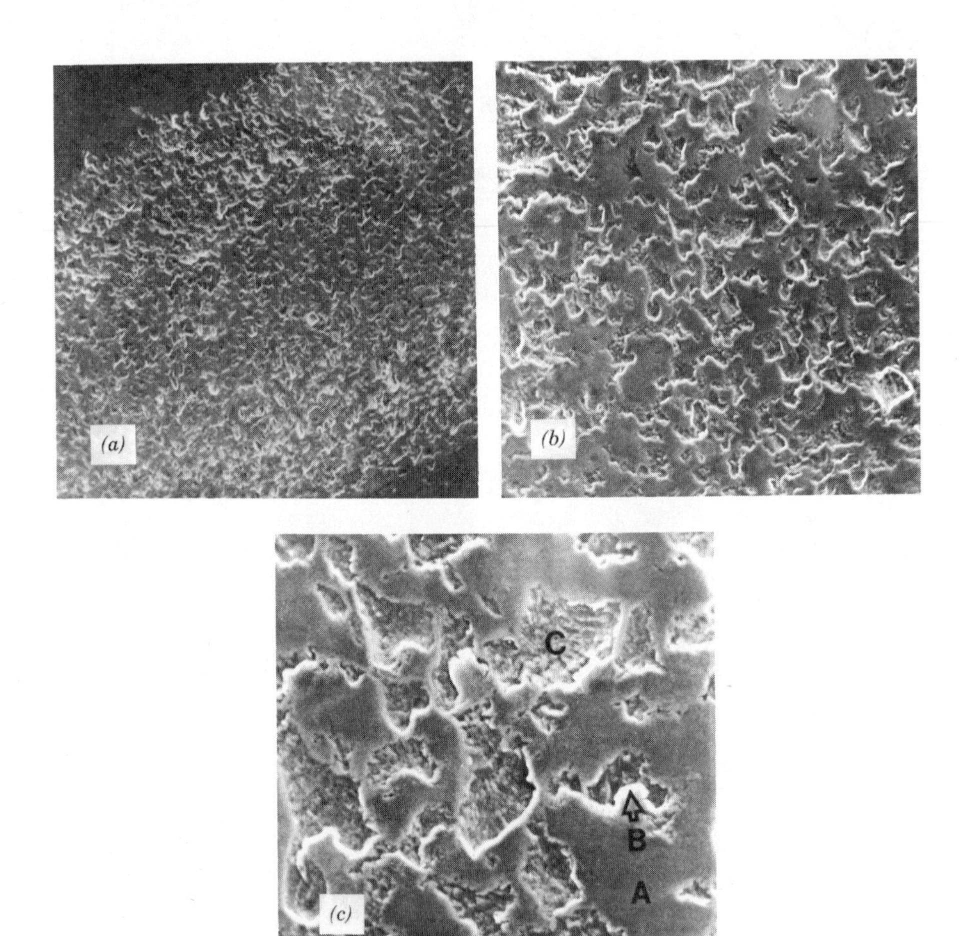

FIGURE 24.14. Final stages of surface distress of a ball bearing inner ring shown at three different magnifications. (*a*) Frosted appearance. (*b*) Multiple spalling of burnished surface. (*c*) Smooth appearance of plastically flowed material [A] (from [24.13]).

Fatigue. The final mechanistic wear process is fatigue. Fatigue is caused by cyclically repeated stresses on the contact surface, which eventually introduce permanent damage within the material. Damage begins as a crack. After repeated stress cycles, cracks can propagate and eventually lead to loss of surface material. Fatigue may initiate and propagate from the macrostresses induced in the subsurface region, resulting in "spalling" characterized by relatively large craters. Fatigue can also be initiated in the near-surface region as a result of microstresses from asperities or surface defects, such as dents, grooves, nicks, and scratches. If the combined micro- and macrostress fields propagate cracks through the quiescent zone and into the subsurface region, surface-initiated fatigue spalling can occur. "Pitting" and "delamination" occur when crack propagation is confined to the near-surface region. These processes are associated with the final stages of surface distress discussed above. The microstructural material changes and theory for spalling fatigue are discussed further in Chapters 18, 22, and 23.

PHENOMENOLOGICAL VIEW OF WEAR

The previous section reviewed the tribological processes associated with lubrication and wear. In a real system these processes interact and compete with one another in a complicated way so that the contribution of the individual processes to the overall picture is not very clear. Understanding the parts of the tribological system is important for selecting bearing materials, lubrication, bearing selection, life prediction, design decisions, and failure analysis. From an engineering standpoint it is essential to have a phenomenological view of wear in addition to an understanding of the constitutive parts of the tribological processes. In this way the complicated tribological processes can be reduced to a description of simpler observed behavior as a result of external operating conditions. For example, the wear rate for a given system can be observed as a function of time, loads, velocities, temperatures, and lubricant film thicknesses. The phenomenological approach is useful if the behavior is orderly with respect to the controlling variables. Figure 24.15 is a schematic representation for unlubricated, boundary lubricated, and fluid film lubricated systems [24.8].

Several interesting phenomenological investigations have been conducted using simple sliding contact test rigs that demonstrate the usefulness of this approach. Begelinger et al. [24.9], using a simple sliding ball-on-ring test assembly, demonstrated the usefulness of the phenomenological approach for establishing a failure map as shown in Fig. 24.16.

The map defines various transitional regions as functions of load, speed, and temperature. These transitions are identified by changes in friction and wear characteristics, and the regions separated by the transitions are characterized by various regimes of lubrication. Region I is

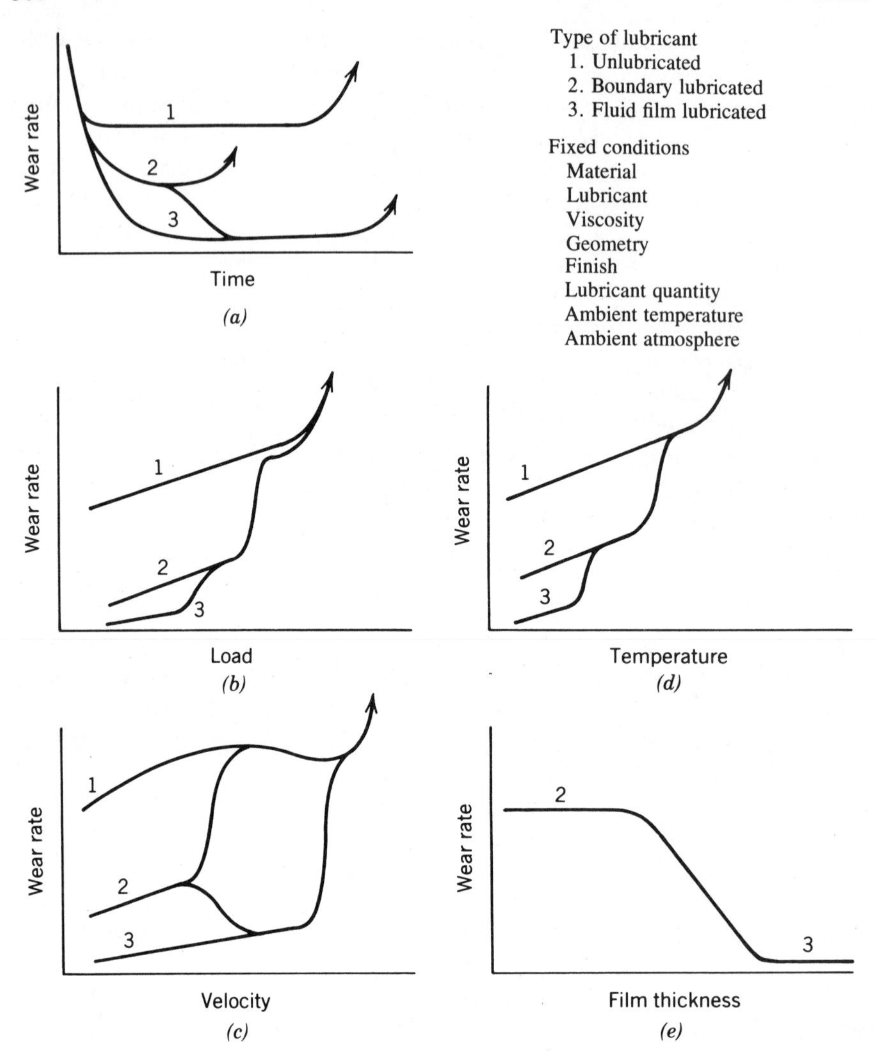

FIGURE 24.15. Phenomenological view of wear (from [24.8]).

associated with EHD and micro-EHD lubrication; region II is characterized by boundary lubrication; and region III reflects unlubricated or smearing (scuffing) behavior.

One of the most useful parameters for characterizing the phenomenological view of lubricated wear is Λ. Chapters 14 and 18 describe how this parameter can be used along with a detailed characterization of surface topography to predict asperity contact severity as a function of the mean plane separation of the surfaces. Detailed studies were conducted in [24.10] to show the wear behavior and fatigue behavior as functions of Λ. The result is shown in Fig. 24.17.

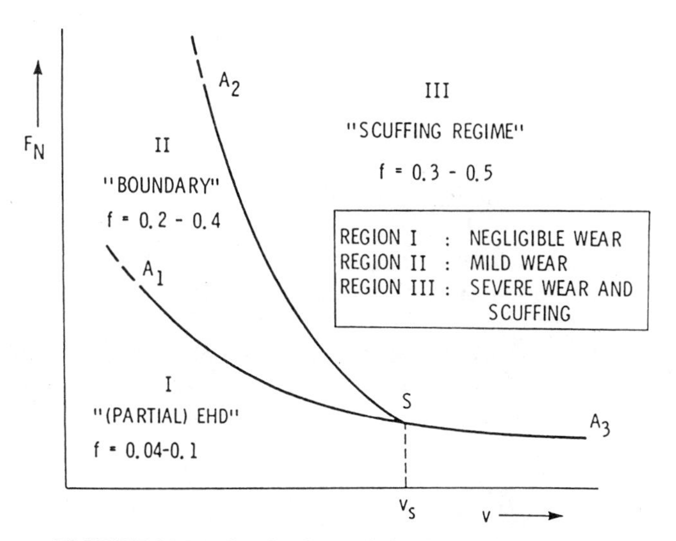

FIGURE 24.16. Load-velocity failure map (from [24.9]).

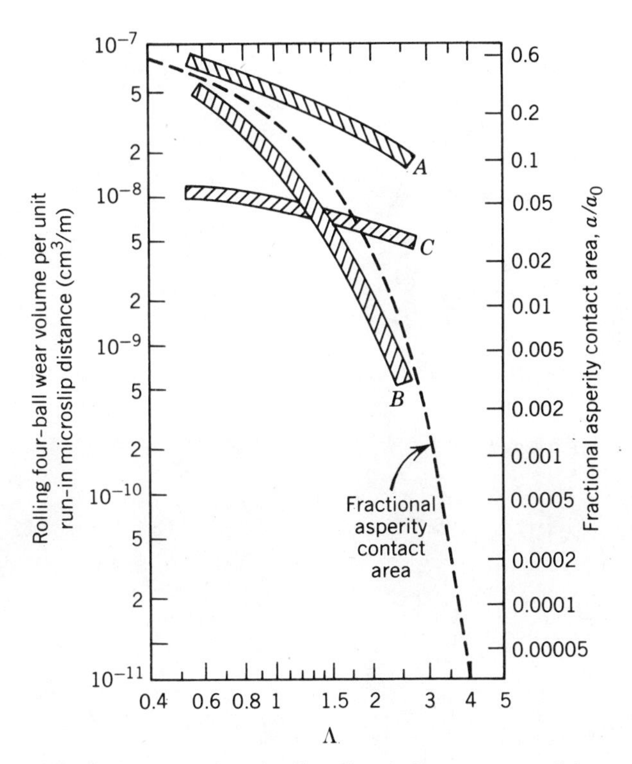

FIGURE 24.17. Radiotracer-measured rolling four-ball wear rate and fractional asperity contact area vs Λ. For $a_0 = 0.13\ \text{mm}^2$, A = Mineral oil, B = Synthetic ester, C = Sodium grease.

INTERACTING TRIBOLOGICAL PROCESSES AND FAILURE MODES

The ultimate failure of surfaces in a rolling contact bearing is the result of a complex sequence of events involving tribological processes of lubrication and wear. Frequently, an event that initiates a wear process can be successfully rescued by a lubrication process. On other occasions a minor wear process proceeding at one contact location can initiate a more devastating wear process at a more critical location. Examples of these will be given.

Figure 24.18*c*,*d* shows debris dents typical of those found in bearings contaminated with hard particles [24.11]. The stress concentrations at the shoulders of defects, similar to debris dents, frequently lead to initiation of spalling fatigue as shown in Fig. 24.18*a*,*b*. The role of EHD lubrication in reducing the stresses at the defect site was studied in [24.12]. If the defect dimensions are small compared with the inlet dimensions where EHD pressure is generated, the stress concentration at the edge of the defect can essentially be eliminated. It was further found that with EHD lubrication, the trailing edge of the dent should have

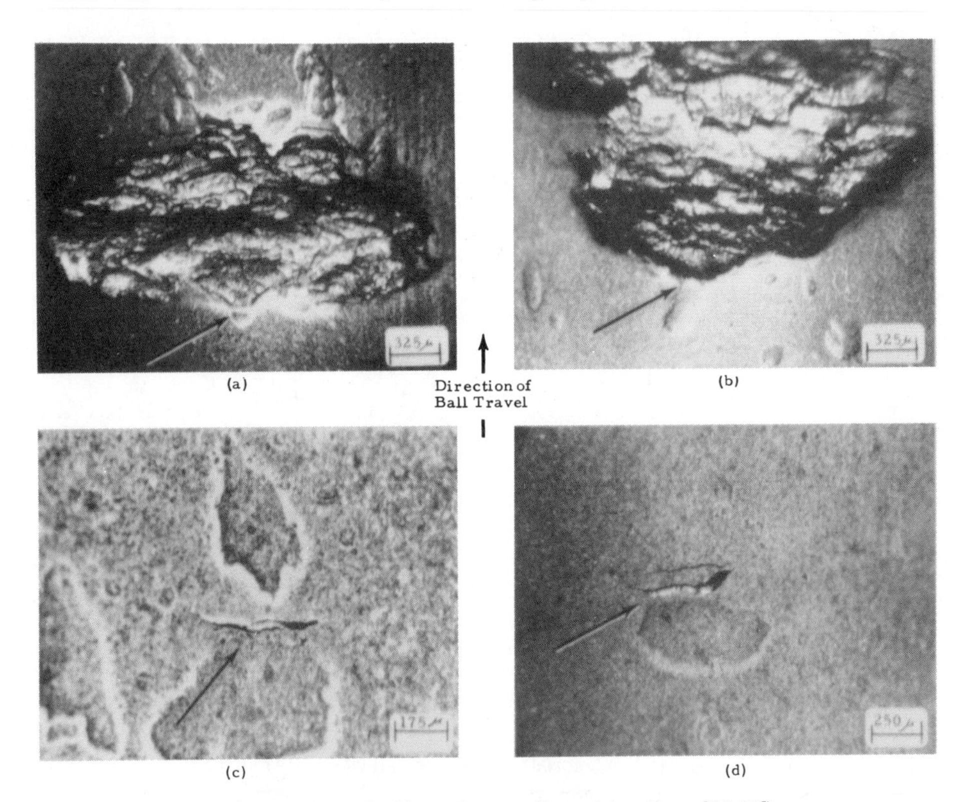

FIGURE 24.18. Spalls and prespall cracking (from [24.11]).

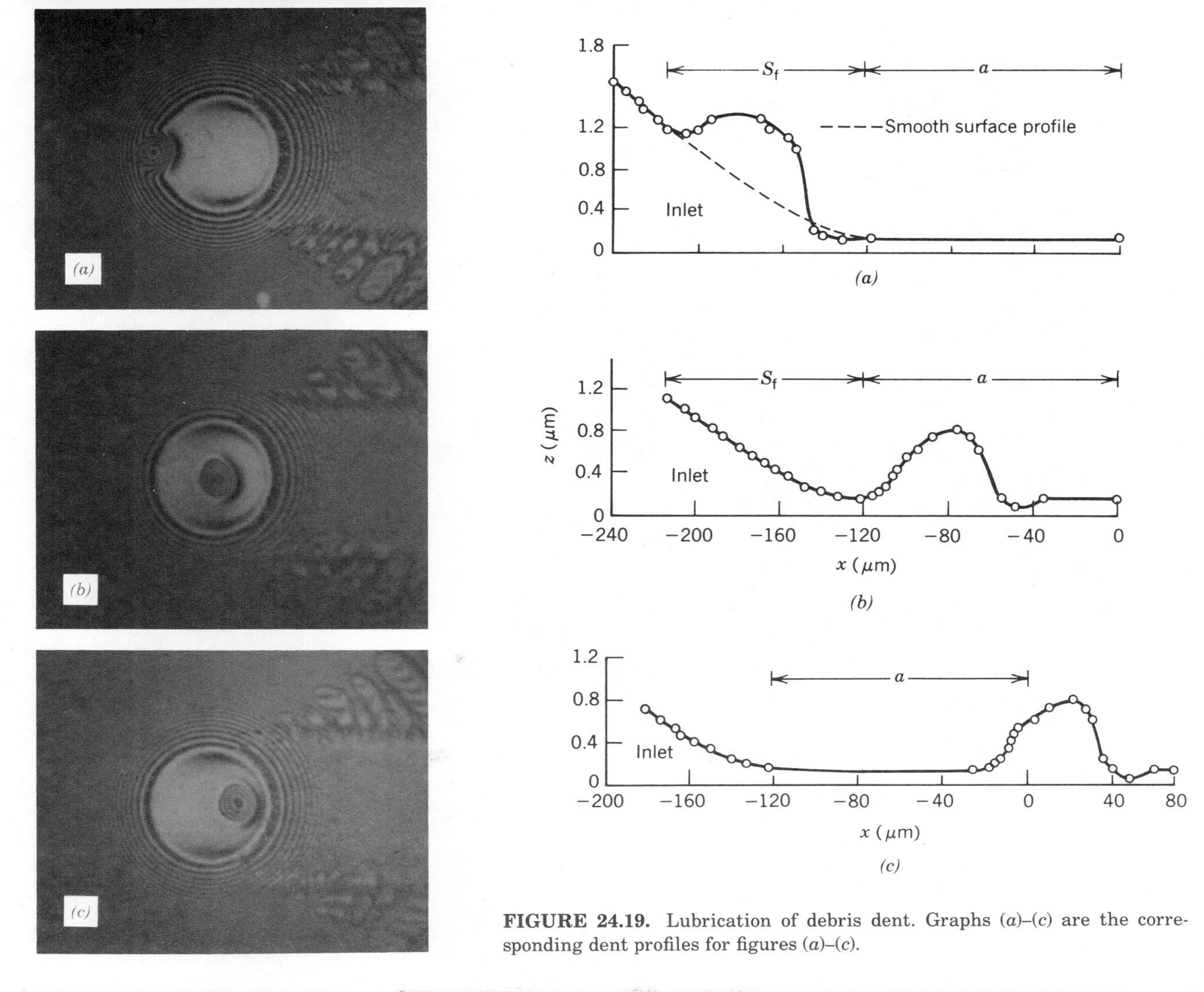

FIGURE 24.19. Lubrication of debris dent. Graphs (*a*)–(*c*) are the corresponding dent profiles for figures (*a*)–(*c*).

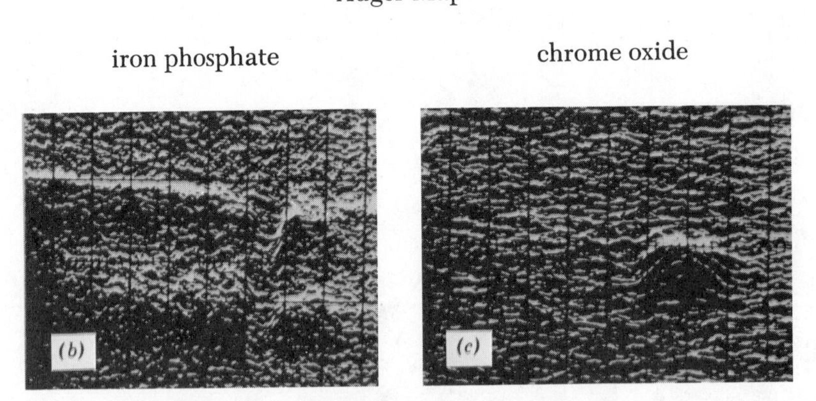

FIGURE 24.20. Boundary films at defect sites.

higher local stress than the leading edge; however, the micro-EHD lubrication behavior at the leading edge gives a much lower film thickness, as shown in Fig. 24.19. This corresponds to frequent observations of fatigue spalls initiating at the trailing edge of defect sites, as shown in Fig. 24.18*c*,*d*, and that surface distress appears more frequently at the leading edge due to the lower film thickness. This can also be seen in the same figure.

When the defects are large, as shown by the groove in Fig. 24.20*a*, the local EHD lubrication is not effective in providing surface separation, which results in surface interaction and the initiation of boundary film formation. The identification of the boundary films is shown in Fig. 24.20*b*,*c*.

An example of interactive wear modes is shown in Fig. 24.21. In Fig.

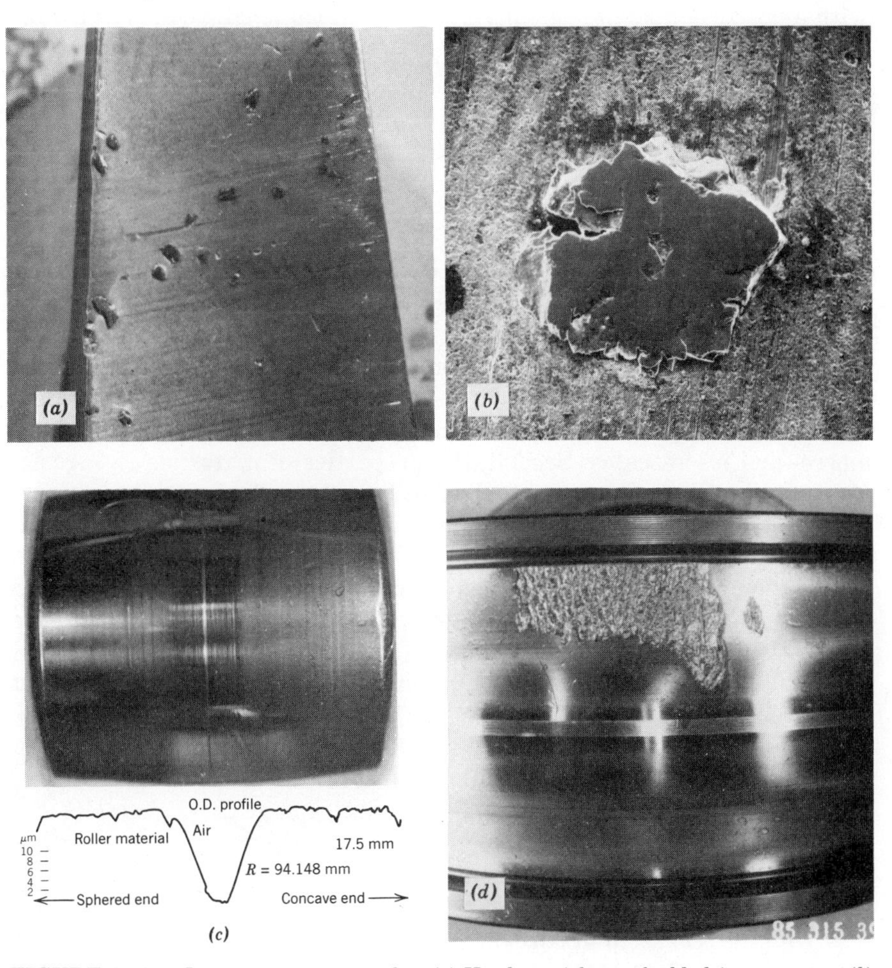

FIGURE 24.21. Interacting wear modes. (*a*) Hard particles embedded in separator. (*b*) SEM of embedded particle. (*c*) Abrasively worn roller and typical surface profile. (*d*) Fatigue spall on inner ring due to stress concentration of worn roller and/or debris denting.

24.21*a*, hard particles are shown imbedded in the cage material of a spherical roller bearing. The hard particles initiated an abrasive wear process on the roller in Fig. 24.21*b*. The abrasive wear has altered the geometry of the roller, as shown in Fig. 24.21*c*. Stress concentrations induced by the altered profile of the roller produced the final failure of the bearing by way of a fatigue spall on the inner ring, as shown in Fig. 24.21*d*.

RECOMMENDATIONS FOR WEAR PROTECTION

Recommendations for wear protection can be derived from a knowledge of the tribological processes associated with lubrication and wear along with a good phenomenological view of wear. Table 24.3 contains a summary of the wear processes and their tribological implications. A detailed overview of the control of rolling contact failure through lubrication is given in [24.10].

CLOSURE

The topic of wear in rolling element bearings is complex, and an attempt to present the subject in simple design criteria and formulas has been avoided. Wear is not an intrinsic property of materials but a complex sequence of events of an entire system. The difficulties of wear prediction and prevention are associated with several factors. First, the stress field that drives the process of wear is not well defined and occurs on both macro- and microscales. Second, the properties of materials that resist wear are usually those associated with the near-surface region generally characterized by insufficiently defined microstructures and mechanical and chemical properties significantly different from the bulk material. Third, chemical reaction processes occur in parallel with stress-induced mechanical deformation and ultimately are interdependent. Fourth, the description of wear modes is generally confused by the proliferation of terminology and a lack of definitive connection between wear mode and wear process.

Consequently, the subject of wear has generally been presented not in quantitative terms with specific materials of construction but in terms of tribological processes associated with specific structural elements of a lubricated contact. Nevertheless, in [24.14] a method to predict wear life in ball and roller bearings was presented. Since it is possible to eliminate significant wear in rolling bearings by effective lubrication, sealing, and/or shielding in most applications, the need to be able to predict wear in a bearing application is far less important than the need to prevent wear.

TABLE 24.3. Wear Processes and Tribological Implication

Observed Wear Behavior	Primary Mechanistic Wear Process	Primary Structural Elements	Tribological Implication
Mild mechanical wear Adhesive wear Smearing	Adhesion	Surface films Near-surface	Interacting surfaces that are not sufficiently protected by lubricating films (surface films or EHD/micro-EHD films) frequently result in adhesive transfer or removal of near surface material. Surface integrity can be maintained by proper selection of material pairs and lubrication to provide protective surface and EHD/micro-EHD films.
Corrosion Tribochemical wear	Chemical reaction	Surface films Near-surface EHD/micro-EHD films	Tribochemical behavior is activated by mechanical interactions of the contacting surfaces producing activated surface sites and local high temperatures for chemical reactions. Tribochemical wear involves a continuous process of surface film formation and removal. Correct balance is required between chemical wear and adhesive wear, requiring appropriate lubricant/additive aggressiveness.

(continued on next page)

TABLE 24.3. Wear Processes and Tribological Implication (*Continued*)

Observed Wear Behavior	Primary Mechanistic Wear Process	Primary Structural Elements	Tribological Implication
Surface indentation Surface distress Abrasive wear	Plastic flow	Near-surface	Most observed wear behavior involves plastic deformation as a result of displacement of material under sliding conditions (plowing, smearing) or material removal (abrasive, adhesive wear). Overrunning of hard particles produces debris denting, giving rise to surface-initiated fatigue spalling. EHD/micro-EHD films reduce or eliminate local surface plastic flow.
Pitting Fatigue spalling	Fatigue	Near-surface Subsurface	EHD/micro-EHD films redistribute surface pressure, thereby modifying the magnitude and location of critical stresses. The induced subsurface shear stress is important with respect to the critical shear stress of the material in the subsurface region. The EHD film/surface roughness ratio influences fatigue life due to high local stress at surface defect sites (asperities, dents, and material inconsistencies).

Consequently, efforts by most major rolling bearing manufacturers have been aimed at prevention, and the approach in [24.14] has not been generally used in industry.

REFERENCES

24.1. H. Czichos, "Importance of Properties of Solids to Friction and Wear Behavior," *Tribology in the 80's*, NASA Conference Publication 2300, Vol. I, Sessions 1 to 4, Proceedings of an International Conference held at NASA-Lewis Research Center, Cleveland, Ohio (April 18–21, 1983).

24.2. L. Wedeven and C. Cusano, "Elastohydrodynamic Contacts—Effects of Dents and Grooves on Traction and Local Film Thickness," NASA TP 2175 (June, 1983).

24.3. A. Dorinson and K. Ludema, *Mechanics and Chemistry in Lubrication*, Tribology Series, 9, Elsevier (1985).

24.4. A. Dyson and L. Wedeven, "Assessment of Lubricated Contacts—Mechanisms of Scuffing and Scoring," NASA TM-83074 (February, 1983).

24.5. J. Lancaster, "The Formation of Surface Films at the Transitions between Mild and Severe Metallic Wear," *Proc. Roy. Soc. A* **273,** 466–483 (1963).

24.6. T. Quinn, "NASA Interdisciplinary Collaboration in Tribology—A Review of Oxidation Wear," NASA Contractor Report 3686.

24.7. C. Rowe, "Lubricated Wear," *Wear Control Handbook*, ASME, pp. 143–160, (1980).

24.8. M. Peterson, "Design Considerations for Effective Wear Control," *Wear Control Handbook*, ASME, pp. 413–473, (1980).

24.9. A. Begelinger, A. deGee, and G. Solomon, "Failure of Thin Film Lubrication—Function Oriented Characterization of Additives and Steels," *ASLE* **23,** 23–34 (1980).

24.10. T. Tallian, "Rolling Contact Failure Control Through Lubrication," *Proc. Inst. Mech. Engrs.* **182,** 205–236 (1967–68).

24.11. R. Parker, "Correlation of Magnetic Perturbation Inspection Data with Rolling Element Bearing Fatigue Results," *ASME Trans. J. Lub. Tech.* **97,** Ser. F, No. 2, 151–158 (Apr. 1975).

24.12. L. Wedeven, "Influence of Debris Dent on EHD Lubrication," *ASLE* **21,** 41–52 (1978).

24.13. T. Tallian, G. Baile, H. Dalal, and O. Gustafsson, *Rolling Bearing Damage Atlas*, SKF Industries, Inc., Revere Press, Philadelphia (1974).

24.14. P. Eschmann, L. Hasbargen, and R. Weigand, *Ball and Roller Bearings-Theory, Design, and Application, 2nd Ed.*, 188–194, Wiley (1985).

25

VIBRATION AND NOISE

LIST OF SYMBOLS

Symbol	Description	Units
A	Peak displacement amplitude	mm (in.)
D	Rolling element diameter	mm (in.)
dB	Decibels, relative logarithmic amplitude	
d_{m}	Pitch diameter	mm (in.)
F_{B}	Reaction force by bearing	N (lb)
f	Frequency	rps, Hz
g	Acceleration due to gravity	mm/sec^2 (in./sec^2)
M	Mass	kg (lb-sec^2/in.)
N	Speed	rpm, rps
R	Radius	mm (in.)
r	Radial deviation	mm (in.)
t	Time	sec
W	Waves per circumference	
Y	Vertical displacement	mm (in.)
Z	Number of balls or rollers	

Symbol	Description	Units
δ_r	Radial deflection	mm (in.)
λ	Wavelength	mm (in.)
θ	Angular measure	rad

SUBSCRIPTS

c	Refers to cage
i	Refers to inner ring, shaft or raceway
o	Refers to outer ring or raceway

GENERAL

This chapter provides a practical overview of bearing vibration. Where relevant, reference is also made to noise, sometimes resulting from excessive bearing vibration. Common bearing applications in which noise and/or vibration are important are described.

Machine vibration or noise levels, whether excessive or not, are affected by bearings in three ways: as a structural element defining in part a machine's stiffness; as a generator of vibration by virtue of the way load distribution within the bearing varies cyclically; as a vibration generator because of geometrical imperfections from manufacturing, installation or wear and damage after continued use.

Illustrations of manufacturing and installation problems are shown, in some cases with the use of vibration measurement taken from machines after bearing installation. Descriptions are also given of methods used in rolling bearing factories to evaluate bearing component quality, control manufacturing processes, and minimize bearing vibration.

Detection of progressive bearing deterioration in operating machinery by vibration measurement has become more economical and reliable in recent years. Some aspects of such machinery monitoring are considered.

VIBRATION AND NOISE-SENSITIVE APPLICATIONS

Significance of Vibration and Noise

In many cases objectionable airborne noise from a machine results from measurable vibration of machine components. Correlation between bearing noise and machine vibration measurements has been reported [25.1, 25.2]. Therefore, with respect to rolling bearings, the terms "noise" and "vibration" usually denote similar and related phenomena. Their relative importance to the bearing user may differ, depending on where and how the machine is used.

Regardless of which seems to be more important in a particular application, noise and vibration may both be indicators in new machines of quality problems with bearings, machine components, or assembly methods. Such problems can limit the functional capabilities of the machine, and they can reduce the potential useful life of the bearings or the machine itself. In cases where a machine has successfully performed its function and is approaching the normal time for repair or replacement, the first indication may come from increased levels of noise or vibration.

Noise-Sensitive Applications

The application that has been the major driving force for reduced noise is that of small and medium electric motors, primarily utilizing deep-groove ball bearings. Figure 25.1 shows such an application. The outer ring of the bearing at the left end of the motor is free to move axially under controlled thrust load of a spring to remove axial clearance from within the bearing. This allows for thermal expansion of the shaft and motor assembly without loss of preload while simultaneously preventing excessive bearing loads or distortion of motor components.

Quiet running characteristics of the electric motor are required in office equipment and household appliances where noise may be an irritant. Noise is also a problem in building heating and air conditioning systems, where motor or fan support bearing noise can be transmitted and amplified through duct work or air columns. Also included in this category are drive systems of elevators, using larger electric motors with deep-

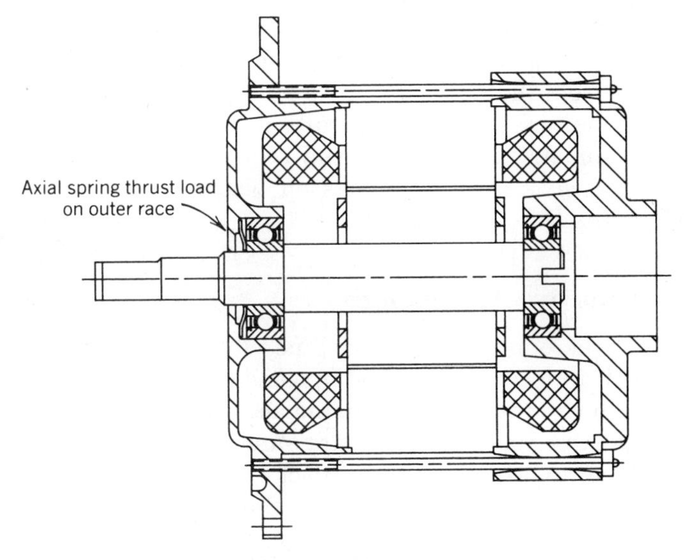

FIGURE 25.1. Electric motor.

groove ball bearings and cylindrical roller bearings and spherical roller bearings in pillow blocks to support cable sheaves. Aside from irritation, excessive noise in the latter application might make passengers concerned. Automotive applications also requiring quiet running performance include alternators (deep-groove ball bearings and needle roller bearings), transmissions, differentials (tapered roller bearings), and fans.

Objectionable noise might be characterized by volume or sound level and pitch or frequency. Sound from a machine may be more irritating if a particular frequency is dominant. Possibly even more objectionable are intermittent or transient sounds that vary with time in either pitch or volume at regular or irregular intervals. Such effects might be more easily heard than measured, since common measuring methods may be acquiring data over time periods that are long compared to short-duration transient sounds or vibrations. In addition, transient sound or vibration is sometimes most significant when a machine is coming up to operating speed or coasting down. Even in vacuum cleaners or dishwashers this effect is sometimes heard. An example is given in "The Role of Bearings in Machine Vibration."

Qualitative audible evaluation of airborne sound from machines such as electric motors is performed as a routine inspection in many cases. Audible evaluation is also performed by processing the output of a vibration measurement transducer through a loudspeaker. This is useful for detecting transients and also in some cases for identifying the cause of excessive measured vibration. The vibration parameter monitored could be velocity or acceleration rather than displacement, which overemphasizes low-frequency vibration.

The United States Navy has made extensive demands on bearing manufacturers with respect to bearing vibration and noise reduction as well as boundary dimension and running accuracy tolerances [25.3]. This stems in part from the requirement to make submarines more difficult to detect by monitoring sound transmitted through water. At the same time, improvements in reliability and reduced maintenance costs are achieved. Extensive research efforts on bearing vibration have been sponsored by the U.S. Navy [25.4].

Vibration-Sensitive Applications

Applications where bearing and machine vibration are more important than noise fall into two categories. In some cases the machine must be capable of high running and positioning accuracies to function properly. In other cases the major concerns are safety, if vibration causes catastrophic failure, and the economic impact of reduced machine utilization and increased repair cost if vibration foreshortens the life of components.

Not only is noise intrinsically less important than vibration in these categories, but it may also be incapable of indicating a significant problem. This would occur if the predominant frequency of high-amplitude vibration falls outside the audible range; for example, rotating imbalance in a machine running at 1800 rpm (30 Hz). In addition, abnormal noise might be undetectable because of ambient noise where the machine operates or because of normal noise from the process the machine performs.

Bearing applications where machine accuracy might be affected by vibration include machine tools. Grinding spindles often must be capable of producing components with size and two- or three-point roundness within a micron (4×10^{-5} in.). Figure 25.2 shows a grinding wheel spindle using precision double-row, cylindrical roller bearings and a double-direction, angular-contact ball thrust bearing to achieve high radial and axial stiffnesses. The cylindrical roller bearings have tapered bores for accurately controlling preload. Precision angular-contact ball bearings in matched sets are also widely applied in spindles.

In addition to size control and roundness, precision spindles must be capable of producing even finer levels of geometrical accuracy such as relatively low levels of surface roughness and circumferential waviness amplitudes of much less than a micron (4×10^{-5} in.). Vibration can contribute to excessive roughness or waviness and can also produce chatter, a more severe form of waviness that can cause permanent metallurgical damage to hardened steel parts. Some examples of components with wavy surfaces are shown in "Nonroundness Effect and Its Measurement." Vibration measurements from two machines that produced such components are illustrated in "Detection of Failing Bearings in Machines."

Other machines in which vibration might prevent the required accuracy from being achieved include rolling mills for sheet steel, paper, and chemical films. Computer disc drives are a further example, requiring

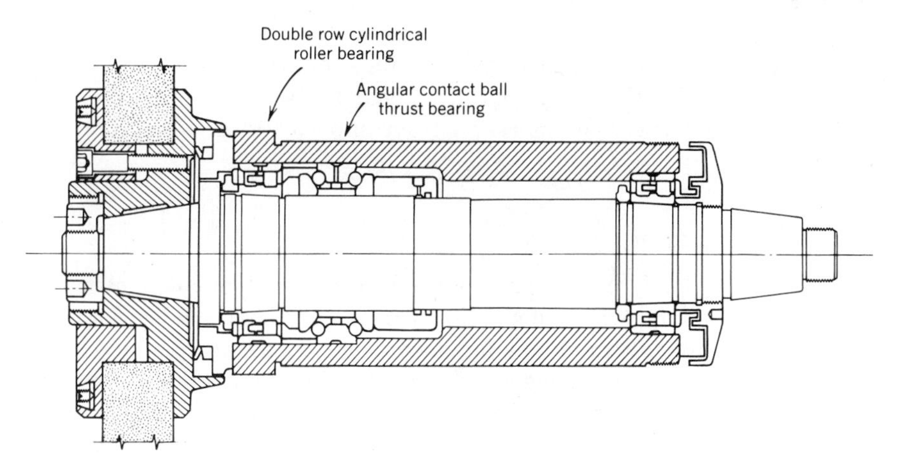

FIGURE 25.2. Machine tool spindle.

nonrepeatable bearing runout accuracy of no more than one quarter to one half of a micron (1-2 $\times$ 10^{-5} in.) for the spindle and head combined. Similarly, gyroscope bearings require good dynamic running accuracy as well as very low torque levels.

Cases where running accuracy is not as important as safety and machine reliability often involve machines that are producing or transmitting high horsepower, have massive rotating components, and are running at high speeds relative to the size of the equipment. Eccentric mass produces large and potentially destructive forces in these applications. As discussed in Chapter 11, such equipment may operate at speeds above resonant frequencies, so large amplification of vibration could occur as equipment is run up to speed. Examples include compressors, pumps, and turbines.

Applications in this section are examples where machine noise or vibration is important. More demanding applications continually arise, requiring greater accuracy, higher speeds and loads, and improved reliability. Therefore, bearing manufacturers have continuously emphasized improvement of bearing quality with respect to noise and vibration through ongoing development of machines and methods for manufacturing and inspection.

THE ROLE OF BEARINGS IN MACHINE VIBRATION

Bearing Effects on Machine Vibration

Rolling bearings have three effects with respect to machine vibration. The first effect is as a structural element that acts as a spring and also adds mass to a system. As such, bearings define, in part, the vibration response of the system to external time-varying forces. The second and third effects occur because bearings act as excitation sources, producing time-varying forces that cause system vibration. In one case this excitation is inherent in the design of rolling bearings and cannot be avoided. In the other case these forces result from imperfections, which usually are avoidable.

Structural Element

With sufficient load, the bearing is a stiff structural member of a machine. It is a spring whose deflection varies nonlinearly with force, in contrast to the usual linear spring characteristics assumed in dynamic models, such as the single-degree-of-freedom spring-mass-damper model discussed in Chapter 11. As a first approximation, it may be adequate to estimate machine vibration response by considering the bearing as a linear spring. In this case a bearing spring constant is determined by taking the slope of the force-deflection curve of the bearing at the normal operating load. The approximation may be insufficient in cases requiring precise knowledge of transient vibration response, particularly near machine resonant frequencies. In these cases extensive mathematical modeling and experimental modal analysis are performed, both of which are beyond the objectives of this chapter. If it is sufficient to consider bearing stiffness as a constant, under a specific set of operating conditions, then this approximation can be derived from equations in Chapter 11.

Bearing stiffness increases with increasing load, a characteristic referred to as a "hardening" spring. Larger nominal operating loads or built-in preload would result in smaller variations in dynamic bearing deflection when subjected to a particular dynamic load variation. Similarly, increased bearing stiffness raises the value of a resonant frequency associated with this spring, since a resonant frequency is inversely proportional to the square root of stiffness. Moreover, radial stiffness decreases with increasing contact angle, whereas the reverse is true for axial stiffness. Therefore, response to dynamic load variation will depend strongly on the direction of such loads relative to that of the nominal load that governs contact angle.

Since the bearing "spring" is nonlinear, it is evident that sinusoidal deviations from the nominal load will not cause sinusoidal bearing deflection. When the load is greatest, the increase to nominal bearing deflection will be less than is the decrease from nominal bearing deflection when the load is at its lowest value. If large dynamic fluctuations in load are experienced, say in a radially loaded bearing, then it is possible for the load zone to alternate from the bottom to the top of the outer raceway. If the bearing has radial internal clearance, there is the possibility of essentially no loading at all on the outer raceway for brief instances. Such conditions could arise because of external loading or conditions within the bearing.

Variable Elastic Compliance

The second effect of bearings on machine vibration occurs because bearings carry load with discrete elements whose angular position, with respect to the line of action of the load, continually changes with time. This mere change of position causes the inner and outer raceways to undergo periodic relative motion even if the bearing is geometrically perfect. Analysis of this motion is described in [25.4]. The following example illustrates how variable elastic compliance vibration occurs.

Example 25.1. Consider a 204 radial ball bearing with eight 7.938-mm (0.3125-in.) balls. The bearing supports a 4450-N (1000-lb) radial load. Figure 25.3 shows the bearing at two different times. In Fig. 25.3*a* ball 1 is located directly under the load, and balls 1, 2, and 8

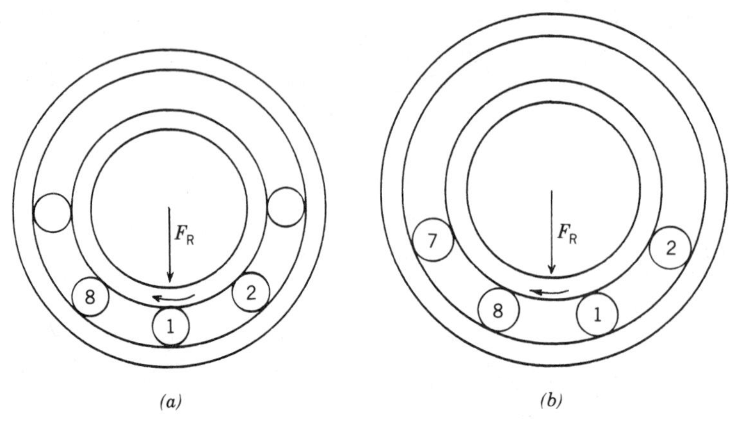

FIGURE 25.3. (*a*) Angular position of ball set, time = 0. (*b*) Angular position, time = $\frac{1}{2}$[1/Z(cage rotation frequency)].

carry the load. In Fig. 25.3*b* balls 1 and 8 straddle the load symmetrically, and balls 1, 2, 7, and 8 carry the load. Obviously, the radial deflection is different in each situation.

With methods from Chapter 5 the radial deflection in the first case is estimated to be 0.04323 mm (0.001702 in.). In the second case the deflection is approximately 0.04353 mm (0.001714 in.). The bearing deflection is less in the former arrangement. The position of the ball set in Fig. 25.3*a* gives a stiffer bearing at that instant. The shaft and inner raceway have approached closer to the outer raceway in the time it takes for one half of the ball spacing to pass a point on the outer raceway to reach the position shown in Fig. 25.3*b*. The shaft will return to its original position as ball 1 comes under the load line. This frequency of vibration is therefore equal to the cage rotational frequency multiplied by the number of balls; that is, the frequency of this vibration occurs at the frequency of balls passing the outer raceway.

This example illustrates vertical elastic compliance vibration. Horizontal motion also occurs, at the same vibrational frequency, as the ball set assumes angular positions that are asymmetrical with respect to the load line. Both vertical and horizontal vibration amplitudes are nonsinusoidal as a result of the nonlinear deflection characteristics. The existence of this type of vibration, which occurs even with a geometrically perfect bearing, is one reason why bearing damage detection is best performed by monitoring frequencies other than the fundamental bearing frequencies.

Geometrical Imperfections

The first effect that bearings have on machine vibration arises from geometrical imperfections. These imperfections are always present to varying degrees in manufactured components. Sayles and Poon [25.6] discuss three mechanisms by which imperfections in bearings cause vibration: waviness (Fig. 25.4) and other form errors causing radial or axial motion of raceways; microslip together with asperity collisions and entrained debris that break through the lubricant film and shocks due to local elastic deformations caused by summits that do not break the lubricant film.

The local elastic contacts are of approximately the same size or smaller than Hertzian contact areas. At any instant there may be only a few such summits in the Hertzian deformation zone, depending on the type of bearing component finishing processes employed; for example, honing and lapping. Elastic deformations of the type discussed occur rapidly, and the time separating one such contact from the next is brief. A major contribution to bearing vibration in the higher frequencies, for example, above 10,000 Hz, is thought to be the result of such deformations. Due to their impulsive nature, however, they are capable of exciting lower-frequency resonances.

Controlling component waviness and other types of errors from manufacturing, distortion, or damage occurring while the bearing is assembled to the machine, is a high priority. The effects of such form errors on machine vibration or noise can be significant.

Waviness Model

Figure 25.4 represents a bearing with waviness on the outer raceway. It is assumed that the bearing supports a mass and that the outer ring is

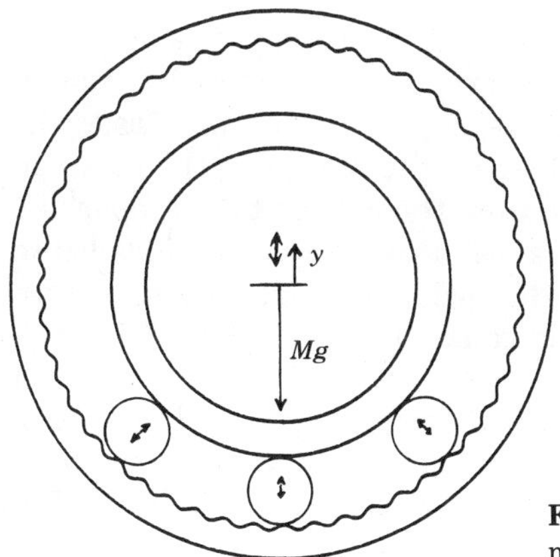

FIGURE 25.4. Vibration from raceway waviness.

rigidly supported by a housing. If no waviness is present on the surface of the bearing raceways, a force balance in the vertical direction is

$$F_B - Mg = 0 \tag{25.1}$$

If waviness is present, then for an approximation it will be assumed that the mass will move up and down as a rigid body, with reaction force produced in the bearing as a result of the acceleration of the mass. In this case the force balance is

$$F_B + \Delta F_B - Mg = M\ddot{y} \tag{25.2}$$

For waviness that can be approximated as sinusoidal the equations are

$$y = A \sin (2\pi ft) \tag{25.3}$$

and

$$\ddot{y} = -A(2\pi f)^2 \sin (2\pi ft) \tag{25.4}$$

The frequency f is the rate at which balls pass over a complete wave cycle. The assumption of only vertical motion of the mass implies two conditions: (1) that the wave peaks are always in phase with balls, and (2) that variation of the ball set angular position within the load zone has no influence on the direction of motion. For illustrative purposes, these simplifying assumptions will suffice to demonstrate the importance of relatively small form errors.

Combining equations (25.2) and (25.4) and rearranging,

$$F_B + \Delta F_B = Mg - MA(2\pi f)^2 \sin (2\pi ft) \tag{25.5}$$

For sufficient waviness amplitude and passage frequency, the right side of the equation can vanish, in which case the bearing force (left side of the equation) vanishes, or it can become negative, in which case the bearing produces a negative force to restrain the motion of the mass. In this case the load zone would alternate from the bottom to the top of the outer raceway. If the bearing has clearance, it could become unloaded in either direction at some instant. The following example gives an estimate of waviness amplitude that would cause this condition.

Example 25.2. For a bearing with 50 waves on the circumference of the outer raceway, estimate the amplitude of the waviness that could cause sufficient acceleration to momentarily unload the bearing. The shaft speed is 1800 rpm (30 Hz), and the cage speed is 11 rps.

The rate at which any ball passes over a wave cycle is the product of the cage speed and the number of waves per circumference of the outer raceway—in this case, 550 wave cycles per second. The condition for bearing unloading would occur when

$$Mg = MA(2\pi f)^2$$

or

$$A = \frac{g}{(2\pi f)^2} = \frac{980}{(2\pi \times 550)^2}$$
$$= 8.208 \times 10^{-4} \text{ mm } (3.231 \times 10^{-5} \text{ in.})$$

Wave amplitude is usually expressed in terms of peak-to-valley amplitude with units of microns (μm) (1 μm $= 10^{-6}$ m). In these terms, the waviness peak-to-valley amplitude is 1.64 μm.

Bearing raceway waviness of this amplitude and frequency are in excess of acceptable levels. Although wavy components of this type rarely occur, they can occur due to improper manufacturing procedures or manufacturing machine malfunctions. Examples of excessively wavy components are presented later on.

Examples of Electric Motor Vibration

Three examples are discussed to reveal sources of noise and vibration problems in small electric motors with newly installed bearings. The first example shows measurements of bearing distortion that, occurring as a result of assembly to faulty machine components, produced waviness on the inner raceway. The second example shows the effect of improper assembly, and the third illustrates the effect of a defective bearing on motor vibration.

Example 25.3. Assembly of bearings in housings or on shafts with poorly controlled geometry can distort the bearing components and produce wavy running surfaces that affect the machine vibration or noise. Johansson [25.6] discusses effects of inner raceway distortion on electric motor vibration.

Figure 25.5*a* shows a circumferential trace of a motor shaft-bearing journal where a 6 mm bore deep-groove ball bearing was mounted. The motor was rejected after assembly for audible noise and for vibration as determined by hand-turning the armature while it was supported by the bearing. The shaft speed in the power tool application was 23,000 rpm. The shaft exhibits a three-point out-of-roundness con-

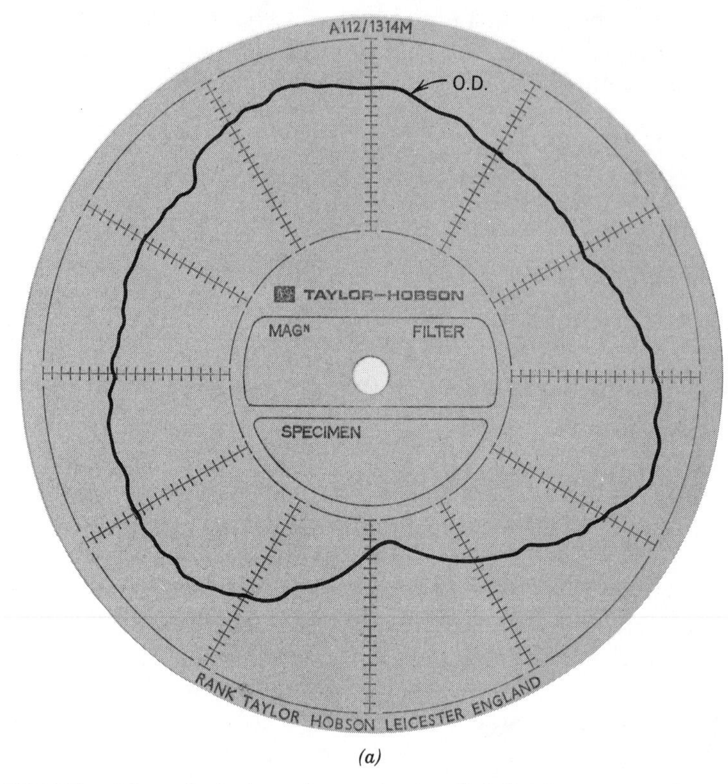

(a)

FIGURE 25.5*a*. Motor shaft circumference (each radial division is 2 μm) (from [25.7]).

dition of approximately 24 μm (0.001 in.). Shaft diameter tolerance for the particular application would normally be held within a total spread, from one shaft to another, of 8 μm (0.0003).

Figure 25.5*b* shows traces of the bore and ball groove after disassembly from the shaft, with both surfaces being less than 1 μm (0.00004 in.) out-of-round. Figure 25.5*c* shows a trace of the ball groove as mounted on the shaft. This indicates that the raceway in the mounted condition exhibited 16 μm (0.0006 in.) of three-point out-of-round. Note the two local imperfections on the raceway. The largest is approximately 2 μm (0.00008 in.) deep and is located on one of the lobes. The origin of this defect was not determined, although it probably occurred during press-fit assembly or from damage during running.

The bearing has six balls, so the three high points on the distorted inner raceway could be in contact with three balls, while the other three balls carry no load. This effect would lower the bearing stiffness, either axial or radial, which depends on the number of balls. On the

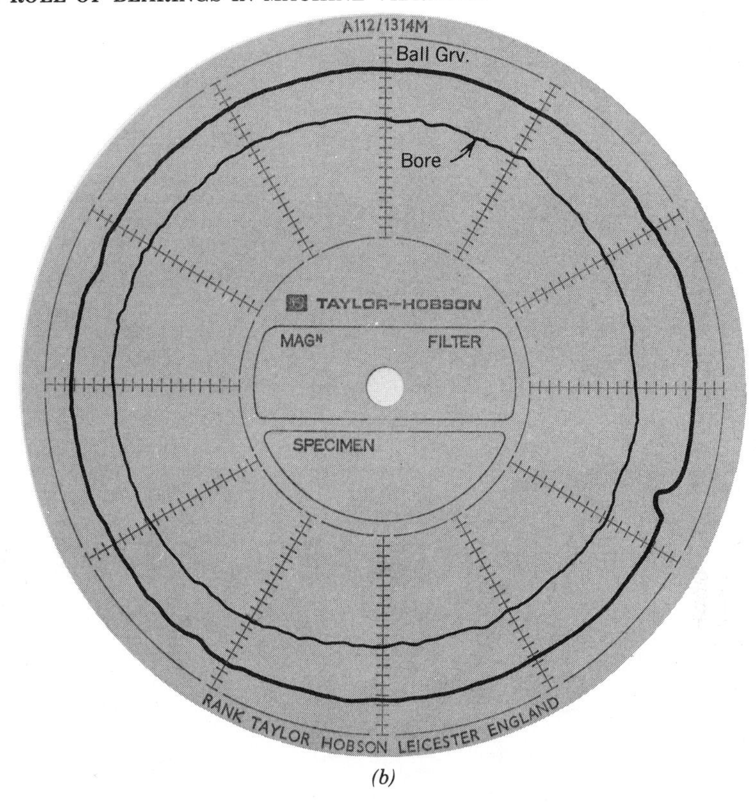

(b)

FIGURE 25.5b. Inner raceway after disassembly (each radial division is 1 μm) (from [25.7]).

other hand, larger individual ball loads are expected to be generated during parts of the rotational cycle of the shaft, tending to raise stiffness, but simultaneously causing large axial vibration.

The cage operating speed can be calculated by using equations of Chapter 7 to be approximately 138 Hz. The cage speed relative to the inner raceway is 245 Hz, so the rate at which a wave cycle passes a ball is 735 Hz. The rate at which any of the high points passes from one ball to the next is equal to the product of the cage speed relative to the inner raceway and the number of balls (1470), harmonically related to the wave passage frequency, because the number of balls is a multiple of the number of waves. Accordingly, there is the potential for large-amplitude vibrations with two fundamental frequencies (735 and 1470 Hz) well into the audible range. In addition, numerous high harmonics of each would be expected, with the potential for excitation of various structural resonances in the motor.

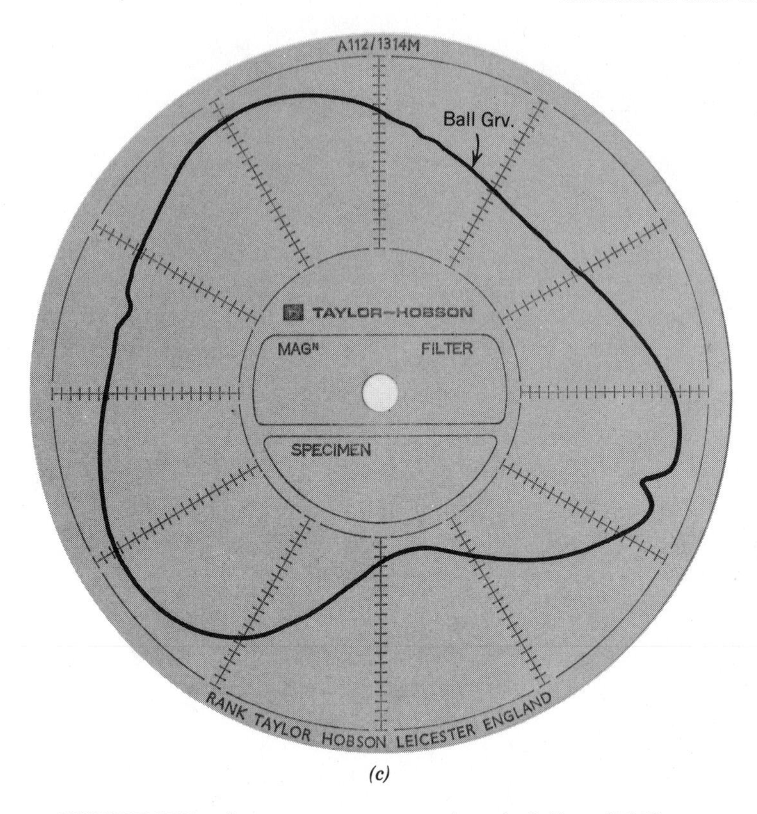

FIGURE 25.5c. Inner raceway mounted on shaft (from [25.8]).

Example 25.4. This example illustrates a loose assembly that contributes to noise and vibration. In this case airborne sound measurement in the form of frequency spectrum analysis is used to identify the source of the problem.

The digital frequency spectrum analyzer is a computer-based instrument that transforms time-sampled data into the frequency domain through Fourier series analysis. Knowledge of dominant frequencies in vibration signals can often reveal sources of a specific problem. An additional benefit of the technique is that storage and documentation of data are facilitated. Figure 25.6*a* shows a simplified time signal, which might be obtained as a voltage signal from a transducer such as an accelerometer or eddy current displacement sensor. This signal might be displayed on an oscilloscope. The same signal can be sampled and Fourier transformed into the frequency domain, as shown in Fig. 25.6*b*, to show the amplitude as a function of frequency. In this case the time signal consists of only a few frequencies, which in the frequency spectrum show up distinctly.

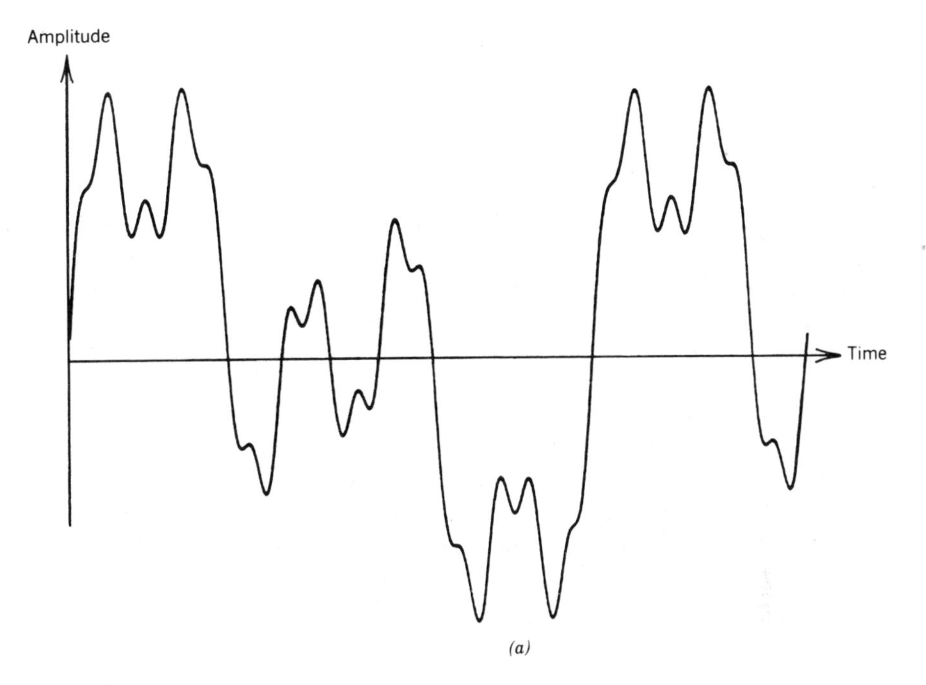

(a)

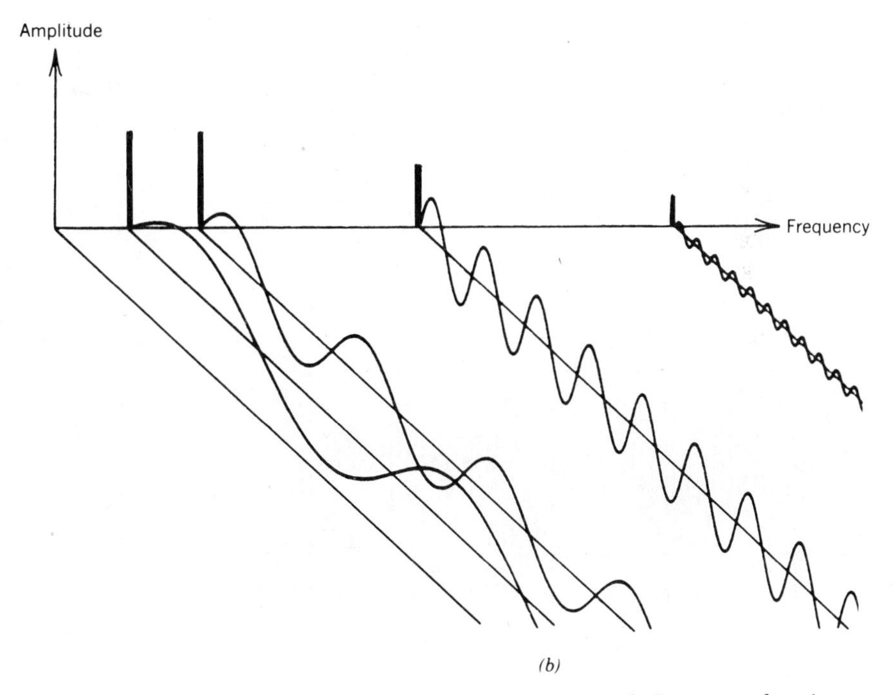

(b)

FIGURE 25.6. Simplified time signal. (*a*) Time domain. (*b*) Frequency domain.

Figure 25.7*a* shows the frequency spectrum of a vacuum cleaner motor rejected for noise after assembly. The spectrum was obtained from airborne noise measurement utilizing a microphone. The amplitude scale is uncalibrated and logarithmic; each vertical division represents an amplitude increase of a factor of 10, with lower frequencies being attenuated, as is common in sound measurement. The normal operating speed of the motor is 20,000 rpm. Neither the sound spectrum nor qualitative audible evaluation revealed anything abnormal at this speed. When the motor was turned off and was coasting down, however, a distinct rumble was heard at a speed subsequently estimated to be around 10,000 rpm. The spectrum of Fig. 25.7*a* was then obtained with the motor running at that speed. It seemed likely that some system resonance was occurring as the motor coasted down and passed a critical speed.

The spectrum shows distinct frequency peaks, determined to be 165.5, 325, 487.5, 650, 975, and 1137.5 Hz. A similar good motor (e.g., Fig. 25.7*b*) shows only a peak at 975. Rotor endplays on the motors

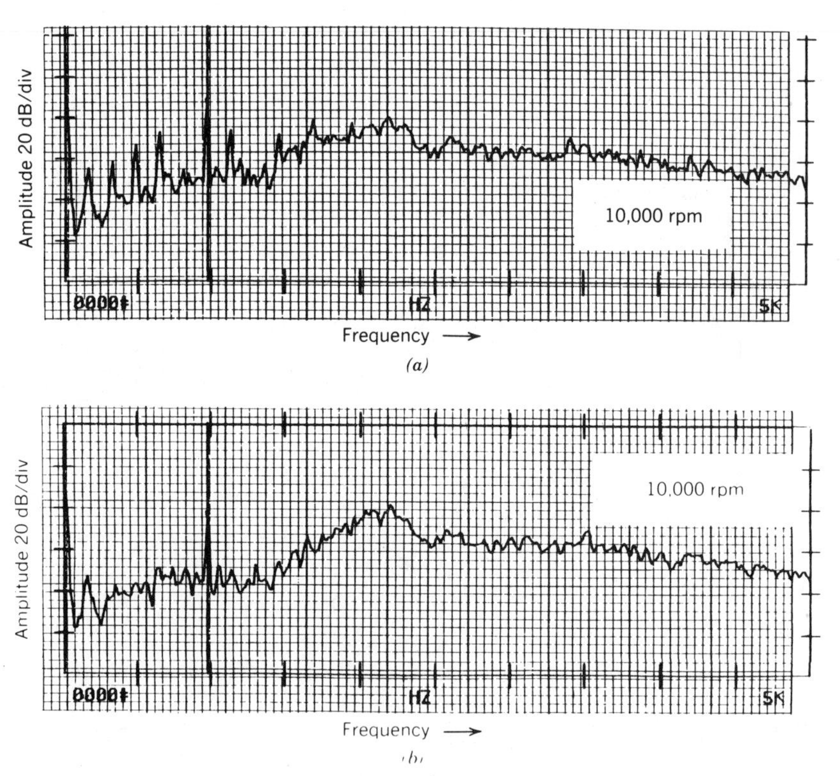

FIGURE 25.7. Electric motor sound spectrum. (*a*) Loose bearing mount. (*b*) Normal motor.

were 0.094 mm (0.0037 in.) for the motor of Fig. 25.7*a* and 0.0508 mm (0.002 in.) for the other motor. The peak at 975 Hz for the good motor was expected to be the blade pass frequency on the motor fan, which has six blades. This means that the actual running speed was 9750 rpm. In this case the measured spectral peaks on the noisy motor were all harmonically related to the running speed. Harmonics of the running speed usually occur when mechanical looseness, which could be caused by improper bearing mounting [25.8], exists.

Looseness would result in low stiffness, lower resonant frequency, and "play" in the system. The imbalance force, rotating at the shaft speed, could then produce significant vibration amplitude at that frequency. The harmonics probably result from either of two effects: (1) directional stiffness variation or (2) shocks occurring if the load zone shifts from one side of the bearing to the other in an unstable manner.

The noisy motor was disassembled. A spring clip, which retains the bearing outer raceway in a plastic housing, had been improperly seated during assembly, resulting in a loose bearing mount. Repair and reassembly reduced the noise to an acceptable level. Diagnosis of this problem could have been made with vibration transducers rather than sound measurement. The audible evaluation of the transient vibration during coast-down also provided a clue to the source of the problem.

Example 25.5. Noise emitted from a small electric motor with a defective bearing is discussed. Figure 25.8 shows vibration spectra of two small electric motors (3600 rpm) on the same plot. These data were obtained by screwing a small accelerometer into a nut glued to the end cover. The frequency range investigated was to 10,000 Hz, covering the most important part of the audible frequency range at usual machine operating speeds.

The rms (root mean square) vibration amplitude at each frequency is plotted in volts dB, where the dB value is equal to 20 $\log_{10}$ (voltage/reference voltage). In this case the reference voltage is 1.0. Each major division on the plot is equal to 10 dB, with the amplitude scale ranging from -40 dB (0.01 V) full scale down to -120 dB (10^{-6} V). The accelerometer output is 0.010 V/g of acceleration.

One motor is seen to have vibration amplitudes from 5 to 25 dB higher than the other motor over much of the frequency range. The motor with the higher vibration also gave torque readings more than a factor of 2 higher than the other motor. Frequency spectra at various points on the motors were taken with the same results. Spectra in other frequency ranges were taken with no conclusive results regarding the origin of the vibration. Qualitative evaluation of audible noise indicated a low-amplitude clicking sound that repeated at a fairly high rate.

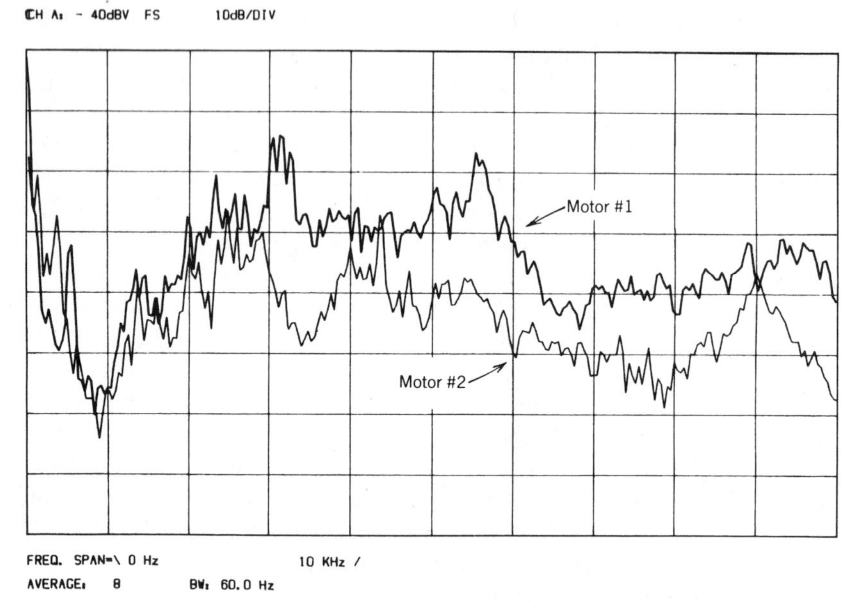

FIGURE 25.8. Motor vibration spectra.

The motors were disassembled, the shafts and housings were checked for geometry and found to be normal, and the bearing torques were measured. The bearings from the motor with high vibration had rubbing seals, whereas the other set had noncontacting seals. The four bearings were vibration tested, with seals removed, on a standard bearing test apparatus discussed in the next section. It was found that one of the bearings from the motor with high vibration also gave high readings on the inspection tester. Spectrum analysis indicated harmonics of the frequency at which balls pass a point on the outer raceway. Examination of the outer raceway revealed a defect that appeared to be related to manufacturing and had escaped detection in final inspection vibration testing.

The discussion and examples of this section have viewed several ways in which bearings can affect or cause machine vibration. The irregularities on bearing surfaces that exist from manufacturing, assembly into the machine, or from deterioration after long-term use provide a source of forced excitation to rotating machine members or structural components that can increase stress, accelerate wear, increase frictional losses, and possibly cause catastrophic machine failure. Other forms of bearing distortion or imperfections occurring as a result of assembly include misalignment (housing centers out of line from each other or not parallel), Brinell damage, contamination by debris, and bell-mouthed housings preventing axial movement of the outer ring (e.g., in small electric motors with spring preload).

NONROUNDNESS EFFECT AND ITS MEASUREMENT

Roundness and Waviness

Generally, a part is said to be round in a specific cross section if there exists a point within that *x*-section from which all other points on the periphery are equidistant. The first-mentioned point is of course the center of the circle, and the *x*-section is a perfect circle as in Fig. 25.9*a*. If the *x*-section is not a perfect circle, as in Fig. 25.9*b*, it is said to be out-of-round with the "out-of-roundness" specified as the difference in distance of points on the periphery from the center. Thus, out-of-roundness in Fig. 25.9*b* is $r_1 - r_2$. In addition to the basic profile of Fig. 25.9*b*, an irregular profile similar to Fig. 25.9*c* is usually present in manufactured machine elements, and this includes rolling bearing raceways and rolling elements. The irregular surface of Fig. 25.9*c* is of substantial importance to bearing frictional performance and endurance; this was discussed in Chapters 13 and 18. The lobed surface of Fig. 25.9*b* is also significant, as it is a causative factor of bearing vibration. The important feature is called *waviness*, that is, the number of lobes per circumference.

Waviness can occur in the machining process. A round bar or ring-type element is compressed at the points of contact in a chuck, three jaw or five jaw, causing stresses in the part. The part is then "turned" or ground perfectly circular; however, when it is released from the chuck, the stresses are released, and the part becomes lobed. Centerless grinding also causes waviness where the original bar stock is irregular; perhaps due to the previous machining operation.

Waviness is measured by equipment such as the Taylor–Hobson

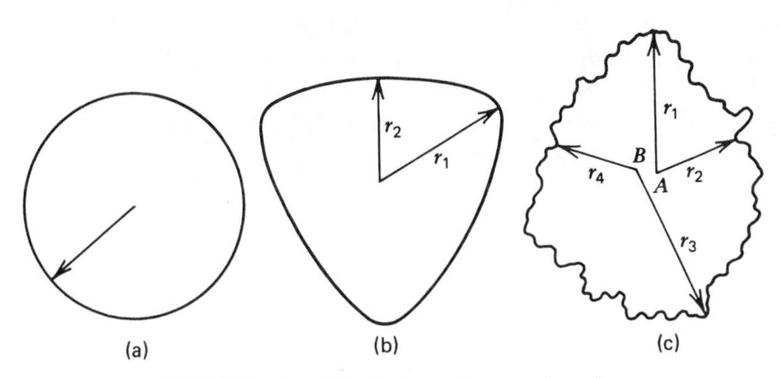

FIGURE 25.9. Illustration of a *round* surface.

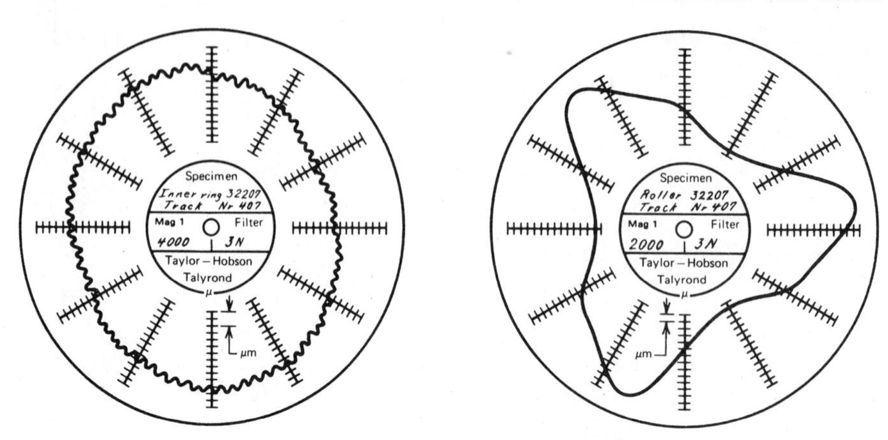

FIGURE 25.10. Talyrond traces of *circular* parts indicating waviness.

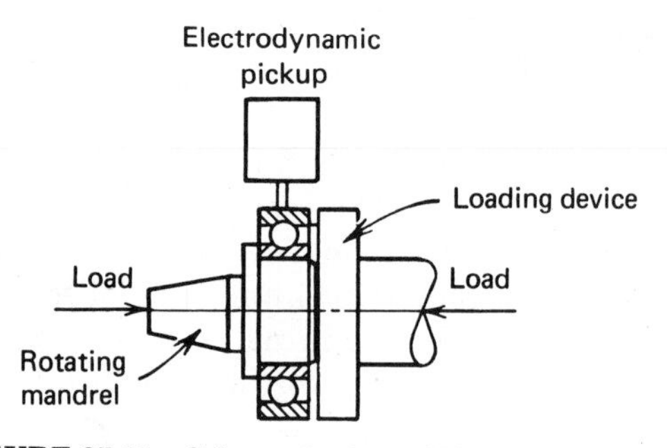

FIGURE 25.11. Schematic view of VKL vibration tester.

Talyrond; Talyrond traces are shown in Fig. 25.10. They were also shown in Fig. 25.5(a)–(c).

For a tapered roller bearing mounted in an SKF VKL tester, shown schematically by Fig. 25.11, Yhland [25.9] examined the correspondence between waviness and the resulting vibration spectrum. For a bearing with Z rolling elements and if p and q are integers equal to or greater than 1 and 0 , respectively, then for vibrations in the radial direction measured at a point on the o.d. of the outer ring, the vibration circular frequencies as functions of inner ring, outer ring, and roller waviness are given in Table 25.1.

In Table 25.1, ω_i, ω_c, and ω_r are the inner ring, cage, and roller angular velocities, respectively. Rigid body vibrations are indicated when $p = 1$; that is, the bearing outer ring moves as a rigid body. For $p > 1$, vibrations are of the flexural type with p equal to the number of lobes per circumference of the outer ring deflection curve. For a waviness spec-

TABLE 25.1. Vibration Frequency vs Waviness

Component	Waviness of Orders	Cause	Vibration with Circular Frequencies
Inner ring	$k = qZ \pm p$		$qZ\,(\omega_i - \omega_c) \pm p\omega_i$
Outer ring	$k = qZ \pm p$		$qZ\omega_c$
Roller	k (even)		$k\omega_r \pm p\omega_c$

trum obtained at an inner ring speed of 900 rpm, for a bearing with an accentuated inner ring waviness, Yhland [25.9] obtained the vibration spectrum at 1800 rpm shown by Fig. 25.12. Also shown by Fig. 25.12 is the Talyrond trace of the inner ring; the tested tapered roller bearing contained very smooth rollers and outer ring.

As compared to the inner ring distortion shown in Fig. 25.5, waviness is a more uniform type of form error. Figure 25.13 shows circumferential traces of two spherical roller bearing inner raceways. One raceway has a peak-to-valley amplitude of approximately 4 μm (0.00016 in.), the other approximately 9 μm (0.00036 in.); each has nine wave cycles per circumference.

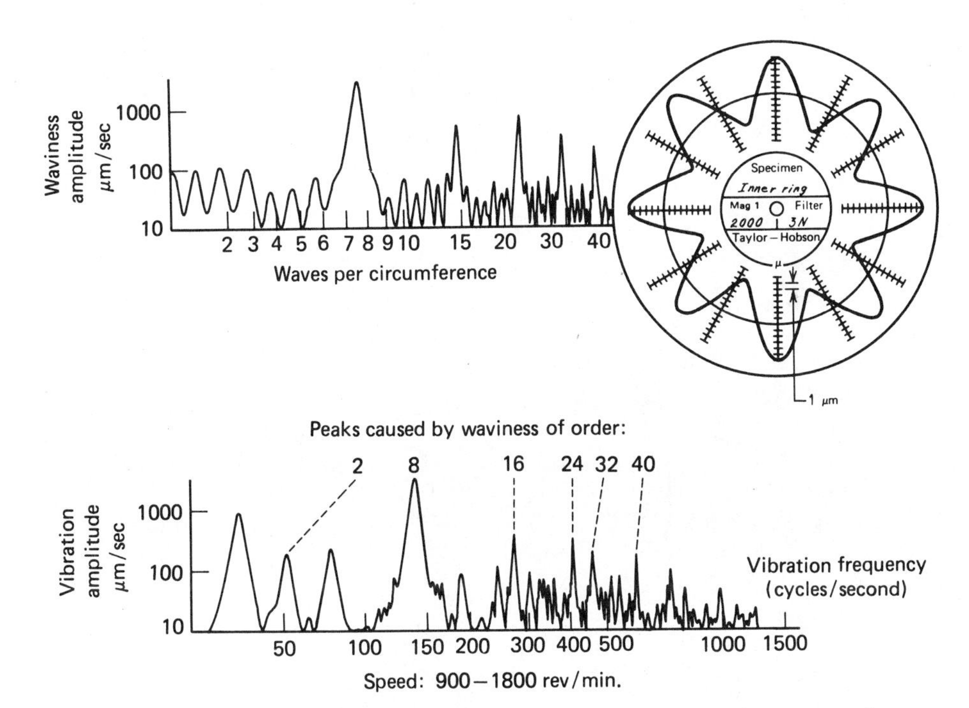

FIGURE 25.12. Waviness and vibration spectra from inner ring with accentuated waviness [25.9].

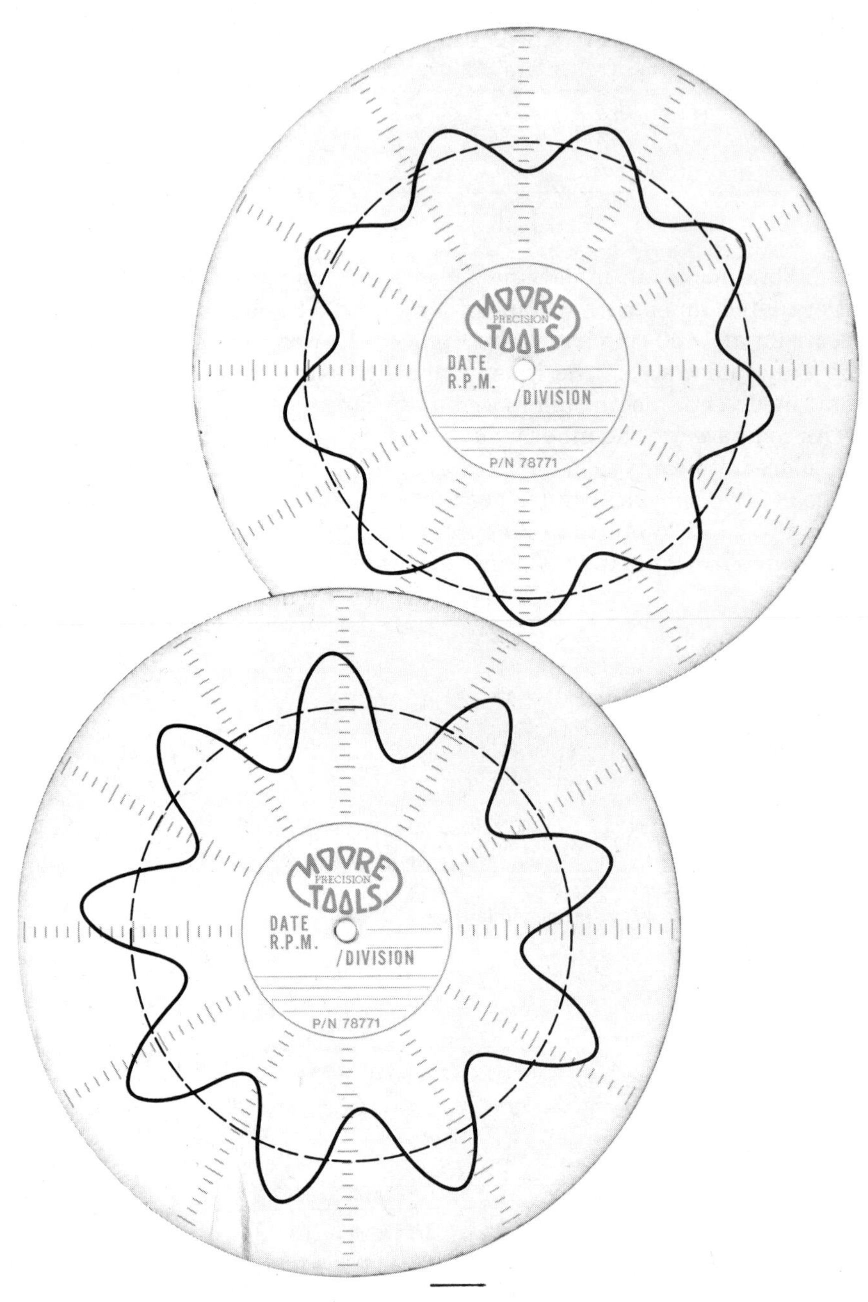

FIGURE 25.13. Spherical roller bearing inner ring waviness, machine setup error. Each radial division equals 1 μm.

Example 25.6. From Fig. 25.13 the radial geometrical deviation of the surface from a true circle (shown on the traces as the dashed line) can be approximated as a function of angular position. The shape is approximately sinusoidal. The radial amplitude variation is then

$$r = A \sin\left(\frac{2\pi R\theta}{\lambda}\right) \tag{25.6}$$

where A = peak amplitude
R = circumferential distance measured from a starting point $\theta = 0$.
λ = wavelength of one cycle or

$$\lambda = \frac{\text{circumference}}{\text{number of waves}} = \frac{2\pi R}{W} \tag{25.7}$$

Therefore, from equations (25.6) and (25.7)

$$r = A \sin W\theta \tag{25.8}$$

and $r = \sin 9\theta$, approximating a sinusoidal wave.

This type of discrete frequency waviness is an excitation source for vibration as well as generation of dynamic force variations on bearing components. Defects of the magnitude shown in Fig. 25.13 are rare and are easily detected when they occur. Waviness of relatively low number (usually odd) of waves per circumference occurs because of inaccuracies in grinding machine tooling or setup involving sliding contact shoe supports for the workpiece as it is ground. Contact of two high points simultaneously on two shoes causes more material to be removed by the wheel at a position opposite the shoes; conversely, low points on the shoes result in less material being removed, producing high points on either side of the wheelwork contact zone. This condition is detectable with conventional in-process gaging used to control diameter, provided that gages are set up for three-point diameter measurement rather than for two point.

Other types of imperfections can result from machine malfunction. See Fig. 25.14, which shows a circumferential trace of a spherical roller. This component was ground "on-centers" and does not show a waviness pattern of the type seen before. A machine control system malfunction, however, allowed the roller to be released from the grinding station before the wheel was retracted, with the result that the roller contacted the wheel and produced a localized flat spot on the roller surface over

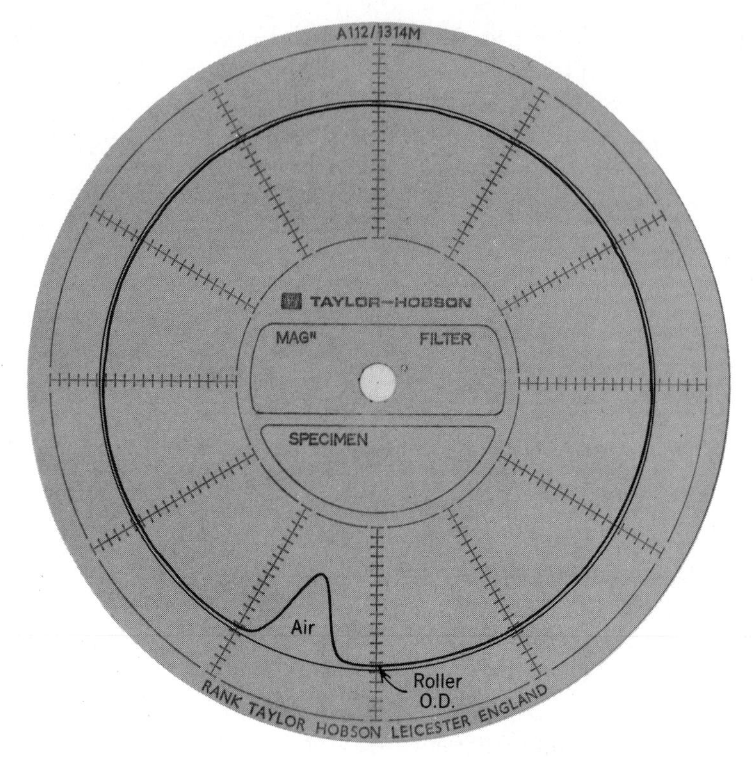

FIGURE 25.14. Roller flat spot, machine malfunction (2 μm per radial division).

approximately 5% of the circumference to a maximum depth of 18 μm (0.00072 in.).

In contrast to waviness from incorrect setup, three-point diameter measurement is not likely to reveal this imperfection unless measurements are made around the entire circumference.

More subtle defects can also occur. They may be characterized by much smaller deviations from true geometrical form and require more detailed component inspection, such as waviness testing or vibration testing of assembled bearings.

Examples of such defects are shown in Figs. 25.15 and 25.16. These traces show, respectively, a spherical roller and a cylindrical roller with lower-amplitude and higher-frequency waviness than the examples of Fig. 25.13. The roller of Fig. 25.15 has 36 waves with an average peak-to-valley amplitude of approximately 1 μm (0.00004 in.). The roller of Fig. 25.16 has over 100 waves with a peak-to-valley amplitude of less than 0.5 μm (0.00002 in.). Both rollers were identified as causes of noise and vibration in assembled bearings. The cylindrical roller bearing corresponding to Fig. 25.16 had been installed in a large electric motor. The

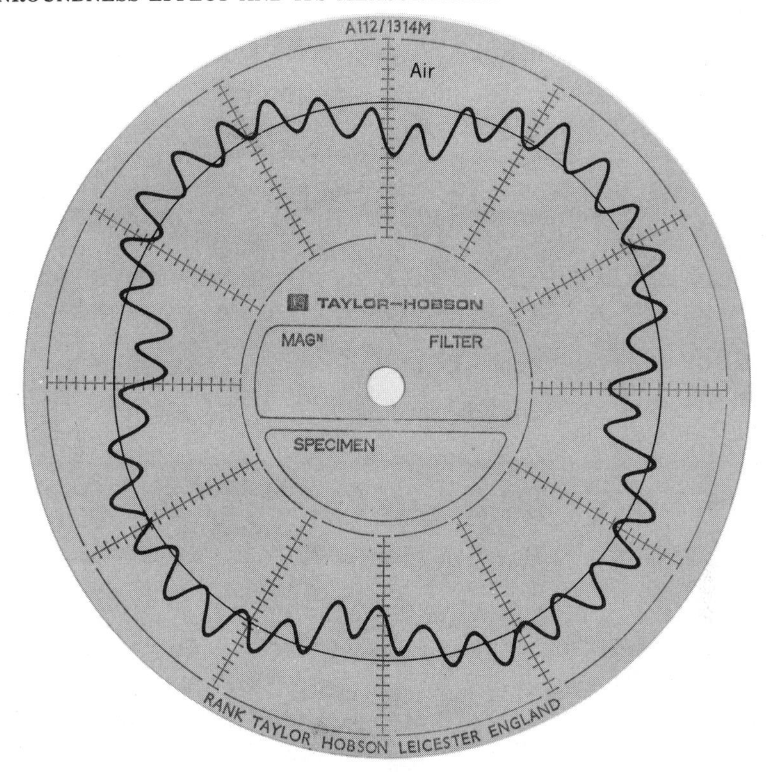

FIGURE 25.15. Roller waviness.

motor emitted a periodic audible noise at slow running speed. With a stopwatch, the repetition rate of the noise could be associated with each revolution of the cage. Subsequent investigation on a test rig traced the noise to this particular roller.

Waviness Testing

Component inspection for waviness has been performed for many years and is described in the literature [25.9, 25.10]. This inspection is used to assess the degree of radial deviations from a true circle on the circumference of a component. This is accomplished by rotating the component on a hydrodynamic spindle and applying a contacting transducer perpendicular to the surface of the component. The transducer is a stylus that follows the radial deviations and produces a voltage output proportional to the instantaneous radial rate of change of the displacement of the stylus; that is, the signal from the transducer is proportional to velocity. This proportionality exists over a wide frequency range, such as 10,000 Hz, which allows reasonably high test speeds to be used.

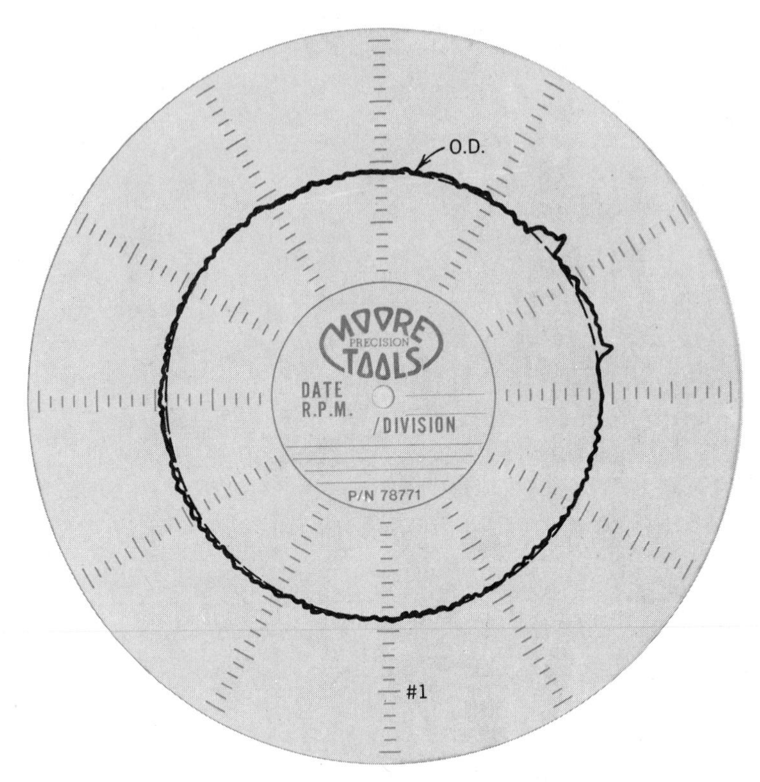

FIGURE 25.16. Low-amplitude, high-frequency waviness.

The voltage signal from the transducer is amplified and bandpass filtered into three or more bands typically 2.5 octaves wide. The filter bands combined with the selected testing speed encompass a broad range of waviness frequencies. Frequencies shown in the previous component traces extend to approximately 100 waves per circumference. Waviness testing equipment and procedures in use cover a range well beyond this.

The rms value of the signal in each filter band is obtained and compared to specifications to determine acceptance or rejection of the lot of components being inspected and to provide information for corrective action on the manufacturing processes. Frequency spectrum analysis is also becoming widely applied as instruments become less expensive and more suitable for use in the factory.

The following discussion illustrates some reasons why velocity measurement has several advantages over displacement for evaluating bearing component quality. The cylindrical roller of Fig. 25.16 was rotated on a waviness testing machine, and the amplified output of the velocity transducer was analyzed with a frequency spectrum analyzer. Results are shown in the plot of Fig. 25.17.

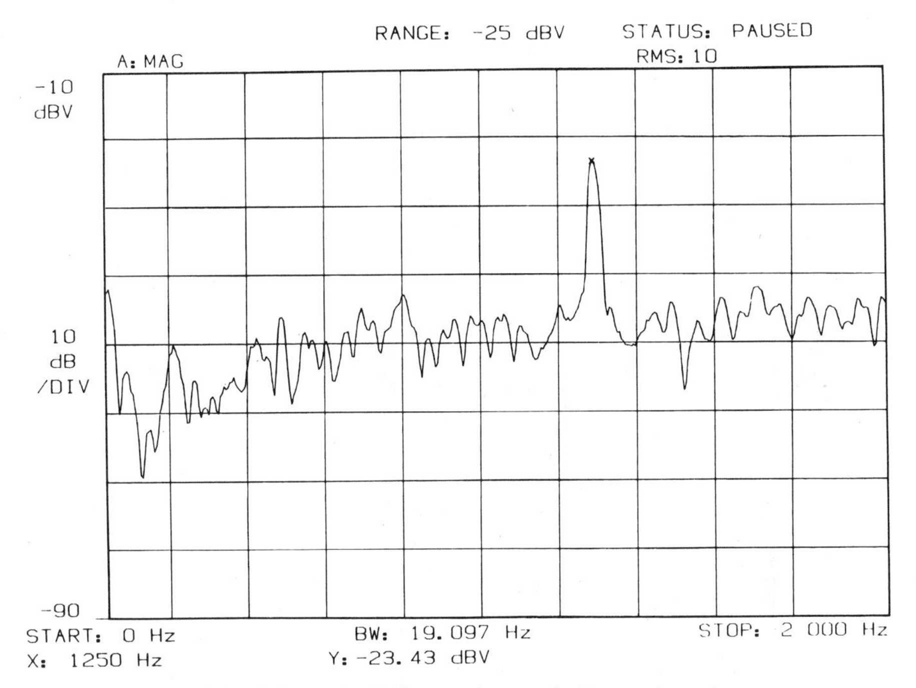

FIGURE 25.17. Roller waviness velocity spectrum.

The abscissa covers a frequency range of 0–2000 Hz, and the roller was tested at 720 rpm (12 Hz), so this frequency range would detect a dominant wave pattern up to 166 waves per circumference (2000/12). The ordinate gives rms voltage in dB referenced to 1 V, ranging from -10 dB full scale (0.3162 V) to -90 dB (3.162×10^{-5} V). Nominal calibration for the velocity transducer and amplifier is 3.0 μV/μin.-sec. A cursor mark is indicated at the peak, with corresponding coordinates printed below the frequency axis. The peak occurs at 1250 Hz, with rms amplitude at -23.43 dB. Since roller test speed was 12 Hz, the frequency at which peak occurs corresponds to approximately 104 waves per circumference.

Example 25.7. For the test arrangement previously indicated, determine the rms radial velocity of the predominant waviness and the number of waves per circumference. Also estimate its average peak-to-valley amplitude in microns.

From Example 25.4, rms voltage is determined from -23.43 dB = $20 \log_{10} V$. Therefore, $V = 0.067375$. Rms velocity is determined by dividing the voltage by the transducer-amplifier conversion of 3 μV/μin.$^{-1}$/sec^{-1}. Therefore,

$$\left|\frac{dr}{dt}\right|_{\text{rms}} = (22{,}458\ \mu\text{in./sec})\ 570\ \mu\text{m/sec}$$

In equation (25.8) the radial deviation r was given as a function of angular position θ on the component and the number of waves W.

In the waviness test apparatus the part is rotated, so the radial deviation r is a function of time. Any angular location on the part also becomes a function of time with respect to the fixed location of the transducer being used to measure the radial deviations:

$$\theta = 2\pi Nt \tag{25.9}$$

where N is the rotational test speed (rps). Therefore,

$$W\theta = 2\pi NtW \tag{25.10}$$

and

$$r = A \sin (2\pi NtW) \tag{25.11}$$

The radial velocity, measured by the transducer, is the change of the radial deviation with respect to time:

$$\dot{r} = 2\pi NWA \cos (2\pi NtW) \tag{25.12}$$

and

$$\dot{r}_{\text{rms}} = 1.414\ \pi NWA \tag{25.13}$$

From the measured rms velocity the peak amplitude A can be estimated:

$$22{,}458 = 1.414\ \pi(720/60)104\ A$$

$$A = 0.1029\ \mu\text{m}\ (4.05 \times 10^{-6}\ \text{in.})$$

The peak-to-valley amplitude is twice that value and is therefore estimated to be 0.206 μm (8.1×10^{-6} in.)

Table 25.2 summarizes estimates of average peak-to-valley displacement and rms velocity values from equation (25.13) of the components of Figs. 25.5*c*, 25.13, 25.15, and 25.16.

Peak-to-valley displacement of these components, one of which was determined as mounted on a shaft with excessive three-point out-of-roundness, varies over a range of approximately 80:1, whereas the number of waves per circumference varies over a range of approximately 35:1. The velocity values, however, are within a range of only 4:1 for components of diverse size and configuration. Additionally, transducers

TABLE 25.2. Waviness Displacement Velocity Amplitudes

Component Trace Figure	Peak-to-Valley Amplitude (μin.)	Peak-to-Valley Amplitude μm	Waves per Circumference	rms[a] Velocity (μin./sec)	rms[a] Velocity μm/sec
25.5*c*	(629.9)	16[b]	3	(50,368)	1,279
25.9*a*	(157.5)	4	9	(37,776)	960
25.9*b*	(354.3)	9	9	(84,997)	2,159
25.11	(9.37)	1	36	(37,776)	960
25.12	(8.11)	0.206[c]	104	(22,458)	570

[a]Displacement assumed sinusoidal; therefore $dr/dt|_{\mathrm{rms}} = (0.707)(2\pi NW/60)$, where N is the rpm of the waviness test spindle.
[b]Value estimated as mounted on out-of-round shaft.
[c]Calculated from measured velocity in Example 25.6.

and instrumentation for velocity measurement systems do not require as great a dynamic range as displacement measurement. Numerical specifications of similar magnitude are also more easily developed and applied.

Some types of defects that can arise on components are very local in nature. Detection of these defects may not be feasible with waviness testing, which may only acquire data from one or two circumferences on components being tested. Balls have numerous potential axes of rotation. Therefore, visual component inspection and vibration testing of assembled bearings provides more definitive assurance of final bearing quality.

Vibration Testing

Aside from defects that are not discovered in waviness testing, vibration testing of the assembly allows the detection of damage occurring in assembly, such as binding or excessively loose cages, Brinell damage to raceways or scuffing of balls, and distortion of raceways from incorrect insertion of seals or shields with bearings tested after grease insertion.

Certain types of geometrical problems may also be detectable in vibration testing. These can include, for example, oversized rolling elements, improper cross-groove form on raceways, or groove runout to side faces of the raceways. In addition, testing can reveal contamination by dirt or inferior grease quality.

Figure 25.18 shows a manually operated vibration testing apparatus for relatively small bearings, for example, up to 100 mm od. Similar equipment is in use for larger diameter bearings, and automatic versions are implemented on production lines. The main elements of the system

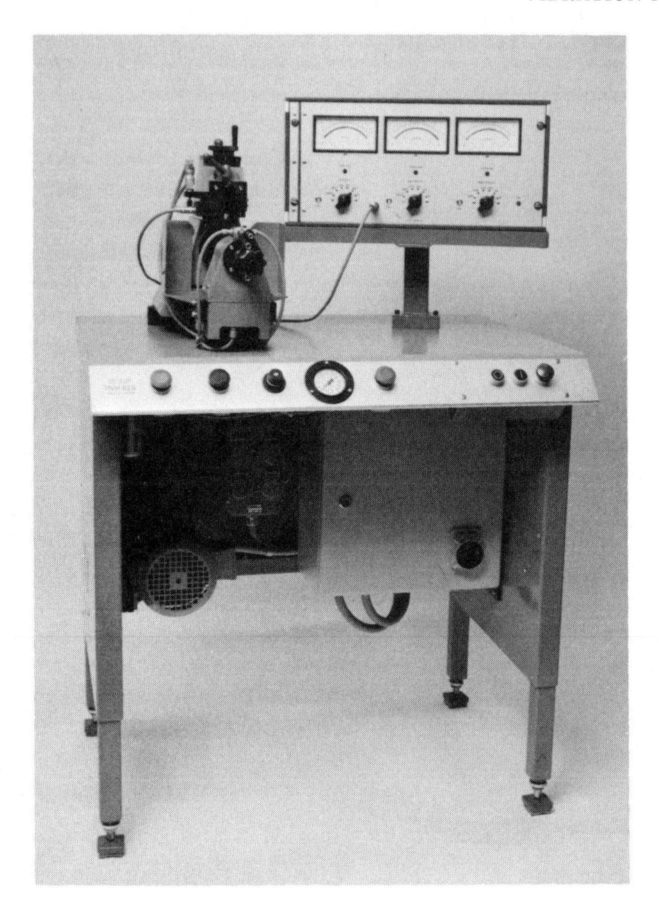

FIGURE 25.18. Bearing vibration tester.

are the test station and the vibration signal analysis instrument. The test station consists of a hydrodynamic spindle, an air cylinder for applying load to the bearing being tested, and an adjustable slide for positioning the velocity transducer. The spindle is belt driven by the motor mounted beneath the stand. A schematic representation of the system is shown in Fig. 25.19.

The inner raceway of the bearing mounts on a precision arbor fastened to the spindle, which rotates at 1800 rpm. A specified thrust load is applied to the side face of the nonrotating outer ring. The tip of the velocity transducer is lightly spring-loaded on the outer diameter of the outer ring. The loading tool (not shown) consists of a thin-walled steel ring molded into a neoprene annulus. The ring contacts the side face of the outer ring. The tool and load combinations are sufficiently compliant to allow radial motion of the outer ring to occur as balls roll over wavy

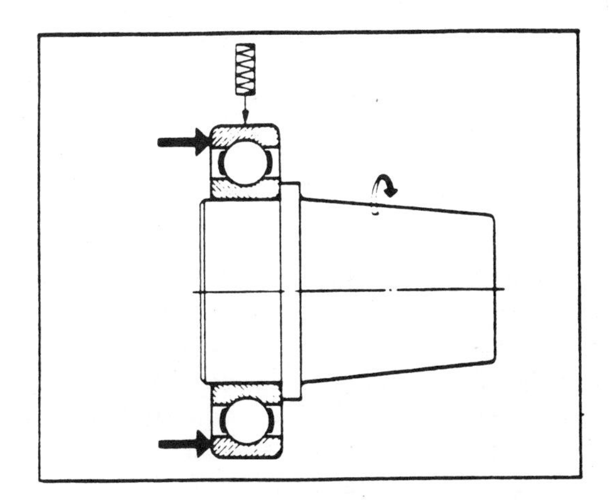

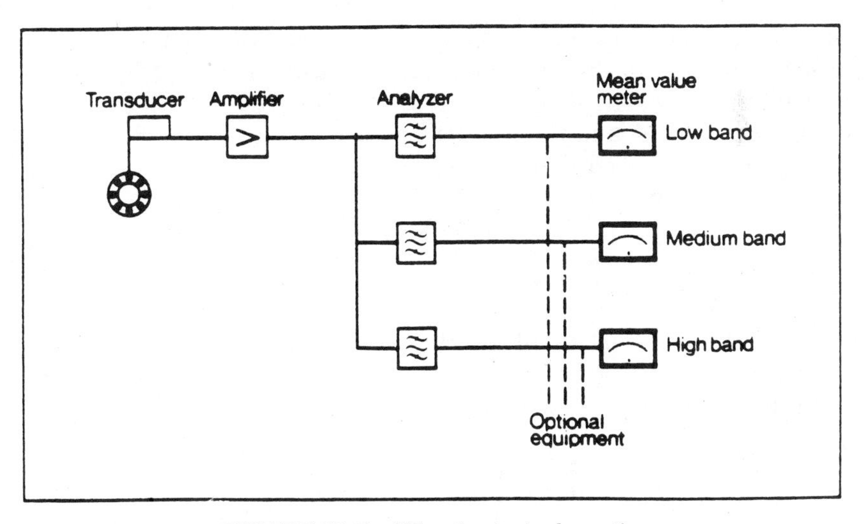

FIGURE 25.19. Vibration test schematic.

surfaces or defects in the ball grooves. The voltage signal from the transducer is input to the analysis instrument, which amplifies the signal, bandpass filters it, and displays the rms velocity values in each band. The three frequency bands are 50–300 Hz, 300–1800 Hz, and 1800–10,000 Hz. Larger bearings are tested at slower rotational speed (700 rpm) with correspondingly lower filter bands: 20–120 Hz, 120–700 Hz, and 700–4000 Hz.

The basic testing technique has been successfully used by bearing manufacturers and customers for many years. Numerous refinements have been made during this time, and development work continues in the area of vibration measurement. Such efforts include investigations of alternative transducer design and system calibration procedures, different methods of applying load to the test bearings, increased application of statistical methods in setting product specifications and analyzing test results, and implementation of supplementary methods of signal analysis.

Bearing Frequencies

The calculation of the fundamental pass frequencies of rolling contact bearings is used to establish component waviness testing speeds and filter bands that coincide with vibration measurement bands. In addition, knowledge of these frequencies is useful, though not always essential, in machinery condition monitoring. Derivation of these equations is presented in Chapter 7. Results are given here for the case of a stationary outer ring and rotating inner ring, as is used in the vibration test system described before and is most often the case in typical bearing applications. Assuming no skidding of rolling elements, the frequencies of interest are related to the rotational speed of the inner ring N, the pitch diameter of the bearing d_m, the rolling element diameter D, the number of balls or rollers Z, and the contact angle α.

The rotational speed of the cage f_c is

$$f_c = \frac{N_i}{2}\left(1 - \frac{D}{d_m}\cos\alpha\right) \tag{25.14}$$

The rotational speed of the inner ring relative to the cage is the rate at which a fixed point on the inner ring passes by a fixed point on the cage. This relative speed is

$$f_{ci} = \frac{N_i}{2}\left(1 + \frac{D}{d_m}\cos\alpha\right) \tag{25.15}$$

The rate at which balls pass a point in the groove of the outer raceway (also called the ball-pass-outer-raceway frequency or outer raceway defect frequency) is

$$f_{bpor} = Zf_c \tag{25.16}$$

f_{bpor} is also that frequency at which variable elastic compliance vibration occurs (see "The Role of Bearings in Machine Vibration").

The rate at which a point in the inner ring groove passes balls (also called the ball-pass-inner-raceway frequency or inner raceway defect frequency) is

$$f_{bpir} = Zf_{ci} = Z(N_i - f_c) \tag{25.17}$$

The rate of rotation of a rolling element about its own axis is

$$f_R = \frac{N_i d_m}{2D}\left[1 - \left(\frac{D}{d_m}\cos\alpha\right)^2\right] \tag{25.18}$$

A single defect on a ball or roller would contact both raceways in one ball or roller revolution so that the defect frequency is $2f_R$. In addition, the defect could contact one or both sides of the cage pocket; however, this usually will have little influence on vibration measured external to the bearing.

Relation of Vibration and Waviness

Within reach of the vibration measurement frequency bands, the number of waves on a component that influence a particular band can be calculated. For outer raceway waviness any ball rolls over all the waves in the outer raceway groove in one cage revolution. Therefore, the ball passage frequency over an individual wave cycle on the outer raceway is $f_c \times$ number of waves per circumference. Consequently, dividing the filter frequencies by f_c determines the number of waves per circumference of the outer raceway producing vibration within a particular band. For balls, the filter band frequencies are divided by f_R to determine the number of waves per circumference of a ball producing vibration within a band.

Similarly, for the inner raceway the band frequencies are divided by f_{ci}. The lobes of low orders of inner raceway waviness, such as two- and three-point out-of-roundness, however, can cause flexure of the outer ring (two- or three-point lobing) and vibration at two or three times N_i, affecting readings in the low-frequency band.

Example 25.8. Compute the cage rotational frequencies, ball rotational frequency, and estimate the waviness orders for each component that fall within vibration testing bands for a 203 ball bearing with an assumed contact angle of 12° and test speed of 1800 rpm (30 Hz). For a 203 ball bearing, ball diameter = 6.747 mm (0.2656 in.) and pitch diameter = 28.5 mm (1.122 in.).

$$f_c = \frac{N_i}{2}\left(1 - \frac{D}{d_m}\cos\alpha\right) \tag{25.14}$$

$$= \frac{30}{2}\left(1 - \frac{6.747\cos 12^\circ}{28.5}\right)$$

$$= 11.53 \text{ Hz}$$

$$f_{ci} = \frac{N_i}{2}\left(1 + \frac{D \cos \alpha}{d_m}\right) \tag{25.15}$$

$$= \frac{30}{2}\left(1 + \frac{6.747 \cos 12°}{28.5}\right)$$

$$= 18.47 \text{ Hz}$$

$$f_R = \frac{N_i d_m}{2D}\left[1 - \left(\frac{D \cos \alpha}{d_m}\right)^2\right] \tag{25.18}$$

$$= \frac{30 \times 28.5}{2 \times 6.747}\left[1 - \left(\frac{6.747 \cos 12°}{28.5}\right)^2\right]$$

$$= 59.97 \text{ Hz}$$

Therefore, calculations indicate that waviness orders of the components fall within the vibration test bands approximately as follows:

	50–300 Hz	300–1800 Hz	1800–10,000 Hz
Outer raceway	4–26	27–156	157–868
Inner raceway	2*–16	17–97	97–541
Balls	2–5	6–30	31–167

*Including two- and three-point out-of-roundness.

Similar values are obtained for the range of ball bearing sizes tested in this manner: for example, deep-groove ball bearings in the 2 and 3 series. Waviness testing procedures are established to correlate with average waviness ranges over a wide range of bearing sizes. In addition, the range of waviness orders tested in either vibration or waviness measurement corresponds approximately to wavelengths of the size of the small axis of the Hertzian contact ellipse in typical applications such as electric motors [25.9].

Other Factors in Vibration Testing

Defects other than waviness can contribute to bearing vibration or noise. Some of them can be difficult to detect with the conventional three-band inspection method. Such defects include local defects on raceways or balls, dirt, grease with improper constituents or properties, and cages with incorrect clearance or geometry. Some of these defect types may produce brief disturbances spaced widely apart in time, which, as a consequence, have only a small effect on the average measured vibration in the inspection bands. For example, a single localized defect on the inner raceway or on an individual ball will be remote from the transducer location during most of the time that it takes for the inner raceway (or

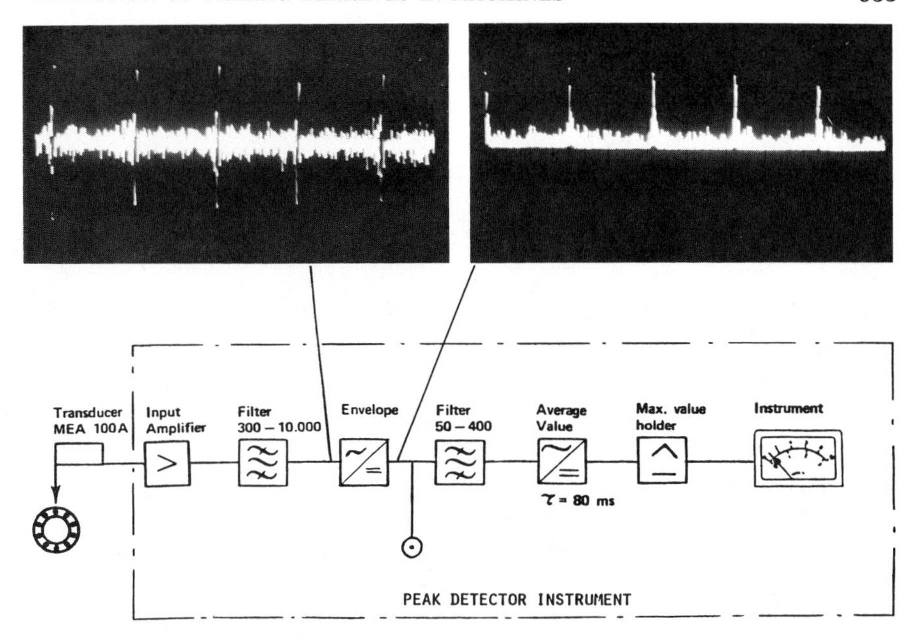

FIGURE 25.20. Peak detection.

cage in the case of a defective ball) to make one revolution. Such defects impacting the outer ring will momentarily excite its various natural frequencies. The lowest of these natural frequencies is a rigid body mode (individual balls act as springs) and higher natural frequencies being modes of outer ring flextural vibration. These modes result from bending of the outer ring into shapes that have an integral number of lobes, as analyzed in [25.4]. Resonant vibration of the outer ring amplifies the effects of these local defects, and the resonant vibration can be used to detect their presence. Therefore vibration measurement is supplemented with the use of a peak detector instrument whose functions are shown schematically in Fig. 25.20. Although the figure shows peaks of relatively constant amplitude, which might be the case for an outer raceway defect, maximum peak values are obtained and evaluated because they can vary with time in the case of ball and inner raceway defects.

DETECTION OF FAILING BEARINGS IN MACHINES

Vibration analysis is one of the most common methods used to evaluate the condition of bearings in an operating machine. As previously shown, such measurements may be used for machines with bearings in new condition as well as for machines whose bearings are deteriorating and approaching the end of their useful lives. If a machine's vibration response to known excitation forces has been determined through techniques such as finite element analysis and modal analysis, then vibration measurements during its in-service operation can define the dynamic characteristics of the forces acting on the machine.

Vibration data can also be used to infer forcing characteristics and condition of machine components, including bearings. General methods for evaluating data include one or more of the following:

1. Comparison of data with guidelines developed empirically on similar types of equipment [25.8, 25.11, 25.12]
2. Comparison of data from similar or identical machines in service within the same factory
3. Trending of data from one machine over time
4. Evaluation of data in an absolute sense with no prior history. For example, by evaluating time signals or frequency spectra to associate vibration with specific machine components

Many machine problems can be traced to faults other than damaged bearings. If a moderately detailed vibration analysis capability is not available, however, bearings are often replaced unnecessarily.

The beginning of progressive bearing damage, which can be called incipient failure, is often characterized by a sizable local defect on one of the components. When this occurs, subsequent rolling over of the damage will produce repetitive shocks or short-duration impulses. It can be surmised that such impulses might appear, if they could be measured, as those in Fig. 25.21*a* and *b*.

Figure 25.21*a* could represent, for example, the effect of successive rolling elements passing over a damaged area on the outer raceway. Similarly, Fig. 25.21*b* might represent the effect of inner raceway damage interacting with several rolling elements in the load zone of a radially loaded bearing without preload. In this case the damage enters the load zone once per revolution of the shaft. The location of the rolling elements with respect to the load zone will vary somewhat from one shaft revolution to the next. If a sensor were placed on the bearing housing to measure the resulting vibration from the series of impacts, it may show a response as in Fig. 25.21*c*. This vibration corresponds to lightly damped oscillation of some system natural frequency greater than the repetition frequency of the train of impacts. It could, for example, be a resonance of the bearing outer ring or of the housing or sensor itself. Such resonant response is excited by harmonics that exist in the periodic nonsinusoidal forcing function. Figure 25.22 shows the time history of an electrical signal representing a pulse train with a fundamental frequency of 160 Hz; Fig. 25.23 is the frequency spectrum of that signal. It contains all harmonics of the fundamental.

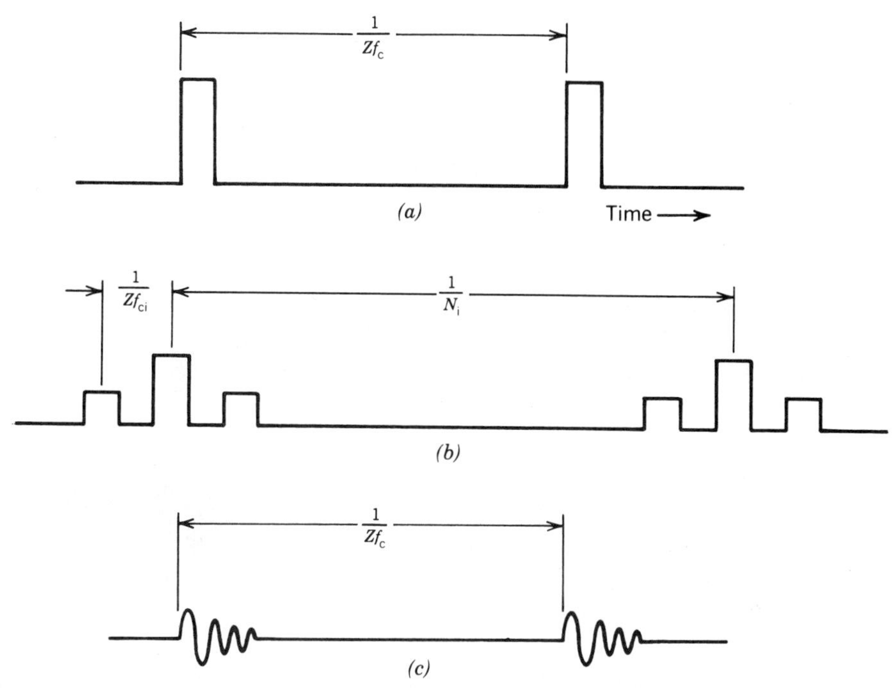

FIGURE 25.21. Impulse train. (*a*) Outer raceway damage. (*b*) Inner raceway damage. (*c*) Resonant vibration.

Impulsive occurrences in bearings, therefore, can cause system vibration at many frequencies that can be harmonically related. Forcing harmonics that are near system resonant frequencies can cause significantly amplified vibration response compared to the vibration at nonresonant frequencies. In the early stages of failure the impulse might have little effect on the amplitude of vibration at the fundamental bear-

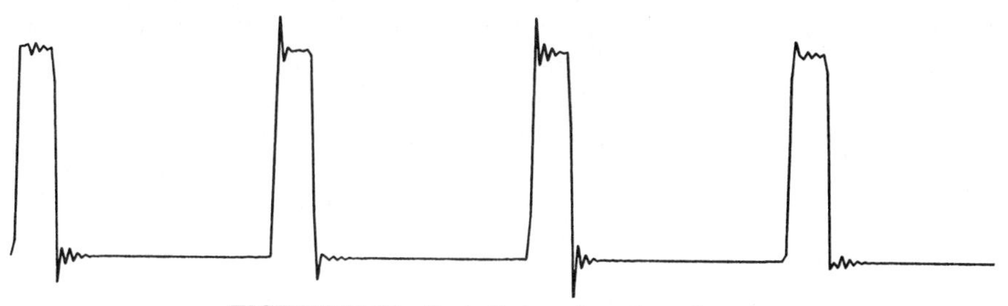

FIGURE 25.22. Periodic impulse—time domain.

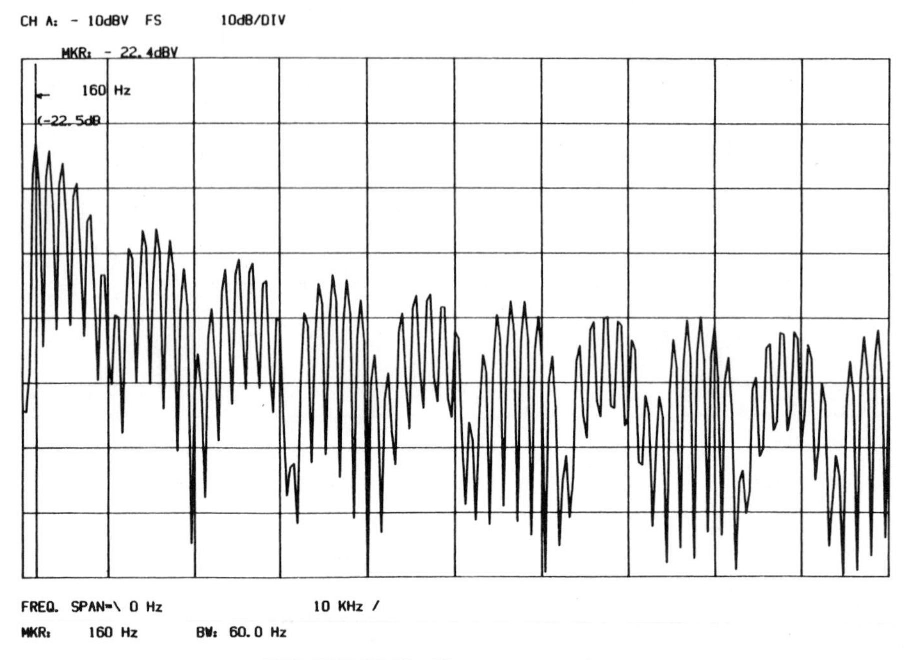

FIGURE 25.23. Frequency spectrum.

ing pass frequencies. In addition, significant normal machine vibration could occur at these lower frequencies, so small changes in vibration amplitude initially may be difficult to detect. Higher-order harmonics, with spacing related to specific component frequencies, however, might be detectable at higher frequencies if the sensor and mounting method provide sufficient response at the higher frequencies. Small accelerometers stud-mounted to electrically isolated nuts and glued to a surface on the machine work satisfactorily. Magnetic mounting is faster but it requires a better surface and frequency is lower. The following example illustrates the effect of local bearing damage on vibration response.

Example 25.9. A test rig was used to run two 205 ball bearings, each mounted on a pillow block at opposite ends of a shaft. The shaft was belt driven at 1690 rpm by a pulley mounted on the shaft between the bearings. The outer raceway of one of the bearings contained localized damage, approximately circular in shape of 1.6 mm (0.0625 in.) diameter, located in the bottom of the groove and in the load zone. The radial bearings each contained nine balls of 7.938 mm (0.3125 in.) diameter on a pitch diameter of 39.04 mm (1.537 in.).

The calculated bearing frequencies are

$$N_i = \frac{1690}{60} = 28.2 \text{ Hz}$$

$$f_c = \frac{N_i}{2}\left(1 - \frac{D \cos \alpha}{d_m}\right) \tag{25.14}$$

$$= \frac{28.2}{2}\left(1 - \frac{7.938 \cos 0°}{39.04}\right)$$

$$= 11.23 \text{ Hz}$$

$$f_{bpor} = Zf_c = 9 \times 11.23 = 101.1 \text{ Hz} \tag{25.16}$$

$$f_{ci} = \frac{N_i}{2}\left(1 + \frac{D \cos \alpha}{d_m}\right) \tag{25.15}$$

$$= \frac{28.2}{2}\left(1 + \frac{7.938 \times 1}{39.04}\right)$$

$$= 16.97 \text{ Hz}$$

$$f_{bpir} = Zf_{ci} = 9 \times 16.97 = 152.7 \text{ Hz} \tag{25.16}$$

Figure 25.24 shows spectra of the two bearings on the same plot.

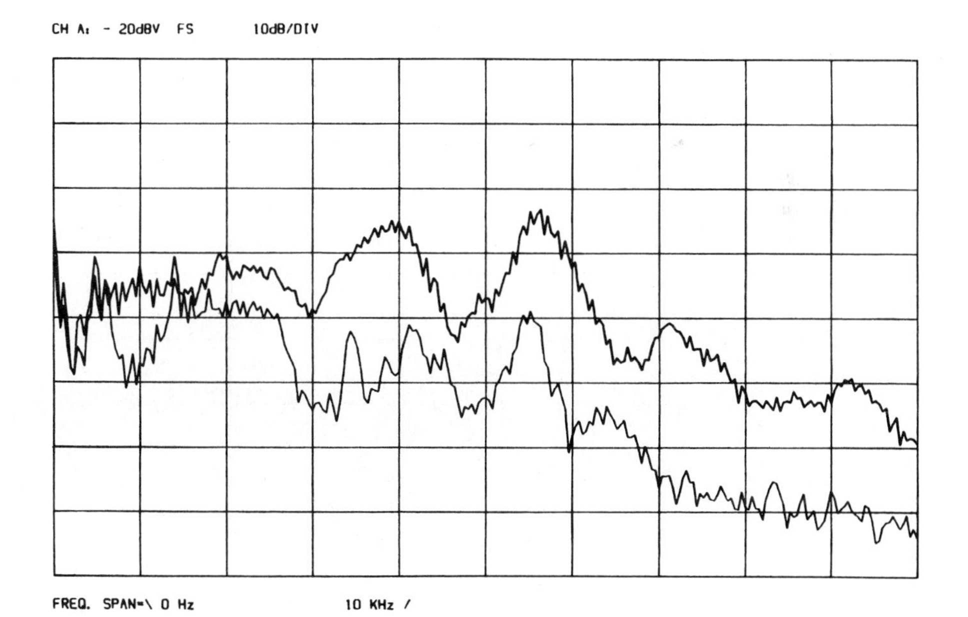

FIGURE 25.24. Vibration of damaged and undamaged bearings.

The frequency span is 0–10,000 Hz and full-scale amplitude is −20 dB. Data were obtained with a stud-mounted accelerometer, 0.010 V/g.

The vibration amplitudes of the damaged bearing are 20 dB greater than those of the undamaged bearing at most frequencies above 3000 Hz. The spectrum of the damaged bearing also shows peaks, about 10 of them in each 1000 segment, whose spacing corresponds to Zf_c. Better resolution of peak spacing would require using a narrower frequency span in regions of the spectrum or an instrument with finer resolution. Nevertheless, the figure illustrates the major effect of local bearing damage on vibration in the higher-frequency regions.

Figure 25.25, taken on the damaged bearing over 2500 Hz, clearly shows harmonics of the ball passage frequency over the outer raceway from 500 to 1250 Hz. Amplitudes of harmonics from 700 to 1200 Hz were approximately 10 dB greater than vibration amplitudes of the undamaged bearings in this range. Below 700 Hz, amplitudes of the two bearings were the same. Depending on the presence of other sources of machine vibration and the magnitude of bearing damage that exists, it might also be possible to successfully identify a problem at low frequencies.

Figure 25.26 shows part of a digitized time sample taken from the damaged bearing, indicating some perturbation at a spacing corre-

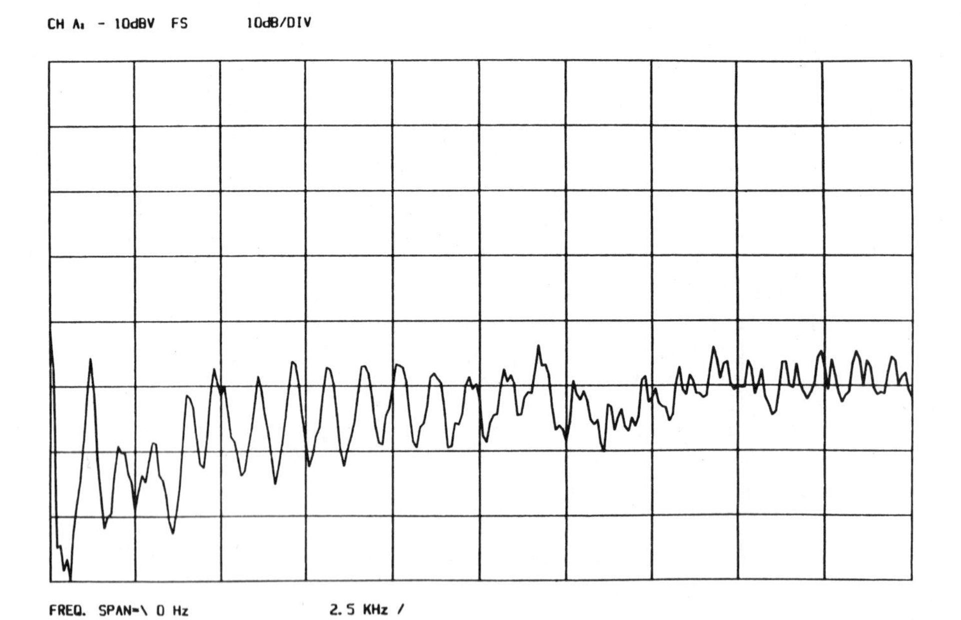

FIGURE 25.25. Low-frequency vibration of damaged bearing.

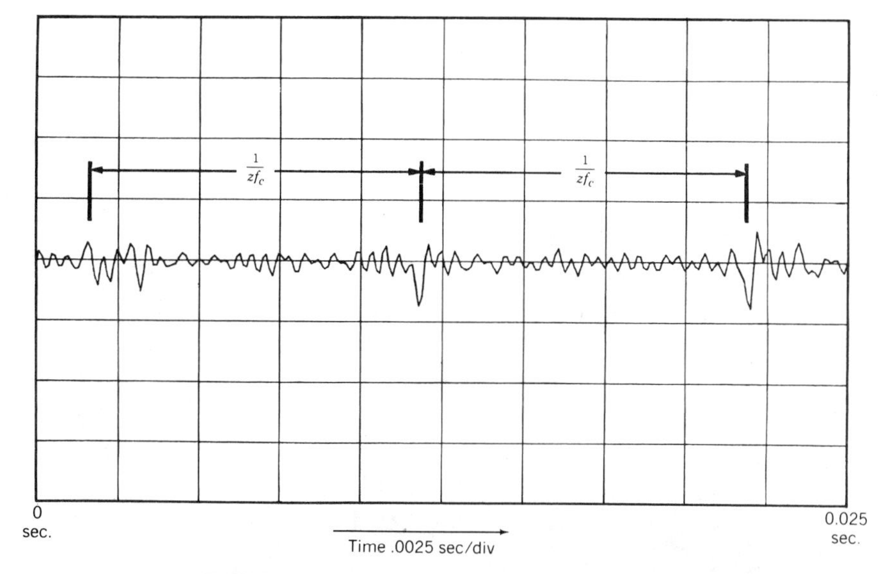

FIGURE 25.26. Damaged bearing—time domain.

sponding to $1/Zf_c$ sec. The previous illustration considers only a single local damage that can be associated with a specific bearing component with frequency spectrum analysis. Numerous cases of such failure detection are reported in the literature [25.13, 25.14].

Most forms of damage preceding bearing failure will result in progressive wear and roughening of component surfaces and irregular running geometry. Such irregularities may produce vibration that can clearly be identified with specific components, or they could produce vibration whose amplitude varies randomly in time and frequency content. In either case machine vibration measurement in one form or another can be used to periodically assess bearing condition.

A particularly damaging form of failure is caused by electrical current passing through bearings in large motors. Arcing erodes the bearing and creates numerous large damaged areas. Sample data show detection of such damage in a 1200-rpm, 1000-hp vertical pump motor using axial vibration on the pump motor base. The top bearing is a 170-mm (6.69-in.) bore, 40° angular-contact ball bearing. In contrast to an 800-hp pump with rotating imbalance, discussed earlier, vibration spectra indicated harmonics, as seen in Fig. 25.27*a*.

Accurate evaluation of harmonics was performed with cepstrum analysis [25.15], which determines periodicity within a spectrum, allowing indentification to "zoom" analysis of frequency bands within the spectrum. Harmonics with a spacing of 101.7 Hz were identified, and they corresponded to the outer raceway ball pass frequency. The

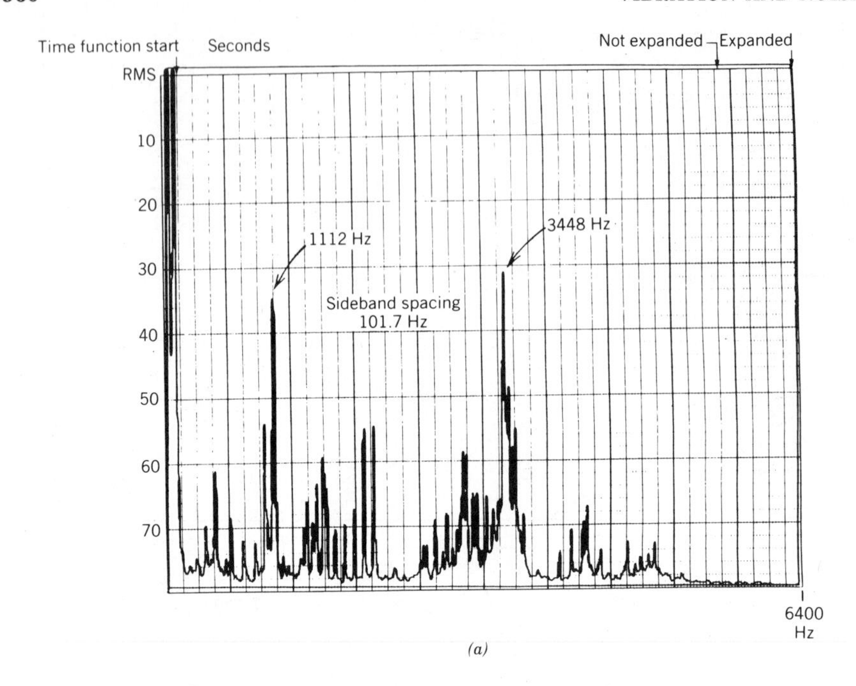

(a)

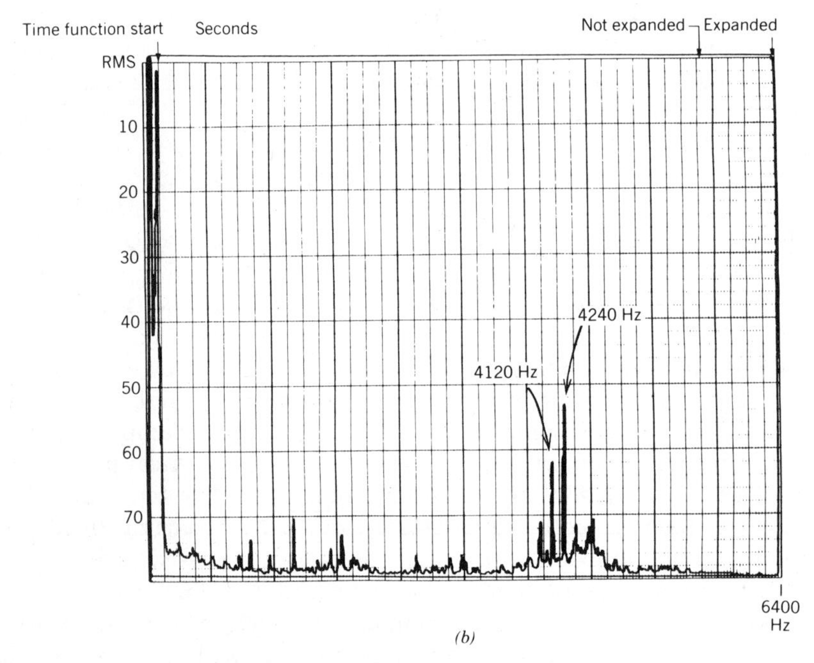

(b)

FIGURE 25.27. (*a*) Pump motor vibration two weeks before bearing replacement. (*b*) Pump motor vibration after bearing replacement.

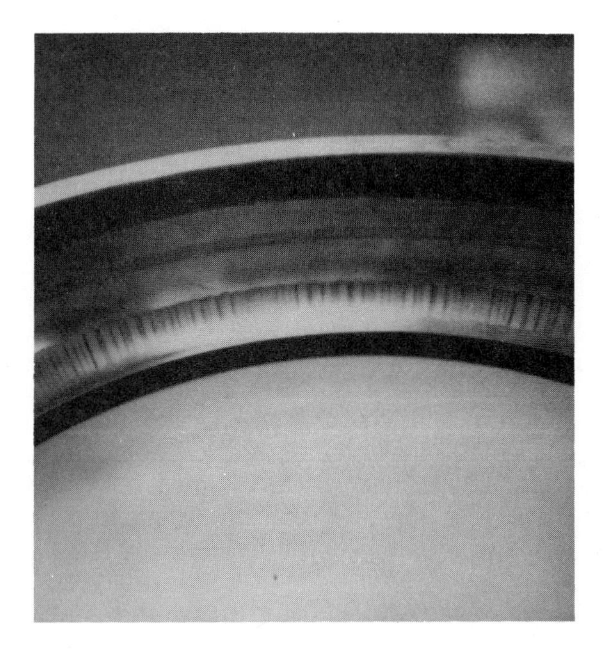

FIGURE 25.28. Pump motor outer raceway failure by electrical arcing.

bearing was removed and replaced after less than nine months of operation. Data from the pump after rebuild are shown in Fig. 25.27*b*, and a photograph (Fig. 25.28) illustrates the nature of damage on the outer raceway. Analysis and means of prevention of this type of damage are presented in [25.16–25.18].

OTHER EVALUATION METHODS

Aside from evaluation of vibration spectra to identify specific machine frequencies, data can be obtained or analyzed by other means to trend the onset of failure. Mathew and Alfredson [25.19] present a comprehensive evaluation of vibration parameters over the life of bearings run to advanced stages of damage progression or failure. Conditions under which bearings were tested include bearing components with initial damage, contaminated lubrication, overload condition leading to cage collapse, and sudden loss of lubrication. Parameters that might be obtained with relatively low cost instrumentation include peak acceleration, rms acceleration over a broad frequency band, and shock pulse data. The last evaluates vibration at a frequency corresponding to the resonant frequency of the accelerometer (32 Hz). Other parameters were calculated by performing arithmetic operations on two frequency spectra, one of which was usually the initial spectrum obtained when tests were begun. The calculated parameters were then trended. Statistical functions, including probability density, skewness, and kurtosis were also evaluated. The results indicated that several parameters evaluated from frequency spectra were successful trend indicators, generally providing a 30-dB increase or more by the time a test run was completed. One such parameter is simply obtained by subtracting the initial spectrum from each new spectrum and computing the rms of the resulting spectrum. This value is then trended over the duration of tests.

Aside from computations of trend parameters from complete vibration spectra, the shock pulse method was reported to provide successful detection for all tests except the case of total lubrication loss. For the successful tests the shock pulse values are estimated to have increased 40 dB. In the test with lubrication loss, seizure occurred in 2 hr. This suggests that lower-frequency vibration may be a better initial indicator than higher-frequency vibration unless the components have time to undergo sufficient gradual distress to be detectable in the high-frequency regime.

CLOSURE

This chapter provides an indication of how bearings can affect machine vibration, as a result of either inherent design characteristics or imperfections and deviations from ideal running geometry within the bearing. Some examples illustrated that such imperfections and geometric deviations can occur during bearing component manufacture, during bearing assembly into a machine, or from bearing deterioration during operation. Each can have a pronounced effect on machine vibration, either by altering stiffness properties or by acting as a source of forces to directly generate vibration.

REFERENCES

25.1. T. Tallian and O. Gustafsson, "Progress in Rolling Bearing Vibration Research and Control," ASLE Paper 64C-27 (October 1964).

25.2. R. Scanlan, "Noise in Rolling-Element Bearings," ASME Paper 65-WA/MD-6 (November 1965).

25.3. Military Specification Mil-B-17931D (Ships), "Bearings, Ball, Annular, for Quiet Operation," (April 15, 1975).

25.4. O. Gustafsson, T. Tallian et al., "Final Report on the Study of Vibration Characteristics of Bearings," U.S. Navy Contract NObs-78552, U.S. Navy Index No. NE 071 200 (December 6, 1963).

25.5. R. Sayles and S. Poon, "Surface Topography and Rolling Element Vibration," *Precision Engineering*, IPC Business Press (1981).

25.6. L. Johansson, "Bearing Noise in Electric Motors," *Ball Bearing J.* **200** (1979).

25.7. J. Hyer and D. Sileo, "Some Practical Considerations in the Selection and Use of Ball Bearings in Small Electric Motors," Small Motor Manufacturers Association (March 1985).

25.8. J. Mitchell, *Machinery Analysis and Monitoring*, PennWell, Tulsa, Chap. 9 and 10 (1981).

25.9. E. M. Yhland, "Waviness Measurement—An Instrument for Quality Control in Rolling Bearing Industry," *Proc. Inst. Mech. Eng.*, **182**, Pt. 3K, 438–445 (1967–68).

25.10. O. Gustafsson and U. Rimrott, "Measurement of Surface Waviness of Rolling-Element Bearing Parts," SAE Paper 195C (June 1960).

25.11. International Organization for Standardization, "Acoustics, Vibration and Shock," ISO Standards Handbook 4 (1980).

25.12. S. Norris, "Suggested Guidelines for Forced Vibration in Machine Tools for Use in Protective Maintenance and Analysis Applications," *Vibration Analysis to Improve Reliability and Reduce Failure*, ASME H00331 (September, 1985).

25.13. J. Taylor, "Identification of Bearing Defects by Spectral Analysis," *ASME J. Mech. Design* **102** (April 1980).

25.14. J. Taylor, "An Update of Determination of Antifriction Bearing Condition by Spectral Analysis," Vibration Institute (April 1981).

25.15. R. Randall, "Cepstrum Analysis and Gearbox Fault Diagnosis," Bruel & Kjaer Application Note 233-80.

25.16. E. Wallin, "Prevention of Bearing Damage Caused by the Passage of Electric Current," *Ball Bearing J.* **153** (1968).

25.17. S. Andreason, "Passage of Electric Current Through Rolling Bearings," *Ball Bearing J.* **153** (1968).

25.18. A. Boto, "Passage of Electric Current Through a Rolling Contact," *Ball Bearing J.* **153** (1968).

25.19. J. Mathew, and R. Alfredson, "The Condition Monitoring of Rolling Element Bearings Using Vibration Analysis," ASME Paper 83-WA/NCA-1 (November 1983).

26

FAILURE INVESTIGATION AND ANALYSIS

GENERAL

Although rolling bearings are extremely reliable, failure can occur by improper operation or manufacture. With early failure–that is, within the anticipated service life requirement—postmortem investigations are frequently conducted to prevent another failure after the component is replaced.

To determine rolling bearing failure, one must first gather data about the alloys from which the bearing was made and their suitability for the application, as well as operating conditions relative to loading, lubrication system, and external environment. Data must then be evaluated with respect to the manner in which the bearing failed. A bearing is deemed to fail when it does not perform as intended; this is frequently much sooner than when it ceases to function. Figure 26.1 is a flowchart of the data gathering process.

PRELIMINARY INVESTIGATION

When possible, preliminary examination of the hardware should include observations of general features on related components before, and dur-

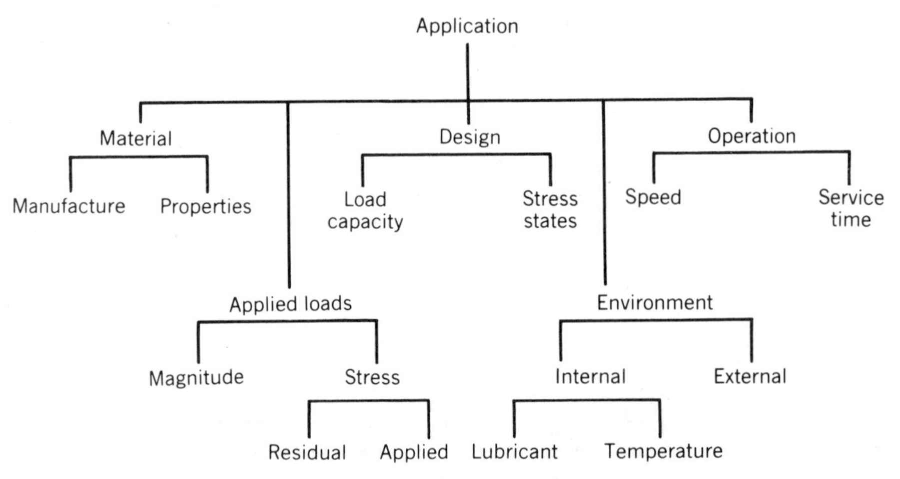

FIGURE 26.1. Flowchart of failure data gathering process.

ing, removal from equipment. Observations should be noted for future reference when more detailed examinations are conducted on the bearing. Similarly, general features should then be noted for the bearing components during disassembly.

DISASSEMBLY OF BEARINGS

Disassembly varies with the types of bearings encountered, so it is beyond the scope of this chapter to deal with specifics for each type. Whatever method is employed, care should be exercised to avoid inadvertent damage, which could be misleading later in the investigation. If destructive methods are required to separate bearing components, the bearing should first be examined to avoid destroying areas that contain regions of interest. Components should be identified with regard to corresponding sides of rings and rolling elements. Samples of grease or residual lubricant should be collected for future reference in the event the failure is found to be lubricant related. If destructive methods have been used to separate components, extraneous debris should be removed, without rotating the rolling elements. Water and other fluids that come into contact with the bearing components during disassembly should be removed immediately to avoid corrosion. Degreasing should be followed by immersion into a preservative.

FAILURE MECHANISMS

Mechanical Damage

Precision bearing components are produced to fine surface finishes on the order of fractions of microns (microinches). When rolling contact surfaces are marred, stress conditions can be imposed that introduce a potential to reduce bearing life significantly. Abusive handling can induce nicks and dents that are harmful, particularly when located in regions tracked by rolling elements. Displaced metal particles generated by nicks and scuffing-type damage introduce secondary effects when they dislodge and indent the raceway.

Permanent indentation created by rolling element overload is called brinelling, a type of damage destined to result in failure. Brinelling may occur by an overload mechanism, such as dropping the bearing or improper mounting techniques. Initial signs of brinelling are signaled by noisy bearing operation.

Raceways may be damaged when they are subjected to vibratory motion while rolling elements are not rotating. This type of damage, called false brinelling, can occur before and after mounting on equipment. False brinelling damage has been observed in bearings that were subjected to vibration during transit as well as on equipment that lay idle for a period of time.

Wear Damage

Wear generally results in gradual deterioration of bearing components, which in turn leads to loss of dimensions and other associated problems. Failure by wear does not mean that bearings will be removed solely because of changes in fit or clearances. Secondary conditions arising from wear can become the predominant failure mechanism. Lubricants may be affected or become contaminated to the degree where lubrication is severely diminished. Stress raisers could be generated that may serve as sites for crack initiation.

Adhesive wear is described in Chapter 24 and is involved in removal of material and possible transfer to mating components. Under properly lubricated conditions, mating components' microscopic asperities could yield and be flattened by cold work. Under these conditions the bearing might function adequately for its projected life. When lubricating conditions become inadequate, however, increased friction results in metal-to-metal contact, giving rise to localized deformation and friction welding. Operating forces cause increased plastic deformation by tearing the locally friction-welded regions from the matrix. One component is now pitted, and the other contains the transferred metal. This condition could be progressive, depending on operating conditions. Generally, lighter adhesive damage is described as scuffing and scoring, whereas more gross damage is described as seizing and galling.

Abrasive wear occurs when hard particles become entrained between bearing components, moving relative to the contacting surfaces. Coarser hard particles can induce microscopic furrows, whereas fine particles may produce a highly polished surface. Abrasive particles may originate internally or externally from adjacent components in the mechanism's system and be transported by the lubricant. Oxidation products and carbides from ferrous components serve as abrasive media. Fretting, a wear mechanism described under corrosion, is a prime example of oxidation by-products.

Lubricant Deficiency

Lubrication problems may be associated with an inadequate lubricant, an inefficient lubricating system, or a combination of both factors. Ideally, rolling elements are separated from the raceways by an elastohydrodynamic (EHD) lubricant film, thereby reducing friction and minimizing mutual wear of bearing components [26.1]. Improperly lubricated bearings produce varying conditions, which lead to progressive contacting surface deterioration and reduced life.

Initial stages of wear involve the plastic deformation of grinding furrow asperities, which, in subsequent cold-working, fracture to produce extremely fine platelets of steel. This stage may also be accompanied by adhesion of microscopic asperities that delaminate and pass on into the lubricant. These cold-worked, hard particles, which also contain carbides, serve as abrasive media. After a time, original grinding furrows in the rolling element tracks are worn smooth to produce a glazed condition. Continued operation will generally lead to a deterioration stage that manifests itself as a frosted condition and, sequentially, to scuffing. Microscopic pits and crevices created by this adhesion mechanism serve as stress raisers for the initiation of microspalls.

In a lubricant system where the quantity of lubricant to vital areas is too low, as a function of either poor design or inadequate flow rates, bearing component temperatures increase. This reaction in turn increases lubricant bulk temperature. As lubricant temperature increases, viscosity decreases, effecting increased frictional forces that make the condition progressively worse. Bearing surface degradation will be accelerated, and the surfaces will be discolored as the process progresses.

Crack Damage

Cracking of bearing components may originate as a function of operating stress conditions via overload or cyclic loading (fatigue). Additionally, manufacturing-related cracks may derive from the steelmaking process and/or working processes. With the exception of cracks arising from inclusions and hydrogen in steels, cracks associated with steelmaking and primary working processes seldom survive through secondary working processes.

Cracks that arise from bearing manufacture secondary working processes frequently are associated with heat treatment and grinding. The nature of the processes are such that sequential operations tend to promote rapid crack propagation in bearing steels, if any are present during manufacture. An obvious exception is cracking that began at the final grinding stages. In view of the foregoing, cracking problems encountered are usually operationally related as compared to manufacturing-related, occurring via cyclic loading. Stress states may be complex, arising from combined effects of component residual stress state, static stresses imposed by mounting, and stresses superimposed by applied loads. Some degree of contribution may be deduced by the direction of crack propagation because propagation tends to proceed normal to the acting stresses.

Corrosion Damage

Corrosion-initiated failures are often difficult to recognize in bearings. Applications involving moisture-laden environments may result in surface oxidation and produce rust particles and pits. These particles are potential media for producing rapid wear via abrasion, and the pits can function as stress raisers providing sites for crack initiation. The series of events that occurs during subsequent operation could alter the conditions initiating the failure by the time a bearing fails and is examined.

Bearings that operate in an environment where water is absorbed in the lubricant may be subjected to pitting corrosion by hydrolysis. Sulfur and chlorine contaminants in lubricants could dissociate from their respective compounds and react with the water to attack the steel on a microscopic scale [26.2]. Pits would then provide sites for spall initiation. Should spalling occur in this environment, it would be assisted by the corroding species and accelerate the rate of crack propagation. Hydrogen dissociated from water is also reported to assist and accelerate crack propagation [26.3]. In the event the contaminating species were to become available in higher concentrations, the lubricant would become acidified, and a more general corrosion could be observed.

Fretting corrosion is an oxidation wear mechanism generated by relative motion between ferrous components that contact bearing surfaces. The by-products are various stable and unstable oxides of iron (Fe), FeO, and Fe_2O_3, which are colored red, and Fe_3O_4, which is black.

EXAMINATION AND EVALUATION OF SPECIFIC CONDITIONS

Rolling Element Tracking

Rolling elements revolve in ring raceways and develop annular tracks that reflect the distribution of applied loads. These track patterns could signal unsuspected problems contributing to or causing premature bearing failures. Alteration of the normal rolling element trajectory might be caused by misalignment associated with assembly or machine operating characteristics. Assembly deficiencies are self-explanatory. Excessive deflection and axial looseness are examples of machinery deficiencies. Since the stress distribution will be adversely altered, fatigue life will be reduced and the propensity for spalling will be increased. The relationship between tracking pattern and failure mode must be evaluated with respect to cause and effect. Characteristic patterns are illustrated in Fig. 26.2.

Brinell Marks

"Brinelling" is a term describing depressions in raceways. For example, rolling element indentations may be induced by overload inadvertently while mounting bearings on machinery. Raceways may deform plastically either by impact or by static forces. The effects stemming from this type of damage are similar regardless of the overload mechanism. Raceways are permanently deformed with the flow of metal corresponding to the conformation of the rolling elements. Usually, multiple indentations are induced, separated by distances equal to the rolling element spacing. Indentations are accompanied by localized stress alterations. When the steel yields, it flows in all directions to accommodate the compressive forces. Microscopic elevations project outward from the raceway, forging crowns around surface regions adjacent to the indenting obstacle(s). Stress states of indented regions are in compression, whereas the

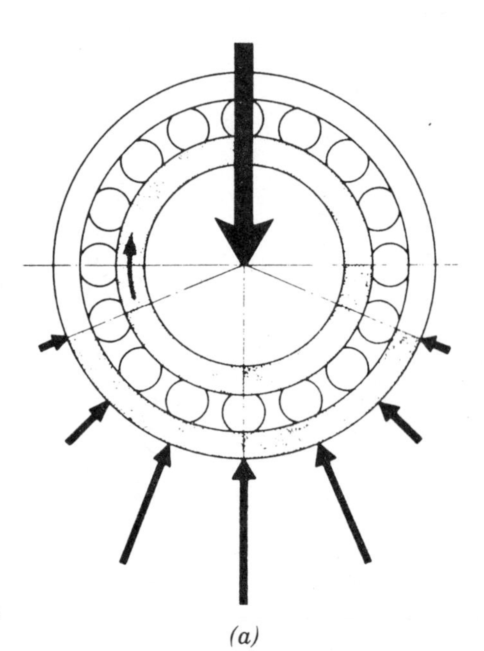

FIGURE 26.2. (*a*) Load distribution within a bearing.

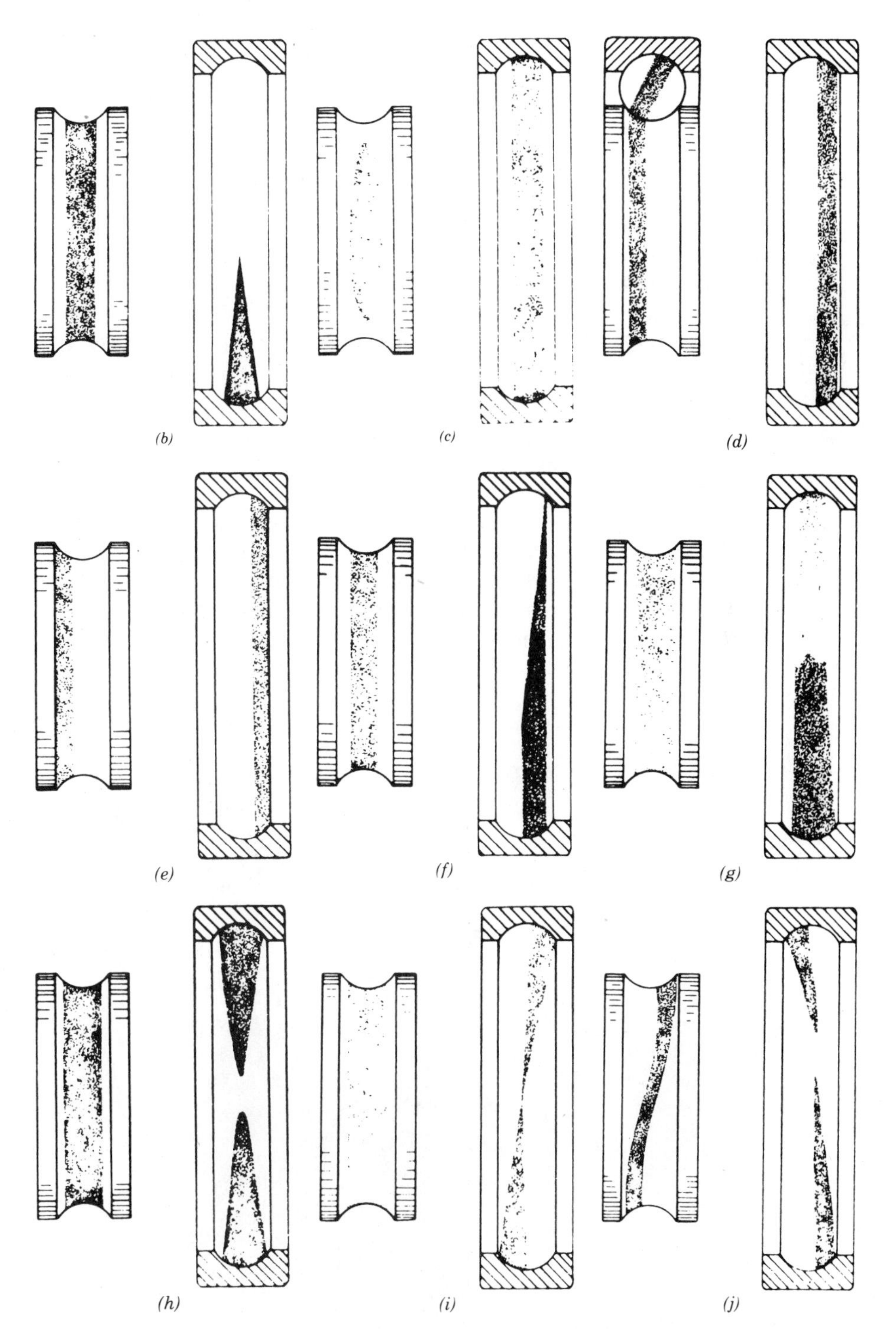

FIGURE 26.2. **(***Continued*) (*b*) Normal load zone inner ring rotating relative to load. (*c*) Normal load zone outer ring rotating relative to load or load rotating in phase with inner ring. (*d*) Normal load zone. Axial load. (*e*) Load zone when thrust loads are excessive. (*f*) Normal load zone combined thrust and radial load. (*g*) Load zone from internally preloaded bearing supporting radial load. (*h*) Load zones produced by out-of-round housing pinching bearing outer ring. (*i*) Load zone produced when outer ring is misaligned relative to shaft. (*j*) Load zones when inner ring is misaligned relative to housing.

crowned regions are in tension. Consequently, crowned regions experience greater cyclic stress amplitudes during bearing operation. The effect is twofold: not only are the crowns subject to stress concentrations, but the separating EHD lubricant film is interrupted by the projections. This imposes increased frictional effects on the crowns. Spall initiation at the crowned regions is imminent.

Identification of brinelled raceways is easiest during early stages of spalling where the distance aspects are readily identified. Multiple occurrences of brinelling tend to make identification more difficult. When spalling progresses to advanced stages and spacing is not apparent, brinelled locations may be detectable by macroscopic fatigue crack propagation features. Indented regions are in compression and therefore tend to spall during later stages of progression. If examined at this stage, these features may be evident and related to the rolling element spacing.

Brinell marks have been identified after spalling has progressed beyond that described here by careful examination of fatigue crack patterns. Spalls generally reflect the influence of the dominating stress field. More severe indentations resulting from rolling elements have exhibited fatigue crack propagation patterns that correspond to the configuration indicative of the roller. An example of this condition is shown in Fig. 26.3.

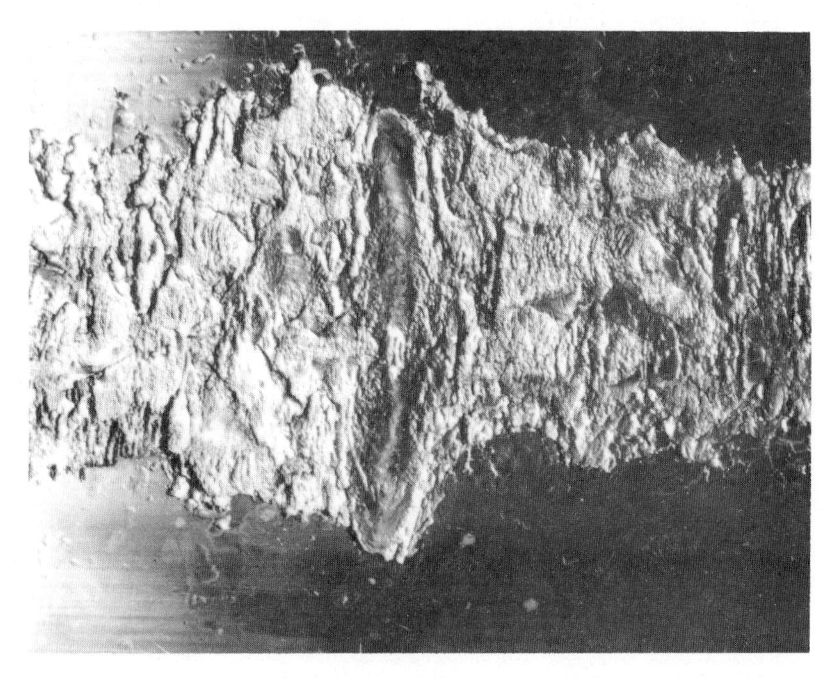

FIGURE 26.3. Spalled area of raceway containing a brinelled region in which fatigue crack propagation is normal to the roller track (magnification 2×).

False Brinell Marks

False brinelling is a condition that resembles true brinelling but is generated by a different mechanism. True brinelling is generated by plastic deformation of the steel, whereas false brinelling is generated by a corrosion-wear mechanism. Stationary nonrotating bearings subjected to vibration, wear by fretting corrosion between rolling elements and raceways. Corrosion products accumulate, which then proceed to accelerate wear by abrasion. Surface irregularities created by wear serve as initiation sites for spalls during subsequent operation. The EHD lubricant film may also be affected locally, effecting marginal lubricating conditions.

Score Marks

Careless handling is the primary cause for scores and digs that mar bearing surfaces. Depressions of this type plastically deform the steel, usually displacing metal that is subsequently cold-worked by the rolling elements. Eventually, the cold-worked slivers fracture, producing fragments in the system that may be coined in the roller path; crowns are generated around the indentations, which are subject to the effects discussed under brinelling.

Notched regions created by the scores function as stress raisers. The sharpness of the notches and their locations with respect to operating forces are important factors affecting the propensity for crack initiation. Stress intensity is greater for sharply notched surface discontinuities.

Adhesion Damage

Adhesion damage manifests itself as a buildup of metal on a component, resulting from metal transfer from the interacting component. In turn, this interacting component, if examined during early stages of progress, would contain corresponding pitted regions where metal had been pulled. Figure 11.4 displays the characteristic features associated with metal transfer.

Lubricant Contamination

Particle Dents. Particulate matter suspended in the lubricant results in mutual deformation of rolling elements and accompanying rings by indentation. The nature of the indentations corresponds to the hardness of the particles. Particle dents initiating from relatively soft materials will display shallower, smoother features. Dents initiating from hard materials will display depths and configurations conforming to the particles. Dents will also be accompanied with the crown features discussed

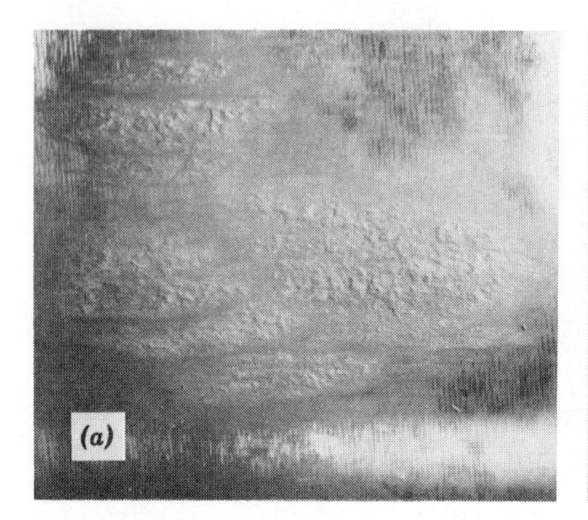

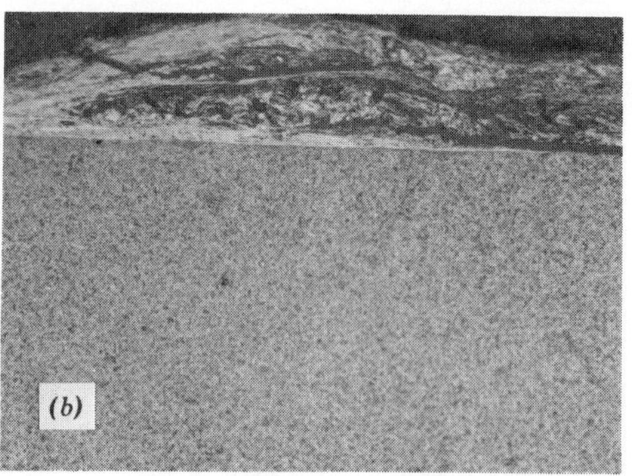

FIGURE 26.4. Adhesive wear. (*a*) Topography of scuffed region displaying metal transfer (magnification 1.5×). (*b*) Cross-sectional view exhibiting deformation associated with scuffing (magnification 10×).

under brinelling. Here again, crown heights will correspond to the volume of metal displaced, and crown surface areas are subject to surface distress and spall initiation, as evidenced in Fig. 26.5. Unless lubricant systems contain filters, spall fragments add to the particulates and generate a self-propagating mechanism.

FIGURE 26.5. Particle dent encircled by distressed crowned region depicted as a darker gray border. Note that spall initiated in the distress region (magnification 90×).

Fine Particle Dispersion. Fine particulate matter dispersed in the lubricant functions as a lapping medium in bearings effectively wearing interacting components by abrasion. Contact regions, such as flanges and ball and roller paths can wear away grinding furrows to highly polished finishes as displayed in Fig. 26.6.

Scanning electron microscope (SEM) examination of affected regions may disclose features to aid in the identification of the particulate matter. Embedded particles may also be detected, which may be analyzed with a microprobe for positive identification. Bearing life is adversely affected as a result of this wear, not only as dimensions are altered but also by progressive deterioration of the lubricant. As the ability to lubricate is diminished, associated new problems arise.

Hydrolysis. Bearing lubricants have limited tolerance for water, and external environmental conditions could provide corroding species that react with bearing steels. Equipment that remains idle for periods of time is susceptible to corrosion by oxidation, which occurs at the lines of contact between rolling elements and raceways. See Fig. 26.7. Affected regions may appear red or black, depending on the state of oxidation. These regions are sites for spall initiation upon subsequent bearing operation.

Bearings that operate at relatively high temperatures appear to be susceptible to localized pitting, as depicted in Fig. 26.8. Pitting in bearing steels has been associated with sulfur and chlorine. Dissolution of their respective ions in a lubricant containing moisture can lead to hydrolysis, whereby localized anodes are established that provide sites for pitting corrosion. In turn, pits provide sites for spall initiation when located in the roller path.

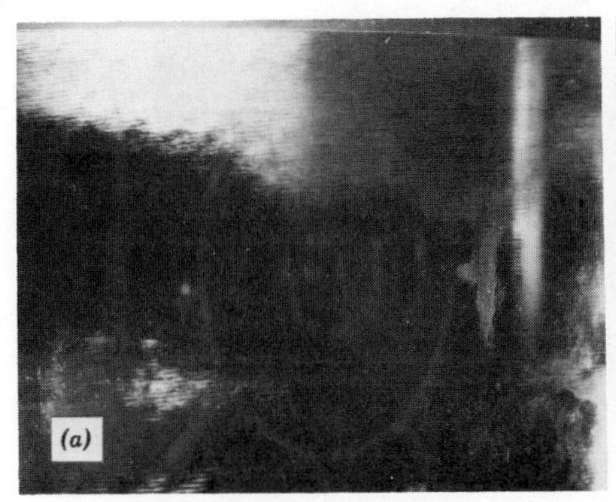

FIGURE 26.6. Abrasive wear. (*a*) Surface area polished by fine coal particle conveyed by the lubricant system (magnification 1.5×). (*b*) SEM photograph showing coal particle indentations (magnification 200×).

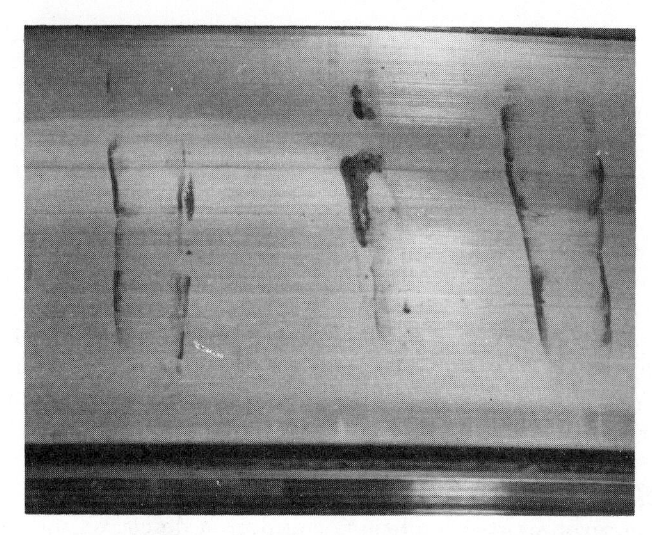

FIGURE 26.7. Static corrosion caused by moisture between rolling elements and raceway.

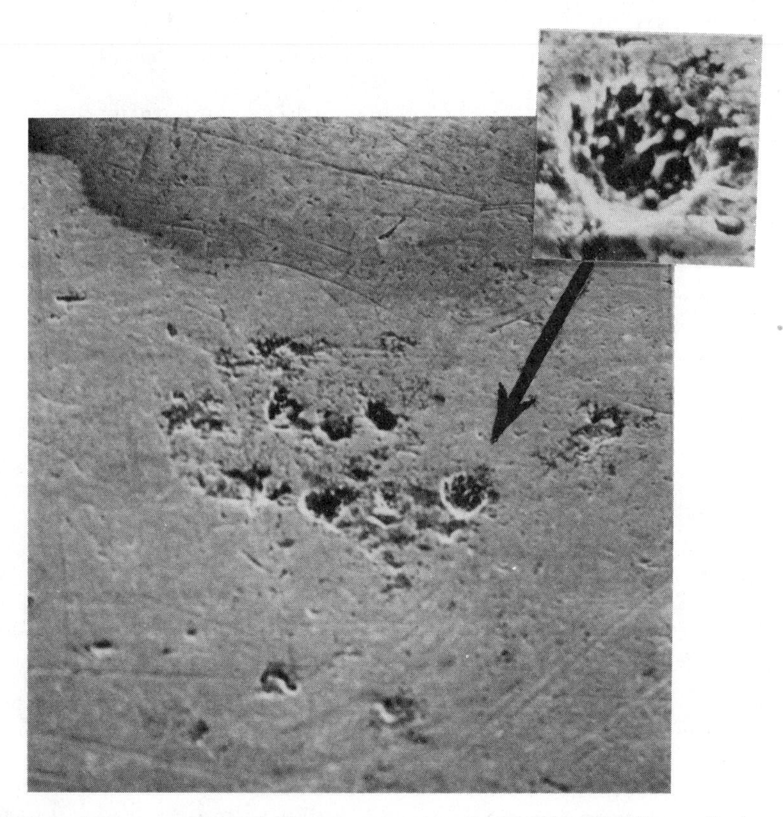

FIGURE 26.8. Pitting corrosion. Sulfur concentrations detected by EDS. Insert displays characteristic dissolution features (magnification 500×).

Discoloration. Generally, when bearings operate under conditions where the lubricant remains unaffected with respect to its composition, the steel remains relatively bright. Most common problems associated with the permanent discoloration of bearing steel surfaces are heat related. For example, frictional heat generated by abrading metal components can oxidize the surfaces, producing oxidation colors ranging from straw color to blue. These oxides are related to temperatures ranging between 177°C (350°F) and 316°C (600°F), respectively. When bearings are overheated, mechanical and associated fatigue strengths are reduced. The color-temperature relationship, however, may be altered by decomposition of the lubricant, which in turn produces a tenacious film that tarnishes the steel. When a lubricant decomposes, lubricating ability is diminished, and, under the conditions discussed in the preceding paragraphs, bearing surfaces may be subjected to corrosion effects.

Fretting Corrosion. Relative movement between a bearing ring and the surface it contacts produces wear on a microscopic scale. Outer and inner ring surfaces interact with internal housing and shaft surfaces. This microscopic interaction produces oxidation wear products, as discussed earlier. Surfaces appear smooth to the unaided eye, but magnification will reveal microscopic crevices indicative of the direction of motion. See Fig. 26.9.

Fretting features on the respective bearing surfaces are indicative of an inadequate fit. This condition can occur when mating surface areas are not intimately supported. Rings are mounted to resist movement by applied loads via friction, which is developed between contacting surfaces by pressure fits. This is usually accomplished by press-fitting or by thermal treatments that expand or contract rings to fit shafts and housings, respectively. Continued wear by fretting results in dimensional changes and loss of fit. When wear is excessive, rings will rotate and overheat. Ring cracking has also been observed to initiate from surface areas subjected to fretting. Cracks initiate by fatigue at cold-worked microscopic crevices.

Electric Arc Damage

Improperly grounded electrical equipment can subject bearings to mechanical and metallurgical damage. When current seeking ground passes through bearings, it tends to arc between noncontacting rolling elements and raceways. Apparent visual effects range from random pitting to fluted patterns composed of numerous pits. Obviously, bearings operating in the latter condition will vibrate noisily.

Other problems arising from electric arc damage are related to metallurgical alterations. Metallurgical properties are affected to signifi-

FIGURE 26.9. Fretting corrosion. Wear pattern indicates direction of motion (magnification 40×).

cant depths. Electric arcs produce intense, highly localized temperatures that have the potential to melt the surface. Heat is rapidly dissipated as it is conducted into the mass. Temperature transitions, as evidenced metallographically, can exhibit remelting, austenitization, and retempering. Surface and near-surface regions are markedly affected. Initial surface zones may consist of brittle untempered martensite followed by high-temperature retempered zones. As temperatures dissipate, the effects are less pronounced. The effects of arcing are more easily identified in the microstructure when arc events are random. A single arc generates a hemispherically shaped, affected region consisting of the zones discussed previously. Succeeding arc events superimpose heat-treating effects on the previously affected region. An example of a fluted condition is presented in Fig. 26.10.

Arcing also alters the previously existing stress field. Rehardened zones exhibit high residual compressive stresses, whereas adjacent retempered zones counteract these stresses with residual tensile stresses. Under cyclic loading this tensile zone is vulnerable to fatigue crack initiation and propagation.

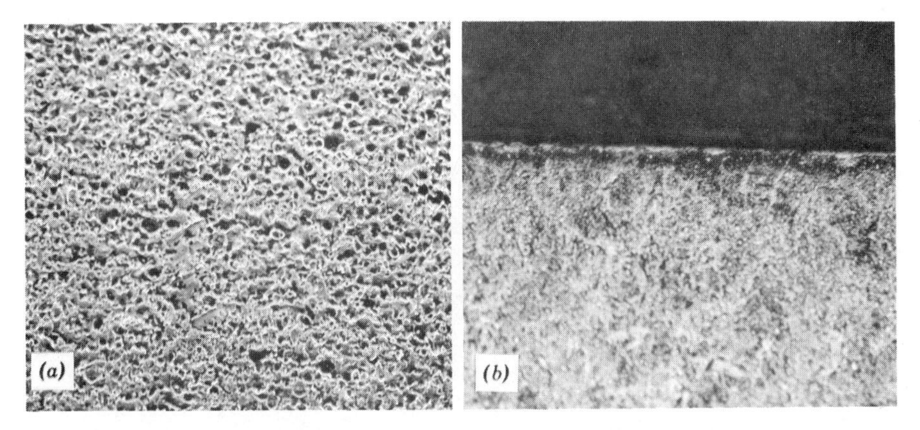

FIGURE 26.10. Electric arc damage. (*a*) Topography of fluted region consisting of pits (magnification 250×). (*b*) Cross-sectional view showing the effect on the microstructure (magnification 500×).

Spalling

Spalling, sometimes called flaking, is a stable stress-related crack mechanism generating from subcritical cyclic loading. Spalls may be surface initiated or subsurface initiated. The processes are called fatigue. Surface-initiated spalling usually occurs by progressive deterioration of rolling contact surfaces resulting when lubrication is inadequate. Several of the previously described mechanisms also exhibited conditions that could potentially culminate in surface-initiated spalling. When inadequately lubricated, rolling contact surfaces become glazed and then microscopically pitted. SEM examination shows these pits to be microscopic spalls. Continued operation results in propagation of the microspalls as well as continued initiation of new spall sites until the surface appears frosted.

Subsurface-initiated spalling is associated with stress concentrations, usually within the depth of the maximum orthogonal shear stress [26.4]. Constituents in the matrix, particularly oxide-type inclusions, intensify the stress. Regions surrounding the inclusion are strained, eventually initiating microscopic cracks. These microscopic regions often manifest themselves as white etched areas commonly called butterflies [26.6]. The direction of crack propagation is dictated by the effects of the applied and residual stresses. When the Hertzian stresses are dominant, fatigue cracks propagate transgranularly to the depth z_{max}, then reverse direction toward the surface, hence the spall. See Fig. 26.11.

Inner rings often sustain tensile hoop stresses resulting from fitting practices on shafts. If excessive, these imposed stresses attain magnitudes that influence the direction of crack propagation. Instead of prop-

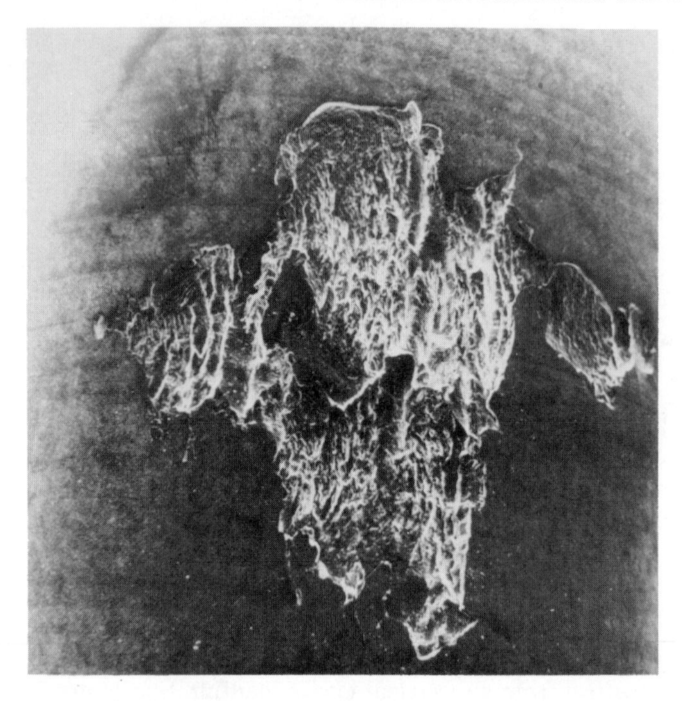

FIGURE 26.11. Spalling. Surface-initiated spall showing characteristic arrowhead growth pattern (magnification 15×).

agating back to the surface, they propagate radially until the ring fractures.

Fatigue crack propagation can also be helped by the lubricant, whereby fluid trapped in the spall exerts high hydrostatic pressures. This condition also results in crack branching and presumably accelerates crack propagation. Moisture also abets fatigue crack propagation by virtually eliminating the endurance limit. Cracking can then occur at any applied stress if the number of cycles is satisfied. This condition is called corrosion fatigue.

FRACTOGRAPHY

Scanning Electron Microscopy

Earlier discussion involved macroscopic features that are identified with the various failure mechanisms that affect bearing life. However, apparent macroscopic features might not represent the initiating mechanism. It is imperative that the failure mode be identified at the origin area, preferably the initiation site itself. This is not always possible with conventional optical microscope techniques due to the inherent loss of focusing ability at higher magnifications. A SEM provides greater depth of focus and the ability to study irregular surfaces. Identifying modes of initiation at crack origins will often provide data leading to the cause of failure, despite damage incurred in other regions. These data will also provide an insight to the types and relative magnitudes of the applied loads. Four basic modes of fracture will be presented with regard to how they appear in bearing steels. Rolling bearings may be manufactured from various grades of steel, but only those that are termed "bearing steels" are considered here. These steels consist of high-carbon and low-alloy compositions.

Microvoid Coalescence

Under abnormal loading conditions in bearings, microvoid coalescence, commonly called dimples, represent the ductile mode of cracking due to overloading. In this mode cracks propagate transgranularly under loads that transmit tensile force components. This mode can be generated by pure tensile, shear, or torsional loading. The last is not common in bearing failure mechanisms. Variations in the microscopic features identify the type of load experienced.

Due to the high hardnesses and corresponding low ductilities to which precision bearings are manufactured, differences in surface texture are less distinguishable at lower magnifications. As SEM magnifications are increased, dimples will be resolved as hemispherical voids. They differ from ductile metals because the tear edges do not elongate significantly. See Fig. 26.12.

Microvoids initiate in the steel matrix where carbides, inclusions, and matrix imperfections reside. In bearing steels dimples are found to be relatively uniform in size, suggesting that the indigenous spherical carbides are the predominant initiation sites. Carbides are sometimes observed within the hemispherical voids.

Dimples indicate the direction of the applied force by their shapes. When dimple formations are equiaxed, the acting force is normal to the fracture. Dimples exhibiting increasingly more oval shapes indicate that the forces exerted increasingly greater shear components. Dimple modes of failure are sometimes observed in bearing flange fracture when thrust loads are excessive.

Cleavage

Cleavage is a rapid overloading mode of failure resulting in bearing fractures. A fracture may initiate by cleavage, or it may initiate by a different mode and propagate by cleavage. In other words, a crack may initiate and propagate by one or more modes until it reaches a critical size.

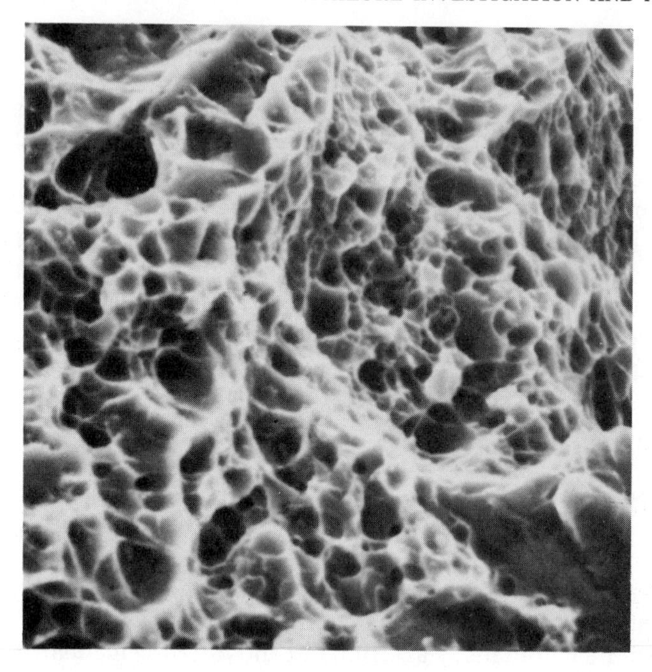

FIGURE 26.12. Microvoidcoalescence. Hemispherical voids are depicted, which are characteristic of bearing steels (magnification 500×).

At this point the crack will expand rapidly; that is, fracture occurs. This unstable crack propagation occurrence is related to the steel's fracture toughness property, which is influenced by composition, microstructure, temperature, and loading rate. If a bearing steel is impacted with sufficient force, cleavage will be observed across the section thickness, including the initiation site.

Cleavage is a low energy fracture that propagates transgranularly along specific crystallographic planes. It appears as flat planes that change orientation from grain to grain. Fan-shaped features are evident on the facets. These features arise from second-order planes, giving the appearance of steps, and are called *river* patterns. These patterns are typically forked, indicating the direction of crack propagation toward the converging feature within a grain.

Matrix carbides interfere with the normal cleavage progression, and bearing steels contain numerous carbides that are precipitated within the prior austenite grains during the temper treatment. Therefore, bearing steels do not exhibit the normal cleavage features. The crack appears to propagate around carbides and develop smaller cleavage facets within grains. This condition is called quasicleavage, which is displayed in Fig. 26.13. Hence, cleavage is more difficult to identify in bearing steels.

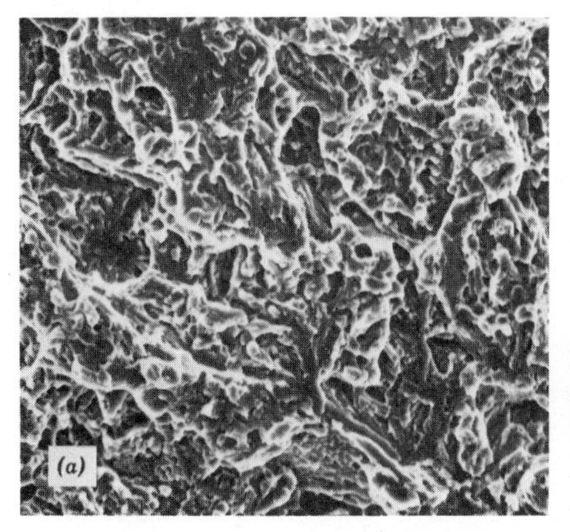

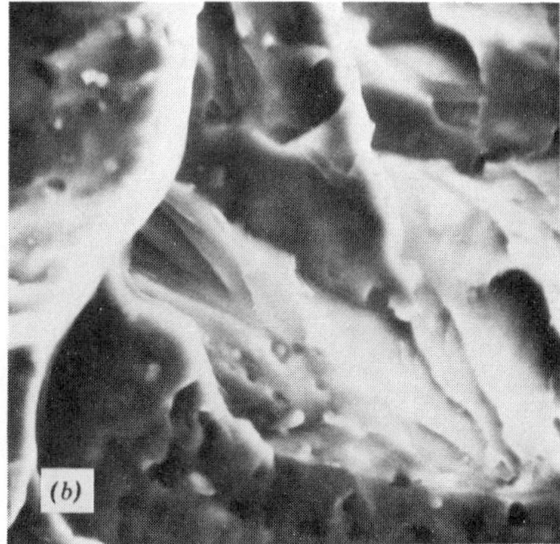

FIGURE 26.13. Quasicleavage. (*a*) Characteristic smaller cleavage facets within crystals (magnification 1500×). (*b*) Fan-shaped features on a cleaved facet (magnification 2500×).

Quasicleavage is indicative of unstable crack propagation in bearing steels occurring due to sudden overload.

Intergranular Fracture

Intergranular fracture is a low energy mode of cracking that starts at grain boundaries. This condition is an embrittlement mechanism, which reduces more tightly bonded grain boundary energy areas. Embrittlement of bearing steels has been caused by improper heat treatment, whereby a brittle phase is precipitated at the grain boundaries. This has been shown to occur by the precipitation of phosphorous at the prior austenite grain boundaries. Quench cracks exhibit this behavior.

Hydrogen gas can also embrittle the steel and cause cracks to progress intergranularly in the affected region. Hydrogen may be dissolved during the melting process and diffuse to form gas pockets during solidification. All processes that provide hydrogen gas environments are potential sources for embrittlement by absorption of hydrogen at the surface. Electroplating is an example.

The intergranular mode of cracking is readily identified by its grainy appearance, as shown by Fig. 26.14. Secondary cracks are often observed. Bearing steels exhibit features similar to those of other steels at lower magnifications. Higher magnification may reveal small spherical pockets where carbides reside.

Fatigue

Microscopic fatigue features are the most difficult to identify in bearing steels. These steels consist of spherical carbides dispersed in a fine mar-

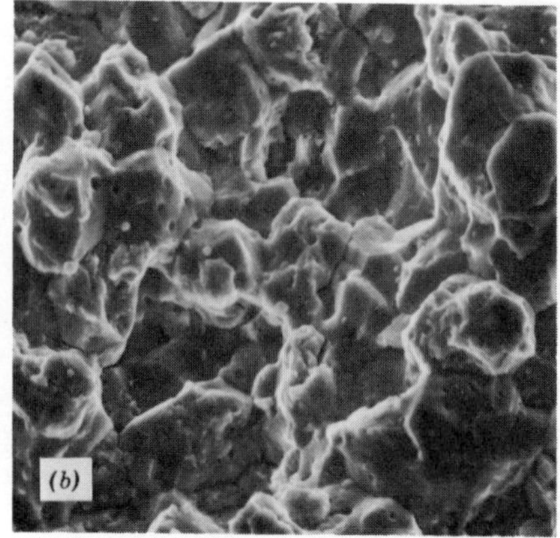

FIGURE 26.14. Intergranular failure. (*a*) General view of quench crack (magnification 250×). (*b*) Higher magnification showing intergranular features (magnification 1000×).

tensitic matrix, constituents which interfere with the normal progression of fatigue striations. Hence, fatigue is usually identified at low and intermediate magnifications.

Fatigue may progress in two stages. The first stage encompasses the crack initiation site and appears relatively smooth and featureless. This stage may only be evident under low stress conditions, often oriented at a slight angle to the adjacent fracture surface. The second stage is characterized by fatigue striations. Ridges oriented normal to the direction of crack propagation may be evident in the transition region. An example of the fatigue features in bearing steels is shown in Fig. 26.15.

Fatigue striation spacing indicates the stress amplitude and, when clearly defined, yields quantitative data about the number of stress cycles. A fatigue striation is the result of a single stress cycle. Bearing steels do not display definitive striations. Consequently, quantitative analysis of fatigue failures in bearing steels does not appear to be feasible at this time.

Mixed Modes

The various mechanisms of crack propagation discussed occur in combinations exhibiting two or more of the identifying features simultaneously. Often these occurrences involve modes associated with material properties and microstructure. An example of this condition observed in bearing steels involves mixtures of quasicleavage and intergranular modes, both of which are low-energy mechanisms. Stress conditions favoring both mechanisms are apparently equal.

Strain rate affects fracture mechanisms. Mixtures of fracture modes are observed in transition regions between stable crack propagation and

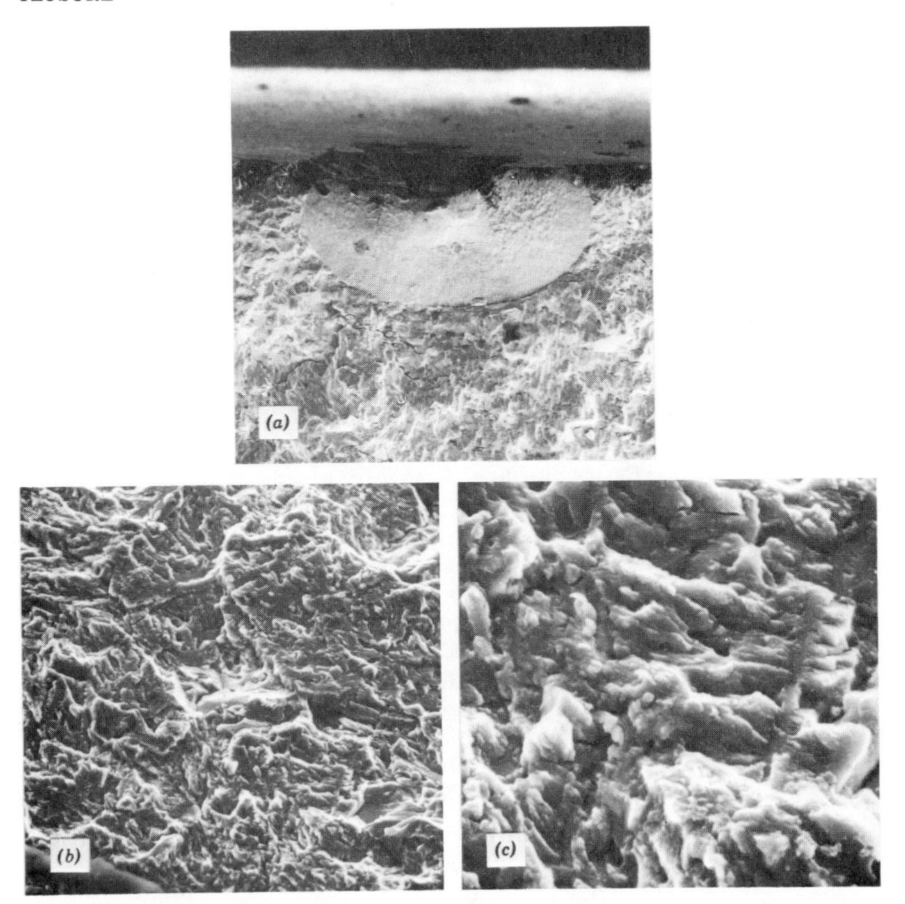

FIGURE 26.15. Fatigue. (*a*) Subsurface-initiated fatigue showing relatively smooth first stage region (magnification 100×). (*b*) Characteristic fatigue features at 1500× magnification. (*c*) Fatigue features at 6000× magnification.

unstable crack propagation. Bearing components that fail by fatigue eventually attain a critical crack size and fracture by quasicleavage. Transition zones between the fatigue and quasicleavage zones sometimes display dimples intermingled in the cleavage facets. These transition zones are relatively narrow.

Evaluation of failure-containing mixed modes of cracking in the origin area depends on which mode is dominant.

CLOSURE

An overview of the more common conditions and damage leading to bearing failures has been presented. Details regarding mechanisms should be referred to in the appropriate chapters. It should not be construed that the examples cited here are all-inclusive; for example, cage problems and wear patterns have not been addressed. Considerable variation may be observed within the examples used. Illustrations and photographs are presented to depict representative features.

Laboratory work, such as metallography, stress determinations, phase identification, microprobe analysis, and so on, should be conducted to verify and support visual observations.

REFERENCES

26.1. T. Tallian, "Rolling Contact Failure Control Through Lubrication," *Proc. Inst. Mech. Eng.* **182,** 205–236 (1967–68).

26.2. J. Mohn, H. Hodgen, H. Munson, and W. Poole, "Improvement of the Corrosion Resistance of Turbine Engine Bearings," AFWAL-TR-84-2014.

26.3. C. Rowe and L. Armstrong, "Lubricant Effects on Rolling-Contact Fatigue," *ASLE Trans.* **23,** 23–39 (Jan. 1982).

26.4. G. Lundberg and A. Palmgren, "Dynamic Capacity of Roller Bearings," *Acta Polytech. Mech. Eng. Ser. 2*, No. 4, 96 (1952).

26.5. J. Martin, S. Borgese, and A. Eberhardt, "Microstructural Alterations of Rolling Bearing Steel Undergoing Cyclic Stressing," ASME Preprint 65-WA/CF-4.

26.6. R. Osterlund, O. Vingsbo, L. Vincent, and P. Guiraldeng, "Butterflies in Fatigued Ball Bearings—Formation Mechanisms and Structures," *Scand. J. Metall.* **11** (1982).

APPENDIX

All equations in the text are written in metric system units. In this appendix, Table A.1 gives factors for conversion of metric system units to English system units. Note that for the former, only millimeters are used for length and square millimeters for area with the exception of viscosity, which being in centistokes is square centimeters per second. Furthermore, the basic unit of power used herein is the watt (as opposed to kilowatt). Consistent with the foregoing, Table A.2 provides the appropriate English system units constant for each equation in the text having a metric system units constant.

TABLE A.1. Unit Conversion Factors[a]

Unit	Metric System	Conversion Factor	English System
Length	mm	0.03937, 0.003281	in., ft
Force	N	0.2247	lb
Torque	mm · N	0.00885	in. · lb
Temperature difference	°C, °K	1.8	°F, °R
Kinematic viscosity	cm^2/sec (centistokes)	0.001076	ft^2/sec
Heat flow, power	W	3.412	Btu/hr
Thermal conductivity	W/mm · °C	577.7	BTu/hr · ft · °F
Heat convection coefficient	W/mm · °C	176,100	BTu/hr · ft^2 · °F
Pressure, stress	N/mm^2	145.0	psi

[a]English system units equal metric system multiplied by conversion factor.

TABLE A.2. Equation Constants for Metric System and English System Units

Chapter Number	Equation Number	Metric System Constant	English System Constant
3	32	4.71×10^{4}	6.83×10^{6}
	33	4.71×10^{4}	6.83×10^{6}
4	41	2.26×10^{-11}	2.11×10^{-6}
	52	3.39×10^{-11}	3.17×10^{-6}
	69	2.26×10^{-11}	2.11×10^{-6}
	70	3.39×10^{-11}	3.17×10^{-6}
	73	4.47×10^{-12}	4.18×10^{-7}
	74	8.37×10^{-12}	7.83×10^{-7}
5	34	0.236	0.0045
	36	0.0236	0.0045
	38	2.79×10^{-4}	1.01×10^{-5}
	47	3.35×10^{-3}	2.78×10^{-4}
	48	3.84×10^{-5}	4.36×10^{-7}
6	8	2.15×10^{5}	3.12×10^{7}
	9	7.86×10^{4}	1.14×10^{7}
	111	3.84×10^{-5}	4.36×10^{-7}
	112	1.24×10^{-5}	8.71×10^{-8}
	115	1.24×10^{-5}	8.71×10^{-8}
	116	1.24×10^{-5}	8.71×10^{-8}
	118	6.20×10^{-6}	4.36×10^{-8}
	121	6.20×10^{-6}	4.36×10^{-8}
	130	1.24×10^{-5}	8.71×10^{-8}
	131	1.24×10^{-5}	8.71×10^{-8}
	133	6.20×10^{-6}	4.36×10^{-8}
	137	1.92×10^{-5}	2.18×10^{-8}
	144	6.20×10^{-6}	4.36×10^{-8}
	146	1.24×10^{-5}	8.71×10^{-8}
7	66	2.24×10^{-12}	2.09×10^{-5}
9	6	4.36×10^{-4}	1.58×10^{-5}
	7	6.98×10^{-4}	2.53×10^{-5}
	8	1.81×10^{-4}	4.33×10^{-6}
	9	7.68×10^{-5}	8.71×10^{-7}
	10	4.36×10^{-4}	1.58×10^{-5}
	11	6.98×10^{-4}	2.53×10^{-5}
	12	5.24×10^{-4}	1.9×10^{-5}
	13	1.81×10^{-4}	4.33×10^{-6}
	14	7.68×10^{-5}	8.71×10^{-7}
	28	0.0472	0.009
	29	0.0472	0.009
	30	0.0472	0.009
	32	1.116×10^{-3}	4.24×10^{-5}
11	50	1.976×10^{-6}	4.62×10^{-5}
	51	6.916×10^{-4}	4.77×10^{-6}
13	78	100,000	8624

TABLE A.2. *(Continued)*

Chapter Number	Equation Number	Metric System Constant	English System Constant
	105	10^{-7}	1.45×10^{-5}
	106	160×10^{-7}	2.32×10^{-3}
15	1	1.05×10^{-4}	0.0404
	7	0.0332	0.332
	8	0.06	0.6
	9	2.3×10^{-5}	0.3
	10	0.03	0.3
	12	5.73	0.173
	14	5.73×10^{-8}	0.173×10^{-8}
18	58	98.1	7,450
	67	98.1	7,450
	99	98.1	7,450
	100	39.9	3,030
	101	98.1	7,450
	102	38.2	2,900
	104	39.9	3,030
	105	39.9	3,030
	106	39.9	3,030
	132	98.1	7,450
	133	98.1	7,450
	134	98.1	7,450
	135	98.1	7,450
	136	88.2	6,700
	138	88.2	6,700
	140	59.1	4,490
	143	552	49,500
	156	207	18,600
	157	207	18,600
	162	552	49,500
	163	469	42,100
	169	207	18,600
	170	552	49,500
	171	469	42,100
21	1	1.3×10^{-7}	6.2×10^{-12}
	2	5.25×10^{-7}	2.5×10^{-11}
	3	2.52×10^{-7}	1.2×10^{-11}
	4	6.03×10^{-11}	1.98×10^{-17}
	6	23.8	3,440
	7	23.8	3,440
	9	44	6,379
	12	220	32,150

INDEX